The Male Gamete

From Basic Science to Clinical Applications

The Male Gamete

From Basic Science to Clinical Applications

Edited by:

Claude Gagnon
McGill University

Cache
River
Press

Library of Congress Catalog Card Number 99-63253

ISBN 1-889899-03-8

1st Edition

Editor:
Claude Gagnon, PhD

Publisher:
Cache River Press
2850 Oak Grove Road
Vienna, IL 62995
USA
1-888-862-2243

Copy edited by Tracey Johnson
Printed in the United States of America

Front Cover Illustrations

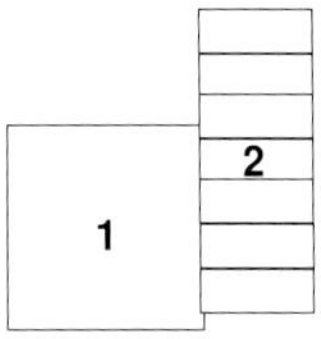

Back Cover Illustrations

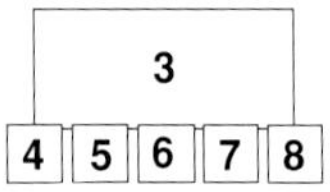

1. A dispermic human mitotic zygote triple-labeled for microtubules (red), gamma tubulin (green) and with Hoechst DNA (blue). Courtesy of C. Simerly and G. Schatten.
2. Bovine fertilization using MitoTraker (green)-tagged sperm labeled with DNA (blue) and fibrous sheath antibody (red). From Sutovsky et al. (Biol Reprod, 55:1195-1205, 1996) used with the permission of the author and The Society for the Study of Reproduction.
3. Sea urchin spermatozoa on the surface of the oocyte. Scanning electron micrograph courtesy of B. Dale.
4. Pseudocolored confocal micrograph of epididymal tubules. Courtesy of Robert N. Zucker, USPA.
5. A human oocyte showing the microtubular organizing center and the centrosome from Simerly et al. (Nat. Med. 1:47-52) used with permission of the publisher and the authors.
6. Pseudocolored confocal micrograph of seminiferous tubules. Courtesy of Robert N. Zucker, USPA.
7. Photograph showing a latex cast of the seminiferous tubules and rete of a rat testis. Courtesy of Martin Dym.
8. Pseudocolored electron micrograph of a cross section through the sperm tail of a moth. See Dallai and Afzelius, this volume.

Contents

Contributors

Afnan, M
Reproductive Biology and Genetics Group, The University of Birmingham, Birmingham Women's Hospital, Birmingham, England

Afzelius, BA
Department of Ultrastructure Research, Stockholm University, Stockholm, Sweden

Allen, M
Digital Instruments, Santa Barbara, CA USA

Arnoult, C
Laboratoire de Biophysique Moleculaire et Cellulaire, CNRS URA, Grenoble, France

Attar, M
Loeb Research Institute, Ottawa Civic Hospital and Reproductive Biology Unit and the Department of Biochemistry, University of Ottawa, Ottawa, Ontario, Canada

Audebert, S
Urology Research Laboratory, McGill University, Royal Victoria Hospital, Montréal, Canada

Auger, J
CECOS, Hopital Cochin, Paris, France

Baba, T
Institute of Applied Biochemistry and Tsukuba Advanced Research Alliance, University of Tsukuba and the National Institute for Advanced Interdisciplinary Research, Tsukuba Science City, Ibaraki, Japan

Baccetti, B
Institute of General Biology, University of Siena, and Center for the Study of Germinal Cells, C.N.R., Siena, Italy.

Balhorn, R
Biology and Biotechnology Research Program, Lawrence Livermore National Laboratory, Livermore, CA USA

Barratt, CLR
Reproductive Biology and Genetics Group, The University of Birmingham, Birmingham Women's Hospital, Birmingham, England

Beagley, K
Discipline in Pathology, University of Newcastle, NSW, Australia

Bedford, JM
Departments of Obstetrics and Gynecology and Cell Biology and Anatomy, Cornell University Medical College, New York, NY USA

Bench, G
Center for Accelerator Mass Spectrometry, Lawrence Livermore National Laboratory, Livermore, CA USA

Billard, R
Laboratoire d'Ichtyologie Générale et Appliquée, Muséum National d'Histoire Naturelle, Paris, France

Bradbury, EM
Department of Biological Chemistry, School of Medicine, University of California, Davis, CA USA

Bradley, L
Department of Biomedical Sciences, Ontario Veterinary College, University of Guelph, Guelph, Ontario, Canada

Braun, RE
Department of Genetics, University of Washington, Seattle, WA USA

Breed, W
Department of Anatomy and Histology, University of Adelaide, Adelaide, South Australia

Brown, JT
Center for X-ray Optics, Lawrence Berkeley National Laboratory, Berkeley, CA USA

Cardullo, RA
Department of Biology, University of California, Riverside, CA USA

Cho, C
Molecular and Cellular Biology, University of California, Davis, CA and Department of Cell Biology and Human Anatomy, School of Medicine, University of California, Davis, CA USA

Cibert, C
CNRS, Institute Jacques Monod, Université de Paris 7, Paris, France

Cohen, DJ
Instituto de Biología y Medicina Experimental, Buenos Aires, Argentina

Corzett, M
Biology and Biotechnology Research Program, Lawrence Livermore National Laboratory, Livermore, CA USA

Cosman, M
Biology and Biotechnology Research Program, Lawrence Livermore National Laboratory, Livermore, CA USA

Cosson, J
URA 671 CNRS, Université de Paris 6, Marine Station, Villefranche-sur -mer, France

Cuasnicú, PS
Instituto de Biología y Medicina Experimental, Buenos Aires, Argentina

Cummins, JM
Anatomy, Division of Veterinary and Biomedical Sciences, Murdoch University, Murdoch, Western Australia

D'Amours, D
Loeb Research Institute, Ottawa Civic Hospital and Reproductive Biology Unit and the Department of Biochemistry, University of Ottawa, Ottawa, Ontario, Canada

Dale, B
Laboratory of Cell and Developmental Biology, Stazione Zoologica "Anton Dohrn," Napoli, Italy

Dallai, R
Dipartimento di Biologia Evolutiva, Università di Siena, Siena, Italy

De Jonge, CJ
Reproductive Medicine Center, Minneapolis, MN USA

de Lamirande, E
Urology Research Laboratory, Royal Victoria Hospital, Montréal, Quebec, Canada

Di Matteo, L
Department of Human Physiology and Biology, Faculty of Medicine and Surgery, Second University of Naples, Naples, Italy

Dominko, T
Oregon Regional Primate Research Center, Beaverton, OR USA

Dréanno, C
Laboratoire de Physiologie des Poissons, IFREMER, Plouzané, France

Eddy, EM
Gamete Biology Group, Laboratory of Reproductive/Developmental Toxicology, National Institute of Environmental Health Sciences, NIH, Research Triangle Park, NC USA

Ellerman, DA
Instituto de Biología y Medicina Experimental, Buenos Aires, Argentina

Fignon, A
Service de Gynécologie-Obstétrique, Hôpital Bretonneau,Tours, France

Florman, HM
Department of Anatomy and Cellular Biology, Tufts University School of Medicine, Boston, MA USA

Fornés, M
Instituto de Histología y Embriología, Universidad Nacional de Cuyo, Mendoza, Argentina

Gagnon, C
Urology Research Laboratory, Royal Victoria Hospital, Montréal, Quebec, Canada

Galantino-Homer, H
Center for Research on Reproduction and Women's Health, University of Pennsylvania Medical Center, Philadelphia, PA USA

Geremia, R
Dipartimento di Sanita' Pubblica e Biologia Cellulare, Sezione di Anatomia, Universita' di Roma "Tor Vergata," Rome, Italy

Hales, BF
Department of Pharmacology and Therapeutics, McGill University, Montréal, Quebec, Canada

Hamamah, S
Service de Biologie et Génétique de la Reproduction, Hôpital A. Béclère, Clamart, France

Hardy, C
CRC for Biological Control of Vertebrate Pest Populations, CSIRO Division of Wildlife and Ecology, Lyneham, Australia

Hausler, CL
Department of Animal Sciences, Food and Nutrition, Southern Illinois University, Carbondale, IL USA

Hecht, NB
Center for Research on Reproduction and Women's Health and Department of Obstetrics and Gynecology, University of Pennsylvania Medical Center, Philadelphia, PA USA

Hewitson, L
Departments of Obstetrics-Gynecology, Oregon Health Science University, Portland, OR and Oregon Regional Primate Research Center, Beaverton, OR USA

Hinds, L
CRC for Biological Control of Vertebrate Pest Populations, CSIRO Division of Wildlife and Ecology, Lyneham, Australia

Holland, MK
CRC for Biological Control of Vertebrate Pest Populations, CSIRO Division of Wildlife & Ecology, Lyneham, Australia

Hoshi, M
Department of Life Science, Tokyo Institute of Technology, Yokohama, Japan

Hud, NV
Department of Chemistry and Biochemistry, Molecular Biology Institute, University of California, Los Angeles, CA USA

Huitorel, P
Biologie Cellulaire Marine, CNRS et Univ. P. and M. Curie, Station Zoologique, Observatoire Océanologique, Villefranche-sur-Mer, France

Huszar, G
Department of Obstetrics and Gynecology, Yale University School of Medicine, New Haven, CT USA

Ikawa, M
Genome Information Research Center, Osaka University, Osaka, Japan

Ikeda, M
Department of Life Science, Tokyo Institute of Technology, Yokohama, Japan

Jamieson, BGM
Zoology Department, University of Queensland, Brisbane, Queensland, Australia

Jégou, B
GERM-INSERM, Université de Rennes, Campus de Beaulieu, Rennes, Bretagne, France

Jones, RC
Department of Biological Sciences, University of Newcastle, NSW, Australia

Jouannet, P
CECOS, Hopital Cochin, Paris, France

Justine, J-L
Laboratoire de Biologie Parasitaire, Protistologie, Helminthologie, Muséum National d'Histoire Naturelle, Paris, France

Kates, M
Department of Biochemistry, University of Ottawa, Ottawa, Ontario, Canada

Kawamura, M
Department of Life Science, Tokyo Institute of Technology, Yokohama, Japan

Kazam, IG
Department of Anatomy and Cellular Biology, Tufts University School of Medicine, Boston, MA USA

Kirz, J
Physics Department, State University of New York at Stony Brook, Stony Brook, NY USA

Kopf, GS
Center for Research on Reproduction and Women's Health, University of Pennsylvania Medical Center, Philadelphia, PA USA

Kramer, C
Biology and Biotechnology Research Program, Lawrence Livermore National Laboratory, Livermore, CA USA

Krishnan, VV
Biology and Biotechnology Research Program, Lawrence Livermore National Laboratory, Livermore, CA USA

Lee IV, J
Biology and Biotechnology Research Program, Lawrence Livermore National Laboratory, Livermore, CA USA

Leibo, SP
Department of Biomedical Sciences, Ontario Veterinary College, University of Guelph, Guelph, Ontario, Canada (currently ACRES, New Orleans, LA USA)

Li, C
Department of Anatomy and Cellular Biology, Tufts University School of Medicine, Boston, MA USA

Litscher, ES
Department of Cell Biology and Anatomy, Mount Sinai School of Medicine, New York, NY USA

Maki, G
Department of Chemistry, University of California, Davis, CA USA

Marino, M
Laboratory of Cell and Developmental Biology, Stazione Zoologica "Anton Dohrn," Napoli, Italy

Maruyama, Y
Department of Life Science, Tokyo Institute of Technology, Yokohama, Japan

Matsumoto, M
Department of Life Science, Tokyo Institute of Technology, Yokohama, Japan

Metz, EC
Center for Marine Biotechnology and Biomedicine, Scripps Institution of Oceanography, University of California, San Diego, La Jolla, CA USA

Meyer-Ilse, W
Center for X-ray Optics, Lawrence Berkeley National Laboratory, Berkeley, CA USA

Miyagawa, I
Department of Urology, Tottori University School of Medicine, Yonago, Japan

Morisawa, M
Misaki Marine Biological Station, Graduate School of Sciences, The University of Tokyo, Tokyo, Japan

Moriyama, H
Department of Beamline Facilities, Japan Synchrotron Radiation Research Institute, Mikazuki, Japan

Mortimer, D
Sydney IVF, Sydney, NSW and Department of Obstetrics and Gynaecology, University of Sydney, NSW, Australia

Multigner, L
INSERM U292, Hopital de Bicêtre, Le Kremlin Bicêtre, France

Myles, DG
Molecular and Cellular Biology, University of California, Davis, CA and Department of Cell Biology and Human Anatomy, School of Medicine, University of California, Davis, CA USA

Nakanishi, T
Genome Information Research Center, Osaka University, Osaka, Japan

Ning, XP
Wyeth Laboratories, Princeton, NJ USA

Nishigaki, T
Instituto de Biotecnologia, Universidad Nacional Autónoma de México, Cuernavaca, Morellos, México

Nishimune, Y
Research Institute for Microbial Diseases, Osaka University, Osaka, Japan

Oda, S
Tokyo Women's Medical University School of Medicine, Tokyo, Japan

Ogiso, M
Department of Life Science, Tokyo Institute of Technology, Yokohama, Japan

Okabe, M
Genome Information Research Center, Osaka University, Osaka, Japan

O'Toole, CMB
Department of Anatomy and Cellular Biology, Tufts University School of Medicine, Boston, MA USA

Pineau, C
GERM-INSERM U435, Université de Rennes, Campus de Beaulieu, Rennes, Bretagne, France

Piomboni, P
Institute of General Biology, University of Siena, and Center for the Study of Germinal Cells, C.N.R., Siena, Italy

Prieto, M
Chemistry and Materials Science, Lawrence Livermore National Laboratory, Livermore, CA USA

Primakoff, P
Molecular and Cellular Biology, University of California, Davis, CA and Department of Cell Biology and Human Anatomy, School of Medicine, University of California, Davis, CA USA

Purdon, M
Center for Research on Reproduction and Women's Health, University of Pennsylvania Medical Center, Philadelphia, PA USA

Rattanachaiyanont, M
Loeb Research Institute, Ottawa Civic Hospital and Reproductive Biology Unit and the Department of Obstetrics and Gynecology, University of Ottawa, Ottawa, Ontario, Canada

Robaire, B
Department of Pharmacology and Therapeutics, McGill University Montréal, Quebec, Canada

Roldan, ERS
Instituto de Bioquímica (CSIC-UCM), Facultad de Farmacia, Universidad Complutense, Madrid, Spain

Rossi, P
Dipartimento di Sanita' Pubblica e Biologia Cellulare, Sezione di Anatomia, Universita' di Roma "Tor Vergata," Rome, Italy

Russell, LD
Department of Physiology, Southern Illinois University, Carbondale, IL USA

Russo, G
Institute of Food Science and Technology, National Research Council, Avellino, Italy

Sakkas, D
Assisted Conception Unit, Birmingham Women's Hospital and Reproductive Biology and Genetics Group, University of Birmingham, Birmingham, England

Schatten, G
Departments of Obstetrics-Gynecology and Cell and Developmental Biology, Oregon Health Science University, Portland, OR and Oregon Regional Primate Research Center, Beaverton, OR USA

Sette, C
Dipartimento di Sanita' Pubblica e Biologia Cellulare, Sezione di Anatomia, Universita' di Roma "Tor Vergata," Rome, Italy

Shoubridge, EA
Montréal Neurological Institute and Department of Human Genetics, McGill University, Montréal, Quebec, Canada

Shur, BD
Department of Cell Biology, Emory University School of Medicine, Atlanta, GA USA

Simerly, C
Cell and Developmental Biology, Oregon Health Science University, Portland, OR and Oregon Regional Primate Research Center, Beaverton, OR USA

Sofikitis, N
Department of Urology, Tottori University School of Medicine, Yonago, Japan

Spira, A
INSERM, Hopital de Bicêtre, Le Kremlin Bicêtre, France

St. John, JC
Reproductive Biology and Genetics Group, The University of Birmingham, Birmingham Women's Hospital, Birmingham, England

Stout, CD
Department of Molecular Biology, The Scripps Research Institute, La Jolla, CA USA

Suarez, S
Department of Biomedical Sciences, College of Veterinary Medicine, Cornell University, Ithaca, NY

Sullivan, R
Ontogeny-reproduction Unit, Research Center, Centre Hospitalier de l'Université Laval, Ste-Foy, Quebec, Canada

Suquet, M
Laboratoire de Physiologie des Poissons, IFREMER, Plouzané, France

Sutovsky, P
Oregon Regional Primate Research Center, Beaverton, OR USA

Suzuki, N
Division of Biological Sciences, Graduate School of Science, Hokkaido University, Sapporo, Japan.

Swanson, WJ
Center for Marine Biotechnology and Biomedicine, Scripps Institution of Oceanography, University of California, San Diego, La Jolla, CA USA

Takahashi, D
Oregon Regional Primate Research Center, Beaverton, OR USA

Takai, H
Department of Geriatric Research, National Institute for Longevity Sciences, Morioka, Obu, Aichi, Japan

Tanphaichitr, N
Loeb Research Institute, Ottawa Civic Hospital and Reproductive Biology Unit and the Departments of Obstetrics and Gynecology, and Biochemistry, University of Ottawa, Ottawa, Ontario, Canada

Taylor, T
Loeb Research Institute, Ottawa Civic Hospital and Reproductive Biology Unit and the Department of Biochemistry, University of Ottawa, Ottawa, Ontario, Canada

Thaler, CD
University of Central Florida, Department of Biology, Orlando, FL USA

Thonneau, P
INSERM U292, Hopital de Bicêtre, Le Kremlin Bicêtre, France

Thornton, K
Biology and Biotechnology Research Program, Lawrence Livermore National Laboratory, Livermore, CA USA

Vacquier, VD
Center for Marine Biotechnology and Biomedicine, Scripps Institution of Oceanography, University of California, San Diego, La Jolla, CA USA

Vigue, L
Department of Obstetrics and Gynecology, Yale University School of Medicine, New Haven, CT USA

Visconti, PE
Center for Research on Reproduction and Women's Health, University of Pennsylvania Medical Center, Philadelphia, PA USA

Wassarman, PM
Department of Cell Biology and Anatomy, Mount Sinai School of Medicine, New York, NY USA

White, Dawn
Loeb Research Institute, Ottawa Civic Hospital and Reproductive Biology Unit and the Department of Biochemistry, University of Ottawa, Ottawa, Ontario, Canada

White, D
Urology Research Laboratory, McGill University, Royal Victoria Hospital, Montréal, Quebec, Canada

Wilding, G
Laboratory of Cell and Developmental Biology, Stazione Zoologica "Anton Dohrn," Napoli, Italy

Yamada, S
Program for Promotion of Basic Research Activities for Innovative Biosciences, Osaka University, Osaka, Japan

Yoshida, E
Department of Life Science, Tokyo Institute of Technology, Yokohama, Japan

Yoshida, M
Misaki Marine Biological Station, Graduate School of Sciences, The University of Tokyo, Tokyo, Japan

Yuan, R
Molecular and Cellular Biology, University of California, Davis, CA and Department of Cell Biology and Human Anatomy, School of Medicine, University of California, Davis, CA USA

Zeyneloglu, HB
Department of Obstetrics and Gynecology, Yale University School of Medicine, New Haven, CT USA

Zhang, X
Physics Department, State University of New York at Stony Brook, Stony Brook, NY USA

Acknowledgements

I would like to express my gratitude to my long-time collaborator, Dr. Eve de Lamirande, for the numerous hours she spent correcting and editing the many chapters of this book. I want to thank the Organizing Committee of the 8th International Symposium on Spermatology, especially Dr. Robert Sullivan, for their effort and time to select the best speakers for the symposium and the contributors of this book. I cannot end these acknowledgments without mentioning the support I received from Lina Ordonselli, both in the organization of the symposium and the assembly of chapters for this book. Finally, I want to thank Dr. Lonnie Russell for providing this opportunity to publish this high quality book at Cache River Press; it is a pleasure to deal with a publisher who is also an excellent reproductive biologist.

Claude Gagnon

Preface

From van Leeuwenhoek's observation of spermatozoa in the 17th century up to now, spermatology has always been at the forefront of experimental research. The articles assembled in this book present the most recent research developments in spermatology.

The following chapters result from the gathering of a group of world-leading scientists and physicians working on a common theme: the spermatozoon, from its assembly to fertilization and up to the management of infertility. Many of these articles originated from the 8th International Symposium on Spermatology, held in Montréal, Canada. However, a few chapters have also been added to complement the book in a few areas not covered during the symposium, and one chapter pays a special tribute to Yves Clermont, the "father of modern mammalian spermatogenesis."

The Male Gamete: From Basic Knowledge to Clinical Applications deals with the most challenging questions raised in this field, from spermatogenesis and epididymal maturation, to interactions between spermatozoa and the female reproductive tract and the oocyte. These chapters are not limited to knowledge gained from mammalian gametes but include information from gametes of less-evolved species which have been the source of so many discoveries that helped pave the way to what is now known in mammals. It is from this cross-fertilization, between evolved and less-evolved model systems, that spermatology will progress through the 21st century. The spermatozoon is definitely a "moving subject."

Claude Gagnon
Montréal, Canada

"Happiness does not rest on what we have
but on what we think we have."

To my sons Jacquelin, Etienne and Alexandre:

Thank you for the joy and unconditional love you have given me.

1 Fifty Years of Insight into the Seminiferous Epithelium: A Tribute to Yves W. Clermont

Bernard Robaire
McGill University

Fifty years ago a young man accepted the challenge from his thesis supervisor to try to make sense of the cell associations found in the seminiferous epithelium of mammals. Yves Clermont successfully responded to the challenge of Charles P. Leblond and completed, in 1953, his Ph.D. thesis at McGill University, entitled *Histology of the Seminiferous Epithelium of the Rat, Hamster and Monkey* (Clermont 1953). These studies constitute the foundation blocks of modern understanding of the functioning of the seminiferous epithelium. Since that time, Yves Clermont and his trainees have continually made a series of insightful contributions regarding complementary aspects of the organization of the seminiferous epithelium, stem cell renewal, the structure and functions of organelles in spermatozoa, and the hormonal control of spermatogenesis. Prior to the official opening of the Eighth International Symposium on Spermatology, a special session was held to pay tribute to the wide range of contributions made by Yves Clermont to this field in the last half century. The invited speakers had either mentored Yves Clermont (C.P. Leblond) or had been his trainees or colleagues (M. Dym, C. Morales, R. Oko, L.D. Russell and L. Hermo).

Clermont likens science to being an artisan, where the love for the work, the drive for perfection and the dedication to quality are overriding. He has always approached complex questions by breaking them down to smaller pieces that were complementary, resolving each, and then reassembling the pieces in order to build up the whole picture. The two major puzzles he has undertaken during his career have been the seminiferous epithelium and the Golgi apparatus; this chapter will focus on his pivotal role regarding the former. His most significant contributions in this field can be subdivided into three major areas: the organization of the seminiferous epithelium, the formation and renewal of spermatogonia and the differentiation of spermatids.

The Organization of the Seminiferous Epithelium

One of the greatest successes of cell biology has been in the domain of male reproduction. During his graduate training, Clermont combined two histochemical stains, the PA-Schiff technique (used to demonstrate the presence of carbohydrates in cells and tissues) with hematoxylin or the Feulgen stain (to analyze the differentiation of spermatids) to resolve how spermatogenesis was organized in a number of mammals (Clermont 1953). To understand spermatogenesis, it is essential to appreciate that, as germ cells differentiate, they move from the basement membrane to the lumen of the tubule. This process takes many weeks and is under remarkably tight control. While germ cells undergo mitotic divisions, they are referred to as spermatogonia, during meiosis they are called spermatocytes, and after becoming haploid, they are referred to as spermatids while they undergo the process of differentiation into spermatozoa.

Clermont subdivided spermiogenesis into a number of steps based on the morphological characteristics of the nucleus and of the PA-Schiff positive acrosome. For example, spermiogenesis of the rat was divided into 19 steps and that of man into 12 steps. Some of these steps of spermiogenesis were then utilized to identify, in the seminiferous epithelium, cell associations composed of spermatogonia, spermatocytes and spermatids that corresponded to stages of a "cycle of the seminiferous epithelium" which takes place repetitively and locally within every region of seminiferous tubules. Thus in the rat, the first 14 steps of spermiogenesis were used to identify 14 stages in one cycle; in man the first six steps of spermiogenesis helped identify six stages in one cycle. Clermont was the first to demonstrate the existence of a cycle in the seminiferous epithelium in man (Clermont 1963). The classifications which he proposed for man, many mammalian species and birds have been universally accepted and are widely utilized in histophysiological, histopathological and toxicological studies of the testis. He also demonstrated that, although germ cells only migrate from the basement membrane toward the lumen over time, there is a coordinate organization along any given seminiferous tubule such that adjacent segments of a tubule are found in contiguous stages of spermatogenesis. This organization thus creates a "wave of spermatogenesis" (Perey et al. 1961). Clermont's ability to conceptualize in a space-time organizational matrix contributed to the scientific resolution of a problem that had appeared insurmountable.

Clermont used these classifications as tools to resolve many questions relating to events taking place in seminiferous tubules. Thus, using tritiated thymidine and a quantitative radioautographic method, he determined the duration of the cycle and derived the duration of spermatogenesis in a number of species including man. For example, in the rat, spermatogenesis lasts 48 days; in the mouse, 35 days; and in man, 64 days. He also showed that the duration of spermatogenesis is fixed for each species and is not modulated by hormones such as androgens or gonadotrophins (Clermont 1972).

The Formation and Renewal of Spermatogonia

Clermont analyzed quantitatively the development of the seminiferous epithelium of growing rats and showed how embryonic gonocytes gave rise to spermatogonia (Clermont, Huckins 1961). Using the subdivision of the cycle of the seminiferous epithelium into stages, he elucidated the mode of differentiation and renewal of spermatogonia in rodents and primates. A distinction was first made between the spermatogonia functioning as stem cells (that is, type A in rodents and types Ap and Ad in primates) and those undergoing differentiation (that is, Intermediate and type B) which are committed to produce spermatocytes. Quantitative data further revealed that spermatogonia did not divide at random throughout the cycle but did undergo mitosis at very precise stages of the cycle. For example, in the rat, all type A cells divide during stages IX, XII, XIV and I, while all Intermediate and Type B cells divide during stages IV and VI of the cycle, respectively. The renewal of the spermatogonia functioning as stem cells takes place during stage I when only some of the type A cells produce new type A cells, while the others yield Intermediate spermatogonia destined to produce type B spermatogonia. It was noted, however, that a fixed proportion of type A spermatogonia degenerated by apoptosis during stage XII of the cycle (Clermont, Perey 1957).

The existence of two main categories of type A spermatogonia that function as stem cells was first described in 1953 (Clermont, Leblond 1953). These stem cells are either 'reserve' stem cells, which in normal adults do not divide significantly, or are 'renewing' stem cells, which proliferate and renew at each cycle. In rat testes exposed to X-rays, all renewing type A and differentiating spermatogonia are destroyed, but the reserve stem cells survive and eventually start to divide, thus yielding new renewing stem cells which reconstitute the spermatogonial population; these in turn restore the entire seminiferous epithelium (Clermont, Bustos-Obregon 1968).

The Differentiation of Spermatids

A remarkable set of changes takes place in germ cells during the transformation of spermatids into spermatozoa. Although the structural changes during this transformation are most striking and were described first, there are also some major changes taking place at the molecular level. This chapter will focus on five facets of this process to which Clermont and his colleagues made fundamental contributions.

The Role of the Golgi Apparatus in the Formation of the Acrosome

The acrosome is a membrane-bound, hydrolytic enzyme-rich structure which arises out of the Golgi apparatus and, after a complex development, becomes closely associated with the nucleus in the spermatozoon and plays a key role in its entry into the oocyte during fertilization.

The structure of the Golgi apparatus has been examined in detail by means of three-dimensional electron microscopy in spermatids and Sertoli cells as well as other cells (Susi et al. 1971; Rambourg et al. 1979). The Golgi apparatus appears as a single organelle consisting of a branching and anastomosing ribbon located around the nucleus or simply next to it. Along this Golgi ribbon, compact regions, made up of the stacked saccules usually described in textbooks, alternate with non-compact, highly fenestrated regions. Each compact region may be subdivided in three compartments.

The Existence, Protein Composition and Formation of the Perinuclear Theca

The perinuclear theca is a rigid capsule-like structure encasing the condensed nucleus of spermatozoa. It has two distinct regions: the "perforatorium," which is in close contact with the acrosome, and the post-acrosomal 'dense lamina,' which partly covers the caudal extremity of the nucleus (Oko, Clermont 1988). The 27 kD outer dense fiber protein of rat spermatozoa has been isolated and sequenced, and its mRNA has been cloned (Morales et al. 1994). This protein falls in the family of lipid binding proteins and serves to bind the acrosomal and cytoplasmic membranes to the nuclear envelope. It has been demonstrated recently that the perinuclear theca is required for fertilization, in the course of which it is internalized by the oocyte (Sutovsky et al. 1997).

The Structure and Evolution of Endoplasmic Reticulum (ER) Cisternae during Spermatogenesis

The endoplasmic reticulum (ER) cisternae were investigated by three-dimensional electron microscopy. ER cisternae were observed to form a single intracytoplasmic membranous system with several regional territories (for example, intra Golgi networks, subplasmalemmal networks, annulate lamellae and radial body) that form, develop and regress in diverse manners during spermiogenesis. Moreover, the ER system was found to extend from cell to cell in the large group of spermatids (about 1000) which are connected by open intercellular bridges. This unusual feature might explain why groups of spermatids evolve synchronously during spermiogenesis (Clermont, Rambourg 1978).

Formation of the Cytoskeletal Components in the Tail of Spermatozoa

The cytoskeletal elements in the tail of spermatozoa include the outer dense fibers (ODF) and the fibrous sheath (FS), both of which are connected to the doublet microtubules of the axoneme and thus may contribute to the helical motion of the spermatozoon's tail. The protein composition of ODFs and FS was determined and the sequencing of some of their major proteins revealed that they were unique to spermatozoa (Oko, Clermont 1989). Their formation during spermiogenesis was investigated using gold-labeled antibodies prepared from purified proteins. The formation of ODF and FS was found to proceed in steps with intermediate storage in cytological sites. Recently, the mRNAs encoding some of these proteins—that is, the 27kD ODF and 75kD FS—have been cloned and their genes identified and analyzed. These investigations open the road to further molecular studies on spermiogenesis.

Role of Sertoli Cells in Spermatogenesis

Since the differentiation of germinal cells requires the support of Sertoli cells, Clermont and his colleagues clarified their functions in relation to spermatids. These studies revealed several unexpected features, three of which are highlighted below. First, Sertoli cell cytoplasmic fragments were found to be delivered into the spermatid cytoplasm. Second, the formation of a novel anchoring device (referred to as tubulobulbar complex) was found; this device prevents the premature release of the late spermatids to the tubular lumen. Finally, Sertoli cells were found to be not only secretory but also to endocytose tubular material by phagocytosis and by receptor mediated and fluid phase pinocytosis; these two processes are integrated to the cycle in such a way that the masses of residual cytoplasm discarded during the last stages of spermatid transformation into spermatozoa can be endocytosed and disposed of by lysosomes (Morales, Clermont 1993).

These investigations on spermatids plus other developmental studies on the chromatoid body, the centrioles and the connecting piece, the manchette and so on have been reviewed by Oko and Clermont (1990) and Clermont et al. (1993).

To achieve his monumental work, Clermont has explored the seminiferous epithelium by a variety of experimental approaches: histochemical, immunocytochemical and morphometric at both the light and electron microscope levels, combined or not with the use of radioactive tracers and radioautography. More recently, he has initiated a molecular biology approach to the analysis of spermatid components. His studies have been done in conjunction with over 30 graduate students, postdoctoral fellows and colleagues who have not only substantially contributed to the outcome of the studies, but who have been infected by the enthusiasm and commitment he exudes for research and teaching.

It is perhaps fitting to conclude by highlighting some of the key qualities that Clermont believes a scientist should possess, as presented in an interview with Lonnie Russell. For Clermont, science is not merely a job but a true love; he dedicates himself to it. He does not perceive the attribute of being a perfectionist as a handicap, but rather as an essential asset that characterizes the commitment and perseverance of the individual.

References

Clermont Y. Histology of the seminiferous epithelium of the rat, hamster and monkey. Ph.D. thesis, Department of Anatomy, McGill University, Canada, 1953.

Clermont Y. The cycle of the seminiferous epithelium in man. Am J Anat 1963; 112:35-51.

Clermont Y. Kinetics of spermatogenesis in mammals: seminiferous epithelium cycle and spermatogonial renewal. Physiol Rev 1972; 52:198-236.

Clermont Y, Bustos-Obregon E. Re-examination of spermatogonial renewal in the rat by means of seminiferous tubules mounted *in toto*. Am J Anat 1968; 122:237-248.

Clermont Y, Huckins C. The microscopic anatomy of the sex cords and seminiferous tubules in growing and adult rats. Am J Anat 1961; 108:79-98.

Clermont Y, Leblond CP. Renewal of spermatogonia in the rat. Am J Anat 1953; 93:475-501.

Clermont Y, Perey B. Quantitative study of the cell population of the seminiferous tubules in immature rats. Am J Anat 1957; 100:241-264.

Clermont Y, Rambourg A. Evolution of the endoplasmic reticulum during rat spermatogenesis. Am J Anat 1978; 151:191-212.

Clermont Y, Oko R, Hermo L. Cell biology of mammalian spermiogenesis In: (Desjardin C, Ewing LL, eds) Cell and Molecular Biology of the Testis. New York/Oxford: Oxford University Press, 1993; pp332-376.

Morales C, Clermont T. Structural changes of the Sertoli cell during the cycle of the seminiferous epithelium. In: (Russell LD, Griswold MD, eds) The Sertoli Cell. Clearwater, Fla: Cache River Press, 1993; pp305-330.

Morales CR, Oko R, Clermont Y. Molecular cloning and developmental expression of an mRNA encoding the 27 kDa outer dense fiber protein of rat spermatozoa. Mol Reprod Dev 1994; 37:229-240.

Oko R, Clermont Y. Isolation, structure and protein composition of the perforatorium of rat spermatozoa. Biol Reprod 1988; 39:673-687.

Oko R, Clermont Y. Immunocytochemical analysis of the formation of the fibrous sheath and outer dense fibers in the seminiferous epithelium of the rat. Anat Rec 1989; 225:46-55.

Oko R, Clermont Y. Mammalian spermatozoa: structure and assembly of the tail. In: (Gagnon C, ed) Control of Sperm Motility. Boca Raton, Fla: CRC Press, 1990, pp3-29.

Perey B, Clermont Y, Leblond CP. The wave of the seminiferous epithelium in the rat. Am J Anat 1961; 108:47-78.

Rambourg A, Clermont Y, Hermo L. Three-dimensional architecture of the Golgi apparatus in the Sertoli cell of the rat. Am J Anat 1979; 154:455-476.

Susi FR, Leblond CP, Clermont Y. Changes in the Golgi apparatus during spermatogenesis in the rat. Am J Anat 1971; 130:251-267.

Sutovsky P, Oko R, Hewitson L, Schatten G. The removal of the sperm perinuclear theca and its association with the bovine oocyte surface during fertilization. Dev Biol 1997; 188:75-84.

2 *Protamine Gene Expression: A Model for Post-Transcriptional Gene Regulation in Male Germ Cells*

Norman B. Hecht
University of Pennsylvania

During spermiogenesis, the haploid phase of spermatogenesis, the male gamete undergoes dramatic morphological and biochemical changes despite the cessation of RNA synthesis (Hecht 1995, 1998). In this time period, the protamines, the predominant arginine-rich proteins of the spermatozoan nucleus, are synthesized. In the mouse there are two protamines encoded by single copy genes on chromosome 16. Protamine 1, a molecule of 50 amino acids, contains three conserved domains, a central basic core and two less basic regions at the amino and carboxyl termini. Although the N-terminus and arginine-rich cores of protamine 1 are highly conserved among all mammals, the amino acid sequences of protamine 1 differ greatly among species. For instance, more than one-third of the amino acids of the mouse and human protamines 1 are different. There are even greater differences between protamines 1 and 2. Mouse protamine 2 is synthesized as a precursor protein of 106 amino acids which is sequentially processed to a mature form of 63 amino acids in the nucleus of late stage spermatids (Yelick et al. 1987). Protamine 1 is not synthesized as a precursor. Although both mouse protamines contain more than 50% arginine, their compositions differ substantially. For example, mouse protamine 2 contains 13 histidines, an amino acid absent in protamine 1. Human sperm also contain multiple protamines, one protamine 1 and two forms of protamine 2. The shorter human protamine 2 lacks the three amino acids of the N-terminus of the 57 amino acid human protamine 2. The reason for more than one protamine in sperm nuclei is unknown and the sperm of the rat and the bull only contain one protamine, protamine 1. The sequence variability of the protamines suggests DNA compaction can be accomplished by many different protein sequences and the mammalian protamines are rapidly evolving.

In this chapter, the protamine genes will be used as a model system to discuss gene regulation in mammalian germ cells. Much of this discussion will focus on post-transcriptional regulation because of the temporally separated periods of protamine transcription and translation. The delayed synthesis of the protamines during the late stages of spermatogenesis offers the investigator a superior system to study translational regulation in differentiating mammalian cells.

The Mouse Protamine 1 and 2 Genes Are Hypermethylated in Expressing Cells

In several mammals, protamine 1 and 2 and a third male germ-cell specific DNA-binding protein, transition protein 2, are clustered on the same chromosome. In the mouse, the protamines and transition protein 2 genes are linked on a DNA fragment of about 8 kb (Choi et al. 1997). Numerous repetitive sequences are also present in this locus of germ cell DNA-binding proteins. Although hypomethylation of DNA is often correlated with gene expression, this appears not to be true for the protamine genes in the mouse. In their expressing cell type, the round spermatid, the protamine 1 and 2 and transition protein 2 genes are located in a large region on chromosome 16 in which an increased pattern of methylation is associated with their transcription. Increased DNA methylation at a time of gene expression is not a general quirk of male germ cells, since other male germ cell specific proteins, such as phosphoglycerate kinase 2 and transition protein 1, show a cell type specific demethylation associated with gene activity. It is likely that the increased methylation of the region of chromosome 16 containing the protamine genes may result from the structural changes in chromatin occurring during spermiogenesis. The changes from the nucleosomal chromatin of earlier stages of germ cells to the form of chromatin in which transcriptional repression of gene expression occurs may require or be influenced by DNA methylation.

The Timing of Protamine Transcription

Protamine transcription is tightly regulated in mammals. The mouse protamine 1 and 2 genes are solely transcribed in round spermatids. This precise control appears to be regulated by a surprisingly small promoter locus. In fact, a mouse protamine 1 promoter of 113 nucleotides is sufficient to direct spermatid-specific expression of a reporter construct in transgenic mice (Zambrowicz et al. 1993). Since no established line of cultured male germ cells exists, investigators seeking to define cis-acting elements of promoters essential for gene expression in germ cells have not been able to utilize traditional transfection assays. To compensate, *in vitro* transcription assays are often used for the analysis of genes coding for proteins such as the angiotensin converting enzyme (Bernstein, Bernstein 1998). This approach has also successfully identified potential positive and negative regulatory regions of the protamine 2 promoter (Bunick et al. 1990a, b). In such studies, a novel testicular protein, protamine activating factor, and a germ cell Y-box protein have been shown to activate protamine 2 transcription in post-meiotic male germ cells (Yiu, Hecht 1997). A number of additional proteins have been reported to bind to the proximal promoters of protamine 1 and 2, including ubiquitous proteins (Zambrowicz, Palmiter 1994) and Tet-1, a putative testis-specific activation factor (Tamura et al. 1992). Gene targeting of the CREM tau gene has also demonstrated a role for the CREM tau protein in protamine 1 and 2 transcription (Blendy et al. 1996; Hummler et al. 1994).

Post-Transcriptional Regulation of the Protamines

Both of the protamine 1 and 2 nuclear transcripts contain an intron which is removed as the protamine mRNAs are processed and become mature functional mRNAs. Following their transcription and processing, the protamine mRNAs are stored in the cytoplasm of the differentiating spermatids for about one week before the protamine protein is synthesized. In contrast to the situation in oocytes where stored maternal mRNAs become polyadenylated upon activation, the stored non-translated protamine mRNAs contain poly (A) tails of about 160 adenines. Upon translation, the poly (A) tails of the protamine mRNAs shorten (Kleene 1989). This partial deadenylation is not specific to protamines 1 and 2, but has also been seen for transition protein 1, a mitochondrial capsule protein, and even a smooth muscle gamma actin. It appears to be specific for mRNAs that are initially transcribed in the post-meiotic germ cells and may represent a mechanism needed to activate stored "paternal" mRNAs in haploid male germ cells.

The delay in translation of protamine mRNA is necessitated by the need to transcribe mRNAs from the haploid genome before the nucleosomal histone-containing chromatin is replaced by transition proteins and transcription ceases. Probably because of the extreme basicity of the protamines, evolution has chosen to store protamine mRNA in the spermatid cytoplasm rather than store the protamine protein. Moreover, many other late stage spermatid and spermatozoal proteins show similar temporal separations of transcription and translation, suggesting a general need for mRNA storage in post-meiotic male germ cells.

Identification of a Trans Acting Factor That Binds Protamine mRNA

RNA-protein interactions play a prominent role in temporally regulating translation as well as in mRNA stability and localization. One protein, Testis Brain RNA-binding protein (TB-RBP) facilitates both translational regulation of stored mRNAs and their transport. TB-RBP specifically binds to highly conserved cis-acting sequences in a number of testicular and brain mRNAs (Kwon, Hecht 1991, 1993). TB-RBP appears to function in the cytoplasm through cytoskeletal associations since TB-RBP binds specific mRNAs to reconstituted microtubules from testis and brain extracts (Han et al. 1995). Both biochemical and immunocytochemical studies have demonstrated that TB-RBP is especially prominent in the cytoplasm of round spermatids, the cell type rich in stored mRNAs.

TB-RBP has been purified from the cytosol of mouse testicular extracts using RNA-affinity chromatography (Wu et al. 1997). Sequences of several peptides were determined from the purified protein TB-RBP, and using reverse genetics, a mouse testis cDNA library was screened with nucleic acid probes whose sequences were predicted from the TB-RBP peptides. Surprisingly, TB-RBP was found to be the mouse homolog of the human single-stranded DNA-binding protein, translin (Wu et al. 1998). Translin has been proposed to bind to specific breakpoint junctions of chromosomal translocations in many lymphoid malignancies (Aoki et al. 1995; Wu et al. 1997). Translin binds to the conserved single stranded DNA sequences, GCAGA [A/T] C and CCCA [C/G] GAC, sequences showing similarity to the RNA-binding recognition Y and H elements of TB-RBP (Kwon, Hecht 1993). Translin is present in both the cytoplasm and nucleus of many cell types and is often found in the nucleus of lymphoid cell lines.

In the testis, TB-RBP is also found in both the nucleus and cytoplasm. Recent immunolocalization studies have demonstrated that TB-RBP is abundant in the nuclei of meiotic pachytene spermatocytes and in the cytoplasm of round spermatids (Morales et al. 1998). This suggests a nuclear role for TB-RBP during meiosis consistent with its binding to consensus sequences of chromosomal breakpoints. The presence of TB-RBP in the cytoplasm of round spermatids argues for an additional function binding to stored mRNAs.

TB-RBP cDNAs encode an open reading frame of 228 amino acids (including the initiating methionine) which contains a leucine zipper in its C terminus (Wu et al. 1997; Fig. 1). TB-RBP has an estimated molecular weight of 26.2 kD. Although TB-RBP does not contain any common RNA binding motifs, TB-RBP contains a leucine zipper (from amino acids 177 to 212) and a transmembrane helix (from amino acids 93 to 114). Mouse TB-RBP shows a 90% identity in nucleotide sequence and 99% amino acid identity to human translin. The three amino acids that differ between translin and TB-RBP at amino acid 49 (alanine to threonine), amino acid 66 (glycine to serine) and amino acid 226 (valine to glycine) represent neutral changes. TB-RBP contains a group of putative consensus phosphorylation sites for protein kinase C (at amino acids 34, 67, 84, 190 and 213) and for tyrosine kinase (at amino acids 85, 200 and 213). TB-RBP also contains motifs with significant homology to other known proteins. Amino acids 9 to 35 and 14 to 47 of TB-RBP reveal a 62% and 55% similarity to regions of a human kinesin heavy chain. Amino acids 82 to 126 exhibit a 51% similarity to a sequence present in the human mitochondrial protein, cytochrome C oxidase polypeptide II.

Expression of TB-RBP

TB-RBP expression is developmentally regulated during male germ cell differentiation with maximal levels of TB-RBP mRNAs detected in pachytene

TB-RBP /TRANSLIN

M**S**VSEIFVELQGFLAAEQDIREEIKVVQSLEQ**T**AREITTLLQGVHQGTG
--A--

FQDIPKRCLKAREHFS**T**VKTHLT**S**LKTKFPAEQ**Y**YRFHEHWRFVLQRL
---------------G--------------------------------

VFLAAFVVYLETETLVTREAVTEILGIEPDREKGFHLDVEDYLSGVLILAS

ELSRLSVNSV**T**AGDYSRPLHISTFINELDSGFRLLNLKND**S**LRKRYDGL

K**Y**DVKKVEEVV**Y**DL**S**IRGFNKETAAACGEK
--------------------------V---

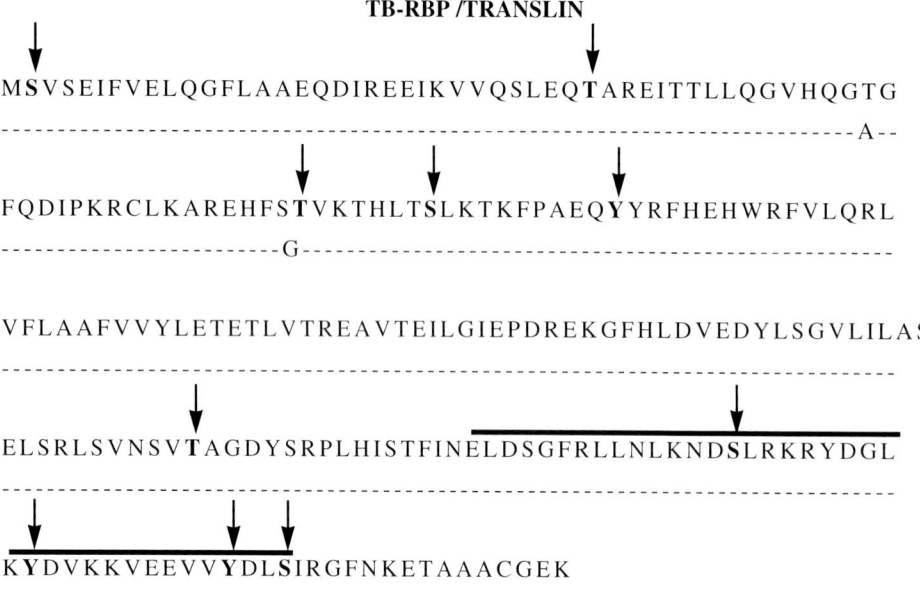

Figure 1. Sequence comparison of TB-RBP and translin. Upper sequence, TB-RBP; lower sequence, translin. Dashes denote identity. The leucine zipper is overlined and putative phosphorylation sites are indicated by arrows.

spermatocytes, a lower amount in round spermatids and a barely detectable amount of TB-RBP mRNA in elongated spermatids (Gu et al. 1998). The high amount of TB-RBP mRNA in meiotic cells precedes the high levels of TB-RBP RNA binding activity seen in the cytoplasm of round spermatids while the near absence of TB-RBP mRNA in elongating spermatids matches the lack of TB-RBP RNA-binding protein activity in late stages of spermatogenesis. Although it is possible that TB-RBP is primarily transcribed in pachytene spermatocytes and its translation is delayed to round spermatids, data indicate that the TB-RBP synthesized during meiosis has an additional role in the nuclei of primary spermatocytes. This is consistent with the single-stranded DNA binding property of TB-RBP/translin.

In the mouse, three TB-RBP mRNAs of about 1.2, 1.7 and 3.0 kb are expressed. The 1.2 kb TB-RBP mRNA is the major species in the testis and the 3.0 kb is the predominant TB-RBP mRNA in the brain. Northern blot analyses of mRNAs after RNase H digestion have revealed that the differences among the three testicular TB-RBP mRNAs reside in their 3´ UTRs, due to alternative polyadenylation site selection and/or splicing. The presence of several TB-RBP transcripts ranging from 1.2 to 3.2 kb in human testis, one major mRNA of about 3.0 kb in rat testis, and one transcript of about 2.8 kb in numerous lymphoid and non-lymphoid human cell lines indicates variability in post-transcriptional processing of the TB-RBP/translin mRNAs. It will be interesting to determine whether the multiple TB-RBP mRNAs have different translational efficiencies or stabilities and why the ratios of the multiple TB-RBP transcripts vary so greatly among tissues.

TB-RBP Functions as a Multimer *in Vitro* and *in Vivo*

To begin to understand the mechanism of action of TB-RBP, structural interactions of TB-RBP with RNA or DNA were investigated. Recombinant full-sized TB-RBP was synthesized from a cloned TB-RBP cDNA and used in *in vitro* protein-binding studies (Wu et al. 1998). TB-RBP dimers were shown to be the minimum structural unit needed for RNA- or DNA-binding (Wu et al. 1998). When one of the two cysteines of TB-RBP, cysteine 225, was replaced by alanine, the stability of the dimer was reduced, but there was not a reduction in DNA- or RNA-binding. Binding studies performed with more truncated forms of TB-RBP revealed that the leucine zipper motif in the C-terminus was essential for both dimer formation and for DNA- or RNA-binding. When the binding properties of recombinant TB-RBP or TB-RBP present in testicular extracts were examined, multimers (dimers, tetramers and octamers) of TB-RBP were detected binding to DNA or RNA. These findings support translin electron microscopic data where an eight-member ring structure of translin was shown to bind single-stranded DNA (Kasai et al. 1997).

To determine whether TB-RBP also functions as a dimer *in vivo,* a full length TB-RBP was used as bait to screen a mouse testis cDNA library in a yeast two-hybrid system assay. The first 10 positive clones isolated encoded TB-RBP, demonstrating that TB-RBP readily dimerizes *in vivo*. When truncated forms of the TB-RBP cDNA were used as bait in the yeast two-hybrid assay, the *in vitro* binding data were confirmed. The leucine zipper was shown to be critical for dimerization of TB-RBP *in vivo* or *in vitro*. These data indicate TB-RBP is likely to function in a complex of proteins. Preliminary immunoprecipitations of testicular and brain cell extracts with an affinity purified antibody to mouse TB-RBP indicate that TB-RBP specifically coprecipitates a number of proteins including an ATPase involved in cellular vesicle transport.

TB-RBP Functions in the Nucleus and Cytoplasm

This characterization of TB-RBP as an RNA-binding protein and the characterization of its human homolog, translin, as a single stranded DNA-binding protein initially appeared contradictory. Using RNA gel shifts as a bioassay for TB-RBP, TB-RBP RNA-binding activity was originally detected in extracts from testes and brain but not in extracts from spleen, kidney, liver, lung or heart (Han et al. 1995). While the studies characterizing the RNA-binding protein, TB-RBP, were in progress, its human homolog, translin, a single-stranded DNA-binding protein, was purified to near homogeneity from TCR β δ T cell leukemia DND-41 nuclear extracts (Aoki et al. 1995). Translin was detected in both the cytosol and nuclei of numerous lymphoid cells and in the cytoplasm of non-lymphoid cell lines.

To determine whether testicular TB-RBP could also function as a DNA-binding protein, biochemical and morphological approaches were taken. First, TB-RBP, isolated from enriched testicular nuclei or cytosol, was assayed for DNA- and RNA-binding activity. The nuclear TB-RBP preferentially bound DNA whereas the cytoplasmic TB-RBP preferred binding to RNA (Morales et al. 1998).

To relate the nucleic acid binding activities of TB-RBP to germ cell type, immunocytochemical studies were undertaken with affinity purified antibody to TB-RBP (Morales et al. 1998). At the light microscope level, TB-RBP is seen scattered in the cytoplasm of diploid pre-meiotic spermatogonia. As the germ cells enter meiosis, high amounts of TB-RBP are detected in the nuclei of the early- and mid-pachytene stage spermatocytes. In later stage primary spermatocytes, TB-RBP is found in both the nuclei and cytoplasm. In round spermatids, the majority of TB-RBP is found in the cytoplasm. Lower levels of TB-RBP are detected in early stages of elongated spermatids, but TB-RBP is not detected in late stage elongated spermatids. TB-RBP binds to RNA sequences that are similar to the single-stranded DNA breakpoint sequences. The RNA sequences are present in many testicular and brain mRNAs that are under translational control and/or are transported. The presence of TB-RBP in the nuclei of meiotic male germ cells argues for an additional function, most likely as a DNA-binding protein in the testis, consistent with the putative functions of translin in lymphoid cells.

When testes sections are examined by electron microscopy, the highest amounts of TB-RBP are detected scattered and in clusters in the nuclei of spermatocytes and in the cytoplasm of round spermatids. In spermatocytes, TB-RBP is primarily located over dense chromatin and is not associated with nucleoli. Although the majority of TB-RBP in round and early elongated spermatids is found in the cytoplasm, low but reproducible amounts of TB-RBP are found in the nuclei of early stage round spermatids. TB-RBP is generally scattered throughout the cytoplasm of the pre-meiotic spermatogonia and the post-meiotic spermatids and is not concentrated in the putative mRNA storage organelle, the chromatoid body, or in granular bodies or reticular bodies.

The abundance of TB-RBP in meiotic nuclei and in the cytoplasm of post-meiotic cells suggests that the intracellular distribution of TB-RBP changes as spermatogenesis proceeds. Clusters of TB-RBP are often seen in the center, periphery, and exiting from pachytene spermatocyte nuclei. In round spermatids, most of the TB-RBP is in the cytoplasm and TB-RBP is often detected in the intercellular bridges connecting germ cells. Thus, TB-RBP moves from the nucleus to the cytoplasm, and once in the cytoplasm it travels between the syncytially connected germ cells.

Biochemical and morphological studies suggest the following model of action for TB-RBP: TB-RBP begins to accumulate in germ cells in the cytoplasm of spermatogonia. During meiosis, TB-RBP enters the nuclei of pachytene spermatocytes, where it functions initially as a DNA-binding protein, perhaps in gene rearrangements as has been postulated for its human homolog, translin, in lymphoid cells. In pachytene spermatocyte nuclei, TB-RBP binds mRNAs which translocate from the nucleus to the cytoplasm. In the cytoplasm, TB-RBP is bound to "paternal" mRNAs, facilitating their storage, translational suppression, and transportation between cells. The transportation of mRNAs through the intercellular bridges between spermatids is likely under specific control to allow each spermatid to receive equal amounts of mRNA. Since colcemid and cytochalasin D release TB-RBP from its cellular associations and TB-RBP binds mRNAs to microtubules, it is likely that TB-RBP interacts with cytoskeletal binding sites such as microtubules and/or the actin bundles near the intercellular bridges. Preliminary data suggest TB-RBP is also present in the nuclei and dendritic processes of cells throughout the brain. The postulated model for TB-RBP function in the testis is presented in Figure 2.

Figure 2. Cellular locations and possible functions of TB-RBP in male germ cells. N= nucleus; C= cytoplasm.

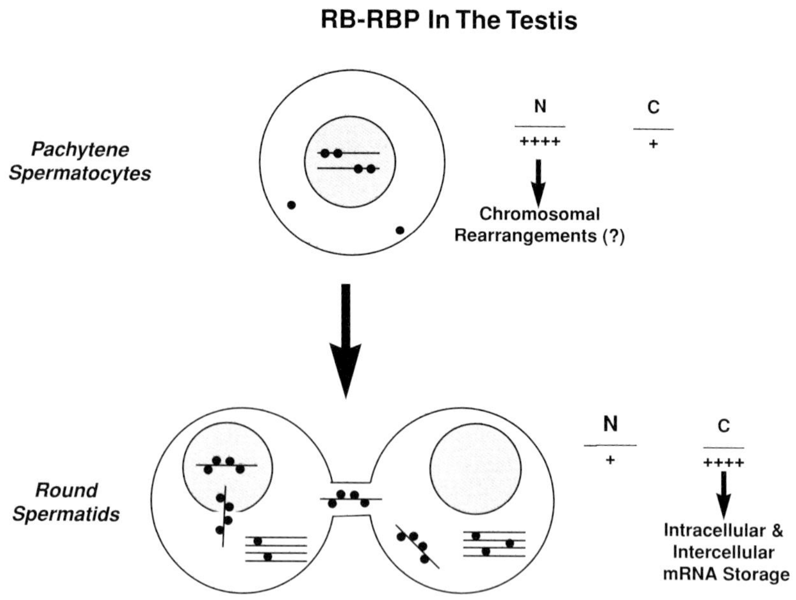

References

Aoki K, Suzuki K, Sugano T, Nakahara K, Kuge O, Omori A, Kasai MA. A novel gene, translin, encodes a recombination hotspot binding protein associated with chromosomal translocations. Nature Genet 1995; 10:167-174.

Bernstein KE, Bernstein E. The control of testis ACE expression. In: (Stefanini M, Boitani C, Galdieri M, Geremia R, Palombi F, eds) Testicular Function: From Gene Expression to Genetic Manipulation. Berlin: Springer, 1998, pp105-114.

Blendy JA, Kaestner KH, Schmid W, Gass P, Schutz G. Targeting of the CREB gene leads to up-regulation of a novel CREB mRNA isoform. EMBO J 1996; 15:1098-1106.

Bunick D, Balhorn R, Stanker LH, Hecht NB. Expression of the rat protamine 2 gene is suppressed at the level of transcription and translation. Exp Cell Res 1990a; 188:147-152.

Bunick D, Johnson PA, Johnson TR, Hecht NB. Transcription of the testis specific mouse protamine 2 gene in a homologous *in vitro* transcription system. Proc Natl Acad Sci USA 1990b; 87:891-895.

Choi Y-C, Aizawa A, Hecht NB. Genomic analysis of the mouse protamine 1, protamine 2, and transition protein 2 genes reveals hypermethylation in expressing cells. Mamm Genome 1997; 8:317-323.

Gu W, Wu X-Q, Meng X-H, Morales C, El-Alfy M, Hecht NB. The RNA- and DNA-binding protein TB-RBP is spatially and developmentally regulated during spermatogenesis. Mol Reprod Dev 1998; 49:219-228.

Han JR, Yiu GKC, Hecht NB. Testis brain-RNA binding protein (TB-RBP) is a microtubule associated protein that attaches translationally repressed and transported mRNAs to microtubules. Proc Natl Acad Sci USA 1995; 92:9550-9554.

Hecht NB. Molecular mechanisms of male germ cell differentiation. BioEssays 1998; 20:555-561.

Hecht NB. The making of a spermatozoon: a molecular perspective. Dev Genet 1995; 16:95-105.

Hummler E, Cole TJ, Blendy JA, Ganss R, Aguzzi A, Schmid W, Beermann F, Schutz G. Targeted mutation of the CREB gene: compensation within the CREB/ATF family of transcription factors. Proc Natl Acad Sci USA 1994; 91:5647-5651.

Kasai M, Matsuzaki T, Katayanagi K, Omori A, Maziarz RT, Strominger JL, Aoki K, Suzuki, K. The translin ring specifically recognizes DNA ends at recombination hot spots in the human genome. J Biol Chem 1997; 272:11402-11407.

Kleene KC. Poly (A) shortening accompanies the activation of translation of five mRNAs during spermatogenesis in the mouse. Development 1989; 106:367-373.

Kwon K, Hecht NB. Cytoplasmic protein binding to highly conserved sequences in the 3' untranslated region of mouse protamine 2 mRNA, a translationally regulated gene of male germ cells. Proc Natl Acad Sci USA 1991; 88:3584-3588.

Kwon K, Hecht NB. Binding of a phosphoprotein to the 3' untranslated region of the mouse protamine 2 mRNA temporally represses its translation. Mol Cell Biol 1993; 13:6547-6557.

Morales CR, Wu X-Q, Hecht NB. The DNA/RNA-binding protein, TB-RBP, moves from the nucleus to the cytoplasm and through intercellular bridges in male germ cells. Dev Biol 1998; 201:113-123.

Tamura T, Makino Y, Mikoshiba K, Muramatsu M. Demonstration of a testis-specific trans-acting factor Tet-1 *in vitro* that binds to the promoter of the mouse protamine 1 gene. J Biol Chem 1992; 267:4327-4332.

Wu X-Q, Gu W, Meng X-H, Hecht NB. The RNA-binding protein, TB-RBP, is the mouse homologue of translin, a recombination protein associated with chromosomal translocations. Proc Natl Acad Sci USA 1997; 94:5640-5645.

Wu X-Q, Xu L, Hecht NB. Dimerization of the Testis Brain RNA-binding protein (translin) is mediated through its C-terminus and is required for DNA- and RNA-binding. Nucleic Acids Res 1998; 26:1675-1680.

Yelick P, Balhorn R, Johnson PA, Cortzett M, Mazrimas JA, Kleene K, Hecht NB. Mouse protamine 2 is synthesized as a precursor whereas mouse protamine 1 is not. Mol Cell Biol 1987; 7:2173-2179.

Yiu GKC, Hecht NB. Novel testis-specific protein-DNA interactions activate transcription of the mouse protamine 2 gene during spermatogenesis. J Biol Chem 1997; 272:26926-26933.

Zambrowicz BP, Palmiter P. Testis-specific and ubiquitous proteins bind to functionally important regions of the mouse protamine 1 promoter. Biol Reprod 1994; 50:65-72.

Zambrowicz BP, Harendza CJ, Zimmermann JW, Brinster RL, Palmiter RD. Analysis of the mouse protamine 1 promoter in transgenic mice. Proc Natl Acad Sci USA 1993; 90:5071-5075.

3

Role of Distinct c-Kit Gene Products in Spermatogenesis and Fertilization

Pellegrino Rossi
Claudio Sette
Raffaele Geremia
University of Rome

The *W* locus in mice is allelic with c-kit, a transmembrane receptor with tyrosine kinase activity (Chabot et al. 1988; Geissler et al. 1988), whereas the *Sl* locus is allelic with stem cell factor (SCF), the physiological ligand of c-kit (Besmer 1991).

The c-Kit Transmembrane Tyrosine Kinase and Its Ligand, Stem Cell Factor

The c-kit tyrosine kinase (the SCF receptor) is a 150 kD protein, consisting of an immunoglobulin-like extracellular domain, a transmembrane domain and an intracellular tyrosine kinase domain which is split into an ATP-binding site and a phosphotransferase catalytic site (Fig. 1). The two domains of the kinase are divided by an intervening sequence, a characteristic shared by other members of a subfamily of receptor tyrosine kinases which are homologous to c-kit; that is, the PDGF receptor and the CSF1 receptor (Qiu et al. 1988). SCF (the c-kit ligand) can exist under both a soluble and a transmembrane form, encoded by distinct mRNAs, normally produced by alternative splicing (Flanagan et al. 1991).

After SCF binding, c-kit dimerization and activation of its tyrosine kinase activity occur, with consequent receptor autophosphorylation, which creates direct or indirect docking sites for signaling proteins containing Src-homology 2 (SH2) domains (Fig. 1), such as phospholipase C γ1 (PLCγ1), p120-Ras-GTPase activating protein (Rottapel et al. 1991; Lev et al. 1991; Herbst et al. 1991, 1995), the GRB2 adaptor protein (Tauchi et al. 1994), the p85 subunit of phosphatidylinositol 3' kinase (PI3K; Serve et al. 1994), Src-related tyrosine kinases (Price et al. 1997) and the SHP-1 and SYP tyrosine phosphatases (Kozlowski et al. 1998). However, some SH2-containing signaling proteins, such as the Tec tyrosine kinase, can also interact with c-kit through SH2-independent mechanisms in the absence of SCF stimulation (Tang et al. 1994).

Many different *W* and *Sl* alleles have been identified, which vary in their degree of severity in affecting the melanoblastic, hematopoietic and germ cell lineages. They are frequently associated to deletions or point mutations in c-kit (Nocka et al. 1990; Tan et al. 1990) or SCF coding sequences (Copeland et al. 1990; Flanagan et al. 1991; Brannan et al. 1991, 1992).

In some cases the phenotype is the consequence of rearrangements of the surrounding genomic region which control c-kit (Geissler et al. 1988) or SCF (Bedell et al. 1995) expression at the transcriptional or post-transcriptional level. In both *W* and *Sl* mice mutants, pigmentation defects are observed in both the heterozygous and the homozygous condition, whereas anemia and sterility are observed mostly in the homozygous condition, which is often lethal (Russell 1979). In Sl^{17h} homozygous mutants, only males are sterile (Brannan et al. 1992), whereas the Sl^{pan} and Sl^{con} mutations affect fertility only in homozygous females (Bedell et al. 1995). Both male and female W^{v} homozygous mutants are viable but sterile (Nocka et al. 1990). In the W^{42} mutants, impairment of fertility is evident even in the heterozygous condition (Tan et al. 1990). In *W* mutants, the defect is intrinsic to stem cells of the affected lineage, whereas in *Sl* mutants the defect

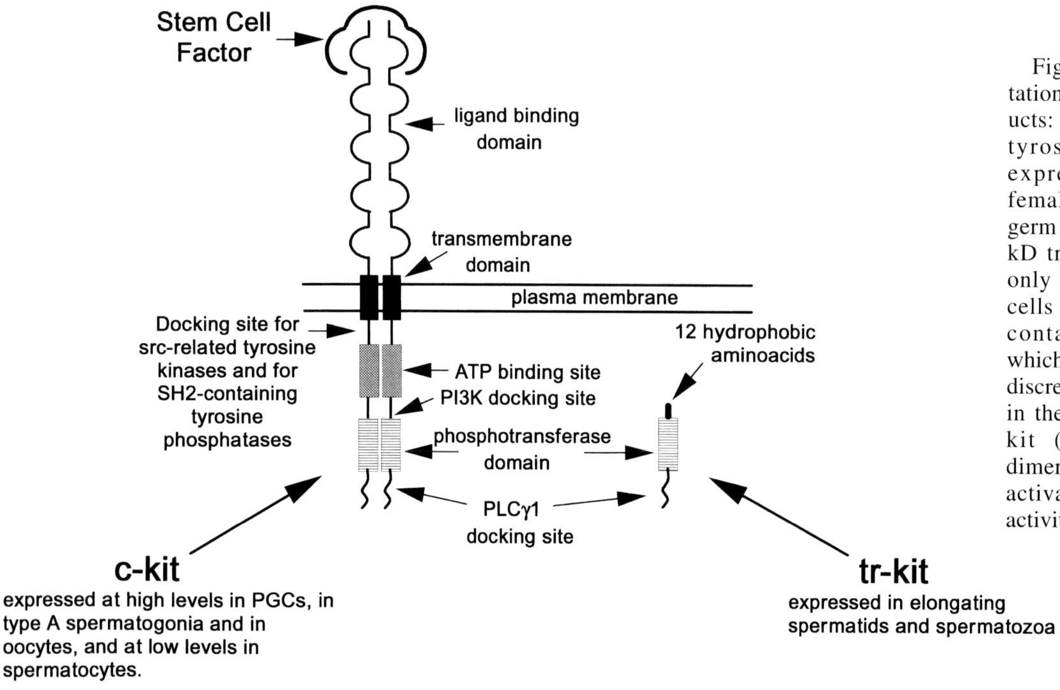

Figure 1: Schematic representation of two distinct gene products: the 150 kD transmembrane tyrosine kinase receptor expressed in both male and female pre-meiotic and meiotic germ cells (c-kit), and the 24-28 kD truncated protein expressed only in post-meiotic male germ cells (tr-kit). Some of the SH2-containing signaling proteins which are known to interact with discrete phosphotyrosine residues in the cytoplasmic domain of c-kit (following SCF-induced dimerization of the receptor and activation of its tyrosine kinase activity) are also indicated.

lies in the microenvironment where stem cells migrate, grow and differentiate (Russell 1979).

Role of SCF and of c-Kit in Survival and Proliferation of Mitotic Germ Cells

The reason underlying sterility in most *W* and *Sl* mutations is that in the homozygous condition an almost complete absence of germ cells is observed within both the postnatal testis and ovary (Coulombre, Russell 1954). Indeed, the c-kit mRNA is expressed in primordial germ cells (PGCs) and it has been shown that the transmembrane form of SCF produced by the surrounding somatic environment of the genital ridge is essential for survival of PGCs in the embryonal gonad (Dolci et al. 1991). These observations explain the sterile phenotype of Sl^{d} mutants, which produce only a soluble form of SCF (Flanagan et al. 1991; Brannan et al. 1991).

After birth, the mRNA encoding the c-kit transmembrane tyrosine kinase receptor is expressed at high levels in spermatogonia of the postnatal mouse testis, as observed by *in situ* hybridization (Manova et al. 1990) and Northern blot analysis (Sorrentino et al. 1991). The 150 kD c-kit protein is readily detectable by immunocytochemistry in mouse spermatogonia (Yoshinaga et al. 1991), and is localized in the cell membrane, as revealed by immunofluorescence analysis (Albanesi et al. 1996). The c-kit protein is also expressed in rat (Dym et al. 1995; Orth et al. 1996) and human (Natali et al. 1992) spermatogonia. SCF mRNA (Rossi et al. 1991) and protein (Tajima et al. 1991) are expressed by the surrounding Sertoli cells. The mRNAs encoding both the soluble and the transmembrane forms of SCF increase dramatically following *in vitro* stimulation of Sertoli cells with follicle stimulating hormone (FSH) or its intracellular mediator, cAMP (Rossi et al. 1991, 1993a). The increase following cAMP stimulation is more evident for the transcript encoding the soluble form of SCF and the ratio between mRNAs for the soluble and the transmembrane form of SCF increases during postnatal testicular development, suggesting that SCF is produced by postnatal Sertoli cells mainly as a soluble factor (Rossi et al. 1993a). Immunohistochemical analysis indicates that the ratio between these two forms of SCF varies in different stages of the seminiferous epithelium (Vincent et al. 1998), probably as a consequence of the different sensitivity of Sertoli cells to FSH stimulation. Addition of the recombinant soluble form of SCF to highly enriched cultures of postnatal germ cells at the mitotic stages stimulates DNA synthesis selectively in type A spermatogonia (Rossi et al. 1993a), in agreement with the presence of a functional c-kit receptor mainly in these cells. The transmembrane form of SCF has been proposed as a survival factor rather than as a proliferative agent for spermatogonia (Packer et al. 1995), a role consistent with its effects on PGCs before birth (Dolci et al. 1991). Furthermore, it has been shown to be important for adhesion of postnatal spermatogonia to cultured Sertoli cell monolayers (Marziali et al. 1993). Researchers in the Rossi laboratory found that the soluble form of SCF also significantly suppresses apoptosis in cultured spermatogonial cells (S Dolci, P Rossi and R Geremia, unpublished data). Possibly, the two forms of the c-kit ligand could have different roles in their interaction with mitotic germ cells, with the transmembrane form being mainly a factor involved in their survival and/or migration and adhesion and the soluble form being an inducer of proliferation during specific stages of spermatogenesis. Further evidence that the c-kit/SCF system plays an important role in mitotic stages of spermatogenesis after birth is given by the observation that injection of antibodies directed against the extracellular domain of c-kit selectively depletes seminiferous tubules from type A spermatogonia (Yoshinaga et al. 1991) and that spermatogenesis can be re-established in *W* mutant mice after transplantation, within the seminiferous tubules, of type A spermatogonia from normal animals (Brinster, Avarbock 1994). FSH induction of the soluble form of SCF in Sertoli cells and stimulation of DNA synthesis in type A spermatogonia by the soluble form of the factor (Rossi et al. 1993a), represent, up to now, the only examples of a link between hormonal stimulation of testicular somatic cells and release by these cells of a growth factor which directly promotes germ cell proliferation. Recently it has been shown that artificial induction of testicular atrophy in rats is accompanied by alterations of the SCF expression pattern and that a gonadotropin-releasing hormone agonist normalizes SCF expression and re-establishes normal spermatogenesis (Blanchard et al. 1998).

Possible Role of c-Kit in Oocyte Maturation

Expression of c-kit mRNA and protein also occur in postnatal oocytes throughout all their maturative stages (Manova et al. 1990; Horie et al. 1991; Yoshinaga et al. 1991). SCF is produced by the surrounding granulosa cells (Manova et al. 1993), is under gonadotropin control (Motro, Bernstein 1993) and has been proposed as one of the factors which promotes oocyte growth and maturation (Manova et al. 1993; Packer et al. 1994). It has also been proposed that either the soluble or the transmembrane form of SCF might be involved in the negative control of meiotic progression of oocytes. Indeed, addition of soluble SCF appears to delay spontaneous germinal vesicle breakdown (GVBD) which is observed *in vitro* in fully grown oocytes; this effect is blocked by anti-c-kit antibodies (Ismail et al. 1996). Moreover, microinjection of antisense c-kit oligonucleotides decreases c-kit expression in oocytes and increases their ability to resume meiosis *in vitro* (Ismail et al. 1997). Stimulation with luteinizing hormone (Motro, Bernstein 1993) or the combined treatment with pregnant mare serum gonadotropin and human chorionic gonadotropin (hCG; Ismail et al. 1996) promotes a dramatic increase in SCF expression in granulosa cells. An increase in the ratio between the transcripts encoding the soluble and the transmembrane form of SCF occurs in follicular granulosa cells after hCG stimulation (Ismail et al. 1997), but a complete loss of expression of both forms of SCF occurs in cumulus cells immediately surrounding the fully grown oocyte (Motro, Bernstein 1993; Ismail et al. 1997). These observations suggest that lack of SCF expression in cumulus cells surrounding the oocyte might allow meiotic resumption and GVBD in cumulus-enclosed fully grown oocytes *in vivo* just before ovulation. Expression of c-kit continues also in ovulated oocytes (Manova et al. 1990; Horie et al. 1991; Yoshinaga et al. 1991), which are arrested in the metaphase of the second meiotic division due to the action of the c-mos proto-oncogene (Colledge et al. 1994; Hashimoto et al. 1994), but the function of the SCF receptor at this stage of development is unknown.

Expression of an Alternative Form of c-Kit (tr-Kit) and Its Localization in the Residual Cytoplasm of Spermatozoa

In the meiotic stages of spermatogenesis, expression of the mRNA encoding 150 kD c-kit transmembrane receptor is strongly reduced, and it ceases completely in the subsequent haploid stages (Sorrentino et al. 1991). However, during spermiogenesis, an alternative shorter c-kit transcript is expressed (Sorrentino et al. 1991), but this transcript encodes a truncated form of the cytoplasmic portion of the receptor. In this c-kit isoform, the extracellular and transmembrane domains, the first box of the split kinase domain (that is, the ATP binding site), and the interkinase domain are missing (Fig. 1). Only the carboxyterminal portion of the c-kit open reading frame (ORF), encoding the second box of the split kinase (that is, the phosphotransferase catalytic domain) and the cytoplasmic tail, is present (Rossi et al. 1992). Immunoblot analysis with a polyclonal antibody directed against the c-kit carboxyterminal portion shows that a 24-28 kD protein, corresponding to the size predicted from the ORF of the alternative c-kit transcript, can be readily identified in elongating spermatids at stages 9-16 of spermiogenesis (Albanesi et al. 1996). The truncated c-kit protein (tr-kit) can be detected at low levels also in round spermatids (stages 1-8 of spermiogenesis), but it is absent both in spermatogonial cells and in spermatocytes.

The 5' end of the alternative c-kit transcript encoding tr-kit maps within the sixteenth intron of the c-kit gene. This intron separates the exon encoding the interkinase domain from the first of the exons encoding the phosphotransferase domain. Twelve novel hydrophobic residues are present at the N-terminus of tr-kit, and they are encoded by intronic sequences in frame with the 190 carboxyterminal residues of the c-kit ORF (Rossi et al. 1992). *In vitro* transcription experiments, using nuclear extracts from round spermatids, suggested that the alternative c-kit transcript during spermiogenesis is generated by a cryptic promoter present within this intron (Albanesi et al. 1996). Moreover, DNA-binding experiments showed that nuclear factors present in extracts from spermatids, but not from spermatocytes or Sertoli cells, bind discrete sequences within the c-kit intron. In order to confirm *in vivo* the activity of the intronic promoter, researchers in the Rossi laboratory have generated transgenic mice in which about 1 Kb of these sequences are linked to the reporter *E. Coli lac-Z* gene. In the transgenic offspring of five out of six separate founders, the reporter gene is specifically expressed only within seminiferous tubules, within haploid germ cells starting from step 8 of spermiogenesis (Albanesi et al. 1996). No expression was found in other organs and cell types (such as brain, liver, kidney, spleen, heart, ovary and oocytes). These data indicate that the intronic promoter is only active in the latest transcriptional stages of the haploid phase, and are in perfect agreement with the finding that the tr-kit protein is mainly accumulated during the elongation steps of spermiogenesis.

Western blot analysis indicated that tr-kit is also present in epidydimal spermatozoa (Albanesi et al. 1996). Immunofluorescence analysis showed that it has a cytoplasmic localization in elongating spermatids and is also present in residual bodies (Albanesi et al. 1996), whereas in the mature sperm, the intracellular distribution of the tr-kit protein is mainly restricted to the midpiece of the flagellum (Sette et al. 1997), which, together with the mitochondrial sheath, contains most of the residual sperm cytoplasm. A weaker positivity to the c-kit antibody could also be detected in the post-acrosomal region at the base of the sperm head, but not in the acrosomal region, the connecting piece, the principal piece of the tail or the terminal tail segment. In immature sperm, tr-kit was mainly accumulated in the cytoplasmic droplet of the midpiece. The presence of tr-kit in the residual bodies of the spermatids and in the midpiece of mature sperm indicates that this protein essentially accumulates in the residual sperm cytoplasm.

Recently, on the basis of immunohistochemical data on isolated spermatozoa obtained with a monoclonal antibody directed against the extracellular domain of c-kit (ACK2), it has also been reported that the c-kit transmembrane protein (or its extracellular portion) might be present in the male gamete and localized in the acrosome (Feng et al. 1997). In support of these data, it has been reported that addition of SCF to spermatozoa *in vitro* stimulates the acrosome reaction (Feng et al. 1998). However, expression of the full-length c-kit transmembrane protein in mature spermatozoa is unlikely, since 1) its coding mRNA is completely absent in the haploid stages of spermatogenesis, as revealed by both Northern blot analysis (Sorrentino et al. 1991; Rossi et al. 1992) and by the much more sensitive RT-PCR technique (Rossi et al. 1993b); 2) the localization of the c-kit transmembrane protein in spermatids and spermatozoa has not been observed by immunohistochemical methods with the ACK2 antibody on testicular sections (Yoshinaga et al. 1991); 3) the 150 kD receptor is not detectable in immunoblots of elongated spermatids or mature spermatozoa using a polyclonal antibody directed against the carboxyterminal portion of the c-kit protein (Albanesi et al. 1996); and 4) immunofluorescence analysis with this antibody does not detect any specific staining in the acrosome of mature sperm (Sette et al. 1997).

Microinjection of Recombinant tr-Kit Parthenogenetically Activates Mouse Eggs and Mimics Egg Activation

Sperm cytosolic factors are released into the egg cytoplasm after sperm-egg fusion and are supposed to trigger the series of events culminating in cell cycle resumption and first mitotic division of the zygote (Swann 1990; Sticc, Robl 1990; Homa, Swann 1994; Dozortsev et al. 1995; Wu et al. 1997; Stricker 1997). A series of Ca^{2+} oscillations is the first event triggered by the sperm at fertilization (Whitaker, Swann 1993), and the increase in intracellular Ca^{2+} is required for at least several of the subsequent events of egg activation (Kline, Kline 1992). Sperm-egg fusion precedes the onset of these Ca^{2+} oscillations in fertilized mouse eggs (Lawrence et al. 1997), suggesting that a factor released by the sperm is responsible for the fertilization-associated Ca^{2+} mobilization. Oscillin, a glucosamine 6-phosphate deaminase localized in the equatorial segment of the hamster sperm head, has been proposed as a possible candidate for such a factor, based on its co-purification with sperm oscillogenic activity (Parrington et al. 1996). However, neither recombinant nor highly purified oscillin have oscillogenic activity and they fail to trigger egg activation, even though they maintain glucosamine 6-phosphate deaminase activity, suggesting that a different protein of the sperm is required for triggering egg activation (Wolosker et al. 1998).

It is known that together with the sperm head, which provides paternal chromosomes, the sperm midpiece also penetrates the mammalian egg and supplies centrosome components responsible for aster formation after fertilization (Yanagimachi 1994). Therefore, the localization of tr-kit in the midpiece is compatible with its entry into the egg and with a possible action inside the oocyte after sperm-egg fusion. The potential action of tr-kit in the egg cytoplasm was investigated by microinjection of a recombinant tr-kit protein expressed in COS cells into mouse oocytes arrested in metaphase II. Microinjection of extracts from COS cells expressing tr-kit reproducibly causes complete parthenogenetic activation of the oocytes (Sette et al. 1997). Activation of MII oocytes triggered by extracts from tr-kit expressing transfected COS cells is coupled to exocytosis of cortical granules and to emission of the second polar body (indicating MII-anaphase transition and completion of the second meiotic division). Activated oocytes also proceed through the first mitotic cycle. Pronuclei appear between four and seven hours after microinjection, a time similar to that observed during natural fertilization (Mori et al. 1988), and most of the activated eggs reach the two-blastomere stage after 24 hours in culture. Furthermore, many of the activated eggs develop to morula stages when cultured for two to four days.

Microinjection of the extracts from mock-transfected COS cells does not induce any of the activation events triggered by tr-kit, indicating that the presence of the truncated c-kit protein in extracts from transfected COS cells is necessary for egg activation (Sette et al. 1997). When the amount of recombinant tr-kit required to activate mouse eggs was compared with the amount carried by a single mouse sperm, it was found that injection of less than one sperm equivalent of recombinant tr-kit is sufficient to exert activation of mouse eggs, suggesting that it can play a physiological role at fertilization (Sette et al. 1998). To determine whether tr-kit is also sufficient to induce the set of events observed after its microinjection into MII oocytes, tr-kit ORF RNA was synthesized *in vitro* and similar microinjection experiments were performed. Synthetic tr-kit RNA reproducibly induced pronuclear formation in the oocytes between four and seven hours after microinjection, whereas microinjection of control RNAs did not trigger activation above background levels observed in non-injected oocytes (Sette et al. 1997). Again, cortical granule exocytosis and second meiotic division preceded pronuclear formation. It can be concluded from this that the microinjected tr-kit protein is directly responsible for MII arrested oocyte activation and that post-translational modifications possibly achieved in a eukaryotic expression environment can be achieved in the MII oocyte cytoplasm after translation of tr-kit RNA.

Intracellular Ca^{2+} mobilization is associated with

sperm-induced egg activation at fertilization and is responsible for the onset of cortical granule exocytosis (Yanagimachi 1994) and for the destruction of the cyclinB/cdc2 complex (the histone-H1 kinase activity known as maturation promoting factor, or MPF) necessary for the completion of meiosis (Lorca et al. 1993). The possibility that tr-kit induced egg activation was also dependent, at least in part, on oocyte intracellular Ca^{2+} ions was investigated. Intracellular Ca^{2+} was chelated by preincubation with the membrane permeable compound BAPTA-AM before microinjection of extracts from COS cells expressing tr-kit. BAPTA-AM completely blocked both second polar body extrusion and pronuclear formation, indicating that egg activation by tr-kit requires calcium ions and that it could involve intracellular Ca^{2+} mobilization (Sette et al. 1997). This last hypothesis is also supported by the evidence that tr-kit is able to trigger cortical granule exocytosis with a pattern similar to that observed in natural fertilization (Sette et al. 1997). MAP kinase activity is stimulated by the c-mos serine-threonine kinase, and both these kinases are involved in prevention of destruction of MPF activity and are essential components of the cytostatic factor (CSF) which prevents spontaneous exit from metaphase II of vertebrate eggs (Haccard et al. 1993; Colledge et al. 1994; Hashimoto et al. 1994; Verlhac et al. 1996). Inhibition of egg MAP kinase activity is observed at fertilization in mouse eggs and is essential for formation of a female pronucleus (Moos et al. 1995). It was found that tr-kit RNA microinjection in MII oocytes is associated with a dramatic decrease in MAP kinase activity coupled to a shift to a slightly higher electrophoretic mobility of p42 MAP kinase (ERK2), due to inactivating dephosphorylation (Sette et al. 1997). Overall, these data demonstrate that tr-kit microinjection into mouse eggs faithfully reproduces many events associated with egg activation occurring after sperm-egg fusion in natural fertilization.

Tr-Kit-Induced Egg Activation is Mediated by Activation of Phospholipase Cγ1

Phospholipase C (PLC) inhibitors suppress Ca^{2+} oscillations at fertilization in mouse eggs (Dupont et al. 1996), suggesting that sperm-egg fusion induces the activation of a PLC isoform, which, in turn, is responsible for inositol 1,4,5 trisphosphate ($InsP_3$) production and consequent Ca^{2+} release from the intracellular stores (Berridge 1993). To investigate whether tr-kit action inside the egg also involves PLC activation, the effect of preincubation of mouse eggs with a specific PLC inhibitor before microinjection experiments was tested. Indeed, tr-kit-induced pronuclear formation was completely suppressed by incubation of eggs with the PLC inhibitor U73122, but not by incubation with its inactive analog U73433 (Sette et al. 1997). These results indicate that egg PLC activity is essential for tr-kit-mediated MII oocyte activation and suggest that tr-kit mediated Ca^{2+} mobilization inside the egg is mediated through PLC activation. This observation reinforces the hypothesis that sperm-derived tr-kit could play a physiological role in natural egg activation at fertilization.

Researchers in the Rossi laboratory reasoned that PLCγ1 is the most likely candidate isoform required for sperm-carried tr-kit action within the egg cytoplasm, since 1) PLCγ1 is expressed in ovulated mouse oocytes (Dupont et al. 1996); 2) PLCγ1 is activated by interaction with tyrosine kinase receptors or with cytoplasmic tyrosine kinases (Lee, Rhee 1995; Kamat, Carpenter 1997; Rhee, Bae 1997); 3) PLCγ1 interacts with c-kit after its stimulation by SCF, the physiological ligand, in several cellular systems (Rottapel et al. 1991; Herbst et al. 1991; Lev et al. 1991); and 4) c-kit-PLCγ1 physical interaction requires a tyrosine residue in the c-kit carboxyterminal portion (Fig. 1; Herbst et al. 1995), which is also present in tr-kit (Rossi et al. 1992).

The hypothesis that PLCγ1 mediates tr-kit effects on egg activation was tested by co-injection into mouse eggs of cell extracts from COS cells transiently expressing a recombinant tr-kit protein, together with glutathione-S-transferase (GST)-fusion proteins containing the Src-homology (SH) region of PLCγ1. The SH region is known to mediate interaction of PLCγ1 with upstream and/or downstream effectors (Lee, Rhee 1995; Kamat, Carpenter 1997; Rhee, Bae 1997). Therefore, researchers in the Rossi laboratory expected that, if PLCγ1 is the PLC isoform involved in tr-kit-induced egg activation, injection of the SH region of PLCγ1 should trigger competition for binding of effectors to the endogenous PLCγ1, thus blocking its enzymatic activity and the consequent egg activation. It was found that co-injection of GST-fusion proteins containing the SH domains of PLCγ1 specifically inhibits tr-kit-induced egg activation (Sette et al. 1998). A GST fusion protein containing only the SH3 domain of PLCγ1 inhibited egg activation as efficiently as the whole SH region, while a GST fusion protein containing the two SH2 domains only slightly competed with tr-kit action in the eggs. SH3 domain inhibition of tr-kit was specific for PLCγ1, since a GST fusion protein containing the Grb2 adaptor protein SH3 domain was ineffective (Sette et al. 1998). Tr-kit-induced egg activation was also suppressed by co-injection of antibodies raised against the PLCγ1 SH domains, but not against the PLCγ1 carboxyterminal domain nor by co-injection of unrelated antibodies. Biochemical experiments in transfected COS cells showed that co-expression of PLCγ1 and tr-kit in these cells significantly increased both diacylglycerol (DAG, a powerful stimulator of some PKC isoforms) and inositol phosphate production with respect to cells expressing PLCγ1 or tr-kit alone. Moreover, co-expression of tr-kit strongly increases tyrosine phosphorylation of PLCγ1 (Sette et al. 1998), which is considered a hallmark of activation of this enzyme (Kim et al. 1991). These data indicate that tr-kit is able to modulate the activity of PLCγ1 and that the SH3 domain of this enzyme is essential for tr-kit-induced activation of mouse eggs.

Recently, it has been reported that activation of PLCγ1 is required for the sperm-induced Ca^{2+} rise observed in starfish eggs at fertilization, and that a GST-PLCγ1-SH2 fusion protein inhibits sperm-induced activation of starfish eggs (Carroll et al. 1997). These observations further support the hypothesis that tr-kit is a sperm factor which plays a physiological role in mammalian egg activation at fertilization and they show that SH-mediated activation of PLCγ1 is an evolutionary conserved route of egg activation. In contrast with the results obtained in starfish eggs, a GST-PLCγ1-SH3 fusion protein is much more effective than a GST-PLCγ1-SH2 fusion protein in inhibiting tr-kit action in mouse eggs. These results were unexpected, since an essential step for tyrosine phosphorylation, translocation, and activation of PLCγ1 is thought to be the interaction of the SH2 domains of PLCγ1 with phosphotyrosine residues present in activated tyrosine kinases (Lee, Rhee 1995; Kamat, Carpenter 1997; Rhee, Bae 1997). However, it must be considered that in addition to tyrosine phosphorylation, other mechanisms of PLCγ1 activation exist and are necessary in some cases for PLCγ1 to fulfill its function. Translocation of PLCγ1 to the membrane and/or the cytoskeleton might bring the enzyme in close proximity to other agents which have been shown to stimulate its hydrolytic activity even in the absence of tyrosine phosphorylation (Jones, Carpenter 1993; Hwang et al. 1996; Bae et al. 1998; Falasca et al. 1998). The SH3 domain of PLCγ1 directs the enzyme to the particulate compartment, via binding at proline-rich sequences of cytoskeletal proteins (Bar-Sagi et al. 1993), thus bringing PLCγ1 in close proximity to its substrate phosphatidylinositol 4,5 bisphosphate (PIP_2). According to this model, the finding that the SH3 domain of PLCγ1 is required for tr-kit action inside the egg cytoplasm suggests that tr-kit triggers activation of PLCγ1 by allowing either the translocation of the enzyme to the particulate compartment or its interaction with effector proteins in this compartment via the SH3 domain. It is also possible that binding of proteins to the SH3 domain is able to derepress PLCγ1 enzyme activity, as shown in the case of src-related tyrosine kinases (Moarefi et al. 1997). Although tr-kit lacks an ATP binding site and should not have tyrosine kinase activity (Rossi et al. 1992), the possibility exists that tr-kit interacts with other tyrosine kinases present in the egg, which in turn phosphorylate tr-kit itself or other proteins mediating PLCγ1 tyrosine phosphorylation and activation. For instance, as schematically proposed in Figure 2, tyrosine phosphorylation of tr-kit by a kinase present in the egg cytoplasm might create docking sites for the SH2 domain of intercalated adaptor proteins containing proline-rich targets, which in turn may stimulate tyrosine phosphorylation and activation of PLCγ1 by interacting with its SH3 domain (Cohen et al. 1995; Pawson 1995). Further experiments are required to identify proteins which mediate the interaction between tr-kit and PLCγ1 inside the egg cytoplasm.

Conclusion

The results discussed earlier indicate that two distinct c-kit gene products play different roles during gametogenesis (Fig. 3). The 150 kD c-kit transmembrane tyrosine kinase receptor and its ligand SCF are essential for survival of PGCs in the embryonal gonad, for proliferation and survival of postnatal spermatogonia, and for regulating maturation of postnatal oocytes. The 24-28 kD truncated c-kit gene product (tr-kit), which is selectively expressed during spermiogenesis and is present in mature spermatozoa, might be important for the final function of germ cells at fertilization.

Several observations make tr-kit a candidate as a cytosolic sperm factor responsible for oocyte activation after sperm-egg fusion: 1) the subcellular localization of tr-kit in mature spermatozoa is compatible with its transfer into the egg cytoplasm; 2) microinjection of less than one sperm equivalent of tr-kit is able to parthenogenetically activate mouse eggs; 3) tr-kit elicits a decrease in egg MAP kinase activity, which is the c-mos dependent component of CSF responsible for the metaphase II arrest of ovulated eggs; 4) tr-kit-induced egg activation is suppressed by preincubation with either chelators of intracellular Ca^{2+} or with specific PLC inhibitors which also block sperm-induced Ca^{2+} spiking at fertilization in mouse eggs; 5) tr-kit-induced egg activation is mediated by stimulation of PLCγ1, and a PLCγ1 isoform has been shown to be responsible for sperm-induced Ca^{2+} mobilization at fertilization in echinoderm eggs; 6) DAG (Colonna et al. 1997; Gallicano et al. 1997) and $InsP_3$ (Miyazaki et al. 1992, 1993; Xu et al. 1994; Berridge 1996) are essential for activation of mammalian eggs at fertilization, and both are produced through PIP_2 hydrolysis consequent to PLCγ1 activation; and 7) tr-kit is accumulated during the cell elongation steps of spermiogenesis and the observation that intracytoplasmic injection of round spermatids fails to activate mouse eggs without additional artificial stimulation, whereas testicular spermatozoa are successful (Kimura, Yanagimachi 1995), indicates that an oocyte activating factor must be

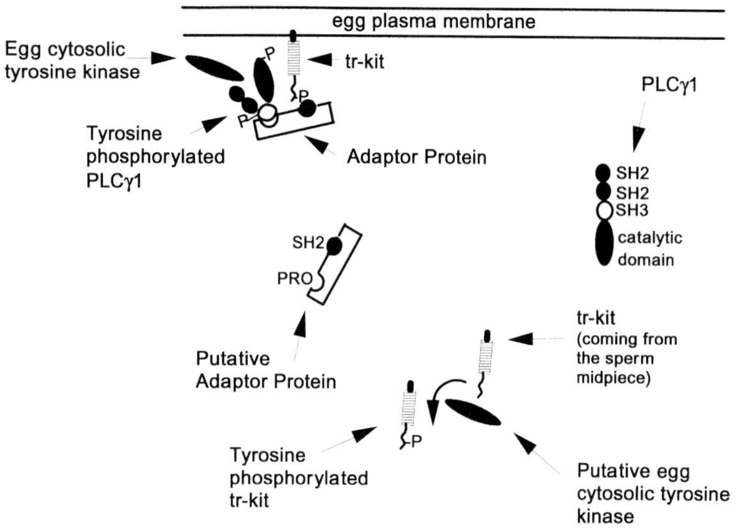

Figure 2: Model for possible interaction between sperm-carried tr-kit and PLCγ1 inside the egg cytoplasm. A hypothetical cytosolic tyrosine kinase is activated by tr-kit. The consequent phosphorylation of tr-kit creates docking sites for a SH2-containing adaptor protein, which, in turn, interacts through proline-rich domains (PRO) with the SH3 domain of PLCγ1. The result is tyrosine phosphorylation and activation of PLCγ1, with stimulation of PIP_2 hydrolysis, and consequent egg activation.

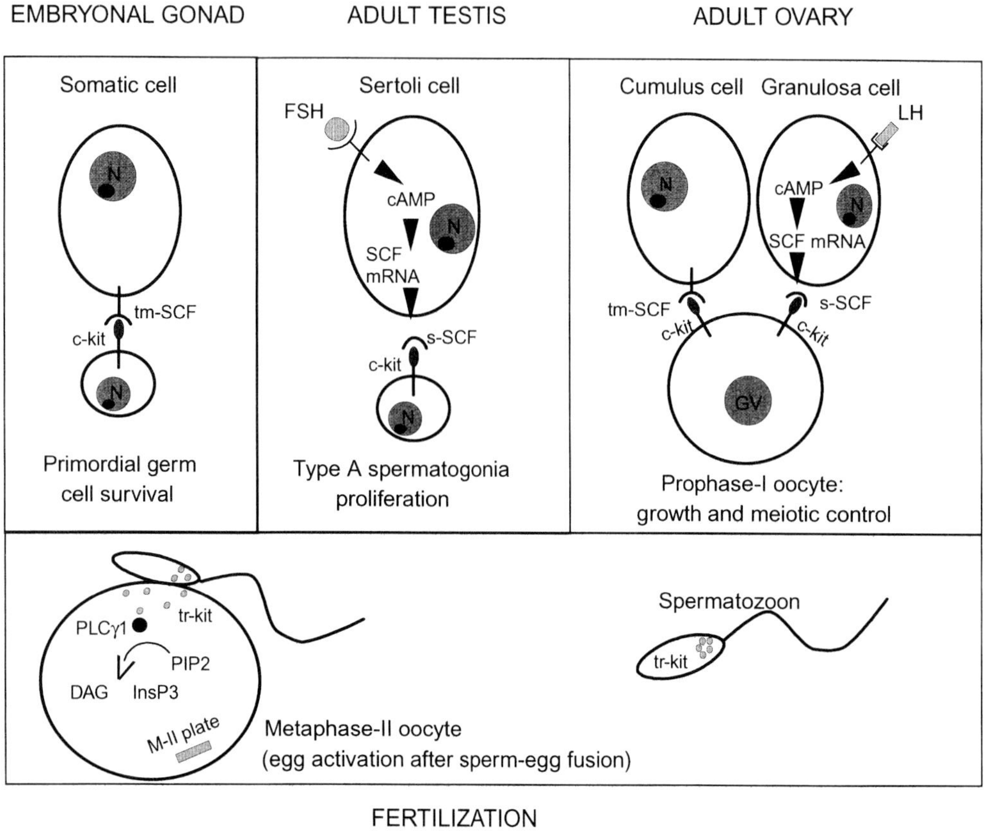

Figure 3: The two distinct c-kit gene products depicted in Figure 1 play different roles during gametogenesis. Activation of c-kit by either transmembrane SCF (tm-SCF) or soluble SCF (s-SCF), produced by the surrounding somatic cells under gonadotropin control, triggers different specific responses in mitotic and meiotic germ cells. After sperm-egg fusion, transfer of sperm-carried tr-kit into metaphase II-arrested oocytes triggers egg activation at fertilization. N= nucleus; GV= germinal vesicle; M-II plate= chromosome aligned on the metaphase-II plate before the second meiotic division.

accumulated during these stages of spermatogenesis.

The evidence that tr-kit actually functions during natural egg activation could come from studies conducted with genetically modified mice. Researchers in the Rossi laboratory are constructing targeting vectors to perform the genetic ablation (Capecchi 1989) of the intronic promoter sequences which drive tr-kit specific expression during spermiogenesis. This approach might reveal whether tr-kit plays a role also in the differentiative events of mouse spermiogenesis. For instance, considering that tr-kit is able to awaken metaphase-II arrested oocytes, it is conceivable that its expression during the haploid stages of spermatogenesis might be responsible for the rapid transition to round spermatids of secondary spermatocytes going through the second meiotic division.

Acknowledgements

The authors thank Drs. Vincenzo Sorrentino, Cristina Albanesi, Susanna Dolci, Laura Pozzi, Paola Grimaldi, Arturo Bevilacqua and Franco Mangia, who contributed to the work from the authors' laboratory described in this chapter. This work was supported by World Health Organization special project for Research Development and Research Training in Human Reproduction, by CNR strategic project "Cell Cycle and Apoptosis," by MURST and by grants from ASI.

References

Albanesi C, Geremia R, Giorgio M, Dolci S, Sette C, Rossi P. A cell- and developmental stage-specific promoter drives the expression of a truncated c-kit protein during mouse spermatid elongation. Development 1996; 122:1291-1302.

Bae YS, Cantley LG, Chen CS, Kim SR, Kwon KS, Rhee SG. Activation of phospholipase C-γ by phosphatydilinositol 3,4,5-trisphosphate. J Biol Chem 1998; 273:4465-4469.

Bar-Sagi D, Rotin D, Batzer A, Mandiyan V, Schlessinger J. SH3 domains direct cellular localization of signaling molecules. Cell 1993; 74:83-91.

Bedell MA, Brannan CI, Evans EP, Copeland NG, Jenkins NA, Donovan PJ. DNA rearrangements located over 100 kb 5' of the Steel (Sl)-coding region in Steel-panda and Steel-contrasted mice deregulate Sl expression and cause female sterility by disrupting ovarian follicle development. Genes Dev 1995; 9:455-470.

Berridge MJ. Inositol triphosphate and calcium signaling. Nature 1993; 361:315-325.

Berridge MJ. Regulation of calcium spiking in mammalian oocytes through a combination of inositol triphosphate-dependent entry and release. Mol Hum Reprod 1996; 2:386-388.

Besmer P. The kit ligand encoded at the murine Steel locus: a pleiotropic growth and differentiation factor. Curr Opin Cell Biol 1991; 3:939-946.

Blanchard KT, Lee J, Boekelheide K. Leuprolide, a gonadotropin-releasing hormone agonist, re-establishes spermatogenesis after 2,5-hexanedione-induced irreversible testicular injury in the rat, resulting in normalized stem cell factor expression. Endocrinology 1998; 139:236-244.

Brannan CI, Lyman DD, Williams DE, Eisenman J, Anderson DM, Cosman D, Bedell MA, Jenkins NA, Copeland NG. Steel-Dickie mutation encodes a c-kit ligand lacking transmembrane and cytoplasmic domains. Proc Natl Acad Sci USA 1991; 88:4671-4674.

Brannan CI, Bedell MA, Resnick JL, Eppig JJ, Handel MA, Williams DE, Lyman SD, Donovan PJ, Jenkins NA, Copeland NG. Developmental abnormalities in Steel17H mice result from a splicing defect in the steel factor cytoplasmic tail. Genes Dev 1992, 6:1832-1842.

Brinster RL, Avarbock MR. Germline transmission of donor haplotype following spermatogonial transplantation. Proc Natl Acad Sci USA 1994; 91:11303-11307.

Capecchi MR. Altering the genome by homologous recombination. Science 1989; 244:1288-1292.

Carroll DJ, Ramarao CS, Mehlmann LM, Roche S, Terasaki M, Jaffe LA. Calcium release at fertilization in starfish eggs is mediated by phospholipase Cγ. J Cell Biol 1997; 138:1303-1311.

Chabot B, Stephenson DA, Chapman VM, Besmer P, Bernstein A. The protooncogene *c-kit* encoding a transmembrane tyrosine kinase receptor maps to the mouse *W* locus. Nature 1988; 335:88-89.

Cohen GB, Ren R, Baltimore D. Modular binding domains in signal transduction proteins. Cell 1995; 80:237-248.

Colledge WH, Carlton MBL, Udy GB, Evans MJ. Disruption of c-mos causes parthenogenetic development of unfertilized mouse eggs. Nature 1994; 370:65-68.

Colonna R, Tatone C, Francione A, Rosati F, Callaini G, Corda D, Di Francesco L. Protein kinase C is required for the disappearance of MPF upon artificial activation in mouse eggs. Mol Reprod Dev 1997; 48:292-299.

Copeland NG, Gilbert DJ, Cho BC, Donovan PJ, Jenkins NA, Cosman D, Anderson D, Lyman SD, Williams DE. Mast cell growth factor maps near the Steel locus on mouse chromosome 10 and is deleted in a number of Steel alleles. Cell 1990; 63:175-183.

Coulombre JL, Russell ES. Analysis of the pleiotropism at the *W*-locus in the mouse. J Exp Zool 1954; 126:277-295.

Dolci S, Williams DE, Ernst MK, Resnick JL, Brannan CI, Lock LF, Lyman SD, Boswell HS, Donovan PJ. Requirement for mast cell growth factor for primordial germ cell survival in culture. Nature 1991; 352:809-811.

Dozortsev D, Rybouchkin A, De Sutter P, Qian C, Dhont M. Human oocyte activation following intracytoplasmic injection: the role of the sperm cell. Hum Reprod 1995; 10:403-407.

Dupont G, McGuinness OM, Johnson MH, Berridge MJ, Borgese F Phospholipase C in mouse oocytes: characterization of β and γ isoforms and their possible involvement in sperm-induced Ca^{2+} spiking. Biochem J 1996; 316:583-591.

Dym M, Jia MC, Dirami G, Price JM, Rabin SJ, Mocchetti I, Ravindranath N. Expression of the c-kit receptor and its autophosphorylation in immature rat type A spermatogonia. Biol Reprod 1995; 52:8-19.

Falasca M, Logan SK, Lehto VP, Baccante G, Lemmon MA, Schlessinger J. Activation of phospholipase C by PI 3-kinase-induced PH domain-mediated membrane targeting. EMBO J 1998; 17:414-422.

Feng H, Sandlow JI, Sandra A. Expression and function of the c-kit proto-oncogene protein in mouse sperm. Biol Reprod 1997; 57:194-203.

Feng H, Sandlow JI, Sandra A. The c-kit receptor and its possible signal transduction pathway in mouse spermatozoa. Mol Reprod Dev 1998; 49:317-326.

Flanagan JF, Chan DC, Leder P. Transmembrane form of the *kit* ligand growth factor is determined by alternative splicing and is missing in the Sl^d mutant. Cell 1991; 64:1025-1035.

Gallicano GI, Yousef MC, Capco DG. PKC: a pivotal regulator of early development. Bioessays 1997; 19:29-36.

Geissler EN, Ryan MA, Housman DE. The dominant-white spotting (W) locus of the mouse encodes the c-kit proto-oncogene. Cell 1988; 55:185-192.

Haccard O, Sarcevic B, Lewellyn A, Hartley R, Roy L, Izumi T, Erikson E, Maller JL. Induction of metaphase arrest in cleaving *Xenopus* embryos by MAP kinase. Science 1993; 262:1262-1265.

Hashimoto N, Watanabe N, Furuta Y, Tamemoto H, Sagata N, Yokoyama M, Okazaki K, Nagayoshi M, Takeda N, Ikawa Y, Aizawa S. Parthenogenetic activation of oocytes in c-mos-deficient mice. Nature 1994; 370:68-71.

Herbst R, Lammers R, Schlessinger J, Ullrich A. Substrate phosphorylation specificity of the human c-kit receptor tyrosine kinase. J Biol Chem 1991; 266:19908-19916.

Herbst R, Shearman MS, Jallal B, Schlessinger J, Ullrich A. Formation of signal transfer complexes between stem cell and platelet-derived growth factor receptors and SH2 domain proteins *in vitro*. Biochemistry 1995; 34:5971-5979.

Homa ST, Swann K. A cytosolic sperm factor triggers calcium oscillations and membrane hyperpolarization in human oocytes. Hum Reprod 1994; 9:2356-2361.

Horie K, Takakura K, Taii S, Narimoto K, Noda Y, Nishikawa S, Nakayama H, Fujita J, Mori T. The expression of the *c-kit* protein during oogenesis and early embryonic development. Biol Reprod 1991; 45:547-552.

Hwang SC, Jhon DY, Bae YS, Kim JH, Rhee SG. Activation of phospholipase C-γ by the concerted action of tau proteins and arachidonic acid. J Biol Chem 1996; 271:18342-18349.

Ismail RS, Okawara Y, Fryer JN, Vanderhyden BC. Hormonal regulation of the ligand of c-kit in the rat ovary and its effects on spontaneous oocyte meiotic maturation. Mol Reprod Dev 1996; 43:458-469.

Ismail RS, Dube M, Vanderhyden BC. Hormonally regulated expression and alternative splicing of kit ligand may regulate kit-induced inhibition of meiosis in rat oocytes. Dev Biol 1997; 184:333-342.

Jones GA, Carpenter G. The regulation of phospholipase C-γl by phosphatidic acid. Assessment of kinetic parameters. J Biol Chem 1993; 268:20845-20850.

Kamat A, Carpenter G. Phospholipase C-γl: regulation of enzyme function and role in growth factor-dependent signal transduction. Cytok & Growth Factor Rev 1997; 8:109-117.

Kim HK, Kim JW, Zilberstein A, Margolis B, Kim JG, Schlessinger J, Rhee SG. PDGF stimulation of inositol phospholipid hydrolysis requires PLC-γl phosphorylation on tyrosine residues 783 and 1254. Cell 1991; 65:435-441.

Kimura Y, Yanagimachi R. Mouse oocytes injected with testicular spermatozoa or round spermatids can develop into normal offspring. Development 1995; 121:2397-2405.

Kozlowski M, Larose L, Lee F, Mingh Le D, Rottapel R, Siminovitch KA. SHP-1 binds and negatively modulates the c-Kit receptor by interaction with tyrosine 569 in the c-Kit juxtamembrane domain. Mol Cell Biol 1998; 18:2089-2099.

Kline D, Kline JT. Repetitive calcium transients and the role of calcium in exocytosis and cell cycle activation in the mouse egg. Dev Biol 1992; 149:80-89.

Lawrence I, Whitaker M, Swann K. Sperm-egg fusion is the prelude to the initial Ca^{2+} increase at fertilization in the mouse. Development 1997; 124:233-241.

Lee SB, Rhee SG. Significance of PIP_2 hydrolysis and regulation of phospholipase C isozymes. Curr Opin Cell Biol 1995; 7:183-189.

Lev S, Givol D, Yarden Y. A specific combination of substrates is involved in signal transduction by the kit-encoded receptor. EMBO J 1991; 10:647-654.

Lorca T, Cruzalegui FH, Fesquet D, Cavadore JC, Mery J, Means A, Doree M. Calmodulin-dependent protein kinase II mediates inactivation of MPF and CSF upon fertilization of *Xenopus* eggs. Nature 1993; 366:270-273.

Manova K, Nocka K, Besmer P, Bachvarova RF. Gonadal expression of *c-kit* encoded at the *W* locus of the mouse. Development 1990; 110:1057-1069.

Manova K, Huang EJ, Angeles M, De Leon V, Sanchez S, Pronovost SM, Besmer P, Bachvarova RF. The expression pattern of the *c-kit* ligand in gonads of mice supports a role for the *c-kit* receptor in oocyte growth and in proliferation of spermatogonia. Dev Biol 1993; 157:85-99.

Marziali G, Lazzaro D, Sorrentino V. Binding of germ cells to mutant Sl^d Sertoli cells is defective and is rescued by expression of the transmembrane form of the *c-kit* ligand. Dev Biol 1993; 157:182-190.

Miyazaki S, Yuzaki M, Nakada K, Shirakawa H, Nakanishi S, Nakade S, Mikoshiba K. Block of Ca^{2+} wave and Ca^{2+} oscillation by antibody to the inositol 1,4,5-triphosphate receptor in fertilized hamster eggs. Science 1992; 257:251-255.

Miyazaki S. Shirakawa H, Nakada K, Honda Y. Essential role of the inositol 1,4,5-triphosphate receptor/Ca^{2+} release channel in Ca^{2+} waves and Ca^{2+} oscillations at fertilization of mammalian eggs. Dev Biol 1993; 158:62-78.

Moarefi I, LaFevre-Bernt M, Sicheri F, Huse M, Lee CH, Kuriyan J, Miller WT. Activation of the Src-family tyrosine kinase Hck by SH3 domain displacement. Nature 1997; 385:650-653.

Moos J, Visconti PE, Moore GD, Schultz RM, Kopf GS. Potential role of mitogen-activated protein kinase in pronuclear envelope assembly and disassembly following fertilization of mouse eggs. Biol Reprod 1995; 53:692-699.

Mori C, Hashimoto H, Hoshino K. Fluorescence microscopy of nuclear DNA in oocytes and zygotes during *in vitro* fertilization and development of early embryos in mice. Biol Reprod 1988; 39:737-742.

Motro B, Bernstein A. Dynamic changes in ovarian c-kit and Steel expression during the estrous reproductive cycle. Dev Dyn 1993; 197:69-79.

Natali PG, Nicotra MR, Sures I, Santoro E, Bigotti A, Ullrich A. Expression of c-kit receptor in normal and transformed human nonlymphoid tissues. Cancer Res 1992; 52:6139-6143.

Nocka K, Tan JC, Chiu E, Chu TY, Ray P, Traktman P, Besmer P. Molecular bases of dominant negative and loss of function mutations at the murine *c-kit*/white spotting locus: W^{37}, W^{v}, W^{41} and *W*. EMBO J 1990; 9:1805-1813.

Orth JM, Jester WF, Qiu J. Gonocytes in testes of neonatal rats express the c-kit gene. Mol Reprod Dev 1996; 45:123-131.

Packer AI, Hsu YC, Besmer P, Bachvarova RF. The ligand of the *c-kit* receptor promotes oocyte growth. Dev Biol 1994; 161:194-205.

Packer AI, Besmer P, Bachvarova RF. Kit ligand mediates survival of type A spermatogonia and dividing spermatocytes in postnatal mouse testes. Mol Reprod Dev 1995; 42:303-310.

Parrington J, Swann K, Shevchenko VI, Sesay AK, Lai FA. Calcium oscillations in mammalian eggs triggered by a soluble sperm protein. Nature 1996; 379:364-368.

Pawson T. Protein modules and signaling networks. Nature 1995; 373:573-580.

Price DJ, Rivnay B, Fu Y, Jiang S, Avraham S, Avraham H. Direct association of Csk homologous kinase (CHK) with the diphosphorylated site Tyr568/570 of the activated c-kit in megakaryocytes. J Biol Chem 1997; 272:5915-5920.

Qiu F, Ray P, Barker PE, Jhanwar S, Ruddle FH, Besmer P. Primary structure of *c-kit*: relationship with the CSF-1/PDGF receptor kinase family - oncogenic activation of *v-kit* involves deletion of extracellular domain and C terminus. EMBO J 1988; 7:1003-1011.

Rhee SG, Bae YS. Regulation of phosphoinositide-specific phospholipase C isozymes. J Biol Chem 1997; 272:15045-15048.

Rossi P, Albanesi C, Grimaldi P, Geremia R. Expression of the mRNA for the ligand of *c-kit* in mouse Sertoli cells. Biochem Biophys Res Commun 1991; 176:910-914.

Rossi P, Marziali G, Albanesi C, Charlesworth A, Geremia R, Sorrentino V. A novel *c-kit* transcript, potentially encoding a truncated receptor, originates within a *kit* gene intron in mouse spermatids. Dev Biol 1992; 152:203-207.

Rossi P, Dolci S, Albanesi C, Grimaldi P, Ricca R, Geremia R. Follicle-stimulating hormone induction of Steel Factor (SLF) mRNA in mouse Sertoli cells and stimulation of DNA synthesis in spermatogonia by soluble SLF. Dev Biol 1993a; 155:68-74.

Rossi P, Dolci S, Albanesi C, Grimaldi P, Geremia R. Direct evidence that the mouse sex-determining gene *Sry* is expressed in the somatic cells of male fetal gonads and in the germ cell line in the adult testis. Mol Reprod Dev 1993b; 34:369-373.

Rottapel R, Reedijk M, Williams DE, Lyman SD, Anderson DM, Pawson T, Bernstein A. The Steel/W transduction pathway: kit autophosphorylation and its association with a unique subset of cytoplasmic signaling proteins is induced by the Steel factor. Mol Cell Biol 1991; 11:3043-3051.

Russell ES. Hereditary anemia of the mouse; a review for geneticists. Adv Genet 1979; 20:357-459.

Serve H, Hsu YC, Besmer P. Tyrosine residue 719 of the c-kit receptor is essential for binding of the P85 subunit of phosphatidylinositol (PI) 3-kinase and for c-kit-associated PI 3-kinase activity in COS-1 cells. J Biol Chem 1994; 269:6026-6030.

Sette C, Bevilacqua A, Bianchini A, Mangia F, Geremia R, Rossi P. Parthenogenetic activation of mouse eggs by microinjection of a truncated c-kit tyrosine kinase present in spermatozoa. Development 1997; 124:2267-2274.

Sette C, Bevilacqua A, Geremia R, Rossi P. Involvement of phospholipase Cγl in mouse egg activation induced by a truncated form of the c-kit tyrosine kinase present in spermatozoa. J Cell Biol 1998; 142:1063-1074.

Sorrentino V, Giorgi M, Geremia R, Besmer P, Rossi P. Expression of the *c-kit* proto-oncogene in the murine male germ cells. Oncogene 1991; 6:149-151.

Stice SL, Robl JM. Activation of mammalian oocytes by a factor obtained from rabbit sperm. Mol Reprod Dev 1990; 25:272-280.

Stricker SA. Intracellular injections of a soluble sperm factor trigger calcium oscillations and meiotic maturation in unfertilized oocytes of a marine worm. Dev Biol 1997; 186:185-201.

Swann K. A cytosolic sperm factor stimulates repetitive calcium increases and mimics fertilization in hamster eggs. Development 1990; 110:1295-1302.

Tajima Y, Onoue H, Kitamura Y, Nishimune Y. Biologically active kit ligand growth factor is produced by mouse Sertoli cells and is defective in Sl/d mutant mice. Development 1991; 113:1031-1035.

Tan JC, Nocka K, Ray P, Traktman P, Besmer P. The dominant W^{42} *spotting* phenotype results from a missense mutation in the *c-kit* receptor kinase. Science 1990; 247:209-212.

Tang B, Mano H, Yi T, Ihle JN. Tec kinase associates with and is tyrosine phosphorylated and activated following stem cell factor binding. Mol Cell Biol 1994; 14:8432-8437.

Tauchi T, Feng GS, Marshall MS, Shen R, Mantel C, Pawson T, Broxmeyer HE. The ubiquitously expressed Syp phosphatase interacts with *c-kit* and Grb2 in hematopoietic cells. J Biol Chem 1994; 269:25206-25211.

Verlhac M-H, Kubiak JZ, Weber M, Geraud G, Colledge WH, Evans MJ, Maro B. Mos is required for MAP kinase activation and is involved in microtubule organization during meiotic maturation in the mouse. Development 1996; 122:815-822.

Vincent S, Grandjean V, Cuzin F, Rassoulzadegan M. Sertoli-germ cell interactions: *in vivo* and *in vitro* studies of Kit/KL and NGF signaling. Tenth European Workshop on Molecular and Cellular Endocrinology of the Testis, Capri, Italy, 1998; abstract C26.

Whitaker M, Swann K. Lighting the fuse at fertilization. Development 1993; 117:1-12.

Wolosker H, Kline D, Bian Y, Blackshaw S, Cameron AM, Fralich TD, Schnaar RL, Snyder SH. Molecularly cloned mammalian glucosamine-6-phosphate deaminase localizes to transporting epithelium and lacks oscillin activity. FASEB J 1998; 12:91-99.

Wu H, He CL, Fissore RA. Injection of a porcine sperm factor triggers calcium oscillations in mouse oocytes and bovine eggs. Mol Reprod Dev 1997; 46:176-189.

Xu Z, Kopf GS, Schultz RM. Involvement of inositol 1,4,5-triphosphate-mediated Ca^{2+} release in early and late events of mouse egg activation. Development 1994; 120:1851-1859.

Yanagimachi R. Mammalian fertilization. In: (Knobil E, Neill JD, eds) The Physiology of Reproduction. New York: Raven Press Ltd, 1994, pp189-317.

Yoshinaga K, Nishikawa S, Ogawa M, Hayashi S, Kunisada T, Fujimoto T, Nishikawa S-I. Role of *c-kit* in mouse spermatogenesis: identification of spermatogonia as a specific site of *c-kit* expression and function. Development 1991; 113:689-699.

4

The Effects of Gene Knockouts on Spermatogenesis

E. M. Eddy
National Institute of Environmental Health Sciences

The development of gene knockout (KO) technology has led to remarkable advances in knowledge in many areas of mammalian biology. It has been less than a decade since it was shown that: 1) mutations can be targeted to specific genes in embryonic stem (ES) cells by homologous recombination, 2) ES cells in which this rare event has occurred can be cloned using positive and negative drug selection strategies, and 3) ES cells containing the mutant gene can be used to produce chimeric mice that give rise to progeny bearing the mutation in every cell. Other essential methods developed previously made this technology possible. These include isolating ES cells from the inner cell mass of preimplantation mouse embryos, maintaining ES cells in culture and keeping them from undergoing differentiation, and producing chimeric mice by injecting ES cells into mouse blastocysts. Although now widely used, the production of gene KO mice requires skills in molecular biology, tissue culture and micromanipulation of mouse embryos, as well as knowledge of mouse reproductive biology and colony management. However, the greatest challenge often occurs after the gene KO mouse is produced, that of determining how the mutation has led to the phenotype found.

Gene KOs have produced effects on male reproduction ranging from a modest reduction in fertility with no obvious changes in spermatogenesis to sterility because of the absence of the testis. Some of these KOs disrupted developmental processes, while others altered endocrine processes or modified the function of somatic cells in the testis. Some of these KOs involved genes expressed in somatic cells, while others involved genes expressed within spermatogenic cells that are responsible for germ cell-autonomous processes and are required for spermatogenesis or sperm function. Some KOs expected to alter male fertility had little or no effect, while others unexpectedly disrupted spermatogenesis or processes essential for normal male fertility.

The focus of this review is on gene KOs with effects on spermatogenesis and male fertility and on the knowledge gained about male reproduction from the use of this relatively new research tool. The gene KO approach has provided novel insights into the process of spermatogenesis and on the role of different proteins involved in sperm function. It has identified diverse genes that are essential for male reproduction, revealed that some aspects of male endocrinology are still not well understood, and demonstrated that genes expressed specifically in germ cells often have unique roles in male reproduction. The information gained from gene KO studies has provided a better understanding of the processes and mechanisms involved in male germ cell development and function and of the genes required for male fertility.

Gene KOs That Directly Disrupt Spermatogenesis

KOs of genes expressed in spermatogenic cells (Table 1) have led to a variety of effects, from subtle to extreme. The KOs have involved genes for growth factors, transcription factors, DNA repair processes, chaperones, proteases, apoptosis regulators and other proteins.

Growth Factors

BMP (bone morphogenetic protein) 8A and 8B are members of the TGF-β superfamily of growth factors. The *Bmp8a* gene is expressed mainly in decidual cells of the uterus between days 7.5 and 10.5 of pregnancy and in the testis at low levels in spermatogonia and primary spermatocytes, and at higher levels in spermatids (Zhao, Hogan 1996). *Bmp8a* gene KO females had normal fertility. In males, the *Bmp8a* gene KO produced no obvious defects during the initiation of spermatogenesis and young gene KO males were fertile, but some males became infertile as they aged and epididymal granulomas were seen occasionally (Zhao et al. 1998). Slightly less than half of 12- to 30-week-old males examined showed varying degrees of germ cell degeneration and apoptotic germ cells were seen in some seminiferous tubules. It was suggested that BMP8A plays a role in the maintenance of spermatogenesis, with the BMP8A protein produced by spermatids acting on primary spermatocytes to regulate their survival and differentiation.

The *Bmp8b* gene has a different pattern of expression in female mice than the *Bmp8a* gene, with the mRNA being found in trophoblast cells in the placenta during days 9.5 to 16.5 of pregnancy. However, in males, the expression of the *Bmp8b* gene is similar to that of the *Bmp8a* gene, with the mRNA found at low levels in spermatogonia and spermatocytes, and at high levels in round spermatids (Zhao, Hogan 1996). *Bmp8b* gene KO females did not show obvious reproductive defects, but the majority of males were infertile and others became infertile after siring two or three litters (Zhao et al. 1996). Histological examination indicated that nearly complete germ cell depletion and seminiferous tubule degeneration had occurred in infertile males, and that partial germ cell depletion and tubule degeneration had occurred in fertile males. BrdU incorporation and TUNEL labeling studies indicated that decreased germ cell proliferation and a subsequent delay in differentiation, rather than increased apoptosis, probably were responsible for the reduction in germ cells in gene KO males. It was suggested that during early puberty BMP8B is required for germ cell proliferation, and during adulthood for the maintenance of spermatogenesis. It was also suggested that the *Madr1* and *Madr2* genes expressed in spermatogonia and spermatocytes might

Table 1: Gene KOs with direct effects on spermatogenesis

Gene[1]	Expression	Effects of KO	References[2]
Growth Factors			
Bmp8a	low in spermatogonia and spermatocytes, higher in spermatids	declining male fertility with age, some germ cell degeneration	Zhao et al. 1998
Bmp8b	low in spermatogonia and spermatocytes, higher in spermatids	majority of males infertile, others become infertile after producing 2-3 litters, germ cell depletion and tubule degeneration	Zhao et al. 1996
Transcription Factors			
Rara	constitutive, present in round spermatids and Sertoli cells	male infertility, some neonatal lethality, severe degeneration of seminiferous epithelium, occasional intact tubules	Lufkin et al. 1993
Crem	several isoforms, some constitutive, CREM tau in spermatids only	male infertility, spermatogenesis disrupted in early spermatid development	Nantel et al. 1996; Blendy et al. 1996
A-myb	low in brain, spleen, ovary, mammary ductal epithelium, high in spermatogonia and spermatocytes	male infertility, degeneration of pachytene spermatocytes, lack of spermatids and sperm development	Toscani et al. 1996
Sprm1	transcript in late pachytene spermatocytes and spermatids, protein in spermatids	reduced male fertility, spermatogenesis unchanged, sperm numbers and motility normal, decreased zona binding	Pearse et al. 1997
Zfx	constitutive	males fertile, some neonatal lethality, low body weight, testis reduced in size, sperm counts reduced	Luoh et al. 1997
DNA Repair			
Hr6b	constitutive, elevated in spermatids	males infertile, spermatogenesis begins in juveniles, fails at 4-5 weeks, sperm have sluggish motion and abnormal heads	Roest et al. 1996
Dmc1	during meiotic prophase only, protein in leptotene and zygotene spermatocytes	males and females infertile, arrest of spermatogenesis at beginning of pachytene spermatocyte development	Yoshida et al. 1998; Pittman et al. 1998
Atm	constitutive, protein in spermatogonia and spermatocytes, on axis of synapsed chromosomes	males and females infertile, growth retardation, lymphomas, testis small, spermatogenesis disrupted at beginning of meiosis, no spermatids or sperm	Barlow et al. 1996; Elson et al. 1996; Xu et al. 1996
Mlh1	constitutive	males and females infertile, small testis, failure of germ cell development beyond pachytene stage	Edelmann et al. 1996; Baker et al. 1996
Pms2	constitutive	males infertile, disruption of pachytene spermatocyte development, abnormal spermatids, some abnormal sperm heads	Baker et al. 1995
Chaperones			
Clgn	pachytene spermatocytes and spermatids	males nearly infertile, spermatogenesis unchanged, sperm numbers and motility normal, sperm fail to adhere to or penetrate zona pellucida	Ikawa et al. 1997
Hsp70-2	begins in leptotene spermatocytes, protein in spermatocytes and spermatids	males infertile, arrest in development and apoptosis at end of pachytene spermatocyte development, no spermatids or sperm	Dix et al. 1996
Proteases			
Ace	germ cell-specific isoform in spermatids	reduced fertility in males, spermatogenesis unchanged, sperm numbers and motility normal, reduced sperm transport in oviduct, reduced sperm binding to zona pellucida and fertilization *in vivo*	Krege et al. 1995; Hageman et al.1998
Psck4	spermatids	reduced fertility, spermatogenesis unchanged, sperm numbers and motility normal, reduced *in vivo* fertilization	Mbikay et al. 1997
Acr	protein in spermatids	males fertile, *in vitro* fertilization delayed	Baba et al. 1994; Adham et al. 1997
Ftnb	protein in pachytene spermatocytes and spermatids	males infertile, testis and spermatogenesis unchanged, poor sperm binding to zona pellucida, reduced binding to egg plasma membrane, poor sperm transport in oviduct	Cho et al. 1998
Apoptosis Regulators			
Bax	constitutive, high in spermatogonia	males infertile, testis atrophic, sperm absent from epididymis, accumulation of spermatogonia	Knudson et al. 1995
Bclw	constitutive, present in Sertoli cells, high in spermatids	males infertile, spermatogenesis blocked in late spermatid development in juveniles, degeneration of tubules later	Ross et al. 1998
Other Genes			
Dazla	type B spermatogonia and early spermatocytes	males and females infertile, testis small, severe disruption of testis organization, reduced fertility in heterozygous males	Ruggiu et al. 1997
Cenpb	constitutive	normal fertility, reduced body weight, testis weights & sperm count	Hudson et al. 1998
Apob	constitutive	homozygous lethal, reduced fertility in heterozygous males with low sperm motility and survival *in vitro*, unable to fertilize eggs with intact zona pellucida	Huang et al. 1996
Ggtb	constitutive	normal fertility, sperm bind less ZP3 and fail to undergo ZP3-induced acrosome reaction, low fertilization *in vitro*	Lu, Shur 1997

[1]Gene symbols used are from the Mouse Genome Database, The Jackson Laboratory.
[2]Only selected references are shown. See listed reference for additional information and additional references.

encode down-stream transducers of BMP8 signaling in the testis (Zhao et al. 1996). However, the nature and location of the receptors for BMP8A and BMP8B in the seminiferous epithelium remain to be determined.

Transcription Factors

The transcription factor genes expressed in male germ cells that have been subjected to KOs include those encoding the retinoic acid receptor alpha (*Rara*), cAMP response-element modulator (*Crem*), myeloblastosis like-1 (A-*myb*), Sperm 1 (*Sprm1*) and X chromosome zinc finger protein (*Zfx*).

The retinoic acid receptor alpha (RARα) is a ligand-activated transcription factor and is a member of the nuclear receptor superfamily. There are two major isoforms of RARα that are produced from alternatively spliced transcripts, RARα1 and RARα2. RARα1 is expressed constitutively in embryonic and adult tissues, including spermatogenic cells, while the less abundant RARα2 is restricted in expression, but present in the testis. RARα is activated by the all-*trans* retinoic acid and 9-*cis* retinoic acid ligands and binds to retinoic acid response elements (RARE) in gene promoters. Mice with a gene KO that disrupted only RARα1 production appeared normal, while 60% of mice with a KO of the entire *Rara* gene had early postnatal lethality. The surviving males were infertile and had severe degeneration of the germinal epithelium, although occasional tubules appeared by histology to be intact (Lufkin et al. 1993). The testicular defect seen in gene KO males was suggested to be similar to that seen in vitamin A deficiency. However, *Rara* transcripts are present both in Sertoli cells and round spermatids and it has not been established if the defects in spermatogenesis seen in gene KO mice are due to disruption of functions of one or both cell types.

The *Crem* gene encodes a family of transcription factor isoforms produced by alternative translational initiation, promoter utilization and splicing events. They bind to the cAMP response element (CRE) in gene promoters to modulate their transcription. The CREM tau isoform is synthesized during the postmeiotic phase of spermatogenesis and is activated by phosphorylation (Delmas, Sassone-Corsi 1994). The *Crem* gene KO resulted in reduced fertility in heterozygous males and infertility in homozygous males, but homozygous females were fertile (Blendy et al. 1996; Nantel et al. 1996). Spermatogenesis was interrupted in the early part of spermatid development and germ cells frequently were seen undergoing apoptosis. The mRNA levels for several genes with putative CRE sites in their promoters expressed during spermatid development were reduced substantially or absent in *Crem* KO mice (Blendy et al. 1996; Nantel et al. 1996). This suggested that the defects in spermatogenesis seen in the *Crem* gene KO were due to lack of activation of genes required for spermatid development and completion of spermiogenesis. However, the KO of the gene for the related CREB (c-AMP response-element binding) protein had no effect on male fertility (Hummler et al. 1994).

The members of the *myb* family of proto-oncogenes are transcription factors that participate in the control of cell growth and differentiation. The A-*myb* gene is located on the X chromosome and expressed in the mouse in mammary gland ductal epithelium during pregnancy and at low levels in the brain, spleen and ovary. It is expressed at high levels in the adult testis in spermatogonia and primary spermatocytes (Mettus et al. 1994). Male A-*myb* KO mice mated but were infertile, their testis size was one-fourth that of wild-type males, and they failed to produce sperm (Toscani et al. 1997). Histological examination of the testes revealed that spermatogonia and preleptotene spermatocytes were normal, most pachytene spermatocytes were degenerating, and spermatids were absent. Spermatogenesis was blocked in early pachytene spermatocytes, suggesting that A-*myb* has an important role in regulating expression of genes required for spermatocyte development. The fertility of female gene KO mice was unaffected, but their mammary glands were under-developed and they were unable to nurse (Toscani et al. 1997).

The *Sprm1* gene is expressed specifically in spermatogenic cells, with the transcript appearing immediately before meiosis and the protein being synthesized in post-meiotic germ cells. The Sperm 1 protein is a POU subclass homeodomain transcription factor (Andersen et al. 1993). The *Sprm1* gene KO produced no detectable changes in the testis or in the morphological differentiation of spermatids, in the number or motility of epididymal sperm, or in mating behavior. There was a statistically significant reduction in the average number of litters sired in two months by 129Sv strain gene KO males, but not by C57BL/6 strain gene KO males (Pearse et al. 1997). There was a modest decrease in the ability of sperm from 129Sv KO males to bind to and penetrate the zona pellucida *in vitro*, but this was not sufficient to explain the reduced fertility. It was suggested that Sperm1 regulates a haploid gene pathway required for optimum fertility in low fecundity strains.

The zinc finger ZFX and ZFY proteins have been suggested to function in sex differentiation and spermatogenesis. The *Zfx* gene is located on the X chromosome and the constitutively-expressed ZFX protein contains a transcription-activating domain, a putative nuclear localization signal and a DNA-binding motif. *Zfx* gene KO mice were small and had low neonatal viability, but there were no obvious defects in sexual differentiation. The number of germ cells was reduced in females and males. Fertility was compromised severely in females, while sperm counts were reduced by half without altering male fertility (Luoh et al. 1997). The number of primordial germ cells in *Zfx* KO embryos was less than half of that in wild-type embryos prior to gonadal sex differentiation. Although these studies indicate that ZFX has a role in the growth and maintenance of germ cells, it has been not been determined if the protein functions within germ cells or if the effect is indirect.

DNA Repair

Bacteria and yeast have been used extensively to study DNA repair and recombination and the same genes are often involved in both processes. The mammalian homologs for several of these genes have been identified and often encode proteins that are abundant in pachytene spermatocytes (Eddy, O'Brien 1998). KOs of some of these genes result in early embryonic lethality, such as in the case of *Rad51* (Lim, Hasty 1996; Tsuzuki et al. 1996), but the effects become apparent only in the adult for others. The genes for DNA repair proteins shown in KO studies to be important in male germ cells include *HR6B*, *Dmc1*, *ATM*, *Mlh1*, and *Pms2*.

The mammalian *HR6A* and *HR6B* genes (homologs of RAD6) are homologs of a yeast gene for an ubiquitin-conjugating DNA repair enzyme. The HR6B protein is expressed constitutively and localized in the nucleus, but is present at elevated levels in mouse spermatids (Koken et al. 1996). The only effect seen in *HR6B* gene KO mice was male infertility. Spermatogenesis was intact in juvenile mice, but failure of germ cell development occurred in four- to five-week-old males as the first wave of spermatogenesis neared completion (Roest et al. 1996). Most adult gene KO males produced sperm in low numbers and the sperm often had abnormal head morphology and sluggish motility. There was considerable variation between mice in the effects seen, but it was suggested that the primary defect occurred in the elongating spermatids, with disruption of the seminiferous epithelium being a secondary effect. It was suggested that the HR6A protein may compensate in gene KO mice for the absence of HR6B in most tissues, but *HR6A* is an X-linked gene and the amount of HR6A protein is too low to compensate (Roest et al. 1996).

The mouse *Dmc1* gene (disrupted meiotic cDNA) is a homolog of the *E. coli RecA* gene, which is involved in initiation of pairing and strand exchange, homologous recombination, and DNA repair. The *Dmc1* gene is expressed only during meiotic prophase in male and female germ cells (Habu et al. 1996) and the DMC1 protein was detected in the nuclei of leptotene and zygotene stage spermatocytes (Yoshida et al. 1998). The *Dmc1* gene KO resulted in male and female infertility. Spermatogenesis was arrested at the beginning of spermatocyte development, post-meiotic germ cells were absent from testes, and ovaries lacked oocytes and follicles (Pittman et al. 1998; Yoshida et al. 1998). Axial elements formed on homologous chromosomes, but synapsis was blocked and synaptonemal complexes failed to develop, resulting in meiotic arrest and apoptosis in spermatocytes and presumably in oocytes. This suggested that DMC1 is required for synapsis of homologous chromosomes during meiosis (Pittman et al. 1998; Yoshida et al. 1998).

The *Atm* gene (ataxia-telangiectasia mutated) encodes a kinase involved in DNA metabolism and cell cycle checkpoint control. Mutations in this gene in humans are associated with increased lymphoid cancer risks, cerebellar ataxia, growth retardation, sensitivity to ionizing radiation, thymic degeneration, immunodeficiency and infertility. The ATM protein is expressed in spermatogonia and spermatocytes and is associated with the axis of synapsed chromosomes in pachytene spermatocytes of the mouse (Keegan et al. 1996). The *Atm* gene KO resulted in male and female infertility, growth retardation and lymphomas by four months of age (Barlow et al. 1996; Elson et al. 1996; Xu et al. 1996). The testes were small and spermatogenesis was disrupted, with meiosis arrested at the beginning of pachytene spermatocyte development, and spermatids and sperm being absent. It appeared that meiotic arrest in male *Atm* KO mice was due to abnormal chromosomal synapsis, fragmentation of chromosomes and apoptosis in spermatocytes.

The *MutH, MutS* and *MutL* genes encode essential components of the bacterial post-replication DNA-mismatch repair system. *Mlh1* (*MutL* homolog 1) and *Pms2* are mouse homologs of *MutL* and the MLH1 and PMS2 proteins can exist as a heterodimer (Li, Modrich 1995), suggesting that they may function cooperatively. The *Mlh1* gene KO resulted in infertility in males and females, although the KO animals mated and females showed normal estrus. Males had small testes, lacked sperm in the epididymis, and their germ cells failed to develop beyond the pachytene spermatocyte stage. Meiosis progressed to the synapsis of homologs, but chiasma formation or stabilization was apparently disrupted in the absence of MLH1 (Baker et al. 1996; Edelmann et al. 1996). Although DNA mismatch repair was severely impaired in fibroblasts from *Mlh1* KO embryos (Edelmann et al. 1996), the mice were not reported to be prone to tumor development or to have a reduced life span.

The *Pms2* gene KO resulted in infertility in males only. There were reduced numbers of sperm in the epididymis, and the sperm often had a misshapen head and a truncated, irregular flagellum (Baker et al. 1995). Vacuolization of pachytene spermatocytes, reduced numbers of round and elongating spermatids, and multinucleated spermatids were present in the testis. Disruption of spermatogenesis occurred in pachytene spermatocytes and abnormalities in chromosome synapsis were present. There was a modest increase in tumor development in *Pms2* gene KO mice, but the frequency of mutation at microsatellite sequences was elevated

~1000-fold, indicating that DNA mismatch repair was severely compromised.

Chaperones

The chaperones are proteins that help other proteins to fold, to avoid aggregation and to assemble into complexes. Two chaperones present only in spermatogenic cells, calmegin and heat-shock protein HSP70-2, have been shown to be important for spermatogenesis and sperm function. The calmegin gene (*Clgn*) encodes a calcium-binding protein located in the membrane of the endoplasmic membrane that is a homolog of the constitutively expressed calnexin gene (*Canx*). Calmegin functions as a calnexin-like molecular chaperone during spermatogenesis. The *Canx* gene is expressed throughout spermatogenesis, while the *Clgn* gene is expressed in pachytene spermatocytes and spermatids (Ikawa et al. 1997; Watanabe et al. 1992). The *Clgn* gene KO had no effects on fertility of female mice, but 120 matings by gene KO males produced only one litter of two offspring. The histology of the testis and the number, morphology and motility of sperm were comparable in gene KO and wild-type males. However, sperm from gene KO males failed to adhere to or penetrate the zona pellucida *in vitro*. It was suggested that calmegin may be necessary for maturation of sperm surface protein(s) required for egg binding (Ikawa et al. 1997).

The HSP70 family of heat-shock proteins is encoded by genes that are expressed constitutively (*Hsc70*, *Grp78*) or in response to stress (*Hsp70-1, Hsp70-3*) in all cell types. In addition, genes for two members of this family are expressed only in spermatogenic cells, with *Hsp70-2* expression beginning in leptotene spermatocytes and *Hsc70t* expression beginning in spermatids (Eddy 1998). The HSP70-2 protein is abundant in pachytene spermatocytes and the gene KO approach was used to determine if the protein has a critical role in meiosis. *Hsp70-2* KO females were fertile, but males were infertile. Spermatocytes were found to arrest in development and to undergo apoptosis at the G2 to M-phase transition of the first meiotic division (Dix et al. 1996). This suggested that HSP70-2 might be a molecular chaperone for a cell-cycle regulatory component. Subsequent studies found that HSP70-2 directly interacts with Cdc2 and that this interaction is required for Cdc2/cyclin B1 complex formation (Zhu et al. 1997). However, HSP70-2 is also associated with the synaptonemal complex (Allen et al. 1996), suggesting that it is also a chaperone for other proteins involved in spermatogenesis.

Proteases

Precursor polypeptides often require cleavage at specific sites to become bioactive. The KOs of genes for two proteases expressed in spermatogenic cells believed to serve this function, angiotensin-converting enzyme (*Ace*) and proprotein convertase 4 (*Psck4*) result in impaired male fertility. KOs of genes for two other proteases unique to spermatogenic cells, acrosin (*Acr*) and fertilin beta (*Ftnb*) also altered sperm function.

The angiotensin-converting enzyme (ACE) increases blood pressure by removing C-terminal dipeptides to activate angiotensin and to inactivate bradykinin. However, a testis form of ACE encoded by an alternative transcript derived from the 3' region of the *Ace* gene is synthesized only in spermatids (Langford et al. 1993). The *Ace* gene KO resulted in greatly reduced fertility in males, but not in females (Krege et al. 1995). Although structures of the testis and the production of sperm were not disrupted, sperm from *Ace* KO mice were less effective at transport through the oviduct, at binding to the zona pellucida and at fertilization *in vivo*, compared to sperm from wild-type mice (Hageman et al. 1998). Male and female mice lacking angiotensin had normal fertility, indicating that angiotensin I is unlikely to be the substrate for testis ACE. However, the role of testis ACE in sperm function remains to be determined.

Proprotein convertase 4 (PC4) is found only in spermatids in the mouse (Torii et al. 1993) and is a serine protease that cleaves precursor polypeptides at specific basic residues to produce activated polypeptides. The KO of the *Psck4* gene encoding PC4 caused reduced litter sizes when either gene KO males or females were mated with heterozygous animals, but the effect was greater for gene KO males than for gene KO females (Mbikay et al. 1997). There were no apparent changes in spermatogenesis, but the ability of sperm from *Psck4* KO mice to fertilize *in vitro* was significantly reduced and there was transmission distortion, with a lower than expected frequency of the mutant allele represented in progeny of heterozygote intercrosses (Mbikay et al. 1997).

Acrosin is a trypsin-like protease present in the acrosome that was thought to be necessary for sperm penetration through the zona pellucida. The *Acr* gene is expressed in late pachytene spermatocytes and the protein is synthesized in spermatids. However, males with an *Acr* gene KO were fertile, with the only effect detected being a modest delay in timing for *in vitro* fertilization (Adham et al. 1997; Baba et al. 1994).

Fertilin is an α:ß dimer encoded by two genes that is present on the sperm surface and hypothesized to act in sperm-egg binding and fusion. The KO of the fertilin beta gene (*Ftnb*) resulted in male infertility (Cho et al. 1998). Sperm failed to bind to the zona pellucida *in vitro*, but after removal of the zona, a few bound to egg plasma membrane and fertilization occurred. In addition, sperm from gene KO males failed to migrate from the uterus into the oviduct. These results indicate that fertilin beta serves an important role in sperm function during the fertilization process.

Regulation of Apoptosis

Different members of the Bcl-2-related family of proteins can have either positive or negative effects on the regulation of apoptosis (Korsmeyer 1995). BAX is a heterodimeric partner of Bcl-2 and counters the apoptosis-repressing role of Bcl-2 to promote apoptosis. BAX is expressed at high levels in spermatogonia (Rodriguez et al. 1997). Male *Bax* KO mice were infertile, the testes were atrophic, and sperm were absent from their epididymis (Knudson et al. 1995). There is an elevated level of apoptosis near the end of spermatogonial development (Allen et al. 1992), which appears to be suppressed in *Bax* KO mice, suggesting that the increase in numbers of spermatogonia caused disruption of subsequent phases of spermatogenesis. This is supported by studies with male transgenic mice that expressed high levels of the apoptosis-inhibiting Bcl2 or Bclx long (BclxL) proteins in spermatogonia. These males were infertile, with the seminiferous tubules containing excessive numbers of spermatogonia, and spermatogenesis being disrupted (Furuchi et al. 1996; Rodriguez et al. 1997). In another related study, a line of mutant mice with male infertility created using a retroviral gene-trap system was found to have a mutation in the *Bclw* gene encoding an apoptosis-inhibiting protein (Ross et al. 1998). Although spermatogenesis was blocked during late spermatid development in juveniles, apoptosis increased substantially in spermatocytes, followed later by the loss of spermatids and then a degeneration of the seminiferous epithelium. However, *Bclw* is expressed both in elongating spermatids and Sertoli cells and the mechanisms involved in the testicular degeneration remain to be determined.

Other Genes

The DAZ (deleted in azoospermia) gene is a candidate for the male sterility gene present on the human Y chromosome. The mouse homolog (DAZ-like) of this gene, *Dazla* is an RNA-binding protein expressed specifically in germ cells of males and females. The transcripts were found predominantly in type B spermatogonia and in preleptotene spermatocytes (Niederberger et al. 1997). The protein is located in the cytoplasm, being detected first in type B spermatogonia and most abundant in early to mid-pachytene spermatocytes (Ruggiu et al. 1997). The *Dazla* gene KO resulted in male and female infertility, with loss of germ cells beginning during fetal development in both sexes. The testes were small and there was severe disruption of testicular histology, with few germ cells present beyond the spermatogonial stage. In addition, a gene dosage effect was seen in heterozygous males with sperm being reduced in numbers and the majority of sperm having abnormal morphology.

The KOs of other genes also have altered the male reproductive system, but the genes are expressed constitutively. It is unclear whether the phenotypes are due to direct or indirect effects of the KO on spermatogenesis and sperm function. One of these is the *Cenpb* gene which encodes the centromere DNA-binding protein B (CENP-B) involved in assembly of arrays of centromeric α-satellite or minor satellite DNA (Muro et al. 1992). The *Cenpb* gene KO resulted in reduced body weight, testis weight and sperm numbers, but litter sizes and reproduction apparently were unchanged (Hudson et al. 1998).

The *Apob* gene encodes apolipoprotein B (apo B), that is synthesized mainly in the liver and intestine and is a structural constituent of several classes of lipoprotein particles. While the *Apob* gene KO was lethal, heterozygous mice had defects in neural tube closure, high density lipoprotein production, and male infertility (Huang et al. 1995). Only an occasional heterozygous male was fertile and sperm *in vitro* had low motility, survived for a short time and failed to fertilize eggs with an intact zona pellucida. However, they were able to fertilize eggs once the zona pellucida was removed (Huang et al. 1996). The *Apob* transcript was found in the testis and epididymis and the fertility defect could be partially corrected with a human *Apob* transgene that was expressed in the testis and epididymis (Huang et al. 1996).

The β1,4-galactosyltransferase enzyme is expressed constitutively, but an isoform associated with the plasma membrane has been suggested to mediate sperm binding to the egg zona pellucida glycoprotein ZP3 (Miller, Shur 1994). Although males with a KO of the *Ggtb* gene were fertile, their sperm *in vitro* bound less ZP3 than sperm from wild-type mice, failed to undergo the ZP3-induced acrosome reaction and penetrated the zona pellucida and fertilized eggs poorly (Lu, Shur 1997). It appears that ZP3 binding and induction of the acrosome reaction are dispensable for fertilization *in vivo*, but that these processes are beneficial to fertilization *in vitro*.

Gene KOs with Indirect Effects on Male Fertility

Some gene KOs have indirect effects on male fertility by disrupting male reproductive tract development, endocrine processes or the function of somatic cells of the testis. These KOs have involved genes for growth factors, growth factor receptors, morphogens, hormone receptors, and transcription factors (Table 2).

Gene KOs with Developmental Effects

Disruption of development of the male reproductive system has been caused by KOs of genes for growth factors or growth factor receptors (*Amh*, *Amhr2*), for transcription factors (*Ftzf1*, *Hoxa10*, *Hoxa11*, *Nhlh2*, *Sp4*,

Pax-2) and an orphan tyrosine kinase receptor (c-*ros*). Although most of these genes are expressed in the developing reproductive tract, some are expressed in the developing brain.

Anti-Müllerian hormone (AMH) is a member of the TGF-β superfamily of growth factors and gene KO studies have shown that AMH has an important role during development of the male reproductive system. AMH is produced by the Sertoli cells of the fetal testis and induces regression of the Müllerian ducts in the male fetus. These ducts give rise to the oviducts, uterus and cranial portion of the vagina in the female. Male *Amh* gene KO mice had scrotal testes and produced sperm, but most were infertile (Behringer et al. 1994). They developed as pseudohermaphrodites with internal male and female reproductive organs, but sperm transfer was obstructed by the presence of the female reproductive organs. However, female *Amh* KO mice were fertile and

Table 2: Gene KOs with indirect effects on male fertility

Gene	Expression	Effects	References
Developmental Effects			
Amh	Müllerian duct in fetus, Sertoli cells	most males infertile, pseudohermaphroditism, testis structure and sperm production normal, sperm transport blocked	Behringer et al. 1994
Amhr2	Müllerian duct in fetus, Sertoli and granulosa cells	most males infertile, pseudohermaphroditism, testis structure and sperm production normal, sperm transport blocked	Mishina et al. 1996
Ftzf1	urogenital ridge, steroidogenic tissues, fetal Sertoli cells	postnatal death, gonads and adrenals fail to develop	Luo et al. 1994
Hoxa10	developing urogenital system	most males infertile, others with declining fertility with age, bilateral cryptorchidism, disruption of spermatogenesis	Satakata et al. 1995
Hoxa11	stromal cells surrounding Müllerian and Wolffian ducts	males and females infertile, failure of testicular descent, malformation of epididymis, disruption of spermatogenesis	Hsieh-Li et al. 1995
Pax-2	developing urogenital system	males and females infertile, testis develops, but epididymis, vas deferens and seminal vesicles absent	Torres et al. 1995
c-*ros*	Wolffian duct and epididymis	male infertility, lack of differentiation of initial segment of epididymis; testis and sperm normal in appearance, sperm able to fertilize eggs *in vitro*, defective sperm transport in oviduct	Sonnenberg-Riethmacher et al.1996
Nhlh2	developing central and peripheral nervous systems	most males infertile, reproductive organs small, spermatogenesis delayed, few spermatids, infrequent matings	Good et al. 1997
Sp4	developing central nervous system, digestive tract, testis	most males infertile, testis structure and sperm production normal, infrequent matings	Supp et al. 1996
Endocrine Function			
Inha	Sertoli cells and granulosa cells	male infertility, spermatogenesis begins and then regresses as interstitial tumors appear and enlarge	Matzuk et al. 1992
Acvr2a	pituitary gonadotrophs	males fertile, sperm production delayed and testis small, low FSH levels	Matzuk et al. 1995
Estra	constitutive, including Sertoli cells, Leydig cells, efferent ducts, epididymis	male and female infertility, male reproductive tract develops normally, spermatogenesis begins and fails, sperm have low motility and are unable to fertilize eggs *in vitro*	Eddy et al. 1996
Sertoli Cells			
Fshb	pituitary gonadotrophs	males fertile, females infertile, small testis and reduced sperm production	Kumar et al. 1997
Cycd2	constitutive, Sertoli cells during perinatal period, spermatogonia	males fertile, females infertile, small testis	Sicinski et al. 1996
Dhh	Sertoli cell precursors, Schwann cells, vascular endothelium, endocardium in fetus	male infertility, testis small, pachytene spermatocytes present, but few spermatids and no sperm	Bitgood et al. 1996
Rxrb	constitutive, only Sertoli cells in testis	male infertility, testis develops normally, progressive degeneration of seminiferous epithelium in adults, few sperm present and immotile and abnormal appearing	Kastner et al. 1996
Fmr1	brain, adrenal, ovary, testis	males and females fertile, testis enlarged, testis structure and spermatogenesis normal	Bakker et al. 1994; Slegtenhorst-Eedgeman et al. 1998

epididymal sperm from male *Amh* KO mice were capable of fertilizing eggs *in vitro*. This indicates that AMH is not required for development of the testis or production of germ cells in males and females. The occurrence of Leydig cell hyperplasia suggested that AMH has an additional role in the regulation of Leydig cell proliferation.

The KO of the gene for the AMH type II receptor (*Amhr2*) also disrupted developmental processes and led to defects in male fertility. The receptor is present on the surface of mesenchymal cells adjacent to the Müllerian duct epithelium during embryogenesis, and on Sertoli cells and granulosa cells in juvenile and adult mice. The effects seen in these gene KO mice were much like those seen in *Amh* KO mice. The gene KO females were fertile and appeared to be unaffected. The *Amhr2* KO males developed as pseudohermaphrodites, with complete male and female reproductive tracts (Mishina et al. 1996). The males produced sperm, but most were infertile due to blockage of sperm transport by the presence of female reproductive organs. The gene KO males also developed Leydig cell hyperplasia, as occurred in *Amr* KO mice.

Steroidogenic factor 1, encoded by the fushi tarazu gene (*Ftzf1*), is expressed in all primary steroidogenic tissues, where it is a transcriptional regulator of steroid hydroxylases. It is also expressed in the urogenital ridge at embryonic day 9-9.5 (E9-9.5) and subsequently in fetal Sertoli cells (Ikeda et al. 1993, 1994). Although primordial germ cells were present in the genital ridges of *Ftzf1* KO fetuses at E10.5 and E11.5, apoptosis and degeneration led to a nearly complete absence of the gonads and adrenals by E12.5 (Luo et al. 1994). Male and female gene KO mice have female internal reproductive organs and die by postnatal day 8 due to adrenal insufficiency. One of the genes regulated by *Ftzf1* is *Amh* (Shen et al. 1994). Thus, gene KO studies have shown that similar effects are produced by disrupting *Amh* gene activation, blocking production of AMH protein or preventing activation of AMH type II receptor-mediated processes. The studies also demonstrate the essential role of AMH in male sexual differentiation.

Homeobox genes encode transcription factors and were originally identified in *Drosophila* by their association with homeotic mutations. The *Abdominal B* homeobox gene is expressed in the posterior region of *Drosophila* and the homologous *Hoxa10* and *Hoxa11* genes are expressed in the developing urogenital system of mammals (Krumlauf 1994). The *Hoxa10* KO produced multiple defects, including bilateral cryptorchidism and disruption of spermatogenesis (Satakata et al. 1995). The majority of gene KO males were infertile and others showed declining fertility with age. The *Hoxa10* gene is expressed strongly in the fetal and postnatal gubernacular cord and bulb and their derivatives. It was suggested that the reproductive defects seen in *Hoxa10* KO males were caused by failure of gubernacular cord shortening, diminished outgrowth of the gubernacular bulb and limited development of the cremasteric muscle. Male infertility was probably due to disruption of spermatogenesis secondary to failure of testicular descent. Females were often infertile as well, with the primary problem being preimplantation death of embryos that was apparently due to an altered uterine environment.

The *Hoxa11* gene is expressed in male and female fetuses in the stromal cells surrounding the Müllerian and Wolffian ducts, but not in the developing gonads. Both male and female *Hoxa11* KO mice were infertile, with the males experiencing failure of testicular descent, malformation of the vas deferens and disruption of spermatogenesis (Hsieh-Li et al. 1995). It was suggested that the main cause of male infertility was the disruption of spermatogenesis due to failure of testicular descent. Females ovulated and mated, and their eggs were fertilized, but the embryos developed only if transferred to wild-type surrogate mothers, suggesting that the uterus failed to accommodate implantation and/or development.

The KO of the *Pax-2* gene that encodes a transcription factor in the paired-box family also disrupted development of the male reproductive system. It is expressed during development in the ducts and mesenchymal components of the urogenital system derived from the intermediate mesoderm. Although the testes and ovaries develop in *Pax-2* KO mice, the Wolffian and Müllerian ducts degenerate and males lack the efferent ducts, epididymis, vas deferens and seminal vesicles, and females lack the oviduct, uterus and vagina (Torres et al. 1995).

The c-*ros* gene KO also disrupted male reproductive tract development. This gene encodes a transmembrane tyrosine kinase receptor for an unknown ligand originally identified as the mutant form of an oncogene. It is expressed in the Wolffian ducts in the embryo and in the epithelial cells of the caput epididymis in the adult, with the highest level of c-*ros* occurring in the initial segment (Sonnenberg-Riethmacher et al. 1996). The c-*ros* gene KO resulted in the absence of the initial segment and male infertility. Although sperm were present in normal numbers in the cauda epididymis and fertilized eggs *in vitro*, no sperm were present in the oviducts following mating between most gene KO males and wild-type females (Sonnenberg-Riethmacher et al. 1996). These results suggest that the initial segment of the epididymis serves an essential role in sperm maturation processes influencing sperm transport in the female reproductive tract.

Two transcription factor genes expressed in the developing nervous system, *Nhlh2* and *Sp4*, have been found to be required for male reproduction. The *Nhlh2* (nescient helix-loop-helix 2) gene encodes a transcription factor expressed in the developing central and peripheral nervous systems, including portions of the brain destined to become the hypothalamus and anterior pituitary. However, there were no overt morphological abnormalities in these regions in adult *Nhlh2* KO mice. Gene KO males at weaning had a small penis, bilateral cryptorchidism, reduced ano-genital distance, vestigial preputial glands and small and convoluted seminal vesicles (Good et al. 1997). The epididymis and vas deferens were slightly reduced in size and the testis was significantly smaller in adults. In gene KO males, the initiation of spermatogenesis was delayed, few spermatids were seen in the testis at 8 weeks, and matings were infrequent. Eleven gene KO males that were caged with wild-type females for over a year produced only one offspring. Androgen levels were low and it was suggested that infertility was due to reduced androgen-regulated sexual response and testicular dysfunction.

Expression of the *Sp4* gene in the fetus occurs in the developing central nervous system, respiratory system, digestive tract and testis. The majority of *Sp4* KO mice died within the first few days after birth from undetermined causes and the surviving gene KO mice were smaller than wild-type littermates (Supp et al. 1996). The testes of gene KO mice appeared normal by histology and produced sperm. However, the males rarely bred and only one of 25 males housed with wild-type females from several weeks to a year sired a litter. It was suggested that the *Sp4* gene is required for development of normal male reproductive behavior. However, the fertility of female mice was unaffected by the *Sp4* gene KO and it apparently was not determined if the gene KO had effects on sperm function.

Gene KOs That Disrupt Endocrine Function

Male fertility has been disrupted by the KO of genes for a growth factor and a growth factor receptor (*Inha*, *Acvr2a*) involved in gonadal-hypophyseal endocrine feedback, and a receptor for a steroid hormone (*Estra*) usually considered a regulator of the female reproductive system.

Inhibin is another member of the TGF-β growth factor superfamily and a dimer of α:β subunits encoded by two genes. The major sites of inhibin synthesis are Sertoli cells in males and granulosa cells in females. Inhibin is secreted and may have intragonadal paracrine or autocrine roles, as well as an inhibitory role in production of follicle-stimulating hormone (FSH) by the pituitary. Male mice with a KO of the inhibin alpha gene (*Inha*) had normal external genitalia and scrotal testes, but were infertile (Matzuk et al. 1992). It was seen by histology that spermatogenesis was occurring by five to seven weeks of age, but that bilateral or unilateral interstitial cell tumors were evident at four weeks. Testicular enlargement was apparent by five weeks and spermatogenesis regressed as the tumor mass enlarged. Most females also were infertile and developed ovarian stromal tumors. FSH levels were elevated two- to threefold in both male and female gene KO mice. In male mice with unilateral tumors, spermatogenesis regressed in the contralateral testis and it was suggested that the tumors might be secreting substances detrimental to germ cells. The specific mechanisms by which this gene KO causes regression of spermatogenesis and the development of interstitial tumors in *Inha* KO mice remain to be determined. However, inhibin alpha appears to be important for male fertility both during development and for regulating endocrine function.

The KO of the gene for the type IIA activin receptor (*Acvr2a*), a cell surface receptor for activin, disrupted endocrine function and led to defects in spermatogenesis and male fertility. Activin is yet another member of the TGF-β superfamily of growth factors. It is produced by Sertoli cells and granulosa cells, stimulates synthesis of FSH by the pituitary and antagonizes the effects of inhibin. Although male *Acvr2a* gene KO mice had delayed sperm production and smaller testes and seminiferous tubules, they were fertile (Matzuk et al. 1995). The effects of this gene KO were thought to be due mainly to reduced FSH production, which caused a lower rate of Sertoli cell proliferation during the perinatal period and resulted in the small testis size.

The estrogen receptor alpha gene *(Estra)* is a member of the nuclear receptor superfamily of ligand-activated transcription factors. It is expressed in Sertoli and Leydig cells of the testis and in the excurrent ducts of the male reproductive system. Male mice with a KO of the *Estra* gene had a fully developed reproductive system, but were infertile (Eddy et al. 1996). Although spermatogenesis began normally, sperm production declined rapidly during the first few weeks of life. The percentage of sperm that were motile was substantially less in gene KO than in wild-type mice, and the sperm were unable to fertilize eggs *in vitro*. The seminiferous tubules began to dilate by 20 days after birth, and spermatogenesis declined and the tubules degenerated over the next few weeks. The highest level of the receptor in the excurrent duct system was reported to be in the efferent ducts (Greco et al. 1992; Schleicher et al. 1984), the region where most of the fluid leaving the testis is reabsorbed (Jones, Jurd 1987). It appears that fluid reabsorption in the efferent ducts is compromised in mice lacking the receptor, resulting in fluid accumulation in the testis and leading to disruption of spermatogenesis (Eddy et al. 1996; Hess et al. 1997). It remains to be determined if the receptor also is required by Sertoli cells, spermatogenic cells and/or the epididymis for the production of sperm capable of fertilization.

KOs That Affect Sertoli Cells

The KOs of several constitutively expressed genes have altered the number or function of Sertoli cells and had effects on male fertility. Included are a gonadotrophin subunit (*Fshb*), a cell cycle component (*Cycd2*), a morphogen (*Dhh*), a member of the nuclear receptor superfamily (*Rxrb*), and a gene associated with mental retardation (*Fmr1*). The effects of some of these KOs (*Fshb, Cycb*) are consistent with well-known FSH-dependent regulation of Sertoli cell proliferation during the perinatal period (Kluin et al. 1984; Orth 1984), whereas the effects of another (*Fmr1*) indicates that additional mechanisms are involved in regulating Sertoli cell numbers.

FSH is a glycoprotein secreted by the anterior pituitary and binds to FSH receptors on the surface of granulosa cells of the ovary and Sertoli cells in the testis. It is a heterodimer with an alpha subunit shared with other hormones and a beta subunit specific to FSH. The KO of the gene for the FSH beta subunit (*Fshb*), resulted in infertility in females due to a block in follicle maturation prior to the antral stage (Kumar et al. 1997). Gene KO males were fertile, but the testes were small and epididymal sperm counts were reduced substantially, compared to those of wild-type males. These results were consistent with a reduced rate of Sertoli cell divisions.

Cyclin D2 is a member of a family of proteins that serves as the regulatory subunit in a heterodimer with a cyclin-dependent kinase to modulate progression of cells through the G1 phase of the cell cycle. The expression of cyclin D2 is widespread and overlaps the expression of other D-type cyclins. Mice with a KO of the *Cycd2* gene appeared normal. However, *Cycd2* KO females were infertile, the number of granulosa cells surrounding each oocyte was reduced, and FSH failed to stimulate granulosa cell proliferation (Sicinski et al. 1996). *Cycd2* expression can be induced in ovarian granulosa cells *in vitro* by FSH via a cAMP-dependent pathway, indicating that cyclin D2 is an FSH-responsive gene (Sicinski et al. 1996). Gene KO males were fertile, but the testes were small and sperm counts were reduced. Cyclin D2 is expressed in dividing Sertoli cells during the perinatal period and in spermatogonia in juvenile and adult mice (Nakayama et al. 1996), suggesting that the effect of FSH on Sertoli cell numbers and testis size may occur through a cyclin D2-dependent mechanism.

Morphogens are secreted proteins implicated in a wide range of activities involving cell interactions, with their best known effects occurring in development. The hedgehog gene was described originally as a key regulator of pattern formation in embryogenic structures in *Drosophila*. The mouse has at least three homologs of hedgehog. One of these, Desert hedgehog (*Dhh*) is expressed in fetal Schwann cells, vascular endothelium and endocardium. It is also expressed in Sertoli cell precursors of the presumptive testis on day 11.5 of embryonic development and in the seminiferous epithelium in the adult, presumably in Sertoli cells, but is not expressed in the fetal or adult ovary. The *Dhh* gene KO had no effects on female fertility, but males were infertile and had small testes containing pachytene spermatocytes, few spermatids and no sperm (Bitgood et al. 1996). Testis size was reduced as early as 10 days after birth. It was suggested that *Dhh* has a role in regulating germ cell proliferation and is required for progression of germ cells through the spermatid stage and completion of spermiogenesis. This may occur indirectly through the homolog of the *Drosophila* patched receptor (*Ptc*) which is present on Leydig cells.

Gene KO studies have shown that the retinoid X receptor β (RXRβ), a ligand-activated transcription faction and an additional member of the nuclear receptor superfamily, is required for effective spermatogenesis. RXRβ binds 9-*cis* retinoic acid, and *Rxrb* transcripts are restricted to Sertoli cells in the testis. Half of the *Rxrb* KO mice died before or at birth, but the survivors appeared normal. However, the males were sterile and the few sperm present in the caudae epididymides had an elongated or sickle-shaped nucleus and a coiled tail and were immotile (Kastner et al. 1996). There was a progressive accumulation with age of lipid in Sertoli cells, degeneration of the germinal epithelium and formation of acellular lipid-filled tubules. It appears that the primary defects caused by the *Rxrb* KO occur in Sertoli cells, but the specific genes in Sertoli cells regulated by this transcription factor remain to be determined.

The fragile X syndrome is a frequent X-linked hereditary form of mental retardation in humans. The gene involved in this disorder encodes an RNA-binding protein and the mouse homolog (*Fmr1*) is expressed in the brain, ovary, adrenal and testis (Bächner et al. 1993; Hinds et al. 1993). The fragile X syndrome results in macroorchidism in humans and the *Fmr1* gene KO produced this effect in male mice (Bakker et al. 1994). It was found that the rate of Sertoli cell proliferation was increased from E12 to 15 days postnatally in *Fmr1* KO mice, compared to wild-type mice. However, serum levels of FSH, FSH receptor mRNA expression, and short-term effects of FSH on Sertoli cells were not changed in the gene KO mice (Slegtenhorst-Eegdeman et al. 1998). This suggested that the increased number of Sertoli cells that results in macroorchidism in *Fmr1* KO mice does not involve a change in FSH action.

Summary

More than 40 KOs of genes are known that have varying effects on spermatogenesis and male reproduction. It is notable that most of the KOs involving genes expressed only in spermatogenic cells (*Sprm1, Clgn, Hsp70-2, Psck4, Ftnb*) resulted in reduced fertility or infertility. Similar results were seen for alternatively spliced transcripts found only in spermatogenic cells (*Crem, Ace, A-myb*). There appear to be many such uniquely expressed genes and transcripts in spermatogenic cells (Eddy, O'Brien 1998), suggesting that these may have been selected during evolution to enhance the success of the reproductive process. It is also interesting that KOs of some more widely expressed genes produced their major effects in spermatogenic cells (*Bmp8a, Bmp8b, Crem,* A-*myb, Hr6b, Pms2, Bax)* or in male and female germ cells (*Zfx, Dmc1, Mlh1, Dazla*). These effects often appeared during meiosis, suggesting that this phase of germ cell development is particularly susceptible to disruption. Several of the KOs were for genes involved in DNA repair and recombination and it is not surprising that their proteins function during meiotic recombination. However, it is less obvious why the products of some other genes (for example, *A-myb, Bax*), are particularly important during germ cell development.

Results of gene KO studies also demonstrate that many other genes are required for the success of spermatogenesis and male reproduction. They encode proteins that participate in development, endocrine function or Sertoli cell activities essential for these processes. Gene KOs have produced developmental effects on sexual differentiation of the reproduction systems (*Amr, Amhr2, Ftzf1*), formation of the urogenital system (*Hoxa10, Hoxa11, Pax-2,* c-*ros*) and on development of mating behavior (*Nhlh2, Sp4*). Some of the gene KOs have produced unanticipated effects on endocrine function in the male, from modest (*Fshb*) to extensive (*Inha, Estra*).It was surprising to many that the *Fshb* gene KO resulted only in reduced testis size without disrupting male fertility, that the *Estra* gene KO resulted in male infertility and that the *Inha* gene KO resulted in interstitial tumors.

Finally, the results of gene KO studies in mice provide information that may be significant for the detection and treatment of infertility in men. Some gene KO studies have found effects on fertility or male reproduction in heterozygous males (*Crem, Dazla, Apob*), indicating that the amount of the protein present is important for germ cell function. This suggests that some forms of infertility in men may occur when only one allele of a gene is mutated. It is unlikely that this would have been learned except through the use of the gene KO method. In addition, several gene KOs were found to reduce *in vivo* or *in vitro* fertility without producing apparent changes in the structure of the male reproductive system, sperm output and motility or endocrine parameters (*Sprm1, Clgn, Ace, Psck4, Acr, Ftnb, Ggtb*). This suggests that some "male factor" causes of infertility in couples will go undetected when physical examination, sperm numbers and motility, and serum hormone levels appear normal. Furthermore, the KOs of some genes expressed in spermatogenic cells (*Ace, Ftnb*) and in the caput epididymis (c-*ros*) resulted in defects in sperm transport in the oviduct of the female reproductive tract. It has been shown for two of these (*Ftnb*, c-*ros*) that the sperm produced were capable of *in vitro* fertilization. These and other gene KO results relate to the concern that gene mutations not causing obvious effects on sperm production may be transmitted to children conceived using *in vitro* fertilization (IVF) or other assisted reproductive technology methods.

References

Adham IM, Nayernia K, Engel W. Spermatozoa lacking acrosin protein show delayed fertilization. Mol Reprod Dev 1997; 46:370-376.

Allen DJ, Harmon BV, Roberts SA. Spermatogonial apoptosis has three morphologically recognizable phases and shows no circadian rhythm during normal spermatogenesis in the rat. Cell Prolif 1992; 25:241-250.

Allen JW, Dix DJ, Collins BW, Merrick BA, He C, Selkirk JK, Poorman-Allen P, Dresser ME, Eddy EM. HSP70-2 is part of the synaptonemal complex in mouse and hamster spermatocytes. Chromosoma 1996; 104:414-421.

Andersen B, Pearse RV II, Schlegel PN, Chichon Z, Shonemann MD, Bardin CW, Rosenfeld MG. Sperm 1: a POU-domain gene transiently expressed immediately before meiosis I in the male germ cell. Proc Natl Acad Sci USA 1993; 90:11084-11088.

Baba T, Azuma S, Kashiwabara S-I, Toyoda Y. Sperm from mice carrying a targeted mutation of the acrosin gene can penetrate the oocyte zona pellucida and effect fertilization. J Biol Chem 1994; 50:31845-31849.

Bächner D, Mance A, Steinbach P, Wohrle D, Just W, Vogel W, Hameister H, Poustka A. Enhanced expression of the murine *FMR1* gene during germ cell proliferation suggests a special function in both the male and the female gonad. Hum Mol Genet 1993; 2:2043-2050.

Baker SM, Bronner CR, Zhang L, Plug AW, Robatzek M, Warren G, Elliott EA, Yu J, Ashley T, Arnheim N, Flavell RA, Liskay RM. Male mice defective in the DNA mismatch repair gene *PMS2* exhibit abnormal chromosome synapsis in meiosis. Cell 1995; 82:309-319.

Baker SM, Plug AW, Prolla TA, Bronner CR, Harris AC, Yao X, Christie D-M, Monell C, Arnheim N, Bradley A, Ashley T, Liskay RM. Involvement of mouse *Mlh1* in DNA mismatch repair and meiotic crossing over. Nature Genet 1996; 14:336-342.

Bakker CE, Verheij C, Willemsen R, van der Helm R, Oerlemans F, Vermey M, Bygrave A, Hoogeveen AT, Oostra BA, Reyniers E, De Boulle K, D'Hooge R, Cras P, Van Velzen D, Nagels G, Martin J-J, De Deyn PP, Darby JK, Willems PJ. *Fmr1* knockout mice: a model to study fragile X mental retardation. Cell 1994; 78:23-33.

Barlow C, Hirotsune S, Paylor R, Liyanage M, Eckhaus M, Collins F, Shiloh Y, Crawley JN, Ried T, Tagle D, Wynshaw-Boris A. *Atm*-deficient mice: a paradigm of ataxia telangiectasia. Cell 1996; 86:159-171.

Behringer RR, Finegold MJ, Cate RL. Müllerian-inhibiting substance function during mammalian sexual development. Cell 1994; 79:415-425.

Bitgood MJ, Shen L, McMahon AP. Sertoli cell signaling by Desert hedgehog regulates the male germline. Current Biol 1996; 5:298-304.

Blendy JA, Kaestner KH, Weinbauer GF, Nieschlag E, Schütz G.

Severe impairment of spermatogenesis in mice lacking the CREM gene. Nature 1996; 380:162-165.

Cho C, Bunch DO, Faure J-E, Goulding EH, Eddy EM, Primakoff P, Myles DG. Male mice lacking fertilin β are infertile and have sperm with multiple defects in the fertilization process. Science 1998; 281:1857-1859.

Delmas V, Sassone-Corsi P. The key role of CREM in the cAMP signaling pathway in the testis. Mol Cell Endocrinol 1994; 100:121-124.

Dix DJ, Allen JW, Collins BW, Mori C, Nakamura N, Poorman-Allen P, Goulding EH, Eddy EM. Targeted gene disruption of Hsp70-2 results in failed meiosis, germ cell apoptosis, and male infertility. Proc Natl Acad Sci USA 1996; 93:3264-3268.

Eddy EM. Regulation of gene expression during spermatogenesis. Sem Cell Develop Biol 1998; 9:451-457.

Eddy EM, O'Brien DA. Gene expression during mammalian meiosis. In: (Handel MA, ed) Meiosis and Gametogenesis. San Diego: Academic Press, 1998, pp141-200.

Eddy EM, Washburn TF, Bunch DO, Goulding EH, Gladen BC, Lubahn DB, Korach KS. Targeted disruption of the estrogen receptor gene in male mice causes alteration of spermatogenesis and infertility. Endocrinology 1996; 137:4796-4805.

Edelmann W, Cohen PC, Kane M, Lau K, Morrow B, Bennett S, Umar A, Kunkel T, Cattoretti G, Chaganti R, Pollard JW, Kolodner RD, Kucherlapati R. Meiotic pachytene arrest in MLH1-deficient mice. Cell 1996; 85:1125-1134.

Elson A, Wang Y, Daugherty CJ, Morton CC, Zhou F, Campos-Torres J, Leder P. Pleiotropic defects in ataxia-telangiectasia protein-deficient mice. Proc Natl Acad Sci USA 1996; 93:13084-13089.

Furuchi T, Masuko K, Nishimune Y, Obinata M, Matsui Y. Inhibition of testicular germ cell apoptosis and differentation in mice misexpressing Bcl-2 in spermatogonia. Development 1996; 122:1703-1709.

Good DJ, Porter FD, Mahon KA, Parlow AF, Westphal H, Kirsch IR. Hypogonadism and obesity in mice with a targeted deletion of the *Nhlh2* gene. Nature Genet 1997; 15:397-401.

Greco TL, Furlow JD, Duello TM, Gorski J. Immunodetection of estrogen receptors in fetal and neonatal male mouse reproductive tracts. Endocrinology 1992; 130:421-429.

Habu T, Taki T, West A, Nishimune Y, Morita T. The mouse and human homologs of *DMC1*, the yeast meiosis-specific homologous recombination gene, have a common unique form of exon-skipped transcripts in meiosis. Nucleic Acids Res 1996; 24:470-477.

Hageman JR, Moyer JS, Bachman ES, Sibony M, Magyar PL, Welch JE, Smithies O, Krege JH, O'Brien DA. Angiotensin-converting enzyme and male fertility. Proc Natl Acad Sci USA 1998; 95:2552-2557.

Hess RA, Bunick D, Lee K-H, Bahr J, Taylor JA, Korach KS, Lubahn DB. A role for oestrogens in the male reproductive system. Nature 1997; 390:509-512.

Hinds HL, Ashley CT, Sutcliff JS, Nelson DL, Warren ST, Houseman DE, Schalling M. Tissue specific expression of FMR-1 provides evidence for a functional role in fragile X syndrome. Nature Genet 1993; 3:36-43.

Hsieh-Li HM, Witte DP, Weinstein M, Branford W, Li H, Small K, Potter SS. *Hoxa 11* structure, extensive antisense transcription, and function in male and female fertility. Development 1995; 121:1373-1385.

Huang L-S, Voyiaziakis E, Markenson DF, Sokol KA, Hayek T, Breslow JL. Apo B gene knockout in mice results in embryonic lethality in homozygotes and neural tube defects, male infertility, and reduced HDL cholesterol ester and apo A-I transport rates in heterozygotes. J Clin Invest 1995; 96:2152-2161.

Huang L-S, Voyiaziakis E, Chen HL, Rubin EM, Gordon JW. A novel functional role for apolipoprotein B in male infertility in heterozygous apolipoprotein B knockout mice. Proc Natl Acad Sci USA 1996; 93:10903-10907.

Hudson DF, Fowler KJ, Earle E, Saffery R, Kalitsis P, Trowell H, Hill J, Wreford NG, de Kretser DM, Cancilla MR, Howman E, Hii L, Cutts SM, Irvine DV, Choo KHA. Centromere protein B null mice are mitotically and meiotically normal but have lower body and testis weights. J Cell Biol 1998; 141:309-319.

Hummler E, Cole TJ, Blendy JA, Ganss R, Aguzzi A, Scmid W, Beermann F, Schütz G. Targeted mutation of the CREB gene: compensation within the CREB/ATF family of transcription factors. Proc Natl Acad Sci USA 1994; 91:5647-5651.

Ikawa M, Wada I, Kominami K, Watanabe D, Toshimori K, Nishimune Y, Okabe M. The putative chaperone calmegin is required for sperm fertility. Nature 1997; 387:607-611.

Ikeda Y, Lala DS, Luo X, Kim E, Moisan M-P, Parker KL. Characterization of the mouse FTZ-F1 gene, which encodes a key regulator of steroid hydroxylases. Mol Endocrinol 1993; 8:852-860.

Ikeda Y, Shen W-H, Ingraham HA, Parker KL. Developmental expression of mouse steroidogenic factor 1, an essential regulator of the steroid hydroxylases. Mol Endocrinol 1994; 8:654-662.

Jones RC, Jurd KM. Structural differentiation and fluid reabsorption in the ductuli efferentes testis of the rat. Aust J Biol Sci 1987; 40:79-90.

Kastner P, Mark M, Leid M, Gansmuller A, Chin W, Grondona JM, Decimo D, Krezel W, Dierich A, Chambon P. Abnormal spermatogenesis in RXRβ mutant mice. Genes Develop 1996; 10:80-92.

Keegan KS, Holtzman DA, Plug AW, Christenson ER, Brainerd EE, Glaggs G, Bentlye NJ, Taylor EM, Meyn MS, Moss SB, Carr AM, Hoekstra MF. The Atr and Atm proteins associate with different sites along meiotically pairing chromosomes. Genes Develop 1996; 10:2423-2437.

Kluin PM, Kramer MF, De Rooij DG. Proliferation of spermatogonia and Sertoli cells in maturing mice. Anat Embryol 1984; 169:73-78.

Knudson CM, Tung KSK, Tourtellotte WG, Brown GA, Korsmeyer SJ. Bax-deficient mice with lymphoid hyperplasia and male germ cell death. Science 1995; 270:96-99.

Koken MHM, Hoogerbrugge JW, Jaspers-Dekker I, de Wit J, Willemsen R, Roest HP, Grootegoed JA, Hoeijmakers JHJ. Expression of the ubiquitin-conjugating DNA repair enzymes HHR6A and B suggests a role in spermatogenesis and chromatin modification. Develop Biol 1996; 173:119-132.

Korsmeyer SJ. Regulators of cell death. Trends Genet 1995; 11:101-105.

Krege JH, John SWM, Langenbach LL, Hodgin JB, Hageman JR, Bachman ES, Jennette JC, O'Brien DA, Smithies O. Male-female differences in fertility and blood pressure in ACE-deficient mice. Nature 1995; 375:146-148.

Krumlauf R. Hox genes in vertebrate development. Cell 1994; 78:191-201.

Kumar TR, Wang Y, Lu N, Matzuk MM. Follicle stimulating hormone is required for ovarian follicle maturation but not male fertility. Nature Genet 1997; 15:201-204.

Langford KG, Zhou Y, Russell LD, Wilcox JN, Bernstein KE. Regulated expression of testis angiotensin-converting enzyme during spermatogenesis in mice. Biol Reprod 1993; 48:1210-1218.

Li G-M, Modrich P. Restoration of mismatch repair to nuclear extracts of H6 colorectal tumor cells by a heterodimer of human MutL homologs. Proc Natl Acad Sci USA 1995; 92:1950-1954.

Lim DS, Hasty P. A mutation in mouse *rad51* results in an early embryonic lethality that is suppressed by a mutation in p53. Mol Cell Biol 1996; 16:7133-7143.

Lu Q, Shur BD. Sperm from β1,4-galactosyltransferase-null mice are refractory to ZP3-induced acrosome reactions and penetrate the zona pellucida poorly. Development 1997; 124:4121-4131.

Lufkin T, Lohnes D, Mark M, Dierich M, Gorry P, Gaub M-P, LeMeur M, Chambon P. High postnatal lethality and testis degeneration in retinoic acid receptor α mutant mice. Proc Natl Acad

Sci USA 1993; 90:7225-7229.
Luo X, Ikeda Y, Parker KL. A cell-specific nuclear receptor is essential for adrenal and gonadal development and sexual development. Cell 1994; 77:481-490.
Luoh S-W, Bain PA, Polakiewicz RD, Goodheart ML, Gardner H, Jaenisch R, Page DC. *Zfx* mutation results in small animal size and reduced germ cell number in male and female mice. Development 1997; 124:2275-2284.
Matzuk MM, Finegold MJ, Su J-GJ, Shueh AJW, Bradley A. α–Inhibin is a tumor-suppressor gene with gonadal specificity in mice. Nature 1992; 360:313-319.
Matzuk MM, Kumar TR, Bradley A. Different phenotypes for mice deficient in either activins or activin receptor type II. Nature 1995; 374:356-360.
Mbikay M, Tadros H, Ishida N, Lerner CP, de Lamirande E, Chen A, El-Alfy M, Clermont Y, Seidah NG, Cretien M, Gagnon C, Simpson EM. Impaired fertility in mice deficient for the testicular germ-cell protease PC4. Proc Natl Acad Sci USA 1997; 94:6842-6846.
Mettus RV, Litvin J, Wali A, Toscani A, Latham K, Hatton K, Reddy EP. Murine A-myb: evidence for differential splicing and tissue-specific expression. Oncogene 1994; 9:3077-3086.
Miller DJ, Shur BD. Molecular basis of fertilization in the mouse. Sem Develop Biol 1994; 5:255-264.
Mishina Y, Rey R, Finegold MJ, Matzuk MM, Josso N, Cate RL, Behringer RR. Genetic analysis of the Müllerian-inhibition substance signal transduction pathway in mammalian sexual differentiation. Genes Develop 1996; 10:2577-2587.
Muro Y, Masumoto H, Yoda K, Nozaki N, Ohashi M, Okazaki T. Centromere protein B assembles human centromeric alpha-satellite DNA at the 17-bp sequence, CENP-B box. J Cell Biol 1992; 116:585-596.
Nakayama H, Nishiyama H, Higuchi T, Kaneko Y, Fukumoto M, Fujita J. Change of cyclin D2 mRNA expression during murine testis development detected by fragmented cDNA subtraction method. Develop Growth Differ 1996; 38:141-151.
Nantel F, Monoco L, Foulkes NS, Masquilier D, LeMeur M, Henricksen K, Dierich A, Parvinen LM, Sassone-Corsi P. Spermiogenesis deficiency and germ-cell apoptosis in CREM-mutant mice. Nature 1996; 380:159-162.
Niederberger C, Agulnik AI, Cho Y, Lamb D, Bishop CE. *In situ* hybridization shows that *Dazla* expression in mouse testis is restricted to premeiotic stages IV-VI of spermatogenesis. Mamm Genome 1997; 8:277-278.
Orth JM. The role of follicle-stimulating hormone in controlling Sertoli cell proliferation in testes of fetal rats. Endocrinology 1984; 115:1248-1255.
Pearse RV II, Drolet EW, Kalla KA, Hooshmand F, Bermingham JRJ, Rosenfeld MG. Reduced fertility in mice deficient for the POU protein sperm-1. Proc Natl Acad Sci USA 1997; 94:7555-7560.
Pittman DL, Cobb J, Schimenti KJ, Wilson LA, Cooper DM, Brugnull E, Handel MA, Schimenti JC. Meiotic prophase arrest with failure of chromosome synapsis in mice deficient for *Dmc1*, a germline-specific RecA homolog. Mol Cell 1998; 1:697-705.
Rodriguez I, Ody C, Araki K, Garcia I, Vassalli P. An early and massive wave of germinal cell apoptosis is required for the development of functional spermatogenesis. EMBO J 1997; 9:2262-2270.
Roest HP, van Klaveren J, de Wit J, van Gurp CG, Koken MHM, Vermey M, van Roijen JH, Hoogerbrugge JW, Vreeberg JTM, Baarends WM, Bootsma D, Grootegoed JA, Hoeijmakers JHJ. Inactivation of the HR6B ubiquitin-conjugating DNA repair enzyme in mice causes male sterility associated with chromatin modifications. Cell 1996; 86:799-810.
Ross AJ, Waymire KG, Moss JE, Parlow AF, Skinner MK, Russell LD, MacGregor GR. Testicular degeneration in *Bclw*-deficient mice. Nature Genet 1998; 18:251-256.
Ruggiu M, Speed R, Taggert M, McKay SJ, Kilanowski F, Saunders P, Dorin J, Cooke HJ. The mouse *Dazla* gene encodes a cytoplasmic protein essential for gametogenesis. Nature 1997; 389:73-77.
Satakata I, Benson G, Maas R. Sexually dimorphic sterility phenotypes in *Hoxa10*-deficient mice. Nature 1995; 374:460-463.
Schleicher G, Drews U, Stumpf WE, Sar M. Differential distribution of dihydrotestosterone and estradiol binding sites in the epididymis of the mouse. Histochemistry 1984; 81:139-147.
Shen W-H, Moore CCD, Ikeda Y, Parker KL, Ingraham HA. Nuclear receptor steroidogenic factor 1 regulates the Müllerian inhibition substance gene: a link to the sex determination cascade. Cell 1994; 77:651-661.
Sicinski P, Donaher JL, Geng Y, Parker SB, Gardner H, Park MY, Robker RL, Richards JS, McGinnis LK, Biggers JD, Eppig JJ, Bronson RT, Elledge SJ, Weinberg RA. Cyclin D2 is an FSH-responsive gene involved in gonadal cell proliferation and oncogenesis. Nature 1996; 384:470-474.
Slegtenhorst-Eegdeman KE, De Rooij DG, Verhoef-Post M, Van de Kant HJG, Bakker CE, Oostra BA, Grootegoed JA, Themmen APN. Macroorchidism in *FMR1* knockout mice is caused by increased Sertoli cell proliferation during testicular development. Endocrinology 1998; 139:156-162.
Sonnenberg-Riethmacher E, Walter B, Riethmacher D, Godecke S, Birchmeier C. The *c-ros* tyrosine kinase receptor controls regionalization and differentiation of epithelial cells in the epididymis. Genes Develop 1996; 10:1184-1193.
Supp DM, Witte DP, Branford WW, Smith EP, Potter SS. Sp4, a member of the Sp1-family of zinc finger transcription factors, is required for normal murine growth, viability, and male fertility. Dev Biol 1996; 176:284-299.
Torii S, Yamagishi T, Murakami K, Nakayama K. Localization of Kex2-like processing endoproteases, furin and PC4, within mouse testis by *in situ* hybridization. FEBS Lett 1993; 316:12-16.
Torres M, Gomez-Pardo E, Dressler GR, Gruss P. *Pax-2* controls multiple steps of urogenital development. Development 1995; 121:4057-4065.
Toscani A, Mettus RV, Coupland R, Simpkins H, Litvin J, Orth J, Hatton KS, Reddy EP. Arrest of spermatogenesis and defective breast development in mice lacking A-myb. Nature 1997; 386:713-717.
Tsuzuki T, Fujii Y, Sakumi K, Tominaga Y, Nakao K, Sekiguchi M, Matusushiro A, Yoshimura Y, Morita T. Targeted disruption of *Rad51* gene leads to lethality in embryonic mice. Proc Natl Acad Sci USA 1996; 93:6236-6240.
Watanabe D, Sawada K, Koshimizu U, Kagawa T, Nishimune Y. Characterization of male meiotic germ cell-specific antigen (Meg 1) by monoclonal antibody TRA 369 in mice. Mol Reprod Develop 1992; 33:307-312.
Xu Y, Ashley T, Brainerd EE, Bronson RT, Meyn MS, Baltimore D. Targeted disruption of *ATM* leads to growth retardation, chromosomal fragmentation during meiosis, immune defects, and thymic lymphomas. Genes Develop 1996; 10:2411-2422.
Yoshida K, Kondoh G, Matsuda Y, Habu T, Nishimune Y, Morita T. The mouse *RecA*-like gene *Dmc1* is required for homologous chromosome synapsis during meiosis. Mol Cell 1998; 1:707-718.
Zhao G-Q, Hogan BLM. Evidence that mouse *Bmp8a* (*Op2*) and *Bmp8b* are duplicated genes that play a role in spermatogenesis and placental development. Mech Develop 1996; 57:159-168.
Zhao G-Q, Deng K, Labosky PA, Liaw L, Hogan BLM. The gene encoding bone morphogenetic protein 8B is required for the initiation and maintenance of spermatogenesis in the mouse. Genes Develop 1996; 10:1657-1669.
Zhao G-Q, Liaw L, Hogan BLM. Bone morphogenetic protein 8A plays a role in the maintenance of spermatogenesis and the integrity of the epididymis. Development 1998; 125:1103-1112.
Zhu D, Dix DJ, Eddy EM. HSP70-2 is required for CDC2 kinase activity in meiosis I of mouse spermatocytes. Development 1997; 124:3007-3014.

5 *Prospects for Spermatogonial Transplantation in Livestock and Endangered Species*

Carl L. Hausler
Lonnie D. Russell
Southern Illinois University

The testis functions to support spermatogenesis and produce androgens. It also supports the transplantation of spermatogonia from one animal (indeed from one species) to another (Brinster et al. 1994; Clouthier et al. 1996). For those who attempt to more fully understand the testis and its function, such inquiry opens new research methodologies and begs answers to an entire new set of questions. Spermatogonial transplantation also offers new solutions to health-related problems.

Arranged in tubules, supported by the secretions from the cells of the interstitium, the mammalian testis produces literally billions of sperm during adulthood. Each individual gamete draws a random sample half of the genes present. The division of the germ cells (spermatogonia) into stem cell spermatogonia and into committed/developing spermatogonia, the process of meiosis to halve the number of chromosomes and form secondary spermatocytes and then spermatids and finally the metamorphosis of the spermatid into a spermatozoon are major events deserving further study. The spermatogenic process is made even more challenging to understand by the intimate relationship of the evolving gametes with the Sertoli cells of the tubule (Russell et al. 1993). Now, with the finding that spermatogonia can be successfully transplanted, there is opportunity to delve further into this process and extend knowledge of the supportive role of Sertoli cells and the timing and control of spermatogenesis. Thus, the development of transplant technology is important to those who seek to understand testis function.

The ability to culture (Nagano et al. 1998) mammalian spermatogonia, and preserve them by freezing (Avarbock et al. 1996), followed by successful transplantation, establishes a potentially important means of producing transgenic animals. The biotechnologists who currently produce transgenics find that the transgene may only transfect somatic cells of the animal, and if the germ cells are not transfected, the desired characteristic will not be passed on to offspring. On the other hand, transfecting the desired gene into spermatogonia followed by successful transplantation would facilitate the production of transgenics. Thus, the molecular biologists gain a potential new tool in their quest to exert genetic control to improve livestock, to save endangered species and to eradicate genetic disease. This review will describe the current status of spermatogonial transplantation and speculate as to how this technology might prove useful in livestock production and preservation of endangered species.

Spermatogonia

Germ cell transplantation must rely on the stem cells of the testis to populate the testis. Spermatogonia are of three distinct types: stem-cell spermatogonia, proliferative spermatogonia and differentiating spermatogonia (Russell et al. 1990). Of these, the stem cell spermatogonia is the only type capable of self-renewal. This is the only type that assures a continuous supply of spermatogonia to repeatedly initiate spermatogenesis during adulthood. There is no certainty as to which spermatogonial cell type is the stem cell, but most investigators believe it is the cell lacking intercellular bridges or the A_S spermatogonia. The logic here is that as cells divide, only the stem cells undergo a normal cell division. The committed cells do not pinch off from each other, but exhibit open continuities of cytoplasm called intercellular bridges (Huckins 1978; Weber et al. 1987). As spermatogonia continue to divide, they form progressively larger clones of cells connected by intercellular bridges. The A_S cells are scattered among these larger clones of spermatogonia lying at the base of the tubule.

Spermatogonia provide the cellular resources needed for spermatogenesis as long as the individual lives. If spermatogonia from an old male were successively transplanted to younger males (or frozen for later transplantation) these cells would, in effect, be immortal. This is a method through which genetic material might be preserved and augmented from a long-ago deceased male by thawing and transplanting his preserved spermatogonia.

It should be noted that spermatogonial transplantation is not cloning and should not engender the negative connotations currently associated with cloning. Cloning preserves the genome of the transplanted nucleus in its entirety. Spermatogonia that have been transplanted must go through meiosis in the recipient, where genetic recombination occurs. Furthermore, the sperm that are produced that fertilize the egg are haploid and contribute only half of their genetic material to the offspring. Thus, genetic diversity among offspring is insured. That is not to say that desirable genetics cannot be inherited.

Collection of Donor Cells

Procedures for the isolation of testes cells were modified by Bellvé et al. (1977) from a procedure originally described by Meistrich et al. (1973). With minor modifications to the Bellvé protocol, Brinster et al. (1994) were able to isolate and collect sufficient quantities of isolated cells to begin transplantation studies. Teased testicular parenchyma was digested sequentially with enzymes and strained through mesh to harvest single, isolated cells. The procedure does not purify spermatogonia but includes all testicular cells, including spermatocytes, spermatids, Sertoli cells and Leydig cells. Only stem-cell spermatogonia—that is, those with the ability to self-renew—will bring about new spermatogenesis upon transplantation (Brinster et al.

1994). The procedures were also effective in harvesting cells from rats (Clouthier et al. 1996) and other species.

The Transplant Recipient

The ideal recipient for transplantation of spermatogonia is a male whose seminiferous tubules contain only normal Sertoli cells. In initial experiments with mouse-to-mouse transplantation of spermatogonia, Brinster et al. (1994) used two types of recipients. One was a compound heterozygous or homozygous mutant W-locus strain (Coulombre et al. 1954) which, due to a mutation, has no spermatogenesis beyond early spermatogonial divisions; the other was a (C5 7BL/6 X SJL) F_1 hybrid adult that was treated once with busulfan at four to six weeks of age and then used as a recipient. Busulfan is known to destroy endogenous spermatogonia in rats, mice, golden hamsters, Indian desert gerbils, guinea pigs, Japanese quail, rabbits and rhesus monkeys (Bishop, Wassom 1986). Busulfan is an alkylating agent that induces chromosomal aberrations and results in loss of proliferative tissues (Viguier et al. 1984). At the doses used for transplantation, busulfan leaves some viable stem cell spermatogonia that will later divide and repopulate some tubules. The major difference between busulfan-treated recipients and W-locus recipients is that the former will usually regenerate some endogenous spermatogenesis.

Transplantation Techniques

In the initial transplantation experiment, Brinster and Avarbock (Brinster et al. 1994) exposed the tunica of the recipient mouse and, using a micromanipulator and a pressure injector, succeeded in filling about 80% of the surface tubules. Later studies demonstrated that mouse testes cells could be transplanted following injection into seminiferous tubules, efferent ducts or into the rete testes with about equal transplantation success (Ogawa et al. 1998). Surprisingly, the injection of germ cells into the rete testes was accomplished without the use of a micromanipulator or pressure injector; simple pressure on the plunger of a syringe was sufficient. Similar techniques were employed by Jiang and Short (Jiang et al. 1995) to transplant rat germ cells into a recipient rat.

Cryopreservation of Spermatogonia

Avarbock et al. (1996) succeeded in transplanting testes stem cells that had been cryopreserved for up to 156 days. These frozen, then thawed, stem cells were shown to give rise to spermatogenesis in the recipient mouse testis. Standard techniques for cell freezing and thawing were used. These results indicate the possibility that stem cells from other species may be similarly preserved, and that, indeed, the genetic potential of an individual male may be made immortal. This opens the possibility to efficiently preserve the genetic potential of endangered species and maintain rare bloodlines of domestic livestock.

Culture of Germ Cells

An important step toward stable transfection of germ cells is to introduce foreign DNA into them as they undergo cell division. Many investigators have attempted to culture germ cells, reporting that most germ cells die in culture within a few days or weeks of initiating cell culture. Indeed, this seems to be the case. Even if they are cultured in an environment of hormones, growth factors and in the presence of Sertoli cells, they appeared not to be maintained for a prolonged period of time. Assessment of the viability of cultured stem cells is difficult because the stem cells may comprise such a small percentage of the cultured germ cells that they may be missed. Transplantation of cultured cells is an effective means to determine if stem cells survive in culture. Indeed, mouse cells cultured up to four months have been successfully transplanted (Nagano, Brinster 1998) into another mouse, proving the extended survival of germinal cells in culture. It is yet to be determined if these stem cells divide in culture.

Interspecies Transplantation of Testis Stem Cells

Perhaps the most surprising, but fascinating, finding in testes stem-cell transplantation efforts is the success of interspecies transplants. Clouthier et al. (1996) reported that rat germ cells transplanted to mouse testes were capable of producing rat spermatozoa in the recipient mouse. The implications that a mouse Sertoli cell, with a spermatogenic cycle 50% shorter than that of a rat, could support the needs of rat spermatogenesis open a whole new means to study the relationships between Sertoli cells and spermatogonia, spermatocytes and spermatids (Russell et al. 1996). A recent study using rat germ cells transplanted into a mouse has shown that rat germ cell genotype determines the cycle length of spermatogenesis (Franca et al. 1998). The somatic cells, in particular the Sertoli cell, do not influence the timing of germ cell development.

Development of Spermatogonial Transplants

A histological study of mouse testes at 10 minutes; nine hours; one week; and one, two and three months following transplantation of spermatogonia demonstrated how spermatogonial transplants develop (Parreira et al. 1998). Following the transplantation procedure, testes weight increased by one day and remained increased in treated animals throughout the remainder of the experiment, suggesting that the presence of intraluminal germ cells increased testis function before evidence of spermatogenesis. Spermatogenesis was first seen at one month after transplantation, and the percentage of tubules supporting spermatogenesis increased dramatically with time. Sperm were first produced two months after transplantation. Most transplanted germ cells were phagocytosed by Sertoli cells, prior to completion of the first week of transplantation.

Rationale for Utilizing Spermatogonial Transplantation in Farm Animals

The success in transplanting spermatogonia in mice suggests that this technology might be extended to serve domestic animal production. Transplanted spermatogonia carry the genome of the donor animal. The recipient male then produces spermatozoa derived genetically from the donor. The offspring of matings to the recipient would be genetically derived from the donor. In theory, then, the genetic benefits of artificial insemination (AI) including the selection of a genetically outstanding male, could be achieved without the drawbacks of AI (labor to detect estrus, corral females and inseminate), processes which are fraught with potential for human error leading to failure to conceive. Transplantation of cryopreserved spermatogonia offers potential for preservation of valuable, but scarce genotypes of animals—genotypes which may prove useful sometime in the future. Furthermore, spermatogonial transplantation also provides a potential to insert desired genes directly into a reproductive stem cell, ensuring the production of transgenic offspring. Currently, production of transgenic livestock is expensive, and carries no assurance that the transgene locates in the reproductive stem cells, making it heritable. This chapter will discuss how spermatogonial transplantation in livestock, if perfected, might contribute to improvement in management of beef and dairy cattle, swine and horses. Progress toward achieving this goal is also described.

Rationale for Spermatogonial Transplantation in Cattle Breeding

Artificial insemination offers producers of beef cattle the opportunity to purchase sires of outstanding genetic potential. Yet this is an opportunity largely missed in beef cattle production. Less than 10% of beef cows are bred artificially each year (Dzuik, Bellows 1983). The principal reason for low incorporation of artificial insemination in beef herds is the impracticality of the extra labor required for proper estrus detection (Godke, Kreider 1977). Successful artificial insemination of a cow depends on completion of each of a series of events described below.

Detection of Estrus

Cows enter estrus on a cycle of approximately 21 days. They are in estrus for an average of 18 hours and ovulate approximately 12 hours after the end of estrus (Jainudeen, Hafez 1993). In order to time artificial insemination to be near the time of ovulation, it is necessary for the producers to check for estrus behavior (standing to be mounted by other cows). This must be done at least twice every day. Observations using videotape to provide continuous observation of dairy cows (King et al. 1976) found 100% of cows in heat, whereas twice-daily observations by herdsmen found only 56%.

Management of Pastures and Ranges

Since the majority of beef cows are managed on pastures and ranges, it takes considerable effort to detect estrus and separate those cows in heat from the rest of the herd in order to be bred. The cows to be bred must be corralled and restrained in a chute to allow for them to be inseminated artificially.

Insemination

The insemination procedure, albeit not difficult, does rely on the skill of the inseminator for ultimate success. Experience of the practitioner, then, is a factor in the ultimate success or failure of each insemination performed (Garner 1991). With a limited breeding season, beef producers may lack the practice necessary to hone these skills. These skills involve not only the deposition of the semen into the cervix, but also the skills of thawing and handling the semen properly and timely insemination following thawing.

Any human error occurring in any of the three steps identified above would result in a failure of the inseminated cow to conceive. This compounds the biological errors within the cow herself which may cause conception failure or embryonic loss. Indeed, some studies demonstrate that a large percentage of failures of cows to conceive following artificial insemination are due to a failure to detect estrus (Barr 1975; King et al. 1976). Compounding the error of failing to detect estrus, 20 to 33% of cows presented by herdsmen as being in estrus and ready to inseminate had high plasma progesterone, indicating they were falsely detected (Appleyard, Cook 1976; Smith 1982). Profitable beef production is predicated on reproductive success. The costs of artificial insemination apparently outweigh the genetic benefits, such that many beef producers opt to breed with natural mating.

Obviously natural mating has its costs. Bulls must be raised or purchased and maintained throughout the year. Bulls can be difficult to handle. Genetically superior bulls of the quality available through artificial insemination are not affordable for the average beef producer. The benefits of natural mating, however, are that the bulls themselves do a better job of estrus detection and, in partnership with the cow in estrus, the optimal time for insemination (mating) is determined and mating is fulfilled.

Spermatogonia transplantation in the bull, then, offers the potential for a recipient bull to produce spermatozoa derived from the donor spermatogonia. The genetic benefits of artificial insemination would be combined with the management benefits of natural mating. Spermatogonia from genetically proven sires could be collected and transplanted to several recipient bulls and the recipients would then be used for mating.

The ability to transplant spermatogonia in beef animals would enable further research into the culture and genetic manipulation of germ cells prior to transplantation. If a DNA fragment were inserted into a cultured spermatogonial stem cell, it should maintain and indeed, multiply, the transgene through the spermatogenic process in all sperm derived from that spermatogonium, once transplanted to a recipient bull. This could prove to be an efficient means to produce transgenic cattle.

Rationale for Spermatogonial Transplantation in Dairy Cattle

Unlike beef producers, dairy producers have utilized artificial insemination to a large extent. A potential market for bulls producing semen from a donor animal at the producer level is unlikely. However, spermatogonial transplantation could serve some potential uses in this industry. Dairy sires used for producing semen for artificial insemination have gone through a lengthy and rigid test to become "proven" sires, where the "proof" refers to their breeding value that they pass on to a large number of offspring. Equation 1 (adapted from Bourdon 1997) is the general equation which can be used to predict genetic change over time. Dairy producers traditionally attempt to increase accuracy of selection (r) and selection intensity (I) to maximize genetic change. This has been done, however, at the cost of increasing the generation interval (L). Increased generation interval (the denominator) decreases genetic change over time. The generation interval is increased, particularly on the male side, because of the need for sire proofs, the process which increases the selection intensity and the accuracy of selection (the numerator). By the time a bull is a "proven" sire he has already reached the age of six to seven years (Bourdon 1997), including two years for the bull to reach maturity and breed, another year for his offspring to be born, two to three years for the female offspring to enter their first lactation and another year to measure the milk production of his female offspring. If it were possible to "prove" a bull at a younger age, it would decrease the generation interval and more rapidly increase genetic change over time.

Equation 1

$$GC/t = \frac{r\,I\,v}{L}$$

where:
GC/t = genetic change over time
r = accuracy of selection
I = selection intensity
v = genetic variation
L = generation interval

Spermatogonial transplantation of spermatogonia collected at birth of a potential young dairy sire could be placed into a mature recipient bull. These spermatogonia would immediately begin spermatogenesis, and within a short time become a continuing flow of spermatozoa genetically derived from an animal that has not yet attained puberty. The donor calf's offspring would be born before he reaches puberty. By the time he is three to four years of age, he should have his initial genetic proofs available. Given the same selection intensity (I) and accuracy of selection (r), along with a generation interval (L) shortened by spermatogonial transplantation, the dairy industry could achieve more rapid genetic change through this potential new technology.

Rationale for Spermatogonial Transplantation in Swine

Artificial insemination is practiced in the swine industry, but not to the extent it is in the dairy industry, largely due to inadequate methods for either semen preservation or synchronization of estrus (Day 1979). Because swine have large litters, short gestation and early puberty compared to cattle, it has largely been the focus of swine breeders to increase the rate of genetic change by decreasing the generation interval (L in Figure 1). Though accuracy of selection (r) may not be ideal, it has proven to be an efficient means of improving swine genetically. Thus, at first glance one might conclude that spermatogonial transplantation will not be useful in swine breeding. There are, however, two possibilities in which this technology could prove useful.

The first possibility is in selecting breeding animals for their carcass value. Market pigs are usually castrated while very young. Their growth patterns, muscle development and fat deposition vary from those of intact males. Carcass measurements are made at slaughter, and intact male litter mates of the best carcass barrows are chosen as potential new sires. While litter mates are full siblings of the slaughtered pigs, their genetic makeup is different due to segregation and recombination of the

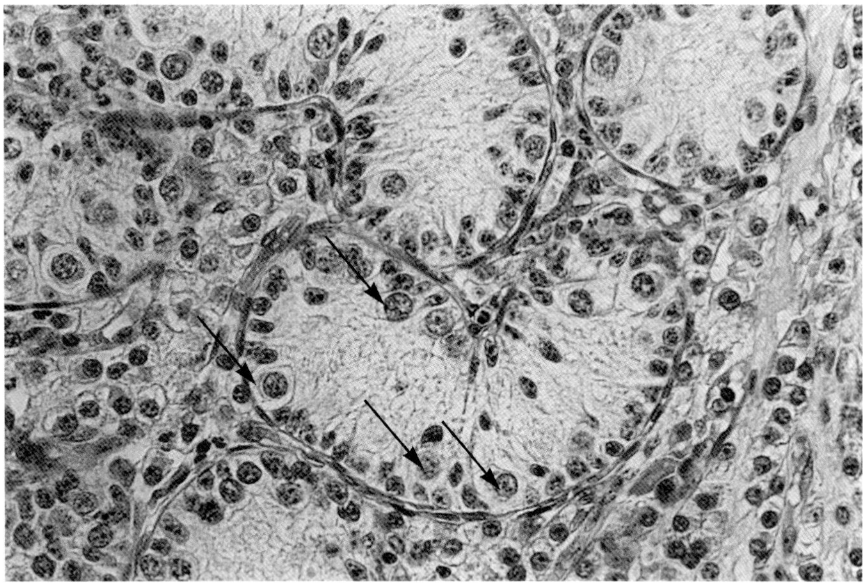

Figure 1A. Developing testis from a normal immature boar. Spermatogonia (arrow) are abundant.

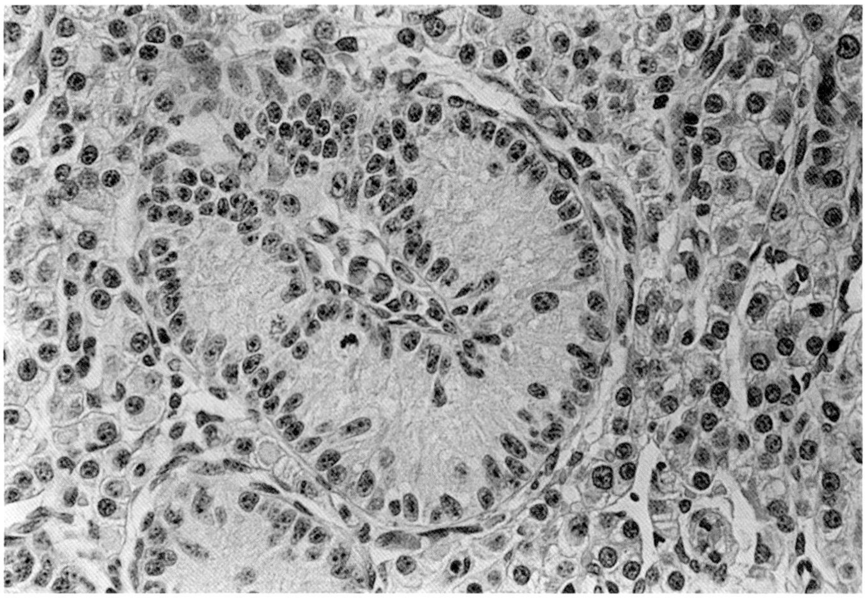

Figure 1B. Developing testis from an immature boar given a single treatment of a sterilizing agent. Spermatogonia are not seen.

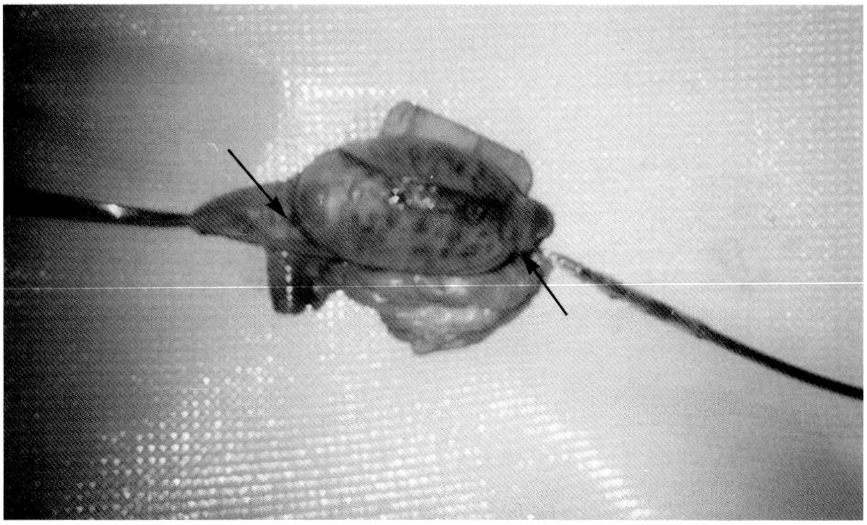

Figure 2. Dye has been infused into the central region of a bull testis through a cannula inserted into the caudal testis (arrow pointing from below). The dye has apparently reached the epididymis via the rete since blue color (arrow pointing from above) is in the region of the efferent ducts and caput.

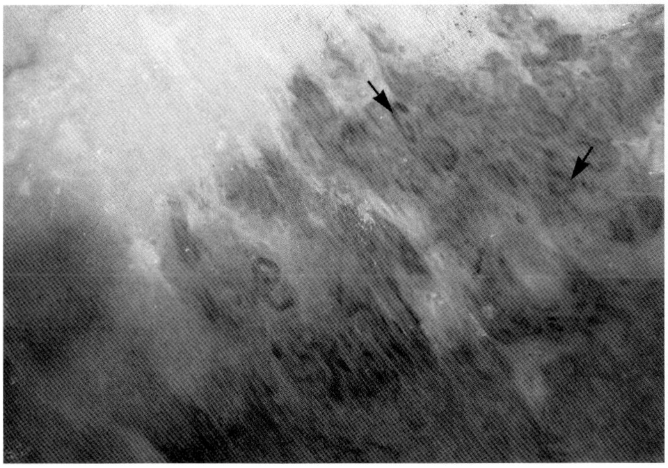

Figure 3. Testis in which dye has been infused into the rete using a method such as that described in Figure 2. The presence of dye within seminiferous tubules is indicated by arrows. The tubules to the lower right are filled and those to the upper left have not filled.

parents' genotype. Thus, there is an inherent loss in selection accuracy (r) and therefore a lowered rate of genetic change (GC/t). If spermatogonia were isolated from the castrated pig's testes and preserved by freezing, then the outstanding carcass pig could, through spermatogonial transplantation, become a sire and pass his outstanding genes for carcass merit to future generations. If this increase in selection intensity (I) and selection accuracy (r) were large enough to compensate for a slightly larger generation interval, then spermatogonial transplantation might find practical usefulness.

A second possibility for spermatogonial transplantation in the swine industry would be in the production of transgenic offspring. Given the possibility that spermatogonia may be transplanted successfully in this species, it would be important to master culturing methods for spermatogonia. With the possibility of culturing spermatogonia comes the possibility of inserting DNA directly into

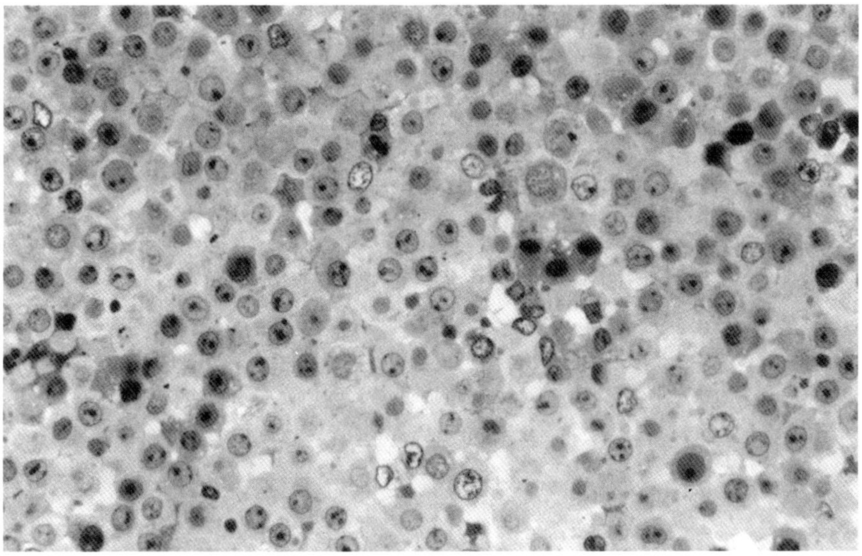

Figure 4. Light micrograph of a cell pellet from an immature bull testis.

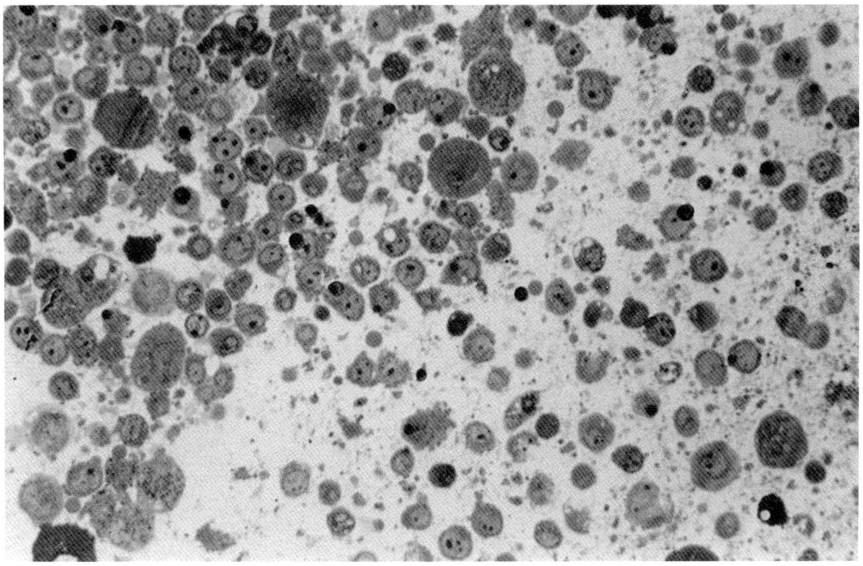

Figure 5. Light micrograph of a cell pellet from an immature boar testis.

the stem cells before transplanting to a recipient. There is currently much interest in production of transgenic pigs. Current methodology is costly, difficult and offers no assurance that the transgene will locate in a germ cell and be passed on through inheritance to offspring.

Rationale for Spermatogonial Transplantation in Horses

Horse breeders do not, generally, accept artificial insemination unless the breeding is done on the premise where the donor stallion resides. It is unknown how the breeders might react to a general practice of spermatogonial transplantation in their industry. However, there is a magnificent example of how spermatogonial transplantation might be useful to horse breeders. Recently, a racehorse named Cigar was retired from racing as the all-time money winner in this sport. He would command much attention as a potential sire of racehorses were it not for the fact that he is a gelding. However, as a stallion, it is unlikely that he would have achieved the success he did on the race track. However, if spermatogonia had been collected from his excised testes at castration time and preserved by freezing, they could be transplanted to a recipient animal to produce his spermatozoa. The possibility of preservation of germ cell lines is only possible if these lines can be transplanted to other animals.

Certainly in favor of spermatogonial transplantation in horses is the presence of an obvious recipient, the mule. The mule may have spermatogonia and even spermatocytogenesis, but lacks the ability to produce spermatozoa (Hernandez J et al. 1977). If horse spermatogonia were transferred to the mule, is it possible that all sperm produced would originate from the transplanted cells and be of the horse genotype?

Spermatogonial Transplantation for Preservation of Endangered Species

Clouthier et al. (1996) successfully transplanted rat spermatogonia into mice, and the mice produced rat spermatozoa. Although this is a cross species transplantation, there is no evidence that such xenotransplants will be successful in other species. However, the demonstration that spermatogonia can be preserved by freezing (Avarbock et al. 1996) prior to transplantation suggests a means to preserve rare germ lines for future use. This would benefit not only attempts to rescue wildlife, but to preserve rare and disappearing genes in domestic livestock. While it seems likely that current technology would allow for the freezing of these cells, the end usefulness is predicated on being able to transplant them at a later time to a breeding animal.

Current Progress in Spermatogonial Transplantation of Domestic Livestock

In the Russell/Hausler laboratories, preliminary experiments which are necessary prerequisites for transplanting spermatogonia in cattle and swine have begun. The immediate challenge is to define methods to destroy endogenous spermatogonia in potential recipient animals. Preliminary evidence that the chemical busulfan will at least partially destroy endogenous spermatogonia in swine is available. Young boars treated with busulfan have smaller testes and considerably fewer spermatogonia than untreated controls (Fig. 1). Further studies have been initiated in this species to optimize dose and appropriate timing of busulfan treatments.

A second challenge is to define a protocol for the transplantation. Dye has been successfully infused into the mediastinum of an excised bull testis and the dye appears in the efferent ducts leading to the caput epididymis (Fig. 2). Examination of the tissue after dissection reveals the dye does indeed enter seminiferous tubules (Fig. 3). More studies are underway to identify appropriate methods for transplantation. Studies in Germany (Schlatt et al. 1998) also indicate that access to the seminiferous tubules may be gained through the rete testis in the bull. The rete of the bull appears to be organized as many centrally positioned ductules.

A third challenge is to adapt the methods for isolating and separating spermatogonia from testes. This has been successfully achieved in both the bull and the boar (Figs. 4, 5). Electron microscopy of the cells isolated from both the bull and the boar confirm that among the cells present are type A spermatogonia (Figs. 6-8). The type A spermatogonia (or perhaps gonocytes) are the kind of cells that serve as stem cells and which will establish spermatogenesis.

Acknowledgements

This work is supported by the Illinois Council on Food and Agricultural Research. The authors acknowledge the support of the Southern Illinois University at Carbondale, University Priorities and Interdisciplinary Initiative Program. The help of Wenuang Cao and Xiong Gu in bull and boar experiments is appreciated.

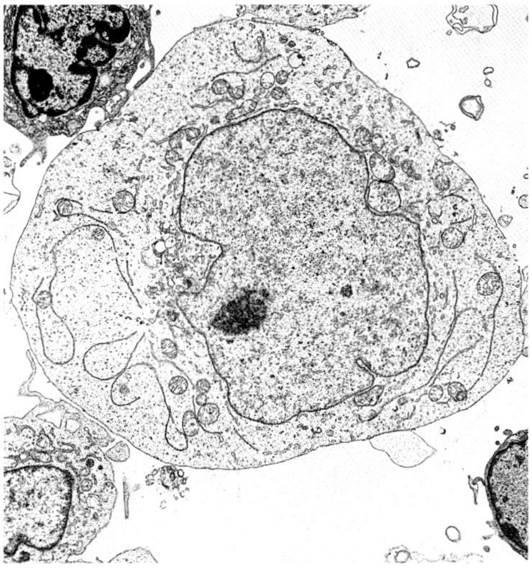

Figure 6. Electron micrograph of a cell pellet from an immature boar testis showing a Type A spermatogonium.

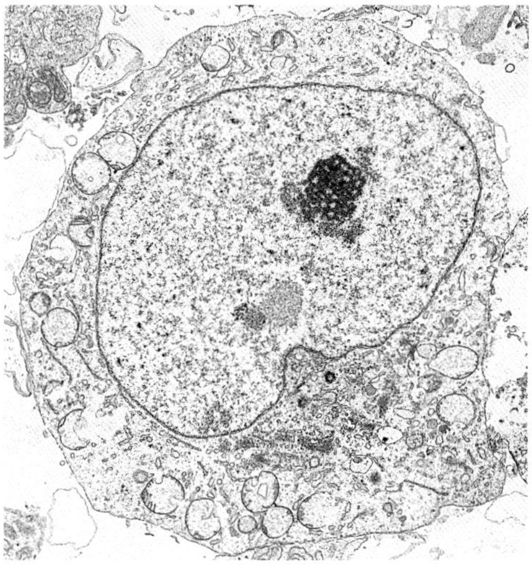

Figure 7. Electron micrograph of a cell pellet from an immature bull testis showing a Type A spermatogonium.

References

Appleyard WT, Cook B. The detection of estrus in dairy cattle. Vet Rec 1976; 99:253-260.

Avarbock MR, Brinster CJ, Brinster RL. Reconstitution of spermatogenesis from frozen spermatogonial stem cells. Nature Med 1996; 2:693-696.

Barr HL. Influence of estrus detection on days open in dairy herds. J Dairy Sci 1975; 58:246-247.

Bellvé AR, Cavicchia JC, Millette CF, O'Brien DA, Bhatnagar YM, Dym M. Spermatogenic cells of the prepuberal mouse. Isolation and morphological characterization. J Cell Biol 1977; 74:68-85.

Bishop JB, Wassom JS. Toxicological review of Busulfan (Myerlan). Mut Res 1986; 168:15-45.

Bourdon RM. Understanding Animal Breeding. Prentice-Hall Inc, Upper Saddle River, NJ, 1997, pp185-211.

Brinster RL, Avarbock M. Germline transmission of donor haplotype following spermatogonial transplantation. Proc Natl Acad Sci USA 1994; 91:11303-11307.

Brinster RL, Zimmerman JW. Spermatogenesis following male germ-cell transplantation. Proc Natl Acad Sci USA 1994; 91:11298-11302.

Clouthier DE, Avarbock MR, Maika SD, Hammer RE, Brinster RL. Rat spermatogenesis in mouse testes following spermatogonial stem cell transplantation. Nature 1996; 381:418-421.

Coulombre JL, Russell ES. Analysis of the pleiotropism at the W-locus in the mouse. J Exp Zool 1954; 126:277-295.

Day BN. Reproductive Problems in Swine. In: (Hawk H, ed) Beltsville Symposium on Agricultural Research. Vol. 3, Animal Reproduction. New York: Allanheld, Osmun and Co Pub Inc, 1979, pp41-51.

Dzuik PJ, Bellows RA. Management of reproduction of beef cattle, sheep and pigs. J Anim Sci 1983; 57 (Suppl 1):355-379.

França L, Ogawa T, Avarbock M, Brinster M, Russell L. Germ cell genotype controls cell cycle during spermatogenesis. Biol Reprod 59:1371-1377.

Garner DL. 1991. Artificial insemination. In: (Cupps PT, ed) Reproduction in Domestic Animals. San Diego: Academic Press, 1998, pp251-274.

Godke RA, Kreider JL. Detecting estrus in beef cattle. Charolais Bull-o-gram. Feb-Mar 1977; p24-39.

Hernandez J, Auregui PM, Marquez MH. Fine structure of mule testes: light and electron microscopy study. Am J Vet Res 1977; 38:443-447.

Huckins C. Spermatogonial intercellular bridges in whole-mounted seminiferous tubules from normal and irradiated rodent testes. Am J Anat 1978; 53:97-122.

Jainuddin MR, Hafez ESE. Cattle and buffalo. In: (Hafez ESE, ed) Reproduction in Farm Animals. Philadelphia, PA: Lea and Febeger, 1993, pp315-329.

Jiang F-X, Short RV. Male germ cell transplantation in rats: apparent synchronization of spermatogenesis between host and donor seminiferous epithelia. Int J Androl 1995; 18:326-330.

King GJ, Humik JF, Robertson HA. Ovarian function and estrus in dairy cows during early lactation. J Anim Sci 1976; 42:688-692.

Meistrich ML, Bruce WR, Clermont Y. Cellular composition of fractions of mouse testis cells following velocity sedimentation separation. Exp Cell Res 1973; 79:213-227.

Nagano M, Brinster RL. Spermatogonial transplantation and reconstitution of donor cell spermatogenesis in recipient mice. APMIS 1998; 106:47-57.

Nagano M, Avarbock M, Leonida E, Brinster C, Brinster RL. Culture of mouse spermatogonial stem cells. Tiss Cell 1998; 30:389-397.

Ogawa T, Arechanga JM, Avarbock MR, Brinster RL. Transplantation of testis germinal cells into mouse seminiferous tubules. Int J Dev Biol 1998; 41:111-122.

Parreira G, Ogawa T, Avarbock M, França LR, Brinster RL, Russell LD. Development of testis cell transplants. Biol Reprod 1998; 59:1360-1370.

Russell LD. Morphological and functional evidence for Sertoli-germ cell relationships. In: (Russell LD, Griswold M, eds) The Sertoli Cell. Clearwater, FL: Cache River Press, 1993, pp315-390.

Russell LD, Brinster RL. Ultrastructural observations of spermatogenesis following transplantation of rat testis cells into mouse seminiferous tubules. J Androl 1996; 17:615-627.

Russell LD, Ettlin RA, Sinha Hikim AP, Clegg ED. Histological and Histopathological Evaluation of the Testis. Cache River Press, Clearwater FL, 1990, pp4-5.

Smith RD. Presenting heat detection as "A" in the A.I. alphabet. NAAB 9th Tech Conf on Artif Insem Reprod 1982; pp104-114.

Viguier MM, Hochereau RM, Barenton B, Perreau C. Effect of prenatal treatment with busulfan on the hypothalamo-pituitary axis, genital tract and testicular histology of prepubertal male rats. J Reprod Fertil 1984; 70:67-73.

Weber JE, Russell LD. A study of intercellular bridges during spermatogenesis in the rat. Am J Anat 1987; 180:1-24.

6 The Putative Chaperone Calmegin and Sperm Fertility

Masaru Okabe
Masahito Ikawa
Shuichi Yamada
Tomoko Nakanishi
Osaka University
Tadashi Baba
University of Tsukuba
Yoshitake Nishimune
Osaka University

Sperm are made to fertilize eggs. However, mammalian sperm are not ready to interact with eggs when ejaculated. They need to undergo a physiological change called "capacitation" before penetrating the eggs (Austin 1951; Chang 1951). Through an understanding of "capacitation," it became possible to fertilize eggs *in vitro*. This very basic phenomenon of mammalian fertilization was found a couple of years before the original report of the double helix structure of DNA (Watson,Crick 1953). Since then, DNA research has progressed far beyond expectations; scientists are now able to cut and paste genes as they design as well as to produce genetically engineered animals. However, these experiments are not the sole accomplishments of DNA research, for they owe much to technological progress in reproductive biology.

Such progress in technology has been rapid. Scientists now know that genetically manipulated somatic cell nuclei can be transplanted to enucleated eggs to produce cloned animals (Wilmut et al. 1997). In the clinical field, sperm are now being injected directly into egg cytoplasm to create life. According to a recent report, sperm are not necessarily "alive" when injected. They can be freeze dried until injection (Wakayama, Yanagimachi 1998).

In spite of this progress in clinical and biotechnology fields, basic reproductive biology awaits further studies that may answer fundamental questions. One of these is, "What do sperm use to interact with eggs?" Several answers have been proposed. Acrosin and galactosyl transferase (GalTase) are believed to play an important role. However, using homologous recombination, mice that were null to these factors were found to be fertile (Baba et al. 1994; Lu, Shur 1997). These results are strikingly important in the sperm/egg recognition research field, as almost all of the other factors postulated for sperm/egg interaction are based up on the same lines of evidence that were used in the cases of acrosin and GalTase. If acrosin and GalTase are not essential to fertilization, the relevance of other candidates may also be in question. If numerous factors are involved in sperm/egg interaction, scientists may be unable to detect a severe detrimental effect by knocking out one of them.

Sperm Membrane Proteins

The molecules involved in sperm/egg fusion must reside on the surface of sperm. Some of them may derive from the epididymis and attach on the surface, while others are produced during spermatogenesis and expressed on the sperm surface. The proper folding of newly synthesized membrane proteins in the endoplasmic reticulum (ER) is required for formation of functional mature proteins. There is a mechanism by which only functional proteins are transported to the plasma membrane and expressed on the surface. Calnexin is an ubiquitous ER chaperone that plays a major role in quality control by retaining incompletely or misfolded proteins (Bergeron et al. 1994; Hebert et al. 1995; Ou et al. 1993; Vassilakos et al. 1996; Wada et al. 1994). In contrast to other known chaperones (Hsps, BiP and calreticulin), calnexin is an integral membrane protein (Bergeron et al. 1994; Wada et al. 1991). Calmegin, on the other hand, is a testis-specific ER protein that is homologous to calnexin (Ohsako et al. 1994a, b; Watanabe et al. 1992; Watanabe et al. 1994).

Calmegin was first identified as a male meiotic germ cell specific antigen (Watanabe et al. 1992) and later proven to be a Ca^{2+}-binding protein localized on the ER membrane. Sequence analysis demonstrated that calmegin is highly homologous to calnexin with 54% identity in amino acid sequence, including two sets of characteristic sequence repeats necessary for Ca^{2+}-binding (Ohsako et al. 1994b; Watanabe et al. 1994), as shown in Figure 1. Researchers in the Okabe laboratory investigated calmegin functions as a calnexin-like molecular chaperone during spermatogenesis. In the testis, calmegin was expressed exclusively in germ cells of pachytene spermatocyte to spermatid stage (Watanabe et al. 1992), while calnexin was expressed throughout the stages. In the ER of somatic cells, calnexin associates transiently with nascent membrane and soluble glycoproteins of the secretory pathway and facilitates their folding (Hebert et al. 1995; Ou et al. 1993; Vassilakos et al. 1996; Wada et al. 1994).

When calmegin was immunoprecipitated from testicular cells pulse-labeled with [^{35}S]-methionine for 30 min, a variety of newly synthesized proteins was found in the immunoprecipitates. Upon chase, these ligands, including p195, p58 and p41, were dissociated from calmegin at various rates. Similarly, the transient interaction of calnexin with nascent proteins was also observed. Castanospermine (CAS) treatment of the cells inhibited the association in either case, suggesting that the glucosidase activity is needed for the binding process as shown for calnexin (Hebert et al. 1995; Wada et al. 1994). It should be noted that both chaperones appeared to interact with distinctive ligands although they may share some. Thus, it was suggested that calmegin as well as calnexin transiently interacts with nascent glycoproteins during spermatogenesis.

To address the physiological role of calmegin *in vivo*, calmegin-deficient mice were generated by homologous recombination. A calmegin-targeting construct, shown in Figure 2, was designed to interrupt the second exon with a neomycin resistance gene (neo). This insertion disrupted the ATG translation start site. Both the targeting event in D3 ES cells and the germ-line transmission of targeted gene were confirmed by Southern

```
hCM    1:M---HFQAFWLCLGLLFISINAEFMDDDVETE-DFEE--NSEE--IDVNESELSSEIK-- 50
mCM    1:M---RFQGVGLCLGLLFITVNADFMDDGVEVE-DFSE--NSDE--SNIKDEPSSGTFK-- 50
mCN    1:MEGKWLLCLLLVLGTAAVEAHDGHDDDAIDIEDDLDDVIEEVEDSKSKSDASTPPSPKVT 60
hCN    1:MEGKWLLCMLLVLGTAIVEAHDGHDDDVIDIEDDLDDVIEEVEDSKPDTTAP-PSSPKVT 59
         *         * **           **     * *        *              *

hCM   51:YKTPQPIGEVYFAETFDSGRLAGWVLSKAKKDDMDEEISIYDGRWEIEELKENQVPGDRG 110
mCM   51:YKTPQPIGEVYFTETFDSGNLAGWVLSKAKKDDMDSEIAIYDGRWEIEELKENQVPGDRG 110
mCN   61:YKAPVPTGEVYFADSFDRGSLSGWILSKAKKDDTDDEIAKYDGKWEVDEMKETKLPGDKG 120
hCN   60:YKAPVPTGEVYFADSFDRGTLSGWILSKAKKDDTDDEIAKYDGKWEVEEMKESKLPGDKG 119
         ** * * *****   ** * * ** ******** * **  *** **  * **   *** *

hCM  111:LVLKSRAKHHAISAVLAKPFIFADKPLIVQYEVNFQDGIDCGGAYIKLLADTDDLILENF 170
mCM  111:LVLKSKAKHHAIAAVLEKPFIFADKPLIVQYEVNFQDGIDCGGAYIKLLADTGDLILENF 170
mCN  121:LVLMSRAKHHAISAKLNKPFLFDTKPLIVQYEVNFQNGIECGGAYVKLLSKTAELSLDQF 180
hCN  120:LVLMSRAKHHAISAKLNKPFLFDTKPLIVQYEVNFQNGIECGGAYVKLLSKTPELNLDQF 179
         *** * ****** * * *** *  ************ ** ***** ***  *  * *  *

hCM  171:YDKTSYIIMFGPDKCGEDYKLHFIFRHKHPKTGVFEEKHAKPPDVDLKKFFTDRKTHLYT 230
mCM  171:YDKTSYTIMFGPDKCGEDYKLHLIFRHKHPKTGVFEEKHAKPPDVDLKEFFTDRKTHLYT 230
mCN  181:HDKTPYTIMFGPDKCGEDYKLHFIFRHKNPKTGVYEEKHAKRPDADLKTYFTDKKTHLYT 240
hCN  180:HDKTPYTIMFGPDKCGEDYKLHFIFRHKNPKTGIYEEKHAKRPDADLKTYFTDKKTHLYT 239
          *** * *************** ***** ****  ****** ** ***  *** ******

hCM  231:LVMNPDDTFEVLVDQTVVNKGSLLEDVVPPIKPPKEIEDPNDKKPEEWDERAKIPDPSAV 290
mCM  231:LVMNPDDTFEVLIDQKVVNQGTLLDDVVPPINPPREIDDPSDKKPEEWDDRAKIPDPTAV 290
mCN  241:LILNPDNSFEILVDQSVVNSGNLLNDMTPPVNPSREIEDPEDRKPEDWDERPKIADPDAV 300
hCN  240:LILNPDNSFEILVDQSVVNSGNLLNDMTPPVNPSREIEDPEDRKPEDWDERPKIPDPEAV 299
         *  ***  ** * ** *** * ** *  **  *  ** ** * *** ** * ** ** **

hCM  291:KPEDWDESEPAQIEDSSVVKPAGWLDDEPKFIPDPNAEKPDDWNEDTDGEWEAPQILNPA 350
mCM  291:KPEDWDENEPAQIEDSSAVKPDGWLDDEPKFIPNPKAEKPEDWSDDMDGEWEAPHIPNPA 350
mCN  301:KPDDWDEDAPSKIPDEEATKPEGWLDDEPEYIPDPDAEKPEDWDEDMDGEWEAPQIANPK 360
hCN  300:KPDDWDEDAPAKIPDEEATKPEGWLDDEPEYVPDPDAEKPEDWDEDMDGEWEAPQIANPR 359
         ** ****  *  * *    ** *******   * * **** **  * ******* * **

hCM  351:C-RI-GCGEWKPPMIDNPKYKGVWRPPLVDNPNYQGIWSPRKIPNPDYFEDDHPFLLTSF 408
mCM  351:C-QI-GCGEWKPPMIDNPKYKGIWRPPMINNPNYQGLWSPQKIPNPDYFEDDHPFLLTSF 408
mCN  361:CESAPGCGVWQRPMIDNPNYKGKWKPPMIDNPNYQGIWKPRKIPNPDFFEDLEPFKMTPF 420
hCN  360:CESAPGCGVWQRPVIDNPNYKGKWKPPMIDNPSYQGIWKPRKIPNPDFFEDLEPFRMTPF 419
         *    *** *  * **** *** * **   ** *** * * ****** ***  **  * *

hCM  409:SALGLELWSMTSDIYFDNFIICSEKEVADHWAADGWRWKIMIANANKPGVLKQLMAAAEG 468
mCM  409:SALGLELWSMTPDIYFDNFIICSEKEVADQWATDGWELKIMVANANEPGVLRQLVIAAEE 468
mCN  421:SAIGLELWSMTSDIFFDNFIISGDRRVVDDWANDGWGLKKAADGAAEPGVVLQMLEAAEE 480
hCN  420:SAIGLELWSMTSDIFFDNFIICADRRIVDDWANDGWGLKKAADGAAEPGVVGQMIEAAEE 479
         ** ******** ** ******       * ** ***  *     *  ***  *   ***

hCM  469:HPWLWLIYLVTAGVPIALITSFCWPRKVKKKHKDTEYKKTDICIPQTKGVLEQEEKEEKA 528
mCM  469:RPWLWLMYLVMAGLPVALVASFCWPRKVKKKYEDTGPKKTELCKLQSKAALEQEAEEEKA 528
mCN  481:RPWLWVVYILTVALPVFLVILFC-C-SGKKQSNAMEYKKTDAPQPDVK--DEEGKEEEKN 536
hCN  480:RPWLWVVYILTVALPVFLVILFC-C-SGKKQTSGMEYKKTDAPQPDVK-EEEEEKEEEKD 536
          ****  *      *  *   **     **       ***       *   *    ***

hCM  529:ALEKPMDLEEEKKQNDGE-M-LEKEEESEPEEKSEEEIEIIEGQEESNQSNKSGSEDEMK 586
mCM  529:P-EKPEDVQEEKKPGEAEVVTVEKEVIGEPEEKSKEDRETLEGQEEVSKLSKSGSEDEMK 587
mCN  537:----------------------KRD-EEEEEEKLEEKQK-SD-AEEDGVTG-SQDEEDSK 570
hCN  537:----------------------KGDEEEEGEEKLEEKQK-SD-AEEDGGTV-SQEEEDRK 572
                                       ***           **      *  *   *

hCM  587:E-ADESTGSGDGPIKSVRKRRVRKD                                    610
mCM  588:D-ADESPGSGDAPLKSLRKRRVRKD                                    611
mCN  571:PKAEEDE----ILNRSPRNRKPRRE                                    591
hCN  573:PKAEEDE----ILNRSPRNRKPRRE                                    592
             *          * * *  *
```

Figure 1. Comparison of the nucleotide sequence and the deduced amino acid sequence of h-calmegin (GenBank accession number: D86322), as seen in Tanaka et al. (1997) with mouse calmegin (GenBank accession number: D86323). Asterisks beneath sequences indicate same sequences in both the human and mouse calmegin. Gaps introduced to maximize the alignment are represented by dashes. The putative hydrophobic leader sequence, transmembrane sequence and ER-retention motif are located at amino acid residue 1-19, 471-492 and 604-610, respectively.

blot analysis. Intercrosses between heterozygous F_1 animals yielded offspring (24 litters), that segregated in a mendelian distribution into these groups: 57 wild-type, 104 heterozygous, and 53 homozygous mutant weaning pups. Homozygous (-/-) mutant mice did not show any overt developmental abnormalities.

Testes derived from adult F_2 males were subjected to analysis of the effect of the neo cassette insertion on calmegin synthesis. The level of full-length (2.4-kb) mRNA is reduced to approximately half of normal in

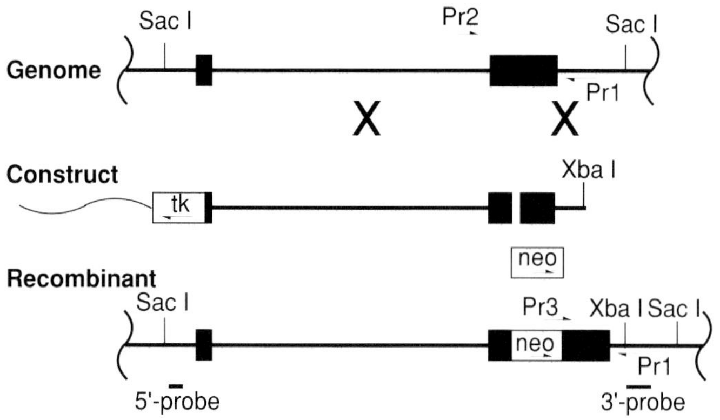

Figure 2. Targeted disruption of calmegin gene. Targeting vector was made by inserting a neomycin resistance (neo) gene into the second exon, which resulted in disruption of the calmegin ATG start codon. For negative selection, a herpes simplex virus thymidine kinase gene (tk) was introduced into the targeting construct.

the heterozygous, while no message is detected in the homozygous mutant mice. When the blots were rehybridyzed to a probe specific for the calnexin or beta-actin gene, constant expressions of both were detected in all F_2 samples. To determine whether calmegin protein was present in the mutant mice, immunoblot experiments were carried out. The anti-calmegin monoclonal antibody (TRA 369) recognized a 93K protein in +/+ and +/- males. Neither normal-sized calmegin protein nor any smaller fragment was detected in the testis from -/- mice. The expression of calnexin in the testis was not significantly affected by calmegin disruption.

The effect of calmegin disruption on spermatogenesis was analyzed histochemically. The adult testis of a -/- mouse was fixed in Bouin's solution and thin sections were made. Testicular sections were stained with HE or PAS and observed under a microscope. Spermatogenesis was normal in the calmegin -/- tubules as indicated by the production of mature spermatozoa and the appropriate proportion of cells in each sub-stage, as shown in Figure 3. There was no difference in the size of the mutant testis or the number of germ cells per section of a tubule counted separately according to the meiotic stage.

However, homozygous mutant males were nearly sterile, although normal copulation and vaginal plug formation could be observed. Even though 11 -/- males were mated with three females each for more than three months, with more than 120 plugs being observed, only one pregnancy, which resulted in two pups, was seen. There was no significant difference in the fecundity between +/+ and +/- males; the average litter size in the wild-type female/heterozygote male cross and the wild-type female/wild-type male cross was (8.2 ± 1.5, n=18) and (8.4 ± 1.1, n=20), respectively. In contrast, all F_2 females (+/+, +/- and -/-) were fertile and showed normal fecundity, consistent with findings that calmegin is not expressed in the ovary (Watanabe et al. 1992; Watanabe et al. 1994).

Sperm was collected from the reproductive tract of females that had been mated with -/- males. The number of recovered sperm and their viability were comparable to that of wild-type males. When incubated *in vitro*, the sperm from both +/+ and -/- mice maintained their motility over 80% and no differences were found between the two during the 120-minute incubation in TYH medium. Epididymal sperm from -/- mice showed normal viability, morphology and motility under microscopic observation (Fig. 4). When sperm from calmegin -/- and +/+ mice were immunostained, using various polyclonal and monoclonal anti-sperm antibodies, no differences in staining pattern or intensity, measured by flow cytometry, were found between the two kinds of sperm. The biotinylation and subsequent SDS-PAGE analysis of solubilized sperm membrane using avidin-peroxidase also showed no significant difference between sperm membrane from +/+ and -/- mice. In spite of their normal appearance in the *in vitro* fertilization medium, sperm from -/- males were generally unable to penetrate the egg extracellular matrix (Ikawa et al. 1997). Removal of cumulus cells revealed that these sperm failed to adhere to the egg despite frequent collisions with the zona pellucida. These results were also applicable to the -/- males that were back crossed to C57BL/6 mice for 10 generations.

Other Cases of Targeted Disruption of Genes Expressed in the Testis

A large number of genes that are known to be expressed in the testis or in maturing spermatocytes have been disrupted in the mouse, to test their function in male fertility. Surprises in this field have been numerous, as the following examples show.

1. Fertilin is reported to function in sperm-egg fusion by binding the sperm to the egg plasma membrane via a binding site in the disintegrin domain of fertilin beta and lead to sperm/egg fusion through its putative "fusion peptide" on alpha subunit. Researchers were surprised to see, however, as one of the phenotypes, mice deficient in fertilin beta gene produce morphologically normal sperm with impaired ability to bind the zona pellucida (Cho et al. 1998).

2. ACE is a membrane-bound dipeptidyl carboxypeptidase that generates the vasoconstricting peptide angiotensin II and that inactivates the vasodilating peptide bradykinin. The gene encoding ACE consists of two homologous regions and encodes both a somatic and a testis-specific isoenzyme. Female mice deficient for both forms of ACE were found to be fertile, but the fertility of homozygous male mutants was greatly reduced (Krege et al. 1995; Hagaman et al. 1998).

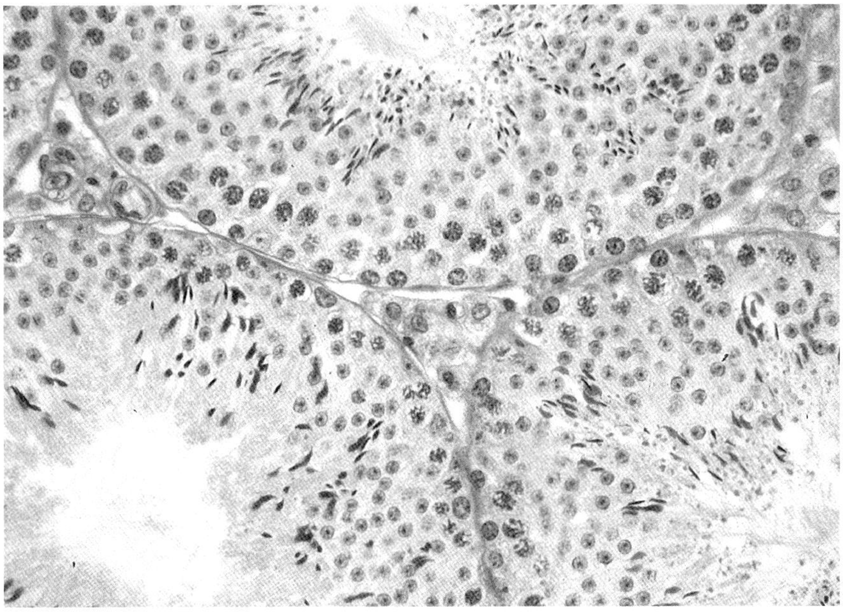

Figure 3 (left). The effect of calmegin disruption on spermatogenesis was analyzed histochemically. No difference in spermatogenesis between wild-type and mutant mice was evident from haematoxylin and eosin-stained (HE) thin sections of adult testis fixed in Bouin's solution. Even in periodic acid Schiff (PAS)-stained sections as shown here, successful accumulation of glycoproteins was observed in the acrosomes of calmegin -/- spermatids.

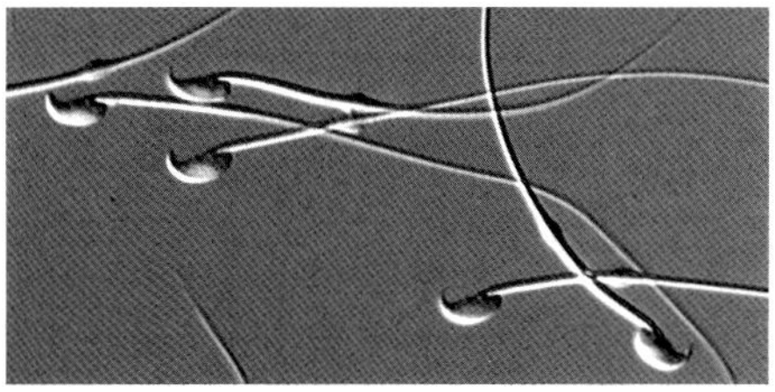

Figure 4 (above). Sperm from calmegin -/- mice showed normal motility. Moreover, no morphological abnormalities were found by observation with Hofman contrast microscope and electron microscope.

The cause of the male sterility was demonstrated to be a defect in sperm migration within the oviducts, as well as a decreased ability to bind to the zona pellucida. This phenotype in the male was not an indirect effect of a blood pressure decrease caused by the loss of somatic ACE expression: mutant males that lacked only the somatic form of ACE were found to be fertile. The male-specific infertility of animals lacking the testis-specific ACE is curious, because angiotensin itself does not seem to be required for spermatogenesis. Indeed, male mice lacking angiotensinogen are reported to have normal fertility (Tanimoto et al. 1994). No other substrate of ACE in the testis has been proposed.

3. PC4 is a member of the proprotein convertase (PC) family of serine proteases, which are implicated in the processing of a variety of prohormones, proneuropeptides and cell surface proteins. In rodents, PC4 transcripts have been detected in spermatocytes and round spermatids exclusively, suggesting a reproductive function for this enzyme. As expected, the *in vivo* fertility of homozygous mutant males was severely impaired (average litter size 0.8, compared with 6.9 in wild-type and +/- mice), but no spermatogenic abnormality was evident (Mbikay et al. 1997). *In vitro*, the fertilizing ability of PC4 null spermatozoa was also significantly reduced. Interestingly, if the average litter size is calculated only on the basis of successful matings, the figure increases to 3.3 in gene knockout mice.

Calmegin-deficient mice are similar in this regard. Their average litter size is 0.02, compared with 8.2 in wild-type mice, but one successful mating resulted in two pups. This indicates that, in some females (or in some ejaculates), the defect in fertility is not as severe as the average value would suggest. Anecdotal evidence suggests that this pattern may occur in humans as well. Many supposedly infertile couples find, to their surprise, that they are suddenly able to conceive.

4. C-ros is a receptor-type tyrosine kinase that is expressed in a small number of epithelial cell types, including those of the caput epididymis. Targeted mutations of c-ros cause male but not female infertility. Sperm isolated from c-ros -/- mice appear normal and can fertilize eggs efficiently *in vitro*. Remarkably, sperm derived from c-ros -/- spermatocytes have their *in vivo* fertilization capacity restored if they pass through the wild-type epididymal epithelium of a chimeric mouse (Sonnenberg Riethmacher et al. 1996). Thus, whereas spermatogenesis does not appear to be affected in c-ros-deficient mice, sperm maturation is impaired, because of a defect in a signaling pathway in epididymal cells. It has been suggested that GPI-anchored proteins are transferred to the sperm surface during epididymal transit (Kirchhoff et al. 1997), and it may be that c-ros is required for such transfer.

However c-ros may act in this tissue, the infertility of these sperm seems to be explained by their failure to reach the oviduct, despite the fact that they are produced in normal numbers. After it was learned that normal-looking sperm failed to ascend into the oviduct in more than two knockout mouse lines, the migrating activity of calmegin knockout mouse sperm was re-examined, and decreased migration into the oviduct was found. These results indicate that there may be many more cases of male infertility in which sperm show normal characteristics but fail to move into the oviduct. There should be some "homing" mechanism for sperm to pass through the uterotubal junction, with the mechanism allowing only "good" sperm to reach the eggs. The ability to pass through the uterotubal junction is hampered by the turbulence of the membrane structure that is caused by disruption of certain genes. However, it was amazing to learn that five of the unrelated gene disruptions resulted in very similar phenotypes. The importance of penetration through the uterotubal junction was clearly demonstrated.

A Non-Invasive Way to Assess Sperm Acrosome Reaction

In all of the infertile sperm from gene-disrupted mice, neither the status of sperm capacitation nor acrosome reaction is clear. The acrosomal status of sperm can be assessed using various probes such as CTC, lectins and acrosome reacted sperm specific antibodies. However, all of these probes are somewhat invasive to sperm function.

In the jellyfish *Aequorea victorea*, GFP is responsible for the green bioluminescence from the margin of its bells with Ca^{2+} mediated activation. *Aequorea* bioluminescence is first activated when Ca^{2+} binds to aequorin following emission of blue fluorescence. Blue fluorescence and subsequent energy transfer from activated aequorin cause GFP to emit green fluorescence. Therefore, unlike enzymes, GFP needs no substrate to emit fluorescence. Rather, it requires only energy to excite the fluorephore.

GFP is a single peptide of 238 amino acids (Prasher et al. 1992). In order to become fluorescent, GFP needs to form its fluorephore by post-translational autocyclization of Ser65, Tyr66 and Gly67 following oxidation. Finally, the chromophore locates in the center of a barrel-like basket composed of eleven beta-sheets (Ormo et al. 1996). GFP becomes fluorescent even when expressed in the heterologous species, probably because the conformational changes require no substrate or cofactor. This could be an ideal feature as a marker protein. To determine if GFP can be used as a vital marker *in vivo*, transgenic mice expressing GFPs were made (Ikawa et al. 1995; Okabe et al. 1997). At first, the polypeptide chain elongation factor 1alpha (EF) promoter was tested, for it is known to have ubiquitous expression (Mizushima, Shigekazu 1990). However, in EF-wtGFP transgenic mice, very weak expression in the liver was observed (unpublished data). When a CAG promoter (combination of beta-actin promoter and hCMV enhancer) was chosen (Niwa et al. 1991), a strong expression was observed in some tissues, especially in muscle and pancreas. Finally, mice expressing wt-GFP, the first application of GFP in mammals *in vivo* (Ikawa et al. 1995), were reported. However, to the researchers' surprise, when EGFP instead of wt-GFP was expressed by a CAG promoter, almost all tissues fluoresced bright green with EGFP (Okabe 1997). These "green mice" were showing that the GFP could be a good transgenic marker protein in mouse cells as well as *C. elegans*.

Green mice produced green eggs but not green sperm. Since the sperm have little cytosol, the sperm from green mice were only faintly green fluorescent. Certainly, cell- or organ-specific expression of GFP is possible if an appropriate promoter is used to drive GFP in transgenic mice and signal peptide leads GFP to target organelle.

From the point of view of making transgenic mice whose acrosomal status could be easily detected non-invasively, the GFP was ligated with acrosin promoter and acrosin signal peptides and injected to pronuclei of fertilized eggs. As shown in Figure 5, sperm from these transgenic mice contained GFP inside the acrosome and under the excitation light, the green fluorescent acrosome could be seen. These mice were useful to investigate the sperm capacitation and acrosome reaction (unpublished data). When the transgenic green mice were bred with calmegin knockout mice, researchers learned that the reactivity of spermatozoa to Ca^{2+} ionophore is not impaired but that their ability to undergo spontaneous acrosome reaction in incubation medium was reduced. The transgenic green sperm mouse

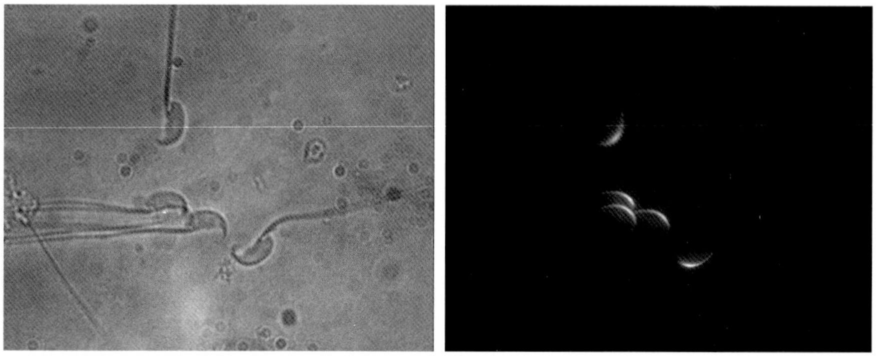

Figure 5. Green sperm. Acrosin promoter-driven green fluorescent protein, which is fused with acrosin signal peptide, was produced in meiotic germ cells and localized only in sperm acrosome. Acrosome reaction was detected by the disappearance of green fluorescence using fluorescent microscopy and flow-cytometry.

line was shown to be a useful tool to investigate sperm capacitation and acrosome reaction *in vitro* and *in vivo*.

Prospects for the Future

It is clear that a man can be diagnosed as infertile if he is azoospermic or oligospermic. However, there are many infertile men who have semen parameters within the normal range but who show diminished binding of sperm to the zona pellucida; still others may be deficient either at one or more of the later steps of sperm maturation or in the expression or activation of stored sperm components.

To date, various proteins have emerged as candidate zona pellucida adhesion molecules on sperm, including sp56 (Bookbinder et al. 1995; Cheng et al. 1994), PH-20 (Hunnicutt et al. 1996), zonadhesin (Hardy, Garbers 1994; Hardy, Garbers 1995), GalTase (Gong et al. 1995; Lopez et al. 1985), p95 (Burks et al. 1995; Saling 1991), sp17 (Richardson et al. 1994; Yamasaki et al. 1995) and so on, and activity of one or more of these molecules is presumably important in sperm-zona pellucida adhesion. Since calmegin transiently interacts with nascent proteins and its gene disruption appeared to impair specifically the sperm binding to the egg, researchers in the Okabe lab suggest that calmegin is expressed during spermatogenesis to ensure the proper maturation of certain sperm surface protein(s) required for the egg binding. However, it was shown by other researchers that calmegin is not the only gene that could cause this interesting phenomenon. The sheer diversity of mouse mutations that cause male infertility may be daunting, but work in this system now provides a number of candidate genes that may be relevant to infertility in men. The prediction of a man's fertility on the basis of his genotype could be possible in the future.

References

Austin CR. Observation on the penetration of the sperm into the mammalian egg. Aust J Sci Res [B] 1951; 4:581-596.

Baba T, Azuma S, Kashiwabara S, Toyoda Y. Sperm from mice carrying a targeted mutation of the acrosin gene can penetrate the oocyte zona pellucida and effect fertilization. J Biol Chem 1994; 269:31845-31849.

Bergeron JJ, Brenner MB, Thomas DY, Williams DB. Calnexin: a membrane-bound chaperone of the endoplasmic reticulum. Trends Biochem Sci 1994; 19:124-128.

Bookbinder LH, Cheng A, Bleil JD. Tissue- and species-specific expression of sp56, a mouse sperm fertilization protein. Science 1995; 269:86-89; erratum in Science 1995; 269:1120.

Burks DJ, Carballada R, Moore HD, Saling PM. Interaction of a tyrosine kinase from human sperm with the zona pellucida at fertilization. Science 1995; 269:83-86.

Chang MC. Fertilizing capacity of spermatozoa deposited in Fallopian tubes. Nature 1951; 168:997-988.

Cheng A, Le T, Palacios M, Bookbinder LH, Wassarman PM, Suzuki F, Bleil JD. Sperm-egg recognition in the mouse: characterization of sp56, a sperm protein having specific affinity for ZP3. J Cell Biol 1994; 125:867-878.

Cho C, Bunch DO, Faure JE, Goulding EH, Eddy EM, Primakoff P, Myles DG. Fertilization defects in sperm from mice lacking fertilin beta. Science 1998; 281:1857-1859.

Gong X, Dubois DH, Miller DJ, Shur BD. Activation of a G protein complex by aggregation of beta-1,4-galactosyltransferase on the surface of sperm. Science 1995; 269:1718-1721.

Hagaman JR, Moyer JS, Bachman ES, Sibony M, Magyar PL, Welch JE, Smithies O, Krege JH, O'Brien DA. Angiotensin-converting enzyme and male fertility. Proc Natl Acad Sci USA 1998; 95(5):2552-2557.

Hardy DM, Garbers DL. Species-specific binding of sperm proteins to the extracellular matrix (zona pellucida) of the egg. J Biol Chem 1994; 269:19000-19004.

Hardy DM, Garbers DL. A sperm membrane protein that binds in a species-specific manner to the egg extracellular matrix is homologous to von Willebrand factor. J Biol Chem 1995; 270:26025-26028.

Hebert DN, Foellmer B, Helenius A. Glucose trimming and reglucosylation determine glycoprotein association with calnexin in the endoplasmic reticulum. Cell 1995; 81:425-433.

Hunnicutt GR, Mahan K, Lathrop WF, Ramarao CS, Myles DG, Primakoff P. Structural relationship of sperm soluble hyaluronidase to the sperm membrane protein PH-20. Biol Reprod 1996; 54:1343-1349.

Ikawa M, Kominami K, Yoshimura Y, Tanaka K, Nishimune Y, Okabe M. Green fluorescent protein as a marker in transgenic mice. Dev Growth Differ 1995; 37:455-459.

Ikawa M, Wada I, Kominami K, Watanabe D, Toshimori K, Nishimune Y, Okabe M. The putative chaperone calmegin is required for sperm fertility. Nature 1997; 387:607-611.

Kirchhoff C, Pera I, Derr P, Yeung C H, Cooper T. The molecular biology of the sperm surface. Post testicular membrane remod elling. Adv Exp Med Biol 1997; 424:221-232.

Krege JH, John SW, Langenbach LL, Hodgin JB, Hagaman JR, Bachman ES, Jennette JC, O'Brien DA, Smithies O. Male-female differences in fertility and blood pressure in ACE-deficent mice. Nature 1995; 375:146-148.

Lopez LC, Bayna EM, Litoff D, Shaper NL, Shaper JH, Shur BD. Receptor function of mouse sperm surface galactosyltransferase during fertilization. J Cell Biol 1985; 101:1501-1510.

Lu Q, Shur B. Sperm from beta 1,4-galactosyltransferase-null mice are refractory to ZP3-induced acrosome reactions and penetrate the zona pellucida poorly. Development 1997; 124:4121-4131.

Mbikay M, Tadros H, Ishida N, Lerner CP, de Lamirande E, Chen A, El Alfy M, Clermont Y, Seidah NG, Chretien M, Gagnon C, Simpson EM. Impaired fertility in mice deficient for the testicular germ-cell protease PC4. Proc Natl Acad Sci USA 1997; 94:6842-6846.

Mizushima S, Shigekazu N. pEF-BOS, a powerful mammalian expression vector. Nucleic Acids Res 1990; 18:5322.

Niwa H, Yamamura K, Miyazaki J. Efficient selection for high-expression transfectants with a novel eukaryotic vector. Gene 1991; 108:193-199.

Ohsako S, Bunick D, Hess RA, Nishida T, Kurohmaru M, Hayashi Y. Characterization of a testis specific protein localized to endoplasmic reticulum of spermatogenic cells. Anat Rec 1994a; 238:335-348.

Ohsako S, Hayashi Y, Bunick D. Molecular cloning and sequencing of calnexin-t. An abundant male germ cell-specific calcium-binding protein of the endoplasmic reticulum. J Biol Chem 1994b; 269:14140-14148.

Okabe M, Ikawa M, Kominami K, Nakanishi T, Nishimune Y. 'Green mice' as a source of ubiquitous green cells. FEBS Lett 1997; 407:313-319.

Ormo M, Cubitt AB, Kallio K, Gross LA, Tsien RY, Remington SJ. Crystal structure of the *Aequorea victoria* green fluorescent protein. Science 1996; 273:1392-1395.

Ou WJ, Cameron PH, Thomas DY, Bergeron JJ. Association of folding intermediates of glycoproteins with calnexin during protein maturation. Nature 1993; 364:771-776.

Prasher DC, Eckenrode VK, Ward WW, Prendergast FG, Cormier MJ. Primary structure of the *Aequorea victoria* green-fluorescent protein. Gene 1992; 111:229-233.

Richardson RT, Yamasaki N, O'Rand MG. Sequence of a rabbit sperm zona pellucida binding protein and localization during the acrosome reaction. Dev Biol 1994; 165:688-701.

Saling PM. How the egg regulates sperm function during gamete interaction: facts and fantasies. Biol Reprod 1991; 44:246-251.

Sonnenberg Riethmacher E, Walter B, Riethmacher D, Godecke S, Birchmeier C. The c-ros tyrosine kinase receptor controls regionalization and differentiation of epithelial cells in the epididymis. Genes Dev 1996; 10:1184-1193.

Tanaka H, Ikawa M, Tsuchida J, Nozaki M, Suzuki M, Fujiwara T, Okabe M, Nishimune Y. Cloning and characterization of the human Calmegin gene encoding putative testis-specific chaperone. Gene 1997; 204:159-163.

Tanimoto K, Sugiyama F, Goto Y, Ishida J, Takimoto E, Yagami K, Fukamizu A, Murakami K. Angiotensinogen-deficient mice with hypotension. J Biol Chem 1994; 269:31334-31337.

Vassilakos A, Cohen Doyle MF, Peterson PA, Jackson MR, Williams DB. The molecular chaperone calnexin facilitates folding and assembly of class I histocompatibility molecules. EMBO J 1996; 15:1495-1506.

Wada I, Rindress D, Cameron PH, Ou WJ, Doherty JJd, Louvard D, Bell AW, Dignard D, Thomas DY, Bergeron JJ. SSR alpha and associated calnexin are major calcium binding proteins of the endoplasmic reticulum membrane. J Biol Chem 1991; 266:19599-19610.

Wada I, Ou WJ, Liu MC, Scheele G. Chaperone function of calnexin for the folding intermediate of gp80, the major secretory protein in MDCK cells. Regulation by redox state and ATP. J Biol Chem 1994; 269:7464-7472.

Wakayama T, Yanagimachi R. Development of normal mice from oocytes injected with freeze-dried spermatozoa. Nat Biotechnol 1998; 16:639-641.

Watanabe D, Sawada K, Koshimizu U, Kagawa T, Nishimune Y. Characterization of male meiotic germ cell-specific antigen (Meg 1) by monoclonal antibody TRA 369 in mice. Mol Reprod Dev 1992; 33:307-312.

Watanabe D, Yamada K, Nishina Y, Tajima Y, Koshimizu U, Nagata A, Nishimune Y. Molecular cloning of a novel $Ca^{(2+)}$-binding protein (calmegin) specifically expressed during male meiotic germ cell development. J Biol Chem 1994; 269:7744-7749.

Watson JD, Crick FHC. Genetic implications of the structure of deoxyribonucleic acid. Nature 1953; 171:964-967.

Wilmut I, Schnieke A, McWhir J, Kind A, Campbell K. Viable offspring derived from fetal and adult mammalian cells. Nature 1997; 385:810-813.

Yamasaki N, Richardson RT, O'Rand MG. Expression of the rabbit sperm protein Sp17 in COS cells and interaction of recombinant Sp17 with the rabbit zona pellucida. Mol Reprod Dev 1995; 40:48-55.

7

Protamine Mediated Condensation of DNA in Mammalian Sperm

R. Balhorn
M. Cosman
K. Thornton
V. V. Krishnan
M. Corzett
Lawrence Livermore National Laboratory
G. Bench
C. Kramer
J. Lee IV
N. V. Hud
M. Allen
M. Prieto
W. Meyer-Ilse
J. T. Brown
J. Kirz
X. Zhang
E. M. Bradbury
G. Maki
R. E. Braun
W. Breed
(see affiliations page)

One of the last structural changes to occur during spermiogenesis in mammals is the repackaging of the spermatid genome. This process is initiated by the synthesis and deposition of protamine in late-step spermatids (Balhorn et al. 1984; Bellvé et al. 1975; Goldberg et al. 1977), and it is accompanied by a dramatic change in chromatin condensation (Dooher, Bennett 1973; Roosen-Runge 1962; Fig. 1). Subtle changes in the condensation state of chromatin are observed prior to this time (Dooher, Bennett 1973; Roosen-Runge 1962), but the final and most dramatic change in DNA compaction is induced when protamines bind to DNA and displace the other chromatin proteins.

While it has been assumed that this process of DNA condensation must play a critical role in reprogramming the sperm genome, there is conflicting evidence supporting this hypothesis. It is not entirely clear, for example, that fertilization and the progression of normal embryonic development require this process to occur. It has been demonstrated by several groups that the nuclei of early stage spermatids can be used to initiate fertilization and embryo development following intracytoplasmic sperm injection (ICSI; Bernabeu et al. 1998; Fishel et al. 1997; Ogura, Yanagimachi 1995; Reubinoff et al. 1998), even though the DNA in the nuclei of these cells have never been packaged and condensed by protamine. In contrast, a number of other studies have shown that changes in the proportion of particular protamines (Balhorn et al. 1988; Belokopytova et al. 1993; Blanchard et al. 1990; de Yebra et al. 1993), deficiencies in zinc in semen and potentially in sperm chromatin (Kjellberg et al. 1992; Kvist et al. 1988), and alterations that disrupt histone removal (Bach et al. 1990) or protamine 2 precursor processing (DP Evenson, LK Jost, M Corzett and R Balhorn, unpublished data) appear to lead to male infertility.

Studies in the Balhorn laboratory have focused on characterizing the structure of sperm chromatin in representative mammals, identifying the molecular processes that occur during the final stages of spermiogenesis, and investigating how defects in these processes relate to the function of the sperm genome

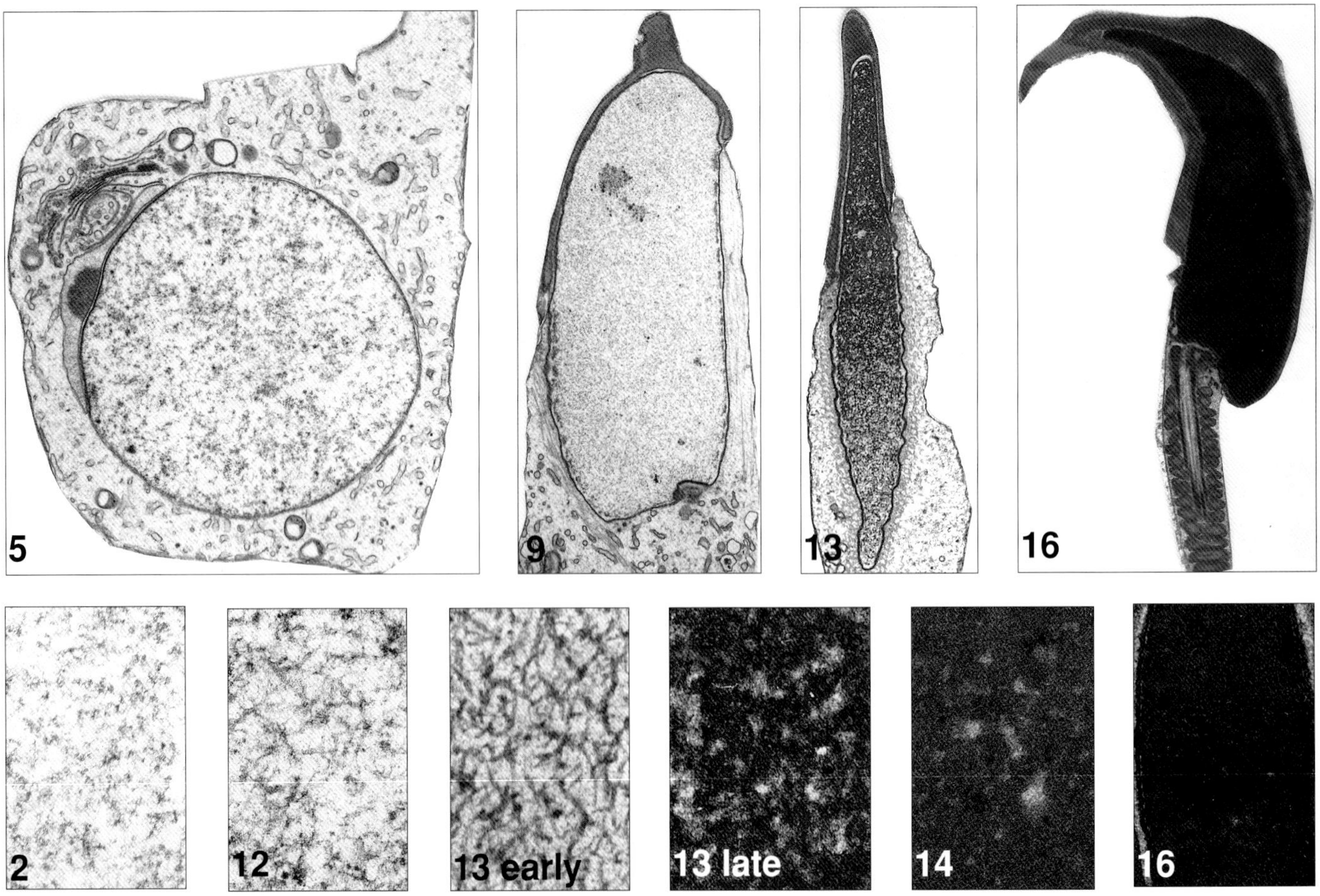

Figure 1. Transmission electron microscopy images of the mouse spermatid nucleus at different steps of spermiogenesis (numbers on figures). The top row depicts low magnifications of spermatids and their nuclei. The bottom row shows high magnification micrographs of spermatid chromatin at different steps of spermiogenesis. Chromatin condensation is initiated at the apical end of the nucleus and progresses back toward the tail as the spermatid matures. Electron micrographs courtesy of LD Russell.

following fertilization. This work on the molecular structure of the DNA-protamine complex and the distribution of protamine 1 (P1) and protamine 2 (P2) along the DNA molecule has begun to provide insight into the mechanism by which DNA is repackaged and the intermolecular interactions (DNA-protamine and P1-P2) that occur in normal sperm chromatin. Analyses of the higher-ordered packing of DNA by protamine both *in vitro* and inside the sperm head using electron microscopy (EM) and scanning probe microscopy (SPM) have identified how the protamines mediate the condensation of DNA through the formation of toroidal subunits. X-ray microscopy analyses of the intact sperm head have also revealed the uniformity with which chromatin is condensed in the head of normal sperm and provided a method for visualizing differences in condensation that occur naturally and as a consequence of defects in the process of spermiogenesis. The results of these studies show that the organization of the genome inside the head of the sperm is very different from that found in any other cell.

Structure of Protamines and Their Distribution along DNA

A combination of gel electrophoresis (Ammer, Henschen 1987; Balhorn et al. 1977, 1987; Bélaïche et al. 1987; Pongsawasdi, Svasti 1976; Tanphaichitr et al. 1978), protein sequencing (Ammer, Henschen 1987; Ammer et al. 1986; Bélaïche et al. 1987; Bellvé et al. 1988; Gusse et al. 1986; Mazrimas et al. 1986; McKay et al. 1985, 1986; Sautiere et al. 1984; Tobita et al. 1982), and protamine gene sequencing (Johnson et al. 1988; Klemm et al. 1989; Krawetz et al. 1987; Lee et al. 1987a, b; Tanhauser, Hecht 1989) has shown that two different protamines are used to package DNA in the sperm of eutherian mammals (Balhorn 1989). The sperm of all mammalian species appear to contain protamine 1 (P1), a small arginine and cysteine rich protein that is characterized by its centralized distribution of polyarginine DNA binding domains (Fig. 2), its lack of proteolytic processing (Klemm et al. 1989; Retief et al. 1993; Yelick et al. 1987), and its amino terminal sequence ARYRCC/X (Balhorn 1989). In addition to P1, the sperm of rodents, primates and selected other species also contain a second protamine. This protein is synthesized as a precursor protein (Yelick et al. 1987), and its amino-terminus is removed through a series of progressive proteolytic cleavages shortly after it binds to DNA in late-step spermatids (Carré-Eusèbe et al. 1991; Elsevier et al. 1991; Martinage et al. 1990; Yelick et al. 1987). The remaining sequence, P2, is characterized by its content of histidine, more uniform distribution of arginine residues (Fig. 2) and its ability to bind zinc (Bianchi et al. 1992; Gatewood et al. 1990).

Differences in the relative proportions of P1 and P2 present in the sperm of different mammals (Balhorn 1989) have made it difficult to reconcile how these two proteins might participate in the packaging of sperm DNA. Unlike the nucleosome subunit of somatic chromatin, in which the proportion of the various histones is universal in all tissues and organisms, the amount of P2 present in sperm chromatin has been found to vary widely among species (Balhorn 1989). In certain species, P2 is the predominant nuclear protein bound to sperm DNA. The sperm produced by many other species do not contain any protamine 2.

To identify the stoichiometry of P1 and P2 bound to DNA in sperm, a quantitative elemental imaging technique, proton induced X-ray emission (PIXE) spectroscopy, was used to measure the amount of DNA and protamine in the sperm of mammals known to contain

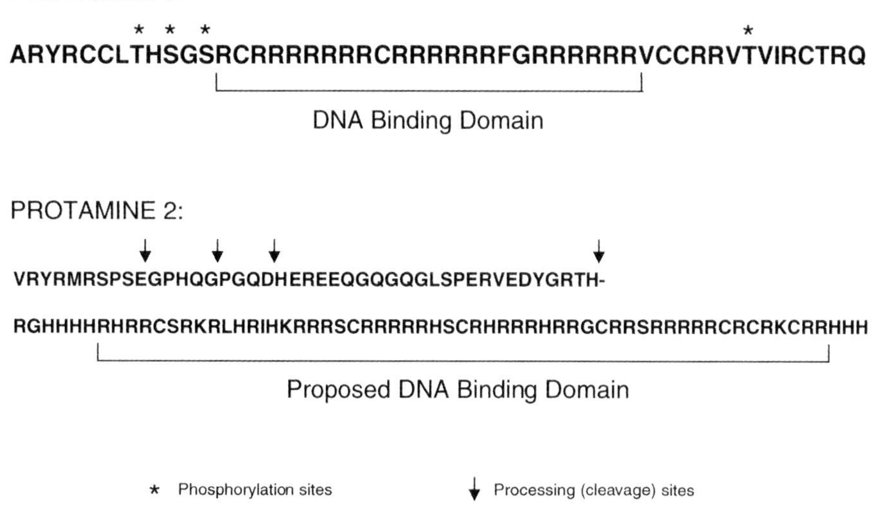

Figure 2. Amino acid sequences of mouse protamine 1 and protamine 2 showing the DNA binding domains and sites of post-translational modification.

different amounts of P2. Individual sperm heads were raster scanned with a 2- to 3-μm proton beam and the phosphorus and sulfur contents of individual sperm nuclei from five different species (Fig. 3) were measured and used to determine the amount of DNA and protamine present in each sperm cell (Bench et al. 1996). The results of these experiments revealed that the total mass of protamine (P1 + P2) bound to DNA in sperm chromatin is constant, irrespective of the relative proportions of the two protamines. Calculations based on this data indicated that the entire length of the DNA molecule must be covered by protamine in the sperm of each species tested, except human. Approximately 15% of the DNA in human sperm is known to be packaged by histones (Gatewood et al. 1987; Pongsawasdi, Svasti 1976; Tanphaichitr et al. 1978), and this was confirmed by these PIXE measurements (Bench et al. 1996).

With this information, the amino acid sequences of the two protamines and the relative proportion of P1 and P2 in the sperm chromatin of each species examined could be used to estimate the length of DNA that must be covered by each P1 and P2 molecule. In bull sperm, which contain only P1, the length of the protamine binding site on DNA (the length of DNA covered by a single protamine molecule) was calculated to be 11 base pairs (bp; Bench et al. 1996). This value is consistent with the length of DNA that could be covered by the central arginine-rich DNA binding domain as suggested previously by modeling studies (Hud et al. 1994). By making the assumption that the size of the binding domain for P1 is constant among species (the similarities in sequence suggest this), this data was also used to estimate the length of DNA covered by each P2 molecule. The size of this binding site, 15 bp, suggests that the entire length of the P2 molecule must bind to DNA in an extended conformation so all the arginine residues in the P2 molecule can interact with the phosphodiester backbone of the helix.

Additional information about the distribution of P1 and P2 along the DNA molecule was obtained from chemical crosslinking experiments performed on sperm chromatin. In these experiments, the bifunctional crosslinker dimethyl suberimidate (DMS) was used to crosslink the protamines in MTAB treated (amembraneous) sperm nuclei. Working with relatively small amounts of crosslinker, only a subset of the total proteins were crosslinked together (crosslinking of all molecules together would prevent protamine extraction from DNA). Gel electrophoretic analyses of the isolated protein and antibody staining of the blotted proteins with a monoclonal antibody specific for P1 (Hup1N) and another for P2 (Hup2B) revealed that P2 molecules could only be crosslinked together as dimers by DMS, while dimers, trimers and higher multimers of P1 were detected in the immunoblots (Fig. 4). These results, albeit preliminary, provide the first direct evidence that P1 and P2 molecules might be clustered in their distribution along the DNA molecule.

Sequences obtained recently for the Syrian hamster P1 and P2 genes (Corzett et al. 1999) have also provided indirect evidence to suggest that the P2 proteins must be clustered and not randomly distributed along DNA. A comparison of the hamster and mouse protamine sequences shows that the number and position of the cysteine residues is identical in the corresponding protein sequences for both P1 and P2 (Fig. 5A). In the sperm chromatin of both of these species, it has also been determined that all the protamine cysteine residues are crosslinked together as disulfides (R Balhorn, unpublished data). Since it is highly unlikely that the P1 or P2 molecules of two closely related species would form different disulfide crosslinks using the same cysteine residues when their sequences are essentially identical, the existence of radically different amounts of P2 in the sperm of hamsters (P2/P1=1/2) and mice (P2/P1=2) could only be explained if all P2 molecules are clustered together as dimers (Fig. 5B). Only under this condition would it be possible to vary the amount of P2 in sperm chromatin over such a wide range and still retain the ability to form the same inter-protamine disulfide bonds.

Figure 3. The DNA and protamine compositions of individual sperm cells are obtained using proton induced X-ray emission spectroscopy by measuring the phosphorus and sulfur contents of each nucleus.

	DNA	Protamine	%P2
Bull	3.42±0.09	2.90±0.12	0
Stallion	3.30±0.10	2.85±0.14	15
Hamster	3.34±0.09	2.98±0.13	34
Human	3.23±0.09	2.36±0.12	50
Mouse	3.36±0.09	2.87±0.15	67

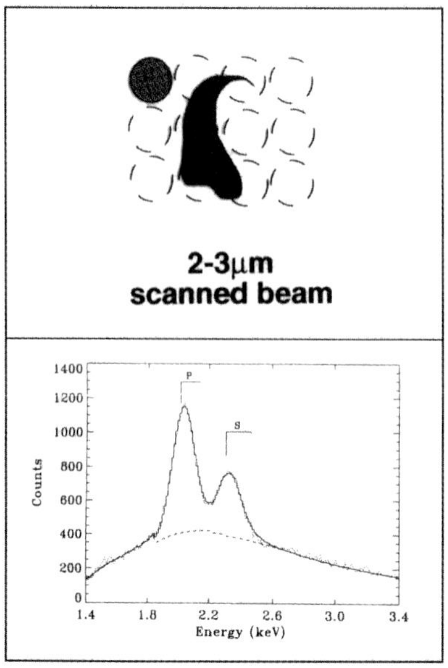

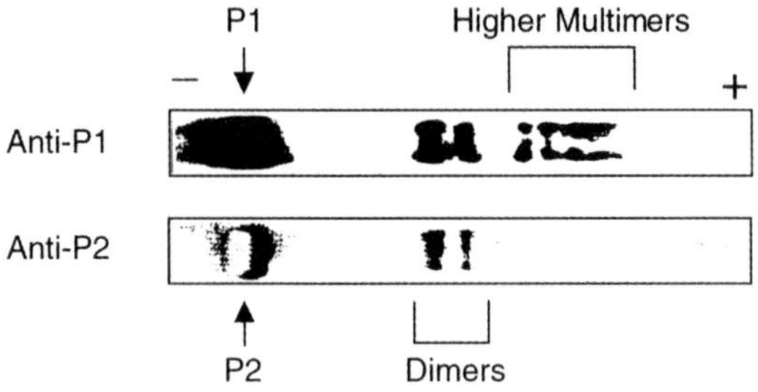

Figure 4. Basic nuclear proteins extracted from hamster sperm chromatin after crosslinking with dimethyl suberimidate are blotted onto Immobilon and probed with antibodies specific for P1 and P2. P1 is crosslinked into dimers and higher multimers while P2 is only crosslinked into dimers.

Conformational Changes Associated with Protamine Binding to DNA and the Site of Binding (Major or Minor Groove)

The insolubility of the native sperm chromatin complex has made it difficult to obtain information about the molecular structure of the DNA-protamine complex. X-ray diffraction studies performed on sperm or DNA-protamine complexes have provided limited information, and the various interpretations of the diffraction data are both confusing and conflicting (Feughelman et al. 1955; Fita et al. 1983; Wilkins 1956). Perhaps the most significant result obtained from these studies has been a measurement of the spacing between adjacent DNA helices packed in sperm chromatin, which has been shown to be ~27Å (Schellman, Parthasarathy 1984).

Raman spectroscopy is a unique form of spectroscopy in that it can be used to identify the conformation of proteins and DNA (Copeland, Spiro 1987; Fodor et al. 1985; Tu 1982) both in solution and in the solid state, and this technique has been used in the Balhorn laboratory to examine the structure of native sperm chromatin and synthetic DNA-protamine and DNA-polyarginine complexes (Hud et al. 1994). Analyses of the stretching vibrations (Fig. 6A) associated with the conformation of the sugar-phosphate backbone and bases (600-840 cm^{-1}) have shown that the DNA remains in the B conformation in the DNA-protamine complex. Comparisons of the amide I bands in free protamine and DNA-protamine and DNA-polyarginine complexes have also indicated that protamine in solution is unstructured, but upon binding to DNA, its α-carbon backbone adopts a unique conformation characteristic of a series of γ-turns containing 1->3 hydrogen bonds (Fig. 6B). Computer modeling studies of the complex show that salt-bridge formation between the guanidinium groups in protamine and the phosphate backbone in DNA would place the arginine side chains in the proper orientation (Fig. 6C) and spacing for the backbone to form a 2_7a-ribbon. The formation of these 1->3 hydrogen bonds would not be the driving force that defines the secondary structure adopted by protamine but rather an interaction favored by a geometry that is promoted when consecutive arginine residues bind to DNA.

Analyses of the excitation state of tryptophan in a synthetic analog of a protamine DNA binding domain bound to bromodeoxyuridine containing duplex DNA have also identified the location of protamine binding on DNA. Optically detected magnetic resonance was used

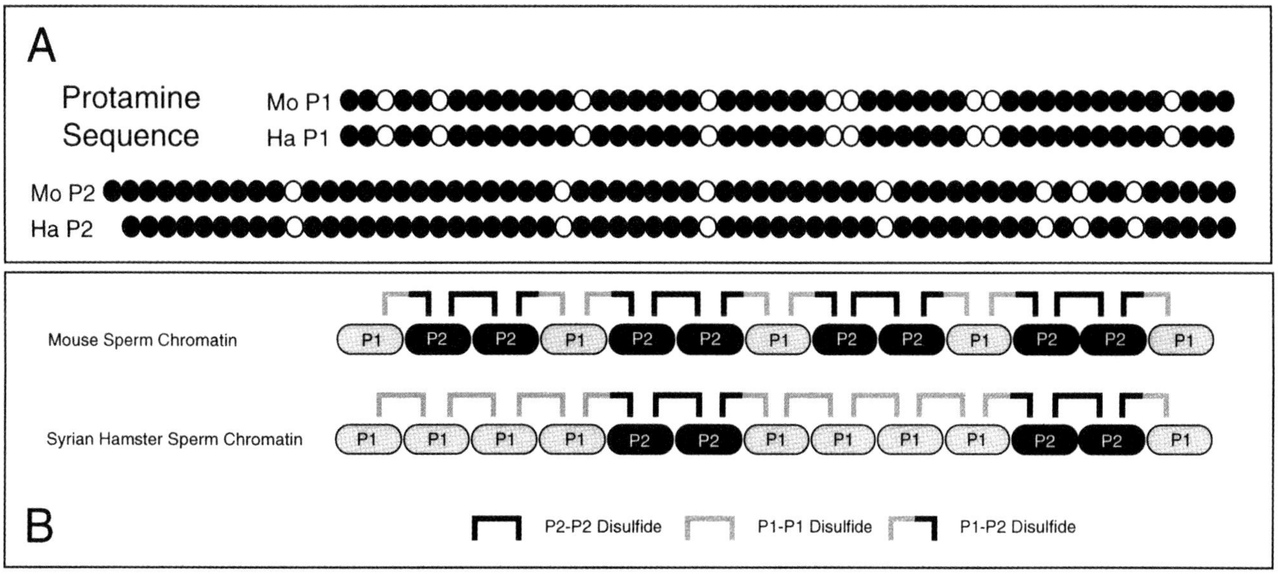

Figure 5. A: Schematic showing location of cysteine residues (white spheres) in the protein sequences of mouse and hamster protamines. B: Hypothetical arrangement of P1 and P2 molecules aligned along DNA in the sperm chromatin of mouse and hamster showing how identical inter-molecular disulfide crosslinks could be made between P2 molecules in chromatin with different amounts of P2 bound to DNA.

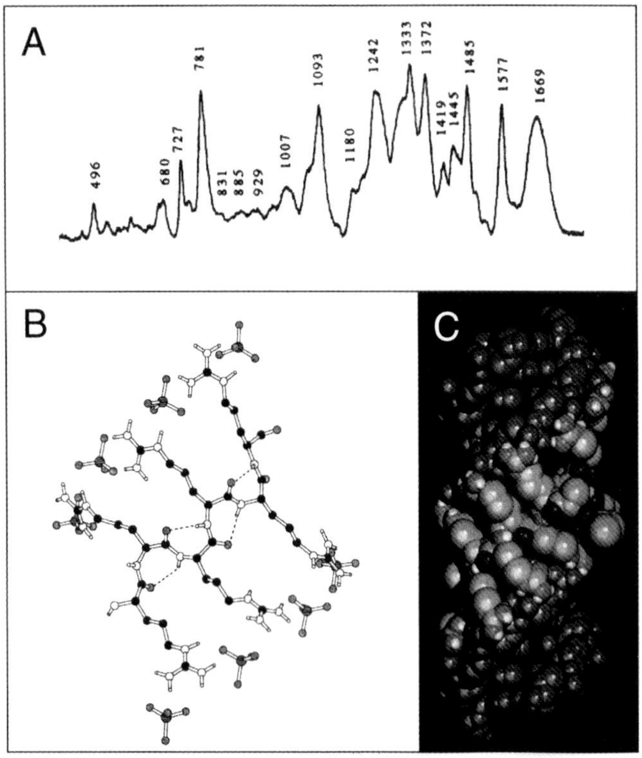

Figure 6. A: Raman spectra of a DNA-protamine (salmine) complex. B: Model of a section of the protamine DNA binding domain (arg6) bound to DNA showing the series of γ-turns that repeat along the α-carbon backbone and the 1->3 hydrogen bonds (dashed lines). C: Computer model of salmine bound to one turn of B-form DNA showing the extension of the arginine side chains and its location in the major groove.

to monitor the kinetics of microwave excitation of the tryptophan residue in the peptide, and the results of these studies confirmed that protamine binds to the major groove and also suggested that aromatic rings on residues like phenylalanine and tyrosine probably are projected in toward the DNA. Only under these conditions would the bromine atoms in the DNA be positioned close enough (within 5Å) to the tryptophan residue in the synthetic complex to accelerate the rate (Fig. 7) of decay of its excited state (Prieto et al. 1997). A similar conclusion was also drawn from a solid-state nuclear magnetic resonance (NMR) spectroscopy study of native bull sperm chromatin. This work demonstrated that the thymine methyl groups located in the major groove of DNA are perturbed in native DNA-protamine complex (Prieto 1995).

NMR Spectroscopy of a Soluble Protamine Binding Domain-DNA Complex

Because it has not been possible to obtain a detailed structure for the native DNA-protamine complex, conditions were developed for preparing a synthetic complex containing the C-terminal half of the bull protamine DNA binding domain and an oligonucleotide containing an eight bp segment of duplex DNA (Fig. 8A). Complexes prepared by titrating the DNA with the protamine DNA binding domain to a final molar ratio of 1:1 remain soluble and are remarkably stable. Unlike higher molecular weight DNA-protamine complexes, the oligonucleotide complexes do not aggregate over time.

Comparisons of the two-dimensional proton-phosphorus correlation spectra and NOESY (proton-proton through space correlation) spectra between the control DNA and the protamine-DNA complex have provided several exciting and unexpected results. It has been assumed that the binding of protamine to DNA involves non-specific salt-bridge interactions between the positively charged guanidinium groups in arginine and the negatively charged DNA backbone phosphates. Surprisingly, the protamine peptide was found to bind specifically to the terminal end of the DNA duplex region (Fig. 8B). None of the peptide bound to the hairpin region, and the preparation did not contain a mixture of complexes with the peptide bound non-specifically along the length of the duplex. Differences in chemical shifts observed for the phosphorus (Fig. 9), carbon and protons in the oligonucleotide and the complex revealed that the binding site spans only five base pairs of DNA sequence, consistent with the 10 arginine residues present in the peptide. In agreement with other studies, the chemical shift differences of the thymine methyls in the duplex region of the oligonucleotide (Fig. 9) show that the peptide is positioned in the major groove of the DNA. Qualitative analysis (relative NOE intensities and sugar coupling patterns) of the NMR data also indicate that the DNA in the complex retains its native B-form conformation.

While the analyses of the NMR data are incomplete, the results have provided additional information about the bound peptide. A comparison of the chemical shifts of the protons and carbons of the protamine in solution alone to those when it is in a complex has shown that the conformation of the peptide backbone in the complex is definitely not helical, but it adopts an extended conformation. The peptide sequence also appears to bind to the DNA with a specific amino to carbonyl orientation, with the phenylalanine residue (located near the N-terminal end of the peptide) intercalating between the d(G1•C21) and d(A2•T20) base pairs located at the end of the DNA duplex.

Another intriguing result, which provides information about the individual relative motions of the nucleotides in the DNA, was obtained by measuring the changes in the intensities of the DNA proton resonances upon protamine binding. As one might expect, significant changes are observed in both the chemical shift

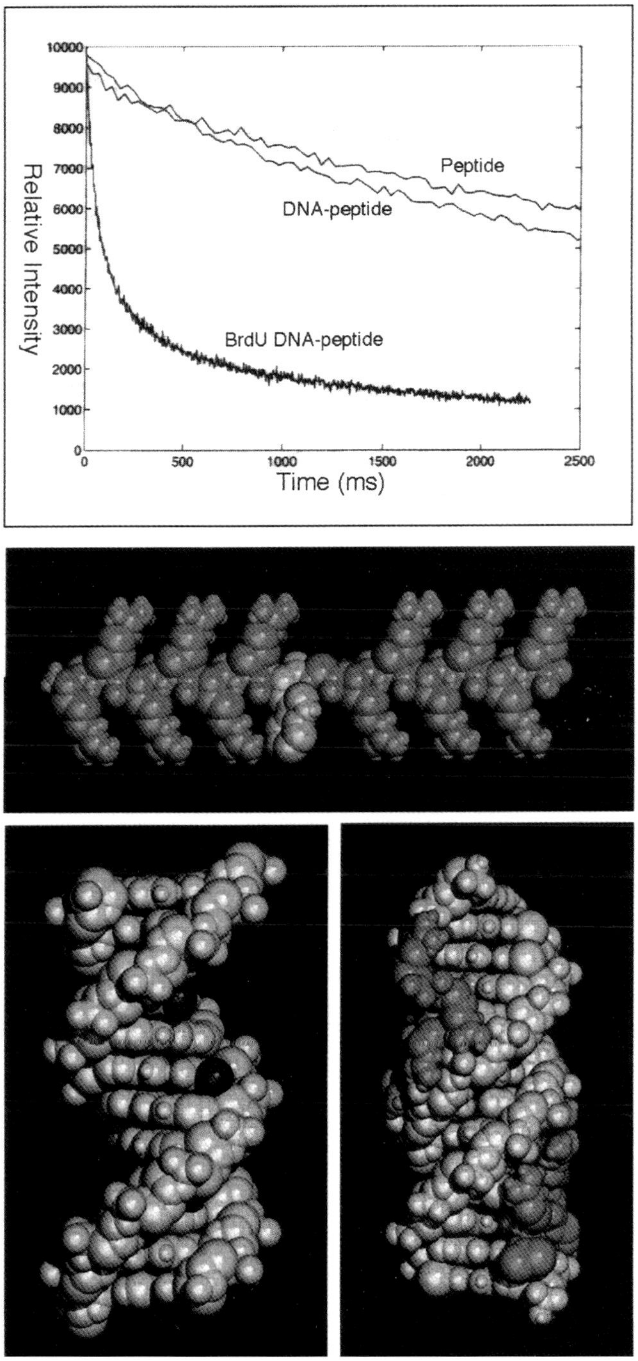

Figure 7. Comparison of decay of excited state of tryptophan residues (graph) in a synthetic protamine DNA binding domain as measured by optically detected magnetic resonance. Rate is similar for peptide (middle panel) in solution and when bound to calf thymus DNA, but rate increases dramatically when the peptide is bound to bromodeoxyuridine containing DNA (lower left structure). Results show protamine binds to the major groove of DNA (lower right structure).

values and the intensities of the DNA resonances located within the peptide binding site upon peptide binding. However, the relative intensities of the DNA resonances located adjacent to this site also change, even though no

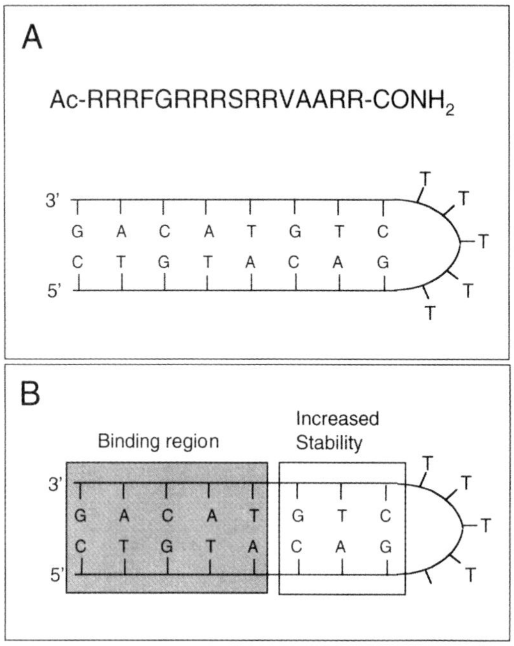

Figure 8. A soluble complex containing the C-terminal half of the bull protamine DNA binding domain and an oligonucleotide (A) was prepared for nuclear magnetic resonance analysis. The analysis of the complex revealed the peptide bound only o the end of the duplex region. These data also revealed that the adjacent three base pairs (B) were also affected by the presence of the peptide, even though they do not interact with it directly.

or very small changes occur in the chemical shift values of the DNA proton resonances in this region. This observation is exciting because these changes in the internal motion of DNA may provide the first structural-based explanation for the cooperativity that has been observed when protamine binds to DNA.

Protamine Binding Induces Higher-Ordered Coiling of the DNA Molecule

It has been known for many years that the binding of protamine to DNA condenses the DNA into a highly compact form. However, both the structure of the condensate and the mechanism of packaging the DNA inside the head of the sperm have never been identified.

The first insight into the higher-ordered packing of DNA by protamine was obtained from electron microscopy studies of DNA condensed by salmine and bull protamine (Hud et al. 1993, 1995). When DNA-protamine complexes were prepared using very dilute concentrations of DNA (to minimize aggregation), these EM studies showed that protamine binding to DNA induced the coiling of the DNA molecule into donut or toroidal structures (Fig. 10A). Measurements performed on a large number of these toroids (Fig. 10B) showed the mean size of the toroidal loop to be ~500 bp—the persis-

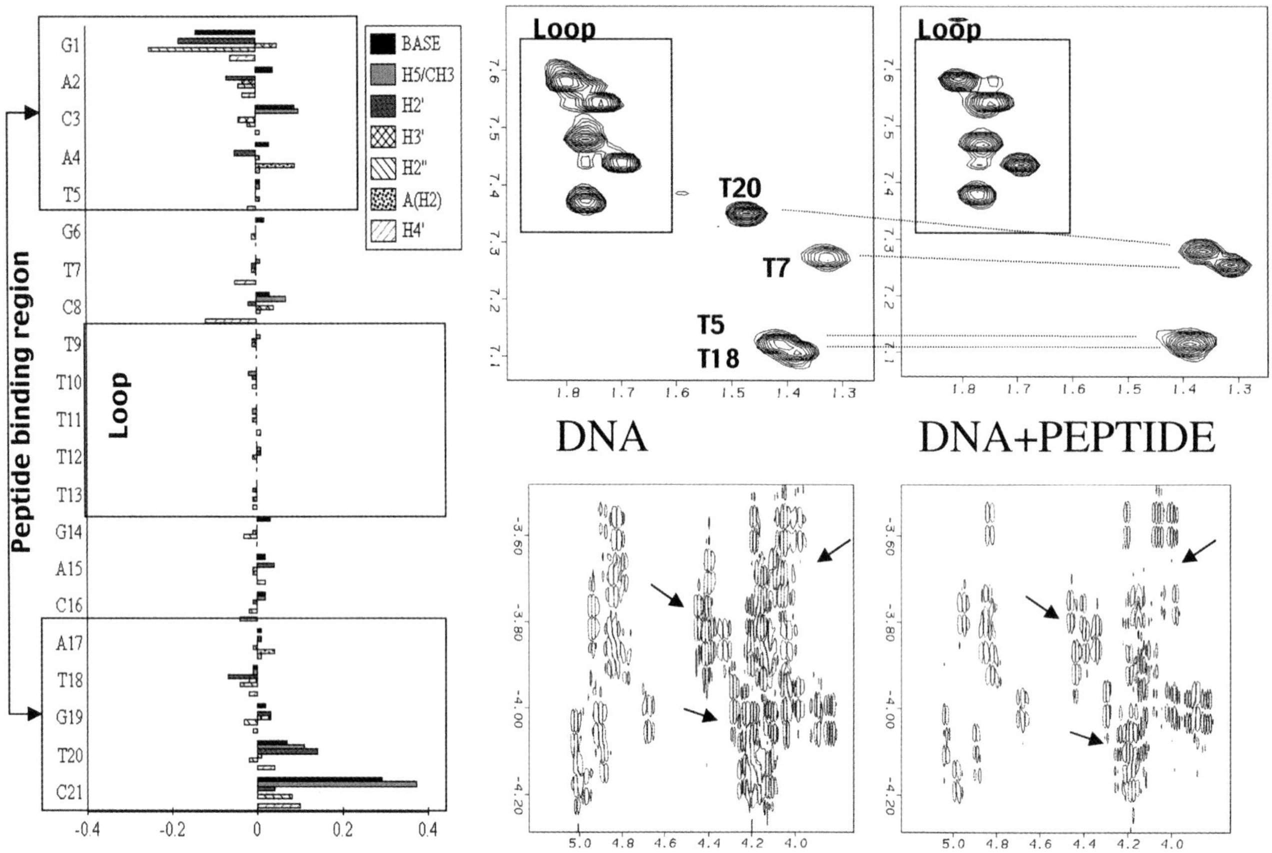

Figure 9. A: The chemical shift differences between the uncomplexed DNA and the complexed DNA is shown for each sugar and base proton for every residue. Positive values indicate upfield shifts, while negative values indicate downfield shifts. The largest differences are observed for the duplex terminal end. B: Expanded region of the 300 ms NOESY spectra for the uncomplexed DNA (left) and the complexed DNA (right) showing through space connectivities between the thymine base H6 proton and its methyl group. While the loop thymines (boxed region) do not exhibit any changes, the duplex thymines show significant shifts, with T20, which is located toward the duplex terminal end, exhibiting the largest change upon peptide binding. C: Proton-phosphorus correlation spectra of the uncomplexed DNA (left) and the complexed DNA (right). Arrows indicate significant changes upon peptide binding in the phosphorus and proton resonances, which have been assigned to residues near the terminal duplex end, reflecting changes in the phosphodiester backbone of the DNA in this region.

tence length of free DNA (Hud et al. 1995). Additional protamine binding experiments using DNA molecules loosely bound to mica (Fig. 10C) confirmed the earlier EM studies which showed that protamine binding to DNA induces the coiling of the DNA molecule into toroidal structures (Allen et al. 1997). Analyses of the diameter of the toroid as a function of the amount of DNA coiled inside it also showed that as the amount of DNA packed inside a single torus increased, the thickness of the edge of the torus increased and the size of the hole decreased. This observation provided the basis for the development of a model that describes how the DNA could be coiled inside the torus (Hud et al. 1995). In this model, the DNA is coiled into a loop defined by the persistence length of DNA, and the torus grows in size as subsequent loops bind adjacent to the previous ones (Fig. 10D).

Subunit Structure of DNA Packaged inside the Sperm Nucleus

Clearly these *in vitro* studies do not prove that sperm DNA is actually coiled by protamines into toroidal like structures. Evidence for the existence of toroidal structures in sperm was obtained by imaging intact and partially decondensed human sperm chromatin with the scanning probe microscope. This technique provides topographical images of surface structures with resolutions approaching that of the electron microscope (Allen et al. 1995). High resolution images of the surface of intact human sperm chromatin after removal of the membranes with a detergent (Fig. 11A) have shown the chromatin to be composed of tightly packed nodular structures 60-100 nm in diameter (Fig. 11B). Occasional toroid-like structures were observed on the glass surface near the edge of the chromatin (Fig. 11C). Smaller

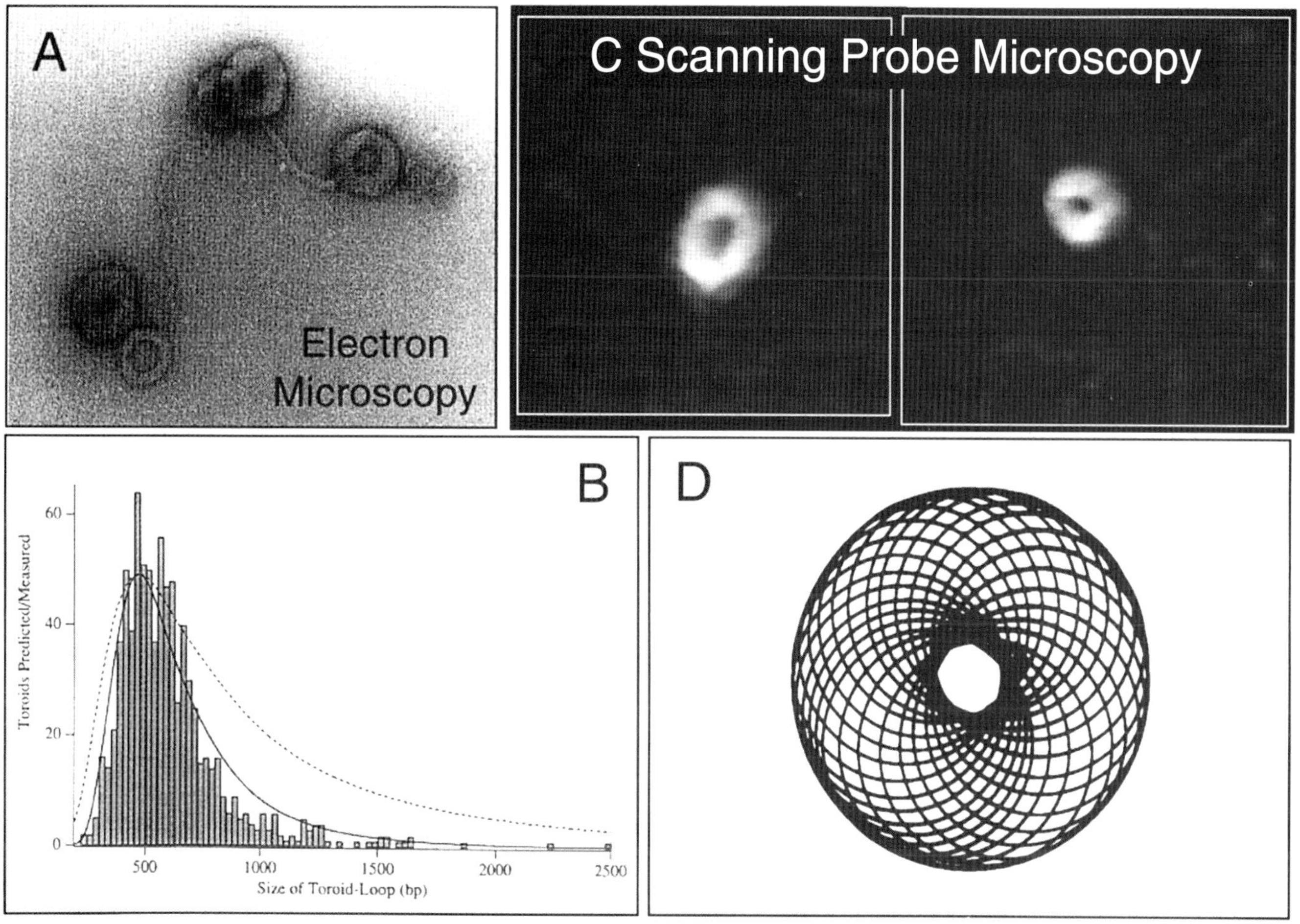

Figure 10. Protamine binding to DNA induces the coiling of DNA into toroids. A: Electron micrograph showing toroids formed by protamine binding to DNA. B: Size-range of toroidal loops measured by electron microscopy. C: Scanning probe images of toroids formed when protamine is added to DNA loosely bound to mica. D: Model of torus showing how coiling involves formation of loops with diameter defined by persistence length and size increases as successive loops are slightly offset.

structures were also seen (Figs. 11C, D), but they were difficult to resolve and measure in the intact chromatin particle.

The presence of two different structural subunits in human sperm chromatin was confirmed by spreading the chromatin on the surface of glass coverslips in distilled water (Fig. 12A) and imaging the samples after air-drying. High-resolution images of the dispersed "nodules" revealed that they are actually donut-shaped (Fig. 12B) and have dimensions (60-100 nm diameter, 20 nm thickness) identical to the toroids produced by binding protamine to DNA *in vitro*. Smaller subunits similar in size to nucleosomes were also observed, distributed in the chromatin spreads as groupings of individual subunits (Figs. 12D, E) and also arranged in bundles (Fig. 12C). Measurements of the thickness of these smaller structures showed there were two populations, one with a thickness of 4.9±0.8 nm and the other with a thickness of 9.4±1.2 nm. These measurements are remarkably close to the width and thickness that have been determined previously for the nucleosome, 5.7 nm x 11 nm (Richmond et al. 1984). These structures are also identical in size to nucleosomal subunits that have been imaged in native (Allen 1997) and synthetic chromatin preparations (Allen et al. 1993).

Higher-Ordered Packing of Chromatin Subunits within the Nucleus

Exactly how these toroids and nucleosomal-like structures are arranged inside the sperm nucleus is not clear. Previous EM studies of rat and rabbit sperm by Koehler (1970) and Koehler et al. (1983) indicated that the chromatin might be arranged as thin layers or sheets stacked on top of each other. Volume measurements obtained for the mouse sperm nucleus by serial section EM also showed that chromatin must be so densely packed inside the nucleus that there was little room for water (Wyrobek et al. 1976). Both of these studies were performed on dehydrated sperm nuclei. Subsequent

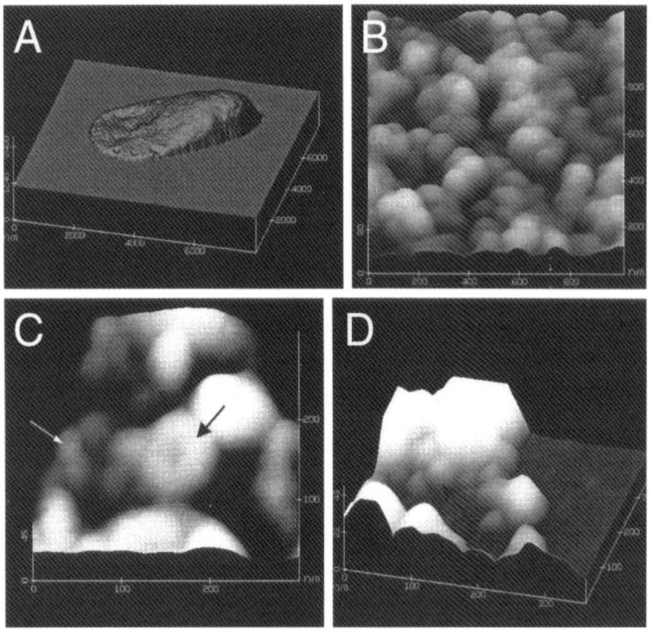

Figure 11 (left). Scanning probe microscopy of native human sperm chromatin. A: Low resolution image of human sperm head treated with detergent to remove acrosome, membranes and tail. B: High resolution image of surface of human sperm chromatin. C: Image of human sperm chromatin near edge showing torus-like structure (black arrow) and nucleosome-like structures (white arrow). D: Chromatin at the edge near the glass surface showing multiple nucleosome-like structures.

Figure 12 (below). Scanning probe microscopy of human sperm chromatin dispersed on the surface of a glass coverslip. A: Low resolution image showing the decondensed chromatin. B: High resolution image showing multiple toroids. C: Clusters of nucleosome-like structures. D, E: Individual nucleosome-like structures distributed on glass surface.

X-ray microscopy (Da Silva et al. 1992) and SPM studies have revealed that the chromatin is more hydrated and less compact than expected.

SPM proved to be a particularly useful technique for measuring the water of hydration of sperm chromatin because it could be used to determine the volume of individual, fully hydrated chromatin particles (or intact sperm nuclei) both in fluid and air, as well as follow changes in their dimensions and volume as each nucleus is dehydrated. Experiments performed on fully hydrated, amembraneous bull and mouse sperm nuclei demonstrated that the volume occupied by sperm chromatin is almost 2.5 times that observed by EM. Data obtained for individual nuclei that were dehydrated by air-drying (Fig. 13A) or by treatment with increasing concentrations of propanol in a fluid cell (Fig. 13B) confirmed that native chromatin is extensively hydrated, and these experiments showed that 50-60% of the volume of sperm chromatin is occupied by water (Allen et al. 1996).

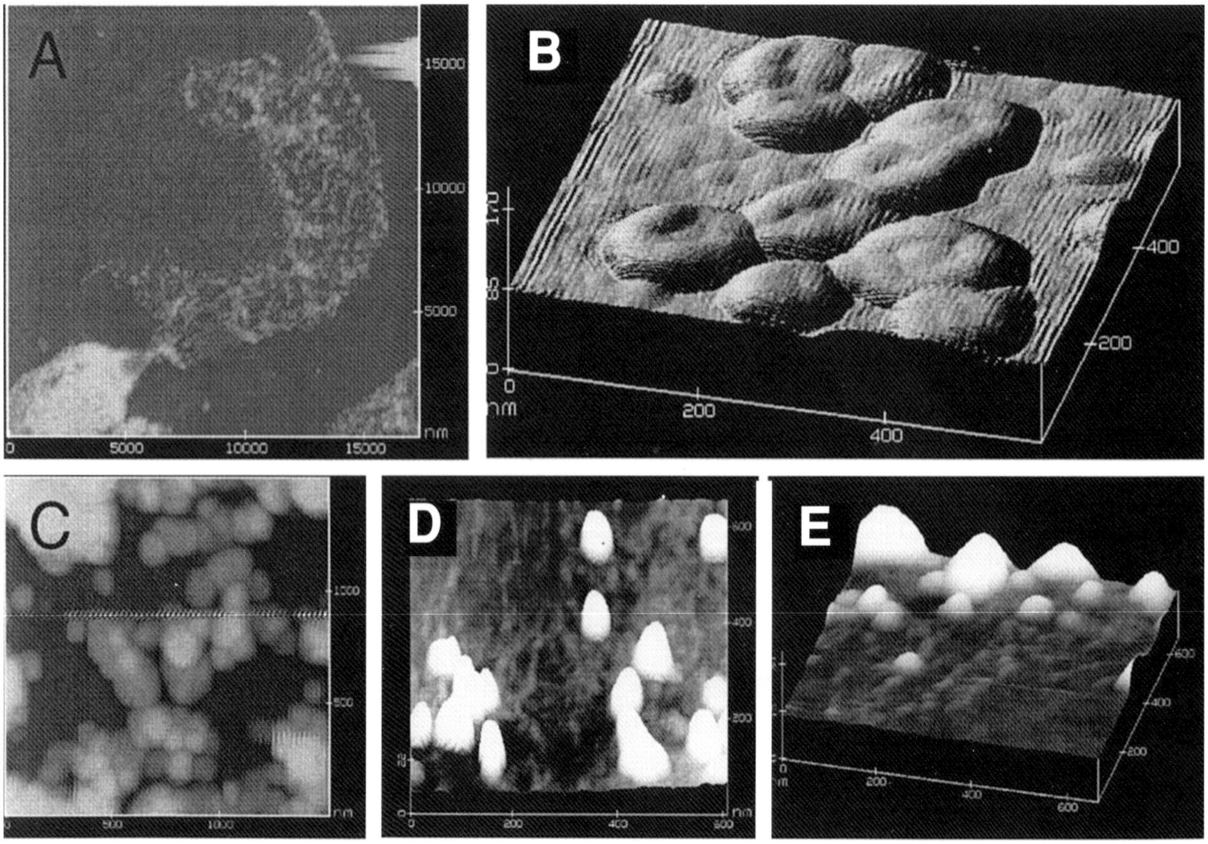

	Volume in μm³	
	Mouse	Bull
Hydrated	10.42±0.45	14.30±1.21
Rehydrated	11.21±0.57	13.25±0.83
Air-dried	6.36±0.47	7.28±0.28
Serial Section EM	4.01±0.03	

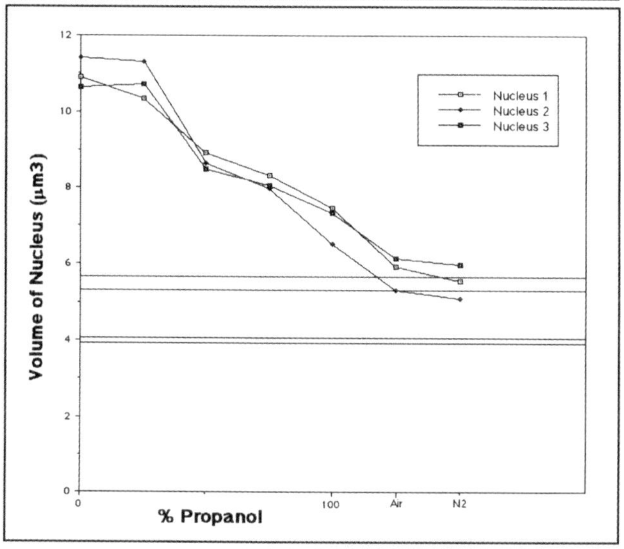

Figure 13. Data showing the volume of hydrated amembraneous mouse and bull sperm nuclei (chromatin) decreases at least two fold when air-dried (upper panel) or dehydrated in increasing concentrations of propanol (lower panel).

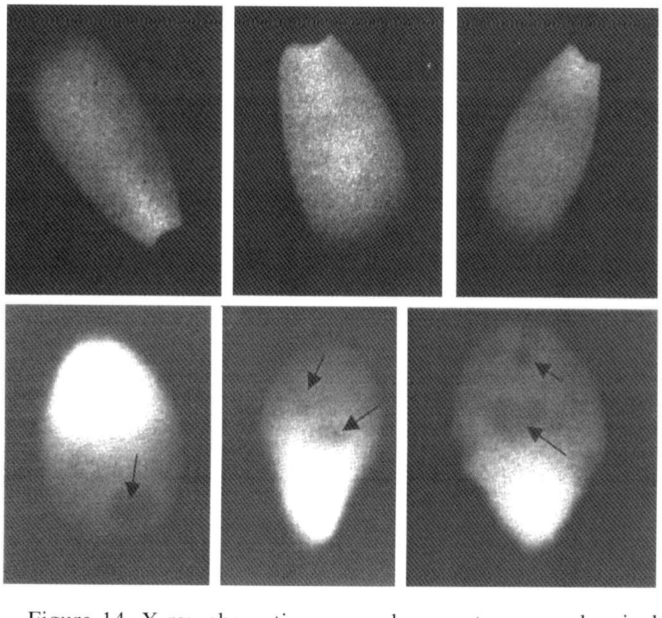

Figure 14. X-ray absorption near edge spectroscopy chemical imaging of bull sperm nuclei (top three panels) and human sperm nuclei (bottom three panels) showing the uniformity of distribution of DNA in the bull sperm head and the presence of vacuoles (arrows) in certain human sperm heads.

While these studies have not provided information about how the chromatin subunits are packed inside the head, changes in the dimensions of the sperm nucleus that were observed during dehydration appear to support Koehler's original idea that the chromatin might be packed as thin layers inside the nucleus. As the sperm chromatin loses water, the largest changes in dimension were changes in thickness of the nucleus. Air-dried nuclei were only half as thick as fully hydrated nuclei. If the layers hypothesized by Koehler were comprised of toroids arranged side by side, a single layer would be approximately 20 nm thick. This thickness would not be expected to change during dehydration because the coiled DNA molecules are tightly packed together. If, on the other hand, 10 to 12 of these layers were separated by a hydrated layer of equivalent size (containing a structural matrix or nucleosomes), the loss of water from these interspersed matrices could easily account for the observed decrease in thickness. This work does not provide any direct evidence for the existence of these layers, but the results obtained are consistent with his hypothesis.

Uniformity of Chromatin Packing inside the Sperm Head

When the organization of sperm chromatin is examined on the scale of the entire nucleus, it has generally been observed that the DNA appears to be uniformly packed throughout the mammalian sperm head. The most notable exceptions are the vacuoles or voids that are often observed by EM to be present in sections of human or other sperm nuclei (el Gothamy, el Samahy 1992; Held et al. 1991; Mundy et al. 1994) or regional differences in chromatin organization that have been observed in a few select rodents (Breed 1997; Breed et al. 1994). X-ray microscopy is well suited to examining the distribution of DNA inside the sperm head because X-rays easily penetrate the sperm and provide transmission images of the intact head. Exploiting known differences in X-ray absorption spectra for molecules such as DNA and protein, it is now possible to obtain quantitative images of cells and nuclei that show the distribution of DNA and certain proteins (Ade et al. 1992; Kirz et al. 1995; Zhang et al. 1996). This technique, called X-ray absorption near edge spectroscopy (XANES), has been used to examine the distribution of DNA inside the nucleus and demonstrate the uniformity of its distribution in bull and hamster sperm (Zhang et al. 1996). Only minor regional differences in DNA content are observed, such as those shown for bull in Figure 14, and this variation simply reflects differences in the thickness of the head. In both bull and human sperm, the nucleus increases in thickness near the implantation fossa.

In contrast, regional differences in DNA concentra-

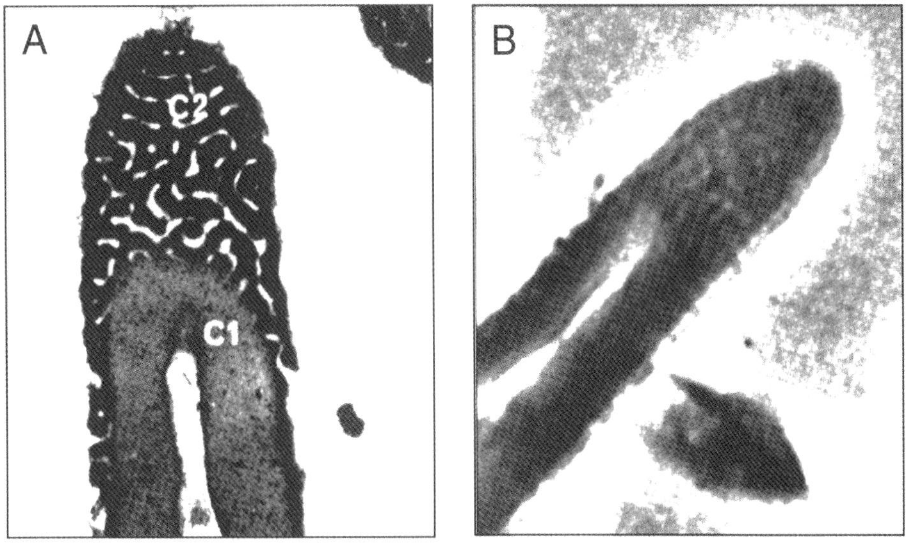

Figure 15. A: Electron micrograph of a marsupial mouse sperm head showing the two different types of chromatin present in the sperm of this rodent. B: Transmission X-ray microscopy image showing the same region of an unstained marsupial mouse sperm head showing the unusual structural features of the two chromatins are not artifacts of dehydration or staining in preparation for electron microscopy.

tion are frequently observed in human sperm heads. XANES images of human sperm taken at wavelengths where DNA absorbs show the presence of vacuoles or voids in a significant fraction of the sperm heads (Fig. 14). The significance of these voids is unclear, since they are frequently found in semen produced by fertile males. Differences in the degree of chromatin compaction have also been observed in sperm produced by the marsupial mouse *Sminthopsis crassicaudata*. Electron micrographs of stained *Sminthopsis* sperm heads revealed the presence of two different types of chromatin, a uniformly condensed type located in the interior of the head and another densely packed region comprised of thick chromatin chords interspersed by DNA free channels (Fig. 15A). Transmission X-ray analyses of unstained nuclei have confirmed the existence of these two different types of chromatin in the head of *Sminthopsis* sperm (Fig. 15B) and provided a second example of sperm containing chromatin that is not uniformly packed throughout the head. Other studies by Breed et al. (W Breed, personal communication) have since shown that in *Sminthopsis*, these differences appear to be related to the presence and spatial segregation of two types of chromatin within the nucleus. One type appears to contain protamines, while the other contains histones.

While X-ray microscopy studies of sperm have been limited to only a few select animals (rats, mice, cattle, horses and hamsters), the results show that the chromatin in these species is, for the most part, uniformly distributed throughout the nucleus. Differences have only been observed in abnormal sperm or when certain steps in the process of spermiogenesis are disrupted. One example is shown in Figure 16. X-ray transmission images of normal mouse sperm nuclei exhibit only subtle differences in DNA density, and these differences simply correspond to regional differences in the thickness of the nucleus. On the other hand, X-ray transmission images of sperm produced by a line of transgenic mice that express the protamine 1 gene at a stage in spermiogenesis much earlier than normal (Lee et al. 1995) show how the uniformity of chromatin condensation is disrupted by the early appearance of protamine 1 in the nucleus. The nuclei of these sperm exhibit a much broader range both in chromatin density and irregularity in packing than observed previously for normal sperm.

Importance of Chromatin Organization to Sperm Function

A great deal has been learned about the structure of the DNA-protamine complex and how it is organized inside the nucleus of mammalian sperm, but researchers have only recently begun to discover how defects in these processes affect sperm function and male fertility. The work of the Balhorn laboratory and that of numerous others has suggested that the reorganization in chromatin structure that takes place in the differentiating spermatid nucleus (Fig. 17) is important both for the production of functional sperm and for ensuring the normal progression of early embryonic development. This process provides a mechanism for "deprogramming" the testicular cell genome so it can be properly "reprogrammed" after fertilization to allow the expression of only those genes required for cell replication and early development.

While it is not yet clear how specific alterations in the composition or structure of sperm chromatin affect its function, several studies have provided enticing hints. One example relates to the importance of protamine 2. Alterations in the relative proportion of P1 and P2 in sperm chromatin have been observed in many

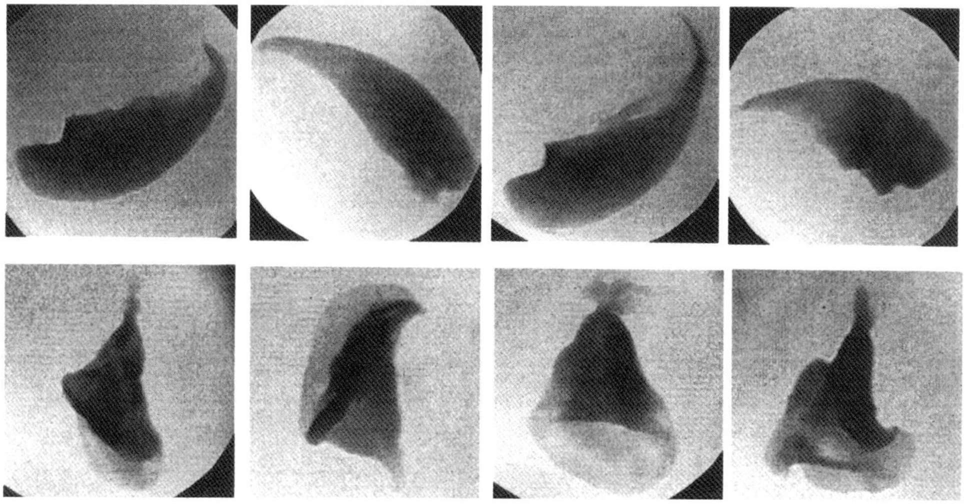

Figure 16. Transmission X-ray microscope images of normal mouse sperm nuclei and nuclei of sperm produced by a transgenic line of mice that turn on expression of the protamine 1 gene early. The condensation of the chromatin in the transgenic line is disrupted, the shape of the head is distorted and the extent of chromatin compaction varies widely throughout the nucleus.

cases of human male infertility (Balhorn et al. 1988; Belokopytova et al. 1993; Blanchard et al. 1990; de Yebra et al. 1993), with some individuals appearing to produce sperm that are completely deficient in protamine 2 (de Yebra et al. 1993). Sperm decondensation studies performed by Perreault et al. (Perreault et al. 1988) have also shown a direct correlation between the amount of protamine 2 present in sperm chromatin and the ease and rate with which the sperm chromatin decondenses in hamster eggs. Other studies have demonstrated the importance of zinc in stabilizing sperm chromatin structure. Zinc, which appears to be bound to protamine 2 (Bianchi et al. 1992; Gatewood et al. 1990) in the sperm cell, is essential for male fertility. Low zinc contents in human semen have been correlated with male infertility (Kjellberg et al. 1992; Kvist et al. 1988), and experiments performed *in vitro* suggest that sperm with low chromatin zinc do not decondense properly (Björndahl, Kvist 1985; Kvist, Björndahl 1985). These combined results suggest that protamine 2 may play an important role in facilitating or regulating the process of chromatin decondensation in the egg.

This example serves to demonstrate one reason it is important that studies be conducted to elucidate and understand the sequence of molecular events that prepare the spermatid genome for fertilization. Without this knowledge, clinicians would not be able to identify the molecular basis for those cases of male infertility that are caused by defects in DNA packaging. It is also critical that researchers understand what happens in cases where the process of reorganization does not occur or is arrested. This is particularly important in light of recent technological advances that now make it possible to perform techniques such as round spermatid nuclei injection (ROSNI; Silber, Johnson 1998) using nuclei containing DNA that has not been properly processed for fertilization.

Acknowledgements

This work was performed at Lawrence Livermore National Laboratory under the auspices of the U.S. Department of Energy and supported by Contract W-7405-ENG-48.

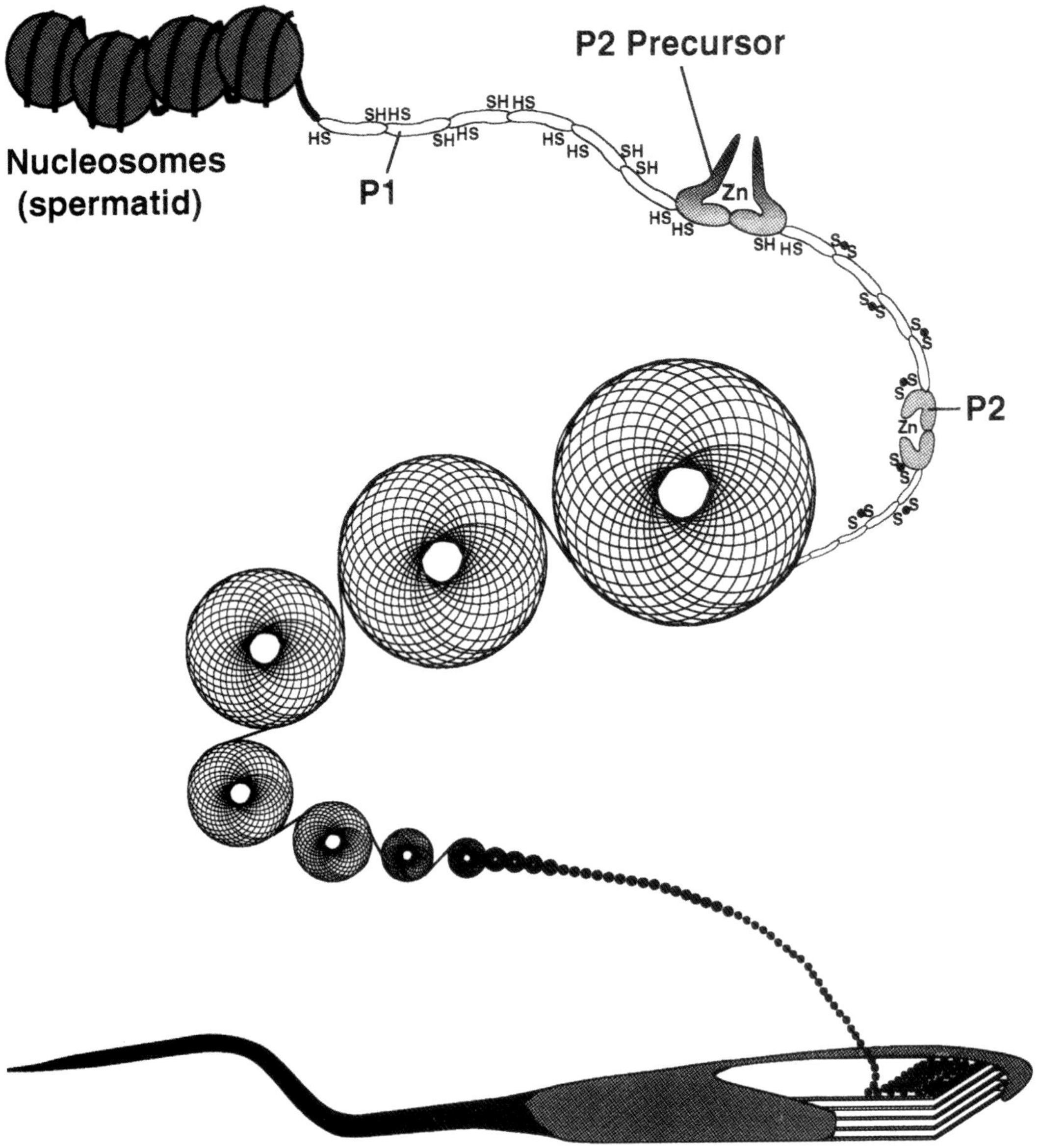

Figure 17. The process of chromatin reorganization in the spermatid is initiated by the synthesis and binding of P1 and P2 to DNA. This displaces all the existing chromatin proteins and coils the DNA into toroidal structures containing ~50Kbp of DNA. Each sperm nucleus contains approximately 50,000 toroids. Defects in the first stage of the process appear to cause infertility. Changes in other processes (for example, alkylation of cysteines and blockage of disulfide formation) can lead to early embryo death. The extensive compaction of DNA that accompanies the change in composition also blocks DNA repair.

References

Ade H, Zhang X, Cameron S, Costello C, Kirz J, Williams S. Chemical contrast in X-ray microscopy and spatially resolved XANES spectroscopy of organic specimens. Science 1992; 258:972-975.

Allen MJ. Atomic force microscopy: a new way to look at chromatin. IEEE Engineer. Med Biol Mag 1997; 16:34-41.

Allen MJ, Dong XF, O'Neill TE, Yau P, Kowalczykowski SC, Gatewood J, Balhorn R, Bradbury EM. Atomic force microscope measurements of nucleosome cores assembled along defined DNA sequences. Biochemistry 1993; 32:8390-8396.

Allen MJ, Bradbury EM, Balhorn R. The natural subcellular surface structure of the bovine sperm cell [published erratum appears in J Struct Biol 1995 Nov-Dec; 115:338-341]. J Struct Biol 1995; 114:197-208.

Allen MJ, Lee JD, Lee C, Balhorn R. Extent of sperm chromatin hydration determined by atomic force microscopy. Mol Reprod Dev 1996; 45:87-92.

Allen MJ, Bradbury EM, Balhorn R. AFM analysis of DNA-protamine complexes bound to mica. Nucleic Acids Res 1997; 25:2221-2226.

Ammer H, Henschen A. The major protamine from stallion sperm. Isolation and amino-acid sequence. Biol Chem Hoppe-Seyler 1987; 368:1619-1626.

Ammer H, Henschen A, Lee CH. Isolation and amino-acid sequence analysis of human sperm protamines P1 and P2. Occurrence of two forms of protamine P2. Biol Chem Hoppe-Seyler 1986; 367:515-522.

Bach O, Glander HJ, Scholz G, Schwarz J. Electrophoretic patterns of spermatozoal nucleoproteins (NP) in fertile men and infertility patients and comparison with NP of somatic cells. Andrologia 1990; 22:217-224.

Balhorn R. Mammalian protamines: structure and molecular interactions. In: (Adolph KW, ed) Molecular Biology of Chromosome Function. New York: Springer-Verlag, 1989, pp366-395.

Balhorn R, Gledhill BL, Wyrobek AJ. Mouse sperm chromatin proteins: quantitative isolation and partial characterization. Biochemistry 1977; 16:4074-4080.

Balhorn R, Weston S, Thomas C, Wyrobek AJ. DNA packaging in mouse spermatids. Synthesis of protamine variants and four transition proteins. Exp Cell Res 1984; 150:298-308.

Balhorn R, Corzett M, Mazrimas J, Stanker LH, Wyrobek A. High-performance liquid chromatographic separation and partial characterization of human protamines 1, 2, and 3. Biotech Appl Biochem 1987; 9:82-88.

Balhorn R, Reed S, Tanphaichitr N. Aberrant protamine 1/protamine 2 ratios in sperm of infertile human males. Experientia 1988; 44:52-55.

Bélaïche D, Loir M, Kruggle W, Sautière P. Isolation and characterization of two protamines St1 and St2 from stallion spermatozoa, and amino-acid sequence of the major protamine St1. Biochim Biophys Acta 1987; 913:145-149.

Bellvé AR, Anderson E, Hanley Bowdoin L. Synthesis and amino acid composition of basic proteins in mammalian sperm nuclei. Dev Biol 1975; 47:349-365.

Bellvé AR, McKay DJ, Renaux BS, Dixon GH. Purification and characterization of mouse protamines P1 and P2. Amino acid sequence of P2. Biochemistry 1988; 27:2890-2897.

Belokopytova IA, Kostyleva EI, Tomilin AN, Vorobev VI. Human male infertility may be due to a decrease of the protamine-P2 content in sperm chromatin. Mol Reprod Dev 1993; 34:53-57.

Bench GS, Friz AM, Corzett MH, Morse DH, Balhorn R. DNA and total protamine masses in individual sperm from fertile mammalian subjects. Cytometry 1996; 23:263-271.

Bernabeu R, Cremades N, Takahashi K, Sousa M. Successful pregnancy after spermatid injection. Hum Reprod 1998; 13:1898-1900.

Bianchi F, Rousseaux-Prevost R, Sautière P, Rousseaux J. P2 protamines from human sperm are zinc-finger proteins with one CYS2/HIS2 motif. Biochem Biophys Res Comm 1992; 182:540-547.

Björndahl L, Kvist U. Loss of an intrinsic capacity for human sperm chromatin decondensation. Acta Physiol Scand 1985; 124:189-194.

Blanchard Y, Lescoat D, Le Lannou D. Anomalous distribution of nuclear basic proteins in round-headed human spermatozoa. Andrologia 1990; 22:549-555.

Breed WG. Unusual chromatin structural organization in the sperm head of a murid rodent from Southern Africa: the red veldt rat, *Aethomys chrysophilus* type B. J Reprod Fert 1997; 111:221-228.

Breed WG, Leigh CM, Washington JM, Soon LL. Unusual nuclear structure of the spermatozoon in a marsupial, *Sminthopsis crassicaudata*. Mol Reprod Dev 1994; 37:78-86.

Carré-Eusèbe D, Lederer F, Lê KH, Elsevier SM. Processing of the precursor of protamine P2 in mouse. Peptide mapping and N-terminal sequence analysis of intermediates. Biochem J 1991; 277:39-45.

Copeland RA, Spiro TG. Secondary structure determination in proteins from deep (192-223-nm) ultraviolet Raman spectroscopy. Biochemistry 1987; 26:2134-2139.

Corzett M, Kramer C, Blacher R, Mazrimas J, Balhorn R. Analysis of hamster protamines: primary sequence and species distribution. Mol Reprod Dev 1999; in press.

Da Silva LB, Trebes JE, Balhorn R, Mrowka S, Anderson E, Attwood DT, Barbee TW Jr, Brase J, Corzett M, Gray J, Koch JA, Lee C, Kern D, London RA, MacGowan BJ, Matthews DL, Stone G. X-ray laser microscopy of rat sperm nuclei. Science 1992; 258:269-271.

de Yebra L, Ballesca JL, Vanrell JA, Bassas L, Oliva R. Complete selective absence of protamine-P2 in humans. J Biol Chem 1993; 268:10553-10557.

Dooher GB, Bennett D. Fine structural observations on the development of the sperm head in the mouse. Am J Anat 1973; 136:339-361.

el-Gothamy Z, el-Samahy M. Ultrastructure sperm defects in addicts. Fertil Steril 1992; 57:699-702.

Elsevier SM, Noiran J, Carré-Eusèbe D. Processing of the precursor of protamine P2 in mouse. Identification of intermediates by their insolubility in the presence of sodium dodecyl sulfate. Eur J Biochem 1991; 196:167-175.

Feughelman M, Langridge R, Seeds WE, Stokes AR, Wilson HR, Hooper CW, Wilkins MHF, Barclay RK, Hamilton LD. Molecular structure of deoxyribonucleic acid and nucleoprotein. Nature 1955; 175:834-838.

Fishel S, Green S, Hunter A, Lisi F, Rinaldi L, Lisi R, McDermott H. Human fertilization with round and elongated spermatids. Hum Reprod 1997; 12:336-340.

Fita I, Campos JL, Puigjaner LC, Subirana JA. X-ray diffraction study of DNA complexes with arginine peptides and their relation to nucleoprotamine structure. J Mol Biol 1983; 167:157-177.

Fodor SP, Starr PA, Spiro TG. Raman spectroscopic elucidation of DNA backbone conformations for poly(dG-dT).poly(dA-dC) and poly(dA-dT).poly(dA-dT) in CsF solution. Biopolymers 1985; 24:1493-1500.

Gatewood JM, Cook GR, Balhorn R, Bradbury EM, Schmid CW. Sequence-specific packaging of DNA in human sperm chromatin. Science 1987; 236:962-964.

Gatewood JM, Schroth GP, Schmid CW, Bradbury EM. Zinc-induced secondary structure transitions in human sperm protamines. J Biol Chem 1990; 265:20667-20672.

Goldberg RB, Geremia R, Bruce WR. Histone synthesis and replacement during spermatogenesis in the mouse. Differentiation 1977; 7:167-180.

Gusse M, Sautière P, Bélaïche D, Martinage A, Roux C, Dadoune JP, Chevaillier P. Purification and characterization of nuclear basic proteins of human sperm. Biochim Biophys Acta 1986; 884:124-134.

Held JP, Prater P, Stettler M. Spermatozoal head defect as a cause of infertility in a stallion. J Am Vet Med Assn 1991; 199:1760-1761.

Hud NV, Allen MJ, Downing KH, Lee J, Balhorn R. Identification of the elemental packing unit of DNA in mammalian sperm cells by atomic force microscopy. Biochem Biophys Res Comm 1993; 193:1347-1354.

Hud NV, Milanovich FP, Balhorn R. Evidence of novel secondary structure in DNA-bound protamine is revealed by raman spectroscopy. Biochemistry 1994; 33:7528-7535.

Hud NV, Downing KH, Balhorn R. A constant radius of curvature model for the organization of DNA in toroidal condensates. Proc Natl Acad Sci USA 1995; 92:3581-3585.

Johnson PA, Peschon JJ, Yelick PC, Palmiter RD, Hecht NB. Sequence homologies in the mouse protamine 1 and 2 genes. Biochim Biophys Acta 1988; 950:45-53.

Kirz J, Jacobsen C, Howells M. Soft X-ray microscopes and their biological applications. Q Rev Biophys 1995; 28:33-130.

Kjellberg S, Björndahl L, Kvist U. Sperm chromatin stability and zinc binding properties in semen from men in barren unions. Int J Androl 1992; 15:103-113.

Klemm U, Lee CH, Burfeind P, Hake S, Engel W. Nucleotide sequence of a cDNA encoding rat protamine and the haploid expression of the gene during rat spermatogenesis. Biol Chem Hoppe-Seyler 1989; 370:293-301.

Koehler JK. A freeze-etching study of rabbit spermatozoa with particular reference to head structures. J Ultrastruct Res 1970; 33:598-614.

Koehler JK, Wurschmidt U, Larsen MP. Nuclear and chromatin structure in rat spermatozoa. Gam Res 1983; 8:357-377.

Krawetz SA, Connor W, Dixon GH. Cloning of bovine P1 protamine cDNA and the evolution of vertebrate P1 protamines. DNA 1987; 6:47-57.

Kvist U, Björndahl L. Zinc preserves an inherent capacity for human sperm chromatin decondensation. Acta Physiol Scand 1985; 124:195-200.

Kvist U, Kjellberg S, Björndahl L, Hammar M, Roomans GM. Zinc in sperm chromatin and chromatin stability in fertile men and men in barren unions. Scand J Urol Nephrol 1988; 22:1-6.

Lee CH, Hoyer-Fender S, Engel W. The nucleotide sequence of a human protamine 1 cDNA. Nucleic Acids Res 1987a; 15:7639.

Lee CH, Mansouri A, Hecht W, Hecht NB, Engel W. Nucleotide sequence of a bovine protamine cDNA. Biol Chem Hoppe-Seyler 1987b; 368:131-135.

Lee K, Haugen HS, Clegg CH, Braun RE. Premature translation of protamine 1 mRNA causes precocious nuclear condensation and arrests spermatid differentiation in mice. Proc Natl Acad Sci USA 1995; 92:12451-12455.

Martinage A, Arkhis A, Alimi E, Sautière P, Chevaillier P. Molecular characterization of nuclear basic protein HPI1, a putative precursor of human sperm protamines HP2 and HP3. Eur J Biochem 1990; 191:449-451.

Mazrimas JA, Corzett M, Campos C, Balhorn R. A corrected primary sequence for bull protamine. Biochim Biophys Acta 1986; 872:11-15.

McKay DJ, Renaux BS, Dixon GH. The amino acid sequence of human sperm protamine P1. Biosci Rep 1985; 5:383-391.

McKay DJ, Renaux BS, Dixon GH. Human sperm protamines. Amino-acid sequences of two forms of protamine P2. Eur J Biochem 1986; 156:5-8.

Mundy AJ, Ryder TA, Edmonds DK. A quantitative study of sperm head ultrastructure in subfertile males with excess sperm precursors. Fertil Steril 1994; 61:751-754.

Ogura A, Yanagimachi R. Spermatids as male gametes. Reprod Fert Dev 1995; 7:155-159.

Perreault SD, Barbee RR, Elstein KH, Zucker RM, Keefer CL. Interspecies differences in the stability of mammalian sperm nuclei assessed *in vivo* by sperm microinjection and *in vitro* by flow cytometry. Biol Reprod 1988; 39:157-167.

Pongsawasdi P, Svasti J. The heterogeneity of the protamines from human spermatozoa. Biochim Biophys Acta 1976; 434:462-473.

Prieto MC. Structural characterization of DNA-protein complexes by optically detected magnetic resonance and nuclear magnetic resonance. Ph.D. thesis, University of California, 1995.

Prieto MC, Maki AH, Balhorn R. Analysis of DNA-protamine interactions by optical detection of magnetic resonance. Biochemistry 1997; 36:11944-11951.

Retief JD, Winkfein RJ, Dixon GH, Adroer R, Queralt R, Ballabriga J, Oliva R. Evolution of protamine P1 genes in primates. J Mol Evolut 1993; 37:426-434.

Reubinoff BE, Abeliovich D, Werner M, Schenker JG, Safran A, Lewin A. A birth in non-mosaic Klinefelter's syndrome after testicular fine needle aspiration, intracytoplasmic sperm injection and preimplantation genetic diagnosis. Hum Reprod 1998; 13:1887-1892.

Richmond TJ, Finch JT, Rushton B, Rhodes D, Klug A. Structure of the nucleosome core particle at 7 A resolution. Nature 1984; 311:532-537.

Roosen-Runge EC. The process of spermiogenesis in mammals. Biol Rev Camb Philos Soc 1962; 37:343-377.

Sautière P, Bélaïche D, Martinage A, Loir M. Primary structure of the ram (*Ovis aries*) protamine. Eur J Biochem 1984; 144:121-125.

Schellman JA, Parthasarathy N. X-ray diffraction studies on cation-collapsed DNA. J Mol Biol 1984; 175:313-329.

Silber SJ, Johnson L. Are spermatid injections of any clinical value? ROSNI and ROSI revisited. Round spermatid nucleus injection and round spermatid injection. Hum Reprod 1998; 13:509-515.

Tanhauser SM, Hecht NB. Nucleotide sequence of the rat protamine 2 gene. Nucleic Acids Res 1989; 17:4395.

Tanphaichitr N, Sobhon P, Taluppeth N, Chalermisarachai P. Basic nuclear proteins in testicular cells and ejaculated spermatozoa in man. Exp Cell Res 1978; 117:347-356.

Tobita T, Nomoto M, Nakano M, Ando T. Isolation and characterization of nuclear basic protein (protamine) from boar spermatozoa. Biochim Biophys Acta 1982; 707:252-258.

Tu AT. Raman Spectroscopy in Biology: Principles and Applications. John Wiley & Sons, New York, 1982.

Wilkins MFH. Physical studies of the molecular structure of deoxyribonucleic acid and nucleoprotein. Cold Spring Harbor Symp Quant Biol 1956; 21:75-90.

Wyrobek AJ, Meistrich ML, Furrer R, Bruce WR. Physical characteristics of mouse sperm nuclei. Biophys J 1976; 16:811-825.

Yelick PC, Balhorn R, Johnson PA, Corzett M, Mazrimas JA, Kleene KC, Hecht NB. Mouse protamine 2 is synthesized as a precursor whereas mouse protamine 1 is not. Mol Cell Biol 1987; 7:2173-2179.

Zhang X, Balhorn R, Mazrimas J, Kirz J. Mapping and measuring DNA to protein ratios in mammalian sperm head by XANES imaging. J Struct Biol 1996; 116:335-344.

8 Regulation of Sperm Transport in the Mammalian Oviduct

Susan S. Suarez
Cornell University

The mammalian oviduct is not merely a simple conduit for sperm. There is evidence that the oviduct regulates admission of sperm and, once admitting sperm, proceeds to regulate their passage to the oocyte.

Of particular importance is the formation of a reservoir of stored sperm in the caudal region of the oviduct. Sperm may be trapped in the reservoir until ovulation is imminent, and then they may be released in a slow trickle to ascend to meet the oocytes. Furthermore, the physiological state of the sperm may be influenced by the oviduct, so that they remain in a preserved state during storage and are promoted to a capacitated state as the time of ovulation approaches.

Sperm Passage through the Uterotubal Junction

Barriers

In eutherian mammals, the uterotubal junction presents sperm with anatomical, physiological and/or biochemical barriers to reaching oocytes in the oviduct. The barriers may serve the dual function of blocking the ascent of microorganisms in the female tract and selecting morphologically normal and vigorously motile sperm for fertilization.

The anatomical barrier is the intricate and baffling luminal space. The lumen of the junction in species as disparate as dairy cattle and mice is tortuous and narrow (Hook, Hafez 1968; Hafez, Black 1969; Beck, Boots 1974; Wrobel et al. 1993; Suarez et al. 1997). There are large and small folds in the mucosa of the wall, creating complex pathways, some of which end blindly. The narrowness of the lumen is especially apparent in living tissue (Suarez 1987) and in frozen sections (Fig. 1), in which tissue does not shrink as it does during standard preparation of paraffin-embedded sections (Suarez et al. 1997).

The physiological barrier consists of smooth muscle tissue surrounding the junction which can contract to reduce the lumen. Muscular compression of the lumen is accentuated in some species, such as cattle, by muscular ligaments attached to the tube that can act to increase flexure in the passageway (Hook, Hafez 1968; Hafez, Black 1969). The action of muscular contraction may be supplemented by that of a vascular plexus in the wall of the junction which resembles erectile tissue and probably swells to further restrict the lumen (Wrobel et al. 1993). The lumen in the intramural portion of the uterotubal junction in mice was observed to shut down within a few hours of mating. Sperm could be seen moving in the lumen above the region of constriction, but not within it (Suarez 1987).

The biochemical barrier, when present, takes the form of a viscous mucus. Mucus has been found in the oviducts of rabbits (Jansen 1978; Jansen, Bajpai 1982), pigs (Suarez et al. 1991), dairy cattle (Fig. 1; Suarez et al. 1990, 1997), and humans (Jansen 1980). At times this mucus may be impenetrable to sperm. *In vitro*, boar sperm added to explants of epithelium became stuck to the surface of mucus sheets secreted in culture by the explants and were not observed to penetrate through the mucus (Suarez et al. 1991; Raychoudhury, Suarez 1991). Jansen has proposed that the mucus may be readily penetrated by morphologically normal sperm under certain hormonal conditions of the female, just as cervical mucus is rapidly penetrated by sperm when the cervix is under the influence of estrogen (Jansen 1980).

The various barriers may function individually or in a concerted effort to prevent passage of sperm and also

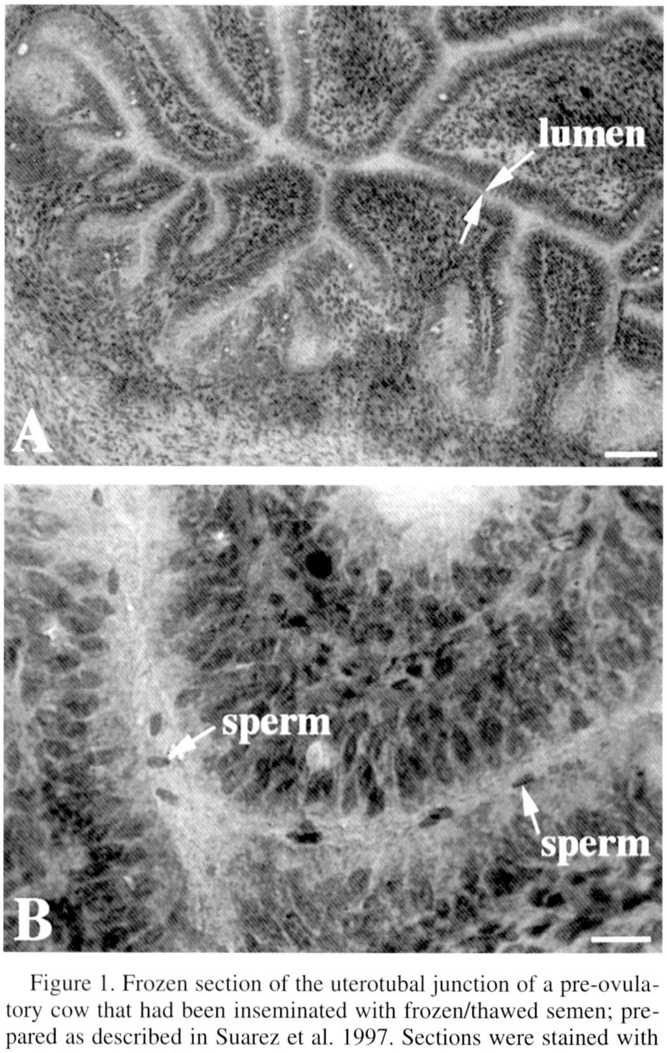

Figure 1. Frozen section of the uterotubal junction of a pre-ovulatory cow that had been inseminated with frozen/thawed semen; prepared as described in Suarez et al. 1997. Sections were stained with periodic acid Schiff to stain mucopolysaccharides and counterstained with hematoxylin. (Prepared by R. Lefebvre, S.S. Suarez, and K. Brockman.) A: Low power photomicrograph of the uterotubal junction, illustrating the complex and narrow luminal space delimited by arrows. Bar = 100 microns. B: High power photomicrograph of the lumen of a similar section, showing the heads of sperm in the mucus-filled lumen. Bar = 20 microns.

microorganisms at various times in the hormonal cycle of the female. At the time when sperm passage through the junction is appropriate, the junction may filter out morphologically abnormal sperm or sperm with poor motility. In pigs (Baker, Degen 1972), rats (Gaddum-Rosse 1981), and hamsters (Smith et al. 1988a), motile sperm pass through the uterotubal junction much more successfully than immotile sperm. Sperm demonstrating activated (progressive) motility are more successful at passing through the uterotubal junction than hyperactivated sperm (Gaddum-Rosse 1981; Shalgi et al. 1992).

Thus, selection for motility and some selection for morphological normality probably occurs at the uterotubal junction. In addition to removing abnormal sperm, the uterotubal junction may filter out seminal plasma. Seminal plasma proteins are left behind in the uterus and are not detected in the oviducts of rats (Carballado, Esponda 1997).

Rapid Sperm Transport

Sperm have been recovered in the cranial reaches of the ampulla only minutes after mating or insemination in several species of mammals (Overstreet, Cooper 1978; Hawk 1983). This phenomenon has been termed rapid sperm transport (or rapid transit, according to Overstreet and Cooper, 1978). Rapid transport of sperm into the oviduct would seem to counter the proposed model of sperm swimming one-by-one through the uterotubal junction. However, when the condition of rabbit sperm recovered from the cranial ampulla shortly after mating was evaluated by Overstreet and Cooper (1978), they found that most of these sperm were immotile and damaged. They proposed that waves of contractions stimulated by insemination cause some sperm to overshoot their destination in the cranial uterus and reach the ampulla of the oviduct, but these sperm are damaged by the associated sheer stress and do not fertilize. Later, motile sperm gradually pass through the uterotubal junction to establish a population that contributes fertilizing sperm.

In the Oviduct: The Oviductal Sperm Reservoir

Functions of the Reservoir

During passage through the uterotubal junction, or soon after passing through to the isthmus, sperm become trapped and form a reservoir. The mammalian sperm reservoir may have been first discovered in hamsters by Yanagimachi and Chang (1963) and has since been reported to exist in a variety of eutherian mammals such as hamsters (Smith et al. 1988b), rabbits (Harper 1973; Overstreet et al. 1978), cows (Hunter, Wilmut 1984), pigs (Hunter 1981), and sheep (Hunter, Nichol 1983).

The oviductal reservoir of sperm may serve a number of functions. First, it may prevent polyspermic fertilization by allowing only a few sperm at a time to reach oocytes in the ampulla. Sperm numbers have been artificially increased at the site of fertilization in the pig by surgical insemination directly into the oviduct (Polge et al. 1970; Hunter 1973); by resecting the oviduct to bypass the reservoir (Hunter, Leglise 1971); and by administering progesterone into the muscularis to inhibit contraction (Day, Polge 1968; Hunter 1972). In all of these cases, the incidence of polyspermy increased.

Second, the oviductal reservoir may maintain the fertility of sperm between the onset of estrus and fertilization. Sperm fertility and motility are maintained longer *in vitro* if the sperm are incubated with oviductal epithelium (porcine: Suarez et al. 1990; bovine: Pollard et al. 1991; equine: Ellington et al. 1993; Chian, Sirard 1994; human: Kervancioglu et al. 1994).

Third, the processes of capacitation and motility hyperactivation may be regulated within the reservoir to provide sperm in the appropriate stage of readiness as the time of ovulation nears. Capacitation is defined as a set of changes in the sperm plasma membrane that enables a sperm to undergo the acrosome reaction (Yanagimachi 1994). Hyperactivation is defined as an increase in flagellar bend amplitude and asymmetry that is observed in sperm recovered from the ampulla of the oviduct near the time of ovulation. Capacitation of bull sperm is enhanced by incubation in medium conditioned by oviductal epithelium (Chian et al. 1995) or in oviduct fluid (Mahmoud, Parrish 1996). Hyperactivation is enhanced when capacitated bull sperm are incubated with explants of isthmic epithelium from preovulatory cows (Lefebvre, Suarez 1996).

With respect to the maintenance of sperm fertility over time, it is interesting that sperm storage structures have developed in other groups of vertebrates. Sperm are stored in folds of ovarian tissue in several species of viviparous fishes (Koya et al. 1997). In several families of turtles, there exist sperm storage tubules in the region of the oviduct that are homologous to the mammalian isthmus (Gist, Jones 1989). Evidently, these storage tubules allow females to fertilize multiple clutches of eggs, sometimes years after mating. In several species of snakes and lizards, sperm storage structures have been described in the anterior vagina and infundibulum of the oviduct (Gist, Jones 1987; Srinivas et al. 1995; Perkins, Palmer 1996; Murphy-Walker, Haley 1996). Neither of these two sites are homologous to the mammalian isthmus, however. Sperm storage tubules have been discovered at the uterovaginal junction in several species of birds (Bakst 1987; Birkhead et al. 1993; Birkhead, Moeller 1993; Bakst et al. 1994), which allow them to lay multiple clutches of eggs after a single mating.

Creation of the Reservoir by Sperm Binding

In several species of eutherian mammals, the oviductal reservoir is evidently created by binding of sperm to oviductal epithelium. Motile sperm have been observed to bind to the apical surface of the oviductal epithelium in mice (Suarez 1987), cattle (Fig. 2)(Suarez et al. 1990), hamsters (Fig. 2)(Smith, Yanagimachi 1991), pigs (Suarez et al. 1991), and horses (Thomas et al. 1994).

The trapping action of the sperm-binding moieties may be enhanced by the narrow, sometimes mucus-filled lumen of the uterotubal junction and isthmus, which would slow the progress of sperm and increase their contact time with the mucosal surface.

Sperm binding to oviductal epithelium involves protein/carbohydrate interaction, like the binding of lectins to their carbohydrate ligands. Hamster sperm were mixed with various glycoproteins, infused into oviducts, and observed by transillumination of the oviducts. Of the molecules tested, fetuin was discovered to inhibit binding of hamster sperm to the epithelium (DeMott et al. 1995). Asialofetuin, which is fetuin with its terminal sialic acids removed, did not inhibit binding. Sialic acid alone did. Fetuin tagged with colloidal gold labeled the heads of hamster sperm and certain glycoprotein bands on Western blots of membrane extracts (DeMott et al. 1995). These data indicate that there is a lectin-like molecule on the head of hamster sperm that binds sialic acid and is responsible for attachment of sperm to the epithelium.

Binding of equine sperm to explants of oviductal epithelium *in vitro* was most effectively inhibited by asialofetuin and its terminal sugar, galactose (Lefebvre et al. 1995b; Dobrinski et al. 1996a). Bovine sperm binding to explants of oviductal epithelium was determined to be specifically blocked by fucoidan and its component fucose (Lefebvre et al. 1997). Pretreatment of bovine epithelium with fucosidase, but not galactosidase, reduced sperm binding (Lefebvre et al. 1997). Fucosylated bovine serum albumin, tagged with fluorescein, labeled the rostral head regions of motile bovine sperm, while fluorescein-tagged plain albumin did not (Revah et al. 1998). So, carbohydrate recognition has been implicated in the binding of three different species of sperm to oviductal epithelium. In each of these species, a different sugar was most effective at inhibiting binding. These species differences may not seem so unusual when one considers that a single amino acid residue can determine the ligand specificity of a lectin (Kogan et al. 1995; Revelle et al. 1996), and that closely-related animal lectins have different carbohydrate specificities (Weiss 1994).

Other forms of heterotypic binding between cells involve carbohydrate recognition. Examples are selectins, which mediate leukocyte binding to endothelium (Varki 1992), and glycolipid ligands on ciliated respiratory cells, which are recognized by mycoplasmas (Zhang et al. 1994). Selectins mediate temporary binding between two cell types, just as binding between sperm and epithelium is temporary. Carbohydrate recognition is also implicated in sperm-zona binding (Yanagimachi 1994; Sinowatz et al. 1997) and sperm-Sertoli cell binding (Raychoudhury, Millet 1997). During the course of evolution, lectins with different specificities could have arisen to regulate sperm attachment to these different surfaces.

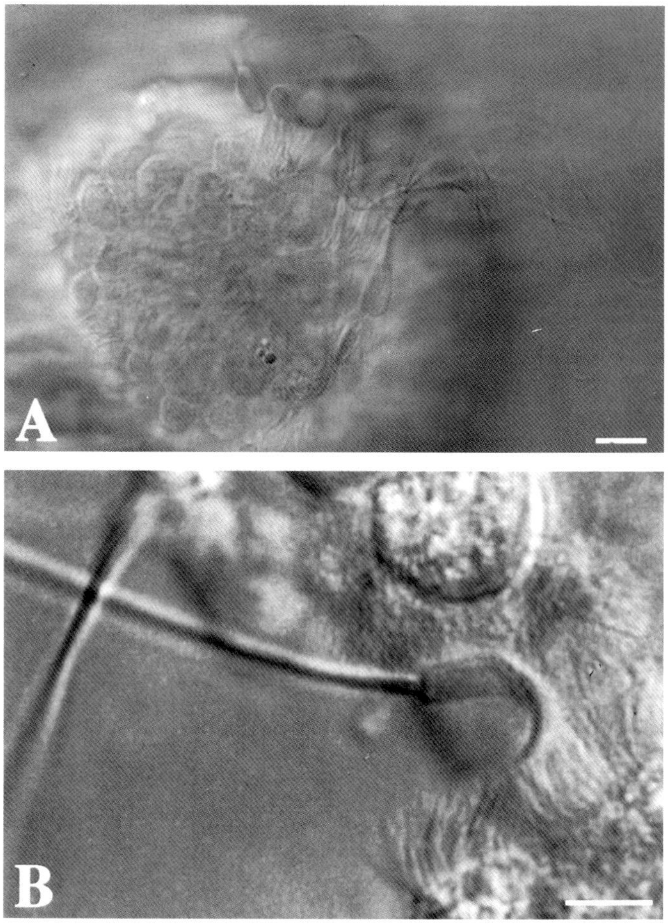

Figure 2. A: Bull sperm binding to an explant of bovine oviductal epithelium. Note the association of the rostral region of the sperm head with cilia of the explant. Bar = 10 microns. (Prepared by R. Lefebvre, as described in Lefebvre et al. 1995a, and photographed by S.S. Suarez; not previously published.) B: Hamster sperm binding to epithelial cells manually stripped from the hamster oviduct as described for cows in Lefebvre et al. 1995a. Bar = 10 microns. (Prepared by M. Lo and photographed by S.S. Suarez.)

In the studies described above, the focus was on identifying a key, specific monosaccharide involved in sperm binding. The next step to characterizing the binding site was to determine how the monosaccharide is linked to a macromolecule in the oviduct. Carbohydrates are usually present on the surface of cells as oligosaccharides attached to proteins or lipids. Because fucose had been determined to be a key monosaccharide for sperm binding in cattle, the next step was to determine which linkage of fucose to other sugars is most effective at blocking bovine sperm binding to oviductal epithelium.

Oligosaccharides containing fucose in various linkages were tested for capacity to inhibit sperm binding to explants *in vitro*. Fucose in an *alpha* 1-4 linkage to N-acetylglucosamine, as found in the trisaccharide Lewis-a, inhibited binding more efficiently than fucose in other linkages (Suarez et al. 1998). It was also more effective than fucose alone. Furthermore, Lewis-a tagged by conjugation to fluorescein-polyacrylamide labeled the heads of motile bovine sperm (Fig. 3; Suarez et al. 1998).

At this point, it can be concluded that there is a lectin-like molecule on the surface of bovine sperm that binds fucose in an *alpha*1-4 linkage to N-acetylglucosamine. The identities of the sperm lectin and corresponding glycoprotein or glycolipd ligand on the oviduct epithelium are yet to be determined.

Binding of bovine sperm to oviductal epithelium is dependent on the presence of Ca^{2+}. In the absence of extracellular Ca^{2+}, bovine sperm fail to bind to explants *in vitro* and they are not labeled with the Lewis-a/polyacrylamide/fluorescein probe (Suarez et al. 1998). Other animal cell lectins are known to be dependent on Ca^{2+} for binding activity (Weiss 1994), so this sperm lectin may be related to the animal "C" or calcium-dependent lectins.

Survival of Sperm during Storage

The oviductal mucosa apparently protects sperm against aging damage during storage. Such damage is a significant problem to sperm compared with most other types of cells, because sperm do not have the machinery for recycling plasma membrane.

Sperm incubated with oviductal epithelium remain viable longer *in vitro* than when they are incubated in medium alone (porcine: Suarez et al. 1990; equine: Ellington et al. 1993; human: Kervancioglu et al. 1994) or with tracheal epithelium (bovine: Pollard et al. 1991). Viability can be extended by incubating sperm with membrane vesicles prepared from the apical membranes of oviductal epithelium (rabbit: Smith, Nothnick 1997; equine: Dobrinski et al. 1997; human: Murray, Smith 1997), indicating that the epithelium can produce at least some of the effect by direct contact rather than by products secreted into the lumen. It was reported that equine sperm binding to epithelium or membrane vesicles maintain low levels of cytoplasmic Ca^{2+}, compared to free-swimming sperm, sperm attached to Matrigel, or sperm incubated with vesicles made from kidney membranes (Dobrinski et al. 1996b, 1997).

Equine and human sperm incubated with oviduct membrane vesicles also capacitated more slowly than sperm incubated in capacitating medium alone, when capacitation was assayed by chlortetracycline fluorescence patterns (Dobrinski et al. 1997; Murray, Smith 1997). Possibly, viability is maintained by preventing capacitation and its concomitant rise in cytoplasmic Ca^{2+}.

It was proposed by Lapointe et al. (1998) that catalase, which is present in the bovine oviduct, serves to protect against peroxidative aging damage to the sperm membranes. Catalase activity was detected in extracts of bovine oviductal epithelium. When bovine sperm were incubated with these extracts, then labeled with a commercial antibody to bovine liver catalase, the sperm were labeled over the acrosomal cap of the head (Lapointe et al. 1998). Hence, one way in which the oviduct may maintain sperm viability during storage in the reservoir could be by protecting them against peroxidative damage.

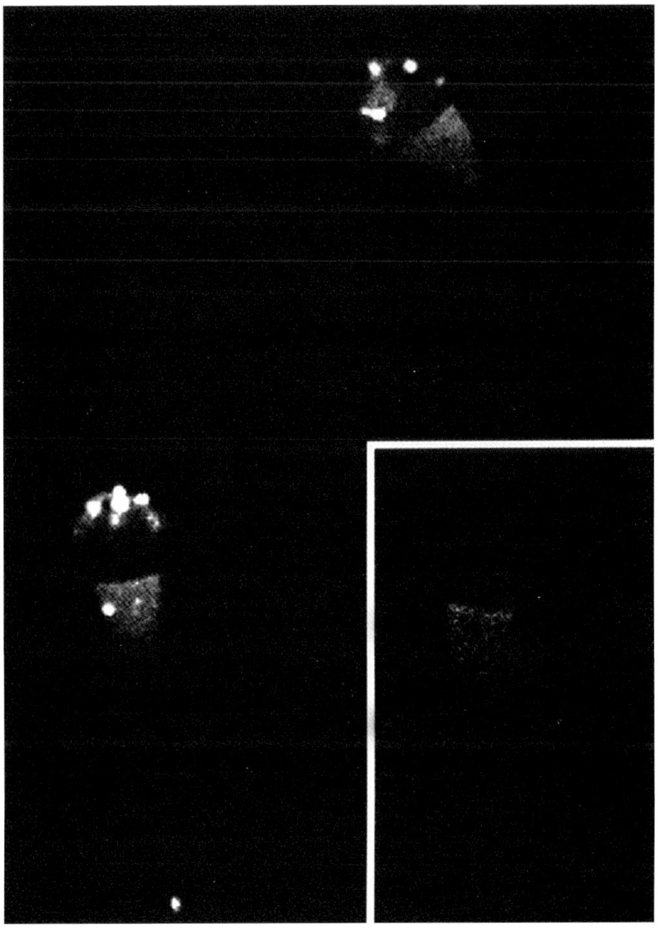

Figure 3. Bovine sperm labeled while living with Lewis-a conjugated to FITC-polyacrylamide (as described in Suarez et al. 1998). Lewis-a is a trisaccharide comprised of fucose, N-acetylglucosamine, and galactose. Inset shows labeling blocked by preincubation with fucose. Used with permission.

Release of Sperm from the Reservoir

Little is known about how sperm are released from the epithelium so that they may fertilize oocytes. Changes in the hormonal state of oviductal epithelium that are related to impending ovulation apparently do not alter the density of binding sites for sperm on the surface of the epithelium (Lefebvre et al. 1995a; Suarez et al. 1991; Thomas et al. 1994).

Thus it appears that the epithelium does not release sperm by enzymatically destroying its binding sites or failing to replenish them. Instead, current evidence indicates that a change in the surface of the sperm brings about its release from the epithelium.

Two types of change are known to occur in sperm in the female reproductive tract prior to fertilization.

One is capacitation, which involves changes in the plasma membrane over the sperm head. Capacitation could effect sperm release by eliminating or modifying binding molecules on the head.

The second is hyperactivation, which could provide the force necessary for overcoming the attraction between sperm and oviductal epithelium. Smith and Yanagimachi (1991) reported that hamster sperm which had undergone both capacitation and hyperactivation *in vitro* did not bind to epithelium when infused into hamster oviducts. While using transillumination to study motile sperm within oviducts removed from mated mice, it was noted that only hyperactivated sperm detach from epithelium (DeMott, Suarez 1992).

Also, when bovine sperm were capacitated by treatment with heparin *in vitro*, initiation of binding to explants of oviductal epithelium was significantly reduced (Lefebvre, Suarez 1996). In this case, the sperm were not hyperactivated. Therefore, it appears that both capacitation and hyperactivation are involved in freeing sperm from the epithelium. Changes in the sperm head surface associated with capacitation could be responsible for loss of binding affinity, while the type of pulling force generated by hyperactivation may enable sperm to overcome the attraction weakened by capacitation.

Although the availability of sperm binding sites on the oviduct epithelium does not change during the hormonal cycle, the oviduct may regulate release of sperm in other ways. One possibility is that hormonal signals initiate secretions that enhance sperm capacitation and initiate hyperactivation, thereby bringing about sperm release. Soluble oviductal factors do enhance capacitation of bull sperm (Chian et al. 1995; Mahmoud, Parrish 1996). Capacitated bovine sperm incubated with fresh explants from preovulatory oviductal epithelium became hyperactivated (Lefebvre, Suarez 1996).

There is evidence in the hamster that capacitation occurs more rapidly when females are mated shortly after ovulation, rather than before (Smith, Yanagimachi 1989). During early estrus, the oviduct may delay capacitation. As the time of ovulation approaches, the oviduct may respond to hormonal messages to secrete substances that initiate or enhance sperm capacitation.

The lectin-like molecule on sperm that appears to be responsible for binding to the epithelium is lost or loses binding affinity in capacitated sperm. Hamster sperm that had been incubated in capacitating medium until they exhibited hyperactivation did not label over the acrosomal region with fetuin, while fresh epididymal sperm did (DeMott et al. 1995). Fetuin bound to certain protein bands on Western blots of electrophoretically separated membrane proteins extracted from fresh epididymal hamster sperm, but binding was reduced on proteins extracted from sperm that were hyperactivated and partially capacitated (DeMott et al. 1995). When bull sperm were capacitated *in vitro*, they no longer labeled with fluorescein-labeled fucosylated bovine serum albumin (Revah et al. 1998).

Summary of Sperm Binding and Release

In summary of what is known about sperm binding in the oviduct to date, the following picture emerges. The sperm reservoir forms in the uterotubal junction and/or isthmus by binding of a lectin-like molecule on sperm to an oligosaccharide ligand on the surface of the oviductal mucosa.

The narrowness of the lumen, and perhaps the mucus within the lumen, must enhance sperm binding by slowing their progress and increasing contact with the epithelial surface. Direct contact with mucosal epithelium prolongs sperm survival and delays capacitation. Capacitation may be initiated by secretions as the time of ovulation approaches. The lectin on the surface of sperm is lost or modified during the complex process of capacitation, thereby allowing sperm to be released. Hyperactivation may provide the force to pull sperm away from their attachment sites. This model is in need of verification for a variety of eutherian mammals.

The Reservoir in Marsupials and Primitive Eutherians

In marsupial mammals (Bedford 1991; Taggart 1994) and birds (Bakst 1992; Bakst et al. 1994), sperm are stored in mucosal crypts in the oviduct called "sperm storage tubules." However, the sperm heads do not attach to the epithelium in the tubules of many species.

Many of the sperm in the tubules of the marsupial *Sminthopsis crassicaudata* were observed to be immotile (Bedford, Breed 1994) and it is thought that the motility of avian sperm is suppressed in the tubules (Bakst et al. 1994). Hence, motility suppression may serve to keep sperm in the tubules until ovulation.

In the primitive eutherian mammals, the shrews,

some species have been reported to possess distinctive bubble-like outpocketings of the oviduct wall in the caudal ampulla. Sperm enter these structures and do not adhere to the epithelium (Bedford et al. 1997a, 1997b).

In more advanced eutherian mammals, the storage structures are less tubular and less distinctive, being organized as grooves created by folds of mucosa. Adhesion may be more effective at trapping sperm in these structures. Motility suppression has been observed in the isthmus of rabbits and has been proposed as a mechanism of storage (Overstreet et al. 1980; Overstreet, Cooper 1975; Burkman et al. 1984). In hamsters (Smith, Yanagimachi 1990) and mice (Suarez 1987), immotile sperm have been observed in the central part of the isthmic lumen; however, in this case, it is thought that these sperm are damaged and may not fertilize (Smith, Yanagimachi 1990). Motility suppression may or may not be operative in sperm storage in species where sperm binding occurs. It is curious that distinctive storage structures would be lost and sperm binding would evolve to replace them.

A Reservoir in Humans?

So far, there has been no conclusive evidence for a distinct oviductal sperm reservoir in humans (Williams et al. 1993). Human sperm have not, for the most part, been observed to bind tightly to oviductal epithelium *in vitro* (SS Suarez and P Morales, personal observations; Yeung et al. 1994; Murray, Smith 1997), although some sperm have been observed to stick to epithelium under certain conditions *in vitro* (Pacey et al. 1995). Despite the lack of obvious strong binding to epithelium, human sperm viability is maintained by incubation with oviductal epithelium (Murray, Smith 1997), as it is in species in which there is strong binding of sperm to epithelium (Pollard et al. 1991; Chian, Sirard 1994).

One possibility is that the human cervix serves as the major site of a sperm reservoir. The lumen of the human cervix is 3 cm in length (Insler et al. 1980). The human uterus is rather small in proportion to body size, compared with those of ruminants, for example, and human sperm must travel only a few cm through the lumen to reach the uterotubal junction.

The entrance to the uterotubal junction in humans is shaped rather like a funnel and is not guarded by mucosal folds (Hafez, Black 1969; Beck, Boots 1974). In comparison, the uterotubal junction of rodents, pigs, dogs, and ruminants presents an elaborate entrance surrounded by mucosal folds. So, human sperm may be guided right into the uterotubal junction, while other species of sperm may be presented with more of a barrier.

The evidence that could be used to argue against a cervical reservoir is that very few sperm have been recovered from human or primate uteri 24 hours after coitus (Rubenstein et al. 1951; Moyer et al. 1970). Furthermore, the leukocytic infiltration of the uterus, which becomes significant several hours after coitus in humans (Thompson et al. 1992), could present a barrier to passage of sperm that had been stored in the cervix. The leukocytes appear to outnumber human sperm in the uterus at 24 hours after coitus (Thompson et al. 1992).

Unless sperm are protected from phagocytosis (and they might be), it is unlikely that they could travel from the cervical reservoir to the oviduct 24 hours post coitus. Alternatively, human sperm could be stored for long periods of time in the oviduct, but not in a distinct reservoir and not by binding tightly to the mucosal surface. The mucosal folds of the human oviductal lumen, which are quite small in the isthmus, increase in size and complexity towards the ovary, possibly offering increasingly greater barriers to the advancement of sperm. Sperm progress could be slowed by the mucus in the lumen (Jansen 1980) and by sticking lightly to the mucosa (Pacey et al. 1995).

So, rather than creating a distinct reservoir, the human oviduct could slow advancement of sperm to the site of fertilization in such a manner so as to increase the chances that a few sperm are present at the site of fertilization when ovulation occurs. At that time, muscular contractions and secretions could move or activate sperm and increase chances of encountering the oocyte.

There is some preliminary evidence that human sperm may be chemotactically drawn to the oocyte by follicular fluid introduced into the oviduct by the cumulus mass at ovulation (Ralt et al. 1991). Data of sperm distribution in the tubes of women have varied so widely that the information cannot be used to construct a useful hypothetical model of the events of sperm transport in humans (Williams et al. 1993). Perhaps sperm transport is a relatively inefficient and unregulated process in humans, because evolutionary selective pressures have worked on the act of mating to promote long-term pair bonding in addition to fertilization success.

Fate of Oviductal Sperm

After fertilization, mammalian sperm may be phagocytosed by isthmic epithelial cells (Chakraborty, Nelson 1975) or may be eliminated into the peritoneal cavity (Mortimer, Templeton 1982) and then phagocytosed. Phagocytosis within the oviduct may be employed by species, such as mice, which have an extensive ovarian bursa that would limit passage of sperm into the peritoneal cavity.

A Model for Sperm Transport

While much remains to be discovered, and some issues remain to be settled, a general model for sperm transport in the oviduct of eutherian mammals can be derived from what is known now.

Sperm are deposited at coitus into the vagina or uterus. Those deposited in the vagina swim through the cervix. Muscular contractions move sperm through the uterine cavity. Eventually a few thousand sperm swim through the uterotubal junction. In the junction or the caudal isthmus, they face a narrow lumen filled with mucus which can slow their progress. Prolonged contact with the wall of the junction or isthmus results in specific binding of sperm to the mucosal epithelium in many species. This serves to create a distinct reservoir. As the time of ovulation approaches, sperm become capacitated and hyperactivated and they release from the epithelium. Meanwhile, the oocyte or oocytes are released from the ovary and transported rapidly into the ampulla and down to the ampullary-isthmic junction.

During this time, a few sperm ascend the oviduct to reach the descending oocytes. Fertilization occurs soon thereafter, as sperm penetrate the cumulus, reach and penetrate the zona pellucida, and finally fuse with the oocyte plasma membrane.

References

Baker RD, Degen AA. Transport of live and dead boar spermatozoa within the reproductive tract of gilts. J Reprod Fertil 1972; 28:369-377.

Bakst MR. Anatomical basis of sperm-storage in the avian oviduct. Scanning Microsc 1987; 1:1257-1266.

Bakst MR. Observations on the turkey oviductal sperm-storage tubule using differential interference contrast microscopy. J Reprod Fertil 1992; 95:877-883.

Bakst MR, Wishart G, Brilliard J-P. Oviductal sperm selection, transport, and storage in poultry. Poultry Sci Rev 1994; 5:117-143.

Beck LR, Boots LR. The comparative anatomy, histology and morphology of the mammalian oviduct. In: (Johnson AD, Foley CW, eds) The Oviduct and Its Functions. New York: Academic Press, 1974, pp2-51.

Bedford JM. The coevolution of mammalian gametes. In: (Dunbar BS, O'Rand MG, eds) A Comparative Overview of Mammalian Fertilization. New York: Plenum Press, 1991, pp3-35.

Bedford JM, Breed WG. Regulated storage and subsequent transformation of spermatozoa in the fallopian tubes of an Australian marsupial, *Sminthopsis crassicaudata*. Biol Reprod 1994; 50:845-854.

Bedford JM, Mock OB, Phillips DM. Unusual ampullary sperm crypts, and behavior and role of the cumulus oophorus, in the oviduct of the least shrew, *Cryptotis parva*. Biol Reprod 1997a; 56:1255-1267.

Bedford JM, Phillips DM, Mover-Lev H. Novel sperm crypts and behavior of gametes in the fallopian tube of the white-toothed shrew, *Crocidura russula* Monacha. J Exp Zool 1997b; 277:262-273.

Birkhead TR, Moeller AP. Sexual selection and the temporal separation of reproductive events: sperm storage data from reptiles, birds, and mammals. Biol J Linnean Soc 1993; 50:295-311.

Birkhead TR, Pellatt EJ, Fletcher F. Selection and utilization of spermatozoa in the reproductive tract of the female zebra finch *Taeniopygia guttata*. J Reprod Fertil 1993; 99:593-600.

Burkman LJ, Overstreet JW, Katz DF. A possible role for potassium and pyruvate in the modulation of sperm motility in the rabbit oviductal isthmus. J Reprod Fertil 1984; 71:367-376.

Carballado R, Esponda P. Fate and distribution of seminal plasma proteins in the genital tract of the female rat after natural mating. J Reprod Fertil 1997; 109:325-335.

Chakraborty J, Nelson L. Fate of surplus sperm in the fallopian tube of the white mouse. Biol Reprod 1975; 12:455-463.

Chian R-C, Sirard M-A. Fertilizing ability of bovine spermatozoa cocultured with oviduct epithelial cells. Biol Reprod 1994; 52:156-162.

Chian R-C, LaPointe S, Sirard M-A. Capacitation *in vitro* of bovine spermatozoa by oviduct cell monolayer conditioned medium. Mol Reprod Dev 1995; 42:318-324.

Day BN, Polge C. Effects of progesterone on fertilization and egg transport in the pig. J Reprod Fertil 1968; 17:227-230.

DeMott RP, Suarez SS. Hyperactivated sperm progress in the mouse oviduct. Biol Reprod 1992; 46:779-785.

DeMott RP, Lefebvre R, Suarez SS. Carbohydrates mediate the adherence of hamster sperm to oviductal epithelium. Biol Reprod 1995; 52:1395-1403.

Dobrinski I, Ignotz GG, Thomas PGA, Ball BA. Role of carbohydrates in the attachment of equine spermatozoa to uterine tubal (oviductal) epithelial cells *in vitro*. Amer J Vet Res 1996a; 57:1635-1639.

Dobrinski I, Suarez SS, Ball BA. Intracellular calcium concentration in equine spermatozoa attached to oviductal epithelial cells *in vitro*. Biol Reprod 1996b; 54:783-788.

Dobrinski I, Smith TT, Suarez SS, Ball BA. Membrane contact with oviductal epithelium modulates the intracellular calcium concentration in equine spermatozoa *in vitro*. Biol Reprod 1997; 56:861-869.

Ellington JE, Ignotz GG, Varner DD, Marcucio RS, Mathison P, Ball BA. *In vitro* interaction between oviduct epithelia and equine sperm. Arch Androl 1993; 31:79-86.

Gaddum-Rosse P. Some observations on sperm transport through the uterotubal junction of the rat. Amer J Anat 1981; 160:333-341.

Gist DH, Jones JM. Storage of sperm in the reptilian oviduct. Scanning Microsc 1987; 1:1839-1849.

Gist DH, Jones JM. Sperm storage within the oviducts of turtles. J Morphol 1989; 199:379-384.

Hafez ESE, Black DL. The mammalian uterotubal junction. In: (Hafez ESE, Blandau RJ, eds) The Mammalian Oviduct: Comparative Biology and Methodology. Chicago: The University of Chicago Press, 1969, pp85-128.

Harper, MJK. Relationship between sperm transport and penetration of eggs in the rabbit oviduct. Biol Reprod 1973; 8:441-450.

Hawk HW. Transport and fate of spermatozoa after insemination of cattle. J Dairy Sci 1983; 70:1487-1503.

Hook SJ, Hafez ESE. A comparative study of the mammalian uterotubal junction. J Morphol 1968; 125:159-184.

Hunter RHF. Local action of progesterone leading to polyspermic fertilization in pigs. J Reprod Fertil 1972; 31:433-444.

Hunter RHF. Polyspermic fertilization in pigs after tubal deposition of excessive numbers of spermatozoa. J Exp Zool 1973; 183:57-64.

Hunter RHF. Sperm transport and reservoirs in the pig oviduct in relation to the time of ovulation. J Reprod Fertil 1981; 63:109-117.

Hunter RHF. Development of the Fallopian tubes and their functional anatomy. In: (Hunter RHF, ed) The Fallopian Tubes: Their Role in Fertility and Infertility. Berlin: Springer-Verlag, 1988, pp12-29.

Hunter RHF, Leglise PC. Polyspermic fertilization following tubal surgery in pigs, with particular reference to the role of the isthmus. J Reprod Fertil 1971; 24:233-246.

Hunter RHF, Nichol R. Transport of spermatozoa in the sheep oviduct: preovulatory sequestering of cells in the caudal isthmus. J Exp Zool 1983; 228:121-128.

Hunter RHF, Wilmut I. Sperm transport in the cow: periovulatory redistribution of viable cells within the oviduct. Reprod Nutr Dev 1984; 24:597-608.

Insler V, Glezerman M, Zeidel L, Bernstein D, Misgav N. Sperm storage in the human cervix: a quantitative study. Fertil Steril 1980; 33:288-294.

Jansen RPS. Fallopian tube isthmic mucus and ovum transport. Science 1978; 201:349-351.

Jansen RPS. Cyclic changes on the human fallopian tubes isthmus and their functional importance. Am J Obstet Gynecol 1980; 136:292-308.

Jansen RPS, Bajpai VK. Oviduct acid mucus glycoproteins in the estrous rabbit: ultrastructure and histochemistry. Biol Reprod 1982; 26:155-168.

Kervancioglu ME, Djahanbakhch O, Aitken RJ. Epithelial cell co-culture and the induction of sperm capacitation. Fertil Steril 1994; 61:1103-1108.

Kogan TP, Revelle BM, Tapp S, Scott D, Beck PJ. A single amino acid residue can determine the ligand specificity of E-selectin. J Biol Chem 1995; 270:14047-14055.

Koya Y, Munehara H, Takano K. Sperm storage and degradation in the ovary of a marine copulating sculpin, *Alcichthys alcicornis* (Teleosti: scorpaeniformes): role of intercellular junctions between inner ovarian epithelial cells. J Morphol 1997; 233:153-163.

Lapointe S, Sullivan R, Sirard M-A. Binding of a bovine oviductal fluid catalase to mammalian spermatozoa. Biol Reprod 1998; 58:747-753.

Lefebvre R, Suarez SS. Effect of capacitation on bull sperm binding to homologous oviductal epithelium. Biol Reprod 1996; 54:575-582.

Lefebvre R, Chenoweth PJ, Drost M, LeClear CT, MacCubbin M, Dutton JT, Suarez SS. Characterization of the oviductal sperm reservoir in cattle. Biol Reprod 1995a; 53:1066-1074.

Lefebvre R, DeMott RP, Suarez SS, Samper JC. Specific inhibition of equine sperm binding to oviductal epithelium. Equine Reproduction VI. Biol Reprod 1995b; Mono 1:689-696.

Lefebvre R, Lo MC, Suarez SS. Bovine sperm binding to oviductal epithelium involves fucose recognition. Biol Reprod 1997; 56:1198-1204.

Mahmoud AI, Parrish JJ. Oviduct fluid and heparin induce similar surface changes in bovine sperm during capacitation. Mol Reprod Dev 1996; 43:554-560.

Moyer DL, Rimdusit S, Mishell DR, Jr. Sperm distribution and degradation in the human female reproductive tract. Obstet Gynecol 1970; 35:831-840.

Mortimer D, Templeton AA. Sperm transport in the human female reproductive tract in relation to semen analysis characteristics and time of ovulation. J Reprod Fertil 1982; 64:401-408.

Murphy-Walker S, Haley SR. Functional sperm storage duration in female *Hemidactylus frenatus* (family gekkonidae). Herpetologica 1996; 52:365-373.

Murray SC, Smith TT. Sperm interaction with Fallopian tube apical plasma membrane enhances sperm motility and delays capacitation. Fertil Steril 1997; 68:352-357.

Overstreet JW, Cooper GW. Reduced sperm motility in the isthmus of the rabbit oviduct. Nature (Lond) 1975; 258:718-719.

Overstreet JW, Cooper GW. Sperm transport in the reproductive tract of the female rabbit: I. The rapid transit phase of transport. Biol Reprod 1978; 19:101-114.

Overstreet JW, Cooper GW, Katz DF. Sperm transport in the reproductive tract of the female rabbit: II. The sustained phase of transport. Biol Reprod 1978; 19:115-132.

Overstreet JW, Katz DF, Johnson LL. Motility of rabbit spermatozoa in the secretions of the oviduct. Biol Reprod 1980; 22:1083-1088.

Pacey AA, Hill CJ, Scudamore IW, Warren MA, Barratt CLR, Cooke ID. The interaction *in vitro* of human spermatozoa with epithelial cells from the human uterine (Fallopian) tube. Hum Reprod 1995; 10:360-366.

Perkins MJ, Palmer BD. Histology and functional morphology of the oviduct of an oviparous snake, *Diadophis punctatus*. J Morphol 1996; 227:67-79.

Polge C, Salamon S, Wilmut I. Fertilizing capacity of frozen boar semen following surgical insemination. Veterinary Rec 1970; 87:424-428.

Pollard JW, Plante C, King WA, Hansen PJ, Betteridge KJ, Suarez SS. Fertilizing capacity of bovine sperm may be maintained by binding to oviductal epithelial cells. Biol Reprod 1991; 44:102-107.

Ralt D, Goldenberg M, Fetterolf P, Thompson D, Dor J, Mashiachi S, Garbers DL, Eisenbach M. Sperm attraction to a follicular fluid factor(s) correlates with human egg fertilizability. Proc Natl Acad Sci USA 1991; 88:2840-2844.

Raychoudhury SS, Millette CF. Multiple fucosyltransferases and their carbohydrate ligands are involved in spermatogenic cell-Sertoli cell adhesion *in vitro* in rats. Biol Reprod 1997; 56:1268-1273.

Raychoudhury SS, Suarez SS. Porcine sperm binding to oviductal explants in culture. Theriogenology 1991; 36:1059-1070.

Revah I, Suarez SS, Flesch FM, Colenbrander B, Gadella BM. Changes in capacity of bull sperm to bind fucose. Biol Reprod 1998; 58 (Suppl 1):117.

Revelle BM, Scott D, Beck PJ. Single amino acid residues in the E- and P-selectin epidermal growth factor domains can determine carbohydrate binding specificity. J Biol Chem 1996; 271:16160-16170.

Rubenstein BB, Strauss H, Lazarus M, Hankin H. Sperm survival in women. Fertil Steril 1951; 2:15-19.

Shalgi R, Smith TT, Yanagimachi R. A quantitative comparison of the passage of capacitated and uncapacitated hamster spermatozoa through the uterotubal junction. Biol Reprod 1992; 46:419-424.

Sinowatz F, Topfer-Petersen E, Calvete JJ. Glycobiology of fertilization. In: (Gabius H-J, Gabius S, eds) Glycosciences. Weinheim: Chapman and Hall, 1997, pp595-610.

Smith TT, Nothnick WB. Role of direct contact between spermatozoa and oviductal epithelial cells in maintaining rabbit sperm viability. Biol Reprod 1997; 56:83-89.

Smith TT, Yanagimachi R. Capacitation status of hamster spermatozoa in the oviduct at various times after mating. J Reprod Fertil 1989; 86:255-261.

Smith TT, Yanagimachi R. The viability of hamster spermatozoa stored in the isthmus of the oviduct: the importance of sperm-epithelium contact for survival. Biol Reprod 1990; 42:450-457.

Smith TT, Yanagimachi R. Attachment and release of spermatozoa from the caudal isthmus of the hamster oviduct. J Reprod Fertil 1991; 91:567-573.

Smith TT, Koyanagi F, Yanagimachi R. Quantitative comparison of the passage of homologous and heterologous spermatozoa through the uterotubal junction of the golden hamster. Gam Res 1988a; 19:227-234.

Smith TT, Koyanagi F, Yanagimachi R. Distribution and number of spermatozoa in the oviduct of the golden hamster after natural mating and artificial insemination. Biol Reprod 1988b; 37:225-234.

Srinivas SR, Shivanandappa T, Hegde SN, Sarkar HBD. Sperm storage in the oviduct of the tropical rock lizard, *Psammophilus dorsalis*. J Morphol 1995; 224:293-301.

Suarez SS. Sperm transport and motility in the mouse oviduct: observations *in situ*. Biol Reprod 1987; 36:203-210.

Suarez SS, Drost M, Redfern K, Gottlieb W. Sperm motility in the oviduct. In: (Bavister BD, Cummins J, Roldan ERS, eds) Fertilization in Mammals. Norwell, MA: Serono Symposia, 1990, pp111-124.

Suarez SS, Redfern K, Raynor P, Martin F, Phillips DM. Attachment of boar sperm to mucosal explants of oviduct *in vitro*: possible role in formation of a sperm reservoir. Biol Reprod 1991; 44:998-1004.

Suarez SS, Brockman K, Lefebvre R. Distribution of mucus and sperm in bovine oviducts after artificial insemination. Biol Reprod 1997; 56:447-453.

Suarez SS, Revah I, Lo M, Kölle S. Bull sperm binding to oviductal

epithelium is mediated by a Ca^{2+}-dependent lectin on sperm which recognizes Lewis-A trisaccharide. Biol Reprod 1998; 58:39-44.
Taggart DA. A comparison of sperm and embryo transport in the female reproductive tract of marsupial and eutherian mammals. Reprod Fertil Dev 1994; 6:451-472.
Thomas PGA, Ball BA, Brinsko SP. Interaction of equine spermatozoa with oviduct epithelial cell explants is affected by estrous cycle and anatomic origin of explant. Biol Reprod 1994; 51:222-228.
Thompson LA, Barratt CLR, Bolton AE, Cooke ID. The leukocytic reaction of the human uterine cervix. Am J Reprod Immunol 1992; 28:85-89.
Varki A. Selectins and other mammalian sialic acid-binding lectins. Curr Opin Cell Biol 1992; 4:257-266.
Weiss WI, Drickamer K. Structural basis of lectin-carbohydrate recognition. Ann Rev Biochem 1996; 65:441-473.
Williams M, Hill CJ, Scudamore I, Dunphy B, Cooke ID, Barratt CLR. Sperm numbers and distribution within the human Fallopian tube around ovulation. Hum Reprod 1993; 8:2019-2026.
Wrobel K-H, Kujat R, Fehle G. The bovine tubouterine junction: general organization and surface morphology. Cell Tissue Res 1993; 271:227-239.
Yanagimachi R. Mammalian fertilization. In: (Knobil E, Neill JD, eds) The Physiology of Reproduction. New York: Raven Press Ltd, 1994, pp189-317.
Yanagimachi R, Chang MC. Sperm ascent through the oviduct of the hamster and rabbit in relation to the time of ovulation. J Reprod Fertil 1963; 6:413-420.
Yeung WSB, Ng VKH, Lau EYL, Ho PC. Human oviductal cells and their conditioned medium maintain the motility and hyperactivation of human spermatozoa *in vitro*. Hum Reprod 1994; 9:656-660.
Zhang Q, Young TF, Ross RF. Glycolipid receptors for attachment of *Mycoplasma hyopneumoniae* to porcine respiratory ciliated cells. Infect Immun 1994; 62:4367-4373.

9 The Implications of Unusual Sperm/Female Relationships in Mammals

J. M. Bedford
Cornell University

The sperm/female relationship as treated here involves not only the interaction of spermatozoa with various regions of the female tract, but also that with the egg leading to fertilization. The first obviously follows from the adoption of internal fertilization, in both the oviparous situation of reptiles, birds and monotremes, and the viviparous state that has evolved independently in some fishes, amphibians and reptiles, as well as all therian mammals. The move to internal fertilization raises several issues, not least an apparent need for the female to cope with what represent foreign antigens presented by spermatozoa. In addition, spermatozoa appear to have undergone functional, metabolic and morphological changes as a consequence of the need to survive in and progress through the female tract.

In that last regard, in external fertilizers (cyclostomes, many teleost fishes and many anuran amphibia), spermatozoa often mature to a functional state within the testis, but where internal fertilization is the rule—as in elasmobranch fishes, reptiles, birds and mammals—sperm maturation tends to be completed in the Wolffian duct or epididymis. Thus, as a common correlate of the adoption of internal fertilization, the male excurrent duct seems to have been coopted for the final phase of sperm maturation, and in many therian mammals this relationship appears to be an essential one. In some reptiles and birds as well as mammals, certain Wolffian/epididymal secretory macromolecules that associate with spermatozoa may possibly function to mask surface antigens, or protect spermatozoa in the female tract in some other way.

In therian mammals at least, certain sperm surface modifications also relate directly to the ability to fertilize and probably to the process of regulated storage in the cauda epididymidis. In view of the fact that internal fertilization often carries the need for spermatozoa to remain motile for 24 hours or more, it seems significant that a third aspect of post-testicular maturation involves the potential for sustained progressive motility. This step may in part reflect the appearance of additional metabolic pathways since bird and mammal spermatozoa derive energy not only from respiratory or oxidative phosphorylation, as do those of external fertilizers, but also from glycolysis of external substrates.

The switch from a relatively short functional life in a freshwater or marine environment to a variably prolonged existence in a female tract has been accompanied also by major developments in sperm organization and design. These include the common appearance of dense fibers and many more mitochondria generally organized as a midpiece sheath in the tail, a general elongation of the sperm head (Franzen 1977; Afzelius 1972; Jamieson et al. 1993; van der Horst et al. 1995), and a more arginine-rich form of nuclear protein (Kazinsky 1989). Further modifications of sperm head design seen in marsupial and eutherian mammals have occurred most probably as a correlate of the prominent character of the zona pellucida in these groups (Bedford 1991). Yet there remain largely unexplained features of the sperm/female relationship in eutherian mammals that include an apparent profligacy often in the number of spermatozoa inseminated, the need to undergo capacitation in the female before they can fertilize, and the unique mode of gamete fusion in this group.

To shed further light on some of these issues, research has turned to more unusual mammals, such as marsupials and sundry members of the Insectivora, with the latter being considered to represent the closest links to the progenitors or ancestral line of eutherian mammals. These groups display interesting deviations from the norm in the way that spermatozoa are disposed in the upper tract, and in the preliminaries to fertilization itself. In the discussion that follows, the commonly studied groups such as primates, ungulates and laboratory mammals are often designated for convenience as "advanced" or "higher" mammals.

Sperm Interactions with the Female Tract

"Normal" Patterns

In discussing "deviant" aspects of the sperm/female relationship, it is necessary first to briefly consider the "norm." This is difficult to define for the lower tract in any general way because the activation, disposition and transport patterns of spermatozoa within the vagina and the uterus may differ widely according to species. There is a more general constancy to events occurring in the oviduct. Regardless of the millions inseminated into the vagina or uterus, only a few thousand spermatozoa pass beyond the uterotubal junction to become adherent in many cases to the epithelium in the initial part of the isthmus. This relationship between the surface of viable spermatozoa and the oviduct epithelium apparently is mediated by cell surface carbohydrates (Smith 1998; Suarez 1998), and the later release of some spermatozoa from it may relate to the final plasmalemma-related step of capacitation (Smith 1998). Only very few then actually ascend to the site of fertilization in the ampulla, the timing of the ascent depending on factors from ovulating follicles of the ipsilateral ovary (Hunter 1995; Sultan, Bedford 1996). Moreover, in polyovular species such as the hamster and rat at least, fewer spermatozoa than eggs may have reached the site of fertilization as this event begins (Cummins, Yanagimachi 1982; Shalgi, Phillips 1988). This arrangement appears to maintain the viability of potential fertilizing spermatozoa over the period between insemination and ovulation (Hunter 1995; Suarez 1998), and provides a means of regulating the final stage of sperm transport in a way that minimizes polyspermy (Hunter 1991).

Unusual Patterns in the Upper Female Tract

In contrast to the isthmic epithelial associations seen in the advanced mammals examined so far, spermatozoa of some elasmobranch fishes are stored in special narrow oviductal crypts (Metten 1941). Reptiles and birds also have special sperm storage tubules, usually in the utero-vaginal junction and/or infundibulum of the oviduct (Bakst et al. 1994; Gist, Jones 1987), or midway between them in turtles, where cryptic spermatozoa can survive for up to four years (Gist, Fischer 1993). Such crypts have not been observed so far among the "higher" mammals studied in this regard. However, it emerges that spermatozoa are maintained in groups of free cells within oviduct crypts in some marsupials, and in certain Insectivora among eutherian mammals.

True sperm storage crypts in the mammalian oviduct were noted first in the opossum, *Didelphis* (Rodger, Bedford 1982), and subsequently in its smaller cousin, *Monodelphis* (HDM Moore, personal communication), in which groups of paired spermatozoa reside as free cells within isthmic crypts until ovulation (Fig. 1).

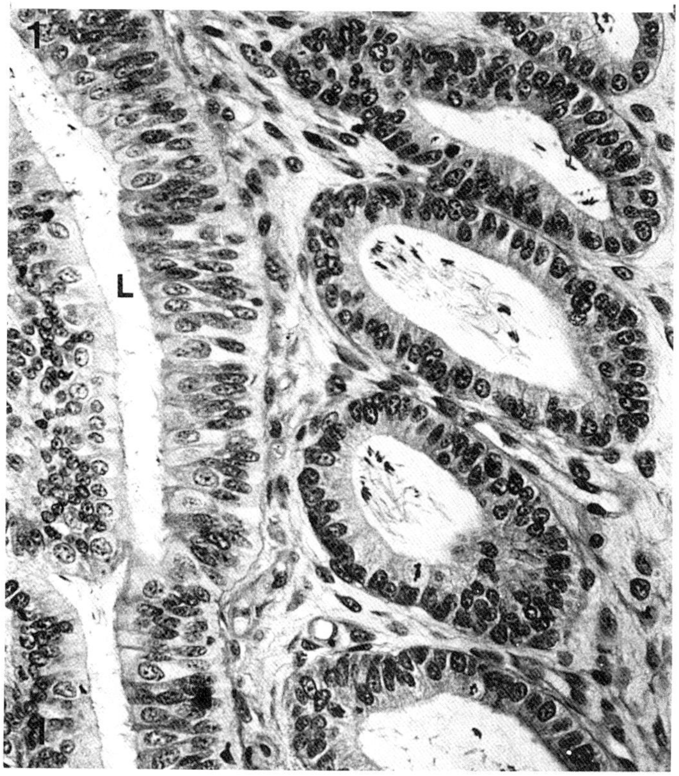

Figure 1. Isthmus of the oviduct in the opossum, *Didelphis virginia*, fixed about 12 hours after natural mating. This section shows groups of spermatozoa within isthmic crypts and illustrates the difference in the morphology of the crypt epithelium from that of the main lumen (L). x270. (From Roger, Bedford 1982.)

A somewhat similar situation has been observed in Australian dasyurid marsupials (Selwood, McCallum 1987; Breed et al. 1989; Taggart, Temple-Smith 1991).

In contrast to those of most eutherian mammals, marsupial spermatozoa maintained within crypts are free of any association with the epithelium (Fig. 2). Since the appearance of the crypt epithelium differs from that of the main lumen, a special environment is probably created within the crypts. After preparation of the oviducts such spermatozoa were sometimes moving actively, but many others were lying in crypts as motionless "sheaves," and became vigorously motile only when released into culture medium (Bedford, Breed 1994). Thus, it appears likely that marsupial spermatozoa stored in isthmic crypts reside there in an immotile state. In *Sminthopsis,* only after ovulation do some hundreds migrate up to the fertilization site in the narrow upper reaches of the fallopian tube, and coincidentally are transformed from a streamlined to a 'T' shape (Bedford, Breed 1994). Those spermatozoa remaining in the crypts after ovulation also eventually adopt a 'T' configuration (Fig. 2), and are cleared later after the zygote has passed through to the uterus.

Preliminary examination of histological sections (JM Bedford, unpublished observations) suggests the presence of true crypts in the oviduct isthmus of another dasyurid, the tuan, and in a glider (*Petaurus breviceps*), whose groups generally produce relatively few spermatozoa and/or have relatively small testes (Taggart et al. 1998). On the other hand, among marsupials which produce relatively large numbers of spermatozoa—

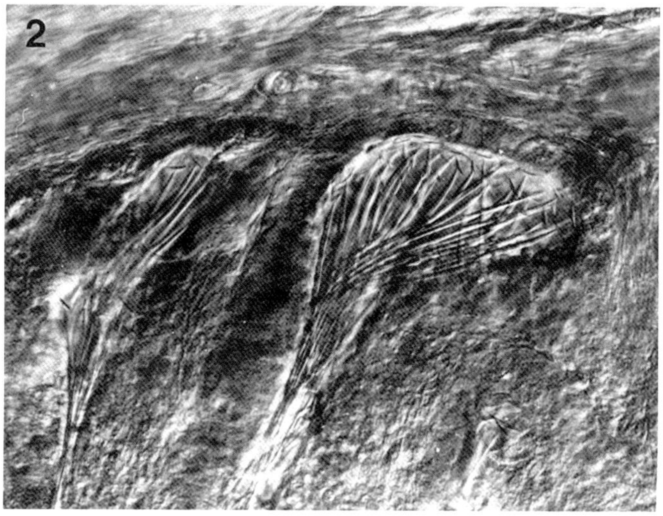

Figure 2. Spermatozoa in crypts in the mid-isthmus of the fallopian tube of the Australian marsupial mouse, *Sminthopsis crassicaudata*, taken about 26 hours after mating, and after ovulation. Visualized in living material with a differential interference contrast microscope, this also illustrates the 'T' shape that develops in spermatozoa after ovulation, even those within isthmic crypts. x300. (From Bedford, Breed 1994.)

macropods and phalangers (Taggart et al. 1998)—true isthmic crypts were not noted in the brushtail possum (Arnold, Shorey 1985), nor in the tammar wallaby (Tyndale-Biscoe, Rodger 1978), which does not seem to possess specialized storage sites nor displays extended sperm storage in the female (Taggart 1994). Neither do preliminary histological examinations reveal differentiated crypts in the hare wallaby, the rock wallaby, or potoroo (JM Bedford, unpublished observations). Thus, while too limited yet to justify any firm conclusions, the evidence begins to suggest that possession of these special oviductal storage sites may be correlated with low sperm production rates in marsupials.

Among eutherian mammals, differentiated crypts were first observed in the musk shrew, *Suncus murinus*, distributed throughout the oviduct isthmus (Fig. 3a). By one hour after mating until approximately 18 to 20 hours later, and so a few hours after ovulation, small groups of slow-moving free spermatozoa, numbering a few hundreds in total, are confined to crypts of the posterior isthmus (Fig. 3b). In the hours after ovulation a very few, sometimes no more than about 10 spermatozoa, then migrate from the isthmic crypts to interact with eggs in a crypt-free ampulla (Bedford et al. 1997b). Spermatozoa have been observed within special oviductal crypts in several other shrews examined, but in unexpectedly varied patterns. In *Crocidura*, another crocidurine shrew, they are first confined to isthmic crypts, but those ascending in the periovulation period become sequestered secondarily within ciliated crypts scattered sparsely throughout the ampulla (Bedford et al. 1997d). In soricine shrews, by contrast, the isthmus appears as a simple straight duct with no resident spermatozoa, these being restricted as active free cells (numbering about 1500 in *Cryptotis*) within discrete ciliated crypts (Fig. 4), distributed around the isthmo/ampullary junction and along the ampulla (Bedford et al. 1997a). The African shrew, *Myosorex varius*, displays both isthmic and ampullary crypts (Fig. 5), but initial observation

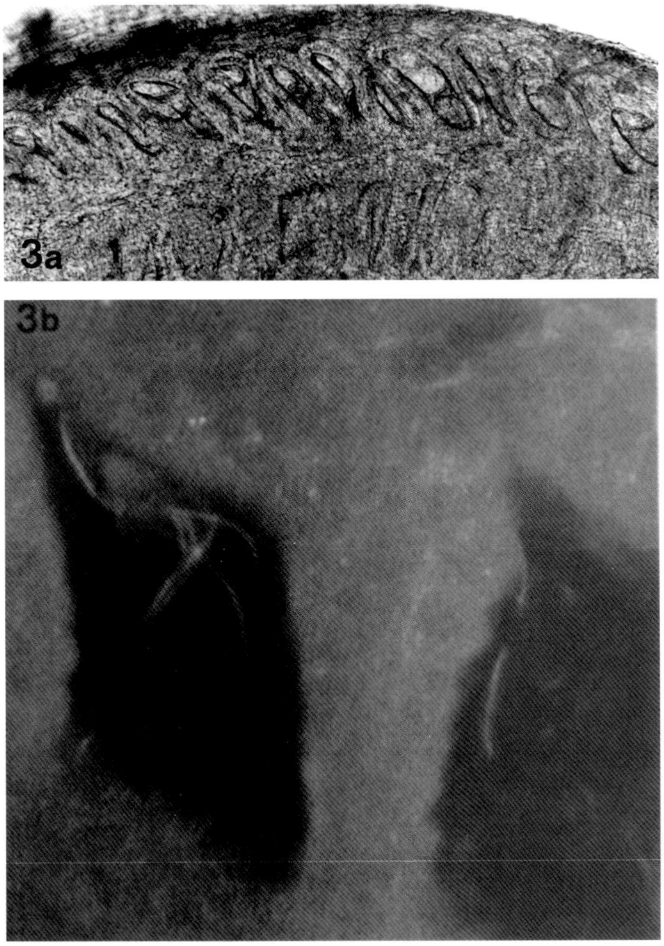

Figure 3. a: Low power view in the DIC microscope of isthmic crypts on one side of the oviduct in the musk shrew, *Suncus murinus*. x 340. b: View obtained with the confocal laser microscope of spermatozoa within isthmic crypts of *Suncus murinus*, 22 hours after coitus. x2,200. (From Bedford et al. 1997b.)

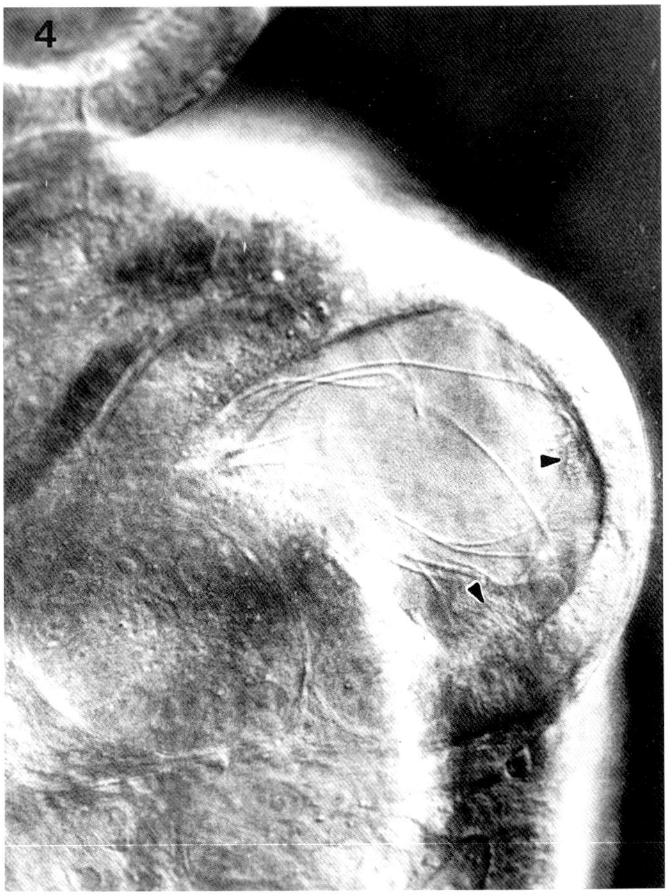

Figure 4. Spermatozoa located within a crypt protruding from the surface of the oviduct ampulla in the least shrew, *Cryptotis parva*, seen in a mated female about two hours after hCG-induced ovulation. Such spermatozoa are actively motile, and do not adhere to the crypt epithelium. Arrowheads point to tufts of cilia within the crypt. x800. (JM Bedford, OB Mock, unpublished.)

suggests that only the latter house spermatozoa (Bedford et al. 1998). The best studied shrews, *Suncus* and *Cryptotis*, retain spermatozoa for only about 30 hours or so in the isthmic and ampullary crypts before being cleared in the postovulatory period.

Among other Insectivora, ampullary crypts are present also in true moles. Histological examination and visualization of fresh material with differential interference contrast (DIC) optics reveals that the oviducts of the eastern mole (*Scalopus aquaticus*) and star-nose mole (*Condylura cristata*) have a very short simple isthmic segment, and a disproportionately long ampulla throughout which ciliated crypts are arrayed (JM Bedford, OB Mock, GE Olson, SK Nagdas and VP Winfrey, unpublished observations). However, such crypts are not necessarily characteristic of all Insectivora. In the tenrec, spermatozoa are shown adhering in large numbers to a simply folded isthmic epithelium, and some then invade a soricid-type cumulus and fertilize eggs within ovulating follicles (Nicoll, Racey 1985; Strauss 1950). The histological description of Deanesly (1934) does not reveal crypts in any region of the hedgehog oviduct, and in a golden mole (*Amblysomus hottentotus*: Chrysochloridae), the isthmic epithelium displays only some irregular pockets and the ampulla some epithelial arborization (RTF Bernard and JM Bedford, unpublished observations).

The Implications of Oviductal Crypts in Mammals

When compared to that in advanced mammals, the interaction of spermatozoa with the oviduct differs strikingly in some marsupials and some Insectivora. The variant situations in these less commonly studied mammals raise several related questions that have to do at least with regulation of sperm transport to the fertilization site, the mechanisms involved in capacitation, and the issue of sperm numbers.

Sperm Regulation. Clearly, different vertebrates have developed diverse strategies to ensure the delivery of appropriate sperm numbers at the fertilization site, and it is evident that these strategies can vary even among

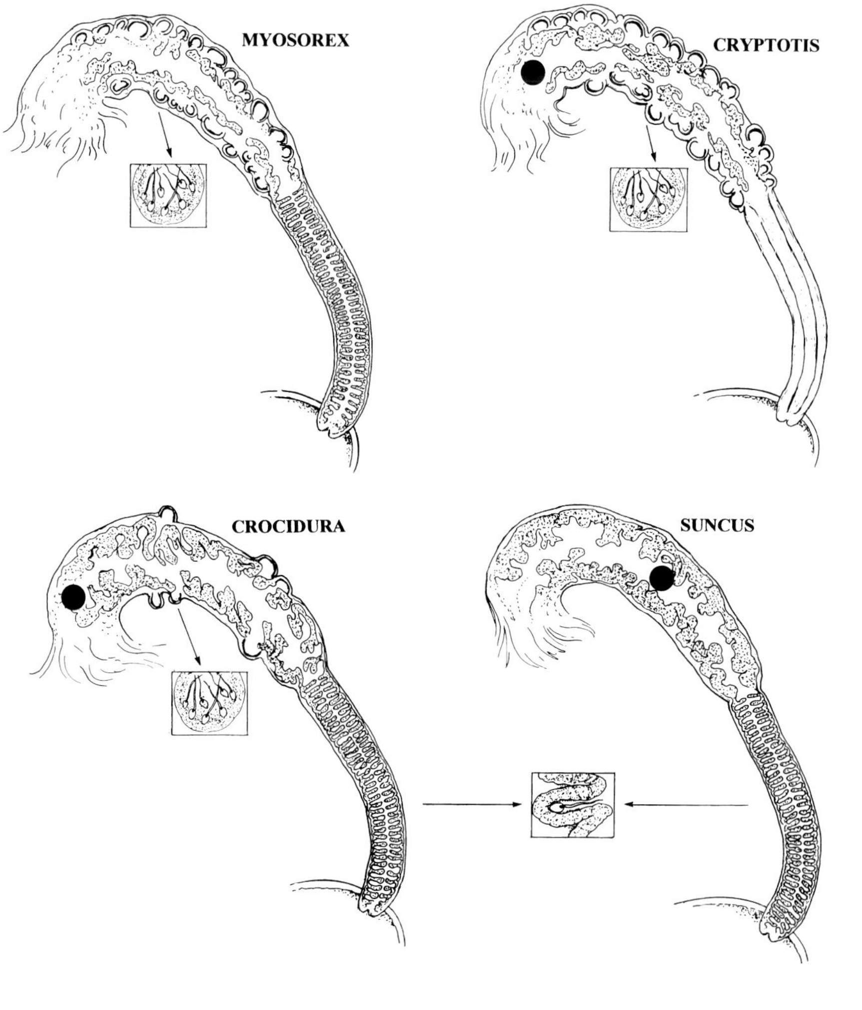

Figure 5. Diagram of the differing distribution of oviduct crypts in four soricid genera. In the typical soricine situation, represented by *Cryptotis*, spermatozoa are found only in ampullary crypts before and after ovulation. In *Crocidura*, some spermatozoa housed in isthmic crypts before ovulation migrate to sparsely distributed ampullary crypts after ovulation. In *Suncus*, spermatozoa are housed within isthmic crypts until ovulation when a very few, often no more than 10 to 20, eventually migrate to the crypt-free ampulla. In *Myosorex*, spermatozoa were observed only in ampullary crypts, notwithstanding the morphology of the isthmus. The black dots indicate the approximate site of fertilization where this has been observed—in the uppermost region where spermatozoa are maintained within and finally released from ampullary crypts, and in the mid-ampulla where this is finally colonized by only a small number of spermatozoa.

therian mammals. From a biological perspective, perhaps the most novel tactic for regulation of sperm numbers reaching the fertilization site is that used by higher mammals, involving sperm head adhesion to and controlled release from the isthmic epithelium (Smith 1998; Suarez 1998).

The situation described here for some marsupials and Insectivora may seem to resemble that in elasmobranch fishes, reptiles and birds. However, oviduct crypts function somewhat differently in these subtherian vertebrates in that they allow fertilization at successive ovulations without additional matings (Howarth 1974; Gist, Jones 1987). Moreover, sperm release from the crypts in marsupials and shrews appears to be coordinated by humoral factors from ovulating follicles, as in higher mammals, rather than through pressures brought by the physical presence of the egg, as postulated for reptiles and birds. In the case of a marsupial mouse, *Sminthopsis*, only after eggs have arrived in the upper oviduct do some spermatozoa ascend to the fertilization site, with a majority remaining in isthmic crypts until they are cleared in the hours after fertilized eggs have passed on into the uterus (Bedford, Breed 1994). Although details of the situation are somewhat variable in shrews (Fig. 5), each system again ensures that very few spermatozoa are present at the site of fertilization as this is occurring. On the other hand, there is no clue yet as to how regulated sperm release from the marsupial and insectivore oviduct crypts is actually brought about.

Although the crypts operate to effect a coordinated release of some sperm up to the fertilization site in both groups, it is important to appreciate that the role of these crypts in marsupials is probably rather different from that in shrews. In the few marsupials studied, isthmic crypts seem to actually "store" spermatozoa; that is, preserve them for future use. The morphology of the crypt epithelium clearly differs from that of the main lumen, suggesting the provision of a special environment there, and sheaves of viable spermatozoa are commonly encrypted in an immotile state (Bedford, Breed 1994), over a period that may last for several days (Selwood, McCallum 1987; Breed et al. 1989).

By contrast, DIC optics reveal that shrew spermatozoa are always motile within oviduct crypts, moderately so in isthmic crypts and very actively in ampullary crypts. Furthermore, although the preliminary mating interactions can be quite prolonged in some soricines, shrews often display a relatively short interval of no more than 13 to 16 hours from ejaculation to ovulation (or from hCG to ovulation), with a limit of 22 hours in *Myosorex*; and spermatozoa are cleared from the oviduct after no more than about 30 hours. This and their variable arrangement gives the impression that such crypts may really be acting in shrews and, by inference from oviduct anatomy, also in moles, not as storage receptacles but as sperm "traps," serving as a means of regulating the number of spermatozoa that reach the site of fertilization until the block to polyspermy is well-established by the fertilizing spermatozoon (Bedford et al. 1997b).

Sperm Capacitation. In higher mammals, the sperm association with the isthmic epithelium has been suggested either to promote (Chian, Sirard 1995) or to inhibit capacitation (Murray, Smith 1997), and capacitation may be required for the release of spermatozoa from the epithelium (Smith, Yanagimachi 1991). In the case of marsupials, *in vitro* studies have suggested that their spermatozoa require some form of capacitation (Mate, Rodger 1991), and a capacitation-like development of the ability to fertilize is demonstrated by the fact that spermatozoa of the dasyurid marsupial, *Sminthopsis*, cannot even adhere to the zona before they move from isthmic crypts to the milieu of the upper oviduct (Bedford, Breed 1994).

Experience from *in vitro* fertilization studies suggests a need for some form of capacitation in shrews as well, yet there is no evidence of sperm adherence to the oviduct epithelium in this group either. Thus, in marsupials and insectivores with oviductal sperm crypts, the process of capacitation, the viability of spermatozoa and their movement to the fertilization site appear to be independent of any direct association with the tubal epithelium.

Sperm Numbers. In recent years, beginning with Cohen (1967), the question has been asked repeatedly why many mammals ejaculate apparently excessive numbers of spermatozoa—for example, as judged by the performance of diluted samples used for artificial insemination. There is increasing interest at present, especially in relation to primates, in the notion of this apparent excess being related to sperm competition, that is, that sperm production rates have been driven upward during evolution by the "raffle" concept first developed for insects by Parker (Stockley, Purvis 1993).

In eutherian mammals, however, this equation is complicated by other very significant factors. As one example, heterospermic inseminations have revealed both an inherent fertilizing superiority of some males' spermatozoa over others, and an effect of the time of insemination in relation to that of ovulation (Dziuk 1996). Furthermore, the storage of a relatively large population in the cauda epididymidis may often relate to the need and ability to be able to effectively impregnate several females within a short period of time, rather than to simply to deliver numbers that will outcompete those of another male.

While the observations are yet too limited to draw strong inferences in regard to sperm numbers, the situation of marsupials suggests that the specific way in which spermatozoa relate to the female tract may have been a significant determinant of sperm production in this group. Marsupials that utilize isthmic crypts produce and ejaculate relatively few spermatozoa, yet a relatively high proportion of those ejaculated (about 10%) and a high absolute numbers (50 to 200 x 10^3) reach the fallopian tube (Bedford et al. 1984; Bedford, Breed 1994). By contrast, according to the very limited observations made so far, those marsupials whose oviducts do not appear to have developed true crypts (such as macropods and phalangers) display much higher sperm production rates, as judged from data compiled by Taggart et al. (1998).

Thus, it may prove that differentiated oviductal storage crypts permit an economy of sperm function in the marsupial female tract, and so a more a modest contribution by the male. Although some genera that exhibit high sperm production rates are promiscuous—for example, tammar wallabies (Rudd 1994)—such group differences in oviduct anatomy hint that intermale competition has not been the only major determinant of sperm production rates in marsupials.

Since some shrew litters are clearly the result of multiple paternity (Searle 1990), it is very interesting in considering male-male competition that shrews nevertheless produce relatively few spermatozoa. Regardless of the precise distribution of their oviductal crypts, small shrews such as *Cryptotis parva, Sorex cinereus, Crocidura russula,* and even the bigger *Suncus murinus* weighing between 35 and 100 g according to strain, house only ~10 x 10^6 spermatozoa in the post-testicular male tract, and, as judged by direct observation in *Suncus* (Bedford et al. 1994, 1997b), many probably ejaculate only ~1 x 10^6 or less. At first, therefore, it might seem that the principles suggested above for marsupials (that is, low numbers ejaculated as a correlate of oviductal crypts) operate in shrews as well. However, not only are rather more spermatozoa (55 to 60 x 10^6) found in the male tract of *Myosorex varius* and *Blarina brevicauda*, in which females house spermatozoa in oviductal ampullary crypts (Bedford et al 1998; OB Mock and JM Bedford, unpublished observations), but the situation presented by true moles, compared to some marsupials, seems to demolish any idea that oviductal sperm crypts may be linked to low sperm numbers in insectivores.

In moles, the ampulla is characterized by differentiated crypts which have a very similar appearance to those in soricine shrews, both in regard to their ciliary tufts and the occasional presence of leukocytes within the crypts, and it can be inferred that these crypts house spermatozoa. Yet, the epididymides of moles can contain huge numbers of spermatozoa. For example, the star-nosed mole, weighing about 50 g, may have about 3000 x 10^6, and the eastern mole, weighing about 100 g, may have some 600 x 10^6 spermatozoa (JM Bedford, OB Mock, GE Olson, SK Nagdas and VP Winfrey, unpublished data). Comparably large sperm numbers are present in the European mole (Racey 1978).

Unusual Sperm Interactions with the Egg

"Normal" Patterns

Over the last 30 years, observations in a small number of higher mammals have given rise to a concept of the "norm" in regard to the way that eutherian spermatozoa approach and penetrate the egg. As judged by fertilizing spermatozoa within eggs still invested by cumulus oophorus, this cell mass usually persists throughout the sperm penetration phase of fertilization, though not necessarily in domestic ruminants. Capacitation often appears to be a requisite for cumulus penetration, but spermatozoa that have already undergone the acrosome reaction generally do not seem able to penetrate the intact cumulus. Whether spermatozoa usually penetrate the cumulus matrix through the agency of a periacrosomal surface hydrolase (Li et al. 1997) is a question that remains to be decided. Moreover, the precise pattern of sperm behavior within the cumulus may vary somewhat.

In the mouse, most spermatozoa are believed to remain intact until they reach the egg, with the acrosome reaction occurring only after they bind to the zona pellucida, but in the rabbit and hamster at least some spermatozoa enter the membrane fusion phase of the acrosome reaction within the cumulus (Bedford 1968; Cummins, Yanagimachi 1982), which appears to produce some activating component in that regard (Boatman, Robins 1991). However, with the apparent exception of the Chinese hamster (Yanagimachi et al. 1983), in all the higher mammals examined cumulus-free oocytes can be fertilized readily *in vitro* and *in vivo*.

In other words, regardless of the precise site of the acrosome reaction for each spermatozoon of each species, among higher mammals the eutherian zona pellucida seems able to induce an appropriate acrosome reaction (Wassarman 1988). As a corollary, this reaction and subsequent penetration of the zona pellucida generally requires that spermatozoa first bind to zona receptors by specific ligands expressed on the periacrosomal plasmalemma (O'Rand 1988).

Variant Patterns

Though established for only a small number of species, it has seemed reasonable to assume that the pattern outlined above for higher mammals reflects the situation in all Eutheria. However, the initial steps of sperm/egg interaction seem to be rather different in shrews, and, if the morphology of the perforatorium is any indication, perhaps in some other mammals as well.

Unlike the cumulus oophorus of most eutherian species, that of shrews presents as a discrete round ball of tightly packed cells (Fig. 6) associated still by junctional complexes and devoid of any visible intercellular matrix (Bedford et al. 1994). In the oviduct, these aggregates remain as individual entities rather than coalescing to form the one large cumulus mass seen in many polyovular mammals. Even so, the postovulatory behavior of the cumulus clearly differs in soricine and crocidurine shrews. In soricine shrews the cumulus enlarges some five to eight hours after ovulation as its cells secrete a matrix in which many ampullary spermatozoa ultimately become enmeshed (Bedford et al. 1997a; OB Mock and JM Bedford, unpublished observations). By contrast, the more stable cumulus in crocidurine shrews (Fig. 6) remains almost unchanged for 10 hours or more after ovulation or fertilization, except that the corona layer soon draws away from the zona to form a distinct 'peri-zonal space' (Bedford et al. 1997c).

In studying the role of the cumulus in the soricine shrew, *Cryptotis*, it was found that 67.5% of eggs with some cumulus remaining were fertilized *in vitro*, whereas in seven parallel experiments no cumulus-free eggs were penetrated, though these had intact spermatozoa adhering to the zona surface (Bedford et al. 1997a). This indication that the soricine cumulus is essential for fertilization, perhaps in induction of the acrosome reaction, is a radical departure from the perceived norm for eutherian mammals. However, this result is complemented by observations in the crocidurine shrews, *Suncus murinus* and *Crocidura russula*, whose giant acrosomes can be assessed easily in the light microscope. In *Suncus*, all spermatozoa seen within cumuli had lost the acrosome, whether moving or not, whereas motile ampullary spermatozoa outside the cumulus were intact (Bedford et al. 1997c). There was no evidence either of an acrosome in single spermatozoa adhering tenaciously to the zona of unfertilized eggs within the peri-zonal space (Fig. 7). Intact spermatozoa adhered to the zona in *Suncus* only where mating was long delayed after hCG-induced ovulation, with consequent prior loss of the cumulus. A similar picture of acrosome loss was evident ultrastructurally in *Crocidura* (Bedford et al. 1997d).

That the acrosome was no longer present in any of a total of some hundreds of spermatozoa seen within crocidurine cumuli again indicates that this cell mass induces the acrosome reaction in such shrews. How then can fertilizing spermatozoa bind to the zona in the absence of the periacrosomal plasmalemmal ligands for zona receptors? In *Suncus*, the perforatorium is decorated by an elaborate array of barbs that are exposed by the acrosome reaction (Phillips, Bedford 1985), and it appears that fertilizing spermatozoa attach to the zona surface by way of these barbs (Fig. 8).

With exceptions such as that in hydromyine rodent spermatozoa (Breed 1997), its anatomy indicates that the eutherian perforatorium falls into one of two groups. Among the "higher mammals" in which it has been described (such as primates, ungulates, rodents, lagomorphs, carnivores and microchiroptera), the

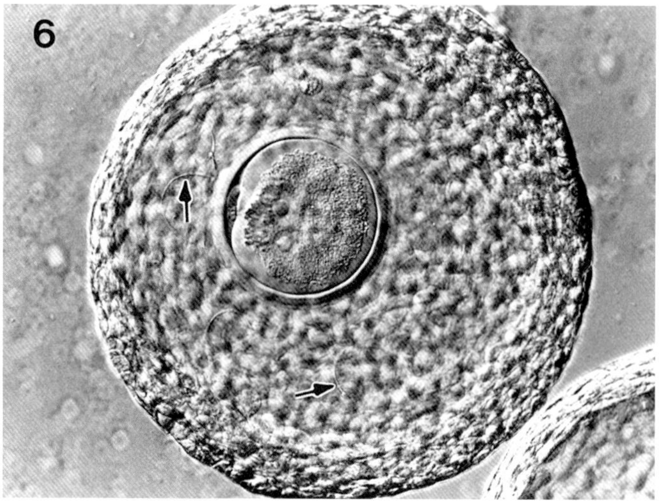

Figure 6. A slightly flattened cumulus/oocyte complex of *Suncus* collected from a mated female about eight hours after ovulation. Although this persists as a discrete matrix-free ball of cells, in the hours after ovulation the inner layers retreat from the zona, forming a peri-zonal space. As seen in a DIC microscope, the egg displays round nucleoli of the male and female pronucleus, and a few spermatozoa (arrows) within the cumulus are located within the peri-zonal space. x250. (T Mori, JM Bedford, unpublished.)

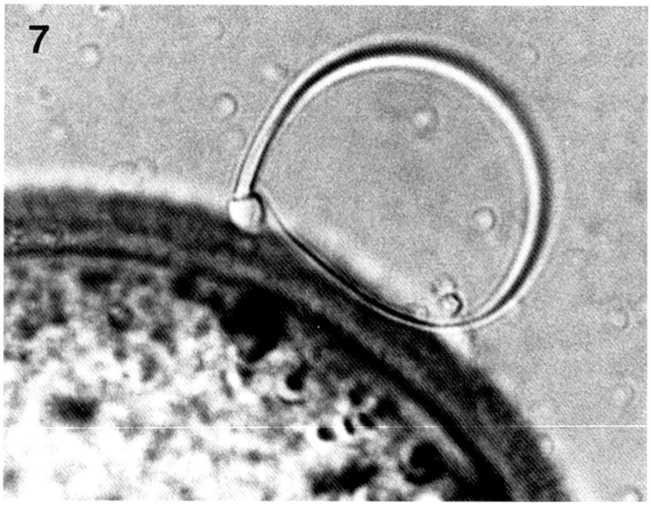

Figure 7. Unfertilized *Suncus* oocyte released from the peri-zonal space of a cumulus taken about 10 hours after ovulation from a mated female. The presumptive fertilizing spermatozoon was adhering tenaciously to the zona, but had no acrosome remnant. x1,200. (From Bedford et al. 1997c.)

perforatorium usually presents as an extension of the inner acrosomal membrane occupied by an -S-S-crosslinked matrix, and in spatulate heads this extension arches smoothly across the rostral surface of the nucleus. The perforatorium proves to be configured differently in what mostly are considered as more primitive mammals. The elaborate array of barbs first noted in *Suncus* spermatozoa occurs in *Crocidura* (Fig. 9), and with fewer barbs in spermatozoa of soricine shrews, too (Mori et al. 1991; Bedford et al. 1997a).

However, the perforatorium proves to be barbed in other Insectivora as well, notably in moles (JM Bedford, OB Mock, GE Olson, SK Nagdas and VP Winfrey, unpublished data), the African hedgehog (*Erinaceus [Atelerix] albiventris*) and a golden mole (*Amblysomus hottentus*), shown in Figure 10. However, this character is not confined to the Insectivora. A barbed perforatorium is present also in elephant shrews (Woodall 1991), and in the fruit bats, *Pteropus poliocephalus* (Rouse, Robson 1986), and *Pteropus dasymallus* (T Mori, personal communication) and apparently in canids, too (wolf: Fig. 1, in Koehler et al. 1998; dog: JM Bedford, unpublished observations).

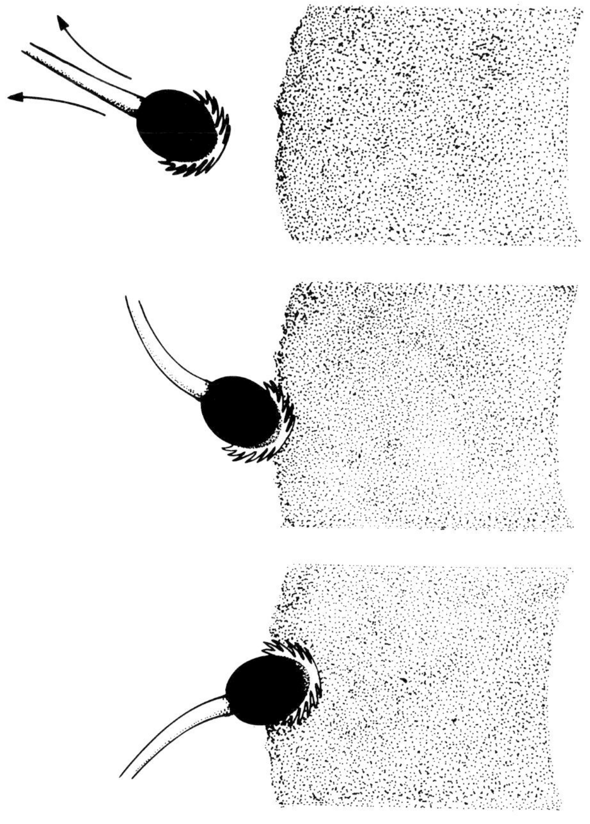

Figure 8. Reacted *Suncus* spermatozoa within the peri-zonal space are believed to attach to the zona surface by way of barbs on the perforatorium. Such barbs have been observed so far occurring in other Insectivora, elephant shrews and megabats. (From Bedford et al. 1997c.)

The Implications of Variant Sperm/Egg Interactions

Why the cumulus oophorus persists at ovulation in eutherian mammals is a matter for debate still. This apparently disperses quite soon after ovulation in some ruminants at least (PJ Dziuk, personal communication). However, in the many animals in which the cumulus remains for a variable period, it may act to promote gamete interaction (Harper 1970), by sequestering the very few spermatozoa that first reach the fertilization site in the ampulla (Bedford, Kim 1993). This is a role that may be optimized by the preovulatory mucification and expansion seen particularly in species characterized by a spacious ampulla (Bedford 1996).

In accord with this, there is no expansion of the persistent cumulus before fertilization in shrews and canines, in both of which the upper oviduct is relatively narrow. In the case of marsupials where the upper oviduct diameter hardly exceeds that of the eggs, the cumulus is shed before ovulation. It therefore seems likely that, while not essential in higher mammals, the cumulus often acts to promote sperm-egg contact in this group, and that its presence and/or particular character may reflect both the dimensions of the fertilization site and the sperm number there.

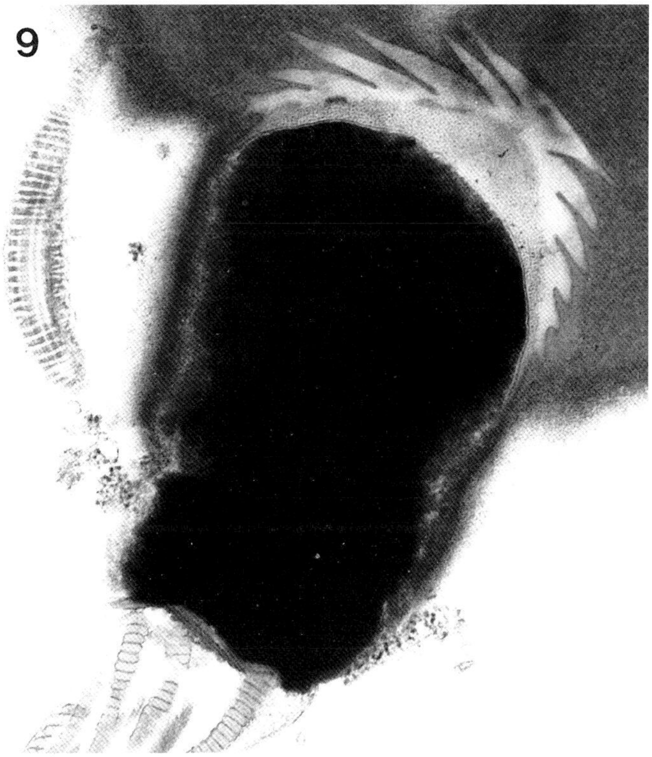

Figure 9. Electron micrograph of the sperm head in the shrew, *Crocidura dzinezumi*. Note the highly barbed configuration of the perforatorium. x15,100. (Used with permission of the publisher from Mori et al. 1991.)

Figure 10. Electron micrograph of a sperm head from the cauda epididymidis of a golden mole, *Amblysomus hottentotus*. Loss of the acrosome has exposed the barbs of the perforatorium, which the plane of section shows here on one side of the head. x33,000. (JM Bedford, RTF Bernard, unpublished.)

In contrast to the situation observed in higher mammals, in shrews and other more primitive mammals the cumulus appears to be involved in more than promotion of sperm-egg contact. *In vitro* fertilization experiments with cumulus-invested versus naked eggs in the soricine, *Cryptotis*, and direct observations of acrosome behavior in two crocidurine shrews indicate that the unusually dense circumscribed cumulus has an essential role in this group, probably as the inducer of the acrosome reaction. As a corollary, it appears that the acrosome reaction is not stimulated by sperm binding to the zona, in soricids at least.

In considering the implications of results obtained with shrews, it must be appreciated that soricids are generally believed to represent existing links to the ancestors of modern Eutheria. Therefore, it is possible that the cumulus evolved first as an essential element, perhaps as a correlate of major changes in the character of the zona, and that this role was later modified in transition to the current situations that present in higher mammals.

Finally, the consistent loss of the acrosome from all *Suncus* spermatozoa within the cumulus, and in those adhering to unfertilized eggs, focuses attention on the barbed configuration of the perforatorium. Originally, this was suggested to function as internal support for the giant acrosome in *Suncus* (Phillips, Bedford 1985), or, in fruit bats, as an adjunct to acrosomal lysins (Cummins et al. 1988). However, a perforatorium with barbs is not necessarily a correlate of the size of the acrosome. Among quite different eutherian species with similar rostral extensions of the acrosome, shrews, moles, hedgehogs, elephant shrews and megabats possess a barbed perforatorium, but others, such as prosimian primates and some microchiroptera, do not.

It seems likely that, in the absence of the surface receptors that mediate sperm-zona binding in higher mammals, the barbs represent another means through which the reacted sperm head can attach to the zona pellucida. Given that the apparently essential role of the soricid cumulus represents a departure from the situation recognized in higher mammals, it will be interesting to discover how critical is the cumulus for fertilization in the other genera, such as moles and hedgehogs, whose spermatozoa display a barbed perforatorium.

Summary

There is no complete uniformity among therian or even eutherian mammals in regard to the behavior of spermatozoa in the oviduct, or in their initial interaction with the egg. Rather than associating with the epithelium of the oviduct isthmus as in higher mammals, in at least two marsupial families and in certain Insectivora, spermatozoa are housed in oviduct crypts as groups of free cells, with some individual spermatozoa being released to advance to the fertilization site in the peri- or post-ovulatory period. In the didelphid and dasyurid marsupials studied, the crypts appear to perform a true storage function that may even allow a conservation of sperm production by the male.

However, the oviductal crypts that house spermatozoa in shrews, and by inference in moles too, probably fill a somewhat different function. In shrews, the arrangement of the crypts can vary significantly according to the genus, and within the crypts the spermatozoa display a persistent motility and have only a relatively short tenure there. The oviduct crypts seen in shrews and moles probably constitute an alternate system whereby sperm transport to the fertilization site is regulated. On the other hand, the fact that moles produce so many spermatozoa speaks against the idea that such crypts are linked to the low sperm numbers that characterize some shrews, as they may prove to be in some marsupials

In regard to the interaction of spermatozoa with the egg, unlike the cumulus in most higher mammals, the characteristically compact matrix-free cumulus in shrews appears essential for fertilization, quite probably in induction of the acrosome reaction. As a corollary, rather than binding to the zona by way of conventional sperm surface ligands, after losing the acrosome, fertilizing spermatozoa of *Suncus* at least appear to attach to the zona by way of barbs arrayed over the perforatorium. Since such barbs decorate the perforatorium also in other insectivores, such as mole, hedgehog and golden mole, and in various elephant shrews and megabats at least, it will be of interest to discover whether the cumulus is critical for fertilization in these other eutherian mammals as well.

Acknowledgements

The author is very grateful to Drs. C.H. Tyndale-Biscoe (Canberra), D.A. Taggart (Melbourne), and W.G. Breed (Adelaide), for the opportunity to examine oviducts of various Australian marsupials; to Dr. O.B. Mock (Kirksville, Missouri) who helped in preparing some of that and also hedgehog material; and to Dr. K. Catania (Vanderbilt University) who collected the eastern and star-nose moles. Spermatozoa of the golden mole were obtained in collaboration with Dr. R.T.F. Bernard (Rhodes University, South Africa). Pauline M. Thomas provided the illustrations.

References

Afzelius BA. Sperm morphology and fertilization biology. In: (Beatty RA, Gluecksohn-Waelsch S, eds) The Genetics of the Spermatozoon. Copenhagen: Bogtrykkeriet Forum, 1972, pp131-143.

Arnold R, Shorey CD. Structure of the oviduct epithelium of the brush-tailed opossum (*Trichosurus vulpecula*). J Reprod Fertil 1985; 73:9-19.

Bakst MR, Wishart G, Brillard JP. Oviducal sperm selection, transport, and storage in poultry. Poultry Sci Rev 1994; 5:117-143.

Bedford JM. Ultrastructural changes in the sperm head during fertilization in the rabbit. Am J Anat 1968; 123:329-358.

Bedford JM. The coevolution of mammalian gametes. In (Dunbar BS, O'Rand MG, eds) A Comparative Overview of Mammalian Fertilization. New York: Plenum Press, 1991, pp3-35.

Bedford JM. What marsupial gametes disclose about gamete function in eutherian mammals. Reprod Fertil Dev 1996; 8:569-580.

Bedford JM, Breed WG. Regulated storage and subsequent transformation of spermatozoa in the Fallopian tubes of an Australian marsupial, *Sminthopsis crassicaudata*. Biol Rcprod 1994; 50:845-854.

Bedford JM, Kim HH. Cumulus oophorus as a sperm sequestering device. J Exp Zool 1993; 265:321-328.

Bedford JM, Rodger JC, Breed WG. Why so many mammalian spermatozoa—a clue from marsupials? Proc Roy Soc B Lond 1984; 221:221-223.

Bedford JM, Cooper GW, Phillips DM, Dryden GL. Distinctive features of the gametes and reproductive tracts of the Asian musk shrew, *Suncus murinus*. Biol Reprod 1994; 50:820-834.

Bedford JM, Mock OB, Phillips DM. Unusual ampullary sperm crypts, and behavior and role of the cumulus oophorus, in the oviduct of the least shrew, *Cryptotis parva*. Biol Reprod 1997a; 56:1255-1267.

Bedford JM, Mori T, Oda S. Ovulation induction and gamete transport in the female tract of the musk shrew, *Suncus murinus*. J Reprod Fertil 1997b; 110:115-125.

Bedford JM, Mori T, Oda S. The unusual state of the cumulus oophorus and of sperm behavior within it, in the musk shrew, *Suncus murinus*. J Reprod Fertil 1997c; 110:127-134.

Bedford JM, Phillips DM, Mover-Lev H. Novel sperm crypts and behavior of gametes in the fallopian tube of the white-toothed shrew, *Crocidura russula monacha*. J Exp Zool 1997d; 277:262-273.

Bedford JM, Bernard RTF, Baxter RM. The 'hybrid' character of the gametes and reproductive tracts of the African shrew, *Myosorex varius*, supports its classification in the Crocidosoricinae. J Reprod Fertil 1998; 112:165-173.

Boatman DE, Robins RS. Detection of a soluble acrosome-reaction-inducing factor, different from serum albumin, associated with the ovulated egg-cumulus complex. Mol Reprod Dev 1991; 30:396-400.

Breed WG. Evolution of the spermatozoon in Australasian rodents. Aust J Zool 1997; 45:459-478.

Breed WG, Leigh CM, Bennett JH. Sperm morphology and storage in the female reproductive tract of the fat-tailed dunnart, *Sminthopsis crassicaudata* (Marsupialia: Dasyuridae). Gam Res 1989; 23:61-75.

Chian RI-C, Sirard M-A. Fertilizing ability of bovine spermatozoa cocultured with oviduct epithelial cells. Biol Reprod 1995; 52:156-162.

Cohen J. Correlation between sperm 'redundancy' and chiasma frequency. Nature (Lond) 1967; 215:862-863.

Cummins JM, Yanagimachi R. Sperm egg ratios and the site of the acrosome reaction during *in vivo* fertilization in the hamster. Gam Rcs 1982; 5:239-256.

Cummins JM, Robson SK, Rouse GW. The acrosome reaction in spermatozoa of the grey-headed flying fox (*Pteropus poliocephalus*: Chiroptera) exposes barbed subacrosomal material. Gam Res 1988; 21:11-22.

Deanesly R. The reproductive processes of certain mammals. Part VI: The reproductive cycle of the female hedgehog. Phil Trans Roy Soc B Lond 1934; 223:240-276.

Dziuk PJ. Factors that influence the proportion of offspring sired by a male following heterospermic insemination. Anim Reprod Sci 1996; 43:65-88.

Franzen A. Sperm structure with regard to fertilization biology and phylogenetics. Verhandl Deutsch Zool Ges 1977:123-138.

Gist DH, Fischer EN. Fine structure of the sperm storage tubules in the box turtle oviduct. J Reprod Fertil 1993; 97:463-468.

Gist DH, Jones FM. Storage of sperm in the reptilian oviduct. Scanning Microsc 1987; 1:1839-1849.

Harper MJK. Factors influencing sperm penetration of rabbit eggs *in vivo*. J Exp Zool 1970; 173:47-62.

Howarth BJ. Sperm storage as function of the female reproductive tract. In: (Johnson AD, Foley CW, eds) The Oviduct and Its Functions. New York: Academic Press, 1974, pp227-237.

Hunter RHF. Oviduct function in pigs, with particular reference to the pathological condition of polyspermy. Mol Reprod Dev 1991; 29:385-391.

Hunter RHF. Ovarian endocrine control of sperm progression in the fallopian tubes. Oxford Rev Reprod Biol 1995; 17:85-124.

Jamieson BGM, Lee MSY, Long K. Ultrastructure of the spermatozoon of the internally fertilizing frog *Ascaphus truei* (Ascaphidae: Anura: Amphibia) with phylogenetic considerations. Herpetol 1993; 49:52-65

Kasinsky HE. Specificity and distribution of sperm basic proteins. In: (Hnilica L, Stein G, Stein J, eds), Histones and Other Basic Nuclear Proteins. Boca Raton, FL: CRC Press, 1989, pp73-163.

Koehler JK, Platz CC, Waddell W, Jones MH, Behrns S. Semen parameters and electron microscope observations of spermatozoa of the red wolf, *Canis rufus*. J Reprod Fertil 1998; 114:95-101.

Li M-W, Yudin AI, VandeVoort CA, Sabeur K, Primakoff P, Overstreet JW. Inhibition of monkey sperm hyaluronidase activity and heterologous cumulus penetration by flavonoids. Biol Reprod 1997; 56:1383-1389.

Mate KE, Rodger JC. Stability of the acrosome of the brush-tailed opossum (*Trichosurus vulpecula*) and tammar wallaby (*Macropus eugenii*) *in vitro* after exposure to conditions and agents known to cause capacitation or acrosome reaction of eutherian spermatozoa. J Reprod Fertil 1991; 91:41-48.

Metten H. Studies on the reproduction of the dogfish. Phil Trans Roy Soc B Lond 1941; 230:217-227.

Mori T, Arai S, Shiraishi S, Uchida TA. Ultrastructural observations on spermatozoa of the Soricidae, with special attention to a subfamily revision of the Japanese water shrew *Chimarrogale himalayica*. J Mammal Soc Japan 1991; 16:1-12.

Murray SC, Smith TT. Sperm interaction with fallopian tube apical plasma membrane enhances sperm motility and delays capacitation. Fertil Steril 1997; 68:351-357.

Nicoll ME, Racey PA. Follicular development, ovulation, fertilization and fetal development in tenrecs (*Tenrec ecaudatus*). J Reprod Fertil 1985; 74:47-55.

O'Rand MG. Sperm-egg recognition and barriers to interspecies fertilization. Gam Res 1988; 19:315-328.

Phillips DM, Bedford JM. Unusual features of sperm ultrastructure in the musk shrew, *Suncus murinus*. J Exp Zool 1985; 235:119-126.

Racey PA. Seasonal changes in testosterone levels and androgen-dependent organs in male moles (*Talpa europaea*). J Reprod Fertil 1978; 52:195-200.

Rodger JC, Bedford JM. Induction of oestrus, recovery of gametes, and the timing of fertilization events in the opossum, *Didelphis virginiana*. J Reprod Fertil 1982; 64:159-169.

Rouse GW, Robson S. An ultrastructural study of megachiropteran (Mammalia: Chiroptera) spermatozoa: implications for chiropteran phylogeny. J Submicrosc Cytol 1986; 18:137-152.

Rudd CD. Sexual behaviour of male and female tammar wallabies (*Macropus eugenii*) at post-partum oestrus. J Zool 1994; 232:151-162.

Searle JP. Evidence for multiple paternity in the common shrew (*Sorex araneus*). J Mammal 1990; 71:139-144.

Selwood L, McCallum F. Relationship between longevity of spermatozoa after insemination and the percentage of normal embryos in brown marsupial mice (*Antechinus stuartii*). J Reprod Fertil 1987; 79:495-503.

Shalgi R, Phillips DM. The motility of rat spermatozoa at the site of fertilization. Biol Reprod 1988; 39:1207-1213.

Smith TT. The modulation of sperm function by the oviduct epithelium. Biol Reprod 1998; 58:1102-1104.

Smith TT, Yanagimachi R. Attachment and release of spermatozoa from the caudal isthmus of the hamster oviduct. J Reprod Fertil 1991; 91:567-573.

Stockley P, Purvis A. Sperm competition in mammals: a comparative study of male roles and relative investment in sperm production. Functional Ecol 1993; 7:560-570.

Strauss F. Ripe follicles without antra and fertilization within the follicle: a normal situation in a mammal. Anat Rec 1950; 106:251-252.

Suarez SS. The oviductal sperm reservoir in mammals: mechanisms of formation. Biol Reprod 1998; 58:1105-1107.

Sultan KM, Bedford JM. Two modifiers of sperm transport within the fallopian tube of the rat. J Reprod Fertil 1996; 108:179-184.

Taggart DA. A comparison of sperm and embryo transport in the female reproductive tract of marsupial and eutherian mammals. Reprod Fertil Dev 1994; 6:451-472.

Taggart DA, Temple-Smith PD. Transport and storage of spermatozoa in the female reproductive tract of the brown marsupial mouse, *Antechinus stuartii* (Dasyuridae). J Reprod Fertil 1991; 93:97-110.

Taggart DA, Breed WG, Temple-Smith PD, Purvis A, Shimmin G. Reproduction, mating strategies and sperm competition in marsupials and monotremes. In: (Birkhead TR, ed) Sperm Competition and Sexual Selection. New York: Academic Press, 1998, pp623-665.

Tyndale-Biscoe CH, Rodger JC. Differential transport of spermatozoa into the two sides of the genital tract of a monovular marsupial, the tammar wallaby (*Macropus eugenii*). J Reprod Fertil 1978; 52:37-43.

van der Horst G, Wilson B, Channing A. Amphibian sperm: phylogeny and fertilization environment. In: (Jamieson BGM, Ausio J, Justine J-L, eds) Advances in Spermatozoal Phylogeny and Taxonomy. Mém Mus Natn Hist Nat 1995; 166:333-342.

Wassarman PM. Zona pellucida glycoproteins. Ann Rev Biochem 1988; 57:415-442.

Woodall PF. An ultrastructural study of the spermatozoa of elephant shrews (Mammalia: Macroscelidea) and its phylogenetic implications. J Submicrosc Cytol 1991; 23:47-58.

Yanagimachi R, Kamiguchi K, Sugawara S, Mikamo K. Gametes and fertilization in the Chinese hamster. Gam Res 1983; 8:569-580.

10

Interaction between Sperm and Epididymal Secretory Proteins

Robert Sullivan
University of Laval

Leaving the testis, mammalian spermatozoa transit along the excurrent duct system formed by the vasa efferentia, epididymis and vas deferens. The epididymis is a single, highly convoluted tubule. Sperm transit along this epithelial cell-lined tubule varies in time—from a few days in the human up to 12 days in the ram. Functions of the epididymis include the absorption of seminiferous fluid and the concentration, transport and storage of spermatozoa. Since the pioneering work of Bedford (1967) and Orgebin-Crist (1967), the epididymis is also recognized for its involvement in sperm maturation. Reviews covering different aspects of the epididymis can be found in Cooper (1986), Moore (1996) and Robaire and Hermo (1988). During epididymal transit, the male gamete undergoes major structural, biochemical and physiological modifications. These are dependent upon epididymal protein synthesis that is under androgen control. Indeed, epididymal development and secretory activity of this tissue are dependent on the presence of testosterone. Moreover, the epididymis contains abundant 5α-reductase enzyme activity that ensures a rich androgenic environment by producing the potent androgen dihydrotestosterone (DHT). Under androgen control, the epididymis ensures the luminal microenvironment necessary for sperm maturation and storage. The luminal fluid of the epididymis modifies the sperm plasma membrane by changing its lipid composition, by adding or post-translational processing surface proteins and by changing membrane fluidity. Taken together, these modifications are responsible for sperm maturation. It is during this process that the spermatozoon acquires its motility and the ability to interact with the zona pellucida surrounding the oocyte.

The epididymis is usually divided into three anatomical regions: the caput, corpus and cauda segments. The epididymal segment where spermatozoa acquire their fertilizing ability varies from one species to another. Generally, it is in the distal caput-proximal corpus region of the epididymis that the first functional male gametes can be found. Interestingly, it is in these regions that epididymal cells display the highest capacity for protein synthesis. In the hamster, the epididymis is highly developed; its segments can be easily identified and the reservoir capacity of the distal cauda epididymidis is considerable. Since large amounts of cauda spermatozoa can be obtained from sexually mature animals, the hamster provides an ideal model for studying epididymal function. The relative ease of performing *in vitro* fertilization using hamster gametes is advantageous for using this species in several experimental protocols. Indeed, many classical works on the epididymis and fertilization have been performed using the hamster (Horan, Bedford 1972; Yanagimachi, Chan 1964). During the last 10 years, this animal model has been used to better understand epididymal sperm maturation. Importantly, results obtained from the hamster allowed researchers to extrapolate their conclusions to human and other mammalian species.

P26h: A Hamster Sperm Antigen

One of the functions acquired by the mammalian spermatozoon during epididymal maturation is the acquisition of the ability to bind to the zona pellucida. The zona pellucida is an acellular, extracellular coat surrounding the mammalian oocyte (Wassarman, Litscher 1995). Sperm-zona pellucida recognition involves a receptor-ligand mechanism (McLeskey et al. 1998). In many species, this step in the process of gamete interaction is a species-specific phenomenon; that is, spermatozoa can interact with the zona pellucida surrounding the oocyte only if they are from the same species (Yanagimachi 1994). This has been particularly well documented in the hamster. Moreover, it has also been demonstrated that, in this species, protein synthesis activity of the epididymis, mainly the corpus segment, is necessary for the acquisition of zona pellucida recognition by maturing spermatozoa (Cuasnicú et al. 1984). In order to document the role of the epididymis in the acquisition of sperm fertilizing ability, researchers in the Sullivan laboratory first identified a zona pellucida ligand on mature hamster spermatozoa (Sullivan, Bleau 1985). Western blots of detergent extracted sperm proteins were probed with iodinated glycoproteins from homologous zonae pellucidae. Autoradiograms of the Western blots revealed that hamster zona pellucida glycoproteins preferentially bind to a 26 kD protein: P26h. This result was functionally significant since, under the same experimental conditions, the hamster I^{125} zona pellucida glycoproteins did not interact with spermatozoa protein extracts from other species. Since the publication of these results, similar experimental strategies have been used by others. In these studies, sperm proteins of different molecular weights have been described. The discrepancies between these studies can be attributed to species differences. Sperm-zona pellucida interactions are probably mediated by a more complex biochemical mechanism than described by the receptor-ligand model. This is supported by the large number of sperm proteins proposed to be involved in zona pellucida binding (McLeskey et al. 1998). Although sperm zona pellucida interactions have been the subject of intense research for more than 20 years, a comprehensive model of this critical step in the process of mammalian fertilization has yet to be described (Aitken, Irvine 1995).

Despite the complexity of sperm-zona pellucida interactions, the species-specific affinity of zona pellucida glycoproteins for P26h suggests that this sperm antigen is likely to be involved in this key event of fertilization, at least in the hamster. In order to test this hypothesis, researchers in the Sullivan laboratory have raised antibodies against P26h purified from hamster cauda epididymal spermatozoa (Coutu et al. 1996; Bérubé et al. 1994). When used to probe hamster sperm proteins separated by two-dimensional gel electrophoresis, these antibodies immunodetected a single protein spot. The P26h antibodies did not detect any proteins when used to probe Western blots of proteins extracted from different hamster somatic tissues such as the spleen, lung, liver, heart, kidney, brain, fat and muscle. Thus, P26h protein is highly specific for spermatozoa and the male reproductive tract (Bérubé, Sullivan 1994). On the other hand, proteins showing common antigenicity with the hamster P26h could be detected in spermatozoa from other mammalian species. Considering the proposed function of this sperm protein in zona pellucida recognition, the tissue specificity of P26h was predictable.

The mammalian spermatozoa is a highly polarized cell with clearly defined membrane domains. This is well illustrated by the restricted distribution of some sperm membrane proteins and antigens. Lipid composition also varies from one membrane domain to another (Eddy, O'Brien 1994). Immunocytochemical studies revealed that the P26h protein is restricted to the plasma membrane covering the acrosomal cap of cauda epididymal hamster spermatozoa. P26h does not undergo membrane relocalization during capacitation. It is clear, however, that it disappears from the sperm surface following the acrosome reaction, an exocytotic event that takes place after sperm-zona pellucida binding. Thus, the localization of P26h is in agreement with the hypothesized involvement of this protein in hamster sperm-zona pellucida interactions. In order to ascertain the role of P26h in fertilization, IgGs purified from either preimmune rabbit serum or P26h antiserum were added to a sperm-zona pellucida binding assay. The presence of preimmune IgGs had no effect on the number of capacitated spermatozoa bound to the egg's zona pellucida. In contrast, anti-P26h IgGs strongly inhibited sperm-zona pellucida interactions in a dose-dependent fashion. Fab fragments generated from the anti-P26h antibodies were as efficient as intact IgGs, thus demonstrating that the inhibition of sperm-zona pellucida binding by the P26h antibodies was not simply due to steric hindrance of some other surface component of the hamster spermatozoon (Bégin et al. 1995; Bérubé, Sullivan 1994). The inhibitory effect of antibodies to document the function of sperm proteins in different steps of fertilization has been extensively used by others (McLeskey et al. 1998; Snell, White 1996).

P26h appears to be highly specific to spermatozoa and apparently plays a key role in fertilization. These two properties suggest that this sperm protein could be a potential target for inducing immunological infertility in hamsters (Boué, Sullivan 1995: Primakoff 1994). P26h was purified from cauda epididymal spermatozoa protein extracts and used to actively immunize fertile male hamsters. Considering the tissue specificity of this protein and the fact that a blood-testis barrier protects the male gamete from an immune attack, a humoral response was expected if a sperm protein was injected outside the reproductive tract. Circulating anti-P26h antibodies were in fact detected in the hamsters immunized against P26h. The circulating antibodies recovered from the serum of immunized animals were able to immunoreact with P26h in Western blots and ELISA as well as in immunocytochemical assays. Interestingly, circulating antibodies resulting from the immunization of hamsters with P26h reached spermatozoa within the epididymal lumen. When high titers of circulating anti-P26h antibodies were reached, cauda epididymal spermatozoa of immunized males were coated with antibodies, as demonstrated by incubating freshly collected spermatozoa with an anti-hamster IgG coupled to fluorescein. This antibody coating was restricted to the acrosomal cap where P26h is localized. Thus, these antibodies crossed the blood-testis barrier, a structure which was thought to be efficient at the level of the epididymis. It is possible that the antibodies reached the epididymal lumen at the level of the vasa efferentia or the proximal caput, where the blood-epididymal barrier has been suggested to be leaky. The fact that circulating antibodies can reach the epididymal lumen raises the possibility that epididymal sperm antigens can be ideal targets for potential immunocontraception in the male.

Interestingly, anti-P26h antibodies associated with the membrane domain covering the acrosome remain on spermatozoa of immunized males after mating. The presence of anti-P26h antibodies inhibited *in vivo* fertilization in a manner similar to that observed in the *in vitro* sperm-zona pellucida binding assay. In fact, immunized males can be mated with superovulated females and motile spermatozoa are found in the reproductive tract of mated females. No fertilization occurs in superovulated females mated with hamsters immunized with a dose as low as 1 μg/animal. This contrasts with a 97% fertilization rate obtained with a control group of male hamsters immunized against purified hamster albumin and mated with superovulated females (Bérubé, Sullivan 1994). The histology of the testis and excurrent duct system of the males immunized against P26h had a normal appearance and circulating testosterone levels were comparable to those measured in sexually mature control animals. Taken together, these data suggest that the P26h sperm protein is a potential candidate for immunocontraception in the male for the following reasons: 1) males respond by producing circulating antibodies, 2) P26h antibodies reach the sperm surface within the epididymal lumen, 3) antibodies remain associated with sperm until they reach the egg surface, 4) antibodies inhibit *in vivo* fertilization without affecting sperm morphology or motility and 5) immunization has no effect on testicular function.

Considering the proposed function of P26h and the fact that hamster spermatozoa acquire their zona pellucida binding ability during epididymal transit, researchers in the Sullivan laboratory investigated the presence of P26h on spermatozoa collected at different levels of the epididymis. Western blots of proteins extracted from spermatozoa collected along the epididymis as well as immunofluorescence studies showed that P26h was barely detectable on spermatozoa collected in the proximal region, whereas P26h accumulated gradually on spermatozoa in the more distal regions of the tissue. During this gradual accumulation along the epididymis, P26h is strictly localized on the acrosomal cap (Fig. 1) and shows the same electrophoretic behavior in sodium dodecyl sulfacte-polyacrylamide gel electrophoresis (SDS-PAGE), as seen in Sullivan, Robitaille (1989). This contrasts with other sperm surface proteins that have been shown to undergo sperm surface relocalization during epididymal transit (Yanagimachi 1994). P26h is also found in soluble form in the luminal fluid collected along the epididymis. In a similar fashion to the accumulation of P26h at the surface of maturing spermatozoa, the concentration of P26 in the luminal fluid decreased during epididymal transit (Robitaille et al. 1991).

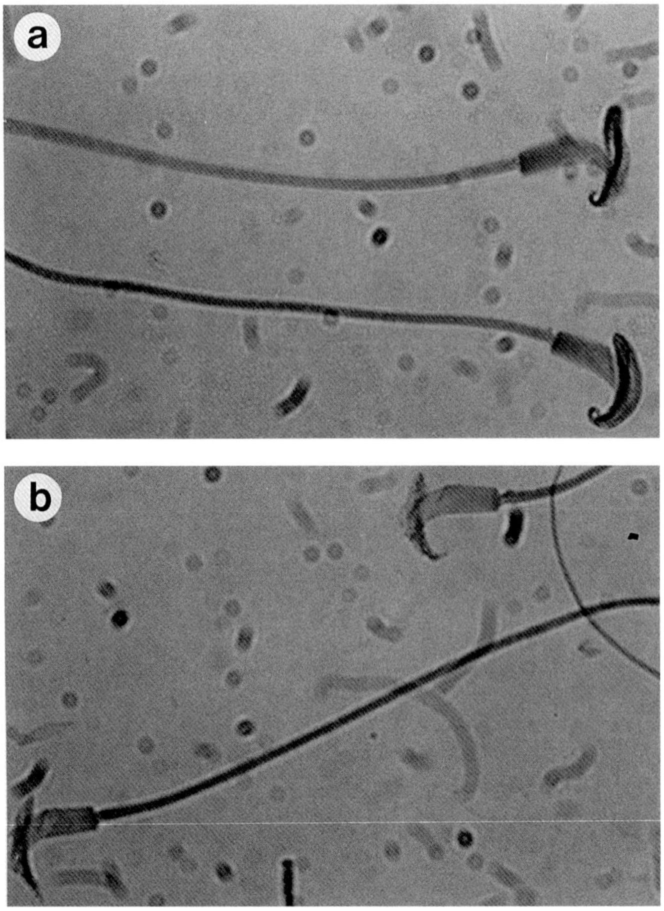

Figure 1. Immunocytochemical localization of P26h on hamster spermatozoa collected from the cauda (a) and caput (b) epididymidis. Positive staining obtained with the P26-antiserum appears dark. P26h is restricted to the acrosomal cap of cauda epididymal spermatozoa. P26h is absent from caput spermatozoa.

The properties of P26h are summarized as follows: 1) P26h is a glycoprotein of 26 kD in SDS-PAGE, 2) P26h is localized on the plasma membrane covering the acrosome, 3) P26h is shows species-specific affinity for zona pellucida glycoproteins, 4) anti-P26h IgGs inhibit sperm-zona pellucida binding *in vitro*, 5) active immunization against P26h inhibits *in vivo* fertilization, 6) P26h is acquired during epididymal transit and 7) P26h is unique to the male reproductive tract.

This protein is added to spermatozoa during post-testicular maturation. Considering the function of this protein in the process of fertilization, P26h can be considered as a marker of epididymal maturation for hamster spermatozoa. Using the highly specific polyclonal antiserum raised against P26h, the presence of an antigenically related sperm protein in other mammalian species, especially in the human, was sought.

P34H: An Analogous Human Protein

When used to probe Western blots of proteins extracted from ejaculated human spermatozoa, the P26h antiserum revealed a minor protein with a molecular weight of 34 kD and an isoelectric point of 6.0-6.2. Depending on the extraction conditions used to prepare human sperm proteins, an additional band with a slightly faster electrophoretic mobility could also be detected. This protein resulted from the proteolytic degradation of the more prominent P34H band: the human sperm protein, with the capital "H" standing for Human (Boué et al. 1994).

P34H immunocytochemical localization was performed on washed ejaculated spermatozoa from fertile donors. Researchers were at first puzzled by the fact that the P26h antiserum did not stain human spermatozoa. Spermatozoa have to be capacitated in order to allow for good immunostaining. Following *in vitro* capacitation, immunostaining revealed that P34H was restricted to the acrosomal region of human spermatozoa (Fig. 2). This observation was in agreement with the fact that mammalian spermatozoa must be capacitated in order to be able to bind to the homologous zona pellucida. Some decapacitating factors added to sperm during ejaculation can mask surface proteins involved in this binding process; P34H is one of these proteins. Following the

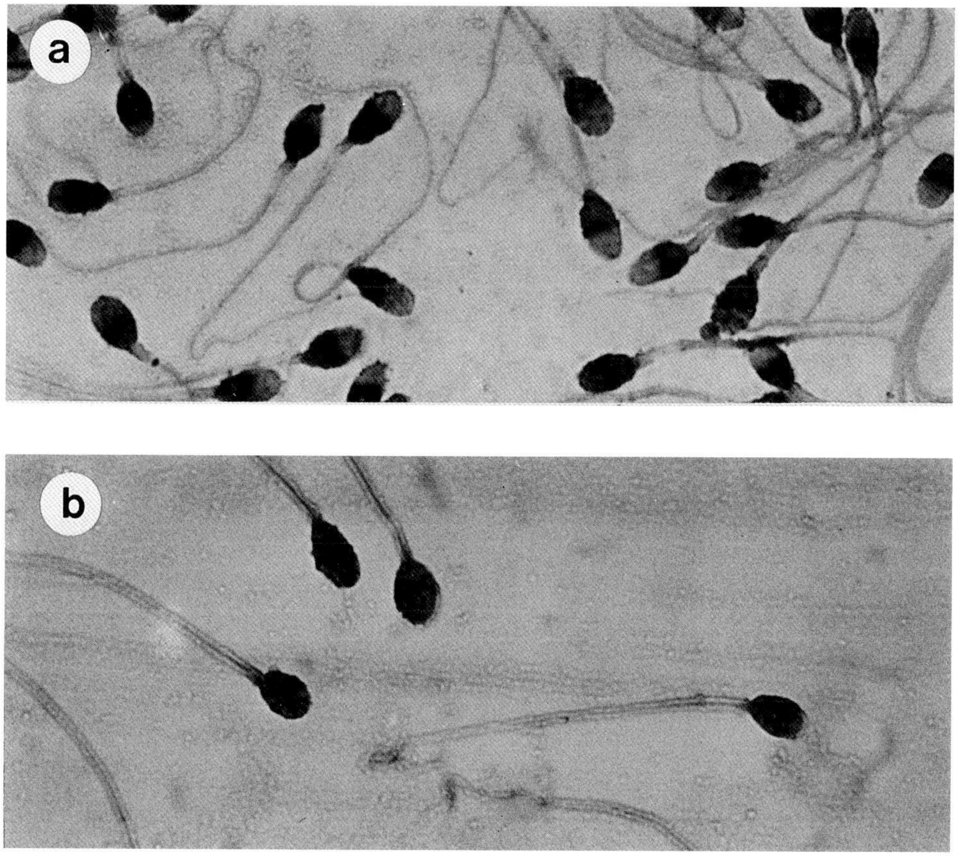

Figure 2. Immunocytochemical localization of P34H on ejaculated human spermatozoa. Immunostaining was performed on washed (a) and capacitated (b) spermatozoa. The P34H labeling is much more evident on capacitated than on washed spermatozoa. Positive staining appears dark.

acrosome reaction, either spontaneous or induced with a calcium ionophore, P34H no longer associates to the sperm surface. Again, this correlates well with the fact that human spermatozoa have to be acrosome-intact in order to bind to the zona pellucida (Boué et al. 1996).

In general, the interaction between spermatozoa and the zona pellucida is considered to be species-specific; that is, only spermatozoa from the same species can efficiently interact with the egg's zona pellucida. In order to document the function of P34H in human fertilization, an *in vitro* sperm-zona pellucida binding assay was developed using human zonae pellucidae obtained from therapeutic *in vitro* fertilization programs. The zonae pellucidae are obtained by aspirating follicles of different diameters. It is important to point out, however, that under these conditions, not all of the cumulus-oocytes complexes obtained are necessarily at the same maturation status. Human spermatozoa can bind to zonae pellucidae of immature oocytes; the efficiency of gamete interactions, however, varies from one oocyte to another. A protocol, modified from Liu et al. (1988), that uses each zona pellucida as its own control was established. This binding assay is performed by incubating the zonae pellucidae with both control spermatozoa and spermatozoa preincubated with an antiserum. The number and the type of spermatozoa bound to each zona pellucida can than be evaluated since each type is differentially stained with vital dyes. Using this protocol, it was demonstrated that the anti-P26h serum, which cross-reacts with human P34H, inhibits human sperm-zona pellucida binding *in vitro* (Fig. 3). This effect was specific to sperm-zona pellucida binding, since the antiserum did not affect other steps of fertilization. P26h antiserum did not interfere with sperm motility parameters as determined by computer assisted sperm analysis (CASA) performed on spermatozoa before and after capacitation. Incubation of spermatozoa with the antiserum had no effect on sperm-egg plasma membrane binding or penetration, as evaluated by the zona-free hamster egg penetration test. Also, the antiserum had no effect on the incidence of spontaneous or induced acrosome reaction, as determined by Pisum sativum-fluorescein staining. Taken together, these results clearly demonstrate that P34H is involved in sperm-zona pellucida binding in humans and that antibodies directed against this sperm surface protein can inhibit this step of fertilization *in vitro* (Boué et al. 1994).

The determination of the origin of P34H along the human reproductive tract was a difficult task. Human tissues that are available for research purposes are frequently of poor quality, especially if the sample is to be used to isolate RNA. Advanced prostatic cancers are usually the cause of surgical orchidectomies.

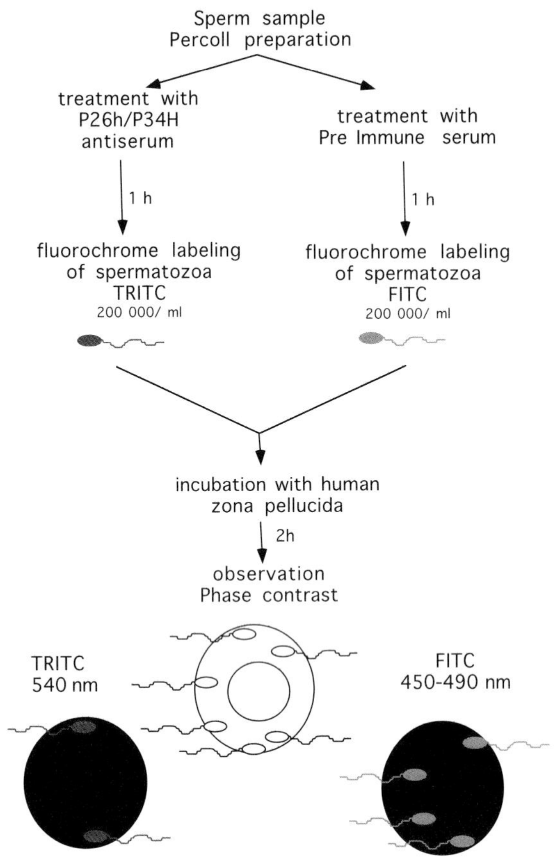

Figure 3. Experimental protocol used to evaluate the inhibitory effect of the P26h/P34H anti-serum on human sperm-zona pellucida binding assay.

Patients are usually advanced in age and have been under anti-androgen therapy, both of which can greatly affect testicular and epididymal physiology. When tissues are obtained from autopsy, the delay for family consent is often too long to ensure good tissue preservation. With the collaboration of the local organ transplantation program, the Sullivan laboratory readily obtains testicular and epididymal tissues from donors aged 20 to 35 years who are victims of accidental death. Importantly, these donors have no medical antecedents that can affect reproductive function and tissues are collected while artificial circulation is used to preserve organs assigned for transplantation. These tissues are processed in optimal conditions for immunocytochemistry, *in situ* hybridization, as well as for protein and RNA extractions.

The antiserum specific for P26h/P34H was used to perform immunohistochemical studies on human epididymal tissues. Results showed that spermatozoa within the seminiferous tubules, as well as in the vasa efferentia, were unreactive for P34H. Spermatozoa within the lumen of caput epididymidis showed very weak staining on the acrosomal cap. P34H labeling increased considerably in the lumen of the proximal corpus segments. P34H staining progressively increased from the proximal corpus to the cauda epididymidis, where spermatozoa were intensively stained. All along the epididymis, the staining was restricted to the acrosomal cap of the human epididymal spermatozoa. Thus, a change in staining intensity was observed as a function of epididymal transit. No changes, however, were noted in the localization of P34H on the spermatozoa itself (Fig. 4; Boué et al. 1994, 1996).

Reproductive tissues obtained from the local organ transplantation program were of satisfactory quality to obtain high quality mRNA. In order to identify a P34H transcript, hamster P26h was purified, submitted to partial proteolysis and N-terminal sequenced by Edman degradation. From these amino acid sequences, oligonucleotide primers were deduced and used in RT-PCR to amplify a cDNA probe. Northern blots of mRNA isolated from human tissues revealed a single transcript of approximately 900 bp. This transcript was abundant in the caput and corpus region, but weaker in the cauda epididymidis. No signal was detected in Northern blots using testicular RNA (Fig. 5), nor RNA isolated from different somatic tissues. *In situ* hybridization experiments, performed on human testicular and epididymal tissues, gave results consistent with the Northern blot studies just described. Moreover, *in situ* hybridization revealed that the P34H transcript was specific to principal cells of the epididymis (Fig. 6). These cells represent more than 80% of the cell population of the epididymal epithelium and play an important role in protein secretion (Hermo et al. 1994). The localization of P34H transcript was in agreement with the immunohistochemical localization of P34H along the male reproductive tract.

Based on its tissue distribution and function in one of the key steps of fertilization, P34H can thus be considered as a marker of epididymal sperm maturation in humans. The properties of P34H can be summarized: 1) P34H is antigenically related to the hamster 26 kD, 2) P34H accumulates during epididymal transit, 3) P34H is localized on the acrosomal cap and 4) an antiserum directed against this human sperm protein inhibits sperm-zona pellucida binding without affecting motility, acrosome reaction and sperm-egg plasma membrane fusion.

The cDNA used to probe Northern blots of human tissues was also used to search for an analogous transcript in the Rhesus monkey. Compared to the results obtained in man, a transcript of a similar length and distribution was detectable in this primate model (Fig. 5). Immunohistochemical studies performed with the anti-P26h/P34H sera revealed that the protein encoded by this epididymal transcript is strictly localized on the acrosomal cap of mature spermatozoa of the Rhesus monkey (data not shown). This result is potentially important if one considers the immunocontraceptive applicability of P34H in humans. Moreover, the

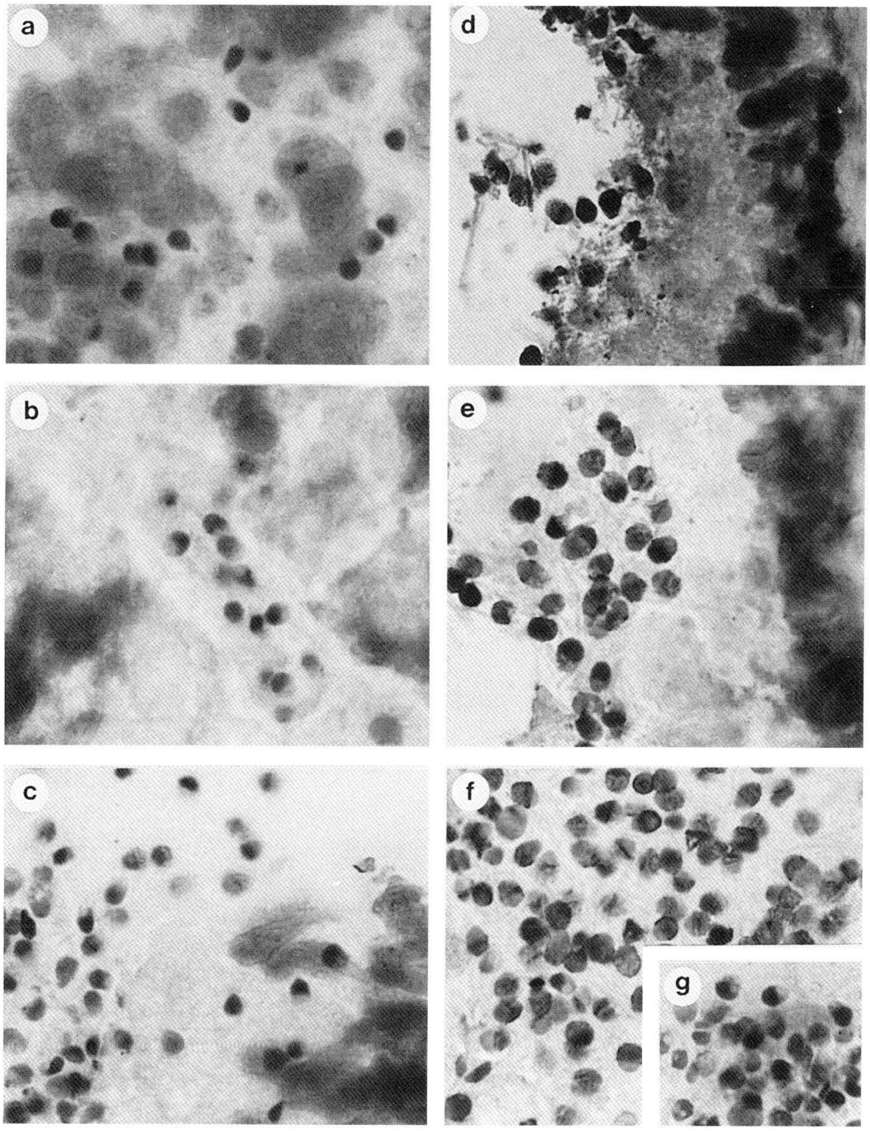

Figure 4. Localization of P34H on histological sections of human seminiferous tubules (a), vasa efferentia (b), caput (c), and proximal (d) and distal (e) corpus and cauda (f, g) segments of the epididymis. Immunostaining using the anti-P26h/P34H antiserum (a-f) or a control preimmune serum (g). Reproduced from Boué et al. 1996 with permission.

presence of similar transcript and protein in the epididymis of a primate model may allow for a preclinical investigation of male contraception based on active immunization against an epididymal antigen.

P34H: A Diagnostic Tool of Male Fertility

According to the World Health Organization (WHO 1992), infertility occurs in 8% of couples worldwide. In about 50% of these cases, male factors are responsible or contribute to this pathological situation. Many causes of male infertility remain unknown. In investigations of male fertility, standard semen analysis is widely used. The latter includes evaluation of sperm concentration, morphology and motility. These parameters, however, fail to explain many cases of male infertility. In order to provide more valuable prognostic values of male fertility, many andrological tests have been proposed. These include sophisticated computer-assisted analyses of sperm motility parameters, evaluation of sperm membrane integrity by a hypoosmotic swelling tests, zona-free hamster egg penetration tests, nuclear condensation and acrosomal status determinations and sperm-zona pellucida binding assays. This latter test requires human zonae pellucidae and involves a relatively complex, time-consuming *in vitro* assay. Although all these assays appear relevant for the estimation of the physiological "efficiency" of the spermatozoon, in general, they correlate poorly with *in vitro* fertilization results. Until now, the human sperm-zona pellucida binding assay represents the andrological test that correlates the best with human sperm fertilizing ability as determined by *in vitro* fertilization (Mortimer, Fraser 1996). Considering the epididymal origin of P34H and its function in zona pellucida binding, it has been proposed by the Sullivan

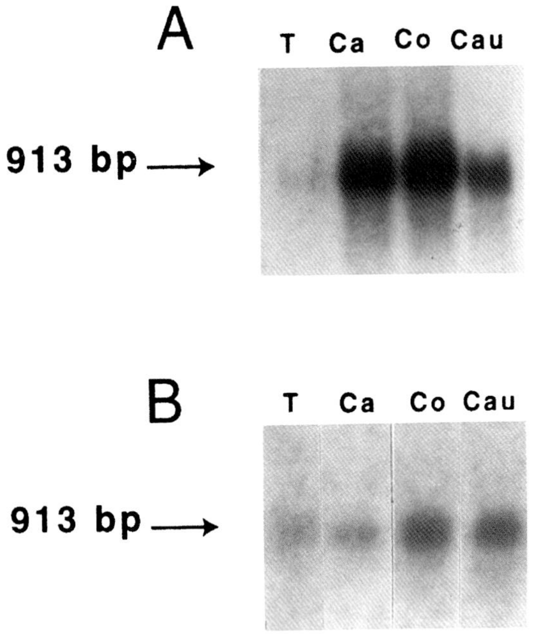

Figure 5. Northern blot analysis of RNA isolated from testis (Te), caput (Ca), corpus (Co), and cauda (Cau) epididymidis of human (A) and Rhesus monkey (B). A 900 bp transcript was detected in the epididymis of both species.

laboratory that some cases of male infertility could be explained by a suboptimal sperm maturation during epididymal transit. Invariably, this would be associated with a decreased P34H levels and consequently, the inability of spermatozoa to bind to the zona pellucida.

Couples consulting an infertility clinic for primary infertility were investigated. They were considered infertile if a period of at least 30 months of unprotected intercourse did not result in a pregnancy. All the cases of infertility were classified as idiopathic following a complete infertility workup. To be included in the study, women had to show normal ovulation as estimated by basal temperature chart, as well as normal plasma hormone concentrations, a normal post-coital test and endometrial biopsy and a negative hysterosalpingography and laparoscopy. Men had to present at least three normal spermograms according to the WHO criteria (WHO 1992) and spermatozoa had to be free of antibodies as determined by the immunobead test. The positive control group of men consisted of healthy volunteers having normal spermogram values. These fertile donors had to have fathered a child less than three years before the study. Sperm donation was performed between three and seven days of sexual abstinence. P34H was determined by densitometric quantification of Western blots performed on proteins extracted from a constant number of spermatozoa from each individual. An internal positive control was always analyzed in parallel in order to allow for comparison. This control consisted of spermatozoa from a fertile donor. Thus, all samples analyzed were expressed as a percentage of this sample.

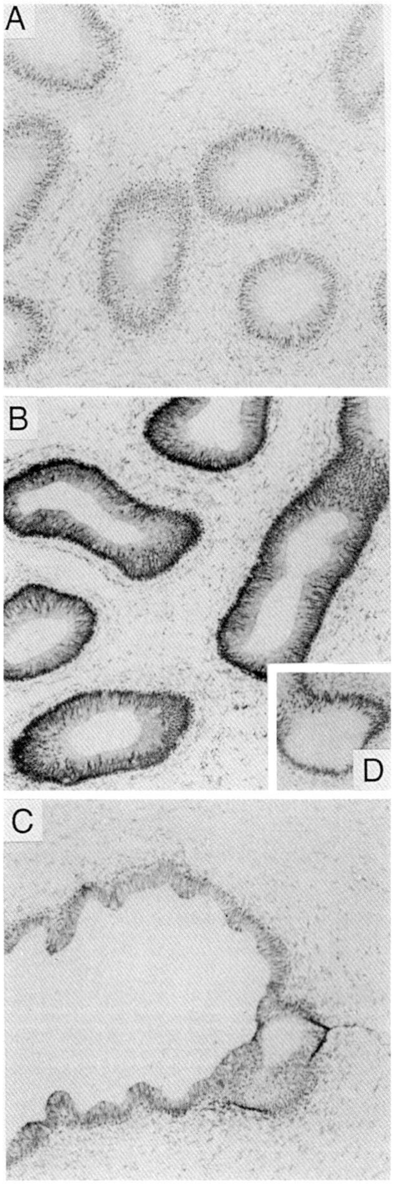

Figure 6. *In situ* hybridization localization of the P34H transcript in the caput (A), corpus (B, D) and cauda (C) human epididymidis. Histological sections were probed with an antisense P34H RNAs probe (A-C) or with the sense RNA probe (D) used as a negative control. P34H mRNA was detected with the digoxigenin-anti-digoxigenin complexes that result in dark staining.

P34H determinations performed on different semen samples from one individual gave comparable results. In contrast, P34H level showed variability from one individual to another. In 16 of the infertile men investigated, nine had spermatozoa with a P34H level less than 30% of the normal values as determined by the levels found in the fertile controls. The other seven infertile men were in the normal range (Fig. 7). If P34H indeed plays a function in human sperm-zona pellucida binding, it can be expected that spermatozoa of infertile men with low P34H levels should be less efficient to bind zonae pellucidae. To ascertain the relationship between the presence of this sperm protein and the ability to interact with the zona pellucida, a double vital staining assay was performed, in which the zona pellucida binding ability of spermatozoa from fertile and infertile men was compared (Fig. 8). Sperm from infertile men with normal levels of P34H were as efficient as spermatozoa of fertile men in a binding assay to homologous zonae pellucidae. In these cases, infertility could not be explained by the inability of the male gametes to undergo this critical step of fertilization. In contrast, spermatozoa of men presenting with low levels of P34H showed a dramatic decrease in their zona pellucida binding capacity (Fig. 9). This result clearly demonstrates the relationship between P34H and human sperm-zona pellucida interaction, a property known to be acquired during the epididymal transit. The low level of P34H associated with cases of infertility in normospermic men suggests that this pathological condition may result from an epididymal defect. Researchers in the Sullivan laboratory are confident that the experimental approach used will provide a better understanding of epididymal physiology in humans. This may also favor a more rational approach to understand the physiopathology of some cases of male infertility.

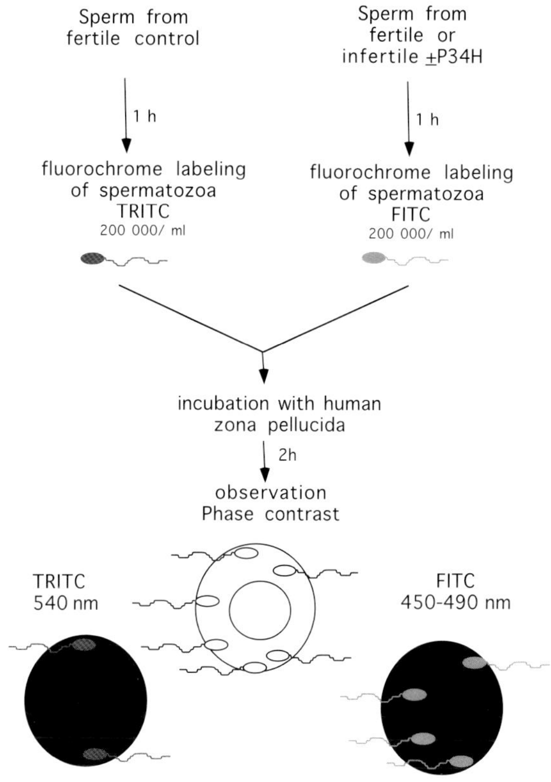

Figure 8. Experimental protocol used to compare the zona pellucida binding ability of spermatozoa from fertile and infertile men with or without P34H.

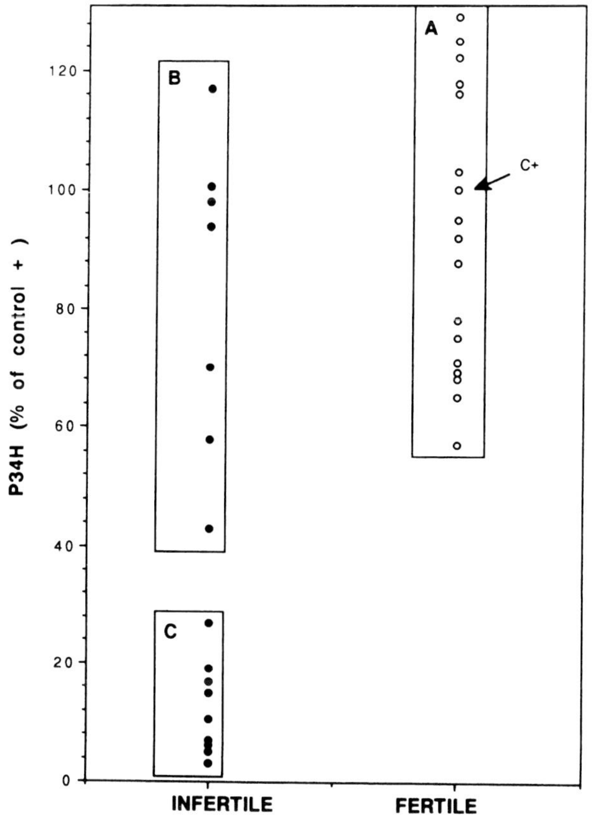

Figure 7. Amount of P34H as determined by densitometric scanning of immunoblots of proteins extracted from $5X10^6$ spermatozoa from fertile (open circles) and infertile (solid circles) men. The results are expressed as a percentage of an internal positive control (C^+=100%). Group A and group B were similar, whereas group C was significantly different from group A ($p< 0.001$). (From Boué, Sullivan 1996, with permission.)

Interestingly, vasectomy may contribute to male infertility by inducing damage to the epididymis in cases of vasovasostomy (surgical vasectomy reversal). Vasectomy has gained increased popularity as a contraceptive method, especially in North America. To be considered as a useful contraceptive method, the induced infertility should be reversible in a relatively short period of time. Vasovasostomy is becoming more and more common. A wide range of vasovasostomy success rates has been reported and controversies exist regarding the post-surgical return of fertility. For this reason, vasectomy is often referred to as a sterilization procedure. The surgical success of vasectomy reversal is higher when evaluated by semen analysis than by the pregnancy rate, and the time between vasectomy and vasovasostomy has an effect on the eventual return to fertility. The failure of patients who have undergone a successful vas deferens reanastomosis to establish a pregnancy has been attributed to partner infertility, antisperm antibodies, epididymal dysfunction or other unknown reasons (Belker et al. 1991).

Considering that P34H is a marker of in sperm maturation and hence, human epididymal function, this

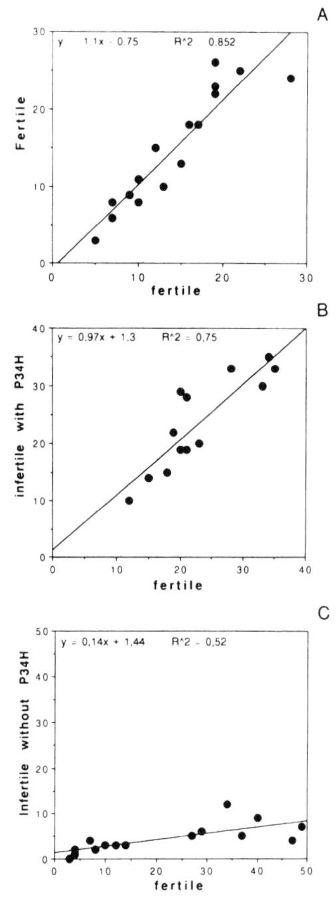

Figure 9. Number of sperm from fertile donor (x axis) or from different men (y axis) that were bound per human zona pellucida. A: fertile men, B: infertile men with P34H, C: infertile men without P34H. The regression line corresponding to each panel is illustrated. From Boué, Sullivan 1996, with permission.

sperm antigen has been sought in semen samples of vasovasostomized men. The results obtained in 25 of these men revealed an interindividual P34H level distribution similar to the one described for idiopathic infertile men. In fact, 18 of the 25 investigated vasovasostomized patients showed P34H levels lower than the values found in fertile men. This proportion of vasovasostomized men with low P34H value may be overestimated due to the fact that this determination was done on semen produced for post-surgical follow-up. This group probably contained more men that did not father children when compared to the population of all men that underwent surgical vasectomy reversal during the same period. Nevertheless, such low P34H levels were never found in spermatozoa from fertile men (Fig. 7). The level of P34H did not correlate with the success of the surgical procedure as measured by spermogram values, nor with the semen neutral α-glucosidase levels. The latter enzyme is used as a marker of epididymal patency. A factor that seemed to affect P34H levels was the time between vasectomy and vasovasostomy. In fact, P34H was low in semen samples from men that had been vasectomized for more than 10 years. This correlated well with the clinical data showing very low chances of paternity in those patients. These P34H determinations suggest that, in at least some men, vasectomy affects the epididymis in a way that sperm maturation cannot occur normally after vasovasostomy. Studies are underway to determine if low P34H levels found in some men that underwent vasectomy reversal can improve with time. If it is the case, this would suggest that the epididymis is able to recover from physiological damage that occurs during the vasectomy period. Thus, P34H could eventually represent a marker of fertility recovery following surgical vasectomy reversal.

The Bull as a Model of Male Subfertility

The pathophysiology of male infertility is poorly understood. In fact, the diagnostic and therapeutic tools available to treat female infertility are much more elaborate than those available for clinical management of infertile men (Adashi et al. 1996). Also, until recently, there was no animal model to study male infertility. Fertility can be defined as a probability, over time, to establish a pregnancy. This biological function can vary greatly from one individual to another. In men, this probability can be associated with spermogram values even though these semen parameters alone cannot explain the interindividual variability of probability, over time, to father a child. The processing of spermatozoa in the epididymis can also be associated with interindividual variability of fertility. In order to investigate this possibility, the anti-P26h/34H polyclonal antiserum was used to search for a related protein on bull spermatozoa.

Bulls used by the artificial insemination industry provide an ideal model to study male fertility interindividual variability. Semen from each bull is used to perform thousands of inseminations. The fertility of these bulls is quantitated as the 60-90 days non-return rate (NRR). The NRR is adjusted by a linear statistical model to include the effects of the month of insemination, the technician, the herd, the age of the inseminated cows and the price of the semen samples (Schaeffer 1993). The NRR can be converted to a scale of 1 to 9, 5 being the average fertility and 1 the less fertile. This provides an interesting animal model to understand the biological causes of male interindividual fertility variability (Parent et al. 1997).

Using the anti-P26h/P34H polyclonal antibodies, researchers have identified on bull spermatozoa an antigenically related protein of 25 kD molecular weight in SDS-PAGE: P25b. Like its human (P34H) and hamster (P26h) counterparts, P25b is associated with plasma membrane covering the acrosomal cap of ejaculated bull spermatozoa. Immunocytochemical studies performed on spermatozoa recovered at different levels of the bull epididymis showed that this protein accumulates on spermatozoa during the maturation process of bull spermatozoa (Lefièvre, Sullivan 1996). This epididymal marker has been quantitated on a constant number of ejaculated spermatozoa from bulls of different NRR. The results showed similarities to the ones described above for P34H levels from idiopathic infertile men as compared to fertile men. There is no linear correlation between the NRR expressed on a scale of 1 to 9 and the amount of P25b associated with spermatozoa. However, low levels of P25b are observed only in low NRR bulls (Fig. 10). Thus, the bull may provide an interesting model of male subfertility associated with variation in epididymal sperm maturation. In the bovine artificial insemination industry, the market demand for a given bull varies with time. As a consequence, bulls with well defined NRRs are often slaughtered. It is then possible to recover the reproductive tract from these animals. With the collaboration of the artificial insemination industry, frozen semen samples from these animals are readily available, as well as their fertility data. This biological material provides an unique opportunity to correlate epididymal gene expression, biochemical modifications occurring during sperm maturation, sperm physiology and interindividual fertility variability. Considering that bull P25b shows antigenetic, ontogenic and cytological localization similarities with human P34H, the bull model may provide information regarding subfertility and infertility in men and how the epididymis contributes to the etiology of male fertility.

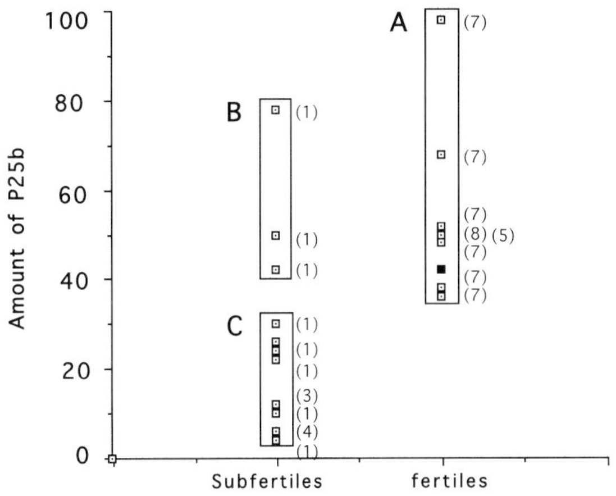

Figure 10. Amount of P25b as determined by densitometric scanning of immunoblots of proteins extracted from spermatozoa of subfertile and infertile bulls. Subfertility was defined as a non-return rate of less than 5 and fertile as a non-return rate of 5 or more. Non-return rate value is expressed on the right of each P25b value. Non-return rate was expressed on a scale of 1 to 9, with 5 being the average fertility and 1 the less fertile. Group A and group B were similar, whereas group C was significantly different from group A ($p < 0.001$).

Conclusions

Assisted reproductive technologies have greatly changed the management of infertile patients and challenged some of the scientific dogma in reproductive biology. One of these has focused on the physiological role of the epididymis in sperm maturation. As stated by Cooper (1993), "More work is required to elucidate specific epididymal proteins that bind to human spermatozoa, their normal temporal appearance on the sperm surface during maturation and their role in fertilization." This could lead to a better understanding of the function of the epididymis, especially in humans. Using different animal models, as well as human material, antigenically related sperm proteins have been identified. These proteins are added to the male gamete during epididymal transit and play a key role in fertilization. These markers of epididymal maturation illustrate the function of post-testicular transit in sperm physiology. Furthermore, these proteins may be used to develop male immunocontraceptive strategies and to understand the physiological causes underlying male infertility and subfertility.

Acknowledgements

Several persons contributed to the work described in this chapter. These are post-doctoral fellows, graduate students, research assistants and technicians. All of them should be acknowledged for their contributions and for sharing the passion for the epididymis and the spermatozoon. They are: B. Bérubé, S. Bégin, J. Blais, F. Boué, L. Coutu, P. Desrosiers, C. Gaudreault, C. Guillemette, N. Lamontagne, L. Lefièvre, C. Légaré, C. Lessard and S. Parent. Drs. J. Bailey, C. Gagnon, M.A. Sirard and S. St-Jacques have also collaborated on this work. Also, the Centre d'Insémination Artificielle du Québec should be thanked, especially Mr. Y. Brindle, for providing bull semen samples and fertility data. Drs. T. Cooper, P. Jouannet, R. Schoysman and M. Thabet are acknowledged for providing semen samples from patients. Also, the staff of the IVF clinic at St-Luc Hospital in Montréal are acknowledged for their help in obtaining human zonae pellucidae. This work was supported by grants from the Medical Research Council of Canada and by the Natural Sciences and Engineering Research Council of Canada.

References

Adashi EY, Ropck JA, Rosenwaks Z. Reproductive endocrinology, surgery, and technology. Lippincott-Raven Publishers, Philadelphia, 1996.

Aitken RJ, Ervine DS. Fertilization without sperm. Science 1995; 269:493-495.

Bedford JM. Effect of duct ligation on the fertilising capacity of spermatozoa in the epididymis. J Exp Zool 1967; 166:271-281.

Bégin S, Bérubé B, Boué F, Sullivan R. Zona pellucida recognition in the mouse and hamster involves both common and specific epitopes of a sperm protein. Mol Reprod Dev 1995; 41:249-256.

Belker AM, Thomas AJ, Fuchs EF, Konnak JW, Sharlip ID. Results of 1,469 microsurgical vasectomy reversals by the vasovasostomy study group. J Urol 1991; 145:505-511.

Bérubé B, Sullivan R. Inhibition of *in vivo* fertilization by active immunization of male hamster against a 26 kDa sperm glycoprotein. Biol Reprod 1994; 51:1255-1263.

Bérubé B, Coutu L, Lefièvre L, Dupont H, Sullivan R. The elimination of keratins artifact in immunoblots probed with polyclonal antibodies. Anal Biochem 1994; 217:331-333.

Bérubé B, Lefièvre L, Coutu L, Sullivan R. Regulation of the epididymal synthesis of P26h, a hamster sperm protein. J Androl 1996; 17:104-110.

Boué F, Sullivan R. Epididymal proteins as targets for contraception in men and women. Ref Gyn Obstet 1995; 3:258-265.

Boué F, Sullivan R. Cases of male infertility are associated with the absence of P34H, an epididymal sperm antigen. Biol Reprod 1996; 54:1018-1024.

Boué F, Bérubé B, de Lamirande E, Gagnon C, Sullivan R. Human sperm-zona pellucida interaction is inhibited by an antibody against a hamster sperm protein. Biol Reprod 1994; 51:577-587.

Boué F, Blais J, Sullivan R. Surface localization of P34H, an epididymal protein, during maturation, capacitation and acrosome reaction of human spermatozoa. Biol Reprod 1996; 54:1009-1017.

Cooper TG. The Epididymis: Sperm Maturation and Fertilisation. Springer-Verlag, Heidelberg, 1986.

Cooper TG. The human epididymis—is it necessary? Int J Androl 1993; 16:245-250.

Coutu L, Desrosiers P, Sullivan R. Purification of P26h: a hamster sperm protein. Biochem Cell Biol 1996; 74:227-231.

Cuasnicú PS, Echeverria FG, Piazza A, Blaquier JA. Addition of androgens to cultured hamster epididymis increases zona recognition by immature spermatozoa. J Reprod Fert 1984; 70:541-547.

Eddy EM, O'Brien DA. The spermatozoon. In: (Knobil E, Neill J, eds) The Physiology of Reproduction. 2nd ed. New York: Raven Press, 1994, pp29-78.

Hermo L, Oko R, Morales CR. Secretion and endocytosis in the male reproductive tract: a role in sperm maturation. Int Rev Cytol 1994; 154:105-189.

Horan AH, Bedford JM. Development of the fertilizing ability of spermatozoa in the epididymis of the Syrian hamster. J Reprod Fert 1972; 30:417-423.

Lefièvre L, Sullivan R. Characterization of a bull epididymal sperm protein involved in fertilization. Biol Reprod 1996; 54 (Suppl 1):70.

Liu DY, Lopata A, Johson WIH, Baker HWG. A human sperm-zona pellucida binding test using oocytes that failed to fertilize *in vitro*. Fertil Steril 1988; 50:782-788.

McLeskey SB, Dowds C, Carballada R, White RR, Saling PM. Molecules involved in mammalian sperm-egg interaction. Int Rev Cytol 1998; 177:57-113.

Moore HDM. The influence of the epididymis on human and animal sperm maturation and storage. Hum Reprod 1996; 11:103-110.

Mortimer D, Fraser LR. Consensus workshop on advanced diagnostic andrology techniques. Hum Reprod 1996; 11:1463-1479.

Orgebin-Crist M-C. Maturation of spermatozoa in the rabbit epididymis: fertilising ability and embryonic mortality in does inseminated with epididymal spermatozoa. Ann Biol Anim Biochim Biophys 1967; 7:373-379.

Parent S, Bousquet D, Brindle Y, Sullivan R. The binding of albumin to bovine spermatozoa affects their relative buoyant density. Theriogenology 1997; 48:1275-1285.

Primakoff P. Sperm proteins being studied for use in a contraceptive vaccine. Am J Reprod Immunol 1994; 31:208-210.

Robaire B, Hermo L. Efferent ducts, epididymis, and vas deferens: structure, functions, and their regulation. In: (Knobil E, Neill J, eds) The Physiology of Reproduction. New York: Raven Press, 1988, pp999-1080.

Robitaille G, Sullivan R, Bleau G. Identification of epididymal proteins associated to hamster sperm. J Exp Zool 1991; 258:69-74.

Schaeffer LR. Evaluation of bulls for non-return rates within artificial insemination organizations. J Dairy Sci 1993; 76:837-842.

Snell WJ, White JM. The molecules of mammalian fertilization. Cell 1996; 85:629-237.

Sullivan R, Bleau G. Interaction between isolated components from mammalian sperm and egg. Gam Res 1985; 12:101-116.

Sullivan R, Robitaille G. The heterogeneity of epididymal spermatozoa in the hamster. Gam Res 1989; 24:229-236.

Sullivan R, Ross P, Bérubé B. Immunodetectable galactosyltransferase is associated only with human spermatozoa of high buoyant density. Biochem Biophys Res Commun 1989; 162:184-188.

Wassarman PM, Litscher ES. Sperm-egg recognition mechanisms in mammals. Curr Top Dev Biol 1995; 30:1-19.

World Health Organization. WHO laboratory manual for the examination of human semen and sperm-cervical mucus interaction. 3rd ed. Cambridge University Press, Cambridge, UK, 1992.

Yanagimachi R. Mammalian fertilization. In: (Knobil E, Neill J, eds) The Physiology of Reproduction. 2nd ed. New York: Raven Press, 1994, pp189-317.

Yanagimachi R, Chang MC. *In vitro* fertilization of golden hamster ova. J Exp Zool 1964; 156:361-376.

11

Signaling Mechanisms Controlling Mammalian Sperm Fertilization Competence and Activation

Gregory S. Kopf
Xiao Ping Ning
Pablo E. Visconti
Marie Purdon
Hannah Galantino-Homer
Miguel Fornés
University of Pennsylvania

Intercellular communication between gametes with the resultant activation of both gametes is essential to the unique event in the life cycle of an organism called fertilization. Successful fertilization results from requisite and reciprocal cell-induced sperm and egg activation events mediated by unique cellular and environmental cues associated with either the gametes or the reproductive tract/environment. The interaction of motile mammalian sperm with the female reproductive tract/environment, as well as with the egg both at a distance and in close proximity, represents a series of integrated processes designed to deliver sperm with optimal fertilizing potential to the site of fertilization. The development of the fertilization-competent state occurs through a poorly understood process called capacitation, which is thought to be integrated with the hyperactivation of sperm motility. Once capacitation has occurred, sperm have the ability to undergo acrosomal exocytosis in response to the egg's unique extracellular matrix, the zona pellucida (ZP). These two activation processes (capacitation and the ZP-induced acrosome reaction) appear to be regulated by intracellular signaling systems that are, in some cases, unique to sperm (capacitation) and, in other cases, similar to those utilized by somatic cells (acrosome reaction). The initiation of transmembrane signaling leading to capacitation appears to occur as a consequence of intrinsically controlled maturational processes lying within the sperm plasma membrane. Intracellular signal transduction involves the integration of both protein kinase A (PK-A) and tyrosine kinase/phosphatase signaling pathways in a manner that, to date, is unique to sperm. In contrast, the ZP-induced acrosome reaction possesses hallmarks of a classical ligand-receptor-effector system leading to regulated secretion. Several candidate sperm proteins have been implicated as binding proteins and/or receptors for the specific ZP ligand, but the nature and function of these candidates is controversial and still unresolved. Transmembrane signal transduction leading to acrosomal exocytosis appears to be regulated via G protein activation with subsequent activation of specific intracellular effectors. An understanding of signal transduction in mammalian sperm will ultimately yield information regarding the "intrinsic" nature of how sperm capacitation is initiated, the nature of the receptors to which these signal transduction pathways are coupled and the intracellular effectors that ultimately regulate sperm function. Moreover, an understanding of these regulatory pathways is essential for the development of clinical approaches designed to enhance or preclude fertilization.

Capacitation

Definition and Endpoints

Historically, capacitation has been defined as the time interval between sperm deposition in the female reproductive tract during natural mating and the time during which fertilization occurs. This interval, therefore, encompasses all of the reciprocal interactions between the sperm and the female tract (Smith 1998; Suarez 1998; Verhage et al. 1998). The definition of capacitation was established following the observation that sperm taken from the female tract immediately following mating did not have the ability to fertilize eggs, and that sperm residence in the female tract in some way conferred fertilization capacity (Austin 1951; Chang 1951). An increased understanding of the biology of sperm and the establishment of more sophisticated assays of sperm function has, over the years, led to a more narrow definition of capacitation reflecting, in part, investigators' biases regarding the physiological importance of this event.

Although fertilization still represents the definitive endpoint that sperm have undergone capacitation, the ability of sperm to undergo a regulated acrosome reaction in response to the ZP can be taken as an earlier, upstream endpoint of capacitation. It must also be emphasized that changes in sperm motility patterns are correlated with capacitation in a number of species; such motility changes are referred collectively as sperm hyperactivation (Yanagimachi 1994; Suarez 1996). Reports demonstrating the dissociation of capacitation and hyperactivation (Neill, Olds-Clarke 1987) have been published, but complete independence of these two events has not been conclusively demonstrated (Suarez 1996). Given these observations, experimental approaches designed to understand the molecular basis of capacitation must consider events that occur both in the head (that is, acrosome reaction) and the tail (that is, motility changes) of the sperm.

Although capacitation *in vivo* must occur to insure successful fertilization, capacitation *in vitro* can also be accomplished in most species in defined media, in various biological fluids (for example, heat-treated serum, oviductal fluid, follicular fluid or vitreous humor) and in a non-species-specific manner (Yanagimachi 1994). Such observations suggest that specific aspects of the capacitation process can be initiated and controlled intrinsically by the sperm itself, and that certain minimal environmental requirements must be met. The intrinsic nature of capacitation is also of interest from a cell regulation standpoint, and must be taken into account when designing and interpreting experiments directed at a molecular understanding of this maturational process. The intrinsic nature of capacitation by no means excludes the physiological relevance of positive and/or negative modulators of capacitation present in the male and female reproductive tracts of various species. It is reasonable to think that the regulation of capacitation lies less in the stimulation of this process and more in the de-repression of inhibitory modulators of capacitation through the removal of decapacitating factors (Hunter, Nornes 1969; Yanagimachi 1994). Regulation of capacitation at this level might be very important, as these modulators may function to extend the fertilizable lifespan of the sperm population in the ejaculate by exerting additional regulation over the intrinsic regulatory aspects of capacitation. Although numerous modulatory factors associated with both the male and female reproductive tracts have been described, their identity and mode of action on sperm have not been elucidated (Kopf et al. 1998). Interactions of sperm with these factors, in conjunction with interactions of the sperm with the cells lining different regions of the reproductive tract (Suarez 1998; Smith 1998), may constitute a concerted regulation of capacitation *in vivo*.

Regulation of sperm function at this level may be extremely important in selecting optimal subpopulations of sperm for fertilization. Sperm composition of the ejaculates from many species may be quite heterogeneous with respect to cellular age, morphology, motility characteristics and ability to undergo capacitation (Bedford 1983). Extrinsic regulation of such a heterogeneous sperm population may serve to select subpopulations of sperm that ultimately could participate in the fertilization process, as well as to extend the fertilizable lifespan of the ejaculate by widening the window of capacitation in these subpopulations (Cohen-Dayag, Eisenbach 1994), given the apparent stochastic nature of capacitation when assessed by *in vitro* assays. Such extrinsic modulation provides a higher order of regulation over the intrinsic properties of capacitation.

Given the poorly understood, but apparently complex nature of the interactions between sperm and the male/female reproductive tracts to regulate capacitation, it is of interest that capacitation can be accomplished *in vitro* in numerous species by incubating cauda and/or ejaculated sperm under a variety of conditions in defined media that mimic the electrolyte composition of the oviductal fluid. If one accepts the theory that capacitation is regulated both by extrinsic and intrinsic factors, capacitation *in vitro* would result in a population of sperm fully competent to fertilize eggs but induced to undergo this maturational event in a very different temporal fashion. Although it is easier to study capacitation *in vitro*, and the sperm capacitated under these conditions could be considered normal, one must always be careful in extending conclusions based on studies carried out *in vitro* to the *in vivo* situation. A molecular understanding of this poorly understood maturational event, however, will have to come initially from studies of capacitation *in vitro*. As stated above, capacitation *in vitro* can be accomplished in media of defined composition. The composition of such media includes energy substrates such as pyruvate, lactate and glucose (depending on the species), a protein source (usually serum albumin), $NaHCO_3$ and Ca^{2+}. Several laboratories, including that of the authors of this chapter, are starting to elucidate the putative mechanism of action by which these media components promote capacitation at the molecular level.

Signal Transduction

Although one must carefully interpret *in vitro* capacitation studies when attempting to extend conclusions from such studies to the *in vivo* situation, *in vitro* approaches still clearly represent the best way to ultimately understand this process *in vivo*. Capacitation in many species can occur *in vitro* spontaneously in defined media without the addition of biological fluids, suggesting that the intrinsic cellular regulation of capacitation involves pre-programmed membrane, transmembrane and/or intracellular signaling events that, once initiated, lead to the capacitated state. Moreover, different defined media support capacitation of sperm from different species, and certain components of these media—that is, serum albumin, Ca^{2+} and HCO_3^-—play critical regulatory roles in promoting capacitation in all species studied thus far. Work, primarily from the Kopf laboratory, has demonstrated that these media components appear to function in a very specific manner to couple membrane, transmembrane and intracellular signaling events leading to the capacitated state. The discussion below will focus on experiments that link the role of these media constituents in regulating membrane, transmembrane and intracellular signaling events that form the basis for a working hypothesis regarding the molecular basis of capacitation. This discussion will draw on the work of others and is intended to integrate these observations into a unifying hypothesis.

Researchers in the Kopf laboratory have established a working model for signal transduction during capacitation (Fig. 1). Several aspects of this model must be tested experimentally and this general model does not take into account other observations regarding the regulation of capacitation seen in other species. Some of the these observations may be unique to a particular species but ultimately must be considered in developing a unifying hypothesis of capacitation. More information along these lines may be found in the work of de Lamirande and Gagnon (1993), Aitken et al. (1995), Leclerc et al. (1997) and Galantino-Homer et al. (1997).

The mechanism by which albumin (in many cases bovine serum albumin, referred to as BSA) supports and promotes capacitation in mammalian sperm is of great interest from a signal transduction standpoint, as it is believed to function during capacitation *in vitro* as a sink for the removal of cholesterol from the sperm plasma membrane (Go, Wolf 1985; Langlais, Roberts 1985; Cross 1998). The association between albumin,

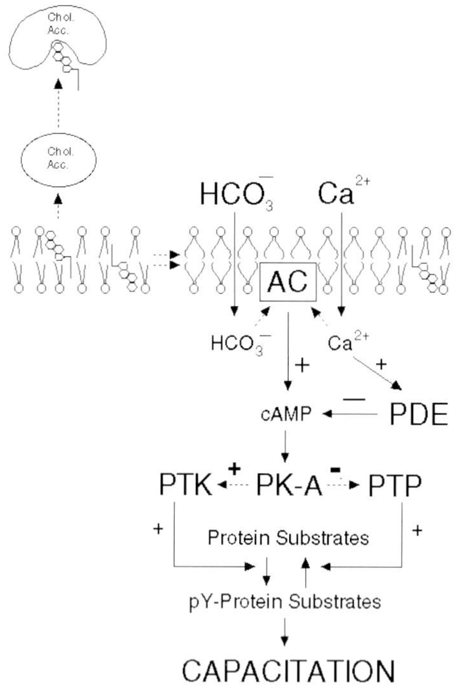

Figure 1. Model summarizing the transmembrane and intracellular signaling pathways hypothesized to play a role in regulating sperm capacitation. This model is based on the work from a number of different laboratories. In this model, cholesterol efflux to an appropriate acceptor initiates changes in membrane architecture due, in part, to a decrease in the orientation order of membrane lipids and phospholipids leading to an increase in bulk membrane fluidity. This change in membrane dynamics results in the activation of signaling pathways leading to capacitation. (-) indicates negative regulation; (+) indicates positive regulation; solid arrows indicate established pathways; dashed arrows indicate hypothesized pathways that are to be experimentally tested. Abbreviations used in this figure: AC, adenylyl cyclase; Chol. Acc., cholesterol acceptor; PTK, protein tyrosine kinase; PTP, phosphotyrosine phosphatase; PDE, cyclic nucleotide phosphodiesterase; PK-A, protein kinase A; pY-protein substrates, phosphotyrosine-containing protein substrates.

cholesterol removal from the sperm plasma membrane and capacitation was first proposed by Davis et al. (1980), and removal of this sterol likely accounts for the membrane fluidity changes observed during this maturational event (Wolf et al. 1986). As a consequence of the removal of this sterol, the cholesterol/phospholipid ratio in the membrane decreases, and such changes in membrane composition could clearly influence transmembrane signaling and cellular function. Such membrane lipid composition changes may underlie the changes in cell surface antigen distribution during capacitation that have been described by several investigators (Moore 1995; Harrison, Gadella 1995; Rochwerger, Cuasnicú 1992). Currently, little is known about the consequences of cholesterol removal on sperm membrane dynamics as it relates to capacitation. Cholesterol movement from the plasma membrane is likely the primary action of BSA, given the fact that other cholesterol binding proteins, such as HDL, can stimulate sperm cholesterol efflux (Langlais et al. 1988) and replace albumin in *in vitro* fertilization assays (Therien et al. 1997). Cross and others (Cross 1998 and references therein) have demonstrated that human semen contains cholesterol and that this sterol can account for the inhibitory effects of seminal plasma on human sperm capacitation, presumably by preventing cholesterol efflux from the sperm plasma membrane. Moreover, the mechanism by which cholesterol movement is initiated and the mechanism by which albumin captures cholesterol are not known.

Although extracellular Ca^{2+} is important for several sperm functions, its role in initiating and/or regulating capacitation is controversial. Capacitation is thought to be dependent on the presence of this divalent cation in mouse sperm (Dasgupta et al. 1993; Visconti et al. 1995a), although Ca^{2+} fluxes or intracellular Ca^{2+} concentrations have not been investigated. Some investigators have demonstrated an increase in intracellular sperm Ca^{2+} during capacitation, while others have shown that no changes occur during this maturational event (Yanagimachi 1994 and references therein). The ambiguity between these studies could be due, in part, to the well-demonstrated action of Ca^{2+} on the acrosome reaction and the inherent difficulties in differentiating both of these events. However, the action of Ca^{2+} at the level of effector enzymes involved in sperm signal transduction (for example, adenylyl cyclase, cyclic nucleotide phosphodiesterase, phosphatases) suggests that this divalent cation is likely to play an important role in this maturational process.

The requirement for HCO_3^- in capacitation has been established in the mouse (Lee, Storey 1986; Neill, Olds-Clarke 1987; Visconti et al. 1995a; Shi, Roldan 1995) and in the hamster (Boatman, Robbins 1991), although it remains to be demonstrated in other mammalian species. It is likely that the movement of this anion across the sperm plasma membrane occurs via specific transporters, since DIDS and SITS, well-known inhibitors of anion transporters, block the actions of HCO_3^- on various sperm functions (Okamura et al. 1988; Visconti et al. 1990; Spira, Breitbart 1992; Parkkila et al. 1993). Sperm contain an immunoreactive protein when probed with an antibody to the AE1 class of anion transporters (Parkkila et al. 1993), but little is known about the identification and function of this protein in these cells. Movement of HCO_3^- has been postulated to play a role in the known increase in intracellular pH that is observed during capacitation (Uguz et al. 1994; Zeng et al. 1996; Cross 1998). In addition, this anion has been demonstrated to modulate cAMP metabolism in sperm through its unique effect to markedly stimulate the adenylyl cyclase of these cells by an as yet unknown mechanism (Okamura et al. 1985; Garty, Salomon 1987; Visconti et al. 1990, 1995b). From a physiological point of view, it is interesting to note that HCO_3^- concentrations are low in the epididymis and high in the seminal plasma and in the oviduct (Harrison 1996). Since HCO_3^- present in the extracellular milieu has also been positively correlated with the motility of pig sperm (Okamura et al. 1985), the HCO_3^- concentrations present in the male and female reproductive tracts could affect the development of the capacitated state. Specifically, low levels of HCO_3^- in the epididymis would be conducive to maintaining sperm in an environment that does not support capacitation, whereas the higher concentrations of this anion in the female tract might contribute to capacitation.

Although little is known about the molecular basis of capacitation, numerous investigators have provided ample evidence that this maturational event is clearly associated with major changes in membrane architecture and composition, the consequence of which is likely to lead to transmembrane signaling and intracellular effector activation. Capacitation-associated changes in the cholesterol/phospholipid ratios of sperm membranes as a consequence of a reduction in membrane cholesterol have been demonstrated using numerous approaches (Davis 1981; Bearer, Friend 1982; Tesarik, Flechon 1986; Ehrenwald et al. 1988; Suzuki, Yanagimachi 1989; Hoshi et al. 1990; Zarintash, Cross 1996), and these changes likely account for the observed alterations in sperm membrane fluidity (Wolf et al. 1986), aggregation of intramembraneous particles and formation of particle-free patches (Koehler, Gaddum-Rosse 1975), and the documented membrane protein redistributions reported with lectins (Cross, Overstreet 1987) and antibodies (Shalgi et al. 1990; Rochwerger, Cuasnicú 1992). These resultant changes in membrane dynamics likely have profound effects on transmembrane signaling, and may constitute specific aspects of the "intrinsic" control of capacitation. For example, changes in ion channel activity and/or the activity of membrane associated enzymatic and non-enzymatic proteins may lead to transmembrane signaling during capacitation and the resultant changes in plasma membrane lipid architecture could also be functionally important, as they may ultimately prime the membrane for fusion with the outer acrosomal membrane during the acrosome reaction. Since cholesterol efflux appears to be the driving force behind these changes in membrane dynamics during capacitation, a clear understanding of the mechanism by which this sterol moves within the plasma membrane and out of the plasma membrane in response to an appropriate acceptor (for example, serum albumin) is critical to a molecular understanding of this maturational event. In this regard, models of reverse cholesterol transport observed in somatic cells might serve as a starting point for designing experiments with the goal of understanding how cholesterol moves out of the sperm plasma membrane during capacitation.

How might changes in cholesterol content of the membrane regulate transmembrane signaling events in the sperm leading to capacitation? It is already known that cholesterol alters the bulk biophysical properties of biological membranes. For example, cholesterol increases the orientation order of the membrane lipid hydrocarbon chains and, as a consequence, reduces the ability of membrane proteins to undergo conformational changes that may control their functions, due to the fact that the membrane is less fluid. As a consequence, high concentrations of cholesterol in the membrane might inhibit membrane protein function. This "indirect" effect of cholesterol on membrane protein function might stabilize membrane and transmembrane events that are part of the "intrinsic" regulatory nature of capacitation. Cholesterol has also been demonstrated to have "direct" effects by binding to and regulating membrane protein function; such binding may serve to exert a positive or negative modulatory effect on the protein in question. In fact, studies of several membrane-associated ion transporters (for example, Na^+, K^+-ATPase, GABA transporter) by cholesterol are consistent with the idea that this sterol can exert direct and indirect effects on enzyme activity (Vemuri, Philipson 1989; Shouffani, Kanner 1990). Such a mode of regulation may have particular relevance to capacitation since the loss of cholesterol from the sperm plasma membrane under conditions conducive to capacitation has been postulated to be coupled to the increase in intracellular pH that accompanies this maturational event (Cross, Razy-Faulkner 1997), suggesting that cholesterol itself and/or its concentration in the membrane could modulate transmembrane ionic movements that ultimately regulate intracellular pH. For the purposes of this review, work from the Kopf laboratory suggests that the release of cholesterol from the sperm membrane leads to the activation of a signal transduction pathway leading to protein tyrosine phosphorylation.

Although Ca^{2+} and HCO_3^- appear to play key roles in regulating aspects of capacitation, the mechanism by which they do so is unclear. Mammalian sperm contain voltage-sensitive Ca^{2+} channels (Florman, Babcock 1991; Florman et al. 1998) that play a role in acrosomal exocytosis induced by the ZP (Arnoult et al. 1996), but the mechanism by which Ca^{2+} influences signal transduction during capacitation is not clear. Elucidating the role of this cation in capacitation is difficult, given the problems in dissociating Ca^{2+} events mediating this maturational process from those involved in the acrosome reaction. Information regarding the role of Ca^{2+} in capacitation will be further clarified once the identity of the Ca^{2+} transport mechanisms present and functioning in sperm are fully characterized.

Capacitation also appears to be regulated by changes in intracellular pH (pH_i). Two acid efflux mechanisms have been identified in mouse sperm and could be involved in this process (Zeng et al. 1996); however, the transport mechanisms that control pH_i in these cells are not fully understood. One of these pathways shares the characteristics of a somatic cell Na^+-dependent Cl^-/HCO_3^- exchanger, and the second pathway does not require extracellular ions to function. These authors described an increase in pH_i during capacitation, consistent with the reports of Vredenburgh-Wilberg and Parrish (1995) describing an increase in pH_i during capacitation of bovine sperm by heparin. Although the increase in pH_i accompanying heparin-induced bovine sperm capacitation is not inhibited by Rp-cAMP (Uguz et al. 1994), this PK-A antagonist can block capacitation, suggesting that a PK-A regulatory pathway(s) functions either parallel to, or downstream of, pathways activated as a consequence of changes in pH_i.

Hyperpolarization of the sperm plasma membrane has also been shown to accompany capacitation in mouse and bovine sperm (Zeng et al. 1995). Enhanced K^+ permeability is at least partially required for this hyperpolarization, and could be related to the release of inhibitory modulation during capacitation (Arnoult et al. 1996). Although little is known about the consequences of this hyperpolarization, it is speculated that such membrane potential changes could recruit Ca^{2+} channels from an inactivated state to a closed, but activatable, state from which they could be subsequently opened by an agonist-induced depolarization, for example, with the ZP (Arnoult et al. 1996; Florman et al. 1998), thus connecting capacitation with the competence of the sperm to undergo acrosomal exocytosis in response to the egg's extracellular matrix. Presently, the role of membrane potential in regulating any of these aspects of capacitation at the molecular level is not known.

In addition to the role for cAMP in regulating sperm motility, recent work supports a role for this second messenger in capacitation (White, Aitken 1989; Parrish et al. 1994; Visconti et al. 1995b; Leclerc et al. 1996). Recently, the Kopf laboratory has demonstrated that an increase in PK-A activity, which represents the most accurate reflection of steady state changes in intracellular cAMP concentrations, accompanies mouse sperm capacitation (Visconti et al. 1997).

The changes in Ca^{2+} and HCO_3^- movements that accompany capacitation provide an intriguing mechanism by which sperm cAMP concentrations are regulated during this maturational event. Both of these ions have been implicated in the regulation of sperm cAMP concentrations through their effects on adenylyl cyclase activity (Hyne, Garbers 1979; Okamura et al. 1985; Garty, Salomon 1987). Although the mammalian sperm adenylyl cyclase possesses unique properties, the sequence and topology of this enzyme has not yet been established and the exact mechanism by which this enzyme is stimulated by these ions is not clear.

In attempts to further understand the signal transduction cascades that regulate capacitation, the Kopf laboratory has recently correlated mouse, human and bovine sperm capacitation with an increase in protein tyrosine phosphorylation of a variety of substrates (Visconti et al. 1995a; Carrera et al. 1996; Galantino-Homer et al. 1997). Many of these results have since been corroborated and extended by other labs (Aitken et al. 1995; Leclerc et al. 1996; Luconi et al. 1996; Emiliozzi, Fenichel 1997). In the Kopf laboratory, the mouse was used as the primary experimental paradigm. It was demonstrated that capacitation *in vitro* of cauda epididymal sperm promotes the tyrosine phosphorylation of a subset of proteins of Mr 40,000-120,000. The presence of BSA, Ca^{2+} and HCO_3^- in the medium is absolutely required for these tyrosine phosphorylations to occur, and the concentrations of these media constituents needed for phosphorylation to occur are correlated with those needed for capacitation (Visconti et al. 1995a). Moreover, caput sperm, which do not possess the ability to undergo capacitation and fertilize eggs (Yanagimachi 1994), do not display these changes in protein tyrosine phosphorylation when incubated under conditions normally conducive to capacitation (Visconti et al. 1995a). In this regard, it is interesting to note that the ability of sperm to display changes in protein tyrosine phosphorylation is first seen during the caput to corpus transition (M Fornes, PE Visconti and GS Kopf, unpublished data), the time during which sperm also gain the ability to undergo capacitation. Taken together, these data suggest that the ability of mouse sperm to become capacitated, as well as their ability to display increases in protein tyrosine phosphorylation, are acquired during epididymal transit and may represent an essential component of epididymal maturation in this species. The mode by which this acquisition occurs has turned out to be an interesting developmental question and is currently being pursued.

The requirement for BSA, Ca^{2+} and HCO_3^- in the extracellular medium to support protein tyrosine phosphorylation represents a novel mode of regulation of the signaling events in sperm leading to these post-translational modifications. Regulation of capacitation *in vitro* by BSA is thought to rely on the ability of this protein to serve as a sink for the removal of cholesterol from the sperm plasma membrane. The interrelationship between BSA and cholesterol movement also appears to be important in the regulation of protein tyrosine phosphorylation, since abolishing the ability of BSA to serve as an extracellular acceptor for cholesterol inhibits protein tyrosine phosphorylation and sperm capacitation (PE Visconti, XP Ning, M Fornes, J Alvarez and GS Kopf, unpublished data). These experiments, in addition to others, suggest that cholesterol release/movement is tightly tied to transmembrane signaling events in the sperm that ultimately regulate protein tyrosine phosphorylation.

These effects of cholesterol efflux on transmembrane signaling and intracellular signal transduction represent a new mode of cellular signaling, and have now generated numerous new and interesting questions. For example, might cholesterol efflux-induced signal transduction occur via changes in membrane fluidity, thereby indirectly modulating membrane-associated signal transduction enzymes and/or ion channels? Or might these signal transduction enzymes and/or ion channels be modulated directly by this sterol?

Since there appears to be a relationship between Ca^{2+}, HCO_3^- and increased adenylyl cyclase activity, experiments were designed to determine whether the action of these ions on protein tyrosine phosphorylation and capacitation involved a cAMP-mediated pathway. It was demonstrated that incubation of sperm in media devoid of BSA, Ca^{2+} or HCO_3^- (which would normally not support protein tyrosine phosphorylation), but in the presence of cAMP agonists (for example, dibutyryl cAMP, 8-bromo cAMP, Sp-cAMPS, isobutyl-methylxanthine), results in an increase in protein tyrosine phosphorylation, as well as capacitation (Visconti et al. 1995b). In addition, protein tyrosine phosphorylation is accelerated by active cAMP agonists in complete media that support capacitation. Results of these and other experiments suggest that 1) the action of cAMP appears to be downstream of the actions of BSA, Ca^{2+} and HCO_3^- but upstream of protein tyrosine phosphorylation, and 2) protein tyrosine phosphorylation and capacitation are regulated through a PK-A pathway. This second conclusion was further reinforced by experiments demonstrating that two inhibitors of PK-A, Rp-cAMPS and H-89, both of which inhibit this enzyme by completely distinct mechanisms, inhibit both protein tyrosine phosphorylation and capacitation of sperm in complete medium (Visconti et al. 1995b), and that PK-A activity increases during capacitation (Visconti et al. 1997). Since the mode of action of BSA appears to be tied to the removal of plasma membrane cholesterol, it is likely that cholesterol release is also upstream of the cAMP-induced protein tyrosine phosphorylation, but whether cholesterol removal is upstream or parallel to the action of Ca^{2+} and/or HCO_3^- is not known.

An attractive hypothesis to be tested is that the removal of cholesterol, with a resultant change in sperm plasma membrane fluidity, could modulate Ca^{2+} and/or HCO_3^- ion fluxes leading to the activation of the adenylyl cyclase. It is interesting to note the observations of Harrison et al. (1996) that HCO_3^- ions can induce major changes in the lipid architecture of boar sperm that appear to be independent of cholesterol removal and changes in intracellular pH. HCO_3^- was demonstrated to increase the binding of the impermeant lipophilic probe merocyanine to the sperm membrane; this probe binds to plasma membranes with enhanced affinity as the lipid components of the membranes become more disordered. This effect was very rapid and could be mimicked in the absence of HCO_3^- by a variety of cyclic nucleotide phosphodiesterase inhibitors, suggesting that these HCO_3^- effects might be mediated by cyclic nucleotides. It is not clear how this effect on membrane structure might be mediated but these observations point to the possibility of multiple effects of HCO_3^- on sperm membrane dynamics, as well as signal transduction. Clearly, additional experiments along this line of investigation are warranted.

In summary, these data suggest that protein tyrosine phosphorylation and capacitation appear to be under regulation of a cAMP/ PK-A pathway. Up-regulation of protein tyrosine phosphorylation by PK-A during sperm capacitation is, to the authors' knowledge, the first demonstration of a connection between these signal transduction pathways at this level. Since similar results have now been reported in sperm of other species (Leclerc et al. 1996; Galantino-Homer et al. 1997), it is likely that this unique mode of signal transduction crosstalk may be universal to mammalian sperm. It is not known whether the resultant increase in protein tyrosine phosphorylation is due to the stimulation of a tyrosine kinase, an inhibition of a phosphotyrosine phosphatase, or to both. Recently, Berruti and Borgonovo (1996) described a male germ cell-specific tyrosine kinase that they have termed sp42. This protein was initially purified from boar sperm, but proteins that cross react with sp42 antibodies are also present in human, mouse and rat sperm. Whether sp42 represents a tyrosine kinase that participates in the signaling cascade leading to capacitation is unknown. Researchers (Carrera et al. 1996; Luconi et al. 1996) have demonstrated in human sperm that extracellular Ca^{2+} can exert an inhibitory effect on protein tyrosine phosphorylation, and Carrera et al. (1996) have provided evidence that this Ca^{2+}-induced dephosphorylation may be regulated by a calmodulin-dependent mechanism, possibly involving calcineurin. Although calcineurin is a major phosphatase in sperm, the role of this particular enzyme in this particular series of dephosphorylations has not been demonstrated. Thus, the identity of the kinases and phosphatases involved in this unique signal transduction pathway is still to be resolved, and remains an area of future experimentation.

Identification and characterization of the substrates that are phosphorylated on tyrosine residues in both a cAMP-dependent and independent manner during capacitation is also of interest, as their identity may yield a great amount of information regarding their roles in capacitation. Researchers have initiated a systematic approach to the identification and characterization of these proteins in sperm, and have identified a major substrate for tyrosine phosphorylation during capacitation in human sperm (Carrera et al. 1996). This protein is associated with the fibrous sheath of the flagellum and is the human homolog of the mouse sperm AKAP82 and pro-AKAP82. AKAPs (A Kinase Anchor Proteins) represent a growing family of scaffolding proteins that function in cells to tether the regulatory subunits of PK-A to organelles or cytoskeletal elements, thus permitting precise control of signal transduction in discrete regions of the cell (Pawson, Scott 1997). It is interesting to note that other members of the AKAP family have been demonstrated to bind a variety of signal transduction enzymes, including calcineurin and protein kinase C (Klauck et al. 1996), suggesting that this family of proteins could potentially serve as scaffolding proteins to anchor entire signal transduction complexes to discrete cellular locations. The role of tyrosine phosphorylation in regulating AKAP82 function, as well as the role of AKAP82, in sperm function is not known. However, it is tempting to speculate that post-translational modifications of this protein might regulate events associated with flagellar bending, such as changes in wave amplitude during hyperactivation of motility.

Although there is no information to date regarding the regulation of a specific protein or process in sperm by protein tyrosine phosphorylation, recent reports implicating protein tyrosine phosphorylation in the modulation of T-type Ca^{2+} channels in mouse spermatogenic cells have potentially important implications with regard to capacitation and its role in preparing sperm to undergo a ZP-induced acrosome reaction. Arnoult et al. (1997) have demonstrated, using whole-cell patch clamp techniques on dissociated mouse spermatogenic cells, that voltage-dependent facilitation of T-currents induced by membrane depolarizations or high frequency stimulations could be mimicked by protein tyrosine kinase inhibitors. In addition, antagonists of protein tyrosine phosphatase activity block this voltage-dependent current facilitation. The authors hypothesize that T channels may be held in a low conductance state by tonic tyrosine phosphorylation and can convert to a high conductance state by dephosphorylation. If T-type Ca^{2+} channels were to play a regulatory role in both capacitation and the ZP-induced acrosome reaction, one could postulate a multi-state model of T-channel conductance controlled by protein tyrosine phosphorylation. T-type Ca^{2+} channels, in such a model, might be partially active in sperm that are not capacitated since the channel is not phosphorylated, and Ca^{2+} movement through channels would initiate signal transduction leading to protein tyrosine phosphorylation and capacitation as described above. One of the substrates that could be phosphorylated during capacitation might be the T-type Ca^{2+} channel which, when phosphorylated, would convert to a low conductance state. This would ensure that changes in Ca^{2+} conductance necessary for capacitation would not continue unabated and result in a precocious induction of the acrosome reaction prior to initiation of exocytosis by a physiological ligand such as the ZP. Binding of capacitated sperm to the ZP would then trigger a rapid dephosphorylation of the T-type Ca^{2+} channel, leading to an increase in conductance and an ensuing acrosome reaction. This ZP-induced dephosphorylation of the T-type Ca^{2+} channel could occur via a phosphoprotein phosphatase. Elements of such a model remain to be tested.

Acrosome Reaction

It is generally accepted that once capacitation has occurred, sperm now possess the ability to fertilize eggs. One of the hallmarks of a capacitated spermatozoon is its ability to undergo a regulated secretory event—the acrosome reaction—in response to the egg's extracellular matrix, the ZP. Researchers from many laboratories, including the Kopf laboratory, have provided evidence that the ZP-induced acrosome reaction has many characteristics of a ligand-receptor-effector activation cascade that has been shown to regulate cellular function in somatic cells. An understanding of the signal transduction events regulating capacitation is likely to help researchers understand signaling events that occur during the acrosome reaction, since it is likely that the identity of phosphorylated substrates and their function may affect the regulatory systems governing the acrosome reaction. One example of such a link may be the phosphorylation of the T- type Ca^{2+} channel. Therefore, just as capacitation and the acrosome reaction are functionally linked, so too are the signal transduction systems regulating these processes. A brief review of events governing sperm-ZP binding and the acrosome reaction will be considered below. The working model for ZP-induced acrosomal exocytosis is shown in Figure 2.

Properties of the Zona Pellucida as a Ligand

The acrosome reaction is essential for fertilization in mammals since it is required for the penetration of the ZP. To date, most information regarding the molecular basis of sperm-ZP interaction has been gleaned from studies in the mouse. Similar results in a variety of other species have allowed researchers to conclude that the ZP is the universal biological trigger of the acrosome reaction in mammalian sperm.

In the mouse, the ZP is synthesized and assembled by the growing oocyte during its growth phase and is comprised of three major glycoproteins, designated as ZP1, ZP2 and ZP3 (Wassarman 1988, 1995). The genes encoding each of these proteins are unique and are under the control of ZP-specific promotors that insure their temporal and tissue-specific expression; the ZP is synthesized only by the oocyte and not by somatic cells. ZP1, ZP2 and ZP3 are all highly glycosylated, and this glycosylation is extremely important for conferring specific biological functions to the ZP. It should be noted that different designations have been given to the components of the ZP in other species; this has led to some confusion in nomenclature. A uniform nomenclature should be carefully considered so as to avoid future confusion. Recent results suggest that the coordinate expression of these individual ZP components is essential for secretion and subsequent assembly of this unique egg-associated extracellular matrix. Following assembly, the ZP is comprised of ZP2/ZP3 heterodimers that are crosslinked in an organized fashion by ZP1 monomers, giving rise to a three-dimensional, relatively insoluble structure that functions biologically as a matrix. It should be noted that the genes encoding the different ZP components in the mouse have been cloned in several other species and the deduced primary polypeptide structures from these other species (including the human) bear remarkable similarity to one another. These data suggest that the primary protein structure of mammalian egg ZP components are similar to one another and that differences observed in biochemical and biological heterogeneity between species may be influenced by the carbohydrate domains. It is notable in this respect that recent studies have demonstrated that the primary structure of the egg vitelline envelopes of certain fish, the extracellular coats of these ancient vertebrates, bear a resemblance to the structure of ZP2 and ZP3 (Wassarman 1995). These data would suggest that specific domains of egg extracellular coats that are involved in sperm recognition and binding are highly conserved in animal evolution.

Binding of Sperm to the Zona Pellucida

Free-swimming, acrosome-intact sperm establish binding to the ZP in a relatively species-specific fashion, and once bound undergo the acrosome reaction. In contrast, acrosome reacted sperm will not initiate binding to the ZP. In the mouse, the sperm-binding and acrosome reaction-inducing activities of the intact ZP are conferred by ZP3, and these biological activities are confined to both the carbohydrate and protein regions of ZP3 (Wassarman 1988; Kopf, Gerton 1991). Moreover, ZP3 binds only to the plasma membrane overlying the acrosome in acrosome-intact sperm, suggesting that this region of the sperm plasma membrane possesses specific ZP3 binding proteins and/or receptors, the identity of which is a subject of much controversy and will be considered below. Presently, most is known about the functional domains involved in sperm binding; these domains are encoded by the O-linked oligosaccharide chains of ZP3 (Wassarman 1995). These functional domains are currently being further refined, and will ultimately be valuable for probing the molecular mechanisms underlying these cell-matrix interactions (Wassarman 1995). Work in other species has demonstrated that the respective ZP3 equivalent also possesses functional equivalence, although detailed functional analysis has yet to be carried out. Although the role of ZP2 has not been directly tested, experiments indicate that this ZP component is involved in anchoring the acrosome reacted sperm on the ZP. In fact, it has been

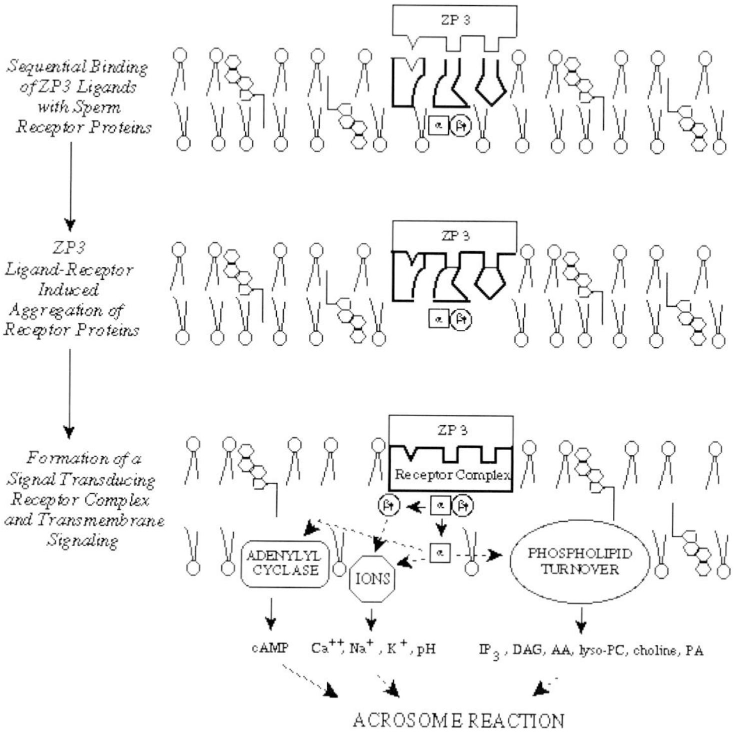

Figure 2. Model for the interaction of the zona pellucida glycoprotein, ZP3, with the plasma membrane overlying the acrosome of mouse spermatozoa to mediate sperm binding and acrosomal exocytosis. In this model, the ZP3 molecule is composed of multiple "functional ligands" ( ) which interact with complementary cell surface binding proteins/receptors present in the sperm plasma membrane. These ZP3-associated ligands are shown as being different from one another in this model, but this does not necessarily have to be the case. Moreover, three ZP3-associated ligands are shown although the actual numbers of ligands involved are not known and are shown in this fashion for illustrative purposes only. The proper interaction of ZP3 with the sperm surface requires the sequential binding of these ligands with sperm-associated binding proteins/receptors. Once these interactions are established, signal transduction is effected by the formation of a functional signal transducing complex which forms in response to ligand-receptor induced aggregation of binding proteins/receptors. Acrosomal exocytosis is then initiated in response to changes in second messengers/ionic conductance that are regulated in response to receptor-mediated signal transduction. In this model a variety of effector systems are shown to be targets for the α and/or $\beta\gamma$ subunits of heterotrimeric G proteins, as indicated by the dashed arrows. cAMP (adenosine-3',5'-cyclic monophosphate); IP_3 (inositol 1,4,5-trisphosphate); AA (arachidonic acid); lyso-PC (lysophosphatidylcholine); PA (phosphatidic acid).

demonstrated that the inner acrosomal membrane, which is retained following the acrosome reaction, is the site for binding of ZP2, suggesting that this particular membrane domain possesses specific ZP2 binding proteins/receptors. It is likely, therefore, that the ZP2/ZP3 heterodimers function together in an integrated manner to regulate sperm-ZP binding and ZP penetration. To date, ZP1 is thought to play solely a structural role in the formation and maintenance of the ZP.

Nature of the Zona Pellucida Binding Proteins/Receptors on Sperm

The identity and function of sperm-associated ZP3 binding proteins/receptors has been quite controversial in recent years, and is still yet to be resolved. Identification of ZP (or ZP3) binding proteins/receptors has been hampered by the lack of information about the precise identity of the active sperm adhesion and acrosome reaction-inducing moieties of the ZP (or ZP3) in any species, as well as the seemingly complex nature of the interactions of these moieties with the sperm surface. These complex interactions do not preclude the possibility that sperm-ZP interactions likely involve multiple interactions between the ZP and the sperm surface, and that sperm-ZP binding and the acrosome reaction result from the formation of functional complexes containing multiple ZP-associated active domains, as well as specific sperm-associated ZP binding proteins and signal transducing receptors (Ward, Kopf 1993). The formation of these functional complexes, when examined at the biochemical level, may manifest themselves as both high- and low-affinity binding interactions. It is premature, therefore, to conclude that there is a single component on the sperm surface that mediates all of these biological events and that this single component has to be a receptor; it is this very perception that has hindered the advancement of this field. For example, although several receptor candidates have been proposed, no one candidate has satisfied all of the properties that one would expect for a specific receptor. These properties include 1) presence in the appropriate region of the sperm cell involved in gamete adhesion and acrosomal exocytosis, 2) specificity and kinetics of ligand binding, 3) presence in numbers on the sperm surface consistent with ligand binding kinetics and biology of the effect, 4) ability to couple to specific signal transduction systems, and 5) appropriate cell and tissue expression. Many candidates have been proposed as binding proteins/receptors. In a majority of cases, these candidates have the ability to bind to carbohydrates and, in some cases, may possess enzymatic activities that may or may not be involved in their putative ZP binding/receptor activity (Table 2, in Tulsiani et al. 1997). It is beyond the scope of this chapter to discuss all of the candidates in detail. Numerous review articles dealing with this topic are available (Kopf, Gerton 1991; Ward, Kopf 1993; Wassarman 1995; Tulsiani et al. 1997). A few candidates that have received the most attention will be discussed briefly here; this does not, in any way, diminish the importance of the other candidates.

Perhaps the most extensively characterized mouse sperm protein that possesses specific binding activity for ZP3 is a protein known as "sp56" (Bookbinder et al. 1995). This protein has lectin-like properties, has properties of a peripheral membrane protein that is associated with the plasma membrane overlying the sperm head, displays appropriate tissue expression and bears homology to the family of complement component 4-binding proteins. The domains on sp56 involved in ZP3 binding and the mechanism by which signal transduction leading to acrosomal exocytosis is integrated with ZP3 binding to this protein are not known. The guinea pig sperm orthologue of sp56 (AM67) has been cloned recently and shown to be present within the acrosomal matrix of the guinea pig sperm. Moreover, re-evaluation of the localization of sp56 in mouse sperm demonstrated that it also was associated with the acrosomal matrix (Foster et al. 1997). This apparent discrepancy in localization of mouse sperm sp56 by two different groups (that is, localization solely to the plasma membrane versus localization to the acrosomal matrix) is of interest and must be resolved by additional experiments. If it is conclusively established that sp56 is a ZP3 binding protein present within the acrosomal matrix, this does not rule out its function in mediating sperm binding to the ZP. It is distinctly possible that acrosome-intact sperm that establish binding to the ZP via ZP3 (or its functional equivalent in other species) may, in fact, be sperm in which the plasma membrane and outer acrosomal membranes have "docked" with one another and have formed intermediate membrane complexes comprising the vesicular face of the outer acrosomal membrane and the extracellular face of the plasma membrane; such membrane docking and formation of intermediate membrane complexes characterize the events of regulated secretion in other secretory model systems. It has been proposed that small fusion pores form that open and close in a dynamic fashion ("flickering pores") may form subsequent to the formation of such intermediate membrane complexes and prior to overt exocytosis. It is entirely possible that stable interactions between an acrosomal matrix-associated sp56 and ZP3 are established during the formation of such a metastable state in which flickering pores precede overt acrosomal exocytosis. The presence of a metastable state following sperm-ZP binding and prior to exocytosis has been documented (Kligman et al. 1991), but has not been characterized in great detail. This new model for sperm-ZP interaction should be examined in greater detail, as it could explain apparent "conflicting" data seen by others using different species regarding the acrosomal status of sperm bound to the ZP. Moreover, such a model might focus attention on a new role for the acrosomal matrix in sperm-ZP interaction and ZP penetration.

A mouse sperm protein designated as p95 was proposed as a ZP3 receptor with properties of a receptor tyrosine kinase (Leyton, Saling 1989). This protein has some characteristics of a membrane protein and it was postulated that p95 possessed intrinsic tyrosine kinase activity that was modulated by ZP3. Subsequent purification of this protein demonstrated it to be a germ cell-specific hexokinase (type I) with unique properties (Kalab et al. 1994; Travis et al. 1998), although its role as a specific ZP3 receptor is now unclear. The reported cloning of the human homolog of mouse p95, which was designated as Hu9 (Burks et al. 1995), has revealed that it is c-mer (Bork et al. 1996), a proto-oncogene of uncertain function and a member of the *axl* family transforming receptor tyrosine kinases. The function of Hu9 as a receptor tyrosine kinase for ZP3 in human sperm is unclear.

A specific form of β-galactosyltransferase (GalTase) has also been implicated as a receptor for ZP3 in mouse sperm (Dubois, Shur 1995). This protein has been postulated to mediate sperm-ZP binding by interacting with oligosaccharide residues specifically on ZP3, and to induce acrosomal exocytosis through GalTase aggregation on the cell surface leading to the activation of sperm heterotrimeric G_i proteins. Targeted over-expression of this form of the enzyme in sperm, predicted in theory to yield sperm that have an enhanced ability to interact with the ZP, yields sperm that, in fact, display a reduced ability to bind to the ZP. This is apparently due to their hypersensitivity to ZP3, such that they undergo acrosome reactions precociously/spontaneously and, therefore, have a reduced avidity of binding. These experiments were interpreted as demonstrating that successful fertilization requires an optimal, rather than a maximal, concentration of GalTase moieties on the sperm surface. In contrast, targeted mutation of the GalTase gene yields null males that are fertile, yet their sperm bind less ZP3 than the sperm from wild-type animals and are unable to undergo a ZP3-induced acrosome reaction (Lu, Shur 1997). The authors conclude that although ZP3 binding and induction of the acrosome reaction are dispensable for fertilization, these properties impart a physiological advantage to sperm for fertilization. It can be concluded that although the sperm-associated GalTase may be important for certain aspects of sperm-ZP interaction, it is not absolutely critical for fertility. These data suggest that GalTase is probably not of primary importance in mediating sperm-ZP3 binding leading to fertilization.

Although a considerable amount of effort by numerous laboratories has identified a number of candidate sperm proteins thought to be involved in sperm-ZP binding and the induction of the acrosome reaction, it is clear that this is a very complex process and that researchers should be rethinking current models for how these interactions occur. It is distinctly possible that some of these candidates discussed here, as well as others not discussed (Table 2 in Tulsiani et al. 1997), may represent only a subset of the complex nature of interactions between the sperm surface and the ZP.

Zona Pellucida-Mediated Signal Transduction

Although there are still many questions regarding the identity of proteins on the sperm surface that mediate ZP binding and acrosomal exocytosis, experiments focusing on mechanisms of transmembrane signal transduction and activation of intracellular effectors have evolved from several laboratories. If one proposes that ZP (or ZP3) effects on sperm are mediated via specific receptors or binding proteins that are co-opted into a functional signaling complex, one could argue that transmembrane signal transduction may have similarities to that seen in somatic cells. Since heterotrimeric G proteins play a key role in signal transduction in response to the activation of several classes of receptors, experiments from several laboratories have focused on the role of these GTP binding proteins in ZP and ZP3-mediated sperm binding and the acrosome reaction. Sperm-associated heterotrimeric G proteins of the G_i class appear to play a key role in regulating ZP or ZP3-mediated signal transduction events leading to the acrosome reaction in mouse, bovine and human sperm (Ward, Kopf 1993). This has been demonstrated by a variety of experimental approaches. First, G_i proteins are present in the membranes overlying the acrosome, that region of the sperm that interacts with ZP3. Second, functional inactivation of G_i proteins in sperm by pertussis toxin treatment does not block the ability of the sperm to establish binding to the ZP, but inhibits the ZP (or ZP3) induced acrosome reaction. Finally, membranes isolated from sperm display G_i protein activation *in vitro* following the addition of ZP3, but not ZP1 or ZP2. Thus, one of the earliest events following ZP or ZP3 interaction with the sperm surface is transmembrane signaling via this particular class of G proteins. Whether other signal transduction events are entrained is unclear. The activation of sperm G proteins, however, appears to be a universal ZP-mediated event.

Intracellular Effectors Mediating the Zona Pellucida-Induced Acrosome Reaction

There is substantial evidence that acrosomal exocytosis is a highly regulated process initiated by the ZP and likely mediated by cell surface binding proteins /receptors and signal transducing G proteins (at least in mammals). It is likely, therefore, that intracellular regulation of acrosomal exocytosis is similar to other receptor-mediated exocytotic events. Such intracellular signals include changes in ionic conductance, changes in cyclic nucleotide metabolism and changes in phospholipid metabolism (Kopf, Gerton 1991). It would also be anticipated that these intracellular effector systems would be modulated in a ZP binding protein/receptor-dependent fashion. Although there have been numerous reports describing ionic and/or second messenger systems purported to play a role in the induction of the acrosome reaction, there is little information linking such effectors to ligands in a receptor-dependent fashion due to the paucity of knowledge of the receptors and the signal transducers. Furthermore, a description of effector systems correlated with the induction of acrosomal exocytosis cannot necessarily be equated with cause and effect.

Studies in both mouse and bull sperm have revealed that elevations in intracellular Ca^{2+}, as well as intracellular pH, represent some of the earliest responses of sperm incubated with ZP or ZP3 (Ward, Kopf 1993; Arnoult et al. 1996). Many of these studies have been performed with Ca^{2+} and pH indicator dyes and have reported localized changes to the acrosomal region. In bovine sperm, ZP-induced Ca^{2+} entry, as well as sperm membrane potential and the acrosome reaction, is dependent upon membrane depolarization, and the Ca^{2+} uptake and acrosome reaction are inhibited by antagonists of voltage-dependent Ca^{2+} channels. Pertussis toxin inhibits the ZP- (and ZP3)-induced pH changes in mouse sperm, as well as the ZP-induced pH and Ca^{2+} changes in bovine sperm, indicating that sperm G_i proteins may regulate such ionic changes. Incubation of bovine sperm under depolarizing conditions, which would activate such voltage-dependent Ca^{2+} channels, bypasses the inhibitory effects of pertussis toxin on the acrosome reaction, suggesting that G_i proteins might regulate such Ca^{2+} channels indirectly. Based on these and other data, a signaling pathway has emerged (Arnoult et al. 1996) whereby ZP3 depolarizes the sperm membrane by activating a pertussis toxin-insensitive pathway with characteristics of a poorly selective cation channel. ZP3 also activates a pertussis toxin-sensitive pathway that produces a transient rise in intracellular pH. Together, the membrane depolarization and rise in intracellular pH open voltage-sensitive Ca^{2+} channels leading to the activation of downstream effectors of the acrosome reaction. These Ca^{2+} channels have characteristics of T-type channels.

Additional studies have suggested that alterations in phospholipid metabolism and/or cyclic nucleotide metabolism may play important intermediary roles in the sperm acrosome reaction (Kopf, Gerton 1991). It has been demonstrated that biologically-active phorbol diesters and diacylglycerols alter the kinetics of the ZP-mediated acrosome reaction in mouse sperm, thus suggesting that this exocytotic event could be regulated in some manner by protein kinase C. This is further supported by the observation that diacylglycerol is formed in mouse sperm in response to the ZP. The products of phospholipase C turnover (for example, IP_3 and *sn*-1,2 diacylglycerol), as well as the role of other phospholipases (A_2 or D), have not yet been examined in sperm challenged with ZP or ZP3. Recently, it has been demonstrated that the mouse ZP stimulates mouse sperm adenylyl cyclase, suggesting that the ZP can alter sperm cAMP metabolism (Leclerc, Kopf 1995). This is consistent with a previous observation that solubilized ZP from mouse eggs can cause transient elevations in sperm cyclic AMP concentrations that are dependent on the presence of extracellular Ca^{2+} (Ward, Kopf 1993). These cyclic AMP elevations precede and are correlated with the induction of the acrosome reaction by the ZP, suggesting that cyclic AMP may be a potential participant in the signaling pathway leading to acrosomal exocytosis. It will be of interest to determine whether such intracellular signaling systems are coupled to sperm G_i proteins, since these second messenger systems are coupled in a receptor-mediated fashion to G proteins in other cell types.

Acknowledgements

The authors thank the members of their lab who have contributed to this work. This work was supported by NIH grants HD-06274, HD-34811, HD-22732 and HD-33052. P.E.V. and X.P.N. were supported by HD-06274 and P.E.V. was supported by the Rockefeller Foundation; H.G.H. was supported by USDA #9502560; M.F. was supported by the Fogarty Foundation.

References

Aitken RJ, Paterso M, Fisher H, Buckingham DW, Van Duin M. Redox regulation of tyrosine phosphorylation in human spermatozoa and its role in the control of human sperm function. J Cell Sci 1995; 108:2017-2025.

Arnoult C, Zeng Y, Florman HM. ZP3-dependent activation of sperm cation channels regulates acrosomal secretion during mammalian fertilization. J Cell Biol 1996; 134:637-645.

Arnoult C, Lemos JR, Florman HM. Voltage-dependent modulation of T-type calcium channels by protein tyrosine phosphorylation. EMBO J 1997; 16:1593-1599.

Austin CR. Observations on the penetration of the sperm into the mammalian egg. Aust J Sci Res 1951; [B]4:581-596.

Bearer EL, Friend DS. Modification of anionic-lipid domains preceding membrane fusion in guinea pig sperm. J Cell Biol 1982; 92:604-615.

Bedford JM. Significance of the need for sperm capacitation before fertilization in eutherian mammals. Biol Reprod 1983; 28:108-120.

Berruti G, Borgonovo B. sp42, the boar sperm tyrosine kinase, is a male germ cell-specific product with a highly conserved tissue expression extending to other mammalian species. J Cell Sci

1996; 109:851-858.
Boatman DE, Robbins RS. Bicarbonate: carbon-dioxide regulation of sperm capacitation, hyperactivated motility, and acrosome reactions. Biol Reprod 1991; 44:806-813.
Bookbinder LH, Cheng A, Bleil JD. Tissue- and species-specific expression of sp56, a mouse sperm fertilization protein. Science 1995; 269:86-89.
Bork P. Sperm-egg binding protein or proto-oncogene? Science 1996; 271:1431-1432.
Burks DJ, Carballada R, Moore HDM, Saling PM. Interaction of a tyrosine kinase from human sperm with the zona pellucida at fertilization. Science 1995; 269:83-86.
Carrera A, Moos J, Ning XP, Gerton GL, Tesarik J, Kopf GS, Moss SB. Regulation of protein tyrosine phosphorylation in human sperm by a calcium/calmodulin-dependent mechanism: identification of A kinase anchor proteins as major substrates for tyrosine phosphorylation. Dev Biol 1996; 180:284-296.
Chang MC. Fertilizing capacity of spermatozoa deposited into the fallopian tubes. Nature 1951; 168:697-698.
Cohen-Dayag A, Eisenbach M. Potential assays for sperm capacitation in mammals. Am J Physiol Cell Physiol 1994; 267:C1167-C1176.
Cross NL. Human seminal plasma prevents sperm from becoming acrosomally responsive to the agonist, progesterone: cholesterol is the major inhibitor. Biol Reprod 1996; 54:138-145.
Cross NL. Role of cholesterol in sperm capacitation. Biol Reprod 1998; 59:7-11.
Cross NL, Overstreet JW. Glycoconjugates of the human sperm surface: distribution and alterations that accompany capacitation *in vitro*. Gam Res 1987; 16:23-35.
Cross NL, Razy-Faulkner P. Control of human sperm intracellular pH by cholesterol and its relationship to the response of the acrosome to progesterone. Biol Reprod 1997; 56:1169-1174.
DasGupta S, Mills CL, Fraser LR. Ca^{2+}-related changes in the capacitation state of human spermatozoa assessed by a chlortetracycline fluorescence assay. J Reprod Fertil 1993; 99:135-143.
Davis BK. Timing of fertilization in mammals: sperm cholesterol/phospholipid ratio as a determinant of the capacitation interval. Proc Natl Acad Sci USA 1981; 78:7560-7564.
Davis BK, Byrne R, Bedigan K. Studies on the mechanism of capacitation: albumin-mediated changes in plasma membrane lipids during *in vitro* incubation of rat sperm cells. Proc Natl Acad Sci USA 1980; 77:1546-1550.
de Lamirande E, Gagnon C. A positive role for the superoxide anion in triggering hyperactivation and capacitation of human spermatozoa. Int J Androl 1993; 16:21-25.
Dubois DH, Shur BD. Cell surface β1,4-galactosyltransferase—a signal transducing receptor? Adv Exp Med Biol 1995; 376:105-114.
Ehrenwald E, Parks JE, Foote RH. Cholesterol efflux from bovine sperm. I. Induction of the acrosome reaction with lysophosphatidylcholine after reducing sperm cholesterol. Gam Res 1988; 20:145-157.
Emiliozzi C, Fenichel P. Protein tyrosine phosphorylation is associated with capacitation of human sperm *in vitro* but is not sufficient for its completion. Biol Reprod 1997; 56:674-679.
Florman HM, Babcock DF. Progress towards understanding the molecular basis of capacitation. In: (Wassarman P, ed) CRC Uniscience-Chemistry of Fertilization. Boca Raton, FL: CRC Press, 1991, pp105-132.
Florman HM, Lemos JR, Arnoult C, Kazam I, Li C, O'Toole CM. A tale of two channels: a perspective on the control of mammalian fertilization by egg-activated ion channels in sperm. Biol Reprod 1998; 59:12-16.
Foster JA, Friday BB, Maulit MT, Blobel C, Winfrey VP, Olson GE, Kim KS, Gerton GL. AM67, a secretory component of the guinea pig sperm acrosomal matrix, is related to mouse sperm protein sp56 and the complement component 4-binding proteins. J Biol Chem 1997; 272:12714-12722.
Galantino-Homer H, Visconti PE, Kopf GS. Regulation of protein tyrosine phosphorylation during bovine sperm capacitation by a cyclic adenosine 3',5'-monophosphate-dependent pathway. Biol Reprod 1997; 56:707-719.
Garty N, Salomon Y. Stimulation of partially purified adenylate cyclase from bull sperm by bicarbonate. FEBS Lett 1987; 218:148-152.
Go KJ, Wolf DP. Albumin-mediated changes in sperm sterol content during capacitation. Biol Reprod 1985; 32:145-153.
Harrison RAP. Capacitation mechanisms, and the role of capacitation as seen in eutherian mammals. Reprod Fertil Dev 1996; 8:581-594.
Harrison RAP, Gadella BM. Membrane changes during capacitation with special reference to lipid architecture. In: (Fenichel P, Parinaud J, eds) Human Sperm Acrosome Reaction. Paris: Colloque INSERM/John Libbey Eurotext Ltd, 1995; 236:45-65.
Harrison RAP, Ashworth PJC, Miller NGA. Bicarbonate/CO_2, an effector of capacitation, induces a rapid and reversible change in the lipid architecture of boar sperm plasma membranes. Mol Reprod Dev 1996; 45:378-391.
Hoshi K, Aita T, Yanagida K, Yoshimatsu N, Sata A. Variation in the cholesterol/phospholipid ratio in human spermatozoa and its relationship with capacitation. Hum Reprod 1990; 5:71-74.
Hunter AG, Nornes HO. Characterization and isolation of a sperm-coating antigen from rabbit seminal plasma with capacity to block fertilization. J Reprod Fertil 1969; 20:419-427.
Hyne RV, Garbers DL. Regulation of guinea pig sperm adenylate cyclase by calcium. Biol Reprod 1979; 21:1135-1142.
Kalab P, Visconti P, Leclerc P, Kopf GS. p95, the major phosphotyrosine-containing protein in mouse spermatozoa, is a form of hexokinase with unique properties. J Biol Chem 1994; 269:3810-3817.
Klauck TM, Faux MC, Labudda K, Langeberg LK, Jaken S, Scott JD. Coordination of three signaling enzymes by AKAP79, a mammalian scaffolding protein. Science 1996; 271:1589-1592.
Kligman I, Glassner M, Storey B, Kopf GS. Zona pellucida-mediated acrosomal exocytosis in mouse spermatozoa: characterization of an intermediate stage prior to the completion of the acrosome reaction. Dev Biol 1991; 145:344-355.
Koehler JK, Gaddum-Rosse P. Media induced alterations of the membrane associated particles of the guinea pig sperm tail. J Ultrastruct Res 1975; 51:106-118.
Kopf GS, Gerton GL. The mammalian sperm acrosome and the acrosome reaction. In: (Wassarman P, ed) Elements of Mammalian Fertilization. Boca Raton, FL: CRC Press Inc, 1991, pp153-203.
Kopf GS, Visconti PE, Galantino-Homer H. Capacitation of the mammalian spermatozoon. In: (Wassarman P, ed) Adv Biochem, JAI Press Inc, 1998; 5:83-107.
Langlais J, Roberts KD. A molecular membrane model of sperm capacitation and the acrosome reaction of mammalian spermatozoa. Gam Res 1985; 12:183-224.
Langlais J, Kan FWK, Granger L, Raymond L, Bleau G, Roberts KD. Identification of sterol acceptors that stimulate cholesterol efflux from human spermatozoa during *in vitro* capacitation. Gam Res 1988; 20:185-201.
Leclerc P, Kopf GS. Mouse sperm adenylyl cyclase: general properties and regulation by the zona pellucida. Biol Reprod 1995; 52:1227-1233.
Leclerc P, de Lamirande E, Gagnon C. Cyclic adenosine 3',5' monophosphate-dependent regulation of protein tyrosine phosphorylation in relation to human sperm capacitation and motility. Biol Reprod 1996; 55:684-692.
Leclerc P, de Lamirande E, Gagnon C. Regulation of protein-tyrosine phosphorylation and human sperm capacitation by reactive oxygen derivatives. Free Rad Biol Med 1997; 22:643-656.
Lee MA, Storey BT. Bicarbonate is essential for fertilization of

mouse eggs; mouse sperm require it to undergo the acrosome reaction. Biol Reprod 1986; 34:349-356.
Leyton L, Saling P. 95 kd sperm proteins bind ZP3 and serve as tyrosine kinase substrates in response to zona binding. Cell 1989; 57:1123-1130.
Lu Q, Shur BD. Sperm from β1,4-galactosyltransferase-null mice are refractory to ZP3-induced acrosome reactions and penetrate the zona pellucida poorly. Development 1997; 124:4121-4131.
Luconi M, Krausz C, Forti G, Baldi E. Extracellular calcium negatively modulates tyrosine phosphorylation and tyrosine kinase activity during capacitation of human spermatozoa. Biol Reprod 1996; 55:207-216.
Moore HDM. Modification of sperm membrane antigens during capacitation. In: (Fenichel P, Parinaud J, eds) Human Sperm Acrosome Reaction. Paris, Colloque INSERM/John Libbey Eurotext Ltd 1995; 236:pp35-43.
Neill J, Olds-Clarke P. A computer-assisted assay for mouse sperm hyperactivation demonstrates that bicarbonate but not bovine serum albumin is required. Gam Res 1987; 18:121-140.
Okamura N, Tajima Y, Soejima A, Masuda H, Sugita Y. Sodium bicarbonate in seminal plasma stimulates the motility of mammalian spermatozoa through the direct activation of adenylate cyclase. J Biol Chem 1985; 260:9699-9705.
Okamura N, Tajima Y, Sugita Y. Decrease in bicarbonate transport activities during epididymal maturation of porcine sperm. Biochem Biophys Res Comm 1988; 157:1280-1287.
Parkkila S, Rajaniemi H, Kellokumpu S. Polarized expression of a band 3-related protein in mammalian sperm cells. Biol Reprod 1993; 49:326-331.
Parrish JJ, Susko-Parrish JL, Uguz C, First NL. Differences in the role of cyclic adenosine 3',5'-monophosphate during capacitation of bovine sperm by heparin or oviduct fluid. Biol Reprod 1994; 51:1099-1108.
Pawson T, Scott JD. Signaling through scaffold, anchoring, and adaptor proteins. Science 1997; 278:2075-2080.
Rochwerger L, Cuasnicú PS. Redistribution of a rat sperm epididymal glycoprotein after *in vitro* and *in vivo* capacitation. Mol Reprod Dev 1992; 31:34-41.
Shalgi R, Matityahn A, Gaunt SJ, Jones R. Antigens on rat spermatozoa with a potential role in fertilization. Mol Reprod Dev 1990; 25:286-296.
Shi Q-X, Roldan ERS. Bicarbonate/CO_2 is not required for zona pellucida- or progesterone-induced acrosomal exocytosis of mouse spermatozoa but is essential for capacitation. Biol Reprod 1995; 52:540-546.
Shouffani A, Kanner BI. Cholesterol is required for reconstitution of the sodium- and chloride-coupled GABA transporter from rat brain. J Biol Chem 1990; 265:6002-6008.
Smith T. The modulation of sperm function by the oviductal epithelium. Biol Reprod 1998; 58:1102-1104.
Spira B. Breitbart H. The role of anion channels in the mechanism of acrosome reaction in bull spermatozoa. Biochim Biophys Acta 1992; 1109:65-73.
Suarez SS. Hyperactivated motility in sperm. J Androl 1996; 17:331-335.
Suarez SS. The oviductal sperm reservoir in mammals: mechanisms of formation. Biol Reprod 1998; 58:1105-1107.
Suzuki F, Yanagimachi R. Changes in the distribution of intramembraneous particles and filipin-reactive membrane sterols during *in vitro* capacitation of golden hamster spermatozoa. Gam Res 1989; 23:335-347.
Tesarik J, Flechon JE. Distribution of sterol and anionic lipids in human sperm plasma membrane: effects of *in vitro* capacitation. J Ultrastruct Res 1986; 97:227-237.
Therien I, Soubeyrand S, Manjunath P. Major proteins of bovine seminal plasma modulate sperm capacitation by high-density lipoprotein. Biol Reprod 1997; 57:1080-1088.
Travis AJ, Foster JA, Rosenbaum NA, Visconti PE, Gerton GL, Kopf GS, Moss SB. Targeting of a germ cell-specific type 1 hexokinase lacking a porin-binding domain to the mitochondria as well as to the head and fibrous sheath of murine spermatozoa. Mol Biol Cell 1998; 9:263-276.
Tulsiani DRP, Yoshida-Komiya H, Araki Y. Mammalian fertilization: a carbohydrate-mediated event. Biol Reprod 1997; 57:487-494.
Uguz C, Vredenburgh WL, Parrish JJ. Heparin-induced capacitation but not intracellular alkalinization of bovine sperm is inhibited by Rp-adenosine-3',5'-cyclic monophosphothioate. Biol Reprod 1994; 51:1031-1039.
Vemuri R, Philipson KD. Influence of sterols and phospholipids on sarcolemmal and sarcoplasmic reticular cation transporters. J Biol Chem 1989; 264:8680-8685.
Verhage HG, Mavrogianis PA, O'Day-Bowman MB, Schmidt A, Arias EB, Donnelly KM, Boomsma RA, Thibodeaux JK, Fazleabas AT, Jaffe RC. Characteristics of an oviductal glycoprotein and its potential role in the fertilization process. Biol Reprod 1998; 58:1098-1101.
Visconti PE, Muschietti JP, Flawia MM, Tezon JG. Bicarbonate dependence of cAMP accumulation induced by phorbol esters in hamster spermatozoa. Biochim Biophys Acta 1990; 1054:231-236.
Visconti PE, Bailey JL, Moore GD, Pan D, Olds-Clarke P, Kopf GS. Capacitation of mouse spermatozoa. I. Correlation between the capacitation state and protein tyrosine phosphorylation. Development 1995a; 121:1129-1137.
Visconti PE, Moore GD, Bailey JL, Leclerc P, Connors SA, Pan D, Olds-Clarke P, Kopf GS. Capacitation of mouse spermatozoa. II. Protein tyrosine phosphorylation and capacitation are regulated by a cAMP-dependent pathway. Development 1995b; 121:1139-1150.
Visconti PE, Johnson L, Oyaski M, Fornés M, Moss SB, Gerton GL, Kopf GS. Regulation, localization, and anchoring of protein kinase A subunits during mouse sperm capacitation. Dev Biol 1997; 192:351-363.
Vredenburgh-Wilberg WL, Parrish JJ. Intracellular pH of bovine sperm increases during capacitation. Mol Reprod Dev 1995; 40:490-502.
Ward CR, Kopf GS. Molecular events mediating sperm activation. Dev Biol 1993; 158:1-26.
Wassarman P. Zona pellucida glycoproteins. Ann Rev Biochem 1988; 57:415-442.
Wassarman P. Towards molecular mechanisms for gamete adhesion and fusion during mammalian fertilization. Curr Biol 1995; 7:658-664.
White DR, Aitken RJ. Relationship between calcium, cyclic AMP, ATP, and intracellular pH and the capacity of hamster spermatozoa to express hyperactivated motility. Gam Res 1989; 22:163-177.
Wolf DE, Hagopian SS, Isogima S. Changes in sperm plasma membrane lipid diffusibility after hyperactivation during *in vitro* capacitation in the mouse. J Cell Biol 1986; 102:1372-1377.
Yanagimachi R. Mammalian fertilization. In: (Knobil E, Neill JD, eds) The Physiology of Reproduction. New York: Raven Press Ltd 1994, pp189-317.
Zarintash RJ, Cross NL. Unesterified cholesterol content of human sperm regulates the response of the acrosome to the agonist, progesterone. Biol Reprod 1996; 55:19-24.
Zeng Y, Clark EN, Florman HM. Sperm membrane potential: hyperpolarization during capacitation regulates zona pellucida-dependent acrosomal secretion. Dev Biol 1995; 171:554-563.
Zeng Y, Oberdorf JA, Florman HM. pH regulation in mouse sperm. Identification of Na^+, Cl^- and HCO_3^- dependent and arylaminobenzoate-dependent regulatory mechanisms and characterization of their role in sperm capacitation. Dev Biol 1996; 173:510-520.

12

How Does the Jelly Coat of Starfish Eggs Trigger the Acrosome Reaction in Homologous Spermatozoa?

Motonori Hoshi
Mayu Kawamura
Yoshinori Maruyama
Eiji Yoshida
Tokyo Institute of Technology

Takuya Nishigaki
Tokyo Institute of Technology
Universidad Nacional Autónoma de México

Masako Ikeda
Manabu Ogiso
Tokyo Institute of Technology

Hideaki Moriyama
Japan Synchrotron Radiation Research Institute

Midori Matsumoto
Tokyo Institute of Technology

The eggs of most animals, though by no means of all, are encased in one or more extracellular investments. These structures protect the egg from biological, chemical and mechanical hazards, and thus sometimes they are called protective coats. Yet they are not simply a protection to the egg, but play crucial roles in sperm-egg interactions for efficient fertilization such as species recognition, as well as self/non-self recognition in self-sterile hermaphrodites, between the gametes. Indeed, egg coats have the specific ligand for sperm binding and specific signal for the induction of acrosome reaction.

In starfish, the outermost egg coat is a relatively thick gelatinous layer called the jelly coat. Upon encountering the jelly coat, the intracellular pH and Ca^{++} concentration of starfish sperm increase in a species-specific manner, which results in the exocytosis of the acrosomal vesicle, namely the acrosome reaction (Dale et al. 1981; Ikadai, Hoshi 1981a; Sase et al. 1995). A long acrosomal process (about 25 μm) thus formed pierces the jelly coat and vitelline coat, and sperm plasma membrane covering the tip of acrosomal process fuses with egg plasma membrane. A jelly component(s) responsible for triggering the acrosome reaction in the starfish *Asterias amurensis* has been sought. Now it is known that the jelly coat of *A. amurensis* eggs contains a high-mannose type neutral glycoprotein, a highly sulfated proteoglycan-like compound(s), peptides and saponins as organic components (Uno, Hoshi 1978; Ikadai, Hoshi 1981a, b; Endo et al. 1987; Nishiyama et al. 1987a; Fujimoto et al. 1987; Nishigaki et al. 1996). It is also known that three of them act in concert on homologous spermatozoa to elicit the acrosome reaction. Those three are a highly sulfated proteoglycan-like compound named acrosome reaction-inducing substance (ARIS), a group of sulfated steroid saponins named Co-ARIS and a group of sperm-activating peptides named asterosap. ARIS alone induces the acrosome reaction only in high Ca^{++} or high pH sea water. In normal sea water, either Co-ARIS or asterosap is also required. Without ARIS, no combination of Co-ARIS and asterosap can induce the acrosome reaction in normal, high calcium or high pH sea water. A mixture of ARIS and Co-ARIS stimulates Ca^{++} uptake by sperm, whereas asterosap increases intracellular pH (Matsui et al. 1986b, c; Nishigaki et al. 1996). A tripartite of ARIS, Co-ARIS and asterosap provides the best condition for the induction of the acrosome reaction in normal sea water; this is reviewed in Hoshi et al. (1994).

This chapter describes the structures of these signal molecules in relation to their biological activities and their mode of action on spermatozoa.

Induction of Acrosome Reaction by Jelly Coat Components

ARIS and Its Receptor

ARIS, P-ARIS and ARIS-Receptor. ARIS is a proteoglycan-like compound of an apparent molecular mass of over 10^7 D, containing fucose, xylose, galactose, glucosamine and galactosamine as constituent monosaccharides. The biological activity of ARIS is species-specific and mostly attributable to sulfated saccharide chains since it is susceptible to periodate oxidation as well as chemical desulfation, but is resistant to pronase digestion (Ikadai, Hoshi 1981a, b; Matsui et al. 1986a; Koyota et al. 1997). A pronase digest of ARIS (P-ARIS) shows an apparent molecular mass of about 10^7 D and an activity similar to that of ARIS.

ARIS, as well as P-ARIS, binds specifically to the anterior tip of the sperm head, suggesting the presence of an ARIS-receptor in that area (Ushiyama et al. 1993). Indeed, ARIS-receptor is specifically localized to a single domain (0.1-0.3 μm in diameter) on the plasma membrane. This site is located near the region occupied by the acrosomal vesicle and the periacrosomal components (Longo et al. 1995).

Structure of Fragment 1. When P-ARIS is sonicated, it is disintegrated into two major fragments, Fragments 1 and 2. Fragment 1 is a protein-free, sulfated polysaccharide consisting of fucose, xylose and galactose in a molar ratio of approximately 3:1:1, with an average molecular mass of 10.4 kD. Fragment 2 contains a much smaller quantity of sugars than Fragment 1, but retains all the five sugars and the protein portion. Fragment 1 is active, but much less than P-ARIS. Fragment 2 is much less active than Fragment 1. Since Fragment 1 retains significant ARIS activity, details of its structure were analyzed using various techniques of glycobiology, such as permethylation analysis, nuclear magnetic resonance (NMR) spectrometry and mass spectrometry, before and after desulfation, periodate oxidation and Smith degradation. Taking all the data of structural analyses into account, it has been concluded that Fragment 1 is a long linear polymer consisting of 10 or 11 pentasaccharide repeating units (Koyota et al. 1997). This structure is confirmed by the isolation of a trisaccharide, Xyl*p*1-3Gal*p*1-(SO_3)3,4Fuc*p*, after mild acid hydrolysis of P-ARIS (Okinaga et al. 1992; Kurono et al. 1998). It is implied from indirect evidence that the main saccharide chain of ARIS is an extremely long linear polymer consisting of one to two thousand of the pentasaccharide repeating units (Fig. 1).

Very recently, the structures of sea urchin ARISs have also been determined: a homopolymer of [3α-L-Gal*p* -2 (OSO_3)-1]$_n$ in *Echinometra lucunter*, linear sulfated α-L-fucans with regular tetrasaccharide repeating units in *Arbacia lixula* and *Lytechinus variegatus*,

A B C

→4)-β-**D**-Xyl*p*-(1→3)-α-**D**-Gal*p*-(1→3)-α-**L**-Fuc*p*-(1→3)-α-**L**-Fuc*p*- (1→4)-α-Fuc*p*-(1→

(residues A and B: 4 ← SO_3^-)

Figure 1. The repeating units of the main saccharide chain of ARIS.

[3α–L-Fuc*p* -2 (OSO_3), 4(±OSO_3)-1]$_n$, and [3α-L-Fuc*p* -2,4(OSO_3)-1→3α-L-Fuc*p* -4(OSO_3)-1→3α–L-Fuc*p* -4(OSO_3)-1]$_n$ in *Strongylocentrotus purpuratus* (Mulloy et al. 1994; Alves et al. 1997; Alves et al. 1998).

Structure-Activity Relationship of Fragment 1. The biological activity of Fragment 1 was determined by means of the acrosome reaction-inducing capacity in high Ca^{++} sea water. With the selective removal of 4-*O*-sulfate of fucosyl residues *A*, the biological activity of Fragment 1 was totally abolished, which demonstrates the importance of that particular sulfate group for the activity. Similarly, modification of the xylosyl residues and fucosyl residues *C* of Fragment 1 by periodate oxidation destroyed the activity even though the sulfated fucosyl residues *A* and *B* were not affected at all. Both the selectively desulfated and the oxidized Fragment 1 lost their specific binding to spermatozoa.

Monoclonal antibodies raised against Fragment 1 inhibited the induction of the acrosome reaction by egg jelly, suggesting that the main saccharide chain consisting of the pentasaccharide repeating unit is indeed crucial for the biological activity of ARIS (M Ikeda, S Koyota, M Kawamura, M Matsumoto, and M Hoshi, unpublished data). These results, together with some other lines of evidence, suggest that the sperm receptor for ARIS primarily recognizes a very specific spatial arrangement of sulfate residues along the main saccharide chain of ARIS.

Molecular modeling of fragment 1. Based on results of the H-nuclear Overhauser effect spectroscopy and fiber diffraction, a three-dimensional modeling of Fragment 1 was attempted by using an in-house software. The plain structure was constructed by means of the Sequence Builder in QUANTA/CHARM system, from CTC Laboratory Systems, with a structure data base for polysaccharides. The model shown in Figure 2 was drawn by a program molscrip (CCP4). The sulfate groups are faced together and are able to coordinate divalent cations such as Ca^{++}. As a result of the super coordination which is likely to happen in sea water, the ARIS barrier traps water more than 100 times in volume. This may explain why ARIS forms a gel.

Co-ARIS

Structure and activity of Co-ARIS. Although saponins are rare in the animal kingdom, two echinoderms are exceptionally rich in saponins: asteroids with steroid saponins called asterosaponins and holothurians with triterpenoid saponins called holothurins (Burnell, ApSimon 1983). The jelly coat of starfish eggs also contains asterosaponins, some of which are known to agglutinate spermatozoa in *A. amurensis* (Uno, Hoshi 1978). Some others are found to serve as a co-factor for ARIS in the induction of the acrosome reaction in normal sea water, and are therefore named Co-ARIS. Among various steroid saponins in the jelly coat, three saponins consisting of a sulfated steroid and a pentasaccharide chain are identified as principal Co-ARIS molecules in *A. amurensis* (Nishiyama et al. 1987a; Fujimoto et al. 1987). The action of Co-ARIS is not species-specific and a Co-ARIS fraction of *Asterina pectinifera* is effective as the co-factor for *A. amurensis* ARIS, and vice versa (Hoshi et al. 1990; Amano et al. 1992b). The sulfate group and the side chain of the steroid are important for the activity of Co-ARIS, while the saccharide chain of Co-ARIS appears to be modified without affecting its activity (Nishiyama et al. 1987b). These results imply that Co-ARIS is taken up into the outer leaflet of sperm plasma membrane and interacts directly or indirectly with a membrane protein(s) involved in the acrosome reaction (for example, ARIS-receptor). It may be not too far-fetched to suggest that the sulfate residue and pentasaccharide moiety regulate the orientation of Co-ARIS in the outer leaflet (Fig. 3).

Asterosap and Its Receptor

Structure and activity of asterosap. Sperm activating peptides were originally found in sea urchins as factors that restore depressed sperm respiration in slightly acidic sea water to the normal level (Ohtake 1976). Since the jelly coat was incorrectly believed to be acidic, their physiological roles had been considered only to maintain sperm motility in the jelly coat (Suzuki et al. 1981). On the contrary, sperm activating peptide is now considered to be a multi-functional peptide essential for efficient fertilization (Ward, Kopf 1993). Eleven glutamine-rich tetratriacontapeptides with an intramolecular disulfide linkage between Cys^8 and Cys^{32} have so far been isolated as asterosaps from the jelly coat of *A. amurensis* eggs (Fig. 4). They are much larger than sea urchin sperm activating peptides and do not show any significant sequence similarities to known proteins. The amino terminal region, where structural diversity of asterosaps is observed, is not important for the activity, but the disulfide linkage is essential. As mentioned before, asterosap does not induce the acrosome reaction by itself, but it is able to induce the acrosome reaction in combination with ARIS. Furthermore, anti-asterosap

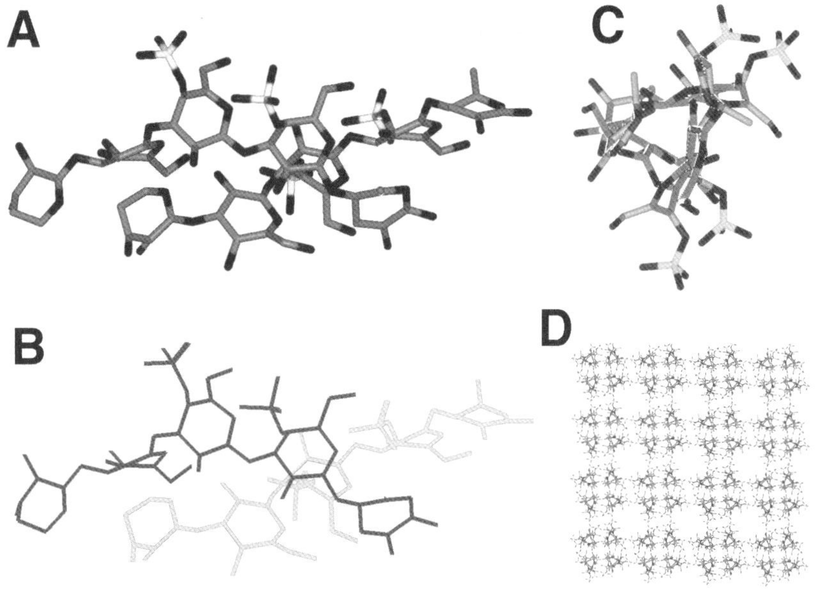

Figure 2. Schematic diagram for the molecular modeling of Fragment 1. In Panels A to C, two chains of the five-sugar repeating units are shown as a double helix. In B, two sugar backbones are differently colored for a contrast. A and B represent side views, whereas C and D are end-on views. A quartet of double helices is shown in D as a possible polymer packing. Hydrogen atoms are not shown.

rabbit antibody significantly decreased the acrosome reaction-inducing activity of the jelly solution, an effect that could be overcome by addition of excess asterosap (Nishigaki et al. 1996). These results support the hypothesis that the main physiological role of sperm activating peptides in the jelly coat is the induction of the acrosome reaction in cooperation with ARIS and Co-ARIS.

Asterosap gene. In sea urchins, it is known that multiple copies of speract (the sperm activating peptide first identified) and its isoforms are encoded in a single mRNA (Ramarao et al. 1990). The question under consideration then was whether or not each female produced multiple isoforms of asterosap. The cDNA clones encoding asterosap contained multiple isoforms and the asterosap mRNA was transcribed only in the oocytes, and not in the follicle cells (Matsumoto et al. 1998). Based upon the deduced amino acid sequences, it was predicted that asterosap precursors are synthesized as a large prepropolypeptides with an unusual "rosary-type" structure made of 10 successive similar stretches of 51-55 residues. Each stretch finishes with a "spacer" of 17-21 residues which is immediately followed by the sequence of an asterosap isoform. The amino terminal of the precursor has 19-21 successive glutamine-rich repeating units. It is also predicted that oocytes may have a unique protease(s) for maturation of asterosaps, since both the N- and C-termini of the spacers are Arg-Lys sequence.

Asterosap-receptor. Asterosap-receptor in *A. amurensis* sperm was then sought (T Nishigaki, K Chiba, M Hoshi, unpublished data; Maruyama et al. 1998). The cell had 1.1 x 10^5 binding sites of high affinity (Kd = 57 pM), and the receptor showed positive cooperativity for asterosap binding. When spermatozoa were treated with fluorophore-labeled asterosaps, sperm flagella were labeled, indicating localization of the receptors in the sperm tail. By using a photo-affinity labeling technique, a 130 kD membrane protein of sperm flagella was identified as the receptor. It was also found that divalent cations are significantly involved in the interaction between asterosap and the receptor.

ARIS and asterosap should act on sperm simultaneously. When starfish spermatozoa are treated in normal sea water with insufficient amounts of jelly to elicit the acrosome reaction, or conversely, with sufficient jelly but in Ca^{++}-deficient sea water, within a minute they lose the capacity to undergo the acrosome reaction and accompanying ionic changes in response to newly added jelly and/or Ca^{++} in excess. Furthermore, they seem to lose their capacity to react to the jelly coat *in situ*. Similarly, if spermatozoa are treated with P-ARIS (or ARIS) and asterosap not simultaneously, but sequentially in either order with an interval over minutes, they do not undergo the acrosome reaction and become incompetent to the jelly (Matsui et al. 1986b). A similar phenomenon is known to occur also in *Asterina pectinifera* (Amano et al. 1992a).

Nevertheless, those spermatozoa undergo the acrosome reaction just like intact spermatozoa if intracellular Ca^{++} concentration and intracellular pH are increased by the treatment with a calcium ionophore (A23187) and monensin (Matsui et al. 1986b; Amano et al. 1992a). Such effects of pretreatments are found with ARIS, P-ARIS and asterosap, but not with Co-ARIS. This effect of asterosap is not simply attributable to its ability to raise intercellular pH, because spermatozoa are still responsive to the jelly even after intercellular pH is significantly increased in alkaline sea water. The spermatozoa appear, therefore, to keep intact the machinery that mediates ionic changes in the cytoplasm and exocytosis of the acrosome. These data suggest that the

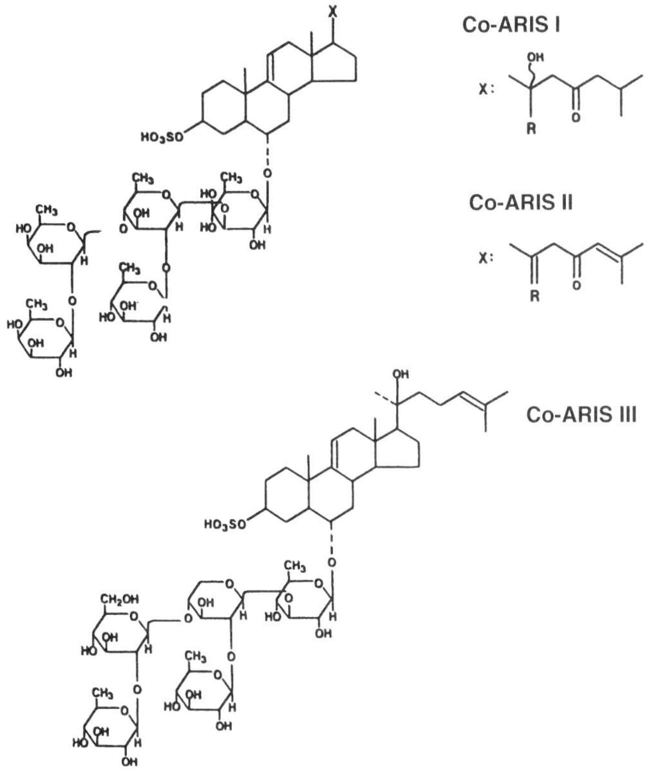

Figure 3. The structures of Co-ARIS I, II and III. Co-ARIS I and II are expressed in the hydrated form at 6-deoxy-xylo-hexos-4-ulose. (From Fujimoto et al. 1987 and Nishiyama et al. 1987b, with a single modification.)

pretreatment effects are due to an irreversible change(s) in the steps not later than the stimulation of Ca^{++} channels and Na^+/H^+ exchangers, presumably the signal transduction machinery mediating the receptors and the channels and exchangers.

The capacity of ARIS/P-ARIS to render spermatozoa incompetent to the jelly by pretreatment is parallel in many features to its capacity to induce the acrosome reaction under favorable conditions (Matsui et al. 1986b). Therefore, it is likely that the two apparently different effects of ARIS upon spermatozoa actually result from a single action of ARIS, presumably to the sperm-membrane receptors. This may be true also for the effects of asterosap.

A Brief Prospect

After more than 20 years of struggling with starfish jelly coat, the cartoon in Figure 5 can be drawn, summarizing the effects of the three signal molecules responsible for triggering the acrosome reaction in starfish spermatozoa: ARIS, Co-ARIS and asterosap. The precise structures of most Co-ARIS saponins and asterosaps are now known, yet knowledge on the chemical structure of ARIS is much limited. Obviously it is urgent to figure out the molecular structures of ARIS protein(s), other saccharide chains besides Fragment 1, and linkage region(s) of saccharide chains to the protein. Naturally, researchers must understand the three-dimensional structures of ARIS, especially Fragment 1 region, as the spermatozoa do. Then much more about spermatozoa must be known, especially the receptor for ARIS, the plasma membrane component(s) that interact with Co-ARIS, ion channels and exchangers, signal transduction systems and so on. Although science is still a long way from understanding the molecular mechanism underlying the acrosome reaction, a door to the next step in elucidating gamete interactions is about to be opened.

Summary

In the starfish, *Asterias amurensis*, three components in the jelly coat of eggs, namely acrosome reaction-inducing substance (ARIS), Co-ARIS and asterosap, act in concert on homologous spermatozoa to elicit the acrosome reaction species-specifically. ARIS is a proteoglycan-like compound which binds to the receptor localized to a small domain in the anterior portion of sperm heads. Its activity is mostly attributable to a sulfated polysaccharide chain composed of one thousand or so repeating units of a sulfated pentasaccharide. Co-ARIS is a group of saponins consisting of a sulfated steroid and a pentasaccharide. In combination with ARIS, it increases intracellular Ca^{++} concentration of sperm. Asterosap is a group of glutamine-rich tetratriacontapeptides with a disulfide linkage. It increases intracellular pH of sperm through binding to its receptor on the plasma membrane of flagella, and thus facilitates the acrosome reaction and restores sperm motility and respiration to a normal level if depressed.

Acknowledgements

The authors are indebted to the directors and staffs of the Akkeshi Marine Biological Station, Asamushi Marine Biological Station, Misaki Marine Biological Station, Otsuchi Marine Research Center, Tateyama Marine Biological Station, Sugashima Marine Biological Station and Ushimado Marine Biological Station for their unfailing courtesy in sharing the stations' facilities. This work was supported in part by Grants-in-Aid for Scientific Research on Priority Areas (No. 09240208; No. 10178102) from the Ministry of Education, Science, Sports and Culture of Japan and a grant from Mizutani Foundation for Glycoscience to M.H.

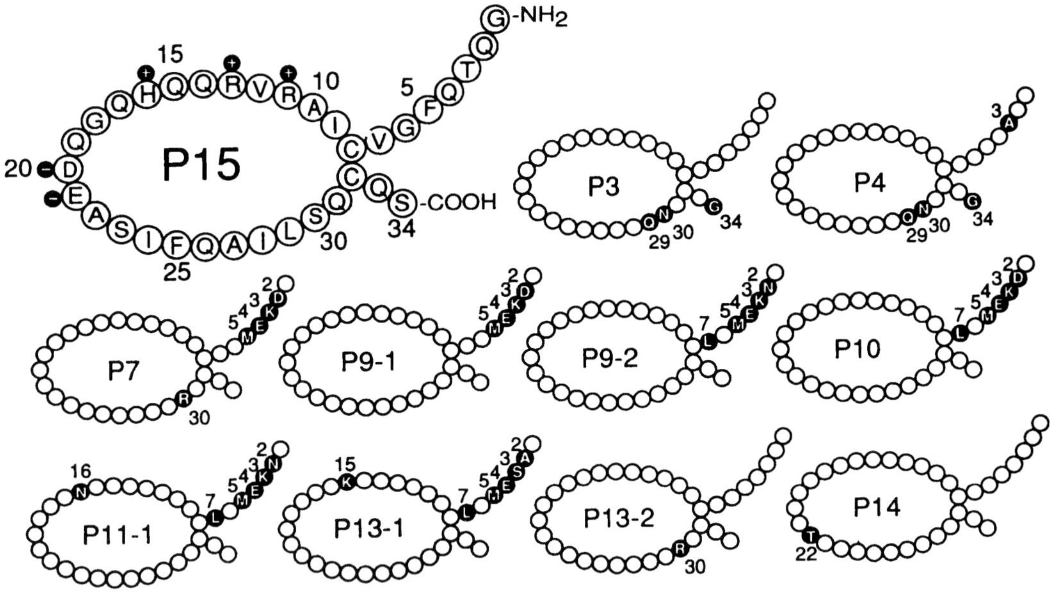

Figure 4. The structures of asterosaps. P15 is shown as the representative. Others are shown as isoform peptides and the filled circles represent amino acid changes compared with P15. The numbers indicate positions counted from the N-terminal. Ionized form of basic and acidic residues are shown in P15.

Figure 5. A cartoon summarizing the effects of ARIS, Co-ARIS and asterosap. ARIS, Co-ARIS and asterosap in the jelly coat of starfish eggs act in concert on the homologous spermatozoa to elicit the acrosome reaction in seconds or so. Combinations of ARIS and either Co-ARIS or asterosap induce the acrosome reaction in normal sea water but the rate is much slower than the tripartite. It also shows that, if sperm are treated only with ARIS or asterosap, they become incompetent to the jelly coat or its components.

References

Alves AP, Mulloy B, Diniz JA, Mourao PA. Sulfated polysaccharides from the egg jelly layer are species-specific inducers of acrosomal reaction in sperms of sea urchins. J Biol Chem 1997; 272:6965-6971.

Alves AP, Mulloy B, Moy GW, Vacquier VD, Mourao PA. Females of the sea urchin *Strongylocentrotus purpuratus* differ in the structures of their egg jelly sulfated fucans. Glycobiology 1998; 8:939-946.

Amano T, Okita Y, Matsui T, Hoshi M. Pretreatment effects of jelly components on the sperm acrosome reaction and histone degradation in the starfish, *Asterina pectinifera*. Biochem Biophys Res Commun 1992a; 187:263-273.

Amano T, Okita Y, Okinaga T, Matsui T, Nishiyama I, Hoshi M. Egg jelly components responsible for histone degradation and acrosome reaction in the starfish, *Asterina pectinifera*. Biochem Biophys Res Commun 1992b; 187:274-278.

Burnell DJ, ApSimon JW. Echinoderm saponins. In: (Scheuer PJ, ed) Marine Natural Products, Chemical and Biological Perspectives. Vol 5. New York: Academic Press, 1983, pp287-389.

Dale B, Dan-Sohkawa M, De Santis A, Hoshi M. Fertilization of the starfish *Astropecten aurantiacus*. Exp Cell Res 1981; 132:505-510.

Endo T, Hoshi M, Endo S, Arata Y, Kobata A. Structures of the sugar chains of a major glycoprotein present in the egg jelly coat of a starfish, *Asterias amurensis*. Arch Biochem Biophys 1987; 252:105-112.

Fujimoto Y, Yamada T, Ikekawa N, Nishiyama I, Matsui T, Hoshi M. Structure of acrosome reaction-inducing steroidal saponins form the egg jelly of the starfish, *Asterias amurensis*. Chem Pharm Bull 1987; 35:1829-1832.

Hoshi M, Amano T, Okita Y, Okinaga T, Matsui T. Egg signals for triggering the acrosome reaction in starfish spermatozoa. J Reprod Fertil 1990; 42:s23-s31.

Hoshi M, Nishigaki T, Ushiyama A, Okinaga T, Chiba K, Matsumoto M. Egg-jelly signal molecules for triggering the acrosome reaction in starfish spermatozoa. Int J Dev Biol 1994; 38:167-174.

Ikadai H, Hoshi M. Biochemical studies on the acrosome reaction of the starfish, *Asterias amurensis*. I. Factors participating in the acrosome reaction. Dev Growth Differ 1981a; 23:73-80.

Ikadai H, Hoshi M. Biochemical studies on the acrosome reaction of the starfish, *Asterias amurensis*. II. Purification and characterization of acrosome reaction-inducing substance. Dev Growth Differ 1981b; 23:81-88.

Koyota S, Wimalasiri KSS, Hoshi M. Structure of the main saccharide chain in the acrosome reaction-inducing substance (ARIS) of the starfish, *Asterias amurensis*. J Biol Chem 1997; 272:10372-10376.

Kurono K, Ohashi Y, Hiruma K, Okinaga T, Hoshi M, Hashimoto H, Nagai Y. Characterization of the sulfated fucose-containing trisaccharides by fast atom bombardment tandem mass spectrometry in the study of the acrosome reaction-inducing substance of the starfish, *Asterias amurensis*. J Mass Spectrometry 1998; 33:35-44.

Longo FJ, Ushiyama A, Chiba K, Hoshi M. Ultrastructural localization of acrosome reaction-inducing substance (ARIS) on sperm of the starfish, *Asterias amurensis*. Mol Rep Dev 1995; 41:91-99.

Maruyama Y, Nishigaki T, Matsumoto M, Hoshi M. Identification and partial characterization of a sperm receptor for sperm-activating peptides of starfish *Asterias amurensis*. Zygote 1998; 6:S108.

Matsui T, Nishiyama I, Hino A, Hoshi M. Induction of the acrosome reaction in starfish. Dev Growth Differ 1986a; 28:339-348.

Matsui T, Nishiyama I, Hino A, Hoshi M. Acrosome reaction-inducing substance purified from the egg jelly inhibits the jelly-induced acrosome reaction in starfish: an apparent contradiction. Dev Growth Differ 1986b; 28:349-357.

Matsui T, Nishiyama I, Hino A, Hoshi M. Intracellular pH changes of starfish sperm upon acrosome reaction. Dev Growth Differ 1986c; 28:358-368.

Matsumoto M, Briones AV, Nishigaki T, Hoshi M. Sequence analysis of cDNA encoding precursor of asterosaps. Zygote 1998; 6:SS109.

Mulloy B, Ribeiro AC, Alves AP, Vieira RP, Mourao PA. Sulfated fucans from echinoderms have a regular tetrasaccharide repeating unit defined by specific patterns of sulfation at the O-2 and O-4 positions. J Biol Chem 1994; 269:22113-22123.

Nishigaki T, Chiba K, Miki W, Hoshi M. Structure and function of asterosaps, sperm-activating peptides from the jelly coat of starfish eggs. Zygote 1996; 4:237-245.

Nishiyama I, Matsui T, Hoshi M. Purification of Co-ARIS, a cofactor for acrosome reaction-inducing substance, from the egg jelly of starfish. Dev Growth Differ 1987a; 29:161-169.

Nishiyama I, Matsui T, Fujimoto Y, Ikekawa N, Hoshi M. Correlation between the molecular structure and the biological activity of Co-ARIS, a cofactor for acrosome reaction-inducing substance. Dev Growth Differ 1987b; 29:171-176.

Ohtake H. Respiratory behavior of sea-urchin spermatozoa. II. Sperm-activating substance obtained from jelly coat of sea-urchin eggs. J Exp Zool 1976; 198:313-322.

Okinaga T, Ohashi Y, Hoshi M. A novel saccharide structure, Xyl1-3Gal1-(SO_3^-)3,4Fuc-, is present in acrosome reaction-inducing substance (ARIS) of the starfish, *Asterias amurensis*. Biochem Biophys Res Commun 1992; 186:405-410.

Ramarao CS, Burks DJ, Garbers DL. A single mRNA encodes multiple copies of the egg peptide speract. Biochemistry 1990; 29:3383-3388.

Sase I, Okinaga T, Hoshi M, Feigenson GW, Kinoshita K Jr. Regulatory mechanisms of the acrosome reaction revealed by multiview microscopy of single starfish sperm. J Cell Biol 1995; 131:963-973.

Suzuki N, Nomura K, Ohtake H, Isaka S. Purification and the primary structure of sperm-activating peptides from the jelly coat of sea urchin eggs. Biochem Biophys Res Commun 1981; 99:1238-1244.

Uno Y, Hoshi M. Separation of the sperm agglutinin and the acrosome reaction-inducing substance in egg jelly of starfish. Science 1978; 200:58-59.

Ushiyama A, Araki T, Chiba K, Hoshi M. Specific binding of acrosome reaction-inducing substance to the head of starfish spermatozoa. Zygote 1993; 1:121-127.

Ward CR, Kopf GS. Molecular events mediating sperm activation. Dev Biol 1993; 158:9-34.

13

Signaling for Exocytosis: Lipid Second Messengers, Phosphorylation Cascades and Cross-Talks

E.R.S. Roldan

Instituto de Bioquímica (CSIC)

At the time of fertilization, the spermatozoon undergoes a process of regulated exocytosis, the "acrosome reaction," which results in the release or exposure of enzymes contained in the acrosomal granule. Completion of exocytosis allows the fertilizing spermatozoon to penetrate oocyte vestment(s) and to fuse with the oolemma (Yanagimachi 1994).

Regulated exocytosis involves three stages: agonist-receptor interaction; intracellular signaling (signal transduction through the plasma membrane, generation of messengers, and activation of phosphorylation cascades); and membrane fusion.

Today, there is a wide variety of agonists (and corresponding receptors) capable of initiating exocytosis in mammalian spermatozoa. It is thought that the main agonists of exocytosis are the glycoprotein zona pellucida 3 (ZP3; Bleil, Wassarman 1980, 1983) and progesterone (Meizel et al. 1990). Progesterone may be trapped in the matrix of the cumulus oophorus and/or could be produced by cumulus cells themselves (Schuetz, Dubin 1981) and, in fact, would be encountered by the spermatozoon when it penetrates through the oocyte coats. Both agonists may interact during initiation of exocytosis, with progesterone priming the sperm cell to respond to ZP3 action (Roldan et al. 1994a). This is in agreement with the observation that the presence of cumulus oophorus or cumulus cells around the oocytes is beneficial for *in vitro* fertilization in several species, especially when the sperm concentration is low or the incubation conditions are sub-optimal; this is reviewed by Yanagimachi (1994).

It has been proposed that other agonists may also have a physiological role in the initiation of acrosomal exocytosis. Some agonists may, indeed, act as modulators or co-factors, but it is difficult to understand how others may act until it is demonstrated that they are actually present in the vicinity of the oocyte at the time of fertilization. Among the various agonists proposed to have a role in the initiation of exocytosis, the following can be included: epidermal growth factor (EGF), as shown in Naz and Ahmad (1992), Lax et al. (1994) and Murase and Roldan (1995); and atrial natriuretic peptide (ANP), as shown in Anderson et al. (1994) and Zamir et al. (1995); but Silvestroni et al. (1992) and Anderson et al. (1995) suggested a role for ANP in motility. Also included in the list are prolactin (Mori et al. 1989), interleukin 6 (Naz, Kaplan 1994), and *c-kit* (Stem Cell Factor), as seen in Feng et al. (1997, 1998). γ-Aminobutyric acid (GABA) can also be included in this list since it has been found that it triggers acrosomal exocytosis (Roldan et al. 1994a; Shi, Roldan 1995; Shi et al. 1997) and is present in oviductal fluid (László et al. 1992). However, it is also possible that GABA effects are just related to activation of receptors usually targeted by progesterone under physiological conditions. This, therefore, means that GABA should be considered as a probe rather than as a physiological agonist.

One of the very early responses generated upon agonist-receptor interaction is the activation of ion fluxes. Among these, Ca^{2+} plays a very important role and it has been demonstrated that this cation is essential for acrosomal exocytosis (Yanagimachi 1994). Ca^{2+} is necessary for the activation of intracellular enzymes and for the actual fusion of membranes, and various Ca^{2+}-dependent steps in the sequence underlying acrosomal exocytosis have been identified (Roldan, Fragio 1993a). The regulation of sperm intracellular Ca^{2+} has been reviewed recently (Yanagimachi 1994; Baldi et al. 1996).

This review will concentrate on intracellular signaling processes leading to generation of lipid messengers and activation of phosphorylation cascades that are triggered in spermatozoa during acrosomal exocytosis and that may be important in eliciting membrane fusion, the final stage of exocytosis. First, it will be addressed which lipid metabolites are generated upon stimulation of spermatozoa, and whether they have a role as messengers. Second, it will be discussed which phosphorylation cascades are activated in spermatozoa, and what interactions may exist between them. Third, it will be argued that other signaling mechanisms (such as the cAMP-protein kinase A pathway, or nitric oxide) may also have important roles in the process of exocytosis. Finally, possible cross-talks between pathways will be summarized.

Lipid Messengers

Diacylglycerol (DAG)

Generation of DAG. Stimulation of capacitated spermatozoa with the natural agonists progesterone or zona pellucida (ZP) causes a rise in diacylglycerol (DAG). This has been observed in mouse spermatozoa treated with progesterone or ZP (Roldan et al. 1994a) and in human spermatozoa stimulated with progesterone (O'Toole et al. 1996a). Treatment with the Ca^{2+} ionophore A23187 also leads to generation of DAG in a variety of species (ram: Roldan, Harrison 1992; mouse: Roldan et al. 1994a; man: O'Toole et al. 1996a; boar: Vazquez, Roldan 1997b). This suggests that generation of DAG takes place after Ca^{2+} entry when sperm cells are stimulated under physiological conditions. In fact, capacitated mouse or human spermatozoa stimulated with natural agonists do not show an elevation of DAG if Ca^{2+} entry is inhibited by inclusion of Ca^{2+} channel blockers (Roldan et al. 1994a; O'Toole et al. 1996b).

Sources of DAG. There are three common sources of DAG (Fig. 1): hydrolysis of phosphoinositides by a specific phospholipase C, hydrolysis of phospholipids other than the phosphoinositides by another type of phospholipase and hydrolysis of phospholipids such as phosphatidylcholine (PC) by phospholipase D (PLD), resulting in generation of phosphatidic acid (PA), followed by conversion of PA to DAG by phosphatidic acid phosphohydrolase (PPH).

Hydrolysis of phosphatidylinositol 4,5-bisphosphate (PIP_2) and phosphatidylinositol 4-phosphate (PIP) by a phosphoinositide (PPI)-specific phospholipase C (PPI-PLC), also known as phosphoinositidase C (PIC), has been demonstrated after treatment with A23187 in spermatozoa of a variety of species (Roldan, Harrison 1989). Similarly, hydrolysis of these phosphoinositides has been demonstrated in mouse (Roldan et al. 1994a) and human (Thomas, Meizel 1989) spermatozoa after stimulation with physiological agonists of exocytosis. It is not well known which types of PIC are involved in generation of DAG in spermatozoa in response to stimulation with natural agonists; this is reviewed in Roldan (1995). Evidence for PIC-γ activation, modulated by tyrosine phosphorylation, has been presented for mouse sperm (Tomes et al. 1996, Feng et al. 1998). On the other hand, there is no clear evidence for the involvement of a PIC-β activated by a pertussis toxin-insensitive GTP-binding protein (Roldan 1995). The presence of other types of PICs has been inferred: spermatozoa may have a PIC activated by a pertussis toxin-sensitive GTP-binding protein (G_o or G_i type), or a Ca^{2+}-dependent isoform, perhaps PIC-δ (Roldan 1995).

It has been argued that in somatic cells, the main route for DAG formation from phospholipids other than the phosphoinositides involves the concerted action of PLD and PPH (Asaoka et al. 1996; Hodgkin et al. 1998) and the same has been assumed for mammalian spermatozoa (Naor, Breitbart 1997). However, the possible role of a phosphatidylcholine (PC)-specific phospholipase C (PC-PLC) has also been recognized in somatic cells (Quest et al. 1996; Bocckino, Exton 1996) and a series of studies has shown that, in fact, this may be the major route for DAG generation in mammalian spermatozoa.

The role of the PLD-PPH pathway in the generation of DAG seems to be limited in spermatozoa, and the DAG generated by this pathway appears to have little importance in the sequence of signaling events culminating in membrane fusion. Evidence gathered using ram, mouse, boar and human spermatozoa, stimulated with A23187, progesterone or ZP, revealed no or very little generation of PA. In no case was PA generated before DAG, and transphosphatidylation in the presence of ethanol (a reaction that unambiguously identifies PLD activity; Hodgkin et al. 1998) was of no or very little quantitative importance (Roldan, Dawes 1993; Roldan et al. 1994a; O'Toole et al. 1996a; Vazquez, Roldan 1997b). Furthermore, stimulation of spermatozoa in the presence of ethanol, or the PPH inhibitor propanolol, did not affect the occurrence of the acrosome reaction (Roldan, Dawes 1993).

On the other hand, spermatozoa stimulated with A23187, progesterone or ZP experienced hydrolysis of diacyl- or alkylacyl-PC, with a concomitant rise in DAG or alkylacylglycerol, indicative of PLC activity (Roldan, Murase 1994; Roldan et al. 1994a; O'Toole et al. 1996a). This agrees well with an earlier study showing that mammalian sperm have a PC-specific PLC localized to the acrosomal region (Sheikhnejad, Srivastava 1986). This PC-PLC has been purified and shown to generate DAG from [1-^{14}C]dioleoyl-PC, but not from similarly labeled phosphatidylinositol or phosphatidylethanolamine (Sheikhnejad, Srivastava 1986); this is coincident with the source of DAG identified in a study in which sperm phospholipids were labeled with radioactive precursors (Roldan, Murase 1994). Thus, it appears that in mammalian spermatozoa DAGs are being generated directly by PLC-mediated hydrolysis of PC.

DAG as substrate for PC re-synthesis. Upon sperm stimulation, a major rise in intracellular DAG has been observed, as summarized above. Nevertheless, using a novel chromatographic separation system (Roldan, Murase 1994), a different pattern of DAG changes was identified in a peculiar pool of sperm DAG.

Spermatozoa have, under resting conditions (that is, non-stimulated cells), a high level of disaturated DAGs (that is, with saturated fatty acids in positions 1 and 2 of the glycerol backbone). Studying the labeling of sperm phospholipids with a variety of radioactive precursors, it was possible to distinguish the kinetics of this pool of disaturated DAG from the kinetics of a pool of saturated/unsaturated-DAG (that is, with saturated fatty acids in position 1 and unsaturated fatty acids in position 2). The latter pool of saturated/unsaturated-DAG was in equilibrium with PC, which also has saturated fatty acids in position 1 and unsaturated fatty acids in position 2 of the glycerol backbone (Vazquez, Roldan 1997a). Interestingly, upon sperm stimulation to undergo acrosomal exocytosis, there was a rapid decrease in PC, followed by re-synthesis of this lipid (Vazquez, Roldan 1997b). This re-synthesis of PC was accompanied by a decrease in the pool of disaturated-DAG, and changes in monacylglycerol and free fatty acids that suggested that there was, in parallel, a process of conversion of disaturated-DAG to saturated/disaturated DAG and use of the latter for synthesis of PC *de novo* (Vazquez, Roldan 1997b). The regulation of this pathway, as well as the one supplying CDP-choline, is unknown and it deserves attention in the future.

DAG as messenger. It has become clear from various studies of signal transduction during acrosomal exocytosis that DAG has a central role as lipid messenger (Roldan 1995). DAG is now known to activate sperm PKC (O'Toole et al. 1996c), and phospholipase A_2

(Roldan, Fragio 1994) and also to have a positive feedback effect on the PC-specific PLC (Roldan, Murase 1994).

Phosphatidic Acid (PA)

Generation of PA. In sperm cells where the PC pool was labeled with fatty acids, little generation of phosphatidic acid (PA) was seen upon stimulation either with A23187 or with natural agonists of exocytosis. This suggests that PLD-mediated hydrolysis of PC is not a major pathway for the generation of this metabolite (Fig. 1). However, when spermatozoa were pre-labeled with $[^{32}P]P_i$, which results in the incorporation of label via the ATP pool, stimulation of spermatozoa led to a considerable formation of $[^{32}P]$PA (Roldan, Harrison 1989, 1990). Under these conditions, $[^{32}P]$PA is formed by the action of DAG kinase on the DAG generated by PIC, as confirmed by results showing that PA formation was blocked by DAG kinase inhibitors (Roldan, Harrison 1990, 1992; O'Toole et al. 1996a).

Very little PA seems to be formed from the DAG that is generated by PC-PLC (Roldan, Murase 1994; Roldan et al. 1994a; O'Toole et al. 1996a). The reason for this may be that the fatty acid composition of PC-derived DAG is different from the PPI-derived DAG and that DAG kinase has a substrate preference for DAG deriving from phosphoinositides.

Role of PA. Several studies have addressed whether PA generated by DAG kinase has any role in events leading to acrosomal exocytosis. It seems that this PA has no special role because prevention of PA formation by inclusion of DAG kinase inhibitors did not reduce the percentage of cells undergoing exocytosis; in fact, exocytosis was enhanced, suggesting that the DAG is the important metabolite (Roldan, Harrison 1992; O'Toole et al. 1996a). Furthermore, addition of exogenous PA had no effect on the time-course of exocytosis (Roldan, Harrison 1992).

Phosphoinositides as Messengers

Generation of 3-phosphorylated phosphoinositides. The principal route for synthesis of phosphatidylinositol 3,4,5-trisphosphate (PIP_3) is by 3-phosphorylation of PIP_2 by a phosphoinositide 3-kinase (PI 3-kinase). Also, 3-phosphorylation of phosphatidylinositol has been documented (Duckworth, Cantley 1996). In somatic cells, formation of PIP_3 takes place by recruitment of cytosolic PI 3-kinase to membranes in response to various agonists, and involves regulation by either GTP-binding proteins or tyrosine phosphorylation (by either cytosolic tyrosine kinase or receptors with intrinsic tyrosine kinase activity) and by Ras and members of the Rho/Rac/CDC42 family.

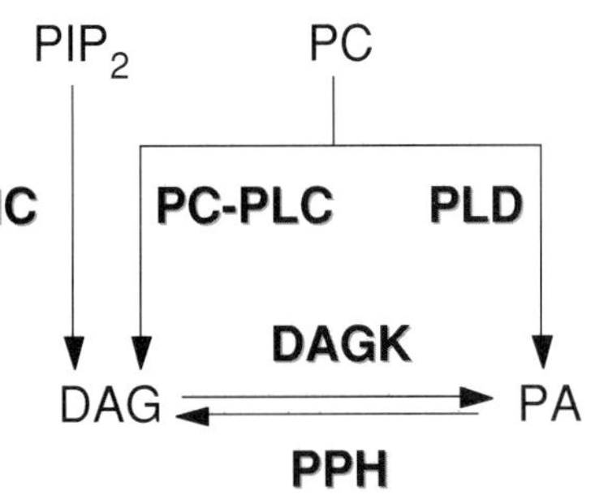

Figure 1. Pathways for the generation of diacylglycerol (DAG) and phosphatidic acid (PA). DAG can be generated via hydrolysis of phosphatidylinositol 4,5-bisphosphate (PIP_2) by phosphoinositidase C (PIC), or from phospholipids such as phosphatidylcholine (PC) by the action of PC-specific phospholipase C (PC-PLC). PA can be generated from PC by phospholipase D (PLD). PA can be converted to DAG by phosphatidic acid phosphohydrolase (PPH). The opposite reaction is mediated by DAG kinase (DAGK).

Role of 3-phosphorylated phosphoinositides. PIP_3 seems to act as messenger underlying processes of cell growth and differentiation (Duckworth, Cantley 1996), as well as exocytosis (Umehara et al. 1997; Chasserot Golaz et al. 1998). Recent work has revealed, by immunodetection, the presence of PI 3-kinase in the acrosomal region of mouse sperm and that wortmannin (a PI 3-kinase inhibitor) blocked agonist-induced acrosomal exocytosis (Feng et al. 1998). It is therefore likely that this pathway would be important in controlling events leading to membrane fusion. A number of targets of, or responses to, PI 3-kinase have been identified, including proteins that directly bind and are modulated by the lipid products including PIP_3, proteins that are phosphorylated by the protein kinase activity of a PI-3 kinase subunit (p110), proteins that bind to PI 3-kinase subunits and are thought to be downstream and other cellular responses downstream of PI 3-kinase with unknown biochemical links; this is reviewed in Duckworth and Cantley (1996) and Duronio et al. (1998). Some of these targets (such as protein kinase C or the mitogen activated protein kinase) are present in spermatozoa and it would therefore be important to determine if they are activated by PI 3-kinase during acrosomal exocytosis.

LysoPC and Arachidonic Acid

LysoPC and arachidonic acid are generated by the action of phospholipase A_2 (PLA_2) on PC. It has to be borne in mind that the action of PLA_2 on other phospholipids such as phosphatidylethanolamine (PE) or phosphatidylinositol (PI) would yield lysoPE or lysoPI, respectively, and that these lysophospholipids may also be important metabolites (Roldan, Fragio 1993b).

Furthermore, arachidonic acid is not the only fatty acid in PC, PE or PI. Other unsaturated fatty acids (such as oleic, linoleic or linolenic acid, or docosahexaenoic acid) are also present in position 2 of these phospholipids and, upon release by the action of PLA_2, could also have a biological function.

Activation of PLA_2 during acrosomal exocytosis. It is now clear that PLA_2 plays an essential role in the release of fatty acids and lysophospholipids involved in sperm membrane fusion during acrosomal exocytosis (Roldan, Fragio 1993a, b). PLA_2 is activated in human (Baldi et al. 1993) and boar spermatozoa (Roldan, Vazquez 1996) in response to progesterone but, to the best of the author's knowledge, no studies have yet reported ZP-induced activation of PLA_2. Interestingly, reagents known to activate GTP-binding proteins, a transduction event involved in ZP-triggered acrosome reaction but not progesterone-induced exocytosis (Murase, Roldan 1996), are capable of activating PLA_2 (Dominguez et al. 1996).

Types of PLA_2 in spermatozoa. PLA_2 constitutes a large superfamily of enzymes (Dennis 1994, 1997) and several groups (I through IX) of PLA_2s have been established based on sequence information. PLA_2s are either secreted, low molecular weight enzymes (13-18 kD) which, in general terms, require millimolar levels of Ca^{2+}, or they are cytosolic enzymes of larger size (two groups of 85 kD forms, and one of 29 kD) and require either no Ca^{2+} or micromolar levels of the cation (Dennis 1997 and references therein).

There is, unfortunately, little information on which are the PLA_2 isoenzymes present in mammalian spermatozoa. Efforts have been directed towards a biochemical characterization of PLA_2 in sperm from various species and information agrees on the fact that sperm PLA_2 requires millimolar levels of Ca^{2+} (Roldan, Mollinedo 1991 and references therein). It should nevertheless be noted that in most studies, protocols of acid extraction of PLA_2 have been used and thus any possibility of detecting high molecular weight isoenzymes has been lost with this procedure.

Partial purification and characterization has revealed the existence of a sperm PLA_2 of about 14-16 kD. Unfortunately, there is only limited sequence data, and it is thus difficult to assign the sperm PLA_2 to any of the groups recognized by Dennis (1997). In any case, partial sequencing of the N-terminal region of a human sperm PLA_2 has revealed some similarities with secretory PLA_2s of groups I and II (from snake venom and porcine/human pancreas), although this human sperm PLA_2 appears to represent a novel sequence (Langlais et al. 1992). Interestingly, antibodies raised against cobra (*Naja naja*) venom recognize a 16 kD protein in SDS-extracts from bull spermatozoa (Weinman et al. 1986). Similarly, antibodies raised against porcine pancreas PLA_2 recognize a protein in hamster and human sperm (Riffo et al. 1992; Riffo, Parraga 1997). The antibodies have also been used in immunolocalization work; PLA_2 has been detected in the acrosomal region, as well as in other sperm compartments. The physiological relevance of these findings is suggested by the fact that Fab fragments of the antibody against porcine pancreas PLA_2 are capable of inhibiting acrosomal exocytosis in hamster spermatozoa (Riffo, Parraga 1996).

In somatic cells, a cytosolic PLA_2 ($cPLA_2$) of larger size (85 kD) has been found to be activated during exocytosis. There is no indication as to whether this isoenzyme is present in mammalian sperm. In any case, recent experiments in which the gene coding for this $cPLA_2$ has been mutated showed that homozygous male mice with the mutations had no impairment in their fertility (Uozumi et al. 1997; Bonventre et al. 1997), suggesting that this $cPLA_2$ may not be essential in events underlying sperm function, including exocytosis.

Roles of lysoPC and fatty acids. These metabolites seem to serve as co-activators of some types of protein kinase C, but their major role appears to relate to perturbation of cell membranes during fusion and this has been extensively investigated in mammalian spermatozoa (Yanagimachi 1994). It is interesting that not all lysophospholipids or fatty acids are capable of exerting this role. Thus, lysoPC and lysoPI were capable of enhancing acrosomal exocytosis, whereas lysophosphatidylserine (lysoPS) was not (Roldan, Fragio 1993b); in fact, lysoPS appears to have an inhibitory role under certain conditions (Roldan, Fleming 1989). Similarly, some fatty acids can enhance exocytosis, whereas others cannot (Roldan, Fragio 1993b).

Further to its role in membranes, one type of lysoPC (alkyl-lysoPC) may serve as substrate for the formation of an important lipid: alkyl-acetyl-PC (also known as "platelet activating factor" or PAF). It has been reported that this lipid can stimulate acrosomal exocytosis (Roldan, Fragio 1993b; Luconi et al. 1995) although it is not yet clear whether PAF acts on a membrane receptor in an autocrine fashion (Luconi et al. 1995). Fatty acids may also have other roles, since they are substrates for cyclooxygenase or lipooxygenase, which generate eicosanoids. These metabolites are also thought to have a role in events leading to membrane fusion.

Phosphorylation Cascades

Protein Kinase C

Protein kinase C (PKC) is a serine/threonine kinase with various isoforms that have been classified into three groups. The first group includes the conventional PKCs (cPKCs: α, βI, βII and γ), which are activated by Ca^{2+}, DAG, phosphatidylserine (PS), free fatty acids and lysoPC. The second group encompasses the novel

PKCs (nPKCs: δ, ε, η, μ and θ) which are activated by DAG and PS, but are Ca^{2+}-independent. The third group consists of the atypical PKCs (aPKCs: ζ, λ, and I) which are DAG- and Ca^{2+}-independent, but require PS and other activators such as free fatty acids (Asaoka et al. 1996).

Early work done on ram spermatozoa stimulated with A23187 and phorbol esters (a standard approach employed to characterize PKC activation in somatic cells) failed to identify PKC activity (Roldan, Harrison 1988). Subsequent efforts by various investigators have resulted in biochemical evidence in favor of PKC activity in spermatozoa from various mammalian species, and have demonstrated that spermatozoa treated with phorbol esters (reagents known to activate PKC) experienced acrosomal exocytosis; this is reviewed by Naor and Breitbart (1997). In addition, circumstantial evidence of PKC activation after ZP stimulation was derived from experiments in which spermatozoa failed to undergo an acrosome reaction if a PKC inhibitor was included (Liu, Baker 1997). Nevertheless, direct evidence of PKC activation and substrate phosphorylation remained elusive for a long time.

Recent work has now shown activation of sperm PKC and phosphorylation of various substrates in response to a natural agonist of exocytosis (O'Toole et al. 1996c). Stimulation of capacitated human spermatozoa with progesterone resulted in phosphorylation of eight different proteins (size from ~20 to 220 kD), and this effect was inhibited by two specific PKC inhibitors. Moreover, phosphorylation of the same substrates was observed after treatment with phorbol esters and the effect was again prevented by the PKC inhibitors. Assessment of completion of exocytosis revealed a tight coupling between phosporylation in response to progesterone and the occurrence of the acrosome reaction under similar conditions, thus providing evidence that PKC is involved in the exocytotic process.

Several immunocytochemical studies have localized PKC both in the head and in the tail of spermatozoa. However, differences were found in the localization of PKCs in bull and human spermatozoa and, furthermore, in the isoenzymes detected in these two species within the acrosomal region. Bull sperm have PKCβI throughout the acrosomal region (Lax et al. 1997), whereas human sperm have PKCβII in the equatorial region (Rotem et al. 1990). With a less specific antibody (recognizing PKC α,β,γ), reactivity was also found in the head region, again with different localization in human (equatorial region) and bull (post-acrosomal region) sperm (Rotem et al. 1990; Breitbart et al. 1992). In mouse spermatozoa, an unidentified PKC isoform has been immunodetected in the acrosomal region (Feng et al. 1998).

Treatment with phorbol esters resulted in translocation of PKCα and PKCβI isoenzymes (in non-capacitated bovine sperm), and this translocation was also found to be Ca^{2+}-dependent (Lax et al. 1997). Nevertheless, it has been claimed that phorbol esters can trigger exocytosis in human sperm in a Ca^{2+}-independent fashion (Rotem et al. 1992) and furthermore, that PKC may modulate Ca^{2+} influx (Naor, Breitbart 1997). This disagrees with findings indicating that PKC activation is, in fact, dependent on a prior increase in intracellular levels of Ca^{2+} (O'Toole et al. 1996c). It has been observed that progesterone-stimulated DAG generation (necessary for PKC activation) and PKC-mediated phosphorylation of substrates were inhibited by inclusion of the Ca^{2+} channel blocker verapamil (O'Toole et al. 1996b, c). Furthermore, the rise in intracellular Ca^{2+} stimulated by progesterone was neither prevented by various PKC inhibitors, nor affected by short- or long-term treatment with phorbol esters (Bonaccorsi et al. 1998). Thus, these results strongly suggest that PKC activation is downstream of Ca^{2+} influx and, in fact, dependent on the increase in intracellular levels of Ca^{2+} (Roldan, Fraser 1998).

As for the substrates phosphorylated by PKC, there is still no evidence regarding their identity but it has been postulated that one of them (in the range of 40 kD) may be a mitogen-activated protein kinase (Luconi et al. 1998b).

MAP Kinase

The Mitogen-Activated Protein (MAP) kinases are a group of protein serine/threonine kinases that are activated in response to a variety of extracellular stimuli and mediate signal transduction from the cell surface to the nucleus (Su, Karin 1996). Together with other signaling pathways they can modulate the phosphorylation status of transcription factors. In this way, MAP kinase cascades are involved in cell proliferation and differentiation. MAP kinase is also involved in the regulation of exocytosis as seen, for example, in chromaffin and natural killer cells (Cox, Parsons 1997; Milella et al. 1997).

Three major types of MAP kinase cascades have been identified in mammalian cells. The most widely studied is that involving the MAP kinases known as Extracellular signal-Regulated Kinases (ERK1 and 2). A general scheme of activation of this cascade (Marshall 1994; Fig. 2) involves the activation of receptor tyrosine kinases by growth factors which provide the binding site of the adapter protein Grb2 which, in turn, localizes Sos to the plasma membrane. Sos activates the small molecular weight G-protein Ras which then binds directly to Raf, a serine/threonine kinase. Binding of Ras to Raf is not sufficient to activate Raf but the additional component needed for its activation is unknown; PKC has been

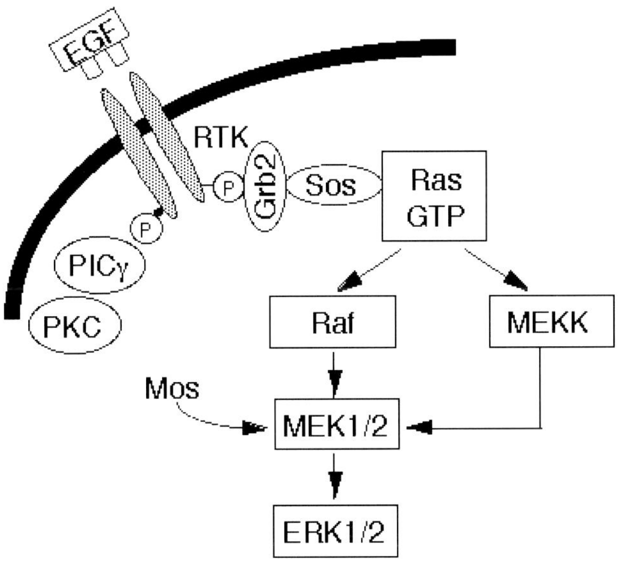

Figure 2. Phosphorylation cascade activated by binding of an agonist such as EGF to its receptor with intrinsic tyrosine kinase activity (RTK). Activation of this cascade leads to phosphorylation and activation of MAP kinases.

implicated as the other component needed to activate Raf. Active Raf phosphorylates MAP kinase/ERK kinase (MEK), a dual-specificity kinase that phosphorylates ERK1/ERK2 on threonine and tyrosine residues to activate it. MEK can also be phosphorylated by the protein kinase Mos and by MEK kinase.

The presence of Ras, which is upstream in the pathway of activation of MAP kinase, has been demonstrated in human spermatozoa by Naz et al. (1992). Furthermore, localization of MAP kinases ($p42^{ERK}$ and $p44^{ERK}$) in human spermatozoa, at the postacrosomal region, but with no detection in the tail or in the acrosome, has also been reported (Luconi et al. 1998a). Thus, human spermatozoa have a Ras/MAP kinase pathway that may underlie changes leading to membrane fusion during acrosomal exocytosis.

Stimulation of human spermatozoa with progesterone, or A23187, resulted in phosphorylation and activation of MAP kinases ($p42^{ERK}$ and $p44^{ERK}$), with redistribution of the proteins from the post-acrosomal region to the equatorial segment of the sperm head (Luconi et al. 1998a, b). Inhibition of the MAP kinase cascade by the inhibitor PD098059 did prevent both progesterone-induced activation (phosphorylation) of MAP kinase, and redistribution of the protein to the equatorial segment, but it did not inhibit the acrosome reaction (Luconi et al. 1998b). Thus, it remains to be established what role, if any, MAP kinases have in acrosomal exocytosis.

Other Signaling Systems

Nitric Oxide Synthase and Nitric Oxide

Nitric oxide synthase (NOS) exists in three isoforms. There are two constitutive, Ca^{2+}-dependent forms, the neuronal NOS (nNOS), also known as brain NOS (bNOS), and the endothelial NOS (eNOS), and a third form which is inducible (iNOS) and Ca^{2+}-independent. Both nNOS and eNOS catalytic activities are triggered by Ca^{2+} entering cells and binding to calmodulin, which stimulates enzyme activity. NOS produces nitric oxide (NO) by deamination of L-arginine, with stoichiometric generation of L-citrulline (Bredt, Snyder 1994). NO binds to the haem group of soluble guanylyl cyclase (sGC), which is the major target for the physiological effects of NO. This stimulates cGMP formation which may activate cyclic nucleotide-gated Ca^{2+} channels or may activate protein kinase G which may, in turn, phosphorylate ion channels (Bredt, Snyder 1994; Mayer, Hemmens 1997).

NO may be an important intracellular messenger in the regulation of exocytosis as seen, for example, in chromaffin cells (Oset-Gasque et al. 1994, 1998). In spermatozoa, NOS has been detected in the acrosome and tail of mouse and human sperm cells using an antibody against nNOS (Herrero et al. 1996a). Evidence gathered using specific NOS inhibitors suggests that this enzyme, and NO, could be involved in progesterone-induced acrosomal exocytosis (Herrero et al. 1996b). However, it is not clear what type of NOS is present in spermatozoa. Antibodies against each of the three isoforms recognized a protein of about 140 kD; this result is intriguing, although it could be due to the fact that the three isoforms have a 50% amino acid identity (Herrero et al. 1997). It is likely that spermatozoa do not have iNOS because of their lack of synthetic capacity; further support for this idea comes from the finding that NOS activity was only affected by inhibitors of the constitutive type of NOS, with no effect exerted by inhibitors of iNOS (Herrero et al. 1997).

The role of NOS in spermatozoa is not well understood. In this context it is important to review work leading to generation of mice with targeted disruptions of nNOS or eNOS. Disruption of nNOS does not appear to affect fertility of mutant homozygous male mice (Huang et al. 1993) and, similarly, homozygous eNOS mutant mice are fertile and indistinguishable from wild-type and heterozygous littermates (Huang et al. 1995). However, it is important to bear in mind that absence of one NOS isoform can be compensated by the presence of the other isoform, so it is difficult to draw firm conclusions from these observations. Thus, the finding that doubly mutant $nNOS^-$ and $eNOS^-$ mice are fertile (Son et al. 1996) suggests that these NOS isoforms may not have an essential role in events leading to completion of acrosomal exocytosis. Alternatively, these results may also indicate a degree of redundancy in pathways underlying exocytosis. In any case, the analysis of signal transduction pathways in spermatozoa lacking NOS would be extremely useful to understand the mechanism leading to membrane fusion.

Adenylyl Cyclase and Cyclic AMP

Stimulation of human sperm with progesterone leads to a Ca^{2+}-dependent increase in cAMP (Parinaud, Milhet 1996) and although there is no direct evidence showing that treatment with ZP causes a generation of cAMP, it has been found that ZP can stimulate adenylyl cyclase in isolated mouse sperm membranes (Leclerc, Kopf 1995). The fact that cAMP elevation is triggered by Ca^{2+} ionophore (Garbers et al. 1982; Parinaud, Milhet 1996) supports the idea that cAMP could be generated after Ca^{2+} influx.

cAMP acts as second messenger activating protein kinase A (PKA). Immunocytochemical studies showing localization of the RIα and RIβ regulatory subunits of PKA in the acrosomal region (Vijayaraghavan et al. 1997) are consistent with the idea that the cAMP/PKA pathway may have a role in acrosomal exocytosis. Furthermore, results obtained using reagents that alter generation and catabolism of cAMP, permeable analogs of cAMP that increase endogenous levels of cAMP, and inhibitors of PKA strongly suggest that this pathway is involved in exocytosis; this is reviewed in Baldi et al. (1996) and Naor and Breitbart (1997).

Nevertheless, it is not clear which is the role of the cAMP/PKA pathway during exocytosis in spermatozoa, since targets for this pathway during signaling for membrane fusion have not been identified. During capacitation, this pathway activates a tyrosine kinase (Visconti et al. 1995; Galantino-Homer et al. 1997), so one possibility is that PKA may target the same substrate during exocytosis. It is possible that cAMP targets an early step in the sequence underlying exocytosis, such as cyclic nucleotide-gated Ca^{2+} channels (Weyand et al. 1994) but this is not entirely in agreement with the finding that cAMP is generated after Ca^{2+} entry.

It is also possible that the cAMP/PKA pathway is involved in the final stages of membrane fusion. It has been recently found that the cAMP/PKA pathway would interact (probably via phosphorylation of rabphilin) with Rab3, a group of Ras-like small GTP-binding proteins, and metabolites generated by phospholipase A_2, to stimulate exocytosis (Garde, Roldan 1996).

"Cross-Talks"

A number of pathways are activated upon sperm stimulation. Some of these pathways serve to generate lipid messengers whereas others involve phosphorylation cascades. It is now clear that there are various interactions between these pathways.

PKC Regulation of PLA_2

DAG may activate a small molecular weight PLA_2 directly or by causing changes in the lipid bilayer, because PLA_2 activity was found to be enhanced by diglycerides in *in vitro* PLA_2 assays (Roldan et al. 1994b) and in labeled sperm cells exposed to diglycerides (Roldan, Fragio 1994). In addition to this role, DAG may also activate a putative cytosolic PLA_2 (85 kD) via a prior activation of PKC. This, however, may not take place directly, since there is no evidence at this time that PKC phosphorylates $cPLA_2$ (Leslie 1997). Phosphorylation of $cPLA_2$ may thus be a two-step process, with PKC first phosphorylating MAP kinase and this enzyme in turn phosphorylating PLA_2 (Lin et al. 1993). No evidence exists for this cross-talk in mammalian sperm; it would be interesting to find out whether it is present, especially considering the apparent lack of a role for MAP kinase in exocytosis (Luconi et al. 1998b).

cAMP/PKA Regulation of Pathways Generating Lipid Messengers

It has been reported that cAMP may activate, inhibit, or have no effect on phosphoinositide-specific PIC (Park et al. 1992; Alava et al. 1992; Bold et al. 1995; Tsai et al. 1995). On the other hand, and to the best of the author's knowledge, there is no clear evidence of cAMP-mediated regulation of PC-PLC.

With regard to interactions with PLD, it has been found that the cAMP/PKA pathway could either activate (Ginsberg et al. 1997) or inhibit PLD (Ruan et al. 1998), depending on the cell system under study. The relevance of this potential cross-talk in mammalian spermatozoa is unknown, and should be explored, taking into account the limited importance of PLD in events leading to membrane fusion during exocytosis.

cAMP/PKA Regulation of Other Phosphorylation Cascades

Evidence has been presented demonstrating the ability of the cAMP/PKA pathway to phosphorylate MAP kinase (Frödin et al. 1994; Young et al. 1994). Activated MAP kinase may then target PLA_2 (Lin et al. 1993), although the significance of MAP kinase in acrosomal exocytosis remains to be established. On the other hand, it is possible that some form of cross-talk between the cAMP/PKA pathway and PKC may take place because it has been found that suboptimal concentrations of reagents known to activate the cAMP/PKA pathway and PKC have an additive effect in stimulating the acrosome reaction (Doherty et al. 1995).

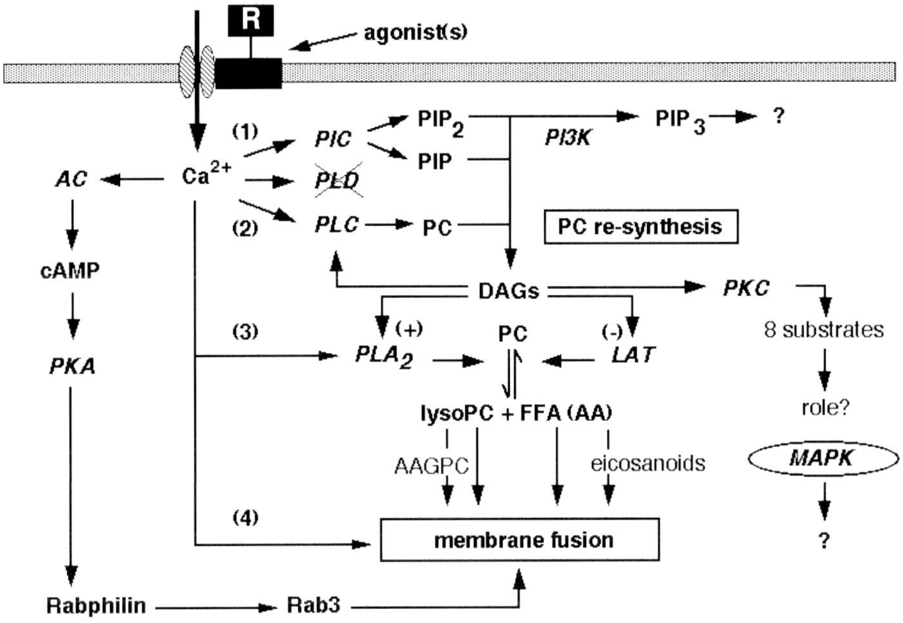

Figure 3. A model of signaling events activated by Ca^{2+} influx in spermatozoa and leading to membrane fusion during acrosomal exocytosis. Pathways for the generation of lipid messengers and for phosphorylation of enzymes and other substrates are shown.

Conclusions

There is now a variety of agonists capable of initiating acrosomal exocytosis, in addition to the two well recognized natural agonists, progesterone and ZP. Receptors for most of these "alternative" agonists have been identified in spermatozoa and one can therefore ask the question, "Are all these receptors physiologically relevant?" In addition, various signaling pathways have now been identified in spermatozoa. It is not clear, however, if all intracellular signals are related to exocytosis, or to the other main sperm function, motility. More detailed studies, including analyses of time-course of events, may help to differentiate between signaling processes underlying each of the two sperm functions. Furthermore, it would also be relevant to ask whether all signaling events activated upon sperm stimulation are really essential, or whether there is a certain degree of redundancy. Finally, it is also worth considering whether some of the signaling events detected so far in mature spermatozoa could actually be relics of signaling pathways important for the last stages of differentiation during spermiogenesis or epididymal maturation.

Pathways for which evidence has been found in mammalian spermatozoa are summarized in Figure 3. It is important to realize that there are differences between this model and hypotheses which draw from evidence obtained in somatic cells, since the latter do not distinguish pathways not yet demonstrated to be activated during acrosomal exocytosis.

Finally, many gaps still exist in the information available on signaling during acrosomal exocytosis and various issues need to be addressed in the future including, perhaps, the following:

- re-synthesis of PC *de novo*, and its regulation,
- role of DAGs in regulating other signal transduction mechanisms (for example, G proteins),
- role of DAGs in the activation of different PKC isoenzymes,
- identity of substrates phosphorylated by PKC,
- role of MAP kinase in the sequence leading to membrane fusion,
- characterization of sperm PLA_2s, and their regulation,
- the regulation of Ca^{2+} at various steps in the sequence underlying exocytosis,
- and, how fusion is triggered and what metabolites (including fusigenic proteins) participate in the fusion between plasma membrane and the membrane of the acrosomal granule.

Acknowledgements

The author wishes to thank colleagues who over the years contributed with discussions and their efforts to develop ideas and work described here: Julian Garde, Cristina Fragio, Lynn Fraser, Robin Harrison, Robin Irvine, Rosa Martinez-Dalmau, Faustino Mollinedo, Tetsuma Murase, Chris O'Toole, Qi-Xian Shi and Juan Maria Vazquez. Financial support from the following organizations made the work possible: Biotechnology and Biological Sciences Research Council (UK), The Wellcome Trust (UK), The Rockefeller Foundation (USA), The Lalor Foundation (USA), JRF Ltd. (UK) and the Ministry of Education and Science (Spain).

References

Alava MA, Debell KE, Conti A, Hoffman T, Bovini E. Increased intracellular cyclic AMP inhibits inositol phospholipid hydrolysis induced by perturbation of the T-cell receptor CD3 complex but not by G-protein stimulation. Biochem J 1992; 284:189-199.

Anderson RA, Feathergill KA, Drisdel RC, Rawlins RG, Mack SR, Zaneveld LJD. Atrial natriuretic peptide (ANP) as a stimulus of the human acrosome reaction and a component of ovarian follicular fluid. Correlation of follicular fluid ANP content with *in vitro* fertilization outcome. J Androl 1994; 15:61-74.

Anderson RA, Feathergill KA, Rawlins RG, Mack SR, Zaneveld LJD. Atrial natriuretic peptide. A chemoattractant of human spermatozoa by a guanylate cyclase-dependent pathway. Mol Reprod Dev 1995; 40:371-378.

Asaoka Y, Tsujishita Y, Nishizuka Y. Lipid signaling for protein kinase C activation. In: (Bell RM, Exton JH, Prescott SM, eds) Lipid Second Messengers. New York: Plenum Press, 1996, pp59-74.

Baldi E, Falsetti C, Krausz C, Gervasi G, Carloni V, Casano R, Forti G. Stimulation of platelet-activating factor synthesis by progesterone and A23187 in human spermatozoa. Biochem J 1993; 292:209-216.

Baldi E, Luconi M, Bonaccorsi L, Krausz C, Forti G. Human sperm activation during capacitation and acrosome reaction: role of calcium, protein phosphorylation and lipid remodeling pathways. Frontiers Biosci 1996; 1:189-205.

Bleil JD, Wassarman PM. Mammalian sperm-egg interaction: identification of a glycoprotein in mouse egg zonae pellucidae possessing receptor activity for sperm. Cell 1980; 20:873-882.

Bleil JD, Wassarman PM. Sperm-egg interactions in the mouse: sequence of events and induction of the acrosome reaction by a zona pellucida glycoprotein. Dev Biol 1983; 95:317-324.

Bocckino SB, Exton JH. Phosphatidic acid. In: (Bell RM, Exton JH, Prescott SM, eds) Lipid Second Messengers. New York: Plenum Press, 1996, pp75-123.

Bold RJ, Ishizuka J, Townsend CM, Tompson JC. Secretin potentiates cholecystokinin-stimulated amylase release by AR4-2J cells via stimulation of phospholipase C. J Cell Physiol 1995; 165:172-176.

Bonaccorsi L, Krausz C, Pecchioli P, Forti G, Baldi E. Progesterone-stimulated intracellular calcium increase in human spermatozoa is protein kinase C-independent. Mol Hum Reprod 1998; 4:259-268.

Bonventre JV, Huang Z, Taheri MR, O'Leary E, Li E, Moskowitz MA, Sapirstein A. Reduced fertility and postischemic brain injury in mice deficient in cytosolic phospholipase A_2. Nature 1997; 390:622-625.

Bredt DS, Snyder SH. Nitric oxide: A physiologic messenger molecule. Ann Rev Biochem 1994; 63:175-195.

Breitbart H, Lax Y, Rotem R, Naor Z. Role of protein kinase C in the acrosome reaction of mammalian spermatozoa. Biochem J 1992; 281:473-476.

Chasserot Golaz S, Hubert P, Thierse D, Dirrig S, Vlahos CJ, Aunis D, Blader MF. Possible involvement of phosphatidylinositol 3-kinase in regulated exocytosis: studies in chromaffin cells with inhibitor LY294002. J Neurochem 1998; 70:2347-2356.

Cox ME, Parsons SJ. Roles for protein kinase C and mitogen-activated protein kinase in nicotine-induced secretion from bovine adrenal chromaffin cells. J Neurochem 1997; 69:1119-1130.

Dennis EA. Diversity of group types, regulation and function of phospholipase A_2. J Biol Chem 1994; 269:13057-13060.

Dennis EA. The growing phospholipase A_2 superfamily of signal transduction enzymes. Trends Biochem Sci 1997; 22:1-2.

Doherty CM, Tarchala SM, Radwanska E, De Jonge CJ. Characterization of two 2nd messenger pathways and their interactions in eliciting the human sperm acrosome reaction. J Androl 1995; 16:36-46.

Dominguez L, Yunes RMF, Fornes MW, Mayorga LS. Acrosome reaction stimulated by the GTP non-hydrolyzable analog GTP-γ-S is blocked by phospholipase A_2 inhibitors in human spermatozoa. Int J Androl 1996; 19:248-252.

Duckworth BC, Cantley LC. PI 3-kinase and receptor-linked signal transduction. In: (Bell RM, Exton JH, Prescott SM, eds) Lipid Second Messengers. New York: Plenum Press, 1996, pp125-175.

Duronio V, Scheid MP, Ettinger S. Downstream signalling events regulated by phosphatidylinositol 3-kinase activity. Cell Signalling 1998; 10:233-239.

Feng H, Sandlow JI, Sandra A. Expression and function of the *c-kit* proto-oncogene protein in mouse sperm. Biol Reprod 1997; 57:194-203.

Feng H, Sandlow JI, Sandra A. The *c-kit* receptor and its possible signaling transduction pathway in mouse spermatozoa. Mol Reprod Dev 1998; 49:317-326.

Frödin M, Peraldi P, Van Obberghen E. Cyclic AMP activates the mitogen activated protein kinase cascade in PC12 cells. J Biol Chem 1994; 269:6207-6214.

Galantino-Homer HL, Visconti PE, Kopf GS. Regulation of protein tyrosine phosphorylation during bovine sperm capacitation by a cyclic adenosine 3',5'-monophosphate-dependent pathway. Biol Reprod 1997; 56:707-719.

Garbers DL, Tubb DJ, Hyne RV. A requirement of bicarbonate for Ca^{2+}-induced elevations of cyclic-AMP in guinea pig spermatozoa. J Biol Chem 1982; 257:8980-8984.

Garde J, Roldan ERS. rab3 peptide stimulates exocytosis of the ram sperm acrosome via interaction with cyclic AMP and phospholipase A_2 metabolites. FEBS Lett 1996; 391:263-268.

Ginsberg J, Gupta S, Matowe WC, Kline L, Brindeley DN. Activation of phospholipase D in FRTL-5 thyroid cells by forskolin and dibutyryl-cyclic adenosine monophosphate. Endocrinology 1997; 138:3645-3651.

Herrero MB, Perez Martinez S, Viggiano JM, Polak JM, Gimeno MF. Localization by indirect immunofluorescence of nitric oxide synthase in mouse and human spermatozoa. Reprod Fertil Dev 1996a; 8:931-934

Herrero MB, Viggiano JM, Perez Martinez S, Gimeno MF. Evidence that nitric oxide synthase is involved in progesterone-induced acrosomal exocytosis in mouse spermatozoa. Reprod Fertil Dev 1996b; 8:301-304.

Herrero MB, Goin JC, Boquet M, Canteros MG, Franchi AM, Perez Martinez S, Polak JM, Viggiano JM, Gimeno MAF. The nitric oxide synthase of mouse spermatozoa. FEBS Lett 1997; 411:39-42.

Hodgkin MN, Pettitt TR, Martin A, Michell RH, Pemberton AJ, Wakelam MJO. Diacylglycerols and phosphatidates: which molecular species are intracellular messengers? Trends Biochem Sci 1998; 23:200-204.

Huang PL, Dawson TM, Bredt DS, Snyder SH, Fishman MC. Targeted disruption of the neuronal nitric oxide synthase gene. Cell 1993; 75:1273-1286.

Huang PL, Huang Z, Mashimo H, Bloch KD, Moskowitz MA, Bevan JA, Fishman MC. Hypertension in mice lacking the gene for endothelial nitric oxide synthase. Nature 1995; 377:239-242.

Langlais J, Chafouleas JG, Ingraham R, Vigneault N, Roberts KD. The phospholipase A_2 of human spermatozoa; purification and partial sequence. Biochem Biophys Res Comm 1992; 182:208-214.

László A, Nádasu GL, Monos E, Zsolnai B, Erdö SL. The GABAergic system in human female genital organs. In: (Erdö SL, ed) GABA Outside the CNS. Berlin: Springer-Verlag, 1992, pp183-197.

Lax Y, Rubinstein S, Breitbart H. Epidermal growth factor induces acrosomal exocytosis in bovine sperm. FEBS Lett 1994; 339:234-238.

Lax Y, Rubinstein S, Breitbart H. Subcellular distribution of protein kinase Cα and βI in bovine spermatozoa, and their regulation by calcium and phorbol esters. Biol Reprod 1997; 56:454-459.

Leclerc P, Kopf GS. Mouse sperm adenylyl cyclase. General properties and regulation by zona pellucida. Biol Reprod 1995; 52:1227-1233.

Leslie CC. Properties and regulation of cytosolic phopholipase A_2. J Biol Chem 1997; 272:16709-16712.

Lin LL, Wartmann M, Lin AY, Knopf JL, Seth A, Davis RJ. $cPLA_2$ is phosphorylated and activated by MAP kinase. Cell 1993; 72:269-278.

Liu DY, Baker HWG. Protein kinase C plays an important role in the human zona pellucida-induced acrosome reaction. Mol Hum Reprod 1997; 3:1037-1043.

Luconi M, Bonaccorsi L, Krausz C, Gervasi G, Forti G, Baldi E. Stimulation of protein tyrosine phosphorylation by platelet activating factor and progesterone in human spermatozoa. Mol Cell Endocrinol 1995; 108:35-42.

Luconi M, Barni T, Vannelli GB, Krausz C, Marra F, Benedetti PA, Evangelista V, Francavilla S, Properzi G, Forti G, Baldi E. Extracellular signal-regulated kinases modulate capacitation of human spermatozoa. Biol Reprod 1998a; 58:1476-1489.

Luconi M, Krausz C, Barni T, Vannelli GB, Forti G, Baldi E. Progesterone stimulates p42 extracellular signal-regulated kinase ($p42^{erk}$) in human spermatozoa. Mol Hum Reprod 1998b; 4:251-258.

Marshall CJ. MAP kinase kinase kinase, MAP kinase kinase, and MAP kinase. Curr Opin Gen Dev 1994; 4:82-89.

Mayer B, Hemmens B. Biosynthesis and action of nitric oxide in mammalian cells. Trends Biochem Sci 1997; 22:477-481.

Meizel S, Pillai MC, Díaz-Pérez E, Thomas P. Initiation of the human sperm acrosome reaction by components of human follicular fluid and cumulus secretions including steroids. In: (Bavister BD, Cummins J, Roldan ERS, eds) Fertilization in Mammals. Norwell, MA: Serono Symposia, 1990, pp205-222.

Milella M, Gismondi A, Roncaioli P, Bisogno L, Palmieri G, Frati L, Cifone MG, Santoni A. CD16 cross-linking induces both secretory and extracellular signal-regulated kinase (ERK)-dependent cytosolic phospholipase A_2 (PLA_2) activity in human natural killer cells. Involvement of ERK, but not PLA_2 in CD16-triggered granule exocytosis. J Immunol 1997; 158:3148-3154.

Mori C, Hashimoto H, Harigaya T, Hoshino K. Biological effects of prolactin on spermatozoa. J Fertil Implant 1989; 6:75-78.

Murase T, Roldan ERS. Epidermal growth factor stimulates hydrolysis of phosphatidylinositol 4,5-bisphosphate, generation of diacylglycerol and exocytosis in mouse spermatozoa. FEBS Lett 1995; 360:242-246.

Murase T, Roldan ERS. Progesterone and the zona pellucida activate different transducing pathways in the sequence of events leading to diacylglycerol generation during mouse sperm acrosomal exocytosis. Biochem J 1996; 320:1017-1023.

Naor Z, Breitbart H. Protein kinase C and mammalian spermatozoa acrosome reaction. Trends Endrocrinol Metab 1997; 8:337-342.

Naz RK, Ahmad K. Presence of expression products of c-erbB-1 and c-erbB-2/HER genes on mammalian sperm cell, and effects of their regulation on fertilization. J Reprod Immunol 1992; 21:223-239.

Naz RK, Kaplan P. Interleukin-6 enhances the fertilizing capacity of human sperm by increasing capacitation and acrosome reaction. J Androl 1994; 15:228-233.

Naz RK, Ahmad K, Kaplan P. Expression and function of ras-protooncogene proteins in human sperm cells. J Cell Sci 1992; 102:487-494.

Oset-Gasque MJ, Parramon M, Hortelano S, Bosca L, Gonzalez MP. Nitric oxide implication in the control of neurosecretion by chromaffin cells. J Neurochem 1994; 63:1693-1700.

Oset-Gasque MJ, Vicente S, Gonzalez MP, Rosario LM, Castro E. Segregation of nitric oxide synthase expression and calcium response to nitric oxide in adrenergic and noradrenergic bovine chromaffin cells. Neuroscience 1998; 83:271-280.

O'Toole CMB, Roldan ERS, Hampton P, Fraser LR. A role for diacylglycerol in human sperm acrosomal exocytosis. Mol Hum Reprod 1996a; 2:317-326.

O'Toole CMB, Roldan ERS, Fraser LR. Role for Ca^{2+} channels in the signal transduction pathway leading to acrosomal exocytosis in human spermatozoa. Mol Reprod Dev 1996b; 45:204-211.

O'Toole CMB, Roldan ERS, Fraser LR. Fraser. Protein kinase C activation during progesterone-stimulated acrosomal exocytosis. Mol Hum Reprod 1996c; 2:921-927.

Parinaud J, Milhet P. Progesterone induces Ca^{++}-dependent 3',5'-cyclic adenosine-monophosphate increase in human sperm. J Clin Endocrinol Metab 1996; 81:1357-1360.

Park DJ, Min HK, Rhee SG. Inhibition of CD3-linked phospholipase C by phorbol ester and by cAMP is associated with decreased phosphotyrosine and increased phosphoserine contents of PLCγ1. J Biol Chem 1992; 267:1496-1501.

Quest AFG, Raben DM , Bell RM. Diacylglycerols: Biosynthetic intermediates and lipid second messengers. In: (Bell RM, Exton JH, Prescott SM, eds) Lipid Second Messengers. New York: Plenum Press, 1996, pp1-58.

Riffo M, Gomez Lahoz E, Esponda P. Immnocytochemical localization of phospholipase A_2 in hamster spermatozoa. Histochemistry 1992; 97:25-31.

Riffo MS, Parraga M. Study of the acrosome reaction and the fertilizing ability of hamster epididymal cauda spermatozoa treated with antibodies against phospholipase A_2 and/or lysophosphatidylcholine. J Exp Zool 1996; 275:459-468.

Riffo MS, Parraga M. Role of phospholipase A_2 in mammalian sperm-egg fusion. development of hamster oolemma fusibility by lysophosphatidylcholine. J Exp Zool 1997; 279:81-88.

Roldan ERS. Role of phosphoinositides in the mammalian sperm acrosome reaction. In: (Fenichel P, Parinaud J, eds) Human Sperm Acrosome Reaction. Montrouge: John-Libbey Eurotext, 1995, pp225-243.

Roldan ERS, Dawes EN. Phospholipase D and exocytosis of the ram sperm acrosome. Biochim Biophys Acta 1993; 1210:48-54.

Roldan ERS, Fleming AD. Is a Ca^{2+}-ATPase involved in Ca^{2+} regulation during capacitation and the acrosome reaction of guinea pig spermatozoa? J Reprod Fert 1989; 85:297-308.

Roldan ERS, Fragio C. Phospholipase A_2 activity and exocytosis of the ram sperm acrosome: regulation by bivalent cations. Biochim Biophys Acta 1993a; 1168:108-114.

Roldan ERS, Fragio C. Phospholipase A_2 activation and subsequent exocytosis in the Ca^{2+}/ionophore-induced acrosome reaction of ram spermatozoa. J Biol Chem 1993b; 268:13962-13970.

Roldan ERS, Fragio C. Diradylglycerols stimulate phospholipase A_2 and subsequent exocytosis in ram spermatozoa. Biochem J 1994; 294:225-232.

Roldan ERS, Fraser LR. Protein kinase C and exocytosis in mammalian spermatozoa. Trends Endocrinol Metab 1998; 9:296-297.

Roldan ERS, Harrison RAP. Absence of active protein kinse C in ram spermatozoa. Biochem Biophys Res Commun 1988; 155:901-906.

Roldan ERS, Harrison RAP. Polyphosphoinositide breakdown and subsequent exocytosis in the Ca^{2+}/ionophore-induced acrosome

reaction of mammalian spermatozoa. Biochem J 1989; 259:397-406.
Roldan ERS, Harrison RAP. Diacylglycerol and phosphatidate production and the exocytosis of the sperm acrosome. Biochem Biophys Res Commun 1990; 172:8-15.
Roldan ERS, Harrison RAP. The role of diacylglycerol in the exocytosis of the sperm acrosome. Studies using diacylglycerol lipase and diacylglycerol kinase inhibitors and exogenous diacylglcyerols. Biochem J 1992; 281:767-773.
Roldan ERS, Mollinedo F. Diacylglycerol stimulates the Ca^{2+}-dependent phospholipase A_2 of ram spermatozoa. Biochem Biophys Res Commun 1991; 176:294-300.
Roldan ERS, Murase T. Polyphosphoinositide-derived diacylglycerol stimulates the hydrolysis of phosphatidylcholine by phospholipase C during exocytosis of the ram sperm acrosome. J Biol Chem 1994; 269:23583-23589.
Roldan ERS, Vazquez JM. Bicarbonate/CO_2 induces rapid activation of phospholipase A_2 and renders boar spermatozoa capable of undergoing acrosomal exocytosis in response to progesterone. FEBS Lett 1996; 396:227-232.
Roldan ERS, Murase T, Shi QX. Exocytosis in spermatozoa in response to progesterone and zona pellucida. Science 1994a; 266:1578-1581.
Roldan ERS, Martinez-Dalmau R, Mollinedo F. Diacylglycerol and alkylacylglycerol stimulate ram sperm phospholipase A_2. Int J Biochem 1994b; 26:951-958.
Rotem R, Paz GF, Homonnai ZT, Kalina M, Naor Z. Protein kinase C is present in human sperm: possible role in flagellar motility. Proc Natl Acad Sci USA 1990; 87:7305-7308.
Rotem R, Paz GF, Homonnai ZT, Kalina M, Lax Y, Breitbart H, Naor Z. Ca^{2+}-independent induction of acrosome reaction by protein kinase C in human sperm. Endocrinology 1992; 131:2235-2243.
Ruan Y, Kan H, Parmentier JH, Fatima S, Allen LF, Malik KU. α1a Adrenergic receptor stimulation with phenylephrine promotes arachidonic acid release by activation of phospholipase D in rat-1 fibroblasts: inhibition by protein kinase A. J Pharmacol Exp Ther 1998; 284:576-585.
Schuetz AW, Dubin NH. Progesterone and prostaglandin secretion by ovulated rat cumulus cell-oocyte complexes. Endocrinology 1981; 108:457-463.
Sheikhnejad RG, Srivastava PN. Isolation and properties of a phosphatidylcholine-specific phospholipase C from bull seminal plasma. J Biol Chem 1986; 261:7544-7549.
Shi QX, Roldan ERS. Evidence that $GABA_A$-like receptor is involved in progesterone-induced acrosomal exocytosis in mouse spermatozoa. Biol Reprod 1995; 52:373-381.
Shi QX, Yuan YY, Roldan ERS. γ-Aminobutyric acid (GABA) induces the acrosome reaction in human spermatozoa. Mol Hum Reprod 1997; 3:677-683.
Silvestroni L, Palleschi S, Guglielmi R, Tosti Croce C. Identification and localization of atrial natriuretic factor receptors in human spermatozoa. Arch Androl 1992; 28:75-82.
Son H, Hawkins RD, Martin K, Kiebler M, Huang PL, Fishman MC, Kandel ER. Long-term potentiation is reduced in mice that are doubly mutant in endothelial and neuronal nitric oxide synthase. Cell 1996; 87:1015-1023.
Su B, Karin M. Mitogen-activated protein kinase cascades and regulation of gene expression. Curr Opin Immunol 1996; 8:402-411.
Thomas P, Meizel S. Phosphatidylinositol 4,5-bisphosphate hydrolysis in human sperm stimulated with follicular fluid or progesterone is dependent upon Ca^{2+} influx. Biochem J 1989; 264:539-546.
Tomes CN, McMaster CR, Saling PM. Activation of mouse sperm phosphatidylinositol-4,5 bisphosphate-phospholipase C by zona pellucida is modulated by tyrosine phosphorylation. Mol Reprod Dev 1996; 43:196-204.
Tsai CH, Hung LM, Chen JK. Perturbation of the platelet-derived growth factor receptor signaling by dibutyryl cAMP in human astrocytoma cells. J Cell Physiol 1995; 164:108-116.
Umehara H, Huang JY, Kono T, Tabassam FH, Okazaki T, Bloom ET, Domae N. Involvement of protein tyrosine kinase $p72^{syk}$ and phosphatidylinositol 3-kinase in CD2-mediated granular exocytosis in the natural killer cell line, NK3.3. J Immunol 1997; 159:1200-1207.
Uozomi N, Kume K, Nagase T, Nakatani N, Ishii S, Tashiro F, Komagata Y, Maki K, Ikua K, Ouchi Y, Miyazaki J, Shimizu T. Role of cytosolic phospholipase A_2 in allergic response and parturition. Nature 1997; 390:618-622.
Vazquez JM, Roldan ERS. Phospholipid metabolism in boar spermatozoa and role of diacylglycerol species in the *de novo* formation of phosphatidylcholine. Mol Reprod Dev 1997a; 47:105-112.
Vazquez JM, Roldan ERS. Diacylglycerol species as messengers and substrates for phosphatidylcholine re-synthesis during Ca^{2+}-dependent exocytosis in boar spermatozoa. Mol Reprod Dev 1997b; 48:95-105.
Visconti PE, Moore GD, Bailey JL, Leclerc, P, Connors SA, Pan DY, Olds-Clarke P, Kopf GS. Capacitation of mouse spermatozoa. 2. Protein tyrosine phosphorylation and capacitation are regulated by a cAMP-dependent pathway. Development 1995; 121:1139-1150.
Vijayaraghavan S, Olson GE, Nag Das S, Winfrey VP, Carr DW. Subcellular localization of the regulatory subunits of cyclic adenosine 3',5'-monophosphate-dependent protein kinase in bovine spermatozoa. Biol Reprod 1997; 57:1517-1523.
Weinman S, Ores-Carton C, Rainteau D, Puszkin S. Immunoelectron microscopic localization of calmodulin and phospholiase A_2 in spermatozoa. J Histochem Cytochem 1986; 34:1171-1179.
Weyand I, Godde M, Frings S, Weiner J, Müller F, Altenhofen W, Hatt H, Kaupp UB. Cloning and functional expression of a cyclic nucleotide-gated channel from mammalian sperm. Nature 1994; 368:859-863.
Yanagimachi R. Mammalian fertilization. In: (Knobil E, Neill JD, eds) The Physiology of Reproduction. 2nd ed. New York: Raven Press, 1994, pp189-317.
Young SW, Dickens M, Tavaré JM. Differentiation of PC12 cells in response to a cAMP analog is accompanied by sustained activation of mitogen-activated protein kinase. Comparison with the effects of insulin, growth factors and phorbol esters. FEBS Lett 1994; 338:212-216.
Zamir N, Barkan D, Keynan N, Naor Z, Breitbart H. Atrial natriuretic peptide induces acrosomal exocytosis in bovine spermatozoa. Am J Physiol 1995; 32:E216-E221.

14

Rapid Evolution of Acrosomal Proteins and Species Specificity of Fertilization in Abalone

Willie J. Swanson

Edward C. Metz
University of California

C. David Stout
The Scripps Research Institute

Victor D. Vacquier
University of California

Elucidation of the molecular mechanisms of gamete interaction has intrinsic interest because fertilization initiates embryogenesis and creation of a new individual. Fertilization also provides a unique system of general value to study molecules mediating intercellular interaction culminating in cell fusion. Sperm-egg interaction during fertilization can comprise the following sequence of events: chemotaxis of sperm toward the egg, attachment of sperm to the egg envelope, induction of the sperm acrosome reaction (AR), penetration of sperm through the egg envelope, and binding and fusion of sperm and egg plasma membranes. In various species, one or more steps may be absent, or their order changed. One or more of the steps can show species-specificity, meaning that sperm and eggs of the same species fuse more efficiently than if the gametes are from different species. Marine invertebrates, such as sea urchins and abalone, broadcast their gametes in seawater, where fertilization and embryogenesis occur. In some cases, strong barriers to sperm binding to the egg envelope have evolved which prevent fertilization between closely related species (Palumbi, Metz 1991; Metz et al. 1994). Although they are internal fertilizers, mammals also exhibit species-specific sperm-egg interaction (O'Rand 1988; Yanagimachi 1988). The goal of research in the Vacquier laboratory is to understand the molecular and evolutionary basis underlying species-specific fertilization in marine invertebrates.

Abalone are broadcast-spawning marine archeogastropod mollusks with external fertilization and embryogenesis. Abalone eggs are inclosed within a vitelline envelope (VE) about 0.6 µm in diameter, composed of glycoprotein filaments, which the sperm must penetrate to reach the egg plasma membrane. Abalone sperm have a giant acrosomal vesicle containing two positively charged proteins which are differentially packaged within the vesicle (Haino-Fukushima, Usui 1986). Sperm bind to the VE by the plasma membrane over the tip of the acrosomal vesicle. The vesicles opens by exocytosis to release lysin (16kD) and the 18kD protein. Lysin dissolves a 3 µm diameter hole in the VE (Lewis et al. 1982; Vacquier et al. 1999) by a non-enzymatic mechanism. The non-enzymatic mechanism was shown by the facts that VE molecules are not degraded by lysin, no new amino-terminal amino acids are created and the kinetics of VE dissolution show a binding stoichiometry as opposed to a catalytic mechanism (Lewis et al. 1982; Swanson, Vacquier 1997). Electron microscopy of the hole in the VE gives the impression that the tightly intertwined filaments of the VE are loosened and splayed apart by lysin (Fig. 1). The 18kD protein preferentially coats the surface membrane of the sperm acrosomal process as it extends to a length of 7 µm (Lewis et al. 1980). Both lysin and the 18kD protein mediate the fusion of artificial phospholipid vesicles. However, the 18kD protein is a much more potent fusagen than is lysin (Swanson, Vacquier 1995a; Fig. 2). Both acrosomal proteins can be purified in hundreds of milligram quantities from the seawater surrounding acrosome reacted abalone sperm.

Lysin and 18kD Arose by a Gene Duplication

Determination of the amino acid sequences of lysin and 18kD proteins from several species showed that both proteins have diverged substantially. Searching GenBank with either protein sequence failed to find the other. However, searching GenBank with a "profile" made of 18kD homologs from five species showed significant matches with only the abalone lysins, suggesting both proteins were related. Further support for this idea came from concordance of the secondary structure predictions of both proteins (Swanson, Vacquier 1995a). Additional evidence that these two proteins evolved by gene duplication was found in genomic DNA

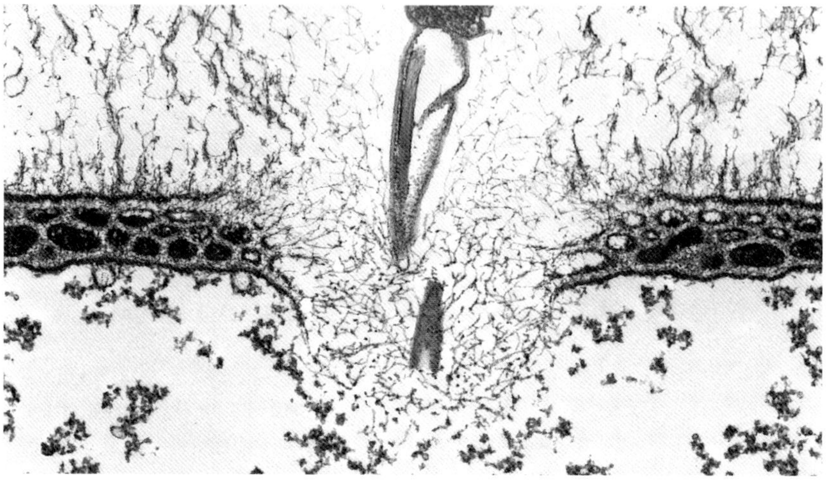

Figure 1. An abalone spermatozoon making a hole (3 µm in diameter) in the egg vitelline envelope as seen by transmission electron microscopy. The sperm is coming in from the top. A grazing section of the wall of the acrosome vesicle appears in the bottom of the hole. A piece of the acrosomal rod appears against the inner edge of the empty vesicle. x21,400. (Lewis et al. 1982.)

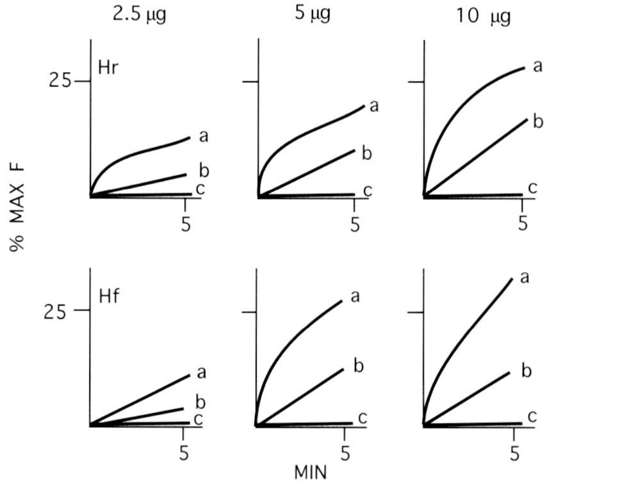

Figure 2. The fusion of phospholipid vesicles induced by lysin (b), the 18kD acrosomal protein (a) and buffer alone (c). Upper three panels are proteins from the red abalone (Hr, *Haliotis rufescens*) and the three lower panels are 18kD from the green abalone (Hf, *H. fulgens*). The 18kD proteins are more potent fusagens than the lysins. Both 18kD proteins have five predicted amphipathic α-helices with high hydrophobic moments. (Swanson, Vacquier 1995a.)

sequences. Both proteins are encoded by five exons and four introns. The introns are in the same positions and interrupt the codon in the same phase in both proteins, providing strong evidence that they arose by gene duplication (Metz et al. 1998). Hypothetically, before the duplication event, one acrosomal protein mediated both VE lysis and plasma membrane fusion. After gene duplication, both proteins diverged substantially, so that today 18kD is a potent fusagen, but has no effect on the integrity of the VE. However, lysin retains the ability to lyse the VE and mediate membrane fusion, although it is a poorer fusagen than 18kD (Swanson, Vacquier 1995a).

Crystal Structures of Lysin

The ease of isolation by CM-cellulose ion exchange chromatography of large quantities of protein led to the solution of lysin's monomeric three-dimensional structure by X-ray crystallography (resolved to 1.9 Å) from the red abalone (*Haliotis rufescens*). This was the first time the structure of a fertilization protein had been solved. Monomeric lysin was revealed as a triple helical bundle with 65% α-helix, no β-sheet and no binding pockets or clefts characteristic of enzymes and lectins (Fig. 3). There are no disulfide bonds and no glycosylation in lysin. Lysin's tertiary structure is implicit in its primary structure, as shown by the fact that it can be renatured to 100% activity following denaturation in boiling guanidinium hydrochloride (VD Vacquier, unpublished data). Three important features of monomeric lysin are noted in the structure. First, the amino-terminal end of residues 1-12 projects away from

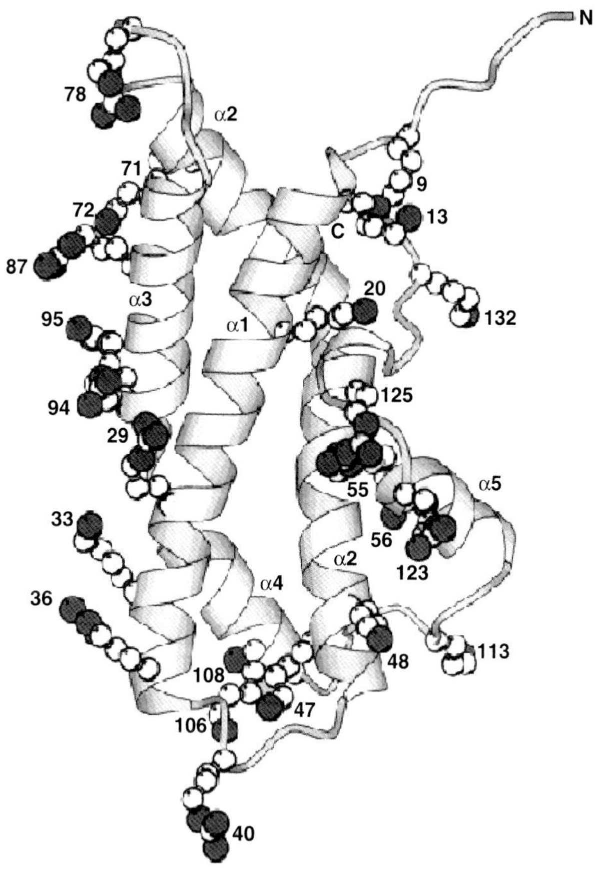

Figure 3. The crystal structure of the monomer of red abalone sperm lysin. This tracing of the positions of all α-carbon atoms shows the tight triple helical bundle with the amino-terminus projecting to the right. The left basic track has nine, and the right track 14, Arg (R) and Lys (K) residues making up the two basic tracks. Arg^1 and Lys^{136} are not seen in the crystal structure. Nitrogen atoms are black. (Shaw et al. 1993.)

the helical bundle (Figs. 3, 5). This domain is always species-specific in its sequence (Lee et al. 1995). Both termini lie on the same face of the monomer. Second, one face of lysin has two nearly parallel tracks of 9 and 14 basic residues running the entire length of the molecule and creating a surface of high positive charge density (Fig. 3). Among the seven California abalone species, the basic tracks are considerably conserved. Third, the surface opposite the basic tracts has 11 solvent-exposed hydrophobic residues which comprise a hydrophobic patch accounting for 10% of the protein's surface (Fig. 4). The hydrophobic patch is invariant in the seven species of California abalone. The hydrophobic patch on one side and the two basic tracks on the other side make lysin an extremely amphipathic protein (Shaw et al. 1993).

Lysin dimerizes by interaction of the hydrophobic patches of monomers. Fluorescence resonance energy transfer (FRET) experiments yield a K_D for dimerization of about 1 μM. The low affinity results from the electrostatic repulsive forces of the four basic tracks of

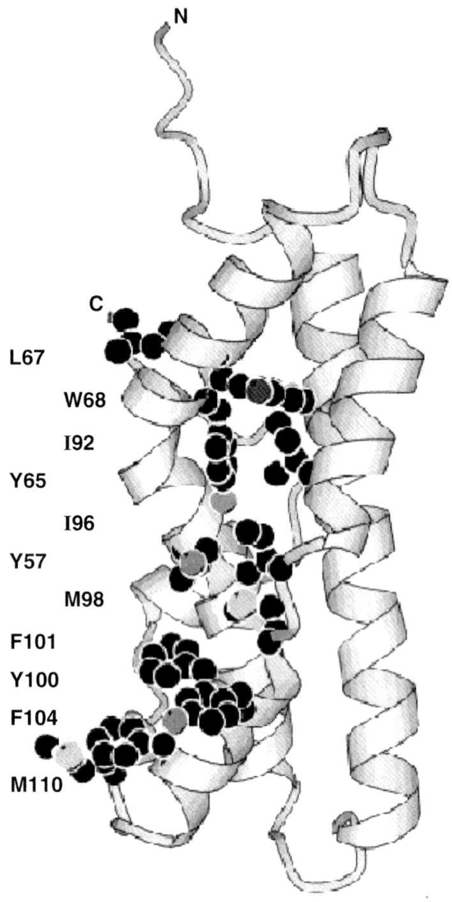

Figure 4. The hydrophobic patch of the lysin monomer is 10% of the molecule's surface area. The 11 hydrophobic residues on one side, balanced by the two basic tracks on the opposite side, make lysin an extremely amphipathic protein. (Shaw et al. 1993, 1995.)

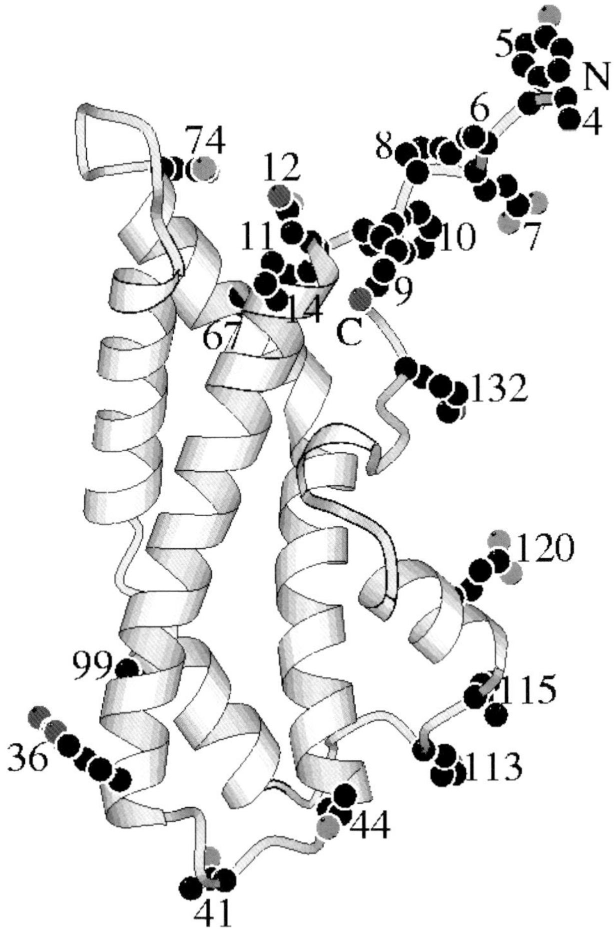

Figure 5. The hypervariable residues of lysins computed from the sequences from 27 species. These residues are involved in the species-specific recognition of the egg vitelline envelope. (Lee et al. 1995.)

the dimer being counteracted by the attractive forces of the two hydrophobic patches. Dimerization occludes only 50% of the hydrophobic patch (Fig. 6). An important feature of the dimer interface is the stacking of aromatic residues His^{61}, Tyr^{57}, Phe^{101}, and Phe^{104} of each monomer. Of the 20 contacts between monomers at the dimer interface, 14 are hydrophobic, two are hydrogen bonds and four are non-specific polar interactions. FRET experiments show that monomers exchange freely between dimers; at 8 μM monomer the half time of exchange is about eight minutes. When isolated VEs (Fig. 7) are added to dimers, the dimer rapidly dissociates, suggesting that the monomer is the active species in dissolving the VE (Shaw et al. 1995).

The basic tracts and the hydrophobic patch of lysin are highly conserved features among species, whereas other regions of lysin, including the amino- and carboxyl- termini, are highly variable (Fig. 5). It is believed that lysin acts by a two-step mechanism. First, the variable features are involved in species-specific recognition of the lysin receptor in the egg VE. This allows the lysin dimer close access to its receptor site. Second, once lysin is in the proper configuration with its receptor, its conserved features mediate the loss of cohesion among the filaments comprising the VE, causing them to splay apart, creating the hole through which the sperm swims.

Lysin and 18kD Protein Evolve Rapidly by Positive Selection

When the lysin sequences of the seven California abalone species were determined, it was surprising to discover how extensively they had diverged (Table 1). In addition, pairwise comparisons of aligned mature sequences of 136-138 residues showed that amino acid altering nucleotide substitutions (non-synonymous substitutions) in codons greatly outnumbered silent substitutions (synonymous substitutions; Lee, Vacquier 1992; Vacquier et al. 1997). Analysis of nucleotide substitutions showed that lysin evolved by positive Darwinian selection, indicating there was adaptive value in altering the amino acid sequence. The mature 18kD sequences

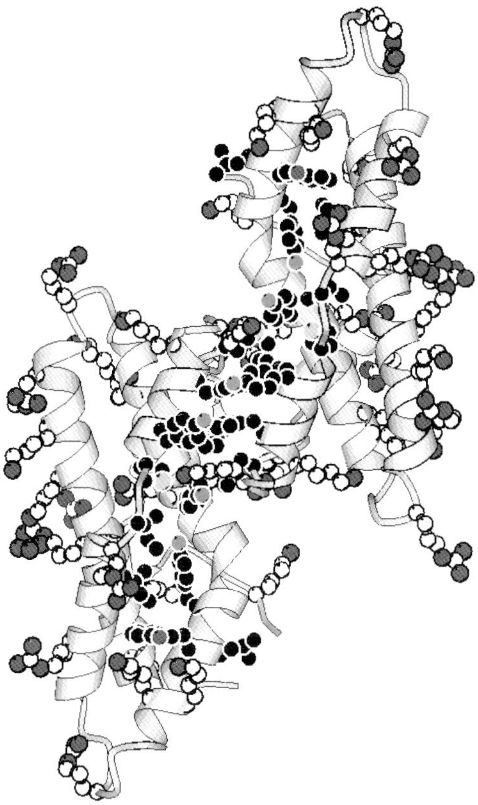

Figure 6. The dimer of red abalone sperm lysin viewed along the local 2-fold axis. Only 50% of the hydrophobic patch is occluded in the dimer. Carbon atoms of the hydrophobic patch residues are black and those of the basic track residues are white, and nitrogens of the basic residues are dark grey. The K_D of dimer formation is ~1 μM; above 8 μM all lysin is present as dimers. (Shaw et al. 1995.)

of 132-146 residues from five species of California abalone showed even greater divergence and signal of positive selection than did lysin (Swanson, Vacquier 1995b; Vacquier et al. 1997, 1999). These data are summarized in Table 1, which represents 10 pairwise comparisons of five California species. In the comparisons of lysins, the percent identities in amino acid position ranged from 63-90%, meaning that 13-50 amino acids varied out of 136. For lysins, the ratios of non-synonymous substitutions (Dn) to synonymous substitutions (Ds) varied from 1.08-3.82.

Likewise, the 18kD protein percent identities varied from 27-87% and the non-synonymous to synonymous ratios from 0.95-4.67. The ratios of Dn:Ds >3 are some of the highest known for the analysis of full-length proteins (Table 1). Comparison of the green (Hf) and red (Hr) 18kD proteins shows they are only 34% identical in sequence. However, comparisons of their five predicted amphipathic α-helices show them to be quite similar, indicating that although the sequences have diverged considerably, the secondary and tertiary structures may be conserved (Swanson, Vacquier 1995a).

A phylogeny for California abalone was determined

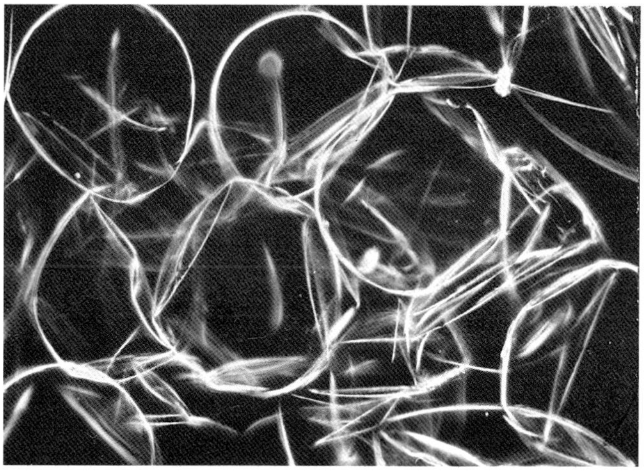

Figure 7 (above). Isolated vitelline envelopes (VE) from red abalone eggs. The VEs are 40% protein and 60% carbohydrate. The VE receptor for lysin is about 30% of the VE protein. x290. (Lewis et al. 1982; Swanson, Vacquier 1997.)

by sequencing 528 base pairs of mitochondrial cytochrome oxidase subunit I (mtCOI). Using a rate of 2% mtCOI divergence per million years, it is calculated that the four most closely related California abalone species diverged within the past one to two million years (Metz et al. 1998). The mtCOI phylogeny showed that about 20 amino acids in lysin, and about 30 amino acids in 18kD, are replaced in the time required for 3% divergence in mtCOI. This means that these two abalone acrosomal proteins are evolving approximately two to 50 times faster than the most rapidly evolving mammalian proteins (Metz et al. 1998).

Rapid evolution is also evidenced by comparison of intron and exon sequences of these two proteins. In most proteins, introns evolve much faster than exons. However, in comparisons of two species, red and green abalone, nucleotide substitution in lysin exons differed by 24% while introns varied 4-6%. With 18kD, exons differed by 81%, whereas introns varied by 3-7%. These low rates of intron nucleotide differentiation between these two species were also found in an intron of a G-alpha protein which differed by 6% (Metz et al. 1998). These data pose two interesting questions: why do these two acrosomal proteins evolve so fast, and could their rapid evolution be involved in the speciation process?

The Species-Specificity of VE Dissolution by Lysin

A light scattering assay (Vacquier, Lee 1993) and a radiometric assay (Vacquier et al. 1990) were developed to test the ability of various lysins to dissolve isolated egg VE. Typical data are shown in Figure 10, where homospecific combinations of lysin and VEs from two

Table 1: Pairwise comparisons of the percent identity of the mature acrosomal proteins: numbers of synonymous (Ds) and non-synonomous (Dn) substitutions per 100 sites.

Species Comparison	% Identity	Ds	Dn	Dn : Ds
LYSIN				
Hr - Hs	90	1.6 ± 1.3	5.8 ± 1.4	3.63*
Hr - Hk	82	2.8 ± 1.8	10.7 ± 1.9	3.82**
Hr - Hc	78	10.6 ± 3.5	14.6 ± 2.3	1.38
Hr - Hf	65	21.3 ± 5.3	24.9 ± 3.2	1.17
Hs - Hk	85	3.3 ± 1.9	8.4 ± 1.7	2.55*
Hs - Hf	64	21.1 ± 5.3	25.4 ± 3.3	1.20
Hs - Hc	80	9.9 ± 3.4	13.7 ± 2.2	1.38
Hk - Hc	72	9.8 ± 3.4	18.1 ± 2.6	1.85*
Hk - Hf	63	22.7 ± 5.5	24.5 ± 3.2	1.08
Hc - Hf	65	16.0 ± 4.4	24.0 ± 3.2	1.50
18 kD				
Hr - Hs	87	1.8 ± 1.7	8.1 ± 1.7	4.50**
Hr - Ha	75	4.9 ± 2.5	18.5 ± 2.7	3.78**
Hr - Hc	31	87.0 ± 17.5	82.6 ± 8.6	0.95
Hr - Hf	34	78.2 ± 15.4	81.1 ± 8.5	1.04
Hs - Ha	77	3.1 ± 1.9	14.5 ± 2.3	4.67**
Hs - Hf	35	83.5 ± 16.8	76.9 ± 8.0	0.92
Hs - Hc	33	78.0 ± 15.5	75.6 ± 7.8	0.97
Ha - Hc	31	92.2 ± 18.6	82.3 ± 8.8	0.89
Ha - Hf	35	84.0 ± 16.6	85.5 ± 9.0	1.02
Hc - Hf	27	114.2 ± 23.8	86.9 ± 8.9	0.76

Significant at the p < *0.025, **0.005 level

species (red and pink abalone) had the same kinetics of VE dissolution, with 50% dissolution occurring at about 7 μg lysin. However, the combination of red abalone VEs and pink lysin required about 17 μg lysin, and pink VEs required 45 μg red lysin to achieve 50% dissolution (Swanson, Vacquier 1997). Species-specificity was most pronounced in the initiation of the dissolution reaction (Fig. 8). Alignment of the sequences of red and pink lysins shows that the signal sequences of -18 to -1 are conserved (Fig. 9). The amino-terminus consisting of residues 1-11 is the most variable region of the two sequences. In addition, there are 19 other replacements between these two species, only three of which are class-conservative changes. Expression of site specific recombinant lysin should clarify which residues are important in the species recognition process and in the mechanism of lysin's action.

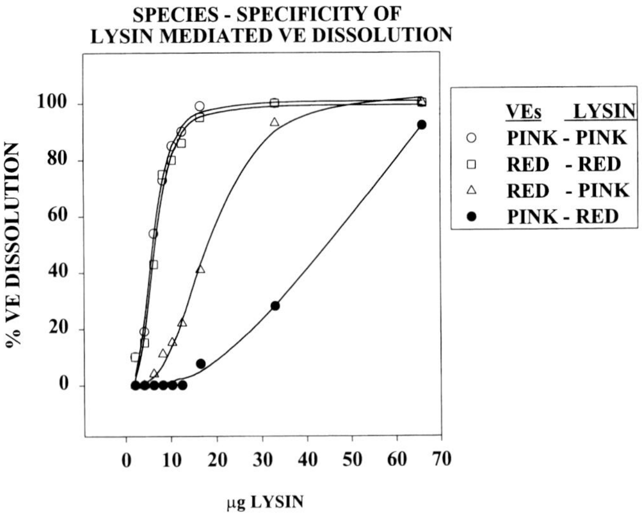

Figure 8. Species-specific dissolution of isolated egg VEs by purified lysins. VEs of pink abalone, *Haliotis corrugata,* and red abalone, *H. rufescens*, show the same kinetics of dissolution with their conspecific lysins, with 50% dissolution occurring at 7 μg lysin. Red VEs require 17 μg pink lysin and pink VEs require 45 μg red lysin to achieve 50% dissolution of VEs. (Swanson, Vacquier 1997.)

Identification and Cloning of Lysin's VE Receptor

The knowledge that lysin binds tightly to the VE and the ability to isolate substantial amounts of VEs allowed the identification of the VE receptor for lysin (VERL). When titrated to lower pHs, VEs dissolve between pH 5.0 and 4.5 (VD Vacquier, unpublished data). Isolated VEs were dissolved by pH 4.0 treatment and then readjusted to pH 7.8 and the solution clarified by low speed centrifugation. The supernatant was applied to a solid support to which lysin was covalently bound. After washing, the column was eluted with pH 3 buffer. Only one high molecular component eluted from the lysin column, but not from a control column. This material was a potent inhibitor of lysin-mediated VE dissolution. VEs were dissolved at acid pH, readjusted to pH 7.8 and a limiting amount of ^{125}I-lysin added. The material was subjected to sucrose density gradient ultracen-

```
                -18                -1 +1        10        20        30        40        50
H.rufescens     MKLLVLCIFAMMATLAMS  R-SWHYVEPKFLNKAFEVALKVQIIAGFDRGLVKWLRVHGRTLSTVQKKALYFVNRRYM
H.corrugata     .......L........V.  .HRFRFIPH.YIR.E.......E.......T.........GR................

                60        70        80        90        100       110       120       130
H.rufescens     QTHWANYMLWINKKIDALGRTPVVGDYTRLGAEIGRRIDMAYFYDFLKDKNMIPKYLPYMEEINRMRPADVPVKYMGK
H.corrugata     ....Q......VR.T.....P...A..S................N..NGR.......................ANR..
```

Figure 9. Alignment of the deduced amino acid sequences of red (*H. rufescens*) and pink (*H. corrugata*) abalone sperm lysins. Dots denote identity to the top sequence and the single dash is inserted for alignment. The signal sequences (positions -18 to -1) are conserved. The amino-terminal domain of residues 1-11 is the most variable part of lysin. (Lee, Vacquier 1992.)

trifugation; all the radioactive lysin migrated with a high molecular weight fraction. An SDS-PAGE system was developed to resolve the lysin binding components which migrated with a relative mass of about one million daltons. Isolated VERL was 50% carbohydrate and contained no unusual sugars.

Fluorescence polarization studies showed strong positive cooperativity in lysin-VERL binding (Hill Coefficient ~3 , EC_{50} ~9.5 nM). Stoichiometry indicated between 120-140 lysin monomers bound each VERL; however, cloning VERL has reduced this number to 60 lysins per VERL (Swanson, Vacquier 1998). These data indicated that VERL probably contained a repetitive lysin binding motif (Swanson, Vacquier 1997).

VERL was isolated, fragmented by CNBr and V8 protease and the fragments subjected to gas phase amino acid sequencing. The same overlapping sequences were obtained from peptides of various sizes, showing that VERL contained repetitive primary structure. An arrayed library of abalone ovary cDNA was made and probed with end-labeled oligonucleotides representing the peptide sequences. Sequencing positive plasmids showed VERL contained a tandemly repeating 153 amino acid motif. VERL repeats could be amplified from either genomic DNA or cDNA, and sequencing showed there were no introns between repeats. Northern blots of ovary cDNA showed the VERL mRNA to be 13.5kb. Assuming the entire open reading frame to be repeats, with short 3' and 5' untranslated regions, one VERL molecular would contain about 28 repeats.

Alignment of VERL repeats from three species (Fig. 10) shows that between species, the repeats are quite similar to each other. This is in contrast to interspecies comparisons of lysins, which show considerable divergence (Table 1). How could there be so little interspecies variation in VERL sequences, yet so much interspecies variation in lysin? An answer to this question could come from the pattern of amino acid substitutions in VERL between species. For example, the 16 positions marked with a star in Figure 10 are ones in which a potential glycosylation site (S or T), or a charged residue (E, D, K or R), is replaced between species by a hydrophobic residue (Fig. 10). This could result in exposure to solvent of a fold in VERL, versus burying that fold within the molecule. Furthermore, a proline rich region (underlined P residues) is surrounded by this type of hydrophilic-hydrophobic replacement pattern, which would allow the VERL repeat greater flexibility in folding. Thus, single amino acid changes in VERL repeats could profoundly affect the folding of VERL and be the basis for its species-specific structure. The calculated isoelectric point of the VERL repeat is 4.7, which is the pH at which VEs dissolve. This is consistent with the idea that VERL maintains the structural integrity of the VE.

Evolution of Species-Specificity of VERL-Lysin Binding and Fertilization

Coding and non-coding tandemly repeated sequences can evolve by concerted evolution, which involves unequal crossing over and gene conversion (Li 1997). Unequal crossing over will expand or contract the number of repeats, or partial repeats, whereas gene conversion will correct repeat sequences. Both processes will homogenize a repeat family so that repeats will be more similar to each other when compared within a species than between any two species. From two to seven individual VERL repeats were randomly sequenced from seven species of California abalone. A neighbor joining tree of these sequences indicated they had been subjected to concerted evolution (Swanson, Vacquier 1998). The ratio of non-

```
             *           *   *   *                                                              * *
Red    VPITQEFGIN MMLIQYTRNE LLDSPGMCVF WGPYSVPKND TVVLYTVTAR LKWSEGPPTN LSIECYMPKS PVAPKPETGP
Red    .......... .......... .......... .......... .......... .......... .......... ..........
Green  ......Y.F. .......... .......... .......... .......... .........D .......... .D.S......
Green  ......Y.F. .......... ....S..... .......... .......... .......... .T........ .D.S......
Pink   ......S... ......SL.D S......... .......... .......... .......... .......... ..........
Pink   ......S... ......SL.D S......... .......... .......... .......... .......... ......K...
Pink   ......S... ......SL.D S......... .......... .......... .......... .......... ..........

       *  *   ***       *       *   *                              *
Red    SSNAPEPETY PTSSAPEKVS SDQPAPSHNQ SKLIDWDVYC SQNESIPAKF ISRLLTSKDQ ALEKTEINCS NGL
Red    T......Q.. .......... .......... .......... .......... ..-.V..... .......... ...
Green  ..STA..G.. .......... .......D.. .......... .......... ....V..... .V...D.... ...
Green  ..STA..G.. .......... .......D.. .......... .......... ....V..... .V...D.... ...
Pink   I.ST....AS ...K..G... .N.....P.. ......E... ...D...... ...IP..... .V....V... ...
Pink   T.ST...DAS ...K..G... ...T...... .......... ...D...... ...IPI.... .V....V... ...
Pink   T.ST....AS ...K..G... .......... .......... ...D...... ...IPI.... .V....V... ...
```

Figure 10. Alignment of the repeating sequence of VERL, the VE receptor for lysin, from three species of abalone. Two red, two green and three pink VERL repeats of 153 residues are shown. Stars indicate positions where a potential site for glycosylation (S or T), or charged residue (E, D, K or R) is replaced with a hydrophobic residue. Clustering of these replacements with proline residues (underlined P) could have profound effects on the species-specific folding of VERL repeats. (Swanson, Vacquier 1998.)

synonymous to synonymous nucleotide substitutions in VERL repeats between species showed that 89% of the pairwise comparisons did not differ significantly from 1, indicating that VERL repeats are subjected to weak purifying selection.

Genes subjected to concerted evolution can produce a selective force on the genes for their cognate protein ligands (Dover 1993). For example, concerted evolution acts on the divergence of promoters to which RNA polymerase-I transcription initiation factors bind (Heix et al. 1997). This causes selective pressure for the transcription initiation factors themselves to change (Heix et al. 1997; Rudloff et al. 1994). Between mice and humans, the transcription initiation factors are 20-34% divergent, a level of divergence similar to rapidly evolving defense proteins (Murphy 1993).

Lysin-VERL co-evolution could be subjected to forces similar to those in the above example of rRNA gene transcription regulation. A mutant in one VERL repeat, which may be unfavorable for lysin binding, could be tolerated since ~27 unchanged repeats are present. Concerted evolution could randomly propagate this repeat within the VERL gene. As the mutant repeat increased in frequency, selective pressure on lysin to adapt to its cognate receptor would be created. The mutant would propagate in the population in a cohesive manner, meaning that the mean ratio of mutant to wild type repeats would have low variance as the mutant spread. Thus, selection could maintain the proper molecular recognition between lysin and VERL. In short, the hypothesis proposed here is that the egg surface component changes first by concerted evolution and the cognate sperm component adapts to the change by positive selection. This means of change in a gamete recognition system could produce and maintain species-specific sperm-egg interaction as species diverge. The process could be continuous, lysin always being under selective pressure to adapt to changes in VERL (Lee et al. 1995). The process could occur within a single population and require no external forces (Swanson, Vacquier 1998).

Whether concerted evolution is involved in the evolution of other gamete recognition systems remains to be determined. Several gamete recognition proteins, such as the *spe-9* gene of the nematode *Caenorhabditis elegans* (Singson et al. 1998), the sea urchin speract gene (Ramarao et al. 1990), sea urchin bindin (Minor et al. 1991), mammalian zonadhesin (Gao, Garbers 1998), and mouse sperm sp56 (Bookbinder et al. 1995), have repetitive elements of primary structure which could be subjected to concerted evolution.

Summary

The abalone egg is inclosed in a resilient, elevated vitelline envelope (VE) made of glycoprotein filaments. During fertilization, spermatozoa attach to the VE and acrosome react. Two proteins, lysin (16kD) and an 18kD protein, are released. Lysin dissolves a hole in the VE by a non-enzymatic process exhibiting species-specificity. The 18kD protein coats the acrosomal process and may mediate fusion of the gametes. Both proteins arose by gene duplication and have since diverged substantially. The crystal structure of lysin reveals a unique, highly amphipathic, surface-active protein which probably dissolves the VE by disrupting hydrogen bonds holding the VE filaments together. Both lysin and 18kD evolve rapidly by positive Darwinian selection, indicating there is adaptive value in altering the sequences of both proteins. The four most closely related California abalone species probably diverged within the past one to two million years. Lysin and 18kD can show rates of evolution which are more rapid than the fastest evolving mammalian proteins. Lysin exhibits species-specificity in dissolving the VE. Several regions of the amino acid sequence could be involved in determining this specificity. The receptor for lysin is a long, unbranched, glycoprotein of about one million daltons which comprises about 30% of the VE. Lysin shows positive cooperativity in binding its receptor; approximately 60 lysin monomers bind each receptor. Sequencing the lysin receptor shows it contains about 28 tandem repeats of a 153 residue sequence. Sequencing the repeat from seven species shows that it evolves by concerted evolution. Concerted evolution of lysin's receptor could be responsible for the positive selection seen in lysin and provide a mechanism for the evolution of species-specific fertilization.

Acknowledgements

This research has supported by NIH grant HD12986 to V.D.V., NSF grant MCB951342 to C.D.S. and a United States-Mexico Foundation for Science grant to V.D.V.

References

Bookbinder LH, Cheng A, Bleil JD. Tissue and species-specific expression of sp56, a mouse sperm fertilization protein. Science 1995; 269:86-89.

Dover GA. Evolution of genetic redundancy for advanced players. Curr Opin Genet Dev 1993; 3:902-910.

Gao Z, Garbers DL. Species diversity in the structure of zonadhesin, a sperm-specific membrane protein containing multiple cell adhesion molecule-like domains. J Biol Chem 1998; 273:3415-3421.

Haino-Fukushima K, Usui N. Purification and immunocytochemical localization of the vitelline coat lysin of abalone spermatozoa. Dev Biol 1986; 115:27-34.

Heix J, Zomerdijk JCBM, Ravanpay A, Grummt I. Cloning of murine RNA polymerase I specific TAF factors: conserved interactions between the subunits of the species-specific transcription initiation factor TIF-IB/SL1. Proc Natl Acad Sci USA 1997; 94:1733-1738.

Lee Y-H, Vacquier VD. The divergence of species-specific abalone sperm lysins is promoted by positive Darwinian selection. Biol Bull 1992; 182:97-104.

Lee Y-H, Ota T, Vacquier VD. Positive selection is a general phenomenon in the evolution of abalone sperm lysin. Mol Biol Evol 1995; 12:231-238.

Lewis CA, Leighton DL, Vacquier VD. Morphology of abalone spermatozoa before and after the acrosome reaction. J Ultrastruct Res 1980; 72:39-47.

Lewis CA, Talbot CF, Vacquier VD. A protein from abalone sperm dissolves the egg vitelline layer by a nonenzymatic mechanism. Dev Biol 1982; 92:227-240.

Li W-H. Molecular Evolution. Sunderland, Mass: Sinauer Associates Inc, 1997.

Metz EC, Kane RE, Yanagimachi H, Palumbi SR. Fertilization between closely related sea urchins is blocked by incompatibilities during sperm-egg attachment and early stages of fusion. Biol Bull 1994; 187:24-34.

Metz EC, Robles-Sikisaki R, Vacquier VD. Nonsynonymous substitution in abalone sperm fertilization genes exceeds substitution in introns and mitochondrial DNA. Proc Natl Acad Sci USA 1998; 95:10676-10681.

Minor JE, Fromson DR, Britten RJ, Davidson EH. Comparison of the bindin proteins of *Strongylocentrotus franciscanus, S. purpuratus* and *Lytechinus variegatus*: sequences involved in the species-specificity of fertilization. Mol Biol Evol 1991; 8:781-795.

Murphy PM. Molecular mimicry and the generation of host defense protein diversity. Cell 1993; 72:823-826.

O'Rand MG. Sperm-egg recognition and barriers to interspecies fertilization. Gam Res 1988; 19:315-328.

Palumbi SR, Metz EC. Strong reproductive isolation between closely related tropical sea urchins (genus *Echinometra*). Mol Biol Evol 1991; 8:227-239.

Ramarao CS, Burks DJ, Garbers DL. A single mRNA encodes multiple copies of the egg peptide speract. Biochemistry 1990; 29:3383-3388.

Rudloff U, Eberhard D, Tora L, Stunnenberg H, Grummt I. TBP-associated factors interact with DNA and govern species-specificity of RNA polymerase I transcription. EMBO J 1994; 13:2611-2616.

Shaw A, McRee DE, Vacquier VD, Stout CD. The crystal structure of abalone sperm lysin. Science 1993; 262:1864-1867.

Shaw A, Fortes PGA, Stout CD, Vacquier VD. Crystal structure and subunit dynamics of the abalone sperm lysin dimer: egg envelopes dissociate dimers, the monomer is the active species. J Cell Biol 1995; 130:1117-1125.

Singson A, Mercer KB, L'Hernault SW. The *C. elegans spe-9* gene encodes a sperm transmembrane protein that contains EGF-like repeats and is required for fertilization. Cell 1998; 93:71-79.

Swanson WJ, Vacquier VD. Liposome fusion induced by a Mr 18 000 protein localized to the acrosomal region of acrosome reacted abalone spermatozoa. Biochemistry 1995a; 34:14202-14208.

Swanson WJ, Vacquier VD. Extraordinary divergence and positive Darwinian selection in a fusagenic protein coating the acrosomal process of abalone spermatozoa. Proc Natl Acad USA 1995b; 92:4957-4961.

Swanson WJ, Vacquier VD. The abalone egg vitelline envelope receptor for sperm lysin is a giant multivalent molecule. Proc Natl Acad Sci USA 1997; 94:6724-6729.

Swanson WJ, Vacquier VD. Concerted evolution in an egg receptor for a rapidly evolving abalone sperm protein. Science 1998; 281:710-712.

Vacquier VD, Lee Y-H. Abalone sperm lysin: unusual mode of evolution of a gamete recognition protein. Zygote 1993; 1:181-196.

Vacquier VD, Carner KR, Stout CD. Species-specific sequences of abalone sperm lysin, the sperm protein that creates a hole in the egg envelope. Proc Natl Acad USA 1990, 87:5792-5796.

Vacquier VD, Swanson WJ, Lee Y-H. Positive Darwinian selection on two homologous fertilization proteins: what is the selective pressure driving their divergence? J Mol Evol 1997; 44 (Suppl 1):S15-S22.

Vacquier VD, Swanson WJ, Metz EC, Stout CD. Acrosomal proteins of abalone spermatozoa. Adv Dev Biochem 1999; 5:49-81.

Yanagimachi R. Sperm-egg fusion. Curr Top Mem Trans 1988; 32:3-43.

15 *Transmembrane Signal Transduction for the Regulation of Sperm Motility in Fishes and Ascidians*

Masaaki Morisawa
University of Tokyo

Shoji Oda
Tokyo Women's Medical University

Manabu Yoshida
University of Tokyo

Hiroyuki Takai
National Institute for Longevity Sciences

Maturation and spawning of gametes are prerequisite processes for the accomplishment of fertilization. Spawning of gametes and the initiation of sperm motility have been subjects of numerous studies. Furthermore, the specificity of interactions between spermatozoa and eggs dictates the chances for a successful fertilization, one of the most important events in the reproduction system. In the process of fertilization, several interactions between the gametes are well established: activation of sperm motility, sperm chemoattraction toward the egg, adhesion and recognition between the sperm cell and the egg, induction of the acrosome reaction, sperm penetration into the egg-associated structures, membrane fusion of the gametes and activation of the egg by the spermatozoon. Studies on the regulatory mechanisms of the initiation of sperm motility have been studied in fishes (Morisawa 1985; Morisawa, Morisawa 1990; Morisawa 1994) since Morisawa and Suzuki (1980) demonstrated that the changes in environmental ionic concentration and osmolality trigger the phenomenon. Activation (Domino, Garbers 1990; Suzuki 1995) and chemotactic behavior of spermatozoa (Miller 1985b; Cosson 1990) as well as motility initiation at spawning in the process of fertilization induced by some signals released from the egg are important aspects in understanding the fertilization mechanisms. Recent studies by Morisawa and his colleagues focused on transmembrane cell signaling mechanisms. This chapter therefore deals with how the environmental factors and factors derived from eggs regulate the movement of spermatozoa at spawning and affect the plasma membrane of spermatozoa.

Factors Initiating Sperm Motility in Salmonid, Marine and Freshwater Fishes

External factors which control the initiation of sperm motility have been examined for decades in fishes (see also the chapter by Cosson et al., this book). In regard to the initiation of sperm motility at spawning, the following experimental conditions are similar to natural conditions. Spermatozoa of salmonid fishes, such as the rainbow trout *Oncorhynchus mykiss*, the chum salmon *Oncorhynchus keta*, the masu salmon *Oncorhynchus masou* and the char *Salvelinus leucomaenis,* are completely immotile when semen is diluted in a medium containing millimolar concentrations of K^+. They show motility when the semen is suspended in K^+-free medium (Morisawa et al. 1983a). The seminal plasma of salmonid fishes contains a high concentration of K^+ which is responsible for the lack of motility of spermatozoa (Morisawa 1985). Motility is activated at spawning into fresh water, when K^+ is diluted below the inhibitory concentration.

Freshwater or seawater spawning fishes also use the change of the natural environment for spawning. In all marine teleosts examined, including the puffers *Takifugu niphobles* and *T. pardalis*, the flounders *Kareius bicoloratus* and *Limanda yokohamae*, the Pacific cod *Gadus macrocephalus* (Morisawa, Suzuki 1980; Oda, Morisawa 1993), the turbot (Chauvaud et al. 1995) and the Atlantic croaker (Detweiler, Thomas 1998), spermatozoa are quiescent in a solution that is isotonic to the seminal plasma (300 mOsm/kg). They become motile when the semen is diluted with a hypertonic solution. In freshwater cyprinid fishes, such as the carp *Cyprinus carpio*, the goldfish *Carassius auratus*, the crucian *Carassius carassius*, the daces *Tribolodon taczanowskii* or *Tribolodon hakonensis* (Morisawa, Suzuki 1980; Morisawa et al. 1983b; Billard et al. 1995) and the pejerrey *Odontesthes bonariensis* (Stüssmann et al. 1994), spermatozoa are also quiescent in isotonic solution and the burst of motility occurs in hypotonic electrolyte and non-electrolyte solutions. The same phenomenon was reported in the cyclostome *Lampetra japonica* (Kobayashi 1993). These results suggest that motility is suppressed by the seminal osmolality and that these sperm initiate motility by exposure to hypertonic sea water or hypotonic freshwater at spawning. These data also suggest that spermatozoa of fish species use the changes of osmolality as the trigger for the intracellular cascade of events leading to the initiation of sperm motility.

K^+-Triggered Transmembrane Cell Signaling in Salmonid Fish Sperm

Activation of adenylyl cyclase is proposed as the key signal transduction mechanism in the initiation of sperm motility in salmonid fishes, rainbow trout and chum salmon, since spermatozoa demembranated with a detergent will reactivate motility in the presence of cAMP (Morisawa, Okuno 1982; Cosson et al. 1995). However, the transmembrane cell signaling underlying the activation of this enzyme is unclear. Several investigators have reported that the first detectable signal, the decrease in external K^+, causes hyperpolarization of the sperm plasma membrane via the efflux of K^+ through K^+ channels (Tanimoto et al. 1988, 1994; Gatti et al. 1990; Boitano, Omoto 1991).

Beltran et al. (1996) have reported that in sea urchin sperm, hyperpolarization of the plasma membrane with valinomycin increases the intracellular cAMP concentration in the absence of external Ca^{2+}, suggesting that adenylyl cyclase in the sperm is modulated by membrane potential. Recently, the Morisawa laboratory obtained data suggesting that the hyperpolarization of the sperm plasma membrane could induce the increase in intracellular cAMP which will cause the initiation of sperm motility in rainbow trout. The increase in cAMP through activation of adenylyl cyclase

(Morisawa, Ishida 1987) would be induced by the K^+ channel-dependent hyperpolarization of sperm membrane and the transmembrane cascade will trigger the initiation of motility in salmonid sperm (Fig. 1).

The issue of the requirement of external Ca^{2+} for the initiation of salmonid sperm motility is controversial. Addition of Ca^{2+} to immotile intact sperm in a medium containing K^+ elevates intracellular Ca^{2+} concentration ($Ca^{2+}{}_i$; Tanimoto et al. 1994) and induces the initiation of sperm motility; thus, Ca^{2+} overcomes the inhibitory effect of K^+ (Scheuring 1925; Baynes et al. 1981; Tanimoto, Morisawa 1988). The $Ca^{2+}{}_i$ increases concomitantly with the decrease in external K^+ (Cosson et al. 1989; Boitano, Omoto 1992). However, the presence of Ca^{2+} and other divalent cations in the medium causes hyperpolarization of the plasma membrane, which results in initiation of sperm motility (Boitano, Omoto 1991). These data raise the question as to whether the Ca^{2+} required for motility initiation of salmonid sperm is derived from external Ca^{2+}.

The intracellular signal transduction mechanism involved in the initiation of salmonid sperm motility has been established. It involves the phosphorylation of tyrosine residues of the movement-initiating phosphoprotein (MIPP) through the activation of a cAMP-dependent protein kinase and, subsequently, of a tyrosine kinase (Morisawa 1994). Recent studies by Inaba et al. (1998) demonstrated that cAMP-dependent phosphorylation of an outer arm dynein light chain is regulated by proteasome to control salmonid sperm motility.

Hyperosmolality-induced Initiation and Termination of Marine Fish Sperm Motility

Changes in external osmolality can modulate the initiation and termination of sperm motility: puffer sperm which initiated and continued their motility in 350 mM NaCl or 700 mM non-electrolyte mannitol solution (about 700 mOsm/kg), became quiescent within 10 seconds when the addition of an appropriate volume of a buffer solution changed the external osmolality to the isotonicity (300 mOsm/kg). Subsequent increases in osmolality in the sperm suspension, caused by the addition of high concentration of NaCl or mannitol solution, induced the reinitiation of sperm motility. The conversion from the immotile to the motile state and vice versa by changing external osmolality is a reproducible phenomenon (Takai, Morisawa 1995). The role of $Ca^{2+}{}_i$ in the hyperosmolality-dependent initiation and termination of sperm motility is clearly established in marine teleosts. Spermatozoa of puffer fish, in which motility is suppressed by osmolality isotonic to that of the seminal plasma, can initiate motility even in an isotonic solution if $Ca^{2+}{}_i$ is allowed to increase by treatment with the Ca^{2+} ionophore A23187 (Oda, Morisawa 1993).

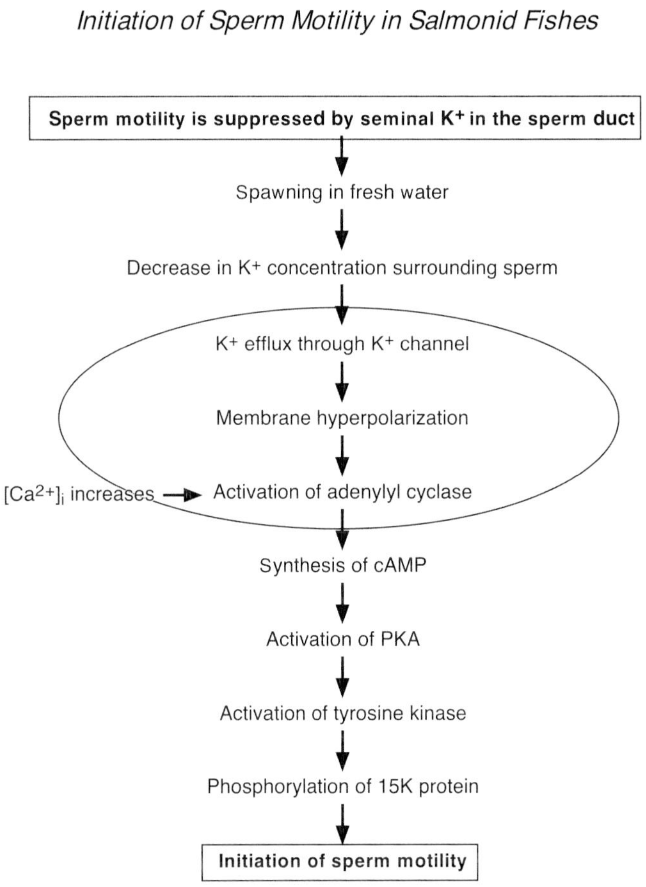

Figure 1. Transmembrane cell signaling for the initiation of sperm motility in salmonid fishes. The experimental conditions for demonstrating the role of K^+ on the initiation of sperm motility parallel the natural condition at spawning in K^+-deficient fresh water. Motility of sperm is suppressed by high seminal K^+ (37 and 87 mM in the rainbow trout and the chum salmon, respectively). A decrease in K^+ concentration surrounding spawned sperm in fresh water causes K^+ efflux, through K^+ channel, resulting in hyperpolarization of the plasma membrane of sperm flagellum. The change of membrane potential could directly activate adenylyl cyclase, which increases intracellular cAMP concentration. Increased $Ca^{2+}{}_i$ may also activate the enzyme in cooperation with membrane hyperpolarization. The transmembrane cascade process described in this chapter is encircled. Following this process, cAMP activates cAMP-dependent protein kinase (PKA), which then activates tyrosine kinase. The resulting phosphorylation of the protein triggers a final step leading to the initiation of sperm motility. (Morisawa, Morisawa 1990; Morisawa 1994.)

This suggests the participation of $Ca^{2+}{}_i$ in the control of axonemal movement. Since spermatozoa of puffer and flounder are motile in a medium containing EGTA, the Ca^{2+} released from intracellular stores may participate in the initiation of motility (Oda, Morisawa 1993).

Addition of 10 μM nigericin, a K^+/H^+ ionophore, to quiescent puffer sperm in an isotonic solution causes the initiation of motility in the same manner as that initiated in the hyperosmotic sea water. The ionophore-induced motility initiation does not occur when KCl was substituted for NaCl, suggesting that increase in K^+ in the sperm cell has a significant role (Takai 1994). Null point analysis using the pH sensitive fluorescent dye BCECF/AM showed that K^+_i of the sperm immotile in isotonic osmolality is 105.5 mM (Takai, Morisawa 1995). It increases to 299.0 mM when sperm become motile in hypertonic conditions, suggesting that the increase in K^+_i is critical for the osmolality-dependent initiation of sperm motility. The demembranated sperm of the puffer, the plasma membrane of which was removed with detergent, are quiescent in the reactivating medium with isotonic 150 mM K^+ and become fully motile in a medium containing 300 mM K^+. The initiation and arrest of sperm motility is repetitive upon changing the K^+ concentration of the reactivating media. Cosson et al. (this book) showed that repetitive changes of turbot sperm cell volume occur concomitantly with the changes of external osmolality. These results suggest that change of ionic strength or K^+ concentration (Takai, Morisawa 1995) in the sperm cell by a change of external osmolality is a factor for triggering the initiation of sperm motility in marine fishes (Fig. 2). Detweiler and Thomas (1998) showed that blockers of K^+-, Ca^{2+}-, Na^+- and Cl^-- channels inhibit sperm motility of a marine fish, the Atlantic croaker, and proposed an hypothesis that influx of Ca^{2+} and K^+ have an important role to play in the activation of sperm motility in marine fishes, although extracellular Ca^{2+} had no influence on the osmolality-dependent initiation of sperm motility in the puffer and the flounder (Oda, Morisawa 1993).

pH_i is a possible factor for the regulation of sperm motility in marine teleosts. Spermatozoa of the flounder and the puffers can initiate their motility under the isotonic osmolality when pH_i is increased by supplementation with NH_4Cl (Oda, Morisawa 1993) or nigericin (Takai 1994). The repetitive initiation and arrest of motility in demembranated spermatozoa also occurs upon changing the pH of the reactivating medium: spermatozoa are immotile or motile in the reactivating medium containing 150mM K-acetate at pH 7.4 or over 7.8 respectively, and vice versa (Takai, Morisawa 1995). However, contradictory results concerning the occurrence of the changes in pH_i during the changes in sperm motility have been obtained in marine teleosts. Using 9-aminoacridine, an increase in pH_i during the initiation of sperm motility in the puffer and the flounder was observed (Oda, Morisawa 1993). However, the increase in pH_i during the hyperosmolality-dependent motility initiation in puffer sperm was not detected clearly when BCECF was used; the change in pH_i that occurs at the initiation of sperm motility in this species is not major. pH_i in immotile sperm in the isotonic solution was 7.32, became 7.33 eight seconds after external osmolality increased, and decreased to 7.03 at 55 seconds, the time

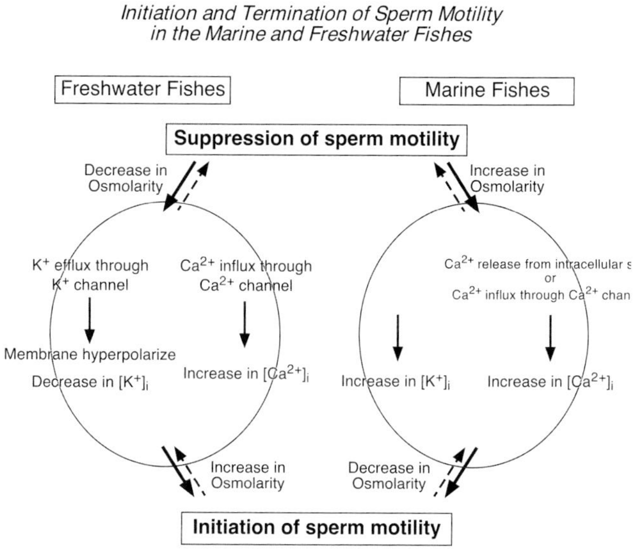

Figure 2. Initiation and termination of sperm motility in the marine and freshwater fishes. The experimental conditions demonstrating the role of osmolality on the initiation of sperm motility (black arrow) parallel the natural spawning conditions in hypotonic freshwater or hypertonic sea water. Generally, duration of sperm motility in teleosts is relatively short (Morisawa, Suzuki 1980; Cosson et al. this book) and mature males approach females and release spermatozoa immediately after oviposition (Stacy, Liley 1974). The released sperm reach the spawned oocytes within a short period. During the approach of sperm to the oocyte, shrinkage or swelling of sperm cells may be caused by the changes in external osmolalities and the events may cause increase or decrease, respectively, in intracellular ion concentration, especially K^+. Increase or decrease in $[K^+]_i$ in the sperm by the changes in external osmolalities may cause the initiation of motility (that is, flagellar axoneme). Participation of K^+ channel in sperm activation was proposed in freshwater fish (for example, the carp) and in marine fish (for example, the Atlantic croaker). Membrane hyperpolarization in the sperm of freshwater fish and increase in $[Ca^{2+}]_i$ of sperm cells in the marine and freshwater fishes have critical roles for the initiation of sperm motility. Termination of sperm motility (dotted arrow) could occur through reversal of the cascade process of the initiation of sperm motility. A change in pH_i does not seem to be a primary factor for the regulation of the initiation of sperm motility in either freshwater or marine teleosts. It may work as a cooperative factor with the intracellular trigger for the phenomenon; for example, K^+ concentration. cAMP is not required for the initiation of sperm motility in freshwater and marine teleosts. (Perchec et al. 1995; Krasznai et al. 1999.)

at which sperm motility has almost ceased (Takai 1994). The following experiments settle the above controversial results on the role of pH_i. When pH_i increased, the K^+ concentration required for initiation of motility became low. The high pH_i induced the initiation of sperm motility even with a low K^+ concentration, suggesting that both K^+_i and pH_i are the factors regulating sperm motility and that considerable change in pH_i is required for the motility initiation by increasing K^+_i; pH_i and K^+_i are both needed in the initiation of flagellar movement in the marine fish.

Hypoosmolality-induced Repetition of Initiation and Termination of Freshwater Fish Sperm Motility

The repetition of initiation and termination of sperm motility in freshwater fishes is quite a mirror image of that of marine fishes (Takai, Morisawa 1995). The conversion of live sperm from the immotile to the motile state and vice versa by changing external osmolality from isotonic to hypotonic was observed in the freshwater teleost fish, zebrafish. The demembranated sperm of the zebrafish were immotile in the reactivating solution containing K-acetate at concentrations over 150 mM at pH 7.4. Spermatozoa became motile when the K-acetate concentration was decreased. Na^+ in the reactivation solution could be substituted for K^+ but the sperm were motile in mannitol solutions at all concentrations of mannitol tested, suggesting that the intracellular environment for the regulation of sperm motility is not osmolality but ionic strength. The analysis of the mechanism concerning conversion of the external signal, the osmolality, to the internal one, ion strength, is the subject of future studies.

Motility of zebrafish demembranated sperm was also sensitive to pH. The demembranated sperm were immotile in the reactivating medium containing 100 mM K-acetate at a pH lower than 6.5, but sperm became motile when pH of the reactivating medium was increased. Initiation and termination of sperm motility in the reactivation medium are also pH dependent. The coordinated effect of K^+ and pH in the initiation of sperm motility is present in freshwater teleosts as well as marine teleosts. When concentrations of K-acetate in the reactivating solution were decreased, the demembranated sperm were motile in the pH range from 6.5-8.5; however, motility did not occur if the reactivating solution contained a high concentration of K^+ even when pH in the reactivating medium was increased. Demembranated sperm of the zebrafish as well as of the puffer could initiate motility in the presence of ATP alone (without cAMP). Many articles have suggested that cAMP is not required for the initiation of activation of sperm motility in freshwater and marine fishes except salmonid fishes (Cosson et al., this volume).

The role of pH_i and K^+_i on the initiation of sperm motility was also examined in the carp (Màriàn et al. 1997). The repetition of initiation and termination of sperm motility also occurs in this species. Hypoosmotic shock induces a fast alkalinization of the sperm cell (Krasznai et al. 1995) concomitantly with fast change of the membrane structure and permeabilization (Màriàn et al. 1993) that could be reversed by restoring physiological osmolality (Màriàn et al. 1997). In the physiological range of both extracellular and intracellular pH, sperm are always motile, but in the extremely unphysiological pH range (both acidic and alkaline), sperm motility is suppressed. These results suggest that pH does not have a primary regulatory role in inducing the initiation of sperm motility, not only in freshwater (Màriàn et al. 1997) but also marine teleosts (Takai, Morisawa 1995). Inhibitors of Na^+ channel, tetrodoxin, and of Cl^- channel, DDIS and ethacrynic acid, have no inhibitory effect on the induction of sperm motility, but K^+ channel inhibitors, 4-AP, quinine and veratrine, decrease the duration of flagellar motion or abolish motility completely (Krasznai et al. 1995). The K^+_i of sperm is 62.4 mM, which is very similar to the K^+ concentration of the seminal plasma, suggesting a depolarization state of the sperm plasma membrane in the semen (Màriàn et al. 1997). Hypoosmotic induction of motility is accompanied by the hyperpolarization of the sperm membrane and the initiation of sperm motility is blocked by K^+ channel blockers, suggesting the critical role of K^+ efflux for the initiation of sperm motility in freshwater teleosts (Krasznai et al. 1995). In a Ca^{2+} deficient hypoosmotic solution, sperm cannot initiate motility, but addition of Ca^{2+} to the immotile sperm causes the initiation of sperm motility. Some Ca^{2+} channel blockers suppress or inhibit sperm motility, suggesting the participation of Ca^{2+} channels in the initiation of sperm motility in the carp (Krasznai et al. 1998). These results are included in the schematic drawing showing the cell signaling mechanism for the initiation of sperm motility in freshwater teleosts (Fig. 2).

The osmolality-induced transmembrane cell signaling for the initiation of sperm motility in marine fishes and the signaling in freshwater fishes seem to be mirror images of each other in terms of regulation. On the other hand, sperm from these two types of fishes also share some similarities (Fig. 2). K^+ efflux through K^+ channels will change the plasma membrane from a depolarized state to a hyperpolarized state in freshwater fishes (Màriàn et al. 1993, 1997; Krasznai et al. 1995, 1998). In contrast, K^+_i increases during the hypertonic osmolality-induced initiation of sperm motility in marine teleosts (Takai, Morisawa 1995). Both hyperosmotic and hypoosmotic shocks cause the increase in Ca^{2+}_i in marine (Oda, Morisawa 1993) and freshwater fishes (Krasznai et al. 1999). It is still obscure how the change of membrane potential causes the common phenomenon of Ca^{2+} elevation in the sperm cell, in the hypotonic or hypertonic mechanisms inducing sperm motility. A two-dimensional reorganization of the lipid structure may be important in the transmembrane cell signaling for the osmotic pressure-induced initiation of sperm motility (Màriàn et al. 1993). Stretch activated ion channels should be activated by structural changes of the sperm cell. Shrinkage of the marine fish spermatozoon in hypertonic solution or the increase in volume and swelling of the freshwater (Perchec et al. 1996) fish sperm cell in hypotonic conditions may play key roles for the initiation of sperm motility in marine and freshwater teleosts.

Egg-derived Factors, HSAPs, Cause Activation of Pacific Herring Sperm Motility

The activation of the sperm motility by the egg factors, a phenomenon ubiquitous in the animal kingdom, is the first communication between the spermatozoa and the egg at fertilization. The molecules by which eggs activate spermatozoa and the underlying molecular mechanisms have been clarified in only a few invertebrate species, such as sea urchins (Ohtake 1976; Hansbrough, Garbers 1981; Suzuki 1995). The peptides from sea urchin egg jelly with unique structures cause changes in cyclic nucleotide concentrations, membrane potential, ion concentrations, pH_i and protein phosphorylation following binding to the receptor. This results in the activation of metabolism and motility (Domino, Garbers 1990), chemotactic behavior (Cook et al. 1994) and potentiation of acrosome reaction (Lievano et al. 1996) in the sperm. Sperm activators were reported in eggs of fishes such as the bitterlings *Acheilognathus lancedata, A. tubira, Rhodeus ocellatus* and the fat minnow *Sarcocheilichthys variegatus* (Suzuki 1958), but the chemical nature of the activators remains unknown.

Spermatozoa of the Pacific herring *Clupea pallasi* are immotile in a solution which is isotonic to the seminal plasma but swimming can be initiated in a hypertonic solution, suggesting that the factor for the initiation of sperm motility could be the change of environmental osmolality. However, sperm motility in hypertonic solution remains very low (Morisawa et al. 1992) or sometimes is absent or very little (Yanagimachi, Kanoh 1953; Yanagimachi et al. 1992). Sperm motility is initiated in the vicinity of the egg (Yanagimachi et al. 1992) or in egg-conditioned medium (Morisawa et al. 1992), suggesting that motility activation by the egg at spawning into the milieu is considered to be indispensable for a successful fertilization (Yanagimachi, Kanoh 1953; Yanagimachi 1957). Eggs of *C. pallasi* release proteinaceous activation factors of sperm motility into the surrounding sea water (Morisawa et al. 1992). These factors were purified from the egg-conditioned sea water by gel filtration and isoelectric focusing and named herring sperm-activating proteins (HSAPs; Oda et al. 1995). There are five isoforms of HSAPs which have similar molecular masses (about 8 kD) and isoelectric points (4.8, 4.9, 5.0, 5.1 and 5.4), very similar sperm-activating-capacity and almost identical N-terminal amino acid sequences, suggesting that the HSAPs are isoforms of water-soluble small proteins. The molecular mass of the HSAP with pI=5.1 was determined to be 8.1 kD by mass spectrometry.

Another activator of sperm motility has been reported in the same species and designated sperm motility initiating factor (SMIF; Yanagimachi et al. 1992; Pillai et al. 1993). SMIF is a water-insoluble glycoprotein with molecular mass of 105 kD that binds to the egg chorion so tightly that the activation of spermatozoa is considered to happen at the time when spermatozoa touch the surface of the egg chorion.

There are two modes for the activation of sperm by egg in *C. pallasii*: the chemokinesis around the egg (Morisawa et al. 1992) and the chemotaxis into the micropylar opening of the egg (Yanagimachi 1957; Yanagimachi et al. 1992). Because SMIF is thought to be localized near the micropylar opening of the herring egg, SMIF was proposed as a candidate for sperm attractant to the micropylar opening (Griffin et al. 1996). The HSAPs are evenly localized in the outermost layer of the egg chorion (Oda et al. 1998). This localization of HSAPs over the egg surface seems to be inconsistent with the proposition that HSAPs are responsible for sperm chemotaxis, but is suggestive of the role of HSAPs in the chemokinesis around the egg as reported by Morisawa et al. (1992). It is possible that the sperm activation by eggs in *C. pallasi* involves two steps. HSAPs released into the surrounding sea water activate the motility of spermatozoa and increase the number of spermatozoa which meet the SMIF, then the SMIF guides spermatozoa into the micropylar opening. A detailed analysis of the flagellar bending pattern of the spermatozoon activated with HSAPs and SMIF will be essential to verify this hypothesis.

Both diffusible acidic small HSAPs and non-diffusible large SMIF cause a depolarization of the sperm plasma membrane and an increase in $Ca^{2+}{}_i$ (Vine et al. 1998; Yoshida et al. 1998), which is known to be prerequisite for herring sperm motility (Yanagimachi, Kanoh 1953; Yanagimachi et al. 1992). However, there are some differences in the mode of Ca^{2+} action between the effects of HSAPs and the effects of SMIF: the increase in $Ca^{2+}{}_i$ induced by SMIF is several times larger than that induced by HSAPs (K Yoshida, personal communication). Nifedipine, a Ca^{2+} channel inhibitor, blocks sperm activation induced by SMIF but has little or no effect on that induced by HSAPs. Concentrations of cAMP and cGMP did not change in the presence of HSAPs, suggesting that the cyclic nucleotides are not factors triggering the cascade of events leading to the activation of herring sperm motility.

Complementary DNA (cDNA) clones of the HSAP were isolated from the ovarian cDNA library of *C. pallasi.* The full-length cDNA sequence of one of the five

HSAPs (the longer clone) is 479 nucleotides long, contains a single open reading frame of 316 nucleotides and encodes for a polypeptide of 94 amino acid residues (Fig. 3). All five HSAPs purified from the egg-conditioned medium lack 21 amino acids from the N-terminal portion of the deduced protein, indicating that the additional N-terminal amino acids obtained by molecular cloning constitute a signal peptide which is characteristic of secreted proteins. The HSAP encoded by the cDNA is therefore a secretory protein consisting of 73 amino acids. Its isoelectric point and molecular mass are estimated at 5.13 and 8,173 D, respectively, and are in good agreement with those of the purified HSAP (Oda et al. 1998).

Molecular cloning of the HSAPs' cDNA revealed that the HSAPs are small secretory proteins which have striking homology with the Kazal-type trypsin inhibitors (Oda et al. 1998). No similarity was found in the amino acid sequence nor nucleotide sequences of cDNA between the HSAPs and the SAPs of sea urchins (Ramarao et al. 1990; Suzuki 1995) or asterosap of starfish (Nishigaki et al. 1996). The findings that eggs of fish and echinoderms release different peptides to activate spermatozoa and that a single mRNA of HSAP encodes a single peptide, whereas a single mRNA of SAPs encodes more than 10 peptides in sea urchin (Ramarao et al. 1990), strongly suggest that the establishment of sperm activation by the egg in fish and echinoderm is independent in each animal and that each underlying mechanism of sperm activation appears to be different. In vertebrates, there is a possibility that the mechanism of sperm activation is common to all species because HSAP homologs are abundant in mammalian seminal plasma and called acrosin inhibitors. The acrosin inhibitors are considered to have significance in the fertilization through the inhibition of the trypsin-like protease, acrosin (Fritz et al. 1975). Their biological roles are still unclear. The finding presented here that an acrosin inhibitor homolog is implicated in the regulation of sperm motility in fish can also suggest a new physiological function for proteinase inhibitors in vertebrate fertilization. A proteinase inhibitor obtained from porcine follicular fluid increases the motility of porcine spermatozoa (Lee et al. 1992; Jeng et al. 1993). Once these mammalian sperm-activating factors are purified and characterized, the relationship between sperm activating factors of mammals and fishes will be clarified, and the mechanism of sperm activation by egg in vertebrates will be more easily addressed. Further studies will also elucidate the relevance of proteinase inhibitors in vertebrate fertilization as well as enhance knowledge of the mechanisms involved in fertilization.

A prolyl endopeptidase from herring the properties of which are similar to those from other tissues or cells, is activated by HSAPs (Yoshida et al. 1998). The activation of sperm motility by HSAPs is inhibited by Z-thioprolinal, a specific inhibitor of prolyl endopeptidase, suggesting that the prolyl endopeptidase is related to herring sperm activation by HSAPs. It is of great interest to identify the HSAPs receptor on the surface of sperm plasma membrane and to elucidate the relationship and role of the protease and the protease inhibitor in the triggering of the intracellular signal transduction underlying the activation of the herring sperm motility.

```
  T CGA CCC ACG CGT CCG GTC CCT TCT GAA GCC ACC ATG AAG  40
                                             M   K
                                             1

CTG AGC ATT GTG ATC TCC ATC TGC GTT CTA CTT TAC TTC TCT  82
 L   S   I   V   I   S   I   C   V   L   L   Y   F   S
                             10

GGT CAC ACT TTG GCC AGG TCA GTC CCG AGG ATC GGG ATT GAT 124
 G   H   T   L   A   R   S   V   P   R   I   G   I   D
             20                                      30

TGT CAA GGC TAC GGT TCT GCT TGC ACT AAG GAG TAC CGC CCT 166
 C   Q   G   Y   G   S   A   C   T   K   E   Y   R   P
                                     40

ATC TGT GGG TCC GAC GAC GTC ACC TAT GAA AAT GAA TGC TTG 208
 I   C   G   S   D   D   V   T   Y   E   N   E   C   L
                     50

TTC TGC GCT GCC AAA CGA GAA AAT AGA TGG GGG ATT TTG GTC 250
 F   C   A   A   K   R   E   N   R   W   G   I   L   V
     60                                      70

GGT CAT CGC GGG GCA TGT ATA GCG TGG GGG GGG ATG GTG GAG 292
 G   H   R   G   A   C   I   A   W   G   G   M   V   E
                             80

GAG TTG AGG GAG TGG AGC TCC GAC TGA GGT GAA CCC TGA CCC 334
 E   L   R   E   W   S   S   D  end
             90

TGA CCC TGA ACC CTG ATC CTA ACC CTA ACC CTA ACC CTA ACC 376

CTA ACC CCA ACC CTG ACC TGA GAT CAG AGG GAA ATG TCA GTA 418

AAG CAT CAA TAA AGC ATC TGT AAA GCA TAC AAA AAA AAA AAA 460
         ** ***  *

AAA AAA AAA AAA AAA AAA A  479
```

Figure 3. DNA sequences of the isolated HSAP-cDNA clone and amino acid sequence of HSAP. The nucleotide sequence is presented above the predicted sequence of amino acids, which are represented by their single-letter code under the appropriate codon. The signal peptide is underlined with a dotted line. The amino acid sequence corresponding to that of N-terminal amino acid residues of HSAP (pI=5.1), determined by Edman degradation method, is underlined with a solid line. The polyadenylation signals are marked by asterisks. The Kozak's consensus initiation sequence for translation is boxed.

An Egg-derived Factor, SAAF, Induces Both Activation and Chemotaxis in Ascidian Sperm

Nature of Sperm Activation and Chemotaxis

The chemotactic behavior and activation of spermatozoa in the process of fertilization induced by some signals released by the egg are important topics to understand. Sperm chemotaxis was first described in a plant, the bracken fern, by Pfeffer (1884) and the sperm attractant was identified as α-malic acid (Brokaw 1958). Spermatozoa of the brown algae also show chemotactic behavior toward the egg, and the sperm attractant was identified as hydrocarbons (Maier, Müller 1986). Chemotaxis was also observed in bacteria, leukocytes, amoeba of cellular slime mold, and more (Hazelbauer 1985; Gerisch 1987; Devreotes, Zigmond 1988). Spermatozoa of the slime mold *Dyctyostelium discoideum*, for example, show chemotactic behavior toward folic acid in the vegetative stage and toward cAMP in starvation. Activation of G proteins is an important signal transduction mechanism by which chemotaxis is regulated (Kesbeke et al. 1990).

The first description of sperm chemotaxis in animals was reported by Dan (1950) in the hydrozoan, *Spirocodon saltatrix*. Since then, the phenomenon has been extensively studied in many animal phyla: Cnidaria, Urochordata, Mollusca (Miller 1985b), Echinodermata (Miller 1985a; Ward et al. 1985; Punnett et al. 1992), hydroids (Cosson et al. 1984) and humans (Ralt et al. 1994). Among small peptides (sperm activating peptides, or SAP) that were isolated from eggs of sea urchins (Ohtake 1976; Hansbrough, Garbers 1981; Suzuki 1990), only resact has both sperm-attracting and sperm-activating activities (Ward et al. 1985). Some alcohols from the coral (Coll, Miller 1990) have also been identified as the sperm attractant. A sperm guidance mechanism at the micropyle area of the rosy barb *Barbus conchonius* egg was reported by Amanze and Iyengar (1990).

Sperm Activation and Chemotaxis in Ascidians

Recent work on the regulation of sperm motility in the ascidian *Ciona* opens a new insight on the relationship between sperm activation and chemotaxis. Sperm activation and agglutination responses to egg or egg sea water in ascidians were reported by Minganti (1951). Miller (1975) demonstrated sperm chemotaxis in *C. intestinalis* and suggested that strong sperm-attracting activity occurred in the follicle cells. Characteristics of the chemotactic behavior of ascidian spermatozoa include activation, followed by the approach to the egg with circular or sudden turning movements (Miller 1982). It was observed that *Ciona* sperm were activated and attracted to the surface of the whole egg (Yoshida et al. 1993). When the egg was separated into follicle cells, chorion, test cells and the egg itself (naked egg), spermatozoa were activated and exhibited chemotaxis toward the naked egg. Such activation and chemotaxis were not observed when isolated follicle cells or chorion with follicle cells and test cells were tested. These results clearly indicate that the sperm-activating and a sperm-attracting substance in *Ciona* is released from the egg itself and not from other accessory cells.

According to Conklin (1905), the spermatozoon of *Styela* enters near the vegetal pole of the egg. However, opposite results were obtained by Speksnijder et al. (1989), who reported that spermatozoa of *Phallusia* enter the animal pole and later are carried to the vegetal pole by ooplasmic segregation. Sawada and Schatten (1989) suggested that spermatozoa of *Molgula* may first bind at any site on the egg surface and then be carried to near the vegetal pole. Lambert (1989) has shown that spermatozoa of the ascidians *Ascidia collosa* and *Ascidia paratropa* pass through the vitelline coat from the site of binding to the vitelline coat and attach to the egg surface underlining this immediate area. In a study by Yoshida et al. (1993), spermatozoa resuspended in sea water were almost quiescent but, upon addition of a naked egg to the suspension, started to swim toward the egg and seemed to accumulate at a specific site, the vegetal pole, of the egg. The first polar body appeared 10 minutes after deformation of the egg at a position opposite to where spermatozoa had accumulated. It seems possible that the sperm-activating substance and attractant are released from the vegetal pole on the egg surface and attract spermatozoa with greater efficiency to the site of vitelline coat that is just above the vegetal pole.

The sperm-activating and sperm-attracting activities of *C. savignyi* eggs disappear at the time of deformation that occurs immediately after fertilization. Miller (1978) has reported a similar observation: the egg of the leptomedusan *Orthopyxis caliculata* no longer has the capacity to attract spermatozoa approximately 11 minutes after insemination, a time during which the egg may be fertilized. In contrast, in the siphonophore *Muggiaea kochi*, the isolated cupule, which is thought to be the portion that supplies the sperm attractant, can maintain sperm-attracting activity for more than a day (Carré, Sardet 1981). During the deformation of the egg, several changes (mitochondrial movement, ooplasmic segregation and so on) occur at the surface of the egg in *Ciona* (Sawada, Osanai 1981). This suggests some functional changes in the plasma membrane. Thus, the disappearance of the sperm attraction during deformation of the egg seems to be caused by a change in permeability of the egg plasma membrane. In *C. intestinalis* and *C. savignyi*, the egg of each species activates and attracts the other's spermatozoa. However, interspecies fertilization does not occur between these two species (Lambert et al. 1990). Therefore, species specificity in sperm attraction and fertilization may be controlled at some other level.

In ascidians, the sperm attractant was extracted with ethanol from the egg of *C. intestinalis* by Miller (1975). Yoshida et al. prepared egg-conditioned sea water (egg sea water, or ESW) with strong sperm-activating and sperm-attracting activities by incubating eggs of the ascidian *Ciona savignyi* with artificial sea water for 14 to 20 hours at 4°C and by subsequent centrifugation. The supernatant had strong activities. ESW was extracted with ethanol and then fractionated by water-chloroform extraction and the activities were obtained in the water rather than chloroform fraction, suggesting that the substances are not non-polar hydrocarbons, as are the sperm attractants in brown algae. The substance with the sperm-activating and sperm-attracting activities was purified with RP columns with ODS groups, which are effective for separation of hydrophilic small organic compounds such as oligopeptides. The substance was finally obtained as a single peak with an absorbance at 205 nm. The sperm-activating and sperm-attracting activities always comigrated during the purification procedures using HPLC. This suggests that both activities, sperm activation and attraction, belong to a single molecule, and thus the molecule was named sperm-activating and -attracting factor (SAAF).

When the purified SAAF was packed in the tip of a glass capillary and put into a sperm suspension, spermatozoa became active and motile and were attracted toward the tip of the capillary and performed the "chemotactic turn." The sperm-activating and sperm-attracting activities remained when SAAF was boiled or treated with proteases, suggesting that the substance is not protein. The fucosyl-containing glycoprotein derived from the vitelline coat (chorion) of the egg of the ascidian *C. intestinalis* activates sperm motility of the same species (De Santis et al. 1983). It was also reported that pNP-β-D-hexosaminide, the substrate of hexosaminase, enhances sperm motility of the ascidian *Halocynthia roretzi* (Hirohashi, Hoshi 1991). Results also suggest that SAAF activity may be independent of sugar moieties because treatment with glycopeptidase and lectins had no effect.

The Role of Second Messengers in Sperm Activation and Chemotaxis

The basic mechanism for activation of *Ciona* sperm motility is considered to be regulated by the decrease in external K^+ concentration (Morisawa et al. 1984) and dependent on cAMP-dependent phosphorylation of protein (Opresko, Brokaw 1983). Yoshida et al. (1994) further showed in *Ciona intestinalis* that the transient increase in intracellular cAMP due to SAAF occurs in the presence of external Ca^{2+} and is a prerequisite process for the activation of sperm motility. Recently, Izumi et al. (1997) observed that a potassium ionophore, valinomycin, initiates sperm motility in the absence of SAAF, even in Ca^{2+}-free sea water. The effect of the ionophore is suppressed by external K^+ at high concentration. Measurement of membrane potential with $DiSC_3$ indicated that sperm plasma membrane has a low K^+ permeability in the absence of SAAF, and that not only valinomycin but also SAAF increases its K^+ permeability, resulting in induction of a sustained hyperpolarization of the sperm plasma membrane. Valinomycin (Izumi et al. 1997)and SAAF (Yoshida et al. 1994) increase cAMP in the sperm cells, suggesting that hyperpolarization of the plasma membrane through an increase in K^+ permeability causes the synthesis of cAMP, which results in the activation of sperm motility.

The requirement of extracellular Ca^{2+} for the activation of *Ciona* sperm motility (Brokaw 1982; Morisawa et al. 1984) was also suggested. SAAF did not activate sperm motility nor cAMP synthesis in Ca^{2+}-free sea water (CaFSW), but activated both these processes in the presence of Ca^{2+}. Sperm activation by SAAF in Ca^{2+}-containing normal sea water was inhibited by flunarizine, a Ca^{2+} channel antagonist, but not by nitrendipine and verapamil, which are L-type Ca^{2+} channel specific antagonists. Theophylline induced both increase in the cAMP and sperm activation in CaFSW without SAAF. These results suggest that Ca^{2+} influx through a Ca^{2+} channel stimulated the synthesis of cAMP, which activated sperm motility (Yoshida et al. 1994). It is also suggested that although Ca^{2+} influx is necessary for the cAMP synthesis, it is not necessary for the hyperpolarization of the plasma membrane of *Ciona* sperm (Fig. 4).

Chemotactic behavior of sperm was recognized in various animal species, but the chemical nature of the attractants is still unknown except in the sea urchin and starfish. Requirement of Ca^{2+} for the sperm chemotactic response toward the egg was found in the ascidian, sea urchin and hydroids (Miller 1982; Cosson et al. 1984; Ward et al. 1985). Analytical studies in siphonophores by Cosson et al. (1984) showed that sperm which are far from the source of the sperm attractant, called the cupule, swim with circular trajectories. Trajectory diameters are reduced as sperm approach the cupule. The studies also indicated that the sperm were not accumulated around the cupule in CaFSW.

In *Ciona,* spermatozoa do not show chemotaxis but are activated when in close contact to the tip of a capillary filled with a solution of theophylline (an inhibitor of phosphodiesterases) which causes an increase in intracellular cAMP. However, spermatozoa which were previously activated with theophylline without the egg factor showed chemotactic behavior to the tip of the capillary in which the egg sea water was packed (Yoshida et al. 1994). This indicates that sperm activation, but not chemotaxis, requires cAMP. Sperm activation and chemotaxis could be controlled by different mechanisms. Although the control mechanism

Activation of Sperm Motility and Sperm Chemotaxis in the Ascidians

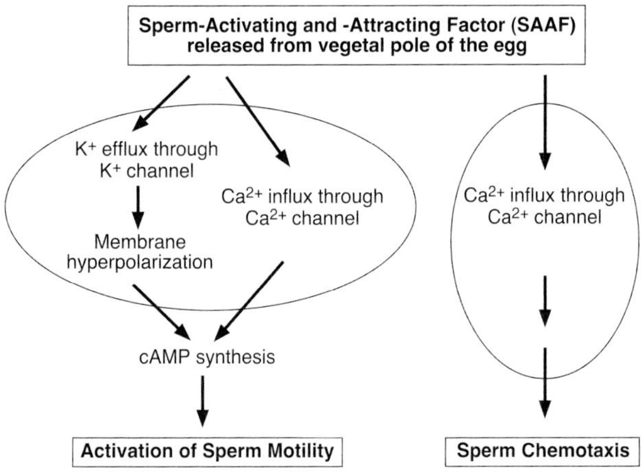

Figure 4. Signal transduction of the activation and chemotaxis of ascidian spermatozoa. Sperm activating- and attracting-factor (SAAF) released from the vicinity of the vegetal pole of the egg activates sperm motility and then attracts sperm toward the egg. One SAAF molecule causes both Ca^{2+} influx into the sperm cell through Ca^{2+} channel and membrane hyperpolarization. Both phenomena cause the increase in intracellular cAMP level, resulting in the activation of sperm motility. Through another route, Ca^{2+} mediates the sperm chemotaxis without participation of cAMP.

of chemotaxis is still unknown, it is possible that a molecule, such as SAAF, released from the egg could control different phenomena, sperm activation and chemotaxis, via different mechanisms (Fig. 4).

The requirement of extracellular Ca^{2+} not only for the activation of *Ciona* sperm but also for sperm chemotaxis was also suggested. Sperm are not attracted toward an egg in Ca^{2+}-free sea water (Miller 1982). On the other hand, the theophylline-activated sperm with a high intracellular cAMP do not exhibit chemotaxis toward the tip of a glass capillary containing SAAF in CaFSW. Upon the addition of Ca^{2+}, they were attracted toward SAAF. These results suggest that sperm activation is induced by cAMP synthesized via a mechanism involving Ca^{2+} influx through Ca^{2+} channel and membrane hyperpolarization, and that Ca^{2+} alone mediates the sperm chemotaxis in *Ciona.*

The diameter of the swimming trajectories of theophylline-activated *Ciona* sperm is reduced in the presence of Ca^{2+}. Spermatozoa are then attracted toward SAAF. Similar modulation of flagellar movement occurs in the demembranated sperm of the sea urchin. Ca^{2+} is indispensable when sea urchin sperm flagellum describe asymmetrical waves (Brokaw 1974). In intact *Ciona* sperm, extracellular Ca^{2+} may be transported into sperm cells through Ca^{2+} channels, suggesting that SAAF, which was released from the eggs, induces a Ca^{2+} influx and increases the asymmetrical flagellar beating. Therefore, it is worth emphasizing again that one molecule mediates two different pathways, Ca^{2+}- and cAMP-dependent sperm activation and sperm attraction (Fig. 4). The latter is regulated only by Ca^{2+}, which causes spiral movement and turn upon approach toward the egg.

Acknowledgements

This work was done with the support of the Ministry of Education, Culture, Sports, and Science of Japan. The authors wish to thank Mrs. Kaoru Yoshida, Miss Hiroko Izumi, Dr. Gary N. Cherr and Dr. Jacky Cosson for critical reading of the manuscript and helpful discussion.

References

Amanze D, Iyengar A. The micropyle: a sperm guidance system in teleost fertilization. Development 1990; 109:495-500.

Baynes SM, Scott AP, Dawson AP. Rainbow trout *Salmo gairdneri* Richardson spermatozoa: effects of cations and pH on motility. J Fish Biol 1981; 19:259-267.

Beltran C, Zapata O, Darszon A. Membrane potential regulates sea urchin sperm adenylyl cyclase. Biochemistry 1996; 35:7591-7598.

Billard R, Cosson J, Perchec G, Linhart O. Biology of sperm and artificial reproduction in carp. Aquaculture 1995; 129:95-112.

Boitano S, Omoto CK. Membrane hyperpolarization activates trout sperm without an increase in intracellular pH. J Cell Sci 1991; 98:343-349.

Boitano S, Omoto CK. Trout sperm swimming patterns and role of intracellular Ca^{2+}. Cell Motil Cytoskel 1992; 21:74-82.

Brokaw CJ. Chemotaxis of bracken spermatozoids.The role of bimalate ions. J Exp Biol 1958; 35:192-196.

Brokaw CJ. Calcium and flagellar response during the chemotaxis of bracken spermatozoids. J Cell Physiol 1974; 83:151-158.

Brokaw CJ. Activation and reactivation of *Ciona* spermatozoa. Cell Motil 1982; 1:185-189.

Carré D, Sardet C. Sperm chemotaxis in siphonophores. Biol Cell 1981; 40:119-128.

Chauvaud L, Cosson J, Suquet M, Billard R. Sperm motility in turbot *Scophthalmus maximus*: initiation of movement and changes with time of spawning characteristics. Env Biol Fish 1995; 43:341-349.

Coll JC, Bowden BF, Clayton MN. Chemistry and coral reproduction. Chem Br 1990; 26:761-763.

Conklin EG. The organization and cell-lineage of the ascidian egg. J Acad Natl Sci Phil 1905; 13:1-126.

Cook SP, Brokaw CJ, Muller CH, Babcock DF. Sperm chemotaxis: egg peptides control cytosolic calcium to regulate flagellar responses. Dev Biol 1994; 165:10-19.

Cosson MP. Sperm chemotaxis. In: (Gagnon C, ed) Controls of Sperm Motility: Biological and Clinical Aspects. Boca Raton, FL: CRC Press, 1990, pp103-135.

Cosson MP, Carré D, Cosson J. Sperm chemotaxis in siphonophores. II. Calcium dependent asymmetrical movement of spermatozoa induced by attractant. J Cell Sci 1984; 68:163-181.
Cosson MP, Billard R, Letellier L. Rise of internal Ca^{2+} accompanies the initiation of trout sperm motility. Cell Motil Cytoskel 1989; 14:424-434.
Cosson MP, Cosson J, Andre F, Billard R. cAMP/ATP relationship in the activation of trout sperm motility: their interaction in membrane-deprived models and in live spermatozoa. Cell Motil Cytoskel 1995; 31:159-176.
Dan JC. Fertilization in the medusan, *Spirocodon saltatrix*. Biol Bull Mar Biol Lab Woods Hole 1950; 99:412-415.
De Santis R, Pinto MR, Cotelli F, Rosati F, Monroy A, D'alessio G. A fucosyl glycoprotein component with sperm receptor and sperm-activating activities from the vitelline coat of *Ciona intestinalis eggs*. Exp Cell Res 1983; 148:508-513.
Detweiler C, Thomas P. Role of ions and ion channels in the regulation of Atlantic croaker sperm motility. J Exp Zool 1998; 281:139-148.
Devreotes PN, Zigmond SH. Chemotaxis in eukaryotic cells: a focus on leucocytes and Dictyostelium. Ann Rev Cell Biol 1988; 4:649.
Domino SE, Garbers DL. Mode of action of egg peptides. In: (Gagnon C, ed) Controls of Sperm Motility: Biological and Clinical Aspects. Boca Raton: CRC Press, 1990, pp91-101.
Fritz H, Schiessler H, Schill WB, Tscheche H, Heimburger N, Wallner D. Low molecular weight protenase (acrosin) inhibitors from human and boar seminal plasma and spermatozoa and human cervical mucus: isolation, properties and biological aspects. In: (Reich E, Rifkin DB, Shaw E, eds) Proteases and Biological Control. Vol 2. Cold Spring Harbor, WA: Cold Spring Harbor Conferences on Cell Proliferation, 1975, pp737-766.
Gatti J-L, Billard R, Christen R. Ionic regulation of the plasma membrane potential of rainbow trout (*Salmo gairdneri*) spermatozoa: role in the initiation of sperm motility. J Cell Physiol 1990; 143:546-554.
Gerisch G. Cyclic AMP and other signals controlling cell development and differentiation in Dictyostelium. Ann Rev Biochem 1987; 56:853-879.
Griffin FJ, Vines CA, Pillai MC, Yanagimachi R, Cherr GN. Sperm motility initiation factor is a minor component of the Pacific herring egg chorion. Dev Growth Differ 1996; 38:193-202.
Hansbrough JR, Garbers D L. Speract. Purification and characterization of a peptide associated with eggs that activates spermatozoa. J Biol Chem 1981; 256:1447-1452.
Hazelbauer GL. Chemotactic migration by bacteria. In: (Metz CB, Monroy A, eds) Biology of Fertilization. New York: Academic Press, 1985, pp237-254.
Hirohashi N, Hoshi M. Role of sperm alpha-L-fucosidase in the ascidian, *H. roretzi*. Zool Sci 1991; 8:1095.
Inaba K, Morisawa S, Morisawa M. Proteasomes regulate the motility of salmonid fish sperm through modulation of cAMP-dependent phosphorylation of an outer arm dynein light chain. J Cell Sci 1998; 111:1105-1115.
Izumi H, Màriàn T, Inaba K, Oka Y, Morisawa M. Hyperpolarization of sperm plasma membrane induces synthesis of cyclic AMP and activation of sperm motility in the ascidians, *Ciona intestinalis* and *C. savignyi*. Zool Sci 1997; 14 (Suppl):119.
Jeng H, Lin K-M, Chang W-C. Purification and characterization of reversible sperm motility inhibitor from porcine seminal plasma. Biochem Biophys Res Commun 1993; 191:435-440.
Kesbeke F, Van Haastert PJM, De Wit RJW, Snaar-Jagalska BE. Chemotaxisis to cyclic AMP and folic acid is mediated by different G proteins in *Dictyostelium discoideum*. J Cell Sci 1990; 96:669-673.
Kobayashi W. Effect of osmolality on the motility of sperm from the lamprey, *Lampetra japonica*. Zool Sci 1993; 10:281-285.
Krasznai Z, Màriàn T, Balkay L, Gaspar R, Tron L. Potassium channels regulate hypo-osmotic shock-induced motility of common carp (*Cyprinus carpio*) sperm. Aquaculture 1995; 129:123-128.
Krasznai Z, Màriàn T, Izumi H, Tron L, Morisawa M. Effect of membrane potential on the activation mechanism of sperm. Zygote 1998; S123.
Krasznai Z, Terez M, Izumi H, Damjanovich S, Farkas T, Morisawa M. Membrane hyperpolarization removes inactivation of Ca^{++} channels leading to Ca^{++} influx and subsequent initiation of sperm motility in common carp. Proc Natl Acad Sci USA 1999; in press.
Lambert, CC. Ascidian sperm penetration and the translocation of a cell surface glycosidase. J Exp Zool 1989; 249:308-315.
Lambert CC, Lafargue F, Lambert G. Preliminary note on the genetic isolation of *Ciona* species (Ascidicea, Urochodata).Vie Milleu 1990; 40:293-295.
Lee S-L, Kuo Y-M, Kao C-C, Hong C-Y, Wei Y-H. Purification of a sperm motility stimulator from porcine follicular fluid. Comp Biochem Physiol 1992; 101:591-594.
Lievano A, Santi CM, Servano CJ, Trevino CL, Bellve AR, Hernandez-Cruz A, Darszon A. T-type Ca^{2+} channels and a_{1E} expression in spermatogenic cells, and their possible relevance to the sperm acrosome reaction. FEBS Lett 1996; 388:150-154.
Maier I, Müller DG. Sexual pheromones in algae. Biol Bull 1986; 170:145-175.
Màriàn T, Krasznai Z, Balkay L, Balazs M, Emri M, Bene L, Tron L. Hypo-osmotic shock induces an osmolality-dependent permiabilization and structural changes in the membrane of carp sperm. J Histochem Cytochem 1993; 41:291-297.
Màriàn T, Krasznai Z, Balkay L, Emiri M, Tron L. Role of extracellular and intracellular pH in carp sperm motility and modifications by hyperosmosis of regulation of the Na^+/H exchanger. Cytometry 1997; 27:374-382.
Miller RL. Chemotaxis of the spermatozoa of *Ciona intestinalis*. Nature 1975; 254:244-245.
Miller RL. Site-specific sperm agglutination and the timed release of a sperm chemo-attractant by the egg of the leptomedusan, *Orthopyxis caliculata*. J Exp Zool 1978; 205:385-392.
Miller RL. Sperm chemotaxis in ascidians. Am Zool 1982; 22:827-840.
Miller RL. Demonstration of sperm chemotaxis in echinodermata: Asteroidea, Holothuroidea, Ophiuroidea. J Exp Zool 1985a; 234:383-414.
Miller RL. Sperm chemo-orientation in metazoa. In: (Metz CB, Monroy A, eds) Biology of Fertilization. New York: Academic Press, 1985b, pp257-337.
Minganti A. Experienze sulle fertilizine nelle Ascidie. Pubbl Stn Zool Napoli 1951; 23:58-65.
Morisawa M. Initiation mechanism of sperm motility at spawning in teleosts. Zool Sci 1985; 2:605-615.
Morisawa M. Cell signaling mechanisms for sperm motility. Zool Sci 1994; 11:647-662.
Morisawa M, Ishida K. Short-term changes in levels of cyclic AMP, adenylate cyclase, and phosphodiesterase during the initiation of sperm motility in rainbow trout. J Exp Zool 1987; 242:199-204, 1987.
Morisawa M, Morisawa S. Acquisition and initiation of sperm motility. In: (Gagnon C, ed) Controls of Sperm Motility: Biological and Clinical Aspects. Boca Raton, FL: CRC Press, 1990, pp137-151.
Morisawa M, Okuno M. Cyclic AMP induces maturation of trout sperm axoneme to initiate motility. Nature 1982; 295:703-704.
Morisawa M, Suzuki K. Osmolality and potassium ions: their roles in initiation of sperm motility in teleosts. Science 1980; 210:1145-1147.
Morisawa M, Suzuki K, Morisawa S. Effects of potassium and osmolality on spermatozoan motility of salmonid fishes. J Exp

Biol 1983a; 107:105-113.
Morisawa M, Suzuki K, Shimizu H, Morisawa S, Yasuda K. Effects of osmolality and potassium on motility of spermatozoa from freshwater cyprinid fishes. J Exp Biol 1983b; 107:95-103.
Morisawa M, Morisawa S, DeSantis R. Initiation of sperm motility in *Ciona intestinalis* by calcium and cyclic AMP. Zool Sci 1984; 1:237-244.
Morisawa M, Tanimoto S, Ohtake H. Characterization and partial purification of sperm-activating substance from eggs of the herring, *Clupea pallasi*. J Exp Zool 1992; 264:225-230.
Morisawa M, Inaba K, Yoshida K, Izumi H, Nomura M. Acquisition, initiation and activation of sperm motility and sperm chemotaxis. Zygote 1998; 6 (Suppl):s14-s15.
Nishigaki T, Chiba K, Miki W, Hoshi M. Structure and function of asterosaps, sperm-activating peptides from the jelly coat of starfish eggs. Zygote 1996; 4:237-245.
Oda S, Morisawa M. Rises of intracellular Ca^{2+} and pH mediate the initiation of sperm motility by hyperosmolality in marine teleosts. Cell Motil Cytoskel 1993; 25:171-178.
Oda S, Igarashi Y, Ohtake H, Sakai K, Shimizu N, Morisawa M. Sperm-activating proteins from unfertilized eggs of the Pacific herring, *Clupea pallasi*. Dev Growth Differ 1995; 37:257-261.
Oda S, Igarashi Y, Manaka K, Koibuchi N, Sakai-Sawada M, Sakai K, Morisawa M, Ohtake H, Shimizu N. Sperm-activating proteins obtained from the herring eggs are homologous to trypsin inhibitors and synthesized in follicle cells. Dev Biol 1998; 204:55-63.
Ohtake H. Respiratory behaviour of sea-urchin spermatozoa. II. Sperm-activating substance obtained from jelly coat of sea-urchin eggs. J Exp Zool 1976; 198:313-322.
Opresko LK, Brokaw CJ. cAMP-dependent phosphorylation associated with activation of motility of *Ciona* sperm flagella. Gam Res 1983; 8:201-218.
Perchec G, Jeulin C, Cosson J, Andre F, Billard R. Relationship between sperm ATP content and motility of carp spermatozoa. J Cell Sci 1995; 108:747-753.
Perchec G, Cosson MP, Cosson J, Jeulin C, Billard R. Morphological and kinetic changes of carp (*Cyprinus carpio*) spermatozoa after initiation of motility in distilled water. Cell Motil Cytoskel 1996; 34:113-120.
Pfeffer W. Locomotorische richtungsbewegungen durch chemische reize. Unters ad bot Institut in Tubingen 1, 363-482, 1884.
Pillai, MC, Shields TS, Yanagimachi R, Cherr GN. Isolation and partial characterization of the sperm motility initiation factor from eggs of the Pacific herring, *Clupea pallasi*. J Exp Zool 1993; 265:336-342.
Punnett T, Miller R L, Yoo B-H. Partial purification and some chemical properties of the sperm chemoattractant from the forcipulate starfish *Pycnopodia helianthoides* (Brandt, 1835). J Exp Zool 1992; 262:87-96.
Ralt D, Manor M, Cohen-Dayag A, Tur-Kaspa I, Ben-Shlomo I, Makler A, Yuli I, Dor J, Blumberg S, Mashiach S, Eisenbach M. Chemotaxis and chemokinesis of human spermatozoa to follicular factors. Biol Reprod 1994; 50:774-785.
Ramarao CS, Burks DJ, Garbers DL. A single mRNA encodes multiple copies of the egg peptide speract. Biochemistry 1990; 29:3383-3388.
Sawada T, Osanai K. The cortical contraction related to the ooplasmic segregation in *Ciona intestinalis* eggs. Wilhelm Roux's Arch. Dev Biol 1981; 190:208-214.
Sawada T, Schatten G. Effects of cyoskeletal inhibitors on ooplasmic segregation and microtubule organization during fertilization and early development in the ascidian *Molgula occidentalis*. Dev Biol 1989; 132:331-342.
Scheuring L. Biologische und physiologische untersuchungen an forellensperma. Arch Hydrobiol 1925; 4(Suppl):181-318.
Speksnijder JE, Jaffe LF, Sardet C. Polarity of sperm entry in the ascidian egg. Dev Biol 1989; 133:180-184.
Stacy NE, Liley NR. Regulation of spawning behavior in the female goldfish. Nature 1974; 247:71-72.
Strüsmann CA, Renard P, Ling H, Takashima F. Motility of pejerrey *Odontesthes bonariensis* spermatozoa. Fish Sci 1994; 60:9-13.
Suzuki N. Structure and function of sea urchin egg jelly molecules. Zool Sci 1990; 7:355-370.
Suzuki N. Structure, function and biosynthesis of sperm-activating peptides and fucose sulfate glycoconjugate in the extracellular coat of sea urchin eggs. Zool Sci 1995; 12:13-27.
Suzuki R. Sperm activation and aggregation during fertilization in some fishes I. Behavior of spermatozoa around the micropyle. Embryologia 1958; 4:93-102.
Takai H. Studies on the regulatory mechanisms for the initiation of sperm motility in marine and freshwater teleosts. PhD thesis, The University of Tokyo, 1994.
Takai H, Morisawa M. Change in intracellular K^+ concentration caused by external osmolality change regulates sperm motility of marine and freshwater teleosts. J Cell Sci 1995; 108:1175-1181.
Tanimoto S, Morisawa M. Roles for potassium and calcium channels in the initiation of sperm motility in rainbow trout. Dev Growth Differ 1988; 30:177-124.
Tanimoto S, Imae Y, Morisawa M. Changes in membrane potential and initiation of sperm motility in rainbow trout. Zool Sci 1988; 5:1218.
Tanimoto S, Kudo Y, Nakazawa T, Morisawa M. Implication that potassium flux and increase in intracellular calcium are necessary for the initiation of sperm motility in salmonid fishes. Mol Reprod Dev 1994; 39:409-414.
Vines CA, Yoshida K, Morisawa M, Cherr GN. Regulation of herring sperm motility initiation by egg-derived molecules. West Coast Regional Developmental Biology Conference, Abstract 20, 1998.
Ward GE, Brokaw CJ, Garbers DL, Vacquier VD. Chemotaxis of *Arbacia punctulata* spermatozoa to resact, a peptide from the egg jelly layer. J Cell Biol 1985; 101:2324-2329.
Yanagimachi R. Studies of fertilization in *Clupea pallasi*. III. Manner of sperm entrance into the egg. Zool Mag (Tokyo) 1957; 66:226-233.
Yanagimachi R, Kanoh Y. Manner of sperm entry in herring egg, with special reference to the role of calcium ions in fertilization. J Fac Sci Hokkaido Univ 1953; 11:487-494.
Yanagimachi R, Cherr GN, Pillai MC, Baldwin JD. Factors controlling sperm entry into the micropyles of salmonid and herring eggs. Dev Growth Differ 1992; 34:447-461.
Yoshida K, Inaba K, Ohtake H, Morisawa M. Purification and characterization of prolyl endopeptidase from the sperm of the Pacific herring, *Clupea pallasi,* and its role in the sperm-egg interaction. Dev Growth Differ 1998; 1999; 41:217-225..
Yoshida M, Inaba K, Morisawa M. Sperm chemotaxis during the process of fertilization in the ascidians *Ciona savignyi* and *Ciona intestinalis*. Dev Biol 1993; 157:497-506.
Yoshida M, Inaba K, Ishida K, Morisawa M. Calcium and cyclic AMP mediate sperm activation, but Ca^{2+} alone contributes sperm chemotaxis in the ascidian, *Ciona savignyi*. Dev Growth Differ 1994; 36:589-595.

16

Ionic Factors Regulating the Motility of Fish Sperm

Jacky Cosson
CNRS
University of Paris 6
Roland Billard
Museum of Natural History
Christian Cibert
CNRS, Institute Jacques Monod
Catherine Dréanno
Marc Suquet
IFREMER

Spermatozoa are unique among cells generated by the metazoans: they are unicells which, in species with external fertilization, are released usually in a hostile external medium where they have to cope with extremely harmful conditions (fresh water, sea or brackish water) even though they appear poorly protected when compared to eggs.

Most of the knowledge on flagellar movement comes from studies on sea urchin sperm (Gibbons 1981a). Nevertheless, some characteristics of sperm cells of other species such as fish show original features. Motility duration (Billard 1978; Billard, Cosson 1988, 1992), motility initiation (Morisawa 1985; Cosson et al. 1995) and motility pattern (Boitano, Omoto 1992; Cosson et al. 1997) are especially well documented for the rainbow trout sperm used as a fresh water species model. In contrast, very little information presently deals with sperm characteristics of marine fish species. Motility lasts for short periods in the case of fresh water teleost fishes. It is in the range of two to 20 minutes for *Dichentrarchus labrax* (Suquet et al. 1993), *Sparus auratus* (Billard 1978), *Scophthalmus maximus* (Chauvaud et al. 1995) and *Hippoglossus hippoglossus* (Billard et al. 1993), but the notion of motility time can become subjective if appreciated as "the time where a few sperm flagella are still seen shaking."

Several advantages favor studies on fish spermatozoa: 1) brood fish are available all year long, in the case of farmed species; 2) fish sperm is easy to collect and to save for a short term; 3) as fish sperm is usually immotile in the seminal fluid, it is easy to fully trigger its motility by transfer into a swimming adequate medium, so sperm motility scores are currently used for selection of genitors in the fish broodstocks or for comparison before and after cryopreservation; 4) fish sperm cells present an homogenous behavior—since all spermatozoa swim with very similar characteristics at a certain time point post-activation, they are a good model for biochemical investigations; 5) in many fish species, the sperm flagellum is 50-60 µm long and up to six curvatures along the length (three sine waves) are observed, plus the ribbon shape (presence of fins, as seen in Figure 1) instead of the usual cylinder shape makes the flagellum brighter when observed by dark-field microscopy and so wave shapes and their three-dimensional distortions are easier to visualize; 6) in several fish species, spermatozoa follow linear tracks and the flagellar bending is symmetrical probably because of the absence of Ca^{2+} sensitivity of the axoneme; 7) in most fish species studied so far, attenuation (so-called dampening) gradually invades the whole length of the flagellum during the short motility period and is a fully reversible phenomenon. However, few detailed studies on fish spermatozoa flagellar motility behavior have been conducted except in trout, probably widely studied because fertilizing ability of trout spermatozoa is easily assessed.

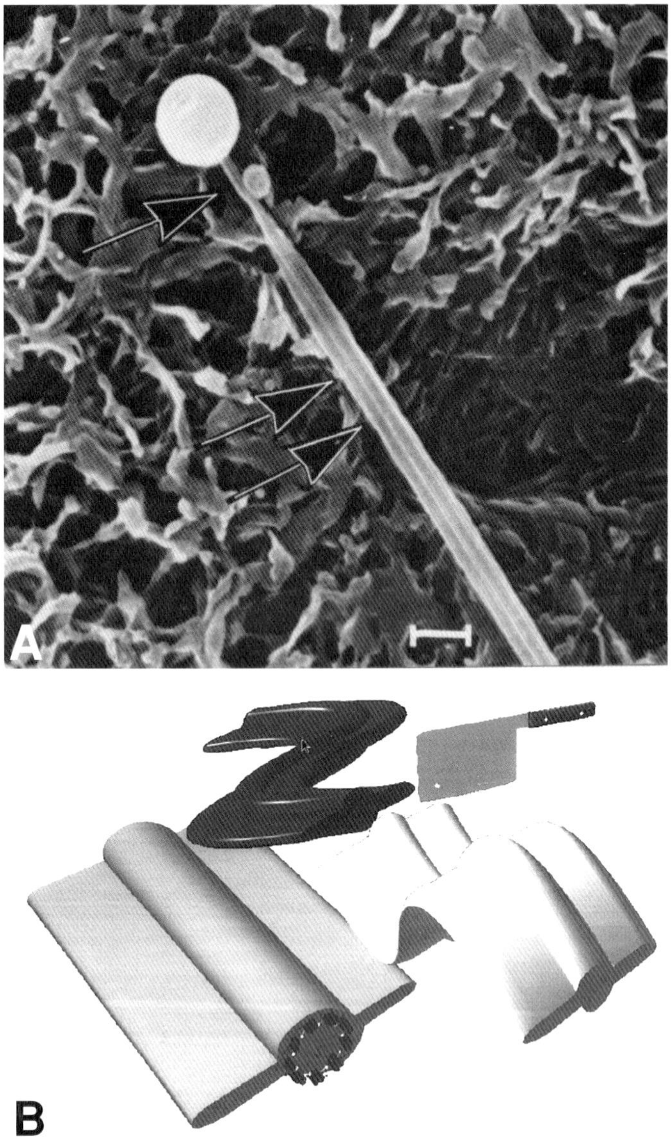

Figure 1. A: Turbot spermatozoon as seen by SEM (reproduced from Suquet et al. 1993). The single arrow shows the head-tail junction devoid of fins, double arrow depicts the fins on both sides of the axoneme. Fins are up to 1 µm long and preferentially oriented along the horizontal axis on both sides of axonemes. B: Schematic representations of the ribbon shape of a fish flagellum. On the right, the waves generated by the axoneme lead to two possibilities of curvature. At the lower left, the axonemal 9+2 structure is seen protruding out of the section of the central cylinder. At the bottom right, the orientation of the plane of the fins is, in cases where it was investigated by electron microscopy, such as in trout spermatozoa (Billard 1970) or in turbot, preferentially orthogonal to the plane passing through doublets 1 and 5-6; that is, parallel to the plane defined by the two singlets of microtubules. In turbot, fins are usually facing groups of doublets 2-3-4 and 7-8-9.

Fish spermatozoa have to adapt to surrounding media having very different ionic/osmotic characteristics: the seminal fluid, the ovarian fluid and the fresh or sea or brackish water. Osmolality is considered in many species as an ubiquitous triggering agent for motility initiation of fish spermatozoa (Morisawa 1985; Morisawa et al., this book); when released in this new environmental condition, fertilization can occur. From the seminal fluid where sperm is stored (usually immotile), osmolality will either rise (sea water) or decrease (fresh water). Usually, an ionic exchange is combined with this osmotic shock. For instance, ionic strength rises drastically at transfer of sperm into sea water in such a way that it changes the sperm membrane potential and as a consequence provokes a rise in the intracellular ionic concentration (specially Ca^{2+}) as well as that of internal pH, both leading to activate flagellar beating (Oda, Morisawa 1993). In some species, such as herring (*Clupea pallasi*), a sperm motility initiation factor (a protein from the egg) could complement the osmolality signal (Pillai et al. 1993, 1994; Morisawa et al., this book).

During the progress of a phase of fish sperm motility, most parameters used to describe motility decrease. This feature leads to a precocious arrest of motility, a handicap for people interested in artificial reproduction. Thus, many attempts to prolong motility were tried, either by helping sperm to sustain a higher energy level (increase of ATP production and/or higher rate of oxidative phosphorylation) or with trials to design "motility enhancing substances," including environment factors such as temperature and swimming medium composition, in efforts to reach the best swimming conditions.

Flagellar wave pattern shows original features during the motile phase of fish spermatozoa: it evolves rapidly from fully motile with waves present along the entire flagellum to intermediate patterns with waves present only in the proximal part of the flagellum. The end of the motility period is identified by the restriction of the waves to one-third or one-quarter of the flagellum length leading to a lower and lower efficiency of the sperm translation (Cosson et al. 1997) rapidly followed by a full stop. This evolution of the wave pattern is paralleled by a decrease in flagellar beat frequency along the swimming period, both contributing to a decrease in the swimming ability.

A possible relationship between ionic effects, osmolality and duration of motility is tentatively established in this chapter: the external osmolality which governs the intracellular ionic concentration would control the wave pattern characteristics. Taking turbot (*Scophthalmus maximus*) spermatozoa as a model and using the axonemal wave pattern as an indicator of the local and temporal ionic concentration, it can be shown that the change of extracellular osmolality perceived by sperm cells when transferred from the seminal fluid to sea water makes the intracellular ionic concentration rapidly evolving during the motility phase; consequently, the axoneme of flagella becomes exposed to a more and more elevated intracellular environment, preventing the development of distal waves and further leading to their full arrest. General information on fish sperm physiology is available in the review by Stoss (1983).

Structure of Fish Spermatozoa

Teleost fish usually lack an acrosome (Afzelius 1978); the lack of this structure is compensated by the presence of a micropyle, which is a hole in the chorion of the fish egg for penetration by the spermatozoon (Ginsburg 1972). The sturgeons and paddlefish are an exception, as their sperm have an acrosome and their eggs have several micropyles. Sperm of teleostean fish have a simple structure (Franzen 1970) called "aqua sperm" (Jamieson 1991). The spermatozoon of the common carp, *Cyprinus carpio*, as described by Billard (1970), which is released at spawning into fresh water, is an example of the primitive type. The head (2.5 µm diameter) is almost spherical with a collar-like midpiece formed by an extrusion of the plasmalemma and an indentation of the nucleus. In the midpiece are located the centrioles plus a few mitochondria. The typical 9+2 arrangement (nine peripheral doublet microtubules and one pair of central singlet tubules) is observed in the flagellum. In other species, the shape of the head may differ slightly; fusion between mitochondria is common (Billard 1970; Mattei, Mattei 1975; Jaspers et al. 1976). Along the flagellar tail, the plasma membrane often forms one or two fin-like ridges (Nicander 1970; Billard 1970; Stein 1981; Suquet et al. 1993); such fins are preferentially oriented along the horizontal plane defined by the central microtubules and on both sides of trout sperm axonemes (Billard 1970). Such a modification of the flagellum (Fig. 1A) could improve the efficiency of the flagellar propulsion, but this was questioned by Afzelius (1978). This general form has been documented in members of Poeciliidae, Jenysiidae, Pantodontidae and Embiotocidae (Billard 1970; Dadone, Narbaitz 1967; Stanley 1969; Van Deurs 1975; Gardiner 1978; Lahnsteiner et al. 1997). Among more than 30 species for which data are available (Linhart et al. 1991), head diameter varies from 2 to 4 µm, with the exceptions of eel spermatozoa (Gibbons et al. 1983) and of sturgeons and paddlefish (O Linhart and J Cosson, unpublished data) which have an elongate sperm head up to 10 µm in length and over 2 µm in width. Flagellar length varies from 20 to 100 µm. From two to nine mitochondria are present per sperm cell with mean dimensions of 200 over 400 nm.

In contrast, spermatozoa from fish employing internal fertilization have both an elongate head deprived of acrosome and a midpiece similar to mammalian sperm and containing extensive mitochondrial structures and intercentriolar material. However, in fish the midpiece remains separated from the flagellum by the cytoplasmic canal. In five fish families, the mitochondria of the spermatozoon is ring shaped (Mattei et al. 1981). Some peculiarities in the tail have been reported for some species: cells with two flagella were found in *Forichtys notatus* (Stanley 1969), and biflagellated spermatozoa were reported for the channel catfish, *Ictalurus punctatus* (Jaspers et al. 1976), and, in a few instances, in the guppy, *Poecilia reticulata* (Billard, Fléchon 1969).

Sperm flagella from Anguilliformes and Elopiformes present a "9+0" pattern with no central microtubules (Billard, Ginsburg 1973; Mattei, Mattei 1975). These cells are motile, even though the outer dynein arms and the structures which connect the peripheral and central microtubules are missing (Baccetti et al. 1979; Gibbons et al. 1983).

The system of microtubules arranged in the flagellum represents the architecture of the motile apparatus in a sperm cell which is called the axoneme (Cosson 1996; Gagnon 1995). One of each of the nine peripheral double microtubules (the A tubule) carries two arms which consist of ATPases (dyneins) and arranged as rows lining the whole length of microtubules. Under hydrolysis of ATP, these dynein arms interact with the protein tubulin of the B tubule from the adjacent double set, causing a sliding process between them. Because of the presence of interconnecting elements between peripheral microtubules, continuous sliding between some of them creates tension and results in oscillation of the flexible flagellum (Satir 1974).

Quite early, Gray and Hancock (1955) postulated an hypothesis in which flagellar beating is due to localized contractions propagated along the doublet tubules. In elaborated models of flagellar functioning, active moments, later attributed to action of the dynein ATPases, are balanced both by viscous (mechanical) and elastic (later supposed passive) moments. Bending waves could propagate in a flagellum if changes in length of contractile elements caused delayed changes in tension: the concept of non-linearity and frame shift are already included in early models. Later, Rikmenspoel (1978) drew attention to the separation of two notions: that of non-propagated active moment which varies sinusoidally with time and that of propagated active moment of adapted phase. These early ideas contained important notions detailed later by others: a local sliding is combined with bending and associated with a propagating mechanism. Conceptually these are the essential content of the notion called "metachronism." They have led to the elaboration of many functional models which, nevertheless, explain only on a theoretical basis using computer models the presence of several curvatures in opposite directions coexisting along a flagellum in movement and the propagation of the curvature along the flagellum. However, the bend initiation mechanism is not fully explained (Gibbons 1981a).

Studies demonstrated that the axoneme of demembranated spermatozoa (from sea urchin origin) can be reinitiated to produce waves if energy is provided in the form of ATP. In rainbow trout or chum salmon, flagella had to be exposed to both cyclic AMP (cAMP) and ATP in order to become functionally motile (Morisawa, Okuno 1982; Morisawa et al. 1983). The environmental factors which activate this ATP-Mg^{2+} system and initiate motility vary among fishes.

Fish Sperm Motility and Its Measurement

Van Leeuwenhoek (1677) was the first to attempt to describe the movement and density of these "bisexual tadpoles, which were agitating more than a thousand in a volume equivalent to a sand grain." If one assumes that the volume of a sand grain is about a microliter, his estimation leads to several million spermatozoa per milliliter, a value in the range of current measurements. Due to the briefness of fish sperm motility, special methods to record spermatozoa motion (Billard, Cosson 1989), and in particular to get high resolution flagellar images (Cosson et al. 1997), have been developed.

Motility of Live Sperm

Motility is a useful parameter for assessing the viability of sperm cells; for instance, to test the preservation of spermatozoa. Observation is easy and rapid and can be done under a simple microscope using a hemocytometer. Assessment of the percentage of motile cells is easy when referring to a simple scale from 0 to 5. Attempts were also made to use the Quasi-Elastic light scattering technique but the first measurement is possible only 10 to 15 seconds after initiation of motility. This is acceptable for guppy, in which motility may last one hour, and in mammals (Adam et al. 1969), but not for trout, with a motility duration of only 20 to 25 seconds (Bergé et al. 1967). Craig et al. (1983) reported motility measurement with trout sperm using this technique, but their results were controversial.

Motility measurements, based on a "yes or no" test (moving or non-moving spermatozoa) were often used. However, information on the type of motility is limited. Thus, more sophisticated techniques, such as CASA (Computer Assisted Sperm Analysis), were applied to fish sperm (Christ et al. 1996; Kime et al. 1996; Lahnsteiner et al. 1997), but the briefness of motility and the high velocity make these techniques difficult to adapt to fish sperm, with the exception of sperm trackers (so-called "Hobson tracker") which copes more easily with such constraints. In general, motility and fertility are well correlated. However, because different organelles of the cell are responsible for motility and fertility, there are numerous examples where motile, gamma-irradiated (Lasher, Rugh 1962) or cryopreserved

(Kossmann 1973; Mounib et al. 1968; Stein, Bayle 1978) cells were not fertile. In turbot, it was shown that it is the percentage of motile sperm which limits the ability of sperm to fertilize, not velocity (Dréanno et al. 1999b). As a general rule for fish sperm, experimental induction of motility usually requires the use of a double dilution protocol: a first dilution in an immobilizing medium (IM) allows dispersion of the spermatozoa clumps frequently encountered; this is followed by a second dilution obtained by mixing a small volume of the diluted sperm in a drop of swimming medium (SM) previously placed on the microscope slide. Such a double dilution procedure allows motility to be triggered for all sperm cells at exactly the same time, and early observation (a few seconds) can start right after triggering. The use of an open drop is supported by three reasons: 1) the absence of coverslip allows observation within a minimal time after motility is triggered, during which time fish sperm is the most active; 2) access to the open drop allows addition of small volumes of "effector" solutions directly during the process of observation; and 3) it is difficult to quantify motility in samples of diluted milt held for long periods in the increasingly hypoxic environment beneath a cover slip, plus sperm cells may stick to the cover slip.

Examples of the solutions used as IM and SM are described in Table I. The following procedure can be adopted for synchronous initiation of motility of live sperm. First, 1 μl of "dry sperm" is added to 50-100 μl of IM. Then movement is initiated by transfer into a drop of SM previously deposited on the glass slide and adjusted ready for observation with the dark field microscope. 1 μl of the intermediate dilution of sperm in IM is deposited into the 20 μl SM drop and mixed by horizontal movement of the tip of the micropipette inside the drop, which is observed right after mixing under the microscope. An efficient mixing is obtained within two to three seconds and therefore observations, video records and frequency measurements always start with a three- to five-second lag period from zero time, while a five- to 10-second average period is needed when using a cover slip, as required for oil immersion lenses.

This protocol was adopted because when sperm motion was initiated by diluting the semen directly into the SM, the dilution ratio of the sperm suspension was difficult to control because of the viscosity of the semen, and it was rather difficult to obtain quickly (within a few seconds) an homogenous suspension of spermatozoa because of the tendency of the cells to remain in small clumps. That resulted into a mixture of spermatozoa with delayed motion phases: a large proportion of spermatozoa (40-60%) started to swim just after dilution, but others remained quiescent. Initially immotile cells usually began to move stochastically within 10 to 30 seconds but some even one minute later. As the characteristics of the movement of any single spermatozoon evolved very rapidly during the first minute after activation, asynchronous initiation of motility resulted in a sperm suspension with very heterogeneous swimming behavior and duration. To overcome this problem, sperm suspensions were routinely diluted five to 100-fold in the IM, which allowed an efficient dispersion of the spermatozoa without initiation of flagellar beating (especially with the precaution of saving diluted sperm at 0-4°C). Under these conditions, very homogeneous initiation of movement was observed for 90-100% of cells within three seconds after mixing with the SM (Billard, Cosson 1992).

The beat frequency (BF) can be measured by adjustment of the strobe frequency set in register with the wave beating until showing apparently non-propagating waves, either on an individual sperm cell or on the whole sperm population. Billard and Cosson (1989) and Cosson et al. (1991) showed a progressive decrease in the flagellar beat frequency in individual fish spermatozoa during the motile phase. Therefore, each evaluation of the BF had to be associated with the precise timing of the measurement. The frequency of the flash illuminator was adjusted before each experiment to the highest value previously observed for flagellar frequency

Table 1: Solutions used for studies of fish sperm motility and the fish species* used for collection of sperm.

Species	Components	Immobilizing Medium (IM)	Swimming Medium (SM)
O. mykiss	NaCl	80mM	125mM
	KCl	40mM	0
	$CaCl_2$	0.1mM	0.1mM
	Tris-HCl pH 9.2	30mM	30mM
A. baeri	Saccharose	400mM	0
	Tris-HCl pH 8.0	20mM	0
S. maximus	Glucose	200mM	0
	Sea water	0	100%
	BSA	5mg cm^{-3}	5mg cm^{-3}
	Tris-HCl pH 8.2	30mM	30mM

*The broodstocks of turbot, *Scophthalmus maximus*, were managed in IFREMER facilities at Brest, France. Turbot sperm was collected by stripping ripe males after cleaning of the genital papilla. In order to avoid any sperm contamination by urine, the urinary bladder was first emptied by gently squeezing the fish abdomen and the ureter was catheterized. After drying of the urogenital pore, sperm was carefully collected into a syringe and stored at 4°C until experimentation. Adult male trout *(Oncorhynchus mykiss)* were from the INRA Laboratory, Rennes, France. They were kept in water tanks and fed with trout pelleted food at 0.5% body weight. The animal chosen for sperm sampling was immersed for one to two minutes in a fresh water solution of 500 mg.dm^{-3} of phenoxy-ethanol (Merck chemical). After the cessation of abrupt tail movement, the genital aperture was dried with tissue paper and a 5 cm^3 plastic tube was positioned close to the genital papilla. By pressure from head to tail, 2-3 cm^3 of "dry sperm" (5 to 15. 10^9 spermatozoa cm^{-3}) was collected; extrusion of fecal material was prevented by placing tissue paper in the rectum and attempts were made to eliminate urine by placing the fish in a dorsal position while pressing the belly. Samples contaminated by urine or feces were discarded. Sperm suspensions were kept in a tube on ice as "dry sperm" for short time experiments (up to 5 hours). Sturgeon broodstocks (*Acipenser baeri*) originated from Ecloserie de Guyenne (Bordeaux, France); they were 6-year-old males weighing 5 to 10 kg, reared in concrete tanks and fed with artificial food. Sperm was collected by stripping after careful cleaning of the genital area in order to avoid any sperm contamination by urine. Sperm was collected into a syringe and stored at 4°C until experimentation (less than 8 hours).

(usually around 50-70 Hz). Then motility was initiated and the flash frequency was permanently adjusted, leading to the observation of an homogeneous picture of the moving sperm population with immobilized bent flagella. To obtain the complete sequence of sperm motion in one experimental condition, five to six successive experiments had to be carried out in order to get overlapping information at various time periods elapsed since activation, especially the earliest time points (five to 10 seconds). Observations were carried out at room temperature, usually 20°C, and eventually down to 13°C for some observations.

To observe the reversibility of the ionic/osmotic effects on turbot sperm, local injection of sea water (SW) or sucrose solution was applied near one spermatozoon and observed under high magnification in an open drop of IM; this was obtained with a microneedle (1 µm opening) filled with the solution to be tested and linked by a tubing to an injection syringe with a micrometric screw, which allowed control of the perfusion process. A single spermatozoon was successively activated by contact with sea water, then, as a result of its intense swim, it would escape out of sea water into IM where it would stop; movement would again be activated in SW, and so on.

The quality of the sperm samples is assessed as the homogeneity of movement in these standard conditions: only sperm samples that show more than 90% of the spermatozoa with synchronously activated movement are usually retained for experiments.

All motility parameters (frequency, velocity, wave characteristics) evolve very rapidly during the motility period. This was documented in trout (Cosson et al. 1985), in silurids (Billard et al. 1997), in carp (Perchec et al. 1993), in *Cottus gobio* (Lahnsteiner et al. 1997), in perch (Lahnsteiner et al. 1995), in sturgeon (Cosson et al. 1995; Tsvetkova et al. 1996), in paddlefish (Linhart et al. 1995; Cosson, Linhart 1996), in sea bass (Dréanno et al. 1999a), in halibut (Billard et al. 1993) and in turbot (Chauvaud et al. 1995). Findings for spermatozoa of these fish contrast with those of eel spermatozoa, which swim for more than 20 minutes without change in their characteristics (Gibbons et al. 1985a, b). This general decrease of the swimming performances is rooted partly in the parallel decrease of the energy stores observed during the motility period.

In spermatozoa of several fish species such as trout (Christen et al. 1987), carp (Perchec et al. 1993), silurids (Billard et al. 1997), sea bass (Dréanno et al. 1999a) or turbot (Dréanno et al. 1999b), ATP stores were shown to shift down to values reaching one-third to one-tenth of the initial content. The interdependence between motility, respiration, ATP production and utilization was investigated in turbot (Dréanno et al. 1999b) and in carp spermatozoa (Perchec et al. 1995b).

Even if mitochondrial respiration plays an important role for the sperm motility of several animal groups (Gosh 1989), only few investigations were carried out on oxygen uptake by fish spermatozoa. A comparison with previous studies is difficult because of the variations in experimental conditions and differences in units used to express oxygen uptake. Data on O_2 uptake by fish spermatozoa are scarce. *Lepomis* and *Catostomus commersonnii* use 110 to 140 µl O_2 per 10^9 cells per hour. Corresponding values for rainbow trout, Atlantic salmon, and Atlantic cod spermatozoa were 20-40 µl O_2 per 10^9 cells per hour (Terner 1962; Terner, Korsh 1963; Mounib et al. 1968). Comparison between species is difficult: differences in cell size among species and high incubation temperatures for salmonid spermatozoa (25°C) may lead to very different conclusions on motility characteristics (Inoda et al. 1988; Vladic, Järvi 1997), so the O_2 uptake was measured in motile sunfish, sucker or cod spermatozoa but measured in immotile salmonid cells. As a general rule, aerobic metabolism of intracellular substrates dominates in fish sperm (Billard, Cosson 1992). Turbot spermatozoa maintained quiescent in IM show a constant and low respiratory rate similar to that in rainbow trout or in the cod *Gadus morhua* (Robitaille et al. 1987). This is similar to what is seen in sea urchin, starfish and bull sperm (Ohtake et al. 1996; Mohri et al. 1990; Halangk et al. 1985). The mitochondrial respiration of turbot spermatozoa is stimulated as a consequence of the initiation of motility and gradually decreases after 30 seconds of movement (Dréanno et al. 1999b). In carp, the ATP content measured after one minute was weaker and ATP from respiration could not compensate for ATP consumption by motility (Perchec et al. 1995b).

The rate of mitochondrial respiration can be limited by the supply of substrate or by that of oxygen (Nicholls 1982). In fish sperm, the mitochondrial oxidative phosphorylation which is highly requested to produce energy required during motion remains insufficient to sustain endogenous ATP stores.

The Use of Membrane Deprived Spermatozoa to Study Flagellar Motility

When demembranated, flagella are directly in contact with chemicals to be tested as potential "effectors" of the motility, without complication due to the presence of a membrane. Demembranation is obtained by application of a mild non-ionic detergent (Triton X 100, for example) and reactivation is initiated by addition of ATP-Mg, the substrate of flagellar dynein-ATPases (Gibbons, Gibbons 1972).

An example of the conditions used for turbot sperm is described in the following section. Two µl of undiluted sperm were mixed at 0°C with 50 µl of demembranation medium (DM) composed of KAcetate (KAc) at various concentrations, 0.5 mM $CaCl_2$, 0.5 mM EDTA,

1 mM DTT, 20 mM Tris HCl pH 8.2, 0.04% Triton X-100 (Cosson, Gagnon 1988). After 30 seconds at 0°C, a 2 μl aliquot was pipetted and mixed at room temperature (18-20°C) on the glass slide with a 50 μl drop of reactivation medium (RM) composed of various concentrations of KAc, 1 mM DTT, 20 mM Tris-HCl pH 8.2, 1 mM $MgCl_2$, 2 mg/ml BSA (fatty acid-free from Sigma) and 1mM ATP (vanadate-free, from Boerhinger). The presence of BSA was required to prevent sticking of the sperm cells to glass or particles. Addition of cAMP either in DM or in RM was not necessary to initiate motility. The presence of Ca^{2+} ions up to 1 mM (as well as that of EGTA up to 0.2 mM) in RM had no effect on the percentage of motile turbot sperm models or on their quality of swim. Triggering of live cell motility was not needed for *in vitro* reactivation in the above conditions.

An additional motility test for the dynein dependent movement is offered by the use of protease induced microtubule sliding, which thus occurs in absence of constraints, the latter leading to usual bend formation in native axonemes. When links between peripheral doublet microtubules are specifically destroyed by a mild trypsin digestion, sliding between adjacent microtubules can be triggered by ATP. As an example, turbot sperm was diluted 1:10 in DM (same as above). After 30 seconds at room temperature, it was further diluted 1:10 in RM including 75 mM KAc and containing trypsin (Type XIII from Sigma) at 0.2 μg/ml and ATP at 1 mM. By dark field microscopy, sperm models were observed to swim during the first minute where they all stopped, and then started to slide at about one minute after addition of trypsin; sliding lasted two to three minutes and appeared stochastically for individual sperm models. Mostly, sliding leads to the induction of one or two local and brief (one to three seconds) but irreversible bends along the axoneme: in the permanently bent region which presents a large curvature, one or several detached microtubular doublets appear to link the borders of the curvature. Sliding efficiency was estimated by the number of curled axonemes appearing as a function of time.

Recording of Flagellar Movement and Analysis of the Motility Parameters

In order to get detailed information about the conformation of the fish sperm flagellum while the waves progress and the whole cell rapidly translates, high speed video recording techniques were developed (Cosson et al. 1997). Spermatozoa were observed with an Olympus BH2 microscope using dry lenses (Zeiss 25X) or oil immersion lenses, either an Olympus 40X D-Apo UV-oil 1.30 with diaphragm or a Reichert 100X Plan Oil 1.25 with iris, combined with an Olympus Dark Field oil condenser DWC 1.4-1.2. Records were obtained using stroboscopic illumination by a Strobex (Chadwick-Helmuth, El Monte, CA, USA). Video records were performed with a Panasonic WV-F 15 E S-VHS video-camera (constant frame rate of the European standard at 50 Hz), connected to a Hamamatsu video image processor (DVS 3000) and to a Panasonic AG 7330 S-VHS video-recorder synchronized to the stroboscopic illumination with a fiber optic "video-sync" module #9630 (Chadwick-Helmuth). The highest resolution images thus obtained allowed images of sperm covering the width of the video format, which corresponds to 250 to 400 pixels per 60 μm (length of the turbot sperm flagellum). The stroboscopic flash illumination with adjustable frequency was set either in automatic register with video frames (50 Hz in Europe) or manually to 150 to 800 Hz depending on the time resolution needed. During the recording, the microscope stage was slowly hand-translated: this allowed the visualization of multiple, well defined successive images of a moving sperm without overlap within every video frame. Eventually two or three successive video frames were associated to document a full beat cycle. In a few instances, direct photographs of sperm cells were also obtained using an Olympus OM2 camera set with a 1/30th second exposure and a flash frequency of 150 to 500 Hz.

For the measurements of shear angle along the flagellum, the angle for each partition either relative to the head axis or to an arbitrary and constant reference axis was plotted versus its curvilinear abscissa. Software was developed for this task. Successive images of the flagellum were scanned (eight bits, 150 dpi) and analyzed without smoothing with a specific program written under the C interpreter from Visilog 4.1 Software (Noesis, France). The skeletons of flagella were divided into 150 partitions. The origin of each partition was considered as the curvilinear origin of a 20-pixel-long segment, and the first derivative of the regression line was calculated. This analysis was repeated for successive images obtained every 1/150 to 1/800 second.

Bend angle and wavelength measurements were obtained on video images captured through an image stabilizer (AT Bus Frame, ATF 101 ForA) and a video-image acquisition card (Spigot/Screen Play 1.2.2); images were transferred for analysis using the N.I.H. image 1.57 software (with the distance and angle measurement facilities) or for montage of images using Adobe Photoshop 3.0 on a MacIntosh G3 computer to control image contrast.

The speed of translation was measured on video frames as the distance covered by the head during 25 successive video images (half a second). The tracks of head position were obtained by accumulation of successive video frames using the DVS Hamamatsu system. The number of waves per flagellum was counted on video still frames as the number of maximal curvatures along the flagellum. The percentage of flagellar length, 60 μm in the case of turbot flagella, presenting waves was estimated on still video frames illuminated by a double flash per video frame (100Hz) which allows visualization of the presence or absence of wave propagation in precise segments of the flagellum.

Wavelength and amplitude were measured on enlarged images from video records. The wave amplitude was measured as the distance between two segments tangent to the wave envelope, with one segment joining two successive principal (P) waves and the opposite segment joining two successive reverse (R) waves, which in the case of turbot spermatozoa is possible because waves are strictly symmetrical. The measurements of bending characteristics of axonemes were according to Brokaw (1991b) with some simplifications. The local bend angle was defined as the angle between tangents to the two straight segments adjacent to the bend. The wavelength was defined as twice the length of the segment between the two inflexion points adjacent to a bend.

Ions and Molecules Regulating Motility

The Activation Process of Fish Spermatozoa: Osmotic or Ionic?

General information on the effects of ions on male gametes is available in reviews by Darszon et al. (1996), Fraser (1995) and Morisawa et al. (this book). Most studies on the regulation of fish sperm motility activation were carried out on trout spermatozoa. These cells become motile as a result of changes in the properties of the plasma membrane potential and its ionic conductance (Cosson et al. 1989; Gatti et al. 1990; Boitano, Omoto 1991, 1992; Tanimoto et al. 1994) due to a decrease in extracellular K^+ concentration (Morisawa et al. 1983; Stoss 1983; Tanimoto et al. 1994). These membrane modifications induce intracellular changes in the second messengers such as cAMP, calcium, pH or protein phosphorylation (Morisawa, Ishida 1987; Cosson et al. 1989; Boitano, Omoto 1991, 1992) which activate axonemal movement.

Spermatozoa of freshwater oviparous teleosts seem well adapted to a brief period of activity upon dilution in a drastically anisoosmotic spawning medium. Spermatozoa are immotile in undiluted milt; proposed stimuli to motility include release from physical constraint in the testis, change in oxygen tension, dilution of an inhibitory factor and the change in osmolality itself (Ginsburg 1972).

Fish spermatozoa are immotile in the testis, and, in most species, in the seminal plasma. During natural reproduction, motility is induced after the delivery of sperm (and thus its dispersion/dilution) from the male genital tract into the aqueous environment. As a general rule, the factors that suppress motility are neutralized at spawning by the environmental conditions. For instance, K^+ ions in the seminal plasma of salmonids or sturgeons prevents sperm motility; this K^+ inhibitory effect is released when sperm is diluted in stream water. Environmental factors, such as pH, ions or osmolality, could stimulate the motility by depolarization of the sperm cell membrane and in synchrony with the egg activation factor released by the female.

Protons (pH)

The pH of an activating solution usually affects motility. Buffered solutions (including ovarian fluid) did not induce motility in rainbow trout spermatozoa when the pH was adjusted to values below 7.8 (Baynes et al. 1981; Schlenk 1933). Alkaline conditions similar to or greater than those of seminal plasma or ovarian fluid (Scott, Baynes 1980) apparently enhance the motility and fertility of salmonid spermatozoa (Petit et al. 1973; Billard et al. 1974; Billard 1981). Gaschott (1928), Inaba et al. (1958), and Pautard (1962) reported that alkaline pH (9) gives better motility in rainbow trout sperm, but buffered electrolyte as well as unbuffered water solutions of acidic pH do not lead to full immotility in salmonids. pH was also shown to influence maturation of salmon sperm (Morisawa et al. 1993). pH between 6 and 8.5 has little effect on sperm motility in northern pike and chain pickerel (Duplinsky 1982), nor in white sucker at pH 6 to 7 (Mohr, Chalanchuk 1985).

A few observations were reported from other fishes. Spermatozoa from the mullet, *Mugil capito*, were active in buffered sea water between pH 5.5 and 10, with a distinct optimum at pH 7 (Hines, Yashouv 1971). Long jawed goby (*Gillihthys mirabilis*) and sea bass (*Dicentrarchus labrax*) spermatozoa were motile in diluted sea water buffered between pH 5 and 10, with the optimum around pH 9 for sea bass (Weisel 1948; Billard 1980; Chambeyron, Zohar 1990), halibut (Billard et al. 1993) and turbot (Chauvaud et al. 1995).

Alteration of the internal pH (pH_i, a possible cellular second messenger) was described in spermatozoa of different species to interfere with motility. It has little impact on carp spermatozoa motility (Perchec-Poupard et al. 1997) except at extreme pH, via an Na^+/H^+ exchanger (Màriàn et al. 1997) and in the latter case, pH_i was measured at room temperature after loading of sperm cells with a fluorescent derivative at 37°C for 30 minutes. Activation of carp spermatozoa appears independent of the pH_i since the addition of NH_4Cl does not trigger sperm motility, in contrast to the case with sea urchin spermatozoa (Christen et al. 1983). Moreover, carp spermatozoa motility can be initiated in medium with external pH (pH_e) between 6 and 9 (Redondo-Müller et al. 1991). Similarly, it has been shown with boar, ram and trout spermatozoa (Gatti et al. 1990) that a change in the pH_e value induces a change in the pH_i. Krasznai et al. (1995) observed that the intracellular alkalinization (0.15 pH units) which accompanied the motility initiation of carp sperm does not play a key role in triggering axonemal movement. Similar results were obtained for trout (Gatti et al. 1990; Boitano, Omoto 1991).

Cations

Scheuring (1925) first stated that induction of motility in *Salmo gairdneri* spermatozoa was not possible after dilution in a diluted K^+ solution. Schlenk and Kahmann (1938) directly related K^+ concentration in the seminal fluid to immotility of sperm. Dilution of potassium or its removal by dialysis from the seminal plasma (Benau, Terner 1980) were shown to induce trout sperm motility. Kusa (1950) showed in chum salmon (*Oncorhynchus keta*) sperm that K^+ inhibits both motility and fertility. The action of K^+ is still apparent at 1 mM, where motility fails to initiate in *Salmo gairdneri* and *Salmo trutta* sperm (Scheuring 1925; Schlenk, Kahmann 1938; Stoss et al. 1977; Baynes et al. 1981). Apart from salmonids and sturgeons, sperm from other fish such as carp (Perchec et al. 1993) and *Esox masquinongy* (Lin, Dabrowski 1996) show much lower sensitivity to K^+.

Because motility is not inhibited in physiological solutions, such as ovarian fluid, which contain more than 1 mM of K^+ (Hwang, Idler 1969; Stoss et al. 1977), interactions with other components must occur. Scheuring (1925) was the first to report that ions such as Na^+, Ca^{2+} and Mg^{2+} reduce the inhibitory action of K^+ ions, the bivalent cations being more effective than Na^+.

When NaCl is added to seminal plasma, motility is induced. Schlenk and Kahmann (1938) observed motility with combined Na^+ and K^+ solutions, provided the Na^+/K^+ ratio was 16:1 or higher. This was confirmed by Scheuring (1925), by Stoss et al. (1977) and by Baynes et al. (1981). The latter demonstrated that the addition of small amounts of Ca^{2+} ions to solutions containing K^+ (or K^+ and Na^+) which did not induce motility, completely activates sperm cells. A similar, but slightly reduced, effect is seen with Mg^{2+}.

Synergistic effects between ions have led researchers to investigate a possible control of motility by the membrane potential which results from the combined effect of several ions (Gatti et al. 1990; Boitano, Omoto 1991; Blaber et al. 1988).

Occasionally, substances in the seminal plasma, termed androgamones, seem to control spermatozoan motility in salmonids (Runnström et al. 1944). However, these androgamones were later characterized as K^+ ions (Baynes et al. 1981). So, the effect of K^+ on the motility of spermatozoa in other teleost fishes is less clear, but it is reported not to inhibit flagellar motility in some species (Terner, Korsh 1963; Morisawa, Suzuki 1980; Chao et al. 1975). In another marine species, the Atlantic croaker, K^+ and Na^+ channels blockers were shown to inhibit sperm motility (Detweiler, Thomas 1998). K^+ ions were also shown to control sturgeon sperm activation at very low concentrations, in the range of 0.01 mM in *Acipenser baeri* (Gallis et al. 1991; Toth et al. 1997), in sterlet, shovelnose sturgeon (J Cosson and O Linhart, unpublished data) and in paddlefish (Cosson, Linhart 1996). Sturgeon spermatozoa has been induced to activate in a non-electrolyte solution composed of 40 mM sucrose and 2 mg/ml BSA (to prevent sticking to the glass slide); these results have shown that Ca^{2+} at 100 μM concentration could reverse the K^+ inhibitory effect, but also that EGTA used as a chelator of divalent ions could abolish the Ca^{2+} effect. In the latter case, readdition of Ca^{2+} in excess to EGTA and combined with the Ca ionophore A23187 completely reestablished the motility. The sensitivity of sturgeon axonemes to Ca^{2+} was confirmed by the use of membrane deprived spermatozoa. Even when using DM and RM deprived of K^+ ions, the flagella of demembranated/ATP reactivated spermatozoa showed only twitching of flagella with Ca^{2+} at 100 μM or below in RM. In contrast, propagating waves with large amplitude developed with Ca^{2+} at 250 μM or above (J Cosson and O Linhart, unpublished data). So it is clear that some antagonism between K^+ and Ca^{2+} ions occurs; the more specific effects of divalent ions are detailed below.

Baynes et al. (1981) demonstrated that solutions of K^+ (or K^+ and Na^+) which did not induce motility of trout sperm cells become activating solutions when small amounts of Ca^{2+} are added. Trout spermatozoa describe circular tracks which become tighter as a function of time elapsed since initiation. Figure 2a illustrates this point, and shows that such circling is the result of the asymmetry between principal and reverse waves developed by the flagellum, due to Ca^{2+} influx occurring during the motility period. First, sperm motility was shown to be blocked by the addition of Ca^{2+} chelators such as EGTA and also by Ca^{2+} channels blockers such as Verapamil (Cosson et al. 1986). Second, it was shown that Ca^{2+} ion influx occurring during the short motility period is responsible of this circling observed with trout spermatozoa (Cosson et al. 1989; Boitano, Omoto 1992) and that Ca^{2+} ions can overcome the K^+ inhibitory effect on motility, probably due to a simultaneous efflux of K^+ and Ca^{2+} (Tanimoto et al. 1994). The effect of Ca^{2+} on circling is easier to control in experiments when using demembranated/reactivated sperm models (Okuno, Morisawa 1989). Other divalent ions such as Cs^{2+}, Sr^{2+} and Ba^{2+} were also shown to induce trout sperm motility (Tanimoto, Morisawa 1988).

In contrast to what is observed in salmonids, carp spermatozoa, either native or demembranated, show no Ca^{2+} sensitivity (Cosson, Gagnon 1988), even though channel blockers such as Verapamil were claimed to decrease sperm motility (Krasznai et al. 1995). More recently it was observed that demembranated carp spermatozoa cannot be activated by ATP in the complete absence of Ca^{2+} but become immediately motile if Ca^{2+} is added (Z Krasznai, unpublished data). No effect of Ca^{2+} is observed with native or demembranated spermatozoa of some teleost fishes; for example, turbot (Chauvaud et al. 1995), halibut (Billard et al. 1993), silurids (Billard et al. 1997) and eels (Gibbons et al. 1985a, b). Nevertheless there are several exceptions where Ca^{2+} ions were shown to act as effectors of sperm motility: in paddlefish and sturgeons (J Cosson and O Linhart, unpublished data), in tilapia (O Linhart, unpublished data), in sea bass (Dréanno et al. 1999a) and in Atlantic croaker (Detweiler, Thomas 1998). It is worth emphasizing that in mammalian spermatozoa, Ca^{2+} ion concentration not only exhibits transient waves along the flagella but also has a crucial role in the hyperactivation process (Suarez et al. 1993); the latter could be similar to the mechanism by which fish sperm activation occurs briefly at fertilization.

Apart from the above mentioned antagonistic effects of Ca^{2+} versus K^+, the Ca^{2+} ions seem to be involved in the initiation of a cascade of events involving membrane signaling followed by cAMP dependent protein phosphorylations. This is not a rule that can be generalized from the results on salmonids. The addition of IBMX, a phosphodiesterase inhibitor, does not stimulate carp sperm motility, in contrast to its effect on bovine spermatozoa (Schoff, First 1995), but prolongs the duration of the motility (Billard 1980). As for intracellular pH, cAMP does not seem to be a second messenger triggering carp sperm motility; this is in contrast with trout spermatozoa (Morisawa, Okuno 1982). Other experiments support this observation since the Mg-ATP reactivation of demembranated carp spermatozoa does not require cAMP (Cosson, Gagnon 1988). Moreover, it has been shown on intact trout sperm that intracellular concentration of cAMP and of Ca^{2+} increases after activation; both compounds are necessary in the demembranation-reactivation experiments (Morisawa, Ishida 1987; Okuno, Morisawa 1989; Cosson et al. 1989; Boitano, Omoto 1991; Cosson et al. 1995). In the case of carp spermatozoa, the intracellular free Ca concentration was observed to increase (Z Krasznai, unpublished data) while the cAMP level does not change after activation (R Billard, unpublished data).

So, the involvement of cAMP and subsequent cascade of protein phosphorylation in the activation process of fish sperm motility seem to be restricted to salmonids. In other groups or species, with representatives such as three species of silurids (J Cosson and R Billard, unpublished data), carp (Cosson, Gagnon 1988), turbot (Dréanno et al. 1999b), sea bass (Dréanno et al. 1999a), halibut (Billard et al. 1993), tilapia, walleye (J Cosson and O Linhart unpublished data), eel (Gibbons et al. 1985a, b), herring (Morisawa et al., this book), demembranated spermatozoa activate upon addition of ATP-Mg without addition of cAMP.

Chemicals referred to as heavy metals have frequently been shown to exhibit deleterious effects on

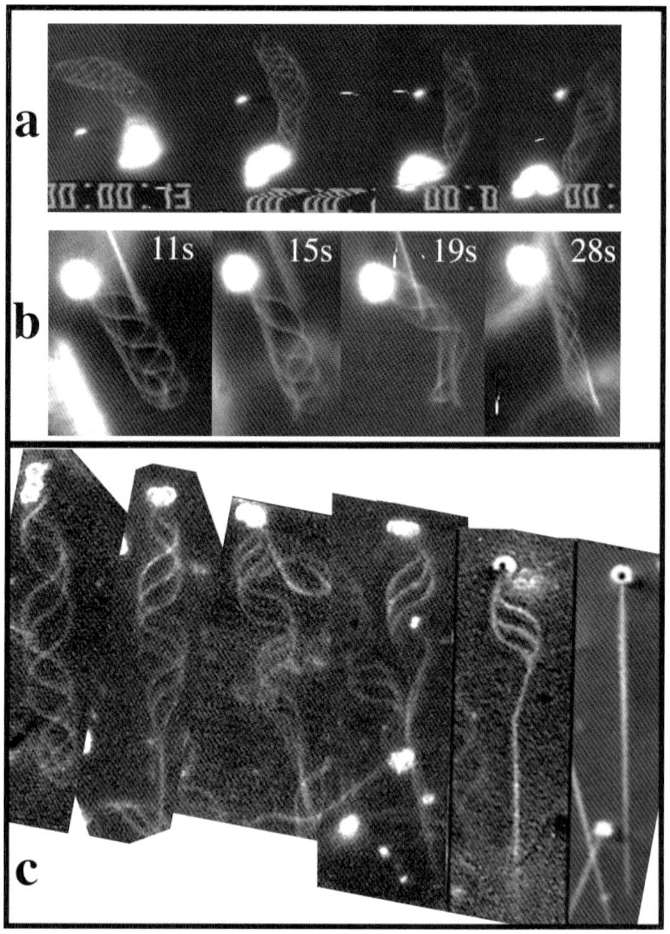

Figure 2. Movement of fish flagella. a, b and c (above): Movement observed at several stages during during the motility phase. Note the wave dampening appearing at the end of the motility phase. a: The same trout spermatozoon (35 µm long) activated since 8 seconds in SM is recorded on successive video frames (every 20 ms) using stroboscopic flashes (every 5 ms). b: Different trout sperm cells are observed at 11, 15, 19 and 28 seconds after activation in SM (respectively, from left to right) using the recording conditions described in a. c: Turbot spermatozoa (about 60 µm long) were recorded and individual video frames are presented at increasing time periods (from left to right) after activation in SM, using stroboscopic illumination (one flash every 7 ms). (2d-2g on opposite page) d: Sturgeon spermatozoa, top series at 15 seconds and bottom series at 50 seconds after activation; same conditions of observations. e: Induction of twists in the axoneme of turbot spermatozoa by incubation with $HgCl_2$. f: The rotation of turbot sperm, from top left, successive images every 20 ms. g: The occasional initiation of waves in the middle portion of the flagellum of turbot sperm with the proximal segment in rigor. Upper panel: one series of waves initiates and is seen progressing as a function of time (left to right); lower panel: three series of waves are successively initiated (single, double and triple arrowheads) with unbent region in between identified by a star (one flash every 3 ms). In both cases, sperm cells were fully motile for several more seconds after sequences were recorded.

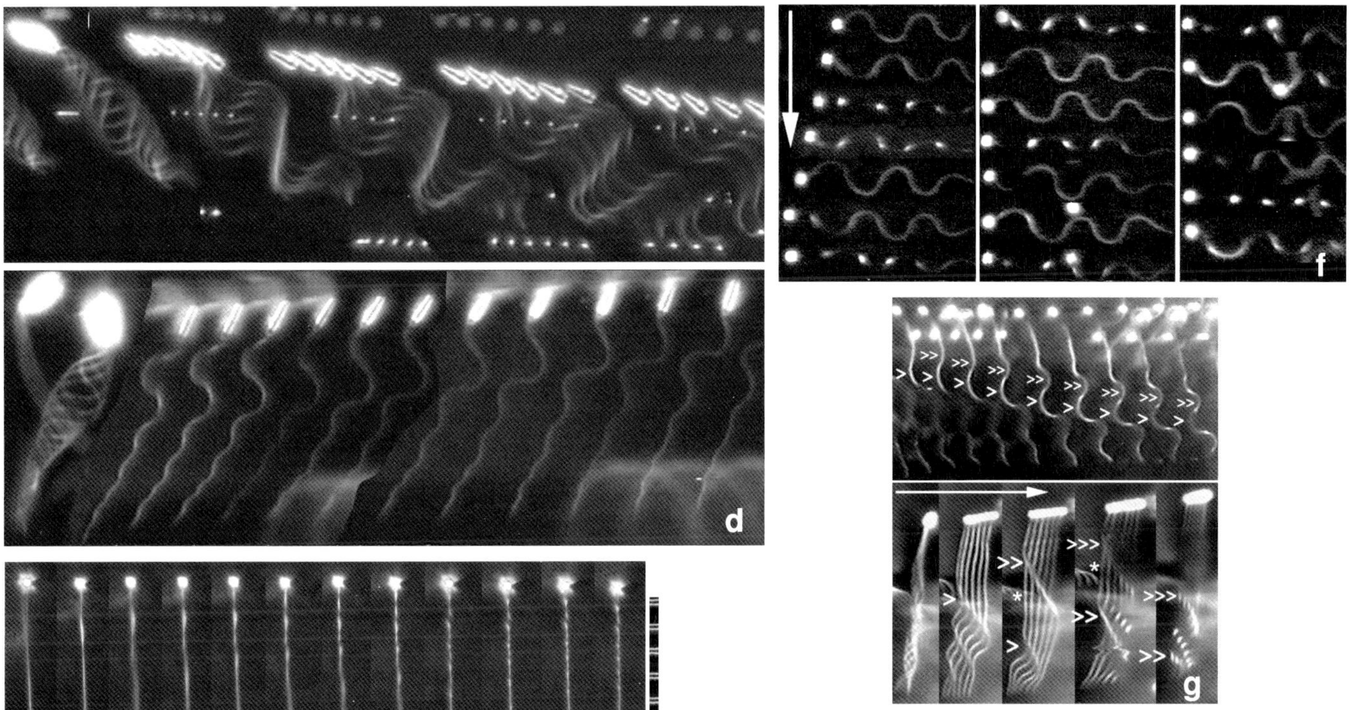

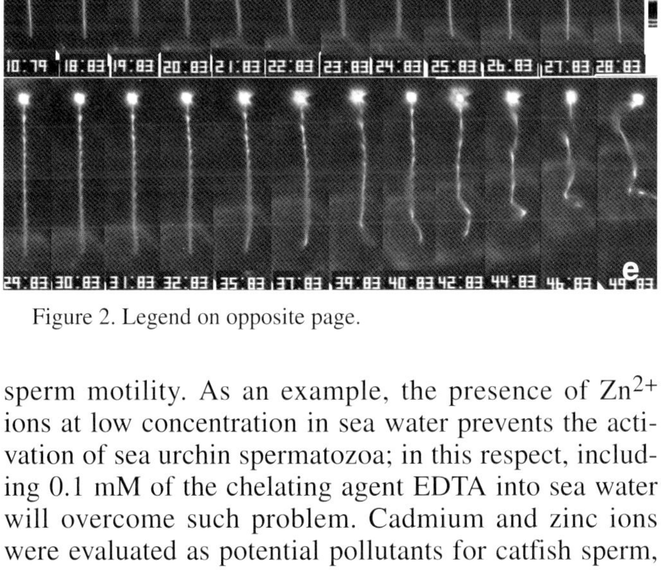

Figure 2. Legend on opposite page.

sperm motility. As an example, the presence of Zn^{2+} ions at low concentration in sea water prevents the activation of sea urchin spermatozoa; in this respect, including 0.1 mM of the chelating agent EDTA into sea water will overcome such problem. Cadmium and zinc ions were evaluated as potential pollutants for catfish sperm, affecting their motility by pre-incubation in extenders containing 100 or 2000 ppm of these chemicals, respectively (Kime et al. 1996).

In experiments with mercury derivatives, intriguing effects of low concentrations of $HgCl_2$ were observed on turbot and sea bass spermatozoa. When applied at 1 mM concentration range in the SM (devoid of BSA to prevent partial fixation of $HgCl_2$ to soluble proteins), turbot spermatozoa were stopped, in a dose- and time-dependent manner. With 200-500 µM $HgCl_2$, periods that were three to four minutes longer were needed for cessation of movement and sperm flagella underwent an unusual curling process, mainly a twisting easily observed by dark field microscopy and taking several seconds to reach completion. Figure 2d shows the successive stages of this twisting. Addition of parachloromercury sulfonate (pCMS) led to similar results.

Such flagellar curling is reminiscent of the naturally occurring curling observed for eel flagella (Gibbons et al. 1985a, b) but during motility in the latter. In turbot, the presence of BSA at 1 mg/ml completely prevented both arrest of motility and curling. Addition of cysteine at 1 to 2 mM prevented the $HgCl_2$ arrest of motility as well as the appearance of the flagellar twisting but did not reverse the Hg induced twists afterward. A deleterious effect of $HgCl_2$ was observed at a much higher concentration (20 mM and above) on demembranated flagella, with motility stopped immediately but no curling observed. A possible effect on specific ion ports in the membrane or on water channels could explain these *in vivo* observations as some water channels, such as aquaporin CHIP28, have been described as especially Hg-sensitive to pCMS or $HgCl_2$ and water channel proteins were characterized in ram and human sperm membranes (Curry et al. 1995). In turbot sperm, the blockage of these water channels could lead to drastic prevention of the osmotic re-equilibration which normally follows transfer of sperm cells into sea water.

Vanadium ions are also known to alter axonemal motility, but as they are poorly permeant through biological membranes, only demembranated flagella show sensitivity to their action. The mechanism of motility blockage was explained by the inhibitory effect of vanadium ions on dynein ATPases (Gibbons et al. 1978). This specific inhibitory effect is true for dyneins of fish sperm flagella in species such as trout (Saudrais et al.

1998), carp, silurids, sturgeons, paddlefish or turbot (J Cosson, unpublished data) in a micromolar concentration range (2 to 10 μM).

Osmolality

Many different studies have emphasized the effect of osmolality on spermatozoan motility in freshwater and marine teleosts other than salmonids and viviparous species (Morisawa et al., this book; Gwo 1995). In the goldfish, *Carassius auratus*, duration of sperm motility peaks between 100 and 200 mOsmol.kg^{-1}, while in the marine puffer, *Fugu niphobles*, sperm cells are immotile at 300 mOsmol.kg^{-1}, quiescent at 1200 mOsmol.kg^{-1} and active for the longest time at 400 mOsmol.kg^{-1}. These findings are in agreement with the sperm quiescence observed in undiluted milt of roughly 300 mOsmol.kg^{-1}. Thus it seems reasonable to assume that activation should occur as osmolality is lowered for freshwater milt and elevated for marine milt.

In freshwater species, the duration of spermatozoan motility in mannitol, sucrose or glucose solutions of varying osmolality is close to the following pattern: forward movement lasted longest between 100 and 200 mOsmol.kg^{-1} and declined at 0 and 300 mOsmol.kg^{-1}. However, a significant number of spermatozoa were briefly activated at osmolalities as high at 450 mOsmol.kg^{-1}, suggesting that osmolality may not be the only factor responsible for immotility in the testis.

One problem with the use of motility duration measurements is that it is based on an interpretation of sperm performance which favors the longest duration of motility in nature. Actually, fertilization in freshwater teleosts can occur at osmolalities much lower than the optimal value of 100-200 mOsmol.kg^{-1} suggested by these experiments; "vigor" may be more important than longevity. In most freshwater teleosts, egg activation begins with water contact (Yamamoto 1961) and selection may not favor the ability to swim long after the micropyle has closed. Ginsburg (1972) found that three to six seconds sufficed for gametic association in rainbow trout.

In many published studies, osmotic sensitivity of teleost spermatozoa has also been assessed by counting the sperm cells that are motile immediately (as soon as possible technically) upon mixing with activating solutions of varying osmolalities. Compared to a "yes or no" response offered by determining the percentage of motile cells, measurements of other parameters such as the velocity of sperm head, displacement and shape of head tracks offer more subtle description of the effects of osmolality. As the latter is a consequence of the diversity of the flagellar movements, the use of the shape of the wave patterns is thus a parameter giving access to more diversified and progressive changes.

Osmolality effects on cells or tissues are difficult to separate from ionic and/or gaseous effects. As an example, interpretation may be complicated in experiments using substitution of ions present in seminal fluid by non-ionic solutions of equivalent osmolality; this may lead to a depletion of the intracellular ionic content. Similarly, dissolved CO_2 is in equilibrium with $NaHCO_3$, which thus contributes to ion concentration but also to osmolality. Nevertheless, ionic effects at the plasma membrane level are not the only trigger of fish sperm motility and another regulation mechanism involves the environmental osmotic variation. Spermatozoa of the puffer, a marine fish, which are quiescent in their seminal plasma (around 300 mOsmol.kg^{-1}), become motile by an increase in the osmolality (1200 mOsmol.kg^{-1}) in the surrounding medium (Oda, Morisawa 1993). This hyperosmotic shock could induce an increase in intracellular K^+ (Takai, Morisawa 1995) and Ca^{2+} concentration, as well as an internal acidification (Oda, Morisawa 1993). These variations have been proposed as the trigger for motility activation. Conversely, carp spermatozoa motility is inhibited by the osmolality found in the seminal plasma at 300 mOsm (Morisawa et al. 1983; Redondo-Müller et al. 1991; Perchec et al. 1995a; Perchec-Poupard et al. 1997). The exposure to hypoosmotic media <200 mOsmol.kg^{-1}, whatever the ionic composition, triggers carp spermatozoa motility. The low osmolality induces also a sperm cell swelling and, in fresh water, the disruption of the plasma membrane occurs due to the large osmotic shock (Billard 1978; Morisawa et al. 1983). It was suggested that the osmotic shock modifies the membrane permeability and the lipid bilayer organization (Màriàn et al. 1993). Some results suggest that a potassium efflux through K^+ channels is involved in the activation (Krasznai et al. 1995) and the decrease in intracellular K^+ concentration has been proposed as the trigger for the activation of the freshwater zebra fish spermatozoa (Takai, Morisawa 1995).

The induction of motility by a change in the osmotic pressure of the medium surrounding sperm cells has been reported repeatedly. According to Morisawa and Suzuki (1980), hypotonic suspension media initiates motility in spermatozoa from freshwater fishes such as *Cyprinus carpio*, *Carassius auratus* and *Tribolodon hakonensis*. Either NaCl, KCl or mannitol solutions were used. Earlier observations by Suzuki (1959) and by Ginsburg (1972) confirm these observations, but there are also examples in freshwater spawners where isotonic media effectively activate motility. Pike, *Esox lucius*, spermatozoa are motile in an isotonic NaCl solution (250 mOsmol.kg^{-1}, pH 9; Billard, Breton 1976), and channel catfish, *Ictalurus punctatus*, spermatozoa are motile in a 0.65% saline (Guest et al. 1976). So, hypotonicity is not the only factor explaining motility induction in freshwater species.

Hypertonicity induces the motility of spermatozoa in marine teleosts, as demonstrated in different species:

cod, *Gadus morhua macrocephalus* (Westin, Nisling 1991), flounders, *Limanda yokohamae* and *Kareius bicoloratus* (Morisawa, Suzuki 1980), sea bass, *Dicentrarchus labrax*, sea bream, *Sparus auratus* (Billard 1978), gray mullet, *Mugil cephalus* (Chao et al. 1975), and the goby, *Gillichthys mirabilis* (Weisel 1948). In the goby, in halibut (Billard et al. 1993) and in turbot (Chauvaud et al. 1995), hypertonic sugar solutions were found efficient to activate motility as well.

In salmonid fishes, motility of spermatozoa can be initiated by hypotonic, isotonic, and, to a certain degree, hypertonic media (Gaschott 1928; Ellis, Jones 1939; Pautard 1962; Stoss et al. 1977; Morisawa, Suzuki 1980). Solutions of NaCl or mannitol (below 400 mOsmol.kg^{-1}) initiate sperm motility in *Salmo gairdneri* and *Oncorhynchus keta* (Stoss et al. 1977; Morisawa, Suzuki 1980). According to Billard (1980), sucrose solutions between 200-300 mOsmol.kg^{-1} at pH 9 do not activate rainbow trout spermatozoa; however, Van der Horst et al. (1980) observed activation at 300 mOsmol.kg^{-1} when the sperm dilution rate was high.

Environmental conditions during spawning activate spermatozoon motility. However, not all the factors involved are well understood. Attempts to change the environmental conditions lead to additional understanding. When Tilapia (*Oreochromis mossambicus*) males were progressively adapted to water of various salinities from fresh water to normal sea water, or even brackish water, successive sperm samples showed an incremental rise of the osmolality activation threshold, relative to each increase of salinity (J Cosson, O Linhart and M Legendre, unpublished data). In this respect, experiments of Harvey and Kelley (1984) emphasize the use of non-electrolyte solutions to control osmolality of testing media. Nevertheless, a drawback of non-electrolyte solutions is that spermatozoa incubated in such media for long periods may experience ion depletion.

Experiments of L. Chauvaud (unpublished data) on turbot show by measurements on sperm gently pelleted in capillary tubes that the volume of sperm cells decreases when previously incubated for brief periods in media with increasing osmolality. The putative role of the osmolality at the axonemal level was also investigated after demembranation/reactivation of carp sperm in various non-ionic solutions containing Mg-ATP; sperm models remain motile up to an osmolality of 550 mOsmol.kg^{-1}, while non-demembranated spermatozoa become immotile from 250 mOsmol.kg^{-1} and above (Perchec-Poupard et al. 1997). The cell volume increases as measured on video images recorded during the motility period and hypo-osmotic saline solutions triggered the swelling of carp sperm. This swelling, obviously resulting from an entrance of water, was affected neither by pCMS, an inhibitor of the aquaporin CHIP28, nor by various inhibitors of the co-transport of water with ions. The activation mechanism may involve specific "stretch activated channels" sensitive to osmotic pressure.

DMSO at 10% in SM triggers transient motility of carp or turbot spermatozoa. Nevertheless, DMSO has previously been extensively and successfully used as an extender for many chemical agents to be tested on sperm cells and in sperm cryopreservation. The use of DMSO for carp sperm cryopreservation could explain the poor capacity for motility initiation after such treatment (Cognie et al. 1989): DMSO removes the inhibition of movement due to the high osmolality (300 to 400 mOsmol.kg^{-1}). In the same way, DMSO has been shown to enhance motility of sea urchin sperm. Because carp spermatozoa swell following addition of DMSO, an influx of water is suggested. However, this swelling is limited and a threshold is obtained for a large range of concentration of DMSO. Conversely to other results, DMSO induced the reduction of intracellular water content in fowl spermatozoa, and a cell shrinking is observed in the same range of DMSO concentration. The delayed response of carp motility activation (up to three to four minutes after mixing), suggests that changes other than the influx of DMSO in spermatozoa are involved. As DMSO can interact with the phospholipid membranes (Anchordoguy et al. 1991), a change in the lipid bilayer structure with DMSO remains possible. Màriàn et al. (1993) propose a reorganization of the lipid bilayer structure of the carp sperm plasma membrane as essential in transmembrane signaling. Such a reorganization of the membrane structure could be a consequence of the osmotic shock; this drastic change is extreme at the end of the motility period in fresh water where the tip of the sperm tail exhibits a curling process (Perchec et al. 1996).

Urea, abundant in urine with other numerous small molecular weight compounds, represents another chemical often in contact with sperm at collection and potentially deleterious to motility. Deleterious effects of urine contamination on sperm quality are observed in carp (Perchec et al. 1995; Perchec-Poupard et al. 1998) and in turbot (Dréanno et al. 1998). Artificial urine contamination induced a delay in the initiation of spermatozoa motility. Velocity, the endogenous ATP store and the fertilizing capabilities of sperm from urine-contaminated samples were significantly decreased.

In viviparous teleosts (topminnows and guppies), sperm initiation occurs in salt solutions with osmolality of 300 mOsmol.kg^{-1}, not in non-ionic solutions (Morisawa, Suzuki 1980). Mechanisms explaining the induction of motility in viviparous fishes are less clear. In fishes such as *Gambiusa affinis* and *Poecilia reticulata*, spermatozoa are released in aggregates (spermatozeugma), and become motile after exposure to NaCl and KCl solutions with osmolality between 50 and 300 mOsmol.kg^{-1} (Morisawa, Suzuki 1980). The breakdown of the spermatozeugma follows and sperm cells are released. Because mannitol was ineffective, Morisawa and Suzuki (1980) concluded that rather than osmolality, a change of ionic concentrations in the environment, particularly in K^+, triggers the initiation of the motility of spermatozoa. A rapid breakdown of spermatozeugma occurs also in Ringer's solution, but no information has been provided on motility (Ginsburg 1972). However, Billard (1978) reported that spermatozoa of *Poecilia reticulata* achieve motility spontaneously after dissociation of the spermatozeugma, without any dilution. The cells then stay motile for an extended period. In the killifish *Fundulus heteroclitus* and in the cichlid *Oreochromis mossambicus*, sperm dilution is not required, possibly due to an unavoidable contamination of urine in sperm. Spermatozoa become motile during stripping or by exposure to air. Motility of sperm without any dilution in a SM was reported in the European catfish *Silurus glanis* (Linhart et al. 1986), in common tench *Tinca tinca* (Linhart, Kvasnika 1992), in asp *Aspius aspius* (Linhart, Benesovsky 1991), in turbot (Dréanno et al. 1999b) and in *Tilapia mossambica* (O Linhart, unpublished data). Strangely enough, it was reported that a very careful collection with prevention of the agitation of samples maintains the immotility in *O. mossambicus* sperm cells (Kuchnow, Foster 1976).

The Duration of Motility: Why So Short?

The duration of motility in the natural environment varies greatly between fish species and coincides in general with the fertile period of spermatozoa. Scott and Baynes (1980) and Ginsburg (1972) summarized data for salmonid fishes and for a number of salt, brackish and fresh water spawners. Chemical characteristics of the motility inducing medium essentially determine the duration of sperm motility but other factors play a role.

Temperature alters the motility period; low temperatures result in prolonged duration of sperm motility but at a reduced velocity (Schlenk, Kahmann 1938; Lindroth 1947; Turdakov 1971; Hines, Yashouv 1971; Billard, Cosson 1988; Vladic, Järvi 1997). For practical reasons, experiments on fish sperm motility were run at room temperature, which is frequently 10 to 15°C above the temperature in natural conditions; for example, 4°C in salmon according to Vladic and Järvi (1997). Large underestimates of the swimming period were made. Estimates were based on a general thermodynamic law called "Q_{10} of 2," which predicts that the swimming period should double when the temperature decreases by 10°C. It is worth noting that fish sperm have to cope with very low temperatures (-2 to +2°C) in the case of Antarctic fishes or Baltic cod *(Gadus morhua)*, for which sperm motility was tested at 7°C (Westin, Nissling 1991). Major proteins such as tubulins (Singer et al. 1994) or microtubule-associated proteins (Detrich et al. 1990) exhibit special adaptations to these low temperatures. More specifically, the dynein-ATPases responsible for motility in sperm axonemes were found to be specially adapted in cod to temperatures around 0°C, when compared to Tetrahymena or trout sperm dyneins (King et al. 1997).

Spermatozoa of freshwater fish species show an immediate burst of motility upon dilution that ceases after 15 to 30 seconds in *Salmo gairdneri* and may last for two to three minutes in *Esox lucius* (Lindroth 1947; Billard, Breton 1976). Swelling and lysis of the cells in the hypotonic water are clearly limiting factors in the duration of motility (Schlenk, Kahmann 1938; Billard 1978), leading to the remarkable observation of Huxley (1930), who emphasized the poor adaptive capacity of trout spermatozoa to cope with the fresh water environment. Motility in salt water may last from two minutes as in *Spariis auratus* (Billard 1978) to exceptionally long periods (two days) in the herring *Clupea harengus pallasi* (Yanagimachi 1957). Ion channels were shown to regulate motility in Atlantic croaker sperm (Detweiler, Thomas 1998). In the viviparous fish *Poecilia reticulata*, motility in the seminal fluid was observed for 48 hours (Billard 1978).

Neither fresh water nor full-strength sea water have ever been shown to be particularly suitable for maintaining spermatozoan motility. In contrast, partly diluted sea water has been used with success (Ellis, Jones 1939; Hines, Yashouv 1971; Billard 1978). Artificial media, which induce good activation of spermatozoa without exposing them to extreme osmotic conditions, will prolong sperm motility and thus their period of fertility. Such media are a prerequisite for the development of incubation or dilution media for short-term and cryopreservation storage of spermatozoa. Trout spermatozoa (*S. trutta* and *S. gairdneri*) were shown to be motile one to five minutes in various isotonic media (Ginsburg 1972; Scheuring 1925; Baynes et al. 1981) and highly fertile for eight minutes when suspended in a buffered NaCl solution (Billard 1977). The phosphodiesterase inhibitor 3-isobutyl-1-methylxanthine (IBMX) extended *Salmo gairdneri* spermatozoan motility to 90 seconds; motility lasts 30 seconds in a saline-lacking IBMX (Benau, Terner 1980). Using 1 to 5 mM IBMX in the $NaHCO_3$, more than 10 minutes of intensive motility in sperm of pink salmon, *Oncorhynchus gorbuscha,* was observed. IBMX inhibits the degradation of cAMP to AMP and causes a distinct increase in cAMP levels in the sperm of *Salmo gairdneri* (Benau, Terner, 1980), as it does in mammalian spermatozoa (Hoskins et al. 1975). In contrast, the addition of cAMP or of ATP to live sperm of rainbow trout (Billard 1980) or to sperm from several other oviparous teleosts (Pautard 1962) had little or no effect on spermatozoan motility. In contrast, addition of ATP to frozen/thawed halibut spermatozoa significantly enhanced their motility (Billard et al. 1993).

Energy production by respiration is slower than its consumption, resulting in energy limitation within the short period of motility. Obviously, ATP is the main energetic store, but phosphocreatine (PCr) was also shown to contribute the maintenance of the ATP level (Robitaille et al. 1987). The simultaneous presence of high levels of PCr and of creatine kinase (CrK) in trout spermatozoa led Saudrais et al. (1998) to suggest the contribution of a PCr/CrK shuttle for such maintenance; such a shuttle would operate similarly to that present in sea urchin spermatozoa (Tombes et al. 1987), which sustains the correct level of ATP all along the flagellar length and prevents its exhaustion in the very distal part. Nevertheless, it is worth noting that the ATP concentration at the end of a motility phase is still high enough to fulfill all needs of the dynein-ATPases. When ATP drops from 7 to 8 mM to 0.5 mM, this decrease appears spectacular, but 0.5 mM is clearly above the K_m ATP of dyneins (150 to 200 μM depending on species). Even ADP accumulated during the motility cannot explain the full motility arrest, even though it was shown to act as a competitive inhibitor of ATP of the dynein-ATPases activity. Thus other factors should be considered to explain motility arrest in fish sperm.

As already mentioned, osmotic damage of cell structures may result in a limitation of the duration of spermatozoa motility. Following dilution in fresh water, the head of carp spermatozoa undergoes considerable swelling while the tip of flagellum exhibits a drastic coiling process (Perchec et al. 1996). In addition, the permeability and the structure of plasma membrane were shown to be modified due to a reorganization of lipid bilayer (Màriàn et al. 1993, 1997).

Carp spermatozoa swim actively within one or two seconds post-activation. While in SM the percentage of motile spermatozoa remains high, it decreases rapidly in distilled water, where a progressive swelling of the head and of the distal part of the tail is followed by a progressive curling of the flagellum. The apparent flagellar length decreases and the spermatozoon stops its progressive motility when the flagellum becomes too short. Dilution in a medium of osmolality 160 $mOsm.kg^{-1}$ prevents such drastic damage (Saad et al. 1988), while the total duration of motility increases (Perchec et al. 1995b; Perchec-Poupard et al. 1998). A similar pattern of sperm behavior in hypotonic solution was observed in trout spermatozoa (Billard 1978) and in pike (Billard, Breton 1976). A swelling phenomenon in hypotonic solution had already been observed on sea urchin and mammalian spermatozoa (Drevius, Eriksson 1966). This emphasizes the large tolerance of sperm membrane to osmolality variations. In human spermatozoa, hypoosmotic shock swelling of the tail is a widely used vitality test in andrology clinics but mammalian sperm appear much more resistant to swelling than fish spermatozoa.

Following incubation into fresh water, drastic morphological changes of trout spermatozoa also occur, such as the disruption of the membrane and mitochondria swelling (Billard 1978). But, in this species, the short duration of motility (25 to 30 seconds) cannot directly result from the hypoosmotic shock since the alteration of the plasma membrane structure occurs later than 30 seconds. After dilution in a hyperosmotic medium, turbot spermatozoa exhibited several structural modifications (Dréanno et al. 1999b). The midpiece became round, probably as a result of the appearance of a vacuole due to an inward water flux. This change of shape of the midpiece region might result from some membrane alterations. At the end of the motility period, turbot spermatozoa did not show any shrinkage of the sperm head, in contrast to puffer spermatozoa (Takai, Morisawa 1995). Turbot spermatozoa might regulate more efficiently their homeostasis despite the hyperosmotic environment of the activating medium. In contrast, images obtained by electron microscopy of sea bass spermatozoa show a swelling when exposed to the high osmolality of sea water; these structural damages attributed to osmotic/ionic shock appear to be morphologically reversible when osmolality is lowered (Dréanno et al. 1999b). In this respect, experiments conducted on sperm from several species show that following a first motility period, spermatozoa can be revived and achieve a second round of motility.

Motility enhancing substances can be of natural origin and represent an integral part of the fertilization process. In the viviparous fishes *Poecilia reticulata* and *Cymatogaster aggregata*, motility is enhanced by the presence of reducible exogenous sugar (Gardiner 1978). While chemotaxis between fish egg and spermatozoa of the same species was not formally demonstrated, it is clear that some of the motility enhancing substances act on sperm motility in a synergetic manner with ions. The eggs from the herring *Clupea harengus* pallasi; several bitterlings, *Acheilognathus lanceolata, A. tabira* and *Rhodeus ocellatus*; and the fat minnow, *Sarcocheilichthys variegatus* (Yanagimachi 1957; Suzuki 1959) are well-known examples. When near an egg or upon physical contact with the micropylar region, the spermatozoa of these species increase their swimming speed and are attracted to the micropyle where they aggregate for a few minutes. Spermatozoa of the herring, unlike those of most other marine fishes, are motionless in sea water. However, they become vigorously motile on contact with the micropyle area of the egg chorion and enter the micropyle rapidly and efficiently (Pillai et al. 1993, 1994). Motility initiation of herring spermatozoa in the micropyle area is dependent on extracellular calcium and potassium. In absence of egg or egg substances, initiation of sperm movement occurs in alkaline sodium-free sea water (Yanagimachi et al. 1992). The motility enhancing factor (Oda et al. 1995) was partly purified (Morisawa et al. 1992) and characterized as a minor component of the egg chorion (Morisawa et al., this book). Its specificity was demonstrated by raising a specific antibody and applying it in an *in vitro* assay. (Griffin et al. 1996). There is no clear-cut demonstration of the chemotactic activity of such motility enhancing compound.

Motility of salmonid spermatozoa is enhanced by ovarian fluid, which is released with the eggs. The duration is at least doubled compared to that in fresh water, and the period of fertility is prolonged (Ginsburg 1972; Billard 1977). This effect has been attributed to the presence of motility enhancing factors in the ovarian fluid such as astaxanthine, beta-carotene (Hartmann et al. 1947), or an unspecified substance of low molecular weight (Yoshida, Nomura 1972). Carotenoid pigments such as astaxanthine or a synthetic cantaxanthine had, according to Quantz (1980), no effect on spermatozoan motility. To explain the good motility of salmonid spermatozoa in ovarian fluid, its isotonicity, combination of ions and alkaline pH are sufficient (Hwang, Idler 1969; Holtz et al. 1977; Baynes et al. 1981).

The secretory product of the seminal vesicles present in some teleosts (Hoar 1969) does not affect spermatozoan motility (Weisel 1948). Its absence, however, reduces the fertilizing ability of spermatozoa in the catfish *Heteropneustes fossilis* (Sundararaj, Nayyar 1969). In cases of internal fertilization, such as in sharks, interaction with the uterine fluid plays an additional role in the regulation of sperm motility (Minamikawa, Morisawa 1996).

Reviving Motility in "Exhausted" Sperm

Once activated, salmonid spermatozoa lose their fertilizing capability very quickly, owing to the brief duration of motility. If activated in a physiological solution, however, motility and fertility can be re-initiated, showing that morphological alterations can be reversed. This was first demonstrated by Schlenk and Kahmann (1938), who immersed activated trout spermatozoa in a K^+ rich solution. After incubation for one-half hour, transfer of the spermatozoa into any solution with a lower K^+ concentration (lower than in seminal plasma) induced motility. A similar finding was made by Nomura (1964) for *Salmo gairdneri* spermatozoa. After a previous dilution of milt in ovarian fluid, motility reactivation with water was efficient from one to 18 hours later. Kusa (1950) reported that *Oncorhynchus keta* milt loses its fertility in Ringer's solution after 90 minutes, but regains it after 24 hours. Diluting *Salmo trutta* milt in Ringer's solution, Ginsburg (1972) noted a loss of fertility after 20 minutes, but a complete restoration after 90 and 120 minutes. The same tendency, but with a lower level of fertility, was also found in ovarian fluid (Billardet al. 1974). Dilution of milt from *Oncorhynchus keta* in Ringer's and redilution with the same medium resulted in a gradual decrease in fertility down to zero at 120 minutes post dilution (Yamamoto 1961). When the second dilution was omitted, fertility was almost zero after two minutes, but fertility was slightly restored between 60 and 80 minutes. Similarly, reactivation has also been induced in spermatozoa from *Ictalurus punctatus* (Guest et al. 1976). Carp spermatozoa, having once been activated, regained motility spontaneously after resting (Sneed, Clemens 1956).

Many examples demonstrate that spermatozoa can become reactivated after a certain period of rest, provided previously activated spermatozoa remain metabolically active; thus, some time is required to restore initial levels of available energy. Such a resting period did not seem to be necessary when phosphodiesterase inhibitors such as IBMX or theophylline were used. IBMX at 5 mM induced immediate reactivation when added in a 0.12 M NaCl solution (pH 7.4) to trout spermatozoa only a few minutes after motility had ceased (Benau, Terner 1980). Similarly reactivated spermatozoa of brook trout, *Salcelinus fontinalis,* were capable of fertilization. Once the motility-suppressing K^+ had been removed during the first activation, other factors such as ATP or cAMP would be responsible for the initiation and maintenance of motility. This was confirmed by Morisawa and Okuno (1982). Billard (1980) maintained fertility most effectively when trout spermatozoa were first diluted with 10 mM theophylline and saline (250 $mOsmol.kg^{-1}$, pH 9) and subsequently rediluted 30 or 60 minutes later with saline only. Reviving could also be aided by a resynthesis of ATP from PCr and ADP via the CrK (Saudrais et al. 1998).

In turbot, sperm rendered immotile by a first incubation in sea water can be revived by letting them settle in an artificial seminal fluid of low osmolality. After spermatozoa are placed back into sea water, they reinitiate a motility, similar to that after the first dilution. If sperm is first activated in diluted sea water, a revival at the end of the motility period can be obtained by transfer into normal sea water. In any case this second transfer needs to be done after a delay, during which cells reload their ATP to a normal level compatible with full motility (Dréanno et al. 1999b). After movement resumes, several morphological changes altering the mitochondria and the midpiece can be observed. Similar observations have been obtained on sea bass (Fauvel et al. 1999).

Several cases have been reported where salmonid spermatozoa maintained motility and fertility for extended periods (hours or days), when sperm were immersed in saline solution, ovarian fluid or diluted sea water (Ellis, Jones 1939; Nomura 1964; Terner, Korsh 1963). Reactivation may have taken place in these cases, leading to the interpretation that motility had never ceased. Since the immersion medium can also induce reactivation (Yamamoto 1961), it may have done so when aliquots of previously diluted milt were transferred onto a slide for microscopic examination. This possibility is further supported by the observation that when these samples are examined microscopically, motility ceases very quickly (Terner, Korsh 1963).

The Wave Shape of Flagella

Wave Shape Changes Drastically during the Motility Phase

Among fish with external fertilization, the traits of the motility behavior of fish sperm flagella are quite similar in several respects (Cosson et al. 1997). As an exception, the eel spermatozoa swim for very long periods with beat frequency reaching 95 Hz, and with original wave pattern. They mainly present a rolling motion at 19 Hz and flagella develop a helicoidal three-dimensional bending, which was recently documented in detail by Wooley (1998). In most other fish species, the evolution of the wave pattern can be summarized according to the scheme proposed in Figure 3. It is presented as five successive representations of a turbot spermatozoon during swim at time periods later and later after triggering of motility. Such wave patterns are not usually present in sea urchin spermatozoa. The statement previously established for *O. mykiss* spermatozoa (Cosson et al. 1991) according to which "the synchronous triggering of trout sperm is followed by an invariable set sequence of movement parameters" holds true for various other fish species: Figures 2a, b, c and d illustrate how wave pattern evolves during the motility phase of trout, turbot and sturgeon sperm.

In the case of turbot sperm, the linearity index remains close to 1 whatever the time after activation and Ca^{2+} has no effect on any of the motility parameters. This includes absence of asymmetric beating; the principal and reverse bends, as defined by Gibbons (1981b), have identical amplitude.

During the motility phase, the wave pattern of turbot spermatozoa (Figs. 2c, 3) rapidly evolves from a fully beating situation where waves proceed through the whole length of the flagellum, to a partially beating situation where waves occupy only the proximal (to the head) part of the flagellum then to absence of wave. When cells are fully stopped, their flagella are in a rigor state, mostly linear; this is the same for sperm flagella prior to their activation. Nevertheless, in both cases, flagella are not completely immotile; transient shivering and local slight bending are observed. The blocking process occurring in the distal part of the flagella is observed for late periods after activation (more than 30 to 40 seconds). In partially beating situations, the distal part of the flagellum appears fully straight, rigid and devoid of any wave propagated in this segment, while a fully developed wave is initiated and propagated in the proximal portion of the flagellum over one-third to one-fourth distance. The proportion of motile sperm cells also decreases as a function of time since activation (Chauvaud et al. 1995). A calculation of the ratio between the initial velocity (230 µm/sec) and the initial beat frequency (70 Hz) leads to a value of 3.3 µm/beat (velocity/frequency) for the so-called "propulsive efficiency." The latter is more and more affected when parameters such as the proportion of the flagellar length occupied by waves as well as the number of bends along the flagellar length decrease as a function of time elapsed since activation. In contrast, length and amplitude of the waves still present in the proximal portion were little affected. Results obtained *in vivo* will be discussed later for comparison with *in vitro* observations obtained with demembranated spermatozoa.

It should be noted in the case of fish sperm, that due to the rotation of the whole cell as seen in Figure 2f, each spermatozoon image appears alternatively with flagella "top view" (waves plane in the observation plane) or "side view" (waves orthogonal to the observation plane). This rotating behavior is reminiscent of the rolling described for eel spermatozoa; for turbot spermatozoa,

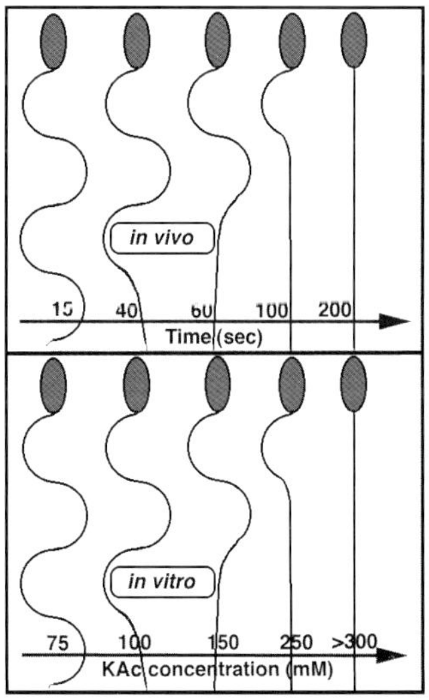

Figure 3. Schematic representation of the evolution of wave shape in fish sperm flagella during the swimming period and *in vitro*. In the upper panel, at 15 seconds after triggering of motility, fully developed waves occupy most of the length of the flagellum and present constant amplitude with high frequency while progressing along the length; at 40-60 seconds (middle of the motility phase) the frequency drops down while at the same time, the wave propagation becomes restricted to the proximal part while the very tip of the flagellum is in rigor; later during the motility phase (60-100 seconds), waves develop only in a very short portion posterior to the head; at 200 seconds, no wave remains: the whole flagellum stops beating and adopts a rigor aspect. In the lower panel, the wave shape of demembranated/ATP reactivated turbot spermatozoa observed at increasing ion concentration, (from 75 to 300 mM K Ac) shows waves more and more restricted to the part of the axoneme most proximal to the head.

the roll frequency was measured as 5 Hz, a low value compared to 19 Hz in the case of eel sperm (Gibbons et al. 1985a, b) and with a large dispersion (±3 Hz).

In order to quantify precisely such wave patterns and be able to compare them quantitatively, a computer assisted method was designed, allowing researchers to measure the local curvature expressed as the shear angle between each elementary segment defined along the flagellum. This gives rise to a mathematical transformation illustrated in Figures 4a, b and c, where the curvature is represented by a measurement of the shear angle and is plotted as a function of the length of the flagellum.

In many instances, examples where waves are present and propagate only in the distal part of the flagellum (Fig. 2g) could be observed. In most cases, this behavior was transient and full wave pattern reappeared within one to two seconds.

Ionic Concentration Affects the Wave Parameters of Permeabilized Axonemes

When exposed to a mild detergent, spermatozoa loose their plasma membrane but their flagellar axoneme remains fully functional after readdition of ATP (Gibbons, Gibbons 1972). When this demembranation process is applied to turbot spermatozoa, ATP reactivates their motility without need of cAMP and sperm models swim for very long periods (10 to 20 minutes). Their wave frequency is proportional to the ATP concentration, which leads to an estimation of K_m ATP of 150 µM, and a plateau value (extrapolated f_{max}) of 80 Hz. This compares to values for eel which were reported by Gibbons et al. (1985) as a K_m ATP of 200 µM and a f_{max} of 83 Hz and for trout axonemes to values of 200 µM and 85 Hz respectively (Saudrais et al. 1998). In eel spermatozoa, the dynein-ATPase activity was measured and found present in inner arms only (Baccetti et al. 1979). Ca^{2+} does not seem to modulate the wave shape of turbot sperm flagella, especially wave asymmetry, in contrast to its known effect on sea urchin spermatozoa (Brokaw 1991a) or on polychaetes spermatozoa (Pacey et al. 1994). In contrast, it has been observed that in turbot axonemes, when the ionic concentration of the reactivating medium is increased, mostly the wave shape is changed dramatically. Figure 5 shows examples of such change as a function of the KAc concentration. Waves become more and more dampened or distally absent when the KAc concentration is increased from 75 to 400 mM. The main features of the KAc effects are a decrease of the percentage of the flagellar length presenting waves and of the number of waves present on the whole length of the axoneme, which contrast with little change of the beat frequency and of the length or amplitude of each remaining individual wave (J Cosson and C Dréanno, unpublished data).

Effects of Ionic Concentration on Wave Shape Are Reversible *in Vivo* and *in Vitro*

Native turbot spermatozoa briefly exposed to conditions where no motility occurs *in vivo*, such as at low osmolality or ionic concentration such as one-to-four diluted sea water, immobilization medium or sucrose solutions of 300 mOsmol.kg^{-1} or lower, showed motility when returned in a SM. Similarly, when membrane deprived immotile sperm models incubated at extreme concentrations of KAc (25 or 250mM or above) were diluted to reactivation solutions containing 75 mM KAc, their motility was fully restored, showing that the blocking effect of extreme ionic strengths is fully reversible.

Ionic Strength Effect Is Not Ion-Specific

In assays with demembranated turbot sperm, KCl as well as K propionate or NaCl gave similar results but K Ac was used instead because motility was more stable in this medium as already observed for sea urchin spermatozoa (Gibbons, Gibbons 1983; Gibbons et al. 1982). This also shows that K^+ ions are not inhibitors of the reactivated movement, even at high concentration. Absence of inhibitory effects of K^+ also holds for trout demembranated sperm models (Saudrais et al. 1998),

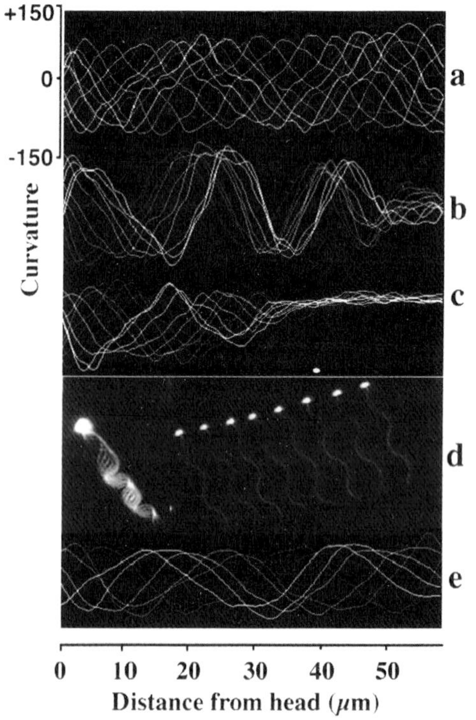

Figure 4. Plots of the bending angle versus the length of the flagellum of native and demembranated turbot spermatozoa. a: 9 seconds; b: 24 seconds; c: 50 seconds after activation in sea water. d, e, demembranated and ATP reactivated turbot sperm axoneme. The local angular curvature (measured as explained in material and methods) is plotted as a function of the distance along the flagellum starting from the head, located on the left of the plot. Successive images every 3 msec are drawn with a gray level decreasing with time elapsed.

which cannot explain its blocking effect observed *in vivo*. With turbot sperm, other ions were also tested; $NaHCO_3$ shows similar results when used at concentrations from 2.5 to 50 mM. When combined with the optimum reactivation medium containing 50-75 mM KAc, no effect was observed up to 5 mM $NaHCO_3$ but a progressive blockage of the axoneme in its distal part was observed when $NaHCO_3$ concentration was increased. These results confirm an effect of ionic strength on distal blockage as $NaHCO_3$ contributes three equivalents per molecule. It was concluded that pH was not significantly affected by addition of $NaHCO_3$. Results obtained *in vitro* with $NaHCO_3$ will be discussed later for comparison with *in vivo* effects.

Effect on Wave Shape in Sperm Models Is under Ionic and Not Osmotic Control

As *in vivo* motility can be triggered in non-electrolyte solutions with osmolality higher than 300 mOsmol.kg^{-1}, a potent effect of osmolality on axonemal machinery was tested. When glucose, sucrose or mannitol were added up to 500 mM to the reactivation solution containing 75 mM KAc as major ions, neither distal blockage nor perturbation of the wave shape or of the frequency were observed. In contrast, media containing glucose from 10 to 500 mM but lacking KAc did not permit any flagellar motility. It is concluded that effects observed *in vitro* are not due to a direct sensitivity of axonemes toward osmolality. In order to identify which part of the axonemal machinery is affected by ionic strength, experiments of microtubule sliding were conducted. Ionic strength had little effect on either the rate of sliding or on the portion of the flagellum where sliding occurred (distal versus proximal) when KAc concentration was varied from 25 to 250 mM (results not shown).

Investigation of the variations of beat frequency of axonemes (or portions of axonemes) reactivated in presence of various ATP-Mg concentrations show that K_m ATP and V_m are little affected when tested at various KAc concentrations from 25 to 250mM. These results tend to show that the axonemal component first blocked by increasing ionic strength could be some dynein subspecies located preferentially in the distal portion of the axoneme, which is reminiscent of the uneven distribution of dynein subspecies along *Chlamydomonas* flagella (Piperno, Ramanis 1991). Nevertheless, *in vivo* observations showing that only distal portions of the flagellum can be active while proximal are fully blocked (Fig. 2g) would not favor such an hypothesis, except if in such case, differences in the local ionic intracellular concentration occur.

Effect of CO_2 on Wave Shape

In view of the potential importance of the oxidative metabolism of sperm cells, CO_2 effects can shed some light on the involvement of respiratory function into motility, due to its putative inhibitory effect on mitochondria. A fortuitous observation of Dr. R. Billard (Dréanno et al. 1995) was that turbot spermatozoa swimming in a drop of sea water on the glass slide beneath a microscope were immediately stopped by expelled air breathed out by the researcher. This simple observation, the "Billard effect," was interpreted as CO_2 inhibition of movement and was confirmed by application of gentle stream of CO_2 on the surface of the drop (Dréanno et al. 1995). The stopping effect of the CO_2 could be video recorded at high magnification even though visibility of flagella was transiently blurred by the CO_2 flux on the drop surface. Figure 6a shows a whole sequence of CO_2 arrest and reinitiation of movement. Arrested state is characterized by a rigid (rigor) shape where flagella appear fully straight. The general methodology is reminiscent of the procedure employed on sea urchin spermatozoa by Brokaw (1977) for CO_2 application, and also of the observations of Gibbons and Gibbons (1980) showing that flagella spontaneously exhibit stochastic arrest and re initiation of beating in sea water. In the latter, this arrest has been related to a transient Ca^{2+} influx. Such observations are rendered possible with turbot spermatozoa because they tend to swim near the upper air/water interface of an open drop (C Cibert and J Cosson, unpublished

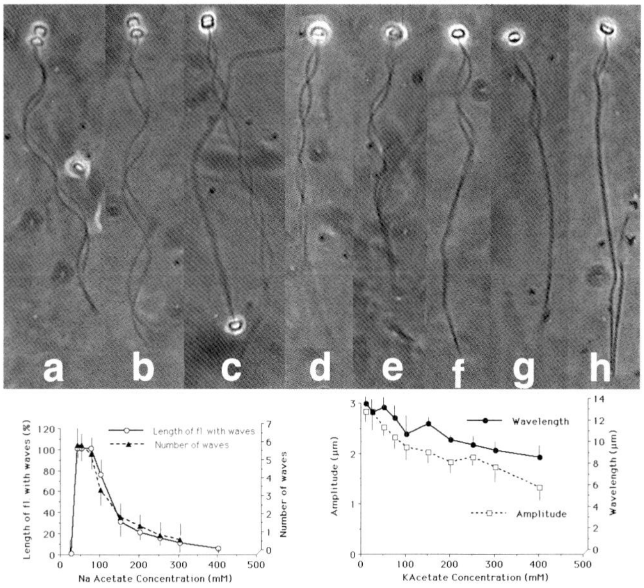

Figure 5. Permeabilized turbot axonemes ATP reactivated at various ionic strengths (varying Na Ac or K Ac concentration). In upper panel, stroboscopic video images obtained by DIC microscopy (two flashes per video frame separated by 10 msec) recorded after ATP reactivation at various K Ac concentrations (increasing from a to h from 50 to 400 mM). In bottom left panel, the graph shows the variations of the number of waves present along flagella and the portion of flagella presenting waves versus the Na Ac concentration. In bottom right panel is a graph of average value of the amplitude and length of waves present along the flagella as a function of K Ac concentration.

data). The CO_2 stream applied to the outer surface diffuses immediately in the few microns located close to this surface, and considering its solubility coefficient, it will generate a local concentration of at least 30 mM from which two-thirds is accounted by the generated $NaHCO_3$. Conversely, when the CO_2 stream is removed, CO_2 is rapidly exchanged with normal air in this thin layer of swimming medium. Observation of the complementary population of sperm cells swimming near the glass slide showed that CO_2 was efficient only after a much longer application, probably because of the delay for its diffusion throughout the entire swimming medium drop.

The CO_2 effect appears fully reversible. In most instances where image quality allowed it, observation of wave pattern before CO_2 application and after CO_2 effect is finished found very similar characteristics (Fig. 6a). A complementary detailed description of wave arrest and re-initiation was obtained by plotting curvature (shear angle) of the flagellum as a function of distance for successive flagellar image collected during CO_2 application (Fig. 6b). Several results argue against a pH shift due to CO_2 dissolution: 1) after intensive bubbling of CO_2 in the SM, direct measurement of the pH in SM was not affected and sperm motility was fully arrested in SM previously CO_2 saturated; 2) arrest of motility by CO_2 was similarly observed when buffering the SM by addition of 50 mM Tris-Cl at pH 8.2; 3) CO_2 could be replaced by bicarbonate ($NaHCO_3$) with the advantage that its chemical concentration can be evaluated precisely. Motility was partly blocked at 50mM $NaHCO_3$ and almost arrested at 100 mM, without effect on the pH of the swimming medium. Swimming parameters, except wavelength and amplitude of waves, were all drastically lowered by increase of the $NaHCO_3$ concentration, with distal blockage similar to that generated by KAc (data not shown). When a CO_2 stream was applied to ATP reactivated demembranated sperm models, the beat propagation immediately stopped, and then resumed after a few seconds, similarly to what was observed *in vivo*. The application of a CO_2 stream on sperm models engaged in the process of trypsin-induced sliding disintegration had no major effect on the completion of sliding. It is worth recalling that bicarbonate ion was shown also to participate in the maturation process of chum salmon spermatozoa (Morisawa, Morisawa 1988).

The simplest conclusion from this set of results is that the CO_2 target resides in the axonemal machinery, affecting primarily the factor(s) responsible for bending and bend propagation, while the sliding process as well as the beat frequency are slightly affected. The inhibitory effects of CO_2 were previously observed on both native (Brokaw, Simonick 1976) and demembranated (Brokaw 1977) sea urchin sperm, but in this case, complications arise from Ca^{2+} interference on flagellar beating asymmetry and from the $CaCO_3$ low solubility. The more general conclusion is that in turbot sperm, CO_2 in equilibrium with $NaHCO_3$ acts mainly as do other ions; that is, they indistinctly affect the behavior of axonemes.

The Mechanism of Wave Dampening

Wave dampening has been observed mostly in sea urchin sperm flagella where such dampening was only partial. It appears that fish sperm flagella present an extreme situation of this feature and give an opportunity to better investigate its regulatory mechanism. In Ciona spermatozoa, Brokaw and Benedict (1968) have shown that thiourea can provoke a partial wave dampening of the distal flagellum. Lithium ions have also been shown to dampen waves of demembranated sea urchin sperm flagella (Brokaw 1987) or of Ciona spermatozoa (Brokaw 1994) without interfering with the microtubule sliding process. Application of pCMS (para-chloromercury sulfonate), iodoacetate or NEM (n-ethylmaleimide) to permeabilized sea urchin sperm also induced a partial wave dampening with little effect on the beat frequency and the beat amplitude (Cosson et al. 1983). In the latter, the Ca^{2+} sensitivity was lost, which could partly explain the

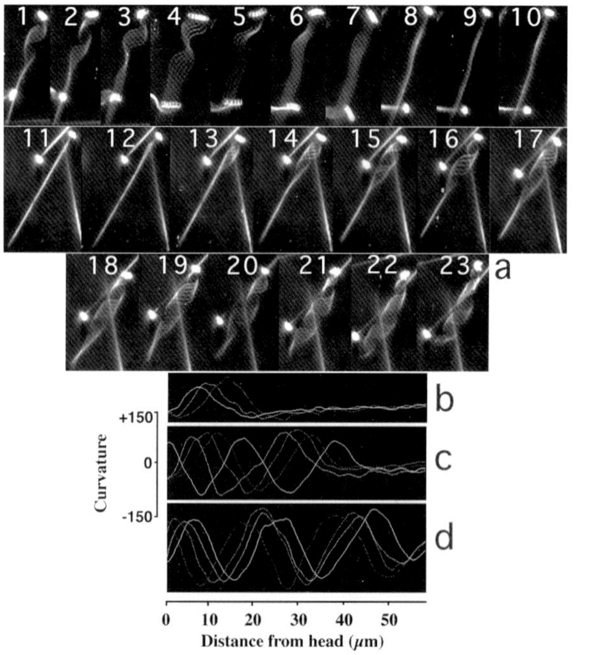

Figure 6. Arrest and recovery from CO_2 applied to swimming turbot spermatozoa. In a panel, video images of spermatozoa recorded while applying a stream of CO_2 to the open drop of SM on the microscope glass slide. The same cell is observed in all panels. Successive video frames (every 20 msec) received six flashes each (every 3 msec), with the exception of images 10 to 11, which are separated by a period of 1.3 seconds. In the lower panels are presented plots of the flagellar curvature as a function of distance from the head as in Figure 4. In b, the first waves generated as seen in the image 13 of panel a; in c, more waves are generated (from image 19); in d, "full" waves pattern (from image 23); successive images every 3 msec are drawn with a gray level increasing with time elapsed.

observed dampening. The use of FDNB (fluorodinitrobenzene) by Tombes et al. (1987) to inhibit the flagellar creatine kinase of sea urchin also leads *in vivo* to a partial dampening of the flagella; this was interpreted by the authors as a blockage of the PCr shuttle which thus becomes inadequate to sustain the ATP concentration needed by dyneins-ATPases, especially the dyneins located in the distal part far away from the production site (in mitochondria located in the perinuclear region). Recent results confirm that all the elements necessary for the functioning of the shuttle are present in trout sperm (Saudrais et al. 1998). Nevertheless, dampening occurs in demembranated trout spermatozoa where no such shuttle can be active. In their studies on the mechanism of sea urchin sperm chemotaxis, Cook et al. (1994) observed wave dampening provoked by speract. Unpublished observations of J. Cosson and C. Gagnon also show distal dampening of waves when NaN_3 is applied at low concentration on demembranated sea urchin sperm. Flagellar dampening also occurs when the axonemal machinery of demembranated sea urchin sperm is partially inhibited by low concentrations of a monoclonal antibody that recognizes the intermediate chain IC1 of sea urchin dynein ATPase (Gagnon et al. 1994). In the case of turbot sperm, the increase of ionic strength could also result in a direct stiffening of the microtubular structure, which would explain the rigor shape of axonemes observed at the end of the motility phase.

In the previous experiments, increasing the ionic strength provoked a flagellar dampening; in contrast, little effect is observed on either trypsin-induced microtubule sliding, on the beat frequency or on the proximal waves amplitude. In turbot, this dampening is not related to Ca^{2+} induction of beat asymmetry as no sensitivity to this ion was observed. Dampening occurs because waves persist preferentially close to the head of flagella. It is striking that the sperm motility activating factor has been localized precisely at the head/tail junction of trout spermatozoa (Morisawa 1994). The dampening observed *in vivo,* which was more and more pronounced as time passed after activation, is directly comparable to the *in vitro* dampening due to increasing ionic strength. The *in vivo* dampening may help explain the *in vitro* dampening.

Fish Sperm Motility: A Result of Osmotic, Ionic and Gaseous Effects

Little information on osmotic response is available on sperm cells, compared to information available on cellular types such as erythrocytes, cultured cells, epithelia, plant cells, bacteria and yeast. In response to hypoosmotic stress, the cytoplasmic volume of these cells swells as a result of the influx of water (Farinas et al. 1997) which is followed by biochemical readjustments to restore the initial cell volume by changing intracellular electrolyte (K^+, Na^+, Ca^{2+}, Cl^-) or non-electrolyte (amino acid, glycerol, sorbitol, inositol) solute concentration. This probably explains how the wave pattern evolves *in vivo* versus time (Fig. 3a) similarly to the way the wave pattern evolves *in vitro* versus ionic strength (Fig. 3b). Intracellular ion concentration probably increases as a reaction to the transfer of seawater fish spermatozoa in the surrounding high osmolality environment.

It has been proposed that the extracellular osmotic signal could act through a change in intracellular K^+ as the factor triggering sperm motility (Takai, Morisawa 1995) in two species, zebra fish (fresh water) and puffer (sea water); this hypothesis was based on measurements of the percentage of motile sperm models exposed to various solute concentrations. Nevertheless, the implication in the activation mechanism of an intracellular effector(s) becoming active by dilution (fresh water) or by concentration (sea water) seems to be ruled out since the entrance of water is a progressive phenomenon (Farinas et al. 1997) taking several seconds, thus occurring late after sperm motility is triggered. For the same reason, it is difficult to imagine how the influx of water alone would trigger motility, following a previous ionic transport across the plasma membrane. Measurements of the ionic flux should give more information. The stretch activated (SA) channels (Sachs, Sokabe 1990) or the mechanically activated (MA) channels (Opsahl, Webb 1994), which are highly sensitive to membrane tension or pressure and which modify the activity of certain membrane proteins (Vandorpe et al. 1994), could be involved in the activation. Gadolinium ions, which are not totally specific for these channels (Sachs, Sokabe 1990), have no effect on carp sperm motility, but 4-AP (a potent blocker of voltage gated potassium channels) blocks carp sperm motility (Krasznai et al. 1995). It was also observed recently that certain Ca^{2+} channel blockers (such as Verapamil and conotoxins) block carp sperm motility, suggesting the involvement of the membrane potential and the Ca^{2+} channels in the signal transduction process (Z Krasznai, unpublished data). Emri et al. (1998) suggest that the hypoosmotic shock induces immediate hyperpolarization and is the regulatory step in the motility activation of common carp sperm. Several groups reported similar effects of transmembrane potential on other species (Boitano, Omoto 1991; Zeng et al. 1995). Specific ion channels (K^+, Na^+ and Ca^{2+}) were also recently localized in mammalian sperm membranes (Chan et al. 1997). The use of electrophysiological techniques such as patch clamp applied to fish spermatozoa and sea urchin spermatozoa (Darszon et al. 1996) should give more information. It is also worth noting that the internal ionic concentration increase (or decrease) occurring during the motility phase could be facilitated due to the following "buffering" mechanism: the intracellular volume could be better modulated thanks to the fin structure of the flagellar membrane; the folding of the latter could thus offer a larger surface allowing a faster water influx (freshwater fishes) or outflux (seawater fishes) and facilitating an intracellular volume adaptation to these water fluxes thanks to a high ratio of surface to internal volume.

In conclusion, it should be realized that fish sperm have to abide by a reproduction strategy in which a very brief period of reaction is needed to achieve their task. They exhibit a hypermotile behavior remarkably similar to the so-called hypermotility of mammalian sperm in the vicinity of eggs before fertilization. The hypermotility of fish sperm is fully triggered at release in the new ionic/osmotic environment and is expressed by an initial high beat frequency, a high velocity and consequently a fast energy consumption. It is possible that such behavior is evolutionarily dictated by a main constraint, which is the short period of competence of the egg for fertilization represented in fish egg by the duration of the micropyle opening, limited to a few seconds in many species.

Acknowledgements

The support of the CNRS, the Muséum National d'Histoire Naturelle and the IFREMER in France is acknowledged. The authors are indebted to colleagues from INRA-Rennes, France; from CEMAGREF Bordeaux, France; from Aquaculture Research Center, Kentucky University, Frankfort, Kentucky, USA; and from University of South Bohemia (Vodany, Czech Republic) for use of their facilities and access to the fish broodstocks. Thanks also go to the Fisheries Society of the British Isles for photographic help. The unpublished data of Dr. Zoltan Krasznai was welcome. The help of Monique Garcia in typing and editing was appreciated.

References

Anchordoguy TJ, Cecchini CA, Crowe JH, Crowe LM. Insights into the cryoprotective mechanism of dimethyl sulfoxide for phospholipid bilayers. Cryobiology 1991; 28:467-473.

Adam M, Hamelin A, Bergé P, Goffaux M. Possibilité d'application de la technique de diffusion inélastique de la lumiére a l'étude de la vitalité des spermatozoïdes de taureaux. Ann Bio Anim Bioch Biophys 1969; 9:651-655.

Afzelius BA. Fine structure of the garfish spermatozoan. J Ultrastruct Res 1978; 64:309-314.

Baccetti B, Burrini AG, Dallai R, Pallini V. The dynein electrophoretic bands in axonemes lacking the inner or the outer arm. J Cell Biol 1979; 80:334-340.

Baynes SM, Scott AP, Dawson AP. Rainbow trout *Salmo gairdneri* Richardson, spermatozoa: effects of cations and pH on motility. J Fish Biol 1981; 19:259-267.

Benau D, Terner C. Initiation, prolongation and reactivation of the motility of salmonid spermatozoa. Gam Res 1980; 3:247-257.

Bergè P, Volochine B, Billard R, Hamelin A. Mise en èvidence du mouvement propre de microorganismes vivants grâce à l'étude de la diffusion inélastique de la lumiére. CR Acad Sci Paris, Sèrie D 1967; 265:889-892.

Billard R. Ultrastructure comparèe de spermatozoïdes de quelques poissons téléostéens. In: (Baccetti B, ed) Spermatologia Comparata, 1970; Quaderno 137:71-80.

Billard R. Utilisation d'un systeme tris-glycolle pour tamponner le diluant d'insémination de la truite. Bull Fr Piscic 1977; 264:102-112.

Billard R. Changes in structure and fertilizing ability of marine and fresh water fish spermatozoa diluted in media of various salinities. Aquaculture 1978; 14:187-198.

Billard R. Prolongation de la durèe de la mobilité et du pouvoir fécondant des spermatozoïdes de truite arc-en-ciel par addition de théophilline au milieu de dilution. CR Acad Sc Paris, Série D. 1980; 291:649-652.

Billard R. Short-term preservation of sperm under oxygen atmosphere in rainbow trout (*Salmo gairdneri*). Aquaculture 1981; 23:287-293.

Billard R, Breton B. Sur quelques problèmes de physiologie du sperme chez les poissons tèlèostèens. Rev Trav Inst Pêches Marit 1976; 40:501-503.

Billard R, Cosson M-P. Sperm motility in rainbow trout *Parasalmo mykiss*: effect of pH and temperature. In: (Breton B, Zohar Y, eds) Reproduction in Fish: Basic and Applied Aspect in Endocrinology and Genetics. Paris: INRA, 1988, pp161-167.

Billard R, Cosson M-P. Measurement of sperm motility in trout and carp. In: (De Pauw N, Jaspers E, Ackefors H, Wilkins N, eds) Aquaculture, a Biotechnology Progress. Bredene: Europ Aquacul Soc, 1989, pp499-503.

Billard R, Cosson M-P. Some problems related to the assessment of sperm motility in freshwater fish. J Exp Zool 1992; 261:122-131.

Billard R, Fléchon J. Particularité de la pièce intermédiaire de quelques Poissons Téléostéens. J Microscopie 1969; 8:36a.

Billard R, Ginsburg AS. La spermiogènése et le spermatozoïde *d'Anguilla anguilla* L, ètude ultrastructurale. Ann Biol Bioch Biophys 1973; 13:523-534.

Billard R, Jalabert B, Breton B. L'insèmination artificielle de la truite (*Salmo gairdnairi* Richardson). III Définition de la nature et de la molarité du tampon à employer avec les dilueurs d'insémination et de conservation. Ann Biol Anim Biochim Biophys 1974; 14:611-621.

Billard R, Cosson J, Crim LW. Motility and survival of halibut sperm during short term storage. Aquat Living Resour 1993; 6:67-75.

Billard R, Linhart O, Fierville F, Cosson J. Motility of *Silurus glanis* spermatozoa in the testicles and in the milt. Pol Arch Hydrobiol 1997; 44/1-1:115-122.

Blaber AP, Hallett J, Ross F. Relationship between transmembrane potential and activation of motility in rainbow trout (*Salmo gairdneri*). Fish Physiol Biochem 1988; 5:21-30.

Boitano S, Omoto CK. Membrane hyperpolarization activates trout sperm without an increase in intracellular pH. J Cell Sci 1991; 98:343-349.

Boitano S, Omoto CK. Trout sperm swimming patterns: role of intracellular Ca^{2+}. Cell Motil Cytoskel 1992; 21:74-82.

Brokaw CJ. CO_2-inhibition of the amplitude of bending of Triton demembranated sea urchin sperm flagella. J Exp Biol 1977; 71:229-240.

Brokaw CJ. A lithium-sensitive regulator of sperm flagellar oscillation is activated by cAMP-dependent phosphorylation. J Cell Biol 1987; 105:1789-1798.

Brokaw CJ. Calcium sensors in sea urchin sperm flagella. Cell Motil Cytoskel 1991a; 18:123-130.

Brokaw CJ. Microtubule sliding in swimming sperm flagella: direct

and indirect measurements on sea urchin and tunicate spermatozoa. J Cell Biol 1991b; 124:1201-1215.
Brokaw CJ. Microtubule sliding in reduced-amplitude bending waves of Ciona sperm flagella: resolution of metachronous and synchronous sliding components of stable bending waves. Cell Motil Cytoskel 1993; 26:144-162.
Brokaw CJ. Microtubule sliding in reduced-amplitude bending waves of Ciona sperm flagella: bending waves attenuated by lithium. Cell Motil Cytoskel 1994; 27:150-160.
Brokaw CJ, Benedict B. Mechanochemical coupling of flagella. II. Effects of viscosity and thiourea on metabolism and motility of Ciona spermatozoa. J Gen Physiol 1968; 52:283-299.
Brokaw CJ, Simonick TF. CO_2 regulation of the amplitude of flagellar bending. In: (Goldman R, Pollard T, Rosenbaum J, eds) Cell Motility. Cold Spring Harbor Conf Cell Prolif, Vol C, 1976; pp933-940.
Chao NH, Chen HP, Liao IC. Study on cryogenic preservation of grey mullet sperm. Aquaculture 1975; 5:389-406.
Chambeyron F, Zohar Y. A diluent for sperm cryopreservation of gilthead seabream, *Sparus auratus*. Aquaculture 1990; 90:345-352.
Chan HC, Zhou TS, Fu WO, Wang WP, Shi YL, Wong PDY. Cation and anion channels in rat and human spermatozoa. Biochim Biophys Acta 1997; 1323:117-129.
Chauvaud L, Cosson J, Suquet M, Billard R. Sperm motility in turbot, *Scophthalmus maximus*, initiation of movement and changes with time of swimming characteristics. Env Biol Fish 1995; 43:341-349.
Christ SA, Toth GP, Mc Carthy HW, Torsella JA, Smith MK. Monthly variations in sperm motility in common carp assessed using computer-assisted sperm analysis (CASA). J Fish Biol 1996; 48:1210-1222.
Christen R, Schackmann RW, Shapiro BM. Metabolism of sea urchin sperm: interrelationships between intracellular pH, ATPase activity and mitochondrial respiration. J Biol Chem 1983; 258:5392-5399.
Christen R, Gatti J-L, Billard R. Trout sperm motility: the transient movement of trout sperm is related to changes in the concentration of ATP following the activation of the flagellar movement. Eur J Biochem 1987; 166:667-671.
Cognie F, Billard R, Chao NH. La cryopreservation de la laitance de carpe *Cyprinus carpio*. J Appl Ichtyol 1989;5:165-176.
Cook SP, Brokaw, CJ, Muller CH, Babcock DF. Sperm chemotaxis: egg peptides control cytosolic calcium to regulate flagellar responses. Dev Biol 1994; 165:10-19.
Cosson J. A moving image of flagella: news and views on the mechanisms involved in axonemal beating. Cell Biol Int 1996; 20:83-94.
Cosson J, Linhart O. Paddlefish, *Polyodon spathula*, spermatozoa: effects of potassium ions and pH on motility. Folia Zoologica 1996; 45:361-370.
Cosson J, Billard R, Cibert C, Dréanno C, Linhart O, Suquet M. Movements of fish sperm flagella studied by high speed videomicroscopy coupled to computer assisted image analysis. Polish Arch Hydrobiol 1997; 44:103-113.
Cosson M-P, Gagnon C. Protease inhibitors and substrates block motility and microtubule sliding of sea urchin and carp spermatozoa. Cell Motil Cytoskel 1988; 10:518-527.
Cosson M-P, Tang W-JY, Gibbons IR. Modification of flagellar waveform and adenosine-triphosphatase activity in reactivated sea urchin sperm treated with N-Ethylmaleimide. J Cell Sci 1983; 60:231-249.
Cosson M-P, Billard R, Gatti J-L, Christen R. Rapid and quantitative assessment of trout spermatozoa motility using stroboscopy. Aquaculture 1985; 46:71-75.
Cosson M-P, Billard R, Carrè D, Letellier L, Christen R, Cosson J, Gatti J-L. Control of flagellar movement in "9+2" spermatozoa: calcium and cyclic AMP dependence, chemotactic behavior. Cell Motil Cytoskel 1986; 6:237.
Cosson M-P, Billard R, Letellier L. Rise of internal Ca^{2+} accompanies the initiation of trout sperm motility. Cell Motil Cytoskel 1989; 14:424-434.
Cosson M-P, Cosson J, Billard R. Synchronous triggering of trout sperm is followed by an invariable set sequence parameters whatever the incubation medium. Cell Motil Cytoskel 1991; 20:55-68.
Cosson M-P, Andre F, Billard R. cAMP/ATP relationship in the activation of trout sperm motility: their interaction in membrane-deprived models and in live spermatozoa. Cell Motil Cytoskel 1995; 31:159-176.
Craig T, Blaber A, Hallett FR. Motility of the spermatozoa of rainbow trout, *Salmo gairdneri*, in solutions of various salinities as studied by quasi-elastic light scattering. Biol Reprod 1983; 29:1189-1193.
Curry MR, Millar JD, Watson PF. The presence of water channel proteins in ram and human sperm membranes. J Reprod Fertil 1995; 104:297-303.
Dadone L, Narbaitz R. Submicroscopic structure of spermatozoa of the Cyprinidiform teleost, *Jenynsia lineata*. F Zellforsch Mikrosk Anat 1967; 80:214-219.
Darszon A, Liévano A, Beltrán C. Ion channels: key elements in gamete signaling. Curr Topics Dev Biol 1996; 34:117-167.
Detweiler C, Thomas P. Role of ions and ion channels in the regulation of Atlantic croaker sperm motility. J Exp Zool 1998; 281:139-148.
Detrich HW, Neighbors BW, Sloboda RD, Williams RC. Microtubule-associated proteins from Antarctic fishes. Cell Motil Cytoskel 1990; 17:174-186.
Dréanno C, Cosson J, Cibert C, André F, Suquet M, Billard R. CO_2 effects on turbot (*Scophthalmus maximus*) spermatozoa motility. In: (Goetz FW, Thomas P, eds) Reproductive Physiology of Fish. Fish Symposium 95, Austin, 1995; pp114.
Dréanno C, Suquet M, Desbruyeres E, Cosson J, Le Delliou H, Billard R. Effect of urine on sperm quality in turbot (*Scophthalmus maximus*). Aquaculture 1998; 169:247-262.
Dréanno C, Cosson J, Suquet M, Dorange G, Christian Fauvel C, Cibert C, Billard R. Effects of osmolality, morphology perturbations and intracellular nucleotid content during the movement of sea bass (*Dicentrarchus labrax*) spermatozoa. J Reprod Fertil 1999a; 115: in press.
Dréanno C, Cosson J, Suquet M, Seguin F, Dorange G, Billard R. Nucleotides content, oxidative phosphorylation, morphology and fertilizing capacity of turbot (*Psetta maxima*) spermatozoa during the motility period. Mol Reprod Dev 1999b; 53:230-243.
Drevius LO, Eriksson H. Osmotic swelling of mammalian spermatozoa. Exp Cell Res 1966; 42:136-156.
Duplinsky PD. Sperm motility of northern pike and chain pickerel at various pH values. Trans Am Fish Soc 1982; 111:768-771.
Ellis WG, Jones JW. The activity of the spermatozoon of *Salmo salar* in relation to osmotic pressure. J Exp Biol 1939; 16:530-534.
Emri M, Màriàn T, Tron L, Balkay L, Krasznai Z. Temperature adaptation changes ion concentrations in spermatozoa and seminal plasma of common carp without affecting sperm motility. Aquaculture 1998;167:85-94.
Farinas J, Kneen M, Moore M, Verksman AS. Plasma membrane water permeability of cultured cells and epithelia measured by light microscopy with spatial filtering. J Gen Physiol 1997; 110:283-296.
Fauvel C, Savoye O, Dréanno C, Billard R, Cosson J, Suquet M. Characteristics of sea bass (*Dicentrarchus labrax* L.) sperm in captivity: sperm concentration, motility and fertilization capacity. J Fish Biol 1999; 54(2):356-369.
Franzen A. Phylogenic aspects of the morphology of spermatozoa and spermiogenesis. In: (Baccetti B, ed) Comparative

Spermatology. Rome: Academia Nazionale Dei Lincei, 1970, pp29-46.
Fraser LR. Ionic control of sperm function. Reprod Fertil Dev 1995; 7:247-267.
Gagnon C. Regulation of sperm motility at the axonemal level. Reprod Fertil Dev 1995; 7:189-198.
Gagnon C, White D, Huitorel P, Cosson J. A monoclonal antibody against the dynein IC1 peptide of sea urchin spermatozoa inhibits the motility of sea urchin, dinoflagellate and human flagellar axonemes. Mol Biol Cell 1994; 5:1051-1063.
Gallis JL, Fedrigo E, Jatteau P, Bonpunt E, Billard R. Siberian sturgeon, *Acipenser baeri*, spermatozoa: effects of dilution, pH, osmotic pressure, sodium, and potassium ions on motility. In: (Willot P, ed) Acipenser. CEMAGREF Publ 1991; pp 143-151.
Gardiner DM. Utilisation of extracellular glucose by spermatozoa of two viviparous fishes. Comp Biochem Physiol 1978; 59A:165.
Gaschott O. Beiträge zur Reizphysiologie des Forellenspermas. I. Die optimalen Konzentrationen einiger Salzlösungen. Arch Hydrobiol 1928; 4 (Suppl):441-478.
Gatti JL, Billard R, Christen R. Ionic regulation of the plasma membrane potential of rainbow trout (*Salmo gairdneri*) sperm: role in the initiation of motility. J Cell Physiol 1990; 143:546-564.
Gibbons BH, Gibbons IR. Flagellar movement and adenosine triphosphatase activity in sea urchin sperm extracted with Triton X-100. J Cell Biol 1972; 54:75-97.
Gibbons BH, Gibbons IR. Calcium-induced quiescence in reactivated sea urchin sperm. J Cell Biol 1980; 84:13-27.
Gibbons BH, Gibbons IR. Certain organic anions improve the reactivated motility of sea urchin sperm flagella. J Cell Biol 1983; 97:5a.
Gibbons BH, Gibbons IR, Baccetti B. Structure and motility of the 9+0 flagellum of eel spermatozoa. J Submicrosc Cytol 1983; 15:15-21.
Gibbons BH, Baccetti B, Gibbons IR. Motility of the 9+0 flagellum of Anguilla sperm. Cell Motil 1985a; 5:333-350.
Gibbons BH, Baccetti B, Gibbons IR. Live and reactivated motility in the 9+0 flagellum of Anguilla sperm. Cell Motil 1985b; 5:333-350.
Gibbons IR. Cilia and flagella of eukaryotes. J Cell Biol 1981a; 91:107s-124s.
Gibbons IR. Transient flagellar waveforms during intermittent swimming in sea urchin sperm. II. Analysis of tubule sliding. J Muscle Res Cell Motil 1981b; 2:83-130.
Gibbons IR, Cosson M-P, Evans JA, Gibbons BH, Houck B, Martinson KH, Sale WS, Tang W-J. Potent inhibition of dynein adenosine triphosphatase and the motility of cilia and sperm flagella by vanadate. Proc Natl Acad Sci USA 1978; 75:2220-2224.
Gibbons IR, Evans JA, Gibbons BH. Acetate anions stabilize the latency of dynein 1 ATPase and increase the velocity of tubule sliding in reactivated sperm flagella. Cell Motil 1982; 1 (Suppl):181-184.
Ginsburg AS. Fertilization in Fishes and the Problem of Polyspermy. Keter Press, Springfield, VA, 1972.
Gosh RI. Energy metabolism of fish spermatozoa, a survey. UDC 1989; 597:62-72.
Griffin FJ, Vines CA, Pillai MC, Yanagimachi R, Cherr GN. Sperm motility initiation factor is a minor component of the Pacific herring egg chorion. Dev Growth Differ 1996; 38:193-202.
Gray J, Hancock GJ. The propulsion of sea-urchin spermatozoa. J Exp Biol 1955; 32:802-814.
Guest WC, Avault JW, Roussel JD. Preservation of channel catfish sperm. Trans Am Fish Soc 1976; 105:469-474.
Gwo JC. Ultrastructural study of osmolality effect on spermatozoa of three marine teleosts. Tissue Cell 1995; 27:491-497.
Halangk W, Bohnensack R, Kunz W. Interdependence of mitochondrial ATP production and extramitochondrial ATP utilization in intact spermatozoa. Biochim Biophys Acta 1985; 808:316-322.
Hartmann M, Medem FG, Kuhn R, Bielig HJ. Untersuchungenüber die Befruchtungsstroffe der Regenbogenforelle. Z Naturforsch B: Anorg Chem, Org Chem Biochem Biophys Biol 1947; 2b:330-349.
Harvey B, Kelley R N. Control of spermatozoan motility in a euryhaline teleost acclimated to different salinities. Can J Zool 1984; 62:2674-2677.
Hines R, Yashouv A. Some environmental factors influencing the activity of spermatozoa of *Mugil capito* Cuvier, a grey mullet. J Fish Biol 1971; 3:123-127.
Hoar WS. Reproduction. In: (Hoar WS, Randall DJ, eds) Fish Physiology. Vol 3. New York: Academic Press, 1969, pp1-72.
Holtz W, Stoss J, Büyükhatipoglu S. Beobachtungen zur Aktivierbarkeit von Forellenspermatozoen mit Fruchtwasser and destillirtem Wasser. Zuchthygiene 1977; 12:82-88.
Hoskins DD, Hall ML, Munsterman D. Induction of motility in immature bovine spermatozoa by cyclic AMP phosphodiesterase inhibitors and seminal plasma. Biol Reprod 1975; 13:168-176.
Huxley JS. The maladaption of trout spermatozoa in fresh water. Nature 1930; 125:494.
Hwang PC, Idler DR. A study of major cations, osmotic pressure and pH in seminal components of Atlantic salmon. J Fish Res Board Can 1969; 26:413-419.
Inaba D, Nomura M, Suyama M. Studies on the improvement of artificial propagation in trout culture. II. On the pH value of eggs, milt, coelomic fluid and others. Bull Japan Soc Sci Fish 1958; 23:762-765.
Inoda T, Ohtake H, Morisawa M. Activation of respiration and initiation of motility in rainbow trout spermatozoa. Zool Sci 1988; 5:939-945.
Jamieson BGM. Fish evolution and systematics: evidence from spermatozoa. Cambridge Univ Press, Cambridge, 1991 pp230-295.
Jaspers EJ, Avault JW, Roussel JD. Spermatozoal morphology and ultrastructure of channel catfish, *Ictalarus punctatus*. Trans Am Fish Soc 1976; 105:475-480.
Kime DE, Ebrahimi M, Nysten K, Roelants I, Rurangwa E, Moore H, Ollevier F. Use of computer assisted sperm analysis (CASA) for monitoring the effects of pollution on sperm quality of fish; application to the effects of heavy metals. Aquatic Toxicol 1996; 36:223-237.
King SM, Marchese-Ragona SP, Parker SK, Detrich H. Inner and outer arm axonemal dyneins from the Antarctic rockcod *Notothenia coriiceps*. Biochemistry 1997; 36:1306-1314.
Kossmann H. Versuche zue Konservierung des Karpfenspermas (*Cyprinus carpio*). Arch Fischereiwiss 1973; 24:125-128.
Krasznai Z, Màriàn T, Balkay L, Gaspar R, Tron L. Potassium channels regulate hypo-osmotic shock-induced motility of common carp (*Cyprinus carpio*) sperm. Aquaculture 1995; 129:123-128.
Kuchnow KP, Foster RS. Thermal tolerance of stored *Fundulus heteroclitus* gametes: fertilizability and survival of embryos. J Fish Res Board Can 1976; 33:676-680.
Kusa M. Physiological analysis of fertilization in the egg of the salmon, *Onchorynchus keta*. I. Why are the eggs not fertilized in isotonic Ringer solution? Annot Zool Japan 1950; 24:22-28.
Lahnsteiner F, Berger B, Weismann T, Patzner R. Fine structure and motility of spermatozoa and composition of the seminal plasma in the perch. J Fish Biol 1995; 47:492-508.
Lahnsteiner F, Berger B, Weismann T, Patzner R. Sperm structure and motility of the freshwater teleost *Cottus gobio*. J Fish Biol 1997; 50:564-574.
Lasher R, Rugh R. The "Hertwig effect" in teleost development. Biol Bull (Woods Hole, MA) 1962; 123:582-588.
Lin F, Dabrowski K. Characterisation of muskellunge spermatozoa II: Effects of ions and osmolality on sperm motility. Trans Am Fish Soc 1996; 125:195-202.
Lindroth A. Time of activity of fresh water fish spermatozoa in relation to temperature. Zool Bidr Uppsala 1947; 25:154-168.
Linhart O, Benesovsky J. Artificial insemination in asp (*Aspius*

aspius L.). Ziv Vur 1991; 36:973-980.

Linhart O, Kvasnicka P. Artificial insemination of tench (*Tinca tinca* L.) Aqua Fish Manag 1992; 23:125-130.

Linhart O, Kouril J, Hamakova J. The motile spermatozoa of wels (*Silurus glanis* L.) and tench (*Tinca tinca* L.) after sperm collection without water activation. Pràce Vúrh Vodnany 1986; 15:28-41.

Linhart O, Slechta V, Slavik T. Fish sperm composition and biochemistry. Bull Inst Zool Academia Sinica 1991; 16:285-311.

Linhart O, Mims SD, Shelton WL. Motility of spermatozoa from shovelnose sturgeon and paddlefish. J Fish Biol 1995; 47:902-909.

Màriàn T, Krasznai Z, Balkay L, Balazs M, Emri M, Bene L, Tron L. Hypo-osmotic shock induces an osmolality dependent permeabilisation and structural changes in the membrane of carp sperm. J Histo Cyto 1993; 41:291-297.

Màriàn T, Krasznai Z, Balkay L, Balazs M, Emri M, Tron L. Role of extracellular and intracellular pH in carp sperm motility and modifications by hyperosmosis of regulation of the Na^+/H^+ exchanger. Cytometry 1997; 27:374-382.

Mattei C, Mattei X. Spermiogenesis and spermatozoa of the Elopomorpha (teleost fish). In: (Afzelius BA, ed) The Functional Anatomy of the Spermatozoon. Pergamon Press: Oxford 1975; pp211-221.

Mattei C, Mattei X, Marchand B, Billard R. Rèinvestigation de la structure des flagelles spermatiques: cas particulier des spermatozoïdes à mitochondrie annulaire. J Ultrastr Res 1981; 74:307-312.

Minamikawa S, Morisawa M. Acquisition, initiation and maintenance of sperm motility in the shark, *Triakis scyllia*. Comp Biochem Physiol 1996; 113A:387-392.

Mohr LC, Chalanchuk SM.The effect of pH on sperm motility of white suckers, *Catostomus commersoni* in the Experimental Lakes Area. Env Biol Fishes 1985; 14:309-314.

Mohri H, Fujiwara A, Daumae M, Yasumasu I. Respiration and motility in starfish spermatozoa at various times in the breeding season. Dev Growth Differ 1990; 32:375-381.

Mounib MS, Hwang PC, Idler DR. Cryogenic preservation of Atlantic cod (*Gadus morua*) sperm. J Res Board Can 1968; 25:2623-2632.

Morisawa M. Initiation mechanism of sperm motility at spawning in teleosts. Zool Sci 1985; 2:605-615.

Morisawa M. Cell signalling mechanism for sperm motility. Zool Sci 1994; 11:647-662.

Morisawa M, Ishida K. Short term changes in level of cyclic AMP, adenylate cyclase and phosphodiesterase during the initiation of sperm motility in rainbow trout. J Exp Zool 1987; 242:199-204.

Morisawa M, Okuno M. Cyclic AMP induces maturation of trout sperm axoneme to initiate motility. Nature 1982; 295:703-704.

Morisawa M, Suzuki K. Osmolality and potassium ion: their role in initiation of sperm motility. Science 1980; 210:1145-1147.

Morisawa M, Suzuki K, Shimizu H, Morisawa S, Yasuda K. Effects of osmolality and potassium on motility of spermatozoa from fresh water cyprinid fishes. J Exp Biol 1983; 107:95-103.

Morisawa M, Tanimoto S, Ohtake H. Characterization and partial purification of sperm-activating substance from eggs of the herring. J Exp Zool 1992; 264:225-230.

Morisawa S, Morisawa M. Induction of potential for sperm motility by bicarbonate and pH in rainbow trout and chum salmon. J Exp Biol 1988; 136:13-22.

Morisawa S, Ishida K, Okuno M, Morisawa M. Role of pH and cyclic adenosine monophosphate in the acquisition of potential for sperm motility during migration from the sea to the river in chum salmon. Mol Reprod Dev 1993; 340:420-426.

Nicander L. Comparative studies on the fine structure of vertebrate spermatozoa. In: (Baccetti B, ed) Comparative Spermatology. Rome: Academia Nazionale Dei Lincei, 1970, pp47-56.

Nicholls DG. Bioenergetics: An Introduction to the Chemiosmotic Theory. Academic Press, London, 1982, pp190.

Nomura M. Studies on reproduction of rainbow trout, *Salmo gairdneri* with special reference to egg taking. VI. The activities of spermatozoa in different diluents and preservation of semen. Bull Japan Soc Sci Fish 1964; 30:723-733.

Oda S, Morisawa M. Rises of intracellular Ca^{2+} and pH mediate the initiation of sperm motility by hyperosmolality in marine teleosts. Cell Motil Cytoskel 1993; 25:171-178.

Oda S, Igarashi Y, Ohtake H, Sakai K, Shimizu N, Morisawa M. Sperm-activating proteins from unfertilized eggs of the Pacific Herring, *Clupea pallasi*. Dev Growth Differ 1995; 37:257-261.

Ohtake T, Mita M, Fujiwara A, Tazawa E, Yasumasu I. Degeneration of respiratory system in sea urchin spermatozoa during incubation in sea water for long duration. Zool Sci 1996; 13:857-863.

Okuno M, Morisawa M. Effect of calcium on motility of rainbow trout sperm flagella demembranated with Triton X-100. Cell Motil Cytoskel 1989; 14:194-200.

Opsahl LR, Webb WW Transduction of membrane tension by the ion channel alamethicin. Biophys J 1994; 66:71-74.

Pacey A, Cosson J, Bentley M G. Intermittent swimming in the spermatozoa of the lugworm *Arenicola marina* (L.) (Annelida: Polychaeta). Cell Motil Cytoskel 1994; 29:186-194.

Pautard FGE. Biomolecular aspects of spermatozoan motility. In: (Bishop DW, ed) Spermatozoan Motility. Publ 72. Washington, DC: Am Assoc Adv Sci, 1962, pp189-232.

Perchec G, Cosson J, Andre F, Billard R. La mobilitè des spermatozoïdes de truite (*Oncorhynchus mykiss*) et de carpe (*Cyprinus carpio*). J Appl Ichthyol 1993; 9:129-149.

Perchec G, Cosson J, Andrè F, Billard R. Degradation of the quality of carp sperm by urine contamination during stripping. Aquaculture 1995a; 129:135-136.

Perchec G, Jeulin C, Cosson J, Andrè F, Billard R. Relationship between sperm ATP content and motility of carp spermatozoa. J Cell Sci 1995b; 108:747-753.

Perchec G, Cosson M-P, Cosson J, Jeulin C, Billard R. Morphological and kinetic sperm changes of carp (*Cyprinus carpio*) spermatozoa after initiation of motility in distilled water. Cell Motil Cytoskel 1996; 35:113-120.

Perchec-Poupard G, Gatti J-L, Cosson J, Jeulin C, Fierville F, Billard R. Effects of extracellular environment on the osmotic signal transduction involved in activation of motility of carp spermatozoa. J Reprod Fertil 1997; 110:315-327.

Perchec-Poupard G, Paxion C, Cosson J, Jeulin C, Fierville F, Billard R. Initiation of carp spermatozoa motility and early ATP reduction after milt contamination by urine. Aquaculture 1998; 160:317-328.

Petit J, Jalabert B, Chevassus B, Billard R. L'insèmination artificielle de la truite (*Salmo gairdneri* Richardson). I. Effets du taux de dilution, du pH et de la pression osmotique du dilueur sur la fècondation. Ann Hydrobiol 1973; 4:201-210.

Pillai MC, Shields T S, Yanagimachi R, Cherr G N. Isolation and partial purification of the sperm motility initiation factor from eggs of the Pacific Herring, *Clupea pallasi*. J Exp Zool 1993; 265:336-342.

Pillai MC, Yanagimachi R, Cherr G. *In vivo* and *in vitro* initiation of sperm motility using fresh and cryopreserved gametes from the pacific herring, *Clupea pallasi*. J Exp Zool 1994; 269:62-68.

Piperno G, Ramanis Z. The proximal portion of *Chlamydomonas* flagella contains a distinct set of inner dynein arms. J Cell Biol 1991; 112:701-709.

Quantz G. Ueber den Einfluss von carotinoidreichem Trockenfutter auf die Eibenfruchtung der Regenbogenforelle (*Salmo gairdneri* R.) Arch Fishereiwiss 1980; 31:29-40.

Redondo-Müller C, Cosson M-P, Cosson J, Billard R. *In vitro* maturation of the potential for movement of carp spermatozoa. Mol Reprod Dev 1991; 29:259-270.

Rikmenspoel R. The equation for motion of sperm flagella. Biophys J 1978; 23:177-206.

Robitaille P M, Munfort K, Brown G. ^{31}P nuclear magnetic resonance study of trout spermatozoa at rest, after motility, and during short-term storage. Biochem Cell Biol 1987; 65:474-485.

Runnström J, Lindvall S, Tiselius A. Gamones from the sperm of sea urchin and salmon. Nature (Lond) 1944; 153:285-286.

Saad A, Hollenbecq MG, Billard R, Pionnier E, Heymann A. Production spermatogenique dilution et conservation du sperme de silure européen, *Silurus glanis*. In: Symposio Sobre Cultivo Intensivo de Peces Continentale. Assoc Espagnola Aquacul, Madrid, 1988; pp1.

Sachs F, Sokabe M. Stretch activated ion channel and membrane mechanics. Neurosci Res 1990; 12:S1-S4.

Satir P. The present status of the sliding microtubule model of ciliary motion. In: (Sleigh MA, ed) Cilia and Flagella. Academic Press, New York, 1974; pp131-141.

Saudrais C, Fierville F, Loir M, Le Rumeur E, Cibert C, Cosson J. The use of phosphocreatine plus ADP as energy source for motility of membrane-deprived trout spermatozoa. Cell Motil Cytoskel 1998; 41:91-106.

Scheuring L. Biologische und physiologishe Untersuchungen an Forellensperma. Arch Hydrobiol Suppl 1925; 4:181-318.

Schlenk W. Spermatozoenbewegung and Wasserstoffionenkonzentration. Versuche mit dem Sperma der Regenboforelle. Biochem Z 1933; 265:29-35.

Schlenk W, Kahmann H. Die chemische Zusammensetzung des Spermaliqors and ihre physiologische Bedeutung. Untersuchung am Forellensperma. Biochem Z 1938; 295:283-301.

Schoff PK, First NL. Manipulation of bovine sperm metabolism and motility using anoxia and phosphodiesterase inhibitors. Cell Motil Cytoskel 1995; 31:140-146.

Scott AP, Baynes SM. A review of the biology, handling and storage of salmonid spermatozoa. J Fish Biol 1980; 17:707-739.

Singer WD, Parker SK, Himes RH, Detrich HW. Polymerisation of antarctic fish tubulins at low temperatures: role of carboxy-terminal domains. Biochemistry 1994; 33:15389-15396.

Sneed KE, Clemens HP. Survival of fish sperm after freezing and storage at low temperature. Prog Fish Cult 1956; 18:99-103.

Stanley HP. An electron microscope study of spermiogenesis in the teleost fish *Oligocottus moculosus*. J Ultrastruct Res 1969; 27:230-243.

Stein H. Licht- und electronenoptishe Untersuchungen an den Spermatozoen verschniedener Süsswasserknochenfishe (Teleostei). Z Angew Zool 1981; 68:183-198.

Stein H, Bayle H. Cryopreservation of the sperm of some freshwater teleosts. Ann Biol Anim Biochem Biophys 1978; 18:1073-1076.

Stoss J. Fish gamete preservation and spermatozoan physiology. In: (Hoar WS, Randall DJ, Donaldson EM, eds) Fish Physiology. New York: Academic Press, 1983, pp305-350.

Stoss J, Büyükhatipoglu S, Holtz W. Der Einfluss bestimmter Elektrolyte auf die Bewegungsauslösung bei Spermatozoen der Regenbogenforelle (*Salmo gairdneri*). Zuchthygiene 1977; 12:178-184.

Suarez SS, Varosi SM, Dai X. Intracellular calcium increases with hyperpolarisation in intact, moving hamster sperm and oscillates with the flagellar beat cycle. Cell Biol 1993; 90:4660-4664.

Sundararaj GI, Nayyer SK. Effect of extirpation of "seminal vesicles" on the reproductive performance of the male catfish *Heteropneustes fossilis* (Bloch). Physiol Zool 1969; 42:429-437.

Suquet M, Dorange G, Omnes MH, Normant Y, Le Roux A, Fauvel C. Composition of the seminal fluid and ultrastructure of the spermatozoon of turbot (*Scophthalmus maximus*). J Fish Biol 1993; 42:509-516.

Suzuki R. Sperm activation and aggregation during fertilization in some fishes. II. Effect of distilled water on the sperm-stimulating capacity and fertilizability of eggs. Embryologia 1959; 4:359-367.

Takai H, Morisawa M. Change in intracellular K^+ concentration caused by external osmolality change regulates sperm motility of marine and freshwater teleosts. J Cell Sci 1995; 108:1175-1181.

Tanimoto S, Morisawa M. Roles of potassium and calcium channels in the initiation of sperm motility in rainbow trout. Dev Growth Differ 1988; 30:117-124.

Tanimoto S, Kudo Y, Nakazawa T, Morisawa M. Implication that potassium flux and increase in intracellular calcium are necessary for the initiation of sperm motility in salmonid fishes. Mol Reprod Dev 1994; 39:409-414.

Terner C. Oxydative and biosynthetic reactions in spermatozoa. In: (Bishop DW, ed) Spermatozoan Motility. Publ 72. Washington, DC: Am Assoc Adv Sci, 1962, pp189-232.

Terner C, Korsh G. The oxidative metabolism of pyruvate, acetate and glucose in isolated fish spermatozoa. J Cell Comp Physiol 1963; 62:243-249.

Tombes RM, Brokaw CJ, Shapiro BM. Creatine kinase-dependent energy transport in sea urchin spermatozoa: flagellar wave attenuation and theoretical analysis of high energy phosphate diffusion. Biophys J 1987; 52:75-86.

Toth GP, Ciereskо A, Christ SA, Dabrowski K. Objective analysis of sperm motility in the lake sturgeon, *Acipenser fulvencens*: activation and inhibition conditions. Aquaculture 1997; 154:337-348.

Tsvetkova LI, Cosson J, Linhart O, Billard R. Motility and fertilizing capacity of fresh and frozen-thawed spermatozoa in sturgeons (*Acipenser baeri* and *A. Ruthenus*). J Appl Ichtyol 1996; 12:107-112.

Turdakov AF. The effects of temperature conditions on the speed and fertilizing capacity of the spermatozoa of some Issyk-kul fishes. J Ichthyol 1971; 11:206-215.

Van der Horst G, Dott HM, Foster GC. Studies on the motility and cryopreservation of rainbow trout *(Salmo gairdneri)* spermatozoa. S Afr J Zool 1980; 15:275-279.

Van Deurs B. The sperm cell of Pantodon (Teleostei) with a note on residual body formation. In: (Afzelius BA, ed) The Functional Anatomy of the Spermatozoon. Oxford: Pergamon Press, 1975, pp311-318.

Vandorpe DH, Small DL, Dabrowski AR, Morris CE. FMRamide and membrane stretch as activators of the Aplysia S-channel. Biophys J 1994; 66:46-58.

Van Leeuwenhoek A. 1677. Quoted in: (Tixier G, ed) "Le sperme." Paris: Presse Univ Fr, 1994, pp15.

Vladic T, Järvi T. Sperm motility and fertilization time span in Atlantic salmon and brown trout—the effect of water temperature. J Fish Biol 1997; 50:1088-1093.

Weisel GF. Relation of salinity to the activity of the spermatozoa of Gillichthys, a marine teleost. Physiol Zool 1948; 21:40-48.

Westin L, Nissling A. Effects of salinity on spermatozoa motility, percentage of fertilised eggs and egg development of Baltic cod (*Gadus morhua*) and implications for cod stock fluctuations in the Baltic. Mar Biol 1991; 108:5-9.

Wooley DM. Studies on the eel sperm flagellum. 2. The kinematic of normal motility. Cell Motil Cytoskel 1998; 39:233-245.

Yamamoto T. Physiology of fertilization in fish eggs. Int Rev Physiol 1961; 12:361-405.

Yanagimachi R. Studies of fertilization in *Clupea pallasi*. Parts I-III. Zool Mag 1957; 66:218-233.

Yanagimachi R, Cherr GN, Muralidharan C, Pillai MC, Baldwin JD, Factors controlling sperm entry into the micropyles of salmonid and herring eggs. Dev Growth Differ 1992; 34:447-461.

Yoshida T, Nomura M. A substance enhancing sperm motility in the ovarian fluid of rainbow trout. Bull Japan Soc Sci Fish 1972; 38:1073-1079.

Zeng Y, Clark EN, Florman HM. Sperm membrane potential: hyperpolarization during capacitation regulates zona pellucida-dependent acrosomal secretion. Dev Biol 1995; 171:554-563.

17 Calcium Channels of Mammalian Sperm: Properties and Role in Fertilization

Harvey M. Florman
Tufts University

Christophe Arnoult
CNRS

Imrana G. Kazam
Chongqing Li
Christine M.B. O'Toole
Tufts University

The acrosome reaction is a secretory event in sperm that is promoted by contact with the egg's extracellular matrix, or zona pellucida, during the early stage of mammalian fertilization. Completion of the acrosome reaction is an essential prerequisite for later events of fertilization, including penetration of the zona pellucida and sperm-egg fusion. The mechanism by which the zona pellucida initiates secretion is thus central to the spatial and temporal coordination of gamete interaction.

A central feature of the acrosome reaction is an elevation of sperm intracellular calcium concentration ($Ca^{2+}{}_i$). It is generally appreciated that $Ca^{2+}{}_i$ is an intracellular mediator of stimulus-secretion coupling in a wide range of somatic cell systems (Katz, Miledi 1967; Douglas 1968; Penner, Neher 1988). Attention focused on the role of this ion in the control of similar events in sperm with the realization that 1) extracellular Ca^{2+} is required for the acrosome reaction that is initiated by eggs in sea urchin sperm and that occurs spontaneously in mammalian sperm (Dan 1954, 1956; Yanagimachi, Usui 1974), 2) that $^{45}Ca^{2+}$ uptake into sperm accompanies these acrosome reactions (Singh et al. 1978; Schackmann et al. 1978; Schackmann, Shapiro 1981), and 3) that Ca^{2+}-transporting ionophores promote acrosome reactions in sperm of mammals and of marine invertebrates (Summers et al. 1976; Green 1978; Tilney et al. 1978) The subsequent identification of ZP3, the acrosome reaction-inducing agonist in the zona pellucida (Bleil, Wassarman 1983), then set the stage for the examination of mechanism of ZP3-dependent Ca^{2+} responses in sperm.

$Ca^{2+}{}_i$ is determined by the net flux through a number of regulatory mechanisms. Studies in somatic cells, where these pathways have been extensively characterized, have revealed the contribution of: 1) Ca^{2+} efflux from the cytosol by ion transporting ATPases in the plasma membrane, in the mitochondrial membrane and in smooth endoplasmic reticulum membrane, as well as by facilitated diffusion through Ca^{2+}/counter ion exchange proteins (Lauger 1991; Carafoli 1992; Clapham 1995); 2) Ca^{2+} entry pathways into the cytosol across the plasma membrane by means of voltage- and ligand-gated channels, across the endoplasmic reticular membrane by means of Ca^{2+} release channels, and from mitochondria (Tsien, Tsien 1990; Hille 1992; Rizzuto et al. 1992; Loew et al. 1994; Clapham 1995; Tsien et al. 1995; Babcock et al.1997; Ichas et al. 1997); and 3) Ca^{2+} buffering by cytosolic proteins (Konishi et al. 1988; Zhou, Neher 1993).

It is likely that all of these regulatory mechanisms are also present in mammalian sperm (Ward, Kopf 1993; Yanagimachi 1994). Nevertheless, a number of observations direct attention towards the role of voltage-sensitive Ca^{2+} channels in the ZP3-evoked signaling pathway of sperm. Key elements of this argument include the observation that depolarization of sperm membrane initiates $Ca^{2+}{}_i$ elevations (Babcock, Pfeiffer 1987; Florman et al. 1992; Arnoult et al. 1996b) and acrosome reaction (Florman et al. 1992; Arnoult et al. 1996b); that these responses are inhibited by non-selective antagonists of Ca^{2+} entry through voltage-sensitive channels, such as Cd^{2+} and Ni^{2+}; and that these same antagonists inhibit the ZP3-induced acrosome reaction and $Ca^{2+}{}_i$ response (Florman et al. 1992; Florman 1994; Arnoult et al. 1996a, b). Here, evidence for the presence of Ca^{2+} channels in mammalian sperm and the mechanism by which they are regulated during ZP3 signal transduction will be considered.

Voltage-Sensitive Calcium Channels Are Activated during the Acrosome Reaction

Two classes of voltage-sensitive Ca^{2+} channels have been described in somatic cells based on their biophysical characteristics. The first is high voltage-activated channels which require depolarizations to >-20 mV for activation, which conduct maximal currents at > +10 mV and may account for the L, N, P, Q and R-type currents. These channels have a heteromeric protein composition consisting of a pore-forming α_1 subunit and auxiliary regulatory $\alpha 2/\delta$ and β subunits (Hille 1992; Dunlap et al. 1995; Tsien et al. 1995). A family of α_1 genes (designated A, B, C, D, E and S) has been described and accounts for the variety of high voltage-activated channels (Hille 1992; Dunlap et al. 1995; Tsien et al. 1995). Many of these channels also can be differentiated from each other, as well as from the T-type low voltage-activated T channels, by pharmacological criteria (Tsien et al. 1995; Ertel, Ertel 1997). The second class contains a low voltage-activated channel that has a voltage threshold of approximately -60 mV, conducts maximal current at -30 to -20 mV and accounts for the T-type current (Tsien et al. 1995; Ertel, Ertel 1997). The α1G and α1H genes encode the neuronal T channel (Perez-Reyes et al. 1998), although the possible participation of other gene products in the generation of T currents cannot be excluded (Soong et al. 1993; Meir, Dolphin 1998). In this regard, pharmacological studies suggest that there may be a diversity of T channels (Ertel, Ertel 1997; Arnoult et al. 1998).

Sperm are transcriptionally and translationally inert. Ion channels that are utilized in sperm must therefore be synthesized during spermatogenesis. A T-type low voltage-activated current is the only Ca^{2+} current that can be detected in mouse spermatogenic cells by whole cell-patch clamp methods (Hagiwara, Kawa 1984; Arnoult et al. 1996a, 1997, 1998; Lievano et al. 1996; Santi et al. 1996). In addition, α1A and α1E genes may be expressed during mouse spermatogenesis, as indicated in RT-PCR experiments using spermatogenic cell RNA (Lievano et al. 1996). The Ca^{2+} currents that are carried by these gene products have not yet been assigned unequivocally. It has been suggested that the α1A gene product mediates both P- and Q-type high voltage-activated currents (Olivera et al. 1994; Tsien et al. 1995), possibly in association with distinct auxiliary subunits (Moreno et al. 1997). Similarly, recombinant α1E protein produces a current in cellular expression systems that was initially described as a low voltage-activated Ca^{2+} current (Soong et al. 1993; Bourinet et al. 1996; Piedras-Renteria et al. 1997). Further analysis has shown that the α1E gene product may participate in the generation of a high voltage-activated current that may be the R-type current (Williams et al. 1994; Varadi et al. 1996; Randall, Tsien 1997). While transcription of these Ca^{2+} channel genes occurs during spermatogenesis, the associated high voltage-activated currents are not observed (Hagiwara, Kawa 1984; Arnoult et al. 1996a, 1997, 1998; Lievano et al. 1996; Santi et al. 1996). Thus, functional expression of one type of voltage-sensitive Ca^{2+} channel can be detected during spermatogenesis. However, other channel types may be present either at low copy number or in an inactive form.

In contrast to spermatogenic cells, it is relatively difficult to form the high resistance (>1GΩ) seals required for patch clamp studies on sperm membranes. However, the presence and functional role of voltage-sensitive Ca^{2+} channels in sperm can be determined using ion-selective fluorescent probes, either in cell populations using spectrofluorometric approaches or in single cells using digital image processing-enhanced fluorescence microscopy. Such studies indicate that sperm maintain a free $Ca^{2+}{}_i$ of 50-100 nM; that, during capacitation, this increases to 125-175 nM with a monotonic time course (~ 0.5 nM min^{-1} in bovine sperm); and that these values then stabilize. Increases in $Ca^{2+}{}_i$ during capacitation occur uniformly throughout both the head and flagellar regions of the sperm (Florman 1994). The mechanisms that account for this physiological alteration in the sperm $Ca^{2+}{}_i$ setpoint have not been determined and the role of Ca^{2+} transport systems, including channels, remains uncertain. What is clear is that these minor increases in $Ca^{2+}{}_i$ do not initiate acrosome reactions.

Addition of ZP3 produces an additional increase in $Ca^{2+}{}_i$ to 300-500 nM within minutes, at which point $Ca^{2+}{}_i$ values either stabilize or slowly decline. Peak rates of $Ca^{2+}{}_i$ elevation during this response are ~150 nM min^{-1} and are associated with the initiation of acrosome reaction. The evidence that T channel activation is an essential component of the ZP3-activated signaling pathway in sperm may be summarized as follows.

T currents in spermatogenic cells are inhibited by several drugs and ions in the following order of potency (IC_{50}): PN200-110 (4×10^{-8} M) > nifedipine (4×10^{-7} M) > pimozide (4.6×10^{-7} M) > Ni^{2+} > (3.4×10^{-5} M) > verapamil (7×10^{-5} M) > amiloride (2.4×10^{-4} M) > Cd^{2+} (2.8×10^{4} M). In sperm both the elevations of $Ca^{2+}{}_i$ and the acrosome reaction induced by ZP3 are inhibited by these same compounds with similar potencies (Florman et al. 1992; Arnoult et al. 1996a, 1997, 1998). These observations suggest that ZP3 stimulation of sperm evokes a T-type Ca^{2+} current that is an essential component of the signal transduction mechanism regulating the acrosome reaction.

Utilization of a T current to mediate Ca^{2+} influx during acrosome reaction can be understood in light of the biophysical characteristics of this class of channels. T channels open following weak depolarizations, when there is a large electrochemical driving force for Ca^{2+} entry (Hille 1992). This provides a mechanism through which non-excitable cells, that do not produce the strong depolarizations that are associated with action potentials and that are required for the opening of high voltage-activated channels, are nevertheless able to transduce external signals into a $Ca^{2+}{}_i$ response.

In summary, T-type low voltage-activated C^{2+} channel genes are synthesized during mouse spermatogenesis, as revealed by direct examination using patch clamp methods. No other functional channel can be detected during spermatogenesis, although transcription of other channel genes has been reported. T channels are retained in sperm following terminal differentiation and play an essential role in the initiation of the acrosome reaction during zona pellucida contact. Since conductance through T channels is regulated by the membrane potential, it is then pertinent to ask how eggs depolarize the sperm membrane.

ZP3-Dependent Cation Channel: Mechanisms of Membrane Depolarization

The discussion to this point has focused on the identification of a T-type voltage-sensitive Ca^{2+} channel and the demonstration of its essential role in the ZP3-activated signal transduction mechanism. This suggests that a depolarization of sperm membrane may be an upstream element of this pathway. To address this question, researchers in the Florman laboratory incubated sperm with the cationic carbocyanine, 3,3'-dipropylthiadicarbocyanine iodide (or $DiSC_3(5)$), and with the anionic oxonol, bis-(1,3-diethylthiobarbituric acid) trimethine oxonol (or $DiSBAC_2(3)$). These compounds are fluorescent probes that redistribute according to the membrane potential. They have been used previously in a variety of cellular systems (Rottenberg 1979; Rink 1982; Ritchie 1984), including sperm (Rink 1977; Lee 1984, 1985; Babcock, Pfeiffer 1987; Gonzalez-Martinez, Darszon 1987; Florman et al. 1992; Espinosa, Darszon 1995).

Capacitated sperm populations from the mouse and the bull have resting membrane potentials of approximately -60 mV, due to permeability of several ions, including K^+ (Zeng et al. 1995) and possibly Ca^{2+} (Espinosa, Darszon 1995). During capacitation *in vitro*, the sperm membrane hyperpolarizes from ~-30 mV to ~-60 mV (Zeng et al. 1995). Approximately 50% of sperm incubated *in vitro* capacitate, as determined from chlortetracycline fluorescence (Zeng et al. 1995) and as estimated from the secretory response to ZP3 (Florman, First 1988; Zeng et al. 1995). The membrane potential of capacitated sperm cells has not yet been determined at the single cell level. However, it is likely that the reported membrane potential of capacitated sperm populations reflects contributions of the relatively depolarized uncapacitated fraction of sperm as well as the contribution of the highly hyperpolarized capacitated sperm.

ZP3 stimulation depolarizes sperm membrane potential to ~-25 mV (Arnoult et al. 1996b). This is a specific response to homologous ZP3 when tested in mouse and bull sperm. It is not observed when sperm are treated with other zona pellucida glycoproteins, with unrelated control glycoproteins or with ZP3 from another species.

Depolarization is due to the activation of a voltage-insensitive ion channel. The characteristics of this channel were determined with the use of potentiometic fluorescent probes (Arnoult et al. 1996b) and with the fura 2-fluorescence quenching probe (Florman 1994). This channel is permeable to Na^+ and to many divalent cations (Ba^{2+}, Ca^{2+}, Co^{2+}, Mn^{2+}, Ni^{2+}), but is impermeable to larger organic cations such as N-methyl-D-glucamine$^+$ and to trivalent cations such as La^{3+}. In addition, the channel appears to be completely impermeable to organic and inorganic anions. Thus, this ZP3-activated pathway has the characteristics of a poorly selective cation channel.

Cation channels gated by cyclic nucleotides (Weyand et al. 1994), by progesterone (Foresta et al. 1993) and by extracellular ATP (Foresta et al. 1996) may be present in mammalian sperm. Moreover, both the progestin- and the ATP-gated channels decrease sperm membrane potential. Comparisons between these channels are difficult due to insufficient data, the physiological state of the sperm cells and the use of different detector systems. Ion selectivity studies suggest that the ZP3-dependent cation channel may conduct certain divalent metal cations (Florman 1994; Arnoult et al. 1996b), whereas other sperm cation channels may be more selective for monovalent ions. Thus, there may be several cation-selective channels in mammalian sperm.

In summary, ZP3 activates a cation channel in sperm during gamete adhesion. This channel conducts monovalent and divalent cations but is not permeable to anions. The role of this channel and of the T-type Ca^{2+} channel are considered in the next section.

Model of ZP3 Signal Transduction

Elevation of $Ca^{2+}{}_i$ is a central event in the control of acrosome reaction. Figure 1 presents a model for the mechanisms by which zona pellucida adhesion regulates sperm $Ca^{2+}{}_i$ levels.

Given the polarization of the membrane maintained by sperm, activation of a cation channel is expected to produce an inward, depolarizing current. Under physiological conditions this cation channel produces a small, transient influx of Ca^{2+} into the sperm head. However, Ca^{2+} entry by this mechanism does not initiate acrosome reaction. The cation channel should also mediate an inward Na^+ current. Researchers in the Florman laboratory have suggested that the major function of these inward currents is to depolarize sperm membrane potential and thereby open voltage-sensitive T channels.

The calculated membrane potential of capacitated sperm is consistent with an essential role of a T-type channel. T-type channels, including that in male germ cells (Arnoult et al. 1996a), exhibit voltage-dependent inactivation at membrane potentials above ~-80 mV (Herrington, Lingle 1992; Hille 1992; McDonald et al. 1994). It is likely that capacitated sperm within a population have a membrane potential considerably more negative than the -60 mV reported for an heterogeneous sperm population (Zeng et al. 1995). Thus, sperm membrane is sufficiently polarized to permit activation of low voltage-activated T channels. ZP3 further depolarizes membrane from resting values to ~-25 mV (Arnoult et al. 1996b), a value that produces a peak T current in germ cells (Arnoult et al. 1996a; Lievano et al. 1996; Santi et al. 1996) and somatic cells (Hille 1992; McDonald et al. 1994). In contrast, the zona pellucida-dependent depolarization would not be expected to activate high voltage-activated Ca^{2+} currents.

Voltage-dependent inactivation of T currents occurs within 50-100 ms during depolarization (Arnoult et al. 1996a), thereby terminating the ZP3-induced Ca^{2+} influx. In contrast, the sustained Ca^{2+} elevation that is required for acrosome reaction is not detected for one to two minutes following zona pellucida stimulation (Florman et al. 1989; Florman 1994; Bailey, Storey 1994; Arnoult et al. 1996a, 1996b). These protracted Ca^{2+} responses require T channel activation and are inhibited by T channel antagonists (Arnoult et al. 1996a). However, they are not a direct measure of T channel function. In this regard, IP3 receptors are present in sperm acrosomes (Walensky, Snyder 1995) and release of sequestered Ca^{2+} triggers acrosome reaction (Meizel, Turner 1993; Walensky, Snyder 1995). A plausible mechanism of action is that T channel activation produces a transient Ca^{2+} influx that, in turn, initiates a downstream process of Ca^{2+}- induced Ca^{2+} release leading to acrosome reaction (Walensky, Snyder 1995;

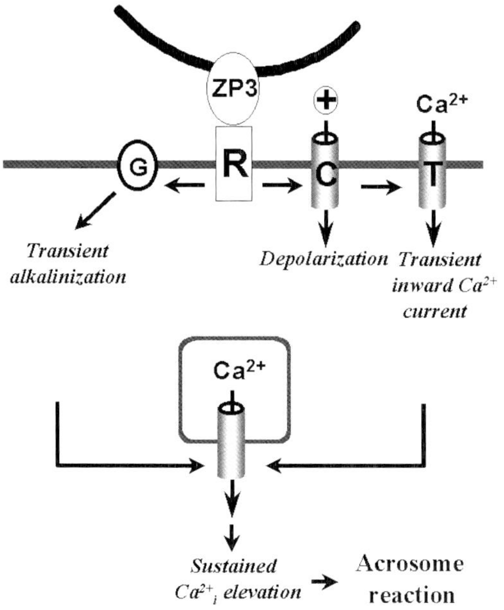

Figure 1. Model of the mechanism of sperm ion channel activation by the zona pellucida. During gamete adhesion, ZP3 in the zona pellucida associates with a sperm surface receptor (R) and produces two separate intracellular signals. One pathway consists of the activation of a cation channel (C) through a pertussis toxin-insensitive and voltage-insensitive mechanism. Sperm maintain an inwardly negative membrane potential and conductance through cation channels produces a depolarizing current. The result of membrane depolarization is the activation of a low voltage-activated T-type Ca^{2+} channel (T). The sperm T channel may additionally be modulated by its tryosine phosphorylation state during capacitation and ZP3 stimulation. The second signaling pathway results from ZP3 activation of a pH regulator, resulting in a transient alkalinization of internal pH. ZP3 activates this pH regulator through a pertussis toxin-sensitive mechanism that likely reflects mediation of the G_i-class of G proteins (G). Transient Ca^{2+} and pH elevations act in synergy to promote a sustained Ca^{2+} elevation, most likely through release from an IP3-sensitive intracellular store. This sustained Ca^{2+} leads to acrosome reaction. This pathway depends on T channel activation, and sustained Ca^{2+} elevations are inhibited by T channel antagonists.

Arnoult et al. 1996a). Other intracellular effectors of ZP3 action, such as elevated pH_i (Florman et al. 1989, 1992; Lee, Storey 1989; Arnoult et al. 1996b) and IP3 (Tomes et al. 1996), may also contribute to Ca^{2+} pool mobilization. In fact, internal alkalinization is required for robust Ca^{2+} elevations in response to ZP3 (Arnoult et al. 1996b) or membrane depolarization (Babcock, Pfeiffer 1987; Arnoult et al. 1996b).

Regulation of ZP3 Signal Transduction during Capacitation

Mammalian sperm require a period of capacitation, or functional reprogramming, to exhibit fertilizing ability (Yanagimachi 1994). Capacitation is also required for the induction of the acrosome reaction by ZP3 (Ward, Storey 1984; Florman, First 1988). This observation has lead to the suggestion that the ZP3 signal transducing mechanisms are regulated during capacitation.

In fact, sperm T channels may be modulated in several ways during capacitation. First, the conductance state is controlled by membrane potential. Voltage-dependent inactivation of T channels occurs with a low voltage threshold. Inactivation of the germ cell channel is first observed at ~-70 mV and is complete by -55 mV (Arnoult et al. 1996b). Importantly, the probability of the T channel opening directly from the inactivated state is low (Hille 1992).

The membrane of uncapacitated sperm is relatively depolarized, with values of ~-30 mV calculated for both mouse and bull sperm *in vitro* (Zeng et al. 1995). Under these conditions there is a maximal voltage-dependent inactivation of T channels (Arnoult et al. 1996a) and Ca^{2+} current through this pathway is minimal. Thus, conditions that may further decrease the membrane potential of uncapacitated sperm will not enhance T currents and are unlikely to initiate acrosomal secretion. During capacitation, sperm membrane becomes hyperpolarized due to an enhanced contribution of K^+ permeability (Zeng et al. 1995). Hyperpolarization relieves steady-state inactivation of T channels. As a result, channels tend to dwell in the closed state, from which opening occurs with high probability following depolarization (Hille 1992). Thus, hyperpolarization may act to prime T channels for subsequent activation by ZP3.

T channel activation is regulated in a second manner during capacitation. A ZP3-activated cation channel produces an inward current involving Ca^{2+} and probably also Na^+. This depolarizing current provides the coupling between gamete contact and T channel activation. Cation channel opening in response to sperm-zona pellucida interaction was assessed in single sperm using Ca^{2+}-selective fluorescent dyes. The sensitivity of this cation channel to evoked opening in response to zona pellucida stimuli is low in uncapacitated sperm and increases during capacitation *in vitro* (Florman 1994). Thus, the ability of ZP3 to depolarize sperm membrane, and hence to activate the voltage-sensitive T channel, is modulated at the level of an upstream cation channel during capacitation.

Finally, and moving into the realm of speculation, T channels may also be modulated during capacitation by tyrosine phosphorylation. Recently, it has been shown that protein tyrosine phosphorylation of either the T channel or of a channel regulator decreases current through the channel. Conversely, protein tyrosine phosphatase action enhances T current (Arnoult et al. 1997). This type of modulation has not yet been observed in somatic cell T channels and may reflect a novel regulatory mechanism. Alternatively, protein tyrosine phosphorylation-dependent modulation may be more readily detectable in the relative simple germ cell model system, where other Ca^{2+} currents are absent.

In this regard, capacitation is associated with a wave of protein tyrosine phosphorylation (Visconti et al. 1995a, b). It is tempting to speculate that the T channel, or a channel regulator, is a substrate for a capacitation-dependent tyrosine kinase. The resultant negative modulation of T currents would reduce Ca^{2+} currents and could minimize spontaneous acrosome reaction. This suggestion implies that ZP3 activates a tyrosine phosphatase activity during the initiation of acrosome reactions. Further studies are required to test this suggested mechanism.

The mechanisms by which K^+ permeability and ZP3-regulated cation channel conductance are regulated during capacitation are not understood. Plausible and intriguing mechanisms may be proposed based on altered protein kinase activity during capacitation (Yanagimachi 1994; Visconti et al. 1995a, b) and on the known effect of post-translational modifications on somatic cell K^+ channel function (Siegelbaum 1994; Jonas, Kaczmarek 1996). Future efforts will focus on testing such proposed mechanisms in sperm. Thus, sperm T channels represent both an essential step in the ZP3 signal transduction pathway and a site of regulation during capacitation.

References

Arnoult C, Cardullo RA, Lemos JR, Florman HM. Egg-activation of sperm T-type Ca^{2+} channels regulates acrosome reactions during mammalian fertilization. Proc Natl Acad Sci USA 1996a; 93:13004-13009.

Arnoult C, Zeng Y, Florman HM. ZP3-dependent activation of sperm cation channels regulates acrosomal secretion during mammalian fertilization. J Cell Biol 1996b; 134:637-645.

Arnoult C, Lemos JR, Florman HM. Voltage-dependent modulation of T-type Ca^{2+} channels by protein tyrosine phosphorylation. EMBO J 1997; 16:1593-1599.

Arnoult C, Villaz M, Florman HM. Pharmacological properties of the T-type calcium current of mouse spermatogenic cells. Mol Pharmacol 1998; 53:1104-1111.

Babcock DF, Pfeiffer DR. Independent elevation of cytosolic $[Ca^{2+}]$ and pH of mammalian sperm by voltage-dependent and pH-sensitive mechanisms. J Biol Chem 1987; 262:15041-15047.

Babcock DF, Herrington J, Goodwin PC, Hille B. Mitochondrial participation in the intracellular Ca^{2+} network. J Cell Biol 1997; 136:833-844.

Bailey JL, Storey BT. Calcium influx into mouse spermatozoa activated by solubilized mouse zona pellucida, monitored with the calcium fluorescent indicator, fluo-3. Inhibition of the influx by three inhibitors of the zona pellucida induced acrosome reaction: tyrphostin A48, pertussis toxin, and 3'-quinuclidinyl benzilate. Mol Reprod Dev 1994; 39:297-308.

Bleil JD, Wassarman PM. Sperm-egg interactions in the mouse: sequence of events and induction of the acrosome reaction by a zona pellucida glycoprotein. Dev Biol 1983; 95:317-324.

Bourinet E, Zamponi GW, Stea A, Soong TW, Lewis BA, Jones LP, Yue DT, Snutch TP. The α_{1E} calcium channel exhibits permeation properties similar to low voltage-activated calcium channels. J Neurosci 1996; 16:4983-4993.

Carafoli E. The Ca^{2+} pump of the plasma membrane. J Biol Chem 1992; 267:2115-2118.

Clapham DE. Calcium signaling. Cell 1995; 80:259-268.

Dan JC. Studies on the acrosome. III. Effect of calcium deficiency. Biol Bull 1954; 107:335-349.

Dan JC. The acrosome reaction. Inter Rev Cytol 1956; 5:365-393.

Douglas WW. Stimulus-secretion coupling: the concept and clues from chromaffin and other cells. Brit J Pharm 1968; 34:451-474.

Dunlap K, Luebke JI, Turner TJ. Exocytotic Ca^{2+} channels in mammalian central neurons. Trends Neurosci 1995; 18:89-98.

Ertel SI, Ertel EA. Low-voltage-activated T-type Ca^{2+} channels. Trends Pharmacol Sci 1997; 18:37-42.

Espinosa F, Darszon A. Mouse sperm membrane potential: changes induced by Ca^{2+}. FEBS Lett 1995; 372:119-125.

Florman HM. Sequential focal and global elevations of sperm intracellular Ca^{2+} are initiated by the zona pellucida during acrosomal exocytosis. Dev Biol 1994; 165:152-164.

Florman HM, First NL. The regulation of acrosomal exocytosis. I. Sperm capacitation is required for the induction of acrosome reactions by the bovine zona pellucida *in vitro*. Dev Biol 1988; 128:453-463.

Florman HM, Tombes RM, First NL, Babcock DF. An adhesion-associated agonist from the zona pellucida activates G protein-promoted elevations of internal Ca^{2+} and pH that mediate mammalian sperm acrosomal exocytosis. Dev Biol 1989; 135:133-146.

Florman HM, Corron ME, Kim TDH, Babcock DF. Activation of voltage-dependent calcium channels of mammalian sperm is required for zona pellucida-induced acrosomal exocytosis. Dev Biol 1992; 152:304-314.

Foresta C, Rossato M, Di Virgilio F. Ion fluxes through the progesterone-activated channel of the sperm plasma membrane. Biochem J 1993; 294:279-283.

Foresta C, Rossato M, Chiozzi P, Di Virgilio F. Mechanism of human sperm activation by extracellular ATP. Am J Physiol 1996; 270:C1709-C1714.

Gonzalez-Martinez M, Darszon A. A fast transient hyperpolarization occurs during the sea urchin sperm acrosome reaction induced by egg jelly. FEBS Lett 1987; 218:247-250.

Green DPL. The induction of the acrosome reaction in guinea pig sperm by the divalent metal cation ionophore A23187. J Cell Sci 1978; 32:137-151.

Hagiwara S, Kawa K. Calcium and potassium currents in spermatogenic cells dissociated from rat seminiferous tubules. J Physiol 1984; 356:135-149.

Herrington J, Lingle CJ. Kinetic and pharmacological properties of low voltage-activated calcium current in rat clonal (GH_3) pituitary cells. J Neurophysiol 1992; 68:213-232.

Hille B. Ionic Channels of Excitable Membranes. 2nd ed. Sinauer Associates Inc, Sunderland MA, 1992.

Ichas F, Jouaville LS, Mazat J-P. Mitochondria are excitable organelles capable of generating and conveying electrical and calcium signals. Cell 1997; 89:1145-1153.

Jonas EA, Kaczmarek LK. Regulation of potassium channels by protein kinases. Curr Opin Neurobiol 1996; 6:318-323.

Katz B, Miledi R. The timing of calcium action during neuromuscular transmission. J Physiol 1967; 189:535-544.

Konishi M, Olson A, Hollingworth S, Baylor SM. Myoplasmic binding of fura-2 investigated by steady-state fluorescence and absorbance measurements. Biophys J 1988; 54:1089-1104.

Lauger P. Electrogenic Ion Pumps. 2nd ed. Sinauer Associates Inc, Sunderland MA, 1991.

Lee HC. Sodium and proton transport in flagella isolated from sea urchin spermatozoa. J Biol Chem 1984; 259:4957-4963.

Lee HC. The voltage-sensitive Na^+/H^+ exchange in sea urchin spermatozoa flagellar membrane vesicles studied with an entrapped pH probe. J Biol Chem 1985; 260:10794-10799.

Lee MA, Storey BT. Endpoint of the first stage of zona pellucida-induced acrosome reaction in mouse spermatozoa characterized by acrosomal H^+ and Ca^{2+} permeability: population and single cell kinetics. Gam Res 1989; 24:303-326.

Lievano A, Santi CM, Serrano J, Trevino CL, Bellve AR, Hernandez-Cruz A, Darszon A. T-type Ca^{2+} channels and α_{1E} expression in spermatogenic cells, and their possible relevance to the sperm acrosome reaction. FEBS Lett 1996; 388:150-154.

Loew LM, Carrington W, Tuft RA, Fay FS. Physiological cytosolic Ca^{2+} transients evoke concurrent mitochondrial depolarizations. Proc Natl Acad Sci USA 1994; 91:12579-12583.

McDonald TF, Pelzer S, Trautwein W, Pelzer DJ. Regulation and modulation of calcium channels in cardiac, skeletal, and smooth muscle cells. Physiol Rev 1994; 74:365-507.

Meir A, Dolphin AC. Known calcium channel α_1 subunits can form low threshold small conductance channels with similarities to native T-type channels. Neuron 1998; 20:341-351.

Meizel S, Turner KO. Initiation of the human sperm acrosome reaction by thapsigargin. J Exp Zool 1993; 267:350-355.

Moreno H, Rudy B, Llinas R. β subunits influence the biophysical and pharmacological differences between P- and Q-type calcium currents expressed in a mammalian cell line. Proc Natl Acad Sci USA 1997; 94:14042-14047.

Olivera BM, Miljanich GP, Ramachandran J, Adams ME. Calcium channel diversity and neurotransmitter release: the omega-conotoxins and omega-agatoxins. Ann Rev Biochem 1994; 63:823-867.

Penner RD, Neher E. The role of calcium in stimulus-secretion coupling in excitable and non-excitable cells. J Exp Biol 1988; 139:329-345.

Perez-Reyes E, Cribbs LL, Daud A, Lacerda AE, Barclay J, Williamson MP, Fox M, Rees M, Lee J-H. Molecular characterization of a neuronal low-voltage-activated T-type calcium channel. Nature 1998; 391:896-900.

Piedras-Renteria ES, Chen C-C, Best PM. Antisense oligonucleotides against rat brain α_{1E} DNA and its atrial homologue decrease T-type calcium current in atrial myocytes. Proc Natl Acad Sci USA 1997; 94:14936-14941.

Randall AD, Tsien RW. Contrasting biophysical and pharmacological properties of T-type and R-type calcium channels. Neuropharm 1997; 36:879-893.

Rink TJ. Membrane potential of guinea pig spermatozoa. J Reprod Fertil 1977; 51:155-157.

Rink TJ. Measurement of membrane potential with chemical probes. Lipid Memb Biochem 1982; B423:1-29.

Ritchie RJ. A critical assessment of the use of lipophilic cations as membrane potential probes. Prog Biophys Molec Biol 1984; 43:1-32.

Rizzuto R, Simpson AWM, Brini M, Pozzan T. Rapid changes of mitochondrial Ca^{2+} revealed by specifically targeted recombinant aequorin. Nature 1992; 358:325-327.

Rottenberg H. The measurement of membrane potential and (d)pH in cells, organelles, and vesicles. Meth Enzymol 1979; 55:547-569.

Santi CM, Darszon A, Hernandez-Cruz A. A dihydropyridine-sensitive T-type Ca^{2+} current is the main Ca^{2+} current carrier in mouse primary spermatocytes. Amer J Physiol 1996; 271:C1583-C1593

Schackmann RW, Shapiro BM. A partial sequence of ionic changes associated with the acrosome reaction of *Strongylocentrotus purpuratus*. Dev Biol 1981; 81:145-154.

Schackmann RW, Eddy EM, Shapiro BM. The acrosome reaction of *Strongylocentrotus purpuratus* sperm. Ion requirements and movements. Dev Biol 1978; 65:483-495.

Siegelbaum S. Ion channel control by tyrosine phosphorylation. Curr Biol 1994; 4:242-245.

Singh JP, Babcock DF, Lardy HA. Increased calcium ion influx is a component of capacitation of spermatozoa. Biochem J 1978; 172:549-556.

Soong TW, Stea A, Hodson CD, Dubel SJ, Vincent SR, Snutch TP. Structure and functional expression of a member of the low voltage-activated calcium channel family. Science 1993; 260:1133-1136.

Summers RG, Talbot P, Keough EM, Hylander BL, Franklin LE. Ionophore A23187 induces acrosome reactions in sea urchin and guinea pig spermatozoa. J Exp Zool 1976; 196:381-385.

Tilney LG, Kiehart D, Sardet C, Tilney M. The polymerization of actin. IV. The role of Ca^{2+} and H^+ in the assembly of actin and in membrane fusion in the acrosomal reaction of echinoderm sperm. J Cell Biol 1978; 77:536-550.

Tomes CN, McMaster CR, Saling PM. Activation of mouse sperm phosphatidylinositol-4,5 bisphosphate-phospholipase C by zona pellucida is modulated by tyrosine phosphorylation. Mol Reprod Dev 1996; 43:196-204.

Tsien RW, Tsien RY. Calcium channels, stores, and oscillations. Ann Rev Cell Biol 1990; 6:715-760.

Tsien RW, Lipscombe D, Madison D, Bley K, Fox A. Reflections of Ca^{2+} channel diversity, 1988-1994. Trends Neurosci 1995; 18:52-54.

Varadi G, Mori Y, Mikala G, Schwartz A. Molecular determinants of Ca^{2+} channel function and drug action. Trends Pharmacol Sci 1996; 16:43-49.

Visconti PE, Bailey JL, Moore GD, Pan D, Olds-Clarke P, Kopf GS. Capacitation of mouse spermatozoa. I. Correlation between the capacitation state and protein tyrosine phosphorylation. Development 1995a; 121:1129-1137.

Visconti PE, Moore GD, Bailey JL, Leclerc P, Connors SA, Pan D, Olds-Clarke P, Kopf GS. Capacitation of mouse spermatozoa. II. Protein tyrosine phosphorylation and capacitation are regulated by a cAMP-dependent pathway. Development 1995b; 121:1139-1150.

Walensky LD, Snyder SH. Inositol 1,4,5-trisphosphate receptors selectively localized to the acrosomes of mammalian sperm. J Cell Biol 1995; 130:857-869.

Ward CR, Kopf GS. Molecular events mediating gamete activation. Dev Biol 1993; 158:9-34.

Ward CR, Storey BT. Determination of the time course of capacitation in mouse spermatozoa using a chlortetracycline fluorescence assay. Dev Biol 1984; 104:287-296.

Weyand I, Godde M, Frings S, Weiner J, Muller F, Altenhofer W, Hatt H, Kaupp UB. Cloning and functional expression of a cyclic-nucleotide-gated channel from mammalian sperm. Nature 1994; 368:859-863.

Williams ME, Marubio LM, Deal CR, Hans M, Brust PF, Philipson LH, Miller RJ, Johnson EC, Harpold MM, Ellis SB. Structure and functional characterization of neuronal α_{1E} calcium channel subtypes. J Biol Chem 1994; 269:22347-22357.

Yanagimachi R. Mammalian fertilization. In: (Knobil E, Neill JD, eds) The Physiology of Reproduction. 2nd ed. New York: Raven Press Ltd, 1994, pp189-317.

Yanagimachi R, Usui N. Calcium dependence of the acrosome reaction and activation of guinea pig spermatozoa. Exp Cell Res 1974; 89:161-174.

Zeng Y, Clark EN, Florman HM. Sperm membrane potential: hyperpolarization during capacitation regulates zona pellucida-dependent acrosomal secretion. Dev Biol 1995; 171:554-563.

Zhou Z, Neher E. Mobile and immobile calcium buffers in bovine adrenal chromaffin cells. J Physiol 1993; 469:245-273.

18 *Defining the Biochemical Mechanisms of Sperm-Zona Pellucida Binding*

Catherine D. Thaler
University of Central Florida

Many complex interactions between gametes are required to complete fertilization. Early observations of mammalian fertilization elucidated the sequence of interactions involved in sperm penetration of the egg extracellular matrices (Gwatkin, Williams 1977; Hartmann et al. 1972; Inoue, Wolf 1975). This discussion will focus on mouse sperm-egg interactions at the zona pellucida (ZP) matrix. Sperm bind to the ZP matrix by their anterior head. Following binding to the ZP, sperm undergo an exocytotic event termed the acrosome reaction (Saling et al. 1979; Florman, Storey 1982). The acrosome reaction releases hydrolytic enzymes from a large secretory vesicle overlying the anterior head of the sperm and also exposes a new membrane surface, formerly the inner acrosomal membrane, as the delimiting membrane of the sperm head. Acrosome reacted sperm proceed through the ZP to arrive at the egg plasma membrane, where they can fuse and complete fertilization. Figure 1 shows capacitated mouse sperm binding to an intact zona pellucida isolated from a mouse egg.

These observations generated a number of questions concerning the interactions of the sperm and the ZP matrix. First, what components are responsible for mediating these interactions? What molecules on the surface of sperm and ZP interact with each other? What components are responsible for adhesion of sperm to this matrix? Is there a signal generated by the ZP that induces acrosomal exocytosis? If so, what ZP molecule is responsible for signaling? What sperm molecules receive the signal? How is it transduced in the cell?

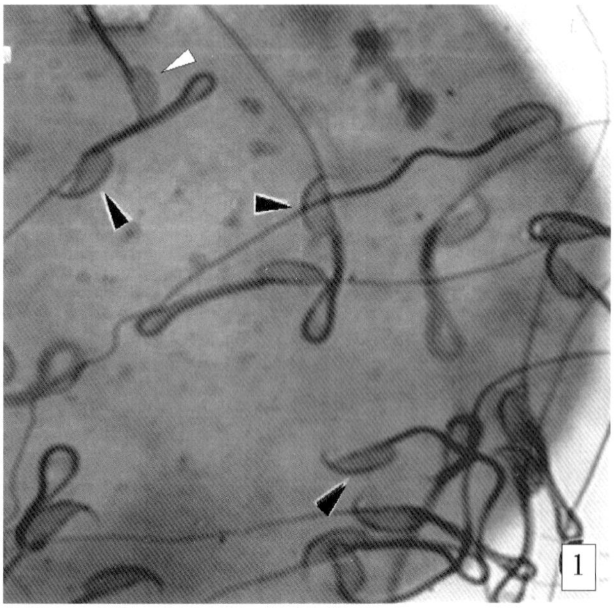

Figure 1. Sperm-zona pellucida binding. Light micrograph showing mouse sperm binding to an intact, isolated mouse zona pellucida. Sperm have been stained with Coomassie Blue to reveal acrosomal status. Sperm containing a darkly stained acrosomal crescent are acrosome intact (black arrowheads). One acrosome reacted sperm can be identified (white arrowhead) due to lack of staining over the acrosomal crescent.

Many of these questions remain unresolved, including the question of sperm proteins which mediate adhesion and signal transduction at the surface of the ZP. Science does know something about the zona pellucida components. The mouse ZP consists of three glycoproteins, ZP1, ZP2 and ZP3 (Bleil, Wassarman 1980a). Sperm that were preincubated with the solubilized ZPs could not bind to intact ZPs of ovulated eggs. Subsequent studies demonstrated that isolated ZP3 could block sperm binding to the intact ZP to the same extent as the whole solubilized ZP and that ZP1 or ZP2 were not able to alter the behavior of sperm in regard to ZP binding (Bleil, Wassarman 1980b). In addition, isolated, solubilized ZP3 was able to stimulate acrosomal exocytosis in capacitated mouse sperm to an extent similar to that with solubilized whole ZP glycoproteins or the calcium ionophore A23187 (Bleil, Wassarman 1983; Florman et al. 1984). Further studies demonstrated that one or more of the O-linked oligosaccharides on ZP3 contained the bioactivity of this molecule (Florman, Wassarman 1985). Studies using the isolated ZP glycoproteins have shown that ZP3 binds only to intact sperm and ZP2 binds only to acrosome reacted sperm (Bleil, Wassarman 1986; Bleil et al. 1988; Mortillo, Wassarman 1991).

Many sperm polypeptides have been proposed to interact with ZP3 and serve as the sperm receptor for ZP3. The studies cited above demonstrated that ZP3 was sufficient to block sperm binding to the ZP as well as induce the sperm acrosome reaction. These data imply that sperm must have a specific molecular mechanism, a receptor, for adhesion and signaling in response to ZP3. It has been generally assumed that a single sperm plasma membrane receptor is responsible for both these activities. However, none of the studies conducted to date have addressed this possibility and it is entirely consistent with current data that adhesion and signaling could be mediated by different sperm proteins. In fact, many sperm polypeptides have been shown to have some ability to interact with ZP3, including the β1,4-galactosyltransferase (Gong et al. 1995; Lu, Shur 1997), the p95/116 hexokinase (Leyton et al. 1992; Kalab et al. 1994), LL95 (Leyton et al. 1995), sp56 (Bookbinder et al. 1995; Foster et al. 1997), zonadhesin (Gao, Garbers 1998; Hardy, Garbers 1995) and ZRK (Burks et al. 1995). Other plasma membrane components have been implicated indirectly; for example, fucosyltransferase (Apter et al. 1988; Thaler, Cardullo 1996b). Currently, it is not possible to make biochemical comparisons of the interactions of these polypeptides with ZP3, since each polypeptide has been characterized independently and by different techniques. Therefore, the understanding of the functional role of any of these polypeptides is limited. In order to understand the role of these polypeptides it must be known how each of them interacts with ZP3. What are the relative affinities of the interactions? Which polypeptides are capable of generating a signal in response to binding ZP3? Could some be involved in adhesion while others are responsible for signaling acrosomal exocytosis?

Current knowledge of sperm-ZP interactions, based primarily on microscopic observations and the assays testing the function of the isolated ZP glycoproteins, has, however, led to the following general model of mouse sperm interactions with the mouse zona pellucida. Acrosome intact sperm bind to the ZP via carbohydrate groups of ZP3, and this binding signals the acrosome reaction. Acrosome reacted sperm cannot bind to ZP3, but interact with ZP2, which maintains their contact with the ZP matrix during their penetration of the matrix. Given the fact that new membrane surfaces are exposed on sperm following the acrosome reaction and that intact and acrosome reacted sperm have different ZP binding properties, there must be separate receptors on mouse sperm mediating the sperm interactions with ZP3 and ZP2. The identity of the polypeptides comprising these receptors is not known.

The three ZP glycoproteins are the products of three related genes that appear to be orthologous throughout mammalian species (Harris et al. 1994). But, it is becoming obvious that there may be differences among mammals in the ZP glycoproteins that are used for adhesion to the zona pellucida and signaling acrosomal exocytosis. Hamster and mouse sperm appear to use ZP3 for both adhesion and signaling (Moller et al. 1990), pig to use both ZP3 and ZP1 (Yurewicz 1998), while in rabbit, r75 (rabbit ZP1) appears to be sufficient for sperm-ZP interactions (Lee et al. 1993). Despite such differences, it is likely that information garnered from one system will be valuable in understanding patterns of fertilization in all mammals. This discussion will focus on work done in the mouse system.

What is the current understanding of the molecular interactions, the receptor ligand interactions, that mediate sperm adhesion to the ZP as well as signal acrosomal exocytosis? How can biochemical approaches help researchers understand the interaction between sperm and ZP molecules, and how can this information help them identify sperm receptors for the zona pellucida?

Modeling Sperm-Zona Pellucida Interactions

It is evident that isolated, soluble mouse ZP3 is sufficient to block sperm binding to intact ZPs and that it is also sufficient to induce acrosomal exocytosis of epididymal mouse sperm (Bleil, Wassarman 1980b; Florman et al. 1984). Only ZP3 is sufficient for these activities and neither isolated ZP1 or ZP2 were able to block binding or stimulate acrosome reactions. Further, it is oligosaccharide groups on ZP3 that are responsible for this interaction (Florman, Wassarman 1985).

ZP3 is the ligand which signals a biochemical response in sperm. This, by definition, postulates the presence of a receptor on the sperm, a protein that is capable of interacting with ZP3 and transducing a signal. Subsequently, the receptor will be referred to as ZP3R, although it should be kept in mind that the functional receptor may contain more than one sperm polypeptide. Many sperm proteins have been proposed to function as the ZP3R and the true composition of this receptor remains controversial. However, the ligand ZP3 can be used to understand and characterize the interactions with its receptor on the sperm by performing traditional biochemical assays to quantify the affinity, number of interactions and so forth. These data can ultimately be used as criteria to identify the ZP3R polypeptides.

First, the interaction must be modeled. The initial model, as illustrated in Figure 2, is based on the question, "Is ZP3-ZP3R binding a simple bimolecular interaction or is it complex?" What is the predicted behavior of the interaction under equilibrium or kinetic analysis for each of these models? A simple bimolecular interaction would mean that one ZP3 ligand binds to one sperm ZP3R polypeptide. A simple bimolecular interaction requires that there be only one species of receptor and ligand, one affinity class of interaction and no cooperativity. This model predicts specific biochemical behaviors in receptor-ligand binding assays. If equilibrium assays are conducted using greater and greater concentrations of ligand in order to saturate the binding, and then the amount of bound ligand at the different concentrations of available ligand is measured, the data would produce a rectangular hyperbola; that is, $B = B_{max}[L]/(K_d + [L])$. If kinetic experiments are conducted to measure the timing of the dissociation of the receptor-ligand complex, and the amount bound over time is plotted, the simple bimolecular interaction predicts a first order exponential decay; that is, $B(t) = B_0 e^{-k_{off}t}$).

If the receptor ligand interaction is complex, there are many possibilities, as far as what would actually be seen in the data obtained from binding studies. These studies can differentiate between "simple" and "not simple"; that is, if the data do not fit the model for a simple bimolecular interaction, the molecular events are not of the simple bimolecular variety, but are by definition complex. Biochemically, the idea of complex interactions incorporates many possibilities, including a multivalent ligand or receptor, as well as heterogeneous receptors or ligands, or cooperativity. Possible sources of complexity will be addressed later in this discussion. For complex interactions, then, the model predicts that equilibrium studies will not yield a simple hyperbola but

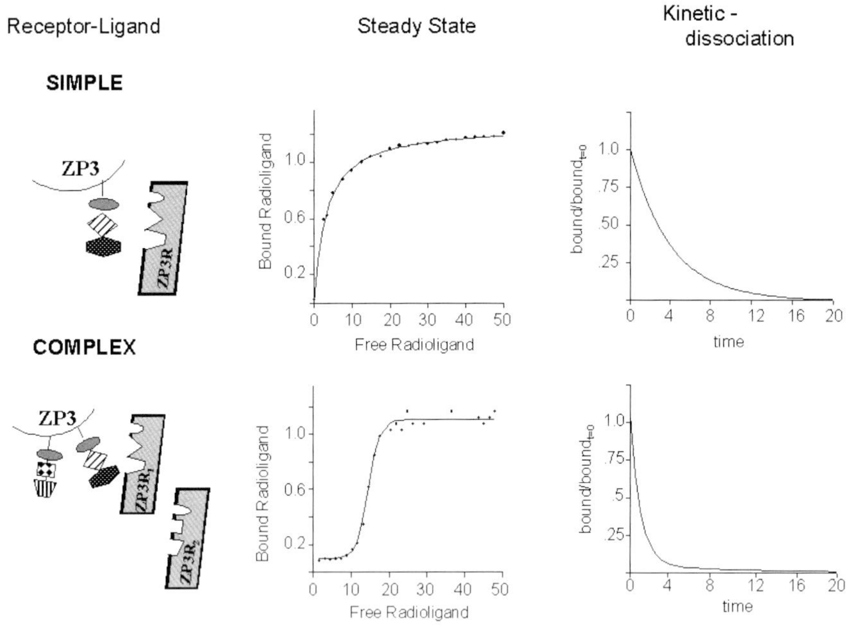

Figure 2. Models of sperm-zona pellucida binding. Simple: a model proposing a simple bimolecular interaction would feature univalent ZP3 and ZP3R. There are no cooperative interactions with other receptor-ligand pairs. Complex: complex interactions could involve many features of receptor and ligand, including multivalence and heterogeneity of receptor and/or ligand. One possibility, shown in this figure, is different ligand oligosaccharides on ZP3, each with a unique ZP3R on the sperm plasma membrane.

some more complex function, which will plateau at a value representing the total number of ZP3 binding sites (B_{max}). Kinetic studies, as well, should produce a more complex function for the dissociation reaction. An initial prediction would be that the dissociation would be a second order or greater exponential decay function.

These models can be tested by actually conducting such receptor ligand studies with sperm and ZP glycoproteins. Such studies have been published (Thaler, Cardullo 1996a) and will be briefly summarized here to illustrate the answer to the question, "Is sperm ZP binding simple or complex?"

Sperm-Zona Pellucida Binding Is Complex

One great advantage in studying the sperm-ZP interactions is that the identity of the ligand is known. The ligand can be isolated, labeled and used as a marker to follow the interaction of the receptor ligand pair. The labeled ligand allows researchers to quantify the interaction and the data thus obtained can be used to evaluate and, potentially, revise the model of this interaction.

The receptor ligand binding studies described here used fixed, capacitated intact mouse sperm to provide a population of the ZP3R and ^{125}I-ZPs to provide a population of the ligand. Receptor and ligand are incubated together under various conditions and subsequently the amount of ligand bound to the receptor population is measured. Given previous work which demonstrated that only ZP3 binds to acrosome intact sperm (Bleil, Wassarman 1986; Mortillo, Wassarman 1991), the radioactivity associated with the sperm is assumed to be due to ZP3-ZP3R binding and the number of ZP3 molecules bound per sperm is calculated accordingly. From these data, the affinity of the interaction (EC_{50} or K_d), the number of binding sites and some information about the complexity of the interactions can be determined.

In equilibrium binding studies, aliquots of sperm were incubated in the presence of ^{125}I-ZPs until the system reached equilibrium. ZP3R samples were incubated with a range of ^{125}I-ZP concentrations. As greater and greater amounts of ^{125}I-ZPs were added, the ZP3R population became saturated; that is, all the receptors were occupied by their ligand. From these data the number of binding sites per sperm was calculated and the average affinity of the ZP3-ZP3R interaction was measured. This experimental approach also allows something about the complexity of the interactions between ZP3 and its receptor ZP3R on the sperm to be determined. The sperm-ZP binding curve obtained from such studies (Fig. 3) shows a distinct inflection at lower concentrations of ZPs. The data do not fit to a rectangular hyperbola, thereby arguing against a simple bimolecular interaction for this receptor-ligand interaction.

In kinetic studies, the amount of receptor-ligand complexes which form are measured over time under different experimental conditions. These types of experiments can be used to determine the intrinsic rate constants, both association rate and dissociation rate, of the ZP3-ZP3R interaction. In order to measure dissociation of ZP3-ZP3R complexes, sperm were incubated with ^{125}I-ZPs to reach equilibrium binding and subsequently the mixture was greatly diluted to favor the dissociation of the complex. Samples were processed at various times after dilution and the amount of ligand remaining bound to the sperm was measured and compared to the initial amount bound at equilibrium. These data were

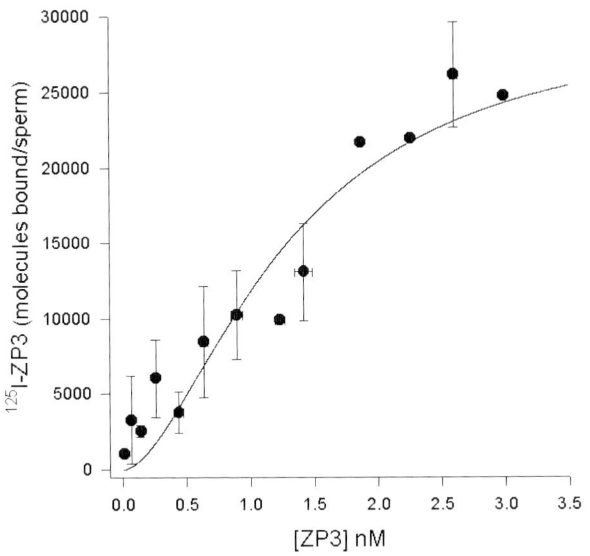

Figure 3. Saturation analysis of sperm-ZP3 binding. Equilibrium binding saturation experiments determined binding affinity, complexity, and B_{max} for sperm-ZP binding. Free ZP3 concentration was calculated from ^{125}I-ZP3 contribution to total ^{125}I-ZPs. Bound radioactivity was assumed to be ZP3 and converted to numbers of ZP3 molecules per sperm. The saturation binding isotherm demonstrates that there are approximately 30,000 ZP3 binding sites per sperm. The binding affinity, EC_{50} (or K_d) is 1.29 nM. The Hill coefficient for the binding isotherm is 1.72, which results in the inflection in the lower part of the curve. A Hill coefficient greater than 1 indicates a deviation from simple bimolecular interactions. The data are binned data points from four independent experiments, plotted +/- SEM for both bound and free ZP3. (r^2=0.92). (Reprinted with permission of the ASBMB, 1996.)

used to follow the dissociation rate of the receptor ligand complexes. The data obtained for sperm-ZP interactions suggested that these interactions are complex. The data displayed a second order exponential decay (Fig. 4), suggesting that there is both a high and low affinity interaction in this system. The low affinity interaction rapidly dissociates following dilution of the complex, with a half time of 4.3 minutes. There is, however, a high affinity interaction which dissociates much more slowly, with a half time of 300.8 minutes. Again, these data are not consistent with a simple bimolecular interaction, but suggest the presence of multiple interactions.

These studies, which quantified the receptor-ligand binding of ZP3 and its receptor on mouse sperm, demonstrated that the interaction is complex (Thaler, Cardullo 1996a). When the data obtained in these binding studies are compared to models, these data do not support the model of a simple bimolecular interaction, but are consistent with a complex binding event.

Possible Sources of Complexity

These initial binding studies demonstrated that sperm-ZP interactions are complex. There are several possible sources of complexity in the sperm-ZP3 system: the ligand, the receptor or both ligand and receptor.

The ligand. ZP3 appears to be the primary ligand for sperm binding in the mouse. ZP3 oligosaccharides provide the ligand groups. Sperm-ZP binding complexity could be the result of a multivalent ligand; that is, the presence of multiple ligand oligosaccharides per ZP3 molecule (Kinloch et al. 1995). The possibility of a multivalent ZP3 has been suggested by experiments demonstrating that monovalent ZP3 glycopeptides were able to bind to acrosome intact sperm but could not induce acrosomal exocytosis unless crosslinked (Leyton, Saling 1989). Complexity could also arise from heterogeneity of ZP3 molecules, either from having more than one species of ligand oligosaccharide present on each ZP3 glycoprotein or from having different species of ligand oligosaccharides present on different ZP3 molecules. In considering this latter possibility, it is important to recognize that if different subpopulations of ZP3 do exist, they may also be differentially distributed in the zona pellucida matrix. This spatial restriction of ligand is not an aspect of sperm-ZP interactions that has been addressed using either the intact or soluble ZP3 binding assays. Investigating the functional role of differential distribution of a heterogeneous ligand would be difficult to address using assays where ligands or competitors are presented to the sperm in soluble form. Therefore, while future studies may be able to determine whether ZP3 ligand oligosaccharides are heterogeneous and the

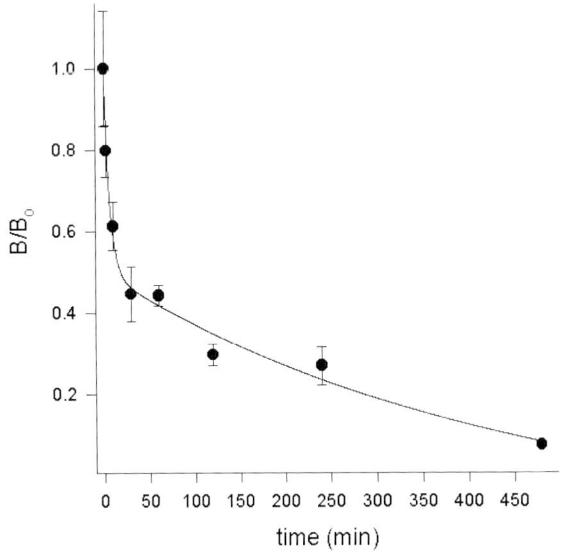

Figure 4. Kinetic analysis of sperm-ZP3 dissociation. Following equilibrium binding of fixed sperm and ^{125}I-ZPs, samples were diluted into a large volume of buffer to favor dissociation of the receptor-ligand complexes. The loss of bound radiolabeled ligand over time (B/B_0) was quantified. Ligand dissociation fit a double exponential decay function. The data points are averages of three independent experiments, each performed in duplicate, +/- SEM. (r^2=0.98). (Reprinted with permission of the ASBMB, 1996.)

distribution of each ligand oligosaccharide within the ZP matrix, understanding the role of each component in such a complex, spatially segregated system will require the development of new experimental approaches. Alternatively, a possibility raised by recent work with porcine ZPs (Yurewicz et al. 1998) is that more than one ZP glycoprotein is involved in high affinity sperm-ZP binding to intact sperm.

The receptor. The identity of sperm polypeptides involved in sperm adhesion and acrosomal exocytosis remains controversial and researchers can currently only speculate about the possible mechanisms by which the ZP3R could contribute to the observed complexity. It is possible that only a single sperm polypeptide interacts with ZP3 and that the complexity in this system is generated entirely by the ligand. However, there are several ways the ZP3R might create a complex interaction. The ZP3R could be multivalent with either two or more identical ZP3 binding sites or with ZP3 binding sites of differing affinities. Alternatively, there may be multiple sperm polypeptides which interact with ZP3. If more than one sperm polypeptide is involved, this raises the additional question of whether adhesion and signaling are segregated to unique interactions with ZP3, mediated by different sperm polypeptides, or are the result of forming a multimeric receptor complex responsible both for adhesion and for signal transduction.

Both receptor and ligand. It is possible that both receptor and ligand contribute to the observed complexity. The possibilities are numerous. Both ZP3 and ZP3R could be multivalent, or heterogeneous or both. So, for example, a multivalent ZP3 could interact with more than one sperm polypeptide (heterogeneous ZP3R), each having a unique affinity for the ligand. Either multivalence or heterogeneity could contribute to the observed complexity in sperm-ZP binding and, again, raise the possibility of segregating adhesion and signaling to different components of the system.

Without more detailed knowledge of the specific composition of the ZP3 oligosaccharides and of the identity and molecular properties of the sperm polypeptides involved in sperm-ZP binding, it is not possible to completely differentiate among these various possible sources of complexity in ZP3-ZP3R binding.

The data from initial binding studies strongly supports the model of complex sperm-ZP interactions. A great deal more work must be done to understand all the factors involved in this system, but biochemical approaches can still be used to further define the interactions and aid in the identification of sperm polypeptides mediating sperm ZP binding in the mouse.

Testing for Sperm-Zona Pellucida Binding Components

Of the possibilities discussed above, recent studies have focused on characterizing the sperm polypeptides which interact with ZP3. The approach that has been used to identify sperm proteins involved in sperm-ZP binding is a competitive binding assay. Competitive assays have been used previously to look at both putative ZP3 ligand moieties (Litscher et al. 1995; Thaler, Cardullo 1996b) and putative sperm receptor polypeptides (Shur, Hall 1982; CD Thaler and RA Cardullo, unpublished results). These assays have used intact ZPs (Litscher et al. 1995; Shur, Hall 1982) or soluble ZPs (Thaler, Cardullo 1996b, and unpublished results) incubated under various conditions with live capacitated sperm and potential competitors, including mimics of the ligand and receptor.

These experiments used soluble ^{125}I-ZPs or isolated ^{125}I-ZP3. The receptor, ZP3R on intact sperm, and ligand, ^{125}I-ZPs or ^{125}I-ZP3, are incubated together briefly, and bound ligand is quantified. The ability of various competitors to block binding of the radiolabeled ligand to sperm can be measured as a decrease in radioactivity bound to the sperm sample relative to control samples in which binding is allowed to occur in the absence of any competitors. Previously this assay was used to investigate sugar residues important in the ZP3 ligand oligosaccharide (Thaler, Cardullo 1996b).

In order to investigate potential sperm polypeptides serving as the ZP3R, sperm membrane protein fractions were used as competitors. Sperm membrane vesicles were generated by sonication of cauda epididymal sperm and the vesicles then fractionated according to density on a sucrose step gradient, similar to previously published work (Millette et al. 1980; Bunch, Saling 1991). Three membrane containing fractions and a pellet of cell debris were obtained.

The accepted model of sperm-ZP binding is that ZP3 binds to acrosome intact sperm and ZP2 binds to acrosome reacted sperm. In an assay where ^{125}I-ZPs are mixed with intact sperm, one would expect to detect radioactivity bound to the sperm and this to be due to binding of ^{125}I-ZP3. Further, addition of sperm proteins that have the ability to interact with ZP3 (that is, ZP3Rs) would result in a net decrease of ^{125}I-ZP3 bound to the intact sperm, relative to control samples incubated in the absence of competitors. Since ZP2 does not interact with intact sperm, the presence of ZP2 binding proteins would not be detected in this assay.

Two experiments were performed, one using ^{125}I-ZPs as the ligand and the second using isolated ^{125}I-ZP3 as the ligand. Each of the three fractions of sperm membrane vesicles were added to the binding assay as competitors. If the isolated membranes contain sperm proteins that bind to ZP3, they will compete for the available ^{125}I-ZP3 and a resulting decrease in ^{125}I-ZP3 bound to the intact sperm would be observed. These experiments should determine which membrane fractions contain sperm proteins that interact with ZP3.

Each of the membrane fractions, termed Band 1, Band 2 and Band 3, were able to significantly reduce the level of ^{125}I-ZPs bound to capacitated, acrosome intact sperm. As shown in Figure 5, the presence of Band 1 membranes resulted in a 47% decrease, Band 2 membranes resulted in a 50% decrease and Band 3 membranes a 63% decrease in the amount of ^{125}I-ZPs bound to intact sperm, relative to control assays measuring binding in the absence of any competitors. Neither Band 4, which contained axonemal remnants devoid of membranes, nor mouse liver membranes fractionated by the protocol used to generate sperm membranes, blocked binding of the ^{125}I-ZPs to the intact sperm (data not shown).

When the competitive assays were repeated, this time with isolated ^{125}I-ZP3, only Band 2 and Band 3 membranes were able to block binding of ^{125}I-ZP3 to the intact mouse sperm (Fig. 6). Band 1 membranes were not able to block the sperm-^{125}I-ZP3 interaction. However, Band 2 membranes reduced binding of ^{125}I-ZP3 to sperm by 63% and Band 3 membranes reduced binding by 50%. Taken together, the data from these two experiments indicate that ZP3 is both necessary and sufficient for binding to sperm polypeptides in Band 2 and Band 3 and that ZP3 is necessary but not sufficient for the interaction with sperm polypeptides in Band 1.

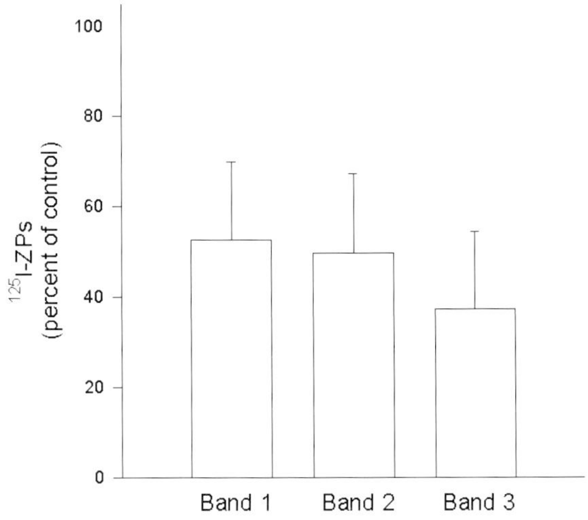

Figure 5. Sperm membrane fractions block binding of ^{125}I-ZPs to intact sperm. Mouse sperm membranes, fractionated by density, were used as competitors in ^{125}I-ZP binding to live, intact sperm. Bands 1-3, which contained membrane vesicles, were able to significantly block ^{125}I-ZP interactions with intact sperm, suggesting that these membranes contain one or more polypeptides which are ZP3Rs ($P<0.05$). In parallel experiments, control fractions of mouse liver membranes did not interfere with ^{125}I-ZP binding to intact sperm.

How do these results fit in with the model? The short answer is that they don't. If only ZP3 is involved in binding to acrosome intact sperm, one would expect to see exactly the same results for competitive assays using ^{125}I-ZPs or using ^{125}I-ZP3 and intact sperm. An alternative possibility is that ZP2 interacts with membrane components present on the acrosome intact sperm and that the decrease in binding of ^{125}I-ZPs to sperm in the presence of Band 1 components is the result of competition for these sites. However, when competitive binding assays were conducted with an ^{125}I-ZP1/2 fraction, none of the membrane fractions interfered with binding and there was very little binding of the ZP1/2 fraction to acrosome intact sperm. These data would suggest that ZP2 alone is not capable of high affinity interactions with acrosome intact sperm. This conclusion is consistent with previous results using the competitive intact ZP assay (Bleil, Wassarman 1980b) or direct microscopic observation of binding of gold-labeled ZP2 to sperm (Mortillo, Wassarman 1991).

It is important to note that although early assays investigating the functional role of the different ZP glycoproteins did not test combinations of the ZP glycoproteins, the design of the assays—competitive binding to intact ZPs or stimulation of acrosomal exocytosis—would not have enabled researchers to detect combinatorial effects. That is, the assays could detect components that were both necessary and sufficient, but not components that were necessary but not sufficient.

What, then, are possible alternatives that are consistent with the data obtained from the competitive binding assays with sperm membrane fractions? First, numerous complex possibilities or scenarios could be proposed that would be consistent with these experimental results. Most of these possibilities could not be differentiated by the few experiments discussed here. Second, if one looks

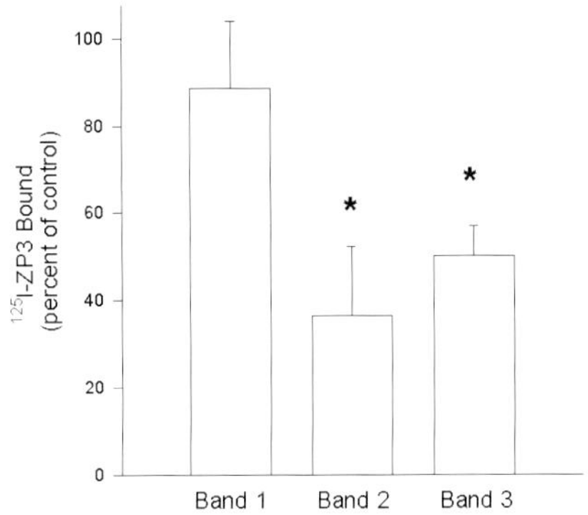

Figure 6. Sperm membranes from Band 2 and Band 3, but not Band 1, block binding of purified ^{125}I-ZP3 to intact sperm. When purified ^{125}I-ZP3 was used as ligand, only membranes from Band 2 and Band 3 were able to interact with the ligand fraction and block its binding to intact sperm ($P<0.05$). Membranes from Band 1 did not block ^{125}I-ZP3 binding to sperm. Control mouse liver membranes had no effect on binding.

beyond the implications for sperm polypeptides which interact with the ZP, one inescapable conclusion is drawn from these experiments: ZP3 in combination with another of the zona glycoproteins (ZP1 or ZP2) is required for interactions with some of the polypeptides present on intact mouse sperm. Third, a few general conclusions about the nature of the sperm polypeptides in the membrane fractions can be drawn. Most importantly, there must be more than one polypeptide capable of interacting with ZP3; that is, the polypeptides responsible for ZP3 binding in Bands 2/3 must be different from the polypeptides responsible for ZP3 binding in Band 1, because of the differential behavior that was observed in the binding assays using the different ligand preparations (^{125}I-ZPs versus ^{125}I-ZP3).

This presents an interesting comparison with recently published work using porcine gametes (Yurewicz et al. 1998). In this system, ZP1 and ZP3 interact with acrosome intact sperm, and neither ZP glycoprotein alone can bind sperm with high affinity. It is only in heterodimeric arrangements that ZP1/ZP3 displays high affinity binding to sperm. In contrast, in mouse some sperm polypeptides appear to interact with ZP3 alone (Band 2/3) while other components (Band 1) appear to require the presence of ZP3 together with ZP1 or ZP2.

It remains to be determined which of these interactions are responsible for signaling and adhesion, or if both interactions are involved in both processes. Additionally, given the previous data on ZP3 and ZP2 interactions with acrosome intact and acrosome reacted sperm, respectively, it is possible that the sperm polypeptides in Band 1 are involved in a transitional interaction. It may be that these polypeptides are involved in maintaining sperm contact with the ZP matrix during and following the acrosome reaction. Perhaps the sperm polypeptides which bind to the ZP3-ZP1/2 complex form a transitional interaction to maintain sperm-ZP contact as sperm are losing the ZP3 binding sites, possessed only by intact sperm, and exposing the ZP2 binding sites, possessed only by acrosome reacted sperm, as they progress through the acrosome reaction.

Summary and Future Directions

Data presented here indicate that binding events between acrosome intact sperm and the zona pellucida matrix may comprise a number of molecular interactions between ZP glycoproteins and several sperm polypeptides. At a minimum, the complexity of the sperm-ZP interactions would be one sperm polypeptide that is a receptor for ZP3 and a different polypeptide that is a receptor for ZP3/ZP2. These data underscore the biochemical complexity of the sperm-ZP interactions.

Certainly, there is much more exciting research to be done in order to understand fully the interactions implied by the biochemical studies discussed here. However, one thing is clear. In asking why is it so difficult to identify zona pellucida receptors, one can now answer, "Because it is a very complex interaction."

Complexity in receptor-ligand interactions is common. In other, more extensively characterized systems, researchers have found many different motifs, but many (if not most) are multimeric receptors. Many cytokine receptors are composed of multiple subunits, one of which binds the ligand and is induced by ligand binding to associate with additional proteins forming an active receptor (Saito et al. 1991; Davis et al. 1993). None of the subunits are capable of signaling, but the complex can interact with cytoplasmic tyrosine kinases to initiate signaling (Hatakeyama et al. 1991; Stahl et al. 1994). The NMDA glutamate receptor is composed of three subunits, each with a number of possible isoforms (Sheng et al. 1994). The NGF receptor is a heterodimer, in which each subunit can bind NGF, at unique affinities, but the heterodimer binds the ligand with greater affinity than either subunit alone (Mahadeo et al. 1994). Even these few examples underscore the trend toward complex interactions in receptor-ligand binding and it should come as no surprise that sperm-ZP interactions appear to follow this trend.

The cell biology of the sperm-ZP interaction is unique. Two crucial events, adhesion and acrosomal exocytosis, occur in response to a single substrate, the zona matrix. In addition, these events are sequentially related and, *in vivo*, at least, appear to be coupled. At the molecular level, there is not sufficient data to determine if these events are intrinsically coupled (one receptor-ligand system) or if they are independent steps (each with its own receptor ligand system) in the progress of sperm toward the egg plasma membrane. The information gathered from biochemical studies of the sperm-zona interaction have preliminarily indicated that both the ZP glycoproteins and the sperm receptors contribute to the observed complexity. Thus, future research will need to address possible sources of complexity in both sperm and zona components in order to elucidate the sequence of molecular interactions during these crucial steps in fertilization.

The zona. It is apparent that a great amount of detail concerning the glycosylation, ligand bioactivity, heterogeneity and functional roles of each of the glycoproteins needs to be addressed in order to understand this system. For example, is ZP3 composed of sub-populations that are heterogeneous in terms of ligand oligosaccharide composition? Does one ZP3 sub-population form a receptor for the Band2/3 polypeptides that are able to interact with ZP3 alone, while the other sub-population forms a complex with ZP2 to interact with Band 1 polypeptides? Could ZP3 ligand heterogeneity be correlated with the differential distribution of lectin binding sites in the intact zona (Aviles et al. 1997)? If so, what are the physiological implications for sperm progression through the zona matrix? Does the spatial heterogeneity of this matrix play a role in controlling not only sperm physiology, such as the acrosome reaction, but also as some type of selective filter for sperm?

The sperm. Currently, there appear to be two classes of zona receptors: those which interact with ZP3 alone and those which interact with ZP3 and ZP1 or ZP2. The extent of complexity within each category of ZP3 receptor will require identifying specific polypeptides that interact with ZP3 or with the proposed ZP3/ZP2 complex. Once specific polypeptides have been identified, future research will need to address many questions, including which of these polypeptides are involved in signaling and which are involved in adhesion.

Such research will need to incorporate multidisciplinary approaches which make use of biochemistry, molecular biology, structural biology and genetics. Only a multidisciplinary approach is likely to further the understanding of these complex interactions. Similarly, future researchers will need to bring fresh ideas and approaches to investigations of sperm-zona binding. Perhaps this may be done by incorporating the techniques and models being developed in the field of receptor biology in order to identify key interactions in the sperm-zona system.

Acknowledgements

The author thanks Rich Cardullo, colleague and collaborator, for stimulating discussions of the work presented here and of its implications for sperm-zona binding. Due to space limitations, only one or two recent references are listed in each of the citations in this review. Parts of the research referred to in this paper were supported by grants from the Lalor Foundation (C.D.T.) and the NIH (R.A.C.).

References

Apter FM, Baltz JM, Millette CF. A possible role for cell surface fucosyltransferase (FT) activity during sperm-zona pellucida binding in the mouse. J Cell Biol 1988; 107:175.

Aviles M, Jaber L, Castells MT, Ballesta J, Kan FW. Modifications of carbohydrate residues and ZP2 and ZP3 glycoproteins in the mouse zona pellucida after fertilization. Biol Reprod 1997; 57:1155-1163.

Bleil JD, Wassarman PM. Structure and function of the zona pellucida: identification and characterization of the proteins of the mouse oocyte's zona pellucida. Dev Biol 1980a; 76:185-202.

Bleil JD, Wassarman PM. Mammalian sperm-egg interaction: identification of a glyco-protein in mouse egg zonae pellucidae possessing receptor activity for sperm. Cell 1980b; 20:873-882.

Bleil JD, Wassarman PM. Sperm-egg interactions in the mouse: sequence of events and induction of the acrosome reaction by a zona pellucida glycoprotein. Dev Biol 1983; 95:317-324.

Bleil JD, Wassarman PM. Autoradiographic visualization of the mouse egg's sperm receptor bound to sperm. J Cell Biol 1986; 102:1363-1371.

Bleil JD, Greve JM, Wassarman PM. Identification of a secondary sperm receptor in the mouse egg zona pellucida: role in maintenance of binding of acrosome reacted sperm to eggs. Dev Biol 1988; 128:376-385.

Bookbinder LH, Cheng A, Bleil JD. Tissue- and species-specific expression of sp56, a mouse sperm fertilization protein. Science 1995; 269:86-89.

Bunch DO, Saling, PM. Generation of a mouse sperm membrane fraction with zona receptor activity. Biol Reprod 1991; 44:672-680.

Burks DJ, Carballada R, Moore HDM, Saling PM. Interaction of a tyrosine kinase from human sperm with the zona pellucida at fertilization. Science 1995; 269:83-86.

Davis S, Aldrich TH, Stahl N, Pan L, Taga T, Kishimoto T, Ip NY, Yancopoulos GD. LIFRβ and gp130 as heterodimerizing signal transducers of the tripartite CNTF receptor. Science 1993; 260:1805-1808.

Florman HM, Storey BT. Mouse gamete interactions: the zona pellucida is the site of the acrosome reaction leading to fertilization *in vitro*. Dev Biol 1982; 91:121-130.

Florman HM, Wassarman PM. O-linked oligosaccharides of mouse egg ZP3 account for its sperm receptor activity. Cell 1985; 41:313-324.

Florman HM, Bechtol KB, Wassarman PM. Enzymatic dissection of the functions of the mouse egg's receptor for sperm. Dev Biol 1984; 106:243-255.

Foster JA, Friday BB, Maulit MT, Blobel C, Winfrey VP, Olson GE, Kim KS, Gerton GL. AM67, a secretory component of the guinea pig sperm acrosomal matrix, is related to mouse sperm protein sp56 and the complement component 4-binding proteins. J Biol Chem 1997; 272:12714-12722.

Gao Z, Garbers DL. Species diversity in the structure of zonadhesin, a sperm-specific membrane protein containing multiple cell adhesion molecule-like domains. J Biol Chem 1998; 273:3415-3421.

Gong X, Dubois DH, Miller DJ, Shur BD. Activation of a G protein complex by aggregation of β-1,4,-galactosyltransferase on the surface of sperm. Science 1995; 269:1718-1721.

Gwatkin RBL, Williams DT. Receptor activity of the hamster and mouse solubilized zona pellucida before and after the zona reaction. J Reprod Fertil 1977; 49:55-59.

Hardy DM, Garbers DL. A sperm membrane protein that binds in a species-specific manner to the egg extracellular matrix is homologous to von Willebrand factor. J Biol Chem 1995; 270:26025-26028.

Harris JD, Hibler DW, Fontenot GK, Hsu KT, Yurewicz EC, Sacco AG. Cloning and characterization of zona pellucida genes and cDNAs from a variety of mammalian species: the ZPA, ZPB and ZPC gene families. DNA Sequence 1994; 4:361-393.

Hartmann, JF, Gwatkin, RBL, Hutchinson, CF. Early contact interactions between mammalian gametes *in vitro*: evidence that the vitellus influences adherence between sperm and zona pellucida. Proc Natl Acad Sci USA 1972; 69:2767-2769.

Hatakeyama M, Kono T, Kobayashi N, Kawahara A, Levin SD, Perlmutter RM, Taniguchi T. Interaction of the IL-2 receptor with the src family kinase $p56^{lck}$: identification of a novel intermolecular association. Science 1991; 252:1523-1528.

Inoue M, Wolf DP. Sperm binding characteristics of the murine zona pellucida. Biol Reprod 1975; 13:340-346.

Kalab P, Visconti P, Leclerc P, Kopf GS. p95, the major phosphotyrosine containing protein in mouse spermatozoa, is a hexokinase with unique properties. J Biol Chem 1994; 269:3810-3817.

Kinloch RA, Sakai Y, Wassarman PM. Mapping the mouse ZP3 combining site for sperm by exon swapping and site-directed mutagenesis. Proc Natl Acad Sci USA 1995; 92:263-267.
Lee VH, Schwoebel E, Prasad S, Cheung P, Timmons TM, Cook R, Dunbar BS. Identification and structural characterization of the 75-kDa rabbit zona pellucida protein. J Biol Chem 1993; 268:12412-12417.
Leyton L, Saling PM. Evidence that aggregation of mouse sperm receptors by ZP3 triggers the acrosome reaction. J Cell Biol 1989; 108:2163-2168.
Leyton L, LeGuen P, Bunch D, Saling PM. Regulation of mouse gamete interaction by a sperm tyrosine kinase. Proc Natl Acad Sci USA 1992; 89:11692-11695.
Leyton L, Tomes C, Saling P. LL95 monoclonal antibody mimics functional effects of ZP3 on mouse sperm: evidence that the antigen recognized is not hexokinase. Mol Reprod Dev 1995; 42:347-358.
Litscher ES, Juntunen K, Seppo A, Penttila L, Niemela R, Renkonen O, Wassarman PM. Oligosaccharide constructs with defined structures that inhibit binding of mouse sperm to unfertilized eggs *in vitro*. Biochem 1995; 34:4662-4669.
Lu Q, Shur BD. Sperm from beta 1,4-galactosyltransferase-null mice are refractory to ZP3-induced acrosome reactions and penetrate the zona pellucida poorly. Development 1997; 124:4121-4131.
Mahadeo D, Kaplan L, Chao MV, Hempstead BL. High affinity nerve growth factor binding displays a faster rate of association than $p140^{trk}$ binding. Implication for multi-subunit polypeptide receptors. J Biol Chem 1994; 269:6884-6891.
Millette CF, O'Brien DA, Moulding CT. Isolation of plasma membranes from purified mouse spermatogenic cells. J Cell Sci 1980; 43:279-299.
Moller CC, Bleil JD, Kinloch RA, Wassarman PM. Structural and functional relationships between mouse and hamster zona pellucida glycoproteins. Dev Biol 1990; 137:276-286.
Mortillo S, Wassarman PM. Differential binding of gold-labeled zona pellucida glycoproteins mZP2 and mZP3 to mouse sperm membrane compartments. Development 1991; 113:141-149.
Saito Y, Tada H, Sabe H, Honjo T. Biochemical evidence for a third chain of the interleukin-2 receptor. J Biol Chem 1991; 266:22186-22191.
Saling PM, Sowinski J, Storey BT. An ultrastructural study of epididymal mouse spermatozoa binding to zonae pellucidae *in vitro*: sequential relationship to the acrosome reaction. J Exp Zool 1979; 209:229-238.
Sheng M, Cummings J, Roldan LA, Jan YN, Jan LY. Changing subunit composition of heterotrimeric NMDA receptors during development of the rat cortex. Nature 1994; 368:144-147.
Shur BD, Hall NG. A role for mouse sperm surface galactosyltransferase in sperm binding to the egg zona pellucida. J Cell Biol 1982; 95:574-579.
Stahl N, Boulton TG, Farruggella T, Ip N, Davis S, Witthuhn BA, Quelle FW, Silvennoinen O, Barbieri G, Pellegrini S, Ihle JN, Yancopoulos GD. Association and activation of Jak-Tyk kinases by CNTF-LIF-OSM-IL-6β receptor components. Science 1994; 263:92-95.
Thaler CD, Cardullo RA. The initial molecular interaction between mouse sperm and the zona pellucida is a complex binding event. J Biol Chem 1996a; 271:23289-23297.
Thaler CD, Cardullo RA. Defining oligosaccharide specificity for initial sperm-zona pellucida adhesion in mouse. Mol Reprod Dev 1996b; 45:535-546.
Yurewicz EC, Sacco AG, Gupta SK, Xu N, Gage DA. Hetero-oligomerization-dependent binding of pig oocyte zona pellucida glycoproteins ZPB and ZPC to boar sperm membrane vesicles. J Biol Chem 1998; 273:7488-7494.

19 *Molecular Dissection of the Sperm Combining-Site of Mouse Egg Zona Pellucida Glycoprotein mZP3, the Sperm Receptor*

Paul M. Wassarman
Eveline S. Litscher
Mount Sinai School of Medicine

Fertilization in mammals takes place within the oviduct where one or more ovulated eggs interact with the hundred or so sperm that have reached the site. All mammalian eggs are surrounded by a thick extracellular coat, called the zona pellucida (Fig. 1), that regulates interactions between sperm and eggs both before and after fertilization (Gwatkin 1977; Wassarman 1987, 1991; Dietl 1989; Yanagimachi 1994). Such regulation is required in order to maintain the characteristics of the species and to prevent polyspermic fertilization; the latter is usually a fatal condition for the embryo. It has been demonstrated that eggs from which the zona pellucida has been removed *in vitro* can fuse with sperm from almost any mammal (that is, no species specificity) and easily become polyspermic.

mZP3 is one of three glycoproteins that constitute the mouse egg zona pellucida (Bleil, Wassarman 1980; Wassarman 1988, 1990, 1995, 1998; Wassarman, Mortillo 1991; Liu et al. 1996; Rankin et al. 1996; Wassarman et al. 1996). It is an essential structural component of zona pellucida filaments and, additionally, acts as a sperm receptor and acrosome reaction-inducer during fertilization. Sperm from most heterologous species are unable to bind to mZP3, thereby ensuring species-specific fertilization. Following fusion of egg and sperm to form a zygote, mZP3 is inactivated as both a sperm receptor and acrosome reaction-inducer ("zona reaction") by degradative enzymes originating from cortical granules underlying the plasma membrane ("cortical reaction").

The complementarity between mammalian sperm and eggs from the same species apparently is based on carbohydrate recognition (Florman et al. 1984; Florman, Wassarman 1985; Wassarman 1990; Litscher et al. 1995; Wassarman, Litscher 1995). In the case of mice, mZP3 polypeptide carries certain oligosaccharides that are recognized by one or more "egg-binding proteins" located on the plasma membrane of the acrosome-intact sperm head. These oligosaccharides are covalently linked to Ser residues (O-linked) located at a region of mZP3 polypeptide referred to as the sperm combining-site. Indeed, there is evidence to suggest that isolated mZP3 oligosaccharides, as well as oligosaccharides from other sources, can prevent binding of sperm to mouse eggs and fertilization *in vitro*. It should be noted that while certain oligosaccharides bind to the sperm head and prevent binding of sperm to eggs, they fail to induce sperm to undergo the acrosome reaction *in vitro*.

During the past few years, several experimental approaches have been employed, including limited proteolysis (Rosiere, Wassarman 1992; Litscher,

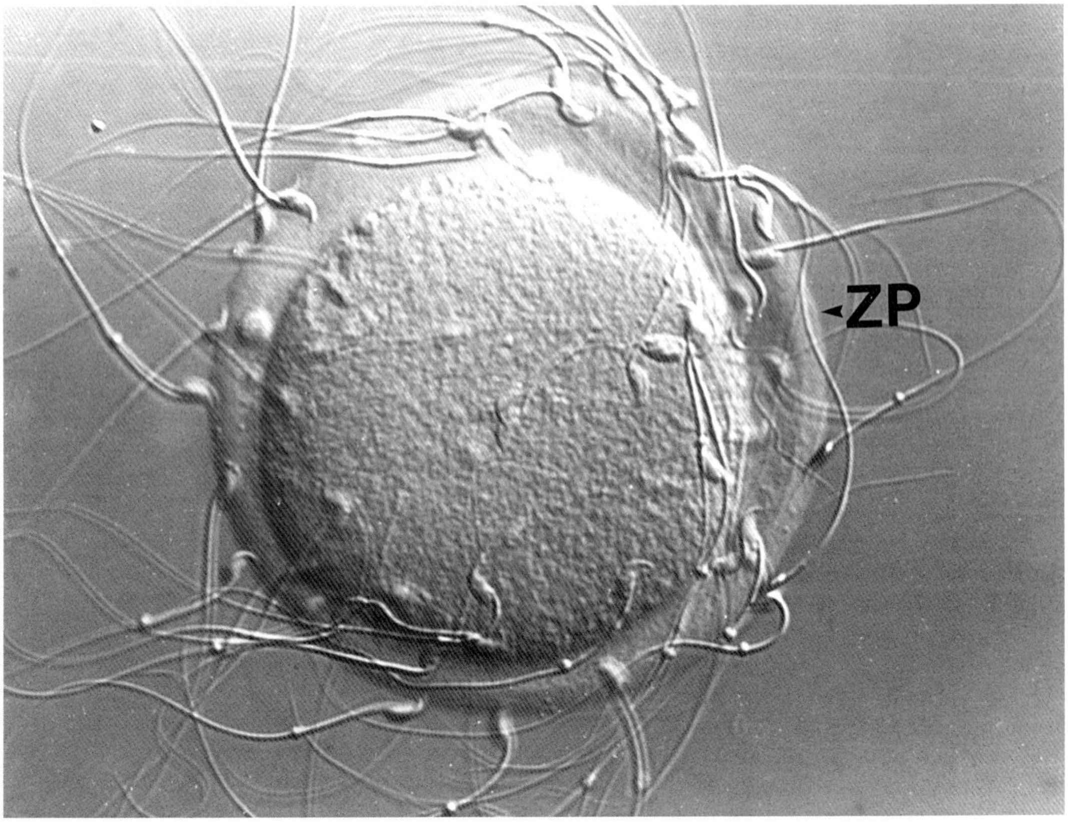

Figure 1. Photomicrograph (Nomarski DIC) of mouse sperm bound to the zona pellucida of an unfertilized mouse egg *in vitro*. ZP = zona pellucida.

Wassarman 1996a), exon swapping (Kinloch et al. 1995), and site-directed mutagenesis (Kinloch et al. 1995; Chen et al. 1998), to locate the sperm combining-site of mZP3 and to characterize some essential features of the site. Each of these approaches is described here and conclusions drawn from the results are presented. Overall, the evidence suggests that sperm recognize and bind to one or two Ser-linked oligosaccharides located at a region of polypeptide encoded by *mZP3* exon-7 near the carboxyl-terminus of the polypeptide. Binding of sperm requires the presence of a correct oligosaccharide structure, but does not require maintenance of the mZP3 polypeptide's secondary and tertiary structure. Such an interaction between mouse sperm and egg is reminiscent of the interactions between certain bacteria, animal viruses and other pathogens and their cellular hosts, as well as of the binding of selectins to their specific oligosaccharide ligands (references are listed later in this chapter).

Mouse Zona Pellucida Glycoprotein mZP3

Each mouse egg zona pellucida (~6.4 μm thick; ~3.0-3.5 ng protein) contains ~1.2-1.4 ng of mZP3. The glycoprotein is encoded by a single-copy gene located on chromosome number 5 that produces a unique nascent polypeptide consisting of 424 amino acids, three or four complex-type asparagine- (N-) linked oligosaccharides, and an undetermined number of O-linked oligosaccharides (Salzmann et al. 1983; Florman, Wassarman 1985; Kinloch et al. 1988; Ringuette et al. 1988; Kinloch, Wassarman 1989; Wassarman 1988, 1995, 1998). The first 22 amino acids at the amino-terminus of the polypeptide constitute a signal sequence that is cleaved during intracellular processing of the nascent glycoprotein. The N- and O-linked oligosaccharides are both sulfated and sialylated, although neither modification is essential for the biological activity of mZP3 (Liu et al. 1997). The polypeptide is particularly rich in Ser plus Thr (~18%) and Pro (~7%) residues, has 13 Cys residues, and possesses a long stretch (22 amino acids) of hydrophobic amino acids near its carboxyl-terminus, 34 residues downstream of a potential furin cleavage site (-Arg-Asn-Arg-Arg-). Solely as a result of N- and O-linked glycosylation, mZP3 has a relatively low pI (<5.0) and a very heterogeneous appearance (that is, a broad band) following one-dimensional SDS-PAGE (~65-100 kD M_r; ave. ~83 kD M_r).

Computer analyses of the amino acid sequence of mZP3 suggest that the polypeptide consists primarily of β-structure, loops, and extended chain conformation, with very little, if any, α-helix. All of the potential N-linked glycosylation sites are predicted to be exposed to solvent. A comparison of ZP3 polypeptides from several mammals, from mice to human beings, suggests a rather high degree of structural similarity (>65% similar). Furthermore, molecular cloning of genes encoding glycoproteins that constitute the vitelline envelope of a variety of non-mammalian animal eggs (for example, from fish, amphibia and birds) has revealed the presence of a polypeptide that resembles ZP3. Thus, ZP3 has been conserved during evolution as a structural component of egg extracellular coats and modified in mammals to accommodate additional biological functions as well.

Limited Proteolysis of mZP3 Polypeptide

The rationale for these experiments was based on the observation that small glycopeptides present in pronase digests of purified egg mZP3 could prevent the binding of sperm to eggs *in vitro* (Florman et al. 1984; Wassarman et al. 1985). This suggested that mZP3 polypeptide, with its native three-dimensional structure intact, was not required for biological activity and, consequently, led to experiments demonstrating that a specific class of mZP3 oligosaccharides, not polypeptide, was involved in binding of sperm to mZP3 (Florman, Wassarman 1985; Wassarman et al. 1985).

To determine which region or regions of mZP3 sperm recognize, mZP3 glycopeptides were generated by limited digestion of purified glycoprotein with either papain or V8 protease (Rosiere, Wassarman 1992; Litscher, Wassarman 1996a). Only a small number of glycopeptides were produced, probably due to masking of potential protease cleavage sites by complex-type N-linked and O-linked oligosaccharides. As expected, this mixture of glycopeptides inhibited binding of sperm to eggs *in vitro*. Papain and V8 protease digests of mZP3 each contained a ~55 kD M_r glycopeptide (called gp55; ~47-62 kD M_r; pI ~4.3-4.5) that was purified by HPLC. gp55 inhibited binding of sperm to ovulated eggs and induced the acrosome reaction about as effectively as intact egg mZP3 (for example, ID_{50} ~50-200 nM). Removal of N-linked oligosaccharides from gp55, by extensive digestion with N-glycanase, reduced its M_r to ~21 kD and increased its pI to ~5.3, but did not significantly affect its biological activity. Immunoblotting and amino acid sequencing results revealed that the ~55 kD M_r glycopeptide was derived from the carboxyl-terminal half of the mZP3 polypeptide and contained amino acid residues 328-343. Within the latter sequence are five clustered Ser residues, Ser -329, -331, -332, -333, and -334.

Exon Swapping between mZP3 and hZP3 Genes

Experiments described here were based on several observations. These include: 1) the sperm receptor and acrosome reaction-inducing activities of mZP3 are dependent on specific O-linked oligosaccharides (Florman, Wassarman 1985; Bleil, Wassarman 1988; Miller et al. 1992); 2) in assays with mouse gametes, mouse embryonal carcinoma (EC) cells stably transfected with either mouse (m) or hamster (h) ZP3 genes secrete active EC-mZP3 and inactive EC-hZP3 (Kinloch et al. 1991; Litscher, Wassarman 1996b); 3) limited digestion of purified mZP3 by certain proteases yields a ~55 kD M_r glycopeptide that originates from the carboxyl-terminal portion of the polypeptide and is active as a sperm receptor and acrosome reaction-inducer (Rosiere, Wassarman 1992; Litscher, Wassarman 1996a); and 4) more extensive digestion of the ~55 kD M_r glycopeptide results in removal of amino acid residues 328-343 and, concomitantly, in loss of biological activity (Rosiere, Wassarman 1992).

To determine whether proper glycosylation of the carboxyl-terminal third of the mZP3 polypeptide is, indeed, essential for sperm binding and acrosome reaction-inducing activity, exon-swapping experiments were carried out. *mZP3* and *hZP3* genes are organized in a very similar manner, with eight exons separated by seven introns (Kinloch et al. 1988, 1990; Ringuette et al. 1988; Kinloch, Wassarman 1989). Exons 6-8 of the *mZP3* gene were replaced with exons 6-8 of the hZP3 gene and exons 6-8 of the *hZP3* gene were replaced with exons 6-8 of the *mZP3* gene, and the hybrid glycoproteins secreted by transfected EC cells (EC-ZP3[m1-5/h6-8] and EC-ZP3[h1-5/m6-8], respectively) were assayed. It was anticipated on the basis of previous results that EC-ZP3[h1-5/m6-8] would be active and EC-ZP3[m1-5/h6-8] would be inactive in assays with mouse sperm and eggs.

Plasmids pPGK/ZP3[h1-5/m6-8] and pPGK/ZP3[m1-5/h6-8], used to generate stably transfected EC cell lines, were constructed. EC-ZP3[h1-5/m6-8] consists of the first 276 amino acids of hZP3 and amino acids 279-424 of mZP3. EC-ZP3[m1-5/h6-8] consists of the first 278 amino acids of mZP3 and amino acids 277-424 of hZP3. PCR was used to identify stably transfected EC cells harboring the hybrid *ZP3* genes and Northern blotting was used to identify cell lines that expressed the hybrid *ZP3* genes (EC-ZP3[h1-5/m6-8], 3 positive lines; EC-ZP3[m1-5/h6-8], 7 positive lines).

To determine whether cell lines that expressed hybrid *ZP3* genes also synthesized and secreted the recombinant glycoproteins, cell lysates and HPLC-fractionated culture medium were analyzed by Western immunoblotting. Cell lysates containing EC-ZP3[h1-5/m6-8] and EC-ZP3[m1-5/h6-8] displayed broad bands that migrated at ~69 kD and ~58 kD M_r, respectively (as compared with EC-hZP3, ~49 kD M_r and EC-mZP3, ~80 kD M_r). Very similar results were obtained using culture medium, freed of EC cells and subjected to HPLC fractionation and Western immunoblotting. The M_r differences between hybrid and wild-type EC-ZP3 probably reflects the fact that, as compared with hZP3, the carboxyl-terminal third of mZP3 possesses more N-linked oligosaccharides.

Two different *in vitro* assays, examining sperm binding and acrosome reaction-inducing activities, were employed to assess secreted EC-ZP3[h1-5/m6-8] and EC-ZP3[m1-5/h6-8] after partial purification by HPLC. Purified mZP3 and hZP3, at nanomolar concentrations, possess both biological activities. EC-mZP3 inhibited binding of mouse sperm to eggs and induced mouse sperm to undergo the acrosome reaction, whereas EC-hZP3 was inactive in both assays. On the other hand, EC-ZP3[h1-5/m6-8] was about as effective as EC-mZP3, and EC-ZP3[m1-5/h6-8] was as ineffective as EC-hZP3 in these assays. These results are consistent with the proposal that O-linked oligosaccharides essential for biological activity are located on the region of polypeptide encoded by exons 6-8 of *mZP3* and *hZP3* genes. While the overall three-dimensional structures of mZP3 and hZP3 probably are very similar, if not identical to each other, differences in the carboxyl-terminal third of the polypeptides apparently determine whether or not transfected EC cells can add essential O-linked oligosaccharide structures to nascent ZP3. It should be noted that, while EC-hZP3 and CHO-hZP3 are inactive with mouse gametes, they are biologically active with hamster gametes (Litscher, Wassarman 1996b).

Site-Directed Mutagenesis of the mZP3 Gene

The portion of mZP3 polypeptide including Ser-329, -331, -332, -333, and -334 was chosen for site-directed mutagenesis since previous experiments revealed that proteolysis of this region resulted in inactivation of mZP3 *in vitro* (Rosiere, Wassarman 1992; Litscher, Wassarman 1996a) and that antibodies directed specifically against this region prevented binding of sperm to eggs both *in vivo* (Millar et al. 1989) and *in vitro* (Rosiere, Wassarman 1992; S Mortillo, PM Wassarman, unpublished results). Apparently, this is the only region of mZP3 polypeptide that carries O-linked oligosaccharides essential for sperm binding and induction of the acrosome reaction. To characterize this region of mZP3 polypeptide in more detail, researchers constructed *mZP3* genes carrying single and multiple mutations in this portion of the polypeptide (amino acids 329-334; Fig. 2.) A description of the eight mutations that were analyzed in stably transfected EC cells is presented in Table 1. The five Ser residues, Ser-329, -331, -332, -333, and -334, were converted as individual mutations to Gly, Ala, or Val residues. In addition, Ser-332, -333, and -334 were converted in a triple mutation to Gly or Ala residues. Finally, Asn-330, a potential N-linked glycosylation site (that is, part of the consensus sequence Asn-X-Ser/Thr, where X is any amino acid other than Pro; Bause 1983; Kornfeld, Kornfeld 1983), was converted to an Ala residue. EC-mZP3 carrying each of these seven mutations, together with a mutant in which all five Ser residues were converted to Gly, Ala, or Val, were tested for biological activity *in vitro* and compared to that of wild-type EC-mZP3 (EC-mZP3-[wt]).

To select cell lines that synthesized and secreted recombinant EC-mZP3, EC cell lysates and culture medium collected from candidate colonies were screened by Western immunoblotting using a goat anti-mZP3 IgG. From the EC cell colonies for each mutation that tested positive for EC-mZP3, the two to three lines that produced the highest levels of secreted EC-mZP3 were chosen for further analysis. Recombinant glycoproteins were partially purified from the culture media by HPLC on a size-exclusion column and were then quantitated and tested for their ability to prevent binding of sperm to eggs *in vitro*.

Stably transfected EC cells secreted relatively large amounts of wild-type and mutated EC-mZP3 into the culture medium. The average apparent M_r (range M_rs) of EC-mZP3 for each of the recombinants follows: EC-mZP3-[wt], 79 kD (67-91); -[Ser-329], 73 kD (62-84); -[Ser-331], 79 kD (67-91); -[Ser-332], 75 kD (62-88); -[Ser-333], 79 kD (67-91); -[Ser-334], 77 kD (66-88); -[Ser-332-334], 74 kD (61-88); -[Ser-329-334], 67 kD (59-76); -[Asn-330], 78 kD (66-91). Since mZP3 is synthesized as an ~44 kD M_r polypeptide (Kinloch et al. 1988; Ringuette et al. 1988), it is apparent that the wild-type and mutant forms of EC-mZP3 are glycosylated. The M_r of EC-mZP3-[Ser-329] is significantly lower than that of EC-mZP3-[wt] and is probably attributable to elimination of the consensus N-linked glycosylation sequence, $Asn_{327}Cys_{328}Ser_{329}$, by conversion of Ser-329 to Ala. A similar case can be made for mutant EC-mZP3-[Ser-329-334] in which Ser-329 was converted to Ala and disrupted the consensus sequence. Overall, the average apparent M_r remains about the same as EC-mZP3-[wt] (~79 kD) for four of the mutations (~77-79 kD) and is decreased significantly for the other four mutations (~67-75 kD).

In vitro competition assays were carried out to assess the ability of different forms of recombinant EC-mZP3 to interfere with binding of sperm to ovulated eggs. Capacitated sperm were treated with HPLC-purified or immunoaffinity-purified wild-type and mutated EC-mZP3, ovulated eggs and two-cell embryos were added, and the extent of sperm binding was determined by light microscopy.

While EC-mZP3-[wt] at a concentration of ~10-15 ng/μl inhibited binding of sperm to eggs, EC-mZP3-[Ser-329-334] at the same concentration had virtually no effect on sperm binding. This result is consistent with the proposal that sperm recognize and bind to O-linked oligosaccharides located in the carboxyl-terminal region of mZP3 polypeptide. On the other hand, the single amino acid mutations EC-mZP3-[Ser-329], -[Ser-331], -[Ser-333], and -[Asn-330], also at a concentration of ~10-15 ng/μl, were just as effective as EC-mZP3-[wt] at inhibiting binding of sperm to eggs *in vitro*. The findings suggest that these four amino acid residues are not essential for mZP3 biological activity.

Only two of the single amino acid mutations had a significant effect on the ability of EC-mZP3 to inhibit sperm binding to eggs. EC-mZP3-[Ser-332] and -[Ser-334], at concentrations of ~10-30 ng/μl, failed to inhibit binding of sperm to eggs as compared to EC-mZP3-[wt] (identical results were obtained at ~50 ng/μl). Since EC-mZP3-[Ser-332] exhibited no biological activity as compared to EC-mZP3-[wt], the mutant glycoprotein was purified by immunoaffinity-chromatography followed by HPLC on a size-exclusion column, and was then tested in the competition assay. As expected, immunoaffinity-purified EC-mZP3-[Ser-332] (~10-20 ng/μl) also exhibited insignificant levels of biological activity (0-6%) as compared to EC-mZP3-[wt] (90-100%). This also was seen with the triple mutation EC-mZP3-[Ser-332-334] at the same concentrations; such a mutant was unable to inhibit sperm binding. Collectively, results with these three mutants strongly suggest that Ser-332 and -334 are essential for mZP3 biological activity.

Discussion

The evidence reviewed here provides support for the proposal that the mZP3 combining-site for sperm (that is, the region of mZP3 polypeptide carrying the O-linked oligosaccharides to which free-swimming sperm bind) is located in the carboxyl-terminal region of the polypeptide. The results strongly suggest that the O-linked oligosaccharides recognized by sperm are located on a region of mZP3 polypeptide that includes amino acids 328-343. There are five Ser residues in this region and no Thr residues. Four of the five Ser residues occur consecutively and the fifth is separated from the others by a single amino acid; an arrangement of hydroxy-amino acids typical of heavily O-glycosylated regions of many glycoproteins (Sadler 1984; Wilson et al. 1991). On the other hand, Pro residues that are often located at positions -1 and +3 relative to glycosylated Ser/Thr residues (Wilson et al. 1991; Yoshida et al. 1997), are not present. It should be noted, however, that the five Ser residues are located adjacent to a stretch of nine amino acids that includes four Cys residues which are conserved in all ZP3 genes characterized thus far (Wassarman 1998). The local three-dimensional structure generated by the Cys residues may render this region of mZP3 polypeptide accessible to N-acetyl-galactosaminyl-transferase, the first enzyme required in the O-linked glycosylation pathway (Sadler 1984).

The sperm combining-site of mZP3 is located relatively close to the carboxyl-terminus of the polypeptide and is encoded by exon-7 of *mZP3*. A relatively small glycopeptide derived from the carboxyl-terminus of mZP3 is about as effective an inhibitor of sperm-egg interaction and inducer of the acrosome reaction as native mZP3 (ID_{50} ~100 nM; Rosiere, Wassarman 1992; Litscher, Wassarman 1996a). Similarly, synthetic oligosaccharides related to O-linked oligosaccharides of mZP3 also inhibit binding of sperm to eggs, albeit at much higher concentrations (ID_{50} ~10 μM), but do not induce sperm to undergo the acrosome reaction *in vitro* (Litscher et al. 1995). Thus, mZP3 polypeptide influences binding of sperm, possibly by affecting the conformation of essential O-linked oligosaccharides, and is necessary for induction of the acrosome reaction. In the latter context, small mZP3 glycopeptides and mZP3 O-linked oligosaccharides bind to the head of acrosome-intact sperm, but do not induce the acrosome reaction (Wassarman 1988, 1990).

It was found that conversion of either of two Ser residues, Ser-332 or -324, to non-hydroxy amino acids, Gly or Ala, respectively, resulted in loss of the ability of EC-mZP3 to inhibit binding of sperm to ovulated eggs *in vitro*. On the other hand, conversion of several neighboring amino acids, Ser-329, Asn-330, Ser-331, and Ser-333 to Ala or Val residues, had no effect on sperm receptor activity of EC-mZP3. Thus, although there are five clustered Ser residues in this region, a situation typical of heavily O-glycosylated regions of many glycoproteins (Sadler 1984; Wilson et al. 1991), mutation of only two of the five affected sperm receptor activity. It is tempting to suggest that Ser-332 and -334 carry oligosaccharides essential for binding of sperm to mZP3. Of course, it is possible that mutation of either Ser-332 or -334 could affect glycosylation of the other Ser residue; such a situation could account for the observed effects on EC-mZP3 sperm receptor activity. Since sperm are induced to undergo the acrosome reaction only after binding to mZP3, as expected, mutation of either Ser-332 or -334 also resulted in the loss of the ability of EC-mZP3 to act as an acrosome reaction-inducer *in vitro*.

It is of interest that Ser-332 and -334, the two residues thought to be essential for mZP3 sperm receptor activity, are conserved residues in mouse, hamster and human ZP3 (Kinloch et al. 1988, 1990; Ringuette et al. 1988; Chamberlin, Dean 1990). In fact, they are conserved residues in a region of polypeptide that has undergone considerable changes during the course of evolution (Kinloch et al. 1995; Wassarman, Litscher 1995). It has been suggested that these changes could direct the addition of alternative oligosaccharide structures to nascent ZP3 and, in this manner, affect species specificity of sperm-egg interaction (Kinloch et al. 1995; Wassarman, Litscher 1995). There is sufficient experimental evidence to suggest that amino acids flanking a Ser or Thr residue influence addition of sugars to the site (Elhammer et al. 1993; Wang et al. 1993; Gooley, Williams 1994; Nehrke et al. 1996).

In this context, it is well documented that mouse sperm bind to hamster eggs and hamster sperm bind to mouse eggs, but that human sperm do not bind to either mouse or hamster eggs *in vitro* (Yanagimachi 1994). If Ser-332 and -334 are glycosylated in hamster and human ZP3, it is likely that the structures of the oligosaccharides differ such that hamster, but not human ZP3, is recognized by mouse sperm *in vitro*.

The putative mZP3 combining-site for sperm is located in a region of polypeptide that has undergone considerable sequence divergence during evolution. For example, whereas mZP3 and human ZP3 (huZP3) polypeptides are ~67% identical overall, the region including amino acids 329-342 is only ~28% identical (that is, four of 14 residues are identical). Two of five Ser residues of mZP3 (Ser-332 and-334) are conserved

Figure 2. Schematic diagram of the *PGK/mZP3* recombinant gene used to generate stably transfected EC-mZP3 cell lines. *pPGK-1* represents the mouse phosphoglycerate kinase-1 promoter region. Restriction enzymes *Cla*I and *Sst*II were used to generate linearized DNA fragments for electroporation of EC cells. Arrow indicates the transcriptional start site on the *PGK-1* promoter. The sites of mutagenesis in exon 7 of the *mZP3* gene are indicated as amino acids 329-334. See Table 1 for a description of the mutations made in *mZP3*.

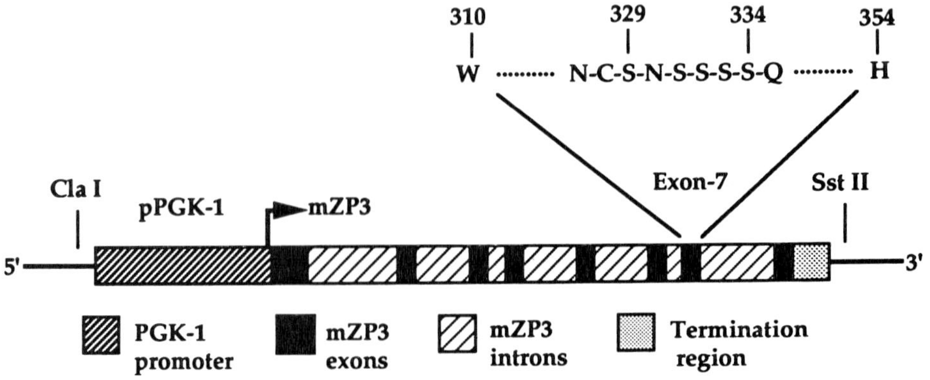

Table 1: Summary of site-directed mutagenesis of EC-mZP3.

	Position of EC-mZP3 Mutation[1]					
EC-mZP3 Designation	329	330	331	332	333	334
EC-mZP3-[wt]-	Ser	Asn	Ser	Ser	Ser	Ser
EC-mZP3-[Ser-329]-	Ala	Asn	Ser	Ser	Ser	Ser
EC-mZP3-[Asn-330]-	Ser	Ala	Ser	Ser	Ser	Ser
EC-mZP3-[Ser-331]-	Ser	Asn	Val	Ser	Ser	Ser
EC-mZP3-[Ser-332]-	Ser	Asn	Ser	Gly	Ser	Ser
EC-mZP3-[Ser-333]-	Ser	Asn	Ser	Ser	Ala	Ser
EC-mZP3-[Ser-334]-	Ser	Asn	Ser	Ser	Ser	Ala
EC-mZP3-[Ser-332-334]-	Ser	Asn	Ser	Gly	Ala	Ala
EC-mZP3-[Ser-329-334]-	Ala	Asn	Val	Gly	Ala	Ala

[1]The mutated amino acids are underlined in this table.

and huZP3 has an additional Ser (Ser-342) and Thr (Thr-330) residue. Furthermore, although mZP3 and hZP3 polypeptides are ~82% identical overall, the region including amino acids 329-342 has only six residues in common (~43% identical). Three of five Ser residues of mZP3 (Ser-331, -332, and -334) are conserved and hZP3 has an additional Ser residue (Ser-342). In fact, a comparison of the primary structures of mZP3 and hZP3 polypeptides, carried out in 40 residue increments from Pro-35 to Leu-394, reveals that the region including amino acids 315-354 has undergone twice as many changes as any other region. In addition, there is evidence to suggest that the combining site of mZP3 is located in one of two structural domains that are separated by a hinge-region located about in the middle of the mZP3 polypeptide (Wassarman, Litscher 1995). While the precise nature of the sperm component that binds to mZP3 remains a controversial issue, it should be noted that some of the principal candidates for egg-binding protein recognize carbohydrate determinants (Litscher, Wassarman 1993; Snell, White 1996).

Differences in ZP3 oligosaccharide structure among species could account for the frequent failure of sperm from one mammalian species to bind to eggs of another. While the rules that govern placement and structure of O-linked oligosaccharides on glycoproteins remain unclear (Sadler 1984; Wilson et al. 1991), it is likely that changes in polypeptide primary structure in and around the ZP3 combining-site influence the location and nature of oligosaccharides added to nascent ZP3 and, in this manner, could determine species specificity of gamete adhesion. In this context, protein-carbohydrate interactions now are thought to be employed in a number of instances of cellular adhesion, in systems as diverse as binding of pathogenic bacteria to animal cells (Lund et al. 1987; Kuehn et al. 1992), neuronal development (Hynes et al. 1989; Schachner 1989), and lymphocyte homing (Lasky et al. 1992; McEver et al. 1995). The potential for enormous variation in oligosaccharide structure, which permits extremely fine tuning of recognition determinants, is an appealing feature of carbohydrate-mediated cellular adhesion in general.

Acknowledgements

The authors are especially grateful to several members of their laboratory, including Drs. Jie Chen, Ross Kinloch, Thomas Rosiere and Yutaka Sakai, for their important contributions to the research reviewed here. The authors are also grateful to all members of their laboratory for good advice and constructive criticism throughout the course of the research. E.L. was initially supported by a post-doctoral fellowship from the Swiss National Science Foundation.

References

Bause E. Structural requirements of N-glycosylation of proteins. Biochem J 1983; 209:331-336.

Bleil JD, Wassarman PM. Mammalian sperm-egg interaction: identification of a glycoprotein in mouse egg zonae pellucidae possessing receptor activity for sperm. Cell 1980; 20:873-882.

Bleil JD, Wassarman PM. Galactose at the nonreducing terminus of O-linked oligosaccharides of mouse zona pellucida glycoprotein ZP3 is essential for the glycoprotein's sperm receptor activity. Proc Natl Acad Sci USA 1988; 85:6778-6782.

Chamberlin ME, Dean J. Human homolog of the mouse sperm receptor. Proc Natl Acad Sci USA 1990; 87:6014-6018.

Chen J, Litscher ES, Wassarman PM. Inactivation of the mouse sperm receptor, mZP3, by site-directed mutagenesis of individual serine residues located at the combining-site for sperm. Proc Natl Acad Sci USA 1998; 95:6193-6197.

Dietl J, ed. The Mammalian Egg Coat: Structure and Function. Springer-Verlag KG, Berlin, 1989.

Elhammer AP, Poorman RA, Brown E, Maggiora LL, Hoogerheide JG, Kézdy FJ. The specificity of UDP-GalNAc:poly-peptide N-acetylgalactosaminyltransferase as inferred from a database of *in vivo* substrates and from the *in vitro* glycosylation of proteins and peptides. J Biol Chem 1993; 268:10029-10038.

Florman HM, Wassarman PM. O-Linked oligosaccharides of mouse egg ZP3 account for its sperm receptor activity. Cell 1985; 41:313-324.

Florman HM, Bechtol KB, Wassarman PM. Enzymatic dissection of the functions of the mouse egg's receptor for sperm. Dev Biol 1984; 106:243-255.

Gwatkin RBL. Fertilization Mechanisms in Man and Mammals. Plenum Press, New York, 1977.

Gooley AA, Williams KL. Towards characterizing O-glycans: the relative merits of *in vivo* and *in vitro* approaches in seeking peptide motifs specifying O-glycosylation sites. Glycobiology 1994; 4:413-417.

Hynes M, Buck L, Gitt M, Barondes S, Dodd M, Jessell T. Carbohydrate recognition in neuronal development: structure and expression of surface oligosaccharides and β-galactoside-binding lectins. In: (Bock G, Harnett S, eds) Carbohydrate Recognition in Cellular Function. Ciba Found Symp 145. Chichester: Wiley, 1989, pp189-223.

Kinloch RA, Wassarman PM. Profile of a mammalian sperm receptor gene. New Biologist 1989; 1:232-238.

Kinloch RA, Roller RJ, Fimiani CM, Wassarman DA, Wassarman PM. Primary structure of the mouse sperm receptor's polypeptide chain determined by genomic cloning. Proc Natl Acad Sci USA 1988; 85:6409-6413.

Kinloch RA, Ruiz-Seiler B, Wassarman PM. Genomic organization and polypeptide primary structure of zona pellucida glycoprotein hZP3, the hamster sperm receptor. Dev Biol 1990; 142:414-421.

Kinloch RA, Mortillo S, Stewart CL, Wassarman PM. Embryonal carcinoma cells transfected with ZP3 genes differentially glycosylate similar polypeptides and secrete active mouse sperm receptor. J Cell Biol 1991; 115:655-663.

Kinloch RA, Sakai Y, Wassarman PM. Mapping the mouse ZP3 combining site for sperm by exon swapping and site-directed mutagenesis. Proc Natl Acad Sci USA 1995; 92:263-267.

Kornfeld R, Kornfeld S. Assembly of asparagine-linked oligosaccharides. Ann Rev Biochem 1983; 54:631-664.

Kuehn MJ, Heuser J, Normark S, Hultgren SJ. P pili in uropathogenic *E. coli* are composite fibres with distinct fibrillar tips. Nature Lond 1992; 356:252-255.

Lasky LA, Singer MS, Dowbenko D, Imai Y, Henzel WJ, Grimley C. Fennie C, Gillett N, Watson SR, Rosen SD. An endothelial ligand for L-selectin is a novel mucin-like molecule. Cell 1992; 69:927-938.

Litscher ES, Wassarman PM. Carbohydrate-mediated adhesion of eggs and sperm during mammalian fertilization. Trends Glycosci Glycotech 1993; 5:369-388.

Litscher ES, Wassarman PM. Characterization of a mouse ZP3-derived glycopeptide, gp55, that exhibits sperm receptor and acrosome reaction-inducing activity *in vitro*. Biochemistry 1996a; 35:3980-3985.

Litscher ES, Wassarman PM. Recombinant hamster sperm receptors that exhibit species-specific binding to sperm. Zygote 1996b; 4:229-236.

Litscher ES, Juntunen K, Seppo A, Penttilä L, Niemelä R, Renkonen O, Wassarman PM. Oligosaccharide constructs with defined structures that inhibit binding of mouse sperm to unfertilized eggs *in vitro*. Biochemistry 1995; 34:4662-4669.

Liu C, Litscher ES, Mortillo S, Sakai Y, Kinloch RA, Stewart CL, Wassarman PM. Targeted disruption of the *mZP3* gene results in production of eggs lacking a zona pellucida and infertility in female mice. Proc Natl Acad Sci USA 1996; 93:5431-5436.

Liu C, Litscher ES, Wassarman PM. Zona pellucida glycoprotein mZP3 bioactivity is not dependent on the extent of glycosylation of its polypeptide or on sulfation and sialylation of its oligosaccharides. J Cell Sci 1997; 110:745-752.

Lund B, Lindberg F, Marklund BI, Normark S. The PapG protein is the α-D-galactopyranosyl-(1-4)-β-D-galactopyranose-binding adhesin of uropathogenic *Escherichia coli*. Proc Natl Acad Sci USA 1987; 84:5898-5902.

McEver RP, Moore KL, Cummings RD. Leukocyte trafficking mediated by selectin-carbohydrate interactions. J Biol Chem 1995; 270:11025-11028.

Millar SE, Chamow SM, Baur AW, Oliver C, Robey F, Dean J. Vaccination with a synthetic zona pellucida peptide produces long-term contraception in female mice. Science 1989; 246:935-938.

Miller DJ, Macek MB, Shur BD. Complementarity between sperm surface β-galactosyltransferase and egg-coat ZP3 mediates sperm-egg binding. Nature Lond 1992; 357:589-593.

Nehrke K, Hagen FK, Tabak LA. Charge distribution of flanking amino acids influences O-glycan acquisition *in vivo*. J Biol Chem 1996; 271:7061-7065.

Rankin T, Familiari M, Lee E, Ginsberg A, Dwyer N, Blanchette-Mackie J, Drago J, Westphal H, Dean J. Mice homozygous for an insertional mutation in the *ZP3* gene lack a zona pellucida and are infertile. Development 1996; 122:2903-2910.

Ringuette MJ, Chamberlin ME, Baur AW, Sobieski DA, Dean J. Molecular analysis of cDNA coding for ZP3, a sperm binding protein of the mouse zona pellucida. Dev Biol 1988; 127:287-295.

Rosiere TG, Wassarman PM. Identification of a region of mouse zona pellucida glycoprotein mZP3 that possesses sperm receptor activity. Dev Biol 1992; 154:309-317.

Sadler JE. Biosynthesis of glycoproteins: formation of O-linked oligosaccharides. In: (Ginsburg V, Robbins PW, eds) Biology of Carbohydrates. New York: Wiley, 1984, pp199-288.

Salzmann GS, Greve JM, Roller RJ, Wassarman PM. Biosynthesis of the sperm receptor during oogenesis in the mouse. EMBO J 1983; 2:1451-1456.

Schachner M. Families of neural adhesion molecules. In: (Bock G, Harnett S, eds) Carbohydrate Recognition in Cellular Function. Ciba Found Symp 145. Chichester: Wiley, 1989, pp156-172.

Snell WJ, White JM. The molecules of mammalian fertilization. Cell 1996; 85:629-637.

Wang Y, Agrwal N, Eckhardt AE, Stevens RD, Hill RL. The acceptor substrate specificity of porcine submaxillary UDP-GalNAc:polypeptide N-acetylgalactosaminyltransferase is dependent on the amino acid sequences adjacent to serine and threonine residues. J Biol Chem 1993; 268:22979-22983.

Wassarman PM. The biology and chemistry of fertilization. Science 1987; 235:553-560.

Wassarman PM. Zona pellucida glycoproteins. Ann Rev Biochem 1988; 57:415-442.

Wassarman PM. Profile of a mammalian sperm receptor. Development 1990; 108:1-17.

Wassarman PM, ed. Elements of Mammalian Fertilization. CRC Press, Boca Raton, FL, 1991.

Wassarman PM. Towards molecular mechanisms for gamete adhesion and fusion during mammalian fertilization. Curr Opin Cell Biol 1995; 7:658-664.

Wassarman PM. Egg zona pellucida glycoproteins. In: (Kreis T, Vale R, eds) Guidebook to the Extracellular Matrix and Adhesion Proteins. Oxford: Oxford University Press, 1999, in press.

Wassarman PM, Litscher ES. Sperm-egg recognition mechanisms in mammals. Curr Top Dev Biol 1995; 30:1-19.

Wassarman PM, Mortillo S. Structure of the mouse egg extracellular coat, the zona pellucida. Intl Rev Cytol 1991; 130:85-109.

Wassarman PM, Bleil, JD, Florman HM, Greve JM, Roller RJ, Salzmann GS, Samuels FG. The mouse egg's receptor for sperm: what is it and how does it work? Cold Spring Harbor Symp Quant Biol 1985; 50:11-19.

Wassarman PM, Liu C, Litscher ES. Constructing the mammalian egg zona pellucida: some new pieces of an old puzzle. J Cell Sci 1996; 109:2001-2004.

Wilson IBH, Gavel Y, von Heijne G. Amino acid distributions around O-linked glycosylation sites. Biochem J 1991; 275:529-534.

Yanagimachi R. Mammalian fertilization. In: (Knobil E, Neill JD, eds) The Physiology of Reproduction. New York: Raven Press, 1994, pp189-317.

Yoshida A, Suzuki M, Ikenaga H, Takeuchi M. Discovery of the shortest sequence motif for high level mucin-type O-glycosylation. J Biol Chem 1997; 272:16884-16888.

20

Zona Pellucida-Induced Signal Transduction Via Sperm Surface β1,4-galactosyltransferase

Barry D. Shur
Emory University

Successful mammalian fertilization reflects the culmination of a wide range of prerequisite events, ranging from differentiation and maturation of the gametes to their transport, activation, binding and fusion with one another. Virtually all of these events have been the subject of considerable study (Yanagimachi 1994).

During spermatogenesis and epididymal maturation, the sperm surface is customized to enable the sperm to recognize and bind to the egg zona pellucida (Eddy 1988; Jones 1989). Some of these sperm surface changes likely result from epididymal secretions that are adsorbed onto the sperm surface and some are intrinsic to the sperm cell itself. Further changes occur when sperm mix with secretions of the accessory glands (Dravland, Joshi 1981). Unfortunately, the physiological significance of many of these surface modifications remains obscure (Storey, Kopf 1991).

In the female reproductive tract, the sperm undergoes a second series of both surface and intracellular transformations during a process collectively called capacitation. Sperm capacitation is associated with changes in the composition, fluidity and permeability of the sperm surface, as well as changes in sperm intermediary metabolism (Carrera et al. 1996; Friend et al. 1977; Hyne, Garbers 1979; Langlais et al. 1981; Leclerc et al. 1996; Llanos et al. 1982; O'Rand 1977; Singh et al. 1978; Talbot, Franklin 1978; Visconti et al. 1997; Vredeburgh-Wilberg, Parrish 1995). Capacitation results in sperm that are capable of completing the acrosome reaction and successfully fertilizing eggs. A limited number of sperm eventually reach the ovulated eggs in the oviduct and must then traverse the surrounding cumulus cells before binding to the outer acellular egg coat, or zona pellucida (Lin et al. 1994; Suarez 1998).

In the mouse, sperm binding to the zona pellucida triggers the completion of the acrosome reaction (Bleil, Wassarman 1983; Florman, Storey 1982; Saling, Storey 1979). In other species, however, the timing of the acrosome reaction, relative to zona binding, probably varies (Huang et al. 1981; Myles et al. 1987). The acrosome-reacted sperm penetrates the zona pellucida to reach the egg plasma membrane, where cell fusion and subsequent zygote formation occur. In the past few years, there has been considerable progress in the understanding of sperm fusion with the egg plasma membrane with the identification of fertilin, the sperm-localized fusigenic protein (Blobel et al. 1992), and its integrin receptor on the egg membrane (Almeida et al. 1995).

Sperm-Egg Recognition

While all of the events that culminate in fertilization are fundamentally important, this chapter focuses on the current understanding of sperm binding to the egg zona pellucida. Studies of a wide range of species, from marine to mammalian, suggest that a common theme is emerging for the molecular basis of gamete recognition (Macek, Shur 1988; Tulsiani et al. 1997). Sperm surface carbohydrate-binding proteins have been identified that have high affinity and specificity for complex glycoconjugate "ligands" on the egg coat. These sperm-derived, carbohydrate-binding proteins may function as lectin-like molecules, as best illustrated in sea urchin (Foltz, Lennarz 1990; Vacquier, Moy 1977), or have enzymatic activities associated with them, as appears the case for ascidians (Hoshi et al. 1983) and mouse (Macek, Shur 1988). Among all species analyzed, the utility of protein-carbohydrate complementarity for mediating sperm-egg binding is best illustrated in the mouse.

Sperm-Egg Binding in Mouse

ZP3 behaves as the zona pellucida receptor for sperm. The mouse egg zona pellucida is composed of three families of glycoproteins, referred to as ZP1, ZP2 and ZP3, with mean molecular weights of 200, 120 and 83 kD, respectively (Bleil, Wassarman 1980a). When these glycoproteins are added back individually to sperm-egg binding assays, only ZP3 inhibits sperm binding to the zona pellucida, and does so in a dose-dependent manner. This suggests that ZP3 is the zona pellucida receptor for sperm (Bleil, Wassarman 1980b). Consistent with this, ZP3 isolated from fertilized eggs, which no longer support sperm binding, no longer competitively inhibits sperm binding to unfertilized eggs.

O-linked oligosaccharides account for ZP3 sperm binding activity. Pronase digestion of ZP3 does not destroy its sperm binding activity, suggesting that the oligosaccharides are the active residues for sperm binding (Florman, Wassarman 1985). Removal of *N*-linked oligosaccharides by endo-H glycosidase digestion does not effect ZP3 sperm binding activity, but alkali release of *O*-linked chains destroys binding activity. The released *O*-linked oligosaccharides bind to sperm and competitively inhibit sperm-zona binding. Thus, sperm-binding oligosaccharides on ZP3 appear to be serine/threonine linked (Florman, Wassarman 1985).

Identification of the ZP3 glycoside epitope recognized by sperm. Although there is good evidence to indicate that *O*-linked oligosaccharide chains on ZP3 display at least one of the binding epitopes recognized by sperm (Florman, Wassarman 1985; Miller et al. 1992), the molecular composition of the sperm-binding oligosaccharide is a matter of active debate. At least four different monosaccharide residues have been implicated as being critical for initial sperm binding: α-galactose, β-*N*-acetylglucosamine (GlcNAc), fucose, and mannose (Cornwall et al. 1991; Johnston et al. 1998). Adding to this dilemma is the suggestion that *N*-linked oligosaccharides are involved in mouse sperm binding, in addition to *O*-linked oligosaccharides, since treatment of intact zona pellucida with *N*-glycanase, as opposed to treating solubilized ZP3, inhibits subsequent sperm binding (Yamagata 1985). Similarly, *N*-linked oligosaccharides have been implicated in the binding of porcine sperm to the zona pellucida (Yonezawa et al. 1995). The α-galactosyl and β-*N*-acetylglucosaminyl residues have been proposed as the recognition determinants on the *O*-linked oligosaccharides of ZP3, whereas the mannosyl residues are thought to be responsible for the sperm binding activity of *N*-linked glycosides (Litscher et al. 1995; Miller, Shur 1994; Tulsiani et al. 1992).

At least some of the confusion stems from the fact that different assays have been used to probe the nature of the sperm-binding glycoside. When synthetic oligosaccharides are used as competitive inhibitors of sperm-egg binding, oligosaccharides with terminal α-galactosyl, fucosyl and mannosyl residues have the strongest competitive inhibitory activity (Cornwall et al. 1991; Johnston et al. 1998; Litscher et al. 1995). One or more of these sugars have been suggested to compete for a "high" affinity binding site on sperm. Unfortunately, different workers have reached different conclusions regarding the involvement of these monosaccharides even when similar assays and reagents have been used (Johnston et al. 1998; Litscher et al. 1995).

Rather than relying upon the use of synthetic oligosaccharides as competitive inhibitors, the nature of the sperm-binding glycoside has been examined using solubilized ZP3 as a source of competitor. These assays have implicated both terminal α-galactosyl and GlcNAc residues in sperm-egg binding (Bleil, Wassarman 1988; Miller et al. 1992).

Even under the best of circumstances, however, the results of using synthetic glycosides or solubilized ZP3 as competitive inhibitors are brought into question by the recent finding that the sugar composition of the zona pellucida is heterogeneous (Avilés et al. 1997). Immunocytochemical analysis at the ultrastructural level reveals that α-galactosyl residues are confined to the inner portions of the zona pellucida, whereas other sugars, such as GlcNAc, are dispersed uniformly throughout the zona. Thus, it is impossible to know whether the inhibitory activity being assayed using soluble glycosides reflects something that is available to the sperm at initial binding or during later aspects of zona penetration, or is accessible to the sperm at all. The spatial heterogeneity of the zona glycosides also indicates a previously unrecognized temporal regulation of some glycosyltransferase activities during secretion of the zona matrix.

Since α-galactosyl residues are not accessible to sperm on the zona surface, they cannot be involved in initial gamete recognition. Consistent with this, eggs from females bearing null mutations in α-galactosyltransferase are fertilized normally, at least *in vivo* (Thall et al. 1995). Furthermore, recombinant ZP3 produced in CV-1 cells, which do not express α-galactosyltransferase, retains its sperm binding activity (Beebe et al. 1992). Nevertheless, α-galactosyl residues do impact sperm-zona interactions and may, therefore, function during later stages of sperm penetration through the zona matrix.

An alternative approach to the use of soluble competitive inhibitors is to biochemically modify the intact zona pellucida to impact its sperm-binding activity. While the use of the intact zona pellucida to probe the biochemical nature of initial sperm-egg interactions is wrought with its own limitations, it does allow one to probe initial aspects of gamete recognition. Using this approach, evidence suggests that sperm recognize a terminal GlcNAc during initial aspects of zona binding. Interestingly, the predominate *O*-linked oligosaccharide in ZP3 has been characterized as a small trisaccharide with a terminal GlcNAc (Nagdas et al. 1994).

ZP3 induces the acrosome reaction by activating a sperm G-protein. Not only is ZP3 responsible for binding sperm, but it also induces the acrosome reaction (Bleil, Wassarman 1983). During the acrosome reaction, the plasma membrane and underlying outer acrosomal membrane fuse together, releasing enzymes that aid sperm penetration through the zona pellucida (Yanagimachi 1994). ZP3 is thought to elicit the acrosome reaction by crosslinking, or aggregating, its receptor within the sperm plasma membrane (Leyton, Saling 1989a), which triggers a pertussis toxin (PTx)-sensitive heterotrimeric G-protein cascade (Endo et al. 1988; Ward, Kopf 1993; Ward et al. 1992). A $G_{i\alpha}$ subunit has been identified in sperm as assessed by PTx-dependent adenosine diphosphate-ribosylation and immunoblotting of a 41-kD sperm protein. ZP3 binding induces GTPase activity and GTP binding, consistent with G_i-protein activation during induction of the acrosome reaction. The binding of ZP3 is also thought to activate other signal transduction pathways critical for completion of the acrosome reaction, including voltage-dependent ion channels and tyrosine kinase activity (Burks et al. 1995; Florman et al. 1989, 1992).

As sperm undergo the acrosome reaction, they must remain transiently attached to the zona prior to initiation of penetration. It has been suggested that the binding of acrosome-reacted sperm to the zona is dependent on ZP2, since acrosome-reacted sperm lose affinity for ZP3 and gain affinity for ZP2 (Bleil, Wassarman 1986; Mortillo, Wassarman 1991). There is evidence that ZP2 binds to a site on sperm that has properties similar to the trypsin-like proteases (Benau, Storey 1987; Bleil et al. 1988).

Acrosome-reacted sperm penetrate the zona pellucida. Eventually, the acrosome-reacted sperm penetrates the zona pellucida, presumably relying upon hydrolytic enzymes released from the acrosome to digest a penetration slit through the zona matrix (Yanagimachi 1994). Most notable among these is the trypsin-like protease acrosin. However, complete elimination of acrosin by homologous recombination still permits sperm to penetrate through the zona, suggesting that other acrosomal enzymes are likely important for penetration (Adham et al. 1997; Baba et al. 1994). Recently, two novel serine proteases have been identified in the mouse acrosome, where they may participate in zona penetration (Kohno et al. 1998). Since the formation of long-term sperm-zona adhesions would impede the progress of sperm penetration through the zona, the acrosome may also contain enzymes that specifically prevent stable molecular adhesions between sperm surface receptors and their zona pellucida ligands.

ZP3 sperm receptor activity is destroyed after egg activation. Soon after egg activation, cortical granules are released to elicit the zona block to polyspermy. Presumably, the contents of the cortical granules inactivate ZP3, resulting in a loss of ZP3's sperm binding activity and accounting for the block to sperm binding. However, there is no observable proteolysis of ZP3 at fertilization (Bleil, Wassarman 1980b). In fact, ZP3 is indistinguishable by SDS-PAGE before and after fertilization. The loss of ZP3's sperm binding activity must, therefore, result from subtle, but critical, modifications in its structure such that it is no longer recognized by the sperm receptor(s).

Several Sperm Proteins Are Thought to Function as Zona Receptors

The relative simplicity of the zona pellucida has allowed for the identification of ZP3 as one of the primary receptors for sperm. However, the complexity of the sperm plasma membrane has made identification of the complementary receptor for the zona pellucida difficult. Several approaches have been applied to this problem, ranging from serological to biochemical to genetic, and consequently, several sperm receptors for the zona pellucida have been proposed and are discussed below. While there are data to suggest many of these sperm proteins may participate in zona binding, the most exhaustively studied gamete receptor is sperm surface β1,4-galactosyltransferase, or GalTase (Benau, Storey 1988; Benau et al. 1990; Cardullo, Wolf 1995; Gong et al. 1995; Lopez, Shur 1987; Lopez et al. 1985; Lu, Shur 1997; Macek et al. 1991; Miller et al. 1992; Scully et al. 1987; Shur, Hall 1982a, b; Shur, Neely 1988; Youakim et al. 1994).

Cell Surface Glycosyltransferases

Glycosyltransferases are traditionally viewed as intracellular enzymes, which synthesize complex oligosaccharides by the sequential transfer of sugar residues from nucleotide or membrane-bound donors to growing polysaccharide chains (Joziasse 1992). When subcellular fractionation became feasible, glycosyltransferase activities were also detected on enriched plasma membrane fractions and on the surface of intact cells. These observations prompted the hypothesis that glycosyltransferases may function as cell adhesion molecules by binding their specific oligosaccharide substrates, or ligands, on adjacent cell surfaces or in the extracellular matrix (Roseman 1971; Roth et al. 1971). Since sugar donor substrates are normally not present in the extracellular fluids, it was presumed that the surface-localized glycosyltransferase-oligosaccharide complex would form a stable adhesive bond.

The initial reports of glycosyltransferase activities on the cell surface were controversial, since it was difficult to eliminate completely all possible contamination from intracellular sources (Evans et al. 1995; Shur 1991, 1992). This controversy has largely been resolved by a wealth of evidence illustrating the presence of a few specific glycosyltransferases on the cell surface, most notably GalTase. GalTase has been demonstrated on the surface of specific cell types at the light, confocal and electron microscopic levels through the use of monospecific polyclonal and monoclonal antibodies raised against affinity-purified and/or bacterially-expressed recombinant proteins, by the presence of GalTase enzyme activity on purified plasma membrane preparations and by a variety of other procedures that have been discussed extensively (Evans et al. 1995; Shur 1991, 1992). A molecular hypothesis accounting for the expression of GalTase on the cell surface has resulted from the cloning of the GalTase gene products (Evans et al. 1995).

Molecular biology of cell surface GalTase. In those species analyzed thus far, the gene for GalTase encodes two similar, but not identical, proteins (Russo et al. 1990; Shaper et al. 1988). Both GalTase proteins have a type II membrane configuration, analogous to all other glycosyltransferases cloned to date (Joziasse 1992), with a relatively short amino terminal cytoplasmic domain, a signal sequence/transmembrane domain, and a large carboxy terminal lumenal or extracellular catalytic domain. The two GalTase proteins have identical catalytic and transmembrane domains, but differ in their cytoplasmic domains due to differential transcription initiation of two distinct RNAs. The shorter protein has a cytoplasmic tail of only 11 amino acids, whereas the longer species has a 24 amino acid cytoplasmic domain.

Whether the two different GalTase isoforms that result from the two different GalTase RNAs have functionally distinct roles within the cell or, rather, reflect tissue-specific proteins with similar, if not identical, biosynthetic functions remains an issue of active debate (Harduin-Leperet et al. 1993; Rajput et al. 1996; Shur et al. 1998). A number of observations suggest that the short GalTase isoform is normally confined to the Golgi complex, where it serves a purely biosynthetic function; it is this isoform that is specifically upregulated during lactation to participate in lactose biosynthesis. Similarly, the long GalTase isoform can function biosynthetically in the Golgi complex, but in addition, there is evidence to indicate that its alternate cytoplasmic domain enables it to function on the cell surface (Evans et al. 1995; Shur et al. 1998). Recent studies of GalTase-deficient mice have largely resolved this controversy and show that the cytoplasmic domain of the long isoform is critical for GalTase to function as a signal transducing receptor on the cell surface (Lu, Shur 1997).

Cell surface GalTase functions as a cell adhesion molecule during fertilization and development. The demonstration that GalTase is present on the cell surface led to an analysis of surface GalTase function. Through the use of a wide range of GalTase-specific perturbants, including competitive GalTase substrates, antibodies raised against affinity-purified bacterially-expressed recombinant GalTase, GalTase modifier proteins, glycosidase pretreatments and others, it became clear that GalTase functions as a cell adhesion molecule by binding to specific oligosaccharide substrates, or ligands, on adjacent cell surfaces or in the extracellular matrix. The ability of a variety of reagents, all of which have different modes of action, but all of which are GalTase-specific, to inhibit selected cell-cell and cell-matrix interactions strongly supports the cell adhesion function of surface GalTase (Shur 1991, 1992). These results have been confirmed by the ability to manipulate, both positively and negatively, GalTase expression on the cell surface in transfected cells as well as in transgenic animals (Shur et al. 1998).

Specificity of surface GalTase function. The surface isoform of GalTase has a much more restricted substrate specificity than does the traditional biosynthetic enzyme in the Golgi complex. In all instances examined (sperm, fibroblasts, PC12 cells, melanoma cells and adrenal carcinoma cells), the surface enzyme usually recognizes only one predominant glycoprotein ligand, although the Golgi enzyme recognizes a broad spectrum on GlcNAc-terminating oligosaccharides (Shur et al. 1998). The presence of a terminal GlcNAc, the terminal monosaccharide recognized by GalTase, is therefore necessary but not sufficient to confer binding by surface GalTase. This may reflect membrane-dependent recognition of other determinants in the oligosaccharide chain or within the protein backbone. In any event, the enzymatic name β1,4-galactosyltransferase is misleading in that it only refers to the linkage catalyzed by this enzyme (that is, galactose-β1,4-GlcNAc) when its UDP-galactose substrate is made available. In this regard, it is now known that GalTase represents one member of a newly defined family of proteins that have β1,4-galactosyltransferase activity. Thus far, six distinct β4-galactosyltransferases have been identified, and it is believed, though not yet proven, that each has a unique substrate specificity and tissue distribution (Almeida et al. 1997; Lo et al. 1998; Sato et al. 1998).

Sperm GalTase Function during Murine Fertilization

Surface GalTase during spermatogenesis and capacitation. GalTase is found on the surface of mouse sperm in the expected location for a zona receptor—a discrete domain on the dorsal, anterior aspect of the sperm head, where it behaves as an integral membrane protein (Shur, Neely 1988; Fig. 1). The suggestion that GalTase is uniformly present on the sperm surface, but is capped onto the dorsal aspect of the sperm head as evidenced from results obtained with anti-GalTase antibodies, is not consistent with the fact that GalTase shows the same dorsal, anterior localization on live and fixed sperm (Cardullo, Wolf 1995; Lopez et al. 1985).

During spermatogenesis, surface GalTase is expressed as early as the pachytene stage of spermatocytes (Scully et al. 1987). There is evidence to suggest that GalTase may facilitate adhesion of the developing germ cell to the Sertoli cell (Pratt et al. 1993). As spermatogenic cells develop, GalTase is redistributed to the anterior aspect of the sperm head, possibly by association with the actin cytoskeleton. Sperm are stored in the cauda epididymis during which time epididymally secreted glycoconjugates mask the sperm GalTase binding site. During capacitation, these competitive glycoconjugates are shed from the sperm surface, making GalTase available to bind its oligosaccharide ligand in the zona pellucida (Shur, Hall 1982a).

Sperm GalTase as a zona receptor. Several laboratories have reported that sperm GalTase functions as a gamete receptor during mouse fertilization by binding to oligosaccharide ligands in the egg zona pellucida (Benau, Storey 1988; Benau et al. 1990; Cardullo, Wolf 1995; Gong et al. 1995; Lopez, Shur 1987; Lopez et al. 1985; Lu, Shur 1997; Macek et al. 1991; Miller et al. 1992; Scully et al. 1987; Shur, Hall 1982a, b; Shur, Neely 1988; Youakim et al. 1994). The importance of GalTase in sperm-zona binding has been demonstrated using a series of reagents that block GalTase or the GalTase recognition site on the zona pellucida, all of which inhibit sperm-zona binding. For example, anti-GalTase antibodies, their Fab fragments and the GalTase substrate modifier α-lactalbumin all inhibit sperm-zona binding in a dose-dependent manner, as do competitive oligosaccharide substrates and affinity-purified sperm GalTase. Alternatively, either blocking or removing the GalTase-binding site on the zona destroys sperm binding activity.

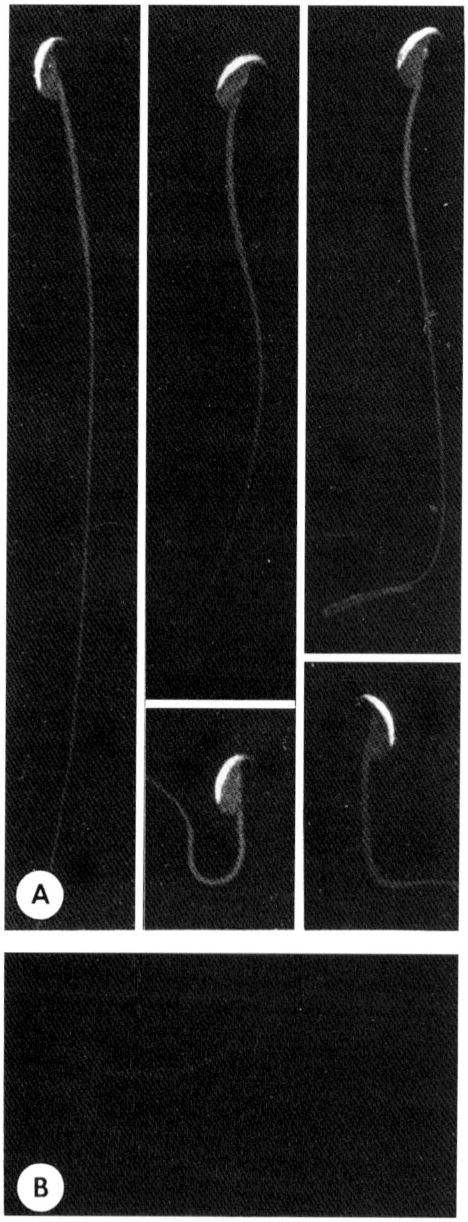

Figure 1. GalTase is localized to the dorsal anterior plasma membrane of the mouse sperm head as assayed by indirect immunofluorescence using antibodies against affinity-purified GalTase (A). Preimmune sera shows no immunoreactivity (B). From Lopez et al. 1985.

Sperm GalTase specifically recognizes O-linked oligosaccharides on ZP3. The sperm GalTase ligand, or substrate, has been shown to be of the same class of *O*-linked oligosaccharides as found on the ZP3 glycoprotein that possess sperm-binding activity (Miller et al. 1992). Sperm GalTase selectively binds *O*-linked oligosaccharides on ZP3; it does not bind other ZP3 oligosaccharides or those on other zona glycoproteins, even though they are recognized by non-sperm GalTase. The interaction between sperm GalTase and ZP3 oligosaccharides is necessary for sperm-zona binding, as shown by two observations. When GalTase recognition sites in soluble ZP3 are blocked by glycosylation, ZP3 can no longer bind GalTase and loses its sperm binding activity (Miller et al. 1992). Similarly, if the GalTase recognition sites of ZP3 are removed by digestion with β-N-acetylglucosaminidase (GlcNAc'dase), ZP3 loses its sperm binding activity (Miller et al. 1992).

GalTase aggregation induces the acrosome reaction. Evidence suggests that ZP3 induces the sperm acrosome reaction by aggregating its sperm-bound receptor, thus activating a PTx-sensitive heterotrimeric G-protein cascade. It is not surprising, therefore, that ZP3 has multiple GalTase-binding oligosaccharides (Miller et al. 1992) and that the acrosome reaction can be induced, in the absence of ZP3, by multivalent, but not by monovalent, anti-GalTase antibodies (Macek et al. 1991; Fig. 2). Crosslinking of GalTase with antibodies, or ZP3, activates a PTx-sensitive, heterotrimeric G-protein complex that is directly associated with the long GalTase cytoplasmic domain, triggering the acrosome reaction (Gong et al. 1995).

These studies predict that the acrosome reaction should be induced by synthetic polymers derivatized with terminal GlcNAc residues. This has recently been shown to be the case using both synthetic polyacrylamide backbones or bovine serum albumin derivatized with GlcNAc. In controls, monovalent GlcNAc is unable to induce the acrosome reaction nor are polymers derivatized with galactosyl residues (Loeser, Tulsiani 1997; E Simanek, Q Lu, BD Shur and G Whitesides, unpublished data).

As a consequence of the acrosome reaction, GalTase is redistributed to the lateral sperm head where its function remains unknown (Lopez, Shur 1987). One possibility is that GalTase may stabilize sperm adhesion to the zona during the initial stages of penetration. In any event, the presence of GalTase on acrosome reacted sperm could conceivably impede sperm penetration through the zona by binding to exposed GlcNAc residues on ZP3 oligosaccharides. This potential problem is eliminated by the release of an acrosomal GlcNAc'dase that removes the GalTase binding site from ZP3 in the vicinity of the penetrating sperm (Miller et al. 1993a).

Destruction of GalTase binding sites after fertilization prevents polyspermic binding. After egg activation, cortical granules are exocytosed at the egg plasma membrane. These vesicles contain various hydrolases that act on the zona pellucida. The resulting zona loses its ability to bind sperm and thereby blocks subsequent fertilization by other sperm. *In vitro* assays demonstrate that ZP3 loses its sperm-receptor activity after fertilization (Bleil, Wassarman 1980b). This is now known to be due to the release of GlcNAc'dase from the egg cortical granules, which destroys the GalTase binding sites on ZP3 and

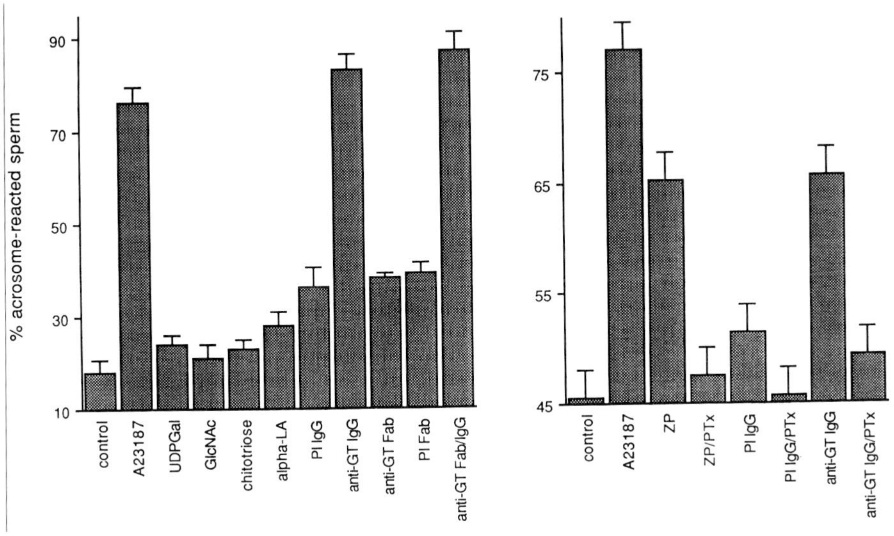

Figure 2 (left). Aggregation of GalTase induces the acrosome reaction. Four different GalTase-specific perturbants all inhibit sperm-egg binding, but do not induce the acrosome reaction as they are monovalent in nature (UDP-galactose, GlcNAc, chitotriose and alpha-lactalbumin). However, anti-GalTase IgG inhibits sperm-egg binding and induces the acrosome reaction by crosslinking GalTase. Anti-GalTase Fab fragments do not induce the acrosome reaction unless they are crosslinked by a secondary anti-rabbit IgG. Pertussis toxin (PTx) inhibits both the zona-induced (ZP) as well as the anti-GalTase-induced acrosome reaction, indicating the involvement of a heterotrimeric G-protein cascade. Maximal acrosome reaction is achieved by the addition of the calcium ionophore A23187. (From Macek et al. 1991; Gong et al. 1995.)

accounts for the loss of ZP3's sperm-binding activity following fertilization (Miller et al. 1993b; Fig. 3). Inhibition of the GlcNAc'dase released at egg activation by either specific substrate analogs or antibodies prevents the loss of the GalTase-binding site on ZP3 and prevents the block to polyspermic binding (Fig. 4).

Altering the Expression of Sperm GalTase Impacts the Efficacy of Sperm-Egg Binding

The use of molecular genetics to manipulate GalTase expression in the germ line allows one to examine the consequences of both elevating and eliminating GalTase expression on sperm fertilizing ability. A simple prediction is that elevating the expression of GalTase on the sperm surface in appropriate transgenic animals would lead to increased binding of ZP3, increased G-protein activation, and increased sensitivity to zona-induced acrosome reaction. Similarly, one might predict that eliminating GalTase from the sperm surface through homologous recombination would lead to reduced ZP3 binding and to sperm that are refractory to zona-induced acrosome reaction. Both of these predictions have been tested and are discussed below.

Overexpressing sperm GalTase makes sperm hypersensitive to ZP3. Sperm from transgenic males that have elevated levels of surface GalTase bind more radiolabeled ZP3 ligand as expected, but unexpectedly, bind much less efficiently to the zona pellucida than do wild-type sperm. The reduced binding of transgenic sperm is due to two GalTase-dependent defects in sperm-egg binding (Youakim et al. 1994).

The first observable defect is that sperm are not able to release their "decapacitation" factors as readily as wild-type sperm. Sperm GalTase is normally cryptic in

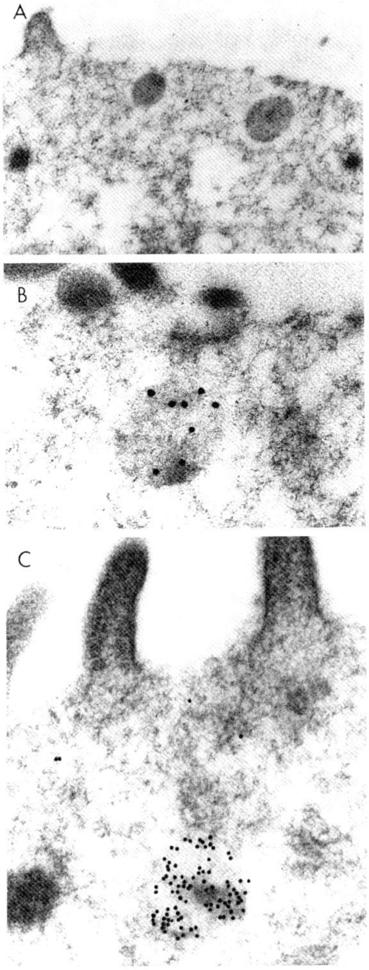

Figure 3 (right). Electron micrograph of unfertilized mouse eggs indicating localization of GlcNAc'dase to the cortical granules. No immunogold labeling is seen with preimmune IgG (A), but anti-GlcNAc'dase IgG readily labels the cortical granules (C). The identity of the cortical granules is confirmed by staining with lens culinaris lectin (B). (From Miller et al. 1993b.)

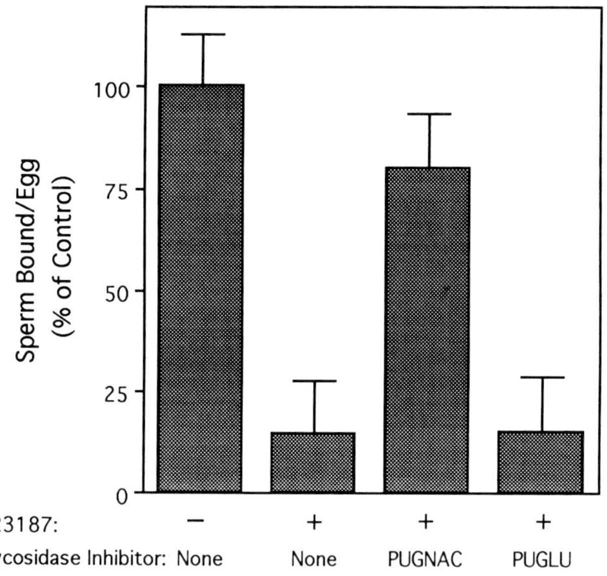

Figure 4. Pretreating eggs with the calcium ionophore A23187 releases cortical granule hydrolases, creating the block to polyspermy as assayed by a reduction in sperm binding after ionophore treatment. Addition of PUGNAC to inhibit the cortical granule GlcNAc'dase inhibits the block to polyspermy by preventing the loss of the GalTase-binding site on ZP3. The structurally related compound PUGLU has no effect on GlcNAc'dase activity or on the block to polyspermy. (From Miller et al. 1993b.)

the epididymis by being masked with epididymal glycoconjugates that are shed from the sperm surface during capacitation, thus exposing the GalTase binding site for ZP3 (Shur, Hall 1982a). Transgenic sperm, due to their greatly increased level of surface GalTase, bind either more epididymal glycosides or bind these with greater affinity, since after capacitation, transgenic sperm remain complexed with higher than normal levels of epididymal glycosides. The end result is that fewer sperm have exposed GalTase available for binding its zona ligand, relative to wild-type sperm.

All sperm populations, however, are heterogeneous, and some transgenic sperm have released sufficient epididymal glycosides to allow zona binding. Those transgenic sperm that do bind show only a transient, low affinity binding, which results from the second, more apparent, GalTase-dependent defect, that of precocious acrosome reactions (Fig. 5). Due to the greatly increased level of sperm GalTase, transgenic sperm that are able to bind the zona are hypersensitive to ZP3, and consequently they undergo accelerated G-protein activation and precocious acrosome reaction and bind to eggs more tenuously than do wild-type sperm (Gong et al. 1995; Youakim et al. 1994).

Eliminating GalTase makes sperm refractory to ZP3. In contrast to that seen with sperm that overexpress GalTase, sperm from GalTase-null males are refractory to zona-induced acrosome reaction (Lu, Shur 1997). When both GalTase isoforms are eliminated by homologous recombination, sperm have negligible levels of GalTase, and as expected, are unable to bind the ZP3 ligand or undergo a ZP3-induced acrosome reaction, as are wild-type sperm (Fig. 6). In contrast, GalTase-null sperm undergo the acrosome reaction normally in response to calcium ionophore, which bypasses the requirement for ZP3 binding. The inability of GalTase-null sperm to undergo a ZP3-induced acrosome reaction renders them physiologically inferior to wild-type sperm, since they are relatively unable to penetrate the egg coat and fertilize the oocyte *in vitro* (Fig. 7).

Since eliminating both GalTase isoforms leads to a loss of surface GalTase expression as well as to a loss of Golgi-based galactosylation, it is possible that the inability of GalTase-null sperm to bind ZP3, undergo an acrosomal reaction and penetrate the zona pellucida is the secondary result of defective galactosylation during spermatogenesis, rather than being the direct result of surface GalTase deficiency. This possibility was tested by creating mice deficient specifically in the long GalTase isoform—the form hypothesized to be responsible for GalTase function as a signal transducing receptor (Lu, Shur 1997).

Sperm from long isoform-null males have near normal levels of GalTase enzyme activity, reflecting the short GalTase isoform, and glycoprotein galactosylation in testis homogenates is similar to that observed in wild-type testis, as assessed by RCA-1 lectin blotting of testicular proteins (Lu, Shur 1997). Although galactosylation appears normal, the sperm are unresponsive to either anti-GalTase antibody- or zona pellucida-induced acrosome reactions; however, they respond normally to

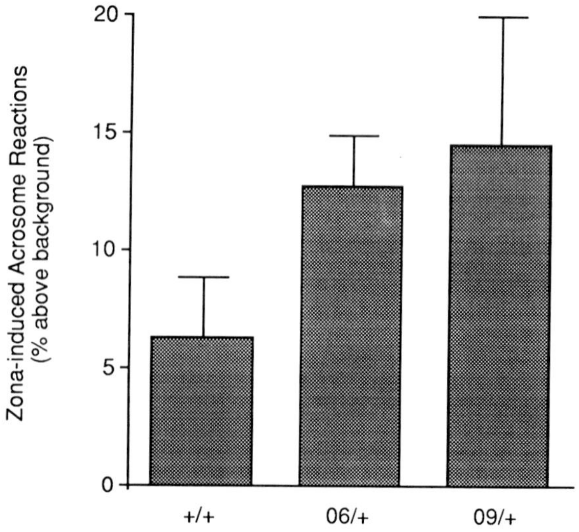

Figure 5. Sperm from two different strains of transgenic mice (06/+, 09/+), which have elevated surface GalTase expression, are hypersensitive to zona-induced acrosome reaction relative to wild-type (+/+) sperm. (From Youakim et al. 1994.)

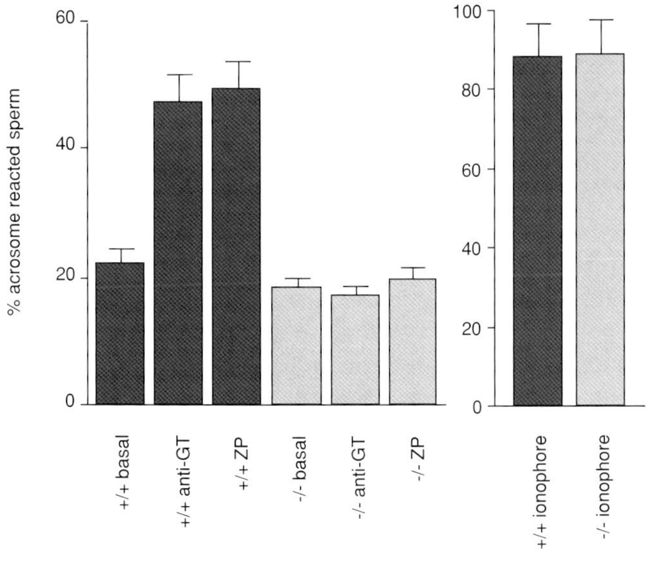

Figure 6. GalTase-null (-/-) sperm are unable to undergo an acrosome reaction in response to either ZP3 or anti-GalTase antibodies. Zona pellucida (ZP) glycoproteins and anti-GalTase antiserum are both able to induce acrosome reactions in wild-type (+/+) sperm, but not in GalTase-null (-/-) sperm. In contrast, the addition of calcium ionophore A23187 induced acrosome reaction to similar degrees in both sperm genotypes. (From Lu, Shur 1997.)

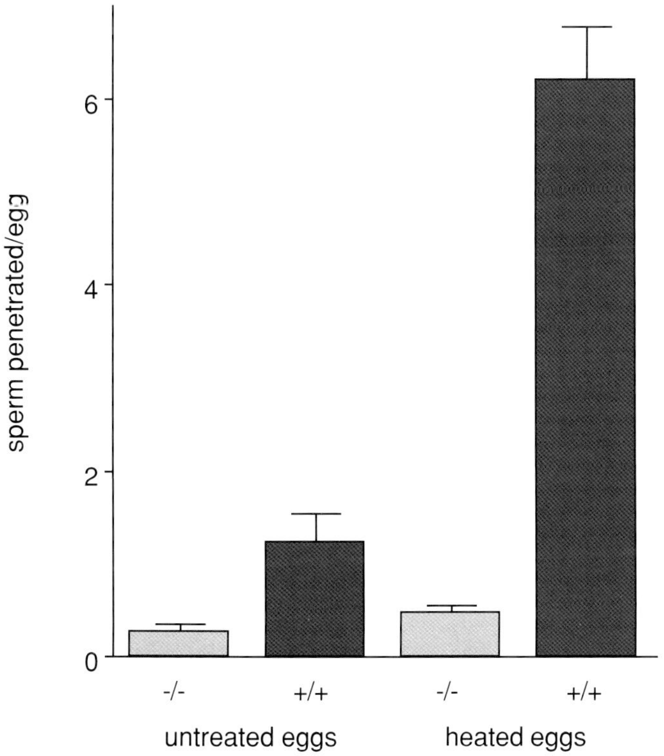

Figure 7. GalTase-null (-/-) sperm are relatively unable to penetrate the zona pellucida. The relative inefficiency of GalTase-null sperm to penetrate the zona is more apparent when the eggs are first mildly heated to destroy the cortical granule enzymes that normally limit the number of sperm that can penetrate the egg. (From Lu, Shur 1997.)

calcium ionophore. Thus, the inability of GalTase-null sperm to undergo a ZP3-induced acrosome reaction is a direct result of the lack of the long GalTase isoform on the sperm surface, and is not the secondary result of defective galactosylation during spermatogenesis. The cytoplasmic domain unique to the long GalTase isoform that is responsible for association of the heterotrimeric G-protein complex (Gong et al. 1995) appears to be absolutely required for ZP3-dependent signal transduction.

Sperm from GalTase-null males still bind to the zona pellucida and are fertile. The acrosome reaction has traditionally been viewed as a prerequisite for sperm penetration through the zona pellucida. However, GalTase-null sperm are fertile, even though they can not undergo ZP3-induced acrosome reaction (Lu, Shur 1997). Similarly, sperm made null for the acrosomal protease, acrosin, are also fertile (Adham et al. 1997; Baba et al. 1994). Both of these observations suggest that the entire ZP3-GalTase-G-protein activated release of acrosomal enzymes is dispensable for fertilization. This then begs the question as to whether the presence of this receptor-ligand complex offers some physiological or competitive advantage to sperm. Consistent with this possibility, GalTase-null sperm are only ~7% as efficient as wild-type sperm in penetrating the zona pellucida and fertilizing the oocyte (Fig. 7). Thus, in a noncompetitive environment where all sperm are phenotypically similar, as in the GalTase-null male, a few sperm eventually find their way through the zona pellucida, by either spontaneous acrosome reaction, breaks in the zona pellucida or other scenarios, and fertilize the egg. However, when compared to what occurs with wild-type sperm, this process is extremely inefficient. This suggests that in a competitive environment where not all sperm may be functionally equivalent, the integrity of the GalTase receptor complex may be advantageous for successful fertilization. A similar explanation has recently been shown to apply to acrosin-null sperm, since when acrosin-null sperm are mixed with wild-type sperm, either by mixed inseminations or in chimeric testes, the wild-type sperm completely outcompeted acrosin-null sperm in the ability to fertilize eggs (Adham et al. 1997).

It is particularly interesting that GalTase-null sperm still bind to the zona pellucida, even though they show low levels of binding ZP3 ligand in solution. This implies that other sperm components cooperate with GalTase to mediate gamete recognition. The most attractive model at present is that GalTase recognizes ZP3 oligosaccharides in the context of another sperm surface component that is responsible for the initial docking of sperm to the egg coat, similar to the concerted action of selectins and integrins in lymphocyte adhesion to the endothelium (Lasky 1992). With this in mind, it would be of interest to determine if other sperm surface components thought to facilitate adhesion to the zona are responsible for the binding of GalTase-null sperm.

GalTase is Present on All Mammalian Sperm Assayed

GalTase is found on all mammalian sperm tested, including human, where it is confined to the plasma membrane consistent with a role in gamete recognition (Fayrer-Hosken et al. 1991; Huszar et al. 1997; Larson, Miller 1997). In this regard, anti-GalTase antibodies and other GalTase-specific perturbants inhibit sperm-egg binding in some species, most notably bovine (M Tengowski, M Wassler and BD Shur, unpublished data). Similarly, a ZP3 homolog is expressed in the ovaries of many mammals (Chamberlin, Dean 1990; Sacco et al. 1989). This then raises the issue of what determines species specificity, if molecules similar to GalTase and ZP3 function in fertilization in other species. One possibility is that the membrane environment influences the substrate specificity of surface GalTase (Shur et al. 1998), which may impact recognition of ZP3 oligosaccharides in a species-specific manner. Alternatively, other sperm surface components may complex with GalTase to impart species specificity to the process. It is noteworthy that GalTase-null sperm are still able to bind to the egg coat (Lu, Shur 1997), suggesting, as before, that GalTase cooperates with other zona-binding proteins to form a multimeric gamete receptor.

Potential GalTase Partners on the Sperm Surface

Three observations support the hypothesis that GalTase functions during zona binding in cooperation with other sperm proteins. First, and most compelling, is that GalTase-null sperm still bind to the zona pellucida, although this binding is unproductive in that it does not lead to an acrosome reaction (Lu, Shur 1997). Second, sperm GalTase specifically recognizes *O*-linked oligosaccharides on ZP3, although soluble non-sperm GalTase recognizes oligosaccharides on all three zona pellucida glycoproteins (Miller et al. 1992). One explanation for this apparent specificity is that GalTase is presented to the zona matrix in concert with other sperm proteins that limit its accessibility to other zona glycosides. Third, GalTase is present on all mammalian sperm assayed and has the proper location expected of a zona-binding protein (Fayrer-Hosken et al. 1991; Huszar et al. 1997; Larson, Miller 1997). While GalTase on different sperm could recognize species-specific differences in oligosaccharide structure, it is equally likely that GalTase serves as a non-species-specific signaling component in response to binding ZP3 oligosaccharides and that the specificity of sperm-zona adhesion is dependent upon other sperm components. This would also account for the ability of GlcNAc-terminating glycosides to inhibit sperm-zona binding, but not necessarily by competing for a high affinity site (Miller, Shur 1994).

Several sperm surface components have been implicated in zona binding. They have been identified by a wide range of approaches, including the biochemical analysis of sperm mutations thought to impact fertilizing ability (GalTase; Shur, Bennett 1979), development of inhibitory monoclonal antibodies (PH-20; Primakoff et al. 1985), analysis of sperm autoantigens (RSA; O'Rand et al. 1988) and identification of sperm proteins that bind the zona pellucida (zonadhesin; Hardy, Garbers 1995). Additional approaches include use of zona pellucida affinity columns (p47; Ensslin et al. 1998), photoaffinity crosslinking to the zona pellucida (sp56; Bleil, Wassarman 1990), application of radiolabeled zona glycoprotein to blots of sperm lysates (p95; Leyton, Saling 1989b), use of competitive glycoside substrates (mannosidase; Cornwall et al. 1991), binding to sulfated glycolipids (SLIP1; Tanphaichitr et al. 1993), and an indirect implication that sperm surface components require the chaperone, calmegin, for proper folding and expression (Ikawa et al. 1997). Of this variety, two proteins other than GalTase have been specifically implicated as sperm receptors for ZP3: sp56 and p95. It is worth considering whether these, or any of the previously identified sperm surface proteins implicated in zona binding, may function in concert with GalTase to mediate species-specific gamete recognition.

sp56 is a sperm peripheral membrane protein that was first identified by ZP3-photoaffinity labeling (Bleil, Wassarman 1990). Although there are data consistent with sp56 having zona binding activity (Bookbinder et al. 1995), the physiological role of sp56 in zona binding remains unclear; sp56 can only be removed from ZP3 affinity columns by strong denaturants (such as urea) and it exists as a peripheral membrane protein on the sperm surface. Recently, sp56 has been shown to be a soluble constituent of the acrosomal matrix and not a cell surface component, which argues that sp56 does not function during initial gamete recognition (Foster et al. 1997). However, this raises the interesting possibility that sp56 may facilitate the binding of acrosome-reacted sperm to the zona prior to penetration.

Another mouse sperm protein thought to physically associate with ZP3 is p95, a phosphotyrosine-containing protein of Mr 95,000 (Leyton, Saling 1989b). This protein has subsequently been demonstrated to be a testis-specific tyrosine phosphorylated form of hexokinase (Kalab et al. 1994). A putative human analog, hu9, has been isolated and shown to have protein kinase activity, and peptides deduced from its sequence inhibit sperm-egg binding (Burks et al. 1995). How these peptides interfere with recognition of sperm-binding oligosaccharides on ZP3 and whether the human clone is, in fact, the mouse homolog rather than the previously identified c-mer proto-oncogene (Bork 1996; Tsai, Silver 1996) is a matter of active debate.

Given that sp56 and p95 are unlikely to function as GalTase-associated proteins during initial binding, attention should be turned towards other sperm proteins implicated in zona binding. There is excellent reason to believe that zonadhesin, p47, PH-20, sperm mannosidase, among others may function as zona-binding proteins. It is now important to determine whether any of them may function during sperm-zona interactions by forming multimeric gamete receptor complexes in association with GalTase. Maybe then researchers will be in a position to formulate a unified molecular understanding of mammalian gamete recognition.

Acknowledgement

Original work in the author's laboratory was supported by grants from the National Institutes of Health.

References

Adham IM, Nayernia K, Engel W. Spermatozoa lacking acrosin protein show delayed fertilization. Mol Reprod Dev 1997; 46:370-376.

Almeida EAC, Huovila A-J, Sutherland AE, Stephens LE, Calarco PG, Shaw LM, Mercurio AM, Sonnenberg A, Primakoff P, Myles DG, White JM. Mouse egg integrin a6β1 functions as a sperm receptor. Cell 1995; 81:1095-1104.

Almeida R, Amado M, David L, Levery SB, Holmes EH, Merkx G, van Kessel AG, Rygaard E, Hassan H, Bennett E, Clausen H. A family of human β4-galactosyltransferases. Cloning and expression of two novel UDP-galactose:β-*N*-acetylglucosamine β1,4-galactosyltransferases, β4Gal-T2 and β4Gal-T3. J Biol Chem 1997; 272:31979-31991.

Avilés M, Jaber L, Castells MT, Ballesta J, Kan FWK. Modifications of carbohydrate residues and ZP2 and ZP3 glycoproteins in the mouse zona pellucida after fertilization. Biol Reprod 1997; 57:1155-1163.

Baba T, Azuma S, Kashiwabara S, Toyoda Y. Sperm from mice carrying a targeted mutation of the acrosin gene can penetrate the oocyte zona pellucida and effect fertilization. J Biol Chem 1994; 269:31845-31849.

Beebe SJ, Leyton L, Burks D, Ishikawa M, Fuerst T, Dean J, Saling P. Recombinant mouse ZP3 inhibits sperm binding and induces the acrosome reaction. Dev Biol 1992; 151:48-54.

Benau DA, Storey BT. Characterization of the mouse sperm plasma membrane zona-binding site sensitive to trypsin-inhibitors. Biol Reprod 1987; 36:282-292.

Benau DA, Storey BT. Relationship between two types of mouse sperm plasma membrane zona-binding sites that mediate binding of sperm to the zona pellucida. Biol Reprod 1988; 39:235-244.

Benau DA, McGuire EJ, Storey BT. Further characterization of the mouse sperm surface zona-binding site with galactosyltransferase activity. Mol Reprod Dev 1990; 25:393-399.

Bleil JD, Wassarman PM. Structure and function of the zona pellucida: identification and characterization of the proteins of the mouse oocyte's zona pellucida. Dev Biol 1980a; 76:185-202.

Bleil JD, Wassarman PM. Mammalian sperm-egg interaction: identification of a glycoprotein in mouse egg zonae pellucidae possessing receptor activity for sperm. Cell 1980b; 20:873-882.

Bleil JD, Wassarman PM. Sperm-egg interactions in the mouse: sequence of events and induction of the acrosome reaction by a zona pellucida glycoprotein. Dev Biol 1983; 95:317-324.

Bleil JD, Wassarman PM. Autoradiographic visualization of the mouse egg's sperm receptor bound to sperm. J Cell Biol 1986; 102:1363-1371.

Bleil JD, Wassarman PM. Galactose at the nonreducing terminus of O-linked oligosaccharides of mouse egg zona pellucida glycoprotein ZP3 is essential for the glycoprotein's sperm receptor activity. Proc Natl Acad Sci USA 1988; 85:6778-6782.

Bleil JD, Wassarman PM. Identification of a ZP3-binding protein on acrosome-intact mouse sperm by photoaffinity crosslinking. Proc Natl Acad Sci USA 1990; 87:5563-5567.

Bleil JD, Greve JM, Wassarman PM. Identification of a secondary sperm receptor in the mouse egg zona pellucida: role in maintenance of binding of acrosome-reacted sperm to eggs. Dev Biol 1988; 128:376-385.

Blobel VP, Wolfsberg TG, Turck CW, Myles DG, Primakoff P, White JM. A potential fusion peptide and an integrin ligand domain in a protein active in sperm-egg fusion. Nature 1992; 356:248-252.

Bookbinder LH, Cheng A, Bleil JD. Tissue- and species-specific expression of sp56, a mouse sperm fertilization protein. Science 1995; 269:86-89.

Bork P. Sperm-egg binding protein or proto-oncogene? Science 1996; 271:1431-1432.

Burks DJ, Carballada R, Moore HDM, Saling PM. Interaction of a tyrosine kinase from human sperm with the zona pellucida at fertilization. Science 1995; 269:83-86.

Cardullo RA, Wolf DE. Distribution and dynamics of mouse sperm surface galactosyltransferase: implications for mammalian fertilization. Biochemistry 1995; 34:10027-10035.

Carrera A, Moos J, Ning XP, Gerton GL, Tesarik J, Kopf GS, Moss SB. Regulation of protein tyrosine phosphorylation in human sperm by a calcium/calmodulin-dependent mechanism: identification of A kinase anchor proteins as major substrates for tyrosine phosphorylation. Dev Biol 1996; 180:284-296.

Chamberlin ME, Dean J. Human homologue of the mouse sperm receptor. Proc Natl Acad Sci USA 1990; 87:6014-6018.

Cornwall GA, Tulsiani DRP, Orgebin-Crist MC. Inhibition of the mouse sperm surface α-D-mannosidase inhibits sperm-egg binding *in vitro*. Biol Reprod 1991; 44:913-921.

Dravland E, Joshi MS. Sperm-coating antigens secreted by the epididymis and seminal vesicle of the rat. Biol Reprod 1981; 25:649-658.

Eddy EM. The spermatozoan. In: (Knobil E, Neill J, eds) The Physiology of Reproduction. New York: Raven Press, 1988, pp189-317.

Endo Y, Lee MA, Kopf GS. Characterization of an islet-activating protein-sensitive site in mouse sperm that is involved in the zona pellucida-induced acrosome reaction. Dev Biol 1988; 129:12-24.

Ensslin M, Vogel T, Calvete JJ, Thole HH, Schmidtke J, Matsuda T, Töpfer-Petersen E. Molecular cloning and characterization of P47, a novel boar sperm-associated zona pellucida-binding protein homologous to a family of mammalian secretory proteins. Biol Reprod 1998; 58:1057-1064.

Evans SC, Youakim A, Shur BD. Biological consequences of targeting β1,4 galactosyltransferase to two different subcellular compartments. BioEssays 1995; 17:261-268.

Fayrer-Hosken RA, Caudle AB, Shur BD. Galactosyltransferase

activity is restricted to the plasma membranes of equine and bovine sperm. Mol Reprod Dev 1991; 28:74-78.

Florman HM, Storey BT. Mouse gamete interactions: the zona pellucida is the site of the acrosome reaction leading to fertilization *in vitro*. Dev Biol 1982; 91:121-130.

Florman HM, Wassarman PM. O-linked oligosaccharides of mouse egg ZP3 account for its sperm receptor activity. Cell 1985; 41:313-324.

Florman HM, Tombes RM, First NL, Babcock DF. An adhesion-associated agonist from the zona pellucida activates G protein-promoted elevations of internal Ca^{2+} and pH that mediate mammalian sperm acrosomal exocytosis. Dev Biol 1989; 135:133-146.

Florman HM, Corron ME, Kim TDH, Babcock DF. Activation of voltage-dependent calcium channels of mammalian sperm is required for zona pellucida-induced acrosomal exocytosis. Dev Biol 1992; 152:304-314.

Foltz KR, Lennarz WJ. Purification and characterization of an extracellular fragment of the sea urchin egg receptor for sperm. J Cell Biol 1990; 111:2951-2959.

Foster JA, Friday BB, Maulit MT, Blobel C, Winfrey VP, Olson G E, Kim K-S, Gerton GL. AM67, a secretory component of the guinea pig sperm acrosomal matrix, is related to mouse sperm protein sp56 and the complement component 4-binding proteins. J Biol Chem 1997; 272:12714-12722.

Friend DS, Orci L, Perrelet A, Yanagimachi R. Membrane particle changes attending the acrosome reaction in guinea pig spermatozoa. J Cell Biol 1977; 74:561-577.

Gong X, Dubois DH, Miller DJ, Shur BD. Activation of a G-protein complex by aggregation of β1,4-galactosyltransferase on the surface of sperm. Science 1995; 269:1718-1721.

Harduin-Leper A, Shaper JH, Shaper NL. Characterization of two *cis*-regulatory regions in the murine β1,4-galactosyltransferase gene. Evidence for a negative regulatory element that controls initiation at the proximal site. J Biol Chem 1993; 268:14348-14359.

Hardy DM, Garbers DL. A sperm membrane protein that binds in a species-specific manner to the egg extracellular matrix is homologous to von Willebrand factor. J Biol Chem 1995; 270:26025-26028.

Hoshi M, De Santis R, Pinto MR, Cotelli F, Rosati F. Is sperm L-fucosidase responsible for sperm-egg binding in *Ciona intestinalis*? In: (Andre J, ed) The Sperm Cell. The Hague: Marinus Nijhoff, 1983.

Huang TTF, Fleming AD, Yanagimachi R. Only acrosome-reacted spermatozoa can bind to and penetrate zona pellucida: a study using the guinea pig. J Exp Zool 1981; 217:287-290.

Huszar G, Sbracia M, Vigue L, Miller DJ, Shur BD. Sperm plasma membrane remodeling during spermiogenetic maturation in men: relationship among plasma membrane β1,4-galactosyltransferase, cytoplasmic creatine phosphokinase, and creatine phosphokinase isoform ratios. Biol Reprod 1997; 56:1020-1024.

Hyne RV, Garbers DL. Calcium-dependent increase in adenosine 3'5'-monophosphate and induction of the acrosome reaction in guinea pig spermatozoa. Proc Natl Acad Sci USA 1979; 76:5699-5703.

Ikawa M, Wada I, Kominami K, Watanabe D, Toshimori K, Nishimune Y, Okabe M. The putative chaperone calmegin is required for sperm fertility. Nature 1997; 387:607-611.

Johnson DS, Wright WW, Shaper JH, Hokke CH, van den Eijnden DH, Joziasse DH. Murine sperm-zona binding, a fucosyl residue is required for a high affinity sperm-binding ligand. A second site on sperm binds a nonfucosylated, β-galactosyl-capped oligosaccharide. J Biol Chem 1998; 273:1888-1895.

Jones R. Membrane remodeling during sperm sperm maturation in the epididymis. In: (Millgan SR, ed) Oxford Reviews of Reproductive Biology. Vol II. Oxford: Oxford University Press, 1989, pp 285-337.

Joziasse DH. Mammalian glycosyltransferases: genomic organization and protein structure. Glycobiology 1992; 2:271-277.

Kalab P, Visconti P, Leclerc P, Kopf GS. p95, the major phosphotyrosine-containing protein in mouse spermatozoa, is a hexokinase with unique properties. J Biol Chem 1994; 269:3810-3817.

Kohno N, Yamagata K, Yamada S, Kashiwabara S, Sakai Y, Baba T. Two novel testicular serine proteases, TESP1 and TESP2, are present in the mouse sperm acrosome. Biochem Biophys Res Commun 1998; 245:658-665.

Langlais J, Zollinger M, Plante L, Chapdelaine A, Bleau G, Roberts KD. Localization of cholesteryl sulfate in human spermatozoa in support of a hypothesis for the mechanism of capacitation. Proc Natl Acad Sci USA 1981; 78:7266-7270.

Larson J L, Miller D J. Sperm from a variety of mammalian species express β1,4-galactosyltransferase on their surface. Biol Reprod 1997; 57:442-453.

Lasky L A. Selectins: interpreters of cell-specific carbohydrate information during inflammation. Science 1992; 258:964-969.

Leclerc P, de Lamirande E, Gagnon C. Cyclic adenosine 3',5'monophosphate-dependent regulation of protein tyrosine phosphorylation in relation to human sperm capacitation and motility. Biol Reprod 1996; 55:684-692.

Leyton L, Saling P. Evidence that aggregation of mouse sperm receptors by ZP3 triggers the acrosome reaction. J Cell Biol 1989a; 108:2163-2168.

Leyton L, Saling P. 95 kD sperm proteins bind ZP3 and serve as tyrosine kinase substrates in response to zona binding. Cell 1989b; 57:1123-1130.

Lin Y, Mahan K, Lathrop WF, Myles DG, Primakoff P. A hyaluronidase activity of the sperm plasma membrane protein PH-20 enables sperm to penetrate the cumulus cell layer surrounding the egg. J Cell Biol 1994; 125:1157-1163.

Litscher ES, Juntunen K, Seppo A, Penttila L, Niemela R, Renkonen O, Wassarman PM. Oligosaccharide constructs with defined structures that inhibit binding of mouse sperm to unfertilized eggs *in vitro*. Biochemistry 1995; 34:4662-4669.

Llanos MN, Lui CW, Meizel S. Studies of phospholipase A2 related to the hamster sperm acrosome reaction. J Exp Zool 1982; 221:107-117.

Lo N-W, Shaper JH, Pevsner J, Shaper NL. The expanding β4-galactosyltransferase gene family: messages from the databanks. Glycobiology 1998; 8:517-526.

Loeser CR, Tulsiani DRP. The acrosome reaction in mouse spermatozoa: induction with neoglycoproteins. Mol Biol Cell 1997; 109a.

Lopez LC, Shur BD. Redistribution of mouse sperm galactosyltransferase after the acrosome reaction. J Cell Biol 1987; 105:1663-1670.

Lopez LC, Bayna EM, Litoff D, Shaper NL, Shaper JH, Shur BD. Receptor function of mouse sperm surface galactosyltransferase during fertilization. J Cell Biol 1985; 101:1501-1510.

Lu Q, Shur BD. Sperm from β1,4-galactosyltransferase-null mice are refractory to ZP3-induced acrosome reactions and penetrate the zona pellucida poorly. Development 1997; 124:4121-4131.

Macek MB, Shur BD. Protein-carbohydrate complementarity in mammalian gamete recognition. Gam Res 1988; 20:93-109.

Macek MB, Lopez LC, Shur BD. Aggregation of β-1,4-galactosyltransferase on mouse sperm induces the acrosome reaction. Dev Biol 1991; 147:440-444.

Miller DJ, Shur BD. Molecular basis of fertilization in the mouse. Sem Develop Biol 1994; 5:255-264.

Miller DJ, Macek MB, Shur BD. Complementarity between sperm surface β-1,4-galactosyltransferase and egg-coat ZP3 mediates sperm-egg binding. Nature (Lond) 1992; 357:589-593.

Miller DJ, Gong X, Shur BD. Sperm require β1,4-acetylglycosaminindase to penetrate through the egg zona pellucida. Development 1993a; 118:1279-1289.

Miller DJ, Gong X, Decker G, Shur BD. Egg cortical granule N-acetylglucosaminidase is required for the mouse zona block to polyspermy. J Cell Biol 1993b; 123:1431-1440.

Mortillo S, Wassarman PM. Differential binding of gold-labeled zona pellucida glycoproteins mZP2 and mZP3 to mouse sperm membrane compartments. Development 1991; 113:141-149.

Myles DG, Hyatt H, Primakoff P. Binding of both acrosome-intact and acrosome-reacted guinea pig sperm to the zona pellucida during *in vitro* fertilization. Dev Biol 1987; 121:559-567.

Nagdas SK, Araki Y, Chayko CA, Orgebin-Crist MC, Tulsiani DRP. *O*-linked trisaccharide and *N*-linked poly-*N*-acetyllactosaminyl glycans are present on mouse ZP2 and ZP3. Biol Reprod 1994; 51:262-272.

O'Rand MG. Restriction of a sperm surface antigen's mobility during capacitation. Dev Biol 1977; 55:260-270.

O'Rand MG, Widgren EE, Fisher SJ. Characterization of the rabbit sperm membrane autoantigen, RSA, as a lectin-like zona binding protein. Dev Biol 1988; 129:231-240.

Pratt SA, Scully NF, Shur BD. Cell surface β1,4galactosyltransferase on primary spermatocytes facilitates their initial adhesion to Sertoli cells *in vitro*. Biol Reprod 1993; 49:470-482.

Primakoff P, Hyatt H, Myles DG. A role for the migrating sperm surface antigen PH-20 in guinea pig sperm binding to the egg zona pellucida. J Cell Biol 1985; 101:2239-2244.

Rajput B, Shaper NL, Shaper JH. Transcriptional regulation of murine β1,4-galactosyltransferase in somatic cells. Analysis of a gene that serves both a housekeeping and a mammary gland-specific function. J Biol Chem 1996; 271:5131-5142.

Roseman S. The synthesis of complex carbohydrates by multiglycosyltransferase systems and their potential function in intercellular adhesion. Chem Phys Lipids 1971; 5:270-297.

Roth S, McGuire EJ, Roseman S. Evidence for cell surface glycosyltransferases. Their potential role in cellular recognition. J Cell Biol 1971; 51:536-547.

Russo RN, Shaper NL, Shaper JH. Bovine β1-4-galactosyltransferase: two sets of mRNA transcripts encode two forms of the protein with different amino-terminal domains. *In vitro* translation experiments demonstrate that both the short and the long forms of the enzyme are type II membrane-bound glycoproteins. J Biol Chem 1990; 265:3324-3331.

Sacco AG, Yurewicz EC, Subramanian MG, Matzat PD. Porcine zona pellucida: association of sperm receptor activity with the a-glycoprotein component of the Mr=55,000 family. Biol Reprod 1989; 41:523-532.

Saling PM, Storey BT. Mouse gamete interactions during fertilization *in vitro*. Chlortetracycline as a fluorescent probe for the mouse sperm acrosome reaction. J Cell Biol 1979; 83:544-555.

Sato T, Furukawa K, Bakker H, van den Eijnden DH, van Die I. Molecular cloning of a human cDNA encoding β-1,4-galactosyltransferase with 37% identity to mammalian UDP-Gal:GlcNAc β-1,4-galactosyltransferase. Proc Natl Acad Sci USA 1998; 95:472-477.

Scully NF, Shaper JH, Shur BD. Spatial and temporal expression of cell surface galactosyltransferase during mouse spermatogenesis and epididymal maturation. Dev Biol 1987; 124:111-124.

Shaper NL, Hollis GF, Douglas JG, Kirsch IR, Shaper JH. Characterization of the full length cDNA for murine β-1,4-galactosyltransferase. Novel features at the 5'-end predict two translation start sites at two in-frame AUGs. J Biol Chem 1988; 263:10420-10428.

Shur BD. Cell surface β1,4 galactosyltransferases: twenty years later. Glycobiology 1991; 1:563-575.

Shur BD. Glycosyltransferases as cell adhesion molecules. Curr Opin Cell Biol 1992; 5:854-863.

Shur BD, Bennett D. A specific defect in galactosyltransferase regulation on sperm bearing mutant alleles of the T/t-locus. Dev Biol 1979; 71:243-259.

Shur BD, Hall NG. Sperm surface galactosyltransferase activities during *in vitro* capacitation. J Cell Biol 1982a; 95:567-573.

Shur BD, Hall NG. A role for mouse sperm surface galactosyltransferase in sperm binding to the egg zona pellucida. J Cell Biol 1982b; 95:574-579.

Shur BD, Neely CA. Plasma membrane association, purification and characterization of mouse sperm β1,4 galactosyltransferase. J Biol Chem 1988; 263:17706-17714.

Shur BD, Evans S, Lu Q. Cell surface galactosyltransferase: current issues. Glycoconjugate J 1998; 15:537-548.

Singh JP, Babcock DJ, Lardy HL. Increased calcium ion influx in a component of capacitation of spermatozoa. Biochem J 1978; 172:548-556.

Storey BT, Kopf GS. Fertilization in the mouse. I. Spermatozoa. In: (Dunbar BS, O'Rand MG, eds) A Comparative Overview of Mammalian Fertilization. New York: Plenum Press, 1991, pp167-216.

Suarez SS. The oviductal sperm reservoir in mammals: mechanisms of formation. Biol Reprod 1998; 58:1105-1107.

Talbot P, Franklin LE. Surface modification of guinea pig sperm during *in vitro* capacitation: an assessment using lectin-induced agglutination of living sperm. J Exp Zool 1978; 203:1-14.

Tanphaichitr N, Smith J, Mongkolsirikieart S, Gradil C, Lingwood CA. Role of a gamete-specific sulfoglycolipid immobilizing protein on mouse sperm-egg binding. Dev Biol 1993; 156:164-175.

Thall AD, Maly P, Lowe JB. Oocyte Galα1,3Gal epitopes implicated in sperm adhesion to the zona pellucida glycoprotein ZP3 are not required for fertilization in the mouse. J Biol Chem 1995; 270:21437-21440.

Tsai J-Y, Silver LM. Sperm-egg binding protein or proto-oncogene? Science 1996; 271:1432-1434.

Tulsiani DRP, Nagdas SK, Cornwall GA, Orgebin-Crist MC. Evidence for the presence of high mannose/hybrid oligosaccharide chain(s) on the mouse ZP2 and ZP3. Biol Reprod 1992; 46:93-100.

Tulsiani DRP, Yoshida-Komiya H, Araki Y. Mammalian fertilization: a carbohydrate-mediated event. Biol Reprod 1997; 57:487-494.

Vacquier VD, Moy GW. Isolation of binding: the protein responsible for adhesion of sperm to sea urchin eggs. Proc Natl Acad Sci USA 1977; 74:2456-2460.

Visconti PE, Johnson LR, Oyaski M, Fornés M, Moss SB, Gerton GL, Kopf GS. Regulation, localization, and anchoring of protein kinase A subunits during mouse sperm capacitation. Dev Biol 1997; 192:351-363.

Vredeburgh-Wilberg WL, Parrish JJ. Intracellular pH of bovine sperm increases during capacitation. Mol Reprod Dev 1995; 40:490-502.

Ward CR, Kopf GS. Molecular events mediating sperm activation. Dev Biol 1993; 158:9-34.

Ward CR, Storey BT, Kopf GS. Activation of a G_i protein in mouse sperm membranes by solubilized proteins of the zona pellucida, the egg's extracellular matrix. J Biol Chem 1992; 267:14061-14067.

Yamagata T. The role of saccharides in fertilization of the mouse. Dev Growth Differ 1985; 27:176-177.

Yanagimachi R. Mammalian fertilization. In: (Knobil E, Neill J, eds) The Physiology of Reproduction. New York: Raven Press, 1994, pp189-317.

Yonezawa N, Aoki H, Hatanaka Y, Nakano M. Involvement of N-linked carbohydrate chains of pig zona pellucida in sperm-egg binding. Eur J Biochem 1995; 233:35-41.

Youakim A, Hathaway HJ, Miller DJ, Gong X, Shur BD. Overexpressing sperm surface β1,4 galactosyltransferase in transgenic mice affects multiple aspects of sperm-egg interactions. J Cell Biol 1994; 126:1573-1584.

21 *Role of Male Germ-Cell Specific Sulfogalactosylglycerolipid (SGG) and Its Binding Protein, SLIP1, on Mammalian Sperm-Egg Interaction*

Nongnuj Tanphaichitr
Dawn White
Tanya Taylor
Mayssa Attar
Manee Rattanachaiyanont
Dominic D'Amours
Morris Kates
Loeb Health Research Institute—Ottawa Hospital
University of Ottawa

Mammalian sperm-egg interaction that results in fertilization is initiated by binding of capacitated sperm to the zona pellucida (ZP) of ovulated mature eggs (Snell, White 1996; Wassarman 1995; Yanagimachi 1994). In all mammals studied, the ZP consists of a few glycoproteins, the functional properties and biochemical nature of which are known mostly in mice and pigs (Harris et al. 1994; Wassarman 1995). In mice, ZP3 and ZP2 sulfoglycoproteins are primary and secondary sperm receptors, respectively (Bleil, Wassarman 1980; Bleil et al. 1988; Wassarman 1995). In pigs, ZP3α is the primary sperm receptor, although its action may be mediated by ZP3β (Yurewicz et al. 1993).

Unlike the egg ZP, where only one or two sperm receptors exist, mammalian sperm appear to possess several surface proteins that are ZP-associated (McLeskey et al. 1998; Snell, White 1996; Tulsiani et al. 1997; Wassarman 1995). The first category of these proteins consists of sperm glycotransferases and glycohydrolases that are not catalytically active, but instead act as lectins conjugating sperm to ZP sugar residues. These include sperm galactosyltransferase (GALT) (Lopez et al. 1985), fucosyltranferase (Apter et al. 1988) and mannosidase (Cornwall et al. 1991). In addition, other mouse sperm surface proteins involved in sperm-ZP binding include trypsin inhibitor-sensitive site (Boettger-Tong et al. 1993), zona receptor kinase (Burks et al. 1995; Leyton, Saling 1989) and sp56 peripheral plasma membrane protein (Bookbinder et al. 1995). In pigs, proacrosin (Jones 1991; Topfer-Petersen, Henschen 1987), zona adhesin (Hardy, Garber 1994), sp38 (Mori et al. 1993) and spermadhesin (Sanz et al. 1993) are sperm surface proteins that have been shown to bind to the ZP. The fact that several sperm surface proteins are involved in ZP binding suggests that their interaction with ZP may be sequential, and/or simultaneous, with back-up mechanisms among one another.

Sulfogalactosylglycerolipid (SGG)

While several laboratories have directed their efforts towards identifying ZP ligands on the sperm surface that are proteins, other sperm surface molecules, which may be potential candidates of ZP binding elements, have been much ignored. One such molecule is the mammalian sperm surface sulfoglycolipid, sulfogalactosylglycerolipid (SGG), commonly known as seminolipid (Murray, Narasimhan 1990), the structure of which was elucidated to be composed of alkylacylglycerol linked to a sulfated galactose polar head group (Murray, Narasimhan 1990). SGG is structurally related to sulfogalactosylceramide (SGC), better known as sulfatide or cerebroside sulfate (Fig. 1), as both sulfoglycolipids possess common features, including the galactose sulfate polar head group and a three carbon chain backbone with two hydrocarbon chains.

SGG is present specifically in mammalian male germ cells (6-10% of total lipids) and to a lesser extent in brain tissue (0.3% of total lipids), while absent in other somatic tissues (Murray, Narasimhan 1990). SGG is synthesized in primary spermatocytes (Lingwood 1985b) and due to its bulky and charged polar head group, it is believed to be present on the outer leaflet of the plasma membrane (Hakomori 1990). Once synthesized, SGG is metabolically stable throughout spermatogenesis, sperm maturation and the initial stage of sperm capacitation (Kornblatt et al. 1974; Tanphaichitr et al. 1990), although 15% of SGG becomes desulfated when sperm are capacitated *in vitro* for more than one hour (Tanphaichitr et al. 1990).

The metabolic stability of SGG suggests that it plays a role in sperm function. Several roles have been postulated based on the known functions of SGC. These include cation trapping/transport and cell-cell/extracellular matrix adhesion (Curatolo 1987). Kidney epithelial cells cultured in high ionic strength medium possess more SGC than those cultured in normal medium (Ishizuka, Nakamura 1991). SGC may be involved in K^+ binding, the intermediate mechanism of (Na^+ /K^+)ATPase activity (Jedlicki, Zambrano 1985). Pressure-tuning Fourier transform infrared spectroscopy (FTIR) performed in Tanphaichitr's laboratory indicates that divalent cations, such as Ca^{2+}, which is pivotal to fertilization (Yanagimachi 1994), bind to both SGG and SGC liposomes with a higher affinity than monovalent cations, such as Na^+, and that this binding results in increased fluidity of the liposomes (Tupper et al. 1992, 1994). FTIR work of the same investigator has been extended using more physiologically representative mixed liposomes containing SGC and a phospholipid, dimyristoylphosphatidylcholine (DMPC). Results indicate that the two lipids fully interact with each other and that their co-existence enhances fluidity of the hydrocarbon chains of both lipids (Attar et al. 1998). It remains to be seen whether SGG can affect the

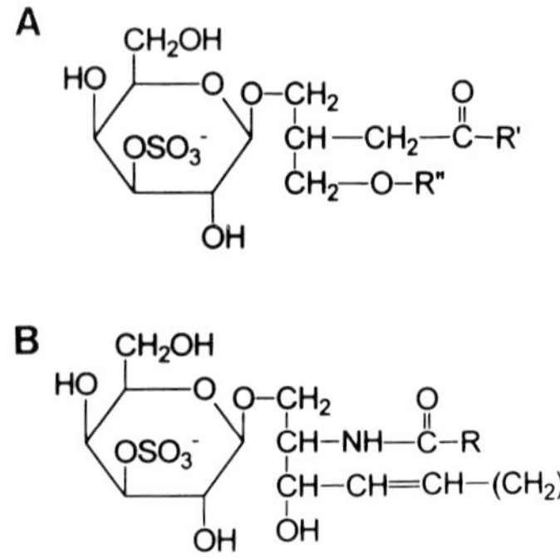

Figure 1. A: Structure of sulfogalactosylglycerolipid (SGG); B: sulfogalactosylceramide (SGC).

fluidity of other phospholipids in the mixed liposomes, and whether Ca^{2+} binding to SGG in mixed liposomes would further enhance membrane fluidity. It is possible that the increased membrane fluidity observed during sperm capacitation (Yanagimachi 1994) may be partially due to the formation of SGG-Ca^{2+} complexes.

The involvement of SGG and SGC, especially their galactosyl sulfate moiety, in cell-cell/extracellular matrix adhesion has been shown in various systems. SGC binds a number of adhesive glycoproteins, including laminin, thrombospondin, von Willebrand factor and vitrogenin (Roberts, Ginsburg 1988). Both SGC and SGG bind L/P selectin (Suzuki et al. 1993), an adhesive protein involved in the lymphocyte homing event (Suzuki et al. 1993). Pathogenic microorganisms, such as mycoplasmas (Lingwood et al. 1990) and malaria sporozites (Pancake et al. 1992), or surface components of virus/microorganisms, such as gp120 coat protein of HIV-1 (Harouse et al. 1991), also interact with SGG and/or SGC.

Role of Sperm SGG in Sperm-ZP Binding

To determine whether sperm SGG was involved in ZP binding, researchers in Tanphaichitr's laboratory planned to mask the sulfoglycolipid with its IgG antibody. However, most antibodies directed against SGG or SGC are multivalent IgM and expected to sterically hinder other sperm surface components besides SGG when employed in *in vitro* functional studies. Therefore, this study was commenced by producing a rabbit polyclonal anti-SGG IgG antibody using the method of Lingwood et al. (1980). Taking advantage of the fact that all antibodies produced against SGC, which is commercially available, also cross-react with SGG, SGC-containing liposomes were used as the antigen. In both rabbits immunized with SGC liposomes, high titer antisera (90,000), as assessed by enzyme linked immunosorbent assay (ELISA), were obtained.

Using thin-layer chromatography overlay technique, the antisera were shown to recognize specifically SGC, SGG and lysoSGG, but not their parental glycolipids, GC, GG, nor a number of other lipids tested (including cholesterol sulfate, monogalactosyldiacylglycerol, phosphatidylcholine, phosphatidylserine, phosphatidylethanolamine, phosphatidic acid and sphingomyelin).

The isolated IgG fraction of anti-SGC/SGG antibody was affinity purified using an SGG affinity matrix and used for localization of mouse sperm SGG by indirect immunofluorescence (IIF). Results shown in Figure 2 revealed that SGG was localized to the postacrosome, and the concave and convex ridges of the mouse sperm head, as well as to the midpiece. The sperm head convex ridge and postacrosome have been shown previously to be ZP binding sites (Chen, Cardullo 1994; Yanagimachi 1994), and pretreatment of sperm with affinity purified anti-SGG IgG prior to sperm-egg co-incubation (30 minutes) inhibited sperm-ZP binding in a dose-dependent manner (Fig. 3). This inhibition was similarly observed when the gamete co-incubation was reduced to only 10 minutes, during which time only primary sperm-ZP binding occurs (Bleil et al. 1988). This treatment of sperm with affinity purified anti-SGG IgG neither changed the motility nor induced the spontaneous acrosome reaction. Therefore, the results suggested that sperm SGG was involved in sperm-ZP binding.

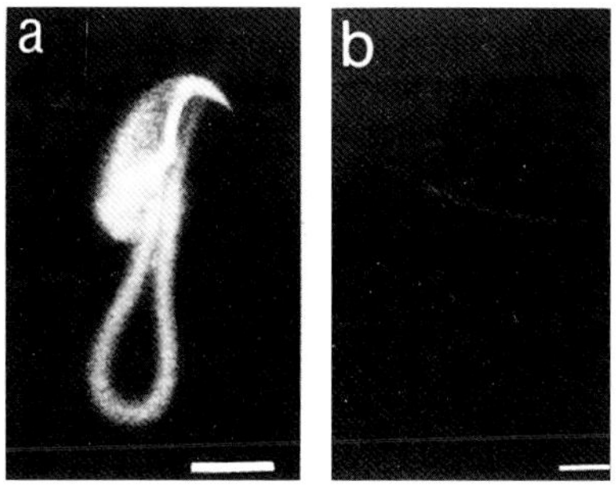

Figure 2. SGG localization in mouse sperm by indirect immunofluorescence. a: Percoll-gradient centrifuged (PGC) mouse sperm were exposed to affinity purified antiSGG IgG (10 μg/ml), followed by FITC-conjugated secondary antibody. b: Negative control showing PGC sperm, which were exposed to preimmune rabbit serum (PRS) IgG (10 μg/ml) instead of antiSGG. Bar = 10 μm.

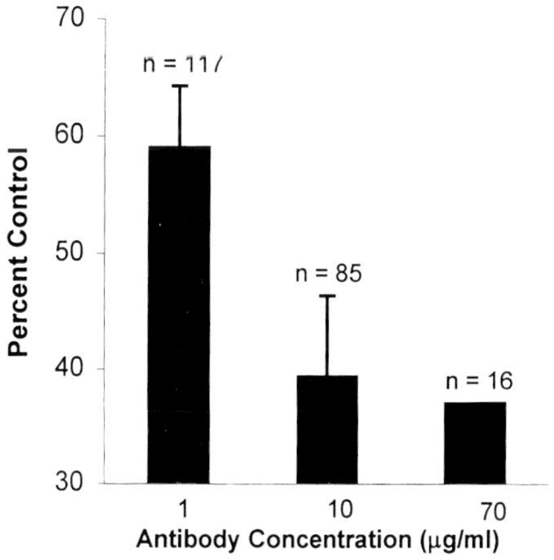

Figure 3. Pretreatment of sperm with anti-SGG IgG inhibited sperm-ZP binding. PGC sperm were treated with various concentrations of affinity purified anti-SGG IgG prior to 30 min co-incubation with zona-intact eggs, as previously described (Tanphaichitr et al. 1993). PGC sperm exposed to 70 μg/ml preimmune rabbit serum (PRS) IgG served as negative controls. Data in this figure were expressed as mean ± SD of percentage of control value from three or more experiments in which 1 and 10 μg/ml anti-SGG IgG were tested. Treatment of sperm with 70 μg/ml anti-SGG IgG was performed only once. However, a higher concentration of anti-SGG IgG did not cause greater inhibition. n = total number of eggs in each sample.

These results corroborated previous findings that SGG and/or its structural analog, SGC, bind to a number of adhesive glycoproteins. In fact, preliminary results revealed that sperm bound to zona-intact eggs pretreated with SGG liposomes at a reduced level, as compared to those pre-exposed to control liposomes constructed with negatively charged phosphatidylserine. In addition, fluorescently labeled SGG-containing liposomes bound to the egg ZP. All these results suggest that SGG interacts directly with the egg extracellular matrix, the ZP, at sperm binding sites.

To demonstrate that the sulfate moiety of SGG was significant in sperm-ZP binding, capacitated mouse sperm were treated with a high amount of arylsulfatase-A, partially purified from pig seminal plasma (pAS-A), provided by Dr. B. Gadella, Utrecht University, The Netherlands (Gadella et al. 1993). This pAS-A is believed to co-purify with its co-activator, saposin B (Fischer et al. 1978; B Gadella, personal communication), and it has been shown previously to desulfate pig sperm SGG (Gadella et al. 1993). Within one hour of treatment, the level of sperm SGG was diminished to only 25% of the level of the control untreated sperm, with a commensurate increase of GG. When these pAS-A-treated sperm were co-incubated with zona-intact eggs, sperm-ZP binding was decreased to only 12%, as compared to that of control untreated sperm. Similar to the results observed with antiSGG-treated sperm, pAS-A treated sperm remained motile and acrosome intact to the same extent as control untreated sperm. These results suggested that the sulfate moiety of SGG was important in sperm-ZP binding.

These results are consistent with the previously documented evidence that suggests the significance of sulfated sugar residues in sperm-ZP binding (Chapman, Barratt 1996). This postulation was derived from the observation that inclusion of sulfated carbohydrate molecules such as fucoidan and dextran sulfate in mammalian gamete coincubates resulted in inhibition of sperm-ZP binding (Oehninger et al. 1991). Further studies revealed that these sulfated carbohydrate molecules bound to sperm inner acrosomal membrane proteins such as proacrosin and sp38 (Brown, Jones 1987; Mori et al. 1993; Topfer-Petersen, Henschen 1987). Therefore, it has been suggested that the sulfated sugar residues of the ZP oligosaccharides play a role in secondary sperm-ZP binding by interacting with sperm proteins on the surface of the inner acrosome membrane (Chapman, Barratt 1996). On the other hand, results described herein indicate that the sulfate moiety of sperm surface SGG is important for primary sperm-ZP binding, and the inhibition caused by fucoidan or dextran sulfate on sperm-ZP binding may implicate the significance of sulfated sugar residues of both sperm SGG and ZP oligosaccharides. It should be noted, however, that Liu et al. (1997) have recently argued against the significance of ZP sulfated sugar residues. Nonetheless, results from Tanphaichitr's laboratory support other findings that sulfated sugar residues are important in cell-cell/extracellular matrix interaction. It is still unclear how SGG interacts with ZP glycoproteins. It is possible that the negatively charged sulfate moiety of SGG may interact electrostatically with positively charged amino acids, such as arginine (Copley, Barton 1994), of ZP glycoprotein(s) similar to that postulated for the interaction between SGG/SGC or the sulfated sugar residues of the GlyCAM-1 O-glycan and L selectin (Rosen, Bertozzi 1996). In addition, carbohydrate-carbohydrate interaction is plausible, based on results from model membrane studies indicating interaction between the galactose sulfate moiety of the SGG-Ca^{2+} complex and the galactose residue of galactosylceramide (Stewart, Boggs 1993). SGG may also act as a complex with its specific binding protein, known as SLIP1, which was shown to be a ZP associated sperm surface protein (Tanphaichitr et al. 1993, 1998), based on the fact that SGG and SLIP1 bind to each other *in vitro* (Boulanger et al. 1995; Lingwood 1985a; Tanphaichitr et al. 1998) and that both molecules are localized to the ZP binding regions on the sperm head (Chen, Cardullo 1994; Yanagimachi 1994). Since the number of ZP binding sites per sperm is ~10^4 to 10^5 (Liu et al. 1995; Mortillo, Wassarman 1991), the majority of SLIP1 molecules on sperm (10^4/sperm) (Tanphaichitr et al. 1993) may be involved in ZP binding. In contrast, it is unlikely that all of the sperm surface SGG (10^8 molecules/sperm) would participate in ZP binding. The remaining SGG, not complexed with SLIP1 (Moase et al. 1997), may bind to other sperm surface proteins, such as zona adhesin, which possesses a domain with high homology to von Willebrand factor, known to have SGG affinity (Hardy, Garber 1994). In addition, SGG may complex with Ca^{2+}, resulting in an increase in membrane fluidity (Tupper et al. 1994). Since SGG is present in sperm of various mammalian species, its role in ZP binding and sperm plasma membrane fluidity would be common to all mammals.

Sulfolipidimmobilizing Protein 1 (SLIP1)

SLIP1, a specific binding protein of SGG,was first isolated from a rat testis homogenate by SGG affinity column chromatography (Lingwood 1985a). It appeared as a 68 kD band on SDS-PAGE from which a rabbit polyclonal antibody, anti-SLIP1, was generated (Lingwood 1985a) for use in biochemical and functional studies. Immunoblotting indicated that SLIP1, like SGG, was present in testicular germ cells and sperm, and to a lesser extent in brain, but not in any somatic tissue tested (Law et al. 1988). SLIP1 is a conserved protein, present in the testis of various vertebrates across the evolutionary scale, from fish to birds to mammals (Law et al. 1988). In addition, SLIP1 exists on the egg plasma membrane without SGG (Ahnonkitpanit et al. 1999).

Testis/sperm SLIP1 binds ATP (Lingwood, Nutikka 1991), allowing for an alternate purification method involving preparation of a crude SLIP1 extract by treating male germ cells with a sucrose solution containing low concentrations of ATP and EDTA (AES). Its extractability from the cells using this simple salt solution strongly suggests that SLIP1 is a peripheral plasma membrane protein (Law et al. 1988; Lingwood, Nutikka 1991). SLIP1 was further purified from this crude extract by DEAE cellulose column chromatography, resulting in a single 68 kD anti-SLIP1-cross-reactive protein band on SDS-PAGE (Lingwood, Nutikka 1991). This purification method of rat testis SLIP1 appears to give a higher yield, as compared to the original method using SGG affinity column chromatography (C Lingwood, Hospital for Sick Children, Toronto, Canada, personal communication). This purified rat testis SLIP1, in addition to anti-SLIP1, was used by researchers in Tanphaichitr's laboratory for functional studies.

SLIP1 was localized, using both immunofluorescent and electron microscopic immunogold labeling techniques, to the plasma membrane of sperm and eggs (Ahnonkitpanit et al. 1999; Tanphaichitr et al. 1993). Sperm SLIP1 is mainly present on the sperm head and its abundance is only 0.0003 pg/mouse sperm (Tanphaichitr et al. 1993). Using the *in vitro* mouse sperm-egg binding assay (Bleil, Wassarman 1980), it was demonstrated that sperm binding to the egg ZP was significantly decreased, when 1) an excess amount of SLIP1 (4 nM) was present in the gamete co-incubate, 2) sperm SLIP1 was masked with anti-SLIP1, and 3) SLIP1 binding sites on the ZP were saturated by egg pretreatment with purified SLIP1 (Tanphaichitr et al. 1993). These results strongly indicate the role of mouse sperm SLIP1 in ZP binding. A similar observation has recently been noted with human sperm SLIP1 (Tanphaichitr laboratory, unpublished results), using the human sperm-human ZP binding assay. Mouse sperm SLIP1 was further demonstrated to participate in fertilization *in vivo*. Superovulated female mice artificially inseminated with anti-SLIP1-IgG-treated sperm allowed the fertilization of only 20% of ovulated eggs, as compared to 80% in mice inseminated with normal rabbit serum (NRS) IgG-treated sperm (Tanphaichitr et al. 1992). It has recently been demonstrated that egg SLIP1 plays a role in mouse sperm-egg plasma membrane binding. Zona-free mouse eggs pre-exposed to anti-SLIP1 IgG antibody, affinity purified using P68, the ZP binding sperm SLIP1 component, showed reduced sperm binding, as compared to zona-free eggs pretreated with NRS IgG. The results suggested that the egg SLIP1 component involved in gamete plasma membrane binding is antigenically related to the ZP-binding sperm SLIP1 component. Since capacitated sperm pretreated with anti-SGG Fab antibody bind to a lesser extent to the egg plasma membrane of zona-free eggs than control sperm (pretreated with NRS Fab), it is postulated that the interaction between egg SLIP1 and sperm SGG may be one of the mechanisms of gamete plasma membrane binding. However, egg SLIP1 appears not to be involved in sperm-egg plasma membrane fusion (Ahnonkitpanit et al. 1999).

The significance of SLIP1 in sperm-egg interaction prompted researchers to characterize its properties. Based on results from Lingwood's and Tanphaichitr's laboratories, researchers in the latter laboratory postulated that purified rat testis SLIP1 (Lingwood, Nutikka 1991) possesses the following properties/components: 1) SGG binding ability (Boulanger et al. 1995; Lingwood 1985a), 2) ZP binding ability (Tanphaichitr et al. 1993), 3) a heat shock protein (HSP) 70 component (Boulanger et al. 1995) and 4) an albumin component (Tanphaichitr et al. 1998; Lingwood's and Tanphaichitr's unpublished peptide sequencing results). Although Lingwood's group has shown that an HSP70 (such as that obtained from *Mycoplasma hyopneumoniae*) binds SGG (Boulanger et al. 1995; Mamelak, Lingwood 1997), Tanphaichitr's unpublished functional studies indicated that this HSP70 did not inhibit mouse sperm-ZP binding, when used at a similar concentration to that of rat testis SLIP1, which caused significant inhibition. Similarly, albumin shows no inhibitory effects on sperm-ZP binding. Therefore, researchers in Tanphaichitr's laboratory believed that the ZP binding component of SLIP1 was a separate entity from the co-purified HSP70 and albumin components, and this prompted them to devise a new method for purifying this ZP binding component.

Instead of the rat testis, they decided to use boar sperm as a source to isolate SLIP1 for the following reasons. First, boar sperm can be obtained in quantity and can be easily washed free of seminal plasma by centrifugation. This lessened the chance of obtaining albumin as part of the crude SLIP1 extract (prepared by treating germ cells with AES), since albumin is present in blood plasma, which would be present during dissection of the testes. Second, sperm SLIP1 was expected to be more biologically active in ZP binding, as compared to SLIP1 purified from less developed testicular germ cells. Using chromatofocusing, protein fractions that appeared as doublets (M_r = 68 and 62 kD) on SDS-PAGE and cross-reacted with anti-SLIP1, but not with anti-rat serum albumin or anti-HSP70, were obtained. This purified protein doublet (named P68/62) binds to SGG affinity matrix and has the ability to inhibit mouse sperm-ZP binding *in vitro* to the background level at 10 nM. Furthermore, it has been shown that P68/62 binds to isolated ZP from various mammalian species (Tanphaichitr et al. 1998). Results demonstrate that SGG-binding P68/62 is the ZP binding component of SLIP1.

Identity of P68, the ZP Binding Component of SLIP1

Since P68 reacts more strongly to anti-SLIP1 than P62, Tanphaichitr's laboratory focused its first effort on peptide sequence P68. Using the peptide sequencing service at Worcester Foundation for Biomedical Research, sequences of three selected peptides revealed high identity (>80%) to human testis arylsulfatase-A (AS-A), described by Stein et al. 1989, and illustrated here in Table 1 and Figure 4. This finding corroborates a number of common properties shared between SLIP1/P68 and AS-A of male germ cells. These include the ability of both proteins to bind SGG and SGC (Boulanger et al. 1995; Lingwood 1985a; Stein et al. 1989; Tanphaichitr et al. 1998); possession of glycosylated moieties (Tanphaichitr's laboratory, unpublished results; Sommerlade et al. 1994); and biosynthesis in primary spermatocytes (Kreysing et al. 1994b; Lingwood 1985b). The apparent molecular weights of the two proteins are also similar (63 kD described for AS-A (Sommerlade et al. 1994), and 68 kD for P68/SLIP1 (Lingwood 1985a; Tanphaichitr et al. 1998); the disparity probably reflects inaccuracy of molecular weight determination by SDS-PAGE (Fig. 5). Furthermore, both SLIP1 and AS-A have been localized to the plasma membrane of the mammalian sperm head (Nikolajczyk, O'Rand 1992; Tanphaichitr et al. 1993). All of these properties strongly suggest that AS-A is a P68 component.

Work in Tanphaichitr's laboratory further confirmed that AS-A is a P68 component by both biochemical and functional studies. A mouse polyclonal anti-P68 IgM antiserum was produced in the laboratory by

Table 1: Homology of P68 peptide sequences to human testis AS-A sequence.

Human Testis AS-A Sequence[1]	Our P68 Sequence	Identity with Human Testis AS-A
FTDFYVPVSLCTP	FTDFYVPVSLXTP	92%
SLFFYPSYPDEVR	TLFFYPAYPDEVR	84%
AQLDAAVTFGPSQVAR	AQFDAAVTFSPSQIAR	81%

[1] Stein et al. 1989

intrasplenic immunization (Tanphaichitr et al. 1998) and used for immunoblotting of P68/62 and recombinant human testis AS-A (rec hAS-A, provided by Dr. V. Gieselmann, Christian-Albrechts Universitat zu Kiel, Germany). Immunodetection of the same blot was also performed using a polyclonal IgG antiserum directed against rec hAS-A (anti-AS-A, also provided by Dr. V. Gieselmann). Both antibodies reacted with both P68 and rec hAS-A (Fig. 5). However, anti-P68 also reacted with another protein band above the P68 band.

IIF of live caudal epididymal mouse sperm using anti-AS-A revealed that in most of the sperm population, AS-A was localized mainly to the convex ridge of the sperm head, and to a lesser extent to the midpiece. This localization pattern of AS-A is similar to that observed for SLIP1 (Tanphaichitr et al. 1993), and prompted these researchers to further determine whether AS-A is involved in sperm-ZP binding. Ovulated zona-intact mouse eggs were pretreated with rec hAS-A and then washed free of unbound protein in culture medium prior to co-incubation with capacitated sperm. Results in Figure 6 demonstrated that pretreatment of eggs with rec hAS-A reduced sperm-ZP binding in a dose-dependent manner, suggesting that AS-A is involved in this process. However, the highest inhibition observed was only 54% (at >15 nM rec hAS-A), in contrast to 70%, when eggs were pretreated with >10 nM P68/62 (Tanphaichitr et al. 1998). This discrepancy may be because 1) human and mouse testis AS-A sequences differ subtly from each other (Kreysing et al. 1994a;

MGAPRSLLLA	LAAGLAVARP	PNIVLIFADD	LGYGDLGCYG	HPSSTTPNLD	50
QLAAGGLR**FT**	**DFYVPVSLCT**	**P**SRAALLTGR	LPVRMGMYPG	VLVPSSRGGL	100
PLEEVTVAEV	LAARGYLTGM	AGKWHLGVGP	EGAFLPPHQG	FHRFLGIPYS	150
HDQGPCQNLT	CFPPATPCDG	GCDQGLVPIP	LLANLSVEAQ	PPWLPGLEAR	200
YMAFAHDLMA	DAQRQDRPFF	LYYASHHTHY	PQFSGQSFAE	RSGRGPFGDS	250
LMELDAAVGT	LMTAIGDLGL	LEETLVIFTA	DNGPETMRMS	RGGCSGLLRC	300
GKGTTYEGGV	REPALAFWPG	HIAPGVTHEL	ASSLDLLPTL	AALAGAPLPN	350
VTLDGFDLSP	LLLGTGKSPR	Q**SLFFYPSYP**	**DEVR**GVFAVR	TGKYKAHFFT	400
QGSAHSDTTA	DPACHASSSL	TAHEPPLLYD	LSKDPGENYN	LLGGVAGATP	450
EVLQALKQLQ	LLK**AQLDAAV**	**TFGPSQVAR**G	EDPALQICCH	PGCTPRPACC	500
HCPDPHA					507

Figure 4. Human testis AS-A sequence (Stein et al. 1989). The bold underlined sequences show great homology to P68 peptide sequences as shown in Table 1.

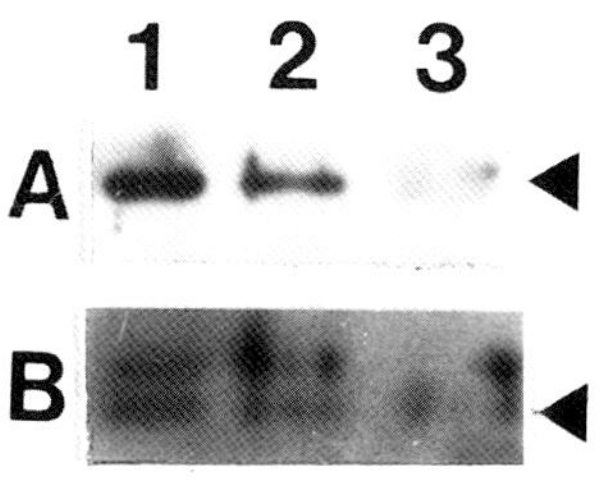

Figure 5. Lane 1 shows cross-reactivity of P68 and rec hAS-A to anti-P68 and anti-AS-A antibodies. Western blot of P68/62 (10 ng), AES extract from pig sperm (~100 ng). In lane 2 rec hAS-A (2.5 ng) (Lane 3) was probed with antiAS-A (panel A) or antiP68 (panel B). Arrowhead indicates the position of 68 kD.

Stein et al. 1989), 2) rec hAS-A does not have the same glycosylated moieties as native hAS-A, and 3) other P68 components that act synergistically with AS-A in ZP binding.

Role of Sperm Surface AS-A

The finding that AS-A is a component of P68, the ZP binding fraction of SLIP1, adds to the list of sperm surface enzymes that act as ZP lectins. However, unlike other enzymes, the substrates of which have been shown or believed to be a component of the ZP oligosaccharides, AS-A's substrate, SGG, is also present on the sperm surface, and at least a population of SGG is co-localized with AS-A . Based on previous *in vitro* studies demonstrating that a co-activator of AS-A, saposin B, is required for the AS-A activity to desulfate SGG (Fischer et al. 1978), it is postulated that a functional form of saposin B is not present on the sperm surface or in the caudal epididymal fluid, thus allowing SGG to remain intact and to act in conjunction with sperm surface AS-A in ZP binding.

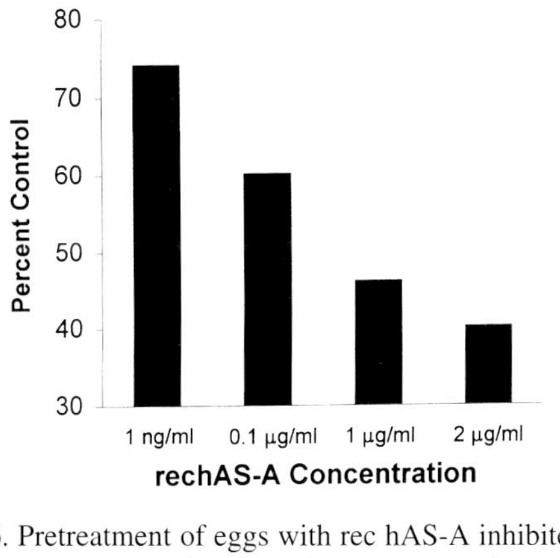

Figure 6. Pretreatment of eggs with rec hAS-A inhibited sperm-ZP binding. Eggs were treated with various concentrations of rec hAS-A prior to co-incubation with sperm. Eggs treated with 2 µg/ml ovalbumin served as negative controls. Data were expressed as average of percent of control value, combined from two replicate experiments.

It is plausible that sperm surface AS-A may be activated following the initial stage of sperm-ZP binding. Since SGG exists in a substantial amount in mammalian sperm, it would be expected that its bulky polar head group would hinder membrane fusion, a requirement for the acrosome reaction. It is therefore postulated that SGG may be desulfated by AS-A and saposin B, generating GG, which may then be further degalactosylated by β-galactosidase, yielding palmitylpalmitoylglycerol (PPG; Fig. 7). Being a diraydylglycerol, PPG may activate sperm phospholipase A-2 (Roldan, Fragio 1994) to hydrolyze phospholipids, yielding fusogenic lysophospholipids and free fatty acids. These end products would enhance membrane fusion essential for initiation of the acrosome reaction. This postulation is based on the facts that hydrolysis of SGG's sulfate group occurs at the same time as a spontaneous acrosome reaction, after mouse sperm are capacitated for a prolonged period of time (Tanphaichitr et al. 1990) and that β-galactosidase has been shown to be present in the epididymal fluid and adsorbed onto the sperm surface (Tulsiani et al. 1995). This postulation will be further confirmed by showing that saposin B, AS-A's co-activator (Fischer et al. 1978), and saposin A or C, β-galactosidase's co-activator (O'Brien, Kishimoto 1991), are formed following sperm-ZP binding. Significantly, prosaposin is present in the male reproductive tract (Hermo et al. 1992) and in quantity in the seminal plasma (Hiraiwa et al. 1993) and is adsorbed onto the sperm head surface of caudal

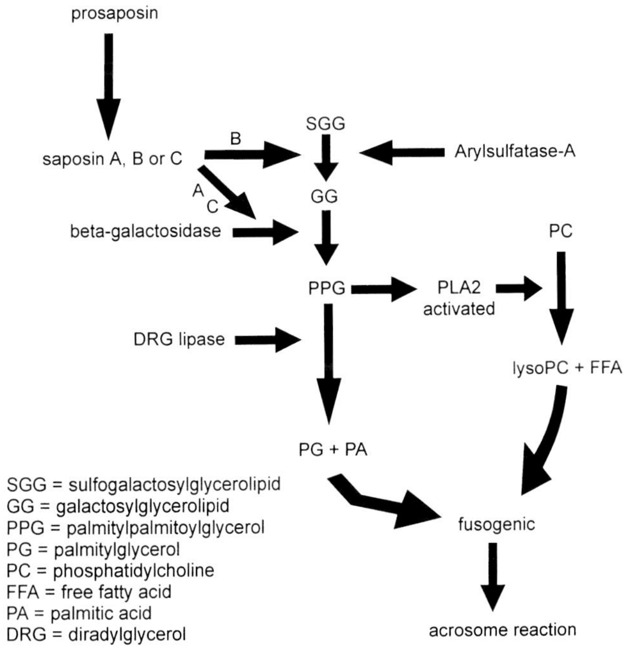

Figure 7. Proposed metabolism of SGG following sperm-ZP binding.

epididymal mouse sperm and ejaculated human sperm (Tanphaichitr's laboratory, unpublished results). Prosaposin may be processed to individual saposins (A, B, C and D) by sperm surface proteases, such as those described previously (Boettger-Tong et al. 1993; Wang et al. 1994), which may be activated upon sperm-ZP binding. Ongoing work in Tanphaichitr's laboratory is focused towards elucidating the metabolic cascade of SGG following sperm-ZP interaction.

Acknowledgements

This work is funded by grants from the Medical Research Council of Canada (MT-10366), the Rockefeller Foundation and the Lalor Foundation. The authors wish to thank Ms. Terri van Gulik for her excellent assistance in manuscript preparation.

References

Ahnonkitpanit V, White D, Suwajanakorn S, Kan F, Namking M, Wells G, Tanphaichitr N. Role of egg sulfolipidimmobilizing protein 1 (SLIP1) on sperm-egg plasma membrane binding. Biol Reprod 1999; in press.

Apter FM, Baltz JM, Millette CF. A possible role for cell surface fucosyl transferase (FT) activity during sperm zona-pellucida binding in the mouse. J Cell Biol 1988; 107:175a.

Attar M, Wong PTT, Kates M, Carrier D, Jaklis P, Tanphaichitr N. Interaction between sulfogalactosylceramide and dimyristoylphosphatidylcholine increases the orientational fluctuations of the lipid hydrocarbon chains. Chem Phys Lipids 1998; 94:227-238.

Bleil JD, Wassarman PM. Mammalian sperm-egg interaction: identification of a glycoprotein in mouse egg zonae pellucidae possessing receptor activity for sperm. Cell 1980; 20:873-882.

Bleil JD, Greve JM, Wassarman PM. Identification of a secondary sperm receptor in the mouse egg zona pellucida: role in maintenance of binding of acrosome-reacted sperm to eggs. Dev Biol 1988; 128:376-385.

Boettger-Tong HL, Aarons DJ, Biegler BE, George B, Poirier GR. Binding of a murine proteinase inhibitor to the acrosome region of the human sperm head. Mol Reprod Dev 1993; 36:346-353.

Bookbinder LH, Cheng A, Bleil JD. Tissue-and species-specific expression of sp56, a mouse sperm fertilization protein. Science 1995; 269:86-89.

Boulanger J, Faulds D, Eddy EM, Lingwood CA. Members of the 70 kDa heat shock protein family specifically recognize sulfoglycolipids: role in gamete recognition and mycoplasma-related infertility. J Cell Physiol 1995; 165:7-17.

Brown CR, Jones R. Binding of zona pellucida proteins to a boar sperm polypeptide of M_r53 000 and identification of zona moieties involved. Development 1987; 99:333-339.

Burks DJ, Carballada R, Moore HDM, Saling PM. Interaction of a tyrosine kinase from human sperm with the zona pellucida at fertilization. Science 1995; 269:83-86.

Chapman NR, Barratt CLR. The role of carbohydrate in sperm-ZP3 adhesion. Mol Hum Reprod 1996; 2:767-774.

Chen S, Cardullo R. Characterization and localization of fluorescent zonae pellucidae on mouse sperm. Mol Biol Cell 1994; 5:224a.

Copley RR, Barton GJ. A structural analysis of phosphate and sulphate binding sites in proteins. J Mol Biol 1994; 242:321-329.

Cornwall GA, Tulsiani DRP, Orgebin-Crist MC. Inhibition of the mouse sperm surface α-D- mannosidase inhibits sperm-egg binding *in vitro*. Biol Reprod 1991; 44:913-921.

Curatolo W. Glycolipid function. Biochem Biophys Acta 1987; 906:137-160.

Fischer G, Reiter S, Jatzkewitz H. Enzymic hydrolysis of sulphosphingolipids and sulphoglycerolipids by sulphatase A in the presence and absence of activator protein. Hoppe-Seyler's Z Physiol Chem 1978; 359:863-866.

Gadella BM, Colenbrander B, Van Golde LMG, Lopes-Cardozo M. Boar seminal vesicles secrete arylsulfatases into seminal plasma: evidence that desulfation of seminolipid occurs only after ejaculation. Biol Reprod 1993; 48:483-489.

Hakomori S. Bifunctional role of glycosphingolipids. J Biol Chem 1990; 265:18713-18716.

Hardy DM, Garber DL. Species-specific binding of sperm proteins to the extracellular matrix (zona pellucida) of the egg. J Biol Chem 1994; 269:19000-19004.

Harouse JM, Bhat S, Spitalnik SL, Laughlin M, Stefano K, Silberberg DH, Gonzalez-Scarano F. Inhibition of entry of HIV-1 in neural cell lines by antibodies against galactosyl ceramide. Science 1991; 253:320-323.

Harris JD, Hibler DW, Fontenot GK, Hsu KT, Yurewicz EC, Sacco AG. Cloning and characterization of zona pellucida genes and cDNAs from a variety of mammalian species: the ZPA, ZPB and ZPC gene families. DNA Sequence 1994; 4:361-393.

Hermo L, Morales C, Oko R. Immunocytochemical localization of sulfated glycoprotein-1 (SGP-1) and identification of its transcripts in epithelial cells of the extratesticular duct system of the rat. Anat Rec 1992; 232:401-422.

Hiraiwa M, O'Brien JS, Kishimoto Y, Galdzicka M, Fluharty AL, Binns EI, Martin BM. Isolation, characterization, and proteolysis of human prosaposin, the precursor of saposins (sphingolipid activator proteins). Arch Biochem Biophys 1993; 304:110-116.

Ishizuka I, Nakamura Y. Accumulation of sulfoglycolipids in hyperosmosis-resistant clones derived from Madin-Darby Canine Kidney Cells (MDCK). Glycoconj J 1991; 8:168a.

Jedlicki A, Zambrano F. Role of sulfatide on phosphoenzyme formation and ouabain binding of the (Na^+/K^+) ATPase. Arch Biochem Biophys 1985; 238:558-564.

Jones R. Interaction of zona pellucida glycoproteins, sulphated carbohydrates and synthetic polymers with proacrosin, the putative egg-binding protein from mammalian spermatozoa. Development 1991; 111:1155-1163.

Kornblatt MJ, Knapp A, Levine M, Schachter H, Murray RK. Studies on the structure and formation during spermatogenesis of the sulfoglycerogalactolipid of rat testis. Can J Biochem 1974; 52:689-697.

Kreysing J, Polten A, Hess B, von Figura K, Menz K, Steiner F, Gieselmann V. Structure of the mouse arylsulfatase A gene and cDNA. Genomics 1994a; 19:249-256.

Kreysing J, Polten A, Lukatela G, Matzner U, van Figura K, Gieselmann V. Translational control of arylsulfatase A expression in mouse testis. J Biol Chem 1994b; 269:23255-23261.

Law H, Itkonnen O, Lingwood CA. Sulfogalactolipid binding protein SLIP 1: a conserved function for a conserved protein. J Cell Physiol 1988; 137:462-468.

Leyton L, Saling P. 95 kDa sperm proteins bind ZP3 and serve as tyrosine kinase substrates in response to zona binding. Cell 1989; 57:1123-1130.

Lingwood CA. Protein-glycolipid interactions during spermatogenesis. Binding of specific germ cell proteins to sulfatoxygalactosylacylalkylglycerol, the major glycolipid of mammalian male germ

cells. Can J Biochem Cell Biol 1985a; 63:1077-1085.
Lingwood CA. Timing of sulphogalactolipid biosynthesis in the rat testis studied by tissue autoradiography. J Cell Sci 1985b; 75:329-338.
Lingwood CA, Nutikka A. Studies on the spermatogenic sulfogalactolipid binding protein SLIP1. J Cell Physiol 1991; 146:258-263.
Lingwood CA, Murray RK, Schachter H. The preparation of rabbit antiserum specific for mammalian testicular sulfogalactoglycerolipid. J Immunol 1980; 124:769-774.
Lingwood CA, Quinn PA, Wilansky S, Nutikka A, Ruhnke HL, Miller RB. Common sulfoglycolipid receptor for mycoplasmas involved in animal and human infertility. Biol Reprod 1990; 43:694-697.
Liu C, Litscher ES, Wassarman PM. Transgenic mice with reduced numbers of functional sperm receptors on their eggs reproduce normally. Mol Biol Cell 1995; 6:577-585.
Liu C, Litscher ES, Wassarman PM. Zona pellucida glycoprotein mZP3 bioactivity is not dependent on the extent of glycosylation of its polypeptide or on sulfation and sialylation of its oligosaccharides. J Cell Sci 1997; 110:745-752.
Lopez LC, Bayna EM, Litoff D, Shaper NL, Shaper JH, Shur BD. Receptor function of mouse sperm surface galactosyltransferase during fertilization. J Cell Biol 1985; 101:1501-1510.
Mamelak D, Lingwood C. Expression and sulfogalactolipid binding specificity of the recombinant testis-specific cognate heat shock protein 70. Glycoconj J 1997; 14:715-722.
McLeskey SB, Dowds C, Carballada R, White RR, Saling PM. Molecules involved in mammalian sperm-egg interaction. Int Rev Cytol 1998; 177:57-113.
Moase CE, Kamolvarin N, Kan FWK, Tanphaichitr N. Localization and role of sulfoglycolipid immobilizing protein 1 on the mouse sperm head. Mol Reprod Dev 1997; 48:1-11.
Mori E, Baba T, Iwanatsu A, Mori T. Purification and characterization of a 38-kD protein, sp38, with zona pellucida-binding property from porcine epididymal sperm. Biochem Biophys Res Commun 1993; 196:196-202.
Mortillo S, Wassarman PM. Differential binding of gold-labeled zona pellucida glycoproteins mZP2 and mZP3 to mouse sperm membrane compartments. Development 1991; 113:141-149.
Murray RK, Narasimhan R. Glycoglycerolipids of animal tissues. In: (Kates M, ed) Glycolipids, Phosphoglycolipids, and Sulfoglycolipids. New York: Plenum Press, 1990, pp321-361.
Nikolajczyk BS, O'Rand MG. Characterization of rabbit testis β-galactosidase and arylsulfatase A: purification and localization in spermatozoa during the acrosome reaction. Biol Reprod 1992; 46:366-378.
O'Brien JS, Kishimoto Y. Saposin proteins: structure, function, and role in human lysosomal storage disorders. FASEB J 1991; 5:301-308.
Oehninger S, Clark GF, Acosta AA, Hodgen GD. Nature of the inhibitory effect of complex saccharide moieties on the tight binding of human spermatozoa to the human zona pellucida. Fertil Steril 1991; 55:165-169.
Pancake S, Holt G, Mellouk S, Hoffman S. Malaria sporozoites and circumsporozoite proteins bind specifically to sulfated glycoconjugates. J Cell Biol 1992; 117:1351-1357.
Roberts DD, Ginsburg V. Sulfated glycolipids and cell adhesion. Arch Biochem Biophys 1988; 267:405-415.
Roldan E, Fragio C. Diradylglycerols stimulate phospholipase A2 and subsequent exocytosis in ram spermatozoa. Biochem J 1994; 297:225-232.
Rosen SD, Bertozzi CR. Leukocyte adhesion: two selectins converge on sulphate. Curr Opin Cell Biol 1996; 6:261-264.
Sanz L, Calvete JJ, Mann K, Gabius HJ, Topfer-Petersen E. Isolation and biochemical characterization of heparin-binding proteins from boar seminal plasma: a dual role for sperm-adhesions in fertilization. Mol Reprod Dev 1993; 35:37-43.
Snell WJ, White JM. The molecules of mammalian fertilization. Cell 1996; 85:629-637.
Sommerlade HJ, Selmer T, Ingendoh A, Gieselmann V, von Figura K, Neifer K, Schmidt B. Glycosylation and phosphorylation of arylsulfatase A. J Biol Chem 1994; 269:20977-20981.
Stein C, Gieselmann V, Kreysing J, Schmidt B, Pohlmann R, Waheed A, Meyer HE, O'Brien JS, von Figura K. Cloning and expression of human arylsulfatase A. J Biol Chem 1989; 264:1252- 1259.
Stewart RJ, Boggs JM. A carbohydrate-carbohydrate interaction between galactosylceramide-containing liposomes and cerebroside sulfate-containing liposomes: dependence on the glycolipid ceramide composition. Biochemistry 1993; 32:10666-10674.
Suzuki Y, Toda Y, Tamatani T, Watanabe T, Suzuki T, Nakao T, Murase K, Kiso M, Hasegawa A, Tadano-Aritomi K, Ishizuka I, Miyasaka M. Sulfated glycolipids are ligands for a lymphocyte homing receptor, L-selectin (LECAM-1), binding epitope in sulfated sugar chain. Biochem Biophys Res Commun 1993; 190:426-434.
Tanphaichitr N, Smith J, Kates M. Levels of sulfogalactosylglycerolipid in capacitated motile and immotile mouse sperm. Biochem Cell Biol 1990; 68:528-535.
Tanphaichitr N, Tayabali A, Gradil C, Juneja S, Leveille MC, Lingwood C. Role of germ cell-specific sulfolipidimmobilizing protein (SLIP1) in mouse *in vivo* fertilization. Mol Reprod Dev 1992; 32:17-22.
Tanphaichitr N, Smith J, Mongkolsirikieart S, Gradil C, Lingwood C. Role of a gamete specific sulfoglycolipid-immobilizing protein on mouse sperm-egg binding. Dev Biol 1993; 156:165-175.
Tanphaichitr N, Moase C, Taylor T, Surewicz K, Hansen C, Namking M, Bérubé B, Kamolvarin N, Lingwood CA, Sullivan R, Rattanachaiyanont M, White D. Isolation of antiSLIP1-reactive boar sperm P68/62 and its binding to mammalian zona pellucida. Mol Reprod Dev 1998; 49:203-216.
Topfer-Petersen E, Henschen A. Acrosin shows zona and fucose binding, novel properties for a serine proteinase. FEBS Lett 1987; 226:38-42.
Tulsiani DRP, Skudlarek MD, Araki Y, Orgebin-Crist MC. Purification and characterization of two forms of β-D-galactosidase from rat epididymal luminal fluid: evidence for their role in the modification of sperm plasma membrane glycoproteins(s). Biochem J 1995; 305:41-50.
Tulsiani DRP, Yoshida-Komiya H, Araki Y. Mammalian fertilization: a carbohydrate-mediated event. Biol Reprod 1997; 57:487-494.
Tupper S, Wong PTT, Tanphaichitr N. Binding of Ca^{2+} to sulfogalactosylceramide and the sequential effects on the lipid dynamics. Biochemistry 1992; 31:11902-11907.
Tupper S, Wong PTT, Kates M, Tanphaichitr N. Interaction of divalent cations with germ cell specific sulfogalactosylglycerolipid and the effects on lipid chain dynamics. Biochemistry 1994; 33:13250-13258.
Wang LF, Wei SG, Miao SY, Liu QY, Koide SS. Calpastatin gene in human testis. Biochem Mol Biol Intl 1994; 33:245-252.
Wassarman PM. Towards molecular mechanisms for gamete adhesion and fusion during mammalian fertilization. Curr Opin Cell Biol 1995; 7:658-664.
Yanagimachi R. Mammalian fertilization. In: (Knobil E, Neill JE, eds) The Physiology of Reproduction. New York: Raven Press Ltd, 1994, pp189-317.
Yurewicz EC, Pack BA, Armant DR, Sacco AG. Porcine zona pellucida ZP3 α glycoprotein mediates binding of the biotin-labeled M_r 55,000 family (ZP3) to boar sperm membrane vesicles. Mol Reprod Dev 1993; 36:382-389.

22

Characteristics of the Sperm-Zona Interaction: Key Issues for Identifying Receptors for ZP3

Richard A. Cardullo
University of California

Following capacitation and passage through the cumulus layer, the mammalian sperm finally reaches the zona pellucida, the innermost extracellular matrix surrounding the mammalian egg. Based on years of research by many investigators, it is clear that this structure plays crucial roles in both fertilization and early development of the zygote and embryo. It is now believed that individual macromolecules within this unique matrix play precise physiological roles, including serving as a species selective adhesive substrate for sperm and as an agonist for regulated exocytosis of the sperm's acrosomal vesicle (the acrosome reaction). In addition, modifications to the zona pellucida following fertilization prevent polyspermy and protect the early embryo in the oviduct prior to hatching and implantation in the uterus. Although biochemical and molecular characterization of the individual glycoproteins that make up the zona pellucida has begun, little is known about the specific molecular interactions that regulate fertilization and early development.

For over 15 years, fertilization biologists have actively tried to characterize and identify the sperm surface molecules that first interact with the zona pellucida. In pursuing these studies, the majority have assumed that this initial interaction, which is responsible for both adhesion and the induction of the acrosome reaction, would be straightforward and relatively simple. However, although a plethora of sperm surface receptors (for example, galactosyltransferase, sp56, zona receptor kinase, zonadhesin, SLIP 1 and so on) has been proposed to interact with a single zona pellucida glycoprotein (ZP3), no consensus has yet been reached and there is increasing evidence that the initial interaction between sperm and egg may involve multiple receptors and/or ligands. Further, it is now clear that this Occam's razor approach should be abandoned and that a complete understanding of sperm-zona interactions requires knowledge about the structural components of the zona pellucida, the nature of the interaction between acrosome intact sperm and the zona pellucida, and the modifications that occur to both the zona pellucida and the sperm following their successful interaction.

This chapter will briefly outline the major issues involved in understanding the initial interaction between the sperm and the zona pellucida. In particular, the chapter will focus on current knowledge of zona pellucida structure, assays used to assess function of the zona pellucida, evidence for the existence of ZP receptors on sperm and modification of the zona pellucida following fertilization. Additional chapters in this volume, by C. Thaler, P. Wassarman, B. Shur and N. Tanphaichitr, focus on specific aspects of this interaction using all of the advantages of the mouse system.

The Molecular Composition and the Structure of the Zona Pellucida

Biochemical and molecular aspects of the zona pellucida have been studied extensively in the mouse, although there is a growing body of literature in other mammals including rabbit (Prasad 1996), pig (Sacco et al. 1989; Yurewicz et al. 1987, 1993a, b, c) and human (Barratt, Hornby 1995). These studies have shown that the proteins which make up the zona pellucida are highly conserved among mammals, perhaps suggesting that the overall organization and function of the zona pellucida is conserved as well.

In the mouse, the zona pellucida is composed of three glycoproteins, designated mZP1, mZP2 and mZP3, with average molecular masses of 200 kD, 120 kD and 83 kD, respectively; this is reviewed in Wassarman (1988) and Wassarman and Litscher (1995). The primary sequence of these glycoproteins was deduced from full length cDNAs with the polypeptide chains of these three macromolecules accounting for only a fraction of the total molecular mass (Ringuette et al. 1988). All three are heavily glycosylated proteins containing both N- and O-linked oligosaccharide chains. Gene sequences reveal a high degree of sequence similarity between members of the ZP3 gene family and to a lesser extent in the ZP1 and ZP2 gene families; this is reviewed in McLeskey et al. (1998).

Studies have shown that synthesis of these glycoproteins is temporally regulated during oogenesis with the ZP genes transcribed exclusively by oocytes. In particular, mZP2 is expressed at low levels in resting oocytes, but mZP1 and mZP3 are expressed exclusively by growing oocytes, with all three transcripts reaching maximal levels in midsized oocytes (Epifano et al. 1995). The translation and secretion of these proteins occurs concomitantly with the transcriptional activation of these genes leading to the assembly of the intact zona pellucida during growth of the developing oocyte (Epifano et al. 1995). A number of recent reports have demonstrated that in some mammals the ZP genes may be expressed by granulosa cells as well (Lee, Dunbar 1993; Grootenhuis et al. 1996; Kolle et al. 1996; Martinez et al. 1996). If this is true, then post-translational modifications to ZP polypeptides may lead to different structural components (for example, glycosylation patterns) between oocyte-derived and granulosa-derived zona pellucida glycoproteins.

The zona pellucida is a compact, highly organized matrix which, in the mouse, is approximately 7 μm thick with an outer diameter of approximately 110 μm. At the level of the electron microscope, the matrix has a lacy appearance. It presents a formidable barrier to sperm. Although little is known concerning the assembly of the different zona pellucida glycoproteins, current evidence suggests that the matrix is a non-covalently assembled structure made up of mZP2 and mZP3 dimers which polymerize into filaments and are crosslinked by mZP1 homodimers to mZP2 (Greve, Wassarman 1985). Since sperm interact with an intact zona pellucida, understanding the nature of this intact structure ultimately may provide important clues about the specific molecular interaction between the sperm receptor and the regions of ZP3 that are available for binding.

In addition to studies done in mammals, there has been some work concerning the composition of analogous structures in other vertebrate models. In *Xenopus laevis*, the vitelline envelope forms a structure which is similar to the zona pellucida and is composed of three glycoproteins (Doren et al. 1997). These three glycoproteins are homologous (30 to 40 percent amino acid identity) to the three mouse zona pellucida glycoproteins. When full length mRNAs from the mouse were injected into stage VI *Xenopus* oocytes, the mouse glycoproteins were expressed and secreted to the extracellular matrix of *Xenopus* eggs, suggesting that the mouse zona proteins have been sufficiently conserved in evolution to be integrated into the vitelline envelope (Doren et al. 1997). In addition, some species of fish express ZP-like proteins which are incorporated into their vitelline envelope (Lyons et al. 1993; Murata et al. 1995; Chang et al. 1996). Interestingly, in the case of white flounder and medaka, these proteins are synthesized in the liver and then transported to the oocyte where they are assembled in the vitelline envelope (Lyons et al. 1993).

The Choice of Binding Assay Depends on the Question Being Addressed

Assays for studying sperm binding to the zona pellucida fall into two categories: those that utilize the intact zona pellucida versus those that use solubilized proteins. The assay chosen depends on particular question being asked. For instance, if information about adhesion at the cellular or biophysical level is needed, then using an intact zona pellucida would be warranted. If, however, molecular details about the interaction between a specific zona pellucida glycoprotein and its complementary receptor on the sperm surface is needed, then it is best to use solubilized and purified glycoproteins.

By far, the most utilized method for studying sperm-zona adhesion is light microscopic adhesion assays using living sperm and intact zonae pellucidae. In these assays, sperm are incubated with eggs in the presence or absence of a competitor for some predetermined amount of time (for example, five to 60 minutes), fixed, and attached sperm are subsequently counted under the light microscope. As a control for non-specific binding, two-cell embryos containing zonae are included, since sperm do not bind to the zona pellucida of embryos. Modifications of this assay include the stop-fix method, in which eggs are centrifuged through fixative to simultaneously stop the assay and remove non-specifically bound sperm (Saling et al. 1978), and another method which analyzes the distribution of sperm bound per zona pellucida in populations of zonae pellucidae (SamAth et al. 1997).

The microscopic assays using intact zonae have proved useful for identifying the key molecules involved in sperm-zona interaction and were instrumental in identifying ZP3 as both the initial adhesion ligand and as the secretagogue for acrosomal exocytosis in mice (Bleil, Wassarman 1980, 1983, 1986; Florman et al. 1984; Florman, Wassarman 1985). In addition, a number of putative receptors for ZP3 on the mouse sperm surface have been identified with this assay by using different antagonists against these sperm surface proteins. In the intact zona pellucida assay, sperm have a limited contact time and contact area for ligand-receptor interactions (Baltz, Cardullo 1989). During this time period (~50-100 ms), sperm must successfully recognize and bind to at least one ZP3 molecule in order for adhesion to occur (Baltz et al. 1988). In competition assays where sperm are coincubated with zonae and an appropriate antagonist, dissociation must occur within 50 ms. Since the lowest affinity constant for ZP3-ZP3 receptor interaction has been quantitatively measured to be around 50 nM (Thaler, Cardullo 1996a), the K_d of a competing molecule would have to be of similar magnitude to successfully displace a sperm from the zona. Further exacerbating the problem is that the density of ZP3 receptors on the sperm surface is extremely high, making competition by simple molecular inhibitors unlikely (Thaler, Cardullo 1996b). Specifically, concentrations of antagonists needed to even moderately inhibit binding most often exceed 1 mM. At these high concentrations, non-specific effects are undoubtedly a significant problem.

The use of solubilized zona pellucida glycoproteins has allowed investigators to study binding interactions in detail. The physiological role of the three glycoproteins has been investigated by a number of fertilization biologists (Bleil, Wassarman 1980, 1983; Leyton, Saling 1989; Miller et al. 1992; Thaler, Cardullo 1996a, b). Early studies by Bleil and Wassarman (1986), used ^{125}I-ZP3 and cellular autoradiography to localize ZP3 receptors to the mouse sperm head. The use of solubilized components to biochemically characterize sperm-zona binding has recently been accomplished using conventional equilibrium, kinetic and competition binding assays (Thaler, Cardullo 1996a, b; Thaler, this volume).

As with the intact zona pellucida assays, the use of solubilized components to study sperm-zona interactions is not without problems. In particular, since molecules are used in isolation, interactions reflecting the filamentous structure of the zona pellucida are lost.

Using solubilized glycoproteins, investigators cannot approach the high density of these proteins in the intact matrix. This obviously has implications in understanding the basis of both the adhesion and the signaling pathways responsible for acrosomal exocytosis.

Identifying Key Cellular Events in Sperm-Zona Interaction

Several laboratories have described a number of discrete steps in sperm zona pellucida binding which can be used to characterize both zonae pellucidae and capacitated sperm *in vitro*. These steps include loose attachment, firm binding, induction of acrosomal exocytosis and penetration of the zona pellucida (Hartmann et al. 1972; Storey 1991). Each step may reveal different molecular interactions between zona pellucida ligands and complementary sperm surface receptors. Understanding the time course of these cellular events is ultimately useful in identifying the molecules involved in primary and secondary binding events.

Combining information at both the cellular and biochemical level has led to a molecular model for the role of the zona pellucida in sperm adhesion before and after acrosomal exocytosis. The sequence of events in this model includes: 1) adhesion of acrosome intact sperm to ZP3, 2) tight binding of acrosome intact sperm to the zona pellucida followed by ZP3-induced activation of signal transduction pathways leading to acrosomal exocytosis, 3) secondary adhesion of acrosome reacted sperm mediated by ZP2, 4) penetration of the zona pellucida and 5) molecular modifications to both ZP2 and ZP3 following sperm-egg fusion and egg activation.

The five steps of the model, in turn, have given fertilization biologists a framework for testing the individual steps of zona binding and penetration by both acrosome intact and acrosome reacted sperm. Hypotheses currently being tested are listed below and are addressed individually in the remainder of this paper. The hypotheses are:

1. ZP3 binds to acrosome intact sperm on the head and should possess a high affinity component to account for adhesion to the matrix. *ZP3 binding to complementary sperm receptors should be detectable using conventional biochemical assays.*
2. ZP3 initiates a signal transduction cascade that leads to acrosomal exocytosis. *In response to ZP3, activation of particular second messengers should be detectable.*
3. Following acrosomal exocytosis, the sperm continues to bind and ultimately penetrate the zona pellucida. *Secondary receptors and/or ligands for sperm binding after the acrosome reaction should be biochemically distinct from the primary molecules.*
4. Following fusion with the egg's plasma membrane, the zona pellucida is chemically modified to prevent subsequent binding by either acrosome intact or acrosome reacted sperm. *Determining which molecular determinants in ZP3 and/or ZP2 are modified after fertilization will provide important clues about what is being recognized by sperm receptors.*

Initial Adhesion between the Zona Pellucida and Sperm

A number of studies using different assays have implicated ZP3 as the primary adhesion molecule in the zona pellucida in most mammals. In the mouse, mZP3 is an 83 kD glycoprotein (Wassarman 1988; Litscher et al. 1995) having three or four N-linked oligosaccharide chains and an undetermined number of O-linked oligosaccharide chains (Wassarman, Litscher 1995). Early experiments suggested that the bioactivity of mZP3 resides solely with one or more of the O-linked oligosaccharides on ZP3 (Bleil, Wassarman 1988; Florman et al. 1984; Florman, Wassarman 1985), in the C-terminal half of the molecule (Rosiere, Wassarman 1992). Among O-linked oligosaccharides from native ZP3, a small oligosaccharide estimated at 3.4-4.6 kD, using size exclusion chromatography, which retained sperm binding activity but did not induce the acrosome reaction, was identified (Florman, Wassarman 1985). This result suggested that the polypeptide chain may play an important role in the organization of oligosaccharides necessary for initiation of signal transduction pathways leading to acrosomal exocytosis.

Early characterization of critical biochemical and cell biological parameters was limited to crude microscopic methods such as binding of sperm to ZP3-conjugated beads (Vazquez et al. 1989), cellular autoradiography (Bleil, Wassarman 1986) and binding of colloidal gold-labeled ZP3 to acrosome intact sperm (Mortillo, Wassarman 1991). These methods showed that ZP3 binding sites localize to the sperm head and the cellular autoradiography estimated the number of ZP3 binding sites to be between 10,000 and 50,000 per sperm (Bleil, Wassarman 1986). Although useful for confirming the localization of ZP3 receptors on the mouse sperm surface, these types of studies have done little to identify the complementary receptors for ZP3 or to establish the precise role of the zona pellucida in adhesion or the initiation of signal transduction pathways necessary for acrosomal exocytosis.

Thaler and Cardullo (1996a) performed a biochemical characterization of ZP binding on sperm using solubilized and ^{125}I-ZP components in a standard biochemical assay. These studies demonstrated that the initial binding event between mZP3 and its complementary receptor on the mouse sperm surface is complex. Both equilibrium and kinetic analyses indicated that these high affinity interactions cannot be explained by a single receptor-ligand interaction. These data suggest the presence of multiple ligands and/or receptors of varying affinities (Thaler, Cardullo 1996a, b). Saturation binding revealed that there are approximately 30,000 mZP3 binding sites per sperm (Thaler, Cardullo 1996a). Along with microscopic evidence that the initial binding event between acrosome intact sperm and the zona pellucida occurs only over the approximately 10-15 μm^2 surface area overlying the acrosomal vesicle, the average mZP3 receptor density is moderately high, at around 2000-3000 molecules/μm^2. Although more detailed experiments need to be performed, these results support the hypothesis that ZP3 is a multivalent ligand, consistent with earlier studies showing that monovalent ZP3 oligosaccharides bind to sperm but do not induce acrosomal exocytosis (Florman, Wassarman 1985).

Experiments using solubilized ZP components may only partially address the interactions that occur on the intact zona pellucida. Theoretical arguments based on biophysical measurements of sperm adhesion have suggested that only a few sperm-zona bonds are needed to tether a sperm to an egg (Baltz et al. 1988). The rate limiting step for sperm adhesion to the zona pellucida is directly proportional to the surface density of ZP3 on the zona pellucida, the receptor density on the surface of sperm, the contact area between sperm and the zona pellucida and the diffusion coefficient of the receptor on the sperm surface (Baltz, Cardullo 1989). Estimates of average ZP3 receptor density on the zona approach 300 molecules/μm^2 and corresponding densities of the ZP3 receptor on the mouse sperm surface are about 2000 molecules/μm^2. These high densities, along with the assertion that few bonds are needed to tether a sperm, virtually ensure that sperm adhesion to the zona pellucida will occur if sperm contact the zona pellucida in the correct orientation.

Successful adhesion of mammalian sperm to the zona pellucida is a combination of achieving the correct molecular specificity between ligands in the zona pellucida and their complementary receptors on the sperm surface along with optimized structural and biophysical characteristics of these surfaces. Zonae pellucidae which do not have the correct chemical modifications to their bioactive ligands would ultimately result in infertility. Similarly, if either the zona pellucida or the sperm surface do not have optimized biophysical parameters to ensure adhesion (such as ligand density on the zona surface, receptor density or mobility on the sperm surface) then fertilization will not occur. In this context it is possible that events upstream from sperm interaction with the zona may play a critical role in the successful interaction of the zona pellucida with the sperm. In particular, capacitating factors in the oviduct or soluble and/or extracellular matrix factors within the cumulus layer may play a role in modifying chemical structures or biophysical parameters (such as increasing receptor diffusion coefficients) which will ensure successful fertilization.

Evidence That the Bioactive Component on ZP3 is an Oligosaccharide

Although it is generally accepted that the bioactive component within ZP3 is related to its carbohydrate composition, there is considerable disagreement over the identity of the oligosaccharide(s) recognized by mouse sperm. Only the identity of the terminal sugar residue has been investigated directly.

Competitive binding assays using intact zonae pellucidae have implicated either a terminal α-linked galactose (Bleil, Wassarman 1988; Litscher et al. 1995) or a terminal N-acetylglucosamine (GlcNAc; Shur, Hall 1982a, b; Lopez et al. 1985; Miller et al. 1992; Youakim et al. 1994) on ZP3. Shur and colleagues have argued that ZP3 must contain terminal GlcNAc residues since substrates for β-1,4 galactosyltransferase (GalTase) decrease the amount of sperm binding to zona intact eggs (Shur, Hall 1982a, b; Lopez et al. 1985), consistent with their hypothesis that a sperm surface GalTase is the mouse sperm ZP3 receptor. This is reviewed in Shur (1991) and Dubois and Shur (1995). More recently, Miller et al. (1992) provided evidence that all ZP glycoproteins contain GlcNAc residues, since all three can be used as substrates for bovine milk soluble GalTase. Further, the mouse sperm GalTase selectively galactosylates ZP3 (and not ZP1 or ZP2), consistent with subsequent work showing that the sperm surface GalTase is distinguishable from Golgi forms of this enzyme (Miller et al. 1992). It has been suggested that GlcNAc residues crosslink GalTase on the mouse sperm surface resulting in the activation of a heterotrimeric G_i protein (Gong et al. 1995). These experiments suggest that a terminal GlcNAc on ZP3 is necessary for sperm zona adhesion and induction of acrosomal exocytosis.

To test for participation of galactose (Gal) or GlcNAc in sperm-zona binding, complex oligosaccharides which varied in composition (terminal Gal or terminal GlcNAc) and in degree of branching (uni-, bi-, tri-, and tetra-antennary structures) were used as competitors in a sperm-zona binding assay. Oligosaccharides which terminated in Gal (in either the α- or β- configuration), but not GlcNAc, significantly inhibited binding (Litscher et al. 1995). Furthermore, the effectiveness of inhibition increased as the degree of branching and the length of terminal Gal containing oligosaccharides increased (Litscher et al. 1995). Studies demonstrating that transgenic mice lacking all α1,3 galactose residues are fertile indicate that a terminal α-galactose may not entirely account for sperm-zona adhesion (Thall et al. 1995). Given the results of Litscher et al. (1995)

showing that both α-Gal and β-Gal are effective inhibitors of binding, it is possible that the important sugar is β-Gal.

Other work suggests that N-linked oligosaccharides are the bioactive components that determine adhesion or that a monosaccharide on O-linked oligosaccharides, other than Gal, may be the ligand for the complementary receptor on the sperm surface. In contrast to earlier reports (Bleil, Wassarman 1988), Nagdas et al. (1994) were unable to detect α-Gal residues on O-linked oligosaccharides of ZP3. These investigators suggested that α-Gal residues are on N-linked rather than O-linked oligosaccharide chains, although this assertion has not yet been demonstrated. When Nagdas et al. (1994) exhaustively treated ZP2 and ZP3 with endo-β-galactosidase, an enzyme which cleaves repeating units of acetyllactosamine (3Galβ1,4GlcNAcβ1), there was a significant reduction in the molecular masses of both ZP2 and ZP3 of 23 kD and 16 kD, respectively. These polylactosaminoglycans were found to be associated solely with N-linked, and not O-linked, oligosaccharides. In addition, these investigators treated de-N-glycosylated ZP3 with mild alkali in the presence of NaB^3H_4, which caused the release of a radiolabeled trisaccharide (GlcNAcβ→ Galβ 1,3 GalNAcol). Significantly, this trisaccharide contains a GlcNAc residue which may serve as the ligand for the sperm surface GalTase as suggested by Shur and colleagues (Shur, Hall 1992a, b; Lopez et al. 1985).

A recent microscopic study using lectins and antibodies against mZP2 and mZP3 has revealed that the intact zona pellucida is heterogenous, possessing both an inner and outer core with ZP2 and ZP3 densities highest in the outer core. Specifically, using a variety of different lectins and electron microscopy, it has been found that the carbohydrate composition in these two regions varies (Avilés et al. 1997). Of particular relevance, the lectin $BSAIB_4$, which recognizes terminal α-Gal residues, was localized only to the inner half of the zona pellucida. This suggests that α-Gal residues would not be available for primary binding events but may be available for binding during sperm penetration of the zona pellucida (Avilés et al. 1997).

It is clear that the initial reaction between acrosome intact sperm and the zona pellucida is complex. This complexity may come from multiple ligands in the zona or from multiple receptors on the sperm surface. It is possible that different mammals may use different molecules or ligands to ensure successful adhesion and onset of acrosomal exocytosis. Alternatively, requiring multiple molecular interactions may serve as a selection filter which allows only competent sperm to undergo a complex binding event and penetrate the zona pellucida. In the context of a complex binding event which displays both high and low affinity states, it will be necessary to determine how these different states relate to adhesion and the initiation of signal transduction pathways leading to acrosomal exocytosis.

Zona Pellucida Glycoproteins Other Than ZP3 May Be Involved in Primary Binding

In some mammals, ZP1 has been suggested to play a role in primary binding between the zona pellucida and acrosome intact sperm. In the current mouse model, mZP1 plays a purely structural role, while mZP2 and mZP3 serve as ligands for receptors on acrosome reacted and acrosome intact sperm, respectively (Wassarman 1988). Evidence for molecules other than mZP3 serving a role in initial binding comes from work done in pig, rabbit and human.

The porcine zona pellucida is composed of three glycoproteins, identified as pZP1, pZP3α and pZP3β, which are homologous to mZP2 (Yamasaki et al. 1996), mZP1 (Yurewicz et al. 1993a) and mZP3 (Harris et al. 1994), respectively. Both pZP3α and pZP3β show identical electrophoretic mobilities on reducing SDS-PAGE and separation of these two molecules can only be achieved following chemical modification (Yurewicz et al. 1987). pZP3α binds to sperm membranes via its N-linked oligosaccharides (Sacco et al. 1989; Yurewicz et al. 1993 a, b; Yonezawa et al. 1995; Nakano et al. 1996). The binding of pZP3α is significantly enhanced by ZP3β. Specifically, pZP3α + pZP3β is approximately 10 times more effective at inhibiting binding of sperm to the zona pellucida than is pZP3α alone (Yurewicz et al. 1993b). Additionally, antibodies against pZP3β inhibit binding of boar sperm to the zona pellucida (Bagavant et al. 1993). In these studies, the antibody was raised against a 25-mer derived against pZP3β, which is putatively rich in O-linked oligosaccharides, perhaps suggesting that like mZP3, pZP3β recognition may involve O-linked carbohydrates. This has yet to be demonstrated, however. Clearly, further studies must be performed to determine the role of pZP3α and pZP3β in sperm-zona binding and to see if pZP3β plays a direct role in adhesion or merely serves to coordinate the binding of pZP3α to its complementary receptor on the sperm surface.

Studies on rabbits and humans have shown that ZP1 may be involved in primary binding events (Prasad et al. 1996). In the rabbit, primary binding is mediated by a 55 kD zona pellucida glycoprotein (R55) that is 51% similar to mZP1 and shows little sequence similarity to mZP3 (Epifano et al. 1995). However, a zona pellucida ligand for sperm binding in humans has greater than 75% sequence homology to pZP3α and R55 and only 50% sequence homology to mZP1 (Epifano, Dean 1994), perhaps suggesting that members of the ZP1 family may have different functions in different mammals. Although further evidence is needed, these data suggest that in these species a homolog of mZP1, and not mZP3, may serve as the primary adhesion ligand for sperm interaction with the zona pellucida.

The Role of the Zona Pellucida following Acrosomal Exocytosis: Secondary Binding

In contrast to data on initial binding events, there is little experimental data available on the role of the zona pellucida following acrosomal exocytosis. Once sperm have undergone acrosomal exocytosis, a new membrane surface is presented and sperm proceed to penetrate the zona matrix. It has been argued that to penetrate the matrix, acrosome reacted sperm must bind to the zona pellucida using either receptors and/or ligands which are distinct from those involved in the primary binding. Monoclonal antibodies against mZP2 do not affect initial binding between acrosome intact sperm and the zona pellucida, but inhibit the continued binding of acrosome reacted sperm to the zona pellucida (Bleil, Wassarman 1988). Later studies using colloidal gold-labeled mZP3 and mZP2 localized these molecules to either acrosome intact or acrosome reacted sperm, respectively, using transmission electron microscopy (Mortillo, Wassarman 1991). Interestingly, low levels of mZP3 binding to the post-acrosomal region of acrosome reacted sperm were also detected.

Unfortunately, no biochemical analysis of mZP2 binding to acrosome reacted sperm has yet been performed. This is somewhat surprising since mZP2 generally presents a stronger signal than the other two glycoproteins, it is easy to obtain enriched populations of acrosome reacted sperm using calcium ionophores or other pharmacological agents such as thapsigargin and no physiological transformations similar to the acrosome reaction are thought to occur as a result of secondary binding. Once key experiments are performed, binding parameters such as K_ds, the number of binding sites and different affinity states can be ascertained.

In the pig, pZP1 has been suggested to be the secondary adhesion ligand for sperm. The pZP1 gene is homologous to that of mZP2, suggesting its role in secondary binding. In addition, fluorescently labeled, recombinant pZP1 (expressed in *E. coli*) binds to the equatorial region on the head of sperm from five different mammals, including boar (Tsubamoto et al. 1996). The study of the fluorescence pattern indicated that this protein was translocated from the equatorial segment to the posterior head over time, suggesting that pZP1 may assist sperm in penetration of the zona pellucida following the acrosome reaction. Further, using affinity blotting, this recombinant pZP1 bound to proacrosin and to an uncharacterized 40 kD protein from sperm (Tsubamoto et al. 1996).

A few mouse sperm surface proteins have been proposed to be involved in secondary binding events. One of these molecules is the GPI-linked hyaluronidase, PH-20, on guinea pig sperm. Some, but not all, antibodies directed against PH-20 block zona pellucida binding of acrosome reacted, but not acrosome intact, guinea pig sperm (Primakoff et al. 1985; Myles et al. 1987). Since PH-20 is localized to the posterior head of acrosome intact sperm and to the anterior head of acrosome reacted sperm, it has been suggested that these two populations play distinct physiological roles: penetration through the cumulus matrix for acrosome intact sperm and secondary binding for acrosome reacted sperm (Hunnicutt et al. 1996).

Soybean trypsin inhibitor (SBTI) also binds to acrosome reacted sperm and blocks secondary binding in a similar manner to anti-ZP2 (Bleil et al. 1981). This data implies that a protease may be involved in secondary binding events. Although the SBTI binding protein has not been identified, proacrosin, the precursor to the sperm's major serine protease, acrosin, has been implicated as a secondary adhesion molecule (Jones 1990; Topfer-Petersen, Calvete 1995; Topfer-Petersen 1996). It has recently been shown that porcine proacrosin recognizes ZP2 (Tsubamoto et al. 1996), supporting the hypothesis that both ZP2 and proacrosin are involved in secondary binding events. Biochemical and molecular analysis of proacrosin has identified potential binding domains in the proacrosin molecule (Topfer-Petersen et al. 1990; Jansen et al. 1995; Richardson, O'Rand 1996). Interestingly, proacrosin may play two important roles in fertilization: first, in secondary adhesion events following acrosomal exocytosis, and second, in penetration of the zona matrix as proacrosin is converted to acrosin, a process which is triggered directly by the zona pellucida (Topfer-Petersen, Cechova 1990). The importance of proacrosin in secondary binding has recently been challenged by experiments showing that proacrosin knockout mice are fertile (Baba et al. 1994). However, a structurally similar, but distinct, molecule known as sp38 has been identified in boar (Mori et al. 1995), perhaps indicating the presence of more than one class of secondary binding molecule in acrosome reacted sperm.

Modifications to the Zona Pellucida following Fertilization

Following fertilization, molecular changes occur in the zona pellucida which prevent additional sperm from binding to, or penetrating, the matrix. This represents a slow block to polyspermy, analogous to the hardening of the vitelline envelope in marine invertebrates. This has been termed the zona reaction (Braden et al. 1954; Gwatkin et al. 1973; Gulyas 1980). Direct and indirect evidence suggest that both mZP2 and mZP3 are modified following egg activation and exocytosis of the cortical granules by the egg. Specific biochemical modifications to mZP2 would render it ineffective as a ligand for secondary adhesion to acrosome reacted sperm, whereas modifications to mZP3 would have a similar effect on acrosome intact sperm. In addition to preventing polyspermy, the zona reaction may provide protection and support for the developing embryo as it passes through the oviduct prior to implantation. It is also possible that these specific molecular changes in the zona pellucida serve as recognition molecules necessary for disruption of the zona pellucida during hatching.

Modification in mZP2 was first detected as a change in electrophoretic mobility and isoelectric point on two-dimensional gels, denoted as a ZP2 to $ZP2_f$ transition (Bleil, Wassarman 1981). The modifications may be due to a protease released from cortical granules following egg activation since mZP2 is readily converted to $mZP2_f$ in the presence of the calcium ionophore, A23187, which induces cortical granule exocytosis in the absence of fertilization (Bleil et al.1981; Kalab et al. 1993). In addition, serine protease inhibitors block the conversion of mZP2 to $mZP2_f$ (Moller, Wassarman 1989).

In addition to a change in mZP2 to $mZP2_f$ following egg activation, it has been shown that eggs incubated *in vitro* will undergo a precocious loss of cortical granules, releasing their contents into the perivitelline space (Kalab et al. 1993). Although this represents only a minor fraction of the entire population of cortical granules within the egg, the spontaneous exocytosis of these vesicles is sufficient to convert mZP2 to $mZP2_f$, making it impossible for sperm to fertilize these eggs. Experiments have shown that this premature conversion of mZP2 to $mZP2_f$ *in vitro* can be prevented by adding a variety of serum components which are found in the oviduct, including fetuin, a known protease inhibitor (Kalab et al. 1993). It has been hypothesized that *in vivo*, spontaneous fusion events do not lead to the conversion of mZP2 to $mZP2_f$ because components in oviductal fluid prevent this conversion. However, these oviductal components are ineffective following egg activation since the massive release of cortical granules is thought to overwhelm those inhibitors leading to the conversion of mZP2 to $mZP2_f$.

In contrast to mZP2, mZP3 undergoes no detectable change in electrophoretic mobility following egg activation (Bleil et al. 1981). However, acrosome intact sperm are unable to bind to fertilized eggs or embryos. mZP3 may be modified by a cortical granule glycosidase which hydrolyzes terminal sugars needed for primary binding. Although researchers who believe that a terminal O-linked galactose is necessary for adhesion might argue that a galactosidase would be sufficient to render ZP3 inactive, Miller et al. (1993) have reported that a cortical granule derived N-acetylglucosaminidase hydrolyzes the GlcNAc residue that is recognized by the sperm surface GalTase. Yet another possibility is that changes in the structure of ZP2 (due to conversion to $ZP2_f$) may lead to a change in conformation of adjacent ZP3 molecules. However, evidence for any structural modifications in mZP3, other than that reported by Miller et al. (1993), still awaits verification.

Recent studies provide additional evidence that both ZP2 and ZP3 are modified after fertilization (Avilés et al. 1997). Immunoreactivity of ZP2 and ZP3 both decreased after fertilization and the binding patterns of a number of lectins changed as well. In particular, lectin binding showed that terminal GlcNAc residues did not decrease following fertilization (Avilés et al. 1997), in contrast to the observation by Miller et al. (1993) that GlcNAc on mZP3 is hydrolyzed by an N-acetylglucosaminidase following fertilization.

Why Has It Been So Difficult to Identify Bona Fide ZP3 Receptors?

The zona pellucida plays critical roles in sperm recognition and adhesion, initiation of acrosomal exocytosis, sperm penetration of the matrix, and subsequently in protection of the fertilized egg and embryo. The structure of the zona pellucida, especially the carbohydrate composition of the individual glycoproteins and the precise arrangement of those molecules, is a key determinant of biological function. Biochemical assays have shown that the interaction between ZP3 and acrosome intact sperm displays both high and low affinity components and suggest that there may be multiple ligands and/or receptors involved in the initial recognition events between sperm and the zona pellucida. Although the molecular determinants are currently being investigated, there is still considerable debate about the role of specific carbohydrates on ZP3 in forming the bioactive ligand of this molecule.

Why has it been so difficult to identify bona fide ZP3 receptors on the surface of acrosome intact sperm? Historically, two major impediments, mainly the small amount of zona pellucida material available for quantitative biochemical characterization and the belief that a molecular determinant on ZP3 interacts with a single complementary receptor on the sperm surface, have prevented researchers from arriving at definitive answers.

In the past, biologists who studied mammalian fertilization were limited by the amount of material available to them for studying sperm-zona interactions in detail. However, the development of biochemical microscale techniques, which are discussed by C. Thaler, in this volume, and the ability to manipulate mouse model systems at genetic and molecular levels, as discussed by B. Shur and P. Wassarman, also in this volume, have now allowed investigators to address specific issues regarding the interaction between the sperm and the zona pellucida. Indeed, whereas in the past the mouse system may have been viewed as being disadvantageous because of the small amount of zona material available for investigation, the development of transgenics and the ability to otherwise manipulate the mouse genome has now perhaps made this the most tractable system for studying mammalian fertilization.

In just the past few years, molecular manipulation of the mouse genome has provided important insights about sperm-zona interactions. Zona pellucida gene families are being constructed with the identification of ZP1, ZP2 and ZP3 genes in a number of different animals including cow, human, mouse and pig. Although these gene families share significant sequence similarities, questions regarding their regulation, secretion, and function still remain. At the molecular level, most ZP genes have the same number of exons and introns although human ZP2 has an additional exon at its C-terminal end (Liang, Dean 1993). In addition, all ZP3 genes are single copy genes except for human ZP3 (Epifano et al. 1995). These similarities and differences may provide important clues about the particular structural components which confer both matrix assembly and function of the zona pellucida in different mammals.

Further information about the role of these zona glycoproteins may be obtained by taking an evolutionary approach. Genes that encode the proteins making up the vitelline envelope from amphibians and fish share significant sequence similarities to the mammalian ZP genes. Additional gene sequences from other vertebrates, especially those which are internal fertilizers, may provide clues about the origins of species specificity conferred by the innermost extracellular matrix. Comparisons with vitelline envelope proteins in marine invertebrates, such as sea urchin and starfish, may also provide additional clues about gamete recognition and modification prior to sperm-egg fusion.

Clearly, biochemical analysis would be greatly aided by expression systems that yield greater amounts of bioactive ZP components. A number of investigators have recently reported that recombinant ZP3 can be expressed in a variety of different cell types. Since it is widely believed that the bioactivity of these glycoproteins is primarily associated with the oligosaccharides, and not the polypeptide chain, choice of expression system and conditions is critical. Given the large amounts of protein that can be obtained from these expression systems, the secreted glycoproteins should allow researchers to determine the critical bioactive components in each of the glycoproteins.

Although the limiting amount of material from the zona pellucida has been a major factor in the ability to identify ZP3 receptors, researchers now must accept that fertilization is a multi-step process which requires intricate molecular machinery. In retrospect, fertilization biologists have for too long assumed that the interaction between acrosome intact sperm and the zona pellucida is a relatively simple one involving a single ligand-receptor interaction. However, researchers are now coming to understand that recognition events between sperm and egg are necessarily complex, involving, perhaps, multiple molecular determinants at both the level of the receptor and the ligand. Further biochemical and biophysical characterization of zona pellucida glycoproteins and complementary receptors before and after fertilization will provide important clues about the role of these molecules in adhesion and the induction of the acrosome reaction.

References

Avilés M, Jaber L, Castells MT, Ballesta J, Kan FWK. Modifications of carbohydrate residues and ZP2 and ZP3 glycoproteins in the mouse zona pellucida after fertilization. Biol Reprod 1997; 57:1155-1163.

Baba T, Azuma S, Kashiwabara S, Toyoda Y. Sperm from mice carrying a targeted mutation of the acrosin gene can penetrate the oocyte zona pellucida and effect fertilization. J Biol Chem 1994; 269:31845-31849.

Bagavant H, Yurewicz EC, Sacco AG, Talwar GP, Gupta SK. Delineation of epitopes on porcine zona pellucida relevant for binding of sperm to oocyte using monoclonal antibodies. J Reprod Immunol 1993; 25:277-283.

Baltz JM, Cardullo RA. On the number and rate of formation of sperm-zona bonds in the mouse. Gam Res 1989; 24:1-8.

Baltz JM, Katz DF, Cone RA. Mechanics of sperm-egg interaction at the zona pellucida. Biophys J 1988; 54:643-654.

Barratt CLR, Hornby DP. Induction of the human acrosome reaction by rhuZP3. In: (Fenichel P, Parinaud J, eds) The Human Acrosome Reaction. Montrouge, France: John Libbey Eurotext, 1995, pp105-122.

Bleil JD, Wassarman PM. Mammalian sperm-egg interaction: identification of a glycoprotein in mouse egg zonae pellucidae possessing receptor activity for sperm. Cell 1980; 20:873-882.

Bleil JD, Wassarman PM. Sperm-egg interactions in the mouse: sequence of events and induction of the acrosome reaction by a zona pellucida glycoprotein. Dev Biol 1983; 76:185-202.

Bleil JD, Wassarman PM. Autoradiographic visualization of the mouse egg's sperm receptor bound to sperm. J Cell Biol 1986; 102:1363-1371.

Bleil JD, Wassarman PM. Galactose at the nonreducing terminus of O-linked oligosaccharides of mouse egg zona pellucida glycoprotein ZP3 is essential for the glycoprotein's sperm receptor activity. Proc Natl Acad Sci USA 1988; 85:6778-6782.

Bleil JD, Beall CF, Wassarman PM. Mammalian sperm-egg interaction: fertilization of mouse eggs triggers modifications of the major zona pellucida glycoprotein, ZP2. Dev Biol 1981; 86:189-197.

Braden AHW, Austin CR, David HA. The reaction of the zona pellucida to sperm penetration. Austr J Biol Sci 1954; 7:391-409.
Chang YS, Wang SC, Tsao CC, Huang FL. Molecular cloning, structural analysis, and expression of carp *Zp3* gene. Mol Reprod Dev 1996; 44:295-304.
Doren S, Landsberger N, Dwyer N, Blanchette-Mackie N, Dean J. Expression of zona pellucida proteins in *Xenopus laevis* oocytes. Mol Biol Cell 1997; 8:438a.
Dubois DH, Shur BD. Cell surface β 1,4-galactosyltransferase. A signal transducing receptor? Adv Exp Med Biol 1995; 376:105-114.
Epifano O, Dean J. Biology and structure of the zona pellucida: a target for immunocontraception. Reprod Fertil Dev 1994; 6:319-330.
Epifano O, Liang LF, Familiari M, Moos ML, Dean J. Coordinate expression of the three zona pellucida genes during mouse oogenesis. Development 1995; 121:1947-1956.
Florman HM, Wassarman PM. O-linked oligosaccharides of mouse egg ZP3 account for its sperm receptor activity. Cell 1985; 41:313-324.
Florman HM, Bechtol KB, Wassarman PM. Enzymatic dissection of the functions of the mouse egg's receptor for sperm. Dev Biol 1984; 106:243-255.
Gong XH, Dubois DH, Miller DJ, Shur BD. Activation of a G protein complex by aggregation of β-1,4-galactosyltransferase on the surface of sperm. Science 1995; 269:718-1721.
Greve JM, Wassarman PM. Mouse egg extracellular coat is a matrix of interconnected filaments possessing a structural repeat. J Mol Biol 1985; 181:253-264.
Grootenhuis AJ, Philipsen HLA, de Breet-Grijsbach JTM, Van Duin M. Immunocytochemical localization of ZP3 in primordial follicles of rabbits, marmoset, rhesus monkey, and human ovaries using antibodies against human ZP3. J Reprod Fertil 1996; 50 (Suppl):43-54.
Gulyas BJ. Cortical granules of mammalian eggs. Int Rev Cytology 1980; 63:357-389.
Gwatkin RBL, Williams DT, Hartmann JF, Kniazuk M. The zona reaction of hamster and mouse eggs: production *in vitro* by a trypsin-like protease from cortical granules. J Reprod Fertil 1973; 32:259-265.
Harris JD, Hibler DW, Fontenot GK, Hsu KT, Yurewicz EC, Sacco AG. Cloning and characterization of zona pellucida genes and cDNAs from a variety of mammalian species: the ZPA, ZPB, and ZPC gene families. DNA Seq 1994; 4:361-393.
Hartmann JF, Gwatkin RBL, Hutchinson CF. Early contact interactions between mammalian gametes *in vitro*: evidence that the vitellus influences adherence between sperm and zona pellucida. Proc Natl Acad Sci USA 1972; 69:2767-2769.
Hunnicutt GR, Primakoff P, Myles DG. Sperm surface protein PH-20 is bifunctional: one activity is hyaluronidase and a second, distinct activity is required in secondary sperm-zona binding. Biol Reprod 1996; 55:80-86.
Jansen S, Quigley M, Reik W, Jones R. Analysis of polysulfate binding domains in porcine proacrosin, a putative zona adhesion protein from mammalian spermatozoa. Int J Dev Biol 1995; 39:501-510.
Jones R. Identification and functions of mammalian sperm-egg recognition molecules during fertilization. J Reprod Fertil 1990; 42(Suppl):89-105.
Kalab P, Schultz RM, Kopf GS. Modifications of the mouse zona-pellucida during oocyte modification - inhibitory effects of follicular fluid, fetuin, and α-2HS-glycoprotein. Biol Reprod 1993; 49:561-567.
Kolle, S, Sinowatz, F, Boie, G, Totzauer, I, Amelsgruber, W, Plendl, J. Localization of the mRNA encoding the zona protein ZP3α in the porcine ovary, oocyte, and embryo by non-radioactive *in situ* hybridization. Histochem J 1996; 28:441-447.
Lee VH, Dunbar BS. Developmental expression of the rabbit 55 kDa zona pellucida protein and messenger RNA in ovarian follicles. Dev Biol 1993; 155:371-382.
Leyton L, Saling PM. Evidence that aggregation of mouse sperm receptors by ZP3 triggers the acrosome reaction. J Cell Biol 1989; 108:2163-2168.
Liang LF, Dean J. Conversion of mammalian secondary sperm receptor genes enables the promoter of the human gene to function in mouse oocytes. Dev Biol 1993; 156:399-408.
Litscher ES, Juntunen K, Seppo A, Pentilla L, Niemala R, Renkonen O, Wassarman PM. Oligosaccharide constructs with defined structures that inhibit binding of mouse sperm to unfertilized eggs *in vitro*. Biochemistry 1995; 34:4662-4669.
Lopez LC, Bayna EM, Litoff D, Shaper NL, Shaper JH, Shur BD. Receptor function of mouse sperm galactosyltransferase during fertilization. J Cell Biol 1985; 101:1501-1510.
Lyons CE, Payette KL, Price JL, Huang RCC. Expression and structural analysis of a teleost homolog of a mammalian zona pellucida gene. J Biol Chem 1993; 268:21351-21358.
Martinez, ML, Fortenot, GK, Harris, JD. The expression and localization of zona pellucida glycoproteins and mRNA in cynomolgus monkeys (*Macaca fascicularis*). J Reprod Fertil 1996; 50 (Suppl):35-41.
McLeskey SB, Dowds C, Carballada R, White RR, Saling PM. Molecules involved in sperm-egg interaction. Int Rev Cytology 1998; 177:57-113.
Miller DJ, Macek MB, Shur BD. Complementarity between sperm surface β-1,4-galactosyltransferase and egg-coat ZP3 mediates sperm-egg binding. Nature 1992; 357:589-593.
Miller DJ, Gong X, Decker G, Shur BD. Egg cortical granule N-acetylglucosaminidase is required for the mouse zona block to polyspermy. J Cell Biol 1993; 123:1431-1440.
Moller CC, Wassarman PM. Characterization of a proteinase that cleaves zona pellucida glycoprotein ZP2 following activation of mouse eggs. Dev Biol 1989; 132:103-112.
Mori E, Kashiwabara S, Baba T, Inagaki Y, Mori T. Amino acid sequences of porcine sp38 and proacrosin required for binding to the zona pellucida. Dev Biol 1995; 168:575-583.
Mortillo S, Wassarman PM. Differential binding of gold-labeled zona pellucida glycoproteins mZP2 and mZP3 to mouse sperm membrane compartments. Development 1991; 113:141-149.
Murata K, Sasaki T, Yasumasu S, Iuchi I, Enami J, Yasumasu I, Yanagamachi K. Cloning of cDNAs for the precursor protein of a low-molecular-weight subunit of the inner layer of the egg envelope (chorion) of the fish *Oryzias latipes*. Dev Biol 1995; 167:9-17.
Myles DG, Hyatt H, Primakoff P. Binding of both acrosome-intact and acrosome-reacted guinea pig sperm to the zona pellucida during *in vitro* fertilization. Dev Biol 1987; 121:559-567.
Nagdas SK, Araki Y, Chayko CA, Orgebin-Crist MC, Tulsiani DR. O-linked trisaccharide and N-linked poly-N-acetyllactosaminyl glycans are present on mouse ZP2 and ZP3. Biol Reprod 1994; 51:262-272.
Nakano M, Yonezawa N, Hatanaka Y, Noguchi S. Structure and function of the N-linked carbohydrate chains of pig zona pellucida glycoproteins. J Reprod Fertil 1996; 50 (Suppl):25-34.
Prasad SV, Wilkins B, Skinner SM, Dunbar BS. Evaluating zona pellucida structure and function using antibodies to rabbit 55 kDa ZP protein expressed in baculovirus expression system. Mol Reprod Dev 1996; 52:1167-1178.
Primakoff P, Hyatt H, Myles DG. A role for the migrating sperm surface antigen PH-20 in guinea pig sperm binding to the egg zona pellucida. J Cell Biol 1985; 101:2239-2244.
Richardson RT, O'Rand MG. Site directed mutagenesis of rabbit proacrosin. Identification of residues involved in zona pellucida

binding. J Biol Chem 1996; 271:24069-24074.

Ringuette MJ, Chamberlin ME, Baur AW, Sobieski DA, Dean J. Molecular analysis of cDNA coding for ZP3, a sperm binding protein of the mouse zona pellucida. Dev Biol 1998; 127:287-295.

Rosiere TK, Wassarman PM. Identification of a region of mouse zona pellucida glycoprotein mZP3 that possesses sperm receptor activity. Dev Biol 1992; 154:309-317.

Sacco AG, Yurewicz EC, Subramanian MG, Matzad PD. Porcine zona pellucida: association of sperm receptor activity with the α-glycoprotein component of the Mr=55,000 family. Biol Reprod 1989; 41:523-532.

Saling PM, Storey BT, Wolf DP. Calcium-dependent binding of mouse epididymal spermatozoa to the zona pellucida. Dev Biol 1978; 65:515-525.

SamAth V, Cardullo RA, Thaler CD. Role of calcium in mouse sperm capacitation and zona pellucida binding. Mol Biol Cell 1997; 8:110a.

Shur BD. Expression and function of cell surface galactosyltransferase. Biochim Biophys Acta 1991; 988:389-409.

Shur BD, Hall NG. Sperm surface galactosyltransferase activities during *in vitro* capacitation. J Cell Biol 1982a; 95:567-573.

Shur BD, Hall NG. A role for mouse sperm surface galactosyltransferase in sperm binding to the egg zona pellucida. J Cell Biol 1982b; 95:574-579.

Storey BT. Sperm capacitation and the acrosome reaction. Ann NY Acad Sci 1991; 637:459-473.

Thaler CD, Cardullo RA. The initial molecular interaction between mouse sperm and the zona pellucida is a complex binding event. J Biol Chem 1996a; 271:23289-23297.

Thaler CD, Cardullo RA. Determining oligosaccharide specificity for initial sperm-zona pellucida adhesion in the mouse. Mol Reprod Dev 1996b; 45:535-546.

Thall AD, Maly P, Lowe JB. Oocyte gal α 1,3gal epitopes implicated in sperm adhesion to the zona pellucida glycoprotein are not required for fertilization in the mouse. J Biol Chem 1995; 270:21437-21440.

Topfer-Petersen E. Molecular mechanism of fertilization in the pig. Reprod Domest Anim 1996; 31:93-100.

Topfer-Petersen E, Calvete JJ. Molecular mechanisms of the interaction between sperm and the zona pellucida in mammals—studies on the pig. Int J Androl 1995; 18:20-26.

Topfer-Petersen E, Cechova D. Zona pellucida induces conversion of proacrosin to acrosin. Int J Androl 1990; 13:190-196.

Topfer-Petersen E, Steinberger M, von Eschenbach CE, Zucker A. Zona pellucida-binding of boar sperm acrosin is associated with the N-terminal peptide of the acrosin β-chain. FEBS Lett 1990; 265:51-54.

Tsubamato H, Hasegawa A, Inoue M, Yamasaki N, Koyama K. Binding of recombinant pig zona pellucida protein 1 (ZP1) to acrosome-reacted spermatozoa. J Reprod Fertil 1996; 50 (Suppl):63-67.

Vazquez MH, Phillips DM, Wassarman PM. Interaction of mouse sperm with purified sperm receptors covalently linked to silica beads. J Cell Sci 1989; 92:713-722.

Wassarman PM. Zona pellucida glycoproteins. Ann Rev Biochem 1988; 57:415-442.

Wassarman PM, Litscher ES. Sperm-egg recognition mechanisms in mammals. Curr Topics Dev Biol 1995; 30:1-19.

Yamasaki T, Tsubamoto H, Hagesawa A, Inoue M, Koyama K. Genomic organization of the gene for the pig zona pellucida glycoprotein ZP1 and its expression in mammalian cells. J Reprod Fertil 1996; 50:19-23.

Yonezawa N, Aoki H, Hatanaka Y, Nakano M. Involvement of N-linked carbohydrate chains of pig zona pellucida in sperm-egg binding. Eur J Biochem 1995; 233:35-41.

Youakim A, Hathaway HJ, Miller DJ, Gong XH, Shur, BD. Overexpressing sperm surface β 1,4-galactosyltransferase in transgenic mice affects multiple aspects of sperm-egg interactions. J Cell Biol 1994; 126:1573-1583.

Yurewicz EC, Sacco EG, Subramanian MG. Structural characterization of the Mr=55,000 antigen (ZP3) of porcine oocyte zona pellucida. J Biol Chem 1987; 262:564-571.

Yurewicz EC, Hibler D, Fontenot GK, Sacco AG, Harris J. Nucleotide sequence of cDNA encoding ZP3α, a sperm-binding glycoprotein from zona pellucida of pig oocyte. Biochem Biophys Acta 1993a; 1174:211-214.

Yurewicz EC, Pack BA, Armant DR, Sacco AG. Porcine zona pellucida ZP3α glycoprotein mediates binding of the biotin-labeled Mr 55,000 family (ZP3) to boar sperm membrane vesicles. Mol Reprod Dev 1993b; 36:382-389.

Yurewicz EC, Zhang S, Sacco EG. Generation and characterization of site-directed antisera against an amino-terminal segment of a 55 kDa sperm adhesive glycoprotein from zona pellucida of pig oocytes. Reprod Fertil 1993c; 98:147-152.

23 *A Current Model for the Role of ADAMs and Integrins in Sperm-Egg Membrane Binding and Fusion in Mammals*

Diana G. Myles
Chunghee Cho
Ruiyong Yuan
Paul Primakoff
University of California

By the time the sperm reaches the egg membrane, it has been altered by its passage through the female reproductive tract. Before its initial contact with the egg, the sperm has been in contact with different reproductive tract fluids and it has undergone capacitation and a change in the pattern of motility.

Sperm also interact with different cells of the female reproductive tract on their way to the egg. In their passage from the uterus to the ampulla of the oviduct, sperm bind to oviductal epithelial cells.

They also must penetrate through the hyaluronic acid-rich extracellular matrix that is formed between the cumulus cells surrounding the ovulated egg in most mammalian species. Upon encountering the egg, sperm bind to the zona pellucida, acrosome react, and penetrate the zona to enter the perivitelline space. Only now do they come in contact with the egg surface.

Sperm-Egg Membrane Interactions

Sperm-egg membrane binding and fusion is a multistep process. This can be seen by following the interaction of live sperm with the egg plasma membrane from the time of initial contact. Some observations have been made *in vitro* on sperm-egg interactions within the perivitelline space of zona-intact eggs (Gaddum-Rosse 1985), but interactions are easier to follow when the zona pellucida has been removed before *in vitro* insemination. During fertilization of zona-free eggs, sperm can be observed approaching the egg and initially adhering to the egg plasma membrane by the tip of the sperm, as shown in Fig. 1A (Gaddum-Rosse 1985; Yoshihara, Hall 1993; Miller B, Primakoff P and Myles D, unpublished observations). It was shown in the golden hamster that this tip attachment can be maintained for an extended period of time in the presence of protease inhibitors that block acrosomal content dispersion. Under these conditions, many sperm remained bound to the egg plasma membrane, pivoting on the tip or central region of their heads, without coming to lie flat on the egg surface or fusing with the egg membrane (Takano et al. 1993).

Under normal conditions, the sperm progress to the second stage of sperm binding to the egg plasma membrane when the sperm turns parallel to the egg surface and there is an attachment between the side or lateral surface of the sperm head and the egg plasma membrane. During this step the sperm lie flat on the surface of the egg plasma membrane and the tail continues to beat vigorously (Fig. 1B). After a variable period of time, the two membranes fuse. A stiffening of the tail has been observed at or near the time of sperm-egg fusion where the tail stops beating and extends out straight from the egg surface (Fig. 1C). In some cases, this "tail stiffening" may be followed by gentle undulations of the tail (Yanagimachi 1988).

The Molecular Basis of Sperm-Egg Membrane Interactions

Several approaches have been used to identify molecules of the gamete plasma membranes that participate in the binding and fusion of sperm and egg plasma membranes. A major method employed has been the

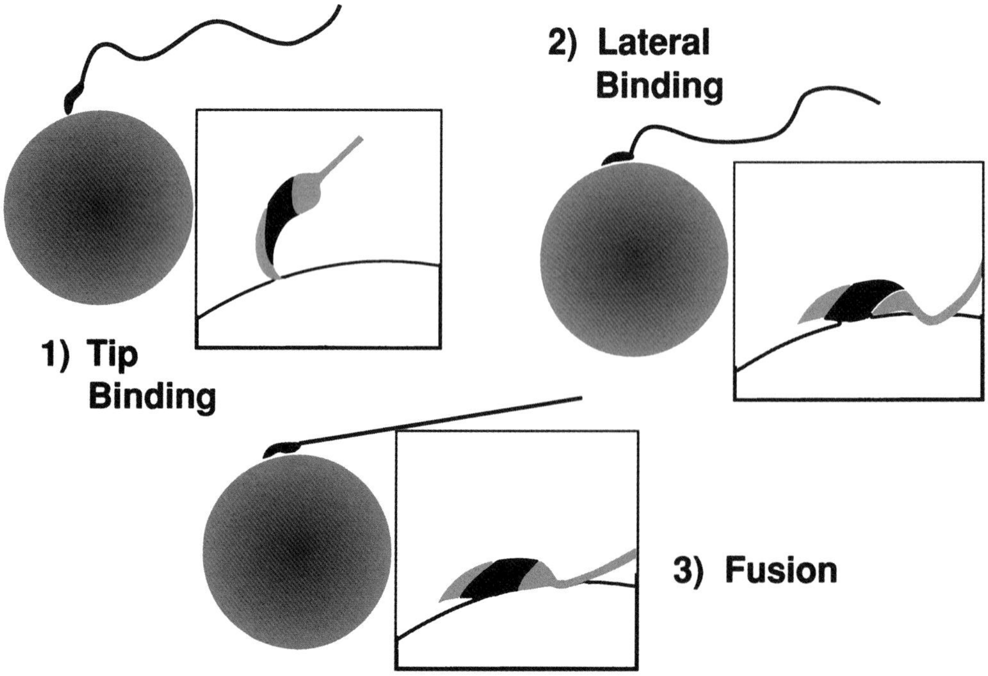

Figure 1. Sperm egg binding and fusion as a multistep process. 1: Sperm bind to the egg plasma membrane by their tips. 2: The tip binding converts to lateral (or flat) binding. 3: Fusion occurs between the equatorial/ posterior head region of the sperm and the egg plasma membrane and the tail stops beating and stiffens.

use of antibody inhibition studies. In these studies, either sperm or eggs are incubated with an antibody that recognizes a specific molecule on the gamete surface. Then *in vitro* assays are used to measure if there is an inhibition of either the number of sperm bound to the egg or the level of membrane fusion.

Other inhibition studies have assayed the effect of proteins or peptides, derived from these proteins, on sperm binding and fusion. In addition, genetic approaches may be used. For example, identification of a protein associated with a particular mutation that reduces sperm-egg membrane binding or fusion could be a direct path to identification of proteins not previously recognized as important in fertilization.

It is also possible to make targeted mutations by homologous recombination (knockouts) of genes that are candidates for a role in binding or fusion.

Currently, fusion is most commonly assayed by incorporating a vital dye that stains chromatin (Hoescht or DAPI) into the cytoplasm of the egg (Hinkley et al. 1986; Longo, Yanagimachi 1993). When the two membranes become confluent, the dye diffuses from the egg cytoplasm into the sperm cytoplasm and gains access to the sperm nucleus so that the sperm nucleus becomes stained.

In fusion assays, it is possible to count both the percentage of eggs fused with at least one sperm (fertilization rate) and the mean number of sperm fused per egg (fertilization index). Both measurements are important as they can reveal information about the results of the assay. For example, if the fertilization index exceeds 1.0, then the assay includes polyspermic eggs. Conditions under which polyspermy occurs could potentially mask the inhibition of fusion in the assay.

A variety of sperm proteins and some egg proteins have been identified as having a potential role in the interaction of sperm and egg plasma membranes. Candidate proteins have been reviewed in previous publications and some are discussed in other chapters of this volume (Myles 1994; Yanagimachi 1988). This chapter focuses on the role of ADAMs and integrins in this process.

The ADAM Family of Transmembrane Proteins

Two of the proteins identified as having a potential role in sperm-egg membrane binding and fusion belong to the ADAM family of membrane proteins. Evidence for the involvement of these two proteins in sperm-egg membrane binding and fusion is detailed below. The ADAM family is a growing family of transmembrane proteins with over 20 members that contain A Disintegrin And Metalloprotease domain (Wolfsberg et al. 1995a; Wolfsberg, White 1998). These are multidomain proteins, as shown in Figure 2.

The ADAM proteins have a least two potential functions. Those with a consensus sequence for an active metalloprotease site potentially have protease activity. Additionally, all members of the family are potentially cell adhesion molecules, because of the sequence similarity of the disintegrin domain of ADAMs to snake disintegrins (soluble peptides found in snake venom that bind to the platelet integrin $\alpha IIb\beta 3$). In many snake venom disintegrin peptides, the active binding site contains an RGD sequence, a frequent recognition site for many integrins (Tomiyama et al. 1992). The RGD-containing region of the snake disintegrin peptides is present in a loop, with two disulfide bonds at the base (Adler et al. 1991; Chen et al. 1991).

However, sequence variations do occur in the loop in some classes of the snake venom peptides (Jia 1996) and in the ADAM family of transmembrane proteins the RGD sequences are usually replaced by other amino acids, thereby potentially changing the specificity of the adhesion partners. When ADAMs function as cell adhesion molecules, the predicted binding partners are integrins, but other adhesion partners could bind to particular ADAMs. The range of specific amino acid sequences that are compatible with adhesion activity has not been defined, and thus, any of the known ADAMs could be an adhesion protein.

Binding of Sperm by the Tip to Egg Plasma Membrane

When sperm first interact with the egg plasma membrane of zona-intact or zona free oocytes, there is an attachment between the tip of the sperm head and the egg plasma membrane (Yanagimachi 1994). This usual-

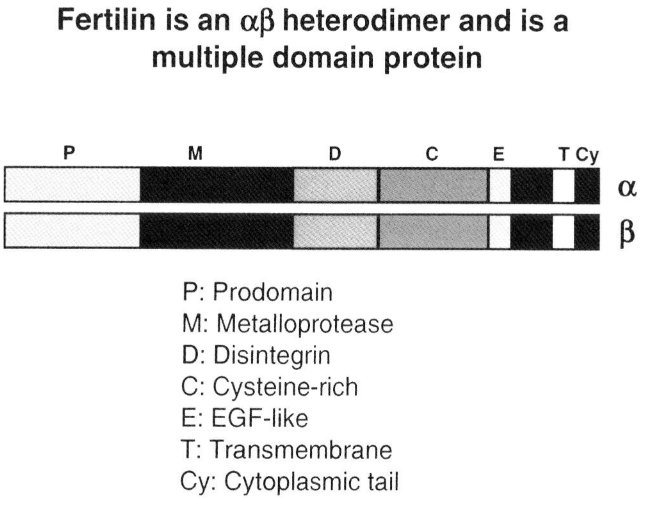

Figure 2. The multidomain organization of fertilin, a member of the ADAM family of proteins.

ly lasts a short time and sperm can be observed to pivot around the point of this attachment (Miller B, Primakoff P and Myles DG, unpublished observations). This attachment via the sperm tip is normally an association between the inner acrosomal membrane (IAM) of the sperm and the egg surface, because sperm at this stage in fertilization are acrosome reacted (Fig. 3).

A candidate adhesion molecule for this step of sperm-egg plasma membrane binding is the sperm membrane protein cyritestin. Cyritestin was first identified as a testis-specific mouse gene belonging to the ADAM family (Heinlein et al. 1994). In the type of *in vitro* fusion assays described above, peptides that included the amino acid sequence of the putative active binding site region of the disintegrin domain were able to block both binding and fusion of sperm with zona-free eggs (Linder, Heinlein 1997; Yuan et al. 1997). Cyritestin has been reported to be associated with either the inner acrosomal membrane (Forsbach, Heinlein 1998; Linder et al. 1995) or equatorial region (Fig. 3) of mouse sperm (Yuan et al. 1997).

Tip localization would indicate that cyritestin is in the correct position to participate in the initial tip binding that is observed *in vitro* between sperm and egg plasma membranes. No adhesion partner for cyritestin has been determined.

After sperm progress to a position where they are flattened against the egg plasma membrane, fusion occurs via the sperm equatorial region or posterior head plasma membrane (Fig. 3). The inner acrosomal membrane is excluded from the membrane fusion and is not incorporated into the membrane that surrounds the developing zygote; instead, the IAM is incorporated into the egg cytoplasm by a process that resembles phagocytosis (Shalgi, Phillips 1980).

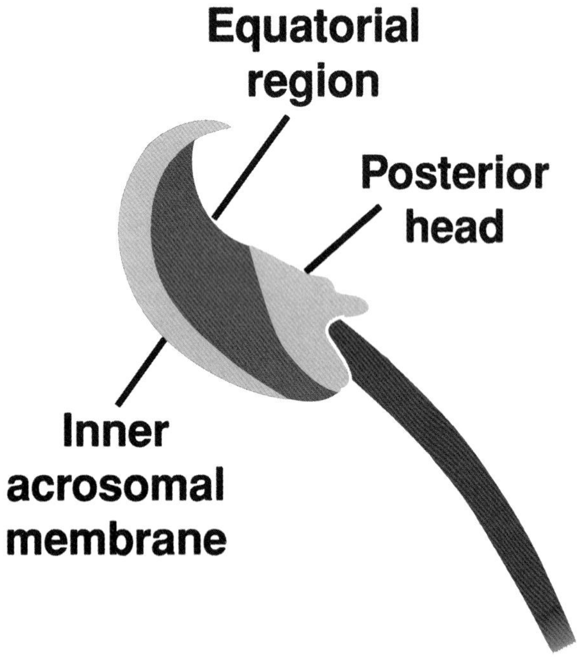

Figure 3. The three surface domains of the sperm head.

Binding of Sperm in Flattened Position to Egg Plasma Membrane: Role of Fertilin β

In most cases, sperm spend only a short time attached to the egg plasma membrane by their tips. The subsequent lateral binding is soon followed by a fusion of sperm and egg plasma membranes. The most extensively studied sperm protein with an apparent role in the process of sperm-egg membrane binding and fusion is fertilin. Fertilin is a heterodimeric protein, first found on guinea pig sperm, where it was originally called PH-30 (Primakoff et al. 1987). Both the α and β subunits of fertilin are members of the ADAM family (Blobel et al. 1992). Immunoprecipation studies indicated that the two subunits are associated non-covalently on guinea pig sperm (Primakoff et al. 1987). Both subunits are made as precursor proteins and are processed to remove the pro- and metalloprotease domains before sperm are mature (Blobel et al. 1990; Wolfsberg et al. 1993). For fertilin α this processing occurs in the testis, but for fertilin β the processing occurs after sperm have left the testis and are in the epididymis (Blobel et al. 1990; Hunnicutt et al. 1997; Lum, Blobel 1997; Phelps et al. 1990). Both subunits have been found in a variety of other species, including the mouse (Wolfsberg et al. 1995b). The localization of fertilin on guinea pig sperm in the posterior head region (Primakoff et al. 1987) and on mouse sperm in the equatorial region (Yuan et al. 1997) makes fertilin a candidate protein for a role in lateral sperm binding. On mouse sperm, fertilin β is additionally observed on the IAM (Yuan et al. 1997); localization on the IAM makes it also a candidate for tip binding.

Many lines of evidence indicate that fertilin functions in sperm-egg binding and fusion. The first evidence obtained was from antibody inhibition studies using a pair of monoclonal antibodies that recognize fertilin β (Primakoff et al. 1987). One of these mAbs (PH-30 mAb) inhibited sperm-egg fusion in a dose-dependent manner with saturating inhibition at 140 μg/ml. A control mAb that recognizes a different epitope on fertilin β (PH-1 mAb) had no effect, even at the highest concentration tested (400 μg/ml). By sequence alignment of fertilin β with the related snake disintegrins where the binding site had been identified, the putative active binding site of fertilin β was predicted.

Mimetic peptides from the guinea pig putative binding site sequence were made. The ability of these peptides to inhibit sperm-egg fusion was tested by *in vitro* fusion assays with zona-intact and zona-free guinea pig eggs (Myles et al. 1994). The mimetic peptides were able to inhibit both the fertilization rate and the fertilization index (80-98%). Similar *in vitro* fusion assays were carried out in mice with fertilin β mouse peptides, where inhibition of the fertilization rate (55%) and fertilization index (59%) was observed. These results in mice were consistent with previous mouse studies where different assay conditions and peptides were used (Evans et al. 1995b).

The role of fertilin β in sperm-egg binding and fusion has also recently been addressed by using homologous recombination to generate mice lacking fertilin β (Cho et al. 1998). When sperm from these fertilin β-/- (knockout) mice were tested for their ability to bind to eggs *in vitro*, a strong inhibition (eight-fold) of sperm-egg binding was found. Sperm from mice lacking fertilin β also showed a reduction (two-fold) in their rate of fusion with eggs. In those cases where sperm-egg fusion did occur, however, no effect on egg activation (assessed by intracellular Ca^{2+} oscillations and polar body formation) was observed. These data are consistent with previous inhibition studies, and, most importantly, would not result from artifactual effects of steric hindrance by antibodies or non-specific effects of peptide inhibitors.

The Role of Integrins

Both fertilin β and cyritestin belong to the ADAM family and are apparently binding via a region of the protein with homology to the snake disintegrins (Myles et al. 1994). The disintegrins are known integrin ligands, suggesting that the binding partner for sperm ADAMs may be egg integrins. The two different sperm proteins (fertilin β and cyritestin) might bind to the same egg integrin or two different integrins. Using a variety of methods, several different integrins have been found on the egg plasma membrane of various mammalian species (Almeida et al. 1995; Campbell et al. 1995; de Nadai et al. 1996; Evans et al. 1995a; Fusi et al. 1992; Fusi et al. 1993; Palombi et al. 1992; Tarone et al. 1993). The reported data indicate that the eggs of mice, hamsters and humans all have at least the integrins $\alpha5\beta1$, $\alpha6\beta1$, and $\alpha_v\beta3$ on their surface.

In vitro assays using mouse eggs suggest that sperm bind to the integrin $\alpha6\beta1$ on the egg plasma membrane (Almeida et al. 1995). Previously the integrin $\alpha6\beta1$ was known to bind laminin. A rat monoclonal antibody that recognizes the $\alpha6$ subunit (GoH3 mAb) was characterized as inhibitory in laminin binding assays. Using *in vitro* fusion assays with zona-free mouse eggs, the inhibition of sperm-egg membrane binding and fusion by GoH3 was compared to the effect of another mAb that also recognizes the $\alpha6$ subunit, but has been used as a non-function-blocking control in laminin binding assays (J1B5 mAb) by Almeida et al. (1995). Half-maximal inhibition of sperm binding was observed at a concentration of between 50-100 μg/ml of GoH3. A concentration of 200 μg/ml GoH3 results in 80-90% inhibition. Sperm-egg fusion, on the other hand, is unaffected at 200 μg/ml GoH3, but is significantly inhibited by 400 μg/ml GoH3. The non-function-blocking rat monoclonal, J1B5, does not affect sperm binding or fusion.

In addition to these antibody inhibition experiments, Almeida and colleagues compared tissue culture cells (particularly P388D mouse macrophages) which do not express $\alpha6\beta1$ to P388D transfected so that they do express $\alpha6\beta1$. The transfected, $\alpha6\beta1$-expressing cells bind sperm at higher levels than mock-transfected cells. However, sperm do not fuse with transfected cells. These data suggest that sperm bind to $\alpha6\beta1$. In addition, there is some evidence that sperm fertilin and egg integrin $\alpha6\beta1$ are adhesion partners (Almeida et al. 1995), but more work needs to be done to investigate this critical point.

Sperm-Egg Membrane Fusion

Currently, there are two basic models of how membrane fusion may occur. One is based on information from membrane fusion between membrane-enveloped viruses and host cell membranes. A variety of experiments indicates that fusion depends upon the presence of a hydrophobic "fusion peptide" occurring in the sequence of the viral fusion protein. At the time of membrane fusion, viral fusion proteins are triggered to undergo a conformational change so that the hydrophobic fusion peptide is revealed and is inserted into the lipid bilayer of the host cell membrane.

Insertion into the host plasma membrane is thought to bring about a destabilization of the membrane and the formation of a "fusion pore" that expands so that the viral and host cell membranes become confluent, with a single surrounding lipid bilayer (Cohen et al. 1996; Xu, Hamilton 1996).

The second model of membrane fusion was recently proposed to explain the fusion that occurs between the transport vesicles in the cytoplasm and target membranes (Weber et al. 1998). In this model, two interacting proteins, v-snare on the vesicle membrane and t-snare on the target membrane, assemble into complexes linking the two membranes in such a way as to make fusion energetically favorable. Evidence that this can occur comes from experiments where two separate preparations of lipid bilayer vesicles were made, with one containing v-snare and the other containing t-snare. These two types of lipid vesicles will fuse with each other but not with themselves.

There is not yet a clear understanding of the molecular nature of membrane fusion between sperm and egg plasma membranes. It has been tempting to speculate that the fertilin α subunit may play a role. Guinea pig fertilin α contains a 22-amino acid peptide in the cysteine-rich domain that is identified as hydrophobic on a hydrophobicity plot and can be modeled as an α helix with the bulky hydrophobic amino acids along one face (Blobel et al. 1992). The occurrence of such a peptide suggests the possibility that it may act as a fusion peptide and insert in the egg plasma membrane after fertilin β binds to the egg plasma membrane.

In support of this theory is the finding that a synthetic peptide, representing the putative fusion domain of guinea pig fertilin, was shown to bind to lipid vesicles and induce fusion between large unilamellar vesicles (Muga et al. 1996).

In species other than the guinea pig, the evidence from protein sequence for a fusion peptide in fertilin α is less convincing, because of the occurrence of helix-breaking amino acids (Hardy, Holland 1996; Perry et al. 1995). The bovine fertilin α sequence contains a hydrophobic peptide, but in a different region of the molecule from the proposed fusion peptide of guinea pig fertilin α (Waters, White 1997).

Furthermore, a substantial rate of fusion of sperm from the fertilin β knockout mice with eggs from wild-type mice was observed. The testicular cells show a reduced amount of fertilin α and there may be no fertilin α on mature mutant β-/- sperm. Thus, there is not a currently valid model in which heterodimeric fertilin, using activities of both fertilin α and β, is absolutely required for fusion.

A Model for Sperm Egg Binding and Fusion

Research results indicate that there are at least three steps involved in sperm-egg plasma membrane interactions. The first step, initial binding via the tip of the sperm, occurs at the inner acrosomal membrane and could be mediated by cyritestin, fertilin β or some other as yet unidentified molecule or molecules. The subsequent lateral binding step, where sperm lie flat on the egg surface, appears to require fertilin β binding.

The best candidate currently for a binding partner for fertilin β on the egg surface is the integrin $\alpha6\beta1$. Fusion of sperm with the egg appears to be facilitated by, but not absolutely require, binding via fertilin β. Whether or not fertilin α has a role in sperm-egg fusion is currently uncertain.

Finally, although fertilin β binding to an integrin in the egg plasma membrane could contribute to some aspect of egg activation, current data indicate that it is not required for two major features of egg activation.

References

Adler M, Lazarus RA, Dennis MS, Wagner G. Solution structure of kistrin, a potent platelet aggregation inhibitor and GP IIb-IIIa antagonist. Science 1991; 253:445-448.

Almeida EA, Huovila AP, Sutherland AE, Stephens LE, Calarco PG, Shaw LM, Mercurio AM, Sonnenber A, Primakoff P, Myles DG, White JM. Mouse egg integrin $\alpha6\beta1$ functions as a sperm receptor. Cell 1995; 81:1095-1104.

Blobel CP, Myles DG, Primakoff P, White JM. Proteolytic processing of a protein involved in sperm-egg fusion correlates with acquisition of fertilization competence. J Cell Biol 1990; 111:69-78.

Blobel CP, Wolfsberg TG, Turck CW, Myles DG, Primakoff P, White JM. A potential fusion peptide and an integrin ligand domain in a protein active in sperm-egg fusion. Nature 1992; 356:248-252.

Campbell S, Swann HR, Seif MW, Kimber SJ, Aplin JD. Cell adhesion molecules on the oocyte and preimplantation human embryo. Hum Reprod 1995; 10:1571-1578.

Chen Y, Pitzenberger SM, Garsky VM, Lumma PK, Sanyal G, Baum J. Proton NMR assignments and secondary structure of the snake venom protein echistatin. Biochemistry 1991; 30:11625-11636.

Cho C, Bunch DO, Faure J-E, Goulding EH, Eddy EM, Primakoff P, Myles DG. Fertilization defects in sperm from mice lacking fertilin β. Science 1998; 281:1857-1859.

Cohen DJ, Munuce MJ, Cuasnicú PS. Mammalian sperm-egg fusion: the development of rat oolemma fusibility during oogenesis involves the appearance of binding sites for sperm protein "DE." Biol Reprod 1996; 55:200-206.

de Nadai C, Fenichel P, Donzeau M, Epel D, Ciapa B. Characterization and role of integrins during gametic interaction and egg activation. Zygote 1996; 4:31-40.

Evans JP, Schultz RM, Kopf GS. Identification and localization of integrin subunits in oocytes and eggs of the mouse. Mol Reprod Dev 1995a; 40:211-220.

Evans JP, Schultz RM, Kopf GS. Mouse sperm-egg plasma membrane interactions: analysis of roles of egg integrins and the mouse sperm homologue of PH-30 (fertilin) β. J Cell Sci 1995b; 108:3267-3278.

Forsbach A, Heinlein UAO. Intratesticular distribution of cyritestin, a protein involved in gamete interaction. J Exp Biol 1998; 201:861-867.

Fusi F, Vignali M, Busacca M, Bronson RA. Evidence for the presence of an integrin cell adhesion receptor on the oolemma of unfertilized human oocytes. Mol Reprod Dev 1992; 31:215-222.

Fusi F, Vignali M, Galit J, Bronson RA. Mammalian oocytes exhibit specific recognition of the RGD (Arg-Gly-Asp) tripeptide and express oolemmal integrins. Mol Reprod Dev 1993; 36:212-219.

Gaddum-Rosse P. Mammalian gamete interactions: what can be gained from observations on living eggs? Am J Anat 1985; 174:347-356.

Hardy CM, Holland MK. Cloning and expression of recombinant rabbit fertilin. Mol Reprod Dev 1996; 45:107-116.

Heinlein UAO, Wallat S, Senftleben A, Lemaire L. Male germ cell-expressed mouse gene TAZ83 encodes a putative, cysteine-rich transmembrane protein (cyritestin) sharing homologies with snake toxins and sperm-egg fusion proteins. Dev Growth Diff 1994; 36:49-58.

Hinkley RE, Wright BD, Lynn JW. Rapid visual detection of

sperm-egg fusion using the DNA specific fluorochrome Hoechst 33342. Dev Biol 1986; 118:148-154.

Hunnicutt GR, Koppel DE, Myles D. Analysis of the process of localization of fertilin to the sperm posterior head plasma membrane domain during sperm maturation in the epididymis. Dev Biol 1997; 191:146-159.

Jia LG, Shimokawa K, Bjarnason JB, Fox JW. Snake venom metalloproteinases: structure, function and relationship to the ADAMs family of proteins. Toxicon 1996; 34:1269-1276.

Linder B, Heinlein UAO. Decreased *in vitro* fertilization efficiencies in the presence of specific cyritestin peptides. Dev Growth Diff 1997; 39:243-247.

Linder B, Bammer S, Heinlein UAO. Delayed translation and post-translational processing of cyritestin, an integral transmembrane protein of the mouse acrosome. Exp Cell Res 1995; 221:66-72.

Longo FJ, Yanagimachi R. Detection of sperm-egg fusion. In: (Duzgunes N, ed) Membrane Fusion Techniques. Part B. San Diego: Academic Press, 1993, pp249-260.

Lum L, Blobel CP. Evidence for distinct serine protease activities with a potential role in processing the sperm protein fertilin. Dev Biol 1997; 191:131-145.

Muga A, Neugebauer W, Hirama T, Surewiez WK. Membrane interaction and conformational properties of the putative fusion peptide of PH-30, a protein active in sperm-egg fusion. Biochemistry 1996; 33:4444-4448.

Myles DG. Molecular mechanism of mammalian sperm-egg fusion. In: (Dufau ML, Fabbri A, Isidori A, eds) Frontiers in Endocrinology: Cell and Molecular Biology of the Testis. Rome: Ares-Serono Symposia, 1994, pp217-220.

Myles DG, Kimmel LH, Blobel CP, White JM, Primakoff P. Identification of a binding site in the disintegrin domain of fertilin required for sperm-egg fusion. Proc Natl Acad Sci USA 1994; 91:4195-4198.

Palombi F, Salanova M, Tarone G, Farini D, Stefanini M. Distribution of β1 integrin subunit in rat seminiferous epithelium. Biol Reprod 1992; 47:1173-1182.

Perry AC, Gichuhi PM, Jones R, Hall L. Cloning and analysis of monkey fertilin reveals novel alpha subunit isoforms. Biochem J 1995; 307:843-850.

Phelps BM, Koppel DE, Primakoff P, Myles DG. Evidence that proteolysis of the surface is an initial step in the mechanism of formation of sperm cell surface domains. J Cell Biol 1990; 111:1839-1847.

Primakoff P, Hyatt H, Tredick-Kline J. Identification and purification of a sperm surface protein with a potential role in sperm-egg membrane fusion. J Cell Biol 1987; 104:141-149.

Shalgi R, Phillips D. Mechanics of sperm entry in cycling hamsters. J Ultrastruct Res 1980; 71:154-161.

Takano H, Yanagimachi R, Urch U. Evidence that acrosin activity is important for the development of fusibility of mammalian spermatozoa with the oolemma: inhibitor studies using the golden hamster. Zygote 1993; 1:79-91.

Tarone G, Russo MA, Hirsch E, Odorisio T, Altruda F, Silengo L, Siracusa G. Expression of β1 integrin complexes on the surface of unfertilized mouse oocyte. Development 1993; 117:1369-1375.

Tomiyama Y, Brojer E, Ruggeri ZM, Shattil SJ, Smiltneck J, Gorski J, Kumar A, Kieber-Emmons T, Kunicki TJ. A molecular model of RGD ligands. J Biol Chem 1992; 267:18085-18092.

Waters SI, White JM. Biochemical and molecular characterization of bovine fertilin alpha and beta (ADAM 1 and ADAM 2): a candidate sperm-egg binding/fusion complex. Biol Reprod 1997; 56:1245-1254.

Weber T, Zemelman BV, McNew JA, Westermann B, Gmachi M, Parlati F, Sollner T, Rothman JE. SNAREpins: minimal machinery for membrane fusion. Cell 1998; 92:759-772.

Wolfsberg TG, White JM. ADAMs in fertilization and development. Dev Biol 1998; 180:389-401.

Wolfsberg TG, Bazan JF, Blobel CP, Myles DG, Primakoff P, White JM. The precursor region of a protein active in sperm-egg fusion contains a metalloprotease and a disintegrin domain: structural, functional, and evolutionary implications. Proc Natl Acad Sci USA 1993; 90:10783-10787.

Wolfsberg TG, Primakoff P, Myles DG, White JM. ADAM, a novel family of membrane proteins containing A Disintegrin And Metalloprotease domain: multipotential functions in cell-cell and cell-matrix interactions. J Cell Biol 1995a; 131:275-278.

Wolfsberg TG, Straight PD, Gerena RL, Huovila AJ, Primakoff P, Myles DG, White JM. ADAM, a widely distributed and developmentally regulated gene family encoding membrane proteins with A Disintegrin And Metalloprotease domain. Dev Biol 1995b; 169:378-383.

Xu W, Hamilton DW. Identification of the rat epididymis-secreted 4E9 antigen as protein E: further biochemical characterization of the highly homologous epididymal secretory proteins D and E. Mol Reprod Dev 1996; 43:347-357.

Yanagimachi R. Sperm-egg fusion. Curr Top Mem Trans 1988; 32:3-43.

Yanagimachi R. Mammalian Fertilization. In: (Knobil E, Neill JD, eds) The Physiology of Reproduction. 2nd ed. New York: Raven Press Ltd, 1994, pp189-317.

Yoshihara CM, Hall ZW. Increased expression of the 43-kD protein disrupts acetylcholine receptor clustering in myotubes. J Cell Biol 1993; 122:169-179.

Yuan R, Primakoff P, Myles DG. A role for the disintegrin domain of cyritestin, a sperm surface protein belonging to the ADAM family, in mouse sperm-egg plasma membrane adhesion and fusion. J Cell Biol 1997; 137:105-112.

24

Diversity of Sperm-Activating Peptide Receptors

Norio Suzuki
Hokkaido University

The spermatozoon, like other cells, responds to specific environmental signals, some of which emanate from the egg and play an important role in fertilization. Fertilization occurs as a result of the interaction between egg and spermatozoa, beginning with spermatozoa reaching the egg and binding to it, and ending with the fusion of the sperm pronucleus and the egg nucleus. In the process, the egg communicates with spermatozoa using molecules in the extracellular matrix. Sea urchin gametes have long been used for analyzing these processes in fertilization. Particularly in the past decade, knowledge of the biochemistry of sea urchin fertilization has increased enormously (Foltz, Lennarz 1993; Garbers 1989; Ward, Kopf 1993). A better understanding of the substances involved in sea urchin gamete interaction would not only lead to an understanding of fertilization in mammals and other animals, but would serve as a valuable guide to mechanisms which might be anticipated in cells other than gametes.

Sea urchin eggs are surrounded by a transparent, gelatinous extracellular matrix called the jelly layer. Sea urchin spermatozoa must pass through the jelly layer before contacting the egg surface. The jelly layer has been shown to induce the sperm acrosome reaction, which was first described in detail by Dan (1952). The acrosome reaction in spermatozoa is an essential requirement for fertilization of eggs in many animals (Dan 1956). In sea urchins, the acrosome reaction occurs within seconds after spermatozoa have come into contact with the jelly layer or the egg surface component, a sperm-binding protein on the vitelline membrane, and is induced by the egg-associated macromolecule, a fucose sulfate glycoconjugate (FSG), as seen in SeGall and Lennarz (1979) and Shimizu et al. (1990).

In the early 1900s, it was shown that a soluble factor associated with the eggs of certain species of sea urchins stimulates the respiration and motility of sea urchin spermatozoa (Lillie 1913). In 1981, researchers in the Suzuki laboratory first isolated a decapeptide (GFDLNGGGVG) from the solubilized jelly layer of the sea urchin *Hemicentrotus pulcherrimus* and demonstrated that this peptide stimulates the respiration and motility of *H. pulcherrimus* spermatozoa (Suzuki et al. 1981). Since then, 74 peptides have been isolated from the solubilized jelly layer of 17 species of sea urchins distributed over five taxonomic orders (Echinoida, Arbacioida, Clypeasteroida, Diadematoida and Spatangoida); this is reviewed in Suzuki (1995). These peptides show essentially the same biological effects on sea urchin spermatozoa, although the biological effects and structures of the peptides are specific at the ordinal level. Therefore, the name sperm-activating peptides (SAPs) was proposed for these peptides. Expanding this nomenclature, peptides from species in the order Echinoida would be called SAP-I (GFDLNGGGVG). Similarly, SAP-II would describe species in the order Arbacioida, and would be divided into subgroups A and B (SAP-IIA: CVTGAPGCVGGGRL-NH_2; SAP-IIB: KLCPGGNCV). Species in the order Clypeasteroida would be SAP-III (DSDSAQNLIG), those in the order Diadematoida, SAP-IV (GCPWGGAVC), and those from species in the order Spatangoida, SAP-V (GCEGLFHGMGNC; Fig. 1).

Interaction of Spermatozoa with SAPs

Respiration and motility of sea urchin spermatozoa are highly dependent on the pH of the suspending medium. Below pH 6.0, sea urchin spermatozoa are practically immotile and do not respire. Decreased motility and respiration rates in slightly acidified sea water (pH 6.6-6.8) can be reversed by the addition of SAPs. The stimulated respiration rates do not exceed those of spermatozoa in normal sea water, which has a pH of 8.0 (Suzuki et al. 1981). Half-maximal stimulation of the respiration rates by SAPs occurs 10-100 pM. The stimulation of respiration rates by SAPs is dependent on the concentration of external Na^+ (Suzuki et al. 1984a). The ability of SAP-I to stimulate H^+ efflux and hence increase intracellular pH (pH_i) correlates well with the ability of SAP-I to stimulate respiration rates. Induction of respiratory stimulation by SAPs can be explained by the hypothesis that SAPs trigger activation of the Na^+/H^+ exchange system across the sperm plasma membrane and raise the pH_i. However, the extents of increases in pH_i induced by the repeated addition of a low concentration of SAP-I (final, 0.59 nM) to *H. pulcherrimus* spermatozoa at pH 6.6 tended to decrease progressively while the

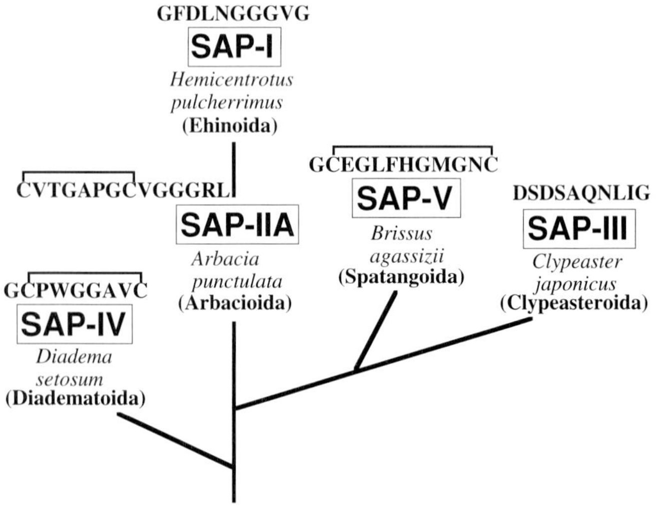

Figure 1. Phylogenetic tree of sea urchins with structures of representative primary structure of sperm-activating peptides.

extents of increases of respiration rates caused by the repeated addition of the same concentration of SAP-I remained unchanged (Fig. 2).

SAP-I increases pH_i and intracellular Ca^{2+} (Ca^{2+}_i) in both acidic (pH 6.6) and normal sea water (pH 8.0-8.2), as seen in Hoshino et al. (1992). The extents of the increase in pH_i in normal sea water are smaller than those in acidic sea water, but the extents of the increase in Ca^{2+}_i in normal sea water are much larger than those observed in acidic sea water. Half-maximal elevations of pH_i and Ca^{2+}_i induced by SAP-I are 45 pM at the external pH of 6.6 and 7.0 nM at the external pH 8.0, respectively. In artificial sea water (ASW) containing 100 mM K^+ (HK^+ASW), the respiration rates and motility of sea urchin spermatozoa are lower than those observed in normal sea water. Under these conditions, SAP-I increases neither the lowered respiration rates and motility nor pH_i and Ca^{2+}_i, although the peptide binds to the spermatozoa as it does in normal sea water (Harumi et al. 1992b). The inhibition of respiration and motility of spermatozoa

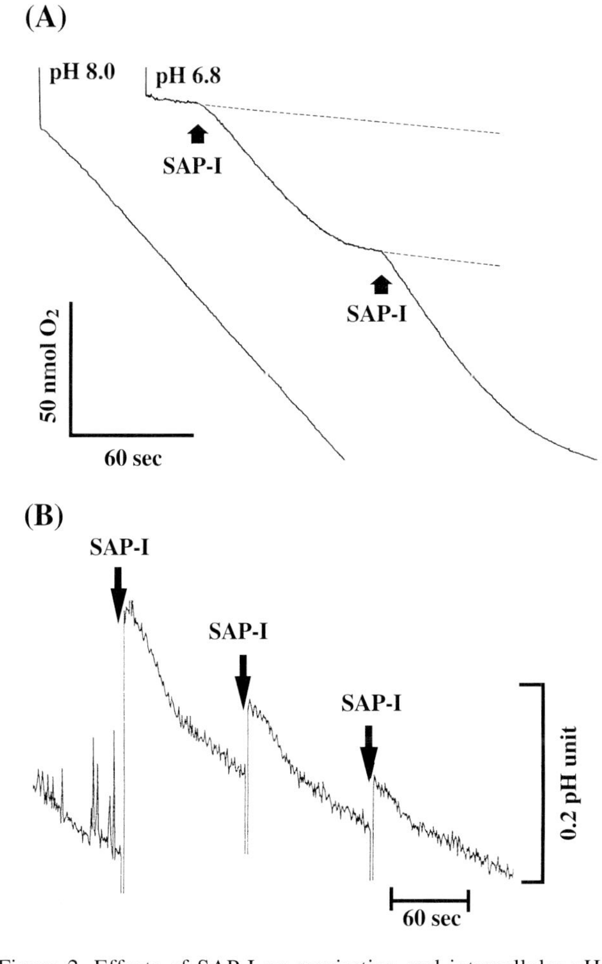

Figure 2. Effects of SAP-I on respiration and intracellular pH of *H. pulcherrimus* spermatozoa. A: At the arrows, SAP-I was added in the final concentration of 3.3 nM. B: At the time indicated by arrows, the same amount of SAP-I (final, 0.59 nM) was added to BCECF/AM-loaded spermatozoa at pH 6.6.

in HK^+ASW may be explained as follows: a high external K^+ depolarizes the sperm plasma membrane and inactivates the Na^+/H^+ exchange system across the sperm plasma membrane, leading to internal acidification (Lee 1985). Because, unlike in the case of SAP-I, the effects of monensin, an ionophore which induces Na^+/H^+ exchange across the plasma membrane, on sea urchin sperm respiration were not affected by a high concentration of external K^+, the inhibition of SAP-I action in HK^+ASW is presumably the result of inhibition in some subsequent steps of signal transduction initiated by SAP-I.

Activation of Membrane-Bound Guanylyl Cyclase

All of the 74 SAPs tested to date caused transient increases in sea urchin sperm cGMP as well as cAMP concentrations in both acidic and normal sea water. Half-maximal elevations of cGMP concentrations occur at about 32 nM of SAP-I. As seen in the cases of respiratory stimulation and elevation of pH_i, the repeated addition of peptide also repeatedly induced increases of cGMP concentrations in both acidic and normal sea water (Fig. 3A). Furthermore, the increases of cGMP concentrations in sperm homogenate with GTP by SAP-I occurred specifically in both acidic and normal sea water (Fig. 3B). SAP-I elevates the sperm cGMP concentrations in HK^+ASW (from 0.37 to 4.81 pmol/mg wet weight spermatozoa) more than in normal sea water (from 0.21 to 0.93 pmol/mg wet weight spermatozoa) (Harumi et al. 1992b). A phosphodiesterase inhibitor, 3-isobutyl-1-methylxanthine (IBMX), and SAP-I synergistically elevated the cGMP concentrations from 0.35 to 33.08 pmol/mg wet weight spermatozoa in HK^+ASW. However, in HK^+ASW, SAP-I did not increase the cAMP concentrations, pH_i, or Ca^{2+}_i, even in the presence of IBMX.

When *Arbacia punctulata* spermatozoa were incubated with [(γ-^{32}P]-ATP, a major sperm plasma membrane protein which has been proven to be a membrane-bound guanylyl cyclase was constitutively labeled (Ward et al. 1985). The ^{32}P-phosphorylated guanylyl cyclase lost its ^{32}P-label and changed its relative mobility in SDS-PAGE by treatment of the spermatozoa with SAP-IIA (Suzuki et al. 1984b; Ward et al. 1985). The increased mobility is thought to result from increased binding of SDS to the dephosphorylated form of the enzyme. Similar mobility changes of a major sperm protein in SDS-PAGE are commonly observed in sea urchin spermatozoa treated with a specific SAP (Suzuki 1995).

H. pulcherrimus sperm guanylyl cyclase contains up to 26 moles phosphates/mole enzyme, but, after treatment of the spermatozoa with SAP-I, the enzyme-bound phosphates decrease to 4 moles/mole enzyme (Harumi

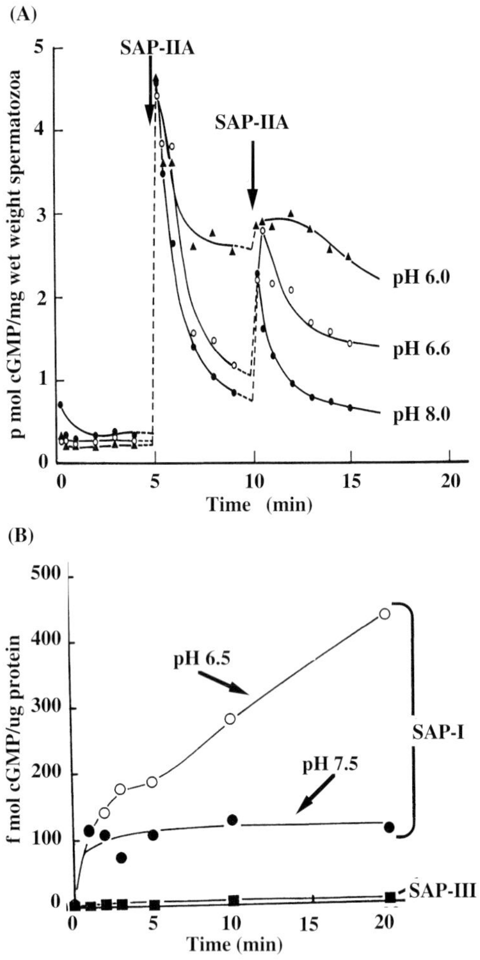

Figure 3. Effects of SAPs on guanylyl cyclase *in vitro* and *in vivo*. A: cGMP levels of *A. punctulata* spermatozoa treated with SAP-IIA in sea water of different pHs. The arrows indicate the time of addition of SAP-IIA (final, 138 nM). B: Specific activation of *Anthocidaris crassispina* sperm homogenate in solution A containing 0.5 M NaCl, 10 mM benzamidine-HCl, and 1 mM DTT buffered with 10 mM MES (pH 6.5) or 40 mM Tris-HCl (pH 7.5). The sea urchins *A. crassispina* and *H. pulcherrimus* are similar, and the jelly layer contains SAP-I and its derivatives. The spermatozoa were treated with 100 nM of SAP-I or SAP-III just before homogenization.

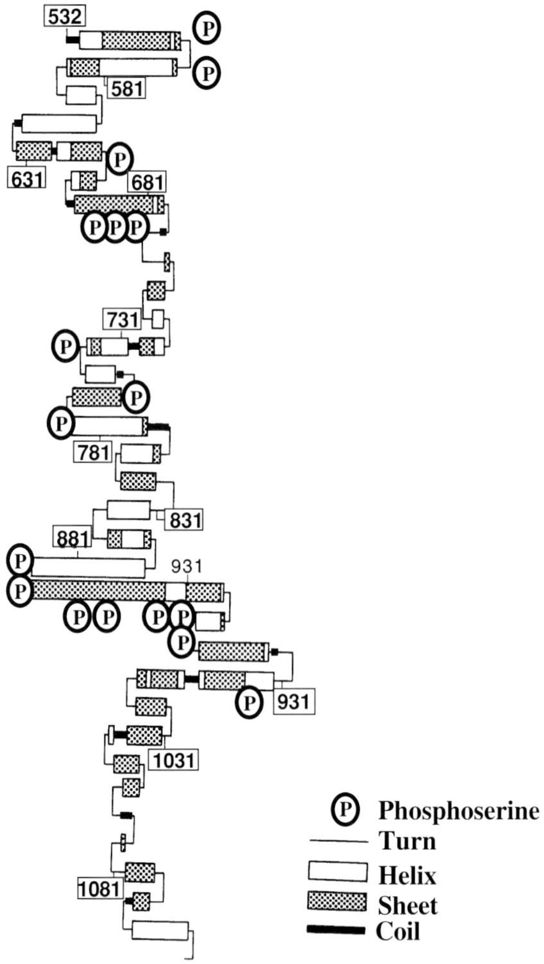

Figure 4. The predicted secondary structure of the intracellular region of *H. pulcherrimus* sperm guanylyl cyclase. The secondary structure was estimated according to the method of Chou and Fasman (1978). The kinase-like domain consists of residues from 600 to 720 and the catalytic domain of residues 860 to 1090.

et al. 1992a; Shimizu et al. 1996). Recently, the positions of 14 phosphoserine residues (serine residues at positions 561, 565, 652, 722, 740, 755, 894, 897, 914, 918, 927, 930, 951 and 985) were assigned, and all of these were located in or near the "turn" region of the predicted secondary structure within the intracellular region (kinase-like and catalytic domains; Fig. 4), as seen in Furuya et al. (1998). Notably, serine residues at positions 894, 918, 927 and 930, which are conserved in the sequence of mammalian and medaka fish membrane-bound guanylyl cyclases, are phosphorylated in *H. pulcherrimus* sperm membrane-bound guanylyl cyclase. The loss of phosphates from guanylyl cyclase and a change in the apparent molecular weight, which is dependent on external Na^+ and Ca^{2+} concentrations, result in a large decrease in the enzyme activity (Ramarao, Garbers 1985, 1988; Ward et al. 1986). Therefore, it is presumed that an SAP causes an initial activation of already phosphorylated guanylyl cyclase and induces dephosphorylation of the guanylyl cyclase. The activity of the dephosphorylated enzyme is much lower than that of the phosphorylated enzyme. The decrease in the activity could be repressed to a large degree by protein phosphatase inhibitors such as calyculin A, or to a lesser degree by microcystin-LR or okadaic acid (Fig. 5). These results suggest that a specific protein phosphatase sensitive to calyculin A dephosphorylates the membrane-bound guanylyl cyclase. On the other hand, in HK^+ASW, SAP-I did not induce the electrophoretic mobility change of the membrane-bound guanylyl cyclase (Harumi et al. 1992a). Thus, it is presumed that SAP-induced activation of the

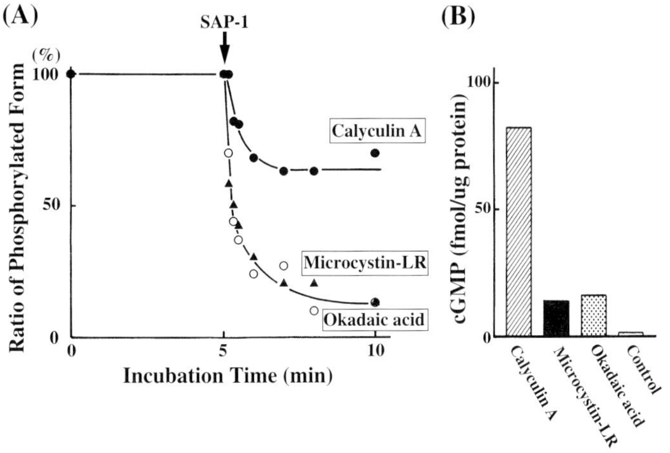

Figure 5. Effects of protein phosphatase inhibitors on the phosphorylation state and activity of SAP-I-treated *A. crassispina* sperm guanylyl cyclase. *A. crassispina* spermatozoa were homogenized in solution A (pH 5.6) containing calyculin A (16 μM), microcystin-LR (1.2 μM) or okadaic acid (12 μM) on ice, and then the pH was adjusted to 7.8 and the homogenate was incubated at 20°C. At the indicated time, 1 μM of SAP-I was added to the homogenate and an aliquot of 50 μl was removed. A: The ratio of phosphorylated form (136 kD)/phosphorylated form plus dephosphorylated form (131 kD). B: Guanylyl cyclase activity at 5 sec after SAP-I addition.

phosphorylated membrane-bound guanylyl cyclase resulting in elevation of the cGMP concentrations in sperm cells occurs before or independently of membrane hyperpolarization induced by the opening of K^+ channels.

SAP Receptors on Spermatozoa

The sea urchin jelly layer also contains a large glycoprotein complex, FSG, which is a major substance inducing the acrosome reaction in sea urchin spermatozoa. However, for full induction of the acrosome reaction in *H. pulcherrimus* spermatozoa, FSG requires SAP-I as a specific co-factor (Yamaguchi et al. 1989). At present, little is known about the binding of FSG to sea urchin spermatozoa, mainly due to technical difficulties in FSG binding experiments. The Suzuki laboratory has shown, however, that when Ca^{2+}_i or pH_i was elevated by a sufficient concentration of FSG, a second addition of the same concentration of FSG did not cause further increase in either Ca^{2+}_i or pH_i. Similarly, when the Ca^{2+}_i or pH_i had been increased by a sufficient concentration of SAP-I, a second addition of the same concentration of SAP-I caused no further increase of either. This may be because all the binding sites on the spermatozoa for FSG or SAP-I became occupied on the first addition of excess FSG or SAP-I, such that a second addition of FSG or SAP-I had no further effect on the Ca^{2+}_i or pH_i. Following SAP-I-induced elevations of Ca^{2+}_i and pH_i, however, subsequent addition of FSG caused further increases in both. Similarly, SAP-I further increased the Ca^{2+}_i and pH_i levels after the initial FSG-induced increases. These results suggest that FSG and SAP-I exert their effects by binding to different sites (receptors) on the spermatozoa (Hoshino et al. 1992; Fig. 6).

Equilibrium Binding of SAP to Spermatozoa

Using a radioiodinated synthetic SAP-I, SAP-IIB or SAP-III analog which exhibits the same respiratory stimulating activity as SAP-I, SAP-IIB or SAP-III, receptors specific for these peptides have been characterized. *Strongylocentrotus purpuratus* spermatozoa possess approximately 6,000-8,000 receptors specific for

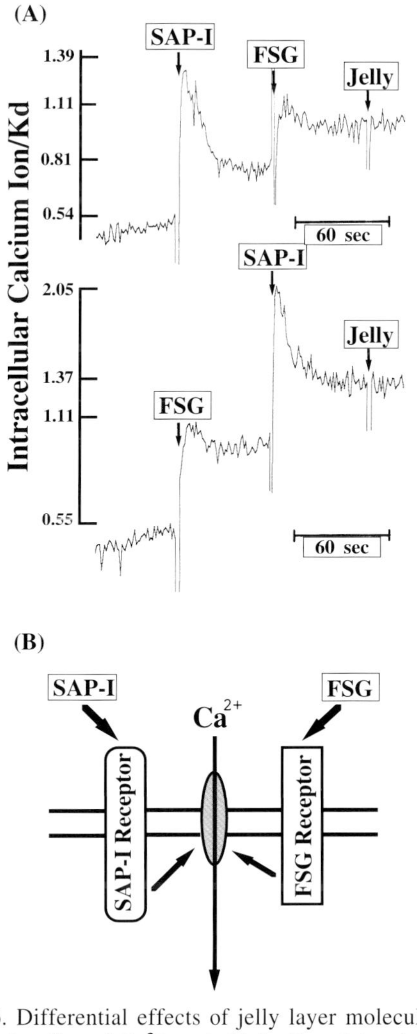

Figure 6. Differential effects of jelly layer molecules, FSG and SAP-I, on elevation of Ca^{2+}_i and pH_i in *H. pulcherrimus* spermatozoa. A: Elevations of Ca^{2+}_i induced by SAP-I, FSG, and solubilized jelly layer. At the times indicated by arrows, SAP-I (final, 0.59 nM), FSG (final, 50 nmole fucose/ml) or solubilized jelly layer (final, 1.24 nmole fucose/ml) was added to Fura 2/AM-loaded spermatozoa at pH 8.0. B: A model for arrangement of receptors for SAP-I, FSG, and a Ca^{2+} channel.

SAP-I per cell (Smith, Garbers 1983). These receptors are exclusively localized on the sperm tail (Suzuki et al. 1987). Analyses of the data obtained from the equilibrium binding of a ^{125}I-SAP-III analog to spermatozoa of the sand dollar *Clypeaster japonicus*, using Klotz, Scatchard and Hill plots, have shown the presence of two classes of receptors specific for SAP-III (Yoshino, Suzuki 1992). One of the receptors (high-affinity) has a Kd of 3.4 nM and 3.4×10^4 binding sites per spermatozoon. Scatchard and Hill plots of the data suggest the existence of positive cooperativity between the high-affinity members. The other receptor (low-affinity) has a Kd of 48 nM, with 6.1×10^4 binding sites/spermatozoon. Similar results were also obtained from a binding experiment using a sperm-membrane fraction prepared from *C. japonicus* spermatozoa and intact *H. pulcherrimus* spermatozoa (Shimizu et al. 1994; Fig. 7, Table 1). The Kd of the high-affinity receptor is comparable to the half-maximal effective concentration of the pH_i-increasing activity of SAP-I or SAP-III, and the Kd of the low-affinity receptor is comparable to the half-maximal effective concentration of the cellular-cGMP-increasing activity of the peptide.

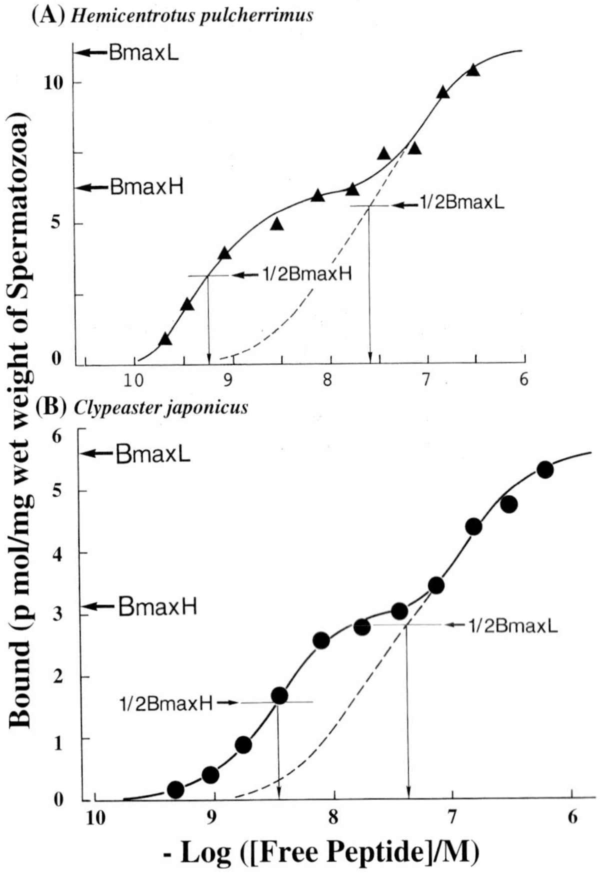

Figure 7. Diphasic binding of radioiodinated SAP analogs to sea urchin spermatozoa. The specific binding was plotted with a logarithmic abscissa according to the method of Klotz (1982). From the values of the plateau in both sygmoidal curves, the Bmax values of the low-affinity (Bmax L) and high-affinity (Bmax H) receptors were estimated, respectively.

SAP-Crosslinking Proteins

It has been reported that SAP-IIA, natriuretic peptides and heat-stable enterotoxins are extracellular peptide ligands for which a membrane-bound guanylyl cyclase serves as a cell-surface receptor (Garbers 1989). These peptide ligands crosslink to the guanylyl cyclase in the respective target tissue in the presence of a chemical crosslinking reagent, disuccinimidyl suberate. Similarly, in the presence of disuccinimidyl suberate, the ^{125}I-SAP-I analog crosslinked specifically to a 77 kD protein in *S. purpuratus* spermatozoa (Dangott, Garbers 1984) and to a 71 kD protein in *H. pulcherrimus* spermatozoa (Shimizu et al. 1994), and the ^{125}I-SAP-IIB analog crosslinked mainly to a 62 kD protein in *Glyptocidaris crenularis* spermatozoa (Harumi et al. 1991; Fig. 8). Unpublished data of Dangott et al. also suggest that a SAP-I-crosslinked protein-like protein exists in *A. punctulata* spermatozoa (Garbers 1989). In this study, these proteins themselves do not show any guanylyl cyclase activity.

A cDNA encoding the 71 kD protein has been isolated from a *H. pulcherrimus* testis cDNA library (Shimizu et al. 1994). An open reading frame of the cDNA predicted a protein of 532 amino acids containing a 30-residue amino-terminal signal peptide, followed by a sequence which corresponds to the N-terminal sequence of the purified 71 kD protein. The amino acid sequence of the mature 71 kD protein is quite similar to that of the 77 kD protein of *S. purpuratus*, at 95.5% identical (Dangott et al. 1989), and has a maximum 48% identity to a bovine and human type-I-specific scavenger receptor cysteine-rich domain (Freeman et al. 1990; Kodama et al. 1990; Matsumoto et al. 1990). On the other hand, the ^{125}I-SAP-III analog has been shown to crosslink to three proteins, 126 kD, 87 kD, and 64 kD, of *C. japonicus* spermatozoa (Yoshino, Suzuki 1992; Fig. 8). Structural analyses of those proteins isolated from *C. japonicus* sperm membranes suggest that the 126 kD protein is a membrane-bound guanylyl cyclase and the 87 kD protein corresponds to the SAP-I-crosslinking protein (Mendoza et al. 1993; Satoh et al. 1996). These analyses also suggest that the 64 kD protein is homologous to the 63 kD GPI-anchored protein present in *S. purpuratus* and *H. pulcherrimus* spermatozoa. In these cases, as well as in the case of SAP-I, however, binding of SAP to these sperm proteins results in a transient activation of the membrane-bound guanylyl cyclase, followed by a large decrease in the activity of this enzyme. The activity appears to be proportional to the amount of enzyme in the phosphorylated form. The dephosphorylation,

Table 1: Equilibrium parameters of radioiodinated SAP analog binding and half-maximal effective concentrations of iodinated or intact SAP on spermatozoa

Binding parameters	**High Affinity**		**Low Affinity**	
	H. pulcherrimus (SAP-I)	*C. japonicus* (SAP-III)	*H. pulcherrimus* (SAP-I)	*C. japonicus* (SAP-III)
Kd (nM)	0.65	3.4	23	63
Bmax (p mol/mg sperm)	6.4	3.1	11.0	5.6
Hill's coefficient (**n**H)	1.24	1.42	0.99	1.05
Co-operativity	positive	positive	none	none
Biological activity*				
Respiratory stimulation	0.056	1.0		
Increase in:				
intracellular pH	0.045	1.8		
intracellular cGMP			32	71

* (nM): half-maximal effective concentration

which is accompanied by a change in the electrophoretic mobility of the enzyme on SDS gels, depends on the pH of the surrounding sea water, and shows an absolute requirement for Na^+. Therefore, the Suzuki laboratory presents a hypothesis that the receptor for SAP consists of three membrane proteins, one of which is a membrane-bound guanylyl cyclase, and the binding of the peptide to all or either of these proteins induces activation of the Na^+ and Ca^{2+}/H^+ exchange systems and the phosphorylated membrane-bound guanylyl cyclase. The protein or proteins to which the peptide binds would be variable.

It should be mentioned that in the purification of the 71 kD protein from *H. pulcherrimus* sperm tails by Shimizu et al. (1994), the protein was always co-purified with a 220 kD wheat germ agglutinin (WGA)-binding protein by gel filtration, ion exchange and affinity chromatography, even in the presence of detergent. The 71 kD protein was separated from the 220 kD protein only after treatment with 2% SDS at 100°C. This suggests that the 71 kD protein is tightly associated with the 220 kD protein. The WGA-binding protein has been reported to be involved in the induction of the acrosome reaction through regulating ion fluxes associated with the acrosome reaction (Podell, Vacquier 1984). SAP-I has also been reported to participate in the induction of

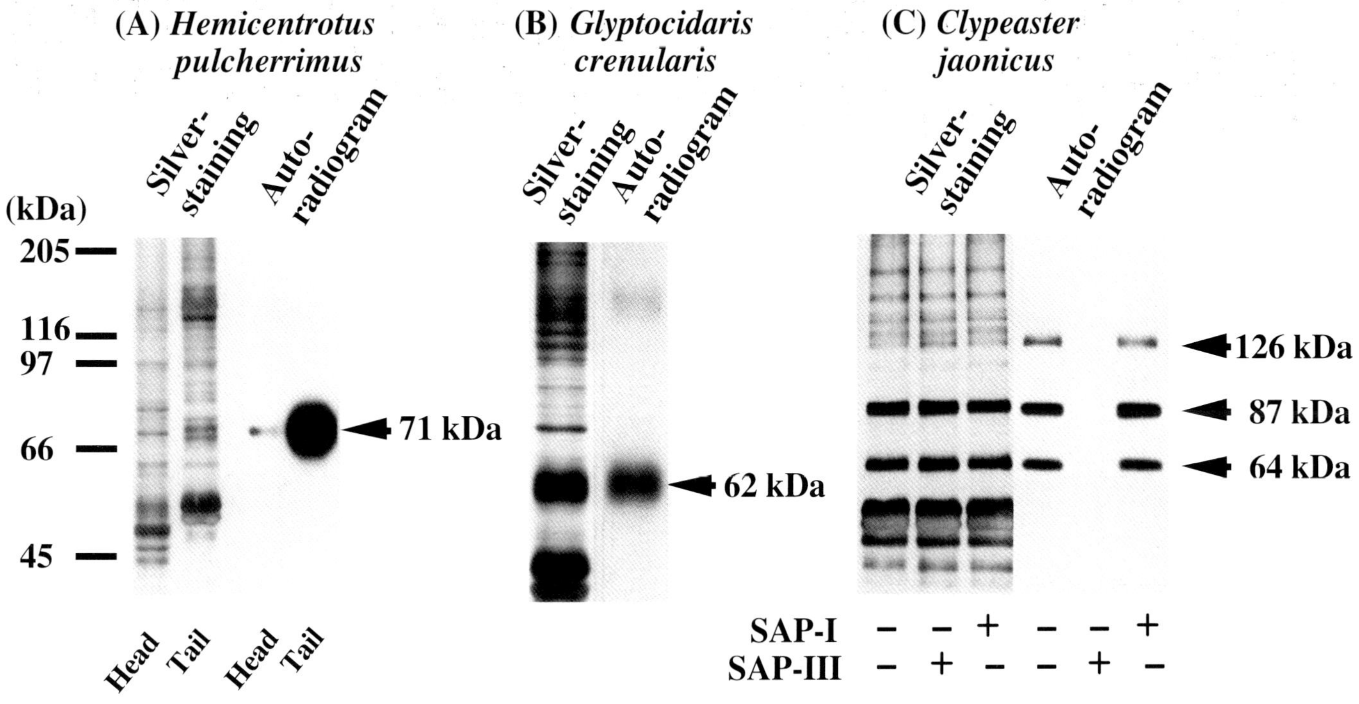

Figure 8. SDS-PAGE of ^{125}I-SAP analog-crosslinked sea urchin sperm proteins.

the acrosome reaction (Yamaguchi et al. 1989). Therefore, SAP-I binding to the 71 kD protein on a spermatozoon may affect the regulatory system of ion fluxes induced by the binding of component(s) in the jelly layer to the WGA-binding protein.

Diversity of Membrane-Bound Guanylyl Cyclase in Echinoderm Gonads

Membrane-bound guanylyl cyclase is the most important protein in SAP receptors, and often the jelly layer of a given species of sea urchin possesses many structurally similar SAPs. For example, the jelly layer of *H. pulcherrimus* contains at least five SAP-I derivatives (Suzuki 1995). In order to exert their effects, these peptides may require a specific receptor consisting of different membrane-bound guanylyl cyclases as well as different peptide-binding proteins. With regard to the membrane-bound guanylyl cyclase, researchers in the Suzuki laboratory found many isoforms in the gonads of sea urchins, such as *H. pulcherrimus* (eight isoforms), *G. crenularis* (six isoforms), *C. japonicus* (six isoforms), *Diadema setosum* (six isoforms) and *Brissus agassizii* (at least three isoforms). Other echinoderm species also possess many isoforms: a starfish, *Asterina pectinifera*, has six isoforms; a brittle star, *Ophioplocus japonicus*, possesses nine isoforms; and a holothurian, *Stichopus japonicus*, contains nine isoforms. At present, however, it is not certain that these isoforms actually consist of SAP receptors in the respective spermatozoa.

Conclusion and Perspective

In the past decade, 74 SAPs have been isolated from the egg jelly of 17 species of sea urchins distributed over five taxonomic orders. These peptides show essentially the same biological effects on sea urchin spermatozoa, although their biological effects and structures are specific at the ordinal level. With regard to that specificity, these peptides can be classified into five groups; that is, SAP-I to SAP-V. Receptors for SAPs seem to be variable and diverse depending on sea urchin species, but appear to consist of three proteins, one of which is a membrane-bound guanylyl cyclase (Fig. 9). In many respects, the mechanism of action of SAPs resembles that of the atrial natriuretic peptides found by de Bold et al. (1981), which are responsible for maintaining body fluid and electrolyte homeostasis through interaction with receptors on a variety of mammalian cell types. In this regard, it should be noted that the structure of a receptor for natriuretic peptides was determined in 1989 as a result of intensive research on the SAP-IIA receptor. This serves as an excellent demonstration that the results of research in one field can be applied to solving the problems in another field. Thus, research on SAPs and their receptors may be useful in helping to clarify the co-evolution of a peptide ligand and its receptor(s), since the biological effects of SAPs are specific at the ordinal level and sea urchins possess diverse structures of SAPs, and since each group of SAPs requires a specific receptor that may have resulted in physiological necessities leading to the development of new orders.

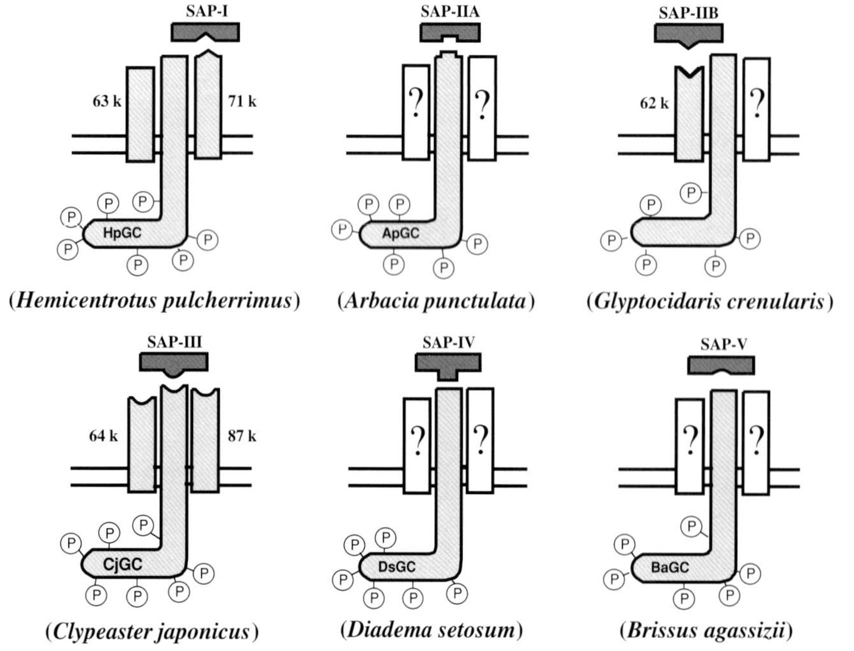

Figure 9. Diverse types of receptors for various sperm-activating peptides.

References

de Bold AJ, Borenstein HB, Veress AT, Sonnenberg H. A rapid and potent natriuretic response to intravenous injection of atrial myocardial extract in rats. Life Sci 1981; 28:89-94.

Chou PY, Fasman GD. Prediction of the secondary structure of proteins from their amino acid sequence. Adv Enzymol 1978; 47:45-148.

Dan JC. Studies of the acrosome. I. Reaction to egg-water and other stimuli. Biol Bull 1952; 103:54-66.

Dan JC. The acrosome reaction. Int Rev Cytol 1956; 5:365-393.

Dangott LJ, Garbers DL. Identification and partial characterization of the receptor for speract. J Biol Chem 1984; 259:13712-13716.

Dangott LJ, Jordan JE, Bellet RA, Garbers DL. Cloning of the mRNA for the protein that crosslinks to the egg peptide speract. Proc Natl Acad Sci USA 1989; 86:2128-2132.

Foltz KR, Lennarz WJ. The molecular basis of sea urchin gamete interaction at the egg plasma membrane. Dev Biol 1993; 158:46-61.

Freeman M, Ashkenas J, Rees DJG, Kingsley DM, Copeland NG, Jenkins NA, Krieger M. An ancient, highly conserved family of cysteine-rich protein domains revealed by cloning type I and type II murine macrophage scavenger receptors. Proc Natl Acad Sci USA 1990; 87:8810-8814.

Furuya H, Yoshino K, Shimizu T, Mantoku T, Takeda T, Nomura K, Suzuki N. Mass spectrometric analysis of phosphoserine residues conserved in the catalytic domain of membrane-bound guanylyl cyclase from the sea urchin spermatozoa. Zool Sci 1998; 15:507-516.

Garbers DL. Molecular basis of fertilization. Ann Rev Biochem 1989; 58:719-742.

Harumi T, Yamaguchi M, Suzuki N. Receptors for sperm-activating peptides, SAP-I and SAP-IIB, on spermatozoa of sea urchins, *Hemicentrotus pulcherrimus* and *Glyptocidaris crenularis*. Dev Growth Differ 1991; 33:67-73.

Harumi T, Kurita M, Suzuki N. Purification and characterization of sperm creatine kinase and guanylate cyclase of the sea urchin *Hemicentrotus pulcherrimus*. Dev Growth Differ 1992a; 34:151-162.

Harumi T, Hoshino K, Suzuki N. Effects of sperm-activating peptide I on *Hemicentrotus pulcherrimus* spermatozoa in high potassium sea water. Dev Growth Differ 1992b; 34:163-172.

Hoshino K, Shimizu T, Sendai Y, Harumi T, Suzuki N. Differential effects of the egg jelly molecules FSG and SAP-I on elevation of intracellular Ca^{2+} and pH in sea urchin spermatozoa. Dev Growth Differ 1992; 34:403-411.

Klotz IM. Numbers of receptors sites from Scatchard graphs: facts and fantasies. Science 1982; 217:1247-1249.

Kodama T, Freeman M, Rohrer L, Zabrecky J, Matsudaira P, Krieger M. Type I macrophage scavenger receptor contains α-helical and collagen-like coiled coils. Nature 1990; 323:531-535.

Lee HC. The voltage-sensitive Na^+/H^+ exchange in sea urchin spermatozoa flagellar membrane vesicles studied with an entrapped pH probe. J Biol Chem 1985; 260:10794-10799.

Lillie FR. Studies on fertilization. V. The behaviour of the spermatozoa of *Nereis* and *Arbacia* with special reference to egg extracts. J Exp Zool 1913; 14:515-574.

Matsumoto A, Naito M, Itakura H, Ikemoto S, Asaoka H, Hayakawa I, Kanamori H, Aburatani H, Takaku F, Suzuki H, Kobari Y, Miyai T, Takahashi K, Cohen E, Wydro R, Housman DE, Kodama T. Human macrophage scavenger receptors: primary structure, expression, and localization in atherosclerotic lesions. Proc Natl Acad Sci USA 1990; 87:9133-9137.

Mendoza LM, Nishioka D, Vacquier VD. A GPI-anchored sea urchin sperm membrane protein containing EGF domains is related to human uromodulin. J Cell Biol 1993; 121:1291-1297.

Podell SB, Vacquier VD. Wheat germ agglutinin blocks the acrosome reaction in *Strongylocentrotus purpuratus* sperm by binding a 210,000 mol-wt membrane protein. J Cell Biol 1984; 99:1598-1604.

Ramarao CS, Garbers DL. Receptor-mediated regulation of guanylate cyclase activity in spermatozoa. J Biol Chem 1985; 260:8390-8396.

Ramarao CS, Garbers DL. Purification and properties of the phosphorylated form of guanylate cyclase. J Biol Chem 1988; 263:1524-1529.

Satoh Y, Shimizu T, Harumi T, Suzuki N. Characterization of sea urchin sperm membrane proteins which interact with a major acrosome reaction-inducing substance, fucose sulfate glycoconjugate. Zool Sci 1996; 13:377-383.

SeGall GK, Lennarz WJ. Chemical characterization of the component of the egg jelly coat from sea urchin eggs responsible for induction of the acrosome reaction. Dev Biol 1979; 71:33-48.

Shimizu T, Kinoh H, Yamaguchi M, Suzuki N. Purification and characterization of the egg jelly macromolecules, sialoglycoprotein and fucose sulfate glycoconjugate, of the sea urchin *Hemicentrotus pulcherrimus*. Dev Growth Differ 1990; 32:473-487.

Shimizu T, Yoshino K, Suzuki N. Identification and characterization of putative receptors for sperm-activating peptide I (SAP-I) in spermatozoa of the sea urchin *Hemicentrotus pulcherrimus*. Dev Growth Differ 1994; 36:209-221.

Shimizu T, Takeda K, Furuya H, Hoshino K, Nomura K, Suzuki N. A mRNA for a membrane form of guanylyl cyclase is expressed exclusively in the testis of the sea urchin *Hemicentrotus pulcherrimus*. Zool Sci 1996; 13:285-294.

Smith AC, Garbers DL. The binding of an ^{125}I-speract analogue to spermatozoa. In: (Lennon DLF, Stratman FW, Zhalten RN, eds) Biochemistry of Metabolic Processes. New York: Elsevier/North-Holland, 1983, pp15-28.

Suzuki N. Structure, function and biosynthesis of sperm-activating peptides and fucose sulfate glycoconjugate in the extracellular coat of sea urchin eggs. Zool Sci 1995; 12:13-27.

Suzuki N, Nomura K, Ohtake H, Isaka S. Purification and the primary structure of sperm-activating peptides from the jelly coat of sea urchin eggs. Biochem Biophys Res Commun 1981; 99:1238-1244.

Suzuki N, Ohizumi Y, Yasumasu I, Isaka S. Respiration of sea urchin spermatozoa in the presence of a synthetic jelly coat peptide and ionophores. Dev Growth Differ 1984a; 26:17-24.

Suzuki N, Shimomura H, Radany EW, Ramarao CS, Bentley JK, Garbers DL. A peptide associated with eggs causes a mobility shift in a major plasma protein of spermatozoa. J Biol Chem 1984b; 259:14874-14879.

Suzuki N, Kurita M, Yoshino K, Yamaguchi M. Speract binds exclusively to sperm tails and causes an electrophoretic mobility shift in a major sperm tail protein of sea urchins. Zool Sci 1987; 4:641-648.

Ward CR, Kopf GS. Molecular events mediating sperm activation. Dev Biol 1993; 158:9-34.

Ward GE, Garbers DL, Vacquier VD. Effects of extracellular egg factors on sperm guanylate cyclase. Science 1985; 227:768-770.

Ward GE, Moy GW, Vacquier VD. Phosphorylation of membrane-bound guanylate cyclase of sea urchin spermatozoa. J Cell Biol 1986; 103:95-101.

Yamaguchi M, Kurita M, Suzuki N. Induction of the acrosome reaction of *Hemicentrotus pulcherrimus* spermatozoa by the egg jelly molecules, fucose sulfate glycoconjugate and sperm-activating peptide I. Dev Growth Differ 1989; 31:233-239.

Yoshino K, Suzuki N. Two classes of receptor specific for sperm-activating peptide III in sand-dollar spermatozoa. Eur J Biochem 1992; 206:887-893.

25 *Participation of Epididymal Protein "DE" and Its Egg Binding Sites in Sperm-Egg Fusion*

Patricia S. Cuasnicú

Débora J. Cohen

Diego A. Ellerman
Institute of Biology and Experimental Medicine

Fusion between gametes is a key event in the fertilization process, involving the interaction of specific domains of the sperm and egg plasma membranes. Although considerable information has been obtained concerning the structural aspects of mammalian sperm-egg fusion (Yanagimachi 1988), only recently have efforts been made towards the identification of the specific molecular components that mediate this event (Myles 1993).

The aim of the present work is to summarize the results obtained on the participation of sperm epididymal protein DE and its egg complementary sites in gamete membrane fusion.

Epididymal Protein DE

Rat epididymal protein DE (MW 37 kD), first described by Cameo and Blaquier (1976), is synthesized and secreted in response to androgens by the epithelium of the proximal segments of the epididymis (Kohane et al. 1980a). Indirect immunofluorescence (IIF) studies using a specific polyclonal antibody against DE (anti-DE) (Garberi et al. 1979, 1982) demonstrated that this protein associates to the surface of the sperm head during epididymal maturation (Kohane et al. 1980a, b) and migrates from the dorsal region to the equatorial segment (Figs. 1A, B, C) concomitantly with the occurrence of the acrosome reaction (AR) (Rochwerger, Cuasnicú 1992).

The fact that a high proportion of this protein is released from the sperm surface by either exposure to high ionic strength (NaCl 0.4 M) or incubation in capacitating media (Kohane et al. 1980b) suggests that DE would be loosely associated to the sperm surface. However, the finding that a remnant of DE remains on sperm after capacitation/AR (Cameo et al. 1986), together with previous evidence indicating the participation of DE in the fertilization process (Cuasnicú et al. 1984), supports the existence of a stronger interaction of DE with the sperm plasma membrane.

In order to study this hypothesis, rat cauda epididymal sperm were incubated with 2 M NaCl and then subjected to IIF. Results indicate that even after this treatment, the majority of the cells were still showing the characteristic fluorescence localized on the dorsal region of the sperm head. Subsequent experiments indicated that this labeling disappears only when sperm were subjected to treatments known to remove integral proteins. These results indicate the existence of two populations of protein DE: one removable by ionic strength and therefore loosely associated to the sperm surface, and a second one, resistant to this treatment, and behaving as an intrinsic protein (Cuasnicú et al. 1996).

In order to examine whether the tightly bound population corresponds to the protein that remains after capacitation and AR, cauda epididymal sperm were first treated with NaCl (2M) to remove the loosely bound protein, then incubated under capacitating conditions for five hours, and finally subjected to IIF for evaluation of the percentage of cells showing migration of protein DE. The high percentage of redistribution observed for these cells confirmed the ability of the strongly bound population to migrate to the equatorial segment of the sperm head (Cuasnicú et al. 1996).

Together, these results indicate that DE would associate to the sperm surface with two different affinities. The tightly associated protein is the one that remains after capacitation and migrates to the equatorial segment.

The relocation of DE over the equatorial segment, the region through which the sperm fuses with the egg plasma membrane, together with the fact that the polyclonal antibody against DE (anti-DE) significantly inhibited penetration of zona-free eggs (Cuasnicú et al.

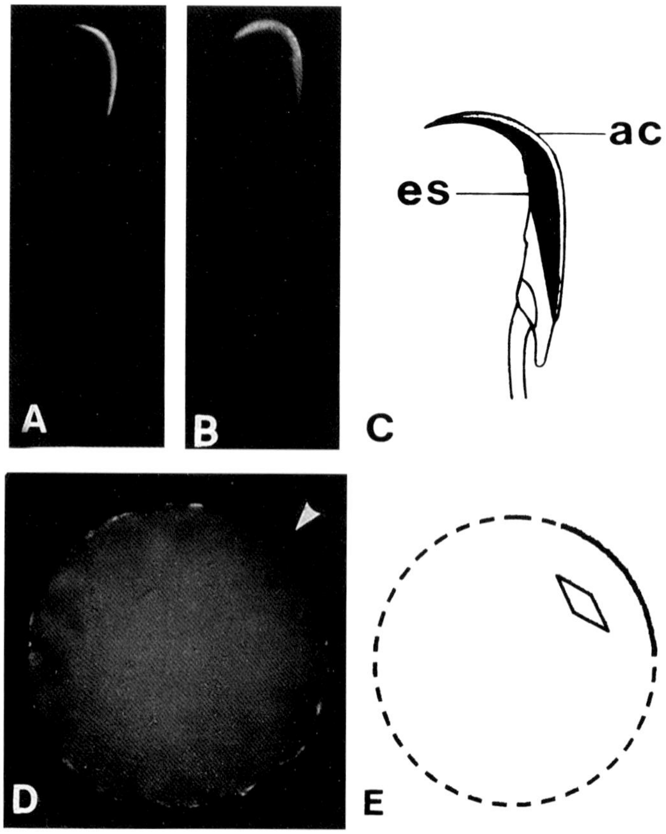

Figure 1. Immunofluorescent localization of rat epididymal protein DE and its egg binding sites. A: fluorescent staining is present over the dorsal region of the sperm head. B: Spermatozoa capacitated *in vitro* for 5 h. Note labeling present over the equatorial region of the head. C: A schematic representation of the rat sperm head showing the acrosomal cap (ac) and equatorial segment (es). D: Fluorescent labeling of zona-free rat eggs. Note the patchy distribution of fluorescence over the egg surface with the presence of a nonstained area (arrowhead). E: Diagram showing the fusogenic (dashed line) and non-fusogenic area of the egg plasma membrane (solid line).

1990), indicated a role for DE in the sperm-egg fusion process. As another approach, zona-free eggs were exposed to purified protein DE and then inseminated with capacitated sperm. The experimental premise was that if complementary sites for DE existed on the egg surface, DE should competitively inhibit the sperm-egg fusion process. Results indicated that exposure of zona-free eggs to purified protein DE produced a concentration-dependent decrease in egg penetration, with almost complete inhibition at 200 μg/ml. This inhibition was not due to an effect of DE on initial sperm binding to the egg plasma membrane, since the presence of this protein did not affect the percentage of oocytes with bound sperm nor the number of bound sperm per egg. In addition, those sperm bound to the egg surface that failed to penetrate the egg in the presence of DE became able to do so after transfer of the eggs to protein- and sperm-free medium, indicating a role for DE in an event subsequent to binding and leading to fusion.

Complementary Sites for DE on the Egg Surface

The inhibition in egg penetration by sperm observed in the presence of DE confirmed the participation of this protein in gamete fusion and indicated the existence of DE-complementary sites on the egg surface (Rochwerger et al. 1992). In order to study the localization of the binding sites on the egg surface, zona-free eggs were incubated with purified protein DE and then subjected to IIF using anti-DE and goat anti-rabbit FITC-conjugated antibodies. Zona-free oocytes incubated with different control proteins (ovalbumin, carbonic anhydrase or α-lactoalbumin) were used as controls. Results indicated that while none of the control oocytes showed evidence of labeling, the oocytes exposed to protein DE exhibited fluorescent labeling localized as patches over the entire egg surface with the exception of the area corresponding to the plasma membrane overlying the meiotic spindle (Fig. 1D), a region through which fusion rarely occurs (Ebensperger, Barros 1984; Johnson et al. 1975). Thus, while protein DE localizes on the fusogenic region of the sperm head, the DE-binding sites are localized on the fusogenic area of the egg surface (Figs. 1C, E). These studies provided the first evidence of the existence of complementary sites for a specific sperm protein on the surface of the mammalian egg.

Presence and Localization of Complementary Sites for DE during Oocyte Growth and Maturation

To study the appearance of DE-binding sites on the rat oolema during oogenesis, zona-free oocytes at different stages of growth and maturation were incubated with purified protein DE and then subjected to IIF as described above. Results indicated that while control oocytes incubated with ovalbumin showed no evidence of fluorescent labeling, oocytes incubated with DE exhibited different patterns of fluorescence. While none of the growing oocytes with a diameter <50 μm showed evidence of fluorescence (Figs. 2A, B), growing oocytes with a diameter >50 μm presented a uniform fluorescent labeling over the entire surface (Figs. 2C, D). This localization of oolema components changed progressively during maturation to the characteristic patchy fluorescence described for ovulated oocytes. Oocytes recovered

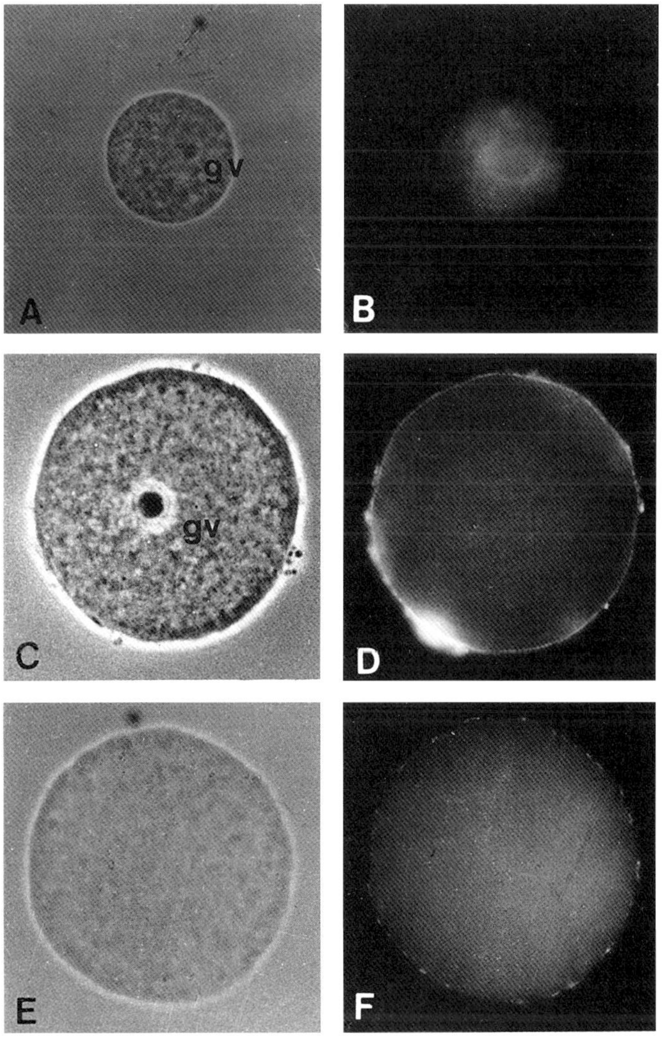

Figure 2. Immunofluorescent localization of DE-complementary sites on the oocyte surface. (left: phase-contrast; right: immunofluorescence). Zona-free oocytes recovered at different stages of growth and maturation were incubated 30 min with purified protein DE (200 μg/ml), fixed and exposed to anti-DE/FITC-goat anti-rabbit IgG. A-B: GV oocyte (40 μm in diameter). Note the absence of labeling on the oocyte surface. C-D: GV oocyte (80 μm in diameter). Note the "uniform" staining over the entire surface. E-F: Oocyte recovered from the ovary 6 h post-hCG. Note the patchy distribution of fluorescence (different from Fig. 2D), over the entire surface. GV: germinal vesicle. x250.

between three and 15 hours post-hCG injection presented decreasing values in the percentages of this "uniform" staining, with a parallel increase in the percentages of patchy labeling. In addition, while those recovered between three and nine hours post-hCG presented fluorescent labeling over the entire surface (Figs. 2E, F), oocytes recovered after this period exhibited the negative area described for ovulated oocytes (Fig. 1D). Considering that the negative area corresponds to the region of the plasma membrane through which fusion rarely occurs, the distribution of DE-binding sites seems to result from an ability of the cell to localize membrane components to specific regions. This ability appears to be essential to membrane function in a number of systems (Johnson et al. 1981; Mc Nutt, Weinstein 1973; Staehelin 1974).

Fusibility of the Oocyte Plasma Membrane during Growth and Maturation

In order to investigate whether the appearance of DE-binding sites on the oocyte surface correlated with the acquisition of fusibility by the rat oolema, the occurrence of sperm-oocyte fusion in oocytes recovered at different stages of oogenesis was determined by the Hoechst 33342 dye transfer technique. As shown in Table 1, it was observed that while growing oocytes with a diameter <50 μm presented a very low incidence of sperm-oocyte fusion (5%), a high percentage (83%) of growing oocytes with a diameter >50 μm already exhibited fusion ability. The percentages of growing oocytes with fused sperm as well as the number of sperm fused per oocyte were not significantly different from those corresponding to maturing oocytes. The fact that growing oocytes showing uniform labeling already exhibited oolema fusibility indicates that the presence of DE-complementary sites, independently of their distribution (uniform or patchy) on the oocyte surface, would be important for fusion with sperm. The acquisition of fusion ability observed in growing oocytes was not due to an increase in their oolema binding ability (Table 1), and occurred concomitantly with the appearance of DE-binding sites on their surface. In order to confirm that the appearance of the sites was involved in the acquisition of oolema fusibility in these oocytes, zona-free growing oocytes with a diameter >50 μm, as well as ovulated oocytes, were incubated with purified protein DE, exposed to capacitated sperm and finally examined for evidence of fusion by the dye transfer technique. Figure 3 indicates that the presence of DE during co-incubation produced a significant ($p < 0.001$) decrease in the percentage of oocytes showing evidence of fusion compared to untreated oocytes.

Together, these observations indicate that the acquisition of fusibility by the rat oolema occurs during the growth period and involves the appearance of DE-binding components on the oocyte surface, providing novel information on the molecular mechanism by which the mammalian egg plasma membrane becomes competent to fuse with sperm during oogenesis.

Fate of DE Binding Sites after Fertilization

Recently, researchers in the Cuasnicú laboratory examined the fate of the DE-binding sites on the egg surface after fertilization. For that purpose, the presence and localization of the sites on both fertilized eggs at the pronuclear stage and two-cell embryos were studied by IIF. Results indicate that while all pronuclear fertilized eggs presented fluorescent staining, only 50% of the

Table 1: Binding ability and fusibility of the oocyte plasma membrane during growth and maturation.

		Binding [a]		Fusion [a]	
Stage	**# of oocytes**	**% oocytes w/ bound sp**	**# of bound sp/ oocyte**	**% oocytes w/ fused sp**	**# of fused sp/ oocyte [b]**
Growing oocytes (diameter <50 μm)	24	91 ± 6	8.4 ± 1.1	5 ± 3	1.3 ± 0.4
Growing oocytes (diameter >50 μm)	59	97 ± 2	12.0 ± 2.0	83 ± 7 *	4.4 ± 0.8
Maturing oocytes (hours post-hCG)					
3	21	100	14.8 ± 3.5	82 ± 8 *	4.6 ± 2.1
6	26	100	14.7 ± 5.8	93 ± 9 *	5.3 ± 2.8
9	24	100	11.8 ± 5.0	90 ± 10*	3.1 ± 1.2
12-15	153	97 ± 3	12.7 ± 1.7	94 ± 2 *	2.3 ± 0.4

[a] Oocytes were incubated with 1 μg/ml Hoechst 33342 for 5 min at 37°C, thoroughly washed in fresh medium, and then inseminated with capacitated sperm (0.5-2.0 x 10^5 cells/ml). After 1 h, oocytes were fixed and both examined under phase contrast and UV microscope for evidence of binding (unlabeled sperm heads) and fusion (bright sperm heads).
[b] calculated as: number of fused sperm / number of oocytes with fused sperm.
sp= spermatozoa.
Values are means ± S.E. of four separate experiments for growing oocytes and maturing oocytes recovered 3-9 h after hCG, and of eight separate experiments for oocytes recovered 12-15 h after hCG.
* $p<0.05$ vs. oocyte diameter <50 μm.

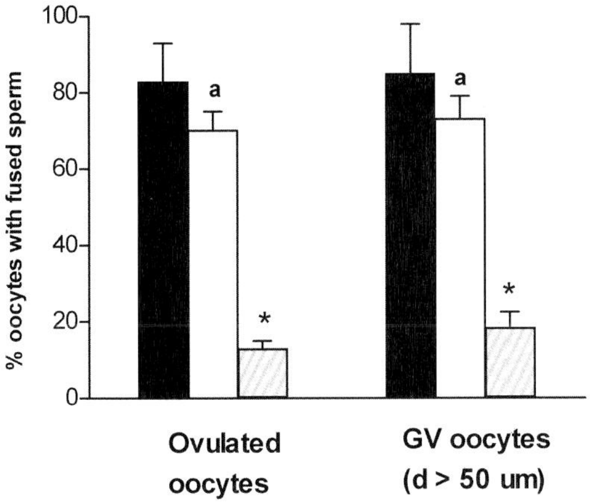

Figure 3. Effect of DE on sperm fusion with GV zona-free oocytes. Zona-free GV oocytes with a diameter >50 μm as well as ovulated oocytes (15 h post-hCG) were preloaded 5 min with Hoechst 33342 (1μg/ml), incubated 30 min in capacitation medium alone (solid bars) or in capacitation medium containing either 200 μg/ml of ovalbumin (open bars) or 200 μg/ml of purified protein DE (cross-hatched bars), and inseminated with capacitated spermatozoa. After 1 h of gamete co-incubation, oocytes were fixed, mounted on slides and examined under a UV microscope for evidence of sperm-oocyte fusion. a: the same results were obtained in medium containing 200 μg/ml of α-lacto-albumin or carbonic anhydrase. * $p < 0.001$.

two-cell embryos showed evidence of labeling. In all the cases the cells presented labeling distributed over the whole surface and the absence of the negative area characteristic of ovulated eggs. At present, the penetrability of pronuclear eggs and embryos is being examined to investigate whether there is any correlation between the fusogenicity of the egg/embryo plasma membrane and the presence of the DE-binding sites on the surface.

Participation of DE in Gamete Fusion in Other Species

Participation of DE in Mouse Gamete Fusion

Considering that two murine homologs of DE have also been identified: cystein-rich secretory protein-1 (CRISP) (Haendler et al. 1993), and mouse acidic epididymal glycoprotein-1 (AEG-1) (Mizuki, Kasahara 1992), the possible participation of DE in mouse gamete fusion was explored. Zona-free mouse eggs were inseminated with capacitated mouse sperm in the presence of purified rat protein DE (200 μg/ml) or a control protein (ovalbumin). Results indicated that while the presence of the same concentration of ovalbumin did not affect egg penetration, the addition of DE during gamete co-incubation produced a significant inhibition ($p<0.001$) in the percentage of penetrated eggs. These results support both the participation of protein DE in the mouse sperm-egg fusion process as well as the existence of complementary sites for DE on the mouse egg surface.

To study the localization of the DE-binding sites, zona-free mouse eggs were incubated with purified protein DE or ovalbumin (control), fixed, and subjected to IIF using anti-DE and goat anti-rabbit FITC-conjugated antibodies. Preliminary results indicated that while none of the zona-free eggs incubated with ovalbumin presented fluorescence, all eggs incubated with DE exhibited fluorescent labeling. As in the rat, this fluorescence was localized as a patchy labeling over the entire egg surface with the exception of a negative area that corresponds to the membrane overlying the meiotic spindle.

Potential Role of the Human Homolog of DE in Sperm-Egg Fusion

Recently, two independent laboratories reported the existence of a human epididymal protein, denominated ARP, with 40% homology to DE (Hayashi et al. 1996; Kratzschmar et al. 1996). Northern and Western blot analysis of various essential and reproductive human organs indicates that the expression of this human protein is strictly limited to the epididymis.

Several characteristics of ARP, such as its epididymal origin, its secretory nature, its molecular weight and its localization on the sperm head, strongly suggest its participation in the human sperm-egg fusion process. However, the removal of ARP from the sperm surface by low ionic strength opened the question as to whether the human homolog of DE would have a role in fertilization (Kratzschmar et al. 1996).

In view of the results obtained in the rat indicating the existence of two populations of DE, the interaction of ARP with the sperm surface was studied to investigate the existence of a second, strongly bound population of ARP which could participate in the fertilization stage. These studies have been conducted using the recombinant ARP protein expressed in bacteria as well as a specific polyclonal antibody against ARP provided by Dr. M. Kasahara's laboratory (University of Hokkaido, Japan). Human ejaculated sperm were subjected to sequential treatments known to remove peripheral and integral proteins, and the presence of ARP in the sperm extracts was evaluated by Western blot.

Preliminary results indicate that while ARP can be removed from the sperm surface by high ionic strength (2M NaCl), a certain amount remains on sperm even after this treatment and can be extracted by Triton X-100 (1%). These results suggest that ARP, as rat epididymal protein DE, would associate to the sperm surface with two different affinities. The existence of a tightly bound population of ARP supports the participation of this protein in fertilization since, at least in the rat, this is the population finally involved in gamete interaction. The possible participation of ARP in sperm-egg fusion as well as the existence of ARP-binding sites on the egg surface are currently being investigated.

Perspectives

Previous results from the Cuasnicú laboratory indicated that active immunization of male and female rats with purified protein DE raised antibodies against the protein in over 90% of the animals, and produced a significant and reversible inhibition of fertility in both sexes (Perez Martinez et al. 1995; Cuasnicú et al. 1990). Recent studies indicated that while no pathological effects were observed on the reproductive organs of DE-immunized males, sperm recovered from these animals exhibited a significant decrease in their ability to fuse with zona-free eggs *in vitro*, with no effects on their motility or viability nor on their ability to undergo capacitation/AR, or to bind to the oolema (Ellerman et al. 1998). These results, together with the finding that a large proportion of cases of human infertility due to defective sperm function involve the failure of acrosome-reacted spermatozoa to fuse with the plasma membrane of the oocyte (Tesarik, Thébault 1993; Aitken 1992), support the study of the participation of epididymal proteins in human sperm-egg fusion for both the development of new and safe male contraceptive methods and a better understanding and treatment of human infertility.

Acknowledgements

This work was supported by OMS grant 87094 and Rockefeller grant 92014 (P.C.). D.C. and D.E. are Research Fellowship recipients from the National Research Council of Argentina. P.C. is a Research Career Award recipient from the National Research Council of Argentina.

References

Aitken RJ. A family of fusion proteins. Nature 1992; 356:196-197.

Cameo MS, Blaquier JA. Androgen-controlled specific proteins in rat epididymis. J Endocrinol 1976; 69:317-324.

Cameo MS, Gonzalez Echeverria MF, Blaquier JA, Burgos MH. Immunochemical localization of epididymal protein DE on rat spermatozoa: its fate after induced acrosome reaction. Gam Res 1986; 15:247-258.

Cuasnicú PS, Gonzalez Echeverria MF, Piazza A, Cameo M, Blaquier JA. Antibodies against epididymal glycoproteins block fertilizing ability in rat. J Reprod Fert 1984; 72:467-471.

Cuasnicú PS, Conesa D, Rochwerger L. Potential contraceptive use of an epididymal protein that participates in fertilization. In: (Alexander NJ, Griffin D, Spieler JM, Waites GMH, eds) Gamete Interaction. Prospects for Immunocontraception. New York: Wiley-Liss, 1990, pp143-153.

Cuasnicú PS, Cohen DJ, Ellerman DA, Rochwerger L. Relationship between the association of an epididymal protein to the sperm surface and its behaviour and function. Biol Reprod 1996; 54 (Suppl 1):140.

Ebensperger C, Barros C. Changes at the hamster oocyte surface from germinal vesicle stage to ovulation. Gamete Res 1984; 9:387-397.

Ellerman DA, Brantúa VS, Pérez Martínez S, Cohen DJ, Conesa D, Cuasnicú PS. Potential contraceptive use of epididymal proteins: immunization of male rats with epididymal protein DE inhibits sperm fusion ability. Biol Reprod 1998; 59:1029-1036.

Garberi JC, Kohane AC, Cameo MS, Blaquier JA. Isolation and characterization of specific rat epididymal proteins. Mol Cell Endocr 1979; 13:73-82.

Garberi JC, Fontana JD, Blaquier JA. Carbohydrate composition of specific rat epididymal protein. Int J Androl 1982; 5:619-626.

Haendler B, Kratzschmar J, Theuring F, Schleuning WD. Transcripts for cysteine-rich secretory protein-1 (CRISP-1; DE/AEG) and the novel related CRISP-3 are expressed under androgen control in the mouse salivary gland. Endocrinology 1993; 133:192-198.

Hayashi M, Fujimoto S, Takano H, Ushiki T, Abe K, Ishikura H, Yoshida M, Kirchhoff C, Ishibashi T, Kasahara M. Characterization of a human glycoprotein with potential role in sperm-egg fusion: cDNA cloning, immunohistochemical localization, and chromosomal assignment of the gene (AEGL1). Genomics 1996; 32:367-374.

Johnson MH, Eager D, Muggleton-Harris A, Grave HM. Mosaicism in organization of concanavalin A receptors on surface membranes of mouse eggs. Nature 1975; 257:321-322.

Johnson MH, Pratt HPM, Handyside AH. The generation and recognition of positional information in the preimplantation mouse embryo. In: (Galsser SR, Bullock DW, eds) Cellular and Molecular Aspects of Implantation. New York: Plenum Press, 1981, pp55-74.

Kohane AC, Cameo MS, Piñeiro L, Garberi JC, Blaquier JA. Distribution and site of production of specific proteins in the rat epididymis. Biol Reprod 1980a; 23:181-187.

Kohane AC, Gonzalez Echeverria F, Piñeiro L, Blaquier JA. Interaction of proteins of epididymal origin with spermatozoa. Biol Reprod 1980b; 23:737-742.

Kratzschmar J, Haendler B, Eberspaecher U, Roosterman D, Donner P, Schleuning WD. The human cysteine-rich secretory protein (CRISP) family. Primary structure and tissue distribution of CRISP-1, CRISP-2 and CRISP-3. Eur J Biochem 1996; 236:827-836.

Mc Nutt NS, Weinstein RS. Membrane ultrastructure at mammalian intracellular junctions. Prog Biophys Mol Biol 1973; 26:45-101.

Mizuki N, Kasahara M. Mouse submandibular glands express an androgen-regulated transcript encoding an acidic epididymal glycoprotein-like molecule. Mol Cell Endocrinol 1992; 89:25-32.

Myles DG. Molecular mechanisms of sperm-egg membrane binding and fusion in mammals. Dev Biol 1993; 158:35-45.

Perez Martinez S, Conesa D, Cuasnicú PS. Potential contraceptive use of epididymal proteins: evidence for the participation of specific antibodies against rat epididymal protein DE in male and female fertility inhibition. J Reprod Immunol 1995; 29:31-45.

Rochwerger L, Cuasnicú PS. Redistribution of a rat sperm epididymal glycoprotein after *in vivo* and *in vitro* capacitation. Mol Reprod Dev 1992; 31:34-41.

Rochwerger L, Cohen DJ, Cuasnicú PS. Mammalian sperm-egg fusion: the rat egg has complementary sites for a sperm protein that mediates gamete fusion. Dev Biol 1992; 153:83-90.

Staehelin LA. Structure and function of intercellular junctions. Int Rev Cytol 1974; 39:191-283.

Tesarik J, Thébault A. Fertilization failure after subzonal sperm insertion associated with defective fusional capacity of acrosome-reacted spermatozoa. Fertil Steril 1993; 60:369-371.

Yanagimachi R. Sperm-egg fusion. In: (Duzgunes N, Bronner F, eds) Current Topics in Membranes and Transport. Orlando, FL: Academic Press, 1988, pp3-43.

26

The Fate of Sperm Components within the Egg during Fertilization: Implications for Infertility

Laura Hewitson
Calvin Simerly
Peter Sutovsky
Tanja Dominko
Diana Takahashi
Gerald Schatten
Oregon Health Sciences University
Oregon Regional Primate Research Center

Understanding the series of events that ultimately leads to fertilization in humans is of tremendous importance for understanding causes of infertility, but due to ethical, political and religious constraints associated with using human sperm and oocytes, basic research is limited. Recent technological advances in single-cell imaging do permit investigators to examine cytoskeletal events with the limited number of donated or discarded human oocytes obtained from consenting patients undergoing infertility treatment, but these may not necessarily reflect normally fertilized human oocytes. Fertilization in the mouse, the best studied animal model, follows a maternal method of centrosome inheritance (Schatten G et al. 1985; Schatten H 1986) that appears to be unique to rodents. Conversely, all other mammals so far studied follow a paternal method of centrosome inheritance, including bovine species (Navara et al. 1994), rhesus monkeys (Wu et al. 1996) and humans (Simerly et al. 1995). This suggests that the mouse may not necessarily be the most applicable model for understanding human fertilization. Studies on bovine fertilization have the advantage that an abundant supply of gametes can be easily obtained, and, more importantly, the reproductive success and genealogy of males used for semen collection is well characterized. However, oocyte micromanipulations, such as intracytoplasmic sperm injection (ICSI), are technically challenging in this species; this is reviewed in Schatten et al. (1998). Non-human primates represent a valuable resource for studying human reproduction since gamete maturation, *in vitro* fertilization (IVF), and embryonic development have been well documented in a variety of subspecies (Bavister et al. 1983, 1984; Lanzendorf et al. 1990; Wu et al. 1996) and all appear to closely follow that of humans.

The advent of ICSI (Palermo et al. 1992; Van Steirteghem et al. 1993) has revolutionized methods of assisted reproduction, demonstrating that the sperm can bypass several important events, such as capacitation, the acrosome reaction and oocyte binding and fusion, yet still result in successful fertilization. Many thousands of babies conceived by ICSI have been born worldwide, although there is mounting concern that ICSI babies may show evidence of increased genetic abnormalities (In't Veld et al. 1995) and even retarded mental development (Bowen et al. 1998). These concerns underscore the need for animal models to be explored thoroughly prior to the clinical introduction of novel assisted reproduction technologies (ART). This is especially relevant when examining new methods of assisted fertilization, such as round spermatid injection (ROSI), as seen in Tesarik et al. (1995) and Fishel et al. (1995) and secondary spermatocyte injection (Sofikitis et al. 1998). This review describes current knowledge on the fate of sperm and egg components during assisted fertilization as well as novel causes of infertility which may be attributed to a defective centrosome in the introduced sperm.

The Role of the Centrosome during Fertilization

Inheritance of the Centrosome

Fertilization in mammals follows a precisely orchestrated series of events which results in irreversible modifications of both the sperm and oocyte, culminating in genomic union at mitosis. The fertilizing sperm first binds to the zona pellucida, inducing the acrosome reaction necessary for the sperm to penetrate through the zona; this is reviewed by Yanagimachi (1994). Fusion of the sperm plasma membrane with that of the oocyte is immediately followed by oocyte activation, characterized by cortical granule exocytosis and depletion of intracellular calcium stores. The accessory structures of the incorporated sperm are disassembled, exposing the sperm centrosome to the oocyte cytoplasm; this is reviewed by Schatten (1994). Restoration of the centrosome at fertilization requires the attraction of maternal centrosomal components, such as γ-tubulin, to the paternal centrosome (Fig. 1). This process ultimately leads to the conversion of the sperm centrosome into the zygotic centrosome, the oocyte's microtubule organizing center (MTOC), which nucleates microtubules responsible for pronuclear movement and apposition (Boveri 1901; Mazia 1987).

Members of every mammalian species examined to date, including primates (Simerly et al. 1995; Wu et al. 1996), domestic species (Le Guen, Crozet 1989; Long et al. 1993; Navara et al. 1994), and even evolutionarily primitive marsupials (Breed et al. 1994), inherit their centrosomes from their fathers; this is reviewed by Schatten (1994). The one exception is rodents, including mice (Maro et al. 1985; Schatten et al. 1985; Schatten et al. 1986), rats (Zernicka-Goetz et al. 1993) and hamsters (Hewitson et al. 1997), which remain under maternal centrosomal control. In unfertilized mouse oocytes, many microtubule asters (cytasters) can be detected throughout the cytoplasm. At fertilization, the incorporated sperm does not nucleate microtubules. Instead, the oocyte-derived cytasters enlarge until the entire cytoplasm is filled with a dense matrix of assembled tubulin protein which ultimately moves the male and female pronuclei into close apposition at the oocyte center (Schatten et al. 1986).

Molecular Characterization of the Centrosome

The centrosome is a complex organelle composed of many different proteins (Kalnins 1992; Kimble, Kuriyama 1992). Its molecular characterization is still in its infancy, although several components that are

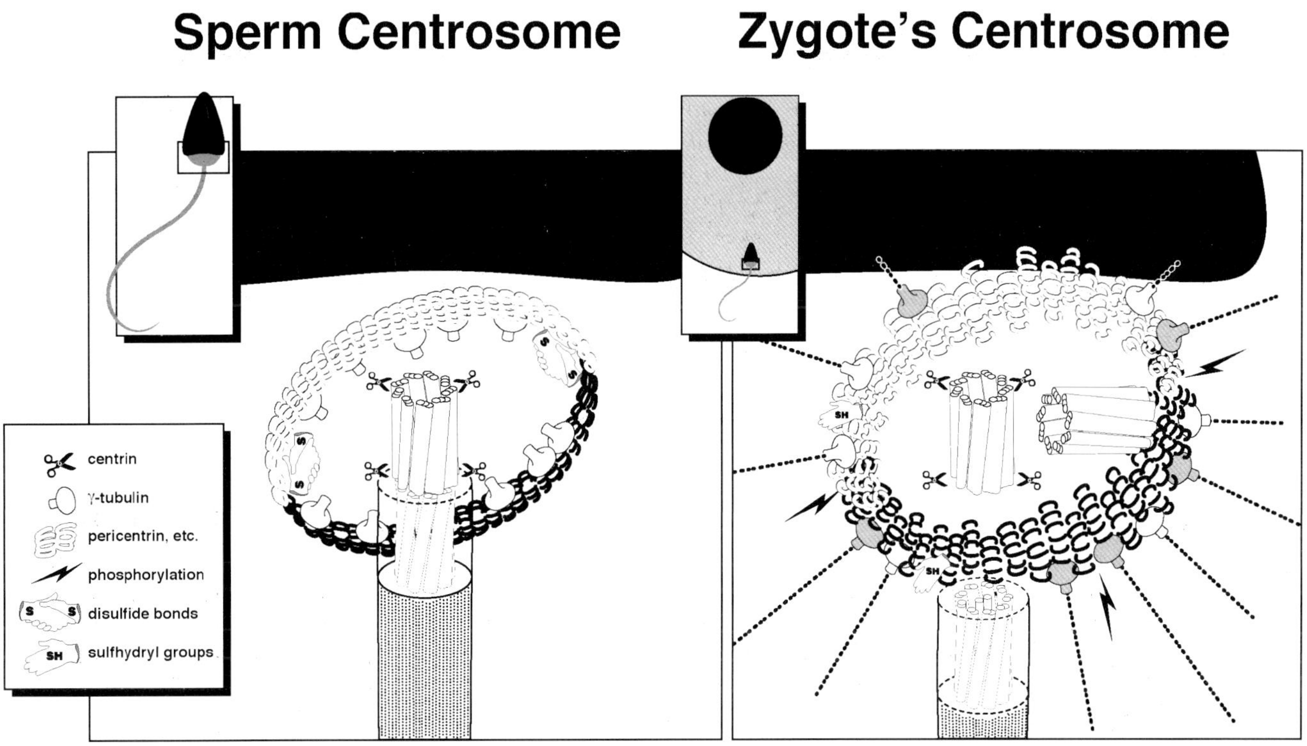

Figure 1. Molecular dissection of the human sperm centrosome and its reconstruction in the zygote. The sperm centrosome (left): The human sperm centrosome has centrin concentrated in one or two focal sites, corresponding to the centrioles. γ-tubulin is not apparent in mature human sperm, but becomes detectable after "centrosomal priming" of the sperm with disulfide reducing agents; this is a novel type of cytoplasmic capacitation. γ-tubulin is also detectable on Western blots with intact or sonicated human sperm. The centrosome is not phosphorylated and the sperm tail microtubules extend from a centriole. The coiled-coil infrastructure of the centrosome probably anchors the centrosome to the sperm nucleus and regulates the exposure of, and binding sites for, γ-tubulin. The zygote's centrosome (right): after permeabilization and incubation in extracts from *Xenopus* oocytes, the human sperm becomes phosphorylated and heavily immunoreactive with antibodies to γ-tubulin. The γ-tubulin found in the human sperm is probably a combination of some paternal and largely maternal protein. The binding of calcium ions, released during the transient increase during egg activation, to centrin is predicted to result later in a centrin-induced severing of the doublet sperm tail microtubules from the triplet microtubules of the centriole. Perhaps the severing of the tail microtubules from the basal body frees the basal body complex so that it can bind additional γ-tubulin and undergo transformation into a centriole. In humans, calcium-mediated centrin excision does not lead to complete sperm tail dissociation from the centriole/centrosome complex. The coiled-coil domains of the centrosome are drawn as unraveling, expanding and everting in the zygote; this exposes paternal γ-tubulin and also exposes binding sites for maternal γ-tubulin. The halo of γ-tubulin nucleates the microtubules which assemble into the sperm aster. (Reprinted with permission from Schatten 1994.)

conserved among centrosomes of many different species have been identified (Oakley, Oakley 1989; Stearns et al. 1992; Horio et al. 1991; Joshi et al. 1992; Palacios et al. 1993; Doxey et al. 1994; Schatten 1994; Salisbury 1995). Included in these conserved constituents is γ-tubulin, a member of the tubulin superfamily that is important in both microtubule nucleation and in defining the polarity of the assembled centrosomal microtubules (Oakley, Oakley 1989; Joshi et al. 1992; Palacios et al. 1993). Mature sperm do not appear to retain appreciable amounts of γ-tubulin at their centrosomes. However, shortly after sperm penetration, maternal γ-tubulin is drawn to the sperm centrosome to assist in formation and subsequent elongation of the sperm aster (Doxsey et al. 1994; Felix et al. 1994; Stearns, Kirschner 1994).

Centrin, a centrosomal, calcium binding protein, has been shown to play a role in the calcium-induced flagellar severing in *Chlamydomonas* (Salisbury 1995). The function of centrin during mammalian fertilization has yet to be determined but it is thought to play a role in centrosome duplication and microtubule severing in yeast (Baum et al. 1986, 1988; Sanders, Salisbury 1994). Centrin is strongly detected in mature human sperm as one or two foci at the base of the sperm head, possibly corresponding to centrioles, and may therefore be involved in sperm tail disassembly. This process is probably facilitated by changes in the phosphorylation status of proteins in the sperm connecting piece and centriole (Pinto-Correia et al. 1994; Zoran et al. 1994) and the binding of calcium ions, released during oocyte activation, to centrosomal proteins such as centrin (Salisbury 1995; Sathananthan et al. 1996). Pericentrin, a 220 kD protein with antigenicity for centrosomes in mammalian gametes (Calarco-Gillam et al. 1983;

Schatten et al. 1986; Doxey et al. 1994) may have a structural role in centrosomal organization. NuMA (nuclear protein that localizes to the mitotic apparatus) is a 240 kD protein which localizes to the centrosome in a cell-specific manner (Compton, Cleveland 1994; Cleveland 1995). NuMA is found exclusively in the male and female pronuclei during interphase but localizes to the spindle poles during mitosis during mouse fertilization (Tang et al. 1995). As additional components of the sperm centrosome are identified, the understanding of centrosome biology and how this pertains to the success of fertilization will be greatly improved.

Correlations between Centrosome Activity and Developmental Success

Bovine fertilization is characterized by a paternally-derived centrosome, necessary for microtubule nucleation and mitotic spindle formation (Navara et al. 1994), events that are nearly identical to those of human fertilization (Simerly et al. 1995). In an IVF study using sperm collected from three bulls of known developmental potential, there was a good correlation between centrosome organization and reproductive success (Navara et al. 1996). The bull with the best breeding potential was found to have the largest, most tightly focused sperm aster. Bulls of lower reproductive potential had smaller, disorganized sperm asters. This demonstrates that microtubule organizing ability of the sperm centrosome varies from male to male and suggests that there may be variations in centrosomal vigor that affect the proficiency and swiftness of fertilization and perhaps the frequency of live births (Navara et al. 1996).

Evidence suggests that some forms of human infertility may be the result of defects in the formation of the sperm aster and in the motility events necessary for gametic union within the activated cytoplasm. Understanding the cell and molecular events which allow the sperm centrosome to be transformed into the active zygotic centrosome necessary for implantation may provide new information for infertility treatments targeted at centrosomal defects. The use of *Xenopus* cell-free cytoplasmic extracts (Tournier et al. 1991; Doxsey et al. 1994; Felix et al. 1994; Stearns, Kirschner 1994) has provided a dynamic way in which to study the transformation of the sperm centrosome into the zygotic centrosome. Experiments in which human sperm are exposed to a *Xenopus* cytoplasmic extract to study microtubule nucleation are currently under investigation (Zoran et al. 1994; Schatten 1994). Human sperm cannot nucleate microtubules following permeabilization and exposure to *Xenopus* extract, but must first undergo a disulfide reduction "priming" step before acquiring the ability to nucleate centrosomal microtubules *in vitro* (Table 1). Once the permeabilized and primed human sperm are exposed to *Xenopus* cytoplasmic extracts, maternal γ-tubulin strongly binds to the paternal centrosome. It is hoped that centrosome function in human sperm may be tested using a cell-free extract assay, which may help to identify human sperm samples which are defective in their microtubule nucleation ability and are therefore likely to lead to fertilization arrest. This would provide an alternative to the more commonly used method of assaying human fertility, the zona-free hamster oocyte sperm penetration assay (Yanagimachi et al. 1976), which is unable to identify sperm centrosomal defects which occur after sperm penetration (Hewitson et al. 1997).

Fate of Sperm Components within the Egg following *in Vitro* Fertilization

In the metaphase-arrested unfertilized rhesus oocyte, the only microtubules present are those of the meiotic spindle, which is barrel-shaped, anastral and oriented radially to the cell cortex (Fig. 2A). The chromosomes are aligned on the metaphase plate. At fertilization, the activated oocyte resumes meiosis separating the female chromosomes resulting in extrusion of the second polar body (Fig. 2B). The incorporated sperm undergoes several steps of disassembly, exposing the sperm centrosome to the oocyte cytoplasm. The replacement of sperm protamines with oocyte-derived histones by the action of glutathione (GSH), as seen in Perreault (1990), is important in pronucleus formation and may also promote the removal of the sperm mitochondria and axonemal striated columns from the connecting piece (Sutovsky et al. 1997; Sutovsky, Schatten 1997) during disassembly. The newly transformed zygotic centrosome nucleates a small aster of microtubules which lengthen as the male pronucleus enlarges (Fig. 2C). It is interesting to note that the adjacent male and female pronuclei are eccentrically positioned at the cortex at this time (Fig. 2D). The sperm centrosome then duplicates and splits, with each centrosome serving as the poles of the mitotic spindle (Fig. 2E). At prophase, the interphase microtubule array disassembles and is replaced by a dense monaster of microtubules still associated with the sperm tail (Sutovsky et al. 1996). The zygotic centrosome duplicates and splits, establishing a bipolar microtubule array. By prometaphase (18 to 20 hours post-insemination), a bipolar spindle emerges as the male and female chromosomes intermix. Mitotic metaphase is marked by an eccentrically positioned, barrel-shaped spindle, upon which the chromosomes become aligned. At mitosis, cleavage results in two equally sized daughter cells each containing a centrosome, one of which is in association with the introduced sperm tail (Wu et al. 1996). The ability of the introduced sperm centrosome to nucleate microtubules demonstrates that during fertilization of rhesus monkey

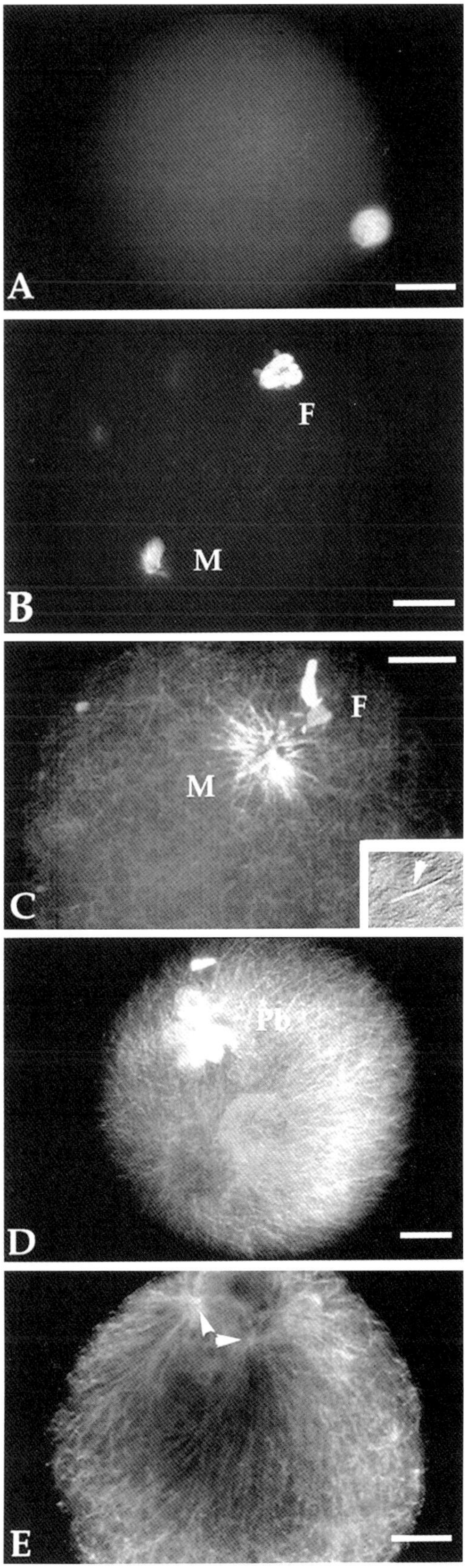

Figure 2. Microtubule patterns in unfertilized and *in vitro*-fertilized rhesus monkey oocytes. A barrel-shaped, anastral, meiotic spindle (A), oriented radially to the cell cortex, is the only microtubule structure present in the mature, unfertilized rhesus oocyte. The chromosomes are aligned on the metaphase plate (A). Between three to five hours post-insemination, a small sperm aster (B; M=male) forms at the base of the sperm head shortly after sperm penetration. The female chromosomes begin to separate on the second meiotic spindle (B; F=female). As development proceeds, the sperm aster microtubules lengthen in the cytoplasm (C; M=male) as the sperm DNA decondenses. The meiotic midbody marks the site of the female chromosomes and the second polar body (C; F=female). By six to eight hours post-insemination, the sperm aster enlarges to completely fill the zygote cytoplasm (D). The centrosome then duplicates and splits to form the two poles of the mitotic spindle (E). All images are double-labeled for microtubules and DNA. Bars: 20μm. (Modified with permission from Wu et al. 1996.)

oocytes a paternally derived centrosome is responsible for microtubule-mediated pronuclear migration. This suggests that the rhesus monkey represents a valuable animal model which closely mimics the events of human fertilization, overcoming the complicated ethical problems associated with using human gametes and embryos. The use of rhesus material in the future will allow researchers to design experiments looking at centrosome biology, which may normally be restricted with human oocytes and embryos, in a non-human primate.

Fate of Sperm Components within the Egg following Intracytoplasmic Sperm Injection

ICSI has been adopted by infertility clinics worldwide as a means of overcoming some severe forms of male factor infertility and has resulted in the birth of thousands of babies (Palermo et al. 1992; Van Steirteghem et al. 1993; Silber 1994). One of the most intriguing aspects of ICSI is that it bypasses many of the upstream events which prime the sperm for egg penetration, such as capacitation and acrosome reaction. However, recent data suggest that some ICSI babies may have increased risk of genetic abnormalities (In't Veld et al. 1995) or show retarded mental development when compared to children born after IVF or conceived naturally (Bowen et al. 1998). There is still clearly a need for basic research devoted to studying ICSI in animal models to address these issues. Rhesus monkey gametes have therefore been used to examine the cell biological basis of ICSI and offer the advantage of exploring primate fertilization without the ethical complexities associated with research on fertilized human oocytes. Rhesus sperm microinjected into rhesus oocytes will nucleate microtubules (Figs. 3A-C) and complete the repositioning of the male and female pronuclei (Fig. 3D) in much the same way as *in vitro* fertilized rhesus (Wu et al. 1996) and human (Simerly et al. 1995) oocytes. Centrosome duplication and bipolarity occur by late interphase (Fig. 3E) culminating in a bipolar mitotic spindle, with a sperm tail attached to the paternally inherited spindle pole (Fig. 3F).

Although these events seem to mimic those of *in vitro* fertilization, a transmission electron microscopy study of ICSI-derived rhesus zygotes demonstrated that sperm decondensation is abnormal after ICSI (Hewitson et al. 1996; Sutovsky et al. 1996). The presence of an intact acrosome over the anterior part of the injected sperm resulted in an asynchrony of sperm chromatin decondensation (Figs. 3G, H), with the chromatin underlying the acrosome cap decondensing at a slower rate than the chromatin in the posterior region of the sperm head. The significance of this asynchronous decondensation is not completely understood, but it may affect the import of intranuclear proteins into the region underlying the acrosome. Additional concerns have also been raised about possible genetic defects resulting from ICSI. Whether certain genetic defects could arise from asynchronous sperm decondensation remains to be addressed.

Despite the general concerns over the technique of ICSI, rhesus zygotes fertilized by ICSI offer a unique way to study new methods of assisted reproduction and to gain insight into the causes of fertilization failure. Unfortunately, many of the types of fertilization failure seen after human IVF were also found after rhesus ICSI, since sperm deficient in microtubule functioning and centrosomal reconstitution are not expected to be remedied by the use of ICSI (Simerly et al. 1995). In fact, many of the documented rhesus ICSI-failures were remarkably similar to those observed in human failed fertilizations (Hewitson et al. 1996) suggesting that these types of failures were centrosomal in origin.

Imaging Sperm Components in Discarded Human Oocytes Fertilized by *in Vitro* Fertilization

Working with clinical collaborators and informed, consenting patients, researchers in the Schatten laboratory have examined the role of the sperm centrosome in donated, unfertilized human oocytes and throughout fertilization, using sophisticated, single-cell imaging techniques. Microtubules were only found in the metaphase-arrested second meiotic spindle (Fig. 4A). Within six hours post-insemination, a small sperm aster could be identified emanating from the introduced sperm centrosome or MTOC (Figs. 4B, C). The activated oocyte extrudes the second polar body, seen attached to the developing female pronucleus by a mid-body (Figs. 4B, C). Sperm astral microtubules elongated throughout the cytoplasm until they came into contact with the developing female pronucleus (Figs. 4D-F). The female pronucleus was translocated towards the decondensing male pronucleus (Figs. 4E, F), possibly with the aid of dynein and kinesin motor proteins. At mitotic prophase, the male and female chromosomes condense separately as the duplicated and split centrosomes form a bipolar array (Figs. 4F, G). Microtubule nucleation from the mitotic spindle poles completes formation of the mitotic spindle, upon which the chromosomes become aligned (Fig. 4H).

These data suggest that humans, unlike rodents, follow a paternal method of centrosome inheritance. Further evidence for paternal centrosome inheritance in humans comes from studies of polyspermy, where the paternal centrosomal contribution is multiplied. During polyspermy, human oocytes develop multiple sperm asters after incorporation, each one associated with a sperm. During parthenogenesis, where there is no paternal centrosomal contribution, no sperm astral microtubules are nucleated. Instead, disarrayed cytoplasmic microtubules are found throughout the oocyte, which may aid in pronuclear migration and apposition. There must therefore be a default mechanism that activates maternal centrosomal components during parthenogenesis, when there is no male contribution. Most species which follow a paternal method of centrosome inheritance can also support parthenogenic activation (Schatten 1994), often leading to the development of cleavage stage embryos.

Imaging Sperm Components in Human Fertilization Failures: Implications for Infertility

Understanding why fertilization sometimes fails is of paramount importance in establishing whether alternative methods of assisted reproduction are necessary. Human oocytes that failed to fertilize after IVF, obtained from consenting patients at collaborating clinics, have been examined by immunocytochemistry to determine the cause or causes of fertilization arrest. A quarter of these oocytes, judged as fertilization failures at 24 hours post-insemination, were actually penetrated by one or more sperm, but had arrested at some point after this event (Simerly et al. 1995). Some of these failures could be attributed to a defective sperm centrosome since some inseminated oocytes failed to support microtubule growth from the sperm centrosome or the sperm centrosome began to nucleate microtubules, but these then failed to elongate (Asch et al. 1995; Simerly et al. 1995). In cases where a large sperm aster had formed, it was often disorganized and not focused, or even prematurely detached from the sperm head. These data demonstrate that there are several points at which fertilization arrest occurs; these arrests seem to be the result of improper centrosome function. These types of defects are probably due to male infertility factors since excess oocytes from some of the same patients were successfully

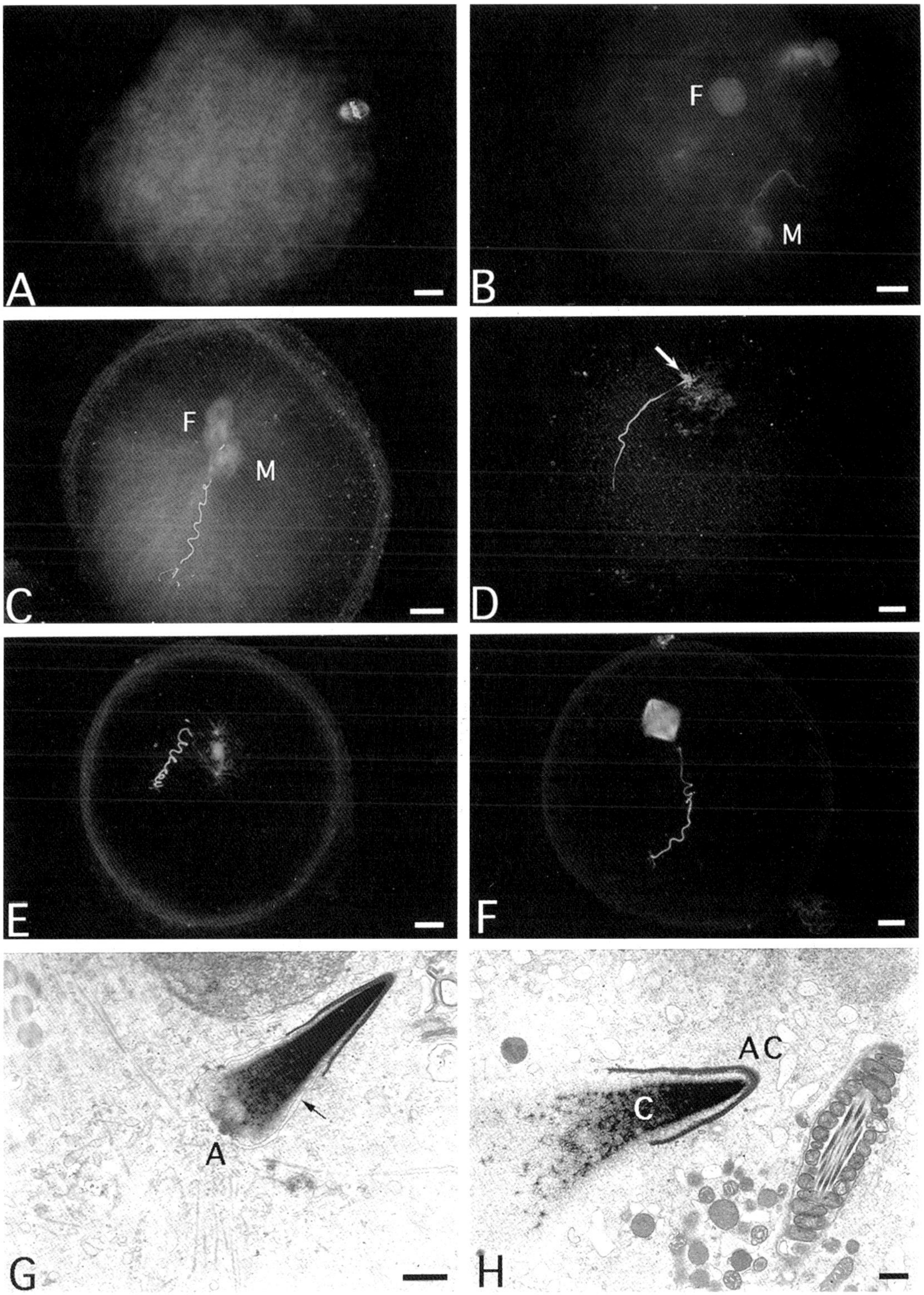

Figure 3. Conventional (A-B) and laser-scanning (C-F) fluorescence microscopy of rhesus oocytes following fertilization by intracytoplasmic sperm injection (ICSI). The second meiotic spindle of a sham injected oocyte shows no apparent damage (A). Following sperm injection, microtubules assemble close to the sperm head, which begins decondensation (B). As the male and female pronuclei decondense, the microtubules elongate to fill the entire cytoplasm (C; M=male; F=female). By prophase, most of the cytoplasmic microtubules disassemble so that just a small aster remains associated with the adjacent male and female pronuclei (D). By prometaphase, microtubules form a bipolar structure opposite the duplicated and split centrosomes (E). A fusiformed anastral spindle slightly eccentric within the cytoplasm forms during metaphase as the condensed chromosomes become aligned at the equator (F). A large aster of microtubules (A) develops around the proximal centriole of the injected sperm despite the asynchronous sperm nuclear decondensation (G). The residual acrosome (AC) caused irregular decondensation of sperm (C) during the early stages of sperm disassembly (H). Note that a new double membrane is being assembled around the base and equatorial region of this sperm. Bar for A-F: 10 µm; bars for G and H: 500 nm. (Modified with permission from Hewitson et al. 1996.)

fertilized with donor sperm, but not spousal sperm. Other types of fertilization failure not directly associated with defects in the sperm centrosome include the failure to incorporate a sperm, failure to exit meiotic metaphase, formation of multiple female karyomeres and the premature condensation of paternal chromosomes, which leads to a paternal meiotic spindle. Despite the recent success of ICSI, men with defective sperm centrosomes are unlikely to benefit from ICSI since centrosomal defects are associated with post-insemination events and oocytes fertilized with defective sperm are still likely to arrest following ICSI.

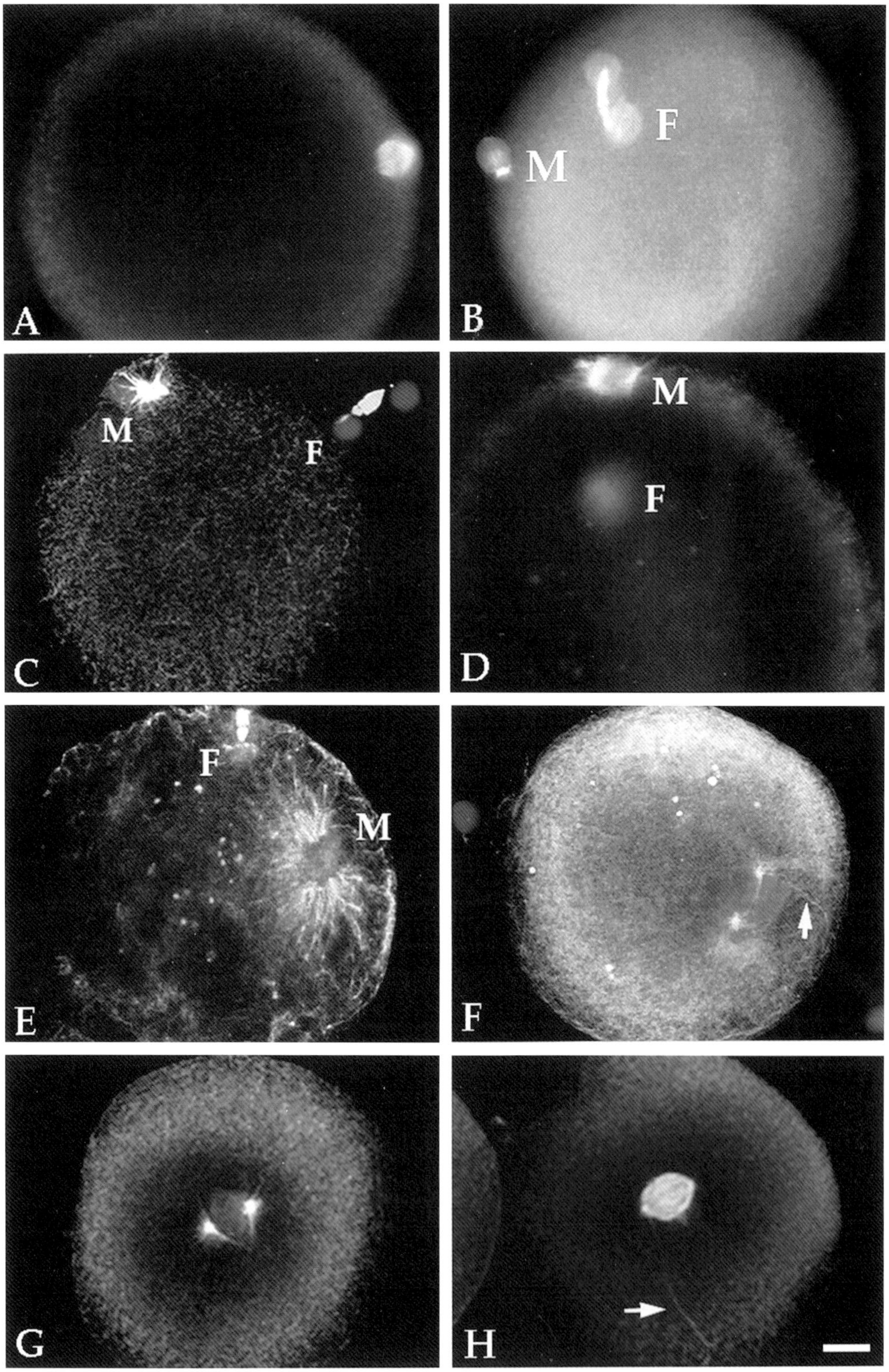

Figure 4. Microtubule and DNA organization in normal inseminated human oocytes. The meiotic spindle in mature, unfertilized human oocytes is anastral, oriented radially to the cell surface, and asymmetric, with a focused pole abutting the cortex and a broader pole facing the cytoplasm (A). No other microtubules are detected in the cytoplasm of the unfertilized human oocyte. Shortly after sperm incorporation (3 to 6.5 hours post-insemination), sperm astral microtubules assemble around the base of the sperm head (M=male), as the inseminated oocytes complete second meiosis and extrude the second polar body (B-D). The close association of the meiotic midbody identifies the female pronucleus (F=female). Short, sparse, disarrayed cytoplasmic microtubules can also be observed in the cytoplasm following confocal microscopic observations of these early activated oocytes (C). As the male pronucleus continues to decondense in the cytoplasm, the microtubules of the sperm aster enlarge, circumscribing the male pronucleus (E). By 15 hours post-insemination, the centrosome splits and organizes a bipolar microtubule array that emanates from the tightly apposed pronuclei (F). The sperm tail is associated with an aster (arrow). At first mitotic prophase (16.5 hours post-insemination), the male and female chromosomes condense separately as a bipolar array of microtubules marks the developing first mitotic spindle poles (G). By prometaphase, when the chromosomes begin to intermix on the metaphase equator, a barrel-shaped, anastral spindle forms in the cytoplasm (H). The sperm axoneme remains associated with a small aster found at one of the spindle poles (arrow). Bars: 10 μm. (Modified with permission from Simerly et al. 1995.)

Conclusions and Future Directions

During fertilization in mammals, the incorporation of one, and only one, sperm into the oocyte results in the introduction of the sperm centrosome, the oocyte's microtubule organizing center (MTOC), thus restoring centrosome function to the oocyte. The restoration and function of the zygotic centrosome is critical for pronuclear apposition and defects in this pathway are likely to lead to fertilization arrest. Fertilization failures associated with defective sperm centrosomes demonstrate that oocytes which inherit a defective sperm centrosome during IVF undergo post-insemination arrest, and may be a novel form of male infertility, not remedied by the use of ICSI. This has also been studied using rhesus monkey oocytes and it has been demonstrated that the types of fertilization failures arising after rhesus IVF and ICSI are very similar to those of humans. These discoveries have important implications for studying centrosome activity and function and may be applied as a diagnostic tool for examining certain types of male and female infertility. However, further characterization of the defects associated with fertilization failure after IVF and ICSI should be addressed. The ability of human sperm to nucleate microtubules *in vitro* can be demonstrated with the use of cell-free extracts, an exciting observation, since phenotypic variations in bovine sperm centrosomes can be correlated with reproductive success. Finally, the success of the rhesus monkey as a relevant and reliable animal model for examining novel forms of fertilization, such as the injection of round spermatids or spermatocytes, will enable a swifter understanding of these emerging methods of assisted reproduction prior to their wide scale clinic adoption.

Acknowledgements

The authors wish to acknowledge colleagues who have contributed to this work: Steve Eisele, Michelle Emme, Vicki Frohlich, Crista Martinovich, Chris Navara, Mark Tengowski, Joseph Wu and Sara Zoran. The authors are grateful to the many collaborators involved in this research. The work would not have been possible without support from the anonymous patients who provided informed consent and agreed to donate excess and discarded oocytes. Funding for animal studies was provided by the National Institute of Health, the United States Department of Agriculture and the Mellon Foundation. Protocols used were approved by the University's Human Subjects Institutional Review Boards and Institutional Animal Care and Use Committee.

References

Asch R, Simerly C, Ord T, Schatten G. The stages at which fertilization arrests in humans: defective sperm centrosomes and sperm asters as causes of human infertility. Hum Reprod 1995; 10:1897-1906.

Baum P, Furlong C, Byers B. Yeast gene required for spindle pole body duplication: homology of its product with Ca^{2+}-binding proteins. Proc Natl Acad Sci USA 1986; 83:5512-5516.

Baum P, Yip C, Goetsch L, Byers B. A yeast gene essential for regulation of spindle pole duplication. Mol Cell Biol 1988; 8:5386-5397.

Bavister BD, Boatman DE, Leibfried L, Loose M, Vernon MW. Fertilization and cleavage of rhesus monkey oocytes *in vitro*. Biol Reprod 1983; 28:983-999.

Bavister BD, Boatman ED, Collins K, Dierschke DH, Eisele SG. Birth of rhesus monkey infant after *in vitro* fertilization and nonsurgical embryo transfer. Proc Natl Acad Sci USA 1984; 81 (7):2218-2222

Boveri T. "Zellen-studieren: Ueber die natur der centrosomen." Fisher, Jena, Germany, 1901.

Bowen JR, Gibson FL, Leslie GI, Saunders DM. Medical and developmental outcome at 1 year for children conceived by intracytoplasmic sperm injection. Lancet 1998; 351:1529-1534.

Breed W, Simerly C, Navara CS, Vanderberg J, Schatten G. Distribution of microtubules in eggs and early embryos of the marsupial, *Monodelphis domestica*. Dev Biol 1994; 164:230-240.

Calarco-Gillam PD, Siebert MC, Hubblc R, Mitchison T, Kirschner M. Centrosome development in early mouse embryos as defined by an autoantibody against pericentriolar material. Cell 1983; 35:621-629.

Cleveland DW. NuMA: a protein involved in nuclear structure spindle assembly, and nuclear reformation. Trends Cell Biol 1995; 5:60-64.

Compton DA, Cleveland DW. NuMA, a nuclear protein involved in mitosis and nuclear reformation. Curr Opin Cell Biol 1994; 6:343-346.

Doxsey SJ, Stein P, Evans L, Calarco P, Kirschner M. Pericentrin, a highly conserved centrosome protein involved in microtubule organization. Cell 1994; 76:639-650.

Félix MA, Antony C, Wright M, Maro B. Centrosome assembly *in vitro*. J Cell Biol 1994; 124:19-31.

Fischel S, Green S, Bishop M. Pregnancy after intracytoplasmic injection of a spermatid. Lancet 1995; 345:1641-1642.

Hewitson LC, Simerly C, Tengowski MW, Sutovsky P, Navara CS, Haavisto AJ, Schatten G. Microtubule and chromatin configurations during rhesus intracytoplasmic sperm injection: successes and failures. Biol Reprod 1996; 55:271-280.

Hewitson LC, Simerly C, Haavisto AJ, Jones J, Schatten G. Microtubule organization and chromatin configurations in hamster oocytes during fertilization, parthenogenesis, and after insemination with human sperm. Biol Reprod 1997; 57:967-975.

Horio T, Uzawa S, Jung MK, Oakley BR, Tanaka K, Yanagida M. The fission yeast γ-tubulin is essential for mitosis and is localized at microtubule organizing centers. J Cell Sci 1991; 99:693-700.

In't Veld P, Brandenburg H, Verhoeff A, Dhont M, Los F. Sex chromosomal abnormalities and intracytoplasmic sperm injection. Lancet 1995; 773:346.

Joshi HC, Palacios MJ, McNamara L, Cleveland DW. γ- tubulin is a centrosomal protein required for cell cycle dependent microtubule nucleation. Nature 1992; 356:80-83.

Kalnins VI. The Centrosome. Academic Press, New York, 1992.

Kimble M, Kuriyama R. Functional components of microtubule organizing centers. Int Rev Cytol 1992; 136:1-50.

Lanzendorf SE, Zelinski-Wooten MB, Stouffer RL, Wolf DP. Maturity at collection and the developmental potential of rhesus monkey oocytes. Biol Reprod 1990; 42:703-711.

Le Guen P, Crozet N. Microtubule and centrosome distribution during sheep fertilization. Eur J Cell Biol 1989; 48:239-249.

Long CR, Pinto-Correia C, Duby RT, Ponce-de Leon FA, Boland MP, Roche JF Robl JM. Chromatin and microtubule morphology during the first cell cycle in bovine zygotes. Mol Reprod Dev

1993; 36:23-32.
Maro B, Howlett SK, Webb M. Non-spindle microtubule organizing centers in metaphase II-arrested mouse oocytes. J Cell Biol 1985; 101:1665-1672.
Mazia, D. The chromosome cycle and the centrosome cycle in the mitosis cycle. Int Rev Cytol 1987; 100:49-92.
Navara C, First NL, Schatten G. Microtubule organization in the cow during fertilization, polyspermy, parthenogenesis and nuclear transfer: the role of the sperm aster. Dev Biol 1994; 162:29-40.
Navara CS, First NL, Schatten G. Phenotypic variations among paternal centrosomes expressed within the zygote as disparate, dissimilar microtubule lengths and sperm aster organization: correlations between centrosome activity and developmental success. Proc Natl Acad Sci USA 1996; 93:5384-5388.
Oakley CD, Oakley BR. Identification of γ-tubulin, a new member of the tubulin superfamily encoded by mipA gene of *Aspergillus nidulans*. Nature 1989; 338:662-664.
Palacios MJ, Joshi HC, Simerly C, Schatten G. Dynamic reorganization of γ-tubulin during murine fertilization. J Cell Sci 1993; 104:383-389.
Palermo G, Joris H, Devroey P, Van Steirteghem AC. Pregnancies after intracytoplasmic sperm injection of a single spermatozoon into an oocyte. Lancet 1992; 340:17-18.
Perreault SD. Regulation of sperm nuclear reactivation during fertilization. In: (Bavister BD, Cummins J, Roldan ERS, eds) Fertilization in Mammals. Norwell, MA: Serono Symposia, 1990, pp285-296.
Pinto-Correia C, Poccia DL, Chang T, Robl JM. Dephosphorylation of sperm mid-piece antigens initiates aster formation in rabbit oocytes. Proc Natl Acad Sci USA 1994; 91:7894-7898.
Salisbury JL. Centrin, centrosomes, and mitotic spindle poles. Cur Opin Cell Biol 1995; 7:39-45.
Sanders MA, Salisbury JL. Centrin plays an essential role in microtubule severing during flagellar excision in *Chlamydomonas reinhardtii*. J Cell Biol 1994; 124:795-805.
Sathananthan AH, Ratnam SS, Ng SC, Tarin JJ, Gianaroli L, Trounson A. The sperm centriole: its inheritance, replication and perpetuation in early human embryos. Hum Reprod 1996; 11:345-356.
Schatten G. The centrosome and its mode of inheritance: the reduction of the centrosome during gametogenesis and its restoration during fertilization. Dev Biol 1994; 165:299-335.
Schatten G, Simerly C, Schatten H. Microtubule configurations during fertilization, mitosis and early development in the mouse and the requirement for egg microtubule-mediated motility during mammalian fertilization. Proc Natl Acad Sci USA 1985; 82:4152-4155.
Schatten G, Hewitson L, Simerly C, Sutovsky P, Huszar G. Cell and molecular biological challenges of ICSI: ART before science? J Law Med Ethics 1998; 26:29-37.
Schatten H, Schatten G, Mazia D, Balczon R, Simerly C. Behavior of centrosomes during fertilization and cell division in mouse oocytes and in sea urchin eggs. Proc Natl Acad Sci USA 1986; 83:105-109.
Silber SJ. The use of epididymal sperm in assisted reproduction. In: (Tesarik J, ed) Male Factor in Human Infertility. Rome: Ares-Serono Symposia, 1994, pp335-68.
Simerly C, Wu G, Zoran S, Ord T, Rawlins R, Jones J, Navara C, Gerrity M, Rinehart J, Binor Z, Asch R, Schatten G. The paternal inheritance of the centrosome, the cell's microtubule-organizing center, in humans and the implications for infertility. Nature Med 1995; 1:47-53.
Sofikitis N, Mantzavinos T, Loutradis D, Yamamoto Y, Tarlatzis V, Miyagawa I. Ooplasmic injections of secondary spermatocytes for non-obstructive azoospermia. Lancet 1998; 351:1177-1178.
Stearns T, Kirschner M. *In vitro* reconstitution of centrosome assembly and function: the central role of γ-tubulin. Cell 1994; 76:623-37.
Stearns T, Evans L, Kirschner M. γ-tubulin is a highly conserved component of the centrosome. Cell 1992; 65:825-836.
Sutovsky P, Schatten G. Depletion of glutathione during bovine oocyte maturation reversibly blocks the decondensation of the male pronucleus and pronuclear apposition during fertilization. Biol Reprod 1997; 56:1503-1512.
Sutovsky P, Hewitson LC, Simerly C, Tengowski MW, Navara CS, Haavisto AJ, Schatten G. Intracytoplasmic sperm injection for Rhesus monkey fertilization results in unusual chromatin, cytoskeletal, and membrane events, but eventually leads to pronuclear development and sperm aster assembly. Hum Reprod 1996; 11:1703-1712.
Sutovsky P, Oko R, Hewitson L, Schatten G. Binding of oocyte microvilli to the perinuclear theca of fertilizing sperm and subsequent theca removal constitute a previously unrecognized step in mammalian fertilization. Dev Biol 1997; 188:75-84.
Tang TK, Tang C-J, Hu H-M. The nuclear mitotic apparatus protein (NuMA) reorganization and function during early mouse development. Mol Biol Cell 1995; 6:422a.
Tesarik J, Mendoza C, Testart J. Viable embryos from injection of round spermatids into oocytes. N Engl J Med 1995; 333:525.
Tournier F, Cyrklaff M, Karsenti E, Bornens M. Centrosomes competent for parthenogenesis in *Xenopus* eggs support pro-centriole budding in cell-free extracts. Proc Natl Acad Sci USA 1991; 88:9929-9933.
Van Steirteghem AC, Nagy Z, Joris H, Liu J, Staessen C, Smitz J, Wisanto A, Devroey P. High fertilization and implantation rates after intracytoplasmic sperm injection. Hum Reprod 1993; 8:1061-1066.
Wu J-G, Simerly C, Zoran S, Funte LR, Schatten G. Microtubule and chromatin configurations during fertilization and early development in rhesus monkeys, and regulation by intracellular calcium ions. Biol Reprod 1996; 55:260-270.
Yanagimachi R. Mammalian fertilization. In: (Knobil E, Neill JD, eds) The Physiology of Reproduction. 2nd ed. New York: Raven Press, 1994, pp189-317.
Yanagimachi R, Yanagimachi H, Rogers BJ. The use of zona free animal ova as a test system for the assessment of the fertilizing capacity of human spermatozoa. Biol Reprod 1976; 15:471-476.
Zernicka-Goetz M, Kubiak JZ, Antony C. Cytoskeletal organization of rat oocytes during metaphase II arrest and following abortive activation: a study by confocal laser scanning microscopy. Mol Reprod Dev 1993; 35:165-175.
Zoran S, Simerly C, Schoff P, Stearns T, Salisbury J, Schatten G. Reconstitution of the human sperm centrosome *in vitro*. Mol Biol Cell 1994; 5:38a.

27 Transmission of Mammalian Mitochondrial DNA

Eric A. Shoubridge
McGill University

Most eukaryotic cells rely on the production of aerobically produced ATP for their normal function. This is produced by the process of oxidative phosphorylation in the enzyme complexes of the mitochondrial respiratory chain, which are located in the inner mitochondrial membrane. There are five such enzyme complexes: complexes I-IV, which make up the electron transfer chain itself, and complex V, the ATP synthase. These multimeric complexes are unique in the cell in that some of the component polypeptide subunits are encoded in the nuclear genome, while others are encoded in mitochondrial DNA (mtDNA).

It has long been established that mtDNA is largely, if not exclusively, maternally inherited in mammals (Giles et al. 1995). Indeed, in the biological world there are only rare exceptions to the rule of uniparental inheritance of either mitochondrial and chloroplast genomes (Birky 1995). This nearly universal phenomenon may have evolved to limit the spread of selfish genetic elements (Hurst et al. 1996). Uniparental inheritance of mtDNA is achieved by a variety of mechanisms in different organisms, including degradation of the organelles in gametes, exclusion of the organelles from one parent from the zygote and selective silencing or degradation of the organelles in the zygote (Birky 1995). In mammals there is evidence for selective elimination of paternal mitochondria in the early embryo (Kaneda et al. 1995), but the molecular basis for this remains a puzzle.

Recently it has become clear that mutations in mtDNA in humans are an important cause of neuromuscular and neurological diseases (Grossman, Shoubridge, 1996; Di Mauro et al. 1998; Shoubridge 1998). The rules governing transmission and segregation of mtDNA in mammals have thus taken on a renewed importance for genetic counselling and clinical management of patients affected with these disorders. Further, the introduction of new reproductive technologies, such as cytoplasmic transfer, direct sperm injection (ICSI) and animal cloning, has forced a re-evaluation of the potential contribution of exogenous or paternally derived mitochondria to the next generation.

This chapter will review current knowledge of the structure and function of mtDNA, transmission and segregation of mtDNA, mechanisms for ensuring maternal transmission and the possibility of leakage of paternal mtDNA.

Mitochondrial DNA: Structure, Replication, Expression

Although there is a great variety in the size and content of mtDNA in the plants and animals, the mtDNA of all mammals investigated has the same basic structure: a double-stranded circular DNA molecule of ~16.5 kb (Fig. 1). The two stands are referred to as heavy (H) and light (L), reflecting their behavior in CsCl density gradients. Mammalian mtDNA codes for 37 genes: 13 polypeptides, all of which are subunits of the mitochondrial respiratory chain; and 22 tRNAs and 2 rRNAs, which constitute part of the mitochondrial translation machinery. The genome is exceedingly compact, containing no introns and only one non-coding region of ~1kb, the D-loop. This is a triple-stranded region that contains one of the replication origins and promoters for transcription of the H- and L-strands. The mtDNA copy number in somatic cells is generally in the range of 10^3-10^4 per cell, organized as two to 10 (mean five) copies per organelle. Gametes are a notable exception to this generalization: mature oocytes have approximately 10^5 mtDNAs (Piko, Taylor 1987) and sperm about 10^2 (Hecht et al. 1984).

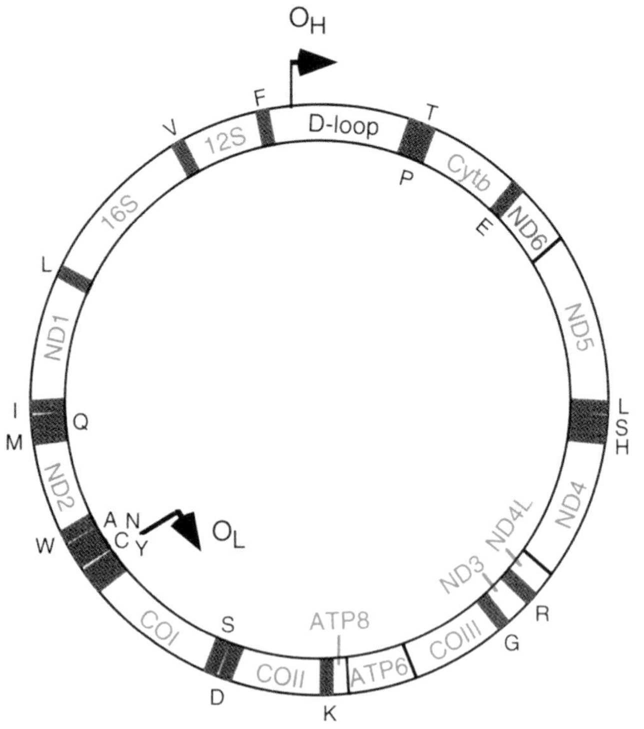

Figure 1. Structure of mammalian mtDNA. The map depicts the structure of the human mt genome. The protein coding genes are: ND1-6, seven subunits of CompleX I; cyt b, subunit of Complex III, COI-III, three subunits of Complex IV and ATP6,*, two subunits of Complex V. 16S and 12S are two rRNA genes. The solid bars represent tRNA genes shown as the single letter amino acid code. The positions of the replication origins (O_H and O_L) and the H- and L-strand promoters (HSP, LSP) are indicated.

Replication of mtDNA occurs asynchronously from two promoters: one for the H-strand in the D-loop and a second for the L-strand about two-thirds of the way around the molecule (Clayton 1982; Shadel, Clayton 1997). Replication is catalyzed by a distinct polymerase, the γ-DNA polymerase, and leading strand synthesis at O_H is primed by a short piece of RNA generated by transcription from the L-strand promoter. Thus, replication of the genome is linked in some way to expression (transcription) of mtDNA genes. Although mtDNA copy number varies widely from cell to cell and is tightly regulated in a cell-specific fashion, replication of mtDNA is not tightly coupled to the cell cycle (Clayton 1982; Birky 1994). Thus, during mitosis some templates may replicate more than once, others not at all. This behavior, coupled with the random distribution of mtDNAs to daughter cells at cytokinesis, provides a mechanism for mitotic segregation of mtDNA sequence variants.

MtDNA is transcribed as three polycistronic units, the entire H- and L-strands and the two rRNAs. The rRNAs are transcribed about 25 times more frequently than the entire H-strand, a process controlled by a specific termination sequence just 3' to the 16S rRNA in the gene coding for $tRNA^{leu(UUR)}$, as seen in Fernandez-Silva et al. (1997). Transcription of mtDNA requires the presence of the only known mitochondrial transcription factor, Tfam, formerly referred to as mtTFA (Shadel, Clayton 1997; Larsson et al. 1998). Maturation of mitochondrial transcripts requires an RNAase P activity and it is thought that the tRNA genes which are interspersed between many of the protein coding genes act as signal sequences in this process. Translation of the mRNAs occurs within the mitochondrial compartment on mitochondrial ribosomes. All of the proteins involved in the replication and expression of mtDNA are encoded in the nuclear genome and must be targeted to the mitochondria.

Transmission of mtDNA in the Female Germ Line

Most mammals have a single mtDNA sequence variant in all of their cells, a condition referred to as mtDNA homoplasmy. There is, however, a great deal of sequence variability between individuals. In human populations, individuals differ by 0.3%, or about 50 nucleotides, in the mtDNA sequence. The rarity of mtDNA heteroplasmy (the occurrence of more than one sequence variant in an individual) and the high degree of population polymorphism suggest that new mtDNA sequence variants are rapidly segregated in maternal lineages. This is paradoxical, given the high genome copy number in mature oocytes (~10^5 copies of mtDNA) and the relatively small number of cell divisions in the development of the female germline. New germline mutations ought to produce significant sequence heterogeneity in the mtDNA population in oocytes and one possible outcome would be a very large number of segregating variants. The fact that this is not seen suggests that the number of segregating units is much less than the copy number of mtDNAs in the oocyte.

These considerations led Hauswirth and Laipis (1982) to hypothesize the existence of a genetic bottleneck for mtDNA in oocyte development or in early embryogenesis. This concept was based on observations of D-loop variants in cows where rapid segregation of the sequence variant was observed in a single generation (Olivio et al. 1983; Ashley et al. 1989). In fact, examining the same sequence variant in a large number of independent Holstein pedigrees, Kohler et al. (1991) observed complete allele switching in a single generation, and suggested that the number of segregating units might be as small as one. Rapid segregation of mtDNA sequence variants has also been observed in a large number of human pedigrees segregating pathogenic mtDNA mutations; however, this process is never as rapid as that observed in the D-loop of cows (Jenuth et al. 1996).

The mechanism responsible for the mtDNA genetic bottleneck has been the subject of recent investigations

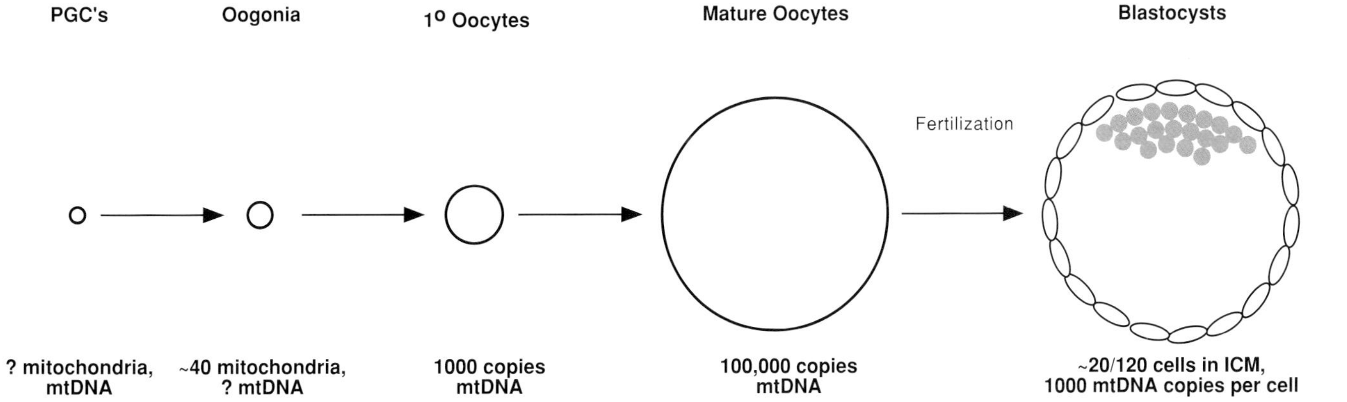

Figure 2. Diagrammatic representation of changes in mtDNA copy number during development of the female germline and preimplantation embryo. PGC: primordial germ cell.

in heteroplasmic mice constructed from two different *M.m. domesticus* strains, NZB and BALB/c (Jenuth et al. 1996). During oocyte maturation and in the early embryo mitochondria and mtDNA copy number change dramatically (Fig. 2). The mtDNA copy number is not known in either primordial germ cells or in oogonia, the precursors of the primary oocyte population. Ultrastructural analyses suggest that oogonia contain approximately 40 mitochondria, which is low compared to most somatic cells (Nogawa et al. 1988). Measurements of mtDNA copy number in immature oocytes (<20µm in diameter) suggest that there are approximately 1000 copies (Piko, Taylor 1987). As the oocyte matures to greater than 80 µm in diameter, the number of mtDNAs increases ~100-fold to approximately 100,000 in the mouse (Piko, Taylor 1987) and close to 200,000 in the cow (Michaels et al. 1982). This is accompanied by a reduction in the copy number per organelle to 1-1.5 mtDNAs per mitochondrion. After fertilization, cell division of the early embryo proceeds without mtDNA replication so that by the time the inner cell mass cells are set aside in the blastocyst, mtDNA copy number is reduced to about 1,000, which is in the same range as in somatic cells.

Two different hypotheses were proposed to explain the bottleneck for mtDNA in female germline development. The first suggested that a limited number of mtDNA templates were used during the 100-fold amplification of mtDNA copy number that accompanies oocyte maturation, reducing the effective number of mtDNAs that contribute to the next generation. The second suggested that perhaps mtDNAs in the ICM were a non-random sample of mtDNAs in the mature oocyte (Olivio et al. 1983; Ashley et al. 1989). These have recently been tested in heteroplasmic mice using single cell PCR on mature and immature oocytes and primordial germ cells (Jenuth et al. 1996). The results of these studies demonstrate that most of the segregation of sequence variants has occurred by the time the primary oocyte population is formed, but that there is little intercellular variation in the degree of mtDNA heteroplasmy in the primordial germ cell population. These data squarely place most of the segregation of mtDNA sequence variants in the mitotic divisions that occur in the oogonial population. Two features of oogonia are important in this context: they contain few mitochondria and they undergo synchronous rounds of mitosis to produce the entire primary oocyte population. Thus one would predict rapid segregation of mtDNA sequence variants in these cells simply by random genetic drift. Studies of homopolymeric tract heteroplasmy in humans are consistent with the mouse studies, showing that segregation occurs by the time oocytes are mature (Marchington et al. 1997).

This model of mtDNA segregation provides a mechanism for the rapid amplification and fixation of new mtDNA mutations in the female germline. It predicts that mutations arising either in the primordial germ cells or in the oogonia will have the possibility of increasing due to the relatively small number of mitochondria in oogonia and the successive rounds of mitosis.

Mitochondrial Function and Selection of Sequence Variants during Embryogenesis and Fetal Life

Early studies by Piko and colleagues established that early development of the preimplantation embryo was unaffected by treatment with chloramphenicol, a compound which inhibits mitochondrial translation (Piko, Chase 1973). Transcription of mtDNA is, however, active from the two-cell stage on, suggesting that the increased capacity in oxidative phosphorylation is preparatory for postimplantation development (Taylor, Piko 1995). Although it is very likely that the expression of mtDNA is required during fetal life, studies of fetal tissues in individuals carrying pathogenic mtDNA mutations show little tissue to tissue variation in the proportion of mutant and wild-type mtDNAs (Matthews et al. 1994; Harding et al. 1992). This suggests that selection for respiratory chain function is not strong during fetal life and that the proportion of mutant mtDNAs at birth is largely determined by the proportion in the oocyte. Similar observations have been made in heteroplasmic mice in which there is little variation in heteroplasmy among tissues at birth, but strong, tissue-specific selection for alternate mtDNA genotypes as the animal ages (Jemuth et al. 1997). Additional evidence for this proposition comes from pedigree analysis of patients segregating mtDNA mutations. Although rapid shifts in the proportions of mutant mtDNAs occur between generations, the shift appears just as likely to occur in the direction of wild-type mtDNA as mutant mtDNA. An apparent exception to the lack of selection for mtDNA sequence variants in embryogenesis was observed in the karyoplast-reconstructed mice (Meirelles, Smith 1998); however, this may represent a highly artificial situation.

Mitotic Segregation of mtDNA after Birth

Although there appears to be little segregation of mtDNA sequence variants during fetal life, mitotic segregation of mtDNA sequence variants occurs in different tissues throughout life. There is good evidence for increases in the proportions of some pathogenic mtDNA mutations, including large-scale deletions (Larsson et al. 1990) and tRNA point mutations (Weber et al. 1997) with age in the skeletal muscle of patients with mitochondrial encephalomyopathies. This apparent selection for a dysfunctional genome may reflect an attempt by the cell to restore oxidative phosphorylation function. Studies of heteroplasmic mice, constructed by karyoplast fusion to enucleated zygotes, have shown marked differences in levels of heteroplasmy among tissues, suggesting the operation of stringent replicative segregation (Meirelles, Smith 1997). Similar studies of mice constructed by cytoplast fusion have shown a different pattern of segregation of mtDNA sequence variants. Little variation was observed except in four tissues—liver/kidney and blood/spleen—where directional selection for opposite mtDNA sequence variants was observed (Jenuth et al. 1997). The molecular basis for this surprising observation remains unknown.

Mechanisms of Exclusion of Paternal mDNA

The basis for the strict maternal transmission of mammalian mtDNA has been the subject of considerable speculation. The fact that mammals are anisogamous and that there exists about a 1000-fold difference in mtDNA copy number between sperm and ovum (10^2 versus 10^5) suggests that simple dilution of male mtDNA might be sufficient to ensure that the vast majority of mtDNAs are inherited maternally. Indeed, the down regulation of mtDNA copy number to about 100, apparently organized at about one per organelle as in the oocyte, during spermatogenesis has been suggested as a mechanism to prevent male transmission of mtDNA (Hecht et al. 1984). Paralleling this decrease is a decrease in Tfam (Larsson et al. 1997); however, whether this is causal remains to be determined. Ultrastructural studies of the fate of hamster sperm post-fertilization have demonstrated so-called microvesicular bodies fusing with mitochondria and leading to their destruction (Hiraoka, Hirao 1988). It has been suggested that sperm mitochondria themselves may not be competent to survive in the zygote. Sperm are known to express a very large number of specific isozymes for important metabolic enzymes in glycolysis (for example, LDH, GAPDH) and for important factors in the mitochondrial respiratory chain, such as cytochrome c (Hecht 1995), and it is possible that metabolic differences select against survival of male sperm.

There is some evidence from somatic cells that the efficiency of repopulation of rho^0 cells with mitochondria (and therefore mtDNA) derived from sperm is dramatically less than the number of cells containing sperm mitochondria after fusion, and less than observed when somatic cell mitochondria are used as donors (Manfredi et al. 1997). The basis for this phenomenon remains unclear and it is possible that the cells containing sperm-derived mtDNA arise only when a rare fusion event occurs between a sperm and a somatic cell mitochondrion.

Early studies suggested that there may be a barrier to transmission of mitochondria from a non-oocyte source (Ebert et al. 1989). However, recent experiments have demonstrated survival of mtDNAs derived from hepatocytes of *M. spretus* mice in *M.m. domesticus* embryos, demonstrating that non-oocyte mitochondria can survive in early embryos (Pinkert et al. 1997). In any case, genetic experiments described below have suggested that none of these factors are critical since male transmission of mtDNA can in fact be observed under some conditions.

Male Transmission of mtDNA in Mammals

Although transmission of mammalian mtDNA was long thought to be strictly maternal, the first indication that paternal mtDNA might be leaked to the next generation came from a study of interspecific crosses in mice involving *M.m. domesticus* and *M. spretus*. On the assumption that any potential male contribution to the mtDNA pool in the offspring of the next generation would be small, Gyllensten and co-workers backcrossed female F_1 hybrids (*domesticus x spretus*) with either *domesticus* or *spretus* males for up to 26 generations in an effort to increase the proportion of paternal mtDNA (Gyllensten et al. 1991). Using PCR, mtDNA specific for *M. spretus* could be detected in all tissues examined in the offspring in these crosses at frequencies of 10^{-3} to 10^{-4}. It was concluded that a small amount of paternal mtDNA was added at each generation and that this accumulated to detectable levels after a large number of generations. This interpretation is, however, clearly untenable in light of what is now known of the mtDNA bottleneck in the female germline. No known mechanism exists for the slow and progressive accumulation of mtDNA in the germline, and the most likely fate of a rare sequence variant would be loss due to genetic drift.

Kaneda et al. (1995) extended these studies by examining the fate of male mtDNA in intraspecific and interspecific crosses in early embryos. They showed that male mtDNA was undetectable in the late pronuclear stage in intraspecific crosses and that this coincided with the loss of rhodamine 123 fluorescence, which had been used as a marker to assess sperm mitochondrial membrane potential. In contrast, *spretus* mtDNA could be detected in the majority of two-cell embryos and in about 50% of neonates, born after implantation of the *in vitro* fertilized embryos into pseudopregnant females. This suggested that sperm from a closely related species could escape the normal surveillance mechanisms that exist to ensure uniparental transmission of mtDNA within a species. Elimination of the paternal mtDNA was also observed in embryos carrying the *spretus* mtDNA on a *domesticus* nuclear background. These results suggested that the molecular target(s) for elimination of paternal mtDNA in intraspecific crosses is a nuclear-encoded molecule(s) associated with the sperm midpiece. Recent experiments, in which sperm were injected directly into the oocyte, demonstrated that male mitochondria, tagged with a Mitotracker dye, were present in some embryos up to the eight-cell stage (Cummins et al. 1997). Ultrastructural investigations, such as those that demonstrated fusion of male mitochondria with multivesicular bodies (Hiraoka, Hirao 1988), have been carried out on interspecific mammalian crosses to look for morphological correlates of male mitochondrial survival.

The group of Yonekawa has studied transmission of paternal mtDNA in somatic tissues and the female germline of F_1 hybrid (*domesticus x spretus*) mice and in the first backcross generation (N_2) between female F_1 hybrids x *M. spretus* males (Shitara et al. 1998). They demonstrated that the presence of paternal mtDNA is random in different tissues of the F_1 animals, with about 40% having some evidence of *spretus* mtDNA. Paternal mtDNA was rare in the ovaries of F_1 hybrid females and undetectable in a large sample of oocytes obtained by superovulation. No paternal mtDNA was detected in any of the N_2 backcross animals, tested by examining embryos obtained by *in vitro* fertilization. This suggested that there was some barrier to transmission even in F_1 hybrid animals. These results are clearly at odds with the observations of Gyllensten et al. (1991) and it is difficult to reconcile the two studies. Although it is possible that stable heteroplasmy for *spretus* mtDNA was established in the F_1 hybrid animals generated by Gyllensten et al. (1991), this would likely be a very rare event.

The fact that leakage of male mtDNA was not observed in the N_2 backcross generation suggests that the genetic factors responsible act as either a dominant or co-dominant maternal trait. Researchers in the Shoubridge laboratory have carried out similar experiments, in collaboration with Dr. Fred Biddle, University of Calgary, with different results. Evidence has been seen of some male leakage of *spretus* mtDNA in N_2 backcross mice (unpublished observations). The pattern of male mtDNA leakage (random tissue distribution, low frequency), does not appear to be remarkably different to that seen in F_1 animals. These observations suggest that prevention of male leakage is inherited as a recessive trait. It is not clear why these results differ from those of Shitara et al. (1998), but it may have to do with the sensitivity of detection of leaked male mtDNA.

In principle it should be possible to map the genetic elements involved in the exclusion of paternal mtDNA in these interspecific backcrosses. However, such an experiment will be complicated by the fact that failure to observe male mtDNA in the tissues of the next generation can result from either exclusion of paternal mitochondria in the early embryo or leakage with subsequent loss due to random factors. This will introduce some uncertainty in the determination of the phenotype (paternal mtDNA leakage or not) in the backcross animals, making genetic analysis difficult.

Conclusions

There is no strong evidence that male mtDNA can ever be transmitted to the next generation within a species in mammals. Although further work needs to be carried out, it appears from studies on mice that intracellular injection of sperm into the ovum does not alter the fate of male mitochondria, which are still actively eliminated from the embryo before the blastocyst stage. The molecular target(s) involved in this surveillance mechanism is nuclear encoded and appears to be associated with the sperm midpiece. In principle it should be possible to identify the gene or genes involved, through appropriate genetic experiments in mice, although this will be complicated by random factors not involved in the targeting mechanism.

Acknowledgements

Research in the author's laboratory is supported by grants from the Medical Research Council of Canada, Muscular Dystrophy Association of Canada and the March of Dimes. E.S. is a Montreal Neurological Institute Killam Scholar. The author is grateful to the those in his laboratory who have contributed to the work and ideas in this paper, particularly Jack Jenuth, Katherine Fu, Tim Johns and Zhiqing Zhu.

References

Ashley MV, Laipis PJ, Hauswirth WW. Rapid segregation of heteroplasmic bovine mitochondria. Nucleic Acids Res 1989; 17:7325-7331.

Birky CW Jr. Relaxed and stringent genomes: why cytoplasmic genes don't obey Mendel's laws. J Hered 1994; 85:355-365.

Birky CW Jr. Uniparental inheritance of mitochondrial and chloroplast genes: mechanisms and evolution. Proc Natl Acad Sci USA 1995; 92:11331-11338.

Clayton DA. Replication of animal mitochondrial DNA. Cell 1982; 28:693-705.

Cummins JM, Wakayama T, Yanagimachi R. Fate of microinjected sperm components in the mouse oocyte and embryo. Zygote 1997; 5:301-308.

DiMauro S, Bonilla E, Davidson M, Hirano M, Schon EA. Mitochondria in neuromuscular disorders. Biochim Biophys Acta 1998; 1366:199-210.

Ebert KM, Alcivar A, Liem H, Goggins R, Hecht NB. Mouse zygotes injected with mitochondria develop normally but the exogenous mitochondria are not detectable in the progeny. Mol Reprod Dev 1989; 1:156-163.

Fernandez-Silva P, Martinez-Azorin F, Micol V, Attardi G. The human mitochondrial transcription termination factor (mTERF) is a multizipper protein but binds to DNA as a monomer, with evidence pointing to intramolecular leucine zipper interactions. EMBO J 1997; 16:1066-1079.

Giles RE, Blanc H, Cann HM, Wallace DC. Maternal inheritance of human mitochondrial DNA. Proc Natl Acad Sci USA 1980; 77:6715-6719.

Grossman LI, Shoubridge EA. Mitochondrial genetics and human disease. Bioessays 1996; 18:983-991.

Gyllensten U, Wharton D, Josefsson A, Wilson AC. Paternal inheritance of mitochondrial DNA in mice. Nature 1991; 352:255-257.

Harding AE, Holt IJ, Sweeney MG, Brockington M, Davis MB. Prenatal diagnosis of mitochondrial DNA8993 T-G disease. Am J Hum Genet 1992; 50:629-633.

Hauswirth WW, Laipis PJ. Mitochondrial DNA polymorphism in a maternal lineage of Holstein cows. Proc Natl Acad Sci USA 1982; 79:4686-4690.

Hecht NB. The making of a spermatozoon: a molecular perspective. Dev Genet 1995; 16:95-103.

Hecht NB, Liem H, Kleene KC, Distel RJ, Ho SM. Maternal inheritance of the mouse mitochondrial genome is not mediated by a loss or gross alteration of the paternal mitochondrial DNA or by methylation of the oocyte mitochondrial DNA. Dev Biol 1984; 102:452-461.

Hiraoka J, Hirao Y. Fate of sperm tail components after incorporation into the hamster egg. Gamete Res 1988; 19:369-380.

Hurst LD, Atlan A, Bengtsson BO. Genetic conflicts. Q Rev Biol 1996; 71:317-364.

Jenuth JP, Peterson AC, Fu K, Shoubridge EA. Random genetic drift in the female germline explains the rapid segregation of mammalian mitochondrial DNA. Nat Genet 1996; 14:146-151.

Jenuth JP, Peterson AC, Shoubridge EA. Tissue-specific selection for different mtDNA genotypes in heteroplasmic mice. Nat Genet 1997; 16:93-95.

Kaneda H, Hayashi J, Takahama S, Taya C, Lindahl KF, Yonekawa H. Elimination of paternal mitochondrial DNA in intraspecific crosses during early mouse embryogenesis. Proc Natl Acad Sci USA 1995; 92:4542-4546.

Koehler CM, Lindberg GL, Brown DR, Beitz DC, Freeman AE, Mayfield JE, Myers AM. Replacement of bovine mitochondrial DNA by a sequence variant within one generation. Genetics 1991; 129:247-255.

Larsson NG, Holme E, Kristiansson B, Oldfors A, Tulinius M. Progressive increase of the mutated mitochondrial DNA fraction in Kearns-Sayre syndrome. Pediatr Res 1990; 28:131-136.

Larsson NG, Oldfors A, Garman JD, Barsh GS, Clayton DA. Down-regulation of mitochondrial transcription factor A during spermatogenesis in humans. Hum Mol Genet 1997; 6:185-191.

Larsson NG, Wang J, Wilhelmsson H, Oldfors A, Rustin P, Lewandoski M, et al. Mitochondrial transcription factor A is necessary for mtDNA maintenance and embryogenesis in mice. Nat Genet 1998; 18:231-236.

Manfredi G, Thyagarajan D, Papadopoulou LC, Pallotti F, Schon EA. The fate of human sperm-derived mtDNA in somatic cells. Am J Hum Genet 1997; 61:953-960.

Marchington DR, Hartshorne GM, Barlow D, Poulton J. Homopolymeric tract heteroplasmy in mtDNA from tissues and single oocytes: support for a genetic bottleneck. Am J Hum Genet 1997; 60:408-416.

Matthews PM, Hopkin J, Brown RM, Stephenson JB, Hilton-Jones D, Brown GK. Comparison of the relative levels of the 3243 (A to G) mtDNA mutation in heteroplasmic adult and fetal tissues. J Med Genet 1994; 31:41-44.

Meirelles FV, Smith LC. Mitochondrial genotype segregation in a mouse heteroplasmic lineage produced by embryonic karyoplast transplantation. Genetics 1997; 145:445-451.

Meirelles FV, Smith LC. Mitochondrial genotype segregation during preimplantation development in mouse heteroplasmic embryos. Genetics 1998; 148:877-883.

Michaels GS, Hauswirth WW, Laipis PJ. Mitochondrial DNA copy number in bovine oocytes and somatic cells. Dev Biol 1982; 94:246-251.

Nogawa T, Sung WK, Jagiello GM, Bowne W. A quantitative analysis of mitochondria during fetal mouse oogenesis. J Morphol 1988; 195:225-234.

Olivo PD, Van de Walle MJ, Laipis PJ, Hauswirth WW. Nucleotide sequence evidence for rapid genotypic shifts in the bovine mitochondrial DNA D-loop. Nature 1983; 306:400-402.

Piko L, Chase DG. Role of the mitochondrial genome during early development in mice. Effects of ethidium bromide and chloramphenicol. J Cell Biol 1973; 58:357-378.

Piko L, Taylor KD. Amounts of mitochondrial DNA and abundance of some mitochondrial gene transcripts in early mouse embryos. Dev Biol 1987; 123:364-374.

Pinkert CA, Irwin MH, Johnson LW, Moffatt RJ. Mitochondria transfer into mouse ova by microinjection. Transgenic Res 1997; 6:379-383.

Shadel GS, Clayton DA. Mitochondrial DNA maintenance in vertebrates. Ann Rev Biochem 1997; 66:409-435.

Shitara H, Hayashi JI, Takahama S, Kaneda H, Yonekawa H. Maternal inheritance of mouse mtDNA in interspecific hybrids: segregation of the leaked paternal mtDNA followed by the prevention of subsequent paternal leakage. Genetics 1998; 148:851-857.

Shoubridge EA. Mitochondrial encephalomyopathies. Curr Opin Neurol 1998; 11:491-496.

Taylor KD, Piko L. Mitochondrial biogenesis in early mouse embryos: expression of the mRNAs for subunits IV, Vb, and VIIc of cytochrome c oxidase and subunit 9 (P1) of H(+)-ATP synthase. Mol Reprod Dev 1995; 40:29-35.

Weber K, Wilson JN, Taylor L, Brierley E, Johnson MA, Turnbull DM, Bindoff LA. A new mtDNA mutation showing accumulation with time and restriction to skeletal muscle. Am J Hum Genet 1997; 60:373-380.

28

Soluble Sperm Activating Factors

B. Dale
Stazione Zoologica "Anton Dohrn"
L. Di Matteo
University of Naples
M. Marino
G. Russo
Stazione Zoologica "Anton Dohrn"
M. Wilding
Stazione Zoologica "Anton Dohrn"

Oocytes are usually blocked twice in meiosis (Dale 1983). The first arrest in prophase I (P-I; G-2 phase) stage may be long, ranging from weeks in lower animals to decades in higher vertebrates. As a general scheme, hormones trigger the release from this first meiotic block, driving the oocyte to a second arrest at meiosis I in many invertebrates, including ascidians, molluscs and insects, or to meiosis II as in vertebrates. Release from the second meiotic block is triggered by the spermatozoon.

There are exceptions to this rule (Fig. 1). At one extreme, in some worms and molluscs, fertilization triggers the release from the first meiotic block. At the other extreme, as in sea urchins and coelenterates, the second meiotic block is at the pronucleus stage, G1, of the first mitotic division, and the spermatozoon enters the oocyte at this late stage. In starfish oocytes, the second meiotic block is not obvious. The hormone 1-methyladenine triggers the P-I blocked oocytes to undergo germinal vesicle breakdown and the spermatozoon usually enters the oocyte shortly after, but in any case, before metaphase I. In ascidians, fertilization drives the oocytes through the first and second meiotic divisions.

Progression from the first to the second meiotic block is usually referred to as oocyte maturation, while the removal of the second meiotic block at fertilization is called oocyte activation. This chapter will look at mechanisms that cause oocytes to arrest in meiosis and how spermatozoa release this second block. Since calcium release is considered to be a fundamental event in oocyte activation, the mechanisms regulating release of this second messenger are described. A central question addressed is whether sperm activating factors are ubiquitous, spanning the phyla, or whether the subcellular targets in P-I, M-I, M-II and G-I mitosis arrested oocytes are different.

The Cell Response—Meiotic Blocks

Meiosis differs from mitosis in terms of checkpoint controls, DNA replication, dependency on external stimuli, and regulation of cell cycle control proteins (Fulka et al. 1994; Page, Orr-Weaver, 1997). Checkpoint controls are mechanisms that assure that one process is completed before another starts (Hartwell, Weinhart 1989; Murray, Kirschner 1989). During meiotic division, it is essential to complete recombination before the beginning of cell division to obtain a correct segregation of homologous chromosomes. Recently, several genes have been identified in yeast that are responsible for the meiotic block when double strand DNA breaks are not repaired (Lydall, Weinert 1996). One of these checkpoint genes is homologous to the mammalian *ATM* gene (coding for human ataxia telangiectasia mutated gene). Disruption of this

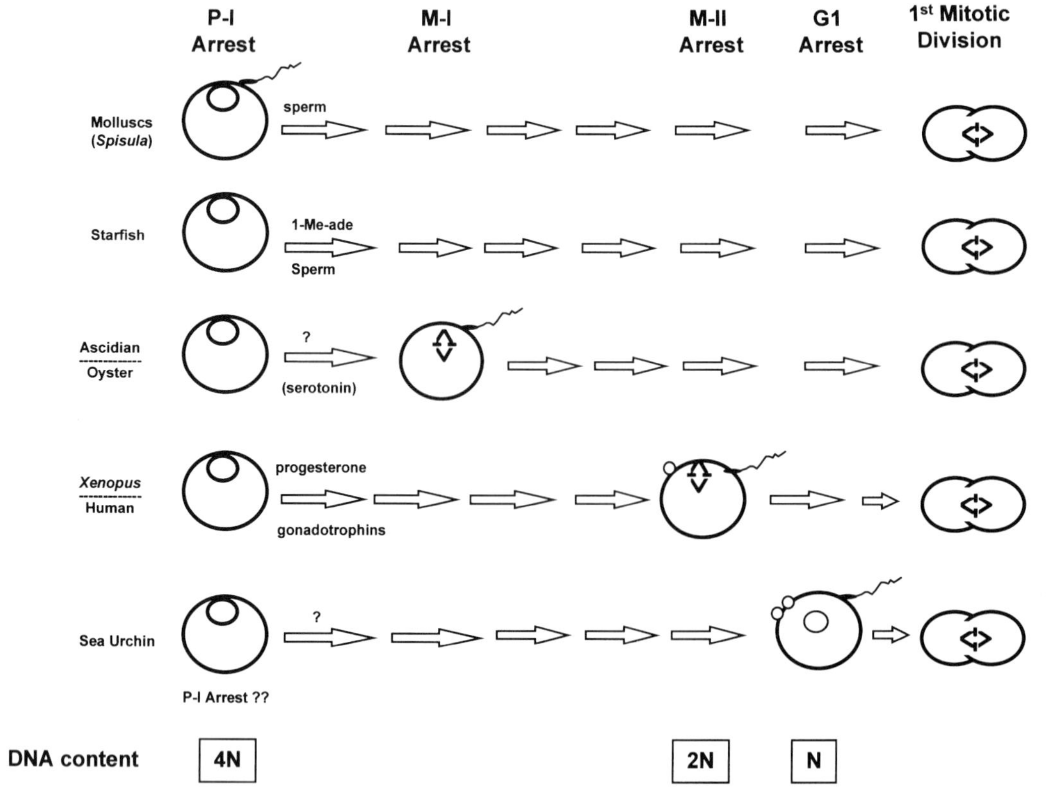

Figure 1. Oocyte meiotic arrest in different species. Generally, there are two meiotic blocks: P-I and a late block at M-I or M-II. Molluscs and echinoderms are exceptions.

gene in the mouse causes sterility and cytological defects during spermatogenesis (Xu et al. 1996). A different checkpoint specific to meiotic cells ensures that anaphase I does not begin until paired chromosomes are correctly attached to the spindle. This control resembles the spindle-assembly checkpoint of mitotic cells (Page, Orr-Weaver 1997). Micromanipulation experiments in spermatocytes show that tension on the spindle generated by attached homologs is a checkpoint. In the experimentally-generated absence of this tension, anaphase is prevented (Li, Nicklas 1995). Mice spermatocytes deficient in the *Mlh1* gene, a DNA mismatch repair gene, are unable to form bivalents and arrest in M-I, due to a defect in this tension-sensitive checkpoint (Baker et al. 1996).

One of the main differences between the mitotic and meiotic cell cycle is the ability in the latter to block the oocyte at precise phases of the cell cycle until a specific stimulus (for example, a hormone or sperm) removes the block. In somatic cells, a state of quiescence or cell cycle block in response to a specific physiological state of the cell is described as the G-0 phase of the cell cycle. However, G-0 differs with respect to the meiotic blocks in terms of cell cycle regulation and the activity of the key kinases that maintain the arrest. Two protein complexes, maturation promoting factor and cytostatic factor, regulate progression through meiosis

The Cell Motors—MPF and CSF

Maturation Promoting Factor (MPF) was discovered in the early '70s (Masui, Markert 1971) by cytoplasmic transfer experiments in amphibian oocytes. The two main components of MPF—a catalytic subunit, the serine/threonine (Ser/Thr) kinase Cdk1 (or Cdc2, or p34^{cdc2}), and a regulatory subunit, cyclin B—have been identified. MPF is recognized to be the universal factor that drives cells (somatic and germinal) into mitosis, from yeast to humans (Nurse 1990). Both components, Cdk1 and cyclin B, are functionally and structurally conserved through evolution. Cdk1 activity is regulated by phosphorylation and dephosphorylation on specific residues, as well as physical binding with cyclins (B and A types). Cyclin B is regulated by its synthesis and degradation, and possibly by phosphorylation (Solomon 1993). MPF activity (measured as the ability of Cdk1 to phosphorylate *in vitro* the specific substrate histone H1 or derived synthetic peptides) is generally maintained at low levels in P-I oocytes. It increases after germinal vesicle breakdown (GVBD) and reaches its maximal levels during M-I and M-II. The activity of MPF decreases in the lag period between M-I and M-II, and at exit from M-II. The majority of data concerning MPF regulation during fertilization is from the amphibians *Rana* and *Xenopus*. However, the same pattern of activation has been described in all vertebrates and invertebrates studied so far (Sagata 1996, 1997). In ascidian oocytes, Cdk1-like activity, as well as cyclin B-like synthesis and degradation, has been reported (Russo et al. 1996). MPF activity oscillates between two maximal levels at M-I and M-II, and two very low levels, at the metaphase-anaphase transition of both meiotic divisions. Fertilized ascidian oocytes show periodic calcium waves that are strictly associated with the MPF oscillations (Fig. 2; Russo et al. 1996).

It is thought that hormonal stimuli increase MPF activity at the exit of the first meiotic block, and that it is maintained at a high level during the second meiotic block by a second factor, Cytostatic Factor (CSF). CSF was also discovered by Masui and Markert (1971) in a bioassay where cytoplasm, which was taken from an unfertilized amphibian oocyte and injected into a blastomere of a two-cell embryo, caused metaphase arrest of the cell. Although the biological activity of CSF has been largely characterized, its biochemical components and its mechanism of action remain elusive (Sagata 1996, 1997; Masui 1991). It is generally accepted that three of the main components of CSF are: the product of c-*mos* oncogene (Sagata et al. 1989), Mitogen Associated Protein Kinase, MAPK (Nebrada, Hunt 1993; Haccard et al. 1993), and possibly Cdk2 kinase (Gabrielli et al. 1993), although the role of the latter is still a matter for conjecture (Furono et al. 1997). Mos is a Ser/Thr kinase originally identified as the transforming component of Moloney murine sarcoma virus (Oskarsson et al. 1980) and specifically expressed in the ovary and testis (Propst et al. 1988). In *Xenopus* oocytes, Mos mRNA increases at the time of GVBD, and remains high until M-II (Sagata 1996, 1997). Its rate of translation, as well as that of cyclin B, is regulated by polyadenylation in response to progesterone (Sheets et al. 1995). Activation by sperm, or parthenogenetic agents, induces CSF inactivation and Mos degradation. In this respect, Mos has two essential roles: during the first meiotic division, in *Xenopus*, it is responsible for the response to progesterone, activation of MPF and MAPK and the suppression of DNA synthesis between the two meiotic divisions (Furono et al. 1994). In meiosis II, Mos acts prevalently in maintaining MPF at a high level of activity, avoiding its inactivation. Synthesis of Mos leads to MAPK activation, via its regulator MAPK kinase (MEK), that is activated by Mos phosphorylation (Posada et al. 1993). In *Xenopus* oocytes, MAPK activity remains high while Mos is present and the activity decreases when activation occurs (Maro et al. 1994) and CSF degradation begins (Watanabe et al. 1989; Weber et al. 1991). The role of MAPK in meiosis and the specific substrates phosphorylated are still unknown. Certainly, MAPK contributes to maintain chromosome condensation and formation of the meiotic

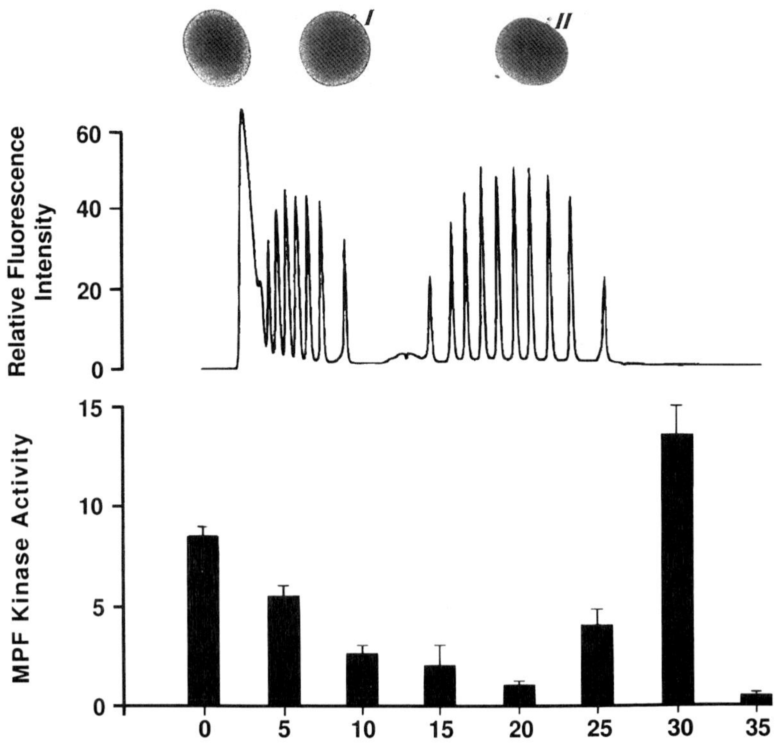

Figure 2. Relationship between intracellular calcium, MPF activity and polar body extrusion in *Ciona intestinalis* oocytes at fertilization and progression through meiosis. The top panel shows the morphological changes during meiosis: contraction, extrusion of first (*I*) and second (*II*) polar body. The trace in the middle panel shows the relative fluorescence intensity in an oocyte loaded with Calcium Green-dextran, while the lower bar diagram shows the Cdk1 kinase activity.

spindle (Verlhac et al. 1994; Choi et al. 1996) and mediates a spindle assembly checkpoint (Minshull et al. 1994). However, several experiments suggest that MAPK is an essential component of CSF in meiosis II, but it is not absolutely necessary in triggering meiosis I (Sagata 1996).

How do CSF and MPF interact in regulating meiosis? Probably Mos maintains the metaphase block and consequently high levels of MPF by two main mechanisms (Sagata 1996, 1997; Murakami, Van Woude 1997). Mos might directly or indirectly activate the pre-MPF pool via activation of Cdc25 (the phosphatase responsible for Cdk1 activation by dephosphorylation of residues Tyr15/Thr14), or inactivation of Wee1, the kinase that phosphorylates and inactivates Cdk1 at the same residues (Freeman et al. 1991; King et al. 1994). Alternatively, Mos may act via the MAPK pathway to inhibit negative regulators of MPF that at the moment are not known (Sagata 1996, 1997).

The Messenger—Intracellular Calcium

Calcium release mechanisms may be divided into three categories, depending on the type of receptor located on the intracellular calcium store. The categories are, first, inositol 1,4,5-trisphosphate (IP_3)-induced calcium release (IICR); second, calcium-induced calcium release (CICR); and third, $NAADP^+$-induced calcium release. Although distinct, these calcium release mechanisms often interact (Fig. 3). The latter pathway has been discovered only recently and is not well defined; in fact, it is highly probable that other pathways related to this family will emerge. IICR is triggered by the binding of IP_3 to its receptor on the endoplasmic reticulum (Terasaki, Sardet 1991). IP_3 is produced by the action of phospholipase C (PLC) on the plasma membrane lipid phosphatidylinositol bisphosphate (PIP_2), as shown in Berridge (1993). CICR is triggered by the opening of the ryanodine receptor on an intracellular store, but can also be triggered in a mechanism involving the IP_3 receptor (Endo 1977; Berridge 1996). This can be triggered by calcium itself, and appears to be modulated by cyclic ADP ribose (Galione, White 1994). Cyclic ADP ribose is in turn produced by metabolism of nicotinamide adenine disphosphate (NAD^+) by ADP ribosyl cyclase or NAD^+ glycohydrolase (Galione, White 1994; Lee et al. 1995; Jacobson et al. 1995). More recently, other calcium-releasing second messengers have been discovered, including cATP ribose and $NAADP^+$ (Lee, Aarhus 1995; Genazzani, Galione 1996; Zhang et al. 1996). Since NAD^+, NADH, $NADP^+$ and NADPH can be metabolized to calcium-releasing second messengers in sea urchin microsomes (Clapper et al. 1987), other calcium-releasing second messengers may be discovered in the nicotinamide nucleotide family.

How are these second messengers formed? This subject has been studied mainly in somatic cells and whether it has any relevance to fertilization is unclear. IP_3 formation is well characterized in somatic cells. It appears to occur by two mechanisms. The first involves a receptor-coupled G protein which stimulates the activity of PLC on the plasma membrane, leading to PIP_2 break-

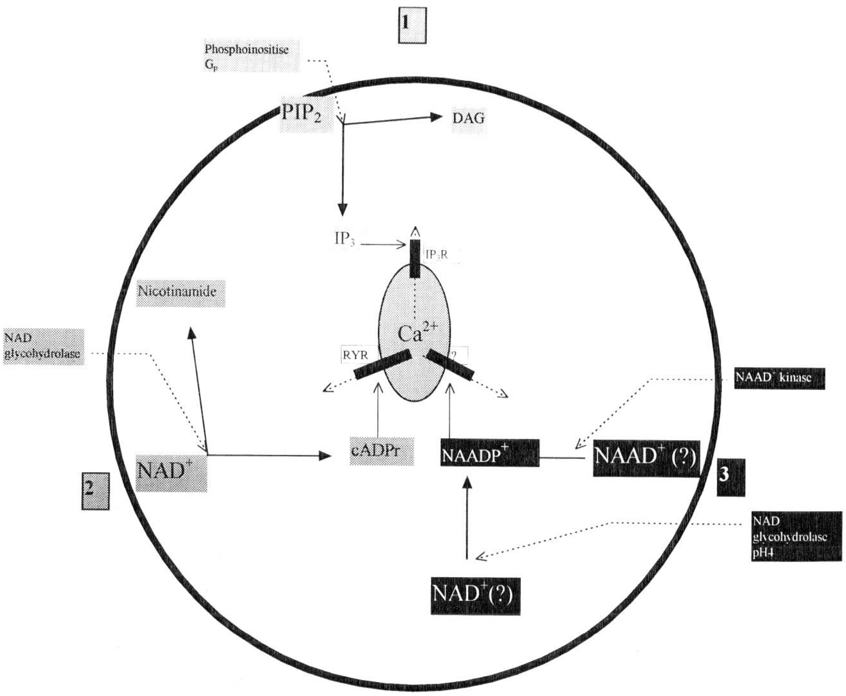

Figure 3. The three categories of intracellular calcium release mechanisms in oocytes. 1) IP_3 -induced calcium release involves the activation of a phosphoinositidase in the plasma membrane, leading to the breakdown of phospholipase biphosphate (PIP_2) and the release of inositol triphosphate (IP_3) and diacylglycerol (DAG). IP_3 binds to its receptor on the endoplasmic reticulum (ER), leading to intracellular calcium release. 2) Calcium-induced calcium release pathway involving the ryanodine receptor. It is thought that cADPr, produced through the breakdown of NAD^+ to cADPr and nicotinamide, acts as a mediator of this channel activity. 3) $NAADP^+$ -induced calcium release. This is a recently discovered pathway and therefore is not totally established, nor does it preclude the presence of other pathways.

down (Berridge 1993). The second involves a receptor tyrosine kinase, with the same result (Berridge 1993). The metabolic pathway leading to the stimulation of cyclic ADP ribose formation is not well established at present. In sea urchins, cyclic ADP ribose formation is stimulated by cGMP (Galione et al. 1993a). This suggests a role for the second messenger nitric oxide, which stimulates guanylate cyclase activity (Bredt, Snyder 1994).

The mode of calcium release at fertilization varies from species to species. However, there are common threads. Both IICR and CICR have been demonstrated in many types of oocyte (Galione et al. 1993b; Whitaker, Swann 1993). However, this does not necessarily mean all oocytes contain both IP_3 and ryanodine receptors, since the IP_3 receptor can support CICR (Galione et al. 1993b; Whitaker, Swann 1993). In sea urchins, both IICR and CICR are triggered at fertilization (Whitaker, Swann 1993). This appears to be due to stimulation of two calcium release pathways by the sperm (Galione et al. 1993b; Lee et al. 1993). In frog oocytes, IICR appears to be uniquely activated at fertilization (Whitaker, Swann 1993); however, what appears to be IICR may in fact be CICR through the IP_3 receptor (Galione et al. 1993b; Whitaker, Swann 1993). The urochordate *Ciona intestinalis* also releases calcium by both IICR and CICR at fertilization, and also generates repetitive calcium transients through meiosis I and II (Speksnijder et al. 1990; McDougall, Sardet 1995; Russo et al. 1996). These transients appear to be triggered by an IP_3-dependent mechanism because they are blocked by heparin (McDougall, Sardet 1995; Russo et al. 1996). In mammalian oocytes, at fertilization there is a large increase in the sensitivity to CICR, together with a series of repetitive calcium spikes (Igusa, Miyazaki 1983; Miyazaki 1988; Cuthbertson, Cobbold 1985; Kline, Kline 1992; Taylor et al. 1993). This again suggests activation of both CICR and IICR at fertilization. The mechanism of repetitive calcium spiking in mammalian oocytes is not clear at present. Mammalian oocytes contain both ryanodine and IP_3 receptors (Rickfords, White 1993; Miyazaki et al. 1992; Yue et al. 1995; Ayabe et al. 1995; Sousa et al. 1996a). However, it is not yet certain whether repetitive calcium transients in mammalian oocytes are propagated by IICR or CICR (Miyazaki et al. 1992, 1993; Carrol, Swann 1992; Swann 1992; Kline, Kline 1994; Fissore et al. 1995; Tesarik, Sousa 1996; Berridge 1996; Tesarik et al. 1995). Sulphydryl reagents regulate the calcium dependent cortical exocytosis in sea urchin oocytes (Dale, Russo 1988); this was later confirmed for mammalian oocytes (Swann 1991). This may involve protein kinase C (Sousa et al. 1996b).

The Primary Trigger: Soluble Sperm Factors

The idea that cortical granule exocytosis and oocyte activation are triggered by a component of the spermatozoon that enters the oocyte cytoplasm was first presented by Robertson (1912) and supported by Loeb (1913). Subsequent models emphasized the effect of the spermatozoon on the external surface of the oocyte membrane (Lillie, Baskerville 1922; Mazia et al. 1975). In the early 1980s two major hypotheses prevailed.

The G-Protein Hypothesis

The G-protein hypothesis for oocyte activation was extrapolated from what was known about the calcium response to hormones in somatic cells. Here, hormone-receptor binding on the outer surface of the plasma membrane signals through a G-protein in the plasma membrane. This signal triggers the activation of PLC, leading to the formation of IP_3 and hence calcium release (Berridge 1993). In the G-protein model of oocyte activation, the sperm behaves as an "honorary hormone," which means that the attachment of sperm to its receptor triggers IP_3 formation through a G-protein linked to this receptor (Kline et al. 1988; Foltz, Schilling 1993). There is limited evidence supporting this hypothesis. First, antisera generated against the sperm receptor on the oocyte plasma membrane or peptides from sperm were found to trigger oocyte activation when added to the outside of oocyte membranes (Perlmann 1954; Gould, Stephano 1987, 1991; Foltz, Lennarz 1992; Moore et al. 1993). Second, receptors expressed in oocyte plasma membranes by microinjection of mRNA can activate various oocytes on application of their respective ligands to the plasma membrane. This demonstrates that receptor-mediated oocyte activation pathways do exist in oocytes (Kline et al. 1988, 1991; Shilling et al. 1991, 1993). Furthermore, *in vitro* G-protein activators such as GTP-γ-S and cholera toxin trigger intracellular calcium release in oocytes with a short delay which parallels the latent period seen after sperm/oocyte fusion (Turner, Jaffe 1989; Crossley et al. 1991). Arguing against this hypothesis is the fact that inhibitors of G-proteins do not block the fertilization calcium wave in sea urchins (Crossley et al. 1991). In summary, although the above data show that the pathway for G-protein-induced oocyte activation does exist, it does not appear to be the physiological pathway to sperm-induced oocyte activation.

The Soluble Sperm Factor Hypothesis

The soluble sperm factor hypothesis is based on a diffusible messenger (or messengers) in the cytoplasm of spermatozoa that enters the oocyte cytoplasm after sperm/oocyte fusion and triggers intracellular calcium release. Evidence for this hypothesis stems from several indirect and direct experiments. Sperm are known to establish cytoplasmic continuity with oocytes several seconds before oocyte activation (Dale, Santella 1985; McCulloh, Chambers 1987). Using immature germinal vesicle stage oocytes, it was shown that only spermatozoa that fused with the oocyte plasma membrane were capable of inducing a fertilization cone (Fig. 4). Immature sea urchin oocytes do not amplify the primary trigger signal from the spermatozoon and thus are useful tools to distinguish primary from secondary activation events. This cytoplasmic protrusion is the first morphological indication of the effect of the spermatozoon on the oocyte cytoplasm. The period between gamete fusion and activation is called the latent period (Ginsberg 1988). The latent period between sperm/oocyte fusion and oocyte activation may allow the diffusion of a messenger between the cytoplasm of the sperm and that of the oocyte (Dale, Santella 1985).

The first direct evidence for a soluble sperm factor was shown by microinjecting the soluble components from spermatozoa into sea urchin and ascidian oocytes (Dale et al. 1985; Dale 1988). Several activation events, including cortical granule exocytosis and gating of plasma membrane currents, were triggered (Fig. 5). This was repeated in mammals, with the same conclusion (Stice, Robl 1990; Swann 1990; Dale et al. 1996).

In the human, during intracytoplasmic sperm injection (ICSI), intracellular microinjection of whole spermatozoa into the cytoplasm also triggers oocyte activation after several hours' delay. This has been suggested to be the time required for sperm plasma membrane breakdown and the release of the cytoplasmic contents of the sperm (Tesarik et al. 1994). Sperm extracts also trigger repetitive calcium transients in mammalian, nemertean and ascidian oocytes which closely mimic the pattern of calcium release seen at fertilization (Miyazaki et al. 1993; Homa, Swann 1994; Stricker 1997; Wilding et al. 1997; Wu et al. 1998). It therefore seems probable that a soluble component of sperm cytoplasm is responsible for the calcium increase that activates oocytes at fertilization.

What Is Sperm Factor?

Spermatozoa are known to contain many molecules capable of releasing intracellular calcium, including cGMP, IP_3, nicotinamide nucleotide metabolites, calcium ions and so on (Whitaker, Crossley 1990; Iwasa et al. 1990; Tosti et al. 1993). More recently, a protein from sperm that triggers repetitive calcium release in hamster oocytes has been sequenced (Parrington et al. 1996) and called oscillin. The supporters of the Oscillin Sperm Factor suggest that this protein is specific to sperm cells (Parrington et al. 1996). It has been suggested that in the mouse a truncated c-kit tyrosine kinase localized in the sperm cytoplasm is the physiological activating factor (Sette et al. 1997).

Soluble extracts of human spermatozoa can activate oocytes from different phyla as well as different species (Homa, Swann 1994; Wilding et al. 1997). Sperm extracts can also trigger calcium oscillations in somatic cells (Currie et al. 1992; Berrie et al. 1996), suggesting that sperm factors are common calcium releasing agents and are not sperm-specific molecules. Furthermore, recent data from several sources point to the role of the

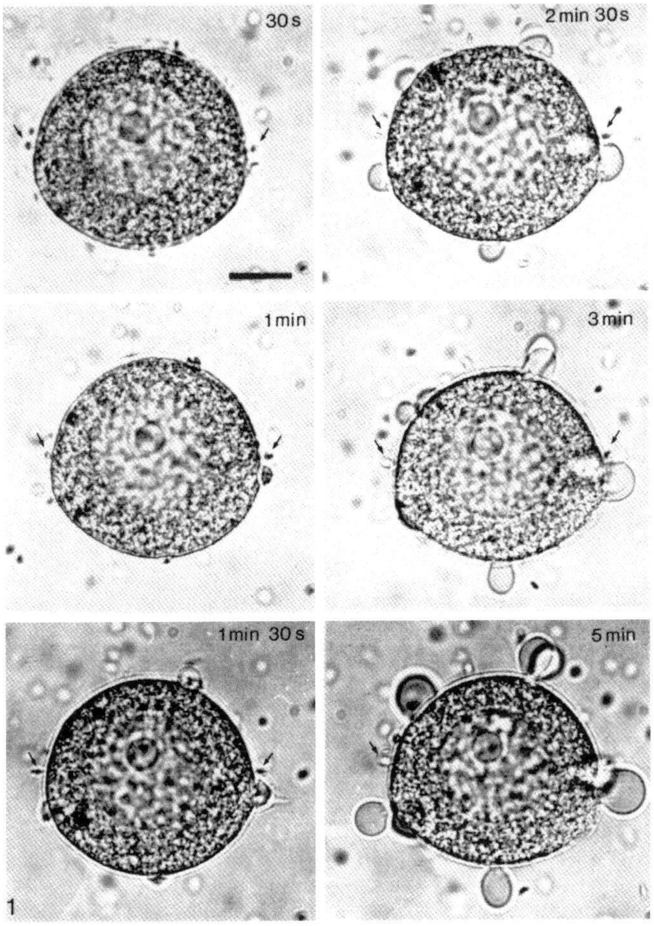

Figure 4. The time course of insemination and activation of immature germinal vesicle stage sea urchin oocytes. Of the many spermatozoa that attach to the surface, only those that fuse (as shown by an increase in electrical conductance) cause a fertilization cone. Many spermatozoa (arrows), although irreversibly attached to the surface, do not fuse or induce fertilization cone formation. The fertilization cone is a localized cytoplasmic disturbance probably caused by the localized diffusion of soluble sperm factor.

sperm as a stimulator of at least two metabolic calcium-releasing systems within the oocyte (Wilding et al. 1997; Osawa 1994; Ciapa, Epel 1996). Unless a common activation pathway can be found for IICR and CICR, it seems probable that the sperm factor is a complex of at least two molecules.

So, the current open debate is whether the sperm factor is a unique sperm protein (Parrington et al. 1996; Sette et al. 1997) or a collection of second messengers that are found in many cell types but are packaged and used differently in spermatozoa (Wilding, Dale 1997; Russo et al. 1998). Data from the Dale laboratory indicate that soluble sperm extracts contain small, temperature-resistant and non-proteic factor(s) able to induce $Ca^{2+}{}_i$ oscillations in ascidian oocytes (Wilding et al. 1997).

Completing the Puzzle—Models for Oocyte Activation

Despite the controversy over the nature of sperm factor, mechanisms regulating activation events are better known and may provide clues to the possible properties of this factor. Fertilization is accompanied by an increase in cytosolic calcium. In mammalian and ascidian oocytes, calcium pulses are essential for initiation and normal embryonic development (McDougall, Sardet 1995; Miyazaki et al. 1993; Homa 1995) and are triggered by the sperm (Cuthbertson, Cobbold 1985). In *Xenopus*, the fertilization calcium transients activate calmodulin (CaM) kinase II, which triggers cyclin B destruction and Mos degradation (Lorca et al. 1991, 1993). However, several aspects of this pathway need to be clarified (Watanabe et al. 1991). In invertebrate oocytes, the role of Ca^{2+}/CaM in M-I exit is not clear, although several lines of evidence indicate a similar mechanism triggering cyclin degradation machinery (Abdelmajid et al. 1993). In many species, the metabolism of nicotinamide nucleotides leads to the production of the second messenger cyclic adenosine disphosphate ribose (cADPr). This molecule has been suggested to be responsible, at least in part, for the calcium release at fertilization (Galione 1993; Galione, White 1994). It has been found that in ascidian oocytes, the nicotinamide nucleotide metabolic pathway is active, although it works differently than in other systems. Here, the major cell mediator appears to be adenosine disphosphate ribose (ADPr). This molecule is responsible for the fertilization current observed in ascidians, which is triggered independently of calcium ions (Dale 1988). In addition, nitric oxide (NO) controls cADPr synthesis in sea urchin eggs via production of cGMP (Galione, White 1994; Galione et al. 1993a). NO is also released at fertilization in ascidian oocytes, and this triggers an inward current similar to the fertilization current; however, NO in *C. intestinalis* does not lead to cADPr production, but possibly ADPr (Grumetto et al. 1997). This information suggests that in ascidians the signal triggered by sperm at fertilization induces the activation of the NO/ADPr pathway.

Two major events at fertilization in ascidian oocytes are induction of calcium oscillations and inactivation of MPF (Russo et al. 1996; McDougall, Sardet 1995). Is there a functional correlation between these two events? Cdk1 activity is maximal at M-I and M-II, and decreases at exit from meiosis I and II (Fig. 2). A series of calcium oscillations occurs simultaneously with the decrease in Cdk1 activity at M-I exit, while a second group of $Ca^{2+}{}_i$ transients precedes the Cdk1 increase at M-II (Russo et al. 1996). A similar temporal correlation

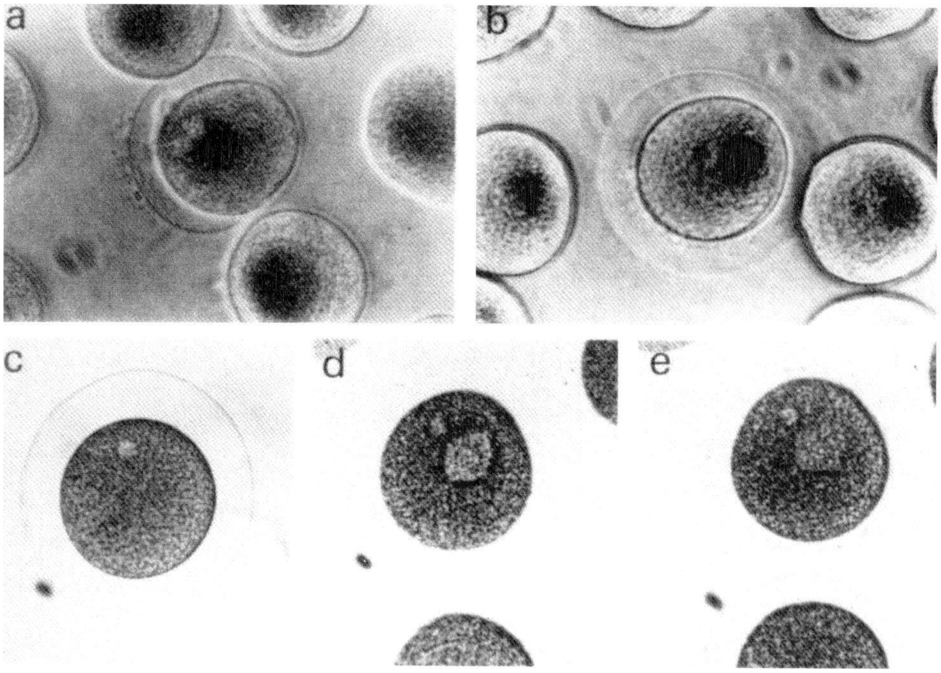

Figure 5. Microinjection of soluble sperm extracts into sea urchin oocytes. Homogenized sea urchin spermatozoa in 0.5M KCl containing EGTA pressure-injected into sea urchin oocytes induces the cortical reaction (a, b). A similar reaction may be induced by injecting hypo-osmotic medium (c), but not by iso-osmotic KCl (d). The oocyte in (d) was subsequently inseminated as a control to show that the cortical granules remained functional following microinjection.

between calcium oscillations and MPF activity has also been reported in the ascidian *Phallusia mammillata* (McDougall, Sardet 1995).

This calcium signaling system, however, does not appear to be the central cell cycle control mechanism during meiosis I. It has been shown that inactivation of the main cell cycle control protein, Cdk1, is $Ca^{2+}{}_i$ independent at this stage (Russo et al. 1996). However, oocytes do not extrude a polar body after fertilization in the presence of calcium chelators, suggesting that calcium is involved in the completion of meiosis I. This suggests that there are a number of signals involved during meiosis I progression. Among them, ADPr probably plays an important role in completion of meiosis I.

In fact, it has been found that nicotinamide, an inhibitor of nicotinamide nucleotide metabolism (Sethi et al. 1996), blocks the large, initial calcium transient triggered at fertilization, suggesting that nicotinamide nucleotides do not directly trigger $Ca^{2+}{}_i$ release, but enhance calcium influx by opening the ADPr channel. In addition, nicotinamide blocks MPF inactivation which normally occurs after fertilization, possibly maintaining high levels of Cdk1 kinase activity. As a consequence, oocytes do not progress from M-I and do not extrude the first polar body, although the calcium transient occurs. This further indicates that phase I calcium oscillations are not essential for MPF inactivation. Attempts to explain the role of nicotinamide on Cdk1 activity showed that unfertilized oocytes of *C. intestinalis* loaded with ADPr abolished their Cdk1-like kinase activity by 20-fold, as compared to that found in control oocytes. It appears that ADPr triggers the inactivation of MPF by inhibiting the kinase activity of the Cdk1-like/cyclin B complex previously characterized in *C. intestinalis* (Russo et al. 1996). Apparently ADPr does not act directly on the enzymatic complex, but exerts its function through a signal transduction pathway, confirming its role as a mediator.

In vertebrate oocytes, fertilization triggers CSF degradation (Watanabe et al. 1989; Weber et al. 1991). Also, MAPK activity decreases after meiosis completion (Maro et al. 1994). A CSF-like biological activity has been described in the ascidian *C. intestinalis* using the classical approach of microinjection into dividing blastomeres (Russo et al. 1998). Little is known about ascidian CSF components. Preliminary data from the Dale laboratory based on antibody cross-reactivity indicate the possible existence of a Mos-like factor. Finally, as expected, MAPK in ascidian *C. intestinalis* and *P. mammilata* decreases at M-I exit and stays low during the remaining part of meiosis completion (Russo et al. 1996). MAPK activity is low in M-II also, when Cdk1 kinase increases, suggesting that meiosis II progression does not require MAPK and that inhibition of DNA synthesis between the two meiotic divisions is not controlled by MAPK. In this respect, it seems that in ascidians MAPK is only required to maintain the M-I arrest, as in vertebrate it is required for the M-II block. How fertilization triggers MAPK inactivation in both systems is presently unknown.

Conclusions

Most oocytes are blocked twice in meiosis: first in prophase I, and later in metaphase I or metaphase II. Generally hormones trigger the release from the first block; spermatozoa from the second block. Two protein complexes, maturation promoting factor (MPF) and cytostatic factor (CSF), regulate progression through meiosis. Generally, MPF activity increases after germinal vesicle breakdown and peaks in metaphase. CSF is thought to maintain the high activity of MPF in metaphase. In M-II arrested oocytes, the spermatozoon induces CSF inactivation and MPF inactivation. In M-I arrested oocytes, the spermatozoon also triggers MPF inactivation; however the role of CSF is still unknown. The major intracellular messenger triggering the activity of these protein complexes is calcium. There are three calcium release mechanisms in oocytes. IP_3-induced calcium release (IICR), calcium-induced calcium release modulated by cyclic ADPribose (cADPr), and $NAADP^+$-induced calcium release. Spermatozoa establish cytoplasmic continuity with oocytes several seconds before oocyte activation, during which time soluble sperm activating factors may diffuse into the oocyte. Spermatozoa contain a variety of molecules that are capable of releasing intracellular calcium. The present debate is whether sperm factor is a unique sperm protein or a collection of second messengers found ubiquitously in cells, but packaged and used differently in spermatozoa. Since there is evidence that spermatozoa triggers at least two independent calcium-releasing systems within the oocyte (Fig. 6), and that a sperm factor triggers calcium release in somatic cells, it seems probable that the sperm factor is a complex of at least two common calcium-releasing molecules.

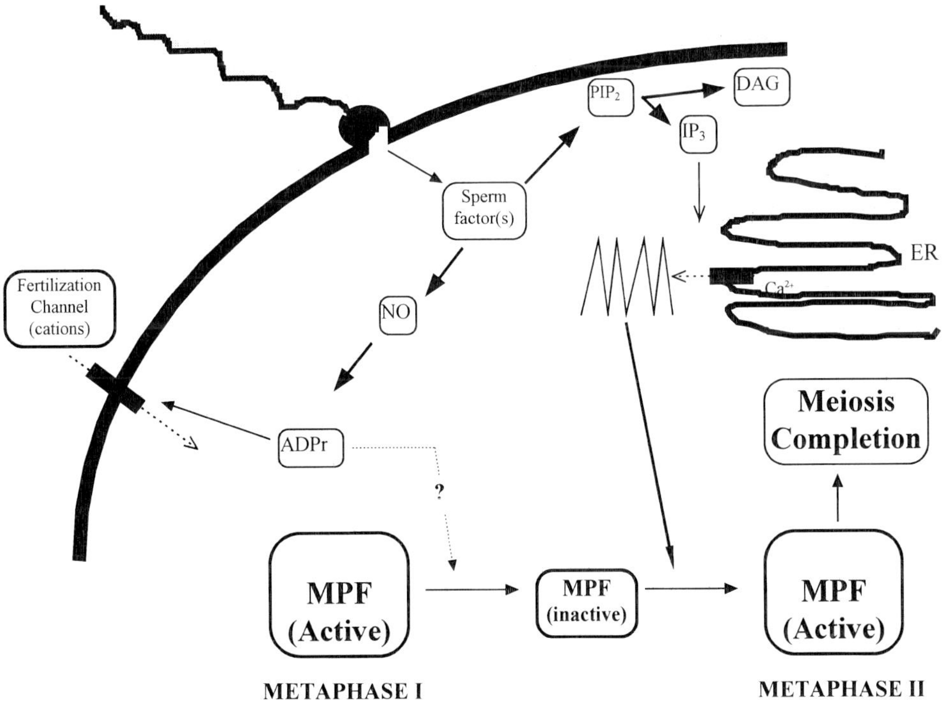

Figure 6. A cartoon summarizing activation events in the ascidian oocyte. The fertilizing spermatozoon fuses to the plasma membrane and releases sperm factors into the oocyte. These factor(s) stimulate both the production of IP_3 and ADPr, the latter probably through the production of NO. ADPr gates the fertilization channels, and may be involved in the inactivation of MPF at meiosis I release. IP_3 gates the release of intracellular calcium, which is required for the activation of MPF and the completion of meiosis II.

References

Abdelmajid H, Leclerc-David C, Moreau M, Guerrier P, Ryazanov A. Release from the metaphase I block in invertebrate oocytes: possible involvement of Ca^{2+}/calmodulin-dependent kinase III. Int J Dev Biol 1993; 37:279-290.

Ayabe T, Kopf G, Schultz R. Regulation of mouse egg activation: presence of ryanodine receptors and effects of microinjected ryanodine and cyclic ADP ribose on uninseminated and inseminated eggs. Development 1995; 121:2233-2244.

Baker SM, Plug AW, Prolla TA, Bronner CE, Harris AC, Yao X, Christie D-M, Monell C, Arnheim N, Bradley A, Ashley T, Liskay RM. Involvement of mouse Mlh1 in DNA mismatch repair and meiotic crossing over. Nature Genet 1996; 13:336-342.

Berridge M. Inositol trisphosphate and cell signalling. Nature 1993; 361:315-325.

Berridge M. Regulation of calcium spiking in mammalian oocytes through a combination of inositol trisphosphate-dependent entry and calcium release. Mol Hum Reprod 1996; 2:386-388.

Berrie C, Cuthbertson K, Parrington J, Lai F, Swann K. A cytosolic sperm factor triggers calcium oscillations in rat hepatocytes. Biochem J 1996; 313:369-372.

Bredt D, Snyder S. Nitric oxide: a physiologic messenger molecule. Ann Rev Biochem 1994; 63:175-195.

Carroll J, Swann K. Spontaneous cytosolic calcium oscillations driven by inositol trisphosphate occur during *in vitro* maturation of mouse oocytes. J Biol Chem 1992; 267:11196-11201.

Choi T, Fukasawa K, Zhou R, Tessarollo L, Borror K, Resau J, Vande Woude GF. The Mos/mitogen-activate protein kinase (MAPK) pathway regulates the size and degradation of the first polar body in maturing mouse oocytes. Proc Natl Acad Sci USA 1996; 93:7032-7035.

Ciapa B, Epel D. An early increase in cGMP follows fertilisation of sea urchin eggs. Biochem Biophys Res Comm 1996; 25:633-636.

Clapper D, Walseth T, Dargie P, Lee H. Pyridine nucleotide metabolites stimulate calcium release from sea urchin egg microsomes sensitized to inositol trisphosphate. J Biol Chem 1987; 262:9561-9568.

Crossley I, Whalley T, Whitaker M. Guanosine 5'-thiotriphosphate may stimulate phosphoinositide messenger production in sea urchin eggs by a different route than the fertilising sperm. Cell Reg 1991; 2:121-133.

Currie K, Swann K, Galione A, Scott R. Activation of calcium-dependent currents in cultured dorsal root ganglion neurons by a sperm factor and cyclic ADP ribose. Mol Biol Cell 1992; 3:1415-1425.

Cuthbertson K, Cobbold P. Phorbol ester and sperm activate mouse oocytes by inducing sustained oscillations in cell calcium. Nature (Lond) 1985; 316:541-542.

Dale B. Fertilization in Animals. Edward Arnold, London, 1983.

Dale B. Primary and secondary messengers in the activation of ascidian eggs. Exp Cell Res 1988; 177:205-211.

Dale B, Russo P. Sulfhydryl groups are involved in the activation of sea urchin eggs. Gam Res 1988; 19:161-168.

Dale B, Santella L. Sperm-oocyte interaction in the sea urchin. J Cell Sci 1985; 74:153-167.

Dale B, DeFelice L, Ehrenstein G. Injection of a soluble sperm extract into sea urchin eggs triggers the cortical reaction. Experientia 1985; 41:1068-1070.

Dale B, Fortunato A, Monfrecola V, Tosti E. A soluble sperm factor gates calcium-activated potassium channels in human oocytes. J Assist Reprod Gen 1996; 13:573-577.

Endo M. Calcium release from the sarcoplasmic reticulum. Phys Rev 1977; 57:71-108.

Fissore R, Pintocorreia C, Robl J. Inositol trisphosphate-induced calcium release in the generation of calcium oscillations in bovine eggs. Biol Reprod 1995; 53:766-774.

Foltz K, Lennarz W. Identification of the sea urchin egg receptor for sperm using an antiserum raised against its extracellular domain. J Cell Biol 1992; 116:647-658.

Foltz K, Schilling F. Receptor-mediated signal transduction and egg activation. Zygote 1993; 1:273-279.

Freeman RS, Ballantyne SM, Donoghue DJ. Meiotic induction by *Xenopus* cyclin B is accelerated by co-expression with mos^{xe}. Mol Cell Biol 1991; 11:1713-1717.

Fulka J Jr, Bradshaw J, Moor R. Meiotic cycle checkpoints in mammalian oocytes. Zygote 1994; 2:351-354.

Furuno N, Nishizawa M, Okazaki K, Tanaka H, Iwashita J, Nakajo N, Ogawa Y, Sagata N. Suppression of DNA replication via Mos function during meiotic divisions in *Xenopus* oocytes. EMBO J 1994; 13:2399-2410.

Furuno N, Ogawa Y, Iwasita J, Nakajo N, Sagata N. Meiotic cell cycle in *Xenopus* is independent of cdk2 kinase. EMBO J 1997; 16:3860-3865.

Gabrielli BG, Roy LM, Maller JL. Requirement for Cdk2 in cytostatic factor-mediated metaphase II arrest. Science 1993; 259:1766-1769.

Galione A. Cyclic ADP-ribose: a new way to control calcium. Science 1993; 259:325-326.

Galione A, White A. Calcium release induced by cyclic ADP ribose. Trends Cell Biol 1994; 4:431-436.

Galione A, White A, Wilmott N, Turner M, Potter B, Watson S. cGMP mobilises intracellular calcium in sea urchin eggs by stimulating cyclic ADP ribose synthesis. Nature 1993a; 365:456-459.

Galione A, McDougall A, Busa W, Wilmott N, Gillot I, Whitaker M. Redundant mechanisms of calcium-induced calcium release underlying calcium waves during fertilisation of sea urchin eggs. Science 1993b; 261:348-352.

Genazzani A, Galione A. Nicotinic acid-adenine dinucleotide phosphate mobilises calcium from a thapsigargin-insensitive pool. Biochem J 1996; 315:721-725.

Ginsberg A. Egg cortical reaction during fertilisation and its role in block to polyspermy. Sov Sci Rev Fert Physiol Gen Biol 1988; 1:307-375.

Gould M, Stephano J. Electrical responses of eggs to acrosomal protein similar to those induced by sperm. Science 1987; 235:1654-1656.

Gould M, Stephano J. Peptides from sperm acrosomal protein that initiate development. Dev Biol 1991; 146:509-518.

Grumetto L, Wilding M, DeSimone ML, Tosti E, Galione A, Dale B. Nitric oxide gates fertilization channels in ascidian oocytes through nicotinamide nucleotide metabolism. Biochem Biophys Res Comm 1997; 239:723-728.

Haccard O, Sarcevic B, Lewellyn A, Hartley R, Roy L, Izumi T, Erikson E, Maller JL. Induction of metaphase arrest in cleaving *Xenopus* embryos by MAP kinase. Science 1993; 262:1262-1265.

Hartwell LH, Weinert TA. Checkpoints: controls that ensure the order of cell cycle events. Science 1989; 246:629-634.

Homa S, Swann K. A cytosolic sperm factor triggers calcium oscillations and membrane hyperpolarisations in human oocytes. Hum Reprod 1994; 9:2356-2361.

Homa ST. Calcium and meiotic maturation of the mammalian oocyte. Mol Reprod Dev 1995; 40:122-134.

Igusa Y, Miyazaki S. Effects of altered extracellular and intracellular calcium concentration on hyperpolarising responses of hamster egg. J Physiol (Lond) 1983; 340:611-632.

Iwasa K, Ehrenstein G, DeFelice L, Russell J. High concentrations of inositol 1,4,5-trisphosphate in sea urchin sperm. Biochem Biophys Res Comm 1990; 172:932-938.

Jacobson M, Cervantes-Laurean D, Strohm M, Coyle D, Bummer P, Jacobson E. NAD glycohydrolases and the metabolism of cyclic ADP ribose. Biochemie 1995; 77:341-344.

King RW, Jackson PK, Kirschner MW. Mitosis in transition. Cell 1994; 79:563-571.

Kline D, Kline J. Repetitive calcium transients and the 145 role of calcium in exocytosis and cell cycle activation in the mouse egg. Dev Biol 1992; 145:80-89.

Kline D, Simoncini L, Mandel G, Maue R, Kado R, Jaffe L. Fertilisation events induced by neurotransmitters after injection of mRNA in *Xenopus* oocytes. Science 1988; 241:464-467.

Kline D, Kopf G, Muncy L, Jaffe L. Evidence for the involvement of a pertussis toxin-insensitive G-protein in egg activation of the frog *Xenopus laevis*. Dev Biol 1991; 143:218-229.

Kline J, Kline D. Regulation of intracellular calcium in the mouse egg: evidence for inositol trisphosphate-induced calcium release, but not calcium-induced calcium release. Biol Reprod 1994; 50:193-203.

Lee H, Aarhus R. A derivative of NADP mobilises calcium stores insensitive to inositol trisphosphate and cyclic ADP ribose. J Biol Chem 1995; 270:2152-2157.

Lee H, Graeff R, Walseth T. Cyclic ADP ribose and its metabolic enzymes. Biochemie 1995; 77:345-355.

Lee HC, Aarhus R, Walseth TF. Calcium mobilization by dual receptors during fertilization of sea urchin eggs. Science 1993; 261:352-355.

Li X, Nicklas RB. Mitotic forces control a cell-cycle checkpoint. Nature 1995; 373:630-632.

Lillie RS, Baskerville M. The action of ultraviolet rays on Arbacia eggs, especially as affecting the response to hypertonic sea water.

Am J Physiol 1922; 61:272-288.
Loeb J. Artificial parthenogenesis and fertilization. University Press, Chicago, 1913.
Lorca T, Galas S, Fesquet D, Devault A, Cavadore J-C, Dorée M. Degradation of the proto-oncogene product p39*mos* is not necessary for cyclin proteolysis and exit from meiotic metaphase: requirement for a Ca^{2+}-calmodulin dependent event. EMBO J 1991; 10:2087-2093.
Lorca T, Cruzalegui FH, Fesquet D, Cavadore JC, Mery J, Means A, Dorée M. Calmodulin-dependent protein kinase II mediates inactivation of MPF and CSF upon fertilization of *Xenopus* eggs. Nature 1993; 366:270-273.
Lydall D, Weinert T. From DNA damage to cell cycle arrest and suicide: a budding yeast perspective. Curr Opin Genet Dev 1996; 6:4-11.
Maró B, Kubiak JZ, Verlhac M-H, Winston NJ. Interplay between the cell cycle control machinery and the microtubule network in mouse oocytes. Semin Dev Biol 1994; 5:191-198.
Masui Y. The role of 'Cytostatic Factor (CSF)' in the control of oocyte cell cycles: a summary of 20 years of study. Dev Growth Differ 1991; 33:543-551.
Masui Y, Markert CL. Cytoplasmic control of nuclear behavior during meiotic maturation of frog oocytes. J Exp Zool 1971; 177:129-146.
Mazia D, Schatten G, Steinhardt R. Turning on of activities in unfertilized sea urchin eggs: correlation with changes of the surface. Proc Natl Acad Sci USA 1975; 72:4469-4473.
McCulloh D, Chambers E. When does the sperm fuse with the egg? J Gen Physiol 1987; 88:384.
McDougall A, Sardet C. Function and characteristics of repetitive calcium waves associated with meiosis. Curr Biol 1995; 5:318-328.
Minshull J, Sun H, Tonks NK, Murray AW. A Map kinase dependent spindle assembly checkpoint in *Xenopus* egg extracts. Cell 1994; 79:475-486.
Miyazaki S. Inositol 1,4,5-trisphosphate-induced calcium release and guanine nucleotide binding protein-mediated periodic calcium rises in golden hamster eggs. J Cell Biol 1988; 106:345-353.
Miyazaki S, Yuzaki M, Nakada K, Shirawaka H, Nakanishi S, Nakade S, Mikoshiba K. Block of calcium wave and calcium oscillation by antibody to the inositol 1,4,5-trisphosphate receptor in fertilised hamster eggs. Science 1992; 257:251-255.
Miyazaki S, Shirakawa H, Nakada K, Honda Y. Essential role of the inositol 1,4,5-trisphosphate receptor calcium release channel in calcium waves and calcium oscillations at fertilisation in mammalian eggs. Dev Biol 1993; 158:62-78.
Moore G, Kopf GS, Schultz R. Complete mouse egg activation in the absence of sperm by stimulation of an exogenous G-protein coupled receptor. Dev Biol 1993; 159:669-678.
Murakami MS, Vande Woude GF. Mechanisms of *Xenopus* oocyte maturation. Meth Enz 1997; 283:584-614.
Murray AW, Kirschner MW. Dominoes and clocks: the union of two views of the cell cycle. Science 1989; 246:614-621.
Nebreda A, Hunt T. The c-*mos* proto-oncogene protein kinase turns on and maintains the activity of MAP kinase, but not MPF, in cell-free extracts of *Xenopus* oocytes. EMBO J 1993; 12:1979-1986.
Nurse P. Universal control mechanism regulating the onset of M-phase. Nature 1990; 344:503-508.
Osawa M. Soluble sperm extract triggers inositol 1,4,5-trisphosphate-induced calcium release in oocytes of the sea urchin *Anthocidaris crassispina*. Cell Struct Funct 1994; 19:73-80.
Oskarsson M, McClements WL, Blair DG, Maizel JV, Vande Woude GF. Properties of a normal mouse cell DNA sequence (sarc) homologous to the src sequence of Moloney sarcoma virus. Science 1980; 207:1222-1224.
Page AW, Orr-Weaver TL. Stopping and starting the meiotic cell cycle. Curr Opin Genet Dev 1997; 7:23-31.
Parrington J, Swann K, Shevchenko V, Sesay A, Lai F. Calcium oscillations in mammalian eggs triggered by a soluble sperm protein. Nature (Lond) 1996; 379:364-368.
Perlmann P. Study on the effect of antisera on unfertilised sea urchin eggs. Exp Cell Res 1954; 6:485-490.
Posada J, Yew N, Ahn NG, Vande Woude G, Cooper JA. Mos stimulates MAP kinase in *Xenopus* oocytes and activates a MAP kinase *in vitro*. Mol Cell Biol 1993; 13:2546-2553.
Propst F, Rosenberg MP, Vande Woude GF. Proto-oncogene expression in germ cell development. Trends Genet 1988; 4:183-187.
Rickfords L, White K. Electroporation of inositol 1,4,5-trisphosphate induces repetitive calcium oscillations in murine oocytes. J Exp Biol 1993; 265:178-184.
Robertson T. Studies on the fertilization of the eggs of a sea urchin *Strongylocentrotus purpuratus* by blood-sera, sperm, sperm extract and other fertilizing agents. Arch Entwick Lungs Mech 1912; 35:64-130.
Russo GL, Kyozuka K, Antonazzo L, Tosti E, Dale B. Maturation Promoting Factor in ascidian oocytes is regulated by different intracellular signals between meiosis I and II. Development 1996; 122:1995-2003.
Russo GL, Kyozuka K, Marino M, Tosti E, Wilding M, De Simone ML, Dale B. Meiotic cell cycle control by Mos in ascidian oocytes. In: (Le Gal Y, Halvorson H, eds) New Developments in Marine Biotechnology. New York: Plenum Publishing Corp, 1998, 115-119.
Sagata N. Meiotic metaphase arrest in animal oocytes: its implication and biological significance. Trends Cell Biol 1996; 6:22-28.
Sagata N. What does Mos do in oocytes and somatic cells? Bio Essay 1997; 19:13-21.
Sagata N, Watanabe N, Vande Woude GF, Ikawa Y. The *c-mos* protooncogene product is a cytostatic factor responsible for meiotic arrest in vertebrate eggs. Nature 1989; 342:512-518.
Sethi JK, Empson RM, Galione A. Nicotinamide inhibits cyclic ADP ribose-mediated calcium signalling in sea urchin eggs. Biochem J 1996; 319:613-617.
Sette C, Bevilacqua A, Bianchini A, Mangia F, Geremia R, Rossi P. Parthenogenetic activation of mouse eggs by microinjection of a truncated c-Kit tyrosine kinase present in spermatozoa. Development 1997; 124:2267-2274.
Sheets MD, Wu M, Wickens M. Polyadenilation of c-*mos* mRNA as a control point in *Xenopus* meiotic maturation. Nature 1995; 374:511-516.
Shilling F, Mandel G, Jaffe L. Activation by serotonin of starfish eggs expressing the rat serotonin 1c receptor. Cell Reg 1991; 1:465-469.
Shilling F, Carroll D, Muslin A, Esobedo J, Williams L, Jaffe L. Evidence for both protein tyrosine kinase and G-protein linked pathways leading to starfish egg activation. Dev Biol 1993; 162:590-599.
Solomon MJ. Activation of the various cyclin/cdc2 protein kinases. Curr Opin Cell Biol 1993; 5:180-186.
Sousa M, Barros A, Tesarik J. The role of ryanodine-sensitive calcium stores in the calcium oscillation machine of human oocytes. Mol Hum Reprod 1996a; 2:265-272.
Sousa M, Barros A, Mendoza C, Tesarik J. Effects of protein kinase C activation and inhibition on sperm, thimerosal and ryanodine-induced calcium responses of human oocytes. Mol Hum Reprod 1996b; 2:699-708.
Speksnijder J, Sardet C, Jaffe L. Periodic calcium waves cross ascidian eggs after fertilisation. Dev Biol 1990; 142:246-249.
Stice S, Robl J. Activation of mammalian oocytes by a factor obtained from rabbit sperm. Mol Reprod Dev 1990; 25:272-280.
Stricker S. Intracellular injections of a soluble sperm factor trigger

calcium oscillations and meiotic maturation in unfertilized oocytes of a marine worm. Dev Biol 1997; 186:185-201.
Swann K. A cytosolic sperm factor stimulates repetitive calcium increase and mimics fertilisation in hamster eggs. Development 1990; 110:1295-1302.
Swann K. Thimerosal causes calcium oscillations and sensitises calcium-induced calcium release in unfertilised hamster eggs. FEBS Lett 1991; 278:175-178.
Swann K. Different triggers for calcium oscillations in mouse eggs involve a ryanodine-sensitive calcium store. Biochem J 1992; 287:79-84.
Taylor C, Lawrence Y, Kingsland C, Biljan M, Cuthbertson K. Oscillations in intracellular free calcium induced by spermatozoa in human oocytes after fertilisation. Hum Reprod 1993; 8:2174-2179.
Terasaki M, Sardet C. Demonstration of calcium uptake and release in sea urchin eggs by cortical endoplasmic reticulum. J Cell Biol 1991; 115:1031-1037.
Tesarik J, Sousa M. Mechanism of calcium oscillations in human oocytes: a two store model. Mol Hum Reprod 1996; 2:383-390.
Tesarik J, Sousa M, Testart J. Human oocyte activation after intracytoplasmic sperm injection. Hum Reprod 1994; 9:511-518.
Tesarik J, Sousa M, Mendoza C. Sperm-induced calcium oscillations in human oocytes show distinct features in oocyte centre and periphery. Mol Reprod Dev 1995; 41:257-263.
Tosti E, Palumbo A, Dale B. Inositol trisphosphate in human and ascidian spermatozoa. Mol Reprod Dev 1993; 35:52-56.
Turner P, Jaffe L. G-proteins and the regulation of oocyte maturation and fertilisation. In: (Schatten H, Schatten G, eds) The Cell Biology of Fertilisation. New York: Academic Press, 1989, pp297-318.
Verlhac MH, Kubiak JZ, Clarke HJ, Maró B. Microtubule and chromatin behavior follow MAP kinase activity but not MPF activity during meiosis in mouse oocytes. Development 1994; 120:1017-1025.
Watanabe N, Vande Woude GF, Ikawa Y, Sagata N. Specific proteolysis of the c-*mos* proto-oncogene product by calpain on fertilization of *Xenopus* eggs. Nature 1989; 342:505-511.
Watanabe N, Hunt T, Ikawa Y, Sagata N. Independent inactivation of MPF and cytostatic factor (Mos) upon fertilization of *Xenopus* eggs. Nature 1991; 352:247-248.
Weber M, Kubiak JZ, Arlinghaus RB, Pines J, Maró B. c-*mos* protooncogene product is partly degraded after release from meiotic arrest and persists during interphase in mouse zygotes. Dev Biol 1991; 148:393-397.
Whitaker M, Crossley I. How does a sperm activate a sea urchin egg? In: (B Dale, ed) Mechanisms of Fertilisation: Plants to Humans. NATO ASI series vol. H45, Cell Biology 45, Springer-Verlag, Berlin, 1990, pp389-417.
Whitaker M, Swann K. Lighting the fuse at fertilization. Development 1993; 117:1-12.
Wilding M, Dale B. Sperm factor: what is it and what does it do? Mol Hum Reprod 1997; 3:269-273.
Wilding M, Kyozuka K, Russo GL, Tosti E, Dale B. A soluble extract from human spermatozoa activates ascidian oocytes. Dev Growth Differ 1997; 39:329-336.
Wu H, Li CH, Fissore RA. Injection of a porcine sperm factor induces activation of mouse eggs. Mol Reprod Dev 1998; 49:37-47.
Xu Y, Ashley T, Brainerd EE, Bronson RT, Meyn SM, Baltimore D. Targeted disruption of *ATM* leads to growth retardation, chromosomal fragmentation during meiosis, immune defects, and thymic lymphoma. Genes Dev 1996; 8:2411-2422.
Yue C, White K, Reed W, Bunch T. The existence of inositol 1,4,5-trisphosphate and ryanodine receptors in mature bovine oocytes. Development 1995; 121:2645-2654.
Zhang F, Yamada S, Gu Q, Sih C. Synthesis and characterisation of cyclic ATP ribose, a potent mediator of calcium release. Bioorg Med Chem Lett 1996; 6:1203-1208.

29 *Spermatozoal Phylogeny of the Vertebrata*

B.G.M. Jamieson
University of Queensland

The ultrastructure of the spermatozoa of fish, from agnathans to Dipnoi, and its phylogenetic significance has been reviewed by Jamieson (1991) and Mattei (1991) and the evolution of tetrapod sperm, with particular reference to amniotes, was discussed by Jamieson (1995a). In this chapter, spermatozoal ultrastructure and phylogeny from the Sarcopterygii (lobed-finned fish and their descendants) will be examined and an attempt will be made to deduce the spermatozoal synapomorphies which distinguish the major constituent groups.

Spermatozoa of Sarcopterygii

Extant sarcopterygians consist of the Actinistia (containing a single species, the coelacanth, *Latimeria chalumnae*), the Dipnoi (containing the three lungfish genera *Neoceratodus*, *Protopterus* and *Lepidosiren*), and the Tetrapoda (containing the Lissamphibia and the Amniota). Alternative classifications and phylogenies of these three major groups are discussed by Jamieson (1991). Despite some equivocal results, recent molecular studies appear to confirm the finding (for example, Meyer, Wilson 1990; Meyer, Dolven 1992) that lungfishes are the sister group of the Tetrapoda though the closest relatives of *Latimeria* (Yokobori et al. 1994; Zardoya, Meyer 1996, 1997a, b). All endorse the sarcopterygian status of *Latimeria* demonstrated by Hillis et al. (1991).

The spermatozoa of the sarcopterygian fish have been described ultrastructurally for the actinistian *Latimeria chalumnae* (Mattei et al. 1988); for the Australian lungfish, *Neoceratodus forsteri* (Jamieson 1995a; Jespersen 1971) and in this account (Figs. 1A-R); for *Protopterus* (Boisson 1963; Boisson et al. 1967; Purkerson et al. 1974); and for *Lepidosiren* (Matos, Azevedo 1989); tetrapod sperm literature is briefly summarized in Jamieson (1995a) and in this chapter.

Symplesiomorphies of Sarcopterygian Sperm

As deduced from the sperm of sarcopterygian fish (*Latimeria* and the Dipnoi) and a survey of tetrapod sperm, the following features appear plesiomorphic for the spermatozoa of Sarcopterygii.

Sarcopterygian fish sperm have a very long, slenderly conical acrosome vesicle (Figs. 1A-E) but lack the subacrosomal cone that is basic to tetrapods. There is only one endonuclear canal in the coelacanth, *Latimeria chalumnae*, but this contains two or three perforatoria (Jamieson 1991, 1995a; Mattei et al. 1988). In *Neoceratodus* there are two or three (Figs. 1C-F, H, I) or sometimes four perforatoria initially in one canal but more posteriorly (Fig. 1H) in as many canals as there are perforatoria (Jamieson 1995a; Jespersen 1971). *Neoceratodus* is exceptional in sarcopterygians in that the perforatoria re-emerge from the nucleus at their posterior ends (Fig. 1I). The number of endonuclear canals and of enclosed perforatoria is one in basal Lissamphibia, in the caiman (though poorly substantiated by micrographs), tinamou, rhea and non-passerines (for example, galliforms), but in the Chelonia and *Crocodylus johnstoni* there are two or three canals and in *Sphenodon* there are two. There are three endonuclear canals in the sperm of the sturgeon, *Acipenser sturio*, in the Actinoperygii (ray-finned fish). It is therefore probable that the presumed common ancestor of Lissamphibia and amniotes possessed more than one perforatorium and possibly more than one endonuclear canal. A single canal occurs in the Lissamphibia, except where lost in more advanced Anura (Jamieson et al. 1993), and appears basic to all amniotes above turtles and *Sphenodon*; that is, in birds, squamates and mammals (Jamieson, Healy 1992). In *Acipenser* the canals are spiralled around each other as they are in turtles, *Sphenodon* and *Crocodylus johnstoni*. The spiral arrangement, or at least the presence of one or more endonuclear canals, may well be a synapomorphy for the Osteichythes, a monophyletic clade including the Actinopterygii, Sarcopterygii and, within the latter, the Tetrapoda. The canals are absent (presumed lost) in the highly simplified sperm of holosteans (a paraphyletic group) and Neopterygii (Jamieson 1991).

The nucleus is long in the sarcopterygian fish (*Neoceratodus*, Fig. 1A). This may be a plesiomorphic retention from osteichthyan fish, as it is also long in *Acipenser*, a basal actinopterygian, and in Chondricthyes (Jamieson 1991, 1995a). Further elongation in sarcopterygian fish appears apomorphic.

A simple midpiece, as in *Neoceratodus* (Figs. 1G, J-L), and in *Acipenser*, with some of the mitochondria in a cytoplasmic collar, is presumably plesiomorphic for the Sarcopterygii. The location of a putatively mitochondrial sleeve, usually incomplete, lateral to the nucleus in *Latimeria* is clearly apomorphic.

A 9+2 axoneme is plesiomorphic for the Sarcopterygii. Whether the lateral fins (Figs. 1M-O) are a plesiomorphy held over from osteichthyan ancestors, and basal to Actinopterygii, or are a new, homoplasic development is debatable.

Spermatozoal Synapomorphies of the Sarcopterygii

If one accepts the validity of the Sarcopterygii as defined above, at least four synapomorphies for the group can be proposed on the basis of sperm ultrastructure. These pertain to the perforatoria, the nucleus, the retronuclear body and the structure of the flagellum.

The great length of the nucleus in Actinistia (acrosome and nucleus 25-26 μm long) and Dipnoi may be an initial synapomorphy of the Sarcopterygii. The nucleus reaches a length of 70 μm in *Neoceratodus forsteri* (Jespersen 1971), the longest recorded in fish sperm (Jamieson 1991).

The extension, anterior to the nucleus, of rod-like structures, the perforatoria, is a new development in the Sarcopterygii and thus constitutes a synapomorphy, and an autapomorphy, for the group. The portions of these within the nucleus lie in one or more endonuclear canals. Perforatoria, or at least endonuclear canals indicating the existence of these, are present in lampreys, in which an acrosomal filament is extruded on reaction (Afzelius et al. 1957); the cladistian *Polypterus senegalus*, in which there is an axial endonuclear canal but a perforatorium remains to be identified (Mattei 1970); the chondrostean *Acipenser stellatus*, in which an acrosome reaction involving subacrosomal material has been demonstrated, although the role of the material in the three endonuclear canals is uncertain (Cherr, Clark 1984; Detlaf, Ginzburg 1963); the dipnoan *Neoceratodus forsteri* (Jamieson 1995a; Jespersen 1971); urodele amphibians; primitive frogs, including *Ascaphus* (Jamieson 1995a; Jamieson et al. 1993); and amniotes, including "reptiles" and non-passerine birds, of which the most basal are the Chelonia (Furieri 1970; Healy, Jamieson 1992; Hess et al. 1991; Jamieson 1995a; Jamieson, Healy 1992) and the sphenodontidan *Sphenodon punctatus* (Healy, Jamieson 1992, 1994; Jamieson, Healy 1992). With the exception of lampreys, *Acipenser* and *Polypterus*, therefore in the sarcopterygians, the perforatorial rods extend anterior to the nucleus (Figs. 1C-F, for *Neoceratodus*). It is probable, in view of the presence of rods and endonuclear canals in *Latimeria* and the ceratodontiform *Neoceratodus*, that their absence in the lepidosireniforms *Protopterus* (Boisson 1963; Mattei 1970) and *Lepidosiren* (Matos, Azevedo 1989) is secondary.

A large dense body, between the nucleus and centrioles and termed the retronuclear body, has been described for *Protopterus annectens* by Boisson (1963; Fig. 2) and *P. aethiopicus* by Purkerson et al. (1974). It has been homologized with a smaller structure which, though postmitochondrial, originates behind the nucleus, in *Neoceratodus* by Jespersen (1971) and Jamieson (1991; Fig. 1G); and a postnuclear structure, termed by Mattei et al. (1988) the "paracentriolar body," in *Latimeria chalumnae* by Jamieson (1991). Its cross striation in *P. aethiopicus* has led to its being compared with the striated columns of mammalian sperm (Purkerson et al. 1974). It is tentatively considered homologous with the neck region of urodele and anuran sperm (Jamieson 1991, 1995a), being, in urodeles, most strongly developed in ambystomatoids, plethodontids (Fig. 6C) and salamandroids (Figs. 6D, E) (Baker 1962, 1963, 1966; Furieri 1962; Jamieson 1995a; Picheral 1967, 1979; Picheral et al. 1966; Werner et al. 1972) and weakly developed in cryptobranchs (Baker 1963). The retronuclear body is here considered to be a synapomorphy, and autapomorphy, of the Sarcopterygii.

In *Neoceratodus*, lateral fin-like prolongations of the sperm flagellum are present at doublets 3 and 8 (Jamieson 1995a; Jespersen 1971), (Figs. 1L-O, Q), except at the endpiece (Fig. 1R). Shortly behind the distal centriole, within the mitochondrial collar (Fig. 1L), and behind this (Fig. 1M), for a short distance, each fin is supported by a large dense juxta-axonemal rod and by a smaller lateral rod within its free extremity (Jamieson 1995a). The fin becomes more extensive behind this short anterior region but the rods are reduced in size (Figs. 1N, O) and soon only the lateral fiber persists (Figs. 1O, Q). Such lateral prolongations, though questionably with supporting rods, in *Latimeria* were appropriately termed undulating membranes by Tuzet and Millet (1959). Lateral fins (also at doublets 3 and 8) in many actinopterygian fish (Jamieson 1991; Mattei 1988) could conceivably have been precursory to actinistian and dipnoan undulating membranes but homoplasy cannot be ruled out as lateral axonemal fins occur also in some echinoderms and protostomes (Jamieson 1995a). Two bilateral elements which also occur at doublets 3 and 8 in Chondrichthyes were presumably convergently acquired.

It is here accepted, as proposed by Jamieson (1995a), that dipnoan axonemal fins are homologous with the undulating membrane of lissamphibian sperm. It is thus proposed that presence of two undulating membranes is a sarcopterygian synapomorphy. As supporting rods are not reported for *Protopterus* (Boisson 1963; Purkerson et al. 1974) or *Lepidosiren* (Matos, Azevedo 1989), nor for *Latimeria* (Mattei et al. 1988), it is possible that presence of such rods is a ceratodontiform-amphibian synapomorphy.

Even if lateral axonemal fins are sarcopterygian symplesiomorphies carried over from an osteichthyan ancestor, their elaboration in dipnoans and amphibians is considered synapomorphic.

Spermatozoal Synapomorphies of the Tetrapoda

A generalized tetrapod spermatozoon manifesting the shared features of basal lissamphibians (for example, *Ascaphus*) and basal amniotes (Chelonia, *Sphenodon*) is illustrated in Figure 3. As it shares features of basal, extant tetrapods, delineation of this hypothetical ancestral tetrapod sperm is not unduly speculative. It should be borne in mind that ancestral tetrapods, in being non-amniote, are classifiable as Amphibia. Their descendants are the Lissamphibia and the Amniota.

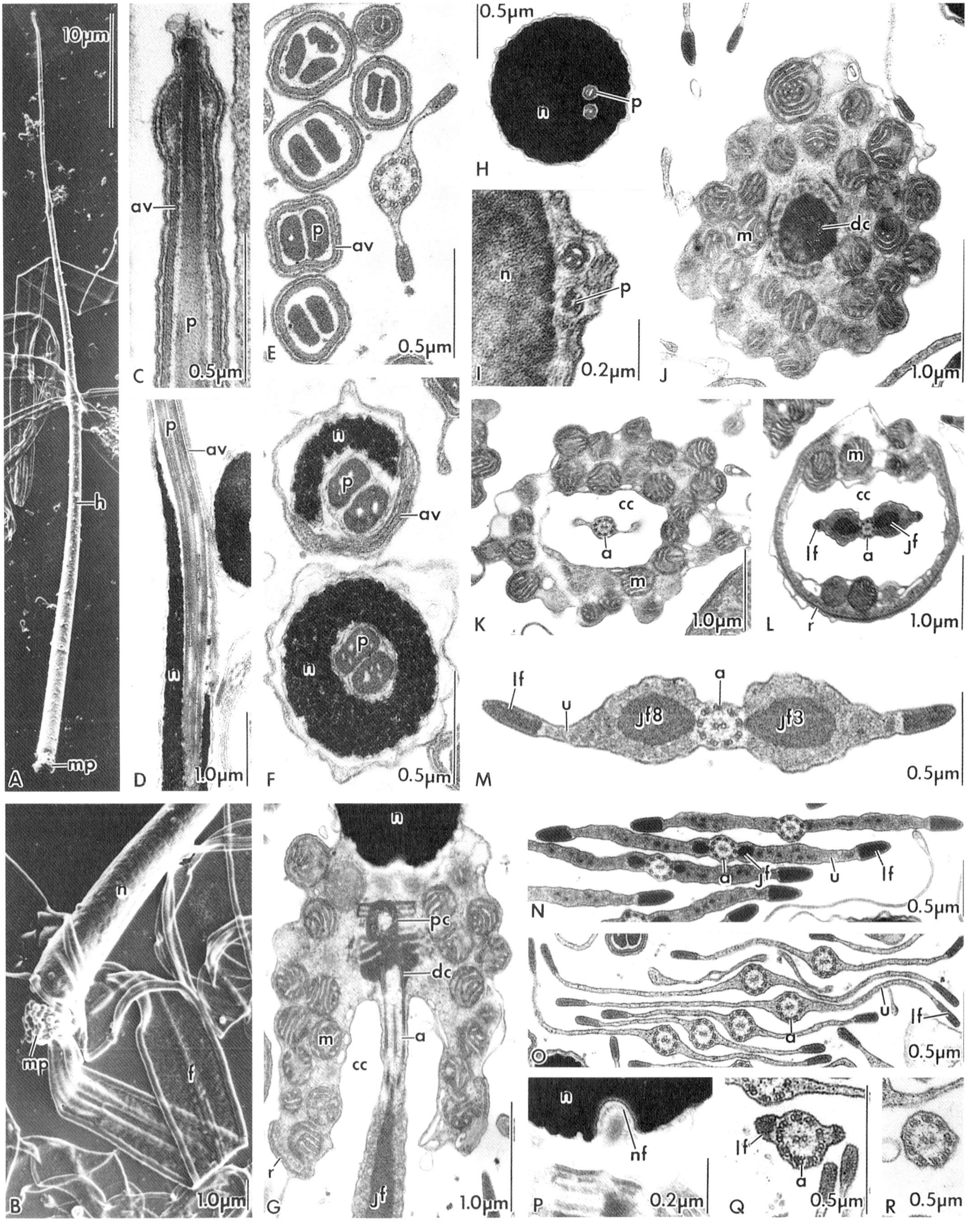
10μm
h
mp
A
av
p
C
0.5μm
p
av
n
D
1.0μm
E
0.5μm
av
n
p
F
0.5μm
0.5μm
n
p
H
n
p
I
0.2μm
dc
m
J
1.0μm
cc
a
m
K
1.0μm
m
cc
lf
a
jf
r
L
1.0μm
lf
u
a
jf8
jf3
M
0.5μm
n
mp
f
B
1.0μm
n
pc
dc
a
m
cc
r
jf
G
1.0μm
a
jf
u
lf
N
0.5μm
u
lf
a
O
0.5μm
n
nf
P
0.2μm
lf
a
Q
0.5μm
R
0.5μm

The tetrapod sperm is derived relative to that of sarcopterygian fish (Actinistia and Dipnoi) in a remodeling of the acrosome complex and nucleus. This involves development of a subacrosomal cone and, presumably to house this, correlated reshaping of the proximal end of the nucleus.

A cone of subacrosomal material, not seen in sarcopterygian fish, is developed, in addition to the basal sarcopterygian perforatorial rods, in the Lissamphibia in *Ascaphus* (Fig. 4) and discoglossoids, and in the amniotes in Chelonia, *Sphenodon* (Fig. 11), crocodiles (Fig. 12), squamates (Fig. 15A) and monotremes. Its function is not known but it may be perforatorial, as suggested by the fact that it is present in the absence of perforatorial rods in monotremes. Birds have lost the subacrosomal cone but plesiomorphically retain a perforatorial rod.

A second synapomorphy of the tetrapod spermatozoon relative to sarcopterygian fish is the development of an abrupt shoulder-like transition from the anterior tapered portion of the nucleus (constituting the nuclear rostrum) within the acrosome complex to the long cylindrical portion of the nucleus. The nuclear shoulders are illustrated for *Ascaphus* (Fig. 4), *Sphenodon* (Fig. 11), *Crocodylus johnstoni* (Fig. 12) and the squamate *Carlia rubrigularis* (Fig. 15A). The shoulders and associated narrowing of the proximal end of the nucleus as the nuclear rostrum, within the acrosome complex, presumably were an adaptation allowing the newly evolving acrosome cone to be housed between the nucleus and the acrosome vesicle.

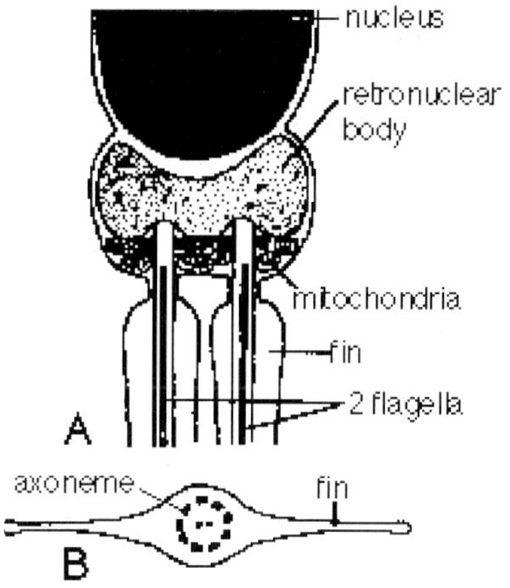

Figure 2. *Protopterus annectens*. Spermatozoon, showing retronuclear body. (From Jamieson 1991, after Boisson 1963.)

Spermatozoa of the Lissamphibia

The extant Amphibia comprise the subclass Lissamphibia. These consist of the Urodela (newts, salamanders and sirenians), the Anura (frogs and toads) and the Gymnophiona (caecilians). It is argued (Jamieson 1995a; Lee, Jamieson 1993) that internal fertilization is basic (plesiomorphic) in the Lissamphibia. It occurs in the great majority of urodeles and in all gymnophionans, but in only primitive frogs.

A survey of the ultrastructure of the spermatozoa of the three orders of Lissamphibia permits delineation of a generalized lissamphibian spermatozoon (Jamieson 1995a; Fig. 5). Many features of this spermatozoon are deduced to be plesiomorphies carried over from the presumed tetrapod ancestral spermatozoon as they are also found in basal amniotes or even in Dipnoi and Actinistia.

Spermatozoal Symplesiomorphies in the Lissamphibia

Features of the ascaphid (Fig. 4) and hypothetical ancestral lissamphibian spermatozoon (Fig. 5) that are

---

Abbreviations used in figures: a, axoneme; af, axial fiber; an, annulus; av, acrosome vesicle; b, barb; cc, cytoplasmic canal; cd, cytoplasmic droplet; cy, cytoplasm; d3, density (juxta-axonemal fiber?) at 3; db, dense, intermitochondrial body; dc, distal centriole; el, electron lucent space; ec, endonuclear canal; f, flagellum; fs, fibrous sheath; h, head; hmt, helical microtubules; jf, juxta-axonemal fiber; jf3, juxta-axonemal fiber at 3; jf8, juxta-axonemal fiber at 8; lc, longitudinal column; lf, lateral fiber; m, mitochondrion; mp, midpiece; mts, sheath of microtubules; n, nucleus; nf, basal nuclear fossa; nk, neck; ni, infolding of nucleus into neckpiece (retronuclear body); nr, nuclear rostrum; nri, nuclear ridge; p, perforatorium; pa, paraxonemal rod; pc, proximal centriole; pf, peripheral fiber; r, retronuclear body; sc, subacrosomal cone; sdb, small dense body; stc, striated (segmented) column; su, subacrosomal material; u, undulating membrane.

Figure 1. *Neoceratodus forsteri*, the Australian lungfish. Ultrastructure of the spermatozoon. A: Scanning electron micrograph (SEM) of the head and midpiece. B: SEM of the base of the nucleus and midpiece and the flagellum with its lateral fins or undulating membranes. C, D: Longitudinal sections (LS) of the perforatoria, showing their extension anterior to the nucleus and, in D, in an endonuclear canal. E: Transverse sections (TS) of the perforatoria within the acrosome. F: TS of perforatoria entering and within the nucleus. G: LS of midpiece, showing mitochondrial collar. H: TS of two perforatoria, each in a separate endonuclear canal. I: TS of two perforatoria posteriorly emergent from the nucleus. J: TS of midpiece through the distal centriole. K: TS of far anterior region of axoneme, within the cytoplasmic canal. L: Same further distally, showing beginning of lateral and juxta-axonemal fibers on each side. M: Same shortly behind the midpiece. N: Further distally, showing more slender undulating membranes still with lateral and greatly reduced juxta-axonemal fibers. O: Still further distally, the slender undulating membranes now lacking the juxta-axonemal fibers. P: LS of basal nuclear fossa. Q: Axoneme far distally, shortly before the endpiece, with greatly reduced undulating membranes. R: TS of endpiece. (Original.)

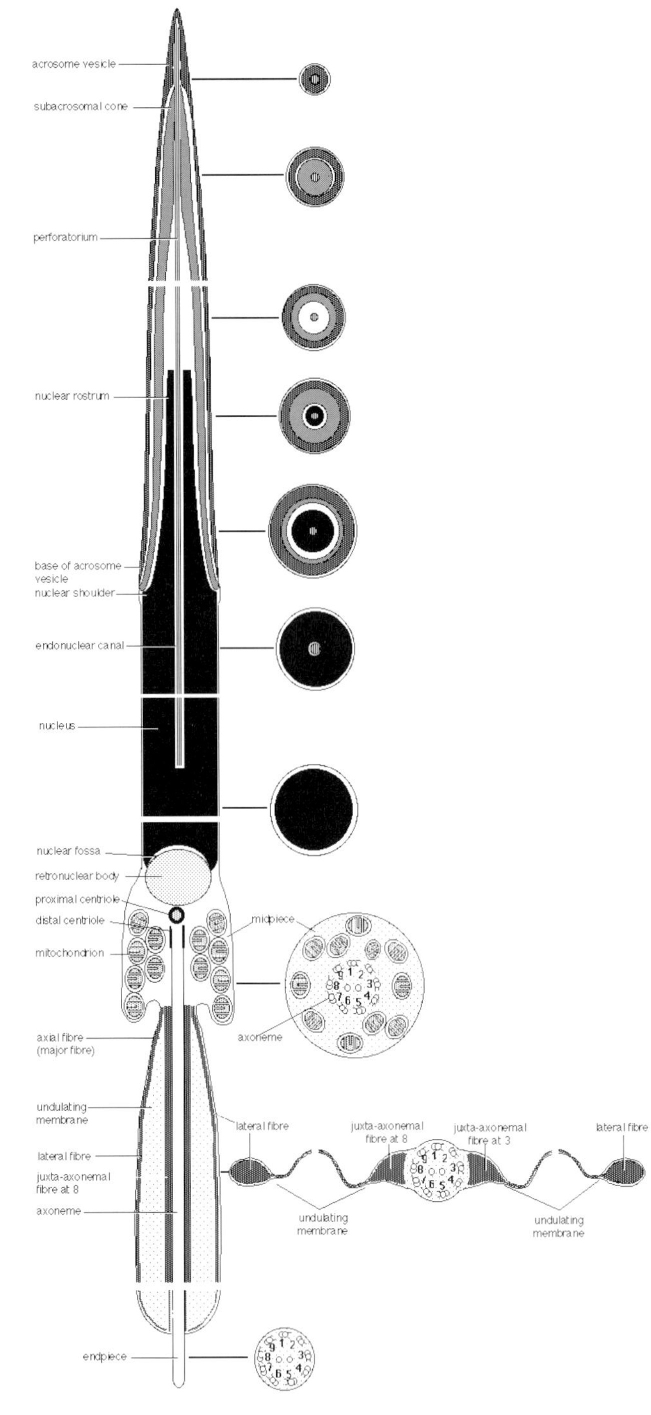

Figure 3. Generalized or hypothetical ancestral tetrapod spermatozoon.

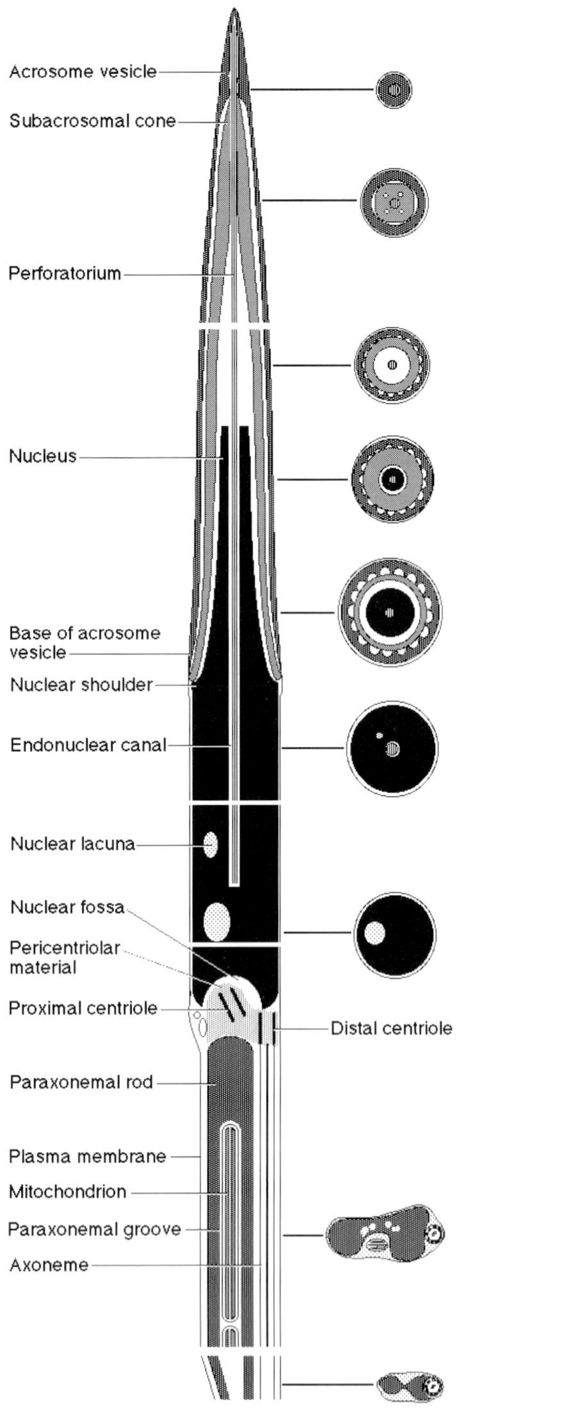

Figure 4. *Ascaphus truei*. Diagrams of spermatozoal ultrastructure as seen in longitudinal section and transverse sections by transmission electron microscopy. (From Jamieson et al. 1993.)

lissamphibian symplesiomorphies are as follows. An anterior acrosomal vesicle forms a hollow cone (sarcopterygian symplesiomorphy) which overlies a cone of subacrosomal material; the subacrosomal cone embraces the tapered anterior end of the nucleus (lissamphibian symplesiomorphy and tetrapod synapomorphy) which is elongate (lissamphibian symplesiomorphy and sarcopterygian synapomorphy). At the posterior end of the acrosome (vesicle and subacrosomal cone), the nucleus forms characteristic "shoulders" (lissamphibian symplesiomorphy and tetrapod synapomorphy) posterior to which its form is cylindrical. Axially, within the acrosome and therefore extending anterior to the nucleus (lissamphibian symplesiomorphy and sarcopterygian synapomorphy), there is a rod, the putative perforatorium, which deeply penetrates the nucleus within an endonuclear canal (lissamphibian, tetrapod and sarcopterygian symplesiomorphy). The basal sarcopterygian feature of a prenuclear, and endonuclear, axial rod (perforatorium), is present not only in *Ascaphus* but also in other primitive frogs, *Discoglossus* and the bombinids *Bombina* and *Alytes*. It is also present in gymnophionids though there the perforatorium lodges posteriorly in a much shorter endonuclear canal.

The base of the nucleus is indented as a basal nuclear fossa (possibly a tetrapod symplesiomorphy as also seen in *Neoceratodus*, Fig. 1P) which, unlike the dipnoan, contains the proximal centriole. Behind this is the distal centriole, which forms the basal body of the axoneme.

Spermatozoal Synapomorphies in the Lissamphibia

Synapomorphies of the lissamphibian sperm relative to an ancestral tetrapod may now be considered.

The presence of an undulating membrane within the flagellar complex has long been thought distinctive of the Lissamphibia but it is proposed here, as in Jamieson (1995a), that it is the loss of the undulating membrane adjacent to doublet 8 which is distinctive and synapomorphic of the Lissamphibia, with retention of that at doublet 3. Evidence for the former existence of an undulating membrane at doublet 8, as in Actinistia and Dipnoi, is the persistence of the juxta-axonemal fiber at doublet 8, in the absence of a membrane on that side, in urodeles (Figs. 6N, O) and exceptionally in Anura.

Concomitant with the loss of one undulating membrane has been the development of a condition in which the axoneme undulates around the remaining lateral fiber which has, therefore, long been termed the axial fiber. This condition is particularly well demonstrated by the Urodela, particularly in the non-cryptobranchs (Figs. 6N, O) as these have a stiffened axial fiber. An axial fiber with undulating membrane is present in caecilians (Fig. 6I), as basically in anurans (Figs. 5, 6P).

Spermatozoal Synapomorphies in the Urodela

Ultrastructural aspects of urodele spermatozoon are illustrated for the salamandrid *Taricha granulosa* (Figs. 6D, E, O) and the plethodontids *Stereochilus marginatus* (Fig. 6C) and *Eurycea quadridigitata* (Figs. 6M, N).

The structure of urodele sperm is uniform relative to the great diversity in anurans though the ground plan has many similarities with that of anuran and caecilian sperm. The more striking, synapomorphic, urodele

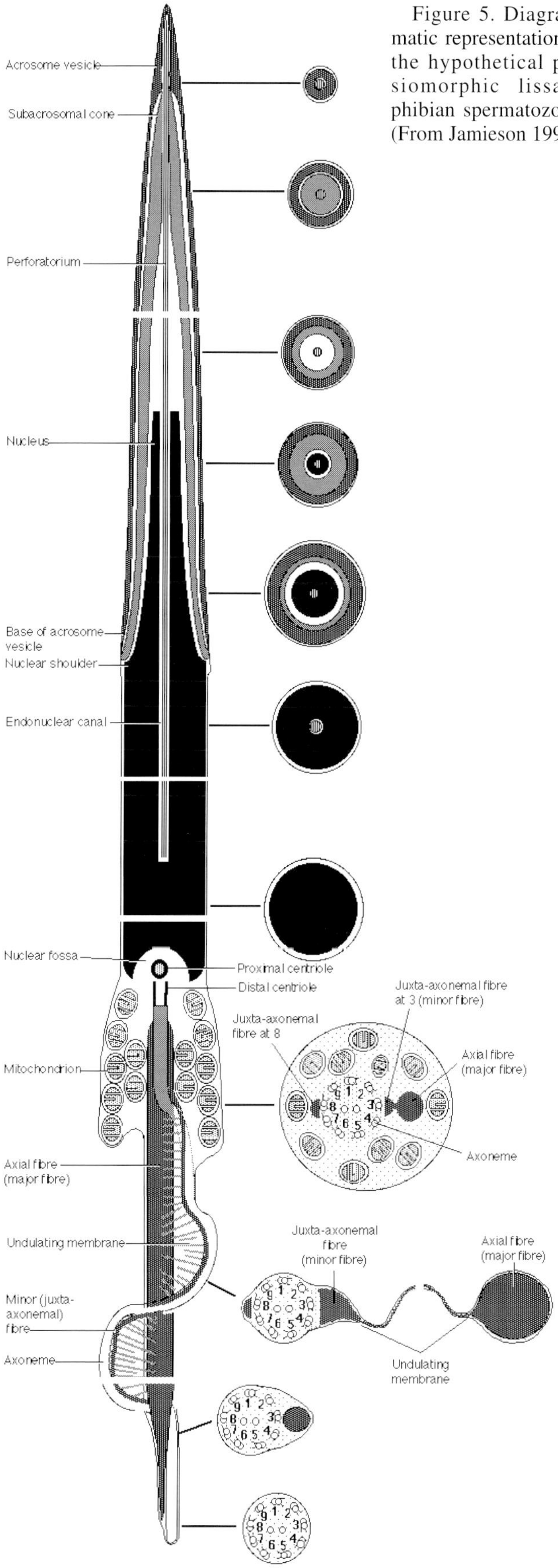

Figure 5. Diagrammatic representation of the hypothetical plesiomorphic lissamphibian spermatozoon. (From Jamieson 1995.)

sperm features will now be discussed, general lissamphibian features having been outlined above.

The acrosome in salamandrids, plethodontids (*Eurycea quadridigitata*, Fig. 6A) and ambystomatids typically ends subapically in a distinct barb (Retzius 1906; Wortham et al. 1982). It is uncertain that a barb is absent in cryptobranchs, amphiumids and sirenids. However, the transverse section of the acrosomal cap (vesicle) in the cryptobranch *Hynobius nebulosus* is strongly trifoliate in a transmission electron micrograph (Picheral 1967, 1979), a condition which is compatible with presence of a barb or at least some unilateral modification of the tip of the vesicle. The presence of a terminal barb to the acrosome of *Amphiuma* claimed by McGregor (1899) was denied by later authors (Baker 1962; Barker, Biesele 1967) and a barb was not described in an optical study of *Pseudobranchus striatus* (Austin, Baker 1964). It is possible that the barb is an internal and not basal synapomorphy of the Urodela.

The barb has no equivalent in anurans. Modification of the tip of the spermatozoon in *Neoceratodus forsteri* (Fig. 1C) is possibly precursory to the urodele condition.

At least the salamandrids and plethodontids (*Eurycea quadridigitata*, Fig. 6A) are distinguished by a stronger development of the subacrosomal cone than that occurring in anurans and gymnophionans. The cone is so large that it projects distally beyond the acrosome vesicle. The condition is uncertain in cryptobranchs and hypermophosis may be a synapomorphy of higher urodeles only.

A nuclear ridge (as in the salamandrid *Taricha granulosa*, Fig. 6M), composed of closely adpressed microtubules internal to the nuclear envelope, is restricted to urodele sperm. It is known for cryptobranchs and salamandrids (Picheral 1967, 1979; Picheral et al. 1966), in addition to plethodontids (Fig. 6M), and can thus be considered a basal synapomorphy and, because unique, an autapomorphy of urodeles.

Behind the nucleus, urodele sperm have a neck region characterized by a structure, considered to be the homolog of the basal sarcopterygian retronuclear body (Jamieson 1991). It is not seen in gymnophionans and is questionably present in anurans. In salamandrids, the neck (also termed the connecting piece) is a long cylinder which fits into the deep fossa at the basal end of the nucleus in such a way that the neck is surrounded by a thin sheath of chromatin (nuclear collar) limited by two nuclear envelopes (Picheral 1967, 1979; Picheral et al. 1966; this study, Fig. 6E). It may present deep indentations into the neckpiece (as in the salamandrid *Taricha granulosa*, Fig. 6D). The length of the neck varies from one-quarter to about one-sixth of the length of the nucleus.

In the ambystomatoids, plethodontids (*Stereochilus marginatus*, Fig. 6C) show a constant and conspicuous difference in the structure of the neckpiece as compared with that of salamandroids. There is no basal nuclear fossa. The posterior end of the nucleus abuts on the neckpiece, which equals the base of the nucleus in width, at a straight transverse border and there is no continuation of nuclear material around the neckpiece (Fig. 6C). The densified nuclear envelope intervenes between nucleus and neckpiece. The form of the neckpiece in Ambystomatidae is unknown.

The sperm of the cryptobranchs *Hynobius* and *Cryptobranchus* have a very short neckpiece, which may be no more than an enlargement of the anterior end of the axial fiber, and resembles the connecting piece of mammalian sperm (Baker 1963). It also resembles the nuclear-centriolar junction of anuran sperm, to which an additional similarity is the presence and subcircular cross section of the major fiber.

The great development of the neck in salamandrids and plethodontids is possibly a synapomorphy of higher urodeles but, alternatively, the modest development in cryptobranchs might be a reduction relative to the intermediate size seen in at least some Dipnoi (*Protopterus*). Although usually considered primitive, cryptobranchs are apomorphic in having reduced spermatophoral glands and therefore external fertilization, which is deduced to be a secondary condition relative to basic internal fertilization of lissamphibians (Lee, Jamieson 1993).

It is possible that the material of the neck in urodeles is homologous with the pericentriolar material of anurans, which is continuous with the major (axial) fiber of the tail.

Figure 6 (right). Spermatozoal ultrastructure in Lissamphibia. A: The plethodontid urodele *Eurycea quadridigitata*. LS of acrosome, showing terminal barb. B: The myobatrachid frog *Limnodynastes peronii*. LS of acrosome and tip of nucleus. C: The plethodontid urodele *Stereochilus marginatus*. LS of anterior region of neck (retronuclear body) and transverse junction with nucleus. D: The salamandrid urodele *Taricha granulosa*. LS of posterior region of neck (retronuclear body) showing that it is ensheathed by the posterior region of the nucleus which intrudes into it. E: Same, showing basal nuclear fossa; F: The microhylid *Cophixalus ornatus*. TS of sperm flagella showing absence of undulating membranes and fibers. G: The rhacophorid frog *Chiromantis xerampelina*. LS (top and left) of five spiral sperm nuclei below which are many cross sections of the paired terminal flagella. H: Same, a TS of the paired flagella embedded in a sheath of microtubules in pseudocrystalline array. I: The gymnophionan *Typhlonectes natans*. TS of axoneme and undulating membrane. J: The hylid frog *Cyclorana alboguttata*. TS of axoneme and thick undulating membrane. K: The hylid frog *Litoria rheocola*. LS of eubufonoid mitochondrial collar. L: The myobatrachid, limnodynastine frog *Limnodynastes convexiusculus*. LS of basal nuclear fossa penetrated by axial fiber. M: *Taricha granulosa*. TS of nucleus showing microtubules of nuclear ridge. N: *E. quadridigitata*. TS of two sperm flagella, showing crescent of mitochondria. O: *Taricha granulosa*. TS of sperm flagella, showing crescent of mitochondria and Y-shaped axial fiber. P: The hylid frog *Litoria eucnemis*. TS of flagella, showing hypermorphosed juxta-axonemal fiber at 3. (Original.)

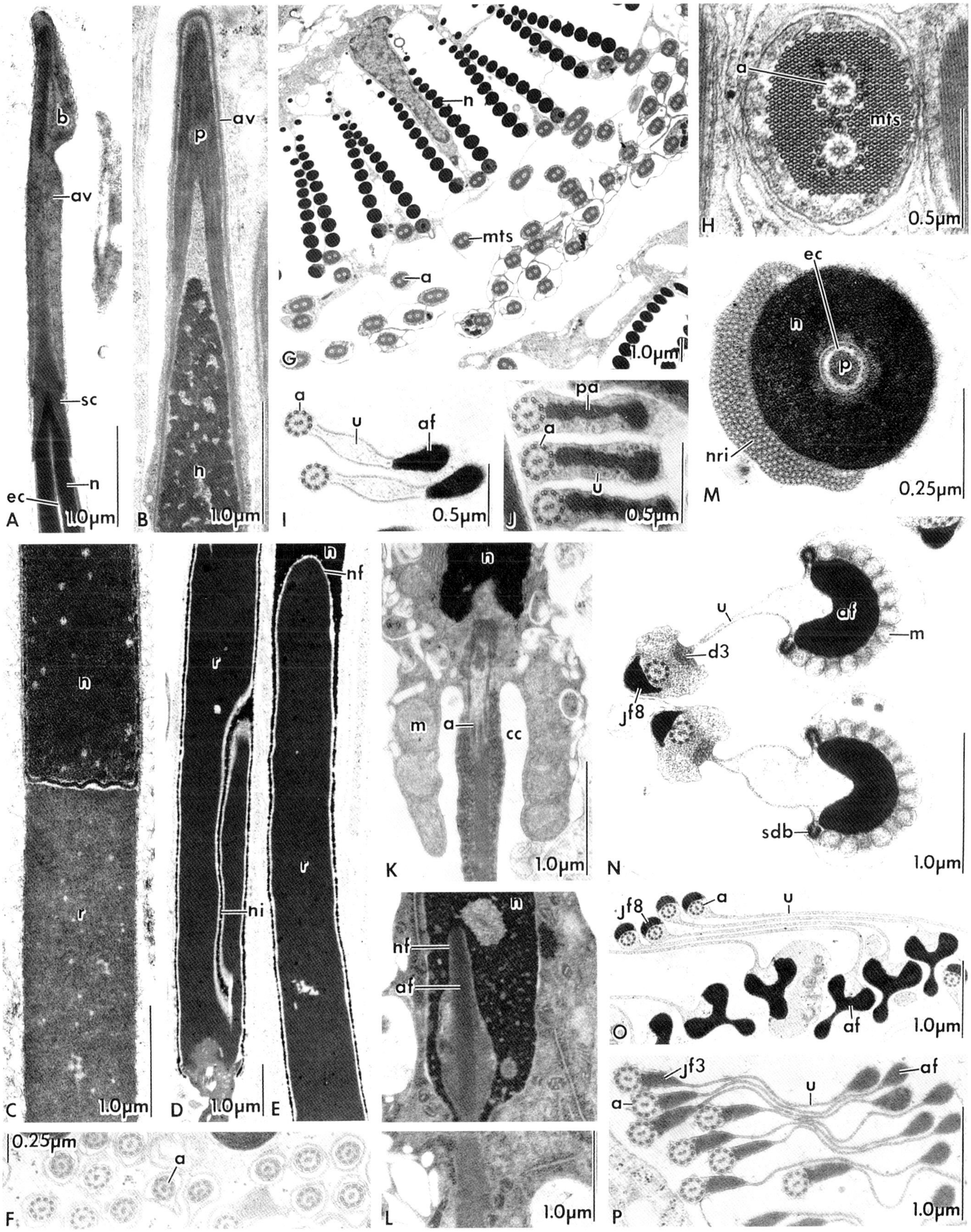
b
av
sc
n
ec
A
1.0μm
p
av
n
B
1.0μm
n
mts
a
G
1.0μm
a
mts
H
0.5μm
a
u
af
I
0.5μm
pa
a
u
J
0.5μm
ec
n
p
nri
M
0.25μm
n
r
C
1.0μm
r
ni
D
1.0μm
n
nf
r
E
n
m
a
cc
K
1.0μm
u
af
m
d3
jf8
sdb
N
1.0μm
jf8
a
u
af
O
1.0μm
n
nf
af
0.25μm
a
F
L
1.0μm
jf3
af
u
a
P
1.0μm

At least some urodele sperm have a dense ring-like structure questionably identified with an annulus (Picheral 1979). The inner wall of a cytoplasmic collar, equivalent to that in many fish sperm (Jamieson 1991), is occupied by two large dense structures: the ring, forming a lining to much of the canal, and opposite the ring, and in *Amphiuma* contiguous with it (Barker, Biesele 1967), the axial fiber. It is uncertain whether the strong development of the ring is a general urodele synapomorphy or is restricted to higher urodeles.

In urodele sperm, the axial fiber, at doublet 3, is connected to the axoneme by the undulating membrane but, typically (*Taricha granulosa*, Fig. 6O), there is no intervening juxta-axonemal fiber. This may represent a synapomorphic loss from a basic lissamphibian condition with juxta-axonemal fibers at 3 and 8. However, the condition in cryptobranchs requires investigation. The plethodontoid ambystomatoids (for example, *Plethodon albagula*, Fig. 7 and *Eurycea quadridigitata*, Fig. 6N) are exceptional in having a density on the adaxonemal end of the undulating membrane but it is questionable, because of its connection to dense bodies near the major fiber rather than to the fiber (Figs. 6N, 7), that this is homologous with the anuran juxta-axonemal fiber (Jamieson 1995a).

The axial fiber has the presumed plesiomorphic subcircular cross section in the cryptobranch *Hynobius nebulosus* (Picheral 1979), as in the Gymnophiona (Fig. 6I) and Anura (Fig. 6P), though in cryptobranchs differing from the typical condition in the latter two orders in also having the smaller fiber on the opposite side of the axoneme at doublet 8.

In the higher urodeles, as an internal synapomorphy, the axial fiber acquires, for much of its length, a Y-shaped or trifoliate cross section (*Taricha granulosa*, Fig. 6O). This sectional profile corresponds with that used in human tools and engineering structures to confer strength and rigidity and is deduced to stiffen the fiber against bending. The trifoliate cross section, and the accompaniment of a fiber on the opposite side of the axoneme, has been demonstrated ultrastructurally in salamandrids (Baker 1966; Furieri 1960, 1962; Picheral 1967, 1972, 1979; *Taricha granulosa*, Fig. 6M); in amphiumids (Baker 1962); in ambystomatids (Russell et al. 1981) and the plethodontids *Eurycea* sp., *Eurycea bislineata bislineata*, *E. quadridigitata*, *E. wilderae*, *Gyrinophilus porphyriticus*, *Plethodon albagula*, *Stereochilus marginatus*, and *Typhlomolge rathbuni* (all in this study).

Presence of an additional juxta-axonemal fiber in many anurans at doublet 3, connected with axial fiber by a thin lamina within the undulating membrane, may well be plesiomorphic, and attributable to the generalized lissamphibian sperm.

In some ambystomatid salamanders (*Ambystoma* and *Rhyacosiredon*) a short tail membrane, absent in the ambystomatid *Rhyacotriton* and unknown in other lissamphibians, has been observed by light microscopy on the opposite side of the major fiber from the undulating membrane at the posterior end of, and extending longitudinally for a fraction of the length of, the membrane (Brandon et al. 1974; Martan, Wortham 1972; Wortham et al. 1982). It does not appear to represent pairing of the undulating membrane, on opposite sides of the axoneme, which is here hypothesized as an ancestral, possibly pre-lissamphibian stage.

Sperm of the sirens, exemplified by *Siren intermedia*, *S. lacertina* and *Pseudobranchus striatus axanthus*, are biflagellate and have two interconnected undulating membranes (Austin, Baker 1964). Such secondary duplication of the axoneme also occurs in some Anura.

The condition of the mitochondria in *Neoceratodus forsteri* in which the mitochondria are located in a short midpiece and in a collar-like posterior extension of this around the base of the axoneme (Figs. 1G, K, L) may represent the plesiomorphic condition for lissamphibia. In the cryptobranch urodeles *Cryptobranchus alleganiensis bishopi* (Baker 1963) and *Hynobius nebulosus* (Picheral 1979), the mitochondria seem to be located in a protoplasmic bead around the nucleus, even in the mature sperm. This possibly represents a reduction from the plesiomorphic condition. The cryptobranch sperm tail lacks mitochondria and there is no remnant of the ring (putative annulus) around the major fiber.

In salamandrids (Fig. 6O), ambystomatids, plethodontids (Figs. 6N, 7) and amphiumids, small, ovoid mitochondria are present in cytoplasm around a long anterior region of the axial (major) fiber where, in cross section of the sperm, they form an arc, the whole constituting the intermediate piece or midpiece (Baker 1966; Fawcett 1970; Grassé 1986; Jamieson et al. 1993; Picheral 1967, 1979; Wortham et al. 1977). This appears to be a derived condition.

This type of midpiece has been said not to occur in anurans but, clearly homoplasically, in *Limnodynastes peronii* an incomplete ring of mitochondria surrounds the major fiber (Fig. 8) much as in salamandrids (Lee et al. 1992); a somewhat similar arrangement is seen in *Bombina variegata* (Folliot 1979; Furieri 1975b; Pugin-Rios 1980) and *Neobatrachus pelobatoides* (Lee, Jamieson 1992) in which the mitochondria lie on each side of the axial fiber.

A major difference of salamandrids, ambystomatids and plethodontids from anurans is the great length of the mitochondrial region, occupying a considerable proportion of the length of the tail. This appears to be an apomorphic elongation as the midpiece is short in dipnoans (Figs. 1B, G) as it is in basal actinopterygians such as *Acipenser* (Jamieson 1991).

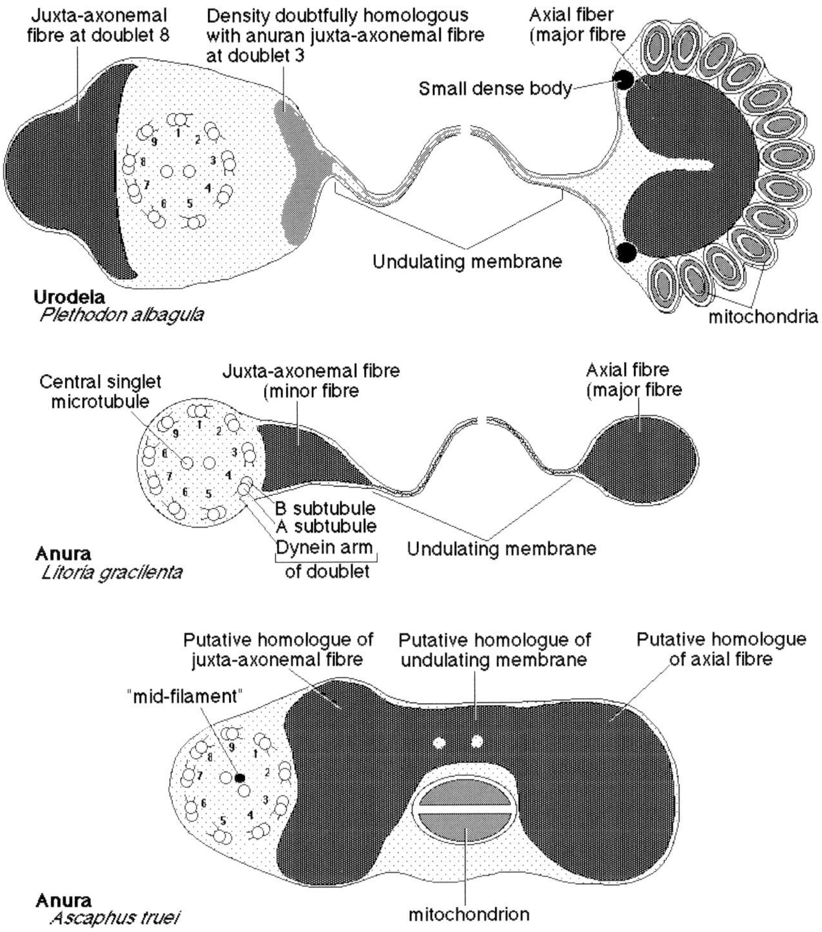

Figure 7. Comparison of the paraxonemal rod in *Ascaphus* with the accessory axonemal fibers and undulating membrane of a urodele and a hylid frog, as seen in transverse sections of the flagellum. Only about one half of the length of the undulating membrane is shown in the latter two species. In the urodele, the major fiber is much larger than indicated relative to the minor fiber. Presence of a fiber on each side of the axoneme, at doublets 3 and 8, in urodeles is probably the plesiomorphic lissamphibian (and sarcopterygian) condition. The juxta-axonemal fiber at 3 is usually absent in urodeles and the density in this position in *Plethodon* may not be homologous with this; the axial fiber is hypertrophied. The two fibers on each side of the axoneme are reduced in Anura to one fiber, at 3, though both may persist as an occasional variation in the same testis. The anuran fiber at 3 is subdivided by the undulating membrane into a juxta-axonemal (minor) fiber and an axial (major) fiber. In the hylid, *Litoria gracilenta*, the minor fiber is particularly well developed, being unusually large relative to the major fiber. It is deduced that in *Ascaphus*, as a paedomorphic condition, the undulating membrane has been greatly reduced in length, and broadened, but that homologs of the juxta-axonemal and axial fibers remain well developed. (From Jamieson et al. 1993.)

The Spermatozoa of Gymnophiona

Caecilian spermatozoa have been examined ultrastructurally in *Typhlonectes natans* (Typhlonectidae) (van der Horst 1991; van der Horst et al. 1991; present study) and by light microscopy for other caecilians (Seshachar 1939, 1940, 1942, 1943, 1945; Wake 1994). The spermatozoon has all the components of the hypothetical lissamphibian sperm which was delineated above (Fig. 5), but there are differences relative to this in the perforatorium and endonuclear canal, the midpiece and the undulating membrane. At least the first two are apomorphic.

Van der Horst et al. (1991) considered the curved tip of the acrosome in *Typhlonectes natans* to be comparable with the hooked shape characteristic of urodeles. However, in urodeles the tip is strongly reflexed and the supposed similarity with gymnophionans cannot be considered a synapomorphy. Furthermore, the tips of the acrosomes in the oviparous *Ichthyophis glutinosus*, *Uraeotyphlus narayani*, *Siphonops annulatus* and *Gegenophis carnosus* are spatulate, not hooked (Seshachar 1940, 1945).

Synapomorphies of Gymnophionan Sperm

The base of the acrosome of the gymnophionan *Typhlonectes natans* conforms with the tip of the nucleus, which is indented medianly for a short distance as an anterior nuclear fossa, into which the base of the perforatorium slightly protrudes. The fossa is here regarded as a short endonuclear canal. The greater part of the width of the tip of the nucleus forms a straight sided bowl-like indentation closely fitting the posterior end of the acrosome and the short endonuclear canal penetrates the center of this indentation (Jamieson, unpublished data; van der Horst et al. 1991). A similar "acrosome seat," where the lance-shaped plug given off at the base of the acrosome fits snugly into a deep pit in the anterior end of the nucleus, has been described for *Uraetyphlus narayani* and *Ichthyophis* (Seshachar 1945) and is here considered a significant synapomorphy and autapomorphy of the Gymnophiona, although the perforatorium and endonuclear canal are themselves symplesiomorphies.

The centrioles and the anterior part of the gymnophionan sperm tail are surrounded by 35-40 spherical mitochondria, which have an extensive array of delicate cristae (van der Horst et al. 1991).

This cylindrical arrangement of mitochondria around the centriolar-axonemal axis, in the absence of a collar, is possibly apomorphic relative to that of finned sarcopterygians and ancestral tetrapods.

The major part of the tail consists of an axial fiber and a 9+2 axoneme enclosed by a plasma membrane, as plesiomorphic features. However, the plasma membranes on the two surfaces of the undulating membrane are not closely apposed but are separated by a considerable amount of cytoplasm (*Typhlonectes natans*, Fig. 6I). It is uncertain whether the wide cytoplasmic band, resulting in a thick undulating membrane, is considered as an apomorphic condition. It is presumably so as, in their widest region, the undulating membranes of Dipnoi are slender (Fig. 1O), and thickening appears to have evolved several times in anurans (though unknown in urodeles), as in *Cyclorana* (Fig. 6J). Nevertheless, the *Typhlonectes* characteristics (Fig. 6I) are not dissimilar to those of some regions of the *Neoceratodus* undulating membranes (Fig. 1O) and could conceivably be plesiomorphic. The unusual and presumably apomorphic structure of the undulating membrane in *Ascaphus* makes determination of the plesiomorphic condition for Anura difficult.

So far as ultrastructural study of *Typhlonectes natans* indicates, the only fiber remaining in the gymnophionan undulating membrane is the non-trifoliate axial fiber (Fig. 6I). Juxta-axonemal fibers are absent. This appears to be an apomorphy, although, again, there is considerable resemblance in this to the more posterior regions of the undulating membranes in *Neoceratodus*.

The Spermatozoa of Anura

Space limitations permit only a brief survey of anuran sperm. A list of the large literature on spermatozoal ultrastructure is available in Lee and Jamieson (1992).

The sperm of the basal families of the Anura (Ascaphidae, Discoglossidae, Bombinidae) have the features described above for the hypothetical lissamphibian sperm, with certain exceptions which are considered apomorphic. Apomorphies of anuran sperm will here be briefly considered within the context of the various basal and more derived families.

The Anura are definable spermatologically on a single, negative autapomorphy, which is the loss of a longitudinal fiber adjacent to axonemal doublets 7, 8 and 9 (juxta-axonemal fiber at 8), a fiber which exists in

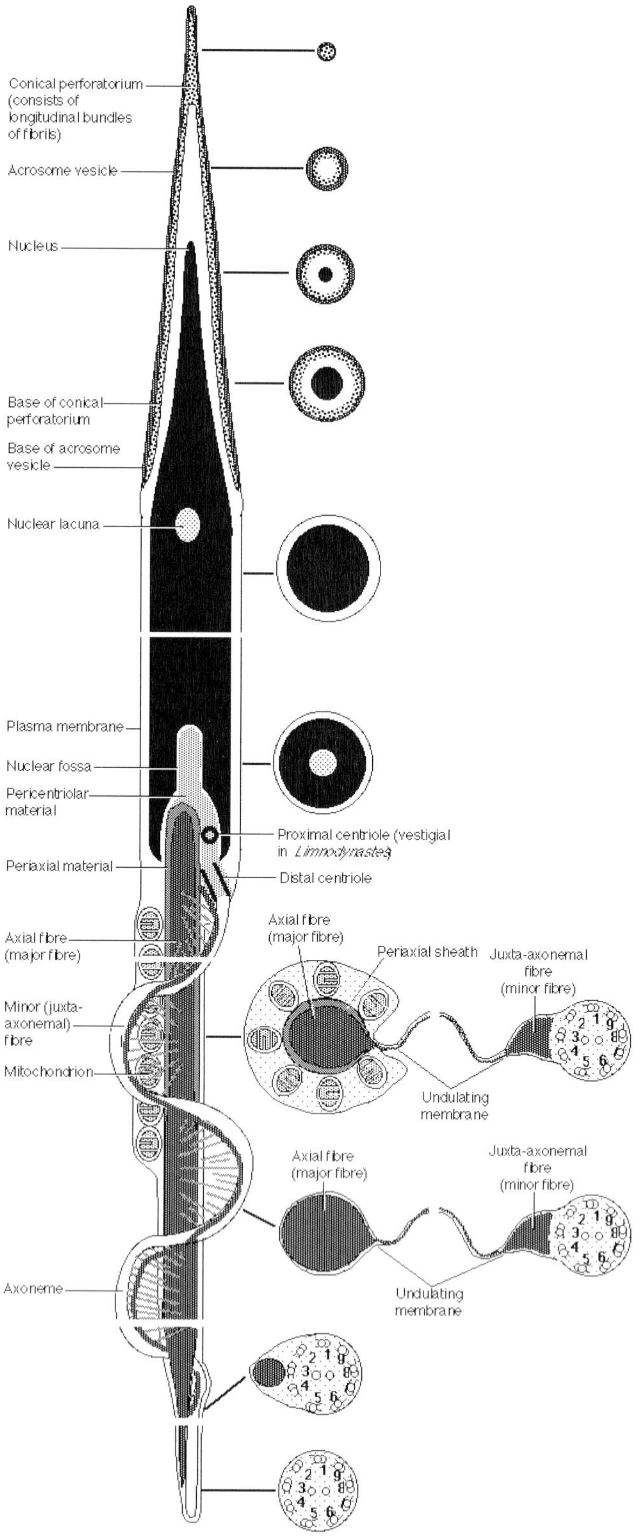

Figure 8. Highly diagrammatic representation of the ultrastructure of a generalized limnodynastine spermatozoon. The conical perforatorium, consisting of bundles of longitudinal fibrils, is a bufonoid autapomorphy. Sperm of the family Myobatrachidae differ from those of other bufonoids in a probable symplesiomorphy, location of the mitochondria bordering the axial fiber and not, as in eubufonoids, in a collar around the anterior axoneme. Extension of the axial fiber into the nuclear fossa characterizing myobatrachids is also known, though less prominent, in one bufonid. The unusually discrete pericentriolar material may be diagnostic of myobatrachids while the periaxial sheath around the proximal region of the axial fiber may be distinctive of limnodynastines. (From Lee, Jamieson 1992.)

urodeles. The fiber is, however, very rarely retained in anurans, as in *Bufo marinus* (Swan et al. 1980), and a transitory fiber is present in this position in the spermatid of *Discoglossus pictus* (Pugin-Rios 1980; Sandoz 1974a). This autapomorphy might, however, be better described as retention of the juxta-axonemal fiber at 3 only, as gymnophionans lack both juxta-axonemal fibers.

It thus appears that anurans are plesiomorphic relative to gymnophionans, which have no juxta-axonemal fibers (if, indeed, this is secondary), and to urodeles, which have a well developed fiber at 8, in retaining the fiber at doublet 3 while being apomorphic in normally losing the fiber at 8.

Major Trends in Spermatozoa of Anura

Ascaphidae. The sole living member of the Ascaphidae, *Ascaphus truei,* is the only anuran with morphological adaptation for internal fertilization. This takes the form of a prominent tail-like intromittent organ. In all other cases of internal fertilization in anurans (for example, *Nectophrynoides*), there are no intromittent organs nor other morphological specializations for internal fertilization, and this occurs simply by cloacal apposition during prolonged amplexus (Townsend et al. 1981). The Ascaphidae are here retained, with the Discoglossidae and Bombinidae, in the Discoglossoidea, although a separate superfamilial status appears justified.

The sperm of *Ascaphus truei* has previously been very briefly described from light microscopy observation (Metter 1964). Its ultrastructure has been described by James (1970) and Jamieson et al. (1993; Fig. 4).

That ultrastructure accords with that of the basic lissamphibian sperm, notably in retaining the basic sarcopterygian synapomorphies of long perforatorial rod which extends anterior to the nucleus but also penetrates a long endonuclear canal, the tetrapod synapomorphy of nuclear shoulders, and the lissamphibian autapomorphy of retention of only one undulating membrane. However, the undulating membrane is of a form, constituting what has been termed a paraxonemal rod (Jamieson et al. 1993), which differs from that shared by the Urodela, Gymnophiona and Anura and must be considered autapomorphic. Putative apomorphies of the *Ascaphus* sperm are as follows.

The paraxonemal rod in *Ascaphus* sperm is considered equivalent to the major and minor fibers (axial fiber and juxta-axonemal fiber at 3) in the sperm tail of other anurans and these structures appear to be jointly homologous with a coarse (peripheral) fiber of the amniote axoneme (Jamieson et al. 1993). The undulating membrane is merely a short, broad bridge between the inner and outer regions of the paraxonemal rod (Fig. 4). The rod carries a longitudinal groove, along this bridge, which houses the mitochondria. This reduction of the undulating membrane cannot be ascribed simply to occurrence of internal fertilization in *Ascaphus* and is tentatively interpreted (Jamieson et al. 1993) as spermatozoal paedomorphism, as an extensive undulating membrane persists in internally fertilizing urodeles and in caecilians. It somewhat resembles the developing undulating membrane complex in the urodele spermatid (Barker, Biesele 1967). In Fig. 7, a cross section of the axoneme of a urodele, *Plethodon albagula,* and of an anuran, *Litoria gracilenta*, both of which have a well developed undulating membrane, is compared with that of *Ascaphus*, with its reduced membrane.Restriction of mitochondria to a groove of the paraxonemal rod (Figs. 4, 7) also appears to be an apomorphy of *Ascaphus*. A fine central "mid-filament" (James 1970) which lies between the central singlets though displaced slightly towards doublet number 1 (Figs. 4, 7) may be an ascaphid apomorphy, though of unknown function.

Discoglossidae—Apomorphies. The spermatozoon of the painted frog, *Discoglossus pictus* (Favard 1955a, b; Furieri 1975a; Pugin-Rios 1980; Sandoz 1970a, b, 1973, 1974a, b), is the longest known in the Amphibia, presumably apomorphically, measuring 2,300 μm to 2,500 μm. The nucleus, 700-800 μm long (Sandoz 1974b), is penetrated for almost its whole length by a narrow endonuclear canal; the perforatorium traverses the canal in the late spermatid (Sandoz 1970b) but is restricted to the prenuclear subacrosomal space in the mature spermatozoon (Pugin-Rios 1980).

The implantation fossa is almost entirely filled by the anterior end of the axial fiber (Pugin-Rios 1980), a putatively apomorphic characteristic which also occurs, independently, in the Myobatrachidae. The short undulating membrane has a thick internal lamina connecting the axial fiber to the juxta-axonemal fiber at 3 (Pugin-Rios 1980). As in many anurans, the mitochondria around the base of the nucleus have disappeared by maturity (Pugin-Rios 1980).

Bombinidae. The spermatozoa of *Alytes obstetricans*, *Bombina variegata* and *Bombina bombina* (=*Bombinator igneus*; Retzius 1906), described by Furieri (1975b), Pugin-Rios (1980) and Retzius (1906), retain the plesiomorphic condition of penetration of the nucleus by an endonuclear canal containing the perforatorial rod. The undulating membrane is also plesiomorphically slender.

A synapomorphy of these three bombinids is apical truncation of the perforatorium (and of the acrosome). *Bombina bombina* (Retzius 1906) and *B. variegata* (Furieri 1975b; Pugin-Rios 1980) are apomorphic in insertion of the axoneme at the anterior end of the nucleus.

Pipoidea—Pipidae. In the Pipidae, only the sperm of *Xenopus laevis* has been examined ultrastructurally (Bernardini et al. 1986, 1988, 1990; Furieri 1972; James

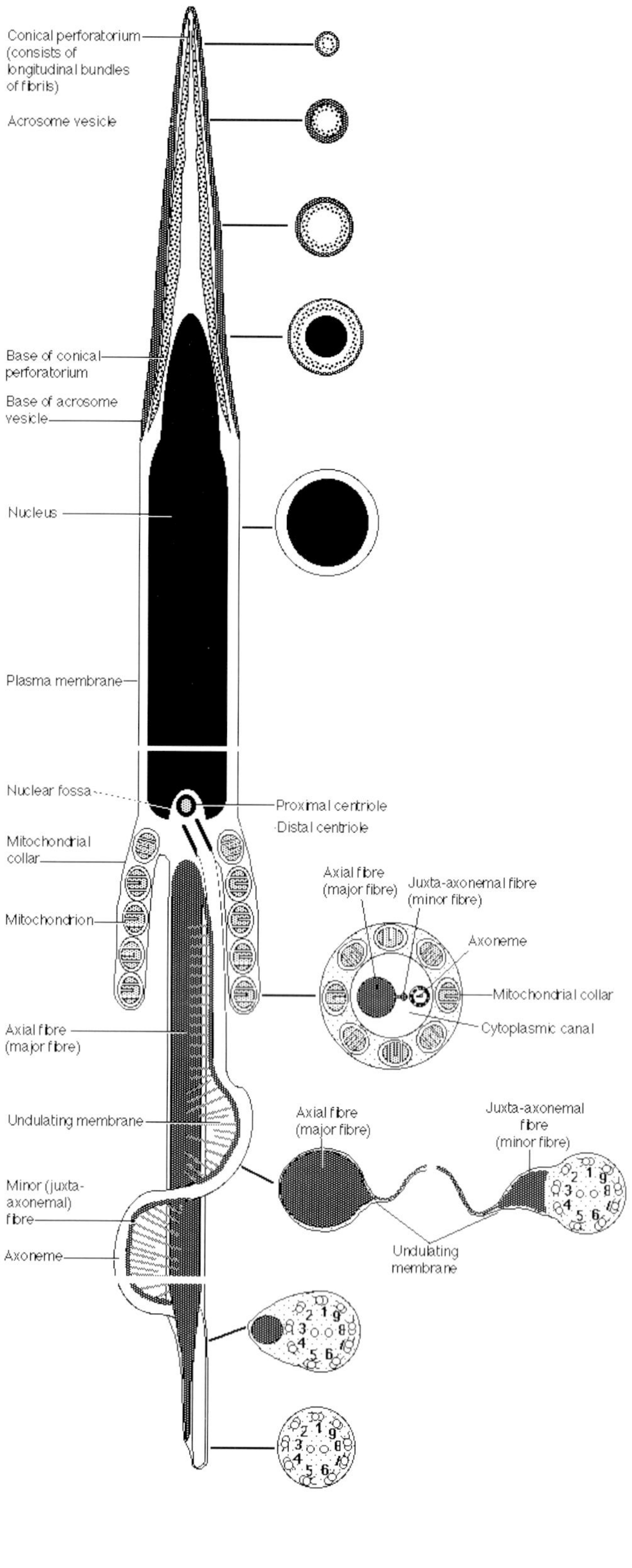

Figure 9. Highly diagrammatic representation of a generalized pelodryadid (Australian hylid)-bufonid spermatozoon. The conical perforatorium is diagnostic of the Bufonoidea. The mitochondrial collar distinguishes eubufonoids, including, *inter alia*, bufonids and hylids, from the Myobatrachidae. (From Lee, Jamieson 1993.)

1970; Pugin-Rios 1980; Reed et al. 1972; van der Horst 1979; Yoshizaki 1987). The spermatozoon is highly apomorphic.

A nuclear fossa is absent. Mitochondria surround the posterior region of the nucleus. There is a long simple flagellum. The undulating membrane, axial and juxta-axonemal fibers have been lost.

The Spermatozoa of the Anura—Neobatrachia

Whereas so-called Archaeobatrachia, constituted by the previous families, clearly comprise a paraphyletic assemblage, the suborder Neobatrachia appears on morphological grounds to be a monophyletic group. This has been supported in parsimony analysis of somatic morphology (Hillis 1991). The Neobatrachia contain the diverse and numerous groups of what may be termed the modern frogs. For brevity the families investigated for sperm ultrastructure will be treated together under their respective superfamilies.

Bufonoidea—Myobatrachidae, Leptodactylidae, Bufonidae, Hylidae, Pelodryadidae

Most of the advanced New World frogs and some frogs in the Old World belong to superfamily Bufonoidea. There appear to be no somatic apomorphies corroborating the monophyly of the bufonoid neobatrachians—indeed, phylogenies based on general morphology have suggested that the group is paraphyletic. However, spermatozoal evidence suggests that the bufonoids are monophyletic, providing a single but convincing synapomorphy.

Bufonoids are united by development of a conical perforatorium (Lee, Jamieson 1992, 1993; *Limnodynastes peronii*, Fig. 6B; also Figs. 8, 9) and loss of the axial, rodlike perforatorium. It consists of separate sheaves or fibers and occupies the location of the subacrosomal cone of plesiomorphic lissamphibian and amniote spermatozoa. It has been argued (Jamieson et al. 1993) that the conical layer of subacrosomal material between the acrosome and the nucleus in urodeles and ascaphids (Jamieson et al. 1993) and in many amniotes including the Chelonia (Healy, Jamieson 1992, 1994; Jamieson, Healy 1992) does not appear to be homologous with the conical perforatorium of bufonoids because the two structures differ ultrastructurally and because a conical perforatorium does not exist in anuran lineages which are more basal than the bufonoids. However, with loss of the central perforatorial rod, it remains possible that the original subacrosomal cone became modified as the bufonoid conical perforatorium.

Within the bufonoids, sperm ultrastructure strongly supports the separation of the myobatrachids (Australasian "leptodactylids") from the hylid-bufonid-

New World leptodactylid assemblage (termed the true or eubufonoids) as sister groups (Lee, Jamieson 1992). This phylogenetic and taxonomic arrangement has been repeatedly proposed in the past.

A generalized myobatrachid (lymnodynastine) spermatozoon is represented diagrammatically in Figure 8.

Myobatrachidae—Synapomorphy

The myobatrachids do not have the mitochondrial collar which is distinctive of eubufonoids, but are united by their own synapomorphy, the extension of the axial fiber up the centriolar fossa (Lee, Jamieson 1992; Figs. 6L, 8; and *Limnodynastes convexiusculus*, Fig. 6L), though this characteristic is approached in the bufonid *Nectophrynoides* and is present homoplasically in *Discoglossus*. Penetration in limnodynastines may be a distinctive apomorphic reversal from non-penetration in lineages intervening between them and the origin of discoglossids. The myobatrachids appear to be the sister group of the eubufonoids.

Eubufonoidae—Synapomorphy

The families Bufonidae, Leptodactylidae (*s. strict.*), and Hylidae (including Australian hylids, the Pelodryadidae) are united, and separated from myobatrachids, by a single synapomorphy: a thick collar-like cytoplasmic sheath (Fig. 9, and the hylid *Litoria rheocola*, Fig. 6K), that emanates from the centriolar region, is separated from the flagellum by a cytoplasmic canal, and contains the mitochondria. This synapomorphy led to erection of the superfamily Eubufonoidea by Lee and Jamieson (1993). A collar is a widespread characteristic of spermatozoa in fish groups, including Dipnoi, and may have been lost in lower anurans to be regained in the eubufonoids.

Eubufonoids have the full complement of conical acrosome (though with a conical rather than rodlike perforatorium, as in myobatrachids), elongate nucleus, and undulating membrane with axial fiber, juxta-axonemal fiber at 3, and rarely (as in *Bufo marinus*) at 8.

Some hylids show reduction of the undulating membrane. This forms a thick lamina with no discrete juxta-axonemal fibers and ends laterally with the major fiber in the Australian *Cyclorana alboguttata* (Fig. 6J), forming a secondary paraxonemal rod. Nevertheless this species exemplifies the bufonoid conical perforatorium and the eubufonoid mitochondrial collar. The form of the undulating membrane confirmed (Meyer et al. 1997) that it should be placed in *Cyclorana* and not in *Litoria,* in which it is commonly included.

Litoria has a well developed, thin undulating membrane with a hypermorphosed juxta-axonemal fiber (Lee, Jamieson 1993; Figs 6P, 9; and *Litoria eucnemis*, Fig. 6P). In *Hyla meridonalis* there is further reduction and the axial fiber directly parallels the axoneme (Pugin-Rios 1980), as is also seen in *Hyla japonica* (Kwon, Lee 1995).

Spermatozoa of the Microhylidae have not previously been investigated ultrastructurally. That of *Cophixalus inornatus* (Fig. 6F) is apomorphic in having lost the undulating membrane and associated fibers, giving a secondarily simple flagellum (BGM Jamieson, DM Scheltinga and KR McDonald, unpublished results). This correlates with terrestrial reproduction in microhylids. However, loss of the undulating membrane also occurs in the lentic Ranidae.

Ranoidea

Ranoid families examined for sperm ultrastructure are the Ranidae, in Mo (1985), Poirier and Spink (1971), Serra and Vincente (1960), Yoshizaki (1987) and Zirkin (1971); and Rhacophoridae (Mainoya 1981; Wilson et al. 1991). A full listing is provided in Lee and Jamieson (1992). The sperm of both families are highly apomorphic.

The sperm of ranids are much modified and simplified, most notably in losing the undulating membrane and axial fiber. The acrosome is caplike, sometimes asymmetrically, and apomorphically lacks a rodlike or conical perforatorium. The mitochondria form a manchette surrounding the base of the nucleus and a considerable region of the axoneme.

The sperm of the Japanese rhacophorid species *Buergeria buergeri*, which lays eggs in streams, has a long head and thin tail, details of which are unknown (Fukuyama et al. 1993). In contrast, the foam-nesting species *Chiromantis xerampelina* (Mainoya 1981; Wilson et al. 1991, DM Scheltinga, BGM Jamieson and AN Hodgson, unpublished results; Figs. 6G, H), *Rhacophorus arboreus* (Mizuhira et al. 1986) and *R. schlegelii* (Mizuhira et al. 1986; Oka 1980) have what appear to be the most modified known amphibian sperm.

Chiromantis spermatozoa have the form of a counterclockwise corkscrew, viewed from the anterior end. The coils involve the acrosome, nucleus and midpiece, which consists of mitochondria around the base of the nucleus. A pair of free parallel flagella comprise the tailpiece. A crystalline matrix composed of many microtubules (Fig. 6H) surrounds these.

The coiled head of the sperm is interpreted as an adaptation to the special microfertilization environment. The coil unwinds in different aqueous media and this lengthens the sperm seven to eight times. The sperm also exhibits a "star-spin" movement, comparable with the hyperactivation of mammalian sperm. The two tails of the sperm seem to enhance this movement, probably facilitating movement in the gelatinous foam and penetration of the outer layers of the egg (Mainoya 1981; Wilson et al. 1991).

The Spermatozoa of the Amniota

From detailed comparative and cladistic considerations of the anatomy of amniote sperm, Jamieson (1995a) recognized the following characteristics of a hypothetical plesiomorphic amniote spermatozoon (Fig. 10). This model is not overly speculative, as it is virtually identical with that of the lowest extant amniotes, the Chelonia and Sphenodontida. Relative to the tetrapod ground plan, deduced from common features of the amniote and lissamphibian sperm, amniotes are seen to have few basal synapomorphies.

Amniote Spermatozoal Symplesiomorphies

Plesiomorphic features of basal amniotes, retained from their tetrapod ancestry, and still seen in Chelonia and Sphenodontida and to varying extents in other amniotes, are as follows. The generalized plesiomorphic amniote spermatozoon (Fig. 10) is elongated and filiform, with a hollow anterior conical acrosome vesicle overlying a simple subacrosomal cone. The base of the acrosome invests the tapered anterior tip (rostrum) of the nucleus and rests on pronounced nuclear "shoulders." The subacrosomal space within the acrosome contains two or three axial rods (putative perforatoria) or, less likely, only one rod. These penetrate the nucleus deeply, almost to its base, in endonuclear canals. The nucleus is plesiomorphically elongated and cylindrical in amniotes from Chelonia through *Sphenodon*, crocodiles, squamates, birds, monotremes and, in therian mammals, the pangolin alone (Leung, Cummins 1988), as in lissamphibians. At the base of the nucleus there is a compact fossa (implantation fossa). Associated with this are two triplet centrioles. The distal centriole forms the basal body of the flagellar axoneme. Whether the presence of an annulus is plesiomorphic or apomorphic is debatable. The terminal portion of the 9+2 axoneme forms a short endpiece distinguished from the principal piece by the absence of the fibrous sheath.

Amniote Spermatozoal Synapomorphies

The Chelonia and Sphenodontida are considered the most basal extant amniotes and have virtually identical spermatozoa (Jamieson 1995a; Jamieson, Healy 1992). The characteristics of these include features considered synapomorphies of the Amniota which are simultaneously symplesiomorphies for these two orders and for the remaining amniotes. The amniote synapomorphies are listed below.

The distal centriole is extremely elongated and extends the entire length of the long midpiece (the latter defined by its mitochondria) in turtles, the tuatara (Fig. 11), crocodiles, and ratites, an apparent basal synapomorphy of amniotes. These elongate centrioles differ from most metazoan basal bodies in being penetrated by two central singlets from the axoneme. Thus, in spermatids of the ratite *Rhea*, the distal centriole elongates and, late in spermiogenesis, becomes penetrated by a central pair of tubules from the developing axoneme (Phillips, Asa 1989). The shorter, though still elongated, distal centriole in the rooster and the somewhat shorter centriole in guinea fowl, at 0.6 μm, and *Geopelia striata*, at 0.5μm (Jamieson 1995a); the short centriole in squamates; and the vestigial centriole in monotremes possibly represent secondary reduction in length of the distal centriole (Healy, Jamieson 1992), culminating in almost total reduction in therian mammals.

In turtles, tuatara (Healy, Jamieson 1992, 1994; Jamieson 1995a; Jamieson, Healy 1992; Fig. 11), *Caiman crocodylus* and *Crocodylus johnstoni* (Jamieson 1995a; Jamieson et al. 1997; Fig. 12), the mitochondria have concentric cristae, known elsewhere in amniotes only in the sperm of some marsupials, notably the Woolly opossum, *Caluromys philander* (Fawcett 1970; Phillips 1970) and the Virginia opossum, *Didelphis virginiana* (Temple-Smith et al. 1980) and also in the macropod *Lagorchestes hirsutus* (Fig. 16). The marsupial condition may be homoplasic but it is here considered possible that it is evidence of an ancient synapsid, and therefore, mammalian link, with the lower, anapsid amniotes. The cristae also tend to a circular arrangement in monotremes (Bedford, Rifkin 1979; Carrick, Hughes 1982).

The mitochondrial cristae in the three "reptilian" taxa usually surround a large central dense body. In all other amniotes studied, the cristae have a "conventional" appearance, being linear or curved, as in Lissamphibia, but never concentric, and do not surround a dense body. In spermatids of *Sphenodon* (Healy, Jamieson 1992; Jamieson, Healy 1992), the cristae have the linear appearance usual for metazoan sperm and the concentric arrangement is a late development. Phylogenetic "reversion" of concentric cristae to the linear condition seen in other amniotes would need only suppression of this final transformation (Jamieson, Healy 1992).

A dense ring, the annulus, at the posterior end of the midpiece is a feature of many metazoan sperm. It is clearly plesiomorphic for amniotes, occurring in all classes (Jamieson, Healy 1992) but its absence in Dipnoi possibly indicates apomorphic re-acquisition in tetrapods. It is well developed in Chelonia, *Sphenodon* (Healy, Jamieson 1992, 1994; Jamieson 1995a; Fig. 11), *Caiman crocodylus* (Saita et al. 1987), the American alligator (Phillips, Asa 1993) and *Crocodylus johnstoni* (Jamieson et al. 1997; Fig. 12).

A dense fibrous sheath (Fig. 10) must, clearly, have developed as an annulated structure in the earliest amniotes, as it is present in all amniote classes. With the exception of squamates (Fig. 15C), it commences immediately behind the midpiece, as in turtles, *Sphenodon*

(Healy, Jamieson 1992; Jamieson, Healy 1992; Fig. 11), *Caiman crocodylus* and *Crocodylus johnstoni* (Jamieson 1995a; Jamieson et al. 1997; Fig. 12), ratites, non-passerines (Fig. 13B) and in mammals (Jamieson 1995a; Figs. 16, 17F, M-Q).

Nine longitudinal dense fibers (coarse fibers) peripheral to the nine axonemal doublets, or to the distal centriole also where this is elongated as in *Emydura*, *Sphenodon* and *Crocodylus*, are a fundamental feature of amniote sperm (Fig. 10), being found in all classes (Jamieson 1995a; Jamieson, Healy 1992, Jamieson, Scheltinga 1993). As nine peripheral fibers are seen in lampreys and *Pantodon* (references in Jamieson 1991) but also in heterobranch and cephalopod molluscs (Healy 1988, 1990), it might be considered that nine is the basic sarcopterygian, rather than merely amniote, number and that amphibians have lost all but those represented by the fibers at doublets 3 and 8. However, there is no evidence in extant Lissamphibia for such a reduction and the presence of only two lateral elements in dipnoans and *Latimeria* suggests that nine fibers were an amniote synapomorphy, albeit homoplasic with the other, non-amniote taxa. The dense fibers are small in turtles, the tuatara (Fig. 11), *Caiman crocodylus* and *Crocodylus johnstoni* (Fig. 12), squamates (Fig. 15B), birds (Figs. 13B, F, N, 14) and monotremes; this is reviewed by Jamieson (1995a).

It is possible that a further basal amniote apomorphy is enlargement and lateral displacement of two fibers, at doublets 3 and 8, and that all fibers in the centriolar region intruded into the inter-triplet radii, as in "lower" amniotes (Chelonians, *Sphenodon* and crocodiles), as seen in Jamieson (1995a).

In the principal piece, two longitudinal keel-like outward projections (longitudinal columns), at doublets 3 and 8, may be present, each aligned with an inward projection of the fibrous sheath, as shown for the mammalian spermatozoon by Fawcett (1975); this is also shown in the elephant shrew *Macroscelides*, in this study (Fig. 17O). At least the two inward projections are present in *Sphenodon* (Fig. 11) and may be a basal synapomorphy for amniotes.

Loss or transformation of the retronuclear body, present in dipnoans and (as the neck structure) in urodeles appears to have occurred as an amniote apomorphy. It is never retained in discrete form seen in Dipnoi. Its putative homolog in urodeles but questionable homology of the striated columns of mammalian sperm (Figs. 17G, H, L) has already been alluded to.

Chelonia and Sphenodontida

The spermatozoa of the Chelonia and of *Sphenodon* (Fig. 11) are indistinguishable from the amniote ground plan.

Crocodilia

The ground plan for the Crocodilia, as exemplified by *Crocodylus johnstoni* (Jamieson 1995a; Jamieson et al. 1997; Fig. 12), is very similar to that of the Chelonia and *Sphenodon*. All three have two or three endonuclear canals and, though requiring further confirmation for crocodiles, concentric cristae with intramitochondrial bodies. In *Crocodylus johnstoni,* the mitochondria are subspheroidal to slightly elongate and possess few septate to (more externally) concentric cristae (Jamieson 1995a; Jamieson et al. 1997); a central dense body reported for *Caiman crocodilus* (Saita et al. 1987) is questionably present.

The spermatozoon of *C. johnstoni* is apomorphic relative to those of Chelonia and *Sphenodon* in reduction of concentric mitochondrial cristae. It is less similar to that of ratites than is that of *Caiman crocodilus*, differing from ratites in having more than one perforatorium.

Synapomorphies of Crocodilian Spermatozoa

In *Caiman crocodylus* (Saita et al. 1987) and *Crocodylus johnstoni,* the investment of the two central singlets of the axoneme or of the distal centriole in a thick dense sheath (Fig. 12) differs from the density, resembling a fiber, associated with the singlets in Chelonia, *Sphenodon* (Healy, Jamieson 1992) and (homoplasically?) snakes (Jamieson, Koehler 1995; Oliver et al. 1996).

Further Apomorphies in *Caiman*

Restriction of the endonuclear canal to the anterior region of the nucleus indicated by Saita et al. (1987) is clearly apomorphic but requires confirmation.

The Spermatozoa of Aves

The conical acrosome, fibrous sheath, elongated centriole and nine dense fibers of ratites are symplesiomorphies not proving avian monophyly. Furthermore, monophyly of ratites cannot be considered proven, as features considered to unify them—conical acrosome, fibrous sheath and elongated centriole (Baccetti et al. 1991)—are all symplesiomorphies.

Spermatozoal Symplesiomorphies of Birds

In birds, a conical acrosome vesicle penetrated almost to its tip by a subacrosomal space which contains a rodlike perforatorium has been demonstrated ultrastructurally in the non-passerines: turkey, *Meleagris gallopavo*; rooster, *Gallus domesticus* (Fig. 13A); guinea fowl, *Numida meleagris* (Thurston et al. 1987); mallard duck, *Anas platyrhynchos* (Humphreys 1972); the quail *Coturnix coturnix* (Humphreys 1972); and parrots

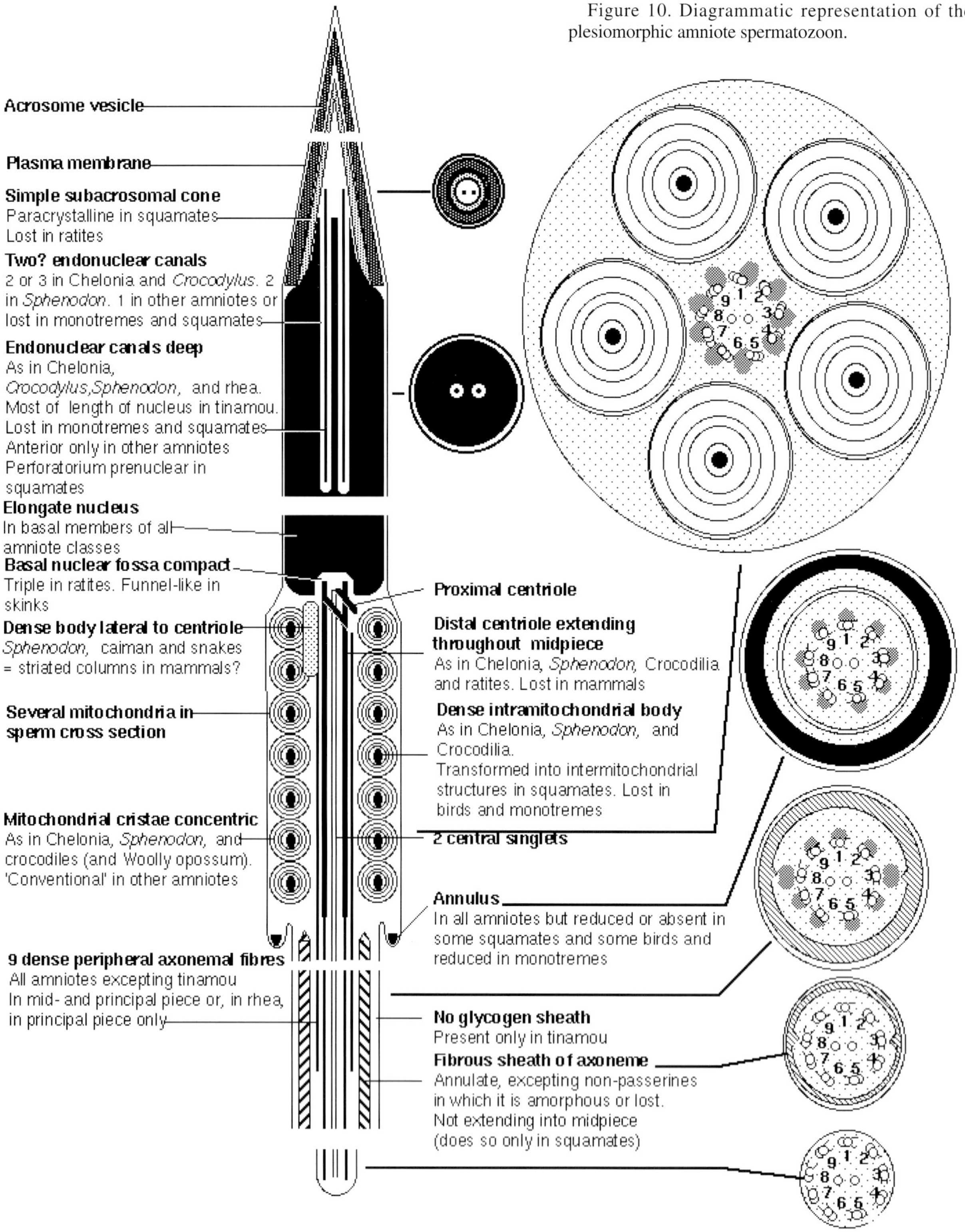

Figure 10. Diagrammatic representation of the hypothetical plesiomorphic amniote spermatozoon.

(Jamieson et al. 1995; Figs. 13J, 14). This has also been demonstrated in the ratites (palaeognaths) tinamou, *Eudromia elegans* (Asa et al. 1986); ostrich, *Struthio camelus* (Baccetti et al. 1991; Soley 1993, 1994); and emu, *Dromaius novaehollandiae* (Baccetti et al. 1991). This is a basic sarcopterygian synapomorphy and therefore avian symplesiomorphy.

The endonuclear canal extends almost to the base of the nucleus as in Chelonia and *Sphenodon* (Healy, Jamieson 1992, 1994; Jamieson, Healy 1992; Fig. 11) in

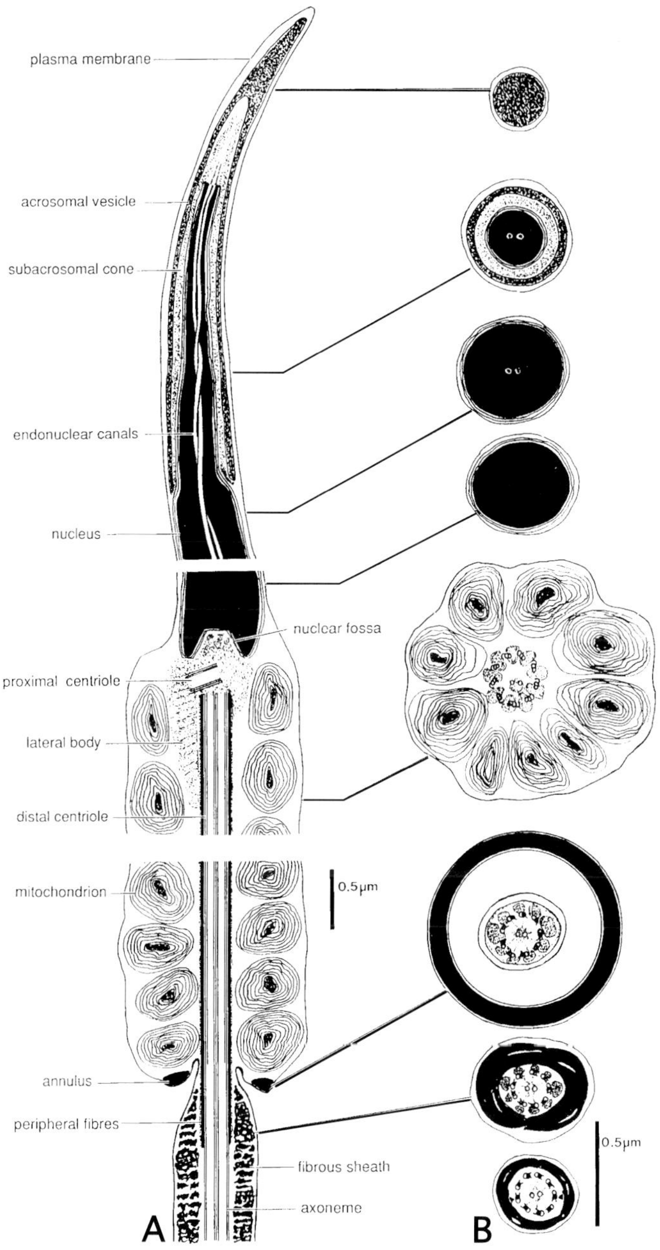

Figure 11. *Sphenodon punctatus*. Semidiagrammatic representation of spermatozoal ultrastructure as seen in longitudinal section and transverse sections by transmission electron microscopy. (From Healy, Jamieson 1992.)

putatively more primitive ratites, such as tinamou (Asa et al. 1986), where it is probably plesiomorphic relative to the shorter condition in other non-passerines and in advanced ratites, the perforatorium being wholly pre-nuclear in the emu (Baccetti et al. 1991). However, there is a possibility that deep penetration in ratites is secondary (Jamieson, Healy 1992).

Nine dense peripheral fibers, a basic amniote synapomorphy, and avian symplesiomorphy, have been observed in turkey, rooster and, though requiring confirmation, in guinea fowl (Thurston, Hess 1987); mallard duck (Humphreys 1972); parrots (Jamieson et al. 1995); doves such as *Geopelia striata* (Jamieson 1995a; Figs. 13E, F, H); passerines such as *Grallina cyanoleuca* (Fig. 13N); and in the anteriormost region of the principal piece of ratite spermatozoa (Asa et al. 1986; Baccetti et al. 1991; Soley 1993, 1994). They are present in suboscine and the more apomorphic oscine passerines (Fig. 13N), being larger in the latter.

Spermatozoal Synapomorphies of Birds

Like the sperm of ratites and other birds, parrot sperm differ from those of "reptiles" in reduction of the subacrosomal material (subacrosomal cone, excluding any perforatorium) to a negligible amount (Jamieson et al. 1995; Figs. 13J, 14). Although avian sperm present several apomorphies, this may be the only basal synapomorphy for bird sperm. It is possible, though, that a further basal synapomorphy has been the loss of concentric cristae.

Secondary Spermatozoal Synapomorphies of Birds

Most of the synapomorphies of bird sperm appear to have been derived within the class rather than basally.

The endonuclear canal is limited to the anterior 1 to 2 µm of the nucleus in rooster (Fig. 13A), guinea fowl, turkey and parrot (Figs. 13J, 14) and the anterior third of the nucleus in the ostrich (Baccetti et al. 1991). Restriction of the endonuclear canal to the anterior region of the nucleus in non-passerines and passerines may be a synapomorphy of these, homoplasic with crocodiles and derived ratites (emu, ostrich).

In the non-passerines rooster (Figs. 13B, C) and guinea fowl, the fibrous sheath has transformed into an amorphous sheath (Thurston, Hess 1987). Less certainly derived is adhesion of all nine dense fibers to their axonemal doublets (Figs. 13F, H, N, 14), a feature also seen in monotremes (Jamieson 1995a).

If birds are, as is commonly held, the sister group of crocodiles, their linear cristae (Figs. 13B, F, G, H, L, M) would indicate loss of the concentric cristae of a common ancestor with crocodiles. However, the sister group relationship with crocodiles was not supported in a cladistic analysis of sperm ultrastructure (Jamieson, Healy 1992). Although a bird-mammal sister group relationship shown in the latter analysis is doubtful, it is remarkable that this relationship was supported by results from a molecular study (Hedges et al. 1990).

Not all non-passerines possess a conical acrosome, and the modifications which occur appear to be secondary, internal apomorphies within the Aves. A small, approximately spherical acrosome has been described for the white-naped crane, *Grus vipio* (Phillips et al. 1987), for *Jacana jacana* (Saita et al. 1983) and most

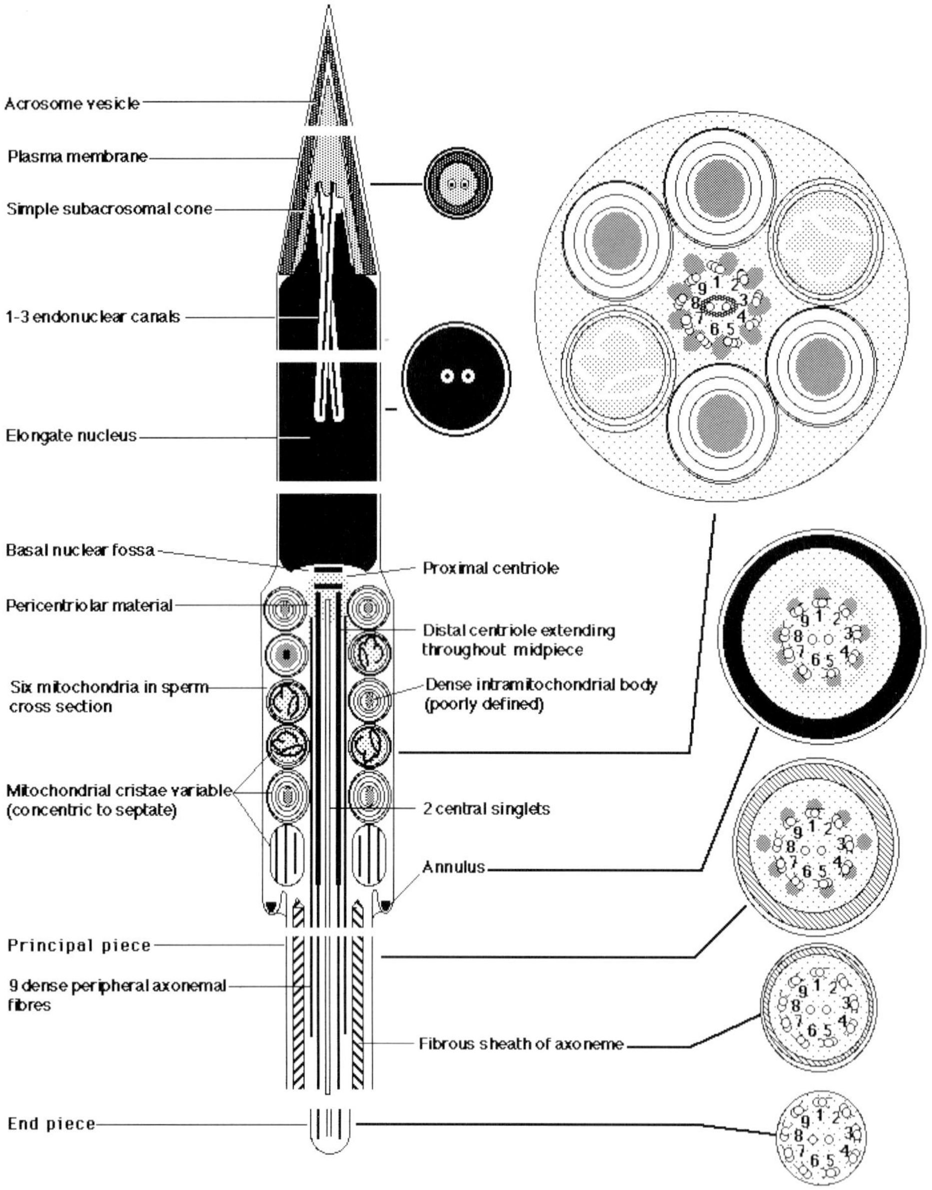

Figure 12. *Crocodylus johnstoni*. Diagram of spermatozoal ultrastructure as seen in longitudinal section and transverse sections by transmission electron microscopy. (From Jamieson et al. 1997.)

Charadriiformes (Fawcett et al. 1971), and for the woodpecker *Melanerpes carolinus* (Henley et al. 1978). These latter avian taxa are considered to be advanced non-passerines, on the basis of DNA hybridization studies (Sibley et al. 1988, 1990).

In the columbiforms, such as *Geopelia striata* (Fig. 13D) and *Ocyphaps lophotes* (Fig. 13K), even a perforatorium is absent although, at least in *Geopelia striata,* some longitudinally orientated subacrosomal material is present and lacunae are present in the nucleus which may represent a vestigial endonuclear canal (Jamieson 1995a). Both of these columbiforms appear plesiomorphic in retaining a nuclear rostrum (Figs. 13D, K), though it is possible that its presence is a reversal.

In parrots (Jamieson et al. 1995; Fig. 14) and doves, such as *Ocyphaps lophotes* and *Geopelia striata*, the fibrous sheath is lost (Fig. 13I). These taxa would therefore provide valuable controls in experimental investigations of the function of the fibrous sheath.

A sheath of putative glycogen external to the fibrous sheath is known only in the tinamou (Phillips, Asa 1989) and cannot be ascribed to the plesiomorphic amniote sperm.

An annulus is basic to bird sperm, being seen in ratites, rooster (Fig. 13B), guinea fowl, and columbiforms (Asa, Phillips 1987), but is apomorphically absent in parrots (Jamieson et al. 1995; Fig. 14).

Dense fibers are described as "tiny" for the rhea, are absent from the tinamou (Asa et al. 1986), and are greatly reduced in columbiforms (Jamieson 1995a). Very small dense fibers are present only in the distal region of the midpiece in the rooster (Fig. 13B) and mallard; dense fibers in turtle dove sperm disappear before maturation is complete (Asa, Phillips 1987), though they persist through a short region of the midpiece in *Geopelia striata* (Jamieson 1995a; Fig. 13F).

Oscine Synapomorphies

Oscine spermatozoa have an extremely large acrosomal complex in contrast to non-passerine and suboscine spermatozoa in which the acrosome is short relative to the nucleus, as in reptiles (Jamieson, Healy 1992; Jamieson, Scheltinga 1993, 1994). In passerines the acrosome vesicle becomes an elongate single-keeled helix, with no evident subacrosomal cone, like that of finches (Kondo et al. 1988; Koehler, 1995).

A helix of densely packed microtubules invests the axoneme in passerines, as seen in *Grallina cyanoleuca* (Fig. 13N).

Spermatozoa of the Squamata

Squamate sperm have all of the basic amniote synapomorphies described above as their plesiomorphies. Their synapomorphies strongly support squamate monophyly.

Synapomorphies of Squamate Sperm

The perforatorial rod in the Squamata is wholly prenuclear. It sits on the tip of a well developed subacrosomal cone and might be termed pre-subacrosomal (*Carlia rubrigularis*, Fig. 15A). Endonuclear canals are absent (Jamieson 1995b; Jamieson et al. 1996; Oliver et al. 1996). This anterior restriction, and reduction to a single perforatorium, are clearly apomorphic relative to the basal amniotes Chelonia and Sphenodontida.

Presence of a well developed epinuclear electron lucent region is a squamate autapomorphy (Jamieson 1995a, b; Jamieson et al. 1996; Oliver et al. 1996; Fig. 15A). The intermitochondrial rings or dense bodies (*Ctenotus inornatus*, Fig. 15B; *C. rawlinsoni*, Fig. 15C) of squamate sperm are regarded as derivations of the intramitochondrial dense bodies (Carcupino et al. 1989; Healy, Jamieson 1992; Jamieson, Healy 1992) seen in basal amniotes (chelonians and sphenodontids). Origin of intermitochondrial material from mitochondria has been confirmed ontogenetically in the sperm of some squamates (Oliver et al. 1996). Extramitochondrial dense bodies are almost limited to squamates but are seen, poorly developed, in the doves *Geopelia striata* (Jamieson 1995a; Figs. 13E, G) and *Ocyphaps lophotes* (Fig. 13L) in which, although appearing homoplasic, they may well indicate persistence of a genetic basis laid down in early amniotes.

In squamates, alone in the Amniota, the fibrous sheath extends anteriorly into the midpiece (*Ctenotus rawlinsoni*, Fig. 15C), a striking squamate autapomorphy (Healy, Jamieson 1992; Jamieson, Healy 1992; Jamieson, Scheltinga 1993).

In the Squamata, the subacrosomal cone has a paracrystalline substructure (Butler, Gabri 1984; Carcupino et al. 1989; Furieri 1970), recently confirmed for Sphenomorphus and Eugongylus group skinks (*Carlia rubrigularis*, Fig. 15A), the gekkonid *Heteronotia binoei*, and in snakes (Oliver et al. 1996). It constitutes a basal synapomorphy of the Squamata (Jamieson, Healy 1992).

Shortening of the distal centriole (Fig. 15C) from the elongated basal amniote character is a basal apomorphy of squamates (Jamieson 1995a). In squamates, there are nine peripheral fibers in the midpiece (Fig. 15B), but at the level of the annulus the only well developed, though small, peripheral fibers are the double fibers at doublets 3 and 8; by the beginning of the principal piece all nine dense fibers are already vestigial (*Ctenotus ornatus*, Fig. 15B) or absent (Jamieson 1995a; Jamieson, Scheltinga 1993).

Secondary Spermatozoal Apomorphies in Squamates

The many internal apomorphic changes within the Squamata are discussed in some detail by Jamieson (1995b). Consideration of these, and the various subgroups, is beyond the scope of this chapter. Only the striking apomorphies of snake sperm will be discussed.

Snake sperm are characterized, apomorphically, by multilaminar membranes in place of the normal plasma membrane of the midpiece and axoneme (Jamieson, Koehler 1995; Oliver et al. 1996); this can be compared to pygopodids (Jamieson 1995b; Jamieson et al. 1996).

Snake sperm are unique in the Squamata in the immense elongation of the midpiece (Jamieson 1995a, b). Snake sperm show reduction of the epinuclear electron-lucent region and reduction or loss of the perforatorial base plate, and greater development of extracellular tubules than is known in any other squamate (Jamieson 1995a, b).

Spermatozoa of Mammalia

A recent paper on the sperm of *Tarsius bancanus* found spermatozoal ultrastructure to be of limited value, in isolation, in reconstructing the phylogeny of primates (Robson et al. 1997). However, other works have shown the utility of spermatozoal ultrastructure for this purpose, notably for marsupials (Harding et al. 1987; Temple-Smith 1987), rodents (Breed 1997), elephant shrews (Woodall et al. 1995) and megachiropterans (Rouse, Robson 1986).

Particularly pertinent to establishing the plesiomorphic sperm type for the Mammalia, and at the same time mammalian synapomorphies, are accounts of the

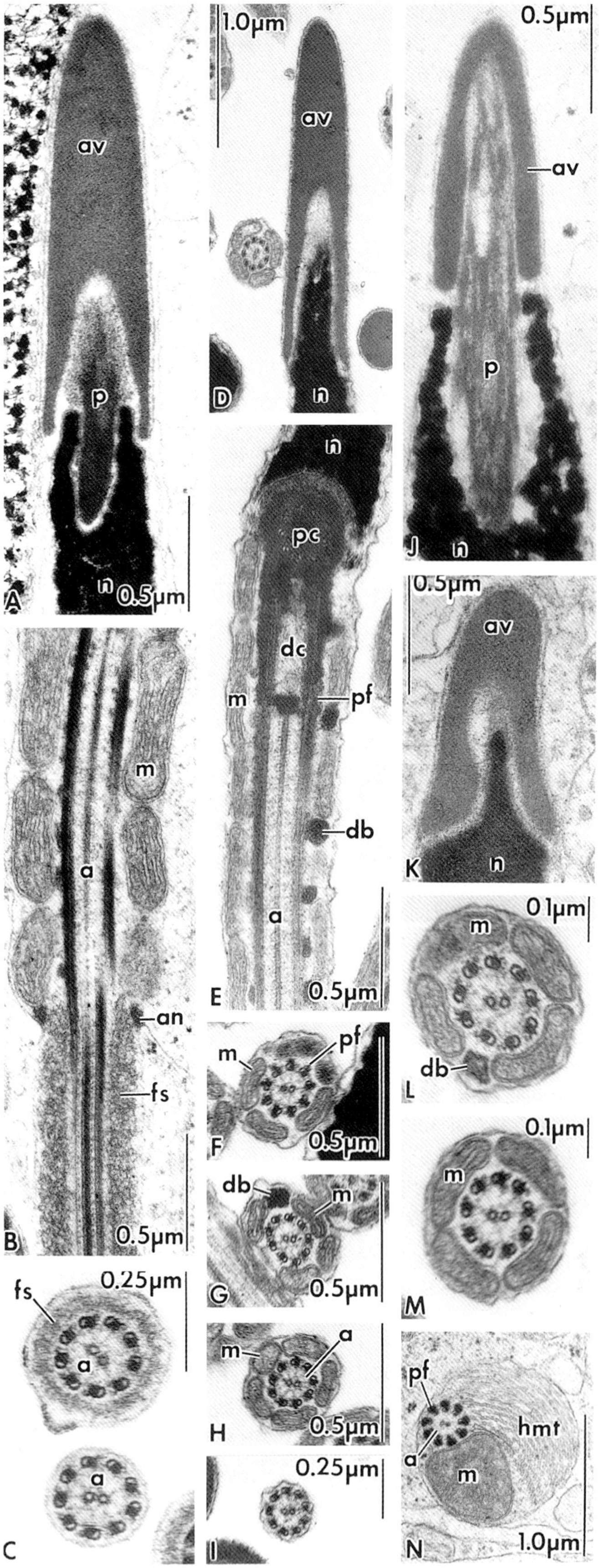

Figure 13. Spermatozoa of Aves. TEM sections. A-C: *Gallus domesticus*, rooster. A: LS of acrosome, perforatorium and endonuclear canal. B: LS of midpiece at junction with principal piece. Note amorphous fibrous sheath distal to annulus. C: TS of principal piece and, below it, an endpiece. D-I: *Geopelia striata*, peaceful dove. D: Acrosome, showing absence of perforatorium. E: Anterior midpiece, showing dense bodies resembling squamate intermitochondrial bodies. F-H: TS of midpiece, showing progressive distalwards reduction of peripheral fibers (in the order F, H, G) and, in H, a dense body. I: TS of principal piece or endpiece; a fibrous sheath is absent from all sections. J: *Platycercus elegans*, crimson rosella. LS acrosome, perforatorium and endonuclear canal. K-M: *Ocyphaps lophotes*, crested pigeon. K: Showing a nuclear rostrum in the absence of a perforatorium. L & M: TS of midpiece, in L, showing a dense body. N: *Grallina cyanoleuca*, magpie lark. Helical sheath of microtubules around the midpiece of the oscine sperm. (Original.)

remarkably reptilian sperm of monotremes (Bedford, Rifkin 1979; Carrick, Hughes 1982) and of pangolins, in the order Pholidota (Ballowitz 1907; Leung 1987).

Symplesiomorphies of Mammalian Sperm

The sperm of monotremes (Bedford, Rifkin 1979; Carrick, Hughes 1982; Jamieson, Healy 1992) are remarkably primitive in retaining the elongated conical structure of the acrosome and the subacrosomal cone. The acrosome retains its plesiomorphic conical form in the pangolin, *Manis pentadactyla* (LKP Leung 1987, and personal communication). The acrosome in *M. pentadactyla* (1.7-2.2 μm long) sits caplike over the rostral fifth of the nucleus, with no substantial subacrosomal material, and does not appear to have the well defined equatorial segment which distinguishes most other Eutheria, which further differ in having flattened sperm heads, from the pangolin sperm (Leung 1987). Thus, monotremes appear to be the only mammals which retain the subacrosomal cone. The nucleus is pointed, elongated and circular in cross section in monotremes (Bedford, Rifkin 1979; Carrick, Hughes 1982) and in pangolins (Ballowitz 1907; Leung 1987) but this condition is unknown in marsupials.

Synapomorphies of Mammalian Sperm

The attempt may be made to distinguish basal synapomorphies of mammalian sperm from secondary synapomorphies of the three chief constituent groups, Prototheria (monotremes), Metatheria (marsupials) and Eutheria (here termed allantoplacental mammals).

Monotreme sperm are in many respects "reptilian." Synapomorphies which they show and which can reasonably be attributed to an ancestral mammal are as follows.

In mammals, a rodlike perforatorium and endonuclear canal do not occur (Jamieson 1995a), a loss which is homoplasic with that in some non-ratite birds.

In the spiral midpiece of mammals, the number of

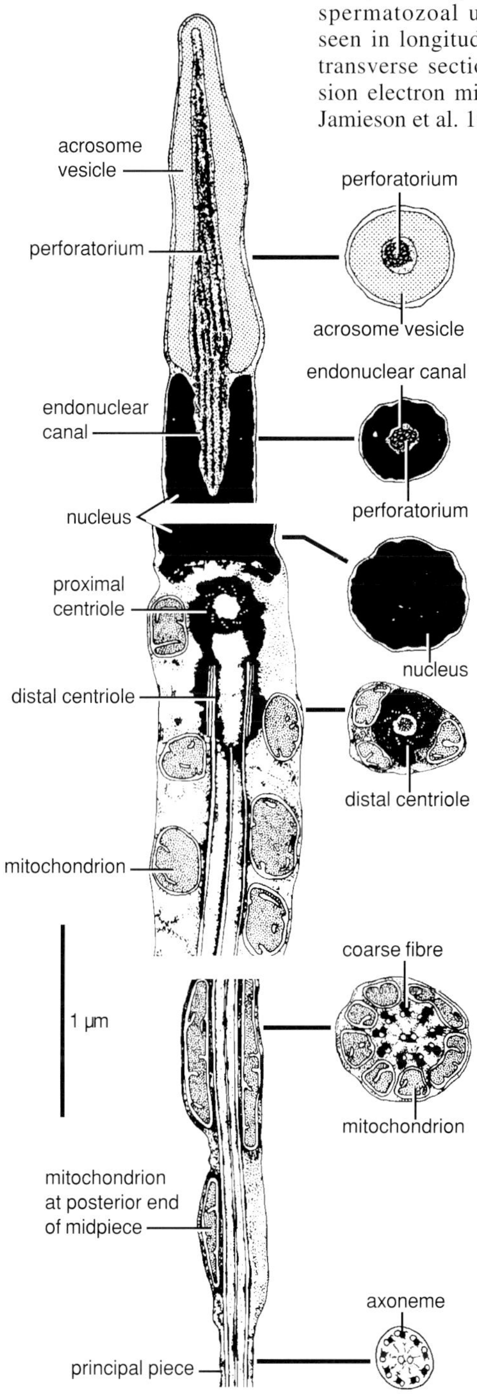

Figure 14. *Melopsittacus undulatus*, the budgerigar. Diagram of spermatozoal ultrastructure as seen in longitudinal section and transverse sections by transmission electron microscopy. (From Jamieson et al. 1995.)

gyres varies from 55 to 300 (Fawcett 1975; Fig. 17F) but is not specified for monotremes (Bedford, Rifkin 1979; Carrick, Hughes 1982). Great reduction of the distal centriole is a mammalian apomorphy. Though normally forming the basal body of flagella, it is at most a vestige in mature mammalian sperm (Baccetti, Afzelius 1976; Fawcett 1975).

Synapomorphies of Extant Monotreme Sperm

The synapomorphies noted for the mammalian ground plan are those attributable to monotremes. As the first mammals were presumably egg-laying monotremes, and the sister group of Metatheria and Eutheria (Rowe 1988), it is questionable that there are any features of the sperm of extant monotremes which may be considered apomorphic departures from the mammalian ground plan.

Synapomorphies of Therian Sperm

No mammalian sperm above the Monotremata is known to possess a subacrosomal cone (*Macroscelides*, Figs. 17A-D). In reptiles, the peripheral fibers at 3 and 8 are detached from their corresponding doublets, while the other seven fibers are attached to their doublets. In birds and monotremes, all of the peripheral fibers are attached to the corresponding doublets (Fawcett, Phillips 1970). It has been said that in marsupial and eutherian sperm, the peripheral fibers are detached from their doublets with the exception of fibers 3 and 8 (Jamieson 1995a), in contrast to what is seen in birds and monotremes. However, in the present study, such differential attachment of fibers 3 and 8 could not be confirmed from a survey of micrographs from the literature on marsupial or eutherian sperm (*Macroscelides*, Figs. 17I-K, O). On the other hand, attachment of these two fibers to the longitudinal columns of the fibrous sheath (Fawcett, Phillips 1970) is demonstrable, but this association is also seen in squamates (Jamieson 1995b). The fibers at 3 and 8 may be lost in at least some therians, as in elephant shrews (Fig. 17O).

Synapomorphies of Marsupial Sperm

The sperm of the rufous hare-wallaby, *Lagorchestes hirsutus*, is shown in Figure 16, to exemplify a marsupial spermatozoon. In marsupials, the acrosome covers, in varying degrees, only one surface of the "anteriorly" flattened (antero-posteriorly compressed) nucleus. This contrasts with the condition in monotreme and eutherian sperm, including the pangolin, in which the acrosome forms a cap which surrounds at least the proximal region of the nucleus and projects for varying distances anterior to its tip (Harding et al. 1979; *Macroscelides*, Figs. 17A, B). The marsupial condition is partly a correlate of distal-proximal flattening of the nucleus.

In marsupials, as exemplified by a member of the primitive family Didelphidae and more derived taxa (Fig. 16), the nucleus is compressed during spermiogenesis in a plane perpendicular to the flagellum (Phillips 1970). The result is that the implantation fossa is situated at the middle of the resultant long axis of the nucleus, giving a T-junction. This condition is typical of marsupials. Only in the koala, *Phascolarctos cinereus*

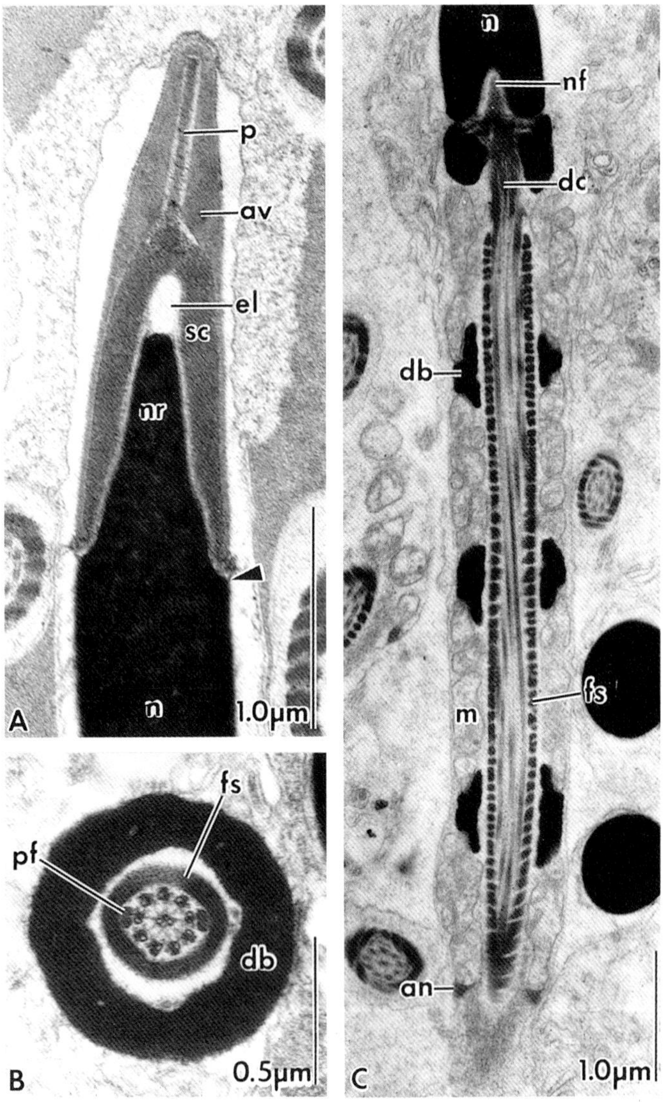

Figure 15. Spermatozoa of Squamata. TEM sections. A: *Carlia rubrigularis*. LS through acrosome and anterior nucleus. Arrow indicates nuclear shoulders. B: *Ctenotus inornatus*. TS through an intermitochondrial dense body, showing that these bodies form large rings in sphenomorph skinks. C: *C. rawlinsoni*. LS through posterior nucleus and complete midpiece, showing series of intermitochondrial rings and fibrous sheath penetrating midpiece. (From DM Scheltinga and BGM Jamieson, unpublished data.)

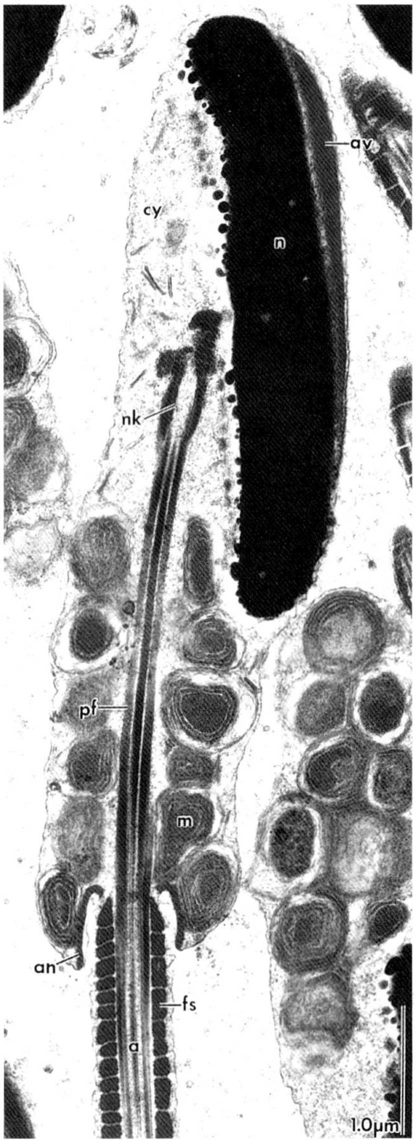

Figure 16. A marsupial spermatozoon exemplified by *Lagorchestes hirsutus*, rufous hare wallaby. LS, showing marsupial feature of compression of the nucleus in a plane at right angles to axoneme and subsequent rotation of the nucleus so that it lies along the midpiece. (From S Johnston, L Smith, F Carrick and BGM Jamieson, unpublished data.)

(Phascolarctidae) (Harding et al. 1979) and, from light microscopy observations, the wombat (Hughes 1965) does the flagellum implant at the short axis of the nucleus, the long axis of the nucleus being continuous with the long axis of the flagellum. Harding et al. (1979) appear to regard the koala sperm type as close to that of the ancestral marsupial, having considered the alternative possibility that the "eutherian" form of the nucleus is a highly specialized product of the long phylogenetic separation of vombatoids from other marsupials. The present writer inclines to the view that one "limb" of the nucleus may have been lost in the koala sperm, giving a

Figure 17 (right). The eutherian spermatozoon exemplified by the Macroscelidae, elephant shrews. All are *Macroscelides proboscideus*, except F, which is *Petrodromus tetradactylus*. A-B: LS of acrosome and nucleus. C-D: proximal and more distal TS of acrosome. E: TS of nucleus, which is enveloped by the acrosome for its entire length. F: LS of entire midpiece, showing cytoplasmic droplet and helical arrangement of mitochondria, with adjacent nucleus and fibrous sheath of principal piece. G-H: LS of base of nucleus and anterior region of midpiece in planes approximately at right angles. I-J: TS of midpiece, showing axoneme with peripheral fibers. K: Same, through cytoplasmic droplet. L: LS of base of nucleus and anterior region of midpiece, showing part of cytoplasmic droplet. M: LS of junction of midpiece and principal piece, showing annulus at the commencement of the fibrous sheath. N: LS of fibrous sheath further distally. O: TS of principal piece where the peripheral fibers at 3 and 8 have each been replaced by an inward keel from the longitudinal column of the fibrous sheath. P: TS of principal pieces of which some, more distal, have lost the peripheral fibers. Q: TS of two endpieces, with disrupted arrangement of microtubules. (Original.)

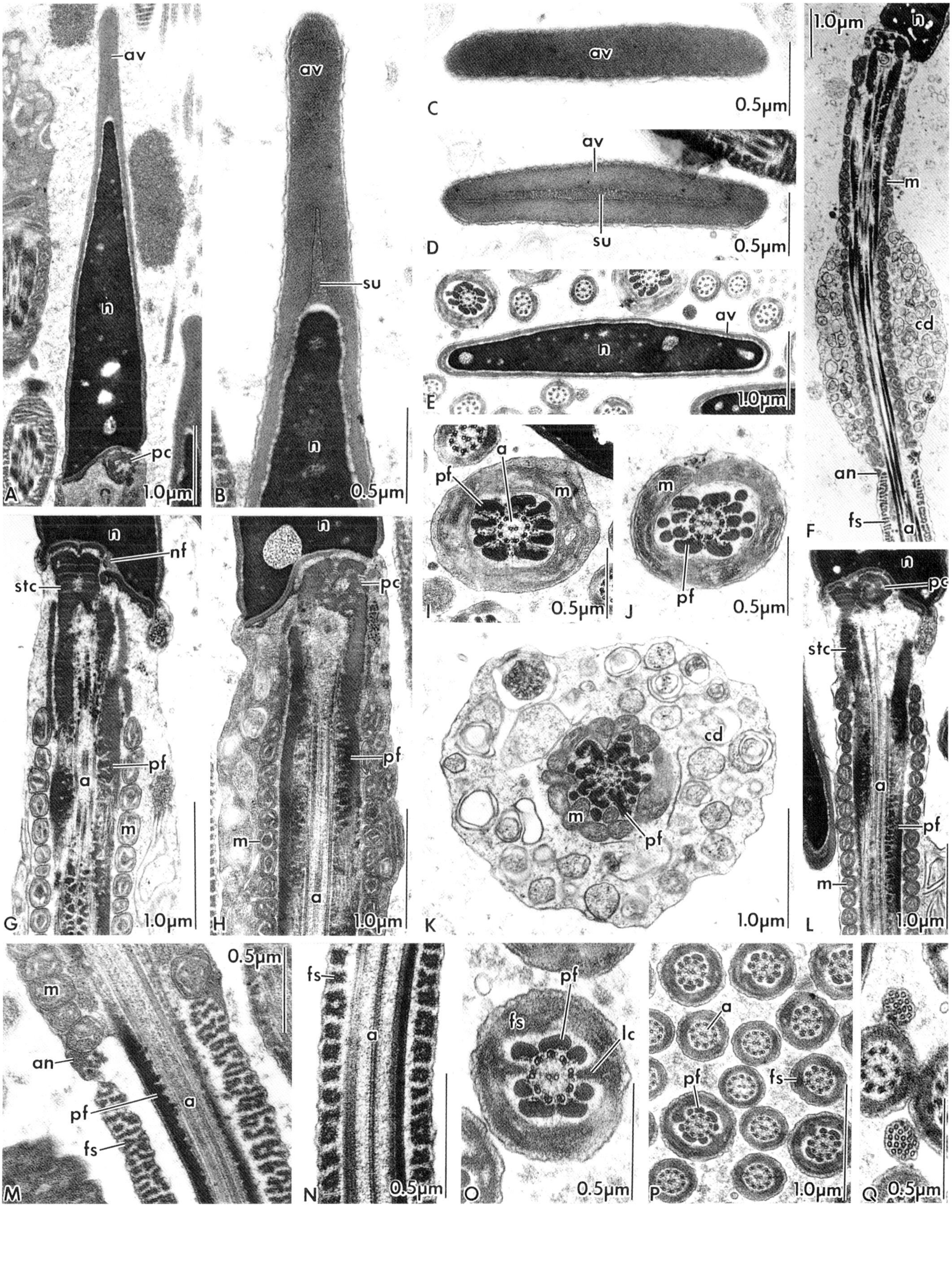
A
B
C
D
E
F
G
H
I
J
K
L
M
N
O
P
Q
av
n
pc
su
m
cd
an
fs
a
nf
stc
pf
lc
1.0µm
0.5µm

secondary appearance of implantation of the axoneme at one end of the nucleus. The fact that, as in other marsupial sperm, the acrosome does not cap the free tip of the nucleus but lies along the abaxonemal face of the nucleus seems to support this view.

Although developmentally the axoneme forms a T-junction with the nucleus in didelphid sperm, the head rotates so that its long axis comes to lie along that of the axoneme. This rotation is reported not only for didelphids but also for phalangerids, petaurids, dasyurids, and peramelids by Harding et al. (1979) and is seen in the macropod *Lagorchestes hirsutus* (Fig. 16).

Synapomorphies of Eutherian Sperm

If the elongated, cylindrical nucleus and caplike acrosome of the pholidotan *Manis* are accepted as plesiomorphic features of therian and eutherian sperm, flattening of the nucleus (here in the plane of the flagellum) (Figs. 17A, E), with the acrosome (Figs. 17C, D), and the many variations of acrosomal structure in the Eutheria must be accepted as internal, secondary eutherian apomorphies. What features, if any, of the basal eutherian sperm were apomorphic relative to the ancestral therians is uncertain. It cannot be determined whether ancestral marsupials had cylindrical sperm nuclei, but if antero-posterior compression were a basal synapomorphy for marsupials, it would have to be concluded that eutherian sperm, as exemplified by *Manis*, are plesiomorphic relative to those of marsupials. This is certainly the situation with regard to extant pholidotan sperm relative to those of marsupials. If the spermatozoal plesiomorphy of eutherians were extrapolated to their phylogenetic position relative to marsupials, researchers would be presented with the profound implication that eutherians with pholidote-like sperm preceded origin of the Marsupalia. This would further imply that yolk sac placentation and premature parturition in marsupials were secondary conditions or that the allantoic placenta of eutherians has originated more than once. This heuristic outcome of spermatozoal studies deserves further consideration. However, the existence of a sperm-type with an elongated, cylindrical nucleus and capping acrosome in ancestral marsupials remains a possibility. The fact that an autosomal marsupial gene appears to have become an X-linked gene in eutherians (Fitzgerald et al. 1993) argues for the derived status of the Eutheria.

Acknowledgements

This work was made possible by a grant from the Australian Research Council. David Scheltinga is thanked for excellent technical and other assistance.

Author Update

Since this chapter was written, a well developed fiber at 8 has been demonstrated for the primitive frog *Leiopelma hochstetteri*. Therefore, absence of a fiber at 8 cannot be considered a synapomorphy of the Anura (DM Scheltinga, BGM Jamieson, K Eggars and DM Green, unpublished data).

References

Afzelius BA, Murray A. The acrosomal reaction of spermatozoa during fertilization or treatment with egg water. Exp Cell Res 1957; 12:325-337.

Asa C, Phillips DM, Stover J. Ultrastructure of spermatozoa of the crested tinamou. J Ultrastruct Res 1986; 94:170-175.

Asa CS, Phillips DM. Ultrastructure of avian spermatozoa: a short review. In: (Mohri H, ed) New Horizons in Sperm Cell Research. Tokyo/New York: Japan Sci Soc Press, Gordon and Breach Sci Publ, 1987, pp365-373.

Austin CR, Baker CL. Spermatozoon of *Pseudobranchus striatus axanthus*. J Reprod Fertil 1964; 7:123-125.

Baccetti B, Afzelius BA. The Biology of the Sperm Cell. Karger, Basel, 1976, pp254.

Baccetti B, Burrini AG, Falchetti E. Spermatozoa and relationships in Palaeognath birds. Biol Cell 1991; 71:209-216.

Baker CL. Spermatozoa of Amphiumae: spermateleosis, helical motility and reversibility. J Tennessee Acad Sci 1962; 37:23-39.

Baker CL. Spermatozoa and spermateleosis in *Cryptobranchus* and *Necturus*. J Tennessee Acad Sci 1963; 38:1-11.

Baker CL. Spermatozoa and spermateleosis in the Salamandridae with electron microscopy of *Diemictylus*. J Tennessee Acad Sci 1966; 41:2-25.

Ballowitz E. Die form und struktur der schuppentierspermien. Zeit Wiss Zool 1907; 86:619-625.

Barker KR, Biesele JJ. Spermateleosis of a salamander *Amphiuma tridactylum* Cuvier: a correlated light and electron microscope study. Cellule 1967; 67:90-118.

Bedford JM, Rifkin JM. An evolutionary view of the male reproductive tract and sperm maturation in a monotreme mammal—the Echidna *Tachyglossus aculeatus*. Am J Anat 1979; 156:207-230.

Bernardini G, Stipani GR, Melone G. The ultrastructure of *Xenopus* spermatozoon. J Ultrastruct Res 1986; 94:188-194.

Bernardini G, Andrietti F, Camatini M, Cosson MP. *Xenopus* spermatozoon: correlation between shape and motility. Gam Res 1988; 20:165-175.

Bernardini G, Podini P, Maci R, Camatini M. Spermiogenesis in *Xenopus laevis*: from late spermatids to spermatozoa. Mol Reprod Dev 1990; 26:347-355.

Boisson C. La spermiogenèse de *Protopterus annectens* (Dipneuste) du Sénégal étudiée au microscope optique et quelques détails au microscope électronique. Ann Fac Sci Univ Dakar 1963; 10:43-72.

Boisson C, Mattei C, Mattei X. Troisième note sur la spermiogenèse de *Protopterus annectens* (Dipneuste) du Sénégal. Institut Fond Afr Noire Bull Sér A (Sci Nat) 1967; 29:1097-1121.

Brandon RA, Martan J, Wortham JWE, Englert DC. The influence of interspecific hybridization on the morphology of the spermatozoa of Ambystoma (Caudata, Ambystomatidae). J Reprod Fertil 1974; 41:275-284.
Breed WG. Evolution of the spermatozoon in Australasian rodents. Austral J Zool 1997; 45:459-478.
Butler RD, Gabri MS. Structure and development of the sperm head in the lizard *Podarcis* (= *Lacerta*) *taurica*. J Ultrastruct Res 1984; 88:261-274.
Carcupino M, Corso G, Pala M. Spermiogenesis in *Chalcides ocellatus tiligugu* (Gmelin) (Squamata, Scincidae): an electron microscope study. Boll Zool 1989; 56:119-124.
Carrick FN, Hughes RL. Aspects of the structure and development of monotreme spermatozoa and their relevance to the evolution of mammalian sperm morphology. Cell Tissue Res 1982; 222:127-141.
Cherr GN, Clark WHJ. An acrosome reaction in sperm from the white sturgeon *Acipenser transmontanus*. J Exp Zool 1984; 232:129-139.
Detlaf TA, Ginzburg AS. Acrosome reaction in sturgeons and the role of calcium ions in the union of gametes. Dokl Akad Nauk SSSR 1963; 153:1461-1464.
Favard P. Mise en évidence d'une sécrétion acrosomique avant la fécondation chez les spermatozoïdes de *Disgoglossus pictus* Otth. et de *Rana temporaria* L. C R Acad Sci (Paris) 1955a; 240:2563-2565.
Favard P. Spermatogenèse de *Discoglossus pictus* Otth.: étude cytologique—maturation du spermatozoïde. Ann Sci Nat Zool 1955b; 11:369-394.
Fawcett DW. A comparative view of sperm ultrastructure. Biol Reprod 1970; 2 (Suppl):90-127.
Fawcett DW. The mammalian spermatozoon. Dev Biol 1975; 44:394-436.
Fawcett DW, Phillips DM. Recent observations on the ultrastructure and development of mammalian spermatozoa. In: (Baccetti B, ed) Comparative Spermatology. Rome: Accademia Nazionale dei Lincei, 1970, pp13-28.
Fawcett DW, Anderson WA, Phillips DM. Morphogenetic factors influencing the shape of the sperm head. Dev Biol 1971; 26:220-251.
Fitzgerald J, Wilcox SA, Graves JAM, Dahl HHM. A Eutherian X-linked gene, PDHA1, is autosomal in marsupials: a model for the evolution of a second, testis-specific variant in Eutherian mammals. Genomics 1993; 18:636-642.
Folliot R. Ultrastructural study of spermiogenesis of the anuran amphibian *Bombina variegata*. In: (Fawcett DW, Bedford JM, eds) The Spermatozoon. Baltimore-Munich: Urban and Schwarzenberg, 1979, pp333-339.
Fukuyama K, Miyazaki K, Kusano T. Spermatozoa and breeding systems in Japanese anuran species with special reference to the spiral shape of sperm in foam-nesting rhacophorid species. 2nd World Congress of Herpetology. Adelaide, South Australia. 1993-6; Abstracts:92-93.
Furieri P. Prime osservazioni al microscopio elettronico sullo spermatozoo di *Triturus cristatus carnifex* (Laurenti): studio al microscopio elettronico. Boll Soc Ital Biol Sperim 1960; 36:1006-1009.
Furieri P. Osservazioni sullo spermatozoo di *Triturus cristatus carnifex* (Laurenti): studio al microscopio elettronico. Monit Zool Ital 1962; 68:90-102.
Furieri P. Sperm morphology of some reptiles: Squamata and Chelonia. In: (Baccetti B, ed) Comparative Spermatology. Rome: Accademia Nazionale dei Lincei, 1970, pp115-131.
Furieri P. La morfologia degli spermi di alcuni anfibi anuri. Boll Zool 1972; 39:618.
Furieri P. La morfologia comparata degli spermi di *Discoglossus pictus* Otth., *Bombina variegata* (L.) e *Alytes obstetricans* (Laurenti). Boll Zool 1975a; 42:458-459.
Furieri P. The peculiar morphology of the spermatozoon of *Bombina variegata* (L.). Monit Zool Ital 1975b; 9:185-201.
Grassé PP. La spermatogénèse. In: (Grassé PP, Delsol M, eds) Traité de Zoologie: Anatomie, Systématique, Biologie. Vol. 14: Batraciens. Apareil uro-genital (suite)—embryogénèse, éthologie, origine, evolution, systematique. Fascicule 1B Paris: Masson, 1986, pp1-20.
Harding HR, Carrick FN, Shorey CD. Special features of sperm structure and function in marsupials. In: (Fawcett DW, Bedford JM, eds) The Spermatozoon. Baltimore-Munich: Urban and Schwarzenberg, 1979, pp289-303.
Harding HR, Aplin K, Shorey CD. Parsimony analysis of marsupial sperm structure: a preliminary report. In: (Mohri H, ed) New Horizons in Sperm Cell Research. Tokyo/New York: Japan Sci Soc Press, Gordon and Breach Sci Publ, 1987, pp375-385.
Healy JM. Sperm morphology and its systematic importance in the Gastropoda. Malacol Rev 1988; 4 (Suppl):251-266.
Healy JM. Euspermatozoa and paraspermatozoa in the trochoid gastropod *Zalipais laseroni* (Trochoidea: Skeneidae). Mar Biol 1990; 105:497-507.
Healy JM, Jamieson BGM. Ultrastructure of the spermatozoon of the tuatara (*Sphenodon punctatus*) and its relevance to the relationships of the Sphenodontida. Phil Trans R Soc Lond Biol Sci B 1992; 335:193-205.
Healy JM, Jamieson BGM. The ultrastructure of spermatogenesis and epididymal spermatozoa of the tuatara *Sphenodon punctatus* (Sphenodontida, Amniota). Phil Trans R Soc Lond B Biol Sci B 1994; 344:187-199.
Hedges SB, Moberg KD, Maxson LR. Tetrapod phylogeny inferred from 18S and 28S ribosomal RNA sequences and a review of the evidence for amniote relationships. Mol Biol Evol 1990; 7:607-633.
Henley C, Feduccia A, Costello DP. Oscine spermatozoa: a light and electron-microscopy study. Condor 1978; 80:41-48.
Hess RA, Thurston RJ, Gist DH. Ultrastructure of the turtle spermatozoon. Anat Rec 1991; 229:473-481.
Hillis DM. The phylogeny of amphibians: current knowledge and the role of cytogenetics. In: (Green DM, Sessions S, eds) Amphibian Cytogenetics and Evolution. New York: Academic Press, 1991, pp7-31.
Hillis DM, Dixon MT, Ammerman LK. The relationships of the coelacanth *Latimeria chalumnae*: evidence from sequences of vertebrate 28S ribosomal RNA genes. Environ Biol Fishes 1991; 32:119-130.
Hughes RL. Comparative morphology of spermatozoa from five marsupial families. Austral J Zool 1965; 13:533-543.
Humphreys PN. Brief observations on the semen and spermatozoa of certain passerine and non-passerine birds. J Reprod Fertil 1972; 29:327-336.
James WS. The Ultrastructure of Anuran Spermatids and Spermatozoa. Ph.D. thesis 1970, University of Tennessee, USA.
Jamieson BGM. Fish Evolution and Systematics: Evidence from Spermatozoa. Cambridge University Press, Cambridge (UK), 1991.
Jamieson BGM. Evolution of tetrapod spermatozoa with particular reference to amniotes. Mém Mus Natl Hist Nat 1995a; 166:343-358.
Jamieson BGM. The ultrastructure of spermatozoa of the Squamata (Reptilia) with phylogenetic considerations. Mém Mus Natl Hist Nat 1995b; 166:359-383.
Jamieson BGM, Healy JM. The phylogenetic position of the tuatara, *Sphenodon* (Sphenodontida, Amniota), as indicated by cladistic analysis of the ultrastructure of spermatozoa. Phil Trans R Soc

Lond Biol Sci B 1992; 335:207-219.

Jamieson BGM, Koehler L. The ultrastructure of the spermatozoon of the Northern Water Snake, *Nerodia sipedon* (Colubridae, Serpentes). Can J Zool 1995; 72:1648-1652.

Jamieson BGM, Scheltinga DM. The ultrastructure of spermatozoa of *Nangura spinosa* (Scincidae, Reptilia). Mem Qld Mus 1993; 34:169-179.

Jamieson BGM, Scheltinga DM. The ultrastructure of spermatozoa of the Australian skinks, *Ctenotus taeniolatus, Carlia pectoralis* and *Tiliqua scincoides scincoides* (Scincidae, Reptilia). Mem Qld Mus 1994; 37:181-193.

Jamieson BGM, Lee MSY, Long K. Ultrastructure of the spermatozoon of the internally fertilizing frog *Ascaphus truei* (Ascaphidae: Anura: Amphibia) with phylogenetic considerations. Herpetologica 1993; 49:52-65.

Jamieson BGM, Koehler L, Todd BJ. Spermatozoal ultrastructure in three species of parrots (Aves, Psittaciformes) and its phylogenetic implications. Anat Rec 1995; 241:461-468.

Jamieson BGM, Oliver SC, Scheltinga DM. The ultrastructure of the spermatozoa of squamata. I. Scincidae, Gekkonidae and Pygopodidae (Reptilia). Acta Zool (Copenhagen) 1996; 77:85-100.

Jamieson BGM, Scheltinga DM, Tucker AD. The ultrastructure of spermatozoa of the Australian freshwater crocodile, *Crocodylus johnstoni* Krefft, 1873 (Crocodylidae, Reptilia). J Submicrosc Cytol Pathol 1997; 29:265-274.

Jespersen Å. Fine structure of the spermatozoon of the Australian Lungfish *Neoceratodus forsteri* (Krefft). J Ultrastruct Res 1971; 37:178-185.

Koehler LD. Diversity of avian spermatozoa ultrastructure with emphasis on the members of the order Passeriformes. Mém Mus Natl Hist Nat 1995; 166:437-444.

Kondo T, Hasegawa K, Uchida T. Formation of the microtubule bundle and helical shaping of the spermatid in the common finch, *lonchura striata* var. *domestica*. J Ultrastruct Res 1988; 98:158-168.

Kwon AS, Lee YH. Comparative spermatology of anurans with special references to phylogeny. In: (Jamieson BGM, Ausio J, Justine J-L, eds) Advances in Spermatozoal Phylogeny and Taxonomy. vol. 166. Paris: Mém Mus Natn His Nat, 1995, pp321-332.

Lee MSY, Jamieson BGM. The ultrastructure of the spermatozoa of three species of myobatrachid frogs (Anura, Amphibia) with phylogenetic considerations. Acta Zool (Stockholm) 1992; 73:213-222.

Lee MSY, Jamieson BGM. The ultrastructure of the spermatozoa of bufonid and hylid frogs (Anura, Amphibia): Implications for phylogeny and fertilization biology. Zool Scr 1993; 22:309-323.

Leung LP. Fish Spermatology: Ultrastructure, phylogeny and cryopreservation. Honours thesis, 1987, University of Queensland.

Leung LKP, Cummins JM. Morphology of immature spermatozoa of the Chinese Pangolin (*Manus pentadactyla*: Pholidota). Proc Aust Soc Reprod Biol (Newcastle, Australia), 20th Ann Conf, 1988; p94.

Mainoya JR. Observations on the ultrastructure of spermatids in the testis of *Chiromantis xerampelina* (Anura: Rhacophoridae). Afric J Ecol 1981; 19:365-368.

Martan J, Wortham E. A tail membrane on the spermatozoa of some ambystomatid salamanders. Anat Rec 1972; 172:460.

Matos E, Azevedo C. Ultrastructural study of the spermatogenesis of *Lepidosiren paradoxa* (Pisces, Dipnoi) in Amazon region. Rev Bras Ciên Morfol 1989; 6:67-71.

Mattei X. Spermiogenèse comparée des poissons. In: (Baccetti B, ed) Comparative Spermatology. New York: Academic Press, 1970, pp57-69.

Mattei X. The flagellar apparatus of spermatozoa in fish: ultrastructure and evolution. Biol Cell 1988; 63:151-158.

Mattei X. Spermatozoon ultrastructure and its systematic implications in fishes. Can J Zool 1991; 69:3038-3055.

Mattei X, Siau Y, Seret B. Étude ultrastructurale du spermatozoïde du coelacanthe: *Latimeria chalumnae*. J Ultrastruct Res 1988; 101:243-251.

McGregor JH. The spermatogenesis of *Amphiuma*. J Morphol 1899; 15:57-104.

Metter DE. On breeding and sperm retention in *Ascaphus*. Copeia 1964; 1964:710-711.

Meyer A, Dolven SI. Molecules, fossils, and the origin of tetrapods. J Mol Evol 1992; 35:102-113.

Meyer A, Wilson AC. Origin of tetrapods inferred from their mitochondrial DNA affiliation to lungfish. J Mol Evol 1990; 31:359-364.

Meyer E, Jamieson BGM, Scheltinga DM. Sperm ultrastructure of six Australian hylid frogs from two genera (*Litoria* and *Cyclorana*): phylogenetic implications. J Submicrosc Cytol Pathol 1997; 29:443-451.

Mizuhira V, Futaesaku Y, Ono M, Ueno M, Yokofujita J, Oka T. The fine structure of the spermatozoa of two species of *Rhacophorus* (*arboreus, schlegelii*). I. Phase-contrast microscope, scanning electron microscope, and cytochemical observations of the head piece. J Ultrastruct Res 1986; 96:41-53.

Mo H. Ultrastructural studies on the spermatozoa of the frog *Rana nigromaculata* and the toad *Bufo bufo asiaticus*. Zool Res 1985; 6:381-390.

Oka T. Ultrastructural observations on the sperm in a frog, *Rhacophorus schlegelii*. Japan J Herpetol 1980; 8:137.

Oliver SC, Jamieson BGM, Scheltinga DM. The ultrastructure of spermatozoa of squamata: II. Agamidae, Varanidae, Colubridae, Elapidae, and Boidae (Reptilia). Herpetologica 1996; 52:216-241.

Phillips DM. Ultrastructure of spermatozoa of the woolly opossum *Caluromys philander*. J Ultrastruct Res 1970; 33:381-397.

Phillips DM, Asa CS. Development of spermatozoa in the Rhea. Anat Rec 1989; 223:276-282.

Phillips DM, Asa CS. Strategies for formation of the midpiece. In: (Baccetti B, ed) Comparative Spermatology 20 Years After. Vol. 75. Serono Symposia Publications. New York: Raven Press, 1993, pp997-1000.

Phillips DM, Asa CS, Stover J. Ultrastructure of spermatozoa of the white-naped crane. J Submicrosc Cytol 1987; 19:489-494.

Picheral B. Structure et organisation du spermatozoïde de *Pleurodeles waltlii* Michah. (Amphibien Urodèle). Arch Biol, (Liège) 1967; 78:193-221.

Picheral B. Les éléments cytoplasmiques au cours de la spermiogenèse du triton *Pleurodeles waltlii* Michah. III. L'évolution des formations caudales. Z Zellforsch 1972; 131:399-416.

Picheral B. Structural, comparative, and functional aspects of spermatozoa in urodeles. In: (Fawcett DW, Bedford JM, eds) The Spermatozoon. Baltimore: Urban and Schwarzenberg, 1979, pp267-287.

Picheral B, Folliot R, Maillet PL. Sur la structure du noyau du spermatozoïde de *Pleurodeles waltlii* Michah. (Amphibien Urodèle). C R Séanc Acad Sci 1966; D 262:1579-1582.

Poirier GR, Spink GC. The ultrastructure of testicular spermatozoa in two species of *Rana*. J Ultrastruct Res 1971; 36:455-465.

Pugin-Rios E. Étude comparative sur la structure du spermatozoïde des Amphibiens Anoures. Comportement des gamètes lors de la fécondation. Thèse 1980, L'Université de Rennes, France.

Purkerson ML, Jarvis JUM, Luse SA, Dempsey EW. X-ray analysis coupled with scanning and transmission electron microscopic observations of spermatozoa of the African lungfish, *Protopterus aethiopicus*. J Zool 1974; 172:1-12.

Reed SC, Stanley HP. Fine structure of spermatogenesis in the South African Clawed Toad, *Xenopus laevis* Daudin. J Ultrastruct Res 1972; 41:277-295.

Retzius G. Die spermien der amphibien. Biol Untersuch, N F 1906; 13:49-70.

Robson SK, Rouse GW, Pettigrew JD. Sperm ultrastructure of *Tarsius bancanus* (Tarsiidae, Primates): implications for primate phylogeny and the use of sperm in systematics. Acta Zool 1997; 78:269-278.

Rouse GW, Robson SK. An ultrastructural study of megachiropteran (Mammalia: Chiroptera) spermatozoa: implications for chiropteran phylogeny. J Submicrosc Cytol 1986; 18:137-152.

Rowe T. Definition, diagnosis, and origin of Mammalia. J Vert Paleontol 1988; 8:241-264.

Russell LD, Brandon RA, Zalisko EJ, Martan J. Spermatophores of the salamander *Ambystoma texanum*. Tissue Cell 1981; 13:609-621.

Saita A, Longo OM, Tripepe S. Osservazioni comparative sulla spermiogenesi. III. Aspetti ultrastrutturali della spermiogenesi di *Jacana jacana* (Caradriformes). Accad Naz Lincei. 1983; 74:417-430.

Saita A, Comazzi M, Perrotta E. Electron microscope study of spermiogenesis in *Caiman crocodylus* L. Boll Zool 1987; 4:307-318.

Sandoz D. Etude cytochimique des polysaccharides au cours de la spermatogénèse d'un amphibien anoure: le discoglosse *Discoglossus pictus* (Otth.). J Microscopie 1970a; 9:243-262.

Sandoz D. Etude ultrastructurale et cytochimique de la formation de l'acrosome du discoglosse (Amphibien Anoure). In: (Baccetti B, ed) Comparative Spermatology. Rome: Accademia Nazionale dei Lincei, 1970b, pp93-113.

Sandoz D. Participation du réticulum endoplasmique a l'élaboration de l'anneau dans les spermatides du discoglosse (Amphibien Anoure). J Microscopie 1973; 17:185-198.

Sandoz D. Development of the neck region and the ring during spermiogenesis of *Discoglossus pictus* (Anuran Amphibia). In: (Afzelius BA, ed) The Functional Anatomy of the Spermatozoon. Oxford: Pergamon, 1974a, pp237-247.

Sandoz D. Modifications in the nuclear envelope during spermiogenesis of *Discoglossus pictus* (Anuran Amphibia). J Submicrosc Cytol 1974b; 6:399-419.

Serra JA, Vicente MJ. New structures of spermatozoa of *Rana* in relation to lipid localization. Revist Portug Zool Biol Gen 1960; 2:223-242.

Seshachar BR. The spermatogenesis of *Uraeotyphlus narayani* Seshachar. Cellule 1939; 48:63-76.

Seshachar BR. The apodan sperm. Current Sci 1940; 10:464-465.

Seshachar BR. Stages in the spermatogenesis of *Siphonops annulatus* Mikan. and *Dermophis gregorii* Blgr. (Amphibia: Apoda). Proc Indian Acad Sci 1942; 15:266-277.

Seshachar BR. The spermatogenesis of *Ichthyophis glutinosus* Linn. Part III. Spermateleosis. Proc Natl Institut Sci, India 1943; 9:271-286.

Seshachar BR. Spermateleosis in *Uraeotyphlus narayani* Seshachar and *Gegenophis carnosus* Beddome (Apoda). Proc Natl Institut Sci, India 1945; 11:336-340.

Sibley CG, Ahlquist JE. Phylogeny and Classification of Birds: A Study in Molecular Evolution. Yale University Press, New Haven, Conn, 1990.

Sibley CG, Ahlquist JE, Monroe BL. A classification of the living birds of the world based on DNA-DNA hybridization studies. Auk 1988; 105:409-423.

Soley JT. Ultrastructure of ostrich (*Struthio camelus*) spermatozoa: 1. Transmission electron microscopy. Onderstepoort J Vet Res 1993; 60:119-130.

Soley JT. Centriole development and formation of the flagellum during spermiogenesis in the ostrich (*Struthio camelus*). J Anat 1994; 185:301-313.

Swan MA, Linck RW, Ito S, Fawcett DW. Structure and function of the undulating membrane in spermatozoan propulsion in the toad *Bufo marinus*. J Cell Biol 1980; 85:866-880.

Temple-Smith P. Sperm structure and marsupial phylogeny. In: (Archer M, ed) Possums and opossums: studies in evolution. Sydney: Surrey Beatty & Sons and the Royal Society of New South Wales, 1987, pp171-193.

Temple-Smith PD, Bedford JM. Sperm maturation and the formation of sperm pairs in the epididymis of the opossum, *Didelphis virginiana*. J Exp Biol 1980; 214:161-171.

Thurston RJ, Hess RA. Ultrastructure of spermatozoa from domesticated birds: comparative study of turkey, chicken and guinea fowl. Scan Microsc 1987; 1:1829-1838.

Townsend DS, Stewart MM, Pough FH, Brussard PF. Internal fertilisation in an oviparous frog. Science 1981; 212:469-470.

Tuzet O, Millot J. La spermatogenèse de *Latimeria chalumnae* Smith (*Crossoptérygien coelacanthidé*). Annal Sci Nat Zool 1959; 1:61-69.

van der Horst G. Spermatozoon structure of three anuran (Amphibia) species. Proc Electron Microsc Soc S Afric 1979; 9:153-154.

van der Horst G. Late spermatid-sperm/Sertoli cell association in the caecilian, *Typhlonectes natans* (Amphibia: Gymnophiona). Electron Microsc Soc S Afric 1991; 21:247-248.

van der Horst G, Visser J, van der Merwe L. The ultrastructure of the spermatozoon of *Typhlonectes natans* (Gymnophiona: Typhlonectidae). J Herpetol 1991; 25:441-447.

Wake MH. Comparative morphology of caecilian sperm (Amphibia: Gymnophiona). J Morphol 1994; 221:261-276.

Werner G, Hübers H, Hübers E, Morgenstern E. Beziehungen zwischen Struktur und chemischer Zussamensetzung der Spermien vom Bergmolch, *Triturus alpestris*. Histochimie 1972; 30:345-358.

Wilson B, van der Horst G, Channing A. Scanning electron microscopy of the unique sperm of *Chiromantis xerampelina* (Amphibia: Anura). Proc Electron Microsc Soc S Afric 1991; 21:255-256.

Woodall PF, Fitzgibbon C. Ultrastructure of spermatozoa of the yellow-rumped elephant shrew *Rhynchocyon chrysopygus* (Mammalia: Macroscelidea) and the phylogeny of elephant shrews. Acta Zool (Copenhagen) 1995; 76:19-24.

Wortham JWEJ, Brandon RA, Martan J. Comparative morphology of some plethodontid salamander spermatozoa. Copeia 1977; 1977:666-680.

Wortham JWEJ, Murphy JA, Martan J, Brandon RA. Scanning electron microscopy of some salamander spermatozoa. Copeia 1982; 1982:52-60.

Yokobori SI, Hasegawa M, Ueda T, Okada N, Nishikawa K, Watanabe K. Relationship among coelacanths, lungfishes and tetrapods: a phylogenetic analysis based on mitochondrial cytochrome oxidase I gene sequences. J Mol Evol 1994; 38:602-609.

Yoshizaki N. Isolation of spermatozoa, their ultrastructure, and their fertilizing capacity in two frogs, *Rana japonica* and *Xenopus laevis*. Zool Sci 1987; 4:193-196.

Zardoya R, Meyer A. Evolutionary relationships of the coelacanth, lungfishes, and tetrapods based on the 28S ribosomal RNA gene. Proc Natl Acad Sci USA 1996; 93:5449-5454.

Zardoya R, Meyer A. The complete DNA sequence of the mitochondrial genome of a "living fossil," the coelacanth (*Latimeria chalumnae*). Genetics 1997a; 146:995-1010.

Zardoya R, Meyer A. Molecular phylogenetic information on the identity of the closest living relative(s) of land vertebrates. Naturwiss 1997b; 84:389-397.

Zirkin BR. The fine structure of nuclei during spermiogenesis in the Leopard Frog, *Rana pipiens*. J Ultrastruct Res 1971; 34:159-174.

30

Accessory Microtubules in Insect Spermatozoa: Structure, Function and Phylogenetic Significance

Romano Dallai
University of Siena

Björn A. Afzelius
Stockholm University

It is well known that insects constitute the largest group in the animal kingdom. Their evolutive radiation consequent to strategies for adaptation to different environments has no counterpart in any animal phylum.

Much less is known about the enormous variation in shape and size of their spermatozoa. There is no other animal group displaying so great a variation in sperm structure as insects, ranging from a conventional flagellated sperm to an aflagellated immotile cell, from minute ones to giant spermatozoa. Sperm variability within insects is independent of the size of the group in which it occurs; thus, a single family of Diptera (Cecidomyiidae) seems to show more sperm models than do large orders such Lepidoptera or Coleoptera.

Insect spermatozoa are generally provided with a flagellar axoneme which can support sperm motility or simply serve as an axial cytoskeleton. There is a tendency in several orders toward a progressive degeneration of the axoneme. Sperm evolution among Homoptera is thus typical in this respect, but similar examples can be found among Protura, Isoptera, Ephemeroptera, Trichoptera, Coleoptera and Diptera (Dallai 1979).

It is often difficult to recognize whether a certain sperm character is due to a primitive position of the group or, alternatively, if it represents an adaptation to new specific functions. Thus, the risk of confusing a primitive, plesiomorphic feature with a derived, apomorphic one is often a possibility. Can, for example, the simple 9+2 flagellar axoneme of the wingless Collembola, one of the oldest (Devonian) groups of insects, be considered to have the same history as that of Mecoptera or Aphaniptera? Or is the 9+2 axoneme in these two latter groups rather due to the secondary loss of the accessory tubules?

The Origin of the Accessory Microtubules

The sperm axoneme of many insects possesses an extra set of nine singlet microtubules, outside the 9+2 complex, and is hence known as a 9+9+2 axoneme. The extra tubules are called accessory microtubules and, sometimes, peripheral singlets (Baccetti, Bairati 1964) or satellites (Kiefer 1970).

These nine accessory microtubules are formed during spermiogenesis as outgrowths on the B-subtubule of each microtubular doublet (Cameron 1965). The early spermatid hence has a simple 9+2 axoneme with each doublet having the "standard" number of tubulin protofilaments: 13 in the A-subtubule and 10 in the B-subtubule (Dallai, Afzelius 1990a; Witman et al. 1972). The inner end of the B-subtubule contains also a particle of smaller diameter (Afzelius et al. 1990), so that the number of units in the B-subtubule wall is sometimes given as 11 (Fujiwara, Tilney 1975; Nojima et al. 1995).

The first step in the formation of the insect accessory tubules takes place as a short projection extending from the B-subtubule of the doublet in the counterclockwise direction, if the dynein arms are clockwise oriented (Fig. 1). In the beginning, this short projection is formed at a precise level corresponding to protofilament n°4 (according to the terminology of Witman et al. 1972) or near this point. The short projection is first formed by three or four elements, then it grows by adding a number of protofilaments and curves to form a complete cylinder that finally detaches from the microtubular doublet (Fig. 2). This growth and detachment is initially somewhat non-synchronous in all nine accessory tubules, but later it proceeds in synchrony with the development of the so-called intertubular material; that is, an electron-dense substance located between adjacent accessory tubules (Dallai, Afzelius 1993a). The extension of such material varies according to the insect orders and in rare cases may be missing, as in Ephemeroptera.

The exact position of accessory tubules relative to the 9+2 axoneme seems to depend on the presence of the intertubular material adhering to the outer side of the microtubules doublets and on the presence of a normally developed outer dynein arm. The *Drosophila melanogaster* mutant HB-223 is characterized by male sterility, due to an autosomal mutation that specifically affects the formation of the Kl-3 loop (Bonaccorsi et al. 1988). These mutants lack outer dynein arms and the corresponding high molecular weight polypeptides. Several axonemes of maturing spermatids of these mutants show aberrant patterns of accessory tubule formation (C Mencarelli, personal communication). In some cases an accessory tubule originates close to protofilament n°4 of the B-subtubule, but its growth occurs into the interdoublet space rather than toward the external side, so that what should have become an accessory tubule results in a third subtubule of each doublet. The final result is a formation that is somewhat similar to a centriolar triplet (Fig. 3).

The Accessory Microtubules Are Unique Structures of Insects

Additional microtubules surrounding the central 9+2 have been described in a few other animal groups other than insects: the oligochaetoid annelid *Questa ersei* (Jamieson 1983), the non-clitellate annelid *Hrabeiella periglandulata* (Rota, Lupetti 1997), the polychaete *Diurodrilus subterraneous* (Kristensen, Eibye-Jacobsen 1995), the priapulid *Tubiluchus* (Storch, Higgins 1989) and all the velvet worms, Onychophora (Baccetti et al. 1976; Jamieson 1987). *Questa* and *Hrabeiella* have 27 singlet microtubules outside the axoneme of the midpiece region. In *Questa* they are

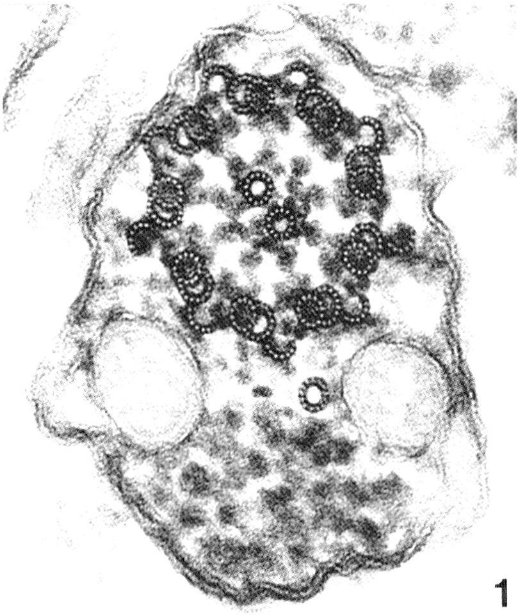

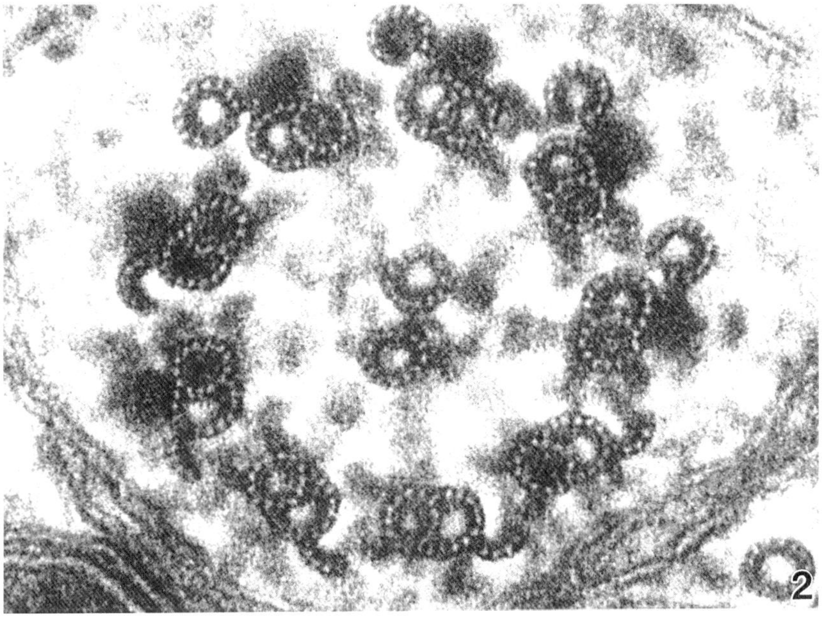

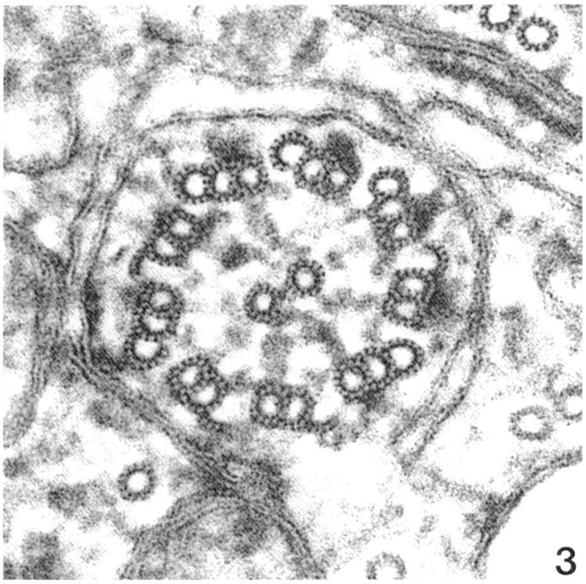

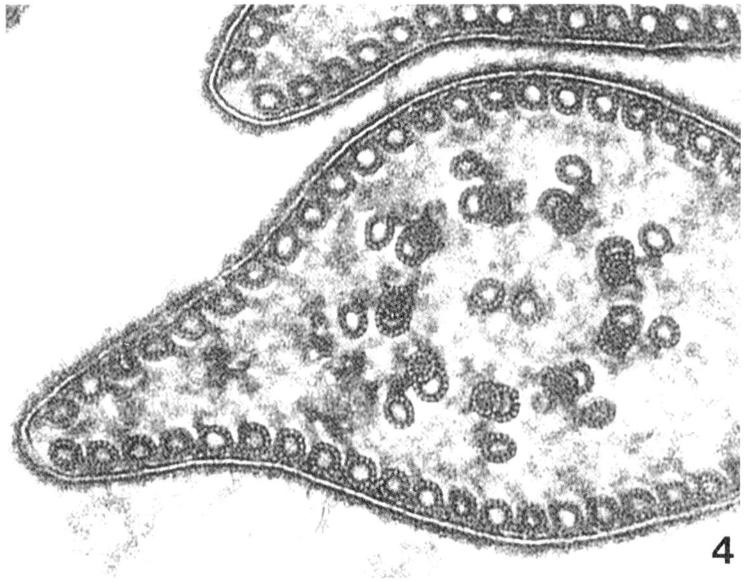

Figure 1. Cross section through the tail flagellum of an early spermatid of *Japyx* sp. The accessory tubules are formed as outgrowths from the B-tubules of the microtubular doublets. x185,000.

Figure 2. Cross section through the flagellum of a *Ceratitis capitata* spermatid. x360,000. (Used with permission from J. Submicrosc. Cytol. and Pathol.)

Figure 3. The sterile mutant HB-223 of *Drosophila melanogaster* has no outer dynein arms and its growing accessory tubules get an abnormal position resembling that of a centriole. x220,000.

Figure 4. Cross section through the spermatid flagellum of the velvet worm *Euperipatoides leuckarti* showing that the nine microtubules external to the 9+2 unit are anchored to the A-subtubule of each doublet. Note also the orderly array of microtubules beneath the plasma membrane. x185,000. (Reprinted with permission from Tiss. Cell.)

arranged in an orderly manner in nine clusters of three microtubules; in the posterior region, however, only nine peripheral singlets are evident. A similar situation occurs in *Tubiluchus*, while in *Diurodrilus* there are nine tubules forming a spiral around the axoneme. It has been suggested that this similarity is due to a close relationship between these groups and insects.

Dallai and Afzelius (1993b) showed that the external microtubules of Onychophora have a different origin from those in insect spermatozoa, thus representing convergent structures and merely indicating that the appearance of this character is due to parallelism. In particular, the accessory microtubules in the spermatozoon of the onychophoran *Euperipatoides leuckarti*, both in spermatophore and in the seminal vesicle, lie outside the axoneme and are separated from it, whereas in the spermatids they are joined to the doublets. These accessory microtubules derive from cytoplasmic microtubules extending posteriorly from the "annulus," rather than being formed as outgrowths from axonemal microtubular doublets; their connection to the 9+2 axoneme hence is secondary rather than primary; moreover, when they adhere to the 9+2 axoneme they do so by means of bridges to the A-subtubules rather than to the B-subtubules, as in insects (Fig. 4).

Accessory Microtubules Are True Microtubules

The origin of accessory tubules as outgrowths from the B-subtubule of microtubular doublets raises some questions as to whether they have the parameters of true microtubules. Are they formed by α- and β-tubulin molecules and other associated proteins (MAPs)? Are these molecules arranged into a defined geometrical surface lattice? Do they have the same polarity and orientation as do cytoplasmic microtubules?

Some data indicate that accessory tubules are true microtubules, irrespective of their diameter: either they measure about 25 nm and consist of 13 protofilaments, as in Diptera, or they have a greater size, up to 34 nm, and consist of 19 protofilaments, as in the caddisfly *Odontocerum*.

The general appearance of the tubular wall of accessory microtubules is comparable to that of cytoplasmic microtubules after they are prepared according to the method by Mizuhira and Futaesaku (1974), although omitting the osmium fixation but employing uranyl acetate block-staining in order to achieve a high contrast (Afzelius 1988). By this technique, in which tannic acid is added to the primary fixative, the cross-sectioned tubular wall of the microtubules consists of a variable number of electron-transparent dots surrounded by dense material. These units seem structurally similar to protofilaments with a somewhat triangular outline, such as it has been described in Afzelius et al. (1990), where they are interpreted as tubulin molecules. The handedness of the individual protofilaments in the accessory tubules of a stick insect was recognized in Afzelius (1988; Figs. 5a, b). This claim was confirmed by Dallai et al. (1993a) and more recently by Hirose et al. (1997) with cryo-electron microscopy after decoration of accessory microtubules of the cricket *Acheta domestica* with the tubulin-motor proteins ncd (nonclaret disjunctional protein) and kinesin. In native accessory microtubules, as well as in microtubules polymerized *in vitro* (Metoz et al. 1997; Sosa, Milligan 1996; Sosa et al. 1997), the tubulin subunits are oriented clockwise, while the motor proteins are oriented anti-clockwise when microtubules

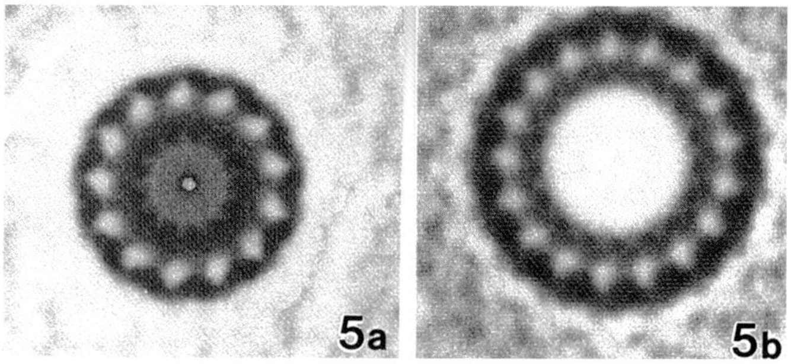

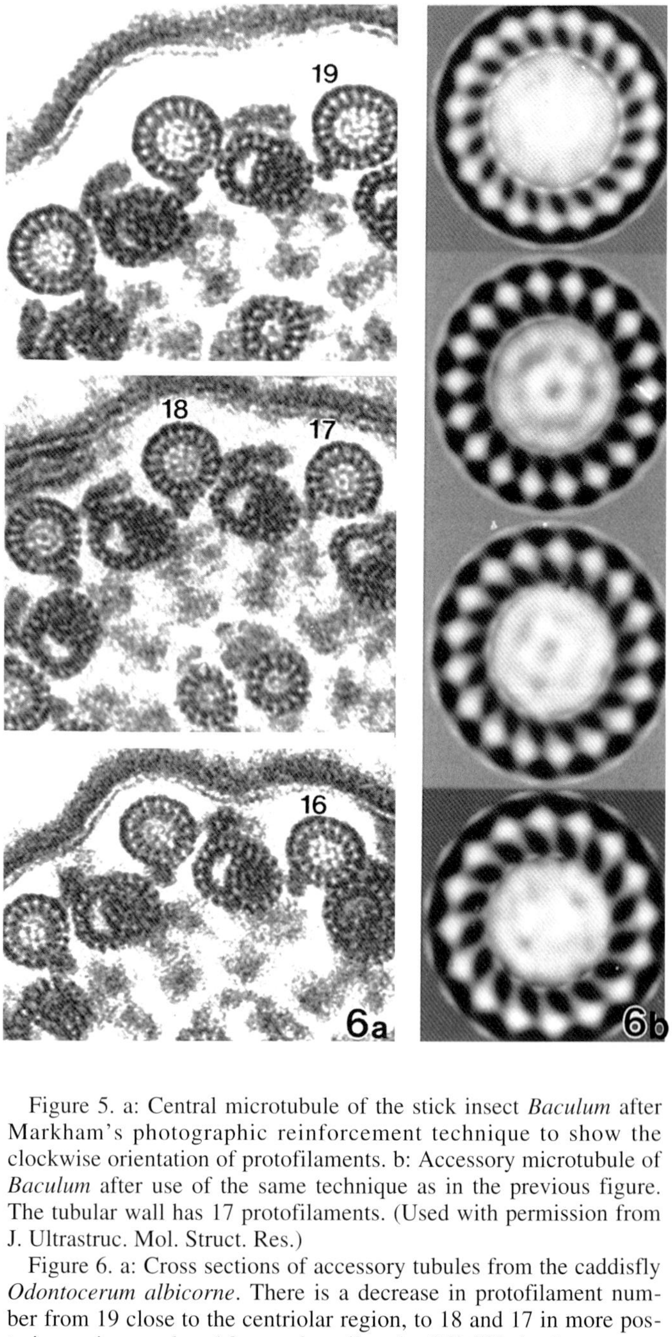

Figure 5. a: Central microtubule of the stick insect *Baculum* after Markham's photographic reinforcement technique to show the clockwise orientation of protofilaments. b: Accessory microtubule of *Baculum* after use of the same technique as in the previous figure. The tubular wall has 17 protofilaments. (Used with permission from J. Ultrastruc. Mol. Struct. Res.)

Figure 6. a: Cross sections of accessory tubules from the caddisfly *Odontocerum albicorne*. There is a decrease in protofilament number from 19 close to the centriolar region, to 18 and 17 in more posterior regions and to 16 near the tail end. x360,000. b: A computer average of the same material showing the decrease from 19 protofilaments proximally to 16 distally. (Reprinted with permission from Cell. Motil. Cytoskel.)

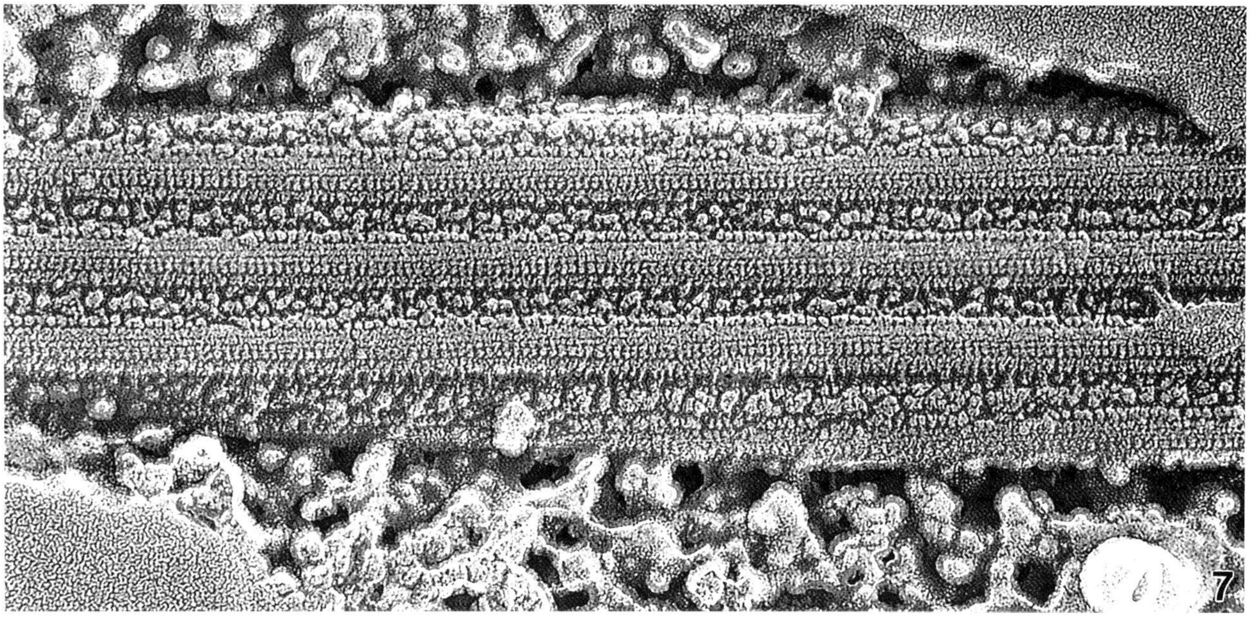

Figure 7. Quick-freeze deep-etch preparation of sperm axoneme of *Apis mellifera*. Note that the tubulin units form protofilaments that run straight in the accessory microtubules. There is an irregular intertubular material between the accessory tubules. x130,000.

are viewed from the minus-end toward the plus-end; that is to say, in a sperm axoneme, from its centriolar region toward the tail tip. Thus, in addition to the polarity of the accessory microtubules, the orientation of the protofilaments can also be recognized (Fig. 6).

These recent data on interaction of tubulin with motor-proteins provide further evidence of the occurrence of α- and β-tubulins in the accessory microtubules. The presence of tubulins in the insect sperm axoneme has also been ascertained with immunoelectron microscopy, using α-antitubulin antibodies (Fernandes, Bào 1996). Accessory microtubules, when negatively stained or prepared with the quick-freeze deep-etching technique, show protofilaments in the tubular wall running parallel to the microtubular long axis, with the same appearance as the axonemal doublets (Fig. 7). Moreover, optical diffraction pattern on micrographs obtained from negative staining preparations reveals spectral lines at 8 and 4 nm, which correspond to the tubulin dimer and monomer repeats, respectively. Studies on insect accessory microtubules with different number of protofilaments have shown that those of the wingless dipluran *Japyx*, provided with 13 protofilaments, gave no equatorial splits in the spectra, thus suggesting that the protofilaments run parallel to the tubule axis (Fig. 8a). The line at 4 nm is related to a three start helix of the monomers; it is consistent with a regular B-lattice model, which is most commonly found, and also described for the two subtubules of the axonemal doublets (Mandelkow et al. 1995; Song, Mandelkow 1995). The strong line at 8 nm indicates the dimer periodicity. The organization of these microtubules with 13 protofilaments is straight rather than helical, in agreement with recent results obtained for cytoplasmic or polymerized microtubules with the same number of protofilaments (Chrétien, Wade 1991; Wade, Hewat 1994). In contrast, accessory microtubules of the nematoceran dipteran *Exechia seriata* (Fig. 8b) and of the caddisfly *Limnephilus bipunctatus*, with 16 and 19 protofilaments, respectively, have skewing protofilaments (Lanzavecchia et al. 1994) as predicted by Langford (1980) and by Wade and Chrétien (1993) for polymerized tubules with other than 13 protofilaments. Consistently, their diffraction patterns clearly reveal spots splitting from the zero line indicating that protofilaments are skewed rather than running parallel to the microtubule axis (Fig. 8b). The twisting allows for accommodating extra molecules of tubulins that must be arranged in each turn (Wade et al. 1990). Phase analysis of spectral peaks also indicates the occurrence of a 4-start helix of dimers, staggered as in the ß-lattice model, in agreement with the predictions by Chrétien and Wade (1991) and Wade and Hewat (1994).

Accessory tubules with more than 16 protofilaments do not retain this number of protofilaments for the entire length of the tubule; the number decreases to 16 posteriorly. Thus the large accessory tubules of the caddisfly *Odontocerum albicorne* are made of 19 protofilaments in the main part of the axoneme, but gradually decrease to 18, 17, and 16 near the distal tip (Dallai et al. 1993a; Fig. 6). The same decrease to 16 was also found in the stick insect *Baculum*, having 17-protofilament accessory

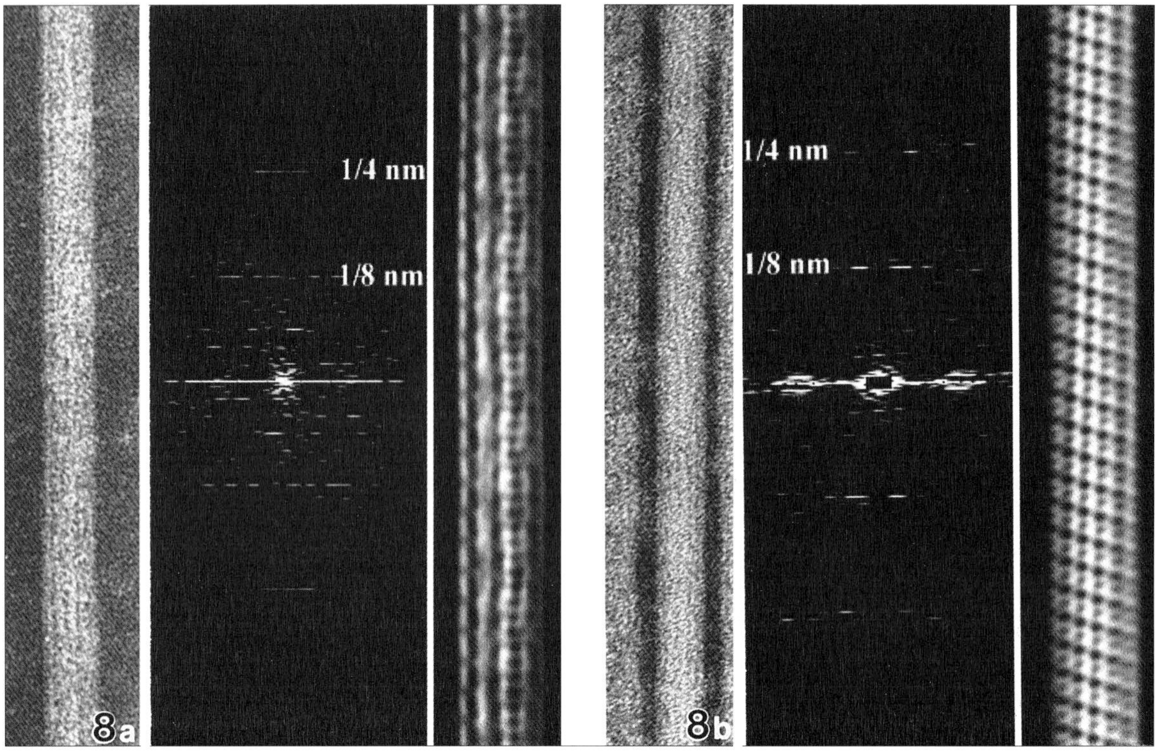

Figure 8. a: Negative staining preparation (1% PTA) of an accessory tubule from *Japyx* sp. (left), optical diffractogram of the same (middle), and computer reconstruction of the straightened microtubule (right). The tubule has 13 protofilaments, which have a straight course as can be appreciated from the fact that there is no equatorial split in the optical diffractogram. x260,000. b: Negative staining preparation (left), optical diffractogram (middle), and computer reconstruction (right) of an accessory tubule from the dipteran *Exechia seriata*, with 16 protofilaments. The equatorial spots splitting from the zero line and the line beyond the one at 4 nm in the diffractogram indicate that the protofilaments are skewed. x260,000. (Used with permission from J. Struct. Biol.)

tubules (Afzelius et al. 1990). A variation along the protofilament number was also recorded from *in vitro* assembled microtubules and interpreted as due to dislocation-like defects (Chrétien et al. 1992).

Another species of Trichoptera, *Oecetis furva*, is provided with accessory tubules having 18 protofilaments in their tubular wall. This number is, however, retained for the whole axonemal length (Dallai, Afzelius 1994a). This is also true for accessory tubules with 16 and with 13 protofilaments. Whether the protofilament number variability along the length has a functional significance in the insect sperm tail or merely represents a simple variant is unknown.

Accessory Microtubules Are Stable Structures of the Insect Axoneme

Unlike the microtubular structures of a conventional 9+2 axoneme, which can be affected by heating (Stephens et al. 1989) or by sarcosyl detergent treatments (Pirner, Link 1995), accessory tubules of insect sperm flagellum are resistant to methods of extraction. Spermatozoa of *Apis mellifera*, treated with 3 M urea, show axonemes which still retain the normal shape, with partially disorganized doublets, while central and accessory tubules have their tubular wall with 13 and 16 protofilaments, respectively (Fig. 9a). A more disruptive treatment with 0.5% sarcosyl + 2.5 M urea, however, is able to induce a greater modification: the axoneme loses its general organization and central tubules and doublets are no longer visible. Accessory tubules and their associated intertubular material, however, are well evident as scattered structures close to the two preserved mitochondrial derivatives; they have 16 protofilaments in their tubular wall (Fig. 9b).

The stability of accessory tubules may be due either to the presence of associated proteins (MAPs), a possibility that, however, has to be demonstrated, and/or to post-translational modifications. During spermatogenesis (or at the late stage of this process), microtubules become enriched in polyglycylated (Bressac et al. 1995), glutamylated and acetylated α-tubulin (Hutchens et al. 1997; Piperno, Fuller 1985; Wilson, Forer 1989; Wolf 1996). An increased degree of α-tubulin acetylation has been observed during male meiosis in the lepidopterans *Ephestia* and *Pieris* (Wolf 1996) using a mouse monoclonal antibody. Data on the greater stability of axonemal microtubules than of short-lived cytoplasmic microtubules are consistent with the observations that the acetylated sperm axoneme of *Drosophila* is still evident after fertilization, while mitotic spindles of the embryo are not (G Callaini, personal communication). In this species, the acetylated cytoplasmic microtubules are seen only at the end of blastoderm cellularization, beyond cycle 14 when a longer cell cycle is observed (Wolf et al. 1988).

Whether or not the process of enhancing stability involves accessory microtubules more than central tubules and peripheral doublets remains to be studied. In this connection, Hutchens et al. (1997) have shown that accessory microtubules in transgenic *Drosophila* constitute an assembly module distinct from both intertubular material and central tubules, which are differently affected by the α85E tubulin isoform; these authors suggested that the tubulin isoform is able to disrupt the initiation mechanism for accessory microtubules.

In the grasshopper *Pezotettix jornai*, which in central Italy passes the wintertime as adult, spermatozoa can be recovered very late in December from male genital ducts. Most of these cells, however, are degenerated with a disordered axoneme. In these axonemes, the microtubular doublets are missing, as are radial spokes. The two central tubules, though still visible, are partially destroyed, their wall having lost most of the protofilaments. However, accessory microtubules, as well as intertubular material and mitochondrial derivatives, are apparently well preserved, even though randomly arranged. Accessory tubules clearly show the expected 16 protofilaments in their tubular wall (Fig. 10).

These observations are reminiscent of what has been described in another orthopteran, *Eyprepocnemis plorans* (Giuffrida, Rosati 1993). In this species, spermatozoa that have reached the spermatheca after mating show a progressive disintegration of the axoneme starting from the microtubular doublets. The accessory tubules, however, have not lost their integrity.

Further evidence of the greater stability of accessory tubules compared to doublets and central tubules can be seen in certain axonemal modifications observed in the sperm tail of a few insects. In the mole cricket *Gryllotalpa gryllotalpa*, as well as in some other orthopterans (Baccetti 1987a) and in fruit flies (Dallai, Afzelius 1991a), the posterior tail region is stiff and its microtubular doublets and central tubules are swollen, apparently due to an excess of material within them (Baccetti et al. 1971). The degeneration along the tail of *Gryllotalpa* begins with doublets 6 and 7 and proceeds to involve all microtubular doublets. The accessory tubules, however, are not affected by the degeneration and retain the 16 protofilaments in their tubular wall (Dallai, Afzelius 1990b; Fig. 11). A somewhat similar degeneration has been also described in the tail end of the dipterans *Dacus* (*Bactrocera*) *oleae* and *Drosophila melanogaster*. In the axoneme of these flies, the end

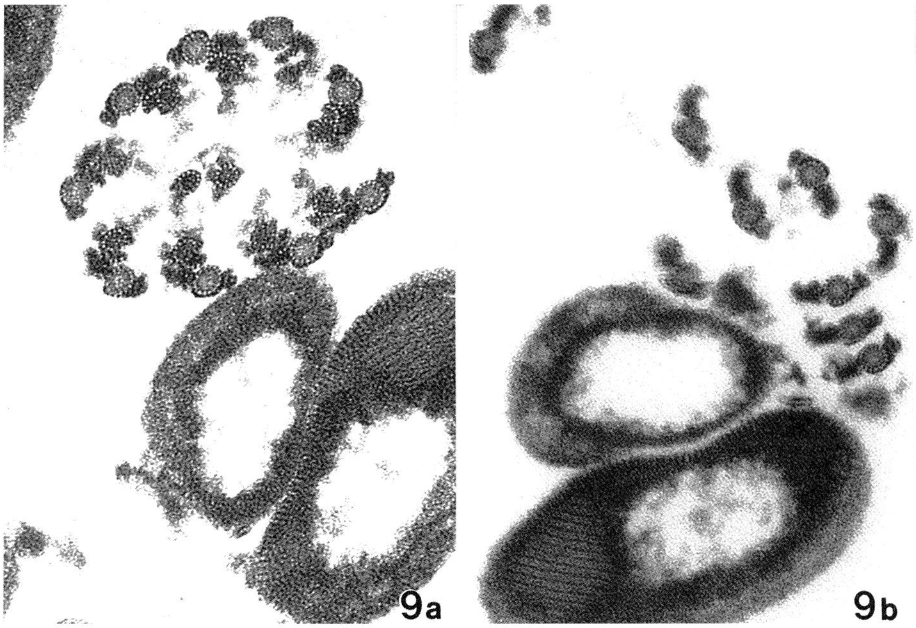

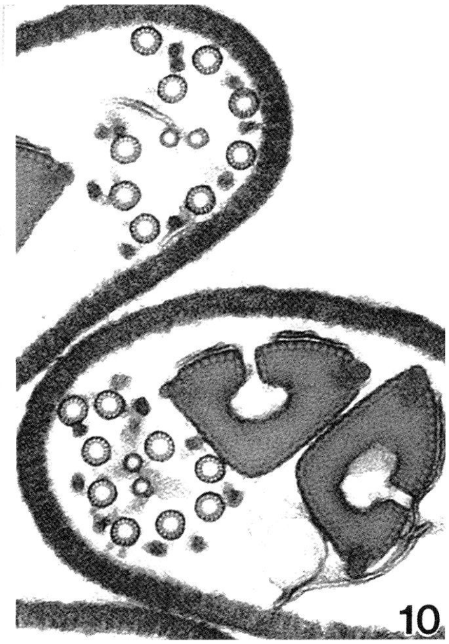

Figure 9. a: Cross section through a urea-treated sperm tail from *Apis mellifera*. The axoneme is somewhat disorganized but accessory tubules remain intact. x185,000. b: Cross section through an *Apis* sperm tail treated with sarcosyl and urea. Only the accessory tubules with their associated intertubular material and the mitochondrial derivatives are visible. The accessory tubules have 16 protofilaments. x185,000.

Figure 10. Cross section through the sperm flagellum of the grasshopper *Pezotettix jornai*. Most cell components are degenerated but the accessory tubules with their 16 protofilaments are retained. x130,000.

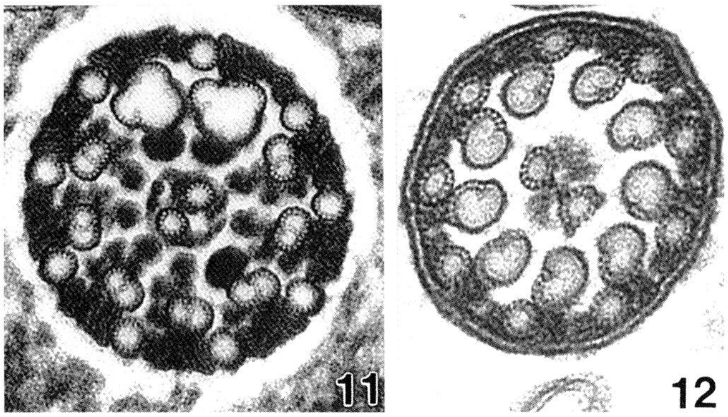

Figure 11. Cross section through the posterior tail portion of a *Gryllotalpa gryllotalpa* spermatozoon in which the degeneration of two microtubular doublets can be seen; the 16 protofilaments of the accessory tubules are retained. x150,000. (Used with permission from Biol. Cell.)

Figure 12. Cross section of a *Dacus* (*Bactrocera*) *oleae* spermatozoon at a distal level where all microtubular doublets are degenerated; the accessory tubules have retained their 13 protofilaments. x180,000.

piece shows modifications of microtubular doublets, while central tubules and accessory tubules retain the conventional appearance, both having 13 protofilaments in their tubular wall (Dallai, Afzelius 1991a; Fig. 12).

Insects in Which Accessory Microtubules Are Missing

There are several insect species, or even whole taxa (that is, family or orders), which have axonemes devoid of accessory tubules. The wingless Collembola and Protura, forming the taxon Ellipura, have no accessory tubules. Collembola are provided with a simple conventional 9+2 axoneme (Dallai 1969) which, at the end of spermiogenesis, rolls around itself. Protura, however, exhibit axonemes consisting of several microtubular doublets (from 12 to 16) arranged in a circle, without central tubules. Axonemal doublets have only inner dynein arms, yet the sperm flagellum displays motility (Dallai et al. 1992a).

Isoptera are the most highly differentiated group of orthopteroid insects and are closely related to blattoids (Kristensen 1981). A flagellate sperm is present only in *Mastotermes darwiniensis*. In this species, the spermatozoon has about 100 flagella (Baccetti, Dallai 1978), each consisting of 9+2 axonemes devoid of accessory tubules.

The rhynchotoid order Thysanoptera is regarded as the sister group of Phthiraptera (Jamieson 1987). They have an odd axonemal pattern consisting of 27 microtubular structures, none of which corresponds to an

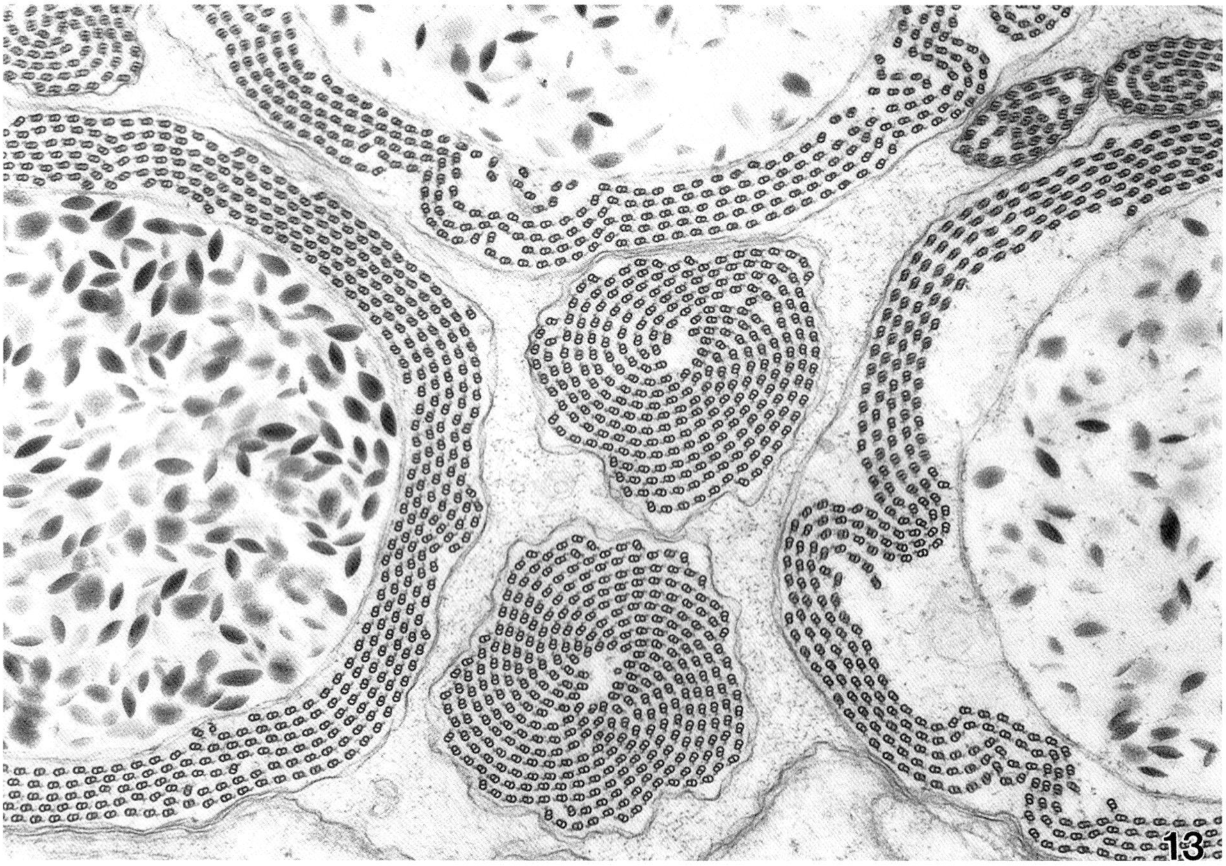

Figure 13. Cross section through sperm tails of the dipteran *Massalongia bachmaieri*. Each axoneme has a great number of microtubular doublets but no accessory tubules. x35,000. (Reprinted with permission from Acta Zool.)

accessory tubule. This aberrant axoneme derives from the fusion, during late spermiogenesis, of three flagellar axonemes of 9+0 type present in the spermatids; the microtubular units are arranged in rows which always begin with a singlet microtubule provided with dynein arms corresponding to the subtubule A of a doublet and end with an armless doublet (Bode 1988; Dallai et al. 1991). The bizarre axoneme of thrips is motile.

Among Holometabola, accessory microtubules are missing in the two orders Mecoptera and Aphaniptera and in a few groups of Diptera, Trichoptera and Lepidoptera.

The presumed sister groups Mecoptera and Aphaniptera share some somatic synapomorphies (Kristensen 1981). They are provided with a 9+2 flagellar axoneme that is coiled around one mitochondrial derivative (Baccetti et al. 1969).

The fungus gnat *Keroplatus reaumurii*, among the dipteran Mycetophilidae, has a simple 9+2 axoneme. In the same family *Exechia seriata* and *E. fusca* have axonemes with only seven, rather than nine, accessory microtubules (Dallai et al. 1995a), doublets 7 and 8 being devoid of such tubules.

In the nematoceran Diptera, the whole family Cecidomyiidae, which is one of the most numerous among insects, exhibits a great axonemal diversity. The sperm axonemes are always devoid of accessory microtubules (Dallai et al. 1996; Fig. 13).

Bibionidae, another nematoceran family, have a sperm axoneme devoid of accessory tubules as described in *Plecia nearctica* (Trimble, Thompson 1974), which has a simple 9+0 axoneme. In the closely related species *Bibio* sp., however, the axoneme does have these structures and the axonemal model is thus 9+9+0 (Dallai, Afzelius 1990a).

Within Trichoptera, two evolutionary trends are identified: members of the suborder Annulipalpia have aberrant sperm cells, while those of the suborder Integripalpia have normal, motile spermatozoa, provided with large accessory microtubules (Dallai, Afzelius 1990a, 1994; Friedländer 1993). Among the former, members of the family Polycentropodidae show sperm axonemes of the 9+7 type; they have immotile spermatozoa, as their microtubular doublets lack dynein arms (Dallai et al. 1995b).

In Lepidoptera, the family Micropterygidae among Zeugloptera has a sperm axoneme that is devoid of accessory tubules and thus it is of a simple 9+2 type (Sonnenschein, Häuser 1990).

Functional Aspects of Accessory Microtubules

Three hypotheses have been forwarded as to the functions of the accessory tubules: a) they may be supporting structures making the axoneme more resistant to mechanical damage; b) they may be part of the motor apparatus; and c) they may be sites for the storage of polysaccharides. The first hypothesis is supported mainly by data from comparative spermatology and by the evidence that accessory tubules are often longer than the 9+2 axoneme. There are several examples where they prolong over the centriolar region, as in Lepidoptera (Medeiros, Silveira 1996), as well as the most posterior tail end, as in Coleoptera (Figs. 12, 25). Accessory tubules of insects (or Onychophora and other animals) are located in the same position relative to the axonemal doublets as the dense fibers of the mammalian (or avian, reptilian, cephalopodan or gastropod) spermatozoa. Fertilization in these latter animal groups is internal and the spermatozoa have to swim in a medium that is rather viscous. By contrast, in species with external fertilization, the spermatozoa are released in seawater or fresh water; these spermatozoa have no accessory tubules or coarse fibers. Spermatozoa from animals with internal fertilization also tend to be longer than those from animals with external fertilization; referred to as introsperm versus ectaquasperm, in the terminology by Jamieson (1991), this is another factor that will make them more fragile. The longest known spermatozoa, with a length of one or several centimeters, are recorded from insects (Afzelius et al. 1976; Joly et al. 1991; Mazzini 1976; Pitnick et al. 1995). Evidence for the opinion that the dense fibers protect mammalian spermatozoa against damage has been given by Baltz et al. (1990). He has, among others, noted that the diameter of dense fibers in mammals increases with the length of the spermatozoon. The stiffness of the tail of the bull sperm has been measured by Lindemann et al. (1973) and that of a simple 9+2 flagellum of the sea urchin tail by Ishijima and Hiramoto (1994). Similar measurements do not seem to have been performed using any insectan 9+9+2 sperm flagella compared with the 9+2 ones. The diameter of the accessory tubules of insect spermatozoa is related to the phylogenetic position of the animal only, not to the sperm size. Regardless of whether sperm length is about 50 mm in certain species of drosophilids and 60,000 mm in others, all investigated species have accessory tubules of a diameter typical of ordinary 13-protofilament microtubules, thus around 25 nm. There might be a difference in the amount of intertubular material, which, for instance, is more abundant in the more derived dipteran suborder Brachycera compared to Nematocera (Figs. 29, 30).

Evidence for the hypothesis that the accessory tubules participate in the generation of propagated sperm bendings is also weak. Rosati (1976) observed connections between accessory tubules and the B-subtubule of doublets and hypothesized their participation in the flagellar motility. According to Stanley et al. (1972), the outer dynein arm in *Drosophila melanogaster* sends thin projections to the accessory tubules and the intertubular material, whereas those of the inner dynein arms go to the neighboring microtubular doublets (as in all cilia and flagella). It would hence seem that the accessory tubules will slide relative to the microtubular doublets in *Drosophila* (and maybe all insects). The micrographs supporting the interpretation of extensions from dynein arm toward the accessory tubules are not very convincing and later investigations of fruit fly spermatozoa have not confirmed the existence of such links (Dallai, Afzelius 1991a).

Another phenomenon that may be of interest in this connection and that has not yet received a satisfactory explanation is the existence of two propagated waves in the swimming sperm tail. It has been suggested that the waves with a small amplitude are produced by the central 9+2 axoneme, whereas those with large amplitudes are produced by accessory bodies in the *Tenebrio* beetle spermatozoon (Baccetti et al. 1973a); or by the laminated bodies in the phasmid spermatozoon (Baccetti et al. 1973b). It would seem more likely that the large-amplitude waves are formed by a structure that is common for both sperm types; for instance, the accessory tubules. Unfortunately it seems that no comparative study has been performed between sperm motility patterns in insects with 9+9+2 axoneme (the majority of insects) and those of 9+2 types.

The third hypothesis is based on the finding that accessory tubules contain a storage of polysaccharides, as seen with the Thiéry technique (Bigliardi et al. 1970; Dallai, Afzelius 1990b). This material might provide the flagellum with nutrients to be used during swimming. This idea can be tested by examining spermatozoa before and after exhaustion by a prolonged swimming.

Diversity of Accessory Microtubules

Whereas the central axis of the axoneme—that is, its 9+2 core—has a fairly invariant ultrastructure in insects (as well as in other animals, plants and protists), there is a considerable diversity in the structure of the accessory tubules and in the intertubular material between them. This diversity seems to be an useful marker of insect relationships. In this context, the data by Raff et al. (1997) are important. These authors have studied spermiogenesis of a sterile transgenic *Drosophila* in which as little as 10% of the β-tubulin pool was replaced by the homolog isoform Hvβt from the moth *Heliothis*. They observed that the organization of some accessory tubules in the fly axoneme is of moth-type: accessory microtubules with 16 rather than 13 protofilaments. The moth protein thus imposes a moth-specific accessory microtubule architecture on the equivalent structure of fruit fly cells, even though only a small amount of the total β-tubulin pool was present.

In the study of phylogenetic relationships, the following characters can be considered: the presence or absence of accessory tubules, their shape, the number of protofilaments and the appearance of the intertubular material.

Diplura are the most archaic insect order in which the sperm tail is unmistakably the same as in the most derived insects. They have nine accessory tubules, which initially flank the nine doublets; later they collect to one side of the axoneme (Figs. 14, 15) and in Japygina, five tubules are visible all along the tail length, while four of them are remarkably shortened and are clustered at the beginning of the axoneme. Each accessory tubule has 13 protofilaments (Dallai et al. 1992b; Fig. 14). There is no intertubular material. The presence of accessory microtubules with 13 protofilaments in both groups (Campodeina and Japygina) supports the monophyly of Diplura and indicates that accessory tubules have been acquired after the diversification of Ellipura (Protura + Collembola). Two other apterygotan orders also have sperm tails with accessory tubules: Zygentoma (silverfish) and Archaeognatha (bristletails). In both orders these tubules consist of 16 protofilaments; those of Archeognatha migrate away from the doublets, whereas those of Zygentoma remain close to the doublets and have a rather prominent intertubular material (Dallai et al. 1992b; Fig. 16). The displacement of accessory microtubules to a position lateral to the axoneme in Archaeognatha is reminiscent of that found in Diplura.

In Ephemeroptera (mayflies) the sperm tail generally has a 9+9+0 axoneme, although it is missing in some species. Its central cylinder has a characteristic appearance resembling a nine-pointed star. The outer dynein arms are missing but the inner ones are present. The accessory tubules are unique in having a "double wall": 13 protofilaments in an outer circle and seven filaments in an inner circle. There is no intertubular material (Afzelius et al. 1991; Dallai, Afzelius 1990a; Fig. 17).

Examined species from the order Odonata (damselflies and dragonflies) have accessory tubules with 16 protofilaments and an intertubular material that does not seem to be in contact with the accessory tubules (Dallai, Afzelius 1990a). The diversity of accessory tubules in the two paleopteran orders mentioned above does not help to elucidate the relationships between them, which remain controversial (Kristensen 1981).

Plecoptera (stone-flies), Blattodea (cockroaches), Dermaptera (earwigs; Fig. 18), Mantodea (mantids) and Embioptera (web-spinners) have accessory tubules with 16 protofilaments and with intertubular material (Dallai, Afzelius 1990a). In the axoneme of Blattodea, accessory tubules and all other microtubules have an electron-dense lumen, whereas the axoneme of Mantodea is characterized by an autoapomorphic feature consisting of a row of nine filaments that connect to the

B-subtubule, somewhat similar to an extra accessory tubule in formation (Dallai, Afzelius 1990a; Fig. 19).

The sperm tail of Phasmida (stick-insects) has a pattern consisting of several unique characteristics. It lacks mitochondria, but instead has the so-called laminated bodies; accessory tubules have 17 protofilaments (Afzelius 1988; Afzelius et al. 1990; Fig. 20). In this respect the phasmids are different from other orthopteroids (Baccetti 1987b); that is, members of the order Orthoptera (grasshoppers), which have a thick glycocalyx, share a 9+9+2 axoneme and accessory microtubules with 16 protofilaments (Dallai, Afzelius 1990a).

The three orders Psocoptera (book lice; Fig. 21), Mallophaga (biting lice; Fig. 22) and Anoplura (sucking lice) share a distinct sperm characteristic: their accessory tubules consist of 13 protofilaments and have an

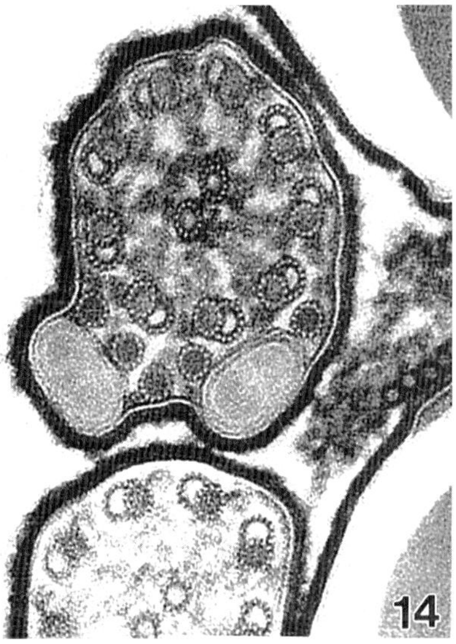

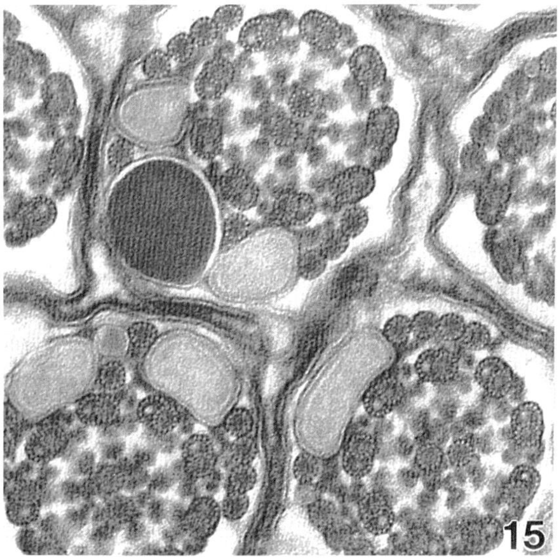

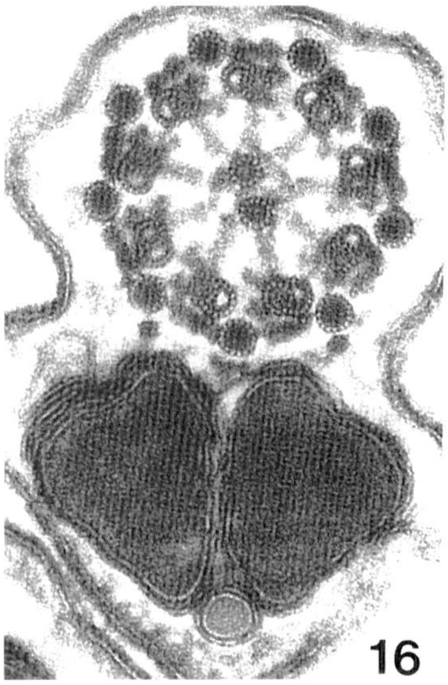

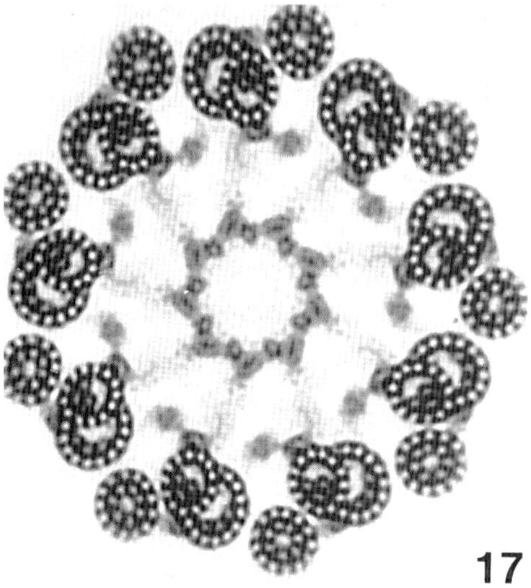

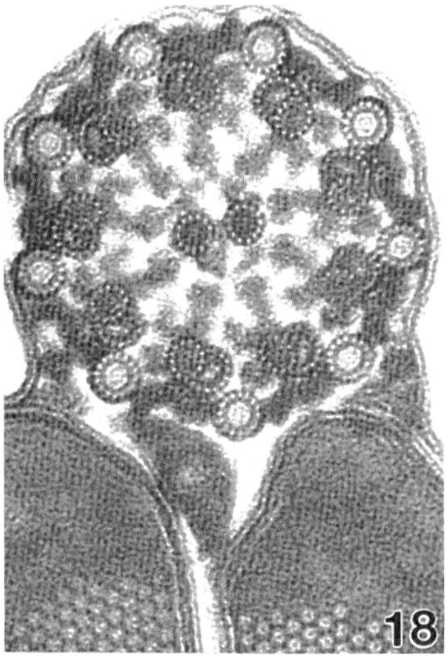

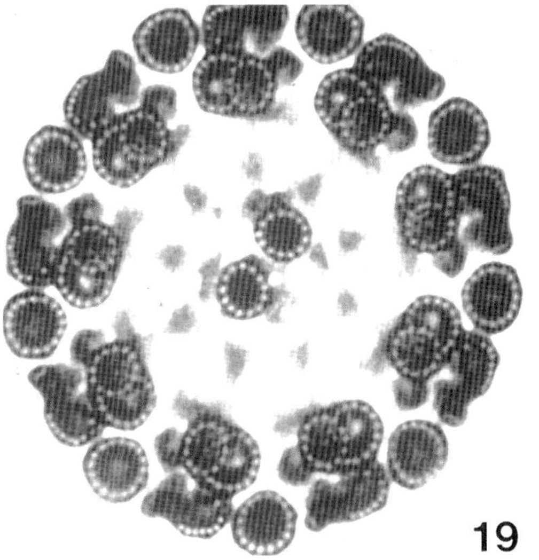

Figure 14. Cross section through a *Japyx* sp. sperm tail. There are only five accessory tubules, each with 13 protofilaments, and they lie on one side of the axoneme intermingled with some mitochondria. x180,000.

Figure 15. Cross section through sperm tails of *Campodea* sp. The accessory tubules are distributed on one side of the axoneme and have 13 protofilaments. x160,000.

Figure 16. Cross section through the sperm tail of the zygentoman insect *Lepismodes inquilinus*. The accessory tubules have 16 protofilaments and intertubular material. x180,000. (Used with permission from Acta Zool.)

Figure 17. Computer reconstruction of the sperm axoneme of the mayfly *Cloëon dipterum*. Note the inner ring of seven units within the accessory tubules, which has 13 protofilaments.

Figure 18. Cross section through the sperm tail of the earwig *Forficula auricularia*. The accessory tubules have 16 protofilaments and are separated by a compact intertubular material. x210,000.

Figure 19. Computer reconstruction of the axoneme of the praying mantid *Mantis religiosa*. The accessory tubules have 16 protofilaments; there is also a row of nine filaments that extends from the B-subtubule of each microtubular doublet.

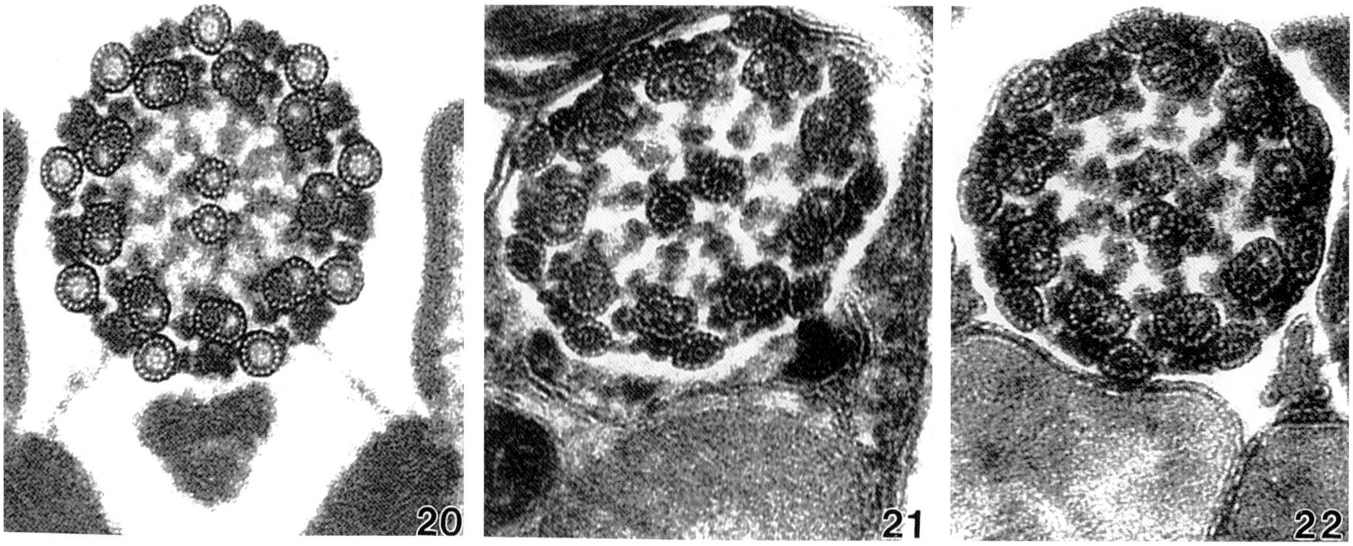

Figure 20. Cross section through the sperm tail of the stick insect *Baculum* sp. The accessory tubules have 17 protofilaments. x200,000.

Figure 21. Cross section through sperm tail of *Dorypteryx palida* (Psocoptera). Note that the accessory tubules have an elliptical shape, have 13 protofilaments only, and that the intertubular material appears as slender bridges. x240,000. (Used with permission from Boll. Zool.)

Figure 22. Cross section through the sperm tail of *Trichodectes bovis* (Mallophaga). The flattened accessory tubules have 13 protofilaments. x240,000. (Reprinted with permission from Boll. Zool.)

elliptic cross section. The intertubular material spans the space between adjacent accessory tubules as thin ropes. These seem to be synapomorphic features shared by the three insect orders (Dallai, Afzelius 1991b), which can be thus recognized as a monophyletic taxon: the Psocodea (Kristensen 1989).

The accessory tubules in the order Heteroptera (true bugs) have a wall consisting of 16 protofilaments and a dense lumen in which a central electron lucid spot is visible, whereas most members of the order Homoptera (leaf hoppers and relatives) have spermatozoa that lack a sperm tail; some of the most archaic members have, however, spermatozoa with a tail of a rather common pattern, among others with accessory tubules consisting

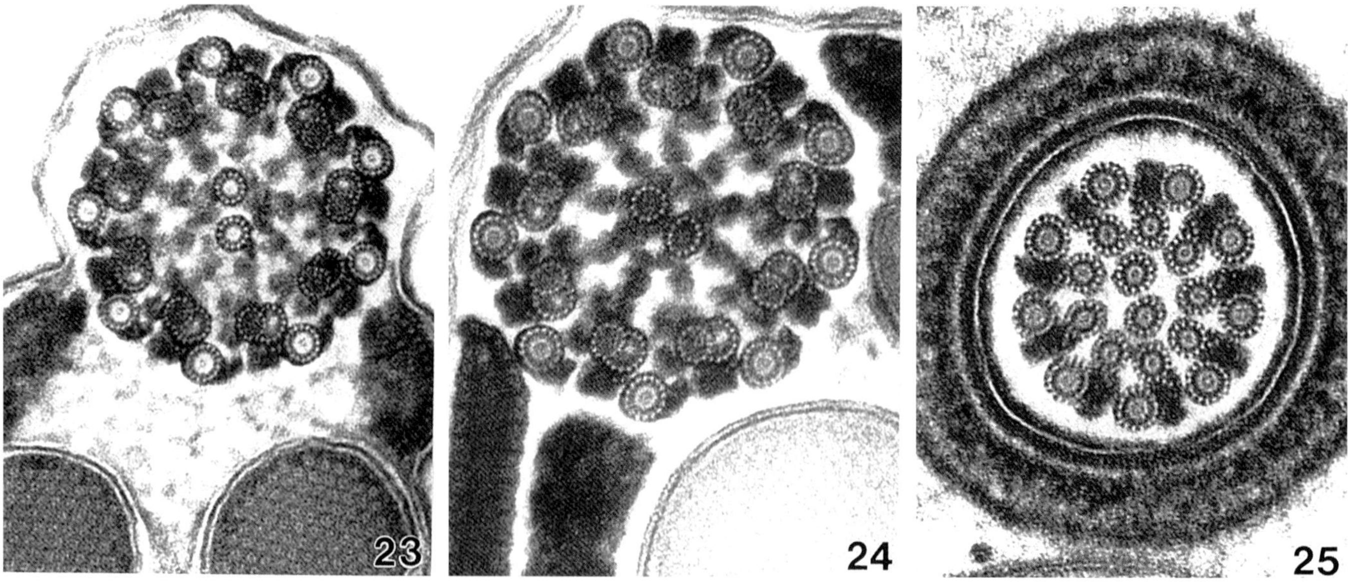

Figure 23. Cross section through the sperm tail of *Ascalaphus* sp. (Neuroptera). The accessory tubules have 16 protofilaments and the intertubular material is divided in two portions. x185,000.

Figure 24. Cross section through the sperm tail of the beetle *Morimus asper*. The accessory tubules have 16 protofilaments and the intertubular material is organized in the same way as in the neuropterans. x210,000.

Figure 25. Cross section through the distal portion of the sperm tail from the curculionid beetle *Protapion dentipes*. At this level, accessory tubules, A-subtubules and central tubules persist, but not the B-subtubules. The accessory tubules have 16 protofilaments and are separated by a prominent intertubular material. x210,000.

of 16 protofilaments (Dallai, Afzelius 1990a).

The four orders Megaloptera (alder-flies), Raphidioptera (snake-flies), Planipennia (lacewings and ant lions; Fig. 23) and Coleoptera (beetles; Figs. 24, 25) have sperm axonemes of approximately the same kind: the accessory tubules thus usually have 16 protofilaments and the intertubular material is divided into two portions: one part in contact with the doublets and one projecting as a beak from the accessory tubules (Afzelius, Dallai 1994; Dallai, Afzelius 1990a). Only a minute percentage of all beetle families have been examined (as is true for most other insect orders). Yet some variability has been recorded, such as microtubules with 17 protofilaments in Coccinellidae and microtubules with 15 in members of Lampyridae and Elateridae (R Dallai and B Afzelius, unpublished observations).

In Strepsiptera, the flagellum has accessory tubules and hence is of the 9+9+2 type (Kathirithamby et al. 1992; Mazzini et al. 1991); their tubular wall consists of 16 protofilaments, although it is not clear whether they form a complete or an incomplete circle. The intertubular material is very reduced (Afzelius, Dallai 1994; Dallai, Afzelius 1990a).

No insect order shows a greater sperm diversity than Diptera (true flies). A 9+9+2 axoneme and accessory tubules with 16 protofilaments, such as seen in most insect orders, have been found in only a few species of the primitive dipterans Mycetophilidae and Keroplatidae (Dallai et al. 1995a). In other families, the accessory tubules have 15 protofilaments; those families are Chironomidae, Dixidae, Culicidae, Bibionidae (Fig. 27) and some species of Mycetophilidae. In Tipulidae, Trichoceridae and Sciaridae, as well as in the more evolved suborders Brachycera and Cyclorrapha, the accessory tubules have 13 protofilaments (Dallai et al. 1993b; Figs. 28-30). These characteristics suggest that the accessory tubules with 16 protofilaments of Mycetophilidae could represent the plesiomorphic status of the character and those with 13 or 15 protofilaments the derived ones. On the basis of this character, it has

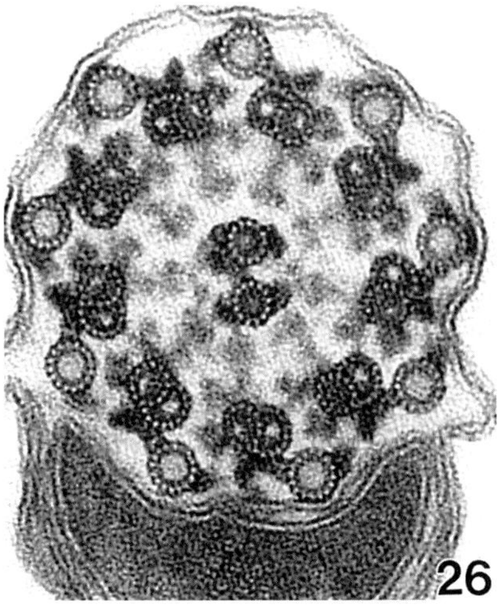
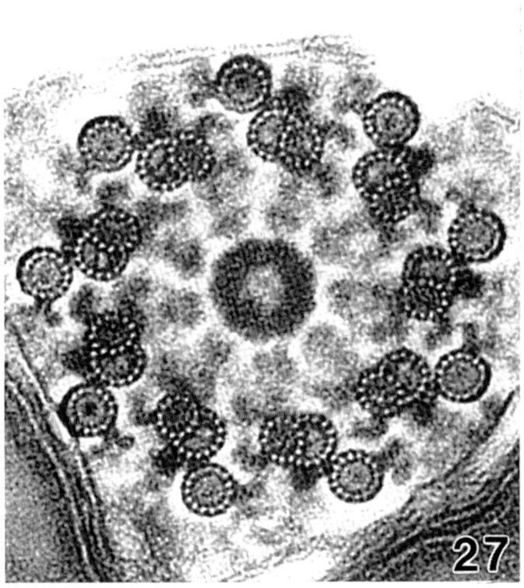
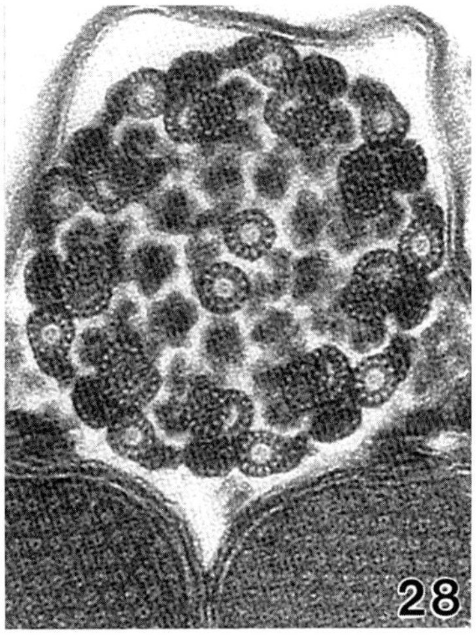
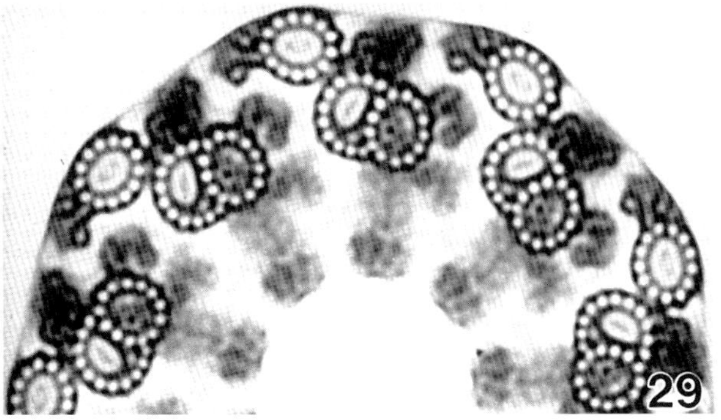
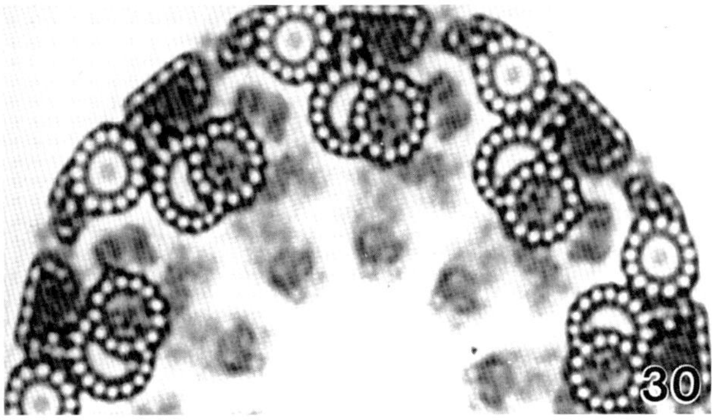

Figure 26. Cross section through the sperm tail of the mycetophilid dipteran *Neoplatyura nigricauda*. Accessory tubules have 16 protofilaments. x215,000.

Figure 27. Cross section through the sperm tail of the dipteran fly *Bibio* sp. The accessory tubules have 15 protofilaments. x215,000.

Figure 28. Cross section through the sperm tail of *Drosophila melanogaster*. Accessory tubules have 13 protofilaments. x185,000.

Figures 29 and 30. Computer reconstructions of the nematoceran dipteran *Tipula* sp. and the brachyceran dipteran *Scatophaga* sp., respectively. Both species have accessory tubules with 13 protofilaments but they differ in that the intertubular material is more highly organized in the brachyceran fly. (Used with permission from Zool. Scripta.)

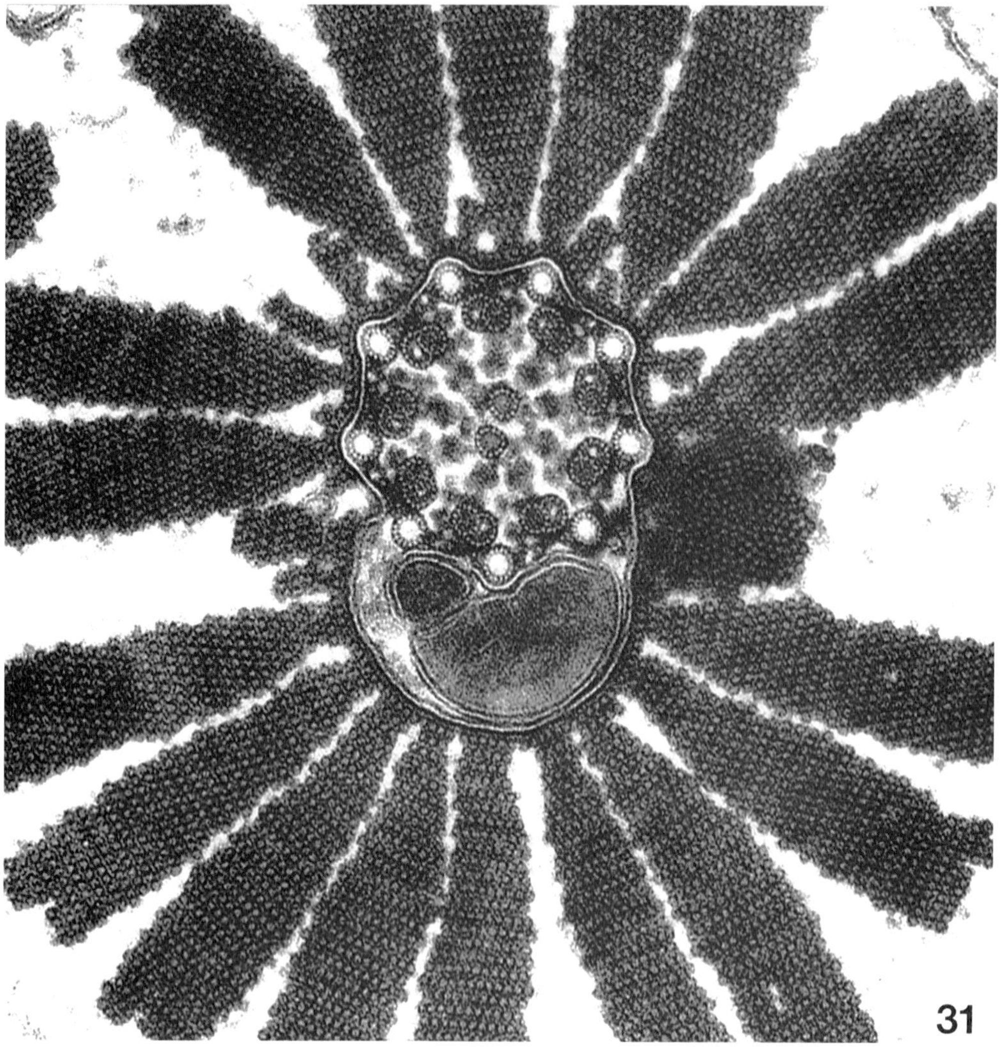

Figure 31. Cross section through the sperm tail of the moth *Apopestes spectrum*. Accessory tubules have 16 protofilaments. Several lacinate and one reticulate appendage radiate from the cell membrane. x180,000.

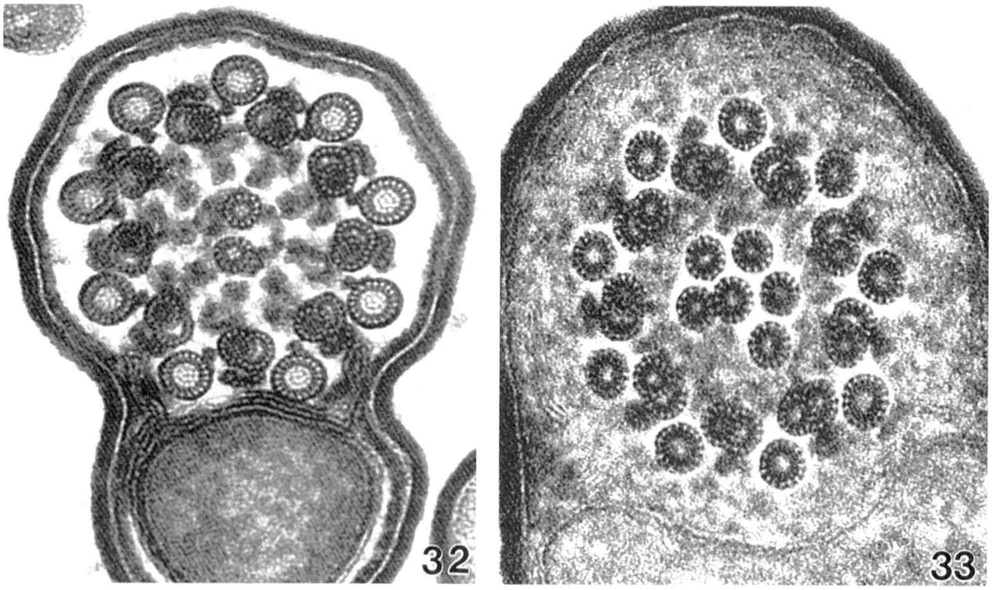

Figure 32. Cross section through the sperm tail of the caddisfly *Stenophylax permistus* (Integripalpia). The spermatozoa are motile in spite of the fact that they have inner dynein arms only. Accessory tubules have 19 protofilaments. x200,000. (Reprinted with permission from Struct. Biol.)

Figure 33. Cross section through the sperm tail of the caddisfly *Philopotamus montanus* (Annulipalpia). The spermatozoa are immotile, since they lack dynein arms. Accessory tubules have 17 or 18 protofilaments. x230,000. (Reprinted with permission from Boll. Zool.)

also been suggested that brachyceran flies originated from Tipulidae-Trichoceridae (Dallai et al. 1993b).

The two sister orders Trichoptera (caddisflies) and Lepidoptera (moths and butterflies) differ in that the lepidopteran sperm axoneme is of the common insectan type, 9+9+2 with 16 protofilaments in the accessory tubules (Dallai, Afzelius 1990a; Medeiros, Silveira 1996; Fig. 31), whereas Trichoptera have sperm axonemes of several different kinds. There is a considerable variability in the trichopteran spermatozoa, although the outer dynein arm is lost in all examined species, a synapomorphic feature shared by all members of the order, and the accessory tubules have a wall with more than 16 protofilaments. The number of protofilaments ranges from 17 to 20, depending on the family.

Among Integripalpia, Rhyacophilidae and Glossosomatidae have 17 and 18 protofilaments, respectively; 18 protofilaments are also found in Leptoceridae, 19 in Limnephilidae (Fig. 32), Goeridae and Odontoceridae (Fig. 5), and 20 in Sericostomatidae, the highest number of protofilaments on record for any microtubule (Dallai, Afzelius 1994; Dallai et al. 1995b). In the suborder Annulipalpia, both inner and outer dynein arms are absent, the spermatozoa are immotile

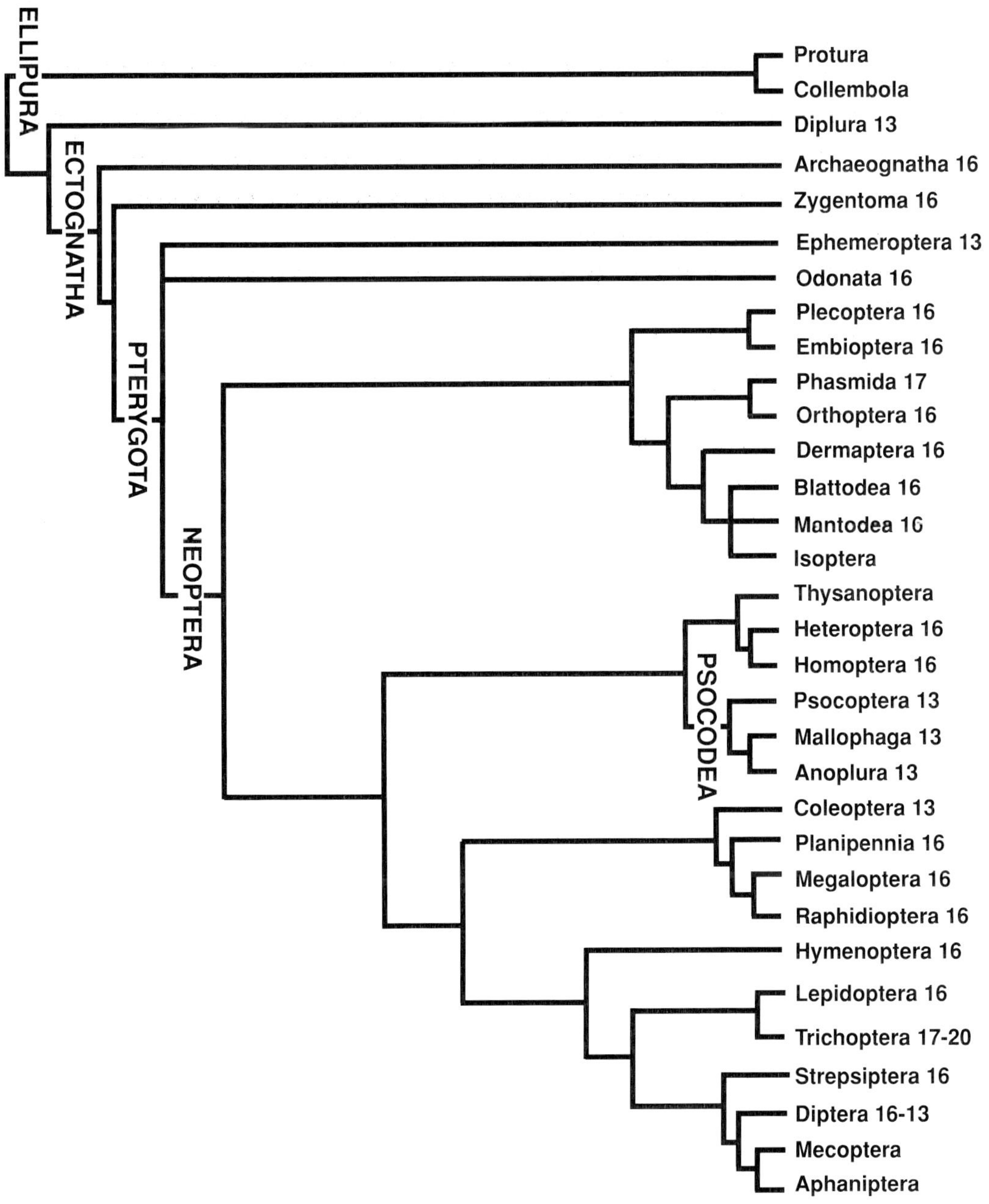

Figure 34. Cladogram showing the presumed insect order relationships. The numbers indicate the number of protofilaments in the accessory tubules. Collembola and Protura (Ellipura); Isoptera, Thysanoptera, Mecoptera and Aphaniptera, among Neoptera, lack accessory tubules.

and the axoneme, when present, has accessory tubules with 16, 17 or 18 protofilaments, as seen in Philopotamidae (Afzelius et al. 1991; Fig. 33). Moreover, in the genus *Wormaldia,* 13 rather than nine doublets and accessory tubules were described (Dallai et al. 1995c).

Spermatozoa from members of the large order Hymenoptera (sawflies, ants, bees and others) are very incompletely studied. However, available data indicate that the sperm tail is of the common insectan type: a 9+9+2 axoneme, accessory tubules with 16 protofilaments and a well-developed intertubular material (Dallai, Afzelius 1990a; Figs. 6, 9).

From this overview on insect accessory tubules, it is reasonable to believe that these structures are modeled, in a few cases, as a cytoplasmic microtubule exhibiting a size of about 25 nm and with a tubular wall consisting of 13 protofilaments (Diplura, Psocodea, Ephemeroptera, some nematoceran Diptera and all brachycerans), or, more commonly, as a microtubule of greater size (about 30 nm) and with 16 or more protofilaments. It is not known why the most common model is a tubule with 16 protofilaments rather than another number. Researchers can only speculate on the fact that the B-tubule of a doublet, from which the accessory tubule grows, would be potentially made of 16 protofilaments if it were closed (Afzelius et al. 1990). The radius of curvature of the axonemal B-tubule is such that it could be completely closed with 16 protofilaments. According to the paper by Hutchens et al. (1997), small changes in tubulin isoforms may have important consequences in the structure and function of accessory microtubules.

Although data are missing for many insect groups, it may be hypothesized that accessory tubules appeared early in insect evolution (Fig. 34). Their presence may well be considered a synapomorphic feature uniting all ectognathan orders plus Diplura, thus excluding Collembola and Protura. This observation could support the placement of Diplura in the ectognathan insect lineage, as suggested by Kukalova-Peck (1987). After the 13-protofilament accessory tubules of Diplura, a perhaps more stable 16-protofilament model might have evolved, prior to the diversification of Ectognathan orders. The fact that the 16-protofilament model is the most common one may suggest that this is the plesiomorphic condition among pterygotan insects, although it needs to be hypothesized that the protofilament number has secondarily changed, as it happens in Phasmida, Diptera, Trichoptera and some Coleoptera. These changes also include either loss of accessory tubules or an independent return to a 13-protofilament condition (as in Ephemeroptera, Psocodea and some Diptera).

Acknowledgements

The authors thank several people of the Department of Evolutionary Biology who helped in various aspects of the study; in particular, G. Callaini, F. Frati, L. Gamberucci, P. Lupetti, E. Malatesta and C. Mencarelli. They thank also Dr. S. Lanzavecchia of the University of Milan for the realization of the optical diffraction patterns. The study was supported by grant to R.D. from Italian Ministry of University, Scientific Research, and Technology (MURST).

References

Afzelius BA. Microtubules in the spermatids of stick insects. J Ultrastr Mol Struct Res 1988; 98:94-102.

Afzelius BA, Dallai R. Characteristics of the flagellar axoneme in Neuroptera, Coleoptera, and Strepsiptera. J Morphol 1994; 219:15-20.

Afzelius BA, Baccetti B, Dallai R. The giant spematozoon of *Notonecta.* J Submicrosc Cytol 1976; 8:1149-1161.

Afzelius BA, Bellon PL, Lanzavecchia S. Microtubules and their protofilaments in the flagellum of an insect spermatozoon. J Cell Sci 1990; 95:207-217.

Afzelius BA, Bellon PL, Dallai R, Lanzavecchia S. Diversity of microtubular doublets in insect sperm tails: a computer-aided image analysis. Cell Motil Cytoskel 1991; 19:282-289.

Baccetti B. Spermatozoa and phylogeny in orthopteroid insects. In: (Baccetti B, ed) Evolutionary Biology in Orthopteroid Insects. Chichester: E. Horwood, 1987a, pp11-112.

Baccetti B. Spermatozoa and stick insects phylogeny. In: (Mazzini M, Scali V, eds) Stick Insects: Phylogeny and Reproduction. Siena: Universities of Siena and Bologna, 1987b, pp117-123.

Baccetti B, Bairati A Jr. Indagini comparative sull'ultrastruttura delle cellule germinali maschili in *Dacus oleae* Gmel ed in *Drosophila melanogaster* Meig. (Ins. Diptera). Redia 1964; 49:1-29.

Baccetti B, Dallai R. The spermatozoon of Arthropopda XXX. The multiflagellate spermatozoon in the termite *Mastotermes darwiniensis.* J Cell Biol 1978; 76:569-576.

Baccetti B, Dallai R, Rosati F. The spermatozoon of Arthropoda III. The lowest holometabolic insects. J Microsc Paris 1969; 8:233-248.

Baccetti B, Rosati F, Selmi G. The spermatozoon of Arthropoda XV. An unmotile "9+2" pattern. J Microsc Paris 1971; 11:133-142.

Baccetti B, Burrini AG, Dallai R, Giusti F, Mazzini M, Renieri T, Rosati F, Selmi G. Structure and function in the spermatozoon of *Tenebrio molitor.* (The spermatozoon of Arthropoda XX). J Mechanochem Cell Motil 1973a; 2:149-161.

Baccetti B, Burrini AG, Dallai R, Pallini, V, Periti P, Piantelli F, Rosati F, Selmi G. Structure and function in the spermatozoon of *Bacillus rossius.* J Ultrastr Res 1973b; 44:1-73.

Baccetti B, Dallai R, Burrini AG, Selmi G. Fine structure of the spermatozoon of an Onychophoran, *Peripatopsis.* Tissue Cell

1976; 8:659-672.
Baltz JM, William PO, Cone RA. Dense fibers protect mammalian sperm against damage. Biol Reprod 1990; 43:485-491.
Bigliardi E, Baccetti B, Burrini AG, Pallini V. The distribution of some enzymes in the insect sperm tail. In: (Baccetti B, ed) Comparative Spermatology. London: Academic Press, 1970, pp451-463.
Bode W. The spermatozoa of Thysanoptera and their relevance for systematics. Acta Physiopathol Entomol Hung 1988; 23:267-273.
Bonaccorsi S, Pisano C, Puoti F, Gatti MY. Chromosome loops in *Drosophila melanogaster*. Genetics 1988; 120:1015-1034.
Bressac C, Bré M-H, Darmanaden-Delorme J, Laurent M, Levilliers N, Fleury A. A massive new posttranslational modification occurs on axonemal tubulin at the step of spermatogenesis in *Drosophila*. Eur J Cell Biol 1995; 67:346-355.
Cameron ML. Some details of ultrastructure in the development of flagellar fibers of the *Tenebrio* sperm. Can J Zool 1965; 43:1005-1010.
Chrétien D, Wade RH. New data on the microtubule surface lattice. Biol Cell 1991; 71:161-164.
Chrétien D, Metoz F, Verde F, Karsenti E, Wade RH. Lattice defects in microtubules: protofilament numbers vary within individual microtubules. J Cell Biol 1992; 117:1031-1040.
Dallai R. The spermatozoon of Arthropoda XI. Further observations on Collembola. In: (Baccetti B, ed) Comparative Spermatology. New York: Academic Press, 1969, pp276-279.
Dallai R. An overview of atypical spermatozoa in insects. In: (Fawcett DW, Bedford JM, eds) The Spermatozoon. Baltimore-Munich: Urban and Schwarzenberg, 1979, pp253-265.
Dallai R, Afzelius BA. Microtubular diversity in insect spermatozoa. Results obtained with a new fixative. J Struct Biol 1990a; 103:164-179.
Dallai R, Afzelius BA. Ultrastructural patterns of the flagellar axoneme in the non-motile part of the mole-cricket sperm. Biol Cell 1990b; 70:19-26.
Dallai R, Afzelius BA. Sperm flagellum of *Dacus oleae* (Gmelin) (Tephritidae) and *Drosophila melanogaster* (Drosophilidae) (Diptera). Int J Insect Morph Embryol 1991a; 20:215-222.
Dallai R, Afzelius BA. Sperm flagellum of insects belonging to insect orders of Psocoptera, Mallophaga and Anoplura. Ultrastructural and phylogenetic aspects. Boll Zool 1991b; 58:211-216.
Dallai R, Afzelius BA. Development of the accessory tubules of insect sperm flagella. J Submicrosc Cytol Pathol 1993a; 25:499-504.
Dallai R, Afzelius BA. Characteristics of the sperm tail of the velvet worm, *Euperipatoides leuckarti* (Onychophora). Tissue Cell 1993b; 25:907-913.
Dallai R, Afzelius BA. Sperm structure of Trichoptera. I. Integripalpia: Limnephiloidea. Int J Insect Morphol Embryol 1994; 23:197-209.
Dallai R, Afzelius BA, Lanzavecchia S, Bellon PL. Bizarre flagellum of thrips spermatozoa (Thysanoptera, Insecta). J Morphol 1991; 209:343-347.
Dallai R, Xue L, Yin WY. Flagellate spermatozoa of Protura (Insecta, Apteygota) are motile. Int J Insect Morphol Embryol 1992a; 21:137-148.
Dallai R, Bellon PL, Lanzavecchia S, Afzelius BA. Sperm axoneme of some apterygote insects examined by computer-aided image analysis. Acta Zool 1992b; 73:109-114.
Dallai R, Afzelius BA, Lanzavecchia S, Bellon PL. Native microtubules with a variable number of protofilaments. Cell Motil Cytoskel 1993a; 24:49-53.
Dallai R, Bellon PL, Lanzavecchia S, Afzelius BA. The dipteran sperm tail: ultrastructural characteristics and phylogenetic considerations. Zool Scr 1993b; 22:193-202.
Dallai R, Lupetti P, Afzelius BA, Mamaev BM. Characteristics of the sperm flagellum in fungus gnats (Mycetophiloidea, Diptera, Insecta). Zoomorphology 1995a; 115:213-219.
Dallai R, Lupetti P, Afzelius BA. Sperm structure of Trichoptera. III. Hydropsychidae, Polycentropodidae and Philopotamidae (Annulipalpia). Int J Insect Morphol Embryol 1995b; 24:171-183.
Dallai R, Lupetti P, Afzelius BA. Sperm structure of Trichoptera. IV: Rhyacophilidae and Glossosomatidae. Int J Insect Morphol Embryol 1995c; 24:185-193.
Dallai R, Lupetti P, Frati F, Afzelius BA, Mamaev BM. Spermatozoa from the supertribes Lasiopteridi and Stomatosematidi (Insecta, Diptera, Cecidomyiidae): ultrastructure data and phylogeny of the subfamily Cecidomyiinae. Zool Scr 1996; 25:51-60.
Fernandes AP, Bào SN. Ultrastructural study of the spermiogenesis and localization of tubulin in spermatid and spermatozoon of *Diabrotica speciosa* (Coleoptera, Chrysomelidae). Cytobios 1996; 86:231-241.
Friedländer M. Phylogenetic position of rhyacophiloid caddisflies (Insecta, Trichoptera): a spermatological analysis of Rhyacophilidae and Glossosomatidae. Zool Scr 1993; 22:299-304.
Fujiwara K, Tilney LG. Substructural analysis of the microtubule and its polymorphic forms. Ann NY Acad Sci 1975; 253:27-50.
Giuffrida A, Rosati F. Changes in sperm tail of *Eyprepocnemis plorans* (Insecta, Orthoptera) as result of *in vitro* incubation in spermathecal extract. Inv Rep Dev 1993; 24:47-52.
Hirose K, Amos WB, Lockhart A, Cross RA, Amos LA. Three-dimensional cryoelectron microscopy of 16-protofilament microtubules: structure, polarity, and interaction with motor proteins. J Struct Biol 1997; 118:140-148.
Hutchens JA, Hoyle HD, Turner FR, Raff EC. Structurally similar *Drosophila* α-tubulins are functionally distinct *in vivo*. Mol Biol Cell 1997; 8:481-500.
Ishijima S, Hiramoto Y. Flexural rigidity of echinoderm sperm flagella. Cell Struct Funct 1994; 19:349-362.
Jamieson BGM. The ultrastructure of the spermatozoon of the oligochaetoid polychaet *Questa* sp. (Questidae, Annelida) and its phylogenetic significance. J Ultrastruct Res 1983; 84:238-251.
Jamieson BGM. The Ultrastructure and Phylogeny of Insect Spermatozoa. Cambridge University Press, Cambridge, 1987.
Jamieson BGM. Fish Evolution and Systematics: Evidence from Spermatozoa. Cambridge University Press, Cambridge, 1991.
Joly D, Bressac C, Devaux J, Lachaise D. Sperm length diversity in Drosophilidae. Drosophila Inform Serv 1991; 70:104-108.
Kathirithamby J, Carcupino M, Mazzini M. Ultrastructure of the spermatozoon of *Elenchus japonicus* and its bearing on the phylogeny of Strepsiptera. Tissue Cell 1992; 24:437-442.
Kiefer BI. Development, organization, and degeneration of the *Drosophila* sperm flagellum. J Cell Sci 1970; 6:177-194.
Kristensen NP. Phylogeny of insect orders. Ann Rev Entomol 1981; 26:135-157.
Kristensen NP. Insect phylogeny based on morphological evidence. In: (Fernholm B, Bremer K, Jörnvall H, eds) The Hierarchy of Life. Amsterdam: Excepta Medica, 1989, pp295-321.
Kristensen RM, Eibye-Jacobsen D. Ultrastructure of spermiogenesis and spermatozoa in *Diurodrilus subterraneus* (Polychaeta, Diurodrilidae). Zoomorphology 1995; 115:117-132.
Kukalova-Peck J. New carboniferous Diplura, Monura, and Thysanura, the hexapod ground plan, and the role of thoracic side lobes in the origin of wings (Insecta). Can J Zool 1987; 65:2327-2345.
Langford GM. Arrangement of subunits in microtubules with 14 protofilaments. J Cell Biol 1980; 87:521-526.
Lanzavecchia S, Bellon PL, Dallai R, Afzelius BA. Three-dimensional reconstructions of accessory tubules observed in the sperm axonemes of two insect species. J Struct Biol 1994; 113:225-237.
Lindemann CB, Rudd WG, Riksmenspoel R. The stiffness of the flagella of impaled bull sperm. Biophys J 1973; 13:437-448.
Mandelkow E, Song Y-H, Mandelkow E-M. The microtubule lattice -

dynamic instability of concepts. Trends Cell Biol 1995; 5:262-266.
Mazzini M. Giant spermatozoa in *Divales bipustulatus* F. (Coleoptera: Cleridae). Int J Insect Morphol Embryol 1976; 5:107-115.
Mazzini M, Carcupino M, Kathirithamby J. Fine structure of the spermatozoon of the strepsipteran *Xenos moutoni*. Tissue Cell 1991; 23:199-207.
Medeiros M, Silveira M. Ultrastructural study of apyrene spermatozoa of *Alabama argillacea* (Insecta, Lepidoptera, Noctuidae) with tannic acid containing fixative. J Submicrosc Cytol Pathol 1996; 28:133-140.
Metoz F, Arnal I, Wade RH. Tomography without tilt: three-dimensional imaging of microtubule/motor complexes. J Struct Biol 1997; 118:159-168.
Mizuhira V, Futaesaku Y. Fine structure of the microtubules by means of the tannic acid fixation. In: Proceedings of the 8th International Congress on Electron Microscopy. Canberra, 1974, pp340-341.
Nojima D, Linck RW, Egelman EH. At least one of the protofilaments in flagellar microtubules is not composed of tubulin. Curr Biol 1995; 5:158-167.
Piperno G, Fuller MT. Monoclonal antibodies specific for an acetylated form of α-tubulin recognize the antigen in cilia and flagella from a variety of organisms. J Cell Biol 1985; 101:2085-2094.
Pirner MA, Linck RW. Methods for the isolation of tektins and sarkosyl-insoluble protofilament ribbons. In: (Dentler W, Witman G, eds) Methods in Cell Biology. Vol 24. San Diego: Academic Press, 1995, pp373-380.
Pitnick S, Spicer GS, Markow TA. How long is a giant sperm? Nature 1995; 375:109.
Raff EC, Fackenthal JD, Hutchens JA, Hoyle HD, Turner FR. Microtubule architecture specified by a β-tubulin isoform. Science 1997; 275:70-73.
Rosati F. Peripheral connections of the axoneme in the sperm cells of *Bacillus rossius* (Rossi) (Phasmida: Bacillidae). Int J Insect Morphol Embryol 1976; 5:223-225.
Rota E, Lupetti P. An ultrastructural investigation of *Hrabeiella* Pizl and Chalupsky, 1984 (Annelida). II. The spermatozoon. Tissue Cell 1997; 29:603-609.
Song Y-H, Mandelkow E. The anatomy of flagellar microtubules: polarity, seam, junctions, and lattice. J Cell Biol 1995; 128:81-94.
Sonnenschein M, Häuser CL. Presence of only eupyrene spermatozoa in adult males of the genus *Micropteryx* Hübner and its phylogenetic significance (Lepidoptera: Zeugloptera, Micropterigidae). Int J Insect Morphol Embryol 1990; 19:269-276.
Sosa H, Milligan RA. Three-dimensional structure of ncd-decorated microtubules obtained by a back projection method. J Mol Biol 1996; 260:743-755.
Sosa H, Hoenger A, Milligan RA. Three different approaches for calculating the three-dimensional structure of microtubules decorated with kinesin motor domains. J Struct Biol 1997; 118:149-158.
Stanley HP, Bowman JT, Romrell L.J, Reed SC, Wilkinson RF. Fine structure of normal spermatid differentiation in *Drosophila melanogaster*. J Ultrastr Res 1972; 41:433-466.
Stephens RE, Oleszko-Szuts S, Linck RW. Retention of ciliary ninefold structure after removal of microtubules. J Cell Sci 1989; 92:391-402.
Storch V, Higgins RP. Ultrastructure of developing and mature spermatozoa of *Tubiluchus corallicola* (Priapulida). Trans Amer Microsc Soc 1989; 108:45-50.
Trimble JJ, Thompson SA. Fine structure of the sperm of the lovebug *Plecia nearctica* Hardy (Diptera: Bibionidae). Int J Insect Morphol Embryol 1974; 3:425-432.
Wade RH, Chrétien D. Cryoelectron microscopy of microtubules. J Struct Biol 1993; 110:1-27.
Wade RH, Hewat EA. Cryoelectron microscopy of macromolecular complexes. Biol Cell 1994; 80:211-220.
Wade RH, Chrétien D, Job D. Characterization of microtubule protofilament numbers. How does the surface lattice accomodate? J Mol Biol 1990; 216:775-786.
Wilson PJ, Forer A. Acetylated α-tubulin in spermatogenic cells of the crane fly *Nephrotoma suturalis*: kinechore microtubules are selectively acetylated. Cell Motil Cytosk 1989; 14:237-250.
Witman GB, Carlson K, Rosenbaum JL. *Chlamydomonas* flagella: II. The distribution of tubulins 1 and 2 in the outer doublet microtubules. J Cell Biol 1972; 54:540-555.
Wolf KW. Cytology of Lepidoptera. VIII. Acetylation of α-tubulin in mitotic and meiotic spindles of two Lepidoptera species, *Ephestia kuehniella* (Pyralidae) and *Pieris brassicae* (Pieridae). Protoplasma 1996; 190:88-98.
Wolf N, Regan CL, Fuller MT. Temporal and spatial pattern of differences in microtubule behaviour during *Drosophila* embryogenesis revealed by distribution of a tubulin isoform. Development 1988; 102:311-324.

31

Spermatozoa of Platyhelminthes: Comparative Ultrastructure, Tubulin Immunocytochemistry and Nuclear Labeling

Jean-Lou Justine
Museum of Natural History, Paris

Spermatozoa of Platyhelminthes show an outstanding variety of structures, including aflagellate, uniflagellate and biflagellate cells. All members of the phylum use internal fertilization, and spermatozoa are generally long and filiform cells. Because of this filiform morphology, description of Platyhelminthes spermatozoa by ultrastructure alone is not an easy task: serial sections are necessary for a thorough description of the cell, and this technique is extremely time consuming, specially for cells 400 μm in length. Therefore, the use of relatively new techniques, such as immunocytochemical labeling of tubulin or labeling of nuclei with specific fluorescent dyes, has recently been tried to enhance the precision of ultrastructural studies.

The first part of this review discusses recent descriptions of Platyhelminthes spermatozoa, with emphasis on parasitic groups, and their use for the understanding of phylogeny. The use of spermatozoa for Platyhelminthes phylogeny has been the subject of numerous reviews in recent years (Bâ, Marchand 1995; Justine 1991a, b, c, 1993, 1995, 1998b; Watson, Rohde 1995). The present review concentrates on recent acquisitions; that is, following the 1995 review by Justine. The second part of this review describes some new results obtained with the use of tubulin and nucleus labeling for the description of Platyhelminthes spermatozoa. Results concerning the occurrence of post-translational modifications of tubulin in various microtubular systems are emphasized.

Structure of Spermatozoa of Parasitic Platyhelminthes

Spermiogenesis in the parasitic Platyhelminthes follows the same pattern in most groups (that is, the Monogenea Polyopisthocotylea, Digenea, Aspidogastrea, Gyrocotylidea, Amphilinidea and Eucestoda), with a very characteristic structure (Justine 1991a, 1995). The spermatids are fused in groups of 32 or 64 cells. Each spermatid produces a conical protruding element, the "zone of differentiation" from which arise three elongating processes: a median cytoplasmic process and two lateral flagella (Fig. 1). Later during spermiogenesis, the three elements fuse and produce the mature spermatozoon, which therefore has two axonemes incorporated within the cytoplasm. The incorporation of the axonemes within the cell occurs as a process known as "proximodistal fusion" (Justine 1991b), because the fusion begins at the proximal extremity (near the common cytoplasmic mass) and ends at the distal extremity of the spermatids. Free flagella are observed at the extremity of spermatids as long as the process is not achieved, but axonemes are completely incorporated in the mature spermatozoa.

This sperm structure, originating from a typical zone of differentiation with two flagella and proximodistal fusion, is found in most Polyopisthocotylea, with the single exception of *Diplozoon*, which is aflagellate (Justine et al. 1985b). This structure is also found in most Digenea, with the single exception of the schistosomes, in which the spermatozoon may be considered as an immature zone of differentiation (Justine 1991d, 1995), and in the Aspidogastrea, the Gyrocotylidea and the Amphilinidea. Variations are found, however, in the Monopisthocotylea and the Eucestoda.

Ultrastructural Studies on Spermatozoa of Parasitic Platyhelminthes

Polyopisthocotylea and Digenea

These groups each display a relatively homogeneous sperm pattern. Polyopisthocotyleans' spermatozoa have two axonemes and cortical microtubules that make a continuous palisade (Justine 1991a). A recent study on *Atriaster heterodus* (Santos et al. 1997) has confirmed this structure.

Digeneans also have a spermatozoon with two axonemes, but the cortical microtubules, in the nuclear region, do not occur in the lateral part of the spermatozoon (Justine 1991a). Recent studies on the digeneans *Allassogonoporus amphoraeformis* (Podvyaznaya 1996), *Dicrocoelium chinensis* (Tang 1996; Tang, Li 1996), *Postorchigenes gymnesicus* (Gracenea et al. 1997), *Ceylonocotyle scoliocoelium* (Li, Wang 1997), *Mesocoelium monas* (Iomini et al. 1997), and *Echinostoma caproni* (Iomini, Justine 1997) have confirmed the homogeneity of sperm structure in the digeneans. Membrane ornamentations are one of the most enigmatic structures of digenean sperm, and their localization along the filiform spermatozoon is difficult to understand with randomly obtained cross sections. A multi-technique approach (SEM, TEM, immunocytochemistry) of the spermatozoon of *E. caproni* has clearly solved the problem, showing that the ornamentations in this species are located in a precise zone which displays reduced permeability to the agents used for immunocytochemistry (Iomini, Justine 1997).

Monopisthocotylea

In the Monopisthocotylea, a true process of fusion is not found: the flagella (or the single flagellum) grow in a process which can be interpreted as homologous to the median cytoplasmic process. Because there are no separate elements, a process of fusion cannot occur. This has been interpreted as derived from the structure found in most parasitic Platyhelminthes (Justine 1991a). However, a recent molecular phylogeny (Mollaret et al. 1997) of the parasitic Platyhelminthes places the

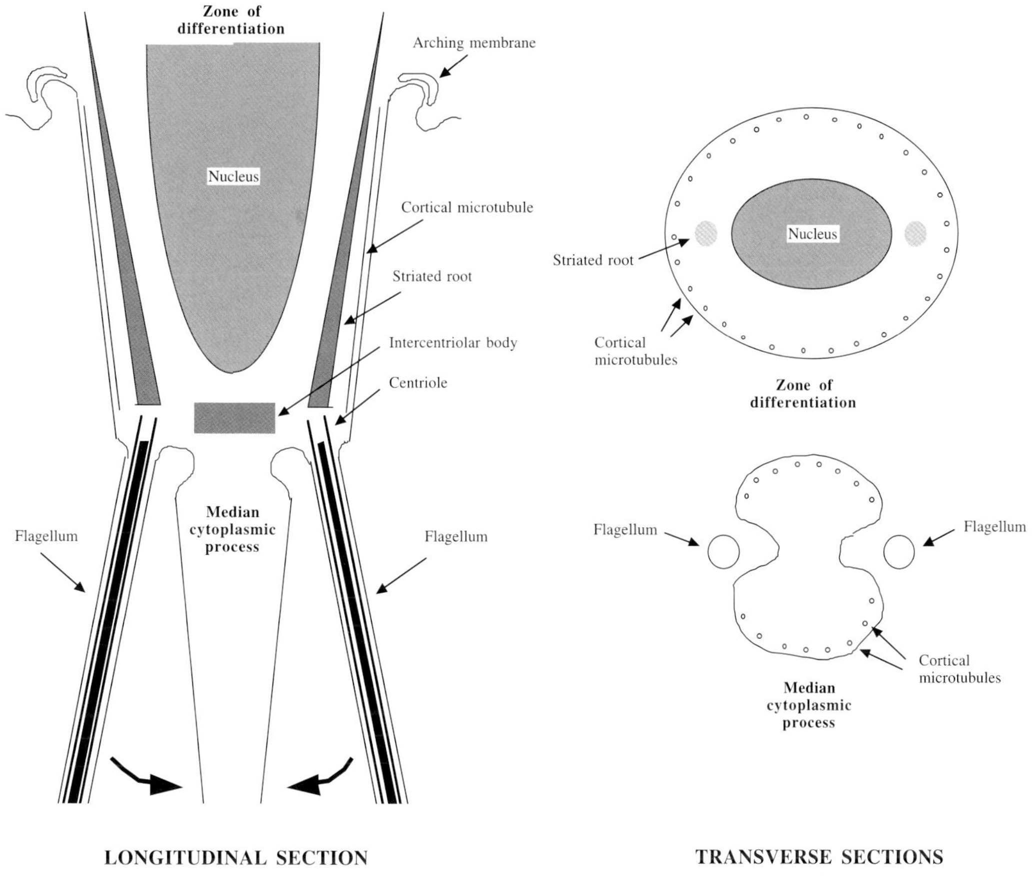

Figure 1. Diagram of the "zone of differentiation" of the spermatid, found in the plesiomorphic pattern of spermiogenesis of the parasitic Platyhelminthes, in the Digenea, Monogenea Polyopisthocotylea, Gyrocotylidea, Amphilinidea and certain Eucestoda. Curved arrows indicate the proximodistal fusion of the flagella with the median cytoplasmic process. (From Justine 1998b; used with permission of the publisher.)

Monopisthocotylea as the sister-group of all other parasitic Platyhelminthes. In this phylogenetic scheme, absence of proximodistal fusion could be a plesiomorphic character for the Monopisthocotylea, with proximodistal fusion as an apomorphy limited to the other groups. This requires further investigation. The question of the monophyly of the Monogenea (that is, a monophylum including the Monopisthocotylea and Polyopisthocotylea) is currently debated (Justine 1998a).

The Monopisthocotylea have been the subject of analyses of sperm structure for phylogenetic purposes (Justine 1991b, 1993; Justine et al. 1985a) and, subsequently, of analyses combining both morphological and spermatological data (Boeger, Kritsky 1993, 1997). Recent results on Monopisthocotylea include a comparative study of spermatozoa of the monocotylids *Troglocephalus rhinobatidis*, *Neoheterocotyle rhinobatidis* and *Merizocotyle australensis* (Watson 1997), showing that sperm structure is not homogeneous in this family, and thus challenging an earlier phylogenetic analysis (Justine 1991b). Interestingly, a recent molecular analysis of the Monopisthocotylea (Mollaret et al. 1997), with one of the different methods of analysis employed, did not support monophyly of the Monocotylidae. Fragmentary results obtained on *Pseudodactylogyroides marmoratae* and *Sundanonchus micropeltis*, two species from Malaysia, showed spermatozoa with a single axoneme (Mollaret et al. 1998). An anecdote interesting for the spermatologist is that a study based only on general morphology had "predicted" the occurrence of sperm with one axoneme in *Sundanonchus* (Kritsky, Lim 1995).

Eucestoda

The Eucestoda is the group of parasitic Platyhelminthes in which comparative ultrastructural studies have been most actively pursued in recent years, by the team of Bernard Marchand (Banyuls, France) and Cheikh Tidiane Bâ (Dakar, Senegal). Previous reviews listed 24 genera (Justine 1991a), then 34-36 genera (Bâ, Marchand 1995; Justine 1995) and 43 genera (Justine 1998b). The total number is now 45 genera and 58 species with the publication of recent studies (Bâ, Marchand 1998; Miquel et al. 1998; Miquel, Marchand 1998a, 1998b).

In the same period, studies of cestode evolution have shown considerable progress. An early cladistic study (Brooks et al. 1991) attempted to identify the relationships among the major cestode groups on the basis of morphological characters. After the Workshop for Tapeworm Systematics (Hoberg et al. 1997a), during which results from morphological, spermatological and molecular characters were compared and discussed, three important papers were published: a major analysis of morphology, with inclusion of some sperm characters (Hoberg et al. 1997b), a molecular analysis based on 18S rDNA (Mariaux 1998), and a review of sperm characters and their use for phylogeny (Justine 1998b).

An illustrative example of the complementary nature of the three analyses is provided by the study of the phylogenetic position of the Tetrabothriidea. This little group of cestodes occurs only in sea birds and mammals. Schematically, various hypotheses proposed their affinities with the Tetraphyllidea, a group of cestodes parasites in marine chondrichthyan fishes, or with the Cyclophyllidea, a group parasitic in land birds and mammals. In the first hypothesis, the marine homeotherms would have acquired the tetrabothriids from their fish prey, but in the second, the tetrabothriids would have followed their host while they were returning to a sea habitat (Hoberg et al. 1997b). An analysis of morphological characters alone favored the first hypothesis (Brooks et al. 1991). In 1991, it was proposed that the twisting of peripheral microtubules be considered a synapomorphy for the Cyclophyllidea (Justine 1991a). Later, the first description of the spermatozoon of a tetrabothriid found this character (Stoitsova et al. 1995). A cladistic analysis based on morphological characters and some spermatological characters finally placed the Tetrabothriidea as the sister-group to the Cyclophyllidea (Hoberg et al. 1997b). At the same time, a molecular analysis of 18S rDNA unequivocally assigned sister-group relationships to the Tetrabothriidea and Cyclophyllidea (Mariaux 1998). This case represents valuable "reciprocal illumination" between various methods.

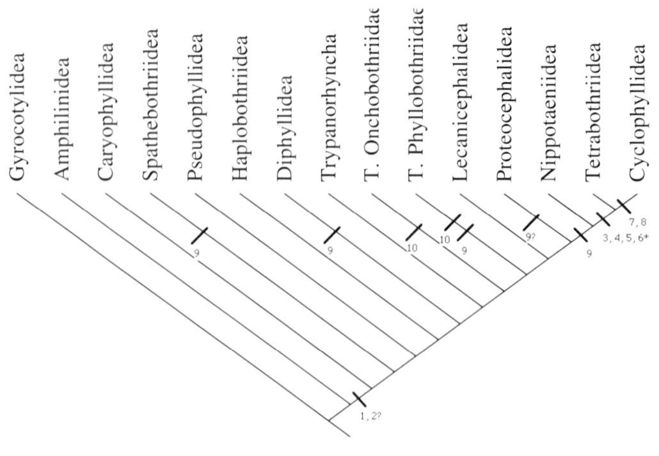

Figure 2. Proposed spermatozoal synapomorphies for the Eucestoda plotted on the phylogenetic tree of Hoberg et al. (1997). No spermatological data are available for the Lecanicephalidea. 1, absence of mitochondrion in mature sperm; 2, presence of crested body; 3, absence of intercentriolar body; 4, absence of striated roots; 5, twisting of peripheral microtubules; 6, presence of periaxonemal sheath (asterisk indicates reversal in certain cyclophyllids); 7, absence of flagellar rotation; 8, a single axoneme in the zone of differentiation; 9, a single axoneme in the mature spermatozoon (question mark for the Proteocephalidea refers to *Sandonella*); 10, two types of peripheral microtubules. (From Justine 1998b; used with permission of the publisher.)

Some spermatozoal synapomorphies are shown on Figure 2, and summarized from Justine (1998b). These are major synapomorphies which are useful for the relationships of cestodes at the ordinal level.

Character 1. Mitochondrion. Although all Platyhelminthes spermatozoa have one or several mitochondria, the mitochondrion is absent in all Eucestodes (Brooks 1989; Ehlers 1985; Justine 1991a, 1995), thus providing a synapomorphy for this group.

Character 2. Crested body. The presence of a crested body has been proposed as a synapomorphy for the Eucestoda (Bâ, Marchand 1995) but in the present state of knowledge, which is meagre for the Caryophyllidea and Spathebothriidea, it is perhaps more prudent to propose it as a synapomorphy for a derived group including the Pseudophyllidea (Justine 1998b).

Character 3. Intercentriolar body. The absence of this structure is a synapomorphy uniting the Tetrabothriidea and Cyclophyllidea (Justine 1998b), but it may also apply to the Nippotaeniidea, which are poorly known in term of sperm structure but are the sister group to this assemblage.

Character 4. Striated roots. The absence of striated roots has been proposed as a synapomorphy for the Tetrabothriidea and Cyclophyllidea (Justine 1998b), and might also apply to the Nippotaeniidea. However, recent publications reported the presence of a striated root in two cyclophyllidean species (Miquel et al. 1998; Miquel, Marchand 1998b). Examination of the micrographs of spermatids of *Dipylidium caninum* (Miquel et al. 1998) reveals that the structures labeled as striated roots do not show the large diameter and the regular striated pattern of the striated roots generally found in other cestodes. It may be that the character "absence of striated root" (Justine 1998b) should be re-coded as "absence of typical striated roots." The presence of the thin striated roots found in some cyclophyllids could be, however, of phylogenetic interest, but at the level of intra-ordinal relationships within the cyclophyllids.

Character 5. Peripheral microtubules. The twisting of peripheral microtubules (in contrast to a parallel arrangement) is a synapomorphy for the Tetrabothriidea and Cyclophyllidea (Justine 1998b).

Character 6. Periaxonemal sheath. This structure consists of a ring of dense material visible in cross section around the axoneme. It has no equivalent in any other Platyhelminthes. The presence of a periaxonemal sheath may be a synapomorphy for the Tetrabothriidea and Cyclophyllidea, but it is absent in some Cyclophyllidea, thus suggesting a reversal in certain cyclophyllids (Justine 1998b).

Character 7. Flagellar rotation. The absence of flagellar rotation is a synapomorphy for the Cyclophyllidea.

Character 8. Number of axonemes in zone of differentiation. The presence of a single axoneme in the zone of differentiation is a synapomorphy for the Cyclophyllidea.

Character 9. Number of axonemes. This is an homoplasic character, with convergence in several unrelated groups. Spermatozoa with two axonemes clearly correspond to the plesiomorphic pattern of the parasitic Platyhelminthes, but spermatozoa with a single axoneme are produced by several distinct spermiogenesis patterns in the Eucestoda (Bâ, Marchand 1995; Justine 1998b).

Character 10. Cortical microtubules. The presence of two types of cortical microtubules, with a usual hollow center and with an electron-dense center, may be a synapomorphy for the Phyllobothriidae and Onchobothriidae (Justine 1998b).

Other characters. Several other characters are potentially useful for the understanding of intra-ordinal relationships, mainly in the Cyclophyllidea, in which the number of sperm studies is now high enough to allow comparisons of phylogenetic value. These characters are: pattern of spermiogenesis, size and numbers of crested bodies, angle of twisted microtubules, shape of nucleus, presence of proteinaceous transverse walls, presence of dense granules, shape of apical cones and posterior structures (Justine 1998b).

Tubulin Immunocytochemistry and Nuclear Labeling

Tubulin is one of the major proteins found in spermatozoa of animals, at least when an axoneme is present. In the Platyhelminthes, the microtubular system of spermatozoa generally consists of one or two axonemes and additional singlet microtubules, thus providing very diverse cell models and tools for the understanding of subcellular sorting of tubulins. Tubulin diversity is generated in eukaryotic cells by the differential expression of several alpha and beta tubulin isogenes, which, in the Platyhelminthes, are known only in the species *Schistosoma mansoni* (Duvaux-Miret et al. 1991). Tubulin diversity is increased by post-translational modifications. Post-translational modifications have been subdivided in two classes by Iomini et al. (1998). The first class corresponds to the addition or removal of one or more amino acids, or the chemical modification of a residue of the polypeptide chain, and includes acetylation (L'Hernault, Rosenbaum 1985; LeDizet, Piperno 1987), detyrosination (Barra et al. 1974; Thompson 1982), excision of the penultimate glutamate residue (Paturle-Lafanechère et al. 1991), phosphorylation (Eipper 1974), and palmitoylation (Caron 1997; Ozols, Caron 1997). The second class, the polymodifications, corresponds to the lateral branching of a chain of variable length, and includes polyglutamylation (Eddé et al. 1990) and polyglycylation (Redeker et al. 1994). In recent years, the Justine laboratory has studied localization of tubulin and several of its post-translational modifications in spermatozoa of several species of Platyhelminthes. Crucial tools for this research are the antibodies developed against tubulin and its variants.

Absence of Tubulin in the Central Core of the 9+"1" Axoneme

The 9+"1" axoneme is a structure typical of the Platyhelminthes. It is found only in a part of the phylum considered as a monophylum under the name of Trepaxonemata (Ehlers 1984, 1985). This name refers to the spiralled structure of the central cylinder, as seen in longitudinal sections. The central cylinder appears, in transverse section, as an electron-dense ring, about 70 nm in diameter (Fig. 3a, c); it is apparently made up of two elements, about 18 nm in diameter, which are coiled in a double spiral (Fig. 3b) with a 65 nm step (Henley et al. 1969; Silveira 1975).

The biochemical nature of the central core has been studied by post-embedding electron microscope immunocytochemistry of spermatids and spermatozoa of the digenean *Echinostoma caproni*, using several antibodies against tubulin: anti-alpha tubulin, anti-beta tubulin and anti-acetylated alpha-tubulin (Iomini et al. 1995). None of the three anti-tubulin antibodies labeled the central core, although the doublet microtubules were, of course, clearly labeled. In contrast, in 9+2 axonemes found in the protonephridia of the same animal, the central microtubules are labeled (Fig. 3). Later, it was shown that antibodies against other variants of tubulin, including glycylated and glutamylated tubulin, do not label the central core (Iomini et al. 1998). This strongly suggests that the central core of the 9+"1" axoneme of the Trepaxonemata does not contain tubulin; this makes a clear difference between the 9+"1" axoneme and the ubiquitous 9+2 pattern which has two microtubules in its center. The chemical nature of the central core of the trepaxonematan 9+"1" axoneme, however, is still unknown.

Tubulin Labeling and Numbers of Spermatids in an Isogenic Group

The number of spermatids in isogenic groups has been proposed as a character of possible phylogenetic importance (Justine 1995). It is 64 in the cestodes, 32 in the digeneans, and 32, 64 and possibly 128 in the monogeneans. In modern literature, this number is generally

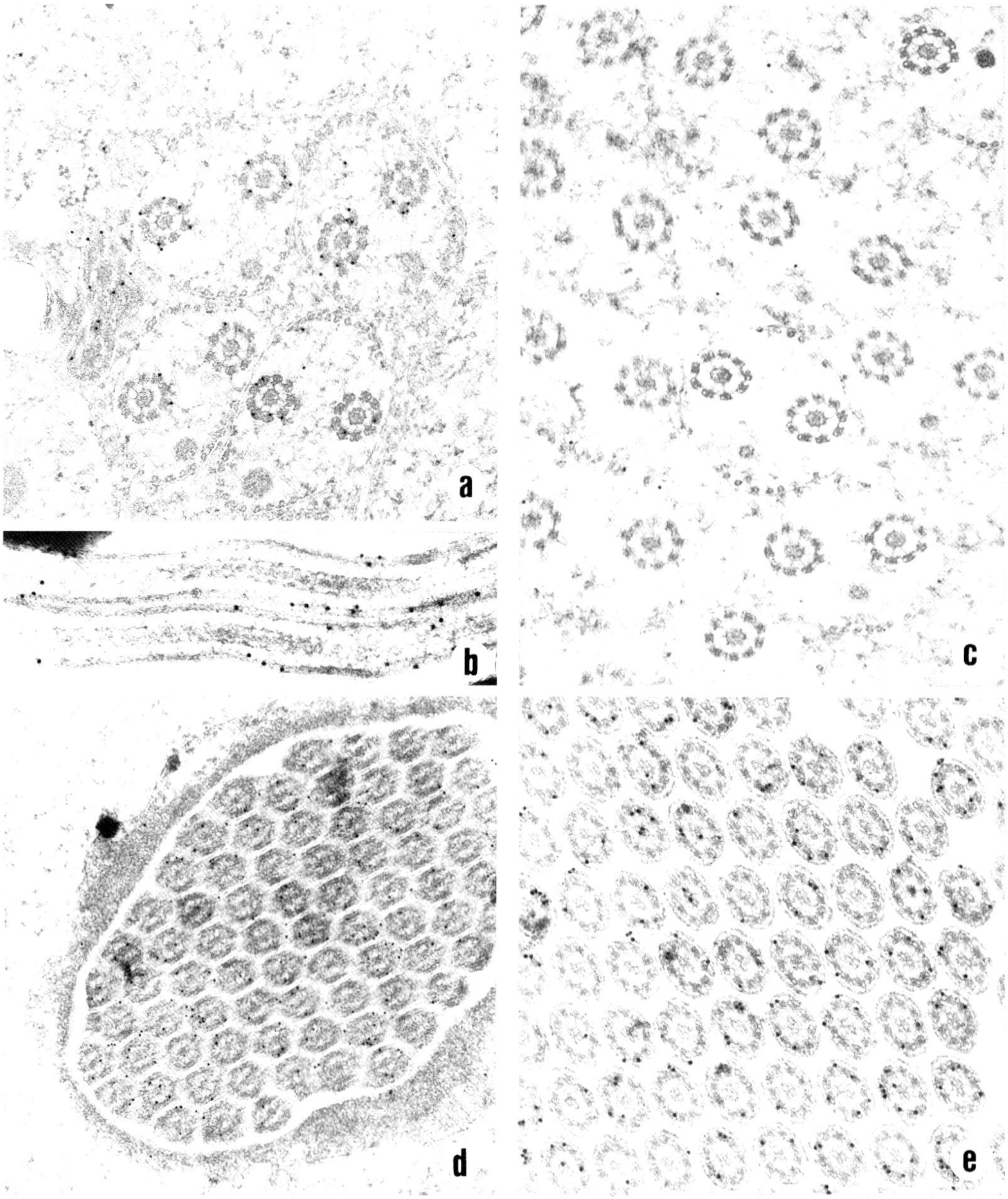

Figure 3. Post-embedding electron microscope immunocytochemistry of spermatids and protonephridia of the digenean *Echinostoma caproni*. a, b, c: Spermatids. a: Transverse section: microtubule doublets, but not the singlet cortical microtubules, are labeled by an anti-acetylated tubulin antibody. b: Longitudinal section of 9+"1" axonemes: the doublets are labeled by an anti-acetylated tubulin antibody, but not the central core. c: Transverse section: no labeling is observed with an anti-glycylated tubulin antibody (AXO 49). d, e: Protonephridia, transverse sections; a labeling of all microtubules is observed (d) with anti-acetylated and (e) anti-glycylated (AXO 49) antibodies.

Tubulin is not detected in the central core of the 9+"1" axoneme. Acetylation is present in protonephridia and in axonemal microtubules of the 9+"1" sperm axonemes, but not in cortical sperm axonemes. High level of polyglycylation is present in protonephridia, but absent in all microtubular systems of the spermatozoon. (Original, by C. Iomini and J.-L. Justine; details of labeling and observation methods in Iomini et al. 1998.)

deduced from the observation of transverse sections of isogenic groups by electron microscopy; this requires well oriented sections. A simple indirect fluorescent labeling of a squash of spermatids provides an easy way to count the number of spermatids (Mollaret, Justine 1997; Fig. 4). A large variety of anti-tubulin antibodies gives excellent results for this kind of observation (Iomini et al. 1997; Mollaret, Justine 1997).

Nuclear Labeling, Migration of the Nucleus and Orientation of Platyhelminthes Sperm

The question of the orientation of Platyhelminthes spermatozoa has long been a confusing question. In the parasitic Platyhelminthes, the centrioles are located in the proximal part (near the cytoplasmic mass of the spermatid) and the nucleus migrates to the distal extremity of the spermatid. This is different from most animal sperm, in which the orientation of the axoneme (centriole forward) and the anterior position of the nucleus coincide. This is the kind of question which is difficult to solve with ultrastructure alone, especially because the great length of Platyhelminthes sperm makes the use of serial sectioning time-consuming. This technique has been used in a few species, however, such as *Gonapodasmius* (Justine 1982; Justine, Mattei 1982). Observations of fertilization have also provided arguments for the posterior position of the nucleus (Justine, Mattei 1984). However, only the use of specific nuclear dyes could definitely solve this question. Observation of spermiogenesis in the digenean *Echinostoma caproni* shows that 32 round nuclei are located in the common cytoplasmic mass at the beginning of spermiogenesis (Fig. 5). They later elongate and migrate to the distal extremity of each spermatid (Fig. 5). These observations in *E. caproni* (Iomini et al. 1998; Iomini, Justine 1997) have been confirmed in another digenean, *Mesocoelium monas* (Iomini et al. 1997). Clearly, the nucleus is in a posterior position in digenean sperm. In a monopisthocotylean, the migration of the nucleus has also been observed, but the nucleus finally occupies a central position in the mature spermatozoon (Mollaret, Justine 1997).

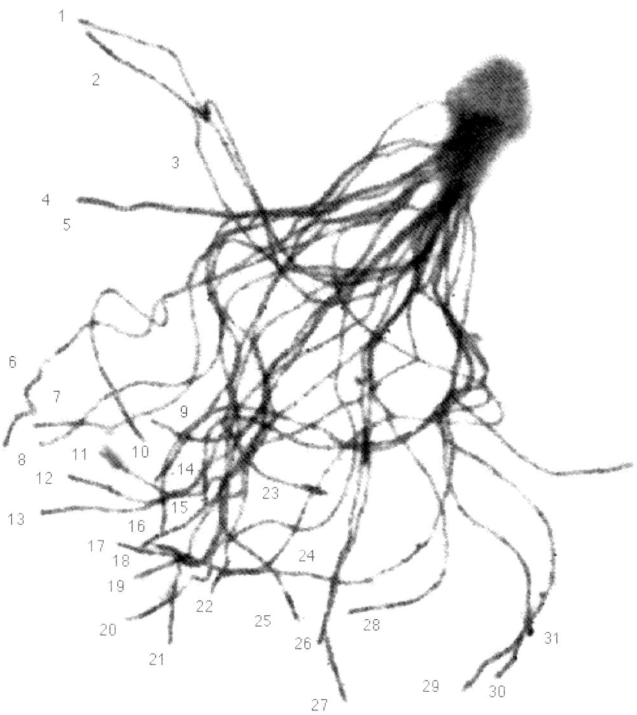

Figure 4. Computer-enhanced image showing 32 spermatids in an isogenic group of spermatids of the monopisthocotylean *Pseudodactylogyrus* sp. Spermatids were labeled by indirect immunofluorescence with a polyclonal anti-tubulin antibody. The image was inverted to produce a negative (details of technique in Mollaret, Justine 1997). This technique allows easy counting of the spermatids. (Used with permission of publisher.)

In the non-parasitic Platyhelminthes ("Turbellaria"), which also have filiform spermatozoa, the process of spermiogenesis is different and the centrioles are located at the distal extremity of the spermatid. This major difference was described early (Hendelberg 1962, 1969). The use of nuclear fluorescent dyes has shown that nuclei do not migrate toward the distal extremity of the spermatid in the scutariellid temnocephalid *Troglocaridicola* sp. (Iomini et al. 1994) and in the acoel *Convoluta saliens* (Raikova, Justine 1999). The shape of the nucleus in mature sperm has also been studied by this technique in the acoel *Paratomella rubra* (Raikova et al. 1997).

Two Subpopulations of Microtubules in Spermatozoa of the Digeneans

Indirect immunofluorescence double labeling experiments with two anti-tubulin antibodies demonstrated that there are two subpopulations of microtubules in digenean spermatozoa, differentiated by their labeling with anti-acetylated tubulin (Fig. 5). Post-embedding electron microscope immunocytochemistry revealed that the doublet microtubules of the axoneme are acetylated, whereas the singlet cortical microtubules are not (Iomini et al. 1995). Further experiments showed that axonemal microtubules contain tubulin which has been submitted to several post-translational modifications, including acetylation, glutamylation, and glycylation (Iomini et al. 1998), but that cortical microtubules do not contain any of these modifications. This demonstrated a subcellular sorting of post-translationally modified tubulin isoforms within the spermatozoon (Iomini et al. 1998). In addition, 9+2 axonemes, found in the same animal in the protonephridia, exhibit a clear labeling with the antibodies against the post-translational modifications of tubulin (Table 1). Results obtained with two antibodies which recognize different levels of glycylation (Bré et al. 1996), namely TAP 952 and AXO 49, are contrasting between the axonemes of the two organs: doublets of the 9+"1" axoneme in spermatozoa exhibit only low levels of glycylation, whereas 9+2 axonemes of protonephridia show high levels of glycylation (Iomini et al. 1998).

The spermatozoon of *Pseudodactylogyrus* sp., a monogenean in which the sperm cell contains only a single axoneme and no cortical microtubule, is a contrasting model for the study of the 9+"1" axoneme. Indirect immunofluorescence studies revealed, as in the 9+"1" of *Echinostoma caproni*, the presence of acetylation and glutamylation (Table 1).

Post-Translational Modifications of Tubulin during Spermiogenesis of a Digenean

Studies of spermiogenesis in a digenean, *E. caproni*, have shown that the post-translational modifications

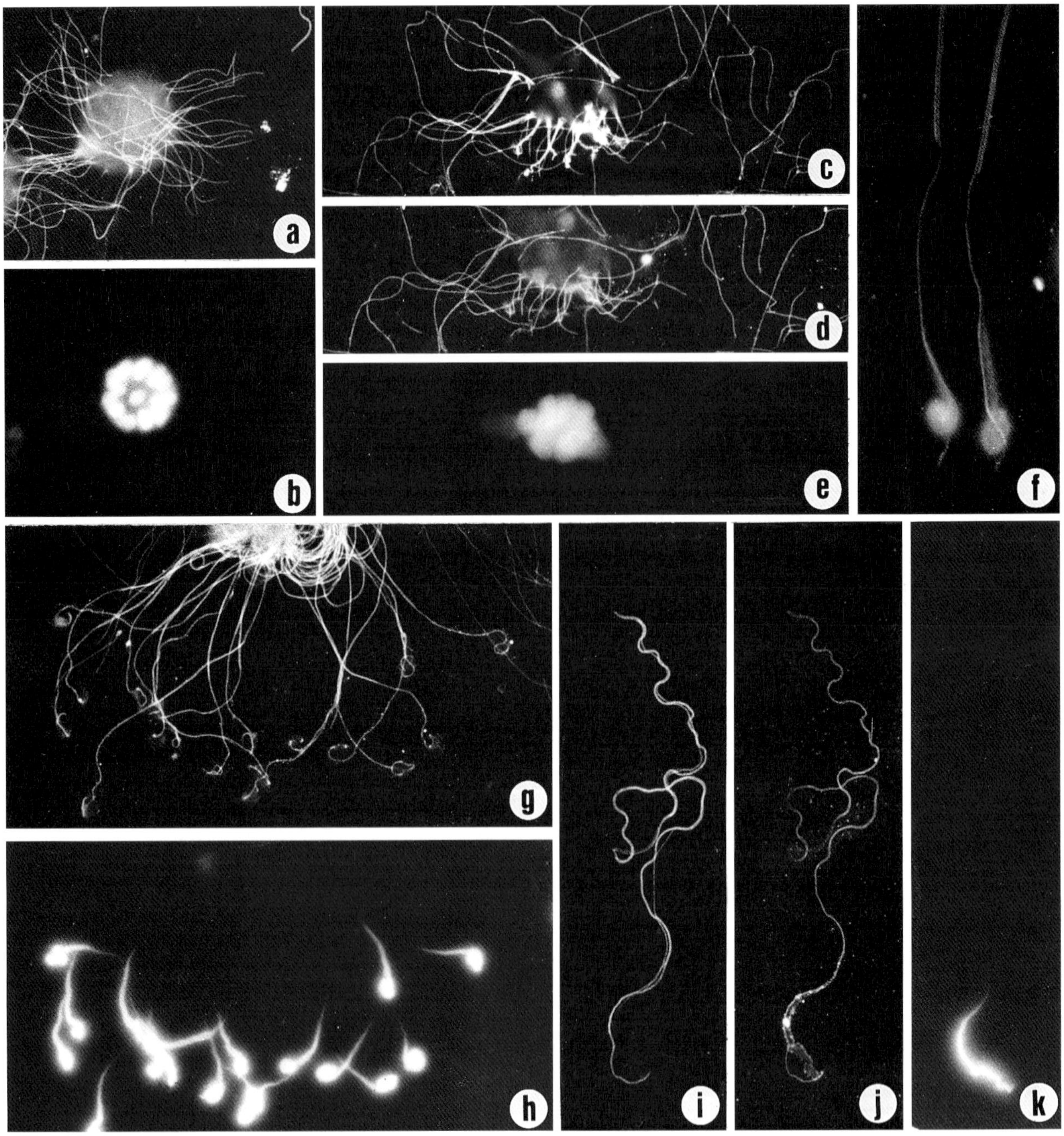

Figure 5. Spermiogenesis and spermatozoon of the digenean *Echinostoma caproni*, immunocytochemistry of tubulin and nuclear labeling. a, b: Early spermatid. a: Tubulin labeling; b: Nuclear labeling, showing the round nuclei grouped in the central cytoplasmic mass. c, d, e: More advanced spermatids, with elongated zones of differentiation. c: Alpha-tubulin labeling; all microtubules are labeled; d: Acetylated alpha tubulin; only axonemal microtubules are labeled; e: Nuclear labeling; the nuclei are still in the common cytoplasmic mass. f: Extremity of elongating spermatid, showing than one axoneme is shorter than the other. The comma-shaped nucleus has migrated at the distal extremity of the spermatid. g, h: Late spermatids, showing migration of nuclei at the distal extremities of elongating spermatids; g: Tubulin labeling; h: Nuclear labeling. i, j, k: Spermatozoon, differential labeling with (i) anti-alpha-tubulin (note cortical microtubules labeled), (j) anti-acetylated alpha tubulin (note that only the axonemes are labeled, as a single line), and (k) nuclear labeling (note distal position of filiform nucleus). (Original, by C. Iomini and J.-L. Justine; details of labeling and observation methods in Iomini et al. 1998.)

appear in a precise order during spermiogenesis: acetylation, then glutamylation, then glycylation (Iomini et al. 1998). Glycylation occurs only at the extreme end of spermiogenesis and in a proximodistal process. This phenomenon could be of general occurrence, because it has already been reported in *Drosophila*

(Bré et al. 1996). Iomini et al. (1998) concluded that glycylation could be involved in a morphogenetic clock for the end of spermiogenesis.

Homology of Microtubules in Acoels and Parasitic Platyhelminthes Spermatozoa

Spermatozoa of certain acoels and spermatozoa of the parasitic Platyhelminthes, although widely separated in phyletic schemes of the Platyhelminthes, show a general morphological resemblance in that they are long cells with two incorporated axonemes and cortical microtubules. Labeling with anti-tubulin antibodies was used to test the possible homology of these cortical microtubules. As mentioned above, cortical microtubules of digenean spermatozoa are not acetylated (Iomini et al. 1995, 1998). Cortical microtubules of the acoel *Actinoposthia beklemischevi* were found to be acetylated (Raikova et al. 1998), therefore suggesting that the sperm cortical microtubules of the acoels are not homologous with similar elements found in spermatozoa of the higher Platyhelminthes (Justine et al. 1998). Recent molecular studies of the Platyhelminthes and related groups confirm this interpretation, because the acoels were found to be independent from the main groups of Platyhelminthes in several analyses (Carranza et al. 1997; Katayama et al. 1996). The question of the relationships of the Platyhelminthes with some groups traditionally included in them, such as the Acoela, Catenulida (Carranza et al. 1997) or Xenoturbellida (Ehlers, Sopott-Ehlers 1997; Norén, Jondelius 1997), is still much debated in the current literature, and sperm structure is providing characters useful for the resolution of Platyhelminthes relationships.

Certain acoel species have no cortical microtubules, but, instead have groups of singlet microtubules in the center of the cytoplasm (Raikova, Justine 1994). In *Convoluta saliens*, central microtubules are present but are not labeled by the anti-acetylated tubulin antibody (Raikova, Justine 1999), therefore suggesting non-homology with the cortical microtubules of other acoel species. In addition, they are not labeled by the widely used DM1A anti-alpha tubulin antibody. This must not be interpreted as the absence of alpha-tubulin, since alpha-tubulin is known to be present in all microtubules (Ludueña 1998); however, this demonstrates a significant difference of the epitope recognized by this antibody between these microtubules and other microtubules of the cell. These findings are summarized in Table 2. The functional significance of this finding is still to be found, but its phylogenetic significance is that studies with various anti-tubulin antibodies can be used to reveal homologies between microtubular elements in diverse species (Raikova, Justine 1999).

Conclusion

Although molecular systematics takes a increasingly prominent place among the methods used for understanding the phylogeny of the Platyhelminthes, characters obtained from the observation of spermatozoa are still important for providing important data sets useful for obtaining a robust phylogeny. The addition to ultrastructure of complementary methods of observation, such as immunocytochemistry of tubulin (or other proteins) and nuclear labeling is important for a better understanding of sperm structure in Platyhelminthes. Comparable results have been obtained on spermatozoa of Crustacea (Tudge et al. 1994; Tudge, Justine 1994) and Nematoda (Mansir, Justine 1996, 1998, 1999; Mansir et al. 1997; Noury-Sraïri et al. 1993). It is hoped that these data will contribute to a robust phylogeny of the Platyhelminthes.

Table 1: Labeling of microtubules with antibodies directed against various tubulin post-modifications. * ** ***

Species, group	Microtubular structure	Acetylation 6-11B-1	Glutamylation GT 335	Glycylation (high levels) AXO 49	Glycylation (low levels) TAP 952
Echinostoma caproni (digenean)	Cortical microtubules (spermatozoa)	0	0/+	0	0
	Axonemal microtubules (spermatozoa)	+	+	0	+
	Axonemal microtubules (protonephridia)	+	+	+	+
Pseudodactylogyrus sp. (monopisthocotylean)	Axonemal microtubules (spermatozoa)	+	+	-	-

* 0, no labeling; +, labeling; -, experiment not done.
** From Iomini et al. 1998; Mollaret, Justine 1997.
*** Acetylation is not found in cortical microtubules of spermatozoa, but is present in axonemal microtubules both in spermatozoa (9+"1" pattern) and protonephridia (9+2 pattern). Glycylation is not found in cortical microtubules of spermatozoa. Only low levels of glycylation occur in spermatozoal axonemes, but low and high levels occur in axonemes of protonephridia.

Acknowledgements

This paper was partly prepared from results recently obtained in the author's team by Carlo Iomini, Isabelle Mollaret, Nezha Noury-Sraïri, Aïcha Mansir and Olga Raikova. Jordi Miquel provided manuscripts of papers in press. Prof. Barrie G.M. Jamieson kindly edited the English. This work was partly funded by an INTAS-EEC grant No. 93-2176, "Ultrastructure and immunocytochemistry of the cytoskeleton of spermatozoa, eggs and fertilization in selected invertebrate species, for the understanding of phylogeny."

Table 2: Immunocytochemical labeling of axonemal, cortical and central microtubules in spermatozoa of various species of Platyhelminthes.* ** ***

	labeling of sperm microtubules								
	axonemal			cortical			central		
Species (group)	alpha	beta	acet.!	alpha	beta	acet.!	alpha	beta	acet !
Echinostoma caproni (digenean)	+	+	+	+	+	0	-	-	-
Pseudodactylogyrus sp. (monopisthocotylean)	+	+	+	-	-	-	-	-	-
Troglocaridicola sp. (temnocephalid)	+	+	+	+	+	0	-	-	-
Actinoposthia beklemischevi (acoel)	+	+	+	+	+	+	-	-	-
Convoluta saliens (acoel)	+	+	+	-	-	-	0	+	0

* 0, no labeling; +, labeling; -, structure not present in this species.
**Tubulin acetylation is found in all axonemes. In cortical microtubules, it is absent in the digenean and the temnocephalid, but is present in the acoel *A. beklemischevi*. In the acoel *C. saliens*, central microtubules are not acetylated and are not labeled by an anti-alpha tubulin antibody which labels all microtubular systems in spermatozoa of other species.
***From Iomini et al. 1998; Justine et al. 1998; Mollaret, Justine 1997; Raikova, Justine 1998.
!, acetylated

References

Bâ CT, Marchand B. Ultrastructure of spermiogenesis and the spermatozoon of *Vampirolepis microstoma* (Cestoda, Hymelolepididae), intestinal parasite of *Rattus rattus*. Microsc Res Tech 1998; 42:218-225.

Bâ CT, Marchand M. Spermiogenesis, spermatozoa and phyletic affinities in the Cestoda. In: (Jamieson BGM, Ausio J, Justine J-L, eds) Advances in Spermatozoal Phylogeny and Taxonomy. Mém Mus Nat Hist Nat 1995; 166:87-95.

Barra HS, Arce CA, Rodriguez JA, Caputto R. Some common properties of the protein that incorporate tyrosine as a single unit and the microtubule proteins. Biochem Biophys Res Commun 1974; 60:1384-1390.

Boeger WA, Kritsky DC. Phylogeny and a revised classification of the Monogenoidea Bychowsky, 1937 (Platyhelminthes). Syst Parasitol 1993; 26:1-32.

Boeger WA, Kritsky DC. Coevolution of the Monogenoidea (Platyhelminthes) based on a revised hypothesis of parasite phylogeny. Int J Parasitol 1997; 27:1495-1511.

Bré M-H, Redeker V, Quibell M, Darmanaden-Delorme J, Bressac C, Cosson J, Huitorel P, Schmitter J-M, Rossier J, Johnson T, Adoutte A, Levilliers N. Axonemal tubulin polyglycylation probed with two monoclonal antibodies: widespread evolutionary distribution, appearance during spermatozoan maturation and possible function in motility. J Cell Sci 1996; 109:727-738.

Brooks DR. A summary of the database pertaining to the phylogeny of the major groups of the parasitic Platyhelminthes, with a revised classification. Can J Zool 1989; 67:714-720.

Brooks DR, Hoberg EP, Weekes PJ. Preliminary phylogenetic systematic analysis of the major lineages of the Eucestoda (Platyhelminthes: Cercomeria). Proc Biol Soc Wash 1991; 104:651-668.

Caron JM. Posttranslational modification of tubulin by palmitoylation: I. *In vivo* and cell-free studies. Mol Biol Cell 1997; 8:621-636.

Carranza S, Baguñà J, Riutort M. Are the Platyhelminthes a monophyletic primitive group? An assessment using 18s rDNA sequences. Mol Biol Evol 1997; 14:485-497.

Duvaux-Miret O, Baratte B, Dissous C, Capron A. Molecular cloning and sequencing of the alpha-tubulin gene from *Schistosoma mansoni*. Mol Biochem Parasitol 1991; 49:337-340.

Eddé B, Rossier J, Le Caer J-P, Desbruyères E, Gros F, Denoulet P. Posttranslational glutamylation of alpha-tubulin. Science 1990; 247:83-85.

Ehlers U. Phylogenetisches System der Plathelminthes. Verh naturwiss Ver Hamburg (NF) 1984; 27:291-294.

Ehlers U. Das Phylogenetische System der Plathelminthes. G Fischer, Stuttgart, 1985.

Ehlers U, Sopott-Ehlers B. Ultrastructure of the subepidermal musculature of *Xenoturbella bocki*, the adelphotaxon of the Bilateria. Zoomorphology 1997; 117:71-79.

Eipper BA. Properties of rat brain tubulin. J Biol Chem 1974; 249:1407-1416.

Gracenea M, Ferrer JR, González-Moreno O, Trullols M. Ultrastructural study of spermatogenesis and spermatozoon in *Postorchigenes gymnesicus* (Trematoda, Lecithodendriidae). J Morphol 1997; 234:223-232.

Hendelberg J. Paired flagella and nucleus migration in the spermiogenesis of *Dicrocoelium* and *Fasciola* (Digenea, Trematoda). Zool Bidr Uppsala 1962; 35:569-587.

Hendelberg J. On the development of different types of spermatozoa from spermatids with two flagella in the Turbellaria with remarks on the ultrastructure of the flagella. Zool Bidr Uppsala 1969; 38:1-50.

Henley C, Costello DP, Thomas MB, Newton WD. The "9+1" pattern of microtubules in spermatozoa of *Mesostoma* (Plathelminthes, Turbellaria). Proc Natl Acad Sci USA 1969; 64:849-856.

Hoberg EP, Gardner SL, Campbell RA. Paradigm shifts and tapeworm systematics. Parasitol Today 1997a; 13:161-162.

Hoberg EP, Mariaux J, Justine J-L, Brooks DR, Weekes PJ. Phylogeny of the orders of the Eucestoda (Cercomeromorphae) based on comparative morphology: historical perspectives and a new working hypothesis. J Parasitol 1997b; 83:1128-1147.

Iomini C, Justine J-L. Spermiogenesis and spermatozoon of *Echinostoma caproni* (Platyhelminthes, Digenea): transmission and scanning electron microscopy, and tubulin immunocytochemistry. Tissue Cell 1997; 29:107-118.

Iomini C, Ferraguti M, Melone G, Justine JL. Spermiogenesis in a scutariellid (Platyhelminthes). Acta Zool (Stockh) 1994; 75:287-295.

Iomini C, Raikova O, Noury-Sraïri N, Justine J-L. Immunocytochemistry of tubulin in spermatozoa of Platyhelminthes. In: (Jamieson BGM, Ausio J, Justine J-L, eds) Advances in Spermatozoal Phylogeny and Taxonomy. Mém Mus Nat Hist Nat 1995; 166:97-104.

Iomini C, Mollaret I, Albaret J-L, Justine J-L. Spermatozoon and spermiogenesis in *Mesocoelium monas* (Platyhelminthes, Digenea): ultrastructure and epifluorescence microscopy of labeling of tubulin and nucleus. Fol Parasitol 1997; 44:26-32.

Iomini C, Bré M-H, Levilliers N, Justine J-L. Tubulin polyglycylation in Platyhelminthes: diversity among stable microtubule networks and very late occurrence during spermiogenesis. Cell Motil Cytoskel 1998; 39:318-330.

Justine J-L. A new look at Monogenea and Digenea spermatozoon. In: (André J, ed) The Sperm Cell. The Hague: Martinus Nijhoff Publishers, 1982, pp454-457.

Justine J-L. Phylogeny of parasitic Platyhelminthes: a critical study of synapomorphies proposed on the basis of the ultrastructure of spermiogenesis and spermatozoa. Can J Zool 1991a; 69:1421-1440.

Justine J-L. Cladistic study in the Monogenea (Platyhelminthes), based upon a parsimony analysis of spermiogenetic and spermatozoal ultrastructural characters. Int J Parasitol 1991b; 21:821-838.

Justine J-L. Spermatozoa as a tool for taxonomy of species and supraspecific taxa in the Platyhelminthes. In: (Baccetti B, ed) Comparative Spermatology 20 Years After. New York: Raven Press, 1991c, pp981-984.

Justine J-L. The spermatozoa of the schistosomes and the concept of progenetic spermiogenesis. In: (Baccetti B, ed) Comparative Spermatology 20 Years After. New York: Raven Press, 1991d, pp977-979.

Justine J-L. Phylogénie des Monogènes basée sur une analyse de parcimonie des caractères de l'ultrastructure de la spermiogenèse et des spermatozoïdes incluant les résultats récents. Bull Fr Pêche Piscic 1993; 328:137-155.

Justine J-L. Spermatozoal ultrastructure and phylogeny of the parasitic Platyhelminthes. In: (Jamieson BGM, Ausio J, Justine J-L, eds) Advances in Spermatozoal Phylogeny and Taxonomy. Mém Mus Nat Hist Nat 1995; 166:55-86.

Justine J-L. Non-monophyly of the monogeneans? Int J Parasitol 1998a; 28:1653-1657.

Justine J-L. Spermatozoa as phylogenetic characters for the Eucestoda. J Parasitol 1998b; 84:385-408.

Justine J-L, Mattei X. Étude ultrastructurale de la spermiogenèse et du spermatozoïde d'un Plathelminthe: *Gonapodasmius* (Trematoda: Didymozoidae). J Ultrastruct Res 1982; 79:350-365.

Justine J-L, Mattei X. Ultrastructural observations on the spermatozoon, ovocyte and fertilization process in *Gonapodasmius*, a gonochoristic Trematode (Trematoda: Digenea: Didymozoidae). Acta Zool (Stockh) 1984; 65:171-177.

Justine J-L, Lambert A, Mattei X. Spermatozoon ultrastructure and phylogenetic relationships in the Monogenea (Platyhelminthes). Int J Parasitol 1985a; 15:601-608.

Justine J-L, Le Brun N, Mattei X. The aflagellate spermatozoon of *Diplozoon* (Platyhelminthe: Monogenea: Polyopisthocotylea). A demonstrative case of relationship between sperm ultrastructure and biology of reproduction. J Ultrastruct Res 1985b; 92:47-54.

Justine J-L, Iomini C, Raikova OI, Mollaret I. The homology of cortical microtubules in platyhelminth spermatozoa: a comparative ultrastructural study of acetylated tubulin. Acta Zool (Stockh) 1998; 79:235-241.

Katayama T, Nishioka M, Yamamoto M. Phylogenetic relationships among turbellarian orders inferred from 18S rDNA sequences. Zool Sci 1996; 13:747-756.

Kritsky DC, Lim SLH. Phylogenetic position of Sundanonchidae (Platyhelminthes: Monogeneoidea: Dactylogyridea), with report of two species of *Sundanonchus* from toman, *Channa micropeltes* (Channiformes: Channidae), in Malaysia. Inv Biol 1995; 114:285-295.

L'Hernault SW, Rosenbaum JL. *Chlamydomonas* alpha-tubulin is post-translationally modified by acetylation on the Sigma-amino group of a lysine. Biochemistry 1985; 24:473-478.

LeDizet M, Piperno G. Identification of an acetylation site of *Chlamydomonas* alpha-tubulin. Proc Natl Acad Sci USA 1987; 84:5720-5724.

Li M-M, Wang X-Y. Spermatogenesis and ultrastructure of the metaphase chromosomes in *Ceylonocotyle scoliocoelium* (Digenea: Paramphistomidae). Acta Zool Sin 1997; 43:1-9.

Ludueña RF. Multiple forms of tubulin: different gene products and covalent modifications. Int Rev Cytol 1998; 178:207 275.

Mansir A, Justine J-L. Actin and major sperm protein in spermatids and spermatozoa of the parasitic nematode *Heligmosomoides polygyrus*. Mol Reprod Dev 1996; 45:332-341.

Mansir A, Justine J-L. The microtubular system and post-translationally modified tubulin during spermatogenesis in a parasitic nematode with amoeboid and aflagellate spermatozoa. Mol Reprod Dev 1998; 49:150-167.

Mansir A, Justine J-L. Actin and major sperm protein in spermatozoa of a nematode, *Graphidium strigosum*. Fol Parasitol 1999; 46:47-51.

Mansir A, Noury-Sraïri N, Cabaret J, Kerboeuf D, Escalier D, Durette-Desset M-C, Justine J-L. Actin in spermatids and spermatozoa of *Teladorsagia circumcincta* and *Trichostrongylus colubriformis* (Nematoda, Trichostrongylida). Parasite 1997; 4:373-376.

Mariaux J. A molecular phylogeny of the Eucestoda. J Parasitol 1998; 84:114-124.

Miquel J, Marchand B. Ultrastructure of the spermatozoon of the bank vole tapeworm, *Paranoplocephala omphalodes* (Cestoda, Cyclophyllidea, Anoplocephalidae). Parasitol Res 1998a; 84:239-245.

Miquel J, Marchand B. Ultrastructure of spermiogenesis and the spermatozoon of *Anoplocephaloides dentata* (Cestoda, Cyclophyllidea, Anoplocephalidae), intestinal parasite of Arvicolidae rodents. J Parasitol 1998b; 84:1128-1136.

Miquel J, Bâ CT, Marchand B. Ultrastructure of spermiogenesis of *Dipylidium caninum* (Cestoda, Cyclophyllidea, Dipylidiidae), an intestinal parasite of *Canis familiaris*. Int J Parasitol 1998; 28:1453-1458.

Mollaret I, Justine J-L. An immunocytochemical study of tubulin in the 9+"1" sperm axoneme of a monogenean (Platyhelminthes), *Pseudodactylogyrus* sp. Tissue Cell 1997; 29:699-706.

Mollaret I, Jamieson BGM, Adlard R, Hugall A, Lecointre G, Chombard C, Justine J-L. Phylogenetic analysis of the Monogenea and their relationships with Digenea and Eucestoda inferred from 28S rDNA sequences. Mol Biochem Parasitol 1997; 90:433-438.

Mollaret I, Lim LHS, Malmberg G, Afzelius B, Justine J-L. Spermatozoon ultrastructure in two monopisthocotylean monogeneans from Malaysia: *Pseudodactylogyroides marmoratae* and *Sundanonchus micropeltis*. Folia Parasitol 1998; 45:75-76.

Norén M, Jondelius U. *Xenoturbella*'s molluscan relatives. Nature 1997; 390:31-32.

Noury-Sraïri N, Gourbault N, Justine J-L. The development and evolution of actin-containing organelles during spermiogenesis of a primitive nematode. Biol Cell 1993; 79:231-241.

Ozols J, Caron JM. Post-translational modification of tubulin by palmitoylation: II. Identification of sites of palmitoylation. Mol Biol Cell 1997; 8:637-645.

Paturle-Lafanechère L, Eddé B, Denoulet P, Van Dorsselaer A, Mazarguil H, Le Caer JP, Wehland J, Job D. Characterization of a major brain tubulin variant wich cannot be tyrosinated. Biochemistry 1991; 30:10523-10528.

Podvyaznaya IM. [The fine structure of the male reproductive system and genital atrium of bat parasite *Allassogonoporus amphoraeformis* (Trematoda: Allassogonoporidae)]. Parazitologiya 1996; 30:229-235. [In Russian.]

Raikova OI, Justine J-L. Ultrastructure of spermiogenesis and spermatozoa in three Acoels (Platyhelminthes). Ann Sci Nat (Zool) 1994; 15:63-75.

Raikova OI, Justine J-L. Microtubular system during spermiogenesis and in the spermatozoon of *Convoluta saliens* (Platyhelminthes, Acoela): tubulin immunocytochemistry and electron microscopy. Mol Reprod Dev 1999; 52:74-85.

Raikova OI, Falleni A, Justine J-L. Spermiogenesis in *Paratomella rubra* (Platyhelminthes, Acoela): ultrastructural, immunocytochemical, cytochemical studies and phylogenetic implications. Acta Zool (Stockh) 1997; 78:295-307.

Raikova OI, Flyatchinskaya LP, Justine J-L. Acoel spermatozoa: ultrastructure and immunocytochemistry of tubulin. Hydrobiologia 1998; 383:207-214.

Redeker V, Levilliers N, Schmitter J-M, Le Caer J-P, Rossier J, Adoutte A, Bré M-H. Polyglycylation of tubulin: a post-translational modification in axonemal microtubules. Science 1994; 266:1688-1691.

Santos CP, Lanfredi RM, Souto-Padrón T. Ultrastructure of spermatogenesis of *Atriaster heterodus* (Platyhelminthes, Monogenea, Polyopisthocotylea). J Parasitol 1997; 83:1007-1014.

Silveira M. The fine structure of 9+1 flagella in turbellarian Flatworms. In: (Afzelius BA, ed) The Functional Anatomy of the Spermatozoon. Oxford: Pergamon Press, 1975, pp289-298.

Stoitsova SR, Georgiev BB, Dacheva RB. Ultrastructure of spermiogenesis and the mature spermatozoon of *Tetrabothrius erostris* Loennberg, 1896 (Cestoda, Tetrabothriidae). Int J Parasitol 1995; 25:1427-1436.

Tang J-Y. Ultrastructural studies on sperm of *Dicrocoelium chinensis* (Trematoda: Digenea). Acta Zool Sin 1996; 42:337-342.

Tang J-Y, Li M-M. Ultrastructural studies on spermatogenesis of *Dicrocoelium chinensis* (Trematoda: Digenea). Acta Zool Sin 1996; 42:225-230.

Thompson WC. The cyclic tyrosination/detyrosination of alpha tubulin. Methods Cell Biol 1982; 24:235-255.

Tudge CC, Justine J-L. The cytoskeletal proteins actin and tubulin in the spermatozoa of four decapod crabs (Crustacea, Decapoda). Acta Zool (Stockh) 1994; 75:277-285.

Tudge CC, Grellier P, Justine J-L. Actin in the acrosome of the spermatozoa of the crab, *Cancer pagurus* L. (Decapoda, Crustacea). Mol Reprod Dev 1994; 38:178-186.

Watson N, Rohde K. Sperm and spermiogenesis in the "Turbellaria" and implications for the phylogeny of the Phylum Platyhelminthes. In: (Jamieson BGM, Ausio J, Justine J-L, eds) Advances in Spermatozoal Phylogeny and Taxonomy. Mém Mus Nat Hist Nat 1995; 166:37-54.

Watson NA. Spermiogenesis and sperm ultrastructure in *Troglocephalus rhinobatidis*, *Neoheterocotyle rhinobatidis* and *Merizocotyle australensis* (Platyhelminthes, Monogenea, Monopisthocotylea, Monocotylidae). Int J Parasitol 1997; 27:389-401.

32 Structured Management as a Basis for Cost-Effective Infertility Care

David Mortimer
Sydney IVF
University of Sydney

Although infertility must always be considered as a problem of a couple, management options are governed by prevailing etiologic factor(s), which may cause treatment to be focused primarily upon one or the other partner. When a comprehensive workup of infertile couples is undertaken, multifactorial infertility is more prevalent than previously recognized, primarily due to the identification of previously unrecognized male factors. In many couples, a problem is identified either in the male partner alone or in both partners, with the result that the prevalence of significant male factor infertility is now being identified in many studies as the most common single diagnostic category, representing perhaps 30%, even 40%, of couples (Irvine 1998; Comhaire et al. 1987; Hargreave 1994; Kamischke, Nieschlag 1998). Nevertheless, there still remains a significant prevalence of idiopathic or "unexplained" infertility but, because of the large number of complex and highly specific steps in the process of gamete approximation (sperm and egg transport to the site of fertilization) and interaction (fertilization and syngamy; Fig. 1), much of this has been suggested as likely being due to an occult male factor caused by sperm dysfunction.

Basic Laboratory Workup of the Infertile Male

In-depth consideration of practical issues in the workup of the male partner is clearly beyond the scope of this chapter and detailed protocols and laboratory methods are available elsewhere, as are critical reviews of the interpretation and clinical significance of these tests (Mortimer 1994b; World Health Organization 1992; Mortimer 1994c).

If results of diagnostic andrology laboratory testing are to be used effectively, they must be accurate and reliable, which requires far greater commitment to providing quality services than has been typical for many such laboratories in the past (Matson 1995). Nevertheless, such goals are achievable if robust methods are used, staff are trained adequately, and the principles of quality

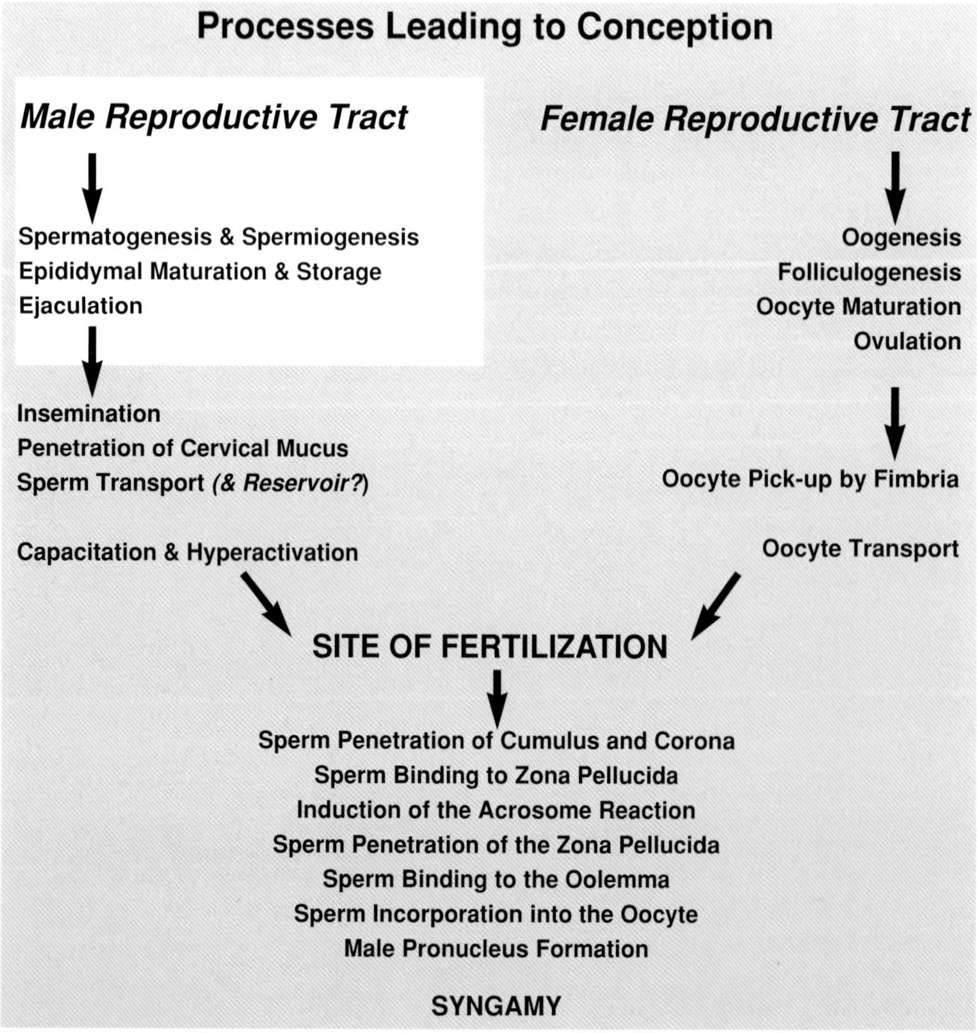

Figure 1. Diagrammatic representation of the various processes leading to conception in Eutheria.

control and quality assurance are inherent in all aspects of laboratory management (Mortimer 1994a; Björndahl, Kvist 1998; De Jonge 1998). Also, in the future, computer-aided sperm analysis (CASA) instrumentation will facilitate not only enhanced analyses of sperm kinematics but more robust determinations of traditional sperm characteristics, provided that these machines are used correctly (ESHRE Andrology Special Interest Group 1998; Mortimer, Mortimer 1998).

Laboratory Testing

When a couple has completed basic infertility investigations according to current World Health Organization (WHO) guidelines, the male partner will have had a careful semen analysis, including at least a screening test for the presence of anti-sperm antibodies (ASABs) on the sperm surface, and participated in an assessment of the interaction between his spermatozoa and his partner's cervical mucus, either as an *in vivo* post-coital test (PCT) or an *in vitro* sperm-mucus interaction test (SMIT). In addition, his partner will have been tested for serum ASABs (World Health Organization 1992; Rowe et al. 1994).

Male Factor Infertility

If an abnormal result has been found that is considered to decrease the man's ability to achieve a pregnancy *in vivo*, then a diagnosis of a male factor in the couple's infertility is made. However, having been able to make a diagnosis does not necessarily mean that a pathophysiological explanation has been identified; indeed, the most common diagnoses of male factor infertility are of one or more idiopathic abnormal semen characteristics.

Idiopathic Infertility

In order for a couple to be diagnosed as having idiopathic infertility, all the basic laboratory tests must have yielded normal results in the absence of endocrinological and physical abnormalities.

Genetic Testing

It is now recognized that problems causing male infertility, even sterility, such as congenital absence of the vasa deferentia (CBAVD) or severe oligozoospermia and secretory azoospermia, can have genetic origins (Mak, Jarvi 1996; Tuerlings et al. 1997; Vogt 1995; Vogt 1997; Schlegel, Shin 1997; Patrizio 1997). Consequently, modern infertility practice must include appropriate testing for defects of the cystic fibrosis gene that cause CBAVD or microdeletions of the Y chromosome that are associated with severely impaired spermatogenesis (see chapter by Barratt et al., this book). The major impact of these problems is not so much in terms of their directing management, but rather in defining the sophistication of assisted conception services that will be required.

Beyond cytogenetic and molecular genetic testing there is also the issue of generalized sperm DNA damage both at the nuclear and mitochondrial levels (Sakkas et al. 1996; Sakkas et al. 1997; Karabinus et al. 1997; also see chapter by Sakkas, this book). The etiology of this damage can be inherent—that is, mitochondrial DNA deletions in the spermatozoa (Cummins 1997; Cummins et al. 1994) causing the production of reactive oxygen species (ROS or free radicals), or due to extrinsic factors, such as smoking, that either cause the widespread generation of ROS or cause damage directly to the sperm nuclear chromatin (Fraga et al. 1996). Apart from intrinsic faults of the mitochondrial genome, many of the factors contributing to this etiology are amenable to remedy through lifestyle alteration and/or systemic therapy or sperm selection.

Sperm Preparation and Selection Issues

The risk of ROS-induced damage to spermatozoa being caused during simple washing procedures and its potential impact upon fertilization has been reviewed previously (Mortimer 1991). However, a further dimension to this problem of iatrogenic impaired fertilization or fertilization failure is now becoming apparent, that of the quality of the sperm chromatin, which impacts not upon fertilization but upon the developmental competence of embryos created by spermatozoa with damaged nuclear DNA (Sakkas et al. 1997).

Since the human embryo does not start expressing its own genome until some time on day three—that is, around the eight-cell stage—assisted reproductive technology (ART) programs transferring embryos on day two, at the two- or four-cell stage, will have no knowledge of their capacity for further development; embryos that have a compromised paternal genome will either fail before implantation or succumb to early embryonic loss post-implantation. This is a primary reason for the move towards transferring embryos at the blastocyst stage, by which time they have gone through this critical stage of development and have established that they have at least some developmental competence (Gardner et al. 1998). Consequently, two issues must be considered. First, ROS-induced damage induced during sperm preparation using sub-optimal procedures will not be evident to ART programs in which embryos are transferred on day two (Janny, Menezo 1994). Second, the opportunity to select spermatozoa with less chromatin damage during semen processing for ART will be expected to provide better clinical outcomes (Pasteur et al. 1994).

For this reason, techniques that just separate spermatozoa from the seminal plasma, other cellular elements such as leukocytes; and dead/immotile spermatozoa, should be avoided in favor of ones that select "better" spermatozoa; for example, swim-up migration direct from liquefied semen or density gradient centrifugation using a product such as PureSperm (since Percoll was withdrawn from clinical use by its manufacturer, Pharmacia Biotech, effective January 1997).

Treatment for Male Factor Infertility

Treatment of a defined cause for male factor infertility may be surgical, medical or involve the application of ART. Although surgical treatment is outside the scope of this review, if indicated it represents a highly effective solution for the couple and therefore must be integral to any modern infertility practice. Medical treatment for male factor infertility is generally considered to be unsuccessful. However, treatment for infections of the male genital tract can lead to restoration of fertility. Furthermore, with increasing recognition of the highly deleterious effects of free radicals upon spermatozoa, there is a growing awareness that lifestyle and diet alterations (for example, decreasing occupational exposure to hazardous physical or chemical agents, taking antioxidants, and so on) can be beneficial, although data from controlled studies are limited.

Assisted Conception Treatment

Nowadays, even at an early point in their management, infertile couples with a contributory male factor are increasingly being referred directly to ART programs for *in vitro* fertilization (IVF) treatment as the most expeditious means of securing a pregnancy. If IVF is unsuccessful, then intracytoplasmic sperm injection (ICSI), in which a spermatozoon is injected directly into the ooplasm, is readily available as the ultimate in assisted conception treatment.

Given the high success rates now being achieved using ICSI (Tarlatzis, Bili 1998), there is an increasingly common trend for ICSI to be used as the treatment of choice. An even more disturbing trend is that infertile couples may be referred directly for ICSI treatment after an initial infertility consultation, in the belief that this is the most cost-effective approach to achieving the desired endpoint of a pregnancy. However, ICSI treatment is the most labor-intensive, invasive and expensive form of assisted conception treatment and is unwarranted for many couples. There is also the issue that many genetic problems in the male will be inherited by sons produced through successful ICSI treatment (Tournaye et al. 1997; Persson et al. 1996). Hence, counseling must be provided to couples before embarking upon such treatment (Meschede, Horst 1997). For example, a man who is functionally azoospermic (even though a few spermatozoa are frequently found in semen or in the testes) because of a deletion in the AZFc region of his Y chromosome has an excellent chance of achieving fertilization by ICSI and then establishing a pregnancy with his partner—but if the child is male, then he will inherit a defective Y chromosome from his father. The child may hence have the same problem of functional sterility, perhaps even worse due to the process of genetic anticipation.

Structured Management

Clearly, a treatment such as ICSI should only be used as a "last resort" when less invasive, lower cost treatments have failed, although it may be the only approach likely to achieve success in couples with extreme male factor infertility, even sterility (for example, azoospermia). A more scientific approach to managing infertile couples with a contributory male factor would be to apply a management strategy that uses information from diagnostic testing in a progressive manner. With such an approach, after initial investigations, couples for whom simple insemination-based treatment would have a good chance of achieving a pregnancy could embark upon such treatment, while couples for whom such treatment would be contraindicated on the grounds of impaired gamete approximation should proceed directly to assisted conception procedures. Couples in whom a severe sperm dysfunction had been identified would proceed directly to ICSI (Mortimer 1994e). This strategy has been termed "structured management."

Structured management protocols for infertile couples would determine the appropriate level of medical intervention required to achieve a reasonable chance of a pregnancy according to available diagnostic information. "Appropriate" would be judged in terms of cost (both health care resource use and to the patient), likelihood of a successful outcome (especially in consideration of the female partner's age and associated risk factors, such as patient morbidity) and be dependent, to a greater or lesser degree, upon local circumstances reflected in the availability and accessibility of the various diagnostic and therapeutic services. Consequently, structured management protocols would allow a more economical use of health care funds by reducing the application of the most invasive techniques until they have been shown to be necessary.

One disadvantage of developing structured management protocols for infertile couples is their requirement for a specialized andrology laboratory service capable of providing accurate and precise diagnostic information. While this requirement is being recognized and remedied by leading laboratories, it is becoming apparent that the way in which such laboratory results are interpreted and applied must also be modernized (Mortimer 1994e; Oehninger et al. 1997; Oehninger et al. 1998). Attempting to use laboratory results to predict a positive outcome (such as fertilization *in vitro*) has poor sensitivity, being prone to a high incidence of false positive predicted outcomes. From the perspective of patient management, such diagnostic tests are better used in the opposing sense; that is, to predict the likelihood or risk of failure of a component in the physiological processes leading to fertilization, and hence of particular therapeutic approaches. This means that what is really needed is tests with good specificity—that is, a high incidence of correctly predicted negative outcomes. These are the tests upon which a structured management protocol should be based.

It is envisaged that the application of structured management protocols will allow physicians to determine the most appropriate (the least invasive and most cost-effective) approach to managing each individual couple's infertility problem in a stepwise manner. However, because the major steps involved in gamete approximation and fertilization occur in the inaccessible domain of the female reproductive tract, tests of sperm functional potential are intrinsic to the development of structured management protocols for infertile couples.

Proposed Structured Management Protocol

Initial Laboratory Testing

This should follow the recommendations of the WHO Manual for Standardized Investigation and Diagnosis of the Infertile Couple (Rowe et al. 1994), although the section on the male partner is currently under revision.

Usually a minimum of two semen analyses, performed to standards at least equivalent to the WHO Laboratory Manual for the Examination of Human Semen and Sperm-Cervical Mucus Interaction (currently in its third edition; World Health Organization 1992), and including tests for ASABs bound to the sperm surface, are required. They should be at least two to three weeks apart. Only if the first analysis shows completely normal results in all respects may the second analysis be omitted.

In addition to tests for ASABs bound to the sperm surface, ASABs must also be tested for in the female partner's serum using indirect methods. For screening purposes, tests such as the Immunobead or SpermMAR test are preferred over functional tests such as the Friberg tray agglutination test or the Isojima sperm-immobilization test (Mortimer 1994b).

Sperm-mucus interaction testing (SMIT) should comprise at least a PCT which, if abnormal, would indicate proceeding to *in vitro* tests. However, *in vitro* SMITs permit more reliable assessment of the ability of a man's spermatozoa to interact with his partner's cervical mucus (Eggert-Kruse et al. 1989a; Eggert-Kruse et al. 1989b) and have been found by several groups to represent better first-line testing. A control for SMIT assessment is the hyaluronate migration test, eliminating the need for donor cervical mucus and semen as required for crossed-hostility testing (Mortimer et al. 1990; Neuwinger et al. 1991; Aitken et al. 1992).

Before a couple proceeds to assisted conception treatment, it is advised that a trial sperm preparation procedure ("trial wash") be undertaken to establish that sufficient progressively motile spermatozoa can be obtained from the man's ejaculate (Mortimer 1994d; Mortimer, Mortimer 1999). This assessment incorporates a comprehensive semen analysis and, ideally, a further screening test for ASABs on the surface of the prepared spermatozoa.

Treatment Options

In the absence of any abnormal findings during the basic workup tests, a couple is considered to have "idiopathic" or "unexplained" infertility. Many such couples then proceed to assisted conception treatment, usually commencing with up to three or four cycles of intrauterine insemination (IUI), typically incorporating mild ovarian stimulation; if unsuccessful they would likely be referred on for IVF. Inherent in any of these therapeutic modalities is the need to prepare a population of selected spermatozoa that have been protected from iatrogenic damage to either their fertilizing ability or chromatin structure.

Severely abnormal semen characteristics will obviously influence treatment options (Mortimer, Mortimer 1999). Low sperm numbers and/or poor sperm motility will preclude IUI, while poor sperm morphology (in terms of low normal forms and elevated teratozoospermia index) may even indicate direct recourse to ICSI. Each unit will need to establish its own cut-offs for decision points using these variables, but an overview of the whole process is shown in Figure 2.

Although defective sperm-mucus interaction is typically treated using IUI, this is not always the best option. When the problem is due to sperm dysfunction, it may be associated with poor fertilizing ability, even at IVF, which would contraindicate IUI (Barratt et al. 1989; Berberoglugil et al. 1993). However, if more careful SMIT assessments are performed before beginning IUI, then patients with dysfunctional spermatozoa, which would have compromised IUI success, could have been diverted towards more invasive treatment such as ICSI, thereby saving wasted time and money. This would also increase the effective success rate of IUI treatment.

Most ART programs offering IUI with controlled ovulation induction report fecundity rates of about 20%,

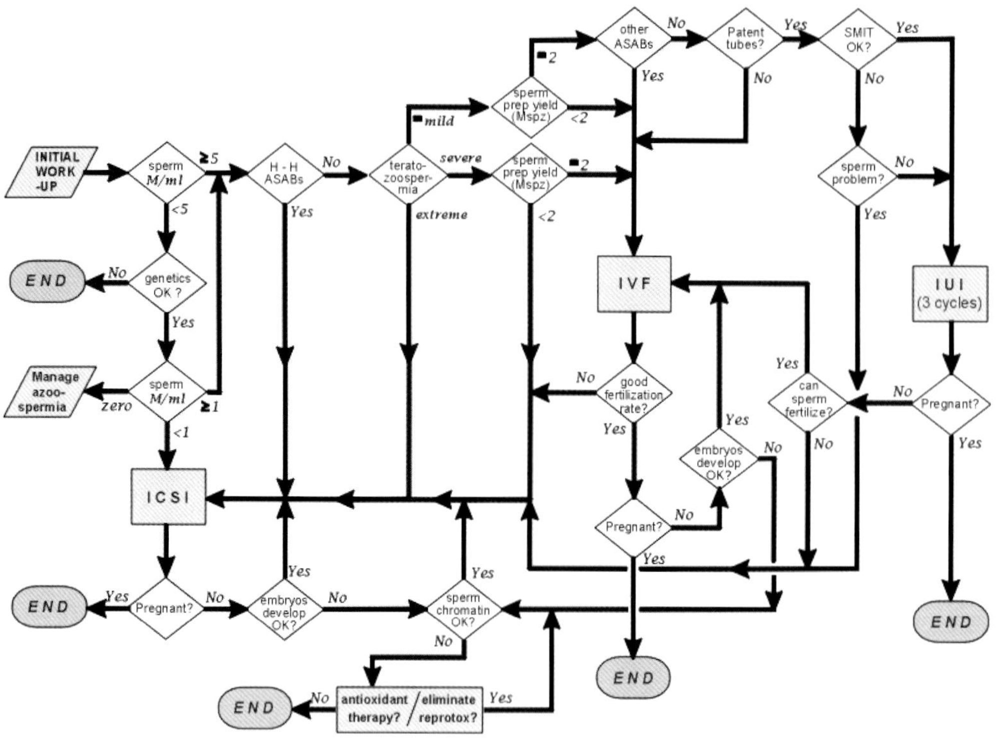

Figure 2. Flow chart illustrating a possible approach to the structured management of couples with male factor infertility.

achieving a cumulative pregnancy rate of perhaps 50% over three treatment cycles. However, pregnancy rates seem to decrease drastically after three or four cycles, so that most units re-assess patients at this time and usually proceed to more elaborate treatment. Consequently, for couples in which the female partner is under 38 years of age and IUI is not contraindicated, for reasons such as blocked fallopian tubes, sperm dysfunction or ASABs, three cycles of IUI should be the recommended first line of treatment. Couples who fail to conceive using IUI should be re-evaluated to determine whether there might be some occult sperm dysfunction. If so, they should be directed towards IVF or ICSI. If all these couples were to go to IVF, some would have an increased risk of fertilization failure, but to direct everyone straight to ICSI would not only be labor intensive for embryology staff, but also unnecessarily expensive for patients who did not actually need ICSI. Assuming it could be provided in a cost-effective manner, some sperm function testing would be appropriate at this juncture.

Couples in which the female partner is over 38 years of age, or where a mild male factor has been identified, would proceed directly to IVF.

In cases where a significant male factor was identified during the initial workup, either a first IVF attempt may be undertaken with recognition of its possible "diagnostic" role, or sperm function testing could be performed to establish whether ICSI might be required in order to achieve a reasonable chance of successful fertilization.

Obviously, in cases where insufficient spermatozoa can be obtained for an IVF attempt, then ICSI becomes the only possible treatment option. This would include all cases where spermatozoa must be obtained from the epididymis or testis.

Sperm Function Testing

Numerous laboratory tests have been proposed for the *in vitro* investigation of human sperm fertilizing ability. The most common tests provide assessment of the expression of hyperactivated motility, the acrosome reaction (AR), sperm-zona pellucida binding ability and the ability of spermatozoa to fuse with zona pellucida-free hamster oocytes and commence nuclear decondensation (see chapter by De Jonge, this book). Detailed reviews of the methodology and interpretation of these tests have been published elsewhere; examples are Mortimer (1994b) and ESHRE Andrology Special Interest Group (1996). Only brief overviews of the most pertinent tests are mentioned here.

Protocols to study acrosome reaction dynamics are well established and incorporate assessment of both spontaneous ARs under capacitating conditions and in response to ionophore challenge (the ARIC test). Acrosome status can be determined by a range of methods using fluorescent lectins, monoclonal antibodies, or even elaborate cytochemical staining. Well-defined cutoffs for the identification of both AR prematurity and AR insufficiency, which can be applied clinically, have been described (ESHRE Andrology Special Interest

Group 1996).

Assessment of sperm hyperactivation requires the use of a CASA machine. While these instruments are not yet universally available, they are becoming more widespread and such a practical application of the technology further justifies their acquisition. An optimized protocol for assessing hyperactivation has been developed and has been shown to provide useful information, enabling prediction of poor IVF fertilization rates (Mortimer et al. 1997).

Although the hemizona assay is an extremely well-established and validated test (ESHRE Andrology Special Interest Group 1996; Oehninger et al. 1998), widespread testing of sperm-zona pellucida binding is not practical until tests using artificial zonae based on rhuZP3 become available. However, it should be noted that some information on this aspect of fertilization is available in "failed fertilization" IVF cycles when the unfertilized oocytes are examined for sperm bound to or penetrating the zona pellucida.

Consequently, while the research value of many of these tests is not in doubt, the cardinal questions remain to be resolved: Who would derive real benefit from these assessments, and at what point in their management? Certainly, extensive, up-front testing of sperm fertilizing ability for infertility patients is neither practical nor cost effective. However, a minimal, yet very informative, package of sperm function tests might only require a combined analysis of AR prematurity/insufficiency and a CASA-based sperm hyperactivation assessment, and the cost for such a procedure, if incorporated into a pre-ART "enhanced trial wash," would be substantially less than the incremental cost for ICSI (ESHRE Andrology Special Interest Group 1996). Infertile couples who would benefit from such testing include those who have failed to achieve pregnancy after three cycles of IUI treatment.

Conclusions

This chapter has introduced and considered the concept of using a structured management approach for couples whose infertility has a contributory male factor. A highly detailed, in-depth treatment of the subject has not been possible in the present forum, and certainly these ideas will develop in the coming years, especially as further insights are gained into the etiology of male infertility. While this approach recognizes the importance of ART in helping these couples, it does not ignore the very real contribution of clinical andrology to their management—indeed, it is founded upon it. Although there remain many skeptics who claim that the advent of ICSI has rendered clinical andrology "a waste of time," more perspicacious individuals have recognized that, in fact, what is really needed is more andrology, not less, in order to provide responsible, cost-effective treatment for infertile couples as humanity enters the next millennium (Cummins, Jequier 1994; Jequier, Cummins 1997).

References

Aitken RJ, Bowie H, Buckingham D, Harkiss D, Richardson DW, West KM. Sperm penetration into a hyaluronic acid polymer as a means of monitoring functional competence. J Androl 1992; 13:44-54.

Barratt CLR, Osborn JC, Harrison PE, Monks N, Dunphy BC, Lenton EA, Cooke ID. The hypo-osmotic swelling test and the sperm mucus penetration test in determining fertilization of the human oocyte. Hum Reprod 1989; 4:430-434.

Berberoglugil P, Englert Y, Van den Bergh M, Rodesch C, Bertrand E, Biramane J. Abnormal sperm-mucus penetration test predicts low *in vitro* fertilization ability of apparently normal semen. Fertil Steril 1993; 59:1228-1232.

Björndahl L, Kvist U. Basic semen analysis courses: experience in Scandinavia. In: (Ombelet W, Bosmans E, Vandeput H, Vereecken A, Renier M, Hoomans E, eds) Modern ART in the 2000s. Carnforth: Parthenon Publishing Group, 1998, pp91-101.

Comhaire FH, de Kretser D, Farley TMM. Towards more objectivity in diagnosis and management of male infertility. Results of a World Health Organization Multicentre Study. Int J Androl 1987; 10:1-53.

Cummins JM. Mitochondrial DNA: implications for the genetics of human male fertility. In: (Barratt C, De Jonge C, Mortimer D, Parinaud J, eds) Genetics of Human Male Fertility. Paris: EDK, 1997, pp287-307.

Cummins JM, Jequier AM. Treating male infertility needs more clinical andrology, not less. Hum Reprod 1994; 9:1214-1219.

Cummins JM, Jequier AM, Kan R. Molecular biology of human male infertility: links with aging, mitochondrial genetics, and oxidative stress. Mol Reprod Dev 1994; 37:345-362.

De Jonge C. Total quality management and the clinical andrology laboratory: essential partners. In: (Ombelet W, Bosmans E, Vandeput H, Vereecken A, Renier M, Hoomans E, eds) Modern ART in the 2000s. Carnforth: Parthenon Publishing Group, 1998, pp55-60.

Eggert-Kruse W, Gerhard I, Tilgen W, Runnebaum B. Clinical significance of crossed *in vitro* sperm-cervical mucus penetration test in infertility investigation. Fertil Steril 1989a; 52:1032-1040.

Eggert-Kruse W, Leinhos G, Gerhard I, Tilgen W, Runnebaum B. Prognostic value of *in vitro* sperm penetration into hormonally standardized human cervical mucus. Fertil Steril 1989b; 51:317-323.

ESHRE Andrology Special Interest Group. Consensus workshop on advanced diagnostic andrology techniques. Hum Reprod 1996; 11:1463-1479.

ESHRE Andrology Special Interest Group. Guidelines on the application of CASA technology in the analysis of spermatozoa. Hum Reprod 1998; 13:142-145.

Fraga CG, Motchnik PA, Wyrobek AJ, Rempel DM, Ames BN. Smoking and low antioxidant levels increase oxidative damage to sperm DNA. Mutat Res Fundam Mol Mech Mutagen 1996; 351:199-203.

Gardner DK, Vella P, Lane M, Wagley L, Schlenker T, Schoolcraft WB. Culture and transfer of human blastocysts increases implantation rates and reduces the need for multiple embryo transfers. Fertil Steril 1998; 69:84-88.

Hargreave TB. Human infertility. In: (Hargreave TB, ed) Male Infertility. 2nd ed. London: Springer-Verlag, 1994, pp1-16.

Irvine DS. Epidemiology and etiology of male infertility. Hum Reprod 1998; 13 (Suppl 1):33-44.

Janny L, Menezo YJR. Evidence for a strong paternal effect on human preimplantation embryo development and blastocyst formation. Mol Reprod Dev 1994; 38:36-42.

Jequier AM, Cummins JM. Attitudes to clinical andrology: a time for change. Hum Reprod 1997; 12:875-876.

Kamischke A, Nieschlag E. Conventional treatments of male infertility in the age of evidence-based andrology. Hum Reprod 1998; 13 (Suppl 1):62-75.

Karabinus DS, Vogler CJ, Saacke RG, Evenson DP. Chromatin structural changes in sperm after scrotal insulation of Holstein bulls. J Androl 1997; 18:549-555.

Mak V, Jarvi KA. The genetics of male infertility. J Urol 1996; 156:1245-1256.

Matson PL. External quality assessment for semen analysis and sperm antibody detection: results of a pilot scheme. Hum Reprod 1995; 10:620-625.

Meschede D, Horst J. Genetic counselling for infertile male patients. Int J Androl 1997; 20 (Suppl 3):20-30.

Mortimer D. Sperm preparation techniques and iatrogenic failures of *in vitro* fertilization. Hum Reprod 1991; 6:173-176.

Mortimer D. Laboratory standards in routine clinical andrology. Reprod Med Rev 1994a; 3:97-111.

Mortimer D. Practical Laboratory Andrology. New York: Oxford University Press, 1994b.

Mortimer D. Semen analysis and other standard laboratory tests. In: (Hargreave TB, ed) Male Infertility. 2nd ed. London: Springer-Verlag, 1994c, pp37-73.

Mortimer D. Sperm recovery techniques to maximize fertilizing capacity. Reprod Fertil Dev 1994d; 6:25-31.

Mortimer D. The essential partnership between diagnostic andrology and modern assisted reproductive technologies. Hum Reprod 1994e; 9:1209-1213.

Mortimer D, Mortimer ST. Value and reliability of CASA systems. In: (Ombelet W, Bosmans E, Vandeput H, Vereecken A, Renier M, Hoomans EH, eds) Modern ART in the 2000s. Carnforth: Parthenon Publishing, 1998, pp73-89.

Mortimer D, Mortimer ST. Laboratory investigation of the infertile male. In: (Brinsden P, ed) A Textbook of *in Vitro* Fertilization and Assisted Reproduction. 2nd ed. Carnforth: Parthenon Publishing Group, 1999, (in press).

Mortimer D, Mortimer ST, Shu MA, Swart R. A simplified approach to sperm-cervical mucus interaction using a hyaluronate migration test. Hum Reprod 1990; 5:835-841.

Mortimer D, Kossakowski J, Mortimer ST. Fussell S. Prediction of fertilizing ability by sperm kinematics. J Assist Reprod Genet 1997; 14 (Suppl):52S.

Neuwinger J, Cooper TG, Knuth UA, Nieschlag E. Hyaluronic acid as a medium for human sperm migration tests. Hum Reprod 1991; 6:396-400.

Oehninger S, Franken D, Kruger T. Approaching the next millennium: how should we manage andrology diagnosis in the intracytoplasmic sperm injection era? Fertil Steril 1997; 67:434-436.

Oehninger S, Sayed E, Kolm P. Role of sperm function tests in subfertility treatment strategies. In: (Ombelet W, Bosmans E, Vandeput H, Vereecken A, Renier M, Hoomans E, eds) Modern ART in the 2000s. Carnforth: Parthenon Publishing Group, 1998, pp25-36.

Pasteur X, Métézeau P, Maubon I, Sabido O, Kiefer H. Identification of two human sperm populations using flow and image cytometry. Mol Reprod Dev 1994; 38:303-309.

Patrizio P. Mapping of the Y chromosome and clinical consequences. In: (Barratt C, De Jonge C, Mortimer D, Parinaud J, eds) Genetics of Human Male Fertility. Paris: EDK, 1997, pp25-42.

Persson JW, Peters GB, Saunders DM. Genetic consequences of ICSI—is ICSI associated with risks of genetic disease? Implications for counselling, practice and research. Hum Reprod 1996; 11:921-924.

Rowe PJ, Comhaire FH, Hargreave TB, Mellows HJ. WHO Manual for the Standardized Investigation and Diagnosis of the Infertile Couple. 1st ed. Cambridge: Cambridge University Press, 1994.

Sakkas D, Urner F, Bianchi PG, Bizzaro D, Wagner I, Jaquenoud N, Manicardi G, Campana A. Sperm chromatin anomalies can influence decondensation after intracytoplasmic sperm injection. Hum Reprod 1996; 11:837-843.

Sakkas D, Bianchi PG, Manicardi G, Bizzaro D, Bianchi U. Chromatin packaging anomalies and DNA damage in human sperm: their possible implications in the treatment of male factor infertility. In: (Barratt C, De Jonge C, Mortimer D, Parinaud J, eds) Genetics of Human Male Fertility. Paris: EDK, 1997, pp205-221.

Schlegel P, Shin D. Urogenital anomalies and genetic defects in men with bilateral congenital absence of the vas deferens. In: (Barratt C, De Jonge C, Mortimer D, Parinaud J, eds) Genetics of Human Male Fertility. Paris: EDK, 1997, pp98-110.

Tarlatzis BC, Bili H. Survey on intracytoplasmic sperm injection: report from the ESHRE ICSI Task Force. Hum Reprod 1998; 13 (Suppl 1):165-177.

Tournaye H, Lissens W, Liebaers I, Van Assche E, Bonduelle M, Fastenaekels V, Van Steirteghem A, Devroey P. Heritability of sterility: clinical implications. In: (Barratt C, De Jonge C, Mortimer D, Parinaud J, eds) Genetics of Human Male Fertility. Paris: EDK, 1997, pp123-144.

Tuerlings JHAM, Kremer JAM, Meuleman EJH. The practical application of genetics in the male infertility clinic. J Androl 1997; 18:576-581.

Vogt PH. Genetic aspects of human infertility. Int J Androl 1995; 18 (Suppl 2):3-6.

Vogt PH. Molecular basis of male (in)fertility. Int J Androl 1997; 20:2-10.

World Health Organization. WHO Laboratory Manual for the Examination of Human Semen and Sperm-Cervical Mucus Interaction. 3rd ed. Cambridge: Cambridge University Press, 1992.

33

The Role of Sperm Function Testing in Infertility Management

Christopher J. De Jonge
University of Minnesota

In the previous chapter, Dr. David Mortimer discussed the basic principles for the work-up of the subfertile male and how the results from andrologic testing are used, in part, to determine the subsequent treatment for the infertile couple. However, the route in getting from the initial semen analysis to therapeutic intervention has not always been straightforward. Often, results from the semen analysis and/or semen-cervical mucus interaction, and even results from *in vitro* fertilization, have directed the clinician to order tests that he or she hopes will offer diagnostic information relative to aspects of sperm function and that will, as a result, help in determining what the next course of therapy will be. The decision-making process surrounding what sperm function test or tests to order often is based on clinical experience and best guess. One reason for this is that there is a paucity of controlled trials of sufficient statistical power with results that provide conclusive evidence to support one or more sperm function tests as providing good prognostic information about the future success of assisted or non-assisted fertilization and pregnancy. This chapter will evaluate several sperm function tests, the role they have in infertility management, and whether the evidence supports their current clinical application.

The third edition of the World Health Organization (WHO) Laboratory Manual for the Examination of Human Semen and Sperm-Cervical Mucus Interaction (1992) provides guidelines to assist in making determinations concerning some of the conditions that must be satisfied in order for fertilization potential to be optimized. The question of whether fertilization potential can be accurately predicted from seminal parameters is of considerable debate and has been the topic of a number of papers (Amann 1989; Amann, Hammerstedt 1993; Duncan et al. 1993). As a result, many studies have been done to determine if correlations exist between seminal parameters and fertility, and if any of these parameters are predictive of fertilization potential. The conclusion is that the semen analysis, in general, lacks the ability to provide definitive diagnostic information about fertilization potential.

Why do basic semen parameters fail to adequately reveal fertility potential? One overwhelming reason is that the parameters detailed in the WHO guidelines address only a few of the conditions that must be satisfied for successful *in vivo* fertilization to take place, and while those items are important, they do not represent the only required conditions. Consequently, one can elaborate on the aforementioned guidelines to include the following additional requirements for successful *in vivo* fertilization. The first condition that must be met is that the fluids contributed by the male accessory glands and the components contained within those fluids should facilitate viability, prevent premature sperm activation by stabilizing the sperm membranes, coat the sperm membranes with elements essential for protection from the hostile environment of the vagina, and have reversibility in the binding of coating proteins to allow for membrane modifications that are required for fertilization to occur. Second, a sufficient number of mature viable spermatozoa must be present in the ejaculate. Third, the morphology of the sperm should be such that the cervical mucus will allow passage into the uterus. In other words, it is unlikely that sperm from a semen sample with a high percentage of morphologically abnormal forms will have a good chance for penetration through the cervical mucus to reach the site of fertilization. Fourth, it is essential that a good percentage of these sperm have forward progressive motion to propel them from the seminal plasma, through the cervical mucus, into the uterine cavity and through the fallopian tube(s) for ultimate encounter with and passage through the cumulus oocyte complex. Fifth, membrane modifications (termed "capacitation") must occur that will allow for a controlled-rate expression of integral proteins that are required for optimized signal transduction between the gametes and within the spermatozoon itself. Sixth, at some time during sperm transport, and presumably close to the time of acrosome reaction and zona penetration, sperm motion should change to a hyperactivated state. Seventh, stimulation of appropriately primed sperm receptors by the cumulus-oocyte complex (and most likely the zona pellucida) must initiate a cascade of biochemical changes to culminate in sperm membrane fusion, vesiculation and expression of acrosomal contents (termed "acrosome reaction") that in conjunction with sperm motion will allow for sperm binding to and penetration through the zona matrix. Eighth, there must be redistribution and/or exposure of sperm receptor proteins that will participate in and facilitate sperm-egg fusion and subsequent incorporation of the spermatozoon into the oocyte. Ninth, complete decondensation of the sperm nucleus into a fully expressed and functional genome capable of competent interaction with the maternal genome must occur. Tenth, activation of the sperm centrosome with subsequent sperm aster assembly to facilitate male and female pronuclear migration must occur. And last, paternal and maternal genomes must co-mingle, resulting in the expression of a normal zygotic genotype and phenotype.

On this basis, it becomes apparent that the ability to predict the *in vivo* fertilization potential of a sperm population using semen parameters is even more questionable, given the multiparametric aspects related to the fertilization process. If one eliminates all factors other than the spermatozoon itself, the task of fertilization potential estimation is not made any easier. One can imagine that the spermatozoon is composed of multiple compartments: the head and its subunits, the midpiece, the tail, and the cell as a whole. Due to the unique anatomical structure of the spermatozoon, each compartment can be thought of as independent from the other, but each part contributes to the whole, and thus the complexity of fertility potential determination is made apparent.

Pitfalls of Sperm Function Tests

In recent years, a number of new sperm function assays have been developed in an attempt to be better able to make fertilization potential estimates, and as a result to facilitate the determination of the most appropriate therapeutic approach, such as intrauterine insemination. However, in reviewing the literature, a number of pitfalls and problems become apparent. First, for many reports in which a new sperm function assay is introduced, one conclusion typically made is that the new assay "correlates significantly" with existing assays. However, the usefulness of a new assay should not necessarily be based on how well it correlates with other assays, but how it is independent from the others—in other words, how it offers an independent measure of function. Second, it is the correlation of the assay with fertility, usually measured in terms of *in vitro* fertilizing ability, that is typically given; however, it is the predictive estimation of fertility, preferably *in vivo*, that is required. Thus, in general, the ability to predict likelihood of success or, more importantly, likelihood of failure from a particular treatment based on results from any given sperm function test is not reliable. A general conclusion that can be drawn is that many new sperm function assays typically do not extend science's ability to determine fertilization potential, but rather may serve only to confound the issue (Barbato et al. 1998; Zaneveld, Jeyendran 1992).

These issues were addressed during a Consensus Workshop on Advanced Diagnostic Andrology Techniques (Excerpts on Human Reproduction Number 3, July 1996) convened at the recent meeting of the European Society for Human Reproduction and Embryology. The workshop was organized to discuss the best known and most widely applied tests of sperm function in relationship to clinical relevance. Tests discussed were: computer-assisted sperm analysis (CASA), acrosome reaction testing, zona-free hamster egg penetration test (SPA) and the zona binding test. It was agreed that most of the tests suffered from a lack of standardized protocols and availability of materials, and were affected by issues of cost and quality control.

In addition to these important deficiencies comes another significant problem, which is that the cutoffs used for making the diagnosis of fertility or infertility are not distinct for many of the sperm function tests. Gray or overlapping areas exist for virtually every test. So, what are the alternatives? Perhaps by measuring function in each of the sperm's compartments and performing multivariate analysis to obtain a discrete number will provide a high level of sensitivity and specificity for predicting *in vivo* fertilization potential.

With this having been said, there are a couple of notable exceptions to the much-maligned spectrum of sperm function tests and also some intriguing "newcomers" to the arena. Each will be discussed separately.

Specific Tests of Sperm Function

Zona-Free Hamster Egg Penetration Test (SPA)

The SPA is perhaps the most widely recognized and used sperm function test. First introduced by Yanagimachi et al. (1976), the SPA measures many aspects of sperm function, specifically the ability of spermatozoa to undergo capacitation, the acrosome reaction, fusion with the oolemma and nuclear decondensation. Unlike many other sperm function tests, the SPA encompasses a majority of the sperm functions required for fertilization, and as such it has a long-standing reputation for being clinically significant. However, the assay does not measure aspects of sperm motion or ligand-receptor interactions between sperm and zona pellucida. It is perhaps the failure of the test to assay the latter constituents of fertilization that contributes to the disparity of data in the literature concerning the diagnostic significance of results, or, more specifically, the negative and positive predictive values obtained using the SPA. Only Aitken and colleagues (1987) have successfully demonstrated predictive value of the assay; others investigators have typically reported correlative results.

Keeping in mind the criticism regarding a lack of standardized or consensus protocol and quality control measures in sperm function assays, it is perhaps the SPA that has been most vulnerable to this criticism. While one laboratory might show statistical and clinical significance of the assay, that same benefit may not be realized by other laboratories due to the deficiencies mentioned earlier. So where does this assay fit in the context of infertility management? The answer to this rather simple question is difficult to come by. If one were to do meta-analysis of SPA reports, it is likely that any conclusions concerning predictive value would be equivocal, due in large part to variability in protocol and study design.

To conclude then, it is inherent that the SPA has the ability for discriminating a broad spectrum of functional defects. However, the SPA first requires protocol standardization; that is, consensus protocol, followed by controlled trials, such as multi-center regional/national trials, plus the application of appropriate statistical analysis, such as receiver-operating characteristic

(ROC plots), before meta-analysis can be conducted to determine true clinical merit (Yusif 1997). If the utopian scenario described can be accomplished even in part, then the circumstances under which the SPA is best applied can be defined and its true role in guiding infertility management therapy can be identified.

The Acrosome Reaction Assay

Acrosome reaction testing provides for a direct measurement of functional activity of the acrosome. In addition, measurement of acrosomal status after no preincubation (basal level), extended incubation (asynchronous/spontaneous reaction) or after incubation and treatment with a stimulator of the acrosome reaction (induced asynchronous or synchronous reaction), may provide diagnostic information about fertilization potential (reviewed in De Jonge 1994 and Tesarik 1996).

As with the SPA, there are many variations in how the acrosome reaction assay can be done. For example, there are different protocols for evaluating acrosomal status with or without treatment with stimulators of the reaction. There are also different protocols for the type of stimulator, such as chemical versus biological, and for the probe used for revealing acrosomal status, such as fluorescent versus non-fluorescent. Thus, when one attempts to evaluate studies in order to discriminate when and for whom the assay is best intended and the subsequent clinical information to be gained, no clear conclusion(s) can be made.

However, unlike the the case of SPA, there is movement in the field to try to standardize an acrosome reaction protocol. Presently, a large multi-center trial is being done. The trial uses a standard protocol which includes the use of two proven and widely available stimulators. It is hoped that when the results from this study are analyzed, a clear indication of whether the acrosome reaction assay, using a consensus protocol, has clinical and diagnostic significance will be found.

A recent report (Franken et al. 1997) evaluated the zona pellucida- and ionophore-induced acrosome reaction in relationship to sperm morphology. After applying ROC analysis, it was found that sperm morphology, at a cutoff of 4% normal forms by strict criteria, was highly predictive of zona pellucida-induced acrosome responsiveness. These results raise the compelling question of whether one can simply use morphology as a sensitive predictor of acrosome responsiveness, thereby obviating the need for an induced acrosome reaction assay in the clinical environment. Perhaps the results from the multicenter study will help to clarify this issue.

Computer-Assisted Sperm Motion Analysis

New instruments for assessing various semen and sperm parameters occupy today's clinical andrology laboratory benches. One notable addition has been the computer-assisted semen analysis (CASA) machine; such machines were introduced into the andrology laboratory more than a decade ago. CASA is a technology that has rapidly advanced, particularly in the last several years. CASA was initially developed to provide accurate determination of sperm concentrations and to discriminate and quantify specific sperm motion (kinematics) parameters.

Specific aspects of sperm movement have been found to correlate with the functional potential of human spermatozoa; specifically their ability to penetrate both human and bovine cervical mucus as well as their intrinsic fertilizing ability as assessed by the zona-free hamster egg penetration test. Initially, there was the indication that assessment of sperm kinematics might provide key prognostic information concerning fertilization (pregnancy) potential (Mortimer 1997). As mentioned earlier, CASA technology has rapidly advanced in response to certain technical aspects that were considered as deficiencies, and as a result reliable and globally-applicable reference values have been slow in coming. However, with improvements in CASA technology have come data that are more reliable and stable and that strongly support the finding that one or more parameters of sperm kinematics have value in regard to providing an indication of fertility potential (Barratt 1993; Mortimer 1997; S Mortimer, personal communication). These exciting but somewhat preliminary data require widespread clinical verification.

The current state-of-the-art status in CASA technology appears very promising, as evidenced by recent publications demonstrating the utility of CASA for assessing key aspects of sperm function (that is, hyperactivated swimming patterns) that are associated with fertilization; this is reviewed in Mortimer (1997). In addition, the most recent generation of CASA machines has strict morphology assessment capabilities included as an added feature. Thus, with one instrument not only can objective measurement of sperm kinematics be done, but also sperm morphology. Perhaps in the future researchers will see implementation into the clinical laboratory of a CASA machine that can do real-time assessment of both parameters simultaneously. Finally, if there is to be global, clinical application of these instruments, then their cost must decrease. Thus, it still remains to be seen whether this promising piece of technology will prove to be useful in providing diagnostic information that can assist in orchestrating clinical management of the infertile couple and if it will be more financially accessible.

The Zona Binding Assay

The zona-binding assay measures the ability of washed spermatozoa to tightly bind to a bisected human zona pellucida (ZP). The hemi-zona assay (HZA) is perhaps the most widely recognized zona-binding assay. It has been reported that the HZA can predict fertilization *in vitro* (Burkman et al. 1990). Indeed, a critical study (Coddington et al. 1994) was reported evaluating 0-100% fertilization *in vitro*. Using ROC analysis, it was calculated that the 100% fertilization group could be separated from the other groups with true positive and negative values of 87% and 83% respectively and the 0% fertilization group separated from all other groups with true positive and negative values of 100% and 71% respectively.

The study noted that there is little within-patient variability over time provided health status remained the same. It should be cautioned that the HZA does not measure post-zona binding events and therefore false positives may result. However, a number of modifications to the assay have been reported whereby acrosomal dyes have been incorporated to provide an indication of the acrosomal status of bound sperm. Furthermore, zona penetration has also been described as an adjunct to the basic assay. Finally, it is important to note the following conclusion as stated by Coddington et al.: "The HZA can be used in a sequential fashion with other bioassays of sperm function to establish accuracy to achieve maximum predictability."

While all of this speaks to the merits of the HZA for assisting in clinical diagnosis, there are some drawbacks. The most significant problem with the HZA is availability of materials, as this assay is dependent on human materials. Oocytes obtained from cadavers, donors and failed fertilization are the few sources for this material. As a result, there is not widespread application of this assay. Thus, while there is very strong data supporting its use in the clinical andrology laboratory, it is unlikely that HZA will ever become a routine assay, and thus it is unlikely to be underwritten by governmental or insurance bodies. What will be helpful is if the proteins that comprise the zona can be reproduced by recombinant technology, fused to synthetic beads and made commercially available. Perhaps this fantasy will become reality before the advent of the next millennium.

Sperm Function Assays for the Next Millennium

It is well accepted that the mission of the fertilizing spermatozoon is to deliver its condensed package of DNA to the mature oocyte. However, as is evidenced by the sperm function assays discussed in this chapter and those not presented, other macromolecular and equally significant micromolecular contributions are made by the spermatozoon to the oocyte, embryo and fetus. Several notable sperm components/structures will be discussed in terms of their role in sperm function, specifically fertilization and post-fertilization events, and how they might be considered as possible sperm function assays for the future.

Human sperm chromosomes are condensed and tightly packaged with a substitution, albeit not in total, of protamines for histones. Thus, defects present in DNA/chromatin prior to packaging will likely persist until eventual nuclear decondensation inside the oocyte. Some of the chromosome/DNA defects can result from endogenous factors, such as failure to repair strand breaks prior to packaging, and/or from exogenous factors, such as free radicals resulting from localized testicular ischemia. So, how might defects be revealed? A review of this subject can be found in Sakkas et al. (1997).

Protamine deposition is critically important to sperm DNA and chromatin stability (Bizzaro et al. 1998). The significance of this becomes apparent when one considers that infertile males are more likely to have chromatin anomalies related to protamine deposition. Furthermore, there is a relationship between under-protamination and damaged sperm DNA. If the DNA has been damaged during packaging, then what impact might this have? Evenson and co-workers (1986, 1990) have established that there is a distinct relationship between susceptibility to DNA denaturation (reflective of altered chromatin structure) and the presence of DNA strand breaks. Adding to this are data from Sakkas et al. (1997) showing that a high percentage of DNA nicks/strand breaks were correlated with decondensation failure. Thus, the way in which sperm DNA has been packaged during spermiogenesis will ultimately influence ability to achieve paternity, regardless of whether one uses advanced assisted reproductive technologies (ART) or not. Based on this information, it is very likely that the assays developed to probe for DNA packaging and chromosome integrity may soon have a place in the clinical andrology laboratory as key sperm functional assays. However, the function or dysfunction in genome expression is not solely restricted to how the nuclear material has been packaged. Equally, if not more covertly, present is the nuclear material proper. The role of the DNA in sperm function is revealed in the following.

There is a higher mutation rate for males in comparison to that of females (Crow 1997). One explanation for this is that there are a far greater number of cell divisions that occur in the male germ cell line to produce the gamete than in the female one. If one were to now consider the subset of the general population that requires fertility assistance, an even greater mutation rate would be calculated for the male. The subject of sperm chromosome abnormalities has received much attention recently, largely in response to the advent of intracytoplasmic sperm injection (ICSI).

Examples of conditions that result from gene mutations in the paternal genome are congenital bilateral absence of the vasa deferentia (CBAVD); microdeletions on the variable region of Yq in the AZF (Azoospermia Factor) locus that can lead to "Sertoli cell only" phenotype, maturation arrest and severe hypospermatogenesis (Vogt et al. 1992; Reijo et al. 1993, 1996). This is significant because scientists are now able to obtain sperm from the extremely subfertile males and, with ICSI, fertilize oocytes and achieve pregnancy. The negative impact of this success may be seen in the offspring where Y-bearing sperm carrying the Deleted in Azoospermia (DAZ) gene can transmit infertility to the male progeny (Kent-First et al. 1996). As such, it is becoming increasingly important to screen genes of these very subfertile males prior to any infertility intervention.

The sperm centrosome is a structure located in the junction between sperm head and tail. The function of the centrosome in embryogenesis is only now beginning to come to light. The paternally-inherited human sperm centrosome, with the assistance of maternal γ-tubulin, nucleates sperm astral microtubules, unites paternal and maternal genomes and forms the mitotic spindle. At the time of fertilization, the human sperm centrosome restores the zygotic centrosome, which is the organizing center for microtubules. In doing so, the polarity and the three-dimensional architecture of the embryo is established. Most of what is known today about sperm centrosome structure and function has come from the work of Dr. Jerry Schatten and co-workers (Schatten 1994; Hewitson et al. 1998).

The significance of the sperm centrosome and its impact on post-fertilization events and early embryo development was revealed in a study that investigated oocytes judged to have failed *in vitro* fertilization (Asch et al. 1995). The results showed that ~50% of oocytes that "failed" to fertilize had actually started fertilization but had arrested for the following reasons: 1) egg activation but no sperm incorporation; 2) post-sperm penetration arrest; 3) no microtubule nucleation; 4) "silent" polyspermy, in which the number of sperm nuclei within the oocyte are not revealed by counting pronuclear number; 5) sperm aster arrests; 6) aster growth defect/detachment; and 7) mitotic arrests.

These abnormalities arose during the time-course of standard *in vitro* fertilization. So, the question can be asked: will ICSI help to overcome some of these deleterious events? On one hand, science can probably say yes. Sperm incorporation problems will be averted. However, what about post-penetration abnormalities? Preliminary data from rhesus monkey ICSI (Hewitson et al. 1998) indicate that the same types of centrosomal dysfunctions occur as were detected in human oocytes. Thus, in the near future, there might be a place in the clinical lab for a sperm function test, whether using the rhesus model or *Xenopus* oocyte extract, that can reveal centrosomal function.

The last clearly identifiable sperm structural component that has significance relative to sperm function is the mitochondrion. The functional significance of the mitochondria lies in their energy generating properties. However, dysfunction in mitochondrial oxidative processes can result in the production of high levels of reactive oxygen molecules that can have permanent destructive effects on cellular and subcellular constituents. More covertly functioning in the mitochondrion is the mitochondrial DNA (mtDNA), which is passed to the oocyte at fertilization. What, if any, influence the paternal mtDNA might have on the oocyte/embryo is largely unknown. Thus, at this time, perhaps the most pertinent issue regarding sperm mitochondria is the impact they may have on nuclear DNA during and after spermiogenesis, when mitochondrial oxidative processes become uncontrollable. The result of this could be irreversible DNA damage.

Conclusions

One or more of the sperm function tests presented have or appear to have not only diagnostic value but also a role in the structured management of the infertile couple. However, there are also substantial questions that must be addressed. First, what, if any, data are there that provide a clear indication of the predictive value for the function test in question? As shown here, there is a paucity of appropriately designed and statistically analyzed studies. If evidence supports the clinical value of the assay, then how widely accessible is the test to the general clinical population? While technology continues to advance, some of that progress has been slowed by biological conundrums inherent to the cell and processes in question. As a result, the appearance of many materials, assays or techniques, such as the recombinant human ZP3-laden agarose beads, is delayed into the unforeseeable future. For the test or tests that can be routinely applied, how will the data influence patient management? The results obtained from some sperm function tests can and do influence subsequent therapy. For example, results from the HZA will likely assist in directing therapy for the infertile couple to ovulation induction with intrauterine insemination or to *in vitro* fertilization with ICSI. Conversely, if a sperm population is judged to be subnormal based on any given sperm function test, such as the detection of under-protaminated DNA, will that diagnosis alter therapy? The answer to that question is not clear. In other words, when functional defects are diagnosed, does the diagnosis help determine a treatment or is the result merely to provide informed consent?

Many of the questions surrounding utilization of sperm function testing in structured infertility management rely on the acceptance or rejection of test validity by the clinician, and in part that is influenced by test cost and whether the patient will receive financial support to diminish or remove that cost. This latter issue depends on the health care industry at large. Presently there is a bit of a cloud that surrounds how the health care industry views the diagnosis and treatment of infertility, particularly the diagnostic aspect.

So, where does that leave the clinical laboratory in regards to contributing to infertility patient management? In the face of technological advances, and specifically ICSI, are the labs relegated to simply providing qualitative information? There answer is a resounding no, as evidenced by recent publications which correctly emphasize the importance of clinical andrology in infertility management (Cummins et al. 1994). Quoting from Dr. Stewart Irvine (Consensus Workshop on Advanced Diagnostic Andrology Techniques): "What the andrology laboratory offers is the ability to select the most appropriate form of treatment for each couple, especially those with male factor infertility, be it overt or occult. Another useful maxim of medical care is the notion that the three basic principles of correct treatment are diagnosis, diagnosis, and diagnosis; it is extremely difficult to provide the correct treatment for a patient whose diagnosis remains obscure." Some sperm function tests already provide diagnosis; others have the potential to do so. It is the responsibility of andrologists to move the many isolated and noncomparable variations of sperm function tests into multi-center trials with consensus protocols to generate data of sufficient statistical power. Upon logistic regression analysis and receiver-operating characteristic analysis, the validity of the assay in question will either be supported or refuted. If support for the test is realized, then prospective meta-analysis can be done, the results from which will likely provide convincing and conclusive evidence of diagnostic value to all interested parties: clinician, third party payor or governmental body. The end result would hopefully be the acceptance and integration of the diagnostic tool—that is, the sperm function test—into the entire gamut of the health care arena, culminating in a direct and visible benefit to the most important person of all. . .the patient.

References

Aitken RJ, Thatcher S, Glasier AF, Clarkson JS, Wu FCW, Baird DT. Relative ability of modified versions of the hamster oocyte penetration test, incorporating hyperosmotic medium or the ionophore A23187, to predict IVF outcome. Hum Reprod 1987; 2:227-231.

Amann RP. Can the fertility potential of a seminal sample be predicted accurately? J Androl 1989; 10:89-98.

Amann RP, Hammerstedt RH. *In vitro* evaluation of sperm quality: an opinion. J Androl 1993; 14:397-406.

Asch R, Simerly C, Ord T, Ord VA, Schatten G. The stages at which fertilization arrests in humans: defective sperm centrosome and sperm asters as causes of human infertility. Hum Reprod 1995; 10:1897-1906.

Barbato GF, Cramer PG, Hammerstedt RH. A practical *in vitro* sperm-egg binding assay that detects subfertile males. Biol Reprod 1998; 58:686-699.

Barratt CLR, Tomlinson MJ, Cooke ID. Prognostic significance of computerized motility analysis for *in vivo* fertility. Fertil Steril 1993; 60:520-525.

Bizzaro D, Manicardi GC, Bianchi PG, Bianchi U, Mariethoz E, Sakkas D. *In situ* competition between protamine and fluorochromes for sperm DNA. Mol Hum Reprod 1998; 4:127-132.

Burkman LJ, Coddington CC, Franken DR, Oehninger SC, Hodgen GD. The hemizona assay (HZA): assessment of fertilization potential by means of human sperm binding to the human zona pellucida. In: (Keel BA, Webster BW, eds) Handbook of the Laboratory Diagnosis and Treatment of Infertility. Boca Raton, FL: CRC Press, 1990, pp212-225.

Coddington CC, Oehninger SC, Olive DL, Franken DR, Kruger TF, Hodgen GD. Hemi-zona index (HZI) demonstrates excellent predictability when evaluating sperm fertilizing capacity in *in vitro* fertilization patients. J Androl 1994; 15:250-254.

Consensus Workshop on Advanced Diagnostic Andrology Techniques. Excerpts on Hum Reprod No. 3, July 1996.

Crow JF. The high spontaneous mutation rate: is it a health risk? Proc Natl Acad Sci USA 1997; 94:8380-8386.

Cummins J, Jequier A. Treating male infertility needs more clinical andrology, not less. Hum Reprod 1994; 9:1214-1220.

De Jonge CJ. Diagnostic significance of the induced acrosome reaction. Reprod Med Rev 1994; 3:159-178.

Duncan WW, Glew MJ, Wang X-J, Flaherty SP, Matthews CD. Prediction of *in vitro* fertilization rates from semen variables. Fertil Steril 1993; 59:1233-1238.

Evenson D, Darzynkiewicz Z, Jost L, Janka F, Ballachey B. Changes in accessibility of DNA to various fluorochromes during spermatogenesis. Cytometry 1986; 7:45-53.

Evenson DP. Flow cytometric analysis of male germ cell quality. Meth Cell Biol 1990; 33:401-10.

Franken DR, Bastiaan HS, Kidson A, Wranz P, Habenicht UF. Zona pellucida mediated acrosome reaction and sperm morphology. Andrologia 1997; 29:311-317.

Hewitson L, Simerly C, Takahashi D, Schatten G. The role of the sperm centrosome during human fertilization and embryonic development: implications for intracytoplasmic sperm injection and other sophisticated ART strategies. In: (Ombelet W, Bosmans E, Vandepput H, Vereecken A, Renier M, Hoomans E, eds) Modern ART in the 2000's: Andrology in the Nineties. Lancaster, UK: Parthenon Publishing, 1998, pp139-156.

Kent-First MG, Kol S, Muallem A, Ofir R, Manor D, Blazer S, First N, Itskovitz-Eldor J. The incidence and possible relevance of Y-linked microdeletions in babies born after intracytoplasmic sperm injection and their infertile fathers. Mol Hum Reprod 1996; 2:943-950.

Mortimer ST. A critical review of the physiological importance and analysis of sperm movement. Hum Reprod Update 1997; 3:403-439.

Reijo R, Lee T, Salo P, Alagappan R, Brown LG, Rosenberg M, Rozen S, Jaffe T, Straus D, Hovatta O, Page D. Diverse spermatogenetic defects in humans caused by Y chromosome deletions encompassing a novel RNA-binding protein. Nature Genet 1993; 10:383-393.

Reijo R, Alagappan RK, Patrizio P, Page DC. Severe oligozoospermia resulting from deletions of the Azoospermia Factor gene on the Y chromosome. Lancet 1996; 347:1290-1293.

Sakkas D, Bianchi PG, Manicardi G, Bizzaro D, Bianchi U. Chromatin packaging anomalies and DNA damage in human sperm: their possible implications in the treatment of male factor infertility. In: (Barratt C, De Jonge CJ, Mortimer D, Parinaud J, eds) Genetics of Human Male Fertility. Paris: Editions EDK, 1997, pp205-221.

Schatten G. The centrosome and its mode of inheritance: the reduction of the centrosome during gametogenesis and its restoration during fertilization. Dev Biol 1994; 165:299-335.

Tesarik J. Acrosome reaction testing. In: Consensus Workshop on Advanced Diagnostic Andrology Techniques. ESHRE Andrology Special Interest Group. Hum Reprod 1996; 11:1467-1470.

Vogt PH, Chandley AC, Hargreave TB, Keil R, Ma K, Sharkey A. Microdeletions in interval 6 of the Y chromosome of males with idiopathic sterility point to disruption of AZF, a human spermatogenesis gene. Hum Genet 1992; 89:491-496.

World Health Organization. WHO laboratory manual for the examination of human semen and sperm-cervical mucus interaction. 3rd ed. Cambridge: Cambridge University Press, 1992.

Yanagimachi R, Yanagimachi H, Rogers BJ. The use of zona free animal ova as a test system for the assessment of the fertilizing capacity of human spermatozoa. Biol Reprod 1976; 15:542-551.

Yusif S. Meta-analysis of randomized trials: looking back and looking ahead. Controlled Clin Trials 1997; 18:594-601.

Zaneveld LJD, Jeyendran RS. Sperm function tests. Infertil Reprod Med Clinics North America 1992; 3:353-371.

34 *The Need to Detect DNA Damage in Human Spermatozoa: Possible Consequences on Embryo Development*

Denny Sakkas
University of Birmingham

It has been clearly established that certain anomalies exist in the DNA of ejaculated spermatozoa, both at the nuclear and mitochondrial levels (Sakkas et al. 1997; St. John et al. 1997). This chapter will deal with the presence of nuclear DNA damage in human sperm. The increased presence of this anomaly in males with abnormal sperm parameters puts the population of patients being treated by assisted reproductive technologies, in particular intracytoplasmic sperm injection (ICSI), at great risk. ICSI forces any selected spermatozoon to enter the oocyte and increases the likelihood that it will participate in fertilization and embryo development.

The question remains, however, as to what may be the consequences of sperm DNA damage during fertilization and embryo development. The greatest impact of faulty nuclear DNA could lead to a defective paternal genome. There are already indications that abnormal sperm can disrupt the process of fertilization, embryo development and fetal development. The role of DNA damage in these processes is, however, poorly understood. If certain evidence indicating that sperm with DNA damage can influence the development of the embryo increases, researchers will need to refine the semen parameters currently assessed. This will lead to a more detailed molecular semen analysis that will provide a risk assessment for the likelihood that a patient's semen sample will lead to defective embryos and will also provide the possibility that sperm with no detectable DNA damage can be selected. The development of sperm assessment techniques based on a molecular analysis will therefore give researchers a greater handle on treatment protocols that should be defined for these patients. This review will attempt to establish the risks of using sperm with DNA damage and outline techniques that will help eliminate these risks.

DNA Damage in Mature Sperm

Numerous techniques have indicated that the mature human spermatozoon can possess nuclear DNA damage. Gorczyca et al. (1993), using the terminal deoxynucleotidyl transferase (TUNEL) assay, indicated that the incidence of DNA fragmentation in human sperm correlated with abnormal sperm chromatin packaging (poorly protaminated sperm). Results have also described the presence of endogenous DNA nicks in ejaculated spermatozoa of men and shown a correlation to poorly protaminated sperm and reduced fertility (Bianchi et al. 1993; Manicardi et al. 1995; Bianchi et al. 1996a; Sakkas et al. 1996). The group of Evenson (Evenson et al. 1980), using the sperm chromatin structure assay (SCSA), which measures the susceptibility of DNA to heat or acid-induced denaturation *in situ*, has also shown that the SCSA test is effective in identifying fertility potential. More recently, the comet assay (single cell gel electrophoresis), which has been recognized in many cell lines to be one of the most sensitive techniques available for measuring DNA strand breaks (Collins et al. 1997), has also been used to highlight the presence of DNA strand breaks in sperm (Hughes et al. 1996; Aravindan et al. 1997). The presence of DNA strand breaks in ejaculated human spermatozoa is therefore an established phenomenon. Two questions arise: first, what is the origin of this DNA damage, and, second, what are the consequences if a DNA-damaged spermatozoon fertilizes an oocyte?

The Source of DNA Damage in Sperm

It is evident from studies in the Sakkas laboratory and others that men with reduced semen parameters (concentration, motility and morphology) have a higher percentage of spermatozoa possessing nuclear DNA damage (Manicardi et al. 1995; Bianchi et al. 1996a, b). How DNA damage arises in mature ejaculated spermatozoa is not precisely understood. There is strong evidence that several mechanisms may be responsible for this damage. Animal models, in particular mouse models, have shown that environmental factors can have a severe influence on spermatogenesis. Using the SCSA, the group of Evenson has shown that both low levels of radiation and heat stress can compromise nuclear chromatin structure of sperm and increase susceptibility of DNA to *in situ* denaturation at low pH (Sailer et al. 1995, 1997).

Another factor implicated in the generation of DNA strand breakage is oxidative stress. Spermatozoa contain high levels of unsaturated lipids in their membranes and are particularly vulnerable to reactive oxygen species (ROS) damage (Alvarez et al. 1987; Aitken, Fisher 1994). Studies by Aitken and Clarkson (1987, 1988) have shown the existence of populations of spermatozoa in an ejaculate that produce ROS that can induce irreversible damage to spermatozoa and impair their fertilizing ability. In various cell types, the level of damage induced by ROS is located at the DNA level (Lafleur, Retel 1993; Gutteridge 1994). Hyperoxia, agents that generate ROS (including chemicals and radiation) or oxidants elaborated by activated neutrophils have been shown to cause DNA damage and mutagenesis in human and animal cells (Parshad, Sanford 1971; Weitberg et al. 1983; Nguyen et al. 1992; Weitzman et al. 1985). It is now widely held that the mutagenic capacity of oxygen is due to the interaction of free oxygen radicals with DNA (Guyton, Kensler 1993). For example, hydroxyl radicals have been detected with electron paramagnetic resonance spectroscopy under conditions of active oxygen induced DNA damage (Inoue, Kawanishi 1987). An unstable radical such as the hydroxyl radical will interact indiscriminately with all components of the DNA molecule, producing a broad spectrum of DNA damage. The forms of DNA damage produced by ROS in experimental systems include modification of all bases as well as the production of base-free sites, deletions, frameshifts, strand breaks, DNA protein crosslinks and chromosomal rearrangements (Halliwell, Aruoma 1991). Indeed, endogenous oxidative DNA damage has been shown in human spermatozoa using the measurement of the oxidized nucleoside 8-hydroxy-2′-deoxyguanosine, or oxo^{8}dG (Fraga et al. 1991). The exposure of spermatozoa to ROS *in vitro* has also been shown to increase the presence of DNA damage in sperm (Twigg et al. 1997; Lopes et al. 1998a). Smoking has also been implicated as a factor responsible for inducing DNA damage in spermatozoa. Smoking can decrease the level of seminal ascorbate, which protects against oxidative DNA damage in human sperm (Fraga et al. 1991). Sorahan et al. (1997) have recently shown that about 15% of all childhood cancers could be attributable to paternal smoking.

A number of factors could therefore be responsible for inducing nuclear DNA damage in the mature spermatozoa of men. These would range from medical events, including fevers, to social behavior such as smoking.

Apoptosis during Spermatogenesis

One theory that may explain the presence of nuclear DNA damage in mature spermatozoa is that it is indicative of apoptosis occurring during spermatogenesis. Classically, cells that have nuclear DNA damage display laddering of their nucleosomes; however, the presence of protamines in mature sperm nuclei does not allow for this classic analysis that is used for somatic cells. The observation that mature ejaculated sperm have nuclear DNA damage strengthens the theory that this is related to apoptosis (Gorczyca et al. 1993). Baccetti et al. (1996) have also described certain ultrastructural features in human sperm that are indicative of apoptosis, while Hadziselimovic et al. (1997) have shown that there was increased apoptosis in the testis of patients with testicular torsion.

An inability to examine spermatozoa for DNA laddering may mean that spermatozoa should be analyzed for other markers of apoptosis. These would include a number of possible molecular markers of apoptosis such as the membrane proteins Fas and Annexin or the Bcl2 family of proteins. Indeed, it has been found that apoptosis is involved in regulating spermatogenesis in the human. In addition, oligozoospermic men have higher levels of Fas expression on their ejaculated spermatozoa when compared to men who have semen concentrations of greater than 20 million sperm per ml (Sakkas et al. 1999). This strengthens the theory that many men with abnormal sperm parameters have underlying problems in spermatogenesis. If this is the case, it is therefore necessary to ascertain at which stage during spermatogenesis the problem arises. If DNA damage arises after the spermatid stage—that is, during spermiogenesis or in the male reproductive tract—spermatids would more likely possess normal DNA. One of the conclusions that may arise from studies on these patients who have a faulty spermatogenesis or defective tract is that it may actually be beneficial for these patients to have a testicular biopsy so that spermatids are used for ICSI.

Consequences for the Developing Embryo

The increased likelihood of DNA damage being present in sperm from men with abnormal sperm parameters places patients being treated by ICSI at the greatest risk of their sperm contributing to an offspring. What indications are there that sperm with nuclear DNA damage may influence abnormal development?

Impact on Fertilization

In the human, it has been observed that oocytes that failed to be fertilized after *in vitro* fertilization (IVF) and are injected with spermatozoa with a high level of endogenous DNA nicks contain more condensed spermatozoa (Sakkas et al. 1996). This indicates that DNA from damaged spermatozoa selected for ICSI may impede the completion or initiation of decondensation, therefore leading to a failure of fertilization. More recently, Lopes et al. (1998b) have also shown that sperm populations containing greater than 25% DNA damage are more likely to fertilize oocytes at low rates (<20%) after ICSI. Although there may be indications that these sperm may fail to fertilize, the greatest risk arises to those embryos that develop and possibly lead to the formation of fetuses.

Development of the Preimplantation Embryos

Sperm defects could impair not only the fertilization process but also the fate of embryos. *In vitro* fertilization is a good model for observing the influence of the male gamete on embryonic quality because it allows the observation of early embryo stages and the evaluation of the embryo's ability to develop by studying the pregnancy per transfer, which rules out fertilization failures.

The effect of various sperm abnormalities on embryo quality is reported by many studies; however, the basis for these observations is not really understood. Some examples of the role of sperm in preimplantation embryo development are:

1) Ron-El et al. (1991) found that delayed fertilization was associated with high levels of morphological abnormalities of spermatozoa or with sperm auto-antibodies. This was thought to be due to a delay in sperm penetration and gamete fusion. Thus an asynchronization occurs between the sequences of events specific to the female and male nuclei. This asynchronization may

cause cleavage disturbances, fragmentations and numeric chromosome abnormalities.

2) Using strict criteria, a relationship between embryo quality and the degree of "normality" of a semen sample regarding morphological features had been observed (Ombelet et al. 1994). The physiological explanation for this finding is unknown. However, it has been proposed that the influence of the paternal genome on the proliferation of extra-embryonic tissue might be important (Hilscher 1991).

3) In human IVF, when the usual parameters of sperm quality are good, there is a strict linear relationship between cleavage and blastocyst formation. For sperm with poor motility and poor morphology parameters, cleavage rate did not correlate with further embryonic development. When fertile abnormal sperm is used in IVF, no correlation exists between the sperm fertility estimated by the cleavage rate and the development potential of the embryos assessed by the blastocyst formation rate. Poor quality sperm may, however, lead to poor rates of blastocyst formation (Janny, Menezo 1994), but once the blastocyst stage is obtained the developmental future is independent of sperm quality at the origin.

4) It has recently been reported that the percentage of embryos that form blastocysts is lower in cases where ICSI was attempted than in cases where routine IVF was attempted (Shoukir et al. 1998).

5) An increased level of chromosome breaks and acentric fragments was reported in sperm of men whose partners have an unexplained history of repeated spontaneous abortions. The embryonic loss could result from a deficiency of genetic material in the embryo secondary to the loss, during the first cleavage, of abnormal male DNA material (Rosenbusch, Sterzik 1991).

Denaturation or redistribution and loss of sperm DNA were reported to occur at higher levels in infertile than fertile men (Peluso et al. 1992).

Impact on Fetal Development and Subsequent Generations

In the above studies, it appears that abnormal spermatozoa can contribute to abnormal development of preimplantation embryos. Whether spermatozoa with damaged DNA can impair the process of fertilization or embryo development is not clear. The alarming studies by the group of Robaire have, however, indicated that damage to sperm DNA may be linked to an increase in early embryo death. They have shown that treatment of male rats with cyclophosphamide had little effect on the male reproductive system but caused single strand DNA breaks in the cauda epididymal spermatozoa and altered the decondensation potential of spermatozoa (Qiu et al. 1995a, b). More disturbingly, similar treatment protocols using cyclophosphamide produce an increase in postimplantation loss and malformations (Trasler et al. 1985, 1986, 1987) and are transmissible to the next generation (Hales et al. 1992).

Another treatment protocol that is known to induce DNA damage in mature spermatozoa in mice is hyperthermia of the testes (Sailer et al. 1997). A number of studies have shown that exposure of mice to a hot environment can lead to abnormalities in embryo growth (Bellvé 1972, 1973). This was verified by Setchell et al. (1998), who showed that embryo growth was retarded in normal female mice made pregnant by males whose testes had been heated.

Treatment Options

The need to detect spermatozoa with abnormal DNA arises because, if a spermatozoon with DNA damage is used to fertilize an egg, the likely effects would not be visible until after the embryo starts to use its own genome. Techniques must be developed to either select the best spermatozoon to use for ICSI or to select which embryo is likely to develop.

The first possibility is to use current techniques that examine nuclear DNA damage in spermatozoa. This could include the TUNEL assay, *in situ* nick translation or the comet assay. Non-toxic markers may, in the future, be used to mark such sperm so that they are not selected for use in ICSI. Unfortunately, current techniques can only give an idea of the percentage of mature sperm with abnormal DNA within the patient's ejaculate. The use of markers, such as Fas or Bcl2, for apoptosis may also give an insight into problems that a specific patient may have during spermatogenesis. A current technique that may be aligned to such assessment is the possibility of using extended embryo culture to the blastocyst stage. This would allow a further selection process at the time of fertilization and during preimplantation embryo development. In ICSI patients, there may be a need to extend the culture period until post-embryonic genome activation, to overcome any errant paternal genome defects. If a patient has a high number of ejaculated spermatozoa with DNA damage, it may be advisable to perform a testicular biopsy to ascertain which stage of spermatogenesis gives rise to abnormal sperm. In doing so, it may be advisable to use spermatids for ICSI. A possible model for selecting these patients is shown in Figure 1.

Conclusion

The use of ICSI has heightened the risk that sperm containing damaged DNA may participate in the development of an infant. Whether sperm possessing damaged DNA will fail in their mission to contribute to a viable offspring at the time of fertilization, embryo development or fetal development is uncertain. There is already a great debate as to whether ICSI may contribute to an increase in the development of abnormal offspring (Bonduelle et al. 1998; Bowen et al. 1998). The above studies show that there is a need for more research into the molecular origin and consequences of sperm DNA damage in patients undergoing ICSI. One key area will be the study of whether or not the oocyte is capable of repairing sperm DNA that has strand breaks post fertilization.

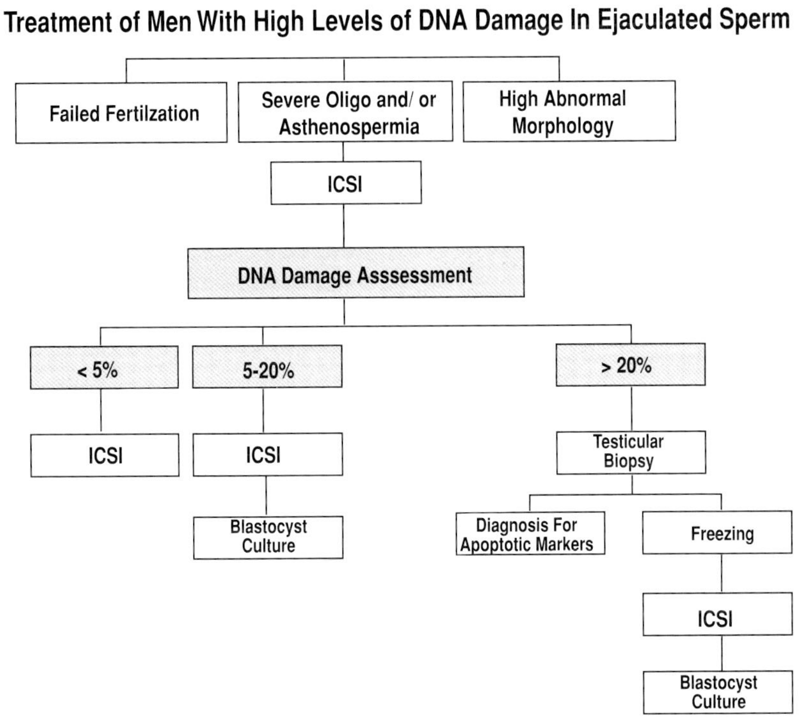

Figure 1. A model for attributing ICSI treatment protocols to male infertility patients, using assessment of sperm nuclear DNA damage as the defining parameter.

References

Aitken J, Fisher H. Reactive oxygen species generation and human spermatozoa: the balance of benefit and risk. Bio Essays 1994; 16:259-267.

Aitken RJ, Clarkson JS. Cellular basis of defective sperm function and its association with the genesis of reactive oxygen species by human spermatozoa. J Reprod Fertil 1987; 81:459-469.

Aitken RJ, Clarkson JS. Significance of reactive oxygen species and antioxidants in defining the efficacy of sperm preparation techniques. J Androl 1988; 9:367-376.

Alvarez JG, Touchstone JC, Blasco L, Storey BT. Spontaneous lipid peroxidation and production of hydrogen peroxide and superoxide in human spermatozoa. Superoxide dismutase as major enzyme protectant against oxygen toxicity. J Androl 1987; 8:338-348.

Aravindan GR, Bjordahl J, Jost LK, Evenson DP. Susceptibility of human sperm to *in situ* DNA denaturation is strongly correlated with DNA strand breaks identified by single-cell electrophoresis. Exp Cell Res 1997; 10:231-237.

Baccetti B, Collodel G, Piomboni P. Apoptosis in human ejaculated sperm cells (notulae seminologicae 9). J Submicrosc Cytol Pathol 1996; 28:587-596.

Bellvé AR. Viability and survival of mouse embryos following parental exposure to high temperatures. J Reprod Fertil 1972; 30:71-81.

Bellvé AR. Development of mouse embryos with abnormalities induced by parental heat stress. J Reprod Fertil 1973; 35:393-403.

Bianchi PG, Manicardi GC, Bizzaro D, Bianchi U, Sakkas D. Effect of deoxyribonucleic acid protamination on fluorochrome staining and *in situ* nick-translation of murine and human mature spermatozoa. Biol Reprod 1993; 49:1083-1088.

Bianchi PG, Manicardi GC, Bizzaro D, Campana A, Bianchi U, Sakkas D. The use of the gc specific fluorochrome chromomycin A_3 (CMA_3) as an indicator of poor sperm quality. J Assist Reprod Genet 1996a; 13:246-250.

Bianchi PG, Manicardi GC, Urner F, Campana A, Sakkas D. Chromatin packaging and morphology in ejaculated human spermatozoa: evidence of hidden anomalies in normal spermatozoa. Mol Hum Reprod 1996b; 2:139-144.

Bonduelle M, Joris H, Hofmans K, Liebars I, Van Steirteghem A. Mental development of 201 ICSI children at 2 years of age. Lancet 1998; 351:1553.

Bowen JR, Gibson FL, Leslie GI, Saunders DM. Medical and developmental outcome at 1 year for children conceived by intracytoplasmic sperm injection. Lancet 1998; 351:1529-1534.

Collins AR, Dobson VL, Dusinska M, Kennedy G, Stetina R. The comet assay: what can it really tell us? Mut Res 1997; 47:1033-1035.

Evenson DP, Darzynkiewicz Z, Melamed MR. Relation of mammalian sperm chromatin heterogeneity to fertility. Science 1980;

240:1131-1133.

Fraga CG, Motchnik PA, Shigenaga MK, Helbock HJ, Jacob RA, Ames BN. Ascorbic acid protects against endogenous oxidative DNA damage in human sperm. Proc Natl Acad Sci USA 1991; 88:11003-11006.

Gorczyca W, Traganos F, Jesionowska H, Darzynkiewicz Z. Presence of strand breaks and increased sensitivity of DNA *in situ* to denaturation in abnormal human sperm cells: analogy to apoptosis of somatic cells. Exp Cell Res 1993; 207:202-205.

Gutteridge JM. Biological origin of free radicals and mechanisms of antioxidant protection. Chem Biol Interact 1994; 91:133-140.

Guyton KZ, Kensler TW. Oxidative mechanisms in carcinogenesis. Br Med Bull 1993; 49:523-544.

Hadziselimovic F, Geneto R, Emmons LR. Increased apoptosis in the contralateral testis in patients with testicular torsion. Lancet 1997; 350:118.

Hales BF, Crosman K, Robaire B. Increased postimplantation loss and malformations among the F_2 progeny of male rats chronically treated with cyclophosphamide. Teratol 1992; 45:671-678.

Halliwell B, Aruoma OI. DNA damage by oxygen derived species: its mechanism and measurement in mammalian systems. FEBS Lett 1991; 281:9-19.

Hilscher W. The genetic control and germ cell kinetics of the female and male germ line in mammals including man. Hum Reprod 1991; 6:1416-1425.

Hughes CM, Lewis SE, McKelvey-Martin VJ, Thompson W. A comparison of baseline and induced DNA damage in human spermatozoa from fertile and infertile men, using a modified comet assay. Mol Hum Reprod 1996; 2:613-619.

Inoue S, Kawanishi S. Hydroxyl radical production and human DNA damage induced by ferric nitriloacetate and hydrogen peroxide. Cancer Res 1987; 47:6522-6527.

Janny L, Menezo YJR. Evidence for a strong paternal effect on human preimplantation embryo development and blastocyst formation. Mol Reprod Dev 1994; 38:36-42.

Lafleur MV, Retel J. Contrasting effects of -SH compounds on oxidative DNA damage: repair and increase of damage. Mutat Res 1993; 295:1-10.

Lopes S, Jurisicova A, Sun JG, Casper RF. Reactive oxygen species: potential cause for DNA fragmentation in human spermatozoa. Hum Reprod 1998a; 13:896-900.

Lopes S, Sun JG, Jurisicova A, Meriano J, Casper RF. Sperm deoxyribonucleic acid fragmentation is increased in poor quality semen samples and correlates with failed fertilization in intracytoplasmic sperm injection. Fertil Steril 1998b; 69:528-532.

Manicardi GC, Bianchi PG, Pantano S, Azzoni P, Bizzaro D, Bianchi U, Sakkas D. Presence of endogenous nicks in DNA of ejaculated human spermatozoa and its relationship to Chromomycin A3 accessibility. Biol Reprod 1995; 52:864-867.

Manicardi GC, Tombacco A, Bizzaro D, Bianchi U, Bianchi PG, Sakkas D. DNA strand breaks in ejaculated human spermatozoa: comparison of susceptibility to the nick translation and terminal transferase assays. Histochem J 1998; 30:33-39.

Nguyen T, Bronson D, Crespi CL, Penman BW, Wishnok JS, Tannenbaum SR. DNA damage and mutation in human cells exposed to nitric oxide *in vitro*. Proc Natl Acad Sci USA 1992; 89:3030-3034.

Ombelet W, Fourie F, Vandeput H, Bosmans E, Cox A, Janssen M, Kruger T. Teratozoospermia and *in vitro* fertilization: a randomised prospective study. Hum Reprod 1994; 9:1479-1484.

Parshad R, Sanford KK. Oxygen supply and stability of chromosomes in mouse embryo cells *in vitro*. J Natl Cancer Inst 1971; 47:1033-1035.

Peluso JJ, Luciano AA, Nulsen JC. The relationship between alterations in spermatozoal deoxyribonucleic acid, heparin binding sites and semen quality. Fertil Steril 1992; 57:665-670.

Qiu J, Hales BF, Robaire B. Damage to rat spermatozoal DNA after chronic cyclophosphamide exposure. Biol Reprod 1995a; 53:1465-1473.

Qiu J, Hales BF, Robaire B. Effects of chronic low dose cyclophosphamide on the nuclei of rat spermatozoa. Biol Reprod 1995b; 52:33-40.

Ron-El R, Nachum H, Herman A, Golan A, Caspi E, Soffer Y. Delayed fertilization and poor embryonic development associated with impaired semen quality. Fertil Steril 1991; 55:338-344.

Rosenbusch B, Sterzik K. Sperm chromosomes and habitual abortion. Fertil Steril 1991; 56:370-372.

Sailer BL, Jost LK, Erickson KR, Tajiran MA, Evenson DP. Effects of x-irradiation on mouse testicular cells and sperm chromatin structure. Environ Mol Mutagen 1995; 25:23-30.

Sailer BL, Sarkar LJ, Bjordahl JA, Jost LK, Evenson DP. Effect of heat stress on mouse testicular cells and sperm chromatin structure. J Androl 1997; 18:294-301.

St. John JC, Cook ID, Barratt CL. The use of long PCR to detect multiple deletions in the mitochondrial DNA of human testicular tissue from azoospermic and severe oligozoospermic patients. In: (Barratt C, De Jonge C, Mortimer D, Parinaud J, eds) Genetics of Human Male Infertility. Paris: Editions EDK, 1997, pp333-347.

Sakkas D, Urner F, Bizzaro D, Bianchi PG, Wagner I, Jacquenoud N, Manicardi GC, Campana A. Sperm chromatin anomalies can influence decondensation after intracytoplasmic sperm injection (ICSI). Hum Reprod 1996; 11:837-843.

Sakkas D, Bianchi PG, Manicardi GC, Bizzaro D, Bianchi U. Chromatin packaging anomalies and DNA damage in human sperm: their possible implications in the treatment of male factor infertility. In: (Barratt C, De Jonge C, Mortimer D, Parinaud J, eds) Genetics of Human Male Infertility. Paris: Editions EDK, 1997, pp205-221.

Sakkas D, Mariethoz E, Manicardi G, Bizzaro D, Bianchi PG, Bianchi U. Origin of DNA damage in ejaculated human spermatozoa. Rev Reprod 1999; 4:31-37.

Setchell BP, Ekpe G, Zupp JL, Surani MAH. Transient retardation in embryo growth in normal female mice made pregnant by males whose testes had been heated. Hum Reprod 1998; 13:342-347.

Shoukir Y, Chardonnens D, Campana A, Sakkas D. Blastocyst development from supernumerary embryos after intracytoplasmic sperm injection: a paternal influence? Hum Reprod 1998; 13:1632-1637.

Sorahan T, Lancashire RJ, Hulten MA, Peck I, Stewart AM. Childhood cancer and parental use of tobacco: deaths from 1953 to 1955. Br J Cancer 1997; 75:134-138.

Trasler JM, Hales BF, Robaire B. Paternal cyclophosphamide treatment causes fetal loss and malformations without affecting male fertility. Nature 1985; 316:144-146.

Trasler JM, Hales BF, Robaire B. Chronic low dose cyclophosphamide treatment of adult male rats: effect on fertility pregnancy outcome and progeny. Biol Reprod 1986; 34:276-283.

Trasler JM, Hales BF, Robaire B. A time course study of chronic paternal cyclophosphamide treatment of rats: effects on pregnancy outcome and the male reproductive and haematologic systems. Biol Reprod 1987; 37:317-326.

Twigg JP, Irvine DS, Aitken RJ. Exposure of human spermatozoa to reactive oxygen species enhances chromatin cross-linking and stimulates DNA strand breakage. ESHRE, June 22-25, Edinburgh, Scotland. Hum Reprod 1997; 12 (Suppl):56-57.

Weitberg AB, Weitzman SA, Destrempes M, Latt SA, Stossel TP. Stimulated human phagocytes produce cytogenetic changes in cultured mammalian cells. N Eng J Med 1983; 308:26-30.

Weitzman SA, Weitberg AB, Clark EP, Stossel TP. Phagocytes are carcinogens: malignant transformation produced by human neutrophils. Science 1985; 227:1231-1233.

35

Cellular Maturity and Fertilizing Potential of Sperm Populations in Natural and Assisted Reproduction

Gabor Huszar

Hulusi B. Zeyneloglu

Lynne Vigue

Yale University

The goal of this article is to develop, based on sperm cell biology and clinical experience, principles for a structured workup of couples with a male factor infertility problem. The basic question is whether semen parameters and sperm function tests are able to provide information that would facilitate the prediction of male fertility, and suggest the most appropriate treatment protocol with respect to natural conception or assisted reproduction. Approximately 95% of non-azoospermic infertile men have either severe male factor infertility requiring *in vitro* fertilization (IVF) or intracytoplasmic sperm injection (ICSI), as found in Palermo et al. (1993), or moderate male infertility characterized by sperm concentrations of at least four to five million sperm/ml semen. About 5% of men in the patient population fall into the class of unexplained male infertility, in which sperm maturity is diminished in spite of normal sperm concentration and motility parameters.

Issues related to the establishment of male fertility have been the subject of many studies in the past two decades. Nevertheless, the progress in this area has been very slow due to some of the following reasons. A) Insistence of finding rigid limits for fertility based on sperm concentration, motility and morphology. B) Failure to recognize that the pathogenesis of diminished male fertility is complex and shows substantial variation among patients. C) Failure to recognize that there are sperm subpopulations in semen that are not equivalent in fertility. D) Application of sperm function tests without understanding the underlying cell biology of human spermatozoa. E) Conducting clinical trials in which the number of patients was too small to reach statistical significance. F) Extremely slow acceptance of the objective biochemical markers of sperm maturity and fertility, which were developed in the past decade by the Huszar laboratory at Yale and by the Aitken-Irvine group in Edinburgh. G) Insufficient consideration of the female factor in infertile couples with oligospermic husbands.

Sperm Concentrations, Sperm Subpopulations and Sperm Function

The ejaculate of men contains seminal fluid, spermatozoa and other cellular elements. Present within the sperm fraction are sperm subpopulations that differ in maturity, in physiological status with respect to the ability to undergo capacitation and acrosome reaction, and in fertilizing potential. The completion of sperm maturation during spermiogenesis is the most important distinguishing factor, because immature spermatozoa with undeveloped structures and cellular signaling systems have diminished functional integrity and a decreased ability to interact with the female reproductive tract and the oocyte.

That the predictive value of sperm concentrations for male fertility is low becomes obvious when one considers the day-to-day variations in sperm concentrations. Although men with concentrations of <10 million sperm per/ml or >30 million sperm/ml semen tend to maintain their sperm concentrations within these ranges, others with concentrations of between 10 to 30 million sperm/ml do not. They have therefore been designated as variablespermic in a previous publication (Huszar et al. 1988b). Moreover, total sperm creatine kinase (CK) activity and the ratio of the CK-M and CK-B isoforms, which are objective biochemical markers of sperm maturity and fertility, indicate that 40% of men in the 20 to 30 million sperm/ml range group have a substantial subpopulation of immature spermatozoa with diminished fertility (Aitken et al. 1992; Huszar et al. 1988b; Lalwani et al. 1996). From this point of view, then, the standard World Health Organization (WHO) sperm concentration ranges for oligospermic men (<20 million sperm/ml) and normospermic men (>20 million sperm/ml) do not reflect male fertility.

Limitation of Sperm Function Studies

The literature contains numerous sperm function tests with correlations around r=0.6 with that of IVF fertilization rates, the gold standard (Consensus Workshop 1996). The most prominent parameter that is predictive for IVF and pregnancies is the concentration of motile sperm (Liu et al. 1991; Fetterolf, Rogers 1990). Other tests with a high predictability for fertility are acrosomal activation, sperm morphology and zona binding. It is not surprising that deficiencies in such sperm functions as sperm motility, acrosomal function and zona binding, all of which are essential for fertilization and pregnancy, would diminish the chances for fertility (Ammann 1989; DeJonge 1994). It is also expected that sperm motility, which controls the incidence of sperm-oocyte encounters, is a major contributor to fertility. However, from the point of view of reproductive physicians, the sperm function tests are not very useful. Since each individual sperm function is part of the sequence of functions necessary for fertilization, no individual function test is able to do what is most important in the clinical setting, which is to identify men with diminished fertility. For instance, assuming that a semen sample shows 40% motility, 15% acrosome reaction rate and 10% zona binding, it is not certain that the sperm subpopulation that is able to undergo acrosome reaction will also be motile or able to bind to the zona.

Other important issues to consider are the stimuli to which the sperm are exposed in testing of sperm function. Because of the redundancy of agents that activate sperm *in vivo*, these stimuli may be very diverse, yet the various agents used to test sperm *in vitro* do not elicit equivalent responses. For instance, one can challenge human sperm with follicular fluid, calcium and/or

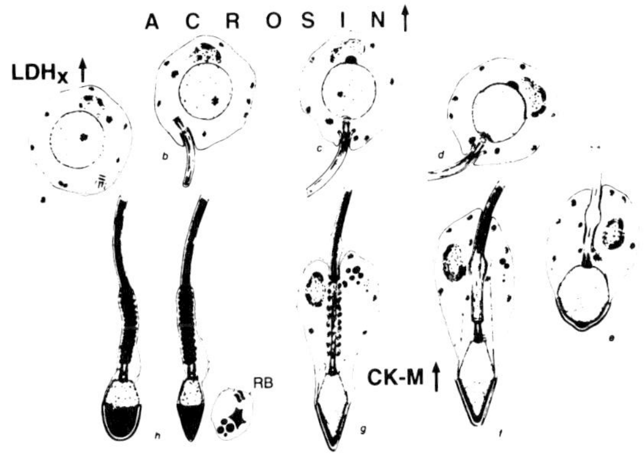

Figure 1. Germ cell development during spermiogenesis. The enzyme LDHx is normally expressed by the formation of the round spermatid, the first haploid male germ cell. During the formation of the elongated spermatid, the sperm tail sprouts, the acrosome is formed and the synthesis of acrosin commences. In terminal spermiogenesis, cytoplasmic extrusion, the synthesis of the CK-M isoform and remodeling of the sperm plasma membrane occur. The extruded cytoplasm remains in the adluminal area as residual bodies (RB).

progesterone to induce acrosome reaction. However, the effects of any or all of these agents would not be as powerful as that of the zona pellucida (Roldan et al. 1994). Thus, the *in vitro* tests may grossly underestimate the performance of sperm *in vivo*. Another aspect to consider is the type of the correlation between the results of sperm function tests and IVF success. For example, in the case of the zona-free hamster oocyte penetration assay, the correlation with IVF data is based mostly on negative events: spermatozoa that do not achieve penetration of zona-free oocytes also fail to fertilize in IVF conditions, and penetration of the hamster oocytes does not predict the IVF outcome. Thus, this test is not useful in the clinical setting (Kuzan et al. 1987).

Because results of sperm function tests show some correlation with fertility, but are of limited usefulness in identifying men with diminished fertility, in the past decade objective biochemical markers of sperm function and maturity, CK activity and CK-M isoform ratio, and the properties of sperm subpopulations in semen specimens have been studied (Huszar et al. 1988a; Huszar 1994). The rationale of this work is based on the fact that an objective parameter is easily measurable and that a biochemical marker indicative of sperm development would reflect the presence of the infrastructures that are necessary for sperm to achieve functional competence. Other goals of this work have been identifying men with unexplained male infertility, who have diminished fertility in spite of adequate motile sperm concentrations, and the assessment of fertilizing potential of sperm subpopulations in view of the various modes of assisted reproduction (Huszar et al. 1990a, 1992).

Sperm Maturity and Sperm Function

In order to discuss the relationship between sperm maturity and sperm function, the concept and work by the Huszar lab will be briefly summarized, focusing upon spermiogenetic maturation (Fig. 1). Once the development of the spermatid (the first haploid male germ cell) occurs, the final stage of spermatogenesis, spermiogenesis, ensues. During spermiogenetic development, the sperm tail sprouts, while the Golgi-apparatus rises to the opposite pole of the developing spermatid and becomes the acrosome. Finally, prior to the release of the sperm to the seminiferous tubuli, the excess cytoplasm, which is collected in the midpiece during spermiogenesis, is extruded and left in the adluminal area as the residual body (Clermont et al. 1963).

In initial studies on CK activity in ejaculated spermatozoa, it was observed that in oligospermic men, the sperm exhibit high CK activity (Huszar et al. 1988a). Through CK-immunocytochemistry of individual spermatozoa, it was demonstrated that the increased CK activity was related to retained cytoplasm in sperm (Huszar, Vigue 1993). It was also found that the degree of cytoplasmic retention was related to abnormal sperm morphology with respect to the sperm head perimeter, area roundness and the incidence of amorphous sperm forms in the specimen (Fig. 2, a-e). Based on the sperm patterns of Fig. 2, the hypothesis was made that increased cytoplasmic retention signified an interruption of spermiogenesis and diminished sperm maturity. Immature sperm are expected to have diminished functional integrity and thus diminished fertility. Subsequent clinical studies demonstrated that CK activity predicts the success of intrauterine insemination for couples in which the male partner is oligospermic (Huszar et al. 1990a). Indeed, in a group of men whose semen contained

Table 1: Semen characteristics and sperm CK activity of oligospermic husbands.

	Men (n)	Samples (n)	Motility (%)	Sperm Concentrations (10^6/ml) Semen	IUI fraction	CK Activities (CK IU 10^6 sperm/ml) Semen	IUI Fraction
Fertile	24	33	23.73 ± 1.7	11.9 ± 0.91	5.6 ± 1.04	0.61 ± 0.12	0.36 ± 0.08
Infertile	18	66	23.0 ± 1.3	11.9 ± 0.5	4.5 ± 0.43	1.29 ± 0.22	0.67 ± 0.09
Fertile vs. Infertile			p = 0.68	p = 0.32	p = 0.22	p = 0.02	p = 0.002

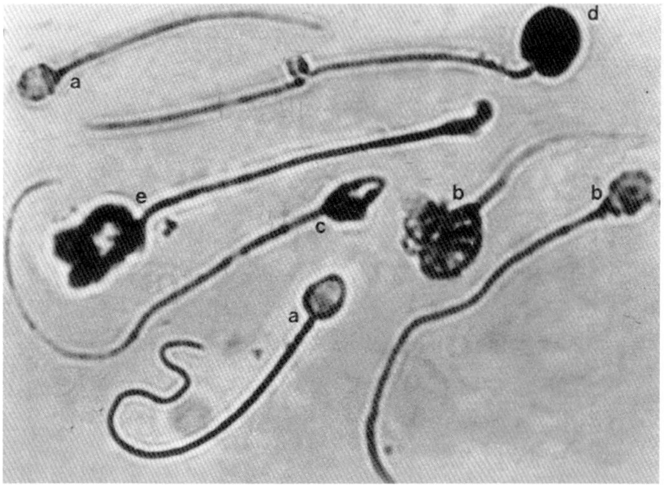

Figure 2. Montage of CK-immunostained sperm of various degrees of cytoplasmic retention. Note the relationship between cytoplasmic retention and abnormal sperm morphology.

5-20 million sperm/ml and who were characterized by similar sperm concentration and motility, the only distinguishing parameter between those who did or did not achieve pregnancy was the sperm CK activity (Table 1). In sperm that had completed cytoplasmic extrusion and had low CK activity, a new sperm CK isoform (in addition to the CK-B) with electrophoretic properties similar to those of the muscle CK-M isoform was found. The ratio of the CK-M and CK-B thus predicted the proportions of mature and immature sperm in the samples (Huszar, Vigue 1990). Three other laboratories have confirmed the value of biochemical measurements with respect to sperm CK, lactate dehydrogenase-X (LHD_X) and glucose-6-phosphate dehydrogenase (Aitken et al. 1994; Orlando et al. 1994; Sidhu et al. 1998).

In further work, the Huszar lab showed that the CK-M ratio [%CK-M/(CK-M+CK-B)] predicts fertility of men in IVF (Huszar et al. 1992). In a blinded study, 84 men were classified based on their CK-M ratios as prospectively fertile or prospectively infertile (Table 2) using 10% CK-M ratio as a cut-off value. The overall pregnancy rate in the 62 prospectively CK-M-fertile men was 24.3%. Among the 62 women with CK-fertile husbands, however, there were 15 cases in which oocyte fertilization was not achieved, indicating potential oocyte defects. When the pregnancy rates were recalculated, considering the couples with CK-fertile husbands and at least one oocyte fertilized, the pregnancy rate was 30.4%, an IVF pregnancy rate which was optimal at the time of the study in 1991. The importance of the sperm CK-M ratio was further supported by the finding that none of the 22 prospectively CK-infertile husbands caused pregnancies, and that nine of the 22 men with diminished sperm CK-M ratios and sperm maturity were normospermic. Thus, the CK-M ratios detected diminished male fertility and unexplained male infertility; that is, men with normal sperm concentration and motility but diminished fertility.

The reason for this predictive value was established in studies demonstrating that immature sperm with cytoplasmic retention failed to bind to the zona pellucida of the oocyte (Huszar et al. 1994). All sperm bound to the zona were the mature type, without cytoplasmic retention (Fig. 3). This suggested that, simultaneously with the spermiogenetic events of cytoplasmic extrusion and expression of the CK-M isoform, the sperm plasma membrane is also remodeled, thereby promoting the formation of the zona-binding site on the sperm plasma membrane. This hypothesis was proved by the study of β1,4,galactosyltransferase which is present exclusively in the plasma membrane of human sperm (Huszar et al. 1997). There was a very close correlation ($r>0.8$) between the density of β1,4,galactosyltransferase on the plasma membrane and the CK activity and CK-M isoform ratio within the cytoplasmic compartment in sperm fractions of various maturities. This demonstrated that a remodeling process does indeed occur within the sperm and on the plasma membrane simultaneously with the process of cytoplasmic extrusion during spermiogenesis and that the formation of the zona-binding site in sperm is part of this remodeling process.

Relation between Sperm Maturity and Morphology

Based on Figure 2, it is clear that the larger and asymmetrical shape of the postacrosomal region of the head and of the midpiece, which is detected by the

Table 2: Blinded study of IVF couples with fertile and diminished fertility husbands.

	CK-M Fertile Group (n=62 couples)	**CK-M Infertile Group** (n=22 couples)
Number of oocytes inseminated (per cycle)	4.9	6.2
Oocyte fertilization	53.4% (2.4/cycle)	14.2% (0.6/cycle)
Number of pregnancies	14	0
Overall pregnancy rate (PR)	16.7%	0%
PR in the CK-M fertile group (n=62)	22.6% (14/62)	
PR in the CK-M fertile group with oocyte fertilization (n=46)	30.4% (14/46)	

Tygerberg-Kruger "strict morphology" and by the "multiple anomalies index" methods, are reflections of cytoplasmic retention in immature sperm (Jouannet et al. 1988; Kruger et al. 1988; Menkveld et al. 1990). In this sense, sperm morphology is related to sperm maturity, although the interpretation of sperm morphology results in the clinical setting is controversial (Barratt et al. 1995; Morganthaler et al. 1995). In an ongoing blinded study, based on CK activity and CK-M isoform determinations and on strict sperm morphology measurements by the Hamilton-Thorne Dimensions program, which is based on the Tygerberg criteria, the Huszar lab found a correlation of about r=0.7 between results of strict sperm morphology and of either CK parameters. In some samples where the normal morphology was >15%, there were CK values indicating diminished maturity (Yamada et al. 1995). It is suggested that this occurs because there is a sperm subpopulation in which, in spite of normal morphology and apparently normal cytoplasmic extrusion, there is an arrest of sperm remodeling in terminal spermiogenesis or there is a small pocket of cytoplasm retained in the sperm which does not affect morphology, as in sperm "c," Figure 2.

A second related morphometrical marker of sperm maturity emerging from the spermiogenetic events illustrated by Figure 1 is the sperm tail length/head long axis ratio. It was hypothesized that incomplete spermiogenetic development will cause shorter sperm tail length and larger head size due to arrested tail development and increased cytoplasmic retention. Indeed, it was found that immature sperm, whether arising from the ejaculate, the epididymis or the testes, have shorter tails, and also that mature sperm have significantly higher tail length/head long axis ratios than immature sperm: the tail length/head long axis ratio of groups of mature and immature sperm was about 12.6 versus 8.9, respectively. The median ratios within the two groups were 7.6 and 5.2, dimensions that are easily recognized under the ICSI microscope by assessing how many sperm head lengths would fit along the length of the tail (Buradagunta et al. 1996; Gergely et al. 1999). These observations are important for the identification of mature sperm in sperm selection for assisted reproduction and especially for testicular sperm, where it is unclear whether the sperm originate from the seminiferous tubuli or from the adluminal area.

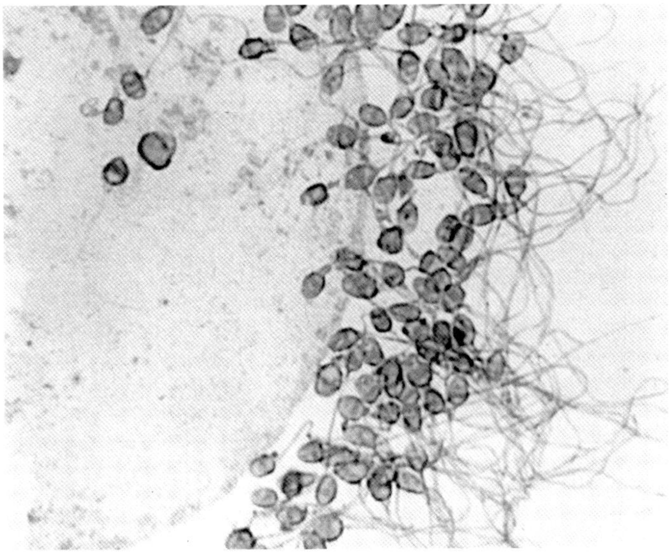

Figure 3. CK-immunostained sperm-hemizona complexes. Note that all sperm bound to the zona are the clear, mature type without cytoplasmic retention.

Characterization of Sperm Subpopulations for Treatment of Male Infertility

In this section, the role of semen analysis results in developing the clinical treatment plan of couples, the diversity of sperm populations, the benefits of sperm function tests in men with various types of male factor infertility, and approaches to the selection of the most appropriate mode of assisted reproduction (Fig. 4) will be considered.

Azoospermia

Examination of the semen begins with the volume. If the volume is <1.0 ml and no sperm are present in the ejaculate, the question of retrograde ejaculation arises, and it should be investigated either by administering drugs that enhance the proper ejaculatory response or by post-ejaculatory catheterization. If the volume is >1.0 ml and the ejaculate contains no sperm, then it is appropriate to measure fructose concentration, which indicates the presence of seminal vesicle excretions in the ejaculate. For men with fructose-negative semen and in whom an obstruction is likely, a urological workup with visualization and imaging techniques is indicated. In cases of epididymal obstruction, it is possible to recover the sperm by means of the microsurgical epididymal sperm aspiration (MESA) procedure and then to proceed with IVF or ICSI. However, since 80% of men with congenital absence of the vas deferens are carriers of the cystic fibrosis (CF) deletion, genetic counseling of the couple is warranted and testing for CF deletions in the wife is mandatory (Patrizio, Zielinski 1996; Pauer et al. 1997). Sperm function tests are not necessary because the MESA, which allows the recovery of limited numbers of motile sperm, predestines the couple for IVF or ICSI. In addition to enhancing the chance for pregnancy, IVF also has a diagnostic utility, because the binding or lack of binding of the sperm to the zona pellucida, the occurrence of fertilization and the development of the zygote to the two-cell stage and beyond may be directly observed.

If fructose is present in the semen, then the patienthas non-obstructive azoospermia. In these circumstances, no sperm function test is applicable. However, the maturity of the few sperm that may be isolated by high-speed centrifugation and used for the ICSI procedure can be studied by CK-immunocytochemistry or by a non-invasive method based on the visual assessment of the tail length/head long axis ratio. If the sperm shows a slight twitching motion, it is considered to be viable, and the likelihood for success by ICSI is higher.

The assumption that the sperm is viable when it displays motility, or even just tail movement, is generally true. Immotile sperm are not necessarily non-viable, however. Distinguishing between non-motile viable or non-viable sperm may be accomplished by means of the hypoosmotic swelling test (HOS) alone or combined with vital stains directed to the head, such as the Hoechst-33258 stain or the FertiLight kit, which is composed of a green and of a red component (Garner et al. 1994; Huszar et al. 1990b; Jeyendran et al. 1992). Because the green nuclear stain permeates the sperm membrane, the viable sperm will appear green. The red stain, propidium iodide, can only pass through sperm membranes with diminished integrity and hence stains non-viable sperm fluorescent red. Obviously, stained sperm cannot be used for assisted reproduction, but viable sperm that in the HOS show curled tails may be selected for ICSI. The HOS, which is used to test the viability of non-motile sperm, is helpful; however, the HOS may have a detrimental effect on the sperm membrane. Also, HOS fails to identify the sperm population that is viable as determined by tail curling, but show diminished membrane integrity and viability in the sperm head. When non-motile but viable sperm are present in semen, the option of ICSI is indicated as in cases of obstructive azoospermia, where non-motile sperm are recovered by MESA. If the HOS test indicates the absence of viable sperm, then testicular sperm extraction (TESE) may be carried out.

Asthenospermia: Sperm Motility in Various Subpopulations

Because sperm motility is essential for *in vivo* fertilization and IVF, the evaluation of asthenospermia is very important. Asthenospermia is defined as sperm motility of <50% (by WHO standards). In the case of men who have no motile sperm, but have viable sperm, the treatment option is ICSI. If the viability of ejaculated sperm is diminished, sperm must be recovered by TESE. In the case of asthenospermic men who have motile sperm in their ejaculate, the total motile sperm concentration must be considered. Somewhat lower motility is not cause for alarm, as long as there is an adequate motile sperm concentration. The key question, however, is whether the man has steady asthenospermia (that is, the motile sperm population remains motile), or if the sperm motility continuously declines. The latter case may be caused by epididymal factors, anti-sperm antibodies or by improper sperm development with respect to defects in sperm energy synthesis or utilization. The results of the sperm migration test, a method developed in the Yale laboratory and used in the daily management of patients, are very useful. The migration test facilitates the establishment of the type of asthenospermia and the sperm migration test may also serve as a sperm preparation method for intrauterine insemination (Makler et al. 1984).

The method used for the migration test is as follows (Fig. 5): A) The semen is diluted with a sperm preparation medium in a ratio of 1:2, and the semen-medium mixture is centrifuged in a tube which has a flat bottom. These tubes are prepared by adding 1.0 ml electron microscopy embedding medium to conical centrifuge tubes, and baking the tubes for 24 hours at 60°C. Following centrifugation of the semen mixture at 450 x g for 10 minutes, all except 1.0 ml of the supernatant is siphoned off. B) All the motile and non-motile sperm and the cellular components of the semen remain on the flat surface. C) The tube is placed in a vertical position into a 37°C incubator, and after 30 minutes, D) the upper part of the supernatant with the "migrated" sperm fraction is carefully removed.

Subsequently, a semen analysis is carried out on both the initial semen and on the migrated sperm fraction. The proportion of motile sperm recovered in the migrated sperm fraction provides the "motile sperm yield." Based on the distribution of the yield in 100 normal patients, a motile sperm yield of >30% has been established as the normal value. If >30% of the motile sperm are recovered, it indicates that the asthenospermia is of the steady type. If the migration test result is <30%, for instance 5% or 15%, this indicates that the asthenospermia is of the declining type, which does not bode well for the success of *in vivo* or *in vitro* fertilization. Sperm motility is determined by computer assisted semen analysis (CASA), and a curvilinear velocity of 8μm/sec is used as the cut-off value for motile sperm. Thus, sperm that twitches and show non-progressive motility are considered to be immotile. In the case of a man who is a candidate for ICSI, the rate of motility decline will also indicate the optimal timeframe between ejaculation and sperm selection for the ICSI procedure.

Motile Sperm Concentration and the Mode of Assisted Reproduction

Based on the motile sperm concentration in the ejaculate, husbands of couples with male factor or unexplained infertility may be divided into four groups. The first group, consisting of men with fewer than

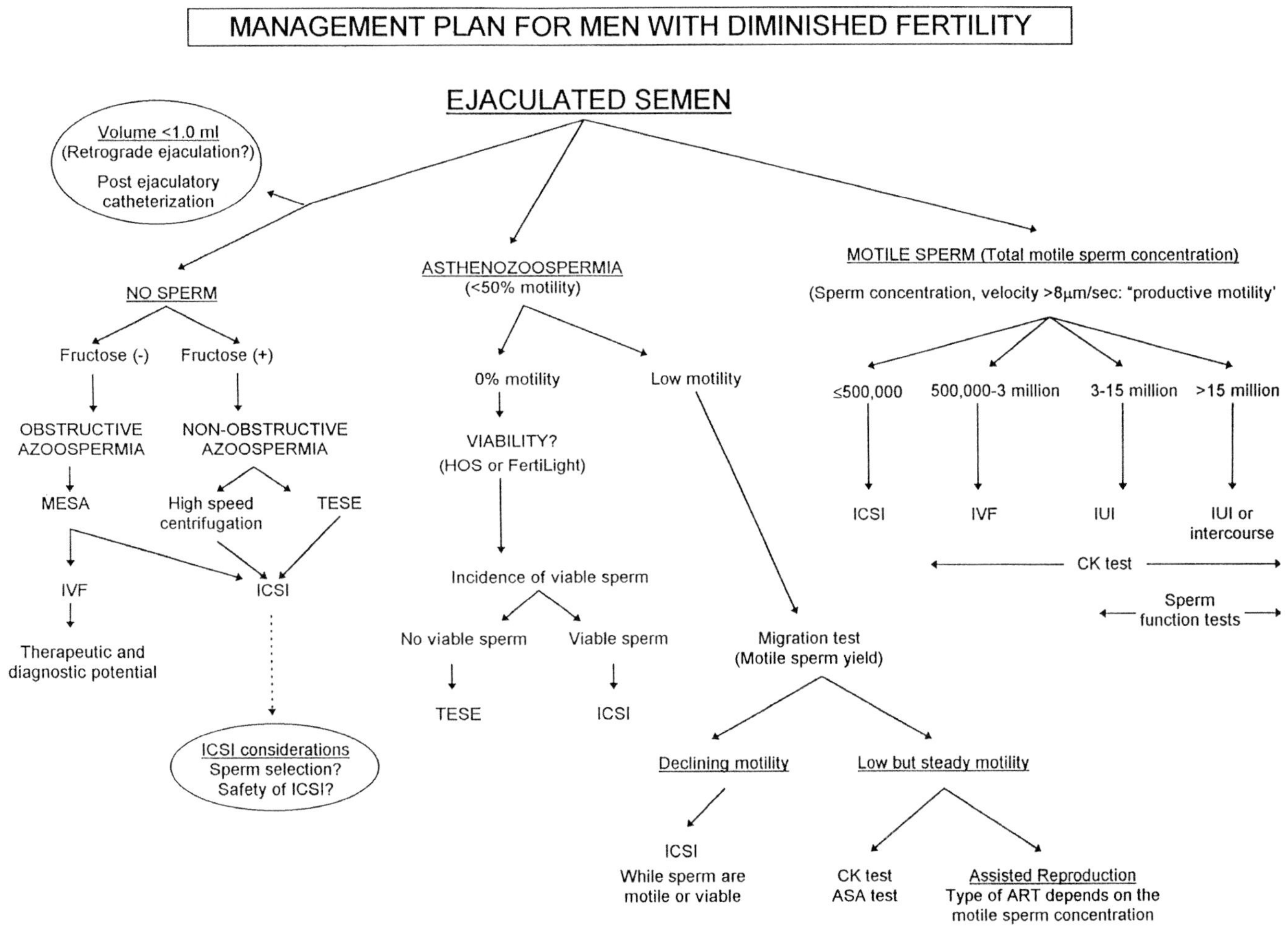

Figure 4. Management plan for men with reduced fertility.

500,000 motile sperm in the ejaculate, is usually restricted to the choice of ICSI. Sperm function tests are not appropriate for these men. For the second group of men, those with motile sperm concentrations between 500,000 to 3 million, IVF is the most efficient treatment. Here, the diagnostic value of IVF also comes into consideration. However, it is also a test of sperm function when the sperm faces the strongest stimulus, the zona pellucida. It must also be remembered that there is an unexplained female contributory factor related to deficiencies of the oocyte's ability to be fertilized, which may occur in about 10-15% of couples who require IVF.

As indicated earlier, about 40% of men whose sperm ranges from 20 to 30 million sperm/ml also have a substantial population of immature sperm, according to the objective biochemical parameters of sperm maturity, which include sperm lipid peroxidation, LDH_X activity, CK-activity and CK-M ratio parameters (Aitken et al. 1992; Alvarez et al. 1987; Huszar, Vigue 1994; Huszar et al. 1992; Lalwani et al. 1996; Orlando et al. 1994). Diminished sperm binding or no fertilization, if evident, can be attributed either to sperm immaturity and the lack of the zona-binding site or to oocyte failure. For this reason, a sperm CK-M isoform ratio test is performed. This test establishes the proportion of mature sperm in the specimen. In case of diminished sperm maturity, ICSI or ICSI and IVF is indicated. Due to the diagnostic benefits and an improved understanding of unexplained infertility, the reproductive physicians' earlier position, that IVF is indicated only as a "last resort" after unsuccessful intrauterine insemination (IUI) cycles, has

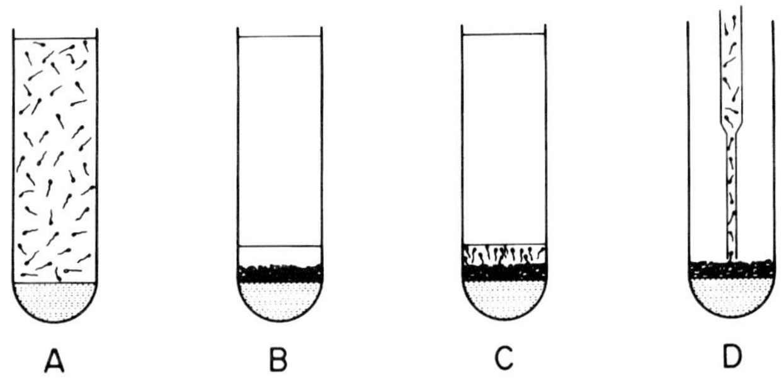

Figure 5: Steps of the sperm migration method.

increasingly given way to the conviction that early IVF is indicated in cases of unexplained infertility. The utility of the gamete intrafallopian transfer (GIFT) procedure is very low, because it does not provide diagnostic information in cases where pregnancy does not occur.

The other two groups are the oligospermic men, with motile sperm concentrations of about 3 to 15 million sperm/ml, and men of low normal sperm concentrations of 15 to 30 million sperm/ml. These men are best treated with intrauterine insemination and would benefit most from tests for anti-sperm antibodies and for sperm function. The CK-M isoform ratio determination, which provides the proportion of mature and immature sperm subpopulations within the specimen, is beneficial in identifying those who should choose between IVF, IUI and natural intercourse. For these men, testing for anti-sperm antibodies is also of interest. The immunobead method is most beneficial because it allows the identification of the type of anti-sperm antibodies (whether they are IgG, IgM or IgA), and the test can also determine whether the anti-sperm antibodies are directed to the head of the sperm. Such antibodies do not diminish sperm motility but may be detrimental from the point of view of fertilization. In the case of asthenospermic patients, it can be ascertained whether there are antibodies directed to the tail. The presence of anti-sperm antibodies in women is also tested by using their serum with the sperm of an antibody-free donor.

From the conceptual point of view, for men whose semen is positive for anti-sperm antibodies, it can be assumed that some sperm subpopulations in the specimen are free, while others carry the antibodies. Because anti-sperm antibodies cannot be removed from sperm, men affected by anti-sperm antibodies have a lower proportion of motile spermatozoa available for fertilization. Thus, for these men the modality of assisted reproduction that is most appropriate according to the motile sperm concentration should be modified by a more active management: when motile sperm concentrations are sufficient to cause natural conception, intrauterine insemination should be attempted, whereas IVF is now indicated for oligospermic samples that would normally be used for intrauterine insemination. Because the ratio of sperm and the titers of anti-sperm antibodies in the semen fluctuate from day to day, in some cycles more sperm are antibody-free, and the couple may achieve pregnancy by means which are in line with the man's actual motile sperm concentration. However, it may not be prudent to wait too long for this "self heal" once the couple seeks treatment.

The Potential Risks of ICSI

It is appropriate to include here a short discussion of the potential risks of ICSI for the offspring. With IVF, the sperm which would fertilize the oocyte must be mature in order to bind to the zona and engage in sperm-oocyte interaction. Indeed, immature sperm with cytoplasmic retention are deficient in zona binding. In the case of ICSI, however, the maturity of the sperm is not an issue. Because ICSI overrides the physiological sperm-zona selection process which only allows fertilization by mature sperm (Huszar et al. 1997), the selection of sperm is of critical importance. Sperm function tests are not applicable here because the test results of $10^6 - 10^9$ sperm would not reflect the properties of a single selected sperm. Fertilization by sperm injection may occur with immature sperm that have higher rates of DNA degradation and chromosomal aneuploidies, which may cause potentially adverse outcomes in the infant. Because most ICSI children are only three to six years old, there are no long-term data on the safety of ICSI. Certainly there are concerns about the four-times higher incidence of sex chromosome abnormalities, the congenital malformation rates, the mental development and the future health of ICSI children (Boundelle et al. 1998a, 1998b; Bowen et al. 1998; Kurinczuk, Bower 1997). For these reasons, IVF should be attempted if possible. Any couple considering ICSI should be informed about the potential risks of the procedure, and made aware that, as of now, very limited information is available on the development, cancer rate and other health aspects of ICSI children. To be prudent, those conducting ICSI should try to inject only those sperm that would likely have fertilized under conventional sperm-zona interaction if only the sperm had been more numerous or more motile. As a means for the selection of mature sperm for ICSI, current approaches focus upon the binding of sperm to substrates for which receptors on the plasma membrane are exclusively present in mature sperm (Huszar et al. 1998).

Sperm Function Tests and Assessment of Fertilizing Potential of Sperm Subpopulations

In spite of the limited ability of sperm function tests to identify men with diminished fertility, there are some tests that are useful in detecting selected deficiencies of fertilizing potential in mature sperm, which in the future can be treated with sperm function enhancement procedures or with early IVF.

The CASA technology may be important in evaluating men with oligospermia, although a more precise determination of the concentration and motility, which are subject to day-to-day changes in all men, may not provide a clinical advantage. Certainly, in addition to the improved precision of CASA, some kinematic parameters may be helpful. For instance, the Hamilton-Thorne instrument, which displays the proportion of a "rapid" sperm subpopulation, can be important when such sperm are absent in a sample. With the utilization of an indirect method, the Hamilton-Thorne instrument will also detect the incidence of hyperactivated sperm, which may offer a new potential for sperm function testing. The measurement of the hyperactivated sperm population itself is not likely to be as valuable as the development of challenge conditions in which one can study the proportion of a functionally active sperm population that is able to convert into the hyperactivated motility pattern (Mortimer, Swan 1995). Along with the measurement of capacitation and plasma membrane changes, to which the hyperactivated motility is likely to be related, such a test would be important because it is non-invasive, and the function of the sperm membrane receptors and of the second messenger pathways could be tested without diminishing the usefulness of the sperm for subsequent assisted reproduction procedures.

The presence of the sperm population with zona-binding ability may be demonstrated by the hemizona assay, which utilizes bisected, unfertilized human oocytes (Coddington et al. 1994). Evaluation of the sperm zona binding is based on a comparison with sperm of donors with proven fertility. There are three potential results. a) If the donor's sperm is able to bind to the zona but the husband's sperm is not, this indicates sperm immaturity and male factor infertility in the husband. b) If both the husband's and donor's sperm bind to the zona, the test is inconclusive but other types of sperm defects causing diminished male fertility may still be present in the husband. c) If neither the husband's nor the donor's sperm bind to the zona, this indicates a female factor infertility related to inadequate development of the zona.

The hemizona assay is informative in the case of IVF with fertilization failure and/or inadequate sperm binding. As discussed above, it has been shown that during spermiogenesis, maturation-related changes take place within the cytoplasmic compartment of the developing sperm while a remodeling process occurs in its plasma membrane, which also includes the formation of the zona binding site or sites (Huszar et al. 1997). Using sperm-hemizona complexes immunostained for CK, it has been found that immature sperm with cytoplasmic retention failed to bind to the zona pellucida (Huszar et al. 1994; Fig. 3). For this reason, the measurements of the sperm CK-M can predict the binding of the sperm to the zona.

Assuming adequate sperm concentration and motility, the sperm functions that may limit zona interaction are the capacitation and acrosome reaction. The capacitation process is poorly defined. It most likely represents demasking changes of the sperm acrosomal membrane and the activation of the signaling system between the surface and interior cell compartments of sperm which improves the response by sperm to the female reproductive tract and the zona. *In vivo*, the sperm is probably capacitated in the fallopian tubes. The concept is based on studies of the bovine fallopian tube, in which, along with ovulation and increased progesterone levels, increased high density lipoprotein (HDL) concentrations are also present, particularly at the utero-tubal junction, which has the optimal blood supply. The hypothesis indicates that the increased HDL would cause the removal of cholesterol and subsequent capacitation, facilitating the acrosome reaction in sperm upon binding to the zona pellucida (Cross 1996; Ehrenwald et al. 1990).

The capacitation rate of sperm *in vitro* may be detected by probes that reflect the integrity of the cellular signaling system and cholesterol content of spermatozoa. Such probes include chlortetracycline and the expression of the mannose-receptor which depend on plasma membrane changes, and fluorescent lectin and antibody markers directed to the acrosomal and post-acrosomal membranes, which indicate whether the sperm have acrosome reacted (Benoff et al. 1993; Cross et al. 1987; Das Gupta et al. 1993; Moore et al. 1987; Moutaffian et al. 1995). These studies can be readily carried out *in vitro* in media conditions with controlled concentrations of calcium, progesterone or other agents. The capacitation and acrosome reaction rates attained under these conditions reflect the relative activability of the sperm populations in *in vitro* conditions, but they do not indicate how the sperm would behave *in vivo* when exposed to the powerful stimulus of the zona pellucida. The results should be interpreted on a relative scale because they apply only to the sample tested, but there are no "standard" rates. In addition, even if the assays give low scores, the underlying dysfunction causing the lack of acrosomal response is not apparent. A sluggish response *in vitro* may not necessarily mean that a particular sperm population will not readily undergo acrosome reaction when facing the zona pellucida.

With respect to assays of acrosome reaction and zona binding, it is important to understand the limitations of methods that are presently used. The gross measurement of acrosin activity has a limited usefulness because it is an average value for millions of spermatozoa that individually contain vastly different amounts of acrosin. In addition, acrosin activity does not reflect the responsiveness of the acrosomal system to biological stimuli. As opposed to sperm activation with physiological agents, the ARIC test utilizing a calcium ionophore-mediated acrosome reaction is not fully valuable in the physiological sense, because the calcium ionophore may bring about acrosome reaction without utilizing the pathways that must be operational for acrosome reaction *in vivo* (Yovich et al. 1994). Also, in addition to the sperm developmental defects, the acrosomal function may be diminished by iatrogenic agents such as calcium channel blockers or environmental toxins such as pesticides (Benoff et al. 1994; Turner et al. 1997).

The zona-free hamster oocyte penetration assay is over 20 years old. While the data indicate that it is a fairly good measure of acrosome reaction, a close and consistent relationship between sperm penetration assay (SPA) and the fusion of the sperm plasma membrane and the oolemma that has been implied by some investigators has not been substantiated. Several papers have reported correlations between the SPA and IVF results, but they are based mostly on the association between SPA failures and lack of fertilization and pregnancies in IVF. There is no solid evidence supporting a positive correlation between SPA success and occurrence of IVF pregnancies (Kuzan et al. 1987; Rogers et al. 1983; Yanagimachi et al. 1976).

The Role of the Female Factor in Male Infertility

The development of objective biochemical markers of sperm function and fertility has facilitated a more specific classification of couples with unexplained infertility by establishing whether a male factor component is present. This is particularly important in the case of couples in which the husband is oligospermic and the wife's ovulatory pattern may receive little attention. In the Yale clinic, each husband is evaluated with a semen analysis, a migration test and the CK test, which together provide an overall indication of the sperm concentration, motility, motile sperm yield, maintenance of motility, and sperm maturity. This yields a good assessment of male fertilizing efficiency. This information also allows a prospective determination of the optimal mode of assisted reproduction. At least two semen analyses are done, two to three weeks apart, and if the CK-M ratio data indicate diminished maturity, that test is repeated in order to confirm the data. For the female partner, ovulation monitoring is initiated with a luteinizing hormone (LH) kit and also progesterone level determination following eight to 10 days after the LH surge. This appears to be a more sensitive measurement of the adequacy of ovulation than the endometrial biopsy. The patency of the tubes is also evaluated, and, if warranted, the shape of the uterine cavity is assessed by hysteroscopy. If the woman has a history of pelvic surgery, inflammation, or endometriosis, laparoscopy is also carried out. Age is also a very important factor, particularly if the woman is over 35 years of age.

Often, the price for sperm maturity and function tests, for example, the CK test, becomes an issue with insurers, and patients are unfairly shifted to clinical laboratories that do not specialize in andrology testing and perform only the old-fashioned manual sperm concentration and motility measurements. Of course, the usefulness of such semen analyses is limited, and the data do not address male factor infertility. In fact, an adequate andrological workup by an expert laboratory is very cost-efficient because the documentation of male factor may substantially reduce the cost of the female factor workup. For this reason it is very unfortunate that now, when science is developing a better understanding of sperm cell biology and would be able to use more specific sperm function tests, the health insurance providers fail to recognize and provide reimbursement for these tests. Table 3 lists the costs of workups in the male and female partners at comparable levels of complexity. It is clear that even at the most sophisticated level of andrology testing, including acrosomal activation, and future tests of zona binding, utilizing genetically engineered zona-protein and hyperactivation challenge assay, the full male workup is a fraction of the cost of that of the female workup. Since 40% of infertile couples have a male factor component and in about 18% of the couples both the wife and the husband are affected, undertaking a thorough andrology testing of the male partner would result in a more effective and cost-efficient workup for about 30% of all infertile couples.

Another point worth emphasizing is that the couple should be seen by an andrologist who is also an infertility physician or by an experienced andrologist-physician team (either an obstetrician or a urologist), rather than being treated, as in the past, separately by a urologist and the woman's obstetrician/gynecologist.

It is imperative that insurers be educated regarding these issues, because if the male factor is identified, then the female workup may be streamlined and the diagnostic and treatment phases of care can be made more cost-efficient. In addition, in oligospermic men who have adequate sperm motility, urological treatment such as varicocelectomy or drug treatment to enhance sperm concentration may be omitted, and the couple can proceed directly to intrauterine insemination or IVF. Conversely, if insufficient sperm maturity is documented in oligospermic or normospermic men with adequate sperm motility, the circumstances of the female will modulate how soon the couple will be treated with IVF or ICSI. In spite of a blinded study, which indicated that men with <10% CK-M ratio did not achieve pregnancy in IVF (Huszar et al. 1992), a combination of IVF and ICSI is worth a try if there are more than four to five oocytes available because of the potential risks of ICSI for the offspring. The IVF embryos, if any develop, may be preferentially transferred and the others cryopreserved.

Conclusions

An outline of structured management of male infertility, based on the classical parameters of sperm concentration and motility and on the recognition of

Table 3: Cost of workup (in U.S. Dollars).

Male		Female	
FSH	42	FSH	42
LH	32	LH	32
Semen analysis	105	Estradiol	46
Migration test	95	Progesterone	70
Sperm CK test	135	Prolactin	85
Anti-sperm antibody testing (male/female)	160	Hysterosalpingogram	580*
Diagnostic testicular biopsy	700*	Diagnostic laparoscopy	5500*

*Including facility fee, anesthesia fee, operating room fee and fees of surgeon, radiologist, anesthesiologist and pathologist.

differences in motility and fertilizing potential of sperm populations within a single specimen, has been presented. Whereas the study of sperm functions such as motility, morphology, capacitation, acrosomal activation and zona binding may find isolated deficiencies affecting sperm fertility, the most important determinant of sperm fertility in conventional fertilization is the completion of the sperm maturation process during spermiogenesis. This is because, in the absence of the remodeling step within the sperm cytoplasmic compartment and plasma membrane that occurs during spermiogenesis, the sperm fail to develop the oocyte recognition sites and their zona-binding ability is deficient. The increasing recognition of these new elements of human sperm cell biology, and the clinical validation of the objective biochemical markers in detecting fertility and the lack of fertility, strongly support the utility of the CK-M isoform determinations which measure the proportion of spermatozoa which have the structural elements necessary for normal sperm function. Now that the specific antibody to CK-M is available, the test will be based on an immunoassay. However, the assessment of sperm maturity is only the laboratory basis for identifying men with diminished fertility, independently from their sperm concentrations. It is important that infertile couples are thoroughly evaluated for both male and female factors, including the possibilities that there are diminished oocyte maturity and/or that mature sperm may have isolated functional defects.

Acknowledgement

This research was supported by the National Institute of Health (HD-19505, HD-32902).

References

Aitken RJ, Buckingham DW, West KM, Wu FC, Zikopoulos K, Richardson DW. On the contribution of leukocytes and spermatozoa to the high levels of reactive oxygen species recorded in the ejaculates of oligozoospermic patients. J Reprod Fertil 1992; 94:451-462.

Aitken RJ, Krausz C, Buckingham D. Relationship between biochemical markers for residual sperm cytoplasm, reactive oxygen species generation, and the presence of leukocytes and precursor germ cells in human sperm suspensions. Mol Reprod Dev 1994; 39:268-279.

Alvarez JG, Touchstone JC, Blasco L, Storey B. Spontaneous lipid peroxidation and production of hydrogen peroxide and superoxide in human spermatozoa: superoxide dismutase as major enzyme protectant against oxygen toxicity. J Androl 1987; 8:338-348.

Amman RP. Can the fertility potential of a seminal sample be predicted accurately? J Androl 1989; 10:89-98.

Barratt CLR, Naeeni M, Clements S, Cooke LD. Clinical value of sperm morphology for *in vivo* fertility: comparison between World Health Organization criteria of 1987 and 1992. Hum Reprod 1995; 10:587-593.

Benoff S, Cooper GW, Hurley I, Napolitano B, Rosenfeld DL, Scholl GM, Hershlag A. Human sperm fertilizing potential *in vitro* is correlated with differential expression of a head-specific mannose-ligand receptor. Fertil Steril 1993; 59:854-862.

Benoff S, Cooper GW, Hurley I, Mandel FS, Rosenfeld DL, Scholl GN, Gilbert BR, Hershlag A. The effect of calcium ion channel blockers on sperm fertilization potential. Fertil Steril 1994; 62:606-617.

Boundelle M, Joris H, Hofmans K, Liebaers I, Van Steirteghem A. Mental Development of 210 ICSI children at 2 years of age. Lancet 1998a; 351:1553-1556.

Boundelle M, Aytoz A, Wilikens A, Buysee A, Van Assche E, Devroy P, Van Steirteghem A, Liebaers I. Prospective follow-up study of 1,987 children born after intracytoplasmic sperm injection. In: (Filicori M, ed) Treatment of Infertility: The New Frontiers. Princeton, NJ: Communication Media for Education Inc, 1998b; pp425-439.

Bowen JB, Gibson FL, Garth IL, Saunders DM. Medical and developmental outcome at 1 year for children conceived by intracytoplasmic sperm injection. Lancet 1998; 351:1529-1534.

Buradagunta, S, Honig, SC, Patrizio, P, Vigue LC, Keefe DL, Huszar GB. The role of morphometrical features in sperm selection from testicular tissue for ICSI. Am Soc Reprod Med, 52nd Ann Meet, Boston, MA, 1996.

Clermont, Y. The cycle of the seminiferous epithelium in man. Am J Anat 1963; 112:35-51.

Coddington CC, Oehninger SC, Olive DL, Franken DR, Kruger TF, Hodgen GD. Hemi-zona index demonstrates excellent predictability when evaluating sperm fertilizing capacity in *in vitro* fertilization patients. J Androl 1994; 15:250-254.

Consensus Workshop on Advanced Diagnostic Andrology Techniques. Hum Reprod 1996; 11:1463-1479.

Cross N. Effect of cholesterol and other sterols on human sperm acrosomal responsiveness. Mol Reprod Dev 1996; 45:212-217.

Cross NL, Morales P, Overstreet JW, Hanson FW. Two simple methods for detecting acrosome-reacted human sperm. Gam Res 1987; 15:213-226.

Das Gupta S, Mills CL, Fraser LR. Ca^{2+}-related changes in the capacitation state of human spermatozoa assessed by a chlortetracycline fluorescence assay. J Reprod Fertil 1993; 99:135-143.

De Jonge CJ. Diagnostic significance of the induced acrosome reaction. Reprod Med Rev 1994; 3:159-178.

Fetterolf PM, Rogers BJ. Prediction of human sperm penetrating ability using computerized motion parameter. Mol Reprod Dev 1990; 27:326-331.

Garner DL, Johnson LA, Yue ST, Roth BL, Haugland RP. Dual DNA staining assessment of bovine sperm viability using SYBR-14 and propidium iodide. J Androl 1994; 15:620-629.

Gergely A, Kovanci E, Senturk L, Cosmi E, Vigue L, Huszar G. Morphometrical assessment of mature and diminished maturity human spermatozoa: sperm regions that reflect in maturity. Hum Reprod 1999; in press.

Ehrenwald E, Foote RH, Parks JE. Bovine oviductal fluid components and their potential role in sperm cholesterol efflux. Mol Reprod Dev 1990; 25:195-204.

Huszar G. The role of sperm creatine kinase in the assessment of male fertility. Reprod Med Rev 1994; 3:179-187.

Huszar G, Vigue L. Spermatogenesis related change in the synthesis of the creatine kinase B-type and M-type isoforms in human spermatozoa. Mol Reprod Dev 1990; 25:258-262.

Huszar G, Vigue L. Incomplete development of human spermatozoa is associated with increased creatine phosphokinase concentrations and abnormal head morphology. Mol Reprod Dev 1993; 34:292-298.

Huszar G, Vigue L. Correlation between the rate of lipid peroxidation and cellular maturity as measured by creatine kinase activity in human spermatozoa. J Androl 1994; 15:71-77.

Huszar G, Corrales M, Vigue L. Correlation between sperm creatine phosphokinase activity and sperm concentrations in normospermic and oligospermic men. Gam Res 1988a; 19:67-75.

Huszar G, Vigue L, Corrales M. Sperm creatine phosphokinase activity as a measure of sperm quality in normospermic, variablespermic and oligospermic men. Biol Reprod 1988b; 38:1061-1066.

Huszar G, Vigue L, Corrales M. Sperm creatine kinase activity in fertile and infertile oligospermic men. J Androl 1990a; 11:40-46.

Huszar G, Willetts M, Corrales M. Hyaluronic acid (Sperm Select) improves sperm motility and viability in normospermic and oligospermic specimens. Fertil Steril 1990b; 54:1127-1134.

Huszar G, Vigue L, Morshedi M. Sperm creatine phosphokinase M-isoform ratios and fertilizing potential of men: a blinded study of 84 couples treated with *in vitro* fertilization. Fertil Steril 1992; 57:882-888.

Huszar G, Vigue L, Oehninger S. Creatine kinase immunocytochemistry of human hemizona-sperm complexes: selective binding of sperm with mature creatine kinase-staining pattern. Fertil Steril 1994; 61:136-142.

Huszar G, Sbracia M, Vigue L, Miller D, Shur B. Sperm plasma membrane remodeling during spermiogenetic maturation in men: relationship among plasma membrane β-1,4,-galactosyltransferase, cytoplasmic creatine phophokinase, and creatine phosophokinase isoform ratios. Biol Reprod 1997; 56:1020-1024.

Huszar GB, Gordon EL, Irvine DS, Aitken RJ. Absence of DNA cleavage in mature human sperm selected by their surface membrane reeceptors. Am Soc Reprod Med, Ann Meet, San Francisco, CA, 1998.

Jeyendran RS, Van der Ven HH, Zaneveld LJ. The hypoosmotic swelling test: an update. Arch Androl 1992; 29:105-116.

Jouannet P, Ducot B, Feneux D, Spira A. Male factors and the likelihood of pregnancy in infertile couples. I. Study of sperm characteristics. Int J Androl 1988; 11:379-394.

Kurinczuk JJ, Bower C. Birth defects conceived by intracytoplasmic sperm injection: an alternative interpretation. Br Med J 1997; 718:1260-1266.

Kuzan FB, Muller CH, Zarutskie PW, Dixon LL, Soules MR. Human sperm penetration assay as an indicator of sperm function in human *in vitro* fertilization. Fertil Steril 1987; 48:282-286.

Kruger TF, Acosta AA, Simmons KF, Swanson RJ, Matta JF, Oehninger S. Predictive value of abnormal sperm morphology in *in vitro* fertilization. Fertil Steril 1988; 49:112-117.

Lalwani S, Sayme N, Vigue L, Huszar G. Biochemical markers of early and late spermatogenesis: relationship between LDH_x and CK-M isoform concentrations in human sperm. Mol Reprod Dev 1996; 43:495-502.

Liu DY, Clarke GN, Baker HW. Relationship between sperm motility assessed with the Hamilton-Thorne motility analyzer and fertilization rates *in vitro*. J Androl 1991; 12:231-239.

Makler A, Murrillo O, Huszar G, DeCherney AH. Improved techniques for collecting motile spermatozoa from human semen. I. A self-migratory method. Int J Androl 1984; 7:61-70.

Menkveld R, Stander FSH, Kotze TJ vW, Kruger TF, Van Zyl JA. The evaluation of morphological characteristics of human spermatozoa according to stricter criteria. Hum Reprod 1990; 5:586-592.

Moore HD, Smith CA, Hartman TD, Bye AP. Visualization and characterization of the acrosome reaction of human spermatozoa by immunolocalization with monoclonal antibody. Gam Res 1987; 17:245-249.

Morgenthaler A, Fung MY, Harris DH, Powers RD, Alper MM. Sperm morphology and *in vitro* fertilization outcome: a direct comparison of World Health Organization and strict criteria methodologies. Fertil Steril 1995; 64:177-178.

Mortimer ST, Swan MA. Variable kinematics of capacitating human spermatozoa. Hum Reprod 1995; 10:3178-3182.

Moutaffian H, Perinaud J. Selection and characterization of human acrosome reacted spermatozoa. Hum Reprod 1995; 10:2948-2951.

Orlando C, Krausz C, Forti G, Casano R. Simultaneous measurement of sperm LDH, LDH_x, CPK activities and ATP content in normospermic and oligospermic men. Int J Androl 1994; 17:13-18.

Palermo GD, Cohen J, Alikani M, Adler A, Rosenwaks Z. Intracytoplasmic sperm injection: a novel treatment for all forms of male factor infertility. Fertil Steril 1993; 63:1231-1240.

Patrizio P, Zielenski J. Congenital absence of the vas deferens: a mild form of cystic fibrosis. Molec Med Today 1996; 7:24-31.

Pauer HU, Hinney B, Michelmann HW, Krasemann EW, Zoll B, Engel W. Relevance of genetic counseling in couples prior to intracytoplasmic sperm injection. Hum Reprod 1997; 12:1909-1912.

Rogers BJ, Perreault SD, Bentwood BJ, McCarville C, Hale RW, Solderdahl DW. Variability in the human-hamster *in vitro* assay for fertility evaluation. Fertil Steril 1983; 39:204-211.

Roldan ER, Murase T, Shi QX. Exocytosis in spermatozoa in response to progesterone and zona pellucida. Science 1994; 266:1578-1581.

Sidhu RS, Sharma RK, Agarwal A. Relationship between creatine kinase activity and semen characteristics in subfertile men. Intern J Fertil 1998; 43:192-197.

Turner KO, Syvanen M, Meizel S. The human reaction is highly sensitive to inhibition by cyclodiene insecticides. J Androl 1997; 18:571-575.

Yamada Y, Vigue L, Huszar G. Sperm creatine kinase parameters and strict sperm morphology in men: relationship between the biochemical and morphological measures of sperm maturity and fertilizing potential. Am Soc Reprod Med, 51st Ann Meet, Seattle, WA, 1995.

Yanagimachi R, Yanagimachi H, Rogers BJ. The use of zona free animal ova as a test system for the assessment of the fertilizing capacity of human spermatozoa. Biol Reprod 1976; 15:542-551.

Yovich JM, Edirisinghe WR, Yovich JL. Use of the acrosome reaction to ionophore challenge test in managing patients in an assisted reproduction program: a prospective, double-blind randomized controlled study. Fertil Steril 1994; 61:902-910.

36 Genetic Testing of the Male

Christopher L.R. Barratt

Justin C. St. John

Masoud Afnan

The University of Birmingham

Two important developments have led to a dramatic increase in the awareness of the genetic components of male infertility and subsequent consequences of using assisted reproductive techniques. Firstly, there is clear evidence to suggest that intracytoplasmic sperm injection (ICSI) results in a significant number of genetic abnormalities, such as a higher incidence of sex chromosome abnormalities (Bonduelle, et al. 1998a) and an increased incidence of Y chromosome deletions in male children (Kent-First et al. 1996). In addition, there are conflicting reports about a higher incidence of congenital abnormalities (Kurinczuk, Bower 1997) and a possible decrease in mental development of the children born as a result of ICSI (Bowen et al. 1998; Bonduelle et al. 1998b, editorial: te Velde et al. 1998).

Secondly, there have been breathtaking advances in molecular biology, including genetic diagnosis, which have transformed understanding of male reproductive function. Knockout mouse experiments, for example, have challenged the role of the follicle-stimulating hormone (FSH), as found in Kumar et al. (1997), and acrosin (Baba et al. 1994) in reproduction.

From this, the obvious question arises: what is the diagnostic and prognostic value of this information? To address this question, the following will be considered: chromosomal analysis, molecular screening, genetic counseling and future molecular/genetic tests. In conclusion, a suggestion for minimal screening is offered.

Chromosome Analysis

There is considerable interest in the etiology, frequency and reproductive consequences of chromosome abnormalities in infertile men. For example, a recent study documented a significant increase in chromosome abnormalities following ICSI. In this study of 1082 prenatal tests, 18 (1.66%) *de novo* aberrations were detected. Nine (0.83%) were sex chromosome aberrations and nine (0.83%) were autosomal aberrations (Bonduelle et al. 1998a). This highlights the critical need to screen for chromosomal aberrations prior to ICSI.

A number of studies have documented the incidence of chromosomal abnormalities in many different infertile male groups, such as those selected for ICSI (van der Ven et al. 1998; Montag et al. 1997; Meschede et al. 1998; Tuerlings et al. 1998a; see Table 1 for summary) and those attending an infertility clinic (Chandley 1997; Pandiyan, Jequier 1996; Micic et al. 1984). The incidence of chromosomal abnormalities in these groups of men varies depending on sample size, diagnosis of infertility, length of infertility, methodological technique, and so on. In practically all studies to date, an increase in the severity of male factor infertility (usually defined by sperm concentration; see studies cited above) is accompanied by a higher incidence of chromosomal abnormality.

For example, Micic et al (1984) reported 39/356 of the azoospermic men (11%) compared to 17/464 of the oligospermic men (3.6%) had a chromosome abnormality. The higher incidence of chromosomal abnormality among men from infertile couples has led to the strong suggestion that all men attending an infertility clinic, particularly those referred for *in vitro* fertilization (IVF) and ICSI, should have a karyotype analysis performed (Montag et al. 1997; van der Ven et al. 1998; Meschede et al. 1998; Tuerlings et al. 1998a).

The reproductive consequences of an abnormal male karyotype vary considerably and cannot always be predicted, particularly with ICSI where an abnormal female karyotype may also be of concern (for example, Meschede et al. 1998; van der Ven et al. 1998). As described by Meschede and colleagues (1998), chromosomal abnormalities can be categorized into several groups which suggest low, moderate or high risk levels of giving rise to an aneuploid pregnancy. The most serious adverse consequence of a chromosomal abnormality may not be a spontaneous pregnancy loss but the birth of a handicapped aneuploid child; in such cases, reciprocal translocations carry a high risk (Meschede et al. 1998). These risk analyses are largely based on studies of natural pregnancies (Gardner, Sutherland 1996). However, because of the lack of selection during ICSI, such risk estimates may not be correct and, in fact, the risks following ICSI may be higher. Due to the limited number of prospective studies determining the consequences to the resulting child when sperm from men with abnormal karyotypes are used for ICSI, it is critically important to extend studies and continue to collect such data. At present, there simply is not enough biological and clinical information about chromosome abnormalities available to provide clear-cut answers.

Table 1: Cytogenetic studies of male patients enrolled in IVF or ICSI programs.

	Treatment planned /performed	# of men	# of abnormal karyotypes (%)
Hens et al. (1988)	IVF, GIFT, ZIFT	500	4 (0.8)
Lange et al. (1993)	IVF	72	2 (2.8)
Baschat et al. (1996)	ICSI	32	2 (6.2)
Mau et al. (1997)	ICSI	150	18 (12.0)
Van de Ven et al. (1997)	ICSI	158	6 (2.6)
Testart et al. (1996)	ICSI	261	11 (4.2)
Meschede et al. (1998)	ICSI	432	9 (2.1)
Tuerlings et al. (1998a)	ICSI	1792	72 (4.0)
Montag et al. (1997)	ICSI	434	8 (1.9)*
van der Ver et al. (1998)	ICSI	305	10 (3.3)

Excludes single-cell aberrations.
Table adapted from Meschede et al. (1998).

Chromosome Abnormalities in Spermatozoa: Use of Fluorescent *in situ* Hybridization (FISH)

With the recent advances in FISH, such as using triple color FISH (Downie et al. 1997), numerous studies document the incidence of chromosome abnormalities (mainly aneuploidy) in spermatozoa from both individual cases where the man has an "interesting karyotype" (Colls et al. 1998; Mercier et al. 1998; Newberg et al. 1998; In't Veld et al. 1997) or a variety of patient and control (fertile) groups (Egozcue et al. 1997; Downie et al. 1997). Not surprisingly, men with severe male factor infertility (oligoasthenoteratozoospermia) have a significantly higher number of spermatozoa carrying chromosome abnormalities (Bernardini et al. 1997).

Bearing in mind the demonstrated increase in sex chromosome abnormalities in prenatal diagnosis following ICSI (Boundelle et al. 1998a), there has been particular interest in the frequency of sex chromosome abnormalities in infertile men. It may not be possible to predict with accuracy the frequency of abnormal chromosomes in spermatozoa from examining peripheral blood. For example, a small number of men with sex chromosome abnormalities may show mosaicism. Such men will have varying degrees of spermatogenesis. Recently, Kruse and colleagues (1998) described an XXY/XXXY/XY mosaic Klinefelter patient in whom 94% of the peripheral blood lymphocytes had sex chromosome aberrations, yet only 7.5% of the sperm had an extra sex chromosome.

These results suggest that mosaic Klinefelter patients who plan to undergo ICSI may benefit from analysis of the sex chromosomes in their spermatozoa. Increasingly, it is apparent that FISH on spermatozoa will allow a more accurate analysis of the chances of transmission of chromosome abnormalities.

Very little is known about chromosome aberrations in the spermatozoa of men with various forms of cancer; likewise, little is known about the effects of chemo/radiotherapy or radiotherapy regimens. Robbins and colleagues performed a FISH study on the spermatozoa of eight men with Hodgkin's disease (HD) before, during and after treatment (Robbins et al. 1997).

There was a five-fold increase in spermatozoa with disomies, diploidies and complex genotypes involving chromosomes X,Y and 8 during treatment. Reassuringly, the effects were transient, declining to pretreatment levels. However, it was of concern that some HD patients had higher proportions of certain sperm aneuploidy types before their first therapy. These results need confirming in a larger series of men with different forms of cancer before definitive conclusions can be drawn.

Molecular Screening

Detection of Deletions on the Y Chromosome

A large number of studies are specifically examining the incidence of Y chromosome deletions in men prior to ICSI (Pryor et al. 1997; Roberts 1998; Cooke et al. 1998). These microdeletions can be detected through sequence-tagged site-polymerase chain reaction (STS-PCR), as described in Kobayashi et al. (1994), Nakahori et al. (1994), Reijo et al. (1995) and Stuppia et al. (1998). This technique, which uses deletion mapping and screening of DNA through multi-primer pairs, was also employed to clone two transcription factors, RBM (RNA binding motif), found in Ma et al. (1993), and DAZ (Deleted in Azoospermia), found in Reijo et al. (1995). STS-PCR allows the assessment of the potential fertility of the offspring and may provide critical information to the patient about the choice of treatment, such as ICSI versus the use of donor gametes (Barratt, St. John 1998).

The incidence of Y chromosome deletions depends on many factors, such as the techniques used to categorize deletions (and number of STS) and patient selection, and so on. It is therefore very difficult to compare studies to obtain an overall frequency. A recent study by Pryor et al. (1997), using PCR analysis, examined the frequency of Y chromosome deletions in 200 unselected men attending an infertility clinic. In total, 7 percent of men harbored microdeletions while 23 percent of the azoospermic men had deletions. There was no obvious correlation between a deletion interval and the degree of spermatogenesis.

Interestingly, 11 of the 12 men with Y chromosome microdeletions had a normal karyotype, strongly suggesting the necessity for PCR analysis. Many deletions detected in interval 6 were outside the DAZ region, clearly suggesting that other genes on the Y chromosome outside the DAZ region are necessary for spermatogenesis (Pryor et al. 1997). Men with deletions in the DAZ region can be fertile following ICSI and under normal circumstances (Mullah et al. 1997; Pryor et al. 1997). To date, many studies documenting the frequency of Y chromosome deletions in infertile men show a higher incidence in men with azoospermia compared to those with oligozoospermia (Stuppia et al 1998; Pryor et al. 1997). In addition, such studies show a significantly higher frequency of deletions in men with oligozoospermia versus those with normal sperm concentrations. The current data do not allow researchers to decide, with certainty, a cut-off sperm concentration which can be used as a tool to determine which men should be screened for Y deletions. However, many studies suggest screening men with sperm concentrations below 10 million per ml (Stuppia et al. 1998). All men with non-obstructive azoospermia should be screened for Y deletions.

Recently, Lahn and Page (1997) reported on 12 novel genes, seven of which demonstrated multiple repeats on the Y chromosome. Clinical studies need to be performed to determine if deletions of one or more of these genes are associated with male infertility.

Straightforward PCR analysis of blood samples to detect microdeletions on the Y chromosome for a small number of men may not be satisfactory. In the literature, the necessity for the screening of spermatozoa in addition to blood is demonstrated. In the first example, Reijo et al. (1996) observed that two severely oligozoospermic men harbored Y-microdeletions that were not detectable in the leukocytes of their fathers. One of these men also showed the same Y-chromosome deletion in his spermatozoa. In this case, the deletion may have arisen *de novo* (Reijo et al. 1996) in the infertile patient. It could also be due to a post-meiotic error in the germline of his fertile father. Such a microdeletion was detected in the patient's blood (Kent-First et al. 1996).

The second example arises from a study of 32 men with male-factor infertility and their ICSI-derived sons. In this study, one common deletion in one father-son pair was diagnosed as the *de novo* deletion arising from the germ-cell lineage of their fertile parental generation (Kent-First et al. 1996). Two additional cases in the same study highlighted microdeletions in the AZF (Azoospermic Factor) region of the Y chromosome in the ICSI-derived sons that were not detected in the blood of their infertile fathers. This indicated the presence of a mosaicism consisting of both intact Y and microdeleted Y copies, resulting in the non-detection in blood of men from the infertile generation. Consequently, it is possible that the sons may inherit a mosaic sperm population, possibly necessitating the use of preimplantation genetic diagnosis to differentiate between Y deleted and Y intact embryos.

Cystic Fibrosis Transmembrane Conductance Regulator (CFTR): Screening of Infertile Men

The biology of the cystic fibrosis transmembrane conductance regulator (CFTR), the protein mutated in cystic fibrosis, is remarkably complex (Prince 1998). CFRT is expressed in epithelial cells (including the epididymis) and is generally regarded to be a chloride channel that regulates sodium, ATP and water transport. Interestingly, recent data suggest that CFTR is a receptor for *Salmonella typhi* and heterozygosity of the CFTR allele may have been selected for in certain populations because it protects against typhoid (Pier et al. 1998).

A large number of studies have examined the role of CFTR in male infertility (reviewed by Schlegel, Shin 1997; De Braekeleer, Ferec 1996; Wong 1998). Among the predominant studies, mutations in CFTR have been documented in men with congenital bilateral absence of the vas deferens (CBAVD). CBAVD is a rare condition affecting approximately 13% of men with obstructive azoospermia (Girgis et al. 1969) and is now considered to be a mild (genital) form of cystic fibrosis.

According to a review by De Braekeleer and Ferec (1996), 65% of the 572 individuals with CABVD had at least one CFTR mutation; therefore, all men attending the infertility clinic with CBAVD must be screened for CFTR mutations. To some degree, the incidence of CF mutations in CABVD depends on the number of mutations screened. Several authors now recommend extending the screening to include the 5T mutation, a variant in the non-coding region of the CFTR gene (Mak et al. 1997; Bienvenu et al. 1997). To date there are limited data on the incidence of the 5T mutation in CBAVD, but one study reported that 26% of men with CBAVD had the 5T allele (Bienvenu et al. 1997). Screening depends on the frequency of mutations in the local population and it is against this background that men from infertile couples should be compared.

The mechanism by which CF mutations "cause" CBAVD is not yet clear, but as some cases of CBAVD are not associated with CF mutations, other developmental genes must be involved. Interestingly, if spermatozoa can be recovered from men with CBAVD who have CF mutations, there does not appear to be any significant difference in fertilization rates as compared to those of CBAVD men without CF mutations (Schlegel et al. 1995).

Studies have suggested a higher than normal level of CFTR mutations in men with unilateral absence of the vas deferens (Mickle et al. 1995). Indeed, it has been suggested that there is a higher incidence of CTFR mutations in men with normal vas deferens but male infertility, such as oligozoospermia (van der Ven et al. 1996), although other studies have failed to confirm this (Tuerlings et al. 1998b).

Based on this information, genetic screening of men with CBAVD for CF mutations and also men with unilateral absence of the vas deferens is recommended. The data to support extending this screening to other groups of infertile men is currently not available.

Screening for Other Less Well-Known Defects

There is a plethora of relatively infrequent genetic defects for which the infertile male could be screened. Individual details of these can be obtained from reviews on the subject (Meschede, Horst 1997; Gagnon 1997; Tournaye et al. 1997; Vogt 1997). Two well-documented examples, FSH mutations and androgen receptor mutations, are worthy of further comment.

Surprisingly, recent data using transgenic mice in which the FSH β subunit was deficient demonstrated that FSH-deficient males were fertile despite having small testes. Interestingly, the females were infertile (Kumar et al. 1997).

Apart from transforming thoughts on the critical role of FSH in spermatogenesis, this highlighted the differing role and importance of hormonal action in the male and female. In general, FSH and FSH receptor (FSHR) mutations are very rare in the infertile male population. For example, Tapanainen and colleagues (1997) studied the incidence of a FSH receptor inactivation mutation (566 C to T) in exon 7 of the FSHR gene. Women homozygous for this mutation had severely disrupted folliculargenesis and were infertile. Yet, males who were homozygous had varying degrees of spermatogenic defects but some were fertile. This mutation is rare in the infertile male population. For example, of 151 men screened with severe or moderate oligozoospermia, none were homozygous, all being heterozygous (Tapanainen et al. 1997). It appears that FSH mutations are less important in the male as compared to those in the female (Huhtaniemi, Aittomaki 1998).

Following the identification and cloning of the androgen receptor (AR), as discussed in Chang et al. (1988), there has been intense research documenting the various mutations of the androgen receptor and the classification of their biological and clinical effects (Quigley et al. 1995; MacLean et al. 1997).

The AR belongs to the family of nuclear transcription factors. Steroid binding induces a conformation change which allows the receptor to act as a transcription factor. Defects in the androgen receptor result in various states of androgen resistance ranging from complete androgen insensitivity (CIS) to partial androgen insensitivity. Currently, there are 309 documented mutations of the androgen receptor (Gottlieb et al. 1998; http://www.mcgill.ca/androgendb/ for details of mutations). This makes the androgen receptor the most frequently mutant human transcription factor. The large number and variety of androgen receptor mutations precludes a correlation between phenotype and genotype (MacLean et al. 1997; Quigley et al. 1995).

In fact, there can be a striking difference in phenotypes between and within kindred with the same mutation. This suggests that other, as yet unidentified, molecular/cellular factors are involved in the expression of the phenotype. There is preliminary evidence that longer polyglutamine tracts in the androgen receptor are associated with male fertility (Tut et al. 1997) but such data needs to be confirmed in a larger series. The incidence of androgen receptor mutations in an infertility clinic population is very low (Tincello et al. 1997) and routine screening of infertile men cannot be recommended. However, in men with physical signs of androgen insensitivity syndrome, such as limited masculinization or azoospermia, screening for the presence of androgen receptor may be justified. At present, the majority of research is concentrated on the identification of the molecular mechanism of the action of the androgen receptor (Langley et al. 1998) and until further advances in understanding are made, clinical management of this heterogeneous group of patients cannot be refined.

Genetic Counseling

The consensus is that a multidisciplinary approach to counseling should be developed for couples who may undergo ICSI. This should include genetic counseling explaining the possible risks of treatment using the information from each couple, including history and physical examination plus any genetic tests that have been performed. To date there is very little information available on this multidisciplinary approach and its effectiveness. Clearly there is a need to document the various models of counseling available and to determine their effectiveness. In the United Kingdom, although it is mandatory to offer all couples undergoing IVF and ICSI the opportunity of counseling, the role of genetic counseling in this forum is poorly formulated.

Very recently, the Human Fertilization and Embryology Authority (HFEA) has issued guidelines on the screening of couples prior to ICSI in view of the recent data concerning an increase in abnormalities. These guidelines state "it is strongly recommended by the HFEA that all patient information on ICSI should stress that it is still a new technique and provide a benefit/risk assessment which reflects the current state of knowledge" (HFEA 1998). The HFEA considers that ICSI children are twice as likely to have a major birth defect and 50% more likely to have a minor birth defect (HFEA 1998).

A recent study in the United States by Schover and colleagues (1998) examined the attitudes to genetic risk among 55 couples starting IVF. The couples possessed a strong desire to have their own genetic children and formal genetic counseling was only considered useful to couples faced by a clearly defined genetic risk. This study confirms the experience of many professionals that couples may not give adequate attention to genetic risks to their children in their desperation to have children.

Clearly there is a critical need to investigate the development, impact and effectiveness of counseling prior to ICSI. Such genetic counseling must incorporate the basic principles of genetic counseling used for other subjects (Lerman et al., 1998, contains a discussion on breast cancer screening) and determine the importance of perceived risk by the patients versus actual risk.

Possible Future Developments in the Screening of the Male

Mitochondrial DNA Mutations and Male Infertility

There is a limited amount of evidence to suggest a link between mitochondrial DNA mutations and sperm function. For example, an investigation of a patient harboring a maternally inherited point mutation demonstrated the heteroplasmic segregation of this point mutation resulting in encephalomyopathy and asthenozoospermia (Folgero et al, 1993). Further evidence results from the segregation of maternally inherited sporadic deletions associated with Kearns Sayre Syndrome and oligoasthenozoospermia (Lestienne et al. 1997). Thus, for male patients suffering from a maternally inherited mitochondrial disease and sperm dysfunction, PCR screening of the sperm sample would indicate whether that particular mutation or deletion is responsible.

Mitochondrial DNA mutations can also arise somatically; that is, through factors such as reactive oxygen species generation which can react with mtDNA and form multiple deletions. There are conflicting reports about the role of somatic mutations and sperm function. Kao and colleagues (1995) found a negative correlation between sperm motility and the proportion of the 5kb deletion (common deletion), while Cummins and colleagues (1998) reported that there was no correlation between the level of this deletion and various semen parameters.

Data from the Barratt laboratory support that of Cummins and colleagues (St. John et al. 1997). These somatic deletions tend to be random and each deletion usually occurs at low levels; therefore, it is the cumulative effect of these deletions that can be considerable (Katsumata et al. 1994). This would account for cellular dysfunction through the failure, for example during early spermatogenesis, to transcribe these vital proteins resulting in the inability of the electron transport chain to generate sufficient ATP to sustain the sperm's metabolic function. This has been observed in some astheno-, oligo- and oligoathenozoospermic patients through the use of Long PCR, which has the capability of amplifying the whole mitochondrial genome. When the products are resolved on an agarose gel, both wild-type and deleted molecules can be identified and then quantified to determine what the wild type-to-deleted ratio might be for a sample (St. John et al. 1997).

Screening Men in the Post-ICSI era

There have been breathtaking advances in molecular biology and genetics. Reproductive biologists need to apply these advances to male reproductive function.

ICSI is now a well-accepted technique. The questions that should be addressed are what new screening methods can be adopted for men to provide information about their chances of success or failure for fertilization and post-fertilization events. Using the work of Schatten as an example, scientists know the human sperm centrosome is critical for successful fertilization by directing the nucleation of the sperm astral microtubules (Simerly et al. 1997). Several clearly identified defects in this process have been documented (Simerly et al. 1995, 1997), and with the significant progress that has been made in identifying the centrosomal proteins, research should shortly be at the stage of screening men prior to ICSI for molecular defects in the centrosome.

The degree of nuclear DNA damage in spermatozoa does have a significant influence on fertilization success. An accurate assessment of this damage can be made and thus should be part of routine screening prior to ICSI (Sakkas, this volume). Simple treatments, such as antioxidant therapy, may be effective in reducing the degree of DNA damage.

The use of transgenic knockout mice has significantly advanced understanding of sperm production and function (Okabe et al. 1998). Such experiments usually show that potentially critical molecules for sperm function are, in fact, not so critical—for example, acrosin (Baba et al. 1994). In contrast, knockout experiments examining other cell systems have revealed surprising results, such as the role of apoptotic/cell survival factors in spermatogenesis (Sassone-Corsi 1997), for example. Such experimental approaches are rapidly changing the understanding of sperm production and function. Scientists need to embrace and capitalize on these developments to use the information in a clinical setting so that the potential usefulness of such information can be determined in a more rapid fashion—that is, moving from discovery to application at a faster rate.

DNA Chip Technology

In order to provide an effective and efficient genetic screening protocol for male infertility, researchers clearly need to enhance the basic understanding significantly. It is likely that current techniques used to screen for genetic diseases, such as PCR and gel analysis, will be redundant in the near future. Remarkable developments in DNA chip technology (which allow the accurate and fast screening of nucleotides/gene markers) are enabling the rapid screening of gene deletions, such as cystic fibrosis. In addition, this technology will allow more effective mapping of genomic libraries, discovery of novel genes and expression studies (Ramsey 1998). It is likely that gene chip technology will be utilized in the management of infertile couples in the not-so-distant future. This will enable the simultaneous, rapid and efficient screening of many patients for several gene defects.

Summary—Who Should be Screened?

Providing that a comprehensive history and examination as well as an accurate semen analysis have been performed, the issue of genetic testing can be considered. Based on the information available the following recommendations can be made:

1. Informed genetic counseling must be integrated with assisted conception. Most importantly, clarification by the prospective parents, after a period of reflection, of an understanding of the potential consequences of ICSI should be achieved.

2. All men with azoospermia and severe oligozoospermia (arbitrary limit 10 x 10^6/ml) and those who are referred for ICSI should have a karyotype analysis performed.

3. All men with CBAVD and unilateral absence of the vas deferens should be screened for CF mutations. In addition, if treatment is to be commenced, their partners should also be screened.

4. All men with non-obstructive azoospermia should be screened for Y chromosome deletions.

These suggestions should be regarded as a minimum. Some centers may wish to perform further tests, such as FISH, on spermatozoa from specific subgroups of men to provide an assessment of the possible risk factor of transmission.

In general, more detailed information is required about the subject of genetic testing of the male before recommendations can be extended. By incorporating the advances in genetic testing and molecular biology into male infertility, a rapid increase in the armamentarium and clearer guidelines for testing should be seen soon.

Acknowledgements

The authors are grateful to the Assisted Conception Unit at the Birmingham Women's Hospital for samples, nursing and laboratory support.

References

Baba T, Azuma S, Kashiwabara S, Toyoda Y. Sperm from mice carrying a targeted mutation of the acrosin gene can penetrate the oocyte zona pellucida and effect fertilization. J Biol Chem 1994; 269:31845-31849.

Barratt CLR, St John JC. Diagnostic tools in male infertility. Hum Reprod 1998; 13 (Suppl 1):51-61.

Baschat AA, Kupker W, al Hasani S, Diedrich K, Schwinger E. Results of cytogenetic analysis in men with severe subfertility prior to intracytoplasmic sperm injection. Hum Reprod 1996; 11:330-333.

Bernardini L, Martini E, Geraedts GPM, Hopman AHN, Lanteri S, Conte N, Capitanio GL. Comparison of gonosomal aneuploidy in spermatozoa of normal fertile men and those with severe male factor detected by *in situ* hybridization. Mol Hum Reprod 1997; 3:431-438.

Bienvenu T, Adjiman M, Thiounn N, Jeanpierre M, Hubert D, Lepercoq J, Francoual C, Wolf J, Izard V, Jouannet P, Kaplan JC, Beldjord C. Molecular diagnosis of congenital bilateral absence of the vas deferens: analysis of the CFTR gene in 64 patients. Ann Genet 1997; 40:5-9.

Bonduelle M, Aytoz A, Van Assche E, Devroey P, Liebaers I, Van Steirteghem A. Incidence of chromosomal aberrations in children born after assisted reproduction through intracytoplasmic sperm injection. Hum Reprod 1998a; 13:781-782.

Bonduelle M, Joris H, Hofmans K, Liebers I, van Steirteghem A. Mental development of 201 ICSI children at 2 years of age. Lancet 1998b; 351:1553.

Bowen JR, Gibson FL, Leslie GI, Saunders D. Medical and developmental outcome at 1 year for children conceived by intracytoplasmic sperm injection. Lancet 1998; 351:1529-1534.

Chandley AC. Karyotype and oligozoospermia. In: (Barratt CLR, De Jonge C, Mortimer D, Parinaud J, eds) The Genetics of Human Male Fertility. Paris: EDK Press, 1997, pp111-122.

Chang C, Kokontis J, Liao S. Molecular cloning of human and rat complementary DNA encoding androgen receptors. Science 1988; 240:324-326.

Colls P, Martinez-Pasarell O, Perez MM, Egozcue J, Templado C. Sperm chromosome analysis in the father of a child with a *de novo* reciprocal translocation t(11;15)(q12;q22) by G-banding and fluorescence *in situ* hybridization. Hum Reprod 1998; 13:60-64.

Cooke HJ, Hargreave TB, Elliot DJ. Understanding the genes involved in spermatogenesis: a progress report. Fertil Steril 1998; 69:989-995.

Cummins JM, Jequier AM, Martin R, Mehmet D, Goldblatt J. Semen levels of mitochondrial DNA deletions in men attending an infertility clinic do not correlate with phenotype. Int J Androl 1998; 21:47-52.

De Braekeleer M, Ferec C. Mutations in the cystic fibrosis gene in men with congenital bilateral absence of the vas deferens. Mol Hum Reprod 1996; 2:669-677.

Downie SE, Flaherty SP, Matthews CD. Detection of chromosomes and estimation of aneuploidy in human spermatozoa using fluorescence *in situ* hybridization. Mol Hum Reprod 1997; 3:585 598.

Egozcue J, Blanco J, Vidal F. Chromosome studies in human sperm nuclei using fluorescence *in situ* hybridization (FISH). Hum Reprod 1997; 3 (Update):441-452.

Folgero T, Bertheussen K, Lindal S, Torbergsen T, Oian P. Mitochondrial disease and reduced sperm motility. Hum Reprod 1993; 8:1863-1868.

Gagnon C. Genetic aspects of flagellar dyskinesis. In: (Barratt CLR, De Jonge C, Mortimer D, Parinaud J, eds) Genetics of Human Male Fertility. Paris: EDK Press, 1997, pp76-98.

Gardner RJM, Sutherland GR. Chromosomal Abnormalities and Genetic Councelling. Oxford University Press, Oxford, 1996, p478.

Girgis SM, Etriby A, Ibrahim AA, Kahil SA. Testicular biopsy in azoospermia. Fertil Steril 1969; 20:467-477.

Gottlieb B, Lehvaslaiho H, Beitel LK, Lumbroso R, Prinsky L, Trifiro M. The androgen receptor gene mutations database. Nucleic Acids Res 1998; 26:234-238.

Human Fertilization and Embryology Authority. Risks associated with ICSI treatment. Letter to All Persons Responsible from HFEA, 11/6/98.

Hens L, Bonduelle M, Liebaers I, Devroey P, Van Steirteghem AC. Chromosome aberrations in 500 couples referred for *in vitro* fertilization or related fertility treatment. Hum Reprod 1988; 3:451-457.

Huhtaniemi IT, Aittomaki K. Mutations of follicle-stimulating hormone and its receptor: effects on gonadal function. Eur J Endocrinol 1998; 138:473-481.

In't Veld PA, Broekmans FJM, France HF, Pearson PL, Pieters MHEC, van Kooij RJ. Intracytoplasmic sperm injection (ICSI) and chromosomally abnormal spermatozoa. Hum Reprod 1997; 12:752-754.

Kao S-H, Chao H-T, Wei Y-H. Mitochondrial deoxyribonucleic acid 4977-bp deletion is associated with diminished fertility and motility of human sperm. Biol Reprod 1995; 52:729-736.

Katsumata K, Hayakawa M, Tanaka M, Sugiyama S, Ozawa T. Fragmentation of human heart mitochondrial DNA associated with premature aging. Biochem Biophys Res Commun 1994; 202(1):102-110.

Kent-First MG, Kol S, Muallem A, Blazer S, Itskovitz-Eldor J. Infertility in intracytoplasmic-sperm-injection-derived sons. Lancet 1996; 348:332.

Kobayashi K, Mizuno K, Hida A, Komaki R, Tomita K, Matsushita I, Namiki M, Iwamoto T, Tamura S, Minowada S, Nakahori Y, Nakagome Y. PCR analysis of the Y chromosome long arm in azoospermic patients: evidence for a second locus required for spermatogenesis. Hum Mol Genet 1994; 3:1965-1967.

Kruse R, Guttenbach M, Schartmann B, Schubert R, van der Ven H, Schmid M, Propping P. Genetic counseling in a patient with XXY/XXXX/XY mosaic Klinefelter's syndrome: estimate of sex chromosome aberrations in sperm before intracytoplasmic sperm injection. Fertil Steril 1998; 69:482-486.

Kumar TR, Wang Y, Lu N, Matzuk MM. Follicle stimulating hormone is required for ovarian follicle maturation but not male fertility. Nature Genet 1997; 15:201-204.

Kurinczuk JJ, Bower C. Birth defects in infants conceived by intracytoplasmic sperm injection: an alternative interpretation. Brit Med J 1997; 315:1260-1265.

Lahn BT, Page DC. Functional coherence of the human Y chromosome. Science 1997; 278:675-680.

Lange R, Johannson G, Engel W. Chromosome studies in *in vitro* fertilization patients. Hum Reprod 1993; 8:572-574.

Langley E, Kemppainen JA, Wilson EM. Intermolecular NH2-/carboxyl-terminal interactions in androgen receptor dimerization revealed by mutations that cause androgen insensitivity. J Biol Chem 1998; 273:92-101.

Lerman C, Hughes C, Lemon SJ, Main D, Snyder C, Durham C, Narod S, Lynch HT. What you don't know can hurt you: adverse psychological effects in members of BRCA1-linked and BRCA2-linked families who decline genetic testing. J Clin Oncol 1998; 16:1650-1654.

Lestienne P, Reynier P, Chretien MF, Penisson-Besnier I, Malthiery Y, Rohmer V. Oligoasthenospermia associated with multiple mitochondrial DNA rearrangements. Mol Hum Reprod 1997; 3:811-814.

Ma K, Inglis JD, Sharkey A, Bickmore WA, Hill RE, Prosser EJ, Speed RM, Thomson EJ, Jobling M, Taylor K. A Y chromosome gene family with RNA-binding protein homology: candidates for the azoospermia factor AZF controlling human spermatogenesis. Cell 1993; 75:1287-1295.

MacLean HE, Warne GL, Zajac JD. Localization of functional domains in the androgen receptor. J Steroids Biochem Mol Biol 1997; 62:233-242.

Mak V, Jarvi KA, Zielenski J, Durie P, Tsui LC. Higher proportion of intact exon 9 CFTR mRNA in nasal epithelium compared with vas deferens. Hum Mol Genet 1997; 6:2099-2107.

Mau UA, Backert IT, Kaiser P, Kiesel L. Chromosomal findings in 150 couples referred for genetic counselling prior to intracytoplasmic sperm injection. Hum Reprod 1997; 12:930-937.

Mercier S, Morel F, Fellman F, Roux C, Bresson JL. Molecular analysis of the chromosomal equipment in spermatozoa of a 46, XY, t(7;8) (q11.21;cen) carrier by using fluorescence *in situ* hybridization. Hum Genet 1998; 102:446-451.

Meschede D, Horst J. The molecular genetics of male infertility. Mol Hum Reprod 1997; 3:419-430.

Meschede D, Lemcke B, Exeler JR, De Geyter C, Behre HM, Nieschlag E, Horst J. Chromosome abnormalities in 447 couples undergoing intracytoplasmic sperm injection—prevalence, types, sex distribution and reproductive relevance. Hum Reprod 1998; 13:576-582.

Micic M, Micic S, Diklie V. Chromosome constitution of infertile men. Clin Genet 1984; 25:33-36.

Mickle J, Milunsky A, Amos JA, Oates RD. Congenital unilateral absence of the vas deferens: a heterogenous disorder with two distinct subpopulations based upon aetiology and mutational status of the cystic fibrosis gene. Hum Reprod 1995; 10:1728-1735.

Montag M, van der Ven K, Ved S, Schmutzler A, Prietl G, Krebs D, Peschka B, Schwanitz G, Albers P, Haidl G, van der Ven H. Success of intracytoplasmic sperm injection in couples with male and/or female chromosome aberrations. Hum Reprod 1997; 12:2635-2640.

Mullhall JP, Reijo R, Alagappan R, Brown L, Page D, Carson R, Oates RD. Azoospermic men with deletion of DAZ gene cluster are capable of completing spermatogenesis: fertilization, normal embryonic development and pregnancy occur when retrieved testicular spermatozoa are used for intracytoplasmic sperm injection. Hum Reprod 1997; 12:503-508.

Nakahori Y, Kobayashi K, Komaki R, Matsushita I, Nakagome Y. A locus of the candidate gene family for azoospermia factor (YRRM2) is polymorphic with a null allele in Japanese males. Hum Mol Gen 1994; 3:1709.

Newberg MT, Francisco RG, Pang MG, Brugo S, Doncel GF, Acosta AA, Hoegerman SF. Cytogenetics of somatic cells and sperm from a 46,XY/45,X mosaic male with moderate oligoasthenoteratozoospermia. Fertil Steril 1998; 69:146-148.

Okabe M, Ikawa M, Ashkenas J. Male infertility and genetics of spermatogenesis. Am J Hum Genet 1998; 62:1247-1281.

Pandiyan N, Jequier AM. Mitotic chromosomal anomalies among 1210 infertile men. Hum Reprod 1996; 11:2604-2608.

Pier GB, Grout M, Zaidi T, Meluleni G, Mueschenborn SS, Banting G, Ratcliff R, Evans MJ, Colledge WH. *Salmonella typhi* uses CFTR to enter intestinal epithelial cells. Nature 1998; 393:79-82.

Prince A. The CFTR advantage—capitalizing on a quirk of fate. Nature Med 1998; 4:663-664.

Pryor JL, Kent-First M, Muallem A, van Bergen AH, Nolten WE, Meisner L, Roberts KP. Microdeletions in the Y chromosome of infertile men. N Engl J Med 1997; 336:534-539.

Quigley CA, DeBellis A, Marsche KB, El-Awady MK, Wilson EM, French FS. Androgen receptor defects: historical, clinical, and molecular perspectives. Endocrine Rev 1995; 16:271-321.

Ramsey G. DNA chips: state-of-the art. Nature Biotechnol 1998; 16:40-44.

Reijo R, Lee TY, Salo P, Alagappan R, Brown LG, Rosenberg M, Rozen S, Jaffe T, Straus D, Hovatta O. Diverse spermatogenic defects in humans caused by Y chromosome deletions encompassing a novel RNA-binding protein gene. Nature Genet 1995; 10:383-393.

Reijo R, Alagappan R, Patrizio P, Page DC. Severe oligospermia resulting from deletions of azoospermia factor gene on Y chromosome. Lancet 1996; 347:1290-1293.

Robbins WA, Meistrich ML, Moore D, Hagemeister FB, Weier HU, Cassel MJ, Wilson G, Eskenazi B, Wyrobek AJ. Chemotherapy induces transient sex chromosomal and autosomal aneuploidy in human sperm. Nature Genet 1997; 16:74-78.

Roberts KP. Y-chromosome deletions and male infertility: state of the art and clinical implications. J Androl 1998; 19:255-259.

St John JC, Cooke ID, Barratt CLR. Mitochondrial mutations and male infertility. Nature Med 1997; 3:124-125.

Sassone-Corsi P. Transcriptional checkpoints determining the fate of male germ cells. Cell 1997; 88:163-166.

Schlegel PN, Shin D. Urogenital anomalies and genetic defects in men with bilateral congenital absence of the vas deferens. In: (Barratt CLR, De Jonge C, Mortimer D, Parinaud J, eds) Genetics of Male Infertility. Paris: EDK Press, 1997, pp98-110.

Schlegel PN, Cohen J, Goldstein M, Alikani M, Adler A, Gilbert BR, Palermo GD, Rosenwaks Z. Cystic fibrosis gene mutations do not affect sperm function during *in vitro* fertilization with micromanipulation for men with bilateral congenital absence of the vas deferens. Fertil Steril 1995; 64:421-426.

Schover IR, Thomas AJ, Falcone T, Attaran M, Goldberg J. Attitudes about genetic risk of couples undergoing *in vitro* fertilization. Hum Reprod 1998; 13:862-866.

Simerly C, Wu GJ, Zoran S, Ord T, Rawlins R, Jones J, Navara C, Gerrity M, Rinehart J, Binor Z. The paternal inheritance of the centrosome, the cell's microtubule-organizing center, in humans, and the implications for infertility. Nature Med 1995; 1:47-52.

Simerly C, Hewitson LC, Sutozsky P, Schatten G. The inheritance, molecular dissection and reconstitution of the human centrosome during fertilization. In: (Barratt CLR, De Jonge C, Mortimer D, Parinaud J, eds) Genetics of Human Male Fertility. Paris: EDK Press, 1997, pp258-286.

Stuppia L, Gatta V, Calabrese G, Franchi PG, Morizo E, Bombieri C, Mingarelli R, Sforza V, Frajese G, Tenaglia R, Palka G. A quarter of men with idiopathic oligoazoospermia display chromosomal abnormalities and microdeletions of different types in interval 6 of Yq11. Hum Genet 1998; 102:566-570.

Tapanainen JS, Aittomaki K, Min J, Vaskivuo T, Huhtaniemi IT. Men homozygous for an inactivating mutation of the follicle-stimulating hormone (FSH) receptor gene present variable suppression of spermatogenesis and fertility. Nature Genet 1997; 15:205-206.

Testart J, Gautier E, Brami C, Rolet F, Sedbon E, Thebault A. Intracytoplasmic sperm injection in infertile patients with structural chromosome abnormalities. Hum Reprod 1996; 11:2609-2612.

Tincello DG, Saunders PTK, Hargreave TB. Preliminary investigations on androgen receptor gene mutations in infertile men. Mol Hum Reprod 1997; 3:941-943.

Tournaye W, Lissens W, Liebaers I, van Assche E, Bounduelle M, Fastenaekels V, van Steirteghem A, Devroey P. In: (Barratt CLR, De Jonge C, Mortimer D, Parinaud J, eds) Genetics of Human Male Fertility. Paris: EDK Press, 1997, pp123-146.

Tuerlings JHAM, de France HF, Hamers A, Hordijk R, van Hemel JO, Hansson K, Hoovers JMN, Madan K, Van Der Blij-Philipsen M, Gerssen-Schoorl KBJ, Kremer JAM, Smeets DFMC. Chromosome studies in 1792 males prior to intracytoplasmic sperm injection: the Dutch experience. Eur J Hum Genet 1998a; 6:194-200.

Tuerlings JHAM, Mol B, Kremer JA, Looman M, Meuleman EJ, te Meerman GJ, Buys CH, Merkus HM, Scheffer H. Mutational frequency of cystic fibrosis transmembrane regulator is not increased in oligozoospermic male candidates for intracytoplasmic sperm injection. Fertil Steril 1998b; 69:899-903.

Tut TG, Ghadessy FJ, Trifiro MA, Pinsky L, Yong EL. Long polyglutamine tracts in the androgen receptor are associated with reduced trans-activation, impaired sperm production, and male infertility. J Clin Endocrinol Metab 1997; 82:3777-3782.

van der Ven K, Messre L, van der Ven H, Jeyendran RS, Ober C. Cystic fibrosis mutation screening in healthy men with reduced semen quality. Hum Reprod 1996; 11:513-517.

van der Ven K, Montag M, Peschka B, Leygraaf J, Schwanitz G, Haidl G, Krebs D, van der Ven H. Combined cytogenetic and Y chromosome microdeletion screening in males undergoing intracytoplasmic sperm injection. Mol Hum Reprod 1997; 3:699-704.

van der Ven K, Peschka B, Montag M, Lange R, Schwanitz G, van der Ven HH. Increased frequency of congenital chromosomal aberrations in female partners of couples undergoing intracytoplasmic sperm injection. Hum Reprod 1998; 13:48-54.

te Velde ER, van Baar AL, van Kooij RJ. Concerns about assisted reproduction. Lancet 1998; 351:1524-1525.

Vogt PH. Genetic disorders of human spermatogenesis. In: (Waites GMH, Frick J, Baker HWG, eds) Current Advances in Andrology. Proceedings of the Sixth International Congress of Andrology. Monduzzi Editorie, Italia, 1997, pp51-73.

Wong PY. CFTR gene and male infertility. Mol Hum Reprod 1998; 4:107-110.

37

Ooplasmic Injections of Round Spermatids and Secondary Spermatocytes for the Treatment of Non-Obstructive Azoospermia

Nikolaos Sofikitis
Ikuo Miyagawa
Tottori University

The evolution of round spermatid nuclear injections (ROSNI) and intact round spermatid injections (ROSI) can be traced to Ogura and Yanagimachi (1993). They showed that round spermatid nuclei injected into hamster oocytes form pronuclei and participate in syngamy. DNA synthesis was found in these pronuclei. However, the developmental potential of the obtained zygotes was not evaluated in this study. In another study, Ogura et al. (1993) injected intact round spermatids into the perivitelline space of mature hamster or mouse oocytes and applied a fusion pulse, attempting to fuse the intact spermatids with the oocytes. It was found that the spermatid nuclei commonly failed to develop into large pronuclei. In one additional study, Ogura et al. (1994) showed that when mouse intact round spermatids are successfully fused with oocytes, some of the resulting zygotes develop into normal offspring. The overall success rate of the electrofusion of intact spermatids with oocytes was low and attributable to the difficulty of fusing large cells such as oocytes with small cells such as spermatids without lysis of the larger cells.

To avoid oocyte damage due to the fusion process, researchers in the Sofikitis lab chose a micro-surgical approach to transfer round spermatid nuclei into rabbit ooplasm. Three pregnancies were achieved with round spermatid nuclei injections into rabbit oocytes (Sofikitis et al. 1994a). In that study, the proportion of implanted embryos to the number of injected oocytes and the ratio of offspring to the number of injected oocytes were low. The low values of these parameters may be attributable to the low developmental potential of the injected oocytes due to inadequate mechanical stimulation applied to activate oocytes prior to round spermatid nuclear injections. For this reason, another study was designed to evaluate the effects of electrical stimulation of oocytes before ooplasmic round spermatid nuclei injections on oocyte activation and subsequent embryonic development (Sofikitis et al. 1996a). It was shown that electrical stimulation of oocytes prior to ooplasmic round spermatid nuclei injections and embryo transfer procedures has beneficial effects on oocyte activation, fertilization and subsequent embryonic development and results in 13% live birth rate per activated oocyte (Sofikitis at al. 1996a).

After these successful trials to produce offspring by microsurgical transfer of round spermatid nuclei into rabbit oocytes, a question was raised: could round spermatid nuclei selected from subjects with various testicular disorders have similar fertilizing capacity? That question was of great concern because in the two previous studies, the round spermatids had been harvested from healthy male rabbits. Furthermore, the probability that humans and animals with primary testicular disorders may not have anatomically and physiologically normal spermatids cannot be excluded. To answer that question, researchers induced an experimental varicocele model in the rabbit, isolated round spermatid nuclei from the testicles of varicocelized rabbits, injected the nuclei into healthy mature oocytes and proved that these nuclei had fertilizing potential. The overall fertilization rate was 23%. However, embryo transfer procedures did not result in pregnancies (Sofikitis et al. 1996b).

Clinical Application of ROSNI/ROSI Techniques

After the encouraging message from the animal investigations, an attractive challenge was to apply ooplasmic injections of round spermatid nuclei selected from testicular biopsy material for the treatment of non-obstructed azoospermic men (Sofikitis et al. 1994a). The first pregnancies via ROSNI techniques reported in the international literature in 1995 were achieved in 1994 (Sofikitis et al. 1995a; Hannay 1995). However, these pregnancies resulted in abortions. A few months later, Tesarik et al. (1995) reported delivery of two healthy children after round spermatid injections into oocytes. The mean fertilization rate was 45% in that study. Fishel et al. (1995) reported a pregnancy and birth after elongated spermatid injections into oocytes.

The first ROSNI procedures in the USA were applied in California, Louisiana and Florida (Sofikitis et al. 1995b). Fertilization and development up to 10-cell stage embryos was achieved in non-obstructed American couples. The overall fertilization rate per injected oocyte was 31% in that study. The peak of the two pronuclei appearance curve in the group of oocytes injected with round spermatid nuclei was nine hours post-injection. At that time, all normally fertilized oocytes revealed two pronuclei, whereas two hours later both pronuclei disappeared in 20% of the oocytes. Considering that the peak of two pronuclei appearance after intracytoplasmic sperm injection (ICSI) procedures is 16 hours post-injection (Nagy et al. 1994), it appears that the speed of human embryo development after ROSNI procedures is faster as compared to that noted after ICSI procedures, and that oocytes injected with round spermatid nuclei should be checked for pronuclei earlier than oocytes injected with spermatozoa (Sofikitis et al. 1997a).

Yamanaka et al. (1997) reported an oocyte cleavage rate equal to 61% after ROSNI procedures and confirmed that the appropriate time for assessment of fertilization after human ROSNI techniques is nine hours post-injection. Additional pregnancies achieved by ROSNI or ROSI techniques were recently reported by Tanaka et al. (1996), Mansour et al. (1996), Antinori et al. (1997a, 1997b), Vanderzwalmen et al. (1997),

Amer et al. (1997) and Sofikitis et al. (1997b). Average fertilization rates were above 25% in all these studies.

Non-Obstructive Azoospermia and Indications for ROSNI/ROSI Procedures

Non-obstructive azoospermia may be due to secondary testicular damage or primary testicular damage. Secondary endocrine and exocrine testicular dysfunction may be due to defects in the hypothalamo-pituitary-testicular axis, or systemic organic disease, such as chronic renal failure, liver insufficiency, sickle cell anemia or diabetes mellitus. Primary testicular damage may be due to chromosomal abnormalities, orchitis, trauma, varicocele, cryptorchidism,gonadotoxins or radiation. The damage may also be congenital, such as Sertoli cell-only syndrome or myotonic dystrophy. Furthermore, genetic abnormalities affecting the function of germ cells or Sertoli cells may be among the causes of animal or human non-obstructive azoospermia. Thus, mutations in the white spotting locus of the mouse (Chabot et al. 1988), the Sl locus encoding the c-kit ligand (Anderson et al. 1990), and genes encoding retinoic acid receptor α (Akmal et al. 1997) may impair spermatogenesis and result in azoospermia. Sex or autosomal chromosomal deletions are also involved in the etiology of non-obstructive azoospermia. Involvement of at least three Y-linked genes in spermatogenesis has been suggested (Chai et al. 1997). Several studies suggest that two gene families, RBM (RNA Binding Motif) and DAZ, are present in Y-chromosomal regions that are deleted in some non-obstructed azoospermic men (Chai et al. 1997). Both gene families show specific testicular expression and encode proteins with RNA binding motifs. There is also increasing evidence for a putative human male infertility DAZ-like autosomal gene (Chai et al. 1997).

Recent studies have shown that a significant percentage of men with non-obstructive azoospermia have testicular foci of active spermatogenesis up to the stage of round spermatid, elongating spermatid or spermatozoon (Sofikitis et al. 1995a; Silbert 1996; Tesarik et al. 1995; Mansour et al. 1996; Amer et al. 1997; Antinori et al. 1997a; Vanderzwalmen et al. 1997; Yamanaka et al. 1997; Sofikitis et al. 1997b, 1998a). Ooplasmic injections of spermatozoa offer a solution for infertile men whose therapeutic testicular biopsy material contains spermatozoa (Silbert 1996). When spermatozoa are not present, ROSNI/ROSI techniques represent the only hope for treatment.

Several studies have shown clearly that in men with spermatogenic arrest at the primary spermatocyte stage or Sertoli cell-only syndrome, a number of germ cells in a limited number of seminiferous tubules can break the barrier of the premeiotic spermatogenic block and differentiate up to the stage of the round or elongating spermatid (Tesarik et al. 1995; Mansour et al. 1996; Antinori et al. 1997a; Yamanaka et al. 1997; Vanderzwalmen et al. 1997; Amer et al. 1997; Sofikitis et al. 1998a). Defects in the secretory function of the Leydig and Sertoli cells or other factors may not allow the round or elongating spermatids to complete the spermiogenesis. Several biochemical mechanisms may be responsible for the inability of the round spermatids to undergo the elongation process. O'Donnel et al. (1996) have shown that decreased intratesticular testosterone concentration (ITC) may be one of these mechanisms. Additional studies are necessary to clarify whether values of ITC below a threshold ITC value cause failure of elongation of round spermatids. If this hypothesis is correct, testicular pathophysiologies affecting optimal ITC may result in complete spermiogenetic failure. It should be emphasized that varicocele, the most frequent cause of male infertility and known to cause azoospermia occasionally, is accompanied by reduced ITC (Rajfer et al. 1987).

Round spermatids are occasionally present in the seminal plasma of non-obstructed azoospermic men. These ejaculated round spermatids can be used for ooplasmic injections. Mendoza and Tesarik (1996) reported that 69% of non-obstructed azoospermic men have round spermatids in the ejaculate. A current study by Tottori University International Research Group, applying confocal scanning laser microscopy (CSLM), fluorescent *in situ* hybridization (FISH) and transmission electron microscopy (TEM) techniques on semen samples of more than 200 non-obstructed azoospermic men, indicates that round spermatids are present in less than 20% of the samples analyzed (Y Yamamoto, N Sofikitis and I Miyagawa, unpublished observations).

Criteria for Identification/Isolation of Human Round Spermatids

The gold standard for identification of round spermatids is TEM. Recently, Mendoza and Tesarik (1996) attempted to identify round spermatids by selective staining of the acrosin contained in the acrosomal granules. Another approach is to visualize proacrosin with the usage of monoclonal antibodies. A drawback to all of these techniques is that application of these methodologies results in cell death. Therefore, observed spermatids cannot be used in assisted reproduction programs.The following approaches are suggested for identification of undisturbed round spermatids in therapeutic testicular biopsy material or in cellular populations isolated from semen samples.

Observation of Samples by CSLM

Live human and rabbit round spermatids are easily identified via lasermicroscopy by the presence of multiple or single (acrosomic) granule(s) adjacent to the nucleus (Sofikitis et al. 1994b; Yamanaka et al. 1997).

Inverted Microscope-Computer Assisted System (IM-CAS)

Application of quantitative criteria based on computer-assisted image analysis allows identification of round spermatids. "Round cells" with minimal diameter between six to 10 µm that satisfy additional specific quantitative and qualitative criteria are considered to be spermatids (Yamanaka et al. 1997).

Qualitative Criteria

IM-CAS and CSLM are not available in most of the IVF centers. Human round spermatids (Tesarik et al. 1995; Yamanaka et al. 1997; Antinori et al. 1997a; Mansour et al. 1996; Vanderzwalmen et al. 1997) can be distinguished from other cell types according to the cellular shape, size, and the form of the nucleus. A developing acrosomal granule can be recognized in the alive round spermatid as a bright/dark spot adjacent to the cell nucleus.

ROSNI Versus ROSI

ROSI procedures ensure the transfer of all the cytoplasmic components of the male gamete into the maternal gamete and are less time-consuming than round spermatid nucleus injection. Furthermore, manipulations of the male gamete nuclear matrix and envelope are avoided when ROSI techniques are applied. In contrast, ROSI procedures have two disadvantages: injecting micropipettes of larger diameter are necessary and consequently the probability to injure oocytes during injections is larger, and the persistence of a large amount of cytoplasm around the round spermatid nucleus may impede its transformation into male pronucleus. In the mouse (Ogura et al. 1993; Kimura, Yanagimachi 1995a, 1995b) transferring the round spermatid nucleus into the oocyte is a far more efficient procedure in achieving fertilization and embryonic development than transferring the intact round spermatid cell.

Contributions of the Round Spermatid to the Zygote

Genetic Material

The deliveries of normal mouse and rabbit offspring (Sofikitis et al. 1994a; Ogura et al. 1994; Kimura, Yanagimachi 1995a) and healthy human newborns (Tesarik et al. 1995; Mansour et al. 1996; Vanderzwalmen et al. 1997; Antinori et al. 1997a; Sofikitis et al. 1997b) after ROSNI/ROSI techniques indicate the maturity of the genetic material of the early haploid male gamete (that is, the chromosomes of the round spermatid are capable of pairing with those of the oocyte and participate in syngamy, fertilization and subsequent embryonic and fetal development).

Oocyte Activating Substance in Spermatid/ Spermatozoon (OASIS)

The male gamete-induced cascade of biochemical ooplasmic events that results in resumption of meiosis of the female gamete is referred to as oocyte activation. Injections of mouse round spermatids into oocytes do not result in oocyte activation, suggesting that the mouse oocyte activating substance in spermatid/spermatozoon (OASIS) has not been expressed at the round spermatid stage (Kimura, Yanagimachi 1995a, 1995b). In contrast, ooplasmic injections of rabbit round spermatids lead to oocyte activation in a significant percentage (Sofikitis et al. 1994a). Electrical stimulation of the rabbit oocyte supporting the functionality of the OASIS has beneficial effects on the activation process (Sofikitis et al. 1996a). Although electrical stimulation usually results in a monophasic ooplasmic Ca^{+2} response, it appears that there is a synergistic action of electrical stimulation and round spermatid OASIS which eventually produces Ca^{+2} oscillations. There is a strong evidence that the human OASIS has been activated at or before the round spermatid stage (Yamanaka et al. 1997; Sofikitis et al. 1998a). The achievement of human pregnancies via ROSNI/ROSI techniques without application of an exogenous electrical or chemical stimulation supports this thesis.

The human round spermatid OASIS should be nucleus-associated since nuclear injections are sufficient to cause activation (Yamanaka et al. 1997).

Centrosomic Components: A Challenge to the Theory of Centrosomes

The zygote's centrosome is a blend of paternal and maternal components. The restoration of the zygotic centrosome at fertilization requires the attraction of maternal centrosomal components to the paternal reproducing element (Schatten 1994).The male gamete contributes to the zygote centrosome by transferring the reproducing element of the centrosome, the microtubule organizing center and a γ-tubulin binding protein. The delivery of healthy babies after human ROSI/ROSNI techniques tends to suggest that the centrosomic components of the human round spermatid are normal, functional and mature. Additional studies are necessary on the development of the aster and the ooplasmic microtubule organization after ROSNI/ROSI procedures. Several studies have suggested that mammalian oocytes lose their centrosomes when they mature and that centrosomic material is introduced into the oocyte by the spermatozoon (Schatten 1994). However, the normal embryonic and fetal development after ROSNI plus embryo transfer procedures in the rabbit (nuclei were proven to be free of cytoplasmic and subsequently centrosomic material; Sofikitis et al. 1994a), the artificial parthenogenesis in several mammalian female gametes (Schatten 1994), and the development of parthenogenetic rabbit fetuses up to day 10 of pregnancy (Ozil 1990) can be interpreted as a challenge to the theory of centrosomes and raise the probability that, when paternal centrosomic material is absent, novel maternal spindle organizing centers can develop and previously denatured/non-functional female centrosomic material can undergo renaturation following oocyte activation.

Nuclear Proteins

Spermiogenesis is characterized by alterations in the protein composition of the nucleus. Testis-specific histones are replaced by spermatid-specific basic proteins. The latter are gradually replaced by protamines. Following ROSNI/ROSI techniques and disintegration of the round spermatid nuclear membrane within the ooplasm, the round spermatid DNA-nuclear protein complex is exposed to ooplasmic factors. Since the histones are proteins containing a reduced amount of disulfide bonds, a question may be raised as to how the round spermatid DNA that is not associated with proteins with large numbers of disulfide bonds can exist within the ooplasm and how it is protected against an immediate action of ooplasmic factors. The answer to this question is that the activation of the oocyte can rescue the chromosomes of the round spermatid from premature condensation (Kimura, Yanagimachi 1995a, 1995b). Therefore, in non-obstructed azoospermic men whose spermatids cannot activate an oocyte, application of an exogenous stimulus for ooplasmic activation is of paramount importance.

Guidelines and Prerequisites for ROSNI/ROSI Techniques

The following issues are important for successful performance of ROSNI/ROSI techniques.

Quality Control for the Identification of Round Spermatids

Several methods for identification of round spermatids have been discussed. It should also be emphasized that training is necessary for the staff of assisted reproduction centers applying ROSNI/ROSI techniques. Even if a center has an excellent ICSI program, ROSNI/ROSI procedures will result in poor outcome if the staff members of that center do not spend many hours observing animal testicular tissue specimens attempting to identify round spermatids via an inverted microscope. Technicians, embryologists and physicians performing ROSNI/ROSI techniques should also confirm via TEM, FISH or CSLM techniques that fractions of the cells that are considered as human round spermatids really are round spermatids.

Quality Control for Viability of Round Spermatids

An occasional finding in ROSNI/ROSI programs is the absence or reduced number of live spermatids. Fractions of round spermatids retrieved from testicular tissue should be processed for assessment of viability (Sofikitis et al. 1996a). Men with a percentage of live round spermatids lower than 10% have a poor ROSNI outcome. Preliminary studies have shown that spermatids from such men cannot fertilize oocytes (Sofikitis et al. 1998b).

Quality Control for the Capacity of Round Spermatids to Activate Oocytes

A previous study has shown that ICSI or ROSNI failure in a selected subpopulation of infertile men is attributable to subnormal OASIS profiles (Sofikitis et al. 1997c). Application of a recently reported quantitative assay to appreciate OASIS functionality is recommended; in this assay, two round spermatids are injected into a hamster oocyte (Sofikitis et al. 1997c). If the percentage of activated hamster oocytes in the assay is less than 8%, fertilization is not anticipated after human ROSNI/ROSI techniques. Alternatively, when OASIS deficiency is suspected, an exogenous stimulus (chemical or electrical) may be applied to support human oocyte activation.

Media for Maintenance of Round Spermatid Viability

Most of the popular media in assisted reproduction programs have been devised to maintain viability of spermatozoa and not of spermatids. There are several anatomical and biochemical differences between the round spermatid and the spermatozoon. A medium (SOF medium) has been developed to prolong the viability of round and elongating spermatids (Sofikitis et al. 1998a). It has been already used for maintenance of human and rabbit round spermatids (Yamanaka et al. 1997; Sofikitis et al. 1997a). It contains lactate and glucose as energy substrates. Previous studies have demonstrated that lactate is the preferable energy substrate for round spermatids (Nakamura et al. 1978). Round spermatids have a larger amount of cytoplasm than the spermatozoa. To protect round spermatids against environmental shock and stabilize the spermatid membrane, cholesterol has been added to the SOF medium at a small concentration. It has also been demonstrated that iron and vitamins influence the spermatid viability, so vitamins and ferric nitrate were chosen as components of the SOF medium (Sofikitis et al. 1998a).

Media for Culture of Oocytes Injected with Round Spermatids

Previous studies have shown that the addition of antioxidants to media used for culture of embryos generated from the fertilization of oocytes by spermatids has beneficial effects on embryonic development (Sofikitis et al. 1996a).

Time to Observe Pronuclei

Pronuclei should be observed at nine hours after human ROSNI techniques (Sofikitis et al. 1995b; Yamanaka et al. 1997). When human ooplasmic injections of elongating spermatids are performed, the appropriate time for pronuclei observation is 13 hours post-injection (Sofikitis et al. 1998a).

Genetic Implications of ROSNI/ROSI Procedures

To evaluate the genetic risks of assisted reproductive technologies, one has to consider the genetic risks inherent to the treatment population and the genetic risks inherent to the procedure performed. Genetic risks inherent to ROSNI/ROSI procedures may involve centrosomic abnormalities resulting in aberrant spindle formation and subsequently in an increased risk of mosaicism; injection of disomic/diploid genetic material which could give rise at fertilization to a trisomic/triploid embryo and fetus; and genomic imprinting abnormalities. Genetic risks of ROSNI/ROSI techniques inherent to a population of men with primary testicular damage are the same as those associated with ICSI procedures (transferring sex chromosomal abnormalities or reciprocal translocations associated with spermatogenic impairment). Inheritance of gene mutations/deletions of DNA sequences in specific regions of the Y-chromosome long arm is an additional risk.

Genomic Imprinting Abnormalities

Most genes are expressed equally from the two parental alleles, but a small subgroup of mammalian genes is differentially expressed depending on whether the genes have been inherited from the mother or the father. The process which differentially marks the DNA in the parental gametes is termed genomic imprinting. Genes whose expression is inhibited after passage through the mother's germline are called maternally imprinted, whereas genes whose expression is inhibited when transmitted by fathers are called paternally imprinted. Imprinted genes have been identified in mice and humans. Mouse insulin-like growth factor-2 (IGF_2) gene is expressed only from the paternal allele. In contrast, the gene encoding a differentiation-related fetal RNA (H19) is expressed only from the maternal alleles. IGF_2 and H19 are also monoallelically expressed in the human. Additional imprinted genes have been characterized (Reik 1989) in the mouse (Xist, Insulin 1, Insulin 2, Mas-oncogene, Znf127, U2afbp-rs, among others) and the human (p57KIP2, CGb, IPW, among others).

There may be some conservation of the imprinted segments between the human and the mouse. It is not known how large the imprinted regions are and how many genes in these segments could potentially show differences in expression according to their parental origin. The evolution of imprinting may be a consequence of a genomic conflict. The conflict theory of imprinting follows as an extension of the classical parent-offspring conflict (Hurst, Moore 1996). In species with multiple paternity, within or between broods, paternally inherited genes are less related to fellow progeny of the same mother than are maternally inherited genes. Thus, paternally derived genes will be under selection to extract more resources from the mother than it is in her interest to give. In contrast, the maternally inherited genes will have an inhibitory role to the action of the paternally derived genes (Hurst, Moore 1996).

It has been suggested that DNA methylation provides a molecular mechanism maintaining the imprinting of some genes. It has been shown that variable penetrance and expressivity of some transgenes in mice can be influenced by modifying genes and that there can be differences according to parental transmission (Reik 1989). In these cases, it has been shown that the expression of the transgene is tightly correlated with DNA methylation. Higher levels of DNA methylation are accompanied by lower levels of expression of the respective genes. However, the possibility that other epigenetic mechanisms affect and/or regulate the process of the imprinting of specific genes cannot be ruled out.

Several studies have shown that imprinted genes regulate the development of the embryo/fetus. Abnormalities in genomic imprinting are associated with genetic diseases. Prader-Willi syndrome, Angelman syndrome and Beckwith-Wiedemann syndrome are examples of abnormal functional imprinting. The status of imprinting of some mutations may affect/regulate the dominance or recessivity of the mutation. Furthermore, abnormal functional imprinting is implicated in tumorogenesis. In recessive tumor pathophysiologies, repression of the normal allele opposite to the mutant allele may initiate growth of the tumor.

Kimura and Yanagimachi (1995a, 1995b) have suggested that genomic imprinting is complete at the mouse round spermatid stage. To extend that conclusion to the human, additional studies are necessary. If genomic imprinting is incomplete in subpopulations of men with primary testicular damage, abnormalities may not become manifest at the early embryonic development but may be detectable in the fetus or during the postnatal life. A question of great clinical importance is whether genomic imprinting has been completed at the human round spermatid stage. To attempt to answer this question the imprinting of a gene should be divided into three stages: a) erasure of the previous imprint, b) re-imprinting and c) consolidation of the new imprint. There is strong evidence that erasure of the previous imprint occurs prior to meiosis and that re-establishment of the new imprint begins prior to the pachytene stage of meiosis (Tycko 1997). In contrast, the fact that DNA methyltransferase enzyme is present in spermatids may be an argument against the thesis that genomic imprinting is complete at the round spermatid stage. However, it should be emphasized that waves of DNA methylation have been demonstrated during early embryonic development, the blastocyst stage, and the time of implantation (Fishel et al. 1996).

These observations tend to suggest that even if genomic imprinting is not complete at the round spermatid stage, genomic imprinting may be completed after the transfer of the round spermatid within the ooplasm. The observations also point out that it it may not be valid or appropriate to characterize genomic imprinting complete in a subpopulation of spermatogenic cells.

Is It Too Early to Perform ROSNI/ROSI Procedures?

Prior to the first human ICSI pregnancy, there was a lack of studies in experimental animals evaluating the health, chromosomes and genes of offspring born after ICSI techniques. In contrast, before human ROSNI/ROSI techniques were performed, the techniques had been applied in rabbits (Sofikitis et al. 1994a) and mice (Ogura et al. 1994) and resulted in delivery of healthy offspring. It appears that human ROSNI/ROSI techniques had been more carefully scheduled than the ICSI procedures prior to their initial application.

ROSNI/ROSI techniques have been criticized by a number of scientists because of genetic risks, low pregnancy rates and inherent technical difficulties (mainly regarding the identification of live round spermatids) that do not allow the majority of the assisted reproduction centers to perform these techniques.

However, these theoretical genetic risks should not be used to exclude men from appropriate infertility treatment. Rather, genetic risks should be extensively discussed with the ROSNI/ROSI or ICSI candidate. To date, there is no evidence of a major or minor abnormality in ROSNI/ROSI human newborns, or in animal offspring. In addition, it is a fundamental human right for every couple to obtain therapy for relief of the disease of infertility. Thus, non-obstructed azoospermic men have the right to choose their treatment after adequate information. Furthermore, although a specialized staff is necessary to perform ROSNI/ROSI, these techniques should be inexpensive. If fertilization is achieved, embryonic biopsy (preimplantation diagnosis) is recommended. Alternatively, couples should be advised to undergo prenatal control after achievement of pregnancy. ROSNI/ROSI may not be criticized because of their low pregnancy rate, because these techniques represent the only hope for non-obstructed azoospermic men without spermatozoa to father their own children.

In the four years following the first application of ROSNI/ROSI techniques in the human (Sofikitis et al. 1995a), more than 20 pregnancies have been achieved worldwide via ooplasmic injections of spermatids by seven different groups working independently (Sofikitis et al. 1995a; Tesarik et al. 1995; Fishel et al. 1995; Mansour et al. 1996; Tanaka et al. 1996; Antinori et al. 1997a, 1997b; Vanderzwalmen et al. 1997; Sofikitis et al. 1997b; Araki et al. 1997; Amer et al. 1997; Sofikitis et al. 1998a, b, c). Additional research efforts are necessary to improve the outcome of ROSNI/ROSI techniques.

These efforts should be directed to the discovery and development of criteria for identification of round spermatids, biochemical media prolonging the viability of spermatids, exogenous stimuli to support oocyte activation in men with OASIS deficiency, methodology to purify human OASIS and methodology to study the metabolism and the implantation process of embryos generated by ooplasmic injections of spermatids. Finally, criticism of ROSNI/ROSI techniques based on their technical difficulties is not justified. Training and basic research in experimental animals is a necessary prerequisite for the staff of assisted reproduction centers applying ooplasmic injections of spermatids.

The Post-ROSNI Period in the World of Assisted Reproduction

ICSI or ROSI/ROSNI procedures offer alternative solutions for men with non-obstructive azoospermia and testicular foci of spermatogenesis up to the spermatozoon or round spermatid stage, respectively. In addition, preliminary trials of human ooplasmic secondary spermatocyte injections (SECSI techniques) recently showed that the human male second meiotic division can be completed within the ooplasm and the derived product of the male second meiotic division can fertilize the human oocyte. However, clinical applications of SECSI procedures may be limited since most of the non-obstructed azoospermic men with secondary spermatocytes in the therapeutic testicular biopsy material have spermatozoa and/or spermatids, as well. In contrast, men in whom the most advanced spermatogenic cells are primary spermatocytes cannot be candidates in assisted reproduction programs, due to the lack of chromosomally haploid male gametes. For men with the diagnosis of spermatogenic arrest at the primary spermatocyte stage, who have neither spermatids nor spermatozoa in the therapeutic testicular biopsy material, three recent achievements in basic research offer new possibilities in assisted reproduction programs in the future.

Artificial Testis/*in Vitro* Culture of Germ Cells

An artificial testis may be considered an *in vitro* culture system where primary spermatocytes or round spermatids will be cultured under biochemical conditions similar to those of the testicular microenvironment with the goal of inducing meiosis or achieving generation of the spermatid flagella. The haploid products of an artificial testis may be used in assisted reproduction programs. Few studies support the idea that utilization of an *in vitro* culture system in assisted reproduction may be possible in the future. Early studies of Gerton and Millette (1984) demonstrated the generation of spermatid flagella *in vitro*. Gritsch et al. (1997) showed that human spermatogenic cells can survive for a long time in *in vitro* culture conditions. In a landmark study, Weiss et al. (1997) demonstrated the generation of round spermatids from primary spermatocytes *in vitro*.

Transplantation of Human Spermatogenic Cells into a Host Testis

Results from transplantation of mouse spermatogenic cells into mouse seminiferous tubules and transplantation of rat spermatogenic cells into mouse seminiferous tubules indicate that the donor germ cells are capable of differentiating to form spermatozoa with the morphological characteristics of the donor species (Russell, Brinster 1996). Furthermore, recent findings by the authors of this chapter (1997) have shown that spermatogenic cells isolated from hamsters with primary testicular damage are capable of transforming into hamster spermatozoa when transplanted into the seminiferous tubules of non-immunosuppressed animals, indicating that the transluminal compartment of the seminiferous tubules is immunologically privileged; and that such cellular transformations of the donor cells are also inducible within the seminiferous tubules of immunosuppressed animals after transfer techniques.

If these techniques are applied successfully in the human, primary spermatocytes of non-obstructed azoospermic men with spermatogenic arrest at the primary spermatocyte stage may be transformed into human spermatids or spermatozoa within a host testis, giving the opportunity to these men to be candidates for ROSNI/ROSI or ICSI techniques. However, even if induction of human meiosis becomes possible within a host testis, application of human ROSNI or ICSI techniques using human haploid male gametes generated in an animal testis is susceptible to genetic and immunological risks.

Gene Transfer for the Treatment of Non-obstructive Azoospermia

Gene therapy is an exciting and powerful technique capable of introducing novel genetic sequences to alter the cell phenotype. Recent data reported by Werthman et al. (1997) confirm successful gene transfer of a reporter gene to murine testicular tissue. This technique may have the potential to reverse the effects of mutations which lead to non-obstructive azoospermia by reconstitution of the wild-type gene.

References

Akmal KA, Dufour JM, Kim K. Retinoic acid receptor α gene expression in the rat testis: potential role during the prophase of meiosis and in the transition from round to elongating spermatids. Biol Reprod 1997; 56:549-556.

Amer M, Soliman E, El-Sadek M, Tesarik J. Is complete spermiogenesis failure a good indication for spermatid conception? Lancet 1997; 350:116.

Anderson DM, Lyman SD, Baird A. Molecular cloning of mast cell growth factor, a hematopoietin that is active in both membrane bound and soluble forms. Cell 1990; 63:235-243.

Antinori S,Versaci C, Dani G, Antinori M, Selman H. Fertilization with human testicular spermatids: four successful pregnancies. Hum Reprod 1997a; 12:285-291.

Antinori S, Versaci C, Dani G, Antinori M, Selman H. Successful fertilization and pregnancy after injections of frozen-thawed round spermatids into human oocytes. Hum Reprod 1997b; 12:554-556.

Araki Y, Motoyama M, Yoshida A, Araki M. Intracytoplasmic injection of late spermatids: a successful procedure in achieving childbirth for couples in which the male partner suffers from azoospermia due to deficient spermatogenesis. Fertil Steril 1997; 67:559-561.

Chai N, Phyllips A, Fernandez A, Yen P. A putative human male infertility gene DAZLA: genomic structure and methylation status. Mol Hum Reprod 1997; 3:705-708.

Chabot B, Stephenson DA, Chapman VM, Chai N. The protooncogene *c-kit* encoding a transmembrane tyrosine kinase receptor maps to the mouse W locus. Nature 1988; 335:88-89.

Fishel S, Green S, Bishop M, Aslam I. Pregnancy after intracytoplasmic injection of spermatid. Lancet 1995; 345:1641-1642.

Fishel S, Aslam I, Tesarik J. Spermatid conception: a stage too early, or a time too soon? Hum Reprod 1996; 11:1371-1375.

Gerton GL, Millette CF. Generation of flagella by cultured mouse spermatids. J Cell Biol 1984; 98:619-628.

Gritch H, Bruning C, Robles M. Long term spermatogenic cell culture. J Urol 1997; 157 (Suppl):169.

Hannay T. New Japanese IVF method finally made available in Japan. Nature Medicine 1995; 1:289-90.

Hurst LD, Moore T. Imprinted genes have few and small introns. Nature Genet 1996; 12:234-237.

Kimura Y, Yanagimachi R. Mouse oocytes injected with testicular spermatozoa or round spermatids can develop into normal offspring. Development 1995a; 121:2397-2405.

Kimura Y, Yanagimachi R. Development of normal mice from oocytes injected with secondary spermatocyte nuclei. Biol Reprod 1995b; 53:855-862.

Mansour RT, Aboulghar MA, Serour K. Pregnancy and delivery after intracytoplasmic injection of spermatids into human oocytes. J Middle East Fertil Soc 1996; 1:223-225.

Mendoza C, Tesarik J. The occurrence and identification of round spermatids in the ejaculate of men with non-obstructive azoospermia. Fertil Steril 1996; 66:826-829.

Nagy Z, Liu J, Joris H, Devroy P, Van Steirteghem A. Time course of oocyte activation, pronucleus formation, and cleavage in human oocytes fertilized by intracytoplasmic sperm injection. Hum Reprod 1994; 9:1743-1748.

Nakamura M, Romrell LJ, Hall P. The effects of glucose and temperature on protein biosynthesis by immature (round) spermatids from rat testis. J Cell Biol 1978; 79:1-9.

O'Donnell L, Mc Lachlan R, Wreford N, De Kretser B. Testosterone withdrawal promotes stage-specific detachment of round spermatids from the rat seminiferous epithelium. Biol Reprod 1996; 55:895-901.

Ogura A, Yanagimachi R. Round spermatid nuclei injected into hamster oocytes form pronuclei and participate in syngamy. Biol Reprod 1993; 48:219-225.

Ogura A, Yanagimachi R, Usui N. Behavior of hamster and mouse round spermatid nuclei incorporated into mature oocytes by electrofusion. Zygote 1993; 1:1-8.

Ogura A, Matsuda J, Yanagimachi R. Birth of normal young after electrofusion of mouse oocytes with round spermatids. Proc Natl Acad Sci USA 1994; 91:7460-7462.

Ozil JP. The parthenogenetic development of rabbit oocytes after pulsatile electrical stimulation. Development 1990; 109:117-127.

Rajfer J, Turner TT, Rivera F, Howards S, Sikka S. Inhibition of testicular testosterone synthesis following experimental varicocele in rats. Biol Reprod 1987; 36:933-937.

Reik W. Genomic imprinting and genetic disorders in man. Trends Genet 1989; 5:331-336.

Russell LD, Brinster RL. Ultrastructural observations of spermatogenesis following transplantation of rat testis cells into mouse seminiferous tubules. J Androl 1996; 17:615-626.

Schatten G. The centrosome and its mode of inheritance: the reduction of centrosome during gametogenesis and its restoration during fertilization. Dev Biol 1994; 165:299-335.

Silbert SJ. Sertoli cell-only syndrome. Hum Reprod 1996; 11:229-233.

Sofikitis NV, Miyagawa I, Agapitos E, Pasyianos P, Toda T, Hellstrom JG. Reproductive capacity of the nucleus of the male gamete after completion of meiosis. J Assist Reprod Genet 1994a; 11:335-341.

Sofikitis N, Miyagawa I, Zavos PM, Toda T, Iino A, Terakawa N. Confocal scanning laser microscopy of morphometric human sperm parameters: correlation with acrosin profiles and fertilizing capacity. Fertil Steril 1994b; 62:376-386.

Sofikitis N, Miyagawa I, Sharlip I, Hellstrom J, Toda T. Human pregnancies achieved by intra-ooplasmic injections of round spermatid nuclei isolated from testicular tissue of azoospermic men. J Urol 1995a; 153 (Suppl):258A.

Sofikitis N, Toda T, Miyagawa I, Hellstrom W, Sikka S, Dickey R. Application of ooplasmic round spermatid nuclear injections for the treatment of azoospermic men in USA. Fertil Steril 1995b; 64 (Suppl):S88-S89.

Sofikitis NV, Toda T, Miyagawa I, Zavos PM, Pasyianos P, Mastelou E. Beneficial effects of electrical stimulation before round spermatid nuclei injections into rabbit oocytes on fertilization and subsequent embryonic development. Fertil Steril 1996a; 65:176-185.

Sofikitis NV, Miyagawa, I, Incze P, Andrighetti S. Detrimental effect of left varicocele on the reproductive capacity of the early haploid male gamete. J Urol 1996b; 156:267-270.

Sofikitis N, Yamamoto Y, Isoyama T, Miyagawa I. The early haploid male gamete develops a capacity for fertilization after the coalescence of the proacrosomal granules. Hum Reprod 1997a; 12:2713-2719.

Sofikitis NV, Mantzavinos T, Loutradis D, Miyagawa I, Tarlatzis B. Treatment of male infertility caused by spermatogenic arrest at the primary spermatocyte stage with ooplasmic injections of round spermatids or secondary spermatocytes isolated from foci of early haploid male gametes. Hum Reprod 1997b; 12 (Suppl):81-82.

Sofikitis NV, Kanakas N, Mantzavinos T, Loutradis D, Kalianidis K, Miyagawa I, Tarlatzis B. Deficiency in the oocyte-activating substance in spermatozoa: a cause of ICSI failure. Hum Reprod 1997c; 12 (Suppl):81.

Sofikitis N, Yamamoto Y, Miyagawa I, Mekras G, Mio Y, Toda T, Antypas S, Kawamura H, Kanakas N, Antoniou N, Loutradis D, Mantzavinos T, Kalianidis K, Agapitos E. Ooplasmic elongating spermatid injections for the treatment of non-obstructive azoospermia. Hum Reprod 1998a; 13:709-714.

Sofikitis N, Mantzavinos T, Loutradis D, Miyagawa I, Trokoudes K, Tarlatzis V. Parameters influencing the outcome of round spermatid injections into oocytes: source of round spermatids, round spermatid viability, and capacity to activate oocytes. Fertil Steril 1998b; 70:S19-S20.

Sofikitis N, Mantzavinos T, Loutradis D, Yamamoto Y, Tarlatzis V, Miyagawa I. Ooplasmic injections of secondary spermatocytes for non-obstructive azoospermia. Lancet 1998c; 351:1177-1178.

Tanaka A, Nagayoshi M, Awata S, Tanaka I. Clinical evaluation of round spermatid injection into human oocytes. Fertil Steril 1996; 66 (Suppl):S99-S100.

Tanaka A, Nagayoshi M, Awata S, Tanaka I, Sofikitis N. Conclusions from transplantation of human or hamster spermatogonia/primary spermatocytes to rat or mouse testis. Fertil Steril 1997; 68 (Suppl):S61-S62.

Tesarik J, Mendoza C, Testart J. Viable embryos from injection of round spermatids into oocyte. N Eng J Med 1995; 333:525.

Tycko B. Post-graduate course program of The American Society of Andrology. J Androl 1997; 18:83-106.

Vanderzwalmen P, Zech H, Birkenfeld, B, Schoysman R. Intracytoplasmic injection of spermatids retrieved from testicular tissue: influence of testicular pathology, type of selected spermatids and oocyte activation. Hum Reprod 1997; 12:1203-1213.

Weiss M, Vigier M, Hue D. Pre- and postmeiotic expression of male germ cell-specific genes throughout 2-week-cocultures of rat germinal and Sertoli cells. Biol Reprod 1997; 57:68-76.

Werthman P, Kaboo R, Peng S. Adenoviral mediated gene transfer to murine testis *in vivo*. J Urol 1997; 157 (Suppl):169.

Yamanaka K, Sofikitis N, Miyagawa I, Yamamoto Y, Toda T, Antypas S, Dimitriadis D, Takenaka M, Tanigouchi K, Takahashi K, Tsukamoto S, Kawamura H, Neil M. Ooplasmic round spermatid nuclear injections as an experimental treatment of non-obstructive azoospermia. J Assist Reprod Genet 1997; 14:55-62.

38 Potential Pitfalls in Male Reproductive Technology

James M. Cummins
Murdoch University

Since Leeuwenhoek discovered the mammalian sperm in 1677 (following a prompt by the student Johan Hamm), the science of spermatology—at least in vertebrates—has largely revolved around questions of basic biology and its application to the breeding of production animals. However, the past 20 years have seen the rapid reapplication of these fundamental research techniques towards the treatment of human infertility. This is being carried out by clinics where the knowledge of evolutionary biology is frequently meagre, and where short-term commercial survival interests may dominate over principles of long-term accountability and the interests of the child. This review will attempt to bring a biological perspective to this sensitive issue.

While the reproductive pioneers Rock and Menken reportedly attempted *in vitro* fertilization (IVF) in 1944 (Valone 1998), the first practical techniques arose in the 50s and 60s as an exercise in applied biological research with the aim of assisting animal production (Yanagimachi 1981). With the development of laparoscopy and reasonably reliable techniques for stimulating and monitoring ovulation, infertility specialists soon applied IVF to human infertility. This culminated in the birth of Louise Brown in 1978. This watershed in applied reproductive technology (ART) was bitterly opposed by many reproductive biologists at the time but led to an explosion in IVF clinics throughout the world (Edwards 1980). The promise of a child of their own versus adoption (itself becoming very difficult) obviously touched a deep chord with infertile couples, even though radical feminist groups such as FINNRAGE argued (perhaps with some cause) that IVF posed unacceptable risks for women (Purdy 1996). Many ART pioneers asserted that the main rationale for IVF was to assist women with tubal disease where sperm and eggs could not normally meet. Perhaps these arguments were aimed at deflecting the storm of criticism that emerged—particularly from the Catholic Church—by showing that IVF was, if anything, "pro-family." By the early 1980s IVF was being advocated for general female and idiopathic male infertility. At that time, donor insemination was the only serious option, as adoption was increasingly more difficult (Trounson 1982).

IVF worked well for men with reasonably normal sperm counts, even though fertilization rates generally plateaued at around 65%. However, for men with very poor sperm counts, poor motility or high levels of abnormal forms (and especially for men with all three sperm defects), IVF frequently resulted in fertilization failure. This was an expensive disaster for couples, a humiliation for the clinics who had advertised hope to the infertile and of course a life-threatening risk for the women. Tentative attempts were made to improve fertilization success rates. Clinics first tried by increasing sperm numbers around the egg, then by weakening or even drilling the zona pellucida (also proposed to assist hatching and implantation) and then by subzonal sperm injection. None of these approaches made more than a minor impact. There were worries about the introduction of infectious organisms into the embryo, and the problems of polyspermic fertilization and embryo loss through genetic imbalance (Ng 1992). Intracytoplasmic sperm injection (ICSI) was initially carried out in sea urchins (Hiramoto 1962). It was then proposed as an experimental tool for studying sperm-egg interactions in the 1970s and 1980s (Yanagimachi 1995) but very rapidly moved into the infertility clinics, starting in the early 90s (Palermo et al. 1992). It now dominates assisted reproductive technology.

ICSI was first seen as a solution to IVF fertilization failure in men with poor semen quality, but it is now proposed as a solution for cases of total azoospermia due to ductal obstruction or spermatogenetic failure. ICSI certainly improves the prognosis for men with very severe semen defects or obstructive azoospermia. However, it never gives a 100% fertilization rate because some sperm fail to decondense or to activate oocytes (Schill et al. 1998; Asch et al. 1995). It is also being proposed for use in cases of meiotic arrest, where only immature germ cells can be recovered from the testis (Edwards et al. 1994; Tesarik, Mendoza 1996a, b; Tesarik et al. 1996, 1998). This is highly controversial, however, and is banned in several countries, including the United Kingdom and Holland. ICSI is also being proposed as a procedure to improve the chances of pregnancy in cases where only a few oocytes can be recovered, due to poor ovarian response or advanced age of the woman (Ludwig et al. 1997).

In Australia, micro-assisted ART (mostly ICSI) accounted for around one-third of procedures in 1996, and probably now exceeds 50% (Hurst et al. 1997). The success of ICSI means that it is now being applied to a majority of cases of infertility. There is a disquieting trend to skip clinical investigations and move directly to ICSI based purely on a semen analysis. This is an insidiously lazy and even dangerous loss of control over the clinical decision-making process (Jequier, Cummins 1997; Cummins, Jequier 1994). One clear disturbing message to spermatologists is that ICSI effectively bypasses all the selective mechanisms that have built up over the hundreds of millions of years over which amphimixis has evolved (Margulis, Sagan 1986). There is thus profound disquiet that science may have opened a Pandora's box of unknown genetic consequences, particularly as researchers do not understand many of the underlying causes of male infertility. In "curing" male infertility, scientists may be going against the trend of natural selection. There is some evidence that infertility (or to be more precise, subfertility) may be integral to the human male reproductive pattern—or at least not necessarily maladaptive. The genetically controlled menopause, for example, imposes a period of infertility in the mature life of women. Some argue that it may have adaptive value by increasing reproductive inclusive fitness through grandmothering (Cummins 1990, 1998a; Diamond 1997; Hawkes 1998). However, it may be that, like the menstrual cycle (Finn 1998), these aspects of reproductive physiology have persisted in modern humans as selectively neutral consequences of other aspects of human evolutionary history. In other words, they are simply genetic "noise" and it may be pointless to attempt to force them into an adaptive mould.

This review will take the stance that ICSI using immature or abnormal germ cells is unacceptable because of science's ignorance about the possible long-term effects. It is easy to make Jeremiad statements about novel medical techniques in a society that increasingly distrusts the nature and ethics of science, but it is better to attempt to be constructive. The genetic and other risks of ICSI have been discussed and possible guidelines for clinical practice have been offered elsewhere (Cummins, Jequier 1995; Cummins 1997), so this review will concentrate on the most recent findings from a biological perspective.

Reproductive medicine seems to be particularly fertile ground for the development of novel jargon and acronyms (Table 1).

Table 1: A list of some of the more novel acronyms in human assisted reproductive technology.

Acronym	Translation	Reference
ROSI	Round spermatid injection	Tesarik et al. (1998)
ROSNI	Round spermatid nuclear injection	Sofikitis et al. (1995)
ELSI	Elongating sperm injection	Tesarik, Mendoza (1996a, b)
PESA	Percutaneous epididymal sperm aspiration	Craft et al. (1995)
TESA	Percutaneous testicular sperm aspiration	Silber et al. (1995a, b)
TESE	Testicular sperm extraction	Silber et al. (1995a, b)
MESA	Microsurgical epididymal sperm aspiration	Silber et al. (1995a, b)
CBAVD	Congenital bilateral absence of the vas deferens	Patrizio et al. (1993)

Genetic Background to Male Infertility

It is surprisingly difficult to diagnose or even define fertility or infertility in a male. By definition it can only refer to a man who is able to impregnate a woman. False paternity rates of around 10-15% in many societies make assurance of paternity far from clear (Cummins et al. 1994; Jequier 1985). This is tacitly acknowledged in the many laws and customs aimed at controlling female (but not male) reproductive behavior. Fertility can only be defined as a trait that can be expressed in a relationship, and individuals who are fertile with one partner may be infertile or subfertile with another (Steinberger, Rodrigues-Rigau 1983).

For the male, fertility at its basis relies on sperm production and delivery. Spermatogenesis is a complex biological process involving the hierarchical interaction of multiple genes and there is thus potential for error at multiple levels (Vogt et al. 1997). In man it starts with the migration of primordial germ cells to the gonadal ridge at around three weeks (Byskov 1982). These invade the medulla to form cords that at puberty will form the spermatogenetic tubules. In most mammals, the SRY gene that causes the differentiation of Sertoli cells from renal somatic cells (Graves 1995) triggers differentiation of the male tract from the underlying female archetype. These in turn secrete Müllerian Inhibitory Substance (MIS) that suppresses the formation of the female archetype. This causes the pronephric and mesonephric ducts to form the efferent ductules, and epididymis and vas deferens, respectively. This process depends in turn on the masculinizing influence of testosterone from the intertubular Leydig cells. There are between 1250 and 1750 genes controlling spermatogenesis in *Drosophila* but the number in humans is not known (Vogt 1998). While the Human Genome project will undoubtedly identify many of the genes for spermatogenesis, it is likely that control is a multi-nested emergent process involving lifestyle factors and hierarchical interactions between somatic and sex chromosomal genes. One interesting recent paper indicates that completion of meiosis may require the expression of cyclic AMP responsive element modulator (Lin et al. 1998). Work on creating gene knockout mice for spermatogenetic control genes is progressing rapidly but science cannot yet claim to have an adequate animal model for genetically caused human male infertility (Ruggiu et al. 1997; Yanagimachi 1995). For other causes of infertility such as trauma, disease or endocrine disorders, the likelihood of finding an appropriate animal model is even more remote.

Developmental and environmental influences can have profound effects on the efficiency of spermatogenesis, and human spermatogenesis is remarkably inefficient even in "normal" individuals when compared to most other mammals (Johnson 1995). Absolute testis size appears to be determined by the duration of Sertoli cell mitotic proliferation during critical phases of perinatal and prepubertal development. This process acts through Sertoli cell tri-iodothyroxin (T3) receptors (Simorangkir et al. 1995, 1997). Testicular development can also be affected by environmental factors such as polychlorinated biphenyls that depress thyroid function. This is interesting given the prevalent preoccupation with falling sperm counts and xenoendocrine factors that may affect the human male (Carlsen et al. 1995; Jensen et al. 1995; Cooke PS et al. 1996; Sharpe, Skakkebæk 1993; Toppari 1996; Swan et al. 1997). Testicular size is an evolutionary labile trait thought to be closely linked with mating systems and with the intensity of sperm competition. There has been marked divergence in these inter-related factors during the 7-10 million years in the evolution of humans and their close relations, the chimpanzee and gorilla. Men, for example, have testes half the size of the much smaller and highly promiscuous chimpanzee and within humans there are marked differences between ethnic groups (Harcourt et al. 1981; Short 1997; Mittwoch 1988).

The notion of a simplistic "control" mechanism for spermatogenesis is thus a chimera. Even the "25th chromosome"—mitochondrial DNA (mtDNA)—is likely to be involved in modulating male reproduction. Because mtDNA is almost exclusively maternally inherited in mammals, long term male "fitness" is irrelevant to the mitochondrial genome as any defects will not get transmitted. At least one mitochondrial disease (Leigh's Hereditary Optic Neuropathy—LHON) is more severe and more prevalent in the male, and one might suspect that some forms of male infertility may also fit this pattern (Cummins et al. 1994; Johns 1996). While sperm even from normal men can show significant levels of mtDNA deletions (Cummins et al. 1998), it is unlikely that these will have significant impact on children conceived by ICSI as the sperm's mitochondria do not survive in the embryo (Houshmand et al. 1997). The mechanism for this is unknown but probably involves tagging of the sperm's midpiece by ubiquitin before lysis at the early morula stage (Sutovsky et al. 1996; Cummins 1998b).

Inherited Fertility Disorders

One unavoidable complication for clinics is that severe male infertility appears to be associated with a 10-fold increased background load of genetic anomalies. These are reflected in increased levels of disomy and other chromosomal anomalies in spermatozoa that would not normally be transmitted to offspring (Jaffe, Oates 1994; Moosani et al. 1995; Vogt 1995). Inherited genetic disorders such as neurofibromatosis are also linked with increasing paternal age (Crow 1995). This seems to be because the spermatogenetic process, where mitotic divisions continue throughout life, is inherently more mutagenic than oogenesis, where mitosis suspends in mid-gestation (Auroux et al. 1989, 1998). Paternal influences on embryonic mortality have been known to reproductive biologists for years (Bishop 1964). Rather surprisingly, elevated levels of chromosome aberrations are found in the wives as well as the husbands in infertile couples (Montag et al. 1997; Vanderven et al. 1998). This again highlights the need for very thorough clinical examination of both partners, rather than simply relying on ICSI as a quick "fix" to infertility.

The terminology of the Y chromosome has been revised recently (Vogt 1997a, b, c). There are a number of specific mutations in the DAZ (Deleted in Azoospermia) family now known to be associated with male infertility. Most notable are deletions of what is now termed the AZFc region on the euchromatic part (Yq11) of the long arm of the human Y chromosome (Vogt et al. 1997; White et al. 1997). This is a polymorphic region highly liable to mutations as it lacks a homologous pairing chromosome for DNA repair (Short 1997). In *Drosophila*, mouse and man, X-Y pairing failure is a potent cause of spermatogenetic disruption (Burgoyne et al. 1992; Mohandas et al. 1992; McKee et al. 1998). In humans, Vogt finds three different *de novo* Yq11 microdeletions associated with male infertility, designated as AZFa, AZFb, and AZFc (Vogt et al. 1996; Vogt 1997a, b, c). Some candidate genes exist for spermatogenetic control—RBM for AZFb; and DAZ and SPGY for AZFc. DAZ is now known to exist in multiple copies and SPGY is a homologous Y copy. Therefore DAZ has now been redesignated as DAZ1 and SPGY as DAZ2 (Vogt 1997a, b, c). The DAZ genes encode testis-specific RNA binding proteins with similar sequence structure. They are derived from an ancient archetype. This exists as an autosomal gene in the mouse (Dazl1) and is also seen in *Xenopus* (XDAZL) and in *Drosophila* (boule), as shown in Houston et al. (1998), Cooke HJ et al. (1996) and Eberhart et al. (1996). The mouse Dazl1 transcription products are seen in both male and female gonads but not germ cell-depleted mutants (Ruggiu et al. 1997; Seligman, Page 1998). In humans, the levels are much higher in the testis than the ovary, suggesting that these genes act at an early stage of meiosis but are much more prevalent in the active testis compared with the ovary, where meiosis suspends in fetal life (Byskov 1982). In humans, DAZ proteins are located apically in the seminiferous epithelium and in the tails of spermatozoa, suggesting a role in RNA metabolism of late spermatids. Habermann et al. (1998) suggest that they may play a role in the storage or transport of testis-specific mRNA, the translation of which is repressed until the formation of mature spermatozoa. Deletion of DAZ genes therefore may not necessarily interfere with human sperm maturation but instead may cause a gradual reduction of mature spermatozoa. The discovery that human sperm carry a range of mRNAs left over from spermatogenesis suggests a novel way to backtrack the genes that control the system and possibly to map out those that affect sperm production efficiency (Miller 1997; Kramer, Krawetz 1997).

Many clinics now screen for Y deletions and counsel couples about the likelihood of infertility in male offspring (Vanderven et al. 1997; Vogt 1998). Some argue that screening for Y deletion status is largely irrelevant as it occurs at low incidence and the knowledge makes little difference to clinical practice (Ludwig et al. 1998). This is a pragmatic position but one wonders how the male offspring will feel about this if they too are infertile. The Y chromosome is highly labile and it is possible that mutations may arise in the germ cell line—or even a subset therein—and thus appear *de novo* in the male offspring of normal men (Kent-First et al. 1996a, b). This makes the philosophical position for routine screening problematic: the use of resources for such purposes is pointless unless it is going to result in practical output.

Researchers have to bear in mind that in evolutionary terms the Y chromosome is relatively unstable. There is evidence that both the DAZL and hnRNPG (an autosomal homolog of RBM—"RNA-binding-motif") are RNA binding motif genes that moved to the Y from autosomes within relatively recent evolutionary history (Saxena et al. 1996; Delbridge et al. 1998). This process involved translocation of the autosomal gene to the Y, amplification and truncation of exons within the transposed gene, and subsequent amplification of the modified gene. For RBMY there are at least 30 genes and pseudogenes found on both arms of the chromosome (Chai et al. 1998). The loss (or failure to acquire) a pairing partner creates inherent instability for the Y chromosome. This conforms to Ohno's rule (Ohno et al. 1977) that chromosomes with a sex-determining trait tend progressively to lose the ability to recombine and are thus destined eventually to extinction, as has happened for some species of *Drosophila* (George 1997). Vogt (personal communication) prefers the view that any gene that jumps to the Y chromosome will be selected for, as long as its function has any benefit for male reproductive fitness. RBM and DAZ may have become amplified on the Y chromosome, because the non-pairing region is rich in tandem Y-specific repeat elements that facilitate intra-chromosomal unequal crossing-over events due to specific chromatin folding structures (Vogt 1990).

At least one other well-defined genetic factor that needs to be considered when dealing with male infertility is the CFTR gene complex causing cystic fibrosis. Carriers can present with a spectrum of reproductive tract disorders ranging from congenital absence of the vas deferens (a clear indication for TESE or TESA) down to mildly impaired epididymal function resulting in secondary semen defects (Patrizio et al. 1995; Patrizio, Zielenski 1996; Kanavakis et al. 1998; Vanderven et al. 1996). The incidence of men with CFTR mutations and with non-obstructive oligozoospermia is quite variable and some groups find no significant increase in their patient population groups (Tuerlings et al. 1998). Many clinics now screen and counsel both partners for CFTR carrier status and in some cases can offer preimplantation genetic screening of embryos at risk of developing cystic fibrosis (Ao et al. 1996). Some consider that follow-up studies of children born to couples at risk of transmission should be mandatory, although it is not clear who should pay for this service (Lissens, Liebaers 1997).

Besides genes causing infertility that could be transmitted to offspring, there is increasing concern that DNA damage to sperm from infertile men may result in increased risk of childhood cancers in offspring. There is good evidence, for example, that smoking (a known cause of sperm DNA breakage) is a significant risk factor for the children (Ji et al. 1997; Vine 1996) There is now a range of experimental techniques aimed at measuring DNA damage that will soon be applied clinically (Evenson, Jost 1994; Manicardi et al. 1998; Hughes et al. 1997). Infertility is also associated with abnormal protamination in sperm DNA. This is another possible cause of ICSI failure though failure of the male pronucleus to form normally (Bench et al. 1998). There is evidence that sperm DNA fragmentation is higher in poor semen samples and that this DNA damage along with oocyte DNA breakage is increased with age and associated with ICSI failure (Lopes et al. 1998a, b). This topic has been covered elsewhere. Suffice it to say that wherever possible, ICSI technicians should aim to select morphologically normal sperm with fully condensed nuclei (Cummins, Jequier 1995).

The Androgen Receptor Gene and Polyglutamination Diseases

While androgens are essential for testicular function, most men with impaired spermatogenesis have normal levels of circulating androgens. Endocrinopathies as defined classically are rare (de Kretser 1990). Recent work from a Singapore group, however, has indicated that subtle changes to the androgen receptor (AR) gene may increase the risk of defective spermatogenesis and infertility (Tut et al. 1997). The AR protein is activated by binding to androgen and translocates to the nucleus where it binds to androgen receptor elements (AREs) in promoter regions of androgen-sensitive genes. This causes specific gene transcription (Quigley et al. 1995). However, the AR gene has two polymorphic and unstable trinucleotide repeat elements. CAG repeats encode for polyglutamine and GGC repeats encode for polyglycine. These elements are highly unstable. In a transgenic mouse model carrying an AR gene with 45 repeats in a yeast artificial chromosome, intergenerational variation in repeat length was 10%. Inheritance was particularly unstable with maternal transmission and increased maternal age (Laspada et al. 1998), These repeats are in exon 1, encoding for the *trans*-activation region of the receptor protein. Optimization of androgen receptor function is critically dependent on the length of the polyglutamine repeat. Such elements are thought to function in a regulatory fashion in a wide range of receptor genes (Karlin, Burge 1996). Relatively short repeats result in androgen hypersensitivity with increased risk of prostate cancer (Irvine et al. 1995; Hardy et al. 1996). Long repeats (40 or more) cause a fatal neurological disorder, associated with infertility and impaired testicular function (spinal bulbar muscular atrophy or Kennedy syndrome). Tut et al. (1997) demonstrated that for intermediate length repeats there was a close tie between CAG repeat number and defective spermatogenesis. Men with 28 or more repeats had a four-fold increased risk of reduced spermatogenesis. Other regions of the AR gene may also be liable to point mutations associated both with androgen insensitivity and prostate cancer, pointing out again the sensitivity of control mechanisms in this region (Yong et al. 1998). In at least one case of idiopathic hypogonadotrophic hypogonadism, the patient showed a good response to exogenous gonadotropins, to the point where ICSI could be attempted with ejaculated sperm. This attempt produced three embryos but no successful pregnancy. The wife later conceived spontaneously (Yong et al. 1997).

There is evidence that variability in other hormone receptors may contribute to male infertility. The follicle stimulating hormone (FSH) receptor, for example, is polymorphic and mutations to it and to the luteinizing hormone (LH) receptor are known causes of infertility and abnormal reproductive development (Simoni et al. 1998).

Clearly this information should be provided to couples seeking treatment. Perhaps men should be investigated for partial androgen insensitivity as part of routine clinical assessment. There are two copies of the AR gene, one on chromosome 11 and one on the X chromosome, responsible for X-linked spinal bulbar muscular atrophy. In this case the implications for transmission would be for the grandsons, 50% of whom would receive the gene defect.

Immature Germ Cells and Genomic Imprinting

Few topics have caused more controversy of late than the use of immature germ cells in ICSI (Tesarik, Mendoza 1996a, b; Tesarik et al. 1996, 1997; Edwards et al. 1994; Elder, Elliott 1998). While normal animals have been born from spermatids (Ogura, Yanagimachi 1995), spermatocytes (Kimura, Yanagimachi 1995) and from abnormal mouse spermatozoa (Burruel et al. 1996; Sasagawa, Yanagimachi 1997), there are serious doubts as to whether these models are appropriate for humans. There are concerns that very immature germ cells may not have completed aspects of genomic imprinting (Tesarik, Mendoza 1996a, b), and that they lack the capacity to organize the centriole-mediated first embryonic spindle apparatus, thus resulting in incomplete syngamy (Asch et al. 1995; Hewitson et al. 1998; Sathananthan 1998). In a rabbit model, spermatids develop full fertilizing competence only after the coalescence of the acrosomal vesicle (Sofikitis et al. 1997). This may be associated with the development of sperm-borne oocyte activating factor in the perinuclear theca (Kimura et al. 1998).

There are serious doubts as to whether clinics can reliably distinguish round spermatids from other cells or even Sertoli cell nuclei without using thin-layer phase contrast microscopy, which is a technique not normally available. Silber and Johnson (1998) argue that if meiosis has progressed to the point of round spermatid formation, then an exhaustive search should reveal elongating spermatids, as arrest at late meiosis is very rare. While pregnancies have been reported from the injection of round and elongating spermatids from both testicular biopsies and ejaculates, the overall success rate is low (Sofikitis et al. 1998; Tesarik, Mendoza 1996a, b; Sousa et al. 1998; Tesarik 1997a, b). In the United Kingdom, the Human Fertility and Embryology Authority (HFEA) has banned the use of non-motile sperm for ICSI and in its 1997 report stated ". . . there is insufficient evidence from animal studies and from research to demonstrate its safety and efficacy" (HFEA 1997). A similar ban exists in Holland and a voluntary moratorium on reporting pregnancies is in place in Japan.

Diagnosis, Training and Decision-Making in Clinical Andrology

The term "andrology" was first coined over 100 years ago by the Congress of American Physicians and Surgeons to differentiate the study of the male from gynecology, and was re-invented in Germany this century (Niemi 1987). Dorland's Medical Dictionary (Friel 1965) defines it as the "scientific study of the masculine constitution and of the diseases of the male sex; especially the study of diseases of male organs of generation." Sadly, it is a medical discipline that badly needs re-shaping and catching up with its scientific counterpart. The past decade has seen significant advances in the andrology laboratory, with much emphasis on developing external and independent quality assurance schemes. Almost every phase of the sperm's life cycle is now open to routine laboratory testing using a variety of molecular biology and other tools (Jeremias, Witkin 1996; Mortimer 1994). This is reflected in the moves by professional societies to impose rigorous standards on all pathology laboratories involved in semenology.

Unfortunately these advances have left clinical andrology floundering, and there is little consensus on suitable training for the clinician even though there have been a number of recent calls to change this (Jequier 1997; Jequier, Cummins 1997; Cummins, Jequier 1994; Patrizio, Kopf 1997; Vanderven, Haidl 1997). Typically these days announcements about training schemes in andrology mean almost exclusively advanced semen analysis, even though a semen analysis is almost meaningless for any diagnosis of causation.

Male infertility is for the most part investigated by male gynecologists. Few of these have ever had any training in any aspect of the examination of the male genital tract, and few clinicians in infertility know how to—or even bother to—take a history from an infertile male, let alone carry out a clinical examination. It is therefore not very surprising that so much less is known about the etiology of infertility in the male than the female. This is a critical matter for clinics, which appear to be in a downward spiral of increasing ignorance about the causes and alternative options for treatment of male infertility. Already there have been calls to redirect resources away from diagnosis to treatment—that is, ICSI will "solve" everything (Hamberger, Janson 1997). It has been suggested that clinical andrologists should receive equal double training in gynecology and urology (Jequier 1997; Jequier, Cummins 1997), and similar moves have been proposed in Germany (Schirren 1996). This will not be popular with clinicians seeking rapid career advancement, but the inevitable alternative for clinics will be messy litigation over misdiagnosis or maldiagnosis. For example, "treating" a couple with ICSI purely on the basis of poor semen quality will not reveal the underlying presence of testicular cancer, as this can only be diagnosed during examination by a trained clinician.

Monitoring Outcomes—Risks and Pitfalls of Follow-Up Studies

The Hippocratic Oath states, "I will prescribe regimen for the good of my patients according to my ability and my judgment and never do harm to anyone." While there is no convincing evidence of serious developmental or psychological abnormalities in IVF children *per se* (Vanbalen 1998), there seems to be consensus with the principle that ICSI and related techniques pose potential problems for the children and that controlled follow-up studies are needed. This is easier to state than to achieve. The parents of children conceived through ART frequently resent and resist intrusion, and even with conventional IVF, generating adequate data is difficult through non-compliance and dropout (Halasz et al. 1993).

The Brussels group has the most comprehensive follow-up of ICSI babies and so far has found no evidence of significantly elevated major birth defects or incidence of mental development problems (Bonduelle et al. 1996, 1998). While their criteria for categorizing major birth defects have been challenged as being over-conservative (Kurinczuk, Bower 1997), these data appear superficially reassuring and do not reinforce early reports of sex chromosome linked anomalies (In't Veld et al. 1995). A recent study found a significant increase in delayed mental development in male ICSI babies at one year of age compared with babies conceived spontaneously or through IVF (Bowen et al. 1998). However, in this study the control groups were unbalanced for socioeconomic group, paternal age and ethnic background. Specifically, the ICSI babies' parent group had significantly more older fathers, lower educational attainment and more came from a non-English-speaking background. As the Bayley score of development used here relies in part on linguistic factors, these demographic factors may well have contributed to the lower scores. Only more studies will allow definite conclusions to be made.

Conclusions

A warning was given several years ago of the potential dangers of ICSI and little has changed to alter the need for that warning (Cummins, Jequier 1995). Humanity is living through an epoch of unparalleled experimentation in human reproduction, with almost no suitable animal models. Tens of thousands of babies have now been born from ICSI and probably less than 1% will ever be followed up in any meaningful way. Clinics have multiplied in every country in the world. Many have no adequate legal, ethical or scientific guidelines, let alone long-term accountability. Many subsist on marginal profits, and intense commercial competition means that the ultimate interests of the child have low priority. Ignorance, poor clinician training and a blind reliance on technology prevail and there seems to be little appreciation of the biological significance of what is being done. There are chilling parallels in the diethylstilbestrol (DES) experience of a generation ago, when clinicians desperate to offer a short term "fix" to what was perceived as a problem perpetuated horrific consequences on the next generation. The only way scientists can proceed in a human and rational way is to insist that clinics acknowledge the potential problems, make full disclosure to infertile patients and above all take responsibility for the lives they are helping to create.

Acknowledgements

The author thanks colleagues Anne Jequier and Peter Vogt for their assistance and advice.

References

Ao A, Ray P, Harper J, Lesko J, Paraschos T, Atkinson G, Soussis I, Taylor D, Handyside A, Hughes M, Winston RML. Clinical experience with preimplantation genetic diagnosis of cystic fibrosis (delta-f508). Prenat Diag 1996; 16:137-142.

Asch R, Simerly C, Ord T, Ord VA, Schatten G. The stages at which human fertilization arrests—microtubule and chromosome configurations in inseminated oocytes which failed to complete fertilization and development in humans. Hum Reprod 1995; 10:1897-1906.

Auroux M, Nawar NNY, Naguib M, Baud M, Lapaquellerie N. Post-pubescent to mature fathers—increase in progeny quality. Hum Reprod 1998; 13:55-59.

Auroux MR, Mayaux MJ, Guihard-Moscato ML, Fromantin M, Barthe J, Schwartz D. Paternal age and mental functions of progeny in man. Hum Reprod 1989; 4:794-797.

Bench G, Corzett MH, Deyebra L, Oliva R, Balhorn R. Protein and DNA contents in sperm from an infertile human male possessing protamine defects that vary over time. Mol Reprod Dev 1998; 50:345-353.

Bishop MWH. Paternal contribution to embryonic death. J Reprod Fertil 1964; 7:383-396.

Bonduelle M, Wilikens A, Buysse A, Vanassche E, Wisanto A, Devroey P, Van Steirteghem AC, Liebaers I. Prospective follow-up study of 877 children born after intracytoplasmic sperm injection (ICSI), with ejaculated epididymal and testicular spermatozoa and after replacement of cryopreserved embryos obtained after ICSI. Hum Reprod 1996; 11:131-155.

Bonduelle M, Joris H, Hofmans K, Liebaers I, Van Steirteghem A. Mental development of 201 ICSI children at 2 years of age. Lancet 1998; 351:115-116.

Bowen JR, Gibson FL, Leslie GI, Saunders DM. Medical and developmental outcome at 1 year for children conceived by intracytoplasmic sperm injection. Lancet 1998; 351:1529-1534.

Burgoyne PS, Mahadevaiah SK, Sutcliffe MJ, Palmer SJ. Fertility in mice requires X-Y pairing and a Y-chromosomal "spermiogenesis" gene mapping to the long arm. Cell 1992; 71:391-398.

Burruel VR, Yanagimachi R, Whitten WK. Normal mice develop from oocytes injected with spermatozoa with grossly misshapen heads. Biol Reprod 1996; 55:709-714.

Byskov AG. Primordial germ cells and regulation of meiosis. In: (CR Austin, RV Short, eds) Germ Cells and Fertilization. Cambridge: Cambridge University Press, 1982, pp1-16.

Carlsen E, Giwercman A, Keiding N, Skakkebæk NE. Declining semen quality and increasing incidence of testicular cancer—is there a common cause? Environ Health Perspect 1995; 103:137-139.

Chai NN, Zhou HY, Hernandez J, Najmabadi H, Bhasin S, Yen PH. Structure and organization of the RBMY genes on the human Y chromosome—transposition and amplification of an ancestral autosomal hnRNPG gene. Genomics 1998; 49:283-289.

Cooke HJ, Lee M, Kerr S, Ruggiu M. A murine homologue of the human DAZ gene is autosomal and expressed only in male and female gonads. Hum Mol Genet 1996a; 5:513-516.

Cooke PS, Zhao YD, Hansen LG. Neonatal polychlorinated biphenyl treatment increases adult testis size and sperm production in the rat. Toxicol Appl Pharmacol 1996b; 136:112-117.

Craft I, Tsirigotis M, Bennett V, Taranissi M, Khalifa Y, Hogewind G, Nicholson N. Percutaneous epididymal sperm aspiration and intracytoplasmic sperm injection in the management of infertility due to obstructive azoospermia. Fertil Steril 1995; 63:1038-1042.

Crow JF. Spontaneous mutation as a risk factor. Exp Clin Immunogenet 1995; 12:121-128.

Cummins JM. Evolution of sperm form: levels of control and competition. In: (Bavister BD, Cummins JM, Roldan ERS, eds) Fertilization in Mammals. Norwell, Massachusetts: Serono Symposia, USA, 1990, pp51-64.

Cummins JM. Controversies in science: ICSI may foster birth defects. J of NIH Res 1997; 9:38-42.

Cummins JM. Genetic aspects of sperm motility disorders. In: (Hamamah S, ed) Ares Serono Symposium on Male Sterility for Motility Disorders: Etiological Factors and Treatment. Paris: Ares Serono Symposia, 1998a, in press.

Cummins JM. Mitochondrial DNA in reproductive biology. Rev Reprod 1998b; 3:172-187.

Cummins JM, Jequier AM. Treating male infertility needs more clinical andrology, not less. Hum Reprod 1994; 9:1214-1219.

Cummins JM, Jequier AM. Concerns and recommendations for intracytoplasmic sperm injection (ICSI) treatment. Hum Reprod 1995; 10:138-143.

Cummins JM, Jequier AM, Kan R. Molecular biology of human male infertility—links with aging, mitochondrial genetics, and oxidative stress? Mol Reprod Dev 1994; 37:345-362.

Cummins JM, Jequier AM, Martin R, Mehmet D, Goldblatt J. Semen levels of mitochondrial DNA deletions in men attending an infertility clinic do not correlate with phenotype. J Androl 1998; 21:47-52.

de Kretser DM. Male infertility. Modern Medicine 1990; 33:98-109.

Delbridge ML, Ma K, Subbarao MN, Cooke HJ, Bhasin S, Graves JA. Evolution of mammalian HNRPG and its relationship with the putative azoospermia factor RBM. Mammalian Genome 1998; 9:168-70.

Diamond J. Why Is Sex Fun? Phoenix Science Masters Series, London, 1997.

Eberhart CG, Maines JZ, Wasserman SA. Meiotic cell cycle requirement for a fly homologue of human Deleted in Azoospermia. Nature 1996; 381:783-785.

Edwards RG, Steptoe PC, Purdy JM. Establishing full-term human pregnancies using cleaving embryos grown *in vitro*. Brit J Obstet Gyn 1980; 87:737-756.

Edwards RG, Tarin JJ, Dean N, Hirsch A, Tan SL. Are spermatid injections into human oocytes now mandatory? Hum Reprod 1994; 9:2217-2219.

Elder K, Elliott T (eds). The Use of Testicular and Epididymal Sperm in IVF. Ladybrook Publishing, Perth, 1998.

Evenson D, Jost L. Sperm chromatin structure assay: DNA denaturability. Meth Cell Biol 1994; 42 (Pt B):159-76.

Finn CA. Menstruation—a nonadaptive consequence of uterine evolution. Q Rev Biol 1998; 73:163-173.

Friel JP, ed. Dorland's Illustrated Medical Dictionary. Philadelphia: W.B. Saunders, 1965.

George CH. Is the Y chromosome behaving badly? J NIH Res 1997; 9:25-27.

Graves JAM. The origin and function of the mammalian Y chromosome and Y-borne genes—an evolving understanding. Bioessays 1995; 17:311-320.

Habermann B, Mi HF, Edelmann A, Bohring C, Backert IT, Kiesewetter F, Aumuller G, Vogt PH. DAZ (Deleted in AZoospermia) genes encode proteins located in human late spermatids and in sperm tails. Hum Reprod 1998; 13:363-369.

Halasz G, Munro J, Saunders K, Astbury J, Spensley J. The Growth and Development of Children Conceived by IVF. Melbourne: Commonwealth Department of Health, 1993.

Hamberger L, Janson PO. Global importance of infertility and its treatment—role of fertility technologies. Int J Gyn Obstet 1997;

58:149-158.
Harcourt AH, Harvey PH, Larson SG, Short RV. Testis weight, body weight and breeding system in primates. Nature 1981; 293:55-57.
Hardy DO, Scher HI, Bogenreider T, Sabbatini P, Zhang ZF, Nanus DM, Catterall JF. Androgen receptor CAG repeat lengths in prostate cancer: correlation with age of onset. J Clin Endo Met 1996; 81:4400-4405.
Hawkes K, O'Connell JF, Jones NGB, Alvarez H, Charnov EL. Grandmothering, menopause, and the evolution of human life histories. Proc Natl Acad Sci USA 1998; 95:1336-1339.
Hewitson L, Simerly C, Takahashi D, Schatten G. The role of the sperm centrosome during human fertilization and embryonic development: implications for intracytoplasmic sperm injection and other sophisticated ART strategies. In: (Ombelet W, Bosmans E, Vandeput H, Vereecken A, Renier M, Hoomans M, eds) Modern ART in the 2000s. Andrology in the Nineties. New York: Parthenon Publishing, 1998, pp139-156.
Hiramoto Y. Microinjection of the live spermatozoa into sea urchin eggs. Exp Cell Res 1962; 27:416-426.
Houshmand M, Holme E, Hanson C, Wennerholm UB, Hamberger L. Paternal mitochondrial DNA transferred to the offspring following intracytoplasmic sperm injection. J Assist Reprod Genet 1997; 14:223-227.
Houston DW, Zhang J, Maines JZ, Wasserman SA, King ML. A *Xenopus* DAZ-like gene encodes an RNA component of germ plasm and is a functional homologue of *Drosophila* boule. Development 1998; 125:171-180.
Hughes CM, Lewis SEM, McKelvey-Martin VJ, Thompson W. Reproducibility of human sperm DNA measurements using the alkaline single cell gel electrophoresis assay. Mut Res—Fund Mol Mech Mutagen 1997; 374:261-268.
Human Fertility and Embryology Authority. Sixth Annual Report. 1997.
Hurst T, Shafir E, Lancaster P. Assisted Conception Australia and New Zealand 1996. Sydney: AIHW National Perinatal Statistics Unit, 1997.
In't Veld P, Brandenburg H, Verhoeff A, Dhort M, Los F. Sex chromosomal abnormalities and intra-cytoplasmic sperm injection. Lancet 1995; 346:773.
Irvine RA, Yu MC, Ross RK, Coetzee GA. The CAG and GGC microsatellites of the androgen receptor gene are in linkage disequilibrium in men with prostate cancer. Cancer Res 1995; 55:1937-40.
Jaffe T, Oates RD. Genetic abnormalities and reproductive failure. Urology Clinics of North America 1994; 21:389-408.
Jensen TK, Toppari J, Keiding N, Skakkebæk NE. Do environmental estrogens contribute to the decline in male reproductive health? Clin Chem 1995; 41:1896-1901.
Jequier AM. Non-therapy related pregnancies in the consorts of a group of men with obstructive azoospermia. Andrologia 1985; 17:6-8.
Jequier AM. Clinical assessment of male infertility in the era of intracytoplasmic sperm injection. Baillieres Clin Obstet Gyn 1997; 11:617-639.
Jequier AM, Cummins JM. Attitudes to clinical andrology—a time for change. Hum Reprod 1997; 12:875-876.
Jeremias J, Witkin SS. Molecular approaches to the diagnosis of male infertility. Mol Hum Reprod 1996; 2:195-202.
Ji BT, Shu XO, Linet MS, Zheng W, Wacholder S, Gao YT, Ying DM, Jin F. Paternal cigarette smoking and the risk of childhood cancer among offspring of non-smoking mothers. J Natl Canc Inst 1997; 89:238-244.
Johns DR. The other human genome—mitochondrial DNA and disease. Nature Med 1996; 2:1065-1068.
Johnson L. Efficiency of spermatogenesis. Micros Res Tech 1995; 32:385-422.
Kanavakis E, Tzetis M, Antoniadi T, Pistofidis G, Milligos S, Kattamis C. Cystic fibrosis mutation screening in CBAVD patients and men with obstructive azoospermia or severe oligozoospermia. Mol Hum Reprod 1998; 4:333-337.
Karlin S, Burge C. Trinucleotide repeats and long homopeptides in genes and proteins associated with nervous system disease and development. Proc Natl Acad Sci USA 1996; 93:1560-1565.
Kent-First MG, Koi S, Muallem A, Blazer S, Itskovitz-Eldor J. Infertility in intracytoplasmic-sperm-injection-derived sons. Lancet 1996a; 348:332.
Kent-First MG, Kol S, Muallem A, Ofir R, Manor D, Blazer S, First N, Eldor JI. The incidence and possible relevance of Y-linked microdeletions in babies born after intracytoplasmic sperm injection and their infertile fathers. Mol Hum Reprod 1996b; 2:943-950.
Kimura Y, Yanagimachi R. Development of normal mice from oocytes injected with secondary spermatocyte nuclei. Biol Reprod 1995; 53:855-862.
Kimura Y, Yanagimachi R, Kuretake S, Bortkiewicz H, Perry ACF, Yanagimachi H. Analysis of mouse oocyte activation suggests the involvement of sperm perinuclear material. Biol Reprod 1998; 58:1407-1415.
Kramer JA, Krawetz SA. RNA in spermatozoa—implications for the alternative haploid genome. Mol Hum Reprod 1997; 3:473-478.
Kurinczuk JJ, Bower C. Birth defects in infants conceived by intracytoplasmic sperm injection—an alternative interpretation. Brit Med J 1997; 315:1260-1265.
Laspada AR, Peterson KR, Meadows SA, McClain ME, Jeng G, Chmelar RS, Haugen HA, Chen K, Singer MJ, Moore D, Trask BJ, Fischbeck KH, Clegg CH, McKnight GS. Androgen receptor yac transgenic mice carrying cag 45 alleles show trinucleotide repeat instability. Hum Mol Gen 1998; 7:959-967.
Lin WW, Lamb DJ, Lipshultz LI, Kim ED. Absence of cyclic adenosine 3'/5' monophosphate responsive element modulator expression at the spermatocyte arrest stage. Fertil Steril 1998; 69:533-538.
Lissens W, Liebaers I. The genetics of male infertility in relation to cystic fibrosis. Baillieres Clin Obstet Gyn 1997; 11:797-817.
Lopes S, Jurisicova A, Casper RF. Gamete-specific DNA fragmentation in unfertilized human oocytes after intracytoplasmic sperm injection. Hum Reprod 1998a; 13:703-708.
Lopes S, Sun JG, Jurisicova A, Meriano J, Casper RF. Sperm deoxyribonucleic acid fragmentation is increased in poor-quality semen samples and correlates with failed fertilization in intracytoplasmic sperm injection. Fertil Steril 1998b; 69:528-532.
Ludwig M, Alhasani S, Kupker W, Bauer O, Diedrich K. A new indication for an intracytoplasmic sperm injection procedure outside the cases of severe male factor infertility. Eur J Obstet Gyn Reprod Biol 1997; 75:207-210.
Ludwig M, Kupker W, Hahn K, Montzka P, Alhasani S, Diedrich K. Clinical significance of Y-chromosomal microdeletions in reproduction genetic routine diagnostics in severe male factor subfertility. Geburtshilfe und Frauenheilkunde 1998; 58:73-78.
Manicardi GC, Tombacco A, Bizzaro D, Bianchi U, Bianchi PG, Sakkas D. DNA strand breaks in ejaculated human spermatozoa—comparison of susceptibility to the nick translation and terminal transferase assays. Histochem J 1998; 30:33-39.
Margulis L, Sagan D. Origins of Sex. Three Billion Years of Genetic Recombination. New Haven: Yale University Press, 1986.
McKee BD, Wilhelm K, Merrill C, Ren XJ. Male sterility and meiotic drive associated with sex chromosome rearrangements in *Drosophila*—role of X-Y pairing. Genetics 1998; 149:143-155.
Miller D. RNA in the ejaculate spermatoozoon—a window into molecular events in spermatogenesis and a record of the unusual

requirements of haploid gene expression and post-meiotic equilibration. Mol Hum Reprod 1997; 3:669-676.

Mittwoch U. Ethnic differences in testis size: a possible link with the cytogenetics of true hermaphroditism. Hum Reprod 1988; 3:445-449.

Mohandas TK, Speed RM, Passage MB, Yen PH, Chandley AC, Shapiro LJ. Role of the pseudoautosomal region in sex-chromosome pairing during male meiosis: meiotic studies in a man with a deletion of distal Xp. Am J Hum Genet 1992; 51:526-533.

Montag M, Vanderven K, Ved S, Schmutzler A, Prietl G, Krebs D, Peschka B, Schwanitz G, Albers P, Haidl G, Vanderven H. Success of intracytoplasmic sperm injection in couples with male and/or female chromosome aberrations. Hum Reprod 1997; 12:2635-2640.

Moosani N, Pattinson HA, Carter MD, Cox DM, Rademaker AW, Martin RH. Chromosomal analysis of sperm from men with idiopathic infertility using sperm karyotyping and fluorescence *in situ* hybridization. Fertil Steril 1995; 64:811-817.

Mortimer D. The essential partnership between diagnostic andrology and modern assisted reproductive technologies. Hum Reprod 1994; 9:1209-1213.

Ng SC, Bongso TA, Liow SL, Edirisinghe R, Tok V, Ratnam SS. Controversies in micro-injection. J Assist Reprod Gen 1992; 9:186-189.

Niemi M. Andrology as a specialty—its origin. J Androl 1987; 8:201-202.

Ogura A, Yanagimachi R. Spermatids as male gametes. Reprod Fertil Dev 1995; 7:155-159.

Ohno S, Nagai Y, Ciccarese S. The X and Y chromosomes: mechanism of sex determination. In: (Sparkes RS, Comings DE, Fox CF, eds). Molecular Human Cytogenetics. New York: Academic Press, 1977, pp294-303.

Palermo G, Joris H, Devroey P, Van Steirteghem AC. Pregnancies after intracytoplasmic injection of single spermatozoon into an oocyte. Lancet 1992; 340:17-18.

Patrizio P, Kopf GS. Molecular biology in the modern work-up of the infertile male—the time to recognize the need for andrologists. Hum Reprod 1997; 12:879-883.

Patrizio P, Zielenski J. Congenital absence of the vas deferens—a mild form of cystic fibrosis. Mol Med Today 1996; 2:24-31.

Patrizio P, Asch RH, Handelin B, Silber SJ. Etiology of congenital absence of vas deferens: genetic study of three generations. Hum Reprod 1993; 8:215-220.

Patrizio P, Ord T, Balmaceda JP, Asch RH. Use of epididymal sperm for assisted reproduction in men with acquired, irreparable obstructive azoospermia. Reprod Fertil Dev 1995; 7:841-845.

Purdy LM. What can progress in reproductive technology mean for women? J Med Phil 1996; 21:499-514.

Quigley CA, De Bellis A, Marschke KB, el-Awady MK, Wilson EM, French FS. Androgen receptor defects: historical, clinical, and molecular perspectives. Endocrinol Rev 1995; 16:271-321.

Ruggiu M, Speed R, Taggart M, McKay SJ, Kilanowski F, Saunders P, Dorin J, Cooke HJ. The mouse DAZLA gene encodes a cytoplasmic protein essential for gametogenesis. Nature 1997; 389:73-77.

Sasagawa I, Yanagimachi R. Spermatids from mice after cryptorchid and reversal operations can initiate normal embryo development. J Androl 1997; 18:203-209.

Sathananthan AH. Paternal centrosomal dynamics in early human development and infertility. J Assist Reprod Gen 1998; 15:129-139.

Saxena R, Brown LG, Hawkins T, Alagappan RK, Skaletsky H, Reeve MP, Reijo R, Rozen S, Dinulos MB, Disteche CM, Page DC. The DAZ gene cluster on the human Y chromosome arose from an autosomal gene that was transposed, repeatedly amplified and pruned. Nature Genet 1996; 14:292-299.

Schill T, Alhasani S, Kupker W, Diedrich K. Light microscopic examination of cytomorphology and cytogenetics in non-fertilised oocytes after *in vitro* fertilization and intracytoplasmatic sperm injection [German]. Geburtshilfe und Frauenheilkunde 1998; 58:79-87.

Schirren C. Andrology—development and future—critical remarks after 45 years of medical practice. Andrologia 1996; 28:137-140.

Seligman J, Page DC. The DAZH gene is expressed in male and female embryonic gonads before germ cell sex differentiation. Biochem Biophys Res Commun 1998; 245:878-882.

Sharpe RM, Skakkebæk NE. Are oestrogens involved in falling sperm counts and disorders of the male reproductive tract? Lancet 1993; 341:1392-1395.

Short RV. The testis—the witness of the mating system, the site of mutation and the engine of desire. Acta Paed 1997; 86:3-7.

Silber SJ, Johnson L. Are spermatid injections of any clinical value? —ROSNI and ROSI revisited. Hum Reprod 1998; 13:509-515.

Silber SJ, Devroey P, Tournaye H, Van Steirteghem AC. Fertilizing capacity of epididymal and testicular sperm using intracytoplasmic sperm injection (ICSI). Reprod Fertil Dev 1995a; 7:281-293.

Silber SJ, Van Steirteghem AC, Liu J, Nagy Z, Tournaye H, Devroey P. High fertilization and pregnancy rate after intracytoplasmic sperm injection with spermatozoa obtained from testicle biopsy. Hum Reprod 1995b; 10:148-152.

Simoni M, Gromoll J, Nieschlag E. Molecular pathophysiology and clinical manifestations of gonadotropin receptor defects. Steroids 1998; 63:288-293.

Simorangkir DR, De Kretser DM, Wreford NG. Increased numbers of Sertoli and germ cells in adult rat testes induced by synergistic action of transient neonatal hypothyroidism and neonatal hemicastration. J Reprod Fertil 1995; 104:207-213.

Simorangkir DR, Wreford NG, de Kretser DM. Impaired germ cell development in the testes of immature rats with neonatal hypothyroidism. J Androl 1997; 18:186-193.

Sofikitis N, Miyagawa I, Sharlip I, Hellstrom W, Mekras G, Mastelou E. Human pregnancies achieved by intra-ooplasmic injections of round spermatid (RS) nuclei isolated from testicular tissue of azoospermic men. In: (American Urological Association, eds). Las Vegas, Nevada: AUA Meeting Abstracts/PRISM Productions, 1995, p0616.

Sofikitis N, Yamamoto Y, Isoyama T, Miyagawa I. The early haploid male gamete develops a capacity for fertilization after the coalescence of the proacrosomal granules. Hum Reprod 1997; 12:2713-2719.

Sofikitis NV, Yamamoto Y, Miyagawa I, Mekras G, Mio Y, Toda T, Antypas S, Kawamura H, Kanakas N, Antoniou N, Loutradis D, Mantzavinos T, Kalianidis K, Agapitos E. Ooplasmic injection of elongating spermatids for the treatment of non-obstructive azoospermia. Hum Reprod 1998; 13:709-714.

Sousa M, Barros A, Tesarik J. Current problems with spermatid conception. Hum Reprod 1998; 13:255-258.

Steinberger E, Rodrigues-Rigau LJ. The infertile couple. J Androl 1983; 4:111-118.

Sutovsky P, Navara CS, Schatten G. Fate of the sperm mitochondria, and the incorporation, conversion, and disassembly of the sperm tail structures during bovine fertilization. Biol Reprod 1996; 55:1195-1205.

Swan SH, Elkin EP, Fenster L. Have sperm densities declined—a reanalysis of global trend data. Environ Health Perspect 1997; 105:1228-1232.

Tesarik J. Fertilization of oocytes by injecting spermatozoa, spermatids and spermatocytes. Rev Reprod 1996; 1:149-52.

Tesarik J. Sperm or spermatid conception? Fertil Steril 1997a; 68:214-216.

Tesarik J. Use of immature germ cells for the treatment of male infertility. Baillieres Clin Obstet Gyn 1997b; 11:763-772.

Tesarik J, Mendoza C. Genomic imprinting abnormalities: a new potential risk of assisted reproduction. Mol Hum Reprod 1996a; 2:295-298.

Tesarik J, Mendoza C. Spermatid injection into human oocytes. 1. Laboratory techniques and special features of zygote development. Hum Reprod 1996b; 11:772-779.

Tesarik J, Rolet F, Brami C, Sedbon E, Thorel J, Tibi C, Thebault A. Spermatid injection into human oocytes. 2. Clinical application in the treatment of infertility due to non-obstructive azoospermia. Hum Reprod 1996; 11:780-783.

Tesarik J, Greco E, Mendoza C. ROSI—instructions for use—1997 update. Hum Reprod 1997; 13:519-523.

Toppari J, Larsen JC, Christiansen P, Giwercman A, Grandjean P, Guillette LJ, Jegou B, Jensen TK, Jouannet P, Keiding N, Leffers H, McLachlan JA, Meyer O, Muller J, Rajpert-Demeyts E, Scheike T, Sharpe R, Sumpter J, Skakkebæk NE. Male reproductive health and environmental xenoestrogens. Environ Health Perspect 1996; 104:741-803.

Trounson AO. Current perspectives of *in vitro* fertilisation and embryo transfer. Clin Reprod Fertil 1982; 1:55-65.

Tuerlings J, Mol B, Kremer JAM, Looman M, Meuleman EJH, Meerman GJT, Buys C, Merkus H, Scheffer H. Mutation frequency of cystic fibrosis transmembrane regulator is not increased in oligozoospermic male candidates for intracytoplasmic sperm injection. Fertil Steril 1998; 69:899-903.

Tut TG, Ghadessy FJ, Trifiro MA, Pinsky L, Yong EL. Long polyglutamine tracts in the androgen receptor are associated with reduced trans-activation, impaired sperm production, and male infertility. J Clin Endocrin Metab 1997; 82:3777-3782.

Valone DA. The changing moral landscape of human reproduction—two moments in the history of *in vitro* fertilization. Mount Sinai J Med 1998; 65:167-172.

Vanbalen F. Development of IVF children. Dev Rev 1998; 18:30-46.

Vanderven H, Haidl G. Clinical andrology is important for treatment of male infertility with ICSI. Hum Reprod 1997; 12:879.

Vanderven K, Messer L, Vanderven H, Jeyendran RS, Ober C. Cystic fibrosis mutation screening in healthy men with reduced sperm quality. Hum Reprod 1996; 11:513-517.

Vanderven K, Montag M, Peschka B, Leygraaf J, Schwanitz G, Haidl G, Krebs D, Vanderven H. Combined cytogenetic and Y chromosome microdeletion screening in males undergoing intracytoplasmic sperm injection. Mol Hum Reprod 1997; 3:699-704.

Vanderven K, Peschka B, Montag M, Lange R, Schwanitz G, Vanderven HH. Increased frequency of congenital chromosomal aberrations in female partners of couples undergoing intracytoplasmic sperm injection. Hum Reprod 1998; 13:48-54.

Vine MF. Smoking and male reproduction—a review. Intl J Androl 1996; 19:323-337.

Vogt P. Potential genetic functions of tandem repeated DNA sequence blocks in the human genome are based on a highly conserved "chromatin folding code." Hum Genet 1990; 84:301-36.

Vogt PH. Genetic aspects of artificial fertilization. Hum Reprod 1995; 10:128-137.

Vogt PH. Genetics of idiopathic male infertility—Y chromosomal azoospermia factors (AZFa, AZFb, AZFc). Baillieres Clin Obstet Gyn 1997a; 11:773-795.

Vogt PH. Human Y chromosome deletions in Yq11 and male fertility. Adv Expl Med Biol 1997b; 424:17-30.

Vogt PH. Molecular basis of male (in)fertility. Intl J Androl 1997c; 20:2-10.

Vogt PH. Genetic aspects of male sterility. In: (Hamamah S, ed) Ares Serono Symposium on Male Sterility for Motility Disorders: Etiological Factors and Treatment. Paris: Ares Serono Symposia, 1998, in press.

Vogt PH, Edelmann A, Kirsch S, Henegariu O, Hirschmann P, Kiesewetter F, Kohn FM, Schill WB, Farah S, Ramos C, Hartmann M, Hartschuh W, Meschede D, Behre HM, Castel A, Nieschlag E, Weidner W, Grone HJ, Jung A, Engel W, Haidl G. Human Y chromosome azoospermia factors (AZF) mapped to different subregions in Yq11. Hum Mol Genet 1996; 5:933-943.

Vogt PH, Affara N, Davey P, Hammer M, Jobling MA, Lau YF, Mitchell M, Schempp W, Tyler-Smith C, Williams G, Yen P, Rappold GA. Report of the Third International Workshop on Y Chromosome Mapping 1997; Heidelberg, Germany, April 13-16, 1997. Cytogenet Cell Genet 1997; 79:1-20.

White JA, McAlpine PJ, Antonarakis S, Cann H, Eppig JT, Frazer K, Frezal J, Lancet D, Nahmias J, Pearson P, Peters J, Scott A, Scott H, Spurr N, Talbot C, Jr., Povey S. Guidelines for human gene nomenclature (1997). HUGO Nomenclature Committee. Genomics 1997; 45:468-471.

Yanagimachi R. Mechanisms of fertilization in mammals. In: (Mastroianni L, Biggers JD, eds) Fertilization and Embryonic Development *in Vitro*. New York: Plenum Press, 1981, pp81-182.

Yanagimachi R. Is an animal model needed for intracytoplasmic sperm injection (ICSI) and other assisted reproduction technologies? Hum Reprod 1995; 10:2525-2526.

Yong EL, Lee KO, Ng SC, Ratnam SS. Induction of spermatogenesis in isolated hypogonadotrophic hypogonadism with gonadotrophins and early intervention with intracytoplasmic sperm injection. Hum Reprod 1997; 12:1230-1232.

Yong EL, Tut TG, Ghadessy FJ, Prins G, Ratnam SS. Partial androgen insensitivity and correlations with the predicted three dimensional structure of the androgen receptor ligand-binding domain. Mol Cell Endocrin 1998; 137:41-50.

39 *The Spermatozoon as a Vehicle for Viral Infection*

B. Baccetti

P. Piomboni

University of Siena

The transmission of viral particles by spermatozoa has become more and more intriguing when studied in the context of Human Immunodeficiency Virus (HIV) infection. Acquired Immune Deficiency Syndrome (AIDS) is basically a sexually transmitted disease. Other modes of transmission can be classified as acquired. A direct involvement of sperm cells in the spread of infection was considered in the early 1980s, but the data were very controversial. It is only recently that proofs of transport of HIV-1 particles by human spermatozoa have been validated by the scientific community (the International Symposium "HIV and Gametes," held in Siena in 1997).

Ashida and Scofield (1987) were the first to suggest that the CD4 receptor (the classic HIV-1 receptor) was expressed on human spermatozoa. Soon after, Wolff et al. (1988) repeated the same type of experiment but failed to confirm the results of Ashida and Scofield (1987), while Borzy et al. (1988) detected by transmission electron microscopy (TEM) HIV-1 like particles in human seminal fluid but not in sperm cells. However, at the same time, Bagasra et al. (1988), using TEM, reported the presence of HIV-1 particles on the surface and in the cytoplasm of spermatozoa incubated with HIV-1, and only extremely low traces of the CD4 receptor on the sperm surface. In 1990, two groups of investigators (Baccetti et al. 1990; Bagasra et al. 1990) presented TEM micrographs of virus-like particles in the cytoplasm of spermatozoa obtained from AIDS patients. Results obtained from this morphological approach were immediately questioned by Pudney (1990), who suggested a possible confusion between viral particles and vesicles generated during acrosome reaction. At the same time, Miller and Scofield (1990) demonstrated that *in vitro* infected spermatozoa were able to carry the virus into macrophages, while Anderson et al. (1990) did not find CD4 receptors on sperm surface nor viral particles in spermatozoa of AIDS patients. In contrast to what Anderson et al. (1990) reported, Gobert et al. (1990) claimed to have found CD4 on healthy human spermatozoa. Moreover, Anderson et al. (1990) reported some preliminary data of Poiesz that failed to demonstrate the presence of HIV-1 proviral DNA in spermatozoa by polymerase chain reaction (PCR). Subsequently, Mermin et al. (1991) and Van Voorhis et al. (1991) also failed to detect, by PCR, HIV-1 proviral DNA in spermatozoa from seropositive men, while Scofield et al. (1992) found HIV-1 proviral DNA in spermatozoa of infected men. Baccetti et al. (1991, 1992) provided new TEM (Figs. 1-4) images of HIV-1 in spermatozoa infected *in vitro* or obtained from AIDS patients and further supported these observations with results obtained by immunogold labeling with polyclonal anti-HIV-1 and monoclonal anti-p24 antibodies in specimens prepared with the techniques of Carlemalm et al. (1985). Dussaix et al. (1993) localized by electron microscopy HIV particles in infected spermatozoa and demonstrated that *in vitro* infected sperm were able to transmit the virus to peripheral blood leukocytes.

New Evidence for HIV-1 Transmission via Spermatozoa

The consensus by the end of 1993 was that the presence of HIV-1 in human spermatozoa was considered as a very attractive possibility, but that the techniques previously used were inadequate to convince the whole scientific community. The following year, results from four laboratories (Baccetti et al. 1994; Scofield et al. 1994; Bagasra et al. 1994; Nuovo et al. 1994) supported the hypothesis of sperm involvement in the transmission

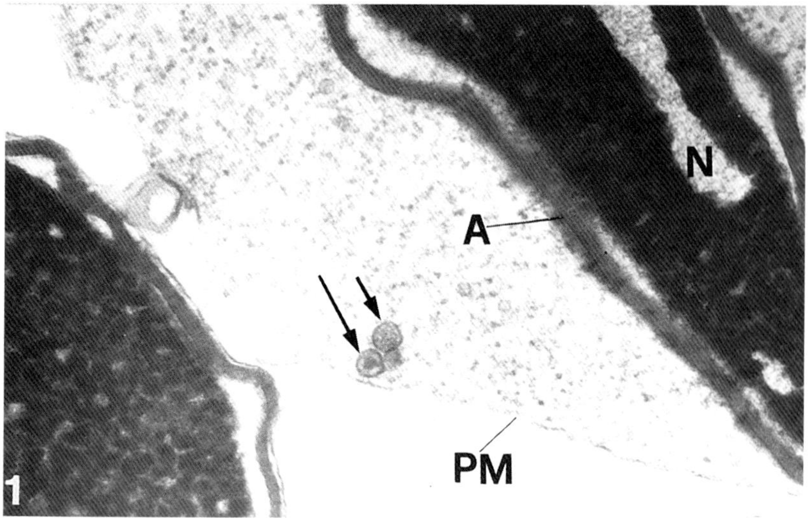

Figure 1. Transmission electron micrograph of HIV-like particles in the cytoplasm of spermatozoa from HIV infected patients. In the cytoplasm embedding the sperm nucleus, the particle indicated by long arrow is surrounded by a membrane-like coat and shows the typical electron-dense nucleoid; the particle indicated by short arrows is full of granular material and devoid of nucleoid. A= acrosome; N= nucleus: PM= plasma membrane. x40,000.

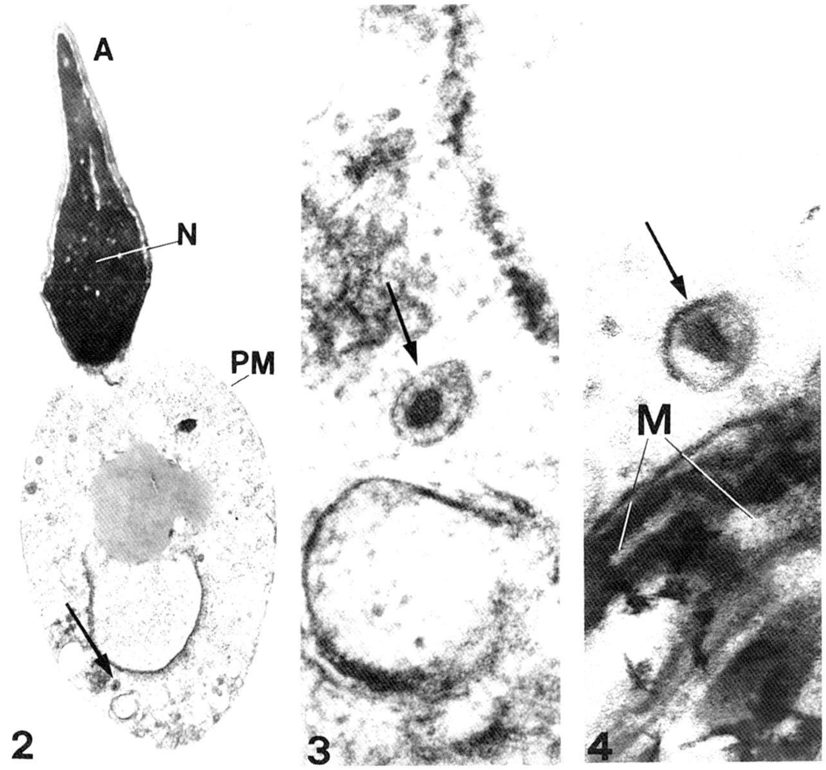

Figure 2. Transmission electron micrograph of HIV-like particle in the cytoplasm of a spermatozoon from HIV infected patient. Viral particle with pyriform coat and eccentric nucleoid (arrow) is present in a large cytoplasmic residue, surrounding the initial region of the midpiece. A= acrosome; N= nucleus; PM= plasma membrane. x13,400.

Figure 3. Transmission electron micrograph of HIV-like particle (arrow) in the cytoplasm of spermatozoa from HIV infected patients. This figure shows an enlargement of Figure 2. x100,000.

Figure 4. Transmission electron micrograph of HIV-like particle in the cytoplasm of spermatozoa from seronegative donors *in vitro*-incubated with HIV-1. A viral particle (arrow) with a dense nucleoid, truncated-cone shaped, and surrounded by a membrane coat, is present around the mitochondrial helix. M= mitochondria. x120,000.

of HIV. More evidence was presented by Baccetti et al. (1994). Using various anti-HIV-1 antibodies (Figs. 5, 6), they provided evidence that human spermatozoa can contain HIV-1. Moreover, they reported results obtained by *in situ* hybridization at the electron microscopy level (Figs. 7, 8) and by PCR analysis that definitely confirmed the presence of HIV-1 in spermatozoa. Baccetti et al. (1993, 1994) also described the transfer of HIV-1 by the spermatozoon into the oocyte (Fig. 9), which underwent normal segmentation. The concept of the membrane HIV-1 receptor was then viewed from a new perspective. Harouse et al. (1991), in fact, had reported that in several CD4 negative neural cells, HIV-1 infection only occurs if galactosylceramide (GalCer), a major membrane component of neural cells, is not blocked by antibodies. Baccetti et al. (1994) detected the presence in human spermatozoa of molecules highly related to GalCer and suggested an alternative hypothesis for the penetration of HIV-1 in these cells in absence of the CD4 receptor. Brogi et al. (1995, 1996, 1998) demonstrated that the new receptor for HIV is a glycolipid molecule, the galactosyl-alkyl-acyl-glycerol (GalAAG), structurally related to galactosylceramide and having strong affinity for the gp120. The GalAAG is preferentially localized in the equatorial segment and in the middle piece of human sperm and it is first expressed during early spermatogenesis. But other confirmations of the presence of HIV-1 in human spermatozoa have been subsequently presented. Using PCR and *in situ* PCR, the presence of HIV-1 proviral DNA was demonstrated in human spermatozoa from infected men (Bagasra et al. 1994; Scofield et al. 1994), as well as in spermatogonia, spermatocytes and rare spermatids from men who died of AIDS (Nuovo et al. 1994; Shevchuk et al. 1998), or in germinal cells at all stages of differentiation in the testis of HIV-1 infected asymptomatic subjects (Muciaccia et al. 1998a, b). Quite recently, the Anderson group (Quayle et al. 1997, 1998) provided a reiterated claim that in the AIDS patient ejaculates tested by PCR, both T cells and macrophages contained HIV-1 DNA, while germ cells, including spermatozoa, were devoid of it. However, in contrast, Dulioust et al. (1998) detected HIV-1 RNA in spermatozoa of seropositive men. Moreover, Scofield (1998) found that spermatozoa penetrated into lymphoid cells enhance their susceptibility to HIV infection.

At this point, it can be concluded that most of the research carried out in the last years demonstrates that HIV-1 virus particles penetrate human spermatozoa *in vitro* and *in vivo*. These particles are recognized to be HIV-1 virions because 1) the particles found in AIDS-patient spermatozoa and in spermatozoa from healthy donors infected *in vitro* exhibit the same morphological features, both in spermatozoa and in CD4 positive lymphocytes (C8166) infected *in vitro* with HIV-1; 2) the particles show the same immunocytochemical characterization in the two types of spermatozoa and in the C8166 T cells, in which the antibody binding is

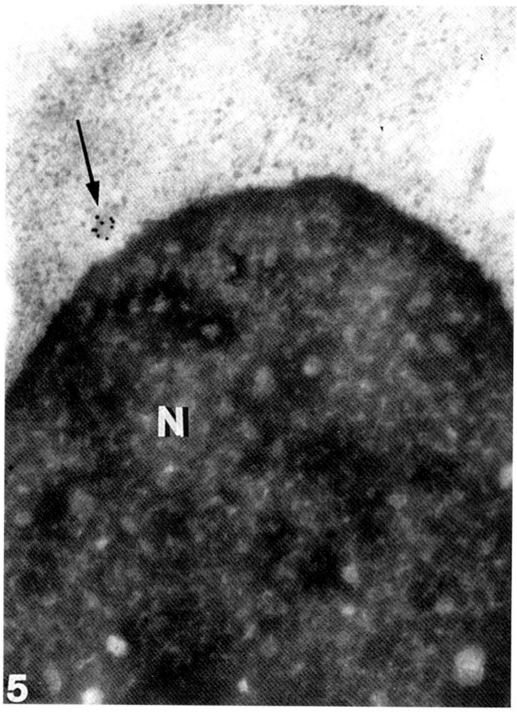

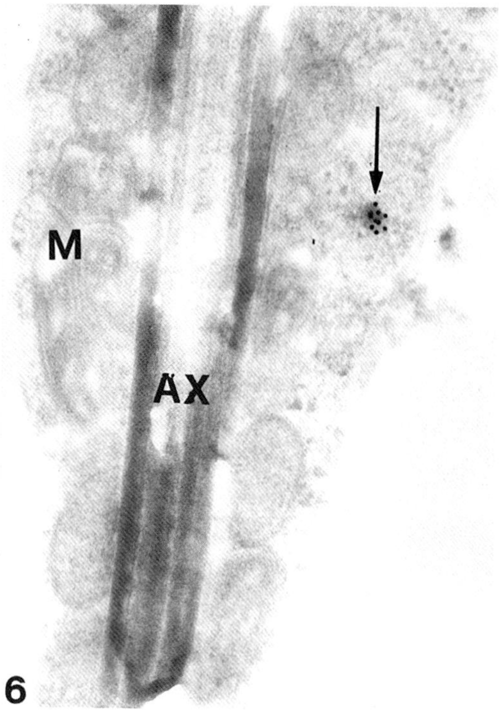

Figure 5. Submicroscopic immunogold detection of HIV-1 p24 antigen by monoclonal antibody in spermatozoa from HIV-1 infected patients. Gold particles are localized on an electron-dense particle (arrow) in the cytoplasm around the sperm head. N= nucleus. x50,000.

Figure 6. Submicroscopic immunogold detection of HIV-1 p24 antigen by monoclonal antibody in spermatozoa from HIV-1 infected patients. Gold granules (arrow) are visible in the cytoplasm around the mitochondrial helix of the sperm tail. Ax= axoneme; M= mitochondria. x71,000.

extremely specific; and 3) this kind of particle has never been detected in uninfected spermatozoa used as controls (Baccetti et al. 1990, 1991, 1994, 1998; Bagasra et al. 1990; Scofield 1992; Dussaix et al. 1993). Only Anderson's group (Anderson 1990, 1992; Pudney 1990; Pudney et al. 1998; Quayle et al. 1997, 1998) did not find HIV-1 particles in spermatozoa.

The HIV particles were in higher number in the healthy sperm infected *in vitro*. This view was further corroborated using immunocytochemistry and *in situ* hybridization. In spermatozoa from infected patients and in *in vitro*-infected spermatozoa from healthy donors, the structure and motility of spermatozoa appear almost normal (Baccetti et al. 1994, 1998).

The mature virus-like particles observed in the HIV infected spermatozoa represent virions penetrated via endocytosis, their uncoating being delayed or prevented by a low hydrolytic activity in the cytosol, as suggested by the absence of lysosomes in spermatozoa (Baccetti et al. 1994). The hypothesis that the virus particles seen in

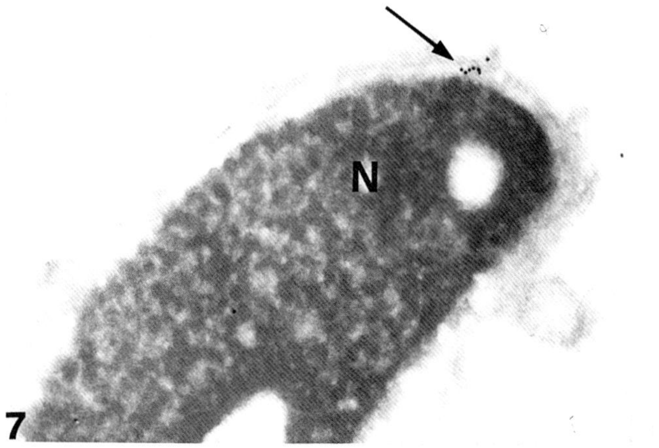

Figure 7. Immunogold labeling of HIV-1 RNA by *in situ* hybridization with a biotinylated DNA probe. Gold particles label the viral RNA (arrow) in the perinuclear region of HIV-1 incubated spermatozoa from seronegative men. N= nucleus. x58,000.

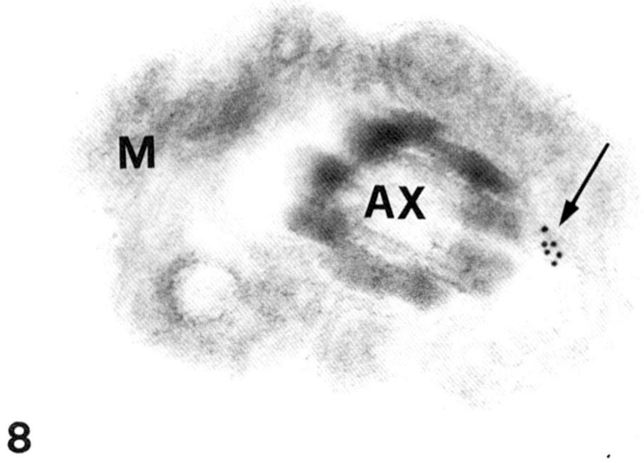

Figure 8. Immunogold labeling of HIV-1 RNA by *in situ* hybridization with a biotinylated DNA probe. Gold particles show the localization of viral RNA (arrow) in the mitochondrial helix of a spermatozoon of an HIV-1 infected patient. Ax= axoneme; M= mitochondria. x65,000.

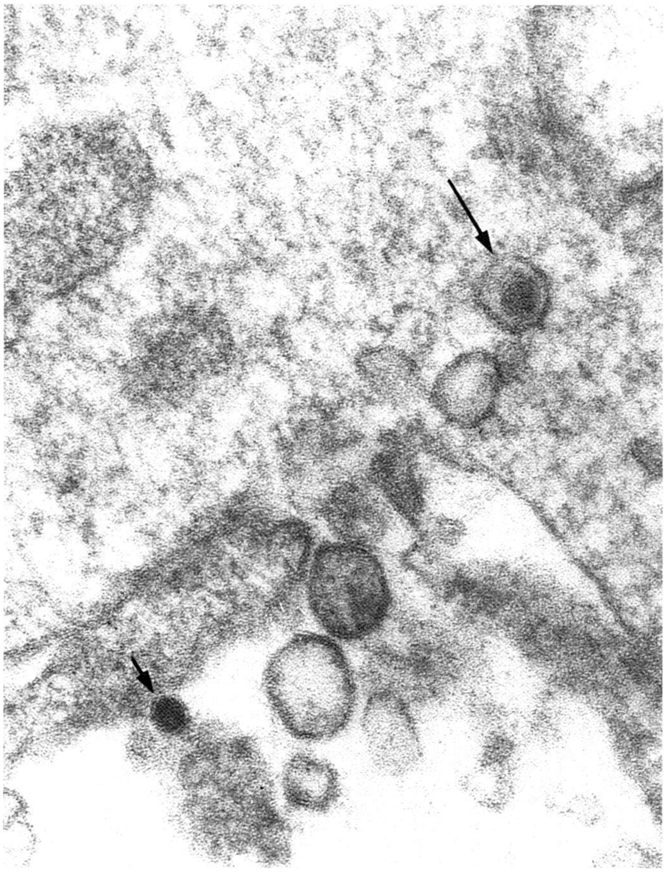

Figure 9. HIV-like particles (arrows) in the cytoplasm of a blastomere of an eight-blastomere human pre-embryo after *in vitro* fertilization with spermatozoa of an HIV-1 infected patient. The particle indicated by long arrow is surrounded by a membrane-like coat and shows the typical electron-dense nucleoid; the particle indicated by short arrow shows a nucleoid devoid of envelope. x115,000.

spermatozoa represent penetrating virions agrees with the rare detection of proviral DNA in the sperm fraction. Mermin et al. (1991), Van Voorhis et al. (1991) and Quayle et al. (1997, 1998) failed to detect HIV-1 DNA in the sperm fraction of the semen from infected patients, while Scofield et al. (1992) identified such sequences in sperm using methods that prevent lipid peroxidation. The last results in some way clarify the intriguing discrepancy. Baccetti et al. (1994) and Dulioust et al. (1998) detected viral RNA in spermatozoa, but no DNA; Nuovo et al. (1994) and Shevchuk et al. (1998) detected viral DNA in spermatogonia and spermatocytes, which decreased during the spermatid stage; Bagasra et al. (1994) detected DNA in 1:100 to 1:800 spermatozoa from HIV-1 infected men, with a precise localization in the center of the nucleus, as visualized by *in situ* PCR. Therefore, it seems (Muciaccia et al. 1998a, b) that the viral DNA content is decreasing during spermatogenesis and seems to be difficult to detect in spermatozoa, where the viral DNA is bound to the extremely concentrated chromatin and to a strong nuclear scaffold.

Summarizing the data presented in the recent Symposium "HIV-1 and Gametes" (Siena 1997), Bagasra (1998) reported that HIV-1 particles bind and enter normal human sperm through a particular receptor (the GalAAG) different from CD4; and that sperm from HIV-1 seropositive men contain antigenically identifiable HIV-1 particles and nucleic acid and that part of such sperm are morphologically normal, motile and can fertilize oocytes and transfer HIV-1 virus to the resulting embryos. Detection of viral nucleic acid by PCR has medical significance for semen-processing technologies now being employed for male-seropositive HIV-disparate couples (Semprini et al. 1992).

In women inseminated with "washed" sperm, no seroconversions were observed by PCR, nor any seropositive child. As a consequence, the authors concluded that spermatozoa do not contain any HIV-1 particles. But the recent data suggest that the absence of HIV-1 infection in the women was probably due to their limited exposure to infected sperm (one insemination per superovulatory cycle) rather than to the absence of HIV in the sperm preparations. In this matter, Hargreave and Ghosh (1998) recently expressed the opinion that to minimize the risk of transmission of HIV to the child, clinicians must consider whether the interest of a future child are best served by using fertility treatments, and conclude "probably not."

Other Viruses and Sperm Transmission

At this point it seems that the possibility of virus transport by spermatozoa is widely accepted by the scientific community. Following is a brief review of what is known about other viruses in humans and in other animal species.

It has been found that some sequences of the Hepatitis B virus (HBV) genome have been identified as integrated in human sperm nuclei, and therefore are vertically transmitted through the germinal line (Hadchouel et al. 1985). In transgenic mice, Baskar et al. (1996) found the expression of the enhancer domain of the major immediate early promoter (MIEP) of Human Cytomegalovirus (HCMV) in many types of cells, including spermatozoa.

The same holds true for Human Papillomaviruses of type 16 and 18 (HPV): some regions of the HPV genome were found in spermatozoa of infertile men (Lai et al. 1996; Pao et al. 1996). The regions E6 and E7 of the viral genome, encoding for important proteins involved in the viral replication and cell transformation, are preferentially localized in spermatozoa, which can, therefore, act as vector of the HPV in the partner, and in the fetus through the fertilization of the oocyte.

Also, Human Herpes Virus type 8 (HHV-8) has been found by *in situ* PCR in the Kaposi's Sarcoma tissues as well as in semen specimens of HIV-1 infected or uninfected individuals (Bobroski et al. 1998).

The Murine Cytomegalovirus (MCMV) was found, by *in situ* hybridization, replicating in mice spermatogenic cells, including spermatozoa (Dutko, Oldstone 1979). The DNA of Murine type C viruses (MuLV) was found in mouse spermatozoa (Levy et al. 1980), which horizontally transmit it to female partners during copulation and in particular to the germ line during penetration of ova. More generally, Kiessling et al. (1989) found that the epididymal epithelium is the principal reservoir for retrovirus expression in the mouse, and in particular viral particles can be associated to spermatozoa. Murine spermatozoa are, in fact, able to integrate foreign DNA sequences in their genome (Magnano et al. 1998).

The enveloped viruses Sendai, Influenza and Semliki Forest have been reported to fuse with bull spermatozoa (Nussbaum et al. 1993). The Feline Immunodeficiency Virus (FIV) DNA has been detected by nested PCR in domestic cat spermatozoa by Jordan et al. (1995, 1998). But this behavior is even more general in the animal kingdom: spermatozoa of salmonid fishes adsorb particles of fish Rhabdoviruses, as well as Hematopoietic Necrosis Virus and Pancreatic Necrosis Virus, and transmit them vertically (Mulcahy, Pascho 1984).

Conclusion

Based on evidence obtained from various methodological approaches, including immunocytochemistry, *in situ* hybridization at the electron microscopy level, PCR and *in vitro* fertilization, an almost unanimous consensus has developed which supports the concept that spermatozoa can act as a vector for HIV-1. Human spermatozoa incorporate HIV-1 via special receptors, different from the usual CD4 receptors, and they remain active and able to transport the viral particles into an oocyte during the fertilization process. Evidence also exists for the presence of viral particles in spermatozoa from animal species. The observation that HIV-1 particles inside spermatozoa can be carried over into oocytes raises concerns for the use of spermatozoa from HIV-1 carriers for achieving a pregnancy.

References

Anderson DJ. Mechanisms of HIV-1 transmission via semen. J NIH Res 1992; 4:106-108.

Anderson DJ, Wolff H, Pudney J, Whenhao Z, Martinez A, Mayer K. Presence of HIV in semen. In: (Alexander NJ, Gebelnick HL, Spieler JM, eds) Heterosexual Transmission of AIDS. New York: Wiley-Liss, 1990, pp167-180.

Ashida RA, Scofield SL. Lymphocyte major histocompatibility complex-coded class II structures may act as sperm receptors. Proc Natl Acad Sci USA 1987; 84:3395-3399.

Baccetti B, Benedetto A, Burrini AG, Collodel G, Elia C, Piomboni P, Renieri T, Zaccarelli M. Retrovirus-like particles detected in spermatozoa of patients with AIDS. First Int Symp on AIDS and Reproduction, 1990, Abstract 4, Genova, Italy.

Baccetti B, Benedetto A, Burrini AG, Collodel G, Elia C, Piomboni P, Renieri T, Sensini C, Zaccarelli M. HIV particles detected in spermatozoa of patients with AIDS. J Submicr Cytol Pathol 1991; 23:339-345.

Baccetti B, Benedetto A, Burrini AG, Collodel G, Elia G, Piomboni P, Renieri T, Zaccarelli M. Spermatozoa of patients with AIDS contain HIV particles. In: (Mélica F, ed) AIDS and Human Reproduction. Basel: Karger, 1992, pp47-54.

Baccetti B, Benedetto A, Burrini AG, Collodel G, Costantino Ceccarini E, Crisà N, Di Caro A, Estenoz M, Garbuglia AR, Massacesi A, Piomboni P, Renieri T, Solazzo D. Atti I Giornata Ital Studio sulla trasmissione eterosessuale dell' HIV. In: (Musicco M, Lazzarin A, Gasparico M, eds) Medico e Paziente, Edifarm S P A Milano, 1993, pp19-22.

Baccetti B, Benedetto A, Burrini AG, Collodel G, Costantino Ceccarini E, Crisà N, Di Caro A, Garbuglia AR, Massacesi A, Piomboni P, Solazzo D. HIV-1 particles in spermatozoa of patients with AIDS and their transfer into the oocytes. J Cell Biol 1994; 127:903-914.

Baccetti B, Benedetto A, Collodel G, Di Caro A, Garbuglia AR, Piomboni P. The debate on the presence of HIV-1 in human gametes. J Reprod Immunol 1998; 41:41-68.

Bagasra O. Summary of the meeting. J Reprod Immunol 1998; 41:373-378.

Bagasra O, Freund M, Weidmann J, Harley G. Interaction of human immunodeficiency virus with human sperm *in vitro*. J Acquir Immunod Syndr 1988; 1:431-435.

Bagasra O, Freund M, Condoluci D, Heins B, Whittle P, Weidmann J, Comito J. Presence of HIV-1 in sperm of patients with HIV/AIDS. Mol Androl 1990; 2:109-125.

Bagasra O, Farzadegan H, Seshamma T, Oakes JW, Saah A, Pomerantz RJ. Detection of HIV-1 proviral DNA in sperm from HIV-1 infected men. AIDS 1994; 8:1669-1674.

Baskar JF, Smith PP, Nilaver G, Jupp RA, Hoffmann S, Peffer NJ, Tenney DJ, Colbert-Poley AM, Ghazal P. The enhancer domain of the human cytomegalovirus major immediate-early promoter determines cell type-specific expression in transgenic mice. J Virol 1996; 70:3207-3214.

Bobroski L, Bagasra AU, Patel D, Saikumari P, Memoli M, Abbey MV, Wood C, Sosa C, Bagasra O. Localization of human herpes virus type 8 (HHV-8) in the Kaposi's Sarcoma tissues and the semen specimens of HIV-1 infected and uninfected individuals by utilizing *in situ* Polymerase Chain Reaction. J Reprod Immunol 1998; 41:149-160.

Borzy MS, Connell RS, Kiessling AA. Detection of human immunodeficiency virus in cell-free seminal fluid. J Acquir Immune Defic Syndr 1988; 1:419-424.

Brogi A, Presentini R, Piomboni P, Collodel G, Solazzo D, Strazza M, Costantino-Ceccarini E. Human sperm and spermatogonia express a galactoglycerolipid which interacts with Gp120. J Submicrosc Cytol Pathol 1995; 27:1-5.

Brogi A, Presentini R, Solazzo D, Piomboni P, Costantino-Ceccarini E. Interaction of human immunodeficiency virus type 1 envelope glycoprotein gp 120 with a galactoglycerolipid associated with human sperm. AIDS Res Hum Retrov 1996; 12:483-489.

Brogi A, Presentini R, Moretti E, Strazza M, Piomboni P, Costantino-Ceccarini E. New insight into the interaction between the gp120 and the HIV receptor in human sperm. J Reprod Immunol 1998; 41:213-232.

Carlemalm E, Villiger W, Hobot JA, Acetarin JD, Kellenberger E. Low temperature embedding with Lowicryl resins: two new formulations and some applications. J Microsc 1985; 140:55-63.

Dulioust E, Tachet A, De Almeida M, Finkielsztejn L, Rivalland S, Salmon D, Sicard D, Rouzioux C, Jouannet P. Detection of HIV-1 in seminal plasma and seminal cells of HIV-1 seropositive men. J Reprod Immunol 1998; 41:27-40.

Dussaix E, Guetard D, Dauguet C, D'Almeida M, Auer J, Ellrodt A, Montagnier L, Auroux M. Spermatozoa as potential carriers of HIV. Res Virol 1993; 144:487-495.

Dutko FJ, Oldstone MBA. Murine cytomegalovirus infects spermatogenic cells. Proc Natl Acad Sci USA 1979; 76:2988-2991.

Gobert B, Amiel C, Tang J, Barbarino P, Béné MC, Faure G. CD4-like molecules in human sperm. FEBS Lett 1990; 261:339-342.

Hadchouel M, Scotto J, Huret JL, Molinie C, Villa E, Degos F, Brechot C. Presence of HBV DNA in spermatozoa: a possible vertical transmission of HBV via the germ line. J Medic Virol 1985; 16:61-66.

Hargreave TB, Ghosh C. The impact of HIV on a fertility problems clinic. J Reprod Immunol 1998, 41:261-270.

Harouse JM, Bhat S, Spitalnik SL, Laughlin M, Stefano K, Silberberg DH, Gonzales-Scarano F. Inhibition of entry of HIV-1 in neural cell lines by antibodies against galactosyl ceramide. Science 1991; 253:320-323.

Kiessling AA, Crowell R, Fox C. Epididymis is a principal site of retrovirus expression in the mouse. Proc Natl Acad Sci USA 1989; 86:5109-5113.

Jordan HL, Howard J, Tompkins WA, Kennedy-Stoskopf S. Detection of Feline Immunodeficiency Virus in semen from seropositive domestic cats. J Virol 1995; 69:7328-7333.

Jordan HL, Howard JG, Bucci JC, Butterworth JL, English R, Kennedy-Stoskopf S, Tompkins MB, Tompkins WA. Horizontal transmission of feline immunodeficiency virus with semen from seropositive cats. J Reprod Immunol 1998; 41:341-358.

Lai YM, Yang FP, Pao CC. Human papillomavirus deoxyribonucleic acid and ribonucleic acid in seminal plasma and sperm cells. Fertil Steril 1996; 65:1026-1030.

Levy JA, Joyner J, Borenfreund E. Mouse sperm can horizontally transmit type C viruses. J Gen Virol 1980; 51:439-443.

Magnano AR, Giordano R, Moscufo N, Baccetti B, Spadafora C. Sperm/DNA interaction: integration of foreign DNA sequences in the mouse sperm genome. J Reprod Immunol 1998; 41:187-196.

Mermin JH, Holodniy M, Katzenstein DA, Merigan TC. Detection of human immunodeficiency virus DNA and RNA in semen by the polymerase chain reaction. J Infect Dis 1991; 164:769-772.

Miller VE, Scofield VL. Transfer of HIV by semen: role of sperm. In: (Alexander NJ, Gebelnick HL, Spieler JM, eds) Heterosexual Transmission of AIDS. New York: Wiley-Liss, 1990, pp147-154.

Muciaccia B, Uccini S, Filippini A, Ziparo E, Paraire F, Baroni CD, Stefanini M. Presence and cellular distribution of HIV in the testes of seropositive subjects: an evaluation by *in situ* PCR hybridization. FASEB J 1998a; 12:151-163.

Muciaccia B, Filippini A, Ziparo E, Colelli F, Baroni CD, Stefanini M. Testicular germ cells of HIV-seropositive asymptomatic men are infected by the virus. J Reprod Immunol 1998b; 41:81-94.

Mulcahy D, Pascho RJ. Adsorption to fish sperm of vertically transmitted fish viruses. Science 1984; 225:333-335.

Nuovo GJ, Becker J, Simsir A, Margiotta M, Khalife M, Shevchuk M. HIV-1 nucleic acids localize to the spermatogonia and their progeny: a study by polymerase chain reaction *in situ* hybridization. Am J Pathol 1994; 144:1142-1148.

Nussbaum O, Laster J, Loyter A. Fusion of enveloped viruses with sperm cells: interaction of Sendai, Influenza, and Semliki Forest viruses with bull spermatozoa. Exp Cell Res 1993; 206:11-15.

Pao CC, Yang FP, Lai YM. Preferential retention of the E6 and E7 regions of the human papilloma virus type 18 genome by human sperm cells. Fertil Steril 1996; 66:630-633.

Pudney J. Caveats associated with identifying HIV using transmission electron microscopy. In: (Alexander NJ, Gebelnick HL, Spieler JM, eds) Heterosexual Transmission of AIDS. New York: Wiley-Liss, 1990, pp197-204.

Pudney J, Nguyen H, Xu C, Anderson DJ. Microscopic evidence against HIV-1 infection of germ cells or attachment to sperm. J Reprod Immunol 1998; 41:105-126.

Quayle AJ, Xu C, Mayer KH, Anderson DJ. T lymphocytes and macrophages, but not motile spermatozoa, are a significant source of human immunodeficiency virus in semen. J Infect Dis 1997; 176:960-968.

Quayle AJ, Xu C, Tucker L, Anderson DJ. The case against an association between HIV-1 and sperm: molecular evidence. J Reprod Immunol 1998; 41:127-136.

Scofield VL. Sperm as vectors and cofactors for HIV-1 transmission. J NIH Res 1992; 4:105:108-111.

Scofield VL. Sperm as infection-potentiating cofactors in HIV transmission. J Reprod Immunol 1998; 41:359-372.

Scofield VL, Poiesz B, Kennedy C, Broder S, Diagne A, Rao B. HIV binds to normal sperm and is present in sperm from infected men. VIII Int Conf on AIDS, Amsterdam, 1992, Abstract 2102.

Scofield VL, Rao B, Broder S, Kennedy C, Wallace M, Graham B, Poiesz BJ. HIV interaction with sperm. AIDS 1994; 8:1733-1736.

Semprini AE, Levi-Setti P, Bozzo M, Ravizza M, Taglioretti A, Sulpizio P, Albani E, Oneta M, Pardi G. Insemination of HIV-negative women with processed semen of HIV-positive partners. Lancet 1992; 340:1317-1319.

Shevchuk MM, Nuovo GJ, Khalife G. HIV in testis: quantitative histology and HIV localization in germ cells. J Reprod Immunol 1998; 41:69-80.

Van Voorhis BD, Martinez A, Mayer K, Anderson DJ. Detection of HIV-1 in semen from seropositive men using culture and polymerase chain reaction deoxyribonucleic acid amplification techniques. Fertil Steril 1991; 55:588-594.

Wolff H, Zhang WH, Anderson DJ. The CD4 antigen (HIV-receptor) is not detectable on human testicular germ cells or spermatozoa In: Program and Abstracts of the IV International Conference on AIDS, Stockholm, 1988, Abstract 2547:112.

40

How Assisted Reproduction Deals with HIV

S. Hamamah
Hôpital A. Béclère

A. Fignon
Hôpital Bretonneau

D. Mortimer
Sydney IVF
University of Sydney

The World Health Organization (WHO) has estimated that 18 million adults and approximately 1.5 million children have been infected by the human immunodeficiency virus (HIV) since the beginning of the pandemic. Entering its second decade, the epidemic of HIV infection continues unabated, now achieving a level of infection that has been estimated at 6000 new cases per day (WHO 1995).

This current global growth is due primarily to heterosexual transmission, which is now responsible for more than 80% of new cases of infections (D'Cruz-Grote 1996). The infection of women, who represent almost half of the individuals infected (Semprini et al. 1997; Marina et al. 1998), and materno-fetal transmission are responsible for this growth.

In Western countries, the level of HIV transmission from mother to child is of the order of 15 to 20%. Recent results have demonstrated the effectiveness of AZT treatment (Center for Disease Control 1994) of pregnant women in reducing the risk of materno-fetal transmission to <5%. This low prevalence allows conception at minimal risk for couples where one of the partners is infected. If only the man is infected, the couple can resort to donor insemination. On the other hand, in both Italy and Spain, for couples in which the man is seropositive but the woman seronegative, the possibility of inseminating women using prepared fractions of normal motile spermatozoa has been proposed (Semprini et al. 1992, 1997; Marina et al. 1998).

The HIV Virus

Epidemiology

The AIDS (Acquired Immunodeficiency Syndrome) virus was discovered in 1983 by Professor Montagnier and colleagues (Pasteur Institute, Paris), and was named the Human Immunodeficiency Virus (HIV). It became HIV-I when, in 1986, HIV-II (which is less pathogenic) was isolated from some African patients. HIV is an RNA retrovirus, in which the viral genome is surrounded by a protein coat, which is itself surrounded by a lipid coat. Because of the presence of reverse transcriptase, which has the ability to reverse transcribe the viral genome (RNA) into complementary DNA, it has also been called a provirus.

This virus, a retrovirus of the slow virus group, has a great affinity for T4 lymphocytes and causes their destruction, inducing lymphadenopathies, opportunistic infections and Kaposi's sarcoma. It is present in all body secretions, notably semen, vaginal secretions, blood, breast milk, cerebrospinal fluid, saliva, tears and urine.

AIDS is a pandemic—that is, it affects the entire world—with more numerous foci of infection in Africa and southeast Asia. WHO has reported an approximate total of 1,300,000 AIDS cases, but estimates the real number at six million; in addition, an estimated 22 million adults and children have been infected with HIV, including about six million women. Of these women, the majority are in the reproductive age range and will give birth to one million children. The majority of those children will be infected by their mothers. The prevalence of HIV-infected women in North America is the same as in Europe, including France. However, in Africa it is far greater, with the proportion of seropositive women varying between 5 and 30%. There are four million HIV-positive women and 1.6 million orphaned children of HIV-infected parents. This is in addition to the 10 million seronegative orphans born of HIV-infected mothers.

Sexual intercourse, either heterosexual or homosexual, is the principal route of HIV transmission. However, HIV can also be transmitted from an infected woman to her fetus, or neonate, before, during or just after delivery. The number of AIDS cases doubles approximately every nine to 12 months: each seropositive individual not using protection against the disease infects a further 25 to 100 others. On the other hand, in developing countries there seems to be a reduction in the number of new cases reported which cannot be due solely to protection and may be related to therapeutic intervention. Nonetheless, projections for the year 2000 indicate that 15 million adults and more than five million children will be infected.

Virology

HIV, being a member of the retrovirus family, is characterized by the ability to destroy the cells that it infects. The target cells of HIV are, essentially, CD4 lymphocytes (helper T cells) and monocytes. Coated virus, between 80 and 120 nm in diameter, leave infected cells by a process of plasma membrane budding. The viral coat is composed of a high molecular weight, heavily glycosylated glycoprotein which is eventually cleaved into three others, gp120, gp110 and gp41. The core of the virus encloses the viral genome, composed of single-stranded RNA and viral enzymes including reverse transcriptase, endonuclease/integrase and protease. This genome of 9200 bases consists of three genes characteristic of retroviruses: *gag*, *pol* and *env*. The *gag* (group antigen) gene synthesizes a polypeptide which is cleaved into the three proteins that constitute the core: the major core protein p24 (24 kD), p17 and p15. The *pol* (polymerase) gene codes for the viral enzymes, and the *env* (envelope) gene codes for the envelope glycoprotein gp120 and the transmembrane glycoprotein gp41.

The second virus, HIV-II, uses modes of replication and transmission identical to those of HIV-I, but differences do exist. The most notable is the HIV-II level of pathogenicity, which is lower than that of HIV-I.

The first stage in the replication cycle of HIV is attachment of the viral particle to the target cell. The surface glycoprotein gp120 recognizes its membrane receptor and binds the virus to the host cell; subsequently, the viral envelope and the cell plasma membrane fuse. The capsid then penetrates the cell and releases the viral genome associated with reverse transcriptase. The viral RNA is transcribed into double-stranded DNA, which then integrates into the nuclear DNA of the host cell.

From a virological standpoint, initial HIV infection is associated with an intense replication, characterized by massive viremia. Six to eight weeks after infection, the immune system attempts to control this viremia by producing anti-HIV antibodies and establishing a specific cell-mediated immunity (activation of cytotoxic lymphocytes).

HIV infects helper T cells (CD4) and monocytes. These cells bear the transmembrane CD4 glycoprotein, facilitating the binding to the viral coat proteins, but alone incapable of ensuring viral entry into the cell. As a consequence, the concept of a co-receptor has been developed. The first candidate to be identified was fusin (or CXCR4), which was renamed CCR5 in 1996. Infected lymphocytes suffer functional modification, compromising the immune system. The cytotoxic effect of HIV leads to the formation of syncytia, leading to cell death. The vulnerability to opportunistic infection is a consequence of the decreasing numbers of CD4 lymphocytes.

Many studies have demonstrated a correlation between viremia titers, proviral DNA, plasma RNA and the severity of immunosuppression. An elevated level of plasma RNA is predictive of clinical progression and of a fall in CD4 lymphocytes. In long-term patients in whom the disease is non-progressing, the viral load is low. In treated patients, the decrease in viral load is correlated with a decrease in clinical progression of the disease. However, if the level of plasma RNA is substantially decreased, changes in proviral DNA are, themselves, of little importance.

HIV and Semen

Which Are the HIV Carrier Cells in Semen?

It is now established that HIV can be found in the semen of infected men, being located either in lymphocytes, CD4 macrophages, or, after reverse transcription, in seminal plasma with non-germ line (round) cells. Of the various semen fractions likely to harbor the virus, monocytes are the prime candidates (Zagury 1984; Borzy et al. 1988; Kiessling 1992); macrophages, whose presence in semen is well established, appear to be the principal vectors of HIV (Gartner et al. 1986; Popovic, Gartner 1987). The role of lymphocytes, which were initially suspected to be the vehicles for the virus, is now uncertain since their presence in semen has been questioned due to their frequent confusion with immature germ line cells. Nevertheless, the ultimate role of infected leukocytes as hosts or vectors for HIV in semen cannot explain certain findings. Indeed, some authors have reported free virus not only in the seminal plasma of contaminated semen (Krieger et al. 1991), but also in semen devoid of infected cells. The current hypothesis for spermatozoa as potential carriers of HIV is therefore justified.

Can Spermatozoa Contain HIV?

The presence or absence of HIV in spermatozoa and the presence or absence of the CD4 receptor on these cells have given rise to an abundant, but contradictory, literature. Although the entry of HIV into spermatozoa has not been proven, *in vitro* incubation of healthy spermatozoa in the presence of a high concentration of HIV particles has shown that the virus can be found on the sperm surface (Bagasra et al. 1988). Similarly, electron microscopic studies have suggested that HIV can enter the sperm cytoplasm (Dussaix et al. 1993). While some *in vivo* studies have not revealed virus particles inside spermatozoa from HIV infected men, others have reported the virus present inside spermatozoa (Bagasra et al. 1990).

If the seminal plasma or non-germ line cells were the only vectors of HIV in semen, insemination using spermatozoa washed from the semen might be considered. On the other hand, if spermatozoa themselves are implicated as vectors, then there is a risk of infection to the woman and to any child resulting from a successful insemination treatment. The possible presence of HIV inside spermatozoa completely discredits the concept of decontaminating the sperm surface by washing. Even if *in vitro* sperm preparation techniques were to permit a 10- to 10,000-fold (1 to 4 log) reduction in semen viral load, some virus particles would still remain. A level of HIV RNA in semen that would be predictive of a potentially infectious viral load therefore needs to be established. But this threshold must be well defined because clearly one cannot consider any possibility of using spermatozoa from a seropositive man for insemination unless the sperm washing techniques (including centrifugation and migration) can reduce the load well below it.

With regard to semen characteristics, Krieger (1991) found no significant difference between seropositive and seronegative individuals, in contrast to Crittenden et al. (1992), who reported a decrease in sperm motility correlated with the fall in the number of circulating CD4 lymphocytes, a reflection of the progression of infection. Dondero et al. (1996) studied the semen of at-risk individuals and their results confirmed the observations of Crittenden et al. (1992) and also revealed an increased proportion of immature germinal cells, atrophic forms of spermatozoa and spermophage cells in the semen of infected men. It seems, therefore, that HIV might activate spermophagic mechanisms.

In addition to this relationship between semen characteristics and clinical features of seropositive individuals, some authors have reported a correlation between seminal viral load, circulating CD4 lymphocyte numbers and clinical symptoms (Borzy et al. 1988; Krieger et al. 1991), with semen viral load increasing during progression of the disease. A recent study has shown a weak but significant correlation between circulating and seminal viral loads (Dussaix et al. 1989). Although PCR positivity appears to be weakly related to the degree of immunosuppression or the presence of a sexually transmitted disease (STD), there does not appear to be a correlation between the number of CD4 lymphocytes and the PCR positivity of semen.

Other authors, including Ashida et al. (1987) and Bagasra et al. (1988), have detected CD4 antigen on some ejaculated spermatozoa, although others have found no evidence of it (Dussaix et al. 1993). Therefore, perhaps the expression of CD4 epitopes on spermatozoa varies quantitatively after ejaculation. Bagasra et al. (1988) and Harouse et al. (1991) have identified another possible receptor for HIV on spermatozoa, galactosylceramide.

According to Baccetti et al. (1994), the entry of HIV into spermatozoa is probably not a result of the binding of the virus to a specific receptor. The sperm membrane appears to be able to incorporate particles, as has been shown during attempts to cause transgenesis by incubating male gametes in a medium containing minigenes, even though these may cross the membrane without being introduced into the zygote genome. HIV may enter cells in a similar manner without requiring the CD4 receptor. Many studies have reported that HIV is capable of infecting cells lacking CD4 epitopes on their surface (Ashida et al. 1987; Harouse et al. 1991).

The Role of Infected Spermatozoa

Initiation of viral replication requires the activation and multiplication of lymphocytes, which may be achieved by mitogenic compounds but also by contact with allogeneic cells (such as spermatozoa) or antigens (such as bacteria, fungi or viruses). Dussaix et al. (1993) demonstrated that HIV on the sperm surface can secondarily infect a culture of lymphocytes. On the other hand, if the virus penetrates spermatozoa, its replication cannot necessarily be established, even if it is integrated into the male gamete's DNA. According to Baccetti et al. (1994), there is no reverse transcription of the viral RNA into DNA in the spermatozoon, which, therefore, acts only as a vector. Bagasra et al. (1992) do not exclude the possibility of proviral RNA, detected integrated into the host cell's DNA. In consideration of this possibility, Dussaix et al. (1993) have foreshadowed the immediate and future consequences for any offspring because as potential carriers of HIV, spermatozoa, in addition to their role in sexual transmission, also have a role in transmission of infection to the zygote.

Several pieces of evidence should be reconsidered and alternative hypotheses perhaps proposed. Experimental studies using electron microscopy (Baccetti et al. 1994; Dussaix et al. 1993; Bagasra et al. 1994) do not necessarily allow demonstration of the presence of HIV inside the spermatozoa. Difficulties of interpreting micrographs also exist so that, even if virus particles are seen using this technique, there remains the question as to whether it may be an artifact. Having detected virus in spermatogonia and spermatids, Bagasra et al. (1994) suggested that HIV infected the testes and that spermatozoa were contaminated as a consequence of this infection; this hypothesis remains unconfirmed. Wolff et al. (1988) have reported the absence of CD4 receptors on the sperm plasma membrane although, as mentioned already, the possibility of HIV infecting cells lacking these receptors warrants further investigation. Similarly, the role of infected spermatozoa in virus transmission to the zygote is unclear and, in fact, no infected neonate has been reported from a seropositive father and seronegative mother.

Management of Conception in HIV Serodiscordant Couples

Possible Transmission of HIV after Artificial Insemination with the Partner's Semen (AIH)

Only seropositive women have produced infected children. Consequently, if a couple in which the woman is seronegative and the man is seropositive seeks a child, artificial insemination (AI) can be considered, so limiting the risk of infecting the female partner. This AI can be performed using donor semen or the partner's semen after sperm washing.

In 1985, Stewart et al. reported four cases of infection of women after donor insemination with frozen semen and, in 1990, the US Center for Disease Control (CDC) reported a case in which a woman was infected after insemination with selected spermatozoa (CDC 1990). According to Baccetti et al. (1992), HIV can attach to spermatozoa and even penetrate into the sperm head, so that following fertilization, viral particles might be transmitted to the zygote. After *in vitro* fertilization of human oocytes using spermatozoa from infected men, Baccetti et al. (1994) found viral particles in or on blastomeres. According to Dussaix et al. (1993), spermatozoa cannot all be contaminated, although the viability and ultimate fertilizing ability of spermatozoa carrying HIV remains unknown. However, would these authors reject the possibility of artificial insemination between a seropositive man and a seronegative woman?

Intrauterine insemination of seronegative women with their seropositive partners' spermatozoa, washed free of HIV, was suggested by Semprini et al. (1992) and by Marina et al. (1998). No case of infection of those women who became pregnant was reported. Furthermore, infection of a woman could also occur from seminal plasma or non-germ line cells during artificial insemination. To date, no seropositive child has been recorded among the more than 200 children born of a seropositive father and a seronegative mother, according to the results presented recently by Semprini et al. (1998). Both Edlin and Holmberg (1993) and Dussaix et al. (1993) emphasized the risk of HIV transmission inherent in such a procedure: if HIV became integrated into the host genome, spermatozoa would therefore be the vectors of altered genetic information. Even if fertilization and embryo viability were not affected, the long-term consequences for any offspring would remain unknown.

In conclusion, the role of spermatozoa in HIV transmission is therefore uncertain and AIH cannot eliminate all risk factors.

Techniques to Reduce the Infectivity of Spermatozoa

The results of Dussaix et al. (1993) showed that spermatozoa of seronegative men incubated with HIV and subjected to repeated washing can still infect receptive cells. To prepare spermatozoa from the semen of a seropositive man for artificial insemination, Chrystie et al. (1998) used a technique involving washing and centrifugation followed by migration on a discontinuous Percoll gradient. The viral load in semen was quantified using PCR and, even if the viral load was decreased, the presence of virus in semen meant that the risk of HIV transmission during insemination of a sample prepared in this way could not be eliminated. Chrystie et al. (1998), after subjecting the semen of seropositive men to repeated washing and centrifugation on a Percoll gradient followed by swim-up migration, used the NASBA technique to determine the viral load in these prepared ejaculates. This method of sperm preparation significantly reduced the number of viral RNA copies by one to four orders of magnitude, but did not completely eliminate it. Nevertheless, the results of Semprini et al. (1992) indicated that *in vitro* sperm preparation followed by swim-up eliminated all risk of infection of the partner when HIV washed semen was used for artificial insemination.

The treatment of semen from HIV-infected men using standard procedures (Percoll gradient, swim-up and so on) for isolating motile spermatozoa can reduce the HIV-RNA concentration in the final sperm sample to low and often undetectable levels (Chrystie et al. 1998). The Percoll gradient and swim-up procedures can efficiently reduce and "eliminate" HIV-RNA from positive semen. The use of these procedures to assist conception in HIV-discordant couples in which the man is HIV positive continues to be used by Semprini et al. (1997) in Italy and by Marina et al. (1998) in Spain.

It might be interesting to determine if the triple therapy, which includes a protease inhibitor, would significantly reduce the viral load in semen. If so, the reduced infectivity of semen used for artificial insemination might limit the risk of infection of the woman and her descendants.

Should HIV Serodiscordant Couples Be Treated Using Assisted Reproductive Technology (ART)?

a) When the woman is HIV positive. The major problem in this situation is the risk of infecting a child born of a mother infected by HIV. Antiretroviral treatment during pregnancy can reduce this risk to below 5%, but promoting a pregnancy with this risk is not acceptable to the medical profession. On the other hand, the fact of not advising a pregnancy under such circumstances should not prevent a doctor doing everything to avoid infection of the male partner. If a couple chooses to ignore such unwanted opinion, the following advice should be given: first, use self-insemination so as to never have unprotected intercourse; or, if self-insemination is unacceptable, engage in unprotected intercourse only once, during the pre-ovulatory period. In this case, in order to avoid a prolonged period of trying, a temperature chart and semen analysis should be the minimum recommended preliminary investigation. Also, all genital infections should be treated, especially if associated with mucosal ulceration.

b) When the man is HIV positive. The principle of managing couples which desire a pregnancy but in which the man is seropositive has been accepted by the medical profession, but this management must be provided by a multi-disciplinary team. There are several primary objectives. First, the man's health must be evaluated and a sufficient life expectancy established. Second, in cases of homologous insemination, semen with the lowest viral load possible should be used. Third, the depth of the couple's desire for a child as well as their psychological state in the face of HIV seropositivity must be ascertained. This aspect is particularly important in order to be able to establish a real desire for a family as distinct from a depression caused by the disease's prognosis. Last, and more important here than in all other requests for reproductive medical assistance, the stability of the couple must be evaluated. Also, the likely success of rehabilitation if the male partner is a drug addict must be considered.

Failure to have protected intercourse in order to

avoid infection of the female partner should be a contraindication for ART treatment of such couples.

What Protocols Should Be Used in Treating Couples?

When a request for assistance from a serodiscordant couple in whom the man is HIV positive is accepted, there is no reason to restrict their management. It is ethically essential that all methods that can minimize or eliminate the risk of infection be used to allow the couple to become parents as quickly as possible. All ART techniques should be considered:

Donor sperm insemination. This represents a technique which carries no risk of infection of the partner or child. However, donor insemination should be performed according to the same principles and procedures as when used for other indications, especially with regard to the waiting period. It is advisable to be particularly aware of psychological reactions. For example, is the man still in the process of coming to terms with his sterility, or has he accepted it?

Homologous insemination. From current knowledge it is impossible to establish a total absence of risk to the partner and, as a consequence, to the offspring, for this form of treatment. Provision of clear and precise information concerning this risk to the couples involved (informed consent) is absolutely essential. Both the study of Semprini et al. (1998) involving more than 1500 AIH cycles, and that of Marina et al. (1998) on 107 AIH cycles, have indicated that the use of semen preparation techniques to separate spermatozoa combined with cryopreservation of semen and virological analysis using validated techniques might reduce the risk to almost zero.

The choice between AIH and donor insemination can only be made after complete evaluation of each couple's request, after clear and precise advice on the techniques and the risks they entail, and after repeated interviews with both partners and a waiting period for reflection.

An Experimental Model

In light of the results reported by Semprini et al. (1992), the efficiency of a Percoll-based sperm preparation method to separate spermatozoa from semen infected with HIV *in vitro* was tested (Hamamah et al. 1998). From HIV cytopathy effect (CPE) analysis, HIV was found to be still present in all semen fractions after processing, even though there had been a decrease in seminal viral load from 10^5 to 10^1 HIV particles = a 4 log reduction (Fig. 1). Percoll processing reduced the semen viral load but did not achieve a total removal of HIV from *in vitro* infected semen, and therefore did not eliminate the risk of HIV transmission using inseminates prepared in this way.

Also investigated was the value of repeated sperm washing after Percoll processing upon the viral load of spermatozoa that had been infected with HIV *in vitro*. Even after many cycles of washing, HIV was still inside the washed spermatozoa, as well as in the supernatants, even though repeated washing significantly reduced the viral load (a 1000-fold reduction after six washing cycles based upon assay of p24 antigen levels: from 160 pg/ml in the first wash supernatant to 11–12 pg/ml in the sixth).

If HIV were completely free in the semen, repeated washing might theoretically allow its complete removal, but in reality, some HIV remains in the final sperm pellet. If HIV entered all the spermatozoa, the level of p24 antigen detected in the wash supernatants would only be reduced by the washing; it could not be eliminated. It therefore seems that HIV, initially free in the semen, progressively enters some spermatozoa, as well as perhaps being adsorbed onto the sperm plasma membrane.

Conclusions

The experimental model indicated that the washing and Percoll gradient procedure used for *in vitro* sperm preparation can reduce, but not eliminate, HIV viral particles. These results have important implications for investigating the HIV infectivity of human semen. Washing of semen from HIV positive men, even by density gradient centrifugation, prior to artificial insemination, can decrease—but not abolish—the risk of HIV transmission.

Regardless of the technique used to detect HIV in semen, there remain many unresolved questions: Does HIV attach to spermatozoa? Is viral genomic material transmitted by spermatozoa? Does the semen quality of seropositive men influence HIV detection? Does HIV replicate inside spermatozoa? What happens to viral particles after their ultimate transmission to oocytes by spermatozoa?

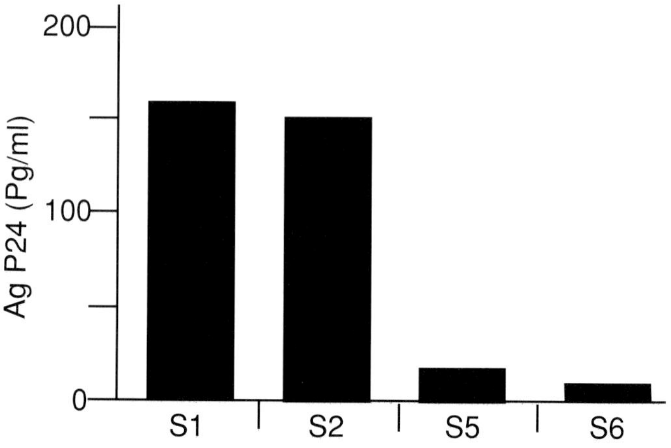

Figure 1. Effects of washing number on HIV p24 antigen of spermatozoa pellet after Percoll preparation (S1, S2, S5, S6: washing number).

For the future, it would be desirable to have a list of centers which treat serodiscordant couples in whom the man is HIV positive, as well as of the procedures employed (homologous or donor gametes). In addition, appropriate follow-up of these different treatment modalities should be ensured; for example in the form of a national registry. These developments would be important so that all medical practitioners could be informed and then able to direct HIV serodiscordant couples to specialist teams for their management.

Acknowledgements

The authors wish to thank G. Dubois and J. Poindron for their technical participation and Pr. F. Barin for kindly providing access to the virology facilities.

References

Ashida ER, Scofield VL. Lymphocyte major histocompatibility complex encoded class II structure may act as sperm receptors. Proc Natl Acad Sci USA 1987; 84:3395-3399.

Baccetti B, Arrigo B, Burrini AG, Collodel G, Costantino Ceccarini E, Cris N, Di Caro A, Estenoz M, Garbuglia AR, Massacesi A. HIV particles in spermatozoa of patients with AIDS and their transfer into the oocyte. J Cell Biol 1994; 127:902-914.

Baccetti B, Benedetto A, Burrini AG, Collodel G, Elia C, Piomboni P, Renier T, Sensini C, Zaccarelli M. Spermatozoa of patients with AIDA contain HIV particles. In: (Mlica F, ed) AIDS and Human Reproduction. Basel: S Karger, 1992, pp47-54.

Bagasra O, Freund M, Weidmann J, Harley G. Interaction of HIV with human sperm *in vitro*. J AIDS 1988; 1:431-435.

Bagasra O, Freund M, Condoluci D, Heins B, Whittle P, Weidmann J, Comito J. Presence of HIV-1 in sperm of patients with HIV:AIDS. Mol Androl 1990; 2:109-125.

Bagasra O, Hauptman SP, Lischner HW, Sachs M, Pomerantz RJ. Detection by *in situ* polymerase chain reaction of provirus in mononuclear cells of individuals infected with HIV1. N Engl J Med 1992; 326:1385-1391.

Bagasra O, Farzadegan H, Seshamma T, Oakes JW, Saah A, Pomerantz RJ. Detection of HIV proviral DNA in sperm from HIV1 infected men. AIDS 1994; 8:1669-1674.

Borzy MS, Connell RS, Kiessling AA. Detection of HIV in cell-free seminal fluid. J AIDS 1988; 1:419-424.

Center for Disease Control. Epidemiology Program Office. HIV1 infection and artificial insemination with processed semen. Mort Morb Week Rep 1990; 39:249-256.

Center for Disease Control. Zidovudine for the prevention of HIV transmission from mother to infant. Mort Morb Week Rep 1994; 43:285-287.

Chrystie IL, Mullen JE, Braude PR, Rowell P, Williams E, Elkington N, de Ruiter A, Rice K, Kennedy J. J Reprod Immun 1998; 41:301-306.

Crittenden JA, Handelsman DJ, Stewart GJ. Semen analysis in human immunodeficiency virus infection. Fertil Steril 1992; 57:1294-1299.

D'Cruz-Grote D. Prevention of HIV infection in developing countries. Lancet 1996; 348:1071-1074.

Dondero F, Rossi T, D'Offizi G, Mazzilli F, Rosso R, Sarandrea N, Pinter E, Aiuti F. Semen analysis in HIV seropositive men and in subjects at high risk for HIV infection. Hum Reprod 1996; 11:765-768.

Dussaix E, Guetard D, Dauguet C, D'Almeida M, Auer J, Ellrodt A, Montagnier L, Auroux M. Potential of spermatozoa in the HIV. Fifth Int Conf on AIDS, June 1989, Montreal, Abstract.

Dussaix E, Guetard D, Dauguet C, D'Almeida M, Auer J, Ellrodt A, Montagnier L, Auroux M. Spermatozoa as potential carriers of HIV. Res Virol 1993; 144:487-495.

Edlin BR, Holmberg SD. Insemination of HIV negative women with processed semen of HIV positive partners. Lancet 1993; 341:570-571.

Gartner S, Markovits P, Markovits DM, Kaplan MH, Gallo RC, Popovic M. The role of mononuclear phagocytes in HTLV III/LAV infection. Science 1986; 233:215-219.

Harouse JM, Bhat S, Spitalnik SL, Laughlin M, Stefano K, Silberberg DH, Gonzalez-Scarano F. Inhibition of entry of HIV1 in neural cell lines by antibodies against galactosyl ceramide. Science 1991; 253:320-323.

Hamamah S, Barin R, Dubois G, Barthelemy C, Lansac J. Spermatozoa as potential carriers of HIV: an experimental model. Abstract. Hum Reprod 1998; 13 (Suppl):76.

Kiessling A. Semen transmission of HIV. Fertil Steril 1992; 58:667-669.

Krieger JN, Coombs RW, Collier AC, Koehler JK, Ross SO, Chaloupka K, Murphy VL, Corey L. Fertility parameters in men infected with human immunodeficiency virus. J Infect Dis 1991; 164:464-469.

Marina S, Marina S, Alcolea R, Exposito R, Huguet J, Nadal J, Verges A. Human immunodeficiency virus type-1 serodiscordant couples can bear healthy children after undergoing intrauterine insemination. Fertil Steril 1998; 70:35-39.

Popovic M, Gartner S. Isolation of HIV1 from monocyte but not T lymphocytes. Lancet 1987; 2:916.

Semprini A, Fiore S, Pardi G. Reproductive counselling for HIV discordant couples. Lancet 1997; 349:1401-1402.

Semprini A, Levy-Setti P, Bozzo M, Ravizza M, Taglioretti A, Sulpizio P. Insemination of HIV negative women with processed semen of HIV positive partners. Lancet 1992; 340:1317-1319.

Semprini A, Fiore S, Oneta M, Castagna C, Savasi V, Guintelli S, Persico T. Assisted reproduction in HIV-discordant couples. Abstract. Hum Reprod 1998; 13 (Suppl):89.

Stewart GJ, Tyler JP, Cunningham AL, Barr JA, Driscoll GL, Gold J, Lamont BJ. Transmission of human T cell lymphotropic virus type M (HTLVIII) by artificial insemination by donor. Lancet 1985; 2:581-584.

Wolff H, Zhang WH, Anderson DJ. The CD4 antigen (HIV-receptor) is not detectable on human testicular germ cells or spermatozoa. Fourth Int Conf on AIDS. Abstract. Stockholm, 1988.

World Health Organization. Global programme on AIDS. The current global situation of the HIV/AIDS pandemic. Geneva, 1995.

Zagury D, Bernard J, Leibowitch J, Safai B, Groopman JE, Feldman M, Sarngadharan MG, Gallo RC. HTLV III in cells cultured from semen of two patients with AIDS. Science 1984; 226:449-451.

41 The Saga of the Sperm Count Decrease in Humans and Wild and Farm Animals

B. Jégou
Université de Rennes

J. Auger
Hopital Cochin

L. Multigner
Hopital de Bicêtre

C. Pineau
Université de Rennes

P. Thonneau
A. Spira
Hopital de Bicêtre

P. Jouannet
Hopital Cochin

Public concern about the changes to the environment and their effects on human and animal health have grown during the last 20 or 30 years to become a major preoccupation in developed societies. Recently, these worries have soared as a result of the publication of data suggesting a decrease in the quality of human sperm, a subject widely reported by the media world-wide. Thus, not only is the health of individuals at risk due to a deteriorating environment, but also the very capacity to reproduce may be threatened.

This article will review the data on which the declining sperm quality hypothesis is based. It will consider the situation of humans and of domesticated and wild animals, and the putative causes of the phenomenon. The possible consequences on human fertility will be addressed.

Is the Deterioration in Sperm Quality a Real Phenomenon?

Some believe that there is no significant decline in sperm quality, whereas others are certain that the problem is real. The debate has been lively for the last 20 years. It started in 1974 when the first article describing a decline in sperm quality, both quantitative and qualitative, was published in the USA. There has since been a constant flow of articles, both epidemiological studies and literature reviews, supporting this hypothesis. However, there have also been an equally large number of papers arguing the opposite (Table 1). It seems surprising that after more than 20 years of debate, and even heated argument, this issue has not been resolved. The uncertainty reflects the difficulties faced by the scientific community in measuring objectively the developments of semen quality and anomalies of the male urogenital tract.

For a long time, the credibility of epidemiological work suggesting the existence of sperm quality decline was limited by numerous biases and also a lesser interest in environmental issues. However, a publication in 1992 by a Danish group, directed by N.E. Skakkebæk, relaunched the debate. The study was a meta-analysis of 61 articles published between 1938 and 1990 including data on 14,947 men on all the world's continents (Carlsen et al. 1992). The conclusion was that the mean volume of the human ejaculate had fallen from 3.4 ml in 1940 to 2.75 ml in 1990, and that simultaneously the mean sperm concentration had fallen from 113 million to 66 million spermatozoa per ml, a decline of 1% per year for 50 years. The conclusions of the meta-analysis were strengthened by the authors' attempts to avoid various common and major biases. For example, data included were obtained as systematically as possible from databases of the Medicus index (1930-1965), Current Lists (1957-1959) and Medline (1966-1991), to try to eliminate any subjective selection of bibliographic sources. The authors only included studies in which sperm concentrations were determined with a haemocytometer to exclude bias due to variability in counting techniques. They also excluded all data for couples consulting for infertility so as to restrict the analysis to supposedly fertile men. Despite these efforts, the study was widely criticized as soon as it was published (Table 2). The major criticisms were that there was necessarily a selection bias as the databases are known not to be exhaustive; data from different laboratories and obtained by different analytical methods could not be compared with reliability; the distribution of the data over the time period was irregular, and particularly that there was little data for the years between 1950 and 1970; the statistical method used (linear regression) was not applicable to this type of non-homogeneous distribution of sperm concentration; and the geographic distribution of the data was not consistent throughout the period studied—the data for the years before 1970 were mostly from the USA where sperm values appear to be high, whereas only a minority of the data for the years after 1970 were

Table 1: Selection of articles on the secular trends in semen quality. Articles in bold and italics support the hypothesis of deterioration in semen quality.

1974	***Nelson and Bunge***	***Fertil Steril 25:503-507***
1975	***Rehan et al.***	***Fertil Steril 26:492-502***
1979	David et al.	Fertil Steril 31:453-455
1979	MacLeod and Wang	Fertil Steril 31:103-116
1980	***James***	***Andrologia 12:381-388***
1981	Jouannet et al.	Int J Androl 4:440-449
1981	***Leto and Frensilli***	***Fertil Steril 36:766-770***
1983	***Bosfolte et al.***	***Int J Fertil 28:91-95***
1984	***Osser et al.***	***Arch Androl 12:113-116***
1986	***Menkveld et al.***	***Arch Androl 17:143-144***
1991	***Bendvold et al.***	***Arch Androl 26:189-194***
1992	***Carlsen et al.***	***Br Med J 305:609-613***
1992	Wittmaack and Shapiro	Wisconsin Med J 91:91-95
1993	Suominen and Vierula	Br Med J 306:1579
1994	Bromwich et al.	Br Med J 309:19-22
1994	Farrow	Br Med J 309:1-2
1995	***Auger et al.***	***N Engl J Med 332:281-285***
1995	Olsen et al.	Fertil Steril 63:887-893
1995	Sherins	N Engl J Med 332:327-328
1996	***Adamopoulos et al.***	***Hum Reprod 11:1936-1941***
1996	Bujan et al.	Br Med J 312:471-472
1996	***De Mouzon et al.***	***Br Med J 313:43***
1996	Fisch et al.	Fertil Steril 65:1009-1014
1996	***Irvine et al.***	***Br Med J 312:467-471***
1996	***Menchini-Fabris et al.***	***Andrologia 28:304***
1996	Paulsen et al.	Fertil Steril 65:1015-1020
1996	***Van Waeleghem et al.***	***Hum Reprod 11:325-329***
1996	Vierula et al.	Int J Androl 19:11-17
1997	***Becker and Berhane***	***Fertil Steril 67:1103-1108***
1997	Handelsman	Hum Reprod 12:2701-2705
1997	***Swan et al.***	***Env Health Perspect 105:1228-1232***

Table 2: Selection of articles discussing the meta-analysis of Carlsen et al. 1992.*

1992	Tumon and Mortimer	Br Med J 305:1228-1229
1992	Brake and Krause	Br Med J 305:1498
1994	Farrows	Br Med J 309:1-2
1994	Bromwich et al.	Br Med J 309:11-22
1994	***Keiding et al.***	***Br Med J 309:22***
1995	Olsen et al.	Fertil Steril 63:887-893
1996	Fisch and Goluboff	Fertil Steril 65:1044-1046
1996	Lerchl and Nieschlag	Exp Clin Endocr Diabetes 104:301-307
1997	***Becker and Berhane***	***Fertil Steril 67:1103-1108***
1997	***Swan et al.***	***Environ Health Perspect 105:228-1232***

*Articles in bold and italics are supportive of Carlsen et al., finding a secular deterioration in human semen quality.

from the USA and a number were from five Third World countries with low values. Keiding et al. (1994) responded to these criticisms and two recent studies, reanalyzing the data reported by Carlsen et al. (1992), confirm that the data reveal a decrease in sperm values. In fact, Becker and Berhane (1997) studied the data from the USA alone, the only country included in the meta-analysis with continuously available data, and reported there has indeed been a decline in the sperm concentration. Moreover, Swan et al. (1997) conducted the most extensive re-analysis of the data and found a decline in the sperm density of 1.5% per year in the USA between 1938 and 1988, and of 3.1% per year in Europe between 1971 and 1990, but no decrease in non-Western countries between 1978 and 1989.

The controversy concerning the meta-analysis of Carlsen et al. (1992) was without precedent, and the least that this study did was to relaunch the debate about sperm quality and incite laboratories world-wide to analyze their own data. Consequently, the hypothesis that sperm quality is declining was reinforced by a study by one of the oldest sperm banks in the world, the Centre d'Etude et de Conservation des Oeufs et du Sperme Humain (CECOS), a human egg and sperm bank in Paris which was founded in 1973 by Georges David. The background of this study is illuminating. In 1979, CECOS, rather than prompted by changes in sperm quality, became interested in their average value: CECOS showed that the mean sperm concentration for fertile Parisian donors was around 90 million/ml (David et al. 1979), a value similar to that published by MacLeod and Gold in 1951 for New Yorkers. This thus gave no evidence for a decline in sperm concentration. In 1983, several epidemiologists working with the CECOS analyzed the sperm concentration of 883 fertile Parisians, with the aim of studying differences according to donor age, rather than changes through time. They again found a mean value, for the period 1973 to 1980, which was high: 103 million/ml (Schwartz et al. 1983). Thus, CECOS greeted the work of Carlsen et al. (1992) with skepticism. Nevertheless CECOS decided to review the sperm characteristics of all the unselected volunteers who came to the CECOS for semen donation. The data available at CECOS had three major strengths. Data included a large number of semen samples donated by a large number of fertile men since 1973, and the methods used for studying sperm have been unchanged for more than 20 years. Furthermore, in the statistical analysis, CECOS researchers could take into account two key confounding factors known to influence sperm concentration values: the age and the delay of sexual abstinence before semen collection. In this study of 1351 healthy men of proven fertility, it was found that there had been a mean decline in sperm concentration of almost 2% per year between 1973 (89 million/ml) and 1992 (60 million/ml), as shown in Auger et al. (1995). The mean percentage of motile spermatozoa had also fallen by 0.6% per year and that of spermatozoa with normal morphology by 0.5% per year. However, unlike Carlsen et al. (1992), CECOS did not find any evidence for a significant decrease in the mean volume of ejaculate (3.8 ml). Auger et al. (1995) went on to show that the year of birth was associated with this decline: the more recently the man was born, the poorer the number and quality of his sperm (motility and morphology).

Although the relationship between indicators of sperm characteristics and age was already known (Schwartz et al. 1983), that between sperm concentration and morphology and year of birth was not. These findings have since been supported by more recent publications reporting a lower sperm concentration the longer after 1950 the man is born, both in a sample of normal Scottish men donating for research (Irvine et al. 1996) and in a French population of partners of women benefiting from *in vitro* fertilization due to tube disorders (De Mouzon et al. 1996).

Since the publication of the study by Auger et al. (1995) at least 12 similar studies have also been published (summarized in Fig. 1). The findings are conflicting. In Paris (France), Ghent (Belgium), Edinburgh (Scotland), Athens (Greece), Pisa (Italy) and Philadelphia (USA), there is evidence of a degradation of sperm characteristics, whereas there is no such evidence in Toulouse (France), Kuopio (Finland), Sydney (Australia) or several cities in the USA.

An interesting issue arising from these epidemiological studies is the enormous geographic diversity in sperm concentration. For example, the concentration found in New York was 131.5 million/ml, but was only 73 million/ml for Los Angeles for the same period (Fisch et al. 1996). The French CECOS network can serve as an observatory for male reproductive health: using the data CECOS supplies, it was recently shown in a study of 4710 fertile candidate sperm donors that

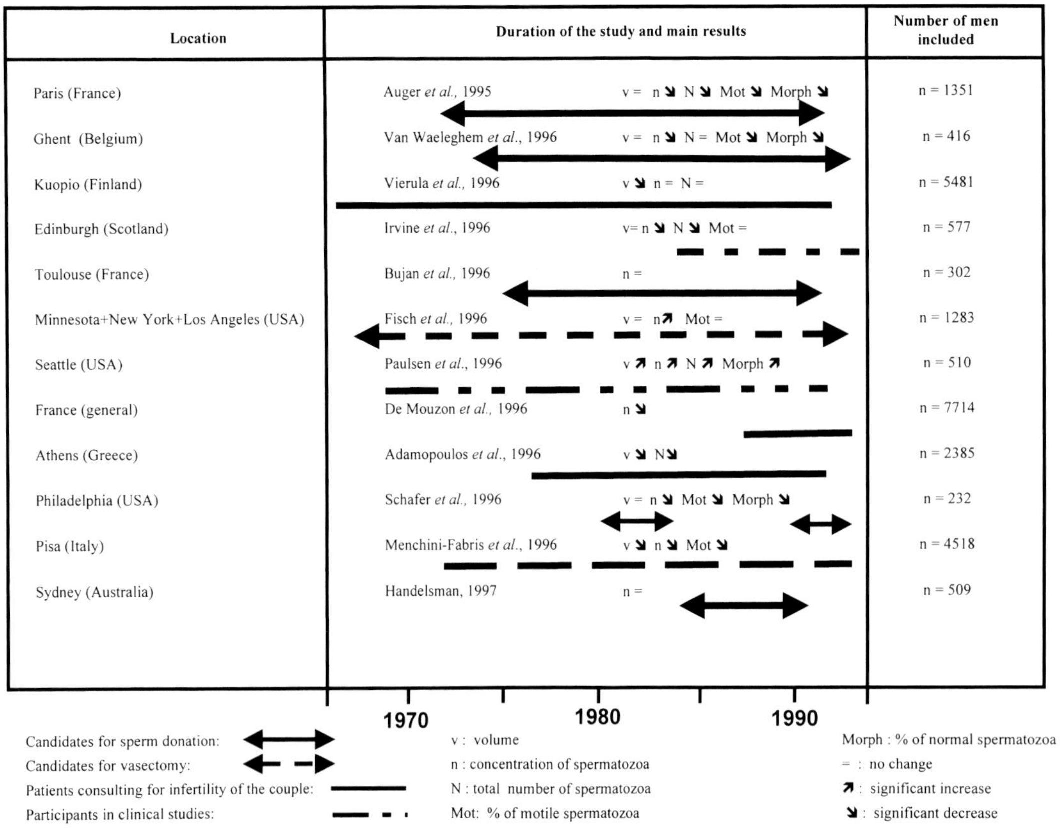
Location
Duration of the study and main results
Number of men included
Paris (France)
Auger et al., 1995
v = n ↘ N ↘ Mot ↘ Morph ↘
n = 1351
Ghent (Belgium)
Van Waeleghem et al., 1996
v = n ↘ N = Mot ↘ Morph ↘
n = 416
Kuopio (Finland)
Vierula et al., 1996
v ↘ n = N =
n = 5481
Edinburgh (Scotland)
Irvine et al., 1996
v= n ↘ N ↘ Mot =
n = 577
Toulouse (France)
Bujan et al., 1996
n =
n = 302
Minnesota+New York+Los Angeles (USA)
Fisch et al., 1996
v = n↗ Mot =
n = 1283
Seattle (USA)
Paulsen et al., 1996
v ↗ n ↗ N ↗ Morph ↗
n = 510
France (general)
De Mouzon et al., 1996
n ↘
n = 7714
Athens (Greece)
Adamopoulos et al., 1996
v ↘ N↘
n = 2385
Philadelphia (USA)
Schafer et al., 1996
v = n ↘ Mot ↘ Morph ↘
n = 232
Pisa (Italy)
Menchini-Fabris et al., 1996
v ↘ n ↘ Mot ↘
n = 4518
Sydney (Australia)
Handelsman, 1997
n =
n = 509
1970
1980
1990
Candidates for sperm donation:
Candidates for vasectomy:
Patients consulting for infertility of the couple:
Participants in clinical studies:
v : volume
n : concentration of spermatozoa
N : total number of spermatozoa
Mot: % of motile spermatozoa
Morph : % of normal spermatozoa
= : no change
↗ : significant increase
↘ : significant decrease

Figure 1: Summary of the recent studies on human semen quality.

there are large differences between eight regions in France (Fédération Française des CECOS et al. 1997). All the CECOS use similar criteria for recruitment and sperm analysis methods. Even after adjustment for age, period of abstinence and year of birth, the differences between regions was very significant. For example, the concentration was one-third lower in Toulouse, in the south, than in Lille, in the north, with the difference being 139 million spermatozoa per ejaculate. These huge differences cannot be due to methodological differences, and are consistent with an environmental effect. Currently available data thus suggest that there is substantial geographic variation in sperm quality, both between distant countries and areas within one country. There is even evidence for differences between districts within a single city; for example, London (Ginsburg et al. 1994) and Paris (Auger, Jouannet 1997). These observations are highly suggestive of local effects, large in some places, and absent in others.

All published studies are retrospective, and are based on data originally collected for reasons other than to assess general or geographic variations in semen characteristics, and which are not representative of the general population. Even if the criticisms of Sherins (1995) are put aside (Table 3), it is nevertheless valid to ask what information and what conclusions can be drawn from these data and studies.

There are several criteria which should be considered before concluding that there is or has been a change in semen characteristics. Most published studies differ with respect to these criteria or only take a small number of them into account (Table 4). One of the major difficulties is the origin and the size of the populations studied. The sperm concentration varies greatly between individuals: among fertile men, there is a tenfold difference between the 10th percentile (~20 million spermatozoa/ml) and the 90th percentile (~200 million spermatozoa/ml), as shown in Jouannet et al. (1981). To measure general or geographic variations, very large groups of men are therefore required (Berman et al. 1996). Unfortunately, in many studies the mean number of men per year or per subgroup used for comparison is very small (Table 4).

The recruitment method also undoubtedly contributes. None of the published studies is representative of the general population (Handelsman 1997). However, it would be almost impossible to conduct such a study on a group of men representative of the general population. In addition to the unwillingness of many men to donate semen collected by masturbation, the men most likely to consent are members of two particular subgroups which would be over-represented: the better educated (Auger et al. 1995), and men with doubts about their own fertility and consequently eager to participate in this type of

Table 3: Sherins' comments (C.), from Sherins (1995) on the work of Auger et al. (1995) and answers (A).

C. "Selected men serving as sperm donors [were included]."
A. No: all candidates for becoming sperm donors were included without selection.

C. "Men attending infertility clinics [were included]."
A. Wrong.

C. "[Men] about to undergo a vasectomy [were included]."
A. No; such men were excluded due to the risk of selection.

C. "Groups of men unrepresentative of the general population [were included]."
A. True, but it is almost impossible to study the general population. Furthermore, what is important is that the criteria characterizing these men corresponding to a single subgroup of the general population have not changed over the 20 years of the study.

C. "Differences in age [and] abstinence before semen analysis... were not controlled."
A. Wrong; these parameters were controlled and included in the statistical analysis.

C. "The number of samples analyzed per person [was] not controlled."
A. Wrong; only one ejaculate per man was included, to avoid the demonstrated risk of selection bias if more than one had been included.

C. A large interindividual variability may mask secular trend of change.
A. True, and this is why a large sample must be studied, as was the case: 1351 men.

C. There is a risk of bias if only one ejaculate per man is included, as a result of the intraindividual variability, which is known to be large.
A. True in theory, but there is no reason to suppose that the samples collected in the early years of the study correspond to high individual values, and those collected later to low values.

C. "Comparisons [which show] statistically significant [differences] do not necessarily reflect biological differences."
A. Quite possibly true, but a decline of 2% per year should not be taken lightly.

C. "There is little evidence that male fertility is declining."
A. True, but the paper does not make this claim. Also, the fact that fertility has not declined in some areas does not mean that there is no problem in others.

C. "Sperm concentration in itself is not the chief determinant of male fertility."
A. Yes, but Auger et al. (1995) also show deterioration in motility and morphology, both much more closely linked to spermatozoa function and fertility.

C. There are sophisticated tests which better measure the ability of spermatozoa to fertilize (acrosomic reaction, hamster test) and they should be used in studies of the general population before drawing conclusions.
A. Extreme naiveté or ignorance of experimental realities; it is simply impractical to apply these to the very large samples necessary to represent as well as possible the general population.

C. Prospective studies should be conducted in the general population.
A. Yes; see comments in the text on the feasibility, limits and advantages of prospective studies, as compared to retrospective studies.

study likely to include analyses which will inform them about their own health status. Analysis of a sample representative of the general population is, however, not necessary to evidence geographic or secular differences in sperm quality. It is sufficient that the various populations included in the analysis are rigorously defined and identical as concerns fertility, socioprofessional status, age and method of recruitment. Unfortunately, the relevant information is absent or unclear in most published studies (Table 4). Consequently, it would be valuable, as well as feasible, to study various well defined categories of men, for example, young army conscripts, university students, fertile partners of pregnant women, voluntary sperm donors, or men giving sperm to a sperm bank prior to vasectomy.

Other biases arise from differences in sperm analysis methods between times or laboratories, and from applying inappropriate statistical methods.

Despite this array of problems, the retrospective analysis of databases can give valuable information, once the biases are taken into account. Indeed, as the differences between regions are so large that they cannot be explained by methodological biases alone, environmental effects are therefore entirely plausible. This possibility needs now to be tested by prospective epidemiological studies designed to minimize potential biases and to collect all information necessary for assessing the factors underlying the putative temporal and geographic variations in semen quality.

Changes in the Male Urogenital Tract

Concerns about the changes in male reproductive function are not limited to semen characteristics. There is various evidence that the incidence of cancers of the prostate and of the testicle have also greatly increased over the last 20 years in almost every country where appropriate studies can be conducted (Adami et al. 1994; Toppari et al. 1996). The case of cancer of the testicle is particularly significant, as, unlike cancer of the prostate, its incidence is not correlated to life expectancy, which has been increasing. It affects mainly younger men, and has become the most common cancer among men in the 25- to 35-year-old group. The incidence of cryptorchidism and hypospadia may have also increased in the last 20 years in several countries (Ansell et al. 1992; Toppari et al. 1996). Although there can be no doubt that the increased incidence of the cancers is real, this is not the case for cryptorchidism and hypospadia because the observed increases could be due to doctors and families being more vigilant and thus identifying a larger proportion of cases of this type of disorder. Note also that the incidence of another disorder, breast cancer, which was previously very rare in men, may also have increased in one country at least: Denmark (Ewertz et al. 1989).

Table 4: Criteria to evaluate the degree of validity of a retrospective study on the secular evolution of semen quality.

Criteria	**Location of the Studies**													
	1*	**2**	**3**	**4**	**5**	**6**	**7**	**8**	**9**	**10**	**11**	**12**	**13**	**14**
Populations homogenous and well-defined	+	+	±	+	+	±	±	±	+	-	-	?	?	±
Mean number of men included/year	**68**	**22**	**196**	**52**	**19**	**26**	**17**	**13**	**24**	**1286**	**140**	**23**	**?**	**34**
Modification of recruitment during the study	**No**	**No**	?	?	**No**	?	?	?	?	?	?	?	?	?
Only one ejaculate analyzed (the first)	+	+	+	+	+	-	-	-	-	-	?	+	-	+
Laboratory methods standardized or well-described	+	+	±	+	±	±	-	-	±	-	+	?	-	+
Statistical analysis**	**L**** **M**	**L NL**	**M**	**L M**	**L M**	**L M**	**L M**	**L M**	**L M**	**L A**	**L NL**	-	-	**L A?**
Year of birth taken into account	+	-	±	+	+	-	-	-	-	+	-	-	-	+
Age taken into account	+	-	+	+	+	+	+	+	-	-	±	-	-	-

*1: Auger et al. 1995; 2: Van Waeleghem et al. 1996; 3: Vierula et al. 1996; 4: Irvine et al. 1996; 5: Bujan et al. 1996; 6: Fisch et al. 1996; 7: Fisch et al. 1996; 8: Fisch et al. 1996; 9: Paulsen et al. 1996; 10: De Mouzon et al. 1996; 11: Adamopoulos et al. 1996; 12: Schafer et al. 1996; 13: Menchini-Fabris et al. 1996; 14: Handelsman 1997.
****A** = Anova; **L** = linear regression; **NL** = non-linear regression; **M** = multiple regression; **+** = yes; **±** = more or less; **-** = no; **?** = not clear.

Are the Characteristics of Farm Animals' Sperm Changing?

If the deterioration in human semen quality results from changes in the environment, it would be expected that similar changes have occurred in animals. To determine whether this is indeed the case, at least three studies have been undertaken recently in farm animals.

The first, by Setchell (1997), was a meta-analysis of data obtained from a systematic review of the *Animal Breeding Abstracts*, between 1932 and 1995. The data were obtained from more than 100 studies on bulls and rams and more than 70 studies on boars, with all studies including samples collected at relatively regular intervals from apparently normal adult animals. Sperm counts were determined using haemocytometers or turbidometric estimations. The results obtained indicated that during this period there has been no change in sperm concentrations of bulls and boars, whereas both the concentrations and the total number of spermatozoa per ejaculate had slightly and significantly risen in the ram.

Van Os et al. (1997) studied the long-term trends in sperm counts of dairy bulls recorded in a single laboratory, at Noordwest, a center for artificial insemination in the Netherlands. The data analyzed corresponded to 75,238 ejaculates collected between 1977 and 1996 from 2,314 bulls. The authors had adjusted their data for effects known to affect semen production in bulls, such as times between semen collections, breed and season. Sperm concentrations in the ejaculates (spermatozoa/ml) had been determined by density measurement using a spectrophotometer. It was found that no significant change had occurred in either the mean sperm output per year of collection or year of birth during the period covered by the study (1978 to 1996 for the year of collection and 1970 to 1995 for the year of birth). Furthermore, this study indicates that data published earlier for 22,120 ejaculates from 3,030 bulls in the same region, tested between 1962 and 1977, showed the same semen characteristics, supporting the conclusion that there had been no decline in bull semen quality in this area of the Netherlands.

The third study in farm animals analyzed 1,830 ejaculates from 476 stallions (Breton draught and Anglo-Arabs) from which semen was collected in France between 1981 and 1996. All stallions were three years old and ejaculates were obtained after the same delay of abstinence. The same techniques used to measure sperm concentration (haemocytometer) were used throughout the period. It was found that there had been no change in the total number of spermatozoa, but that there had been a significant decline in gel-free seminal volume (L Multigner, M Magistrini and A Spira, unpublished data).

Thus, in contrast to the data obtained in man, no study has shown any negative trend in farm animal sperm counts. It is however, difficult to extrapolate these observations to the man for various reasons. The first reason is that the Setchell's meta-analysis (1997), though very useful, suffers from various sources of bias and confounding factors. For example, the animals included in the study were selected (apparently normal adult animals) and were a mixture of different breeds. Furthermore, the sperm counts were obtained by two different techniques, and the delays between collection were irregular. The second reason is that, over the last decades, the breeding conditions of farm animals have changed (for example, quality of the nutrition), and this may have had some positive influence on reproductive efficiency (Setchell 1997). The third reason is that extensive selection of farm animals (particularly bulls) may also have marked effects with sperm production. The last reason is that the studies by Van Os et al. (1997) and Multigner et al. (L Multigner, M Magistrini and A Spira, unpublished data) were performed in areas where human semen quality has not been followed. Therefore, it may also be that in these areas there has been no change in human semen either. Note that there may be large differences in sperm production between humans living in urban and rural environments (Saaranen et al. 1986).

It is interesting, however, that Multigner et al. demonstrate a decline in the volume of stallion gel-free ejaculate collected in constant and standardized conditions. According to these authors, environmental chemicals with anti-androgenic properties could play a role in this phenomenon by acting on the development of accessory sex gland function.

The Situation in Wild Animals

In parallel with the growing debate over human sperm quality and disorders of the human male urogenital tract, workers in another field have been reporting worrying developments in various aspects of the reproductive functions of the wild fauna (gastropods, reptiles, fish, birds and mammals) in numerous parts of the world (Toppari et al. 1996). These concerns about the reproduction of the wild fauna are not new. In her book *Silent Spring,* published in 1962, R. Carson pointed out the severe reproductive problems affecting many species of birds, and suggested that pesticides were the culprits. T. Colborn, working for the World Wildlife Fund, discovered during the 1980s that 16 species of predators (including fish, birds, reptiles and mammals) in the Great Lakes region in the USA were suffering serious reproduction problems. This work led to the hormonal pollution concept, which suggests that certain pollutants act as decoy hormones (Colborn et al. 1993). However, even rigorous analysis of the literature only reveals rare data describing the sperm quality or testicular function in wild animals.

One of the best studied species is the endangered Florida panther (*Felis concolor coryi*), which exhibits developmental and reproductive abnormalities (Facemire et al. 1995; Roelke 1990). In particular, males display low sperm concentrations and ejaculate volumes as well as poor sperm motility, and more than 90% of the sperm present morphological abnormalities. This proportion of abnormalities is 24 to 50% higher than that found in the Texas cougar (*Felis concolor*), as shown in Wildt et al. (1988). Furthermore, an exponential increase in cryptorchidism has been observed in male cubs since 1975, such that more than 90% of the male population born since 1985 displays this major abnormality (Facemire et al. 1995). Problems associated with inbreeding were put forward initially to explain this dramatic trend (Roelke 1990). However, the presence of high body burdens of various chemical contaminants (pesticides and mercury) in this species led Facemire et al. (1995) to propose that this phenomenon is due to contamination of gestational mothers via their major food, the raccoon.

Male Florida alligators have also been studied. Populations of this species of alligator were declining due to reproductive disorders caused by agricultural sewage dumping in Lake Apopka. Males exhibited abnormal germ cells in their testes, abnormally small phalli, low basal and luteinizing hormone (LH) stimulated plasma testosterone levels, but very high plasma estradiol concentrations (Guillette et al. 1996).

There have been studies which show that several wild fish species present, for example, abnormally small gonads, absence of secondary sex characteristics and reduced serum testosterone and 11-ketotestosterone concentrations, consecutive to pollution of their ecosystems (Toppari et al. 1996). A recent study (S Jobling, M Nolan, CR Tyler, G Brighty and JP Sumpter, unpublished data) describes the occurrence of a high incidence of intersexuality in wild populations of riverine fish (roach, *Rutilus rutilus)* throughout the United Kingdom. A large proportion of "males" were intersex, as defined by the simultaneous presence of both male and female gonadal characteristics. The incidence of intersexuality in "male" fish was much higher in rivers polluted by sewage effluents compared to control sites. Analysis of plasma vitellogenin concentrations in these intersex fish revealed that vitellogenin levels were intermediate between the levels found in males and those found in females, leading the authors to conclude that the reproductive disturbances observed are consistent with exposure to endocrine active substances (estrogenic contaminants) present on river stretches downstream from large sewage treatment works. This conclusion was reinforced by another observation demonstrating that the gonadosomatic index (GSI) of the intersex "males" was very significantly different to that in males and females at all types of river sites.

In summary, solid data on a possible negative trend in testicular function in wild animals are scarce, although there are clear indications of a severe deterioration in some species and in some areas. The reproductive function of aquatic species or species feeding on aquatic animals may be more affected due to permanent exposure to the pollutants and a physiology that favors high body concentrations of these pollutants (for example, fish via the continuous flow of water through the gills).

Are Semen Quality Decline and Abnormalities of the Urogenital Tract Linked?

The issue is whether or not there are etiological links between these phenomena: the declining quality of sperm and increasing incidence of urogenital tract abnormalities both in man and animals. In the absence of sufficient relevant data, any answer must be tentative. However, it is a reasonable working hypothesis that there is such a link (Toppari et al. 1996). Consistent with this view, the incidence of both hypospasia and testicular cancer is very low in Finland, the country with the highest known sperm concentrations, and where no decline in sperm quality has been observed since the Second World War (Suominen, Vierula 1993). In Denmark, the incidence of these abnormalities is very high, and the sperm concentration is half that in Finland (Carlsen et al. 1992). Furthermore, cryptorchidism is associated with a halt in spermatogenesis in both man and animals, as well as a higher risk of testicular cancer. Finally, a secular decline in sperm concentration was found in men with a testicular cancer who were born in the 1950s (Auger et al. 1997). The fact that the slope of this decline is higher in this population compared to that found in fertile normal men in the same geographic area further supports the existence of a link between changes in sperm production and the risk of development of a testicular cancer.

Can Environmental Changes Be Blamed?

Considering the increased incidence of urogenital abnormalities and assuming there is indeed a decline in sperm characteristics, it is essential to determine the causes. The direct responsibility of genetic factors seems unlikely in view of the rate of decline (1% to 2% decrease in sperm concentration per year) and the appearance of the phenomenon in diverse parts of the world. However, ethnic differences are possible, as recently demonstrated (Johnson et al. 1998), or genetically based differences in susceptibility to particular changes in the environment may be involved (Jégou 1996). Nevertheless, genetics can be excluded for the overall phenomenon, and it is therefore likely that the causes can be found among the modifications in the environment occurring over the last 50 years. There have been considerable changes in the physical, chemical, biological and sociocultural environments during this period. Note that the testicle is one of the organs most vulnerable to radiation (physical agents) and xenobiotic (chemical agents) exposure, which has substantially increased since the 1950s. During the same period, lifestyle changes have also occurred; for example, smoking (sociocultural agent) or stress, as shown in Fukuda et al. (1996). Note that sexually transmitted diseases known to affect male reproductive function now have a world-wide distribution and affect young age groups (biological agents).

Among this diversity of possibilities, most attention has been paid to chemical agents. Although there are tens of thousands of chemical agents in the environment, the finger has been pointed at a class of compounds known as "endocrine disrupting compounds" (or "endocrine active compounds" or "xenohormones"). They are so called because when in an individual's environment, they may interfere with the actions of the body's hormones. This is based upon the fact that in the laboratory, they present activities that mimic more or less those of steroid sex hormones, and are thus possible candidates as the causative agents of the deterioration of semen quality and of the disorders of the male reproductive tract (Sharpe, Skakkebæk 1993). The first activities of this type to be studied were the estrogenic activities, and to a lesser extent, the anti-androgenic activities. The model xeno-estrogenic compound is diethylstylbestrol, known as DES (Toppari et al. 1996; Golden et al. 1998). The involvement of these agents in the decline in human and animal sperm concentrations remains to be proved, but nevertheless there are intense international programs to detect industrial xenohormonal products, and to assess their toxicity at concentrations found in the environment, both alone and in combination. Only a very few of the tens of thousands of man-made products have been tested, and synthetic xeno-estrogens are only a fraction of all xeno-estrogens; the others are natural, particularly plant products, the phyto-estrogens.

Is Male Fertility in Peril?

If the decline in sperm quality and its regional differences are indeed real, what are the consequences for fertility? This is another difficult question to answer, because fertility is affected by numerous factors other than the quality of sperm (Table 5). Furthermore, the semen characteristics generally studied (concentration, motility and morphology of spermatozoa) are probably not the best indicators of the potential to fertilize. Although tests are now available to measure the ability of sperm to fertilize (for example, evaluation of the acrosomic reaction, the zona pellucida fixation test and the heterospecific fecundation test), they are painstaking, slow and expensive. It is naive to envisage their systematic use for studies of the general population, as suggested by Sherins (1995). Note also that changes in semen do not necessarily have repercussions on fertility: the administration of DES to pregnant women between the 1950s and 1970s does not seem to have had an effect on the fertility of their sons, despite the unusually high incidence of genital tract abnormalities and a large decrease in the number and motility of the spermatozoa they produce (Golden et al. 1998). Furthermore, the sperm concentration can fluctuate by up to 75% from its mean value for a fertile man in perfect health (World Health Organization 1992), even if the period of abstinence is strictly controlled. The mean sperm concentration is around 60 million/ml, but it is difficult to detect any lower fertility until the concentration is below 5 million/ml (Jouannet et al. 1988). It is therefore unlikely that the decrease in concentration of 20% to 50% over the last few decades has caused a decrease in fertility of the population as a whole. However, studies of Carlsen

Table 5: Summary of the factors contributing to the great variability of human semen characteristics.

Factors of Variation Related to:		
People	**Methodology of Sperm Analysis**	**Other**
Degree and type of sexual arousal	materials and methods used	genetic (intra-family)
Delay of abstinence	intra-and inter-technician variability	season
Age		occupational
Sexual activity		health (e.g., venereal diseases, fever)
Body weight, size and body mass		medical (use of pharmaceutical drugs)
		socio-cultural (stress, residency area, nutrition, use of recreational drugs . . .)

et al. (1992) and Auger et al. (1995) suggest that the subgroups of men with a concentration below 20 million/ml, and even 5 million/ml, are growing. If the quality (motility and morphological quality) of the spermatozoa is also affected, as indicated by several studies (Fig. 1), some couples may have to try longer before obtaining a pregnancy, as is the case for couples where the man is exposed to pesticides or heat (de Cock et al. 1994; Thonneau et al. 1996). Other possible consequences are increased medical interventions in human procreation and the disappearance of some already endangered wild species.

Conclusion

The debate over the decline in semen quality is exceptional: it has lasted a long time, is very lively and has had repercussions on public opinion, public authorities, regulatory bodies and industry. One reason for this is the extreme sensitivity of the public at the end of the twentieth century to environmental issues, combined with the potential severity of the consequences of a decline in semen characteristics. All of this strongly suggests that further research is essential. Any such research must consider the probably multifactorial nature of the observed phenomena and involve epidemiological, clinical and basic approaches. Furthermore, humans, domesticated animals and wild animals must be studied simultaneously. These studies should give answers to the major questions which are currently the subject of debate. They will also lead to a better understanding of male reproductive function, an outcome of benefit for individuals; the elucidation of the etiology of various forms of infertility and male contraception are two examples.

Acknowledgements

Supported by INSERM, Université de Rennes 1, the Ministère de Recherche et Technologie, the Ministère de l'Environnement (grant EN96C1) and an European Economic Community grant (Biomed BMH4-CT96-0314).

References

Adamopoulos DA, Pappa A, Nicipoulou S, Andreou E, Kramertzanis M, Michopoulos J, Deliglianni V, Simou M. Seminal volume and total sperm number trends in men attending subfertility clinics in the greater Athens area during the period 1977-1993. Hum Reprod 1996; 11:1936-1941.

Adami H, Bergström R, Möhner M, Zatonski W, Storm H, Ekbom A, Tretli S, Teppo L, Ziegler H, Rahu M, Gurevicius R, Stengrevics A. Testicular cancer in nine northern european countries. Int J Cancer 1994; 59:33-38.

Ansell PE, Bennett V, Bull D, Jackson MB, Pike LA, Pike MC, Chilvers CED, Dudley NE, Gough MH, Griffiths DM, Redman C, Wilkinson AR, Macfarlane A, Coupland CAC. Cryptorchidism: a prospective study of 7500 consecutive male births: 1984-1988. Arch Dis Child 1992; 67:892-899.

Auger J, Jouannet P. Baisse de la production et de la qualité spermatique chez l'homme: facteurs de variation et problèmes méthodologiques. Andrologie 1998; 8:9-24.

Auger J, Kunstmann JM, Czyglik F, Jouannet P. Decline in semen quality among fertile men in Paris during the past 20 years. New Engl J Med 1995; 332:281-285.

Auger J, Czyglik F, Kunstmann JM, Jouannet P. Sperm production of French men with testicular cancer. Fourth Copenhagen workshop, Carcinoma *in situ* and Cancer of the Testis, Molecular and Endocrine Aspects. 1997. Abstract.

Becker S, Berhane K. A meta-analysis of 61 sperm count studies revisited. Fertil Steril 1997; 67:1103-1108.

Berman NG, Wang C, Paulsen CA. Methodological issues in the analysis of human sperm concentration data. J Androl 1996; 17:68-73.

Bosfolte E, Serup J, Rebbe H. Has the fertility of Danish men declined through the years in terms of semen quality? A comparison of semen qualities between 1952 and 1972. Int J Fertil 1983; 28:91-95.

Brake A, Krause W. Decreasing quality of semen. Br Med J 1992; 305:1498.

Bromwich P, Cohen J, Stewart I, Walker A. Decline in sperm counts: an artefact of changed reference range of "normal"? Br Med J 1994; 309:19-22.

Bujan L, Mansat A, Pontonnier F Mieusset R. Time series analysis of sperm concentration in fertile men in Toulouse, France between 1977 and 1992. Br Med J 1996; 312:471-472.

Carlsen E, Giwercman A, Keiding N, Skakkebæk NE. Evidence for decreasing quality of semen during past 50 years. Br Med J 1992; 305:609-613.

Colborn T, Von Saal FS, Soto AM. Developmental effects of endocrine disrupting chemicals in wildlife and humans. Environ Health Perspect 1993; 101:378-384.

David G, Jouannet P, Martin-Boyce A, Spira A, Schwartz D. Sperm counts in fertile and infertile men. Fertil Steril 1979; 31:453-455.

de Cock J, Westveer K, Heederik D, Velde E, Van Kooij R. Time to pregnancy and occupational exposure to pesticides in fruit growers in the Netherlands. Occup Environ Med 1994; 51:693-699.

De Mouzon J, Thonneau P, Spira A Multigner L. Semen quality has declined among men born in France since 1950. Br Med J 1996; 313:43.

Ewertz M, Holmberg L, Karialainen S, Tretli S Adami HO. Incidence of male breast cancer in Scandinavia: 1943-1982. Int J Cancer 1989; 43:27-31.

Facemire CF, Gross TS, Guillette LJJ. Reproductive impairment in the Florida panther: nature or nurture. Environ Health Perspect 1995; 103:79-86.

Farrow S. Falling sperm quality: fact or fiction? Br Med J 1994; 309:1-2.

Fédération Française des CECOS, Auger J, Jouannet P. Evidence for regional differences of semen quality among fertile French men. Hum Reprod 1997; 12:740-745.

Fisch H, Goluboff ET. Geographic variations in sperm counts: a potential cause of bias in studies of semen quality. Fertil Steril 1996; 65:1044-1046.

Fisch H, Goluboff ET, Olson JH, Feldshuh J, Broder SJ, Barad DH. Semen analysis in 1283 men from the United States over a 25-year period: no decline in quality. Fertil Steril 1996; 65:1009-1014.

Fukuda M, Fukuda K, Shimizu T, Yomura W, Shimizu S. Kobe earthquake and reduced sperm motility. Hum Reprod 1996; 11:1244-1246.

Ginsburg J, Okolo S, Prelevic G, Hardiman P. Residence in the London area and sperm density. Lancet 1994; 343:230.

Guillette LJJ, Pickford DB, Crain DA, Rooney AA, Percival HF. Reduction in penis size and plasma testosterone concentrations in juvenile alligators living in a contaminated environment. Gen Comp Endocrinol 1996; 101:32-42.

Golden RJ, Noller KL, Titus-Ernstoff L, Kaufman RH, Mittendorf R, Stillman R, Reese EA. Environmental endocrine modulators and human health: an assessment of the biological evidence. Crit Rev Toxicol 1998; 28:109-227.

Handelsman DJ. Sperm output of healthy men in Australia: magnitude of bias due to self-selected volunteers. Hum Reprod 1997; 12:2701-2705.

Irvine DS, Cawood E, Richardson D, MacDonald E, Aitken J. Evidence of deteriorating semen quality in the United Kingdom: birth cohort study of 577 men in Scotland over 11 years. Br Med J 1996; 312:467-470.

James WH. Secular trend in reported sperm counts. Andrologia 1980; 12:381-388.

Jégou B. Les hommes deviennent-ils moins fertiles? La Recherche 1996; 288:60-65.

Johnson L, Barnard JJ, Rodriguez L, Smith EC, Swerdloff RS, Wang H, Wang C. Ethnic differences in testicular structure and spermatogenic potential may predispose testis of Asian men to a heightened sensitivity to steroidal contraceptives. J Androl 1998; 19:348-357.

Jouannet P, Czyglick F, David G, Mayaux MJ, Spira A, Moscato ML, Schwartz D. Study of a group of 484 fertile men. Part I: Distribution of semen characteristics. Int J Androl 1981; 4:440-449.

Jouannet P, Ducot B, Feneux D, Spira A. Male factors and the likelihood of pregnancy. I. Study of sperm characteristics. Int J Androl 1988; 11:379-394.

Keiding N, Giwercman A, Carlsen E, Skakkebæk NE. Importance of empirical evidence (commentary). Br Med J 1994; 309:22.

Lerchl A, Nieschlag E. Decreasing sperm counts? A critical review. Exp Clin Endocrinol Diabet 1996; 104:301-307.

Leto S, Frensilli FJ. Changing parameters of donor semen. Fertil Steril 1981; 36:766-770.

MacLeod J, Gold RZ. The male factor in fertility and infertility. II. Spermatozoon counts in 1000 men of known fertility and in 1000 cases of infertile marriage. J Urol 1951; 66:436-449.

MacLeod J, Wang Y. Male fertility potential in terms of semen quality: a review of the past, a study of the present. Fertil Steril 1979; 31:103-116

Menchini-Fabris F, Rossi P, Palego P, Simi S, Turchi P. Declining sperm counts in Italy during the past 20 years. Andrologia 1996; 28:304.

Menkveld R, Van Zyl JA, Kotze TJW, Joubert G. Possible changes in male fertility over a 15-year period. Arch Androl 1986; 17:143-144.

Nelson CMK, Bunge RG. Semen analysis: evidence for changing parameters of male fertility potential. Fertil Steril 1974; 25:503-507.

Olsen GW, Bodner KM, Ramlow JM, Ross CE, Lipshultz LI. Have sperm counts been reduced 50% in 50 years? A statistical model revisited. Fertil Steril 1995; 63:887-893.

Osser S, Liedholm P, Rastam J. Depressed semen quality: a study over two decades. Arch Androl 1984; 12:113-116.

Paulsen CA, Berman NG, Wang C. Data from men in greater Seattle area reveals no downward trend: further evidence that deterioration of semen quality is not geographically uniform. Fertil Steril 1996; 65:1015-1020.

Rehan N-E, Sobrero AJ, Fertig JW. The semen of fertile men. Statistical analysis of 1300 men. Fertil Steril 1975; 26:492-502.

Roelke ME. Florida panther biomedical investigations: health and reproduction. Final report. Endangered species project E-III-E-6 7506 Gainsville, FL; Florida Game and Freshwater Fish Commission, 1990.

Saaranen M, Vierula M, Saarikoski S. Semen quality in men of infertile couples and in men of reproductive ages. In: (Aravantinos D, Creatsas G, eds) Proceedings of 5th World Congress on Human Reproduction. Athens: Hellenic Society on the Study of Reproduction, 1986, pp298-300.

Setchell BP. Sperm counts in semen of farm animals 1932-1995. Int J Androl 1997; 20:209-214.

Schafer D, Arredondo-Soberon F, Loret De Mola JR, Blasco L. Has sperm quality declined in the 1990's: analysis from a donor program. Am Soc Reprod Med 52nd Ann Meet, 1996, Abstract, p286.

Schwartz D, Mayaux MJ, Spira A, Moscato M-L, Jouannet P, Czyglik F, David G. Semen characteristics as a function of age in 833 fertile men. Fertil Steril 1983; 39:530-535.

Sharpe RM, Skakkebæk NE. Are œstrogens involved in falling sperm counts and disorders of the male reproductive tract? Lancet 1993; 35:1392-1395.

Sherins RJ. Are semen quality and male fertility changing? New Engl J Med 1995; 332:327-328.

Suominen J, Vierula M. Semen quality of Finnish men. Br Med J 1993; 306:1579.

Swan SH, Elkin EP, Fenster L. Have sperm densities declined? A reanalysis of global trend data. Environ Health Perspect 1997; 105:1228-1232.

Thonneau P, Ducot B, Bujan L, Mieusset R, Spira A. Heat exposure as a hazard to male fertility. Lancet 1996; 347:204-205.

Toppari J, Larsen JC, Christiansen P, Giwercman A, Grandjean P, Guillette LJ, Jégou B, Jensen TK, Jouannet P, Keiding N, Leffers H, McLachlan JA, Meyer O, Müller J, Rejpert-De Meyts E, Scheike T, Sharpe R, Sumpter J, Skakkebæk NE. Male reproductive health and environmental xenœstrogens. Environ Health Perspect 1996; 104:741-803.

Tumon JS, Mortimer D. Decreasing quality of semen. Br Med J 1992; 305:1228-1229.

Van Os JL, De Vries MJ, Den Daas NH, Kaal Lansbergen LM. Long-term trends in sperm counts of dairy bulls. J Androl 1997; 18:725-731.

Van Waeleghem K, De Clercq N, Vermeulen E, Schoonjans F Combaire F. Deterioration of sperm quality in young healthy Belgian men. Hum Reprod 1996; 11:325-329.

Vierula M, Keiski A, Saaranen M, Saarikoski S, Suominen J. High and unchanged sperm counts of Finnish men. Int J Androl 1996; 19:11-17.

Wildt DE, Phillips LG, Simmons LG, Chakroborty PK, Brown JL, Howard JG, Teare A, Bush M. A comparative analysis of ejaculate and hormonal characteristics of the captive male cheetah, tiger, leopard, and puma. Biol Reprod 1988; 38:245-255.

World Health Organization. Laboratory manual for the examination of human semen and sperm–cervical mucus interaction. 3rd ed. Cambridge University Press, Cambridge, 1992, pp1-107.

42

The Dark and Bright Sides of Reactive Oxygen Species on Sperm Function

Eve de Lamirande

Claude Gagnon

McGill University

The impact of reactive oxygen species (ROS) in male reproduction has been the topic of numerous studies (reviews: Aitken 1994; Aitken, Fisher 1994; de Lamirande, Gagnon 1995b; Sikka et al. 1995; de Lamirande et al.1997a; Griveau, Le Lannou 1997b). Excessive ROS levels in semen, which are detected in 25% of semen samples from men consulting for infertility (Iwasaki, Gagnon 1992; Zini et al. 1993; Plante et al. 1994; Okada et al. 1997) and in almost all cases of spinal cord injury (de Lamirande et al. 1995; Padron et al. 1997) were inversely correlated with sperm motility (Iwasaki, Gagnon 1992; de Lamirande et al. 1995) and the outcome of fertility *in vivo* (Aitken et al. 1991). High sperm ROS production is associated most often with the presence of abnormal spermatozoa and neutrophils in semen samples and induces lipid peroxidation, which in turn modifies membrane fluidity and permeability (Aitken, Clarkson 1987; Alvarez et al. 1987; Halliwell, Gutteridge 1989). Despite these known deleterious effects, there is a growing body of evidence indicating that ROS, being small, diffusible, short-lived and ubiquitous molecules that affect cellular responses, are also involved in physiological functions as signalling molecules (reviews: Baeuerle et al. 1996; Suzuki et al. 1997). This seems to be the case for spermatozoa, as the controlled and well-timed generation of very low amounts of ROS appears to be needed for these cells to acquire fertilizing ability (reviews: de Lamirande et al. 1997a, b). This article will attempt to highlight the dark (toxic) and bright (physiological) aspects of ROS on sperm function.

ROS originate from oxygen metabolism and enzymatic reactions (Halliwell, Gutteridge 1989). The ROS most often reported to affect sperm function are the superoxide anion ($O_2 \bullet^-$), hydrogen peroxide (H_2O_2, from the dismutation of $O_2 \bullet^-$) and the hydroxyl radical (•OH, from an iron-catalyzed reaction involving H_2O_2 or H_2O_2 and $O_2 \bullet^-$), as shown in Aitken and Clarkson (1987), Alvarez et al. (1987), de Lamirande and Gagnon (1992a, b), Aitken et al. (1993a), and Griveau et al. (1995a). However, recent evidence indicates that sperm function is also modified by nitric oxide (NO•), which is generated by the nitric oxide synthase (NOS) catalyzed conversion of L-arginine to L-citrulline (Roselli 1997) and by singlet oxygen (^1O), (Griveau et al. 1995a). Furthermore, lipid peroxyl radicals (LOO•) and peroxides (LOOH) generated by ROS attack on the cell membrane phospholipids (Jones et al. 1979; Halliwell, Gutteridge 1989; Kodama et al. 1996; Storey 1997), as well as their residual degradation products such as hydroxy alkenyls (Windsor et al. 1993) and malonaldehyde (Jones et al. 1979; Alvarez et al. 1987) also affect sperm function.

Cellular and Enzymatic Origins of ROS Generation

A variety of semen components, including immotile and/or morphologically abnormal spermatozoa, precursor germ cells, leukocytes and morphologically normal but functionally abnormal spermatozoa, are responsible for the generation of ROS in human semen (Aitken, Clarkson 1988; Rao et al. 1989; Iwasaki, Gagnon 1992; Plante et al. 1994; Aitken et al. 1994). Neutrophils probably represent the most important source of ROS in semen. There is still a controversy concerning the importance of leukocytospermia in male infertility as some authors consider that it may affect fertility, especially if leukocytes are present in sperm preparations used for assisted conception, but others do not (Kovalski et al. 1992; Aitken et al. 1994; Thomlinson et al. 1993; Plante et al. 1994; Ford et al. 1997). This controversy may arise from different factors. An important point is that the state of neutrophil activation in semen and sperm preparations is unknown. Activated neutrophils cause a time- and concentration-dependent decrease in the motility of Percoll-washed spermatozoa but unstimulated neutrophils, which produce 100-fold lower amounts of ROS, do not have such an effect (Kovalski et al. 1992; Plante et al. 1994). Another point is that seminal plasma contains large amounts of ROS scavengers (Zini et al. 1993) but confers a very variable (10-100%) protection against ROS generated by neutrophils (Kovalski et al. 1992). Therefore, the presence of neutrophils would be important mostly in the context of assisted conception because neutrophils are sometimes carried over during sperm preparation (Aitken et al. 1994) and in cases of acute testicular or epididymal infection in which spermatozoa are expected to be in close contact with potentially activated neutrophils and in the absence of the protective action of seminal plasma (Wolf 1995; Wang et al. 1997).

Morphologically abnormal spermatozoa, mainly those with defects in the midpiece region, and precursor germ cells also represent an important source of ROS in semen (Rao et al. 1989; Iwasaki et al. 1992; Plante et al. 1994; Aitken et al. 1994). In animal models (rat, hamster, guinea pig and mouse), the generation of $O_2 \bullet^-$ by precursor germ cells triggered by NADPH appears inversely related to the stage of cell differentiation, the highest levels being measured in pachytene spermatocytes (Fisher, Aitken 1994). On the other hand, epididymal maturation is associated with an increased generation of $O_2 \bullet^-$ and a concomitant decrease in SOD activity of spermatozoa (Kumar et al. 1991). The increased ROS generation of sperm cells that retain residual cytoplasm, due to incomplete extrusion of precursor germ cell cytoplasm during maturation, may be related to higher activities of some enzymes, such as creatine phosphokinase and glucose-6-phosphate dehydrogenase, observed in these cells (Aitken et al. 1994; Gomez et al. 1996). An increased glucose-6-phosphate dehydrogenase activity would enhance the formation of NADPH, which is hypothesized as cofactor in the subsequent triggering of higher ROS production and oxidative stress (Aitken et al. 1994). Morphologically normal spermatozoa from infertile men are probably biochemically abnormal as they also generate significant levels of ROS (Iwasaki, Gagnon 1992; Plante et al. 1994; Ford et al. 1997). However, perhaps because only a third of these ROS are released extracellularly, and these amounts of ROS are very low compared to those generated by activated neutrophils, the presence of ROS generating spermatozoa does not appear to affect normal sperm motility (Plante et al. 1994). Some environmental factors, such as lead exposure, could be responsible for an increased ROS generation and the associated functional impairment of spermatozoa as it is observed in the rat model (Hsu et al. 1997). Fertile spermatozoa produce extremely low levels of ROS in semen or after a Percoll wash but will initiate ROS generation at the time of capacitation (de Lamirande, Gagnon 1993a, b, 1995a; Aitken et al. 1995).

At least two enzymatic systems are possibly responsible for the $O_2\bullet^-$ generation in spermatozoa (Fig. 1). A diaphorase (NADH-dependent oxido-reductase) of mitochondrial origin (Gavella, Lipovac 1992; Gavella et al. 1995) is probably the major source of ROS produced by spermatozoa from infertile patients since two-thirds of these ROS are detected intracellularly (Plante et al. 1994). Furthermore, the activity of this diaphorase and the production of $O_2\bullet^-$ by spermatozoa are directly correlated (Gavella et al. 1995). On the other hand, due to some similarities between events occurring during sperm capacitation, such as calcium influx, increase in intracellular cAMP, protein phosphorylation and so on (Yanagimachi 1994), and during the oxidative burst of neutrophils, it is hypothesized that spermatozoa also possess an oxidase at the level of the plasma membrane (Aitken, Clarkson 1987; de Lamirande, Gagnon 1995a; Griveau et al. 1997a). The observation that NADPH (20 mM) causes an increase in $O_2\bullet^-$ production and causes decreases in sperm motility, hyperactivation, and capacitation, which are prevented by the combination of SOD and catalase (Griveau et al. 1997a), substantiates this hypothesis. However, this membrane oxidase would normally be activated only at the time of capacitation and/or acrosome reaction (de Lamirande, Gagnon 1995a; de Lamirande et al. 1997a, b; Leclerc et al. 1997; Aitken et al. 1995, 1996b, 1997).

Recent studies indicate that NO• may also be toxic to spermatozoa since NO•-releasing substances at high concentrations decrease sperm motility (Hellstrom et al. 1994; Zini et al. 1995; Rosselli et al. 1995) and, conversely, that NOS inhibitors preserve it (Perera et al. 1996). However, even though mouse spermatozoa possess NOS activity (Herrero et al. 1997a, b), there is a controversy regarding the presence of NOS and the generation of NO• in human spermatozoa. Some reports indicate that human spermatozoa possess NOS isoenzymes similar to constitutive brain NOS (bNOS) and endothelial NOS (ecNOS) and generate NO• in a concentration-dependent fashion upon stimulation with the calcium ionophore A23187 (Lewis et al. 1996; Donnelly et al. 1997). However, possibly because of an extremely low enzymatic activity, the presence of NOS in human

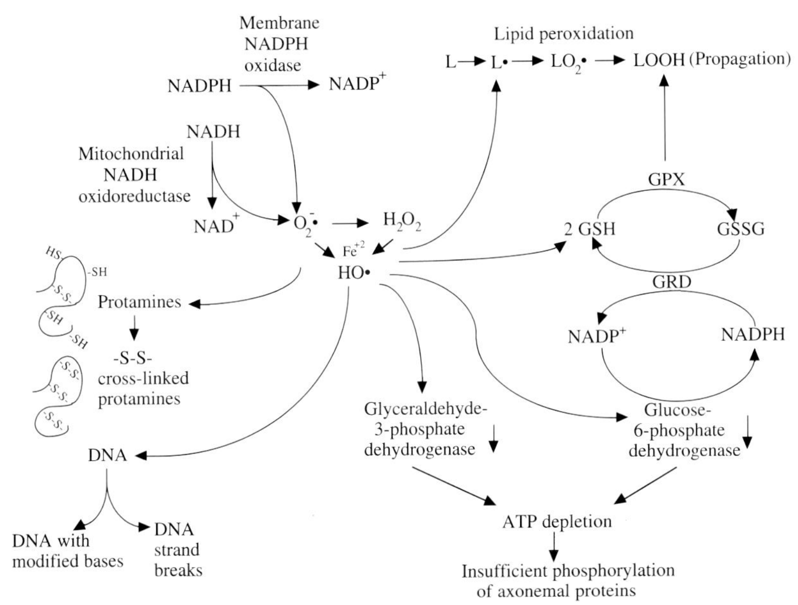

Figure 1. Toxic effects of ROS. A membrane NADPH oxidase, or more likely a mitochondrial NADH oxidoreductase, would be responsible for the generation of high levels of ROS ($O_2\bullet^-$, its dismutation product H_2O_2, and •OH from the iron-catalyzed transformation of H_2O_2 or $O_2\bullet^-$ and H_2O_2) in spermatozoa. ROS attack membrane lipids causing a chain reaction of lipid peroxidation. These peroxidized lipids can be repaired by glutathione peroxidase (GPX) in a reaction that uses glutathione (GSH). The resulting oxidized glutathione (GSSG) is recycled back to GSH by glutathione reductase (GRD) in a reaction that requires NADPH, a cofactor supplied by glucose-6-phosphate dehydrogenase. ROS cause a depletion of GSH and inhibit enzymes such as glyceraldehyde-3-phosphate dehydrogenase and glucose-6-phosphate dehydrogenase. The inhibition of these enzymes results in depletion of intracellular ATP and arrest of sperm motility due to insufficient phosphorylation of axonemal proteins. ROS can also cause DNA oxidation and hyperstability of DNA due to too-extensive cross-linking of protamines.

spermatozoa could not be demonstrated in other studies, even though effects of NO• on sperm function were reported (Zini et al. 1995; Schaad et al. 1996).

ROS Scavenging

The effects of ROS on living organisms depend not only on the production of ROS by cells and their environment, but also on the scavenging capacity of cellular and extracellular components surrounding the cells (Halliwell, Gutteridge 1989). Spermatozoa and seminal plasma possess scavenging systems to counteract or regulate the effects of ROS. They include enzymes such as SOD (Nissen, Kreysel 1983; Alvarez et al. 1987; Aitken et al. 1996a; Peeker et al. 1997), catalase (Jeulin et al. 1989), the glutathione peroxidase/reductase system (Alvarez, Storey 1989; Storey et al. 1998), and phospholipid hydroperoxide glutathione peroxidase (Godeas et al. 1997), but also a variety of substances with SOD- or catalase-like activities (Zini et al. 1993) or peroxyl scavenging capacity (Józwik et al. 1997), such as vitamins E and C (Chow 1991; Dawson et al. 1992), ubiquinol-10 (Alleva et al. 1997), glutathione (Halliwell, Gutteridge 1989; Storey et al. 1998), and taurine and hypotaurine (Alvarez, Storey 1983; Holmes et al. 1992).

The complex relationship between ROS generation and scavenging systems in spermatozoa and paradoxes arising from these processes are emphasized in many studies. The positive relationship between SOD activity in spermatozoa and the percentage and duration of sperm motility in semen suggests that SOD is especially important to these cells (Alvarez et al. 1987). However, another study reports that the higher levels of SOD activity found in spermatozoa retaining excess residual cytoplasm and producing more ROS are negatively correlated with the movement characteristics of human spermatozoa and their capacity for oocyte fusion (Aitken et al. 1996a). Finally, overexpression of SOD and poor $O_2\bullet^-$ generation by spermatozoa are also found in some cases of oligozoospermia (Sinha et al. 1991). Other apparent contradictions are that catalase levels are extremely low in spermatozoa (Jeulin et al. 1989) even though H_2O_2 is the ROS most often responsible for the loss of sperm motility (de Lamirande, Gagnon 1992a, b; Aitken et al. 1993a; Griveau et al. 1995a), and that the key enzymes of the antioxidant defense of spermatozoa, SOD, catalase and glutathione peroxidase can also be inhibited by ROS (Halliwell, Gutteridge 1989; Salo et al. 1990; Griveau et al. 1995a). Finally, NADPH appears as a possible intracellular cofactor needed both for the enzymatic generation of ROS by spermatozoa (Aitken, Clarkson 1987) and for the scavenging of peroxides by the glutathione peroxidase/reductase system (Storey et al. 1998). The peroxidase portion of this system uses glutathione, which is oxidized to glutathione disulfide, to scavenge peroxides; the reductase portion of the system regenerates reduced glutathione, a reaction involving NADPH (Fig. 1). Although exogenous addition of micromolar concentrations of cumene hydroperoxide triggers an increased activity of the pentose phosphate pathway (Ford et al. 1997), one of the possible targets for ROS in spermatozoa is the glucose-6-phosphate dehydrogenase, the enzyme responsible for the formation of NADPH (Griveau et al. 1995b).

ROS scavengers are present throughout the male reproductive system and factors such as nutrition and oxidative stress influence their levels. For example, an epididymis-specific glutathione peroxidase-like protein that is secreted in the porcine epididymal fluid binds to the acrosomal region of spermatozoa and disappears during the acrosome reaction (Okamura et al. 1997). The mRNAs for SOD and the glutathione peroxidase/reductase system are differently expressed in the various regions of the rat epididymis, suggesting that the need for antioxidant may vary during epididymal maturation (Zini, Schlegel 1997). SOD is present both in cells and lumen of human genital organs (prostate, seminal vesicles, testis), as shown in Nonogaki et al. (1992). Both undernutrition and zinc deficiency cause high rates of oxidative damage to testes lipids, proteins, and DNA, for which cells appear to compensate by increasing SOD and glutathione reductase activities (Oteiza et al. 1996). Age also influences the level of ROS scavengers as activity of glutathione peroxidase (in fowls and bulls) and SOD (in bulls) in semen decreased with aging (Kelso et al. 1996, 1997).

Recently, studies have been conducted to assess the effects of *in vitro* and *in vivo* treatment with ROS scavengers in an attempt to increase male fertility. For example, the decrease in washed sperm motility due to ROS generated by spermatozoa themselves or by activated neutrophils can be prevented by the presence of antioxidants such as catalase, dimethylsulfoxide, glutathione, N-acetylcysteine, hypotaurine, pentoxifylline, and so on (Kovalski et al. 1992; Baker et al. 1996; Oeda et al. 1997; Okada et al. 1997; Parinaud et al. 1997). Oral intake of ascorbic acid reduces sperm agglutination and abnormal morphology but only at a high dose (1g/day) and after at least four weeks of treatment (Dawson et al. 1992). Treatment with vitamin E of patients with sperm ROS-associated infertility allows improvement in only one of the sperm function tests, the zona pellucida binding assay (Kessopoulou et al. 1995). Parenteral treatment of infertile males with glutathione results in higher grades of sperm motility and morphology (Irvine 1997). Even though studies cited above describe positive effects of antioxidants on spermatozoa, no data on actual improvement of fertility in humans was yet reported. On the other hand, a pretreatment of mouse spermatozoa with catalase or thioredoxin (an enzyme that recycles oxidized disulfide of proteins back to their reduced sulfhydryl form) improves sperm potential to support embryo development *in vitro* without affecting other parameters such as sperm motility, hyperactivation and fertilization rate (Kuribayashi, Gagnon 1996). Further studies are needed to determine whether this treatment is applicable in human assisted conception technologies.

Mechanisms of ROS Toxicity

All cellular components, lipids, proteins, nucleic acids, and sugars are potential targets for ROS (Halliwell, Gutteridge 1989). The type and extent of the modifications induced by ROS depend not only on the nature and amount of ROS involved (Alvarez, Storey 1989; de Lamirande, Gagnon 1992a, b; Aitken et al. 1993a; Griveau et al. 1995a) but also on the moment and duration of ROS exposure (de Lamirande, Gagnon 1992a) and of extracellular factors such as temperature, oxygen tension and composition of the incubation medium, including ions, proteins, ROS scavengers and so on (Alvarez et al. 1983; Griveau, Le Lannou 1997a). In spermatozoa, the first effect of ROS reported is lipid peroxidation (review: Storey 1997) but recent data indicate that ROS also oxidize DNA (Kodama et al. 1997; Shen et al. 1997), affect ATP and NADPH production (de Lamirande, Gagnon 1992b; Griveau et al. 1995a), and DNA condensation (Rufas et al. 1991; Rodriguez et al. 1985), as shown in Figure 1.

Because of their high content of polyunsaturated fatty acids, spermatozoa are very sensitive to lipid peroxidation, which results in differential loss of phospholipids, mainly phosphatidylcholine and phosphatidylethanolamine (Storey 1997). Lipid peroxidation is detrimental to spermatozoa because it decreases fluidity and increases permeability of their membranes, but also because of the inherent toxicity of lipid peroxides and of their degradation products that accumulate in sperm membranes (Jones et al. 1979; Aitken et al. 1993b; Windsor et al. 1993). High levels of lipid peroxidation are associated with decreased sperm motility and viability (Aitken et al. 1989; Alvarez et al. 1987). Oxidative modification of sperm DNA may be a more elusive cause of male infertility (Shen et al. 1997) since it is not reflected by a decrease in sperm motility (Kodama et al. 1997) but could potentially affect other sperm function such as the competence of these cells to support embryo development (Kuribayashi, Gagnon 1996). Mild and undetectable oxidative stress for extensive periods of time could also alter the condensation state of sperm DNA, an effect that would not be detected before fertilization. Spermatozoa from oligospermic men are significantly less reduced by dithiothreitol than those of fertile men (Rufas et al. 1991). This effect could possibly be explained by an excess formation of disulfide bridges due to oxidation of sulfhydryl groups on protamines. The extreme DNA stability that ensues could prevent glutathione-induced DNA decondensation inside the oocyte (Rodriguez et al. 1985).

Studies in which spermatozoa were subjected to exogenous addition of ROS indicate that H_2O_2 is the ROS responsible for the loss of motility in human (de Lamirande, Gagnon 1992a, b; Aitken et al. 1993a; Griveau et al. 1995a), murine (Kodama et al. 1995) and bovine (Blondin et al. 1997) species. ROS treatment causes a rapid loss of intracellular ATP that triggers a cascade of events leading to a decrease in cAMP-dependent phosphorylation of axonemal proteins and sperm immobilization (de Lamirande, Gagnon 1992b; Leclerc, Gagnon 1996). Oxidative inhibition of enzymes such as glyceraldehyde-3-phosphate dehydrogenase and glucose-6-phosphate dehydrogenase (involved in the generation of ATP and NADPH, respectively) could be responsible for this ATP depletion (de Lamirande, Gagnon 1992b; Griveau et al. 1995a), as seen in Figure 1. Even though exposure to low ROS levels does not affect sperm motility, it may decrease the capacity of spermatozoa to undergo the acrosome reaction when challenged with A23187 (Griveau et al. 1995a), but the underlying mechanism responsible for this action is not yet known. In mice, exogenous addition of ROS (by the combination of xanthine and xanthine oxidase) at concentrations below those that cause detectable levels of lipid peroxidation or loss of sperm motility suppresses sperm fusion with zona-free eggs by a mechanism that appears to involve oxidation of the sulfhydryl groups on sperm membrane proteins (Mammoto et al. 1996).

ROS and the Induction of Sperm Capacitation and Hyperactivation

Recent evidence indicates that the controlled and well-timed generation of very low amounts of ROS participates in signal transduction cascades (Baeuerle et al. 1996; Suzuki et al. 1997) similar to those involved in the acquisition of fertilizing ability by spermatozoa (Yanagimachi 1994; de Lamirande et al. 1997a). Sperm capacitation and the associated hyperactivated motility also appear to be modulated by ROS (Bize et al. 1991; de Lamirande, Gagnon 1993a, b, 1995a; Griveau et al. 1994, 1995b; Aitken et al. 1995, 1996b, 1997; Zhang, Zheng 1996; Leclerc et al. 1997; de Lamirande et al. 1998a). The importance of $O_2\bullet^-$ in human sperm capacitation is evidenced by the observations that exogenously added $O_2\bullet^-$ (xanthine and xanthine oxidase in the presence of catalase) promotes this process and that SOD prevents sperm capacitation triggered by $O_2\bullet^-$, progesterone and biological fluids (de Lamirande, Gagnon 1993a, b; de Lamirande et al. 1993, 1998a). Spermatozoa themselves are the source of the $O_2\bullet^-$

associated with capacitation (de Lamirande, Gagnon 1995a). Even though very low levels of $O_2\bullet^-$ are required for sperm capacitation, higher levels may be needed for the acquisition and maintenance of hyperactivated motility (de Lamirande et al. 1998a). Ultrafiltrates from fetal cord serum (FCSu), follicular fluid (FFu) and seminal plasma (SPu) all induce sperm capacitation but only FCSu, which triggers a three-fold higher $O_2\bullet^-$ generation than that of the two other fluids, promotes sperm hyperactivation (de Lamirande et al. 1998a). The $O_2\bullet^-$ production by spermatozoa is initiated immediately at the beginning of the capacitation period, peaks 15 to 25 minutes later, and slowly decreases over the course of the next hours (de Lamirande, Gagnon 1995a; de Lamirande et al. 1998a). Therefore, the increase in $O_2\bullet^-$ production is one of the first manifestations of capacitation detected and precedes other phenomena such as sperm hyperactivation (observed between one and three hours), protein tyrosine phosphorylation (from one hour) and capacitation (progressive increase over six hours). Interestingly, the sperm $O_2\bullet^-$ generation is required only for the first 30 minutes of incubation under capacitating conditions, which suggests that $O_2\bullet^-$ initiates an early chain of events, but is not needed for the subsequent steps leading to capacitation (de Lamirande et al. 1998a).

The involvement of ROS in capacitation is not limited to $O_2\bullet^-$ and human spermatozoa. In humans, capacitation is also prevented by catalase, and, conversely, is stimulated by a direct addition of H_2O_2 (Griveau et al. 1995b; Aitken et al. 1995; Leclerc et al. 1997); this observation suggests the involvement of H_2O_2 in capacitation and $O_2\bullet^-$ as the source of the endogenous H_2O_2. In addition, the H_2O_2-scavenging capacity of

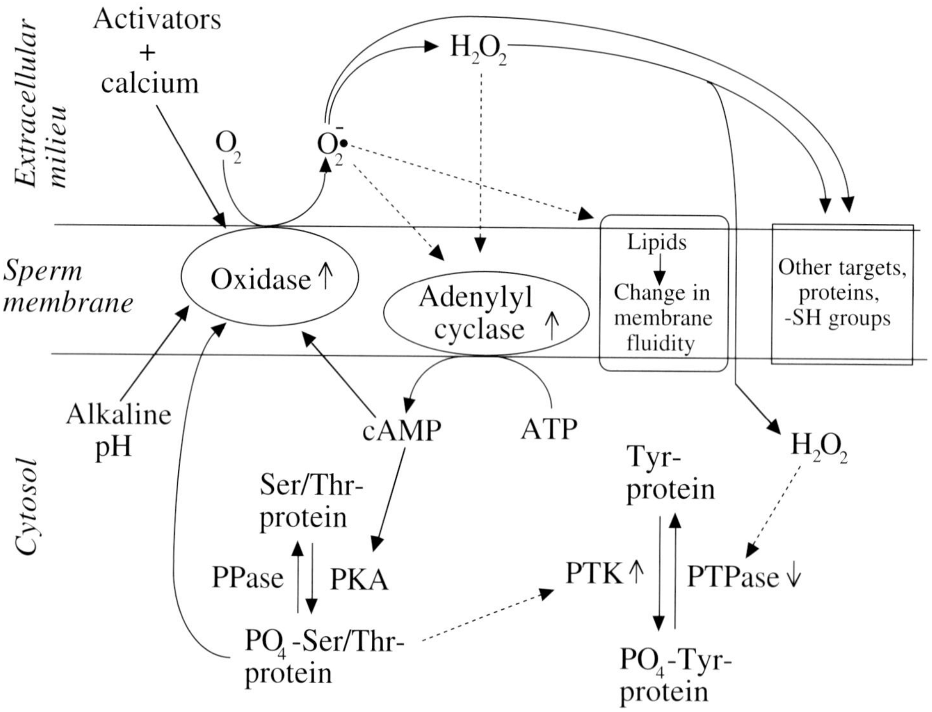

Figure 2. Hypothetical ROS-related signaling pathways during human sperm capacitation. ROS generation, calcium influx, and increase in cAMP concentration appear as early events of sperm capacitation. Extracellular calcium as well as a rise of the intracellular pH to about 7.4 would be needed for the initiation of $O_2\bullet^-$ generation by the sperm membrane oxidase. The observations that an increase in intracellular cAMP promotes the production of $O_2\bullet^-$ and that exogenous ROS (from the combination of xanthine and xanthine oxidase) causes a rise in intracellular cAMP, suggest the involvement of an amplification cycle between these two signaling pathways. Oxidation of specific sulfhydryl (-SH) groups on sperm membrane proteins could be the result of ROS action as well as a triggering event leading to oxidase activation, which would suggest again the presence of an amplification cycle. ROS could modify lipids and cause an increase in membrane fluidity or oxidize proteins and change their structure and functions. Both $O_2\bullet^-$ and H_2O_2 could activate sperm adenylyl cyclase as observed in other cell systems (dashed lines indicate an hypothetical phenomenon). Elevation of intracellular cAMP concentration activates protein kinase A (PKA) and causes an increase in the serine/threonine (Ser/Thr) phosphorylation of numerous proteins, therefore modulating their functions to elicit an overall cellular response. The increase in Ser/Thr phosphorylation could also be caused by inhibition of protein phosphatase activity. Protein tyrosine kinases (PTK) are potential targets for PKA and their phosphorylated substrates; the consequent increase in PTK activity would result in the phosphorylation of sperm proteins. Moreover, the increase in protein tyrosine phosphorylation could be due to inhibition of protein tyrosine phosphatases (PTPases), since these enzymes are very sensitive to oxidation.

spermatozoa prepared by the swim-up technique decreases stepwise in groups of patients whose spermatozoa achieve higher fertilization in an *in vitro* fertilization program (Yeung et al. 1996). Low concentrations of NO• also trigger human sperm capacitation, by a mechanism that may involve the generation of H_2O_2 (Zini et al. 1995). Sperm capacitation in hamsters (Bize et al. 1991), mice (de Lamirande et al. 1997a) and cattle (Blondin et al. 1997; O'Flaherty et al. 1997; Goyette et al. 1998) is also modulated by ROS. Taken together, these data further tend to demonstrate the oxidative nature of the capacitation process. There are presently no data available on the $O_2^{\bullet -}$ generation by spermatozoa during *in vivo* capacitation. However, the oxygen concentration in the female reproductive tract rises sharply at the time of ovulation (Maas et al. 1976), which lends support for such a process. It is also possible that cells and fluids from the female reproductive tract participate in the ROS formation that leads to sperm capacitation.

Origin of ROS during Sperm Capacitation

An oxidase located at the level of the sperm plasma membrane and stimulated during capacitation would be responsible for the increased extracellular $O_2^{\bullet -}$ generation observed during this process (Fig. 2). However, the nature of this oxidase remains elusive. The sperm oxidase would share some similarities with the NADPH oxidase of neutrophils because it appears to be located at the level of the plasma membrane and to release $O_2^{\bullet -}$ in the extracellular milieu; its activation is rapid and transient; the level of its activity is determined by the nature and concentration of the stimulant; and its activity is reduced in the presence of inhibitors of the neutrophil NADPH oxidase, such as diphenyliodonium and lapachol (Leclerc et al. 1997; de Lamirande et al. 1997b; Aitken et al. 1997). However, the sperm oxidase activity is, by at least three orders of magnitude, lower than that of neutrophils (de Lamirande, Gagnon 1995a; de Lamirande et al. 1998a). Furthermore the signal transduction mechanisms leading to oxidase activation appear to be different in spermatozoa and neutrophils, as tyrosine phosphorylation of specific proteins is a prerequisite for the oxidative burst in neutrophils (Fialkow et al. 1993), but a consequence of ROS generation in spermatozoa (Leclerc et al. 1997). There is presently a controversy on the location, mechanism of activation and cofactors of the sperm membrane oxidase. Aitken and colleagues (1995, 1996b, 1997) report that NADPH, but not NADH, when used at millimolar concentrations, stimulates sperm $O_2^{\bullet -}$ generation and capacitation and propose that the sperm oxidase is oriented towards the intracellular space. On the other hand, data presented by de Lamirande et al. (1998a) indicate that capacitation of spermatozoa incubated with biological fluid ultrafiltrates or progesterone is dependent on the release of $O_2^{\bullet -}$ in the extracellular milieu, but that exogenous NADH or NADPH promote capacitation by a mechanism independent of an increased $O_2^{\bullet -}$ production by spermatozoa. These discrepancies are probably due to the use of different experimental conditions and may be resolved soon.

Few of the requirements for activation of the sperm membrane oxidase are known (Fig. 2). Extra- but not intracellular calcium is needed for the induction of $O_2^{\bullet -}$ generation and, once initiated, calcium is not required for the maintenance of $O_2^{\bullet -}$ production (Leclerc et al. 1998). Sperm $O_2^{\bullet -}$ generation is also regulated by intracellular pH. The concentration-dependent effect of bicarbonate ions on the $O_2^{\bullet -}$ production of spermatozoa appears to be related to alkalinization of intracellular pH at about 7.4, an effect that can also be reproduced by the use of buffers such as TAPS, pH 8.4 (Aitken et al. 1998b) or HEPES, pH 8.0 (de Lamirande, Gagnon 1993a, b, 1995a).

Finally, the stimulation of sperm capacitation and $O_2^{\bullet -}$ generation by inhibitors of cyclic phosphodiesterases (isobutyl methylxanthine) and of serine/threonine phosphatases (calyculin A), suggests that increases in cAMP and/or serine/threonine phosphorylation are prerequisites for the sperm oxidase activation (Leclerc et al. 1998). Treatment of spermatozoa with caffeine, another cyclic phosphodiesterase inhibitor, stimulates sperm $O_2^{\bullet -}$ generation and increases membrane fluidity and rotational mobility of thiol-containing membrane proteins, as measured by electron spin resonance (Sinha et al. 1993). This last observation may be related to recent data indicating that a fine balance between oxidation of sulfhydryl groups and reduction of disulfide bridges on sperm proteins may be important for the stimulation of sperm $O_2^{\bullet -}$ production and capacitation (de Lamirande, Gagnon 1998). Reagents targeted at sulfhydryl groups, such as N-ethylmaleimide, phenylarsine oxide, *p*-chloromercuribenzoic acid, and diamide, used at micromolar concentrations, promote sperm $O_2^{\bullet -}$ generation and capacitation that is inhibited by SOD (de Lamirande, Gagnon 1998).

As sulfhydryl groups are potential targets for $O_2^{\bullet -}$ (Halliwell, Gutteridge 1989) and oxidation of sulfhydryl groups may be involved in initiation of sperm $O_2^{\bullet -}$ production, it is tempting to speculate that an amplification loop exists between these two pathways during sperm capacitation. Activation of the sperm oxidase, or mechanisms leading to this process, may also involve the removal of some sialoglycoconjugates from the sperm membrane since a sialic acid binding glycoprotein of endometrial origin increases the $O_2^{\bullet -}$ generation by spermatozoa (Banerjee, Chowdhury 1997). Finally, there is also an indication that the sperm oxidase may preferentially use NADPH as intracellular cofactor since, in a cell-free system, NADPH induces a three-fold higher stimulation of $O_2•^-$ generation than NADH (Aitken et al. 1997).

Mechanisms of ROS Action during Sperm Capacitation

The specific targets for ROS and their role in the ensuing signal transduction cascade leading to sperm capacitation are not yet known. However, there are indications that extracellular ROS may affect the intracellular cAMP concentration (Fig. 2) since treatment of spermatozoa with exogenous ROS (low concentrations of xanthine and xanthine oxidase) causes an increase in sperm intracellular cAMP concentration and capacitation that are both prevented by SOD (Zhang, Zheng 1996), and since stimulation of sperm $O_2•^-$ production with 2.5 mM NADPH is associated with a three-fold increase in intracellular cAMP concentration (Aitken et al. 1998a). Whether the ROS-induced increase in cAMP is due to a stimulation of adenylyl cyclase and/or an inhibition of cyclic phosphodiesterase activities of spermatozoa and how cAMP would stimulate $O_2•^-$ generation remain to be established. On the other hand, sperm capacitation induced by isobutyl methylxanthine is not prevented by SOD, indicating that the intracellular increase in cAMP occurs downstream of ROS production (de Lamirande et al. 1997b). These data may appear contradictory to those presented above which indicated that phosphodiesterase inhibitors, which presumably increase sperm cAMP, also stimulate the generation of $O_2•^-$ by spermatozoa (Sinha et al. 1993; Leclerc et al. 1998). However, these results may also suggest that an amplification loop between the two signaling pathways, cAMP and $O_2•^-$, may occur during sperm capacitation.

Extracellular calcium is needed for the induction of $O_2•^-$ generation and, once initiated, is not required for the maintenance of $O_2•^-$ production (Leclerc et al. 1998); on the other hand, extra- and intracellular calcium are required for sperm capacitation (Leclerc et al. 1998). These observations could suggest that $O_2•^-$ is involved in the calcium influx associated with capacitation. However, flow cytometry experiments performed with Fluo 3-loaded human spermatozoa indicate that the calcium influx associated with FCSu-induced sperm capacitation is not prevented by SOD (E de Lamirande and C Gagnon, unpublished data). Therefore, calcium influx and $O_2•^-$ generation are both essential, but probably occur in parallel and subsequently act on a common pathway during sperm capacitation.

The capacitation-related protein tyrosine phosphorylation appears to be regulated by calcium, cAMP, and ROS (Leclerc et al. 1996, 1997, 1998; de Lamirande et al. 1997b, 1998a; Aitken et al. 1998a), as shown in Figure 2. Sperm capacitation and the associated time-dependent increase in tyrosine phosphorylation of two proteins, p105/p81 (named according to their molecular masses), are both prevented by antioxidants (SOD and catalase) and stimulated by exogenously added ROS, in the form of very low concentrations of H_2O_2 or of xanthine and xanthine oxidase (Aitken et al. 1995; Leclerc et al. 1997). Phosphorylation of p105/p81 occurs through a herbimycin A-sensitive protein tyrosine kinase and incubation of spermatozoa with non-selective protein tyrosine phosphatase inhibitors such as vanadate and phenylarsine oxide results in a marked increase in phosphotyrosine content of these two proteins (Leclerc et al. 1997). The two tyrosine phosphorylated proteins, p105/p81, are components of the fibrous sheath of the sperm flagellum (Leclerc et al. 1997) and are antigenetically related to mouse A kinase anchoring proteins (AKAPs) which sequester protein kinase A through its regulatory subunits to subcellular locations (Carrera et al. 1996). Even though it is hypothesized that phosphorylation of p105/p81 during sperm capacitation may be related to the acquisition of hyperactivated motility (Leclerc et al. 1997), it is also proposed that binding of PKA regulatory units to AKAPs regulates enzymes or biochemical pathways other than, or in addition to, the PKA catalytic subunit (Vijayaraghavan et al. 1997). In other systems, AKAPs are known to bind calmodulin (Rubin 1994), a protein present in spermatozoa (Jones et al. 1980) and which modulates capacitation (Leclerc et al. 1998).

Sperm capacitation is associated with increased membrane fluidity (Yanagimachi 1994; de Lamirande et al. 1997b). Although high levels of ROS cause lipid peroxidation which subsequently decreases membrane fluidity (Halliwell, Gutteridge 1989), caffeine, at concentrations known to stimulate capacitation, increases $O_2•^-$ generation, decreases SOD activity and increases membrane fluidity of spermatozoa (Sinha et al. 1993). The mechanism by which $O_2•^-$ would affect membrane fluidity remains to be elucidated.

Sulfhydryl groups are potential targets for ROS (Halliwell, Gutteridge 1989) and may be important for sperm capacitation since this process is associated with an elevation in the rate of rotational mobility of sulfhydryl-containing sperm proteins, with induction by caffeine (Sinha et al. 1993), and a strong time-dependent increase of sperm membrane sulfhydryl groups exposed to the extracellular space, as observed in spermatozoa incubated with FCSu (de Lamirande, Gagnon 1998). Oxidation of protein sulfhydryl groups by ROS offers cells an easy and reversible means for switching on or off signal transduction enzymes, such as protein kinase C (on) (Gopalakrishna, Anderson 1989) and tyrosine phosphatases (off) (Hecht, Zick 1992). Reagents targeted at sulfhydryl groups, such as N-ethylmaleimide, phenylarsine oxide, *p*-chloromercuribenzoic acid and diamide, promote sperm capacitation and $O_2\bullet^-$ generation. SOD only partially prevents capacitation induced by these substances, which suggests that sulfhydryl groups may also be targets for $O_2\bullet^-$ generated by spermatozoa (de Lamirande, Gagnon 1998). Conversely, disulfide reactants of various structure, such as dithiothreitol (a disulfide compound), thioredoxin (an enzyme that recycles the sulfhydryl groups from the disulfide bridges), and tris-(2-carboxyethyl) phosphine (TCEP, a reductant without sulfhydryl group), used at micromolar concentrations, partly to totally decrease FCSu-induced sperm capacitation and the associated $O_2\bullet^-$ generation (de Lamirande, Gagnon 1998). However, these same disulfide reductants cause sperm capacitation and $O_2\bullet^-$ generation in the absence of FCSu, which indicates that the mechanisms regulating the status of the sulfhydryl-disulfide pair present in spermatozoa and related to sperm capacitation and $O_2\bullet^-$ production are not simple but involve a complex and multifactorial process (de Lamirande, Gagnon 1998).

ROS and Sperm-Zona Binding

The concept that mild oxidative conditions which induce low levels of lipid peroxidation promote or are even required for binding of spermatozoa to the zona pellucida is recent. Even though high levels of lipid peroxidation decrease sperm function and viability, low levels of lipid peroxidation enhance the binding of human spermatozoa to both homologous and heterologous (hamster) zonae pellucidae (Aitken et al. 1989). Treatment of mouse spermatozoa with low concentrations of ferrous iron (Fe^{2+}) and ascorbic acid, which causes lipid peroxidation, improves the fertilization potential of these cells by 50%. This phenomenon appears to be related to a marked increase in binding of spermatozoa to mouse zona pellucida rather than to a change in sperm motility parameters (including hyperactivation) or capacitation (Kodama et al. 1996). Thus, the formation of lipid peroxides in the cell membrane is not necessarily detrimental to sperm function.

ROS and Sperm Acrosome Reaction

Although the involvement of ROS generated by spermatozoa in the capacitating process is now recognized, only three studies dealt with the role of ROS in the acrosome reaction. One study suggests that H_2O_2 is needed for the A23187-induced acrosome reaction, as well as the binding of spermatozoa to zona-free hamster oocytes, because catalase, and not SOD, prevents these processes, and that H_2O_2 initiates them (Aitken et al. 1995). However, no direct evidence is presented for the generation of H_2O_2 by spermatozoa during the acrosome reaction. A second study proposes that $O_2\bullet^-$ is involved in the A23187-induced acrosome reaction because it is associated with a four- to five-fold increase in sperm $O_2\bullet^-$ generation and is inhibited by SOD but not by catalase (Griveau et al. 1995c). However, treatment of capacitated spermatozoa with potassium superoxide (a source of $O_2\bullet^-$) is not by itself sufficient to trigger the acrosome reaction, indicating that this free radical may be essential, but not sufficient by itself, for the induction of this process (Griveau et al. 1995c). The most recent study reports that human sperm acrosome reaction, induced by stimuli that act through very different mechanisms—A23187 (calcium ionophore), lysophosphatidylcholine (membrane disturbing agent), and biological fluids ultrafiltrates (unknown mechanism)—is associated with the production of $O_2\bullet^-$ by spermatozoa and is inhibited by SOD and catalase (de Lamirande et al. 1998b). Furthermore, exogenously added ROS (by the combination of xanthine and xanthine oxidase or direct addition of H_2O_2) promote the acrosome reaction in capacitated spermatozoa (de Lamirande et al. 1998b). The involvement of ROS in the acrosome reaction is not limited to $O_2\bullet^-$ and H_2O_2 and human spermatozoa as NOS plays a role in progesterone-induced acrosomal exocytosis in mouse spermatozoa (Herrero et al. 1997b). Even though these data may appear contradictory, it is possible that, as in the case of sperm capacitation, the involvement of a specific ROS in the acrosome reaction may depend on the incubation conditions. These data converge to emphasize the concept that the acrosome reaction is also part of an oxidative process.

The mechanism by which ROS promote the acrosome reaction appears to involve tyrosine phosphorylation of p105/p81, the same two proteins that were tyrosine phosphorylated during sperm capacitation (Aitken et al. 1995; de Lamirande et al. 1998b). The increase in tyrosine phosphorylation of p105/p81 associated with the acrosome reaction induced by A23187 and by lysophosphatidylcholine is partly reversed by SOD in the case of lysophosphatidylcholine and A23187, or catalase in the case of A23187, and totally abolished in the presence of both antioxidants in the case of A23187 (de Lamirande et al. 1998b). ROS do not appear to influence the tyrosine phosphorylation of Triton soluble sperm proteins during the acrosome reaction (de Lamirande et al. 1998b). These data may be surprising because one would expect to observe changes at the level of the sperm membrane during the acrosome reaction rather than on major fibrous sheath proteins such as p105/p81 (Leclerc et al. 1997). Elucidation of the mechanisms by which ROS modulate phosphorylation of p105/p81 will offer important clues on the mechanisms of the acrosome reaction.

The observations that the same ROS are involved in both sperm capacitation and acrosome reaction may be surprising since these two processes are very different with respect to their kinetics, stimulating agents, signal transduction pathways and so on (Yanagimachi 1994). It is possible that cellular targets for $O_2^{\bullet -}$ and H_2O_2 are different during the two processes, which would imply that the acrosome reaction is associated with modifications, such as membrane conformation, protein translocation and so on, that unmask hidden targets for ROS. It is also possible that two bursts of ROS are needed or that the targets for ROS become available at two different time points, one during sperm capacitation and the other during the acrosome reaction.

Conclusion

In recent years numerous studies reported the effects, both pathologic and physiologic, of ROS on spermatozoa. The balance between the production (time, amount, rate, location and so on) and scavenging of ROS will determine the effects of ROS on spermatozoa. High levels of ROS will endanger sperm function and viability, but the controlled generation of very low amounts of ROS appears to regulate the acquisition of sperm fertilizing ability.

Although most of the mechanisms regulating ROS production and actions in both pathologic and physiologic cases remain to be elucidated, partial answers are now emerging. Elucidation of these mechanisms will allow better options to counteract the toxic effects of high levels of ROS and/or to improve sperm fertility potential in assisted reproduction technologies.

Acknowledgements

The authors acknowledge the support of the Medical Research Council of Canada to C.G.

References

Aitken RJ. A free radical theory of male infertility. Reprod Fertil Dev 1994; 6:19-24.

Aitken RJ, Clarkson JS. Cellular basis of defective sperm function and its association with the genesis of reactive oxygen species by human spermatozoa. J Reprod Fertil 1987; 81:459-469.

Aitken RJ, Clarkson JS. Significance of reactive oxygen species and antioxidants in defining the efficacy of sperm preparation techniques. J Androl 1988; 9:367-376.

Aitken RJ, Fisher H. Reactive oxygen species generation and human spermatozoa: the balance of benefit and risk. Bio Essays 1994; 16:259-267.

Aitken RJ, Clarkson JS, Fishel S. Generation of reactive oxygen species, lipid peroxidation and human sperm function. Biol Reprod 1989; 40:183-197.

Aitken RJ, Irvine DS, Wu FC. Prospective analysis of sperm-oocyte fusion and reactive oxygen species generation as criteria for the diagnosis of male infertility. Am J Obstet Gynecol 1991; 164:542-551.

Aitken RJ, Buckingham D, Harkiss D. Use of a xanthine oxidase free radical generating system to investigate the cytotoxic effects of reactive oxygen species on human spermatozoa. J Reprod Fert 1993a; 97:451-462.

Aitken RJ, Harkiss D, Buckingham DW. Analysis of lipid peroxidation mechanisms in human spermatozoa. Mol Reprod Dev 1993b; 35:302-315.

Aitken RJ, Krausz C, Buckingham D. Relationship between biochemical markers for residual sperm cytoplasm, reactive oxygen species generation, and the presence of leukocytes and precursor germ cells in human sperm suspensions. Mol Reprod Dev 1994; 39:268-279.

Aitken RJ, Paterson M, Fisher H, Buckingham D, van Duin M. Redox regulation of tyrosine phosphorylation in human spermatozoa and its role in human sperm function. J Cell Sci 1995; 108:2017-2035.

Aitken RJ, Buckingham D, Carreras A, Irvine DS. Superoxide dismutase in human sperm suspensions: relationship with cellular composition, oxidative stress, and sperm function. Free Radic Biol Med 1996a; 21:495-504.

Aitken RJ, Buckingham D, Harkiss D. Paterson M, Fisher H, Irvine DS. The extragenomic action of progesterone on human spermatozoa is influenced by redox regulated changes in tyrosine phosphorylation during capacitation. Mol Cell Endocrinol 1996b; 117:83-93.

Aitken RJ, Fisher HM, Fulton N, Gomez E, Knox W, Lewis B, Irvine S. Reactive oxygen species generation by human spermatozoa is induced by exogenous NADPH and inhibited by the flavoprotein inhibitors diphenylene iodonium and quinacrine. Mol Reprod Dev 1997; 47:468-482.

Aitken RJ, Harkiss D, Knox W, Paterson M, Irvine DS. A novel signal transduction cascade in capacitating human spermatozoa characterized by redox-regulated, cAMP-mediated induction of tyrosine phosphorylation. J Cell Sci 1998a; 111:645-656.

Aitken RJ, Harkiss D, Knox W, Paterson M, Irvine DS. On the cellular mechanisms by which the bicarbonate ion mediates the extragenomic action of progesterone on human spermatozoa. Biol Reprod 1998b; 58:186-196.

Alleva R, Scararmucci A, Mantero F, Bompadre S, Leoni L, Littarru GP. The protective role of ubiquinol-10 against formation of lipid peroxides in human seminal fluid. Mol Aspects Med 1997; 18 (Suppl):S221-S228.

Alvarez JG, Storey BT. Taurine, hypotaurine, epinephrine and albumin inhibit lipid peroxidation in rabbit spermatozoa and protect against loss of motility. Biol Reprod 1983; 29:548-555.

Alvarez JG, Storey BT. Role of glutathione peroxidase in protecting mammalian spermatozoa from loss of motility caused by spontaneous lipid peroxidation. Gam Res 1989; 23:77-90.

Alvarez JG, Touchstone JC, Blasco L, Storey BT. Spontaneous lipid peroxidation and production of lipid peroxide and superoxide in human spermatozoa. Superoxide dismutase as a major enzyme protectant against oxygen toxicity. J Androl 1987; 8:338-348.

Baeuerle PA, Rupec RA, Pahl HL. Reactive oxygen intermediates as second messengers of a general pathogen response. Pathol Biol 1996; 44:29-35.

Baker HWG, Brindle J, Irvine DS, Aitken RJ. Protective effect of antioxidants on the impairment of sperm motility by activated polymorphonuclear leukocytes. Fertil Steril 1996; 65:411-419.

Banerjee M, Chowdhury M. Localization of a 25 kDa human sperm surface protein: its role in *in vitro* human sperm capacitation. Mol Hum Reprod 1997; 3:109-114.

Bize I, Santander G, Cabello P, Driscoll D, Sharpe C. Hydrogen peroxide is involved in hamster sperm capacitation *in vitro*. Biol Reprod 1991; 44:398-403.

Blondin P, Coenen K, Sirard M-A. The impact of reactive oxygen species on bovine sperm fertilizing ability and oocyte maturation. J Androl 1997; 18:454-460.

Carrera A, Moos J, Ning XP, Gertol GL, Tesarik J, Kopf GS, Moos SB. Regulation of protein tyrosine phosphorylation in human sperm by a calcium/calmodulin-dependent mechanism: identification of A kinase anchor proteins as major substrates for tyrosine phosphorylation. Dev Biol 1996; 180:284-296.

Chow CK. Vitamine E and oxidative stress. Free Radic Biol Med 1991; 11:215-232.

Dawson EB, Harris WA, Teter MC, Powell LC. Effect of ascorbic acid supplementation on the sperm quality of smokers. Fertil Steril 1992; 58:1034-1039.

de Lamirande E, Gagnon C. Reactive oxygen species and human spermatozoa I. Effects on the motility of intact spermatozoa and on sperm axonemes. J Androl 1992a; 13:368-378.

de Lamirande E, Gagnon C. Reactive oxygen species and human spermatozoa II. Depletion of adenosine triphosphate plays an important role in the inhibition of sperm motility. J Androl 1992b; 13:379-386.

de Lamirande E, Gagnon C. A positive role for the superoxide anion in the triggering of human sperm hyperactivation and capacitation. Int J Androl 1993a; 16:21-25.

de Lamirande E, Gagnon C. Human sperm hyperactivation and capacitation as parts of an oxidative process. Free Radic Biol Med 1993b; 14:157-163.

de Lamirande E, Gagnon C. Capacitation-associated production of superoxide anion by human spermatozoa. Free Radic Biol Med 1995a; 18:487-495.

de Lamirande E, Gagnon C. Impact of reactive oxygen species on spermatozoa: a balancing act between beneficial and detrimental effects. Hum Reprod 1995b; 10:15-21.

de Lamirande E, Gagnon C. Paradoxical effect of reagents for sulfhydryl and disulfide groups on human capacitation and superoxide production. Free Radic Biol Med 1998; 25:803-817.

de Lamirande E, Eiley D, Gagnon C. Inverse relationship between the induction of human sperm capacitation and spontaneous acrosome reaction by various biological fluids and the superoxide scavenging capacity of these fluids. Int J Androl 1993; 16:258-266.

de Lamirande E, Leduc B, Iwasaki A, Hassouna M, Gagnon C. Increased reactive oxygen species formation in semen of patients with spinal cord injury. Fertil Steril 1995; 63:637-642.

de Lamirande E, Jiang H, Zini A, Kodama H, Gagnon. Reactive oxygen species and sperm physiology. Rev Reprod 1997a; 2:48-54.

de Lamirande E, Leclerc P, Gagnon C. Capacitation as a regulatory event that primes spermatozoa for the acrosome reaction and fertilization. Mol Hum Reprod 1997b; 3:175-194.

de Lamirande E, Harakat A, Gagnon C. Human sperm capacitation induced by biological fluids and progesterone, but not by NADH or NADPH, is associated with the production of superoxide anion. J Androl 1998a; 19:215-225.

de Lamirande E, Tsai C, Harakat A, Gagnon C. Involvement of reactive oxygen species in human sperm acrosome reaction induced by A23187, lysophosphatidylcholine, and biological fluid ultrafiltrates. J Androl 1998b; 19:585-594.

Donnelly ET, Lewis SEM, Thompson W, Chakravarty U. Sperm nitric oxide and motility: the effects of nitric oxide synthase stimulation and inhibition. Mol Hum Reprod 1997; 3:755-762.

Fialkow L, Chan CK, Grinstein S, Downey GP. Regulation of tyrosine phosphorylation in neutrophils by the NADPH oxidase. J Biol Chem 1993; 268:17131-17137.

Fisher HM, Aitken RJ. Comparative analysis of the ability of precursor germ cells and epididymal spermatozoa to generate reactive oxygen metabolites. J Exp Zool 1997; 277:390-400.

Ford WC, Whittington K, Williams AC. Reactive oxygen species in human sperm suspensions: production by leukocytes and the generation of NADPH to protect sperm against their effects. Int J Androl 1997; 20 (Suppl 3):44-49.

Gavella M, Lipovac V. NADH-dependent oxido-reductase (diaphorase) activity and isozyme pattern of sperm in infertile men. Arch Androl 1992; 28:135-141.

Gavella M, Lipovac V, Sverko V. Superoxide anion production and some sperm-specific enzyme activities in infertile men. Andrologia 1995; 27:7-12.

Godeas C, Tramer F, Micali F, Soranzo M, Sandri G, Panfili E. Distribution and possible role of phospholipid hydroperoxide glutathione peroxidase in rat epididymal spermatozoa. Biol Reprod 1997; 57:1502-1508.

Gomez E, Buckingham DW, Brindle J, Lanzafame F, Irvine DS, Aitken RJ. Development of an image analysis system to monitor the retention of residual cytoplasm by human spermatozoa: correlation with biochemical markers of the cytoplasmic space, oxidative stress, and sperm function. J Androl 1996; 17:276-287.

Gopalakrishna R, Anderson WB. Ca^{2+}-and phospholipid-independent activation of protein kinase C by selective oxidation of the regulatory domain. Proc Natl Acad Sci USA 1989; 86:6758-6762.

Goyette C, de Lamirande E, Gagnon C. Involvement of reactive oxygen species in heparin-induced capacitation in bull spermatozoa. Can Fert Androl Soc Ann Meet, 1998.

Griveau JF, Le Lannou D. Influence of oxygen tension on reactive oxygen species production and human sperm function. Int J Androl 1997a; 20:195-200.

Griveau JF, Le Lannou D. Reactive oxygen species and human spermatozoa: physiology and pathology. Int J Androl 1997b; 20:61-69.

Griveau JF, Dumont E, Renard P, Le Lannou D. Reactive oxygen species, lipid peroxidation and enzymatic defense systems in human spermatozoa. J Reprod Fertil 1995a; 103:17-26.

Griveau JF, Renard P, Le Lannou D. An *in vitro* promoting role for hydrogen peroxide in human sperm capacitation. Int J Androl 1995b; 17:300-307.

Griveau JF, Renard P, Le Lannou D. Superoxide anion production by human spermatozoa as a part of the ionophore-induced acrosome reaction process. Int J Androl 1995c; 18:67-74.

Halliwell B, Gutteridge JMC. Free Radicals in Biology and Medicine. 2nd ed. Clarendon Press, Oxford, 1989.

Hecht D, Zick Y. Selective inhibition of protein tyrosine phosphatase activities by H_2O_2 and vanadate *in vitro*. Biochem Biophys Res Comm 1992; 188:773-779.

Hellstrom WJG, Bell M, Wang R, Sikka SC. Effect of sodium nitroprusside on sperm motility, viability, and lipid peroxidation. Fertil Steril 1994; 61:1117-1122.

Herrero MB, Goin JC, Canteros MG, Franchi AM, Martinez SP, Polak JM, Viggiano JM, Gimeno MAF. The nitric oxide synthase of mouse spermatozoa. FEBS Lett 1997a; 411:39-42.

Herrero MB, Viggiano JM, Martinez SP, Gimeno MAF. Evidence that nitric oxide synthase is involved in progesterone-induced acrosomal exocytosis in mouse spermatozoa. Reprod Fertil Dev 1997b; 9:433-439.

Holmes RP, Goodman HO, Shihabi ZK, Jarow JP. The taurine and hypotaurine content of human semen. J Androl 1992; 13:289-292.

Hsu PC, Liu MY, Hau CC, Chen Ly, Leon Guo Y. Lead exposure causes generation of reactive oxygen species and functional impairment in rat sperm. Toxicology 1997; 122:133-143.

Irvine DS. Glutathione as a treatment for male infertility. Rev Reprod 1996; 1:6-12.

Iwasaki A, Gagnon C. Formation of reactive oxygen species in spermatozoa of infertile men. Fertil Steril 1992; 57:409-416.

Jeulin C, Soufir JC, Weber P, Laval-Martin D, Calvayrac R. Catalase activity in human spermatozoa and seminal plasma. Gam Res 1989; 24:185-196.

Jones HP, Lenz RW, Palevitz BA, Cormier MJ. Calmodulin localization in mammalian spermatozoa. Proc Natl Acad Sci USA 1980; 77:2772-2776.

Jones R, Mann T, Sherins R. Peroxidative breakdown of phospholipids in human spermatozoa, spermicidal properties of fatty acid peroxides, and protective action of seminal plasma. Fertil Steril 1979; 31:531-537.

Józwik M, Kuczynski W, Szamatowicz M. Non-enzymatic antioxidant activity of human seminal plasma. Fertil Steril 1997; 68:154-157.

Kelso KA, Cerolini S, Noble RC, Sparks NHC, Speake BK. Lipid and antioxidant changes in semen of broiler fowl from 25 to 60 weeks of age. J Reprod Fertil 1996; 106:201-206.

Kelso KA, Redpath A, Noble RC, Speake BK. Lipid and antioxidant changes in spermatozoa and seminal plasma throughout the reproductive period of bulls. J Reprod Fertil 1997; 109:1-6.

Kessopoulou E, Powers HJ, Sharma KK, Pearson MJ, Russell JM, Cooke ID, Barratt CLR. A double-blind randomized placebo cross-over controlled trial using the antioxidant vitamin E to treat reactive oxygen species associated male infertility. Fertil Steril 1995; 64:825-831.

Kodama H, Fukuda J, Karube H Shimizu Y, Ikeda M, Tanaka T. Investigation of cytotoxic effects of reactive oxygen species on mouse spermatozoa. Japan J Fertil Steril 1995; 40:66-72.

Kodama H, Kuribayashi Y, Gagnon C. Effect of sperm lipid peroxidation on fertilization. J Androl 1996; 17:151-157.

Kodama H, Yamaguchi R, Fukada J, Kasai H, Tanaka T. Increased oxidative deoxyribonucleic acid damage in the spermatozoa of infertile male patients. Fertil Steril 1997; 68:519-524.

Kovalski NN, de Lamirande E, Gagnon C. Reactive oxygen species generated by human neutrophils inhibit sperm motility: protective effect of seminal plasma and scavengers. Fertil Steril 1992; 58:809-816.

Kumar PG, Laloraya M, Laloraya MM. Superoxide radical level and superoxide dismutase activity changes in maturing mammalian spermatozoa. Andrologia 1991; 23:171-175.

Kuribayashi I, Gagnon C. Effect of catalase and thioredoxin addition to sperm incubation medium before *in vitro* fertilization on sperm capacity to support embryo development. Fertil Steril 1996; 66:1012-1017.

Leclerc P, Gagnon C. Phosphorylation of Triton X-100 soluble and insoluble protein substrates in a demembranated/reactivated human sperm model. Mol Reprod Dev 1996; 44:200-211.

Leclerc P, de Lamirande E, Gagnon C. Cyclic adenosine 3'5'monophosphate-dependent regulation of tyrosine phosphorylation in relation to sperm capacitation and motility. Biol Reprod 1996; 55:684-695.

Leclerc P, de Lamirande E, Gagnon C. Regulation of protein tyrosine phosphorylation and human sperm capacitation by reactive oxygen species. Free Radic Biol Med 1997; 22:643-656.

Leclerc P, de Lamirande E, Gagnon C. Interaction between Ca^{2+}, cyclic 3'5'adenosine monophosphate, the superoxide anion, and tyrosine phosphorylation pathways in the regulation of human sperm capacitation. J Androl 1998; 19:434-443.

Lewis SEM, Donnelly ET, Sterling ESL. Nitric oxide synthase and nitrite production by human sperm: evidence that endogenous nitric oxide is beneficial to sperm motility. Mol Hum Reprod 1996; 2:873-878.

Maas DHA, Storey BT, Mastroianni L. Hydrogen ion and carbon dioxide content of the oviductal fluid of the Rhesus monkey (Macaca mulatta). Fertil Steril 1977; 28:281-285.

Mammoto A, Masumoto N, Tahara M, Ikebuchi Y, Ohmichi M, Tasaka K, Miyake A. Reactive oxygen species block sperm-egg fusion via oxidation of sperm sulfhydryl proteins in mice. Biol Reprod 1996; 55:1063-1069.

Nissen HP, Kreysel HW. Superoxide dismutase in human semen. Klin Wochenschr 1983; 61:63-65.

Nonogaki T, Yoda Y, Narimoto K, Shiotani M, Mori T, Matsuda T, Yoshida O. Localization of Cu,Zn-superoxide dismutase in the human male genital tract. Hum Reprod 1992; 7:81-85.

Oeda T, Henkel R, Ohmori H, Schill WB. Scavenging effect of N-acetyl-L-cysteine against reactive oxygen species in human semen: a possible therapeutic modality for male factor infertility? Andrologia 1997; 29:125-131.

O'Flaherty C, Beconi M, Beorlegui N. Effect of natural antioxidants, superoxide dismutase and hydrogen peroxide on capacitation of frozen-thawed bull spermatozoa. Andrologia 1997; 29:269-275.

Okada H, Tatsumi N, Kanzaki M, Fujisawa M, Arakawa S, Kamidono S. Formation of reactive oxygen species by spermatozoa from asthenospermic patients: response to treatment with pentoxifylline. J Urol 1997; 157:2140-2146.

Okamura N, Iwaki Y, Hiramoto S, Tamba M, Bannai S, Sugita Y, Syntin P, Dacheux F, Dacheux J-L. Molecular cloning and characterization of the epididymis-specific glutathione peroxidase-like protein secreted in the porcine epididymal fluids. Biochim Biophys Acta 1997; 1336:99-109.

Oteiza PL, Olin KL, Fraga CG, Keen CL. Oxidant defense systems in testes from zinc-deficient rats. Proc Soc Exp Biol Med 1996; 213:85-91.

Padron OF, Brackett NL, Sharma RK, Lynne CM, Thomas AJ, Agarwal A. Seminal reactive oxygen species and sperm motility and morphology in men with spinal cord injury. Fertil Steril 1997; 67:1115-1120.

Parinaud J, Le Lannou D, Vieitez G, Griveau J-F, Milhet P, Richoilley G. Enhancement of motility by treating spermatozoa with an antioxidant solution (Sperm-Fit™) following ejaculation. Hum Reprod 1997; 12:2434-2436.

Peeker R, Abramson L, Marklund SL. Superoxide dismutase isoenzymes in human seminal plasma and spermatozoa. Mol Hum Reprod 1997; 3:1061-1068.

Perera D, Katz M, Heenbanda SR, Marchant S. Nitric oxide synthase inhibitor N^G-monomethyl-L-arginine preserves sperm motility after swim up. Fertil Steril 1996; 66:830-833.

Plante M, de Lamirande E, Gagnon C. Reactive oxygen species released by activated neutrophils, but not by deficient spermatozoa, are sufficient to affect normal sperm motility. Fertil Steril 1994; 62:387-393.

Rao B, Soufir JC, Martin M, David G. Lipid peroxidation in human spermatozoa as related to midpiece abnormalities and motility. Gam Res 1989; 24:127-134.

Rodriguez H, Obanian C, Bustus-Obregon E. Nuclear chromatin decondensation of spermatozoa *in vitro*: a method for evaluating the fertilizing ability of ovine sperm. Int J Androl 1985; 8:147-158.

Rosselli M. Nitric oxide and reproduction. Mol Hum Reprod 1997; 3:639-641.

Rosselli M, Dubey RK, Imthurn B, Macas E, Keller PJ. Effect of nitric oxide on human spermatozoa: evidence that nitric oxide decreases sperm motility and increases sperm toxicity. Hum Reprod 1995; 10:1786-1790.

Rubin CS. A anchor proteins and the intracellular targeting of signals carried by cyclic AMP. Biochim Biophys Acta 1994; 1224:467-479.

Rufas O, Fish B, Seligman J, Tadir Y, Ovadia J, Shalgi R. Thiol status in human sperm. Mol Reprod Dev 1991; 29:282-288.

Salo DC, Pacifici RE, Lin SW, Giulivi, Davies KJ. A superoxide dismutase undergoes proteolysis and fragmentation following oxidative modification and inactivation. J Biol Chem 1990; 265:11919-11927.

Schaad NC, Zhang XQ, Campana A, Schorderet-Slatkine S. Human seminal plasma inhibits brain nitric oxide synthase activity. Hum Reprod 1996; 11:561-565.

Shen H-M, Chia S-E, Ni Z-Y, New A-L, Lee B-L, Ong C-N. Detection of oxidative DNA damage in human sperm and the association with cigarette smoking. Reprod Endocrinol 1997; 11:675-680.

Sikka SC, Rajasekaran M, Hellstrom WJG. Role of oxidative stress and antioxidants in male infertility. J Androl 1995; 16:464-468.

Sinha S, Pradeep KG, Lolaraya M, Warikoo D. Over-expression of superoxide dismutase and lack of surface-thiols in spermatozoa: inherent defects in oligospermia. Biochem Biophys Res Comm 1991; 173:510-517.

Sinha S, Kumar P, Laloraya M. Methyl xanthine and altered biomembrane dynamics: demonstration of protein mobility and enzyme inhibition by caffeine in sperm model system. Biochem Mol Biol Int 1993; 31:1141-1148.

Storey BT. Biochemistry of the induction and prevention of lipoperoxidative damage in human spermatozoa. Mol Hum Reprod 1997; 3:203-213.

Storey BT, Alvarez JG, Thompson KA. Human sperm glutathione reductase activity *in situ* reveals limitations in the glutathione antioxidant defense system due to supply of NADPH. Mol Reprod Dev 1998; 49:400-407.

Suzuki YJ, Florman HJ, Sevanian A. Oxidants as stimulators of signal transduction. Free Radic Biol Med 1997; 22:269-285.

Thomlinson MJ, Barratt CL, Cooke ID. Prospective study of leukocytes and leukocyte subpopulations in semen suggests they are not a cause of male infertility. Fertil Steril 1993; 60:1069-1075.

Vijayaraghavan S, Goueli SA, Davey MP, Carr DW. Protein kinase A-anchoring inhibitor peptides arrest mammalian sperm motility. J Biol Chem 1997; 272:4747-4752.

Wang A, Fanning L, Anderson DJ, Loughlin KR. Generation of reactive oxygen species by leukocytes and sperm following exposure to urogenital tract infection. Arch Androl 1997; 39:11-17.

Windsor DP, White IG, Selley ML, Swan MA. Effects of the lipid peroxidation product (E)-4-hydroxy-2-nonenal on ram sperm function. J Reprod Fertil 1993; 99:359-366.

Wolf H. The biologic significance of white blood cells in semen. Fertil Steril 1995; 63:1143-1157.

Yanagimachi R. Mammalian fertilization. In: (Knobil E, Neill JD, eds) The Physiology of Reproduction. 2nd ed. New York: Raven Press, 1994, pp189-317.

Yeung CH, De Geyter C, De Geyter M, Nieschlag E. Production of reactive oxygen species and hydrogen peroxide scavenging activity of spermatozoa in an IVF program. J Assist Reprod Genetics 1996; 13:495-500.

Zhang H, Zheng R-L. Promotion of human sperm capacitation by superoxide anion. Free Radic Res 1996; 24:261-268.

Zini A, Schlegel PN. Identification and characterization of antioxidant enzymes mRNAs in the rat epididymis. Int J Androl 1997; 20:86-91.

Zini A, de Lamirande E, Gagnon C. Reactive oxygen species in semen of infertile patients: levels of superoxide dismutase- and catalase-like activities in seminal plasma and spermatozoa. Int J Androl 1993; 16:183-188.

Zini A, de Lamirande E, Gagnon C. Low levels of nitric oxide promote human sperm capacitation *in vitro*. J Androl 1995; 16:424-431.

43 *The Male Germ Cell as a Target for Drug and Toxicant Action*

Bernard Robaire
Barbara F. Hales
McGill University

One of the most active sites for cell division in the body is the seminiferous epithelium, where, in man, approximately one hundred million spermatozoa are synthesized daily. In this process, the germ stem cells (spermatogonia) undergo several mitotic divisions to become spermatocytes, which then undergo two meiotic divisions (Clermont 1972). Once the phase of cell division is completed, germ cells, now designated as spermatids, enter a phase of complex differentiation (spermiogenesis) to become spermatozoa. Several major changes take place during spermiogenesis, including the dramatic condensation of the haploid nucleus as a consequence of the replacement of histones with highly basic protamines; the formation of a modified lysosome (acrosome) that is essential for the penetration of an oocyte; the shedding of most of the cytoplasm; and the formation of a tail, to propel the sperm forward, energized by a modified mitochondrial complex (Clermont 1972). Once released from the seminiferous epithelium, spermatozoa undergo a large array of further modifications during epididymal transit; it is in the epididymis that they acquire the ability to swim and to fertilize oocytes (Robaire, Hermo 1988).

Drugs, environmental agents or other toxicants can affect the reproductive function of the testis in different ways. The actions of xenobiotics can be divided into three categories. The first is to suppress the process of spermatogenesis by interfering with the environment necessary to sustain spermatogenesis; an example of such action is the administration of a steroid hormone, such as testosterone. The second is to interfere with the complex nuclear events taking place during mitosis, meiosis and spermiogenesis; an example of such action is exposure to an alkylating agent that causes an increased rate of DNA strand breaks or crosslinks. The third is to prevent spermatozoa from reaching the egg and proceeding with normal fertilization and pronuclear formation; such action is caused, for example, by an agent that blocks the maturation or transport of spermatozoa through the epididymis. A few examples of each of these types of actions are provided to illustrate the wide range of potential actions of xenobiotics on fertility and progeny outcome.

Suppression of Spermatogenesis

Many drugs and environmental chemicals that suppress spermatogenesis act by blocking the communication between the hypothalamic-pituitary axis and the testis. As the development of male contraceptives advances, two concerns arise. The first is the extent to which spermatogenesis needs to be suppressed in order to arrest fertility; the second is whether there are any untoward effects on progeny outcome when the number of spermatozoa is reduced only partially and pregnancy ensues.

Testosterone, or one of its esters, has been advocated for many years as part of a contraceptive formulation (Robaire et al. 1977; Behre et al. 1995; Handelsman et al. 1996). Using the rat as an animal model and sustained administration of testosterone as a means of suppressing spermatogenesis in a stepwise manner, the two issues outlined above were investigated (Robaire et al. 1984). From Figure 1, it can be seen that in spite of substantial decreases in epididymal sperm reserves after treatment for three months with sustained release testosterone implants, pregnancies were not eliminated until sperm reserves reached 5% of those of control. Among the progeny of males with reduced sperm counts, there was no increased incidence of pre- or post-implantation loss, nor was there an increased rate of malformed fetuses (Robaire et al. 1984). These data led to the conclusions that a reduction in spermatogenesis by up to 95% may not be reflected by a loss of fertility, and that simply suppressing sperm count is not associated with an increased risk of an adverse progeny outcome.

Interference with the Nuclear Events Occurring during Spermatogenesis

When the genome of a male germ cell is damaged by a xenobiotic, one of three courses may be followed.

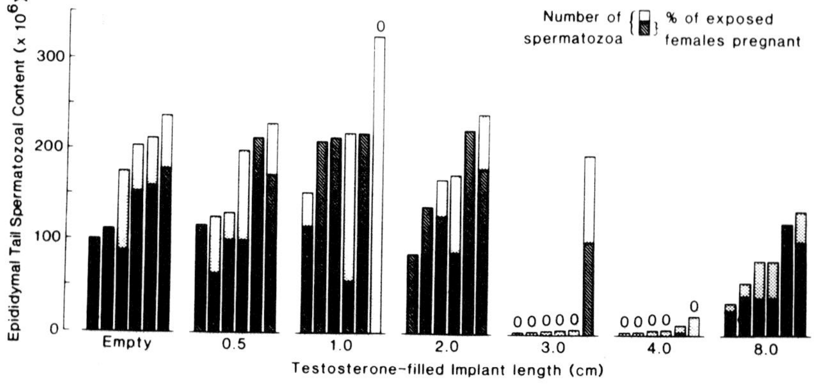

Figure 1. Effect of testosterone-filled polydimethylsiloxane (PDS) capsules implanted subcutaneously for three months in adult male rats (n = 6/group) on the number of spermatozoa in epididymal tail segments. The percent of exposed females (n = 4/male) that became pregnant is indicated by the fraction of each bar that is cross hatched. Zeros indicate that none of the exposed females were pregnant. (Reprinted with permission from Robaire et al. 1984.)

The first is that the damage is relatively minor and the normal DNA repair processes are still functional, allowing for complete repair and thus no further sequela. The second is that the damage is so extensive that the cell cannot repair the damage, and yet it still has the ability to trigger the apoptotic cascade in order to commit suicide. The third is that the damage remains partially or completely unrepaired and is sufficiently minor to allow the spermatogenic process to be completed and fertilization to occur. While the first two scenarios result in little danger to the progeny or future generations, the third scenario presents significant risks.

Exposure to Xenobiotics and DNA Damage

The damage caused by drugs to male germ cell DNA has been assessed using several tests, including alkaline elution, dominant lethality and the specific locus mutation tests, as well as cytogenetics (Hales, Robaire 1997). Alkaline elution has been used to assess overall DNA damage in male germ cells after exposure of rats to cyclophosphamide (Qiu et al. 1995). Under alkaline conditions, DNA unwinds and is eluted through a filter at a rate reflecting the extent of DNA single stand breaks or crosslinks (Kohn 1976). The spermatozoa from male rats treated for one week with cyclophosphamide had DNA single strand breaks which could be detected with proteinase K in the lysis solution; no DNA crosslinks were observed. In contrast, six weeks of treatment with cyclophosphamide induced a significant increase in DNA single strand breaks and crosslinks in the nuclei of spermatozoa; the crosslinks were due primarily to DNA-DNA linkages (Qiu et al. 1995).

There are "hot spots" or loci in the genome which are more susceptible than others to mutations. This specificity has also been observed for the visible specific locus mutations test (Favor 1994). The results of specific locus mutation studies have suggested that exposure of spermatogonia to chemicals or radiation yields few large lesions, while such lesions are common after exposure of post-spermatogonial germ cells. The importance of "specific" genes as targets in mediating the adverse effects of chemicals on the male germ cells is not known.

At the chromosome level, it has proven difficult to detect even bulky deletions, aneuploidy or chromosomal duplications in spermatozoa using cytogenetic approaches (Allen et al. 1994). Mature spermatozoa do not undergo mitosis. Hence, chromosomal structures in the male pronucleus could be analyzed only by allowing denuded hamster eggs to be fertilized by spermatozoa. This approach has been used to identify the effects of age, X-irradiation and drugs on the chromosomal banding pattern of human sperm (Martin, Rademaker 1987). More recently, fluorescent *in situ* hybridization (FISH) has been developed for the analysis of aneuploidy in the male genome (Wyrobek et al. 1994; Martin et al. 1995). Cytogenetic studies of the blastocysts sired by cyclophosphamide-treated males have shown that they have a diploid complement of chromosomes (Austin et al. 1994; Jenkinson, Anderson 1990). Though chromosomes are not identifiable in spermatids or spermatozoa, the position of specific genes on DNA loops attached to the nuclear matrix is constant (Ward, Coffey 1990). Hence, it is reasonable to propose that specific gene loci may be targeted by "non-specific" drugs, even in the absence of transcription. Thus, the paternally mediated effects of cyclophosphamide and other drugs on progeny outcome may be the consequence of gene specific DNA damage.

Repair of DNA Damage

It is possible to take advantage of the tight control of the timing of spermatogenesis to demonstrate that exposure of males to X-rays or to a DNA alkylating agent, such as cyclophosphamide, does not result in uniform damage of all germ cells (Trasler et al. 1985; Russell, Russell 1991; Qiu et al. 1992; Russell 1994; Hasegawa et al. 1997). Indeed, as assessed by the effects of chronic paternal drug exposure on progeny outcome, spermatids have been found to be most dramatically affected, in comparison to spermatogonia and spermatocytes (Trasler et al. 1986; Hales, Robaire 1990); this applies to both the onset and the reversal of drug action. One likely hypothesis to explain this differential germ cell specific effect is that the damage by the agent is similar for all cell types, but early germ cells have the ability to repair the damage or to initiate programmed cell death if the damage is too extensive, whereas spermatids have lost the ability to perform such repairs or to trigger an apoptotic response.

Interestingly, unscheduled DNA synthesis, an assay of DNA repair in post-mitotic male germ cells, has not been found in later stage spermatids and spermatozoa (Sotomayor et al. 1978), both of which are susceptible to damage leading to developmental toxicity. The knockout of some of the DNA repair systems (mismatch repair genes such as *Mlh1* and *PMS2)* results in male infertility (Edelmann et al. 1996; Baker et al. 1995, 1996). There is also evidence suggesting that the presence and/or expression of DNA repair mechanisms are reduced as germ cells proceed through spermiogenesis. For example, the mismatch repair gene *Mlh1* is present in mouse spermatocytes but not spermatids (Baker et al. 1996). Similarly, as spermatogenesis proceeds, there is an increase in the inactive storage form of chromatin associated poly(ADP-ribosyl)transferase that parallels nuclear condensation (Mosgoeller et al. 1996)

Apoptosis

Several studies have established that there is a low spontaneous incidence of apoptosis in the seminiferous tubules of testes from control animals (Cai et al. 1997; Nakagawa 1997; Blanco-Rodriguez, Martinez-Garcia 1998; Strandgaard, Miller 1998). In rats exposed to the anti-tumor agent cyclophosphamide, the incidence of apoptosis increased to a level 3.5-fold above control by 12 hours after treatment (Cai et al. 1997). Drug-induced apoptosis was most pronounced in pre-meiotic germ cells (spermatogonia and spermatocytes) in stages I-IV and XI-XIV of the seminiferous epithelium (Cai et al. 1997; Fig. 2). Two other chemotherapeutic agents, etoposide and doxorubicin/adriamycin, are potent inducers of apoptosis but have somewhat different actions (Sjoblom et al. 1998). The former induces apoptosis in all types of spermatogonia and spermatocytes undergoing meiosis, while the latter causes an increase in apoptosis in a stage-specific manner with type A3-4 spermatogonia, preleptotene, zygotene and early pachytene spermatocytes being the most sensitive. Therefore, apoptosis of damaged mitotic and early meiotic germ cells may serve a critical role in protecting subsequent generations from the diverse untoward effects of toxicants. However, as spermatogenesis proceeds, germ cells lose first the ability to transcribe mRNAs and then to translate proteins, and thus lose the capacity to mount an apoptotic response. If spermatids exposed to DNA damaging agents can neither apoptose nor repair the damage, but can still differentiate into mature spermatozoa and fertilize oocytes, they will carry damaged DNA into the zygote.

Damaged Germ Cells May Have Impaired Function

During chromatin transition processes, the male genome may be in an open dynamic state with many exposed sites which are vulnerable to alkylating agents. Since there is no DNA repair during late spermiogenesis, damage to the genome by alkylation during this phase of spermatogenesis may be cumulative, resulting in the production of dysfunctional germ cells. After entering the egg and decondensing, the spermatozoon must undergo chromatin remodeling to exchange protamines with histones (Perreault 1992). It has been suggested that this chromatin remodeling may explain why the male pronucleus is turned on earlier than the female pronucleus (Worrad et al. 1994). Transcriptional activity was detected in the one-cell murine embryo by G2 when the male, but not when the female, pronucleus was injected with a Sp1 dependent luciferase reporter gene (Ram, Schultz 1993). Coincidentally, there was a higher concentration of transcription factors (Sp1 and TATA-binding protein TBP) in the male pronucleus than in the female (Worrad et al. 1994).

The DNA template function of spermatozoa from cyclophosphamide-treated male rats was determined by using an *in vitro* DNA synthesis system. The availability of spermatozoal DNA for template function was not affected after one week of treatment with cyclophosphamide, but was markedly affected after six weeks of treatment with this drug (Qiu et al. 1995). At the zero time point there was a significant amount of thymidine incorporation in the chromatin from only drug-treated rat spermatozoa; this incorporation may indicate that the drug caused sufficient breaks or nicks in these nuclei to allow DNA polymerase to access the chromatin and to initiate thymidine incorporation. By 90 minutes, the labeled thymidine incorporation into sperm from cyclophosphamide treated males was nearly half that of control.

Consequently, it is hypothesized that chronic exposure to an alkylating agent such as cyclophosphamide results in DNA damage in male germ cells at all phases

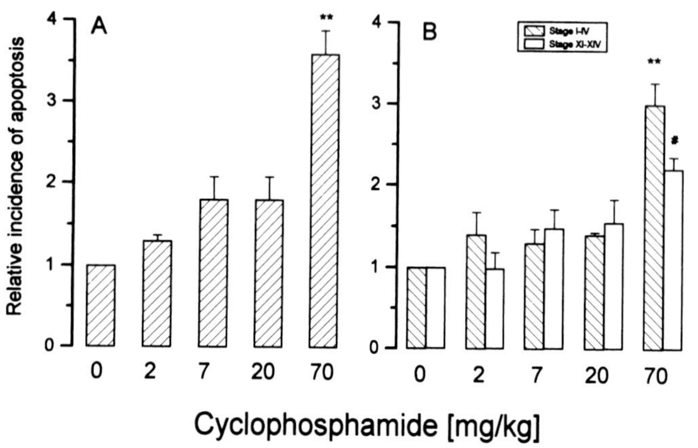

Figure 2. Dose-response of cyclophosphamide-induced apoptosis in male germ cells. A: The apoptosis in male germ cells induced by different doses of cyclophosphamide. Induction of apoptosis is quantified by recording the total number of apoptotic cells in cross sections of 300 seminiferous tubules from three separate sections per male (n=3). The relative incidence is the ratio of apoptotic cells/300 tubules from each treated male to the 0 group (control, 45 ± 8.5 apoptotic cells/300 tubules). The results are presented as mean ±SEM. ** is P<0.01 compared with control. B: Comparison of the relative incidence of apoptotic cells at stages I-IV and XI-XIV after exposure to different doses of cyclophosphamide. The induction of apoptosis is quantified by recording the total number of apoptotic cells in cross sections of stages I-IV and XI-XIV, respectively. The results are presented as mean ±SEM (n=3). $^{\#}$,** are P<0.05 and P<0.01, respectively, compared with corresponding control values (ANOVA and Tukey HSD multiple comparison); # is P<0.05 compared with corresponding values, obtained for stages XI-XIV (t-test). (Reprinted with permission from Cai et al. 1997.)

of spermatogenesis. While repair of this damaged DNA occurs principally in those germ cells that are undergoing mitosis and meiosis, there is evidence for repair in post-meiotic germ cells, such as round spermatids. Apoptosis occurs principally in germ cells capable of transcription and translation; that is, spermatogonia and spermatocytes. Thus, the maximal adverse effects of such drugs on progeny outcome may be mediated via their effects on post-meiotic germ cells (elongating spermatids and spermatozoa), because these cells have lost the ability to repair DNA damage or to undergo apoptosis (Fig. 3).

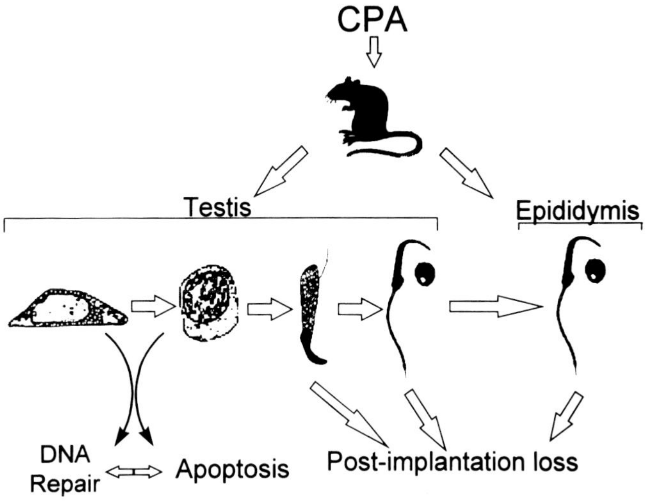

Figure 3. Schematic representation of germ cell phase specificity of the adverse effects of cyclophosphamide treatment of male rats on progeny outcome. (Reprinted with permission from Robaire, Hales 1998.)

Preventing Spermatozoa from Reaching the Egg

Beyond the effects of xenobiotics on sperm chromatin in the testis, there are several means by which such agents can interfere with normal sperm function. It is possible, in theory, to damage chromatin after spermatozoa leave the testis, to block the acquisition of sperm motility or of fertilizing ability in the epididymis, to induce an inflammatory reaction in the epididymal duct or in the vas deferens so as to block the passage of spermatozoa, to prevent spermatozoa from binding to the zona pellucida or recognizing the oocyte, or to prevent the formation of a pronucleus once fertilization has occurred.

An example of a chemical that produces some of these effects is α-chlorohydrin, a monochloro derivative of glycerol (Jones 1983). This agent can reversibly inhibit fertility in an array of species. At low doses, in the cauda epididymidis, it inhibits spermatozoal glyceraldehyde 3-phosphate dehydrogenase, whereas at higher doses, it causes the formation of spermatoceles in rodent caput epididymidis; this results in the blockage of sperm transport through the tissue and consequent atrophy of the seminiferous tubules due to increases in back pressure.

Spermatozoa transiting through the epididymis are still susceptible to damage by xenobiotics. Two such examples are the effects of methyl chloride (Chellman et al. 1986) and cyclophosphamide (Qiu et al. 1992). Methyl chloride (MeCl) induces epididymal inflammation; dominant lethal mutations result from exposure of rodent spermatozoa while they reside in the epididymis. Treatment of rats with an anti-inflammatory agent can reverse both the epididymal inflammation and the dominant lethal mutations induced by this chemical. Short-term treatment with cyclophosphamide in efferent duct ligated rats can increase post-implantation loss several fold, indicating that spermatozoa remain sensitive to xenobiotic attack after spermiation.

Summary

The multiple ways by which xenobiotics can affect male germ cells provide some insight into the molecular and cellular mechanisms underlying the growing literature on the effects of paternal exposure to drugs, environmental chemicals or work conditions that result in adverse effects on progeny outcome (Robaire, Hales 1993).

Acknowledgements

Studies presented above were supported by the Medical Research Council of Canada. The authors express thanks to Robert Vinson for his assistance with the preparation of Figure 3.

References

Allen JW, Collins BW, Cannon RE, McGregor PW, Afshari A, Fuscoe JC. Aneuploidy tests: cytogenetic analysis of mammalian male germ cells. In: (Mattison DR, Olshan AF, eds) Male-Mediated Developmental Toxicity. New York: Plenum Press, 1994, pp59-69.

Austin (Kelly) SM, Robaire B, Hales BF. Paternal cyclophosphamide exposure causes decreased cell proliferation in cleavage-stage embryos. Biol Reprod 1994; 50:55-64.

Baker SM, Bronner CE, Zhang L, Plug AW, Robatzek M, Warren G, Elliot EL, Yu J, Ashley T, Arnhein N, Flavell RA, Liskay RM. Male mice defective in the DNA mismatch repair gene PMS2 exhibit abnormal chromosome synapsis in meiosis. Cell 1995;

82:303-319.
Baker SM, Plug AW, Prolla TA. Involvement of the mouse *Mlh1* in DNA mismatch repair and meiotic crossing over. Nature Genet 1996; 13:261-262.
Behre HM, Baus S, Kliesch S, Keck C, Simoni M, Nieschlag E. Potential of testosterone buciclate for male contraception: endocrine differences between responders and nonresponders. J Clin Endocrinol Metab 1995; 80:2394-2403.
Blanco-Rodriguez J, Martinez-Garcia C. Apoptosis precedes detachment of germ cells from the seminiferous epithelium after hormone suppression by short-term oestradiol treatment of rats. Int J Androl 1998; 21:109-115.
Cai L, Hales BF, Robaire B. Induction of apoptosis in the germ cells of adult male rats after exposure to cyclophosphamide. Biol Reprod 1997; 56:1490-1497.
Chellman GJ, Bus JS, Working PK. Role of epididymal inflammation in the induction of dominant lethal mutations in Fischer 344 rat sperm by methyl chloride. Proc Natl Acad Sci USA 1986; 83:8087-8091.
Clermont Y. Kinetics of spermatogenesis in mammals: seminiferous epithelium cycle and spermatogonial renewal. Physiol Rev 1972; 52:198-236.
Edelmann W, Cohen PE, Kane M, Lau K, Morrow B, Bennett S, Umar A, Kunkel T, Cattoretti G, Changanti R, Pollard JW, Kolodner RD, Kucherlapati R. Meiotic pachytene arrest in MLH1-deficient mice. Cell 1996; 85:1125-1134.
Favor J. Specific-locus mutation tests in germ cells of the mouse: an assessment of the screening procedures and the mutational events detected. In: (Mattison DR, Olshan AF, eds) Male-Mediated Developmental Toxicity. New York: Plenum Press, 1994, pp23-36.
Hales BF, Robaire B. Reversibility of the effects of chronic paternal exposure to cyclophosphamide on pregnancy outcome in rats. Mutat Res 1990; 229:129-134.
Hales BF, Robaire B. Paternally-mediated effects on development. In: (Hood RD, ed) Handbook of Developmental Toxicology. Boca Raton, Fla: CRC Press, 1997, pp91-107.
Hasegawa M, Wilson G, Russell LD, Meistrich ML. Radiation-induced cell death in the mouse testis: relationship to apoptosis. Radiat Res 1997; 147:457-67.
Handlesman DJ, Conway AJ, Howe CJ, Turner L, Mackey MA. Establishing the minimum effective dose and additive effects of depot progestin in suppression of human spermatogenesis by a testosterone depot. J Clin Endocrinol Metab 1996; 81:4113-4121.
Jenkinson PC, Anderson D. Malformed foetuses and karyotype abnormalities in the offspring of cyclophosphamide and allyl alcohol-treated male rats. Mutat Res 1990; 229:173-184.
Jones AR. Antifertility actions of alpha-chlorohydrin in the male. Aus J Biol Sci 1983; 36:333-350.
Kohn KW, Erikson LC, Ewig RAG, Friedman CA. Fractionation of DNA from mammalian cells by alkaline elution. Biochemistry 1976; 15:4629-4637.
Martin RH, Rademaker A. The effect of age on the frequency of sperm chromosomal abnormalities in normal men. Am J Hum Genet 1987; 41:484-492.
Martin RH, Rademaker AW, Leonard NJ. Analysis of chromosomal abnormalities in human sperm after chemotherapy by karyotyping and fluorescence *in situ* hybridization (FISH). Cancer Genet Cytogenet 1995; 80:29-32.
Mosgoeller W, Steiner M, Hosäk P, Penner E, Wesierska-Gadek J. Nuclear architecture and ultrastructural distribution of poly(ADP-ribosyl)transferase, a multifunctional enzyme. J Cell Science 1996; 109:409-418.
Nakagawa S, Nakamura N, Fujioka M, Mori C. Spermatogenic cell apoptosis induced by mitomycin C in the mouse testis. Toxicol Appl Pharmacol 1997; 147:204-213.
Perreault SD. Chromatin remodeling in mammalian zygotes. Mutat Res 1992; 296:43-55.
Qiu J, Hales BF, Robaire B. Adverse effects of cyclophosphamide on progeny outcome can be mediated through post-testicular mechanisms in the rat. Biol Reprod 1992; 46:926-931.
Qiu J, Hales BF, Robaire B. Damage to rat spermatozoal DNA after chronic cyclophosphamide exposure. Biol Reprod 1995; 53:1465-1473.
Ram PT, Schultz RM. Reporter gene expression in G2 of the 1-cell embryo. Dev Biol 1993; 156:552-556.
Robaire B, Hales BF. Paternal exposure to chemicals before conception. Br Med J 1993; 307:341-342.
Robaire B, Hales BF. Target of chemotherapeutic drug action on testis and epididymis. In: (Zirkin BR, ed) Germ Cell Development, Division, Disruption and Death. New York: Springer-Verlag, 1998, pp190-201.
Robaire B, Hermo L. Efferent ducts, epididymis and vas deferens: structure, functions and their regulation. In: (Knobil E, Neill J, eds) Physiology of Reproduction. New York: Raven Press, 1988, pp999-1080.
Robaire B, Ewing LL, Irby DC, Desjardins C. Interactions of testosterone and estradiol-17β on the reproductive tract of the male rat. Biol Reprod 1979; 21:455-463.
Robaire B, Smith S, Hales BF. Suppression of spermatogenesis by testosterone in adult male rats: effect on fertility, progeny outcome and pregnancy. Biol Reprod 1984; 31:221-230.
Russell LD. Effects of spermatogenic cell type on quantity and quality of mutations. In: (Mattison DR, Olshan AF, eds) Male-Mediated Developmental Toxicity. New York: Plenum Press, 1994, pp37-48.
Russell LD, Russell JA. Short-term morphological response of the rat testis to administration of five chemotherapeutic agents. Am J Anat 1991; 192:142-168.
Sjoblom T, West A, Lahdetie J. Apoptotic response of spermatogenic cells to the germ cell mutagens etoposide, adriamycin, and diepoxybutane. Environ Molec Muta 1998; 31:133-148.
Sotomayor RE, Sega GA, Cumming RB. Unscheduled DNA synthesis in spermatogenic cells of mice treated *in vivo* with the indirect alkylating agents cyclophosphamide and mitomen. Mutat Res 1978; 50:229-240.
Strandgaard C, Miller MG. Germ cell apoptosis in rat testis after administration of 1,3-dinitrobenzene. Reprod Toxicol 1998; 12:97-103.
Trasler JM, Hales BF, Robaire B. Paternal cyclophosphamide treatment of rats causes fetal loss and malformations without affecting male fertility. Nature 1985; 316:144-146.
Trasler JM, Hales BF, Robaire B. Chronic low dose cyclophosphamide treatment of adult male rats: effect on fertility, pregnancy outcome and progeny. Biol Reprod 1986; 34:275-283.
Ward WS, Coffey DS. 1990 Specific organization of genes in relation to the sperm nuclear matrix. Biochem Biophys Res Commun 1990; 173:20-25.
Worrad DM, Ram PT, Schultz RM. Regulation of gene expression in the mouse oocyte and early preimplantation embryo: developmental changes in Sp1 and TATA box-binding protein, TBP. Development 1994; 120:2347-2357.
Wyrobeck AJ, Robbins AW, Mehraein Y, Pinkel D, Weier HU. Detection of sex chromosomal aneuploidies X-X, Y-Y, and X-Y in human sperm using two-chromosome fluorescence *in situ* hybridization. Am J Med Genet 1994; 53:1-7.

44 *Role of Tubulin Epitopes in the Regulation of Flagellar Motility*

P. Huitorel
CNRS
P. & M. Curie University

S. Audebert
D. White
McGill University

J. Cosson
CNRS
P. & M. Curie University

C. Gagnon
McGill University

Cilia and flagella are universal organelles present throughout evolution of eukaryotes (Gibbons 1981; Haimo, Rosenbaum 1981; Warner et al. 1989). They have been subjects of numerous studies that turned out to be important not only for themselves, but also for a whole variety of important microtubule-dependent biological functions, such as intracellular transports and mitosis. It is worth emphasizing that dynein was first discovered in axonemes (Gibbons, Rowe 1965) and that its biochemical characterization allowed the development of the more general concept of molecular motors (Huitorel 1988; Warner et al. 1989; Warner, McIntosh 1989), together with myosin, its actin-dependent counterpart. Even though tubulin was well characterized in mammalian brain extracts, it was first discovered in axonemes, where it was identified as the building block of outer doublets and central pair microtubules (Renaud et al. 1968; Shelanski, Taylor 1968), before it was named tubulin (Mohri 1968). Cilia and flagella are all propelled by a complex organelle, the axoneme, whose assembly and function are now better understood because of major advances due to genetics, structural and molecular techniques which complement studies on reactivated systems *in vitro*. Cilia and flagella are still model systems of general value (Thaler, Haimo 1996) and they are universal at three different levels.

Introduction to the Axonemal Machinery at the Molecular Level

The axoneme contains a highly conserved microtubular structure usually made of nine outer doublet microtubules and two central single microtubules, known as the 9+2 axoneme (Afzelius 1995). A number of species show variations around the main theme, with either no central pair, an unusual number of outer doublets, or extra singlet or doublet microtubules (Jamieson et al. 1995). In any case, microtubules represent the main structural feature of the axoneme, anchored at its base on the distal centriole of the basal body. These microtubules are stabilized and interconnected by other protein structures such as radial spokes, nexin links and outer and inner dynein arms that represent the ATP-dependent motor of the whole structure. Both classes of dynein arms are permanently attached to sites located all along A microtubules, outer arms with a periodicity of 24 nm and inner arms with an irregular spacing between three species of arms in a repeated motif of three arms every 96 nm (Piperno et al. 1990). Radial spokes are also repeated as a pair every 96 nm along the axoneme and represent a link between the A microtubule of outer doublets (close to the permanent attachment site of inner dynein arms) and the central apparatus. They not only crosslink the outer doublets together but participate in a complex regulation of inner dynein arms by the central structures (Smith, Sale 1994). Nexin links are repeated every 96 nm and connect outer doublets together (Warner 1983). This whole complex structure is assembled primarily by distal addition of precursors organized by extension of the distal basal body (Dutcher 1995; Johnson, Rosenbaum 1992).

The axoneme is also conserved at the molecular level (Warner et al. 1989). Even though some variations do occur in some species, the axonemal structure is always made up of more than 200 different types of polypeptide chains (Luck 1984). Only 20 to 30% of them are presently characterized; among those are tubulins, radial spoke proteins, nexins, central sheath proteins, the ATP-dependent motors dyneins and kinesins, and a series of regulatory proteins such as calmodulin, protein kinases and phosphatases (Warner et al. 1989). Dyneins, for instance, are composed of 10 to 15 polypeptide chains (Witman et al. 1994): two or three heavy chains of about 500kD each, two to five intermediate chains (ICs) between 150 and 50kD, and two to 10 low chains (LCs) lower than 30kD, forming a huge complex of 1,500 to 2,000 kD, even though molecular variations, such as the number of chains and precise sequences and so on, obviously exist between species. However, at least some of the LCs seem to be highly conserved and shared by both axonemal and cytoplasmic dyneins (King et al. 1996; Wang, Satir 1998). Each dynein heavy chain forms a globular head of about 12 nm in diameter, containing the ATPase catalytic site, a long stem with a coiled-coil structure associating with the equivalent coiled-coil domain of one or two other dynein heavy chains, and a base associated with the ICs and permanently attached to A microtubules. Part of the head is involved in the transient ATP-dependent interaction with the facing adjacent B microtubule lattice. Whereas all outer dynein arms present the same molecular arrangement, inner arms are composed of a combination of different polypeptide chains, belonging to a multigenic family (Asai 1995), along the axoneme (Piperno et al. 1990; Piperno, Ramanis 1991). In addition, they are associated to another macromolecular complex, the dynein regulatory complex (DRC), as seen in Gardner et al. (1994). Similarly, radial spokes are made up of 17 different polypeptides (Curry, Rosenbaum 1993; Piperno et al. 1981) and the central apparatus of at least 23 polypeptides (Dutcher et al. 1984; Smith, Lefebvre 1997). It is worth noting that some dynein regulatory chains (light chains) or the DRC have also been found to be essential components or regulators of non-axonemal dyneins, the so-called cytoplasmic dyneins, which are essential for retrograde axonal transport in neurons and mitosis in all proliferating cell types, leading to the concept that flagellar and cytoplasmic dyneins share similarities in structure, composition, function and regulation (Porter, Johnson 1989).

The basic mechanism of motility residing in axonemes is very ancient and highly conserved throughout evolution (Sleigh 1974). Dynein arms, permanently attached at sites located all along the outer doublet microtubules, bind and hydrolyze ATP according to a mechano-chemical cycle coupling ATP hydrolysis to binding of dynein heads to the neighbor doublet B-microtubule. A conformational change of the dynein arm induces the sliding of doublets relative to each other, followed by the release of the interaction (Brokaw 1989; Cosson 1996; Johnson 1985; Porter, Johnson 1989; Satir et al. 1981). The active sliding due to dynein is unidirectional so that dynein arms "walk" along the neighbor B microtubule towards its minus end or proximal end of the axoneme (Summers, Gibbons 1971; Sale, Satir 1977), a general property of all dynein motors. This whole cycle of force production must be finely regulated so that outer and inner dynein arms play their own specific roles (Brokaw, Kamiya 1987) while cooperating in harmony. In addition, dynein arms must cooperate all along the axonemal length and around the nine outer doublets with sufficient metachrony so that waves are propagated instead of local force produced without propagation. The radial spokes and central apparatus may serve as a component of such a global regulator (Brokaw 1989; Smith, Sale 1994) of not only inner dynein arms but also groups of outer doublets (Lindemann, Kanous 1997). In some studies, the rotation of the central pair was demonstrated, suggesting that it could act as a rotating regulator, activating sequentially the inner dynein arms (Omoto, Kung 1979). The recent discovery of kinesin, a class of motor moving along microtubules towards their plus end and bound to one microtubule (C2) of the central apparatus, has brought new evidence that supports the concept of the rotation of the central pair of microtubules (Bernstein et al. 1994; Fox et al. 1994). All these processes are obviously regulated by factors such as calcium, phosphorylations and proteolysis (Gagnon 1995). However, the mechanisms by which the axoneme regulates its function remain a mystery, and the study of different systems will hopefully allow researchers to distinguish general mechanisms from characteristics of particular systems.

Biochemical and genetic analyses (Dutcher 1995; Luck 1984; Huang 1986) have established the relationship between the presence or absence of important axonemal structures, such as outer and inner dynein arms (Kamiya et al. 1989; Witman 1992; Witman et al. 1994), radial spokes (Luck et al. 1977; Piperno et al. 1981) or the central apparatus (Dutcher et al. 1984), and the polypeptides they contain. This approach has established the multimeric nature of all these substructures. Their absence in mutants induces either a change in motility parameters, such as beat frequency, beating amplitude or curvature, or a total arrest of motility. For instance, the genetic deletion of outer dynein arms in *Chlamydomonas* or their biochemical extraction from sea urchin sperm flagella reduces the beat frequency (Gibbons, Gibbons 1973), whereas the absence of inner arms strongly reduces the beating amplitude and has little effect on the beat frequency (Brokaw, Kamiya 1987). The absence of the radial spokes and central apparatus totally prevents motility (Huang 1986), even though microtubules have the capacity of sliding *in vitro* (Smith, Sale 1992), which shows that these structures regulate the force production rather than contribute to force production itself. Some species, such as eel spermatozoa, naturally do not have a central pair of microtubules in functional "9+0" flagella (Wooley 1997). They tend to show a three-dimensional type of flagellar beating, suggesting that the central apparatus may also be involved in the control of planarity of the flagellar beating (Lindemann, Kanous 1997). Absence of dynein arms on axonemes is associated in all model systems studied (sea urchin spermatozoa, *Chlamydomonas* and so on) with a complete absence of flagellar or ciliary motility. Similarly, in humans, the absence of dyneins is associated with bronchiectasis, situs inversus and infertility. This genetic defect is called Kartagener's syndrome and is reviewed by Afzelius (1995).

Elegant *in vitro* experiments on reactivated models have established that outer dynein arms can be extracted, which results in a reduced axonemal beat frequency. Their reassociation to extracted axonemes restores a normal beat frequency, due to the binding of the extracted dynein arms to their original binding sites on outer doublets (Gibbons, Gibbons 1973). This not only demonstrates that the motor responsible for axonemal motility resides in dynein molecules located all along the length of the axoneme, but also that the axonemal machinery is a highly ordered structure where each component certainly has a very well defined binding site and role to play. Therefore, while the axoneme is being assembled during cell differentiation, the proper binding of the various components is a prerequisite to normal motility, which in turn suggests that the microtubular frame does not allow too many spatial possibilities to organize motors and crosslinking elements within the axoneme. Another example of such a fine and intriguing sorting of axonemal elements is the case, mentioned above, of a kinesin located along one single microtubule of the central pair, or the case of the various kinds of molecular arrangements found in inner dynein arms along the length of axonemes. The mechanisms by which these binding sites are defined on axonemal microtubules are of major interest but remain obscure.

Variations around a universal theme must mean a co-evolution of sequences. For instance, the slow evolution of tubulin and dynein must have produced gene products always able to build up microtubule doublets and singlets, and dynein arms must have absolutely conserved their capacity for permanent attachment onto the tubulin molecules of the A tubule on one side, the transient ATP-dependent interaction with tubulin of the neighboring B tubule on the other side, and the proper distance and flexibility/rigidity in between. For similar reasons, dyneins, radial spokes and the central apparatus must have evolved together. Genetic studies have established that some paralyzed mutants deficient in radial spokes can be partially rescued by suppressor mutations in dynein arms (Huang et al. 1982), as reviewed by Huang (1986) and discussed by Smith and Sale (1994). These results established that the radial spokes-central apparatus positively regulates dynein arm activity, but also suggest that molecules involved in such macromolecular complexes must have evolved simultaneously so that a mutation in one component is corrected by mutations in other components involved in the given interaction in order to preserve axonemal motility and its regulatory mechanisms.

So far, little is known about tubulin-dynein interactions at the molecular level (Gagnon 1995), mostly because of the huge mass of dynein heavy chains that did not allow until recently their cloning and sequencing (Gibbons et al. 1991; Ogawa 1991). Their size is still a major obstacle to their crystallization. Therefore, researchers are far from understanding the dynein-microtubule interactions at the molecular level, whereas other ATP-dependent motors such as myosin, kinesin and ncd (a kinesin-like motor) have been sequenced and their head domains have been crystallized and visualized by cryoelectron microscopy, as reviewed by Downing and Nogales (1998). Cryoelectron microscopy has provided new light in the understanding of kinesin and ncd functions on the surface of microtubules. In particular, the hinge between the head and tail domains plays an important part in the conformational change coupled to ATP hydrolysis and in the polarity of translocation of both dimeric motors along microtubules (towards the plus end for kinesin and towards the minus end for ncd). For both motors, one head is bound and one head is unbound per tubulin dimer and the unbound subunit points towards the plus end in the case of kinesin, and the minus end in the case of ncd (Hirose et al. 1996). Negative staining data suggest that both motors bind to the crest of the protofilaments and establish contacts with both alpha- and beta-tubulin (Hoenger, Milligan 1997). In another approach, limited proteolysis of the head domain of kinesin and ncd by a variety of proteases, either in the presence or in the absence of microtubules, has allowed researchers to produce a map of the cleavage sites protected by the interaction motor-microtubule and locate them on three-dimensional (3D) models (Alonso et al. 1998). All the information acquired in the case of kinesin may be helpful to understand basic rules that may apply to several kinds of motor-microtubule interactions. Before this level of 3D resolution is reached in the case of dynein, indirect evidence may help to elaborate working hypotheses that, unfortunately, must await the availability of crystal structures for confirmation. High resolution electron micrographs reveal the head and stem domains of dynein arms in the "bouquet" configuration both *in situ* and after their isolation from axonemes, and an intermediate chain (IC69) has been localized at the base of the stem domain by direct immuno-localization, as seen in a review by Witman et al. (1994). The use of limited proteolysis has established that the N-terminal domain of dynein heavy chains is involved in the permanent attachment of dynein to microtubules (Mocz, Gibbons 1993). A third approach involved the use of zero-length bifunctional reagents to crosslink neighboring proteins within the axoneme. For example, one regulatory chain of dynein (IC78) was shown to be located at the base of the stem part (coiled-coil region) of the dynein heavy chains where it interacts with alpha-tubulin (King et al. 1991). However, it is not clear whether dyneins may interact with both alpha- and beta-tubulins, either at the permanent attachment sites or at the transient ATP-dependent sites.

Another approach that has been used to contribute to this understanding is a functional approach that involves the use of a series of antibodies, specific of the appropriate targets, including tubulin epitopes, to inhibit *in vitro* reactivated axonemal motility, just as many groups microinject such antibodies in live cells or introduce them in extracts to identify the functional role of various epitopes *in vivo* or in cell-free systems. Before such data can be described, the main tubulin biochemical characteristics must be introduced, in particular sequences and post-translational modifications (PTMs), which are at the origin of a considerable diversity of tubulin isoforms and epitopes.

Tubulin

Tubulin is present in all eukaryotes, where it constitutes the building block of all classes of microtubules: the interphasic network of microtubules structuring all cytoplasmic transports of organelles, mitotic spindles, the nine outer doublets and two central singlets of axonemes and the basal bodies and centriolar triplets (Alberts et al. 1994; Dustin 1984; Hyams, Lloyd 1994). Tubulin belongs to an expending multigenic family, made of at least three major classes of gene products: alpha-, beta- and gamma-tubulins.

The most recently discovered gamma-tubulin has a partial homology with both alpha- and beta-tubulins, represents only about 1 percent of total tubulin and seems to be involved only in the early steps of microtubule assembly (nucleation) at the centrosome (Oakley

1994). Even though present in a particulate form throughout the cytoplasm, it would be functional only in the vicinity of the two centrioles which form the "core" structure of centrosomes. Gamma-tubulin would be responsible for the assembly of the so-called pericentriolar material essential for microtubule nucleation. It is worth noting that basal bodies, also made of centrioles at the base of cilia and flagella, do not show any gamma-tubulin or pericentriolar material, which explains why basal bodies are not competent by themselves to nucleate microtubule assembly: they only assemble axonemal microtubules as an extension of the distal centriole for the outer doublets and by a still unknown mechanism for the central pair (Navara et al. 1995). After fertilization, or after *in vitro* incubation in the appropriate cytoplasmic egg extract, basal bodies recruit gamma-tubulin and other pericentriolar proteins, such as pericentrin, and become competent to nucleate the assembly of microtubules from soluble tubulins, forming the asters that later give rise to the mitotic spindle (Félix 1994; Schatten 1994; Holy, Schatten 1997). More subtle functions of gamma-tubulin may be described in the near future, following the discovery of several genes coding for gamma-tubulins, with additional diversity at the post-translational level, and different functional genes coding for similar but different gamma-tubulins with specific expression patterns (Wilson et al. 1997). A new divergent tubulin (delta) has just been discovered in *Chlamydomonas*; this tubulin is not essential but could participate in the maturation of basal bodies and in the control of flagella number (Dutcher, Trabuco 1998). Whether this is a highly divergent gamma-tubulin or the first member of a new (delta) tubulin family remains to be confirmed.

Alpha- and beta-tubulins are among the most abundant soluble proteins in many tissues. First discovered in flagellar axonemes (Renaud et al. 1968; Shelanski, Taylor 1968), tubulin was extensively studied in the mammalian brain, where it is highly abundant and can be easily purified to homogeneity, as reviewed by Dustin (1984). Alpha- and beta-tubulins are both highly conserved globular proteins of 50 kD (about 451 residues in alpha- and 445 in beta-tubulin) that exist in a functional state only as a heterodimer (100 kD) made of one alpha and one beta polypeptide chains. Both show a GTP binding site, but only the beta subunit hydrolyzes it to GDP during the linear polymerization of tubulin dimers into microtubules (Alberts et al. 1994). Each subunit is a sphere of about 4 nm in diameter. Polymerization proceeds through longitudinal association of dimers building up the so-called protofilaments showing a perfect alternation of alpha and beta subunits, and through the lateral association of 13 (in the vast majority of cases) protofilaments. The formation of such a ribbon is associated with its lateral curvature so that when it contains 13 protofilaments (except in some organisms which may have between 10 and 15 protofilaments), the ribbon closes as a hollow cylinder of 25 nm of outer diameter. All dimers have the same orientation within a microtubule, which is the source of the structural and functional polarity of microtubules. The plus end (growing faster) is distal to the organizing center, with the beta subunit being probably exposed at that end. Two adjacent protofilaments show a lateral tubulin shift so that the microtubule is made of three left-handed helices along which alpha and beta subunits alternate, except along the closure of the cylinder (seam) where a discontinuity shows alpha-alpha or beta-beta lateral interactions (Song, Mandelkow 1995). The presence of this seam could have consequences in terms of motor binding sites on the surface of microtubules. Microtubule doublets present in axonemes are made according to the same rules by elongation of the distal centriole present in basal bodies by addition of tubulin dimers at the distal (plus) end into the A fiber (13 protofilaments) and the B fiber (11 protofilaments). Axonemal microtubule doublets show major longitudinal periodicities: eight nm (tubulin dimer), while outer dynein arms show a 24 nm periodicity (three dimers of tubulin) and inner dynein arms show more complex periodicities with three non-equivalent arms irregularly spaced every 96 nm (12 dimers of tubulin, which is one arm per four dimers as an average).

Tubulin Sequences

Alpha- and beta-tubulins are highly conserved globular proteins that may have slowly evolved since about two billion years, after duplication of a bacterial ancestor gene thought to be FtsZ (Faguy, Doolittle 1998). FtsZ is a GTP-binding protein able to polymerize into filaments and tubes *in vitro*, and is important for bacterial division (Erickson 1997).

According to species, between one and seven genes express alpha-tubulin and between one and nine genes express beta-tubulin, while vertebrates have six active genes coding for six different types of polypeptides (called isotypes) for each subunit. The conservation of tubulin sequences is certainly a reflection of the number of crucial interactions tubulin has to establish for proper function: one GTP binding site in each subunit, four surfaces of tubulin-tubulin interactions between any subunit and its neighbors (two longitudinal and two lateral), and an outer surface devoted to microtubule-associated proteins (MAPs) and motor binding. Apart from a few sites, this sequence conservation (between 70 and 98%) spans the entire N-terminal domain, that shows motifs involved in GTP binding and hydrolysis, the entire central domain and part of the C-terminal domain up to position 436 in alpha- and 430 in beta-tubulin that present "signature" sequences at positions 422-436 and 412-426, respectively (Fig. 1). The rest of the C-terminal domain is highly variable but always extremely rich in Glu residues (Sullivan 1988; Luduena 1993). This part of the polypeptide sequence has allowed researchers to distinguish six well defined classes of beta-tubulins (Sullivan, Cleveland 1986) and several less defined classes of alpha-tubulins (Villasante et al. 1986), which show either a constitutive or a tissue-specific and developmentally regulated expression. Of particular interest is the fact that testis-specific tubulins are among the most divergent in their C-terminal sequences and that the major axonemal beta-tubulin belongs to class IV, which is minor in the brain, whereas minor forms belong to classes II and III, which are major in the brain (Renthal et al. 1993). The binding site for MAPs and motors, such as dynein or kinesin, lies in the C-terminal tubulin sequences.

Tubulin Structure

Although many tubulin sequences have been determined, no one has been able to crystallize tubulin, a protein that self-assembles in all sorts of macromolecular complexes (rings, filaments, sheets and microtubules) but not crystals. However, recent progress made in producing high resolution images by electron microscopy (EM) has allowed researchers to obtain optical diffraction patterns from those EM pictures, as reviewed in Downing and Nogales (1998). As a consequence, 3D tubulin models have been proposed with a resolution gradually increasing during the past 20 years from about 40 angströms (the size of the tubulin monomer itself) to 3.7 angströms (Nogales et al. 1998), which reveals that alpha- and beta-tubulins have almost identical 3D structures. All alpha helices and beta sheets are now localized, as well as the GTP binding site, in the N-terminal domain of both chains and the paclitaxel (taxol) binding site in the central domain of beta-tubulin. The C-terminal domain (residues 385 to 451 in alpha- and 375 to 445 in beta-tubulin) contains two alpha helices. Both helices make the crest of the protofilaments on the outside surface of microtubules and are probably involved in the binding of MAPs and motors, whereas the loop connecting them is important for the monomer-monomer interaction. The C-terminus itself (from residue 440 in alpha- and 427 in beta-tubulin), which is the most acidic and contains the most variable part in tubulins, is absent from the 3D models because it is so flexible that it has no defined structure, producing no crystallographic signal (Nogales et al. 1998). This is unfortunate because this region is of major interest since the precise tubulin sequence involved in MAP or motor binding remains controversial. It is not clear whether the binding site is located within the highly conserved or the variable domain, or at their junction. Therefore the properties of this region have to be approached by indirect methods.

Post-translational Modifications

The determination of tubulin structure and functions is complicated by the existence of an extraordinary diversity within the tubulin molecules due to not less than seven types of PTMs, five of which take place in the C-terminal already variable domain of tubulin (Laferrière et al. 1997; MacRae 1997; Luduena 1998). Different tubulin gene products belonging to the same family are called tubulin *isotypes*, but PTMs of a given gene product give rise to variants called tubulin *isoforms*.

After many years of experimentation, it may be worth saying straight away that these modifications generally occur on tubulin molecules after their polymerization into microtubules and their stabilization by MAPs or by their interaction with other cellular organelles, so that the corresponding enzymes (presumably not abundant and showing a low enzymatic activity) have sufficient time to catalyze the modifications. In other words, the PTMs of tubulin are not a cause but a consequence of microtubule stability (Bulinski, Gundersen 1991), and may serve in the long-term stabilization of microtubule populations during cell differentiation and/or in the regulation of motor binding to the surface of microtubules. This is a particularly important point to bear in mind in the case of axonemal microtubules, which are the most stable microtubules known, with a highly ordered array of microtubular motors on their surface.

The PTMs of tubulin fall into seven types, some of which are classical PTMs of proteins, such as acetylation, palmitoylation and phosphorylation, while others are unique to tubulin.

One type of PTM is located in the N-terminal domain: the acetylation of alpha-tubulin (on Lys 40) occurs on cytoplasmic and axonemal microtubules (L'Hernault, Rosenbaum 1985; LeDizet, Piperno 1987). The acetylation of tubulin occurs soon after its incorporation in axonemes, but its function remains elusive. In *Tetrahymena*, the expression of a non-acetylable tubulin does not produce any new phenotype (Gaertig et al. 1995). However, a human case has been described in which unstable and shorter axonemes, together with retinal degeneration, were associated with a reduced level of tubulin acetylation and axonemal acetylase activity (Gentleman et al. 1996).

The most recently discovered PTM consists in the palmitoylation of both alpha- and beta-tubulins, Cys 376 being a major alpha-tubulin palmitoylation site (Ozols, Caron 1997). Because this type of PTM is generally associated with membrane proteins, it could be involved in tubulin binding to membranes. It is not clear whether this modification is present in many cell types and it would be interesting to know whether it is present in cilia and flagella.

The other five PTMs of tubulin occur in the C-terminal sequence of either alpha or beta or both polypeptide chains (Fig. 1). The C-terminal Tyr 451 of alpha-tubulin can be removed (detyrosination) by a specific carboxypeptidase, and re-added by a specific ligase (Barra et al. 1988; Raybin, Flavin 1977). This occurs after microtubule polymerization, preferentially on stable microtubules. In axonemal structures, it occurs soon after tubulin polymerization into microtubules. Its function is not clear, but interestingly, recent work showed that this modification is more abundant in microtubules interacting with intermediate filaments (Gurland, Gundersen 1995) and that this interaction is mediated by a kinesin type of motor (Liao, Gundersen 1998). Whether this specific interaction with motors is a general case or specific to the transport of intermediate filaments remains to be established.

The penultimate Glu 450 preceding the terminal Tyr can also be removed through an unknown reaction (dipeptide removal), producing an alpha-tubulin which cannot be tyrosinated (Paturle-Lafanechère et al. 1991). This modified tubulin is very abundant in the brain, but also in cilia and flagella, and occurs in highly stable microtubules during differentiation (Paturle-Lafanechère et al. 1994).

Phosphorylation of beta-tubulin has been described at different sites: on Ser 444 or Tyr 437 of beta III tubulin or Ser 441 of beta VI tubulin (Alexander et al. 1991; Khan, Luduena 1996). Its function is unknown, but it accompanies microtubule polymerization during neurite outgrowth (Gard, Kirschner 1985). Whether it is present in axonemes remains to be established.

Polyglutamylation of both alpha- and beta-tubulins of different classes is very abundant in neurons (Alexander et al. 1991; Eddé et al. 1990; Redeker et al. 1992; Wolff et al. 1992). It occurs on up to 90% of total tubulin and is developmentally regulated (Audebert et al. 1994). It consists in the lateral addition of between one to six Glu residues in general, and sometimes more, on the lateral carboxyl of a Glu residue belonging to the primary chain (Eddé et al. 1990) at only one position, in a given isotype of tubulin, that varies around position 440 with the class of tubulin being considered: Glu 445 in alpha , Glu 441 in beta I, Glu 435 in beta II, Glu 438 in beta III and Glu 434 in beta IVa (Mary et al. 1994). In the brain, it is thought to be involved in the regulation of the interaction between tubulin and MAPs (Boucher et al. 1994) or motors (Larcher et al. 1996). It is a major modification of neuronal and axonemal tubulins that occurs in axonemes shortly after the time of acetylation (Iomini et al. 1998). While brain microtubules and the core of centrosomes and basal bodies are made of both alpha- and beta-polyglutamylated tubulins (Bobinnec et al. 1998), only alpha-tubulin seems to be polyglutamylated to a significant level in axonemes (Mary et al. 1996; Plessman, Weber 1997; Fig. 1).

Polyglycylation of both alpha- and beta-tubulins consists in the lateral addition of up to 35 Gly residues on the lateral carboxyl of a Glu residue of the primary chain (Redeker et al. 1994) in the C-terminus of tubulins at a site very close to the polyglutamylation site: Glu 445 in alpha- and Glu 437 in beta-axonemal tubulin. It is highly abundant in ciliary and flagellar axonemes of a variety of species. During spermatogenesis, it occurs as the last PTM of tubulin, after acetylation and polyglutamylation, before the initiation of motility (Bressac et al. 1995; Iomini et al. 1998), and is also present in other classes of stable microtubules associated with membranes.

Tubulin Diversity

Due to the existence of several gene products of alpha- and of beta-tubulins that can combine into a variety of heterodimers, and due to the seven known types of PTMs, some of which are polymodifications of variable length, one can calculate the theoretical existence of hundreds of tubulin isoforms, at least in tissues such as brain where several isotypes are expressed. In neuronal tissues, seven alpha- and 14 beta-tubulin spots can be separated by two-dimensional gel electrophoresis, but each tubulin spot corresponds to several isoforms differing by a combination of PTMs (Audebert et al. 1994, Eddé et al. 1991, 1992). In sperm flagella, where only one alpha- and one beta-tubulin gene products are expressed, about 30 and 15 different isoforms have been evidenced, respectively (Fouquet et al. 1994; Plessman, Weber 1997).

What is the need for such a diversity of tubulin isoforms? At the present time, there is no definitive answer to this question, because there are usually only minor differences in biochemical properties between tubulin isotypes and isoforms and because, *in vivo*, a given isoform usually participates in several classes of microtubules, microtubules contain several isoforms, and in some cases a given tubulin isotype can be replaced by another (Murphy 1991). However, several laboratories have already established that some isotypes of tubulin are involved in a specific function and cannot be replaced. These results support the multi-tubulin hypothesis as originally proposed for flagellar tubulins (Fulton, Simpson 1976). This is the case, for instance, in bird erythrocytes, mechanosensory neurons in *C. elegans* and *Drosophila* testis (Raff 1994; Raff et al. 1997; Wilson, Borisy 1997). Besides the convincing example of the beta-2 tubulin necessary to assemble normal axonemes during spermatogenesis in *Drosophila* (Raff et al. 1997), it has been shown that a highly divergent testis-specific alpha-tubulin isotype is found only in the manchette in mammalian spermatids (Hecht et al. 1988). Another example occurs in a protist in which microtubules from flagellar axonemes and from cytoplasmic origin are composed of distinct tubulins within a single cell (McKeithan, Rosenbaum 1981). Finally, acetylated alpha-tubulin and polyglutamylated tubulins are found only in the flagellar microtubules, whereas tyrosinated and detyrosinated alpha-tubulins are also present in cytoplasmic microtubules (Piperno, Fuller 1985; Fouquet et al. 1994). It can be concluded that sperm axonemes contain a whole diversity of tubulins, just as neurons do. Why are so many different isoforms of tubulin present in cells and in axonemes? Could a combination of tubulin isoforms allow optimal interactions between microtubules and other proteins such as MAPs or motors, at least in differentiated microtubular systems? If so, how can this hypothesis be investigated? Because the axoneme is stable and can be isolated and dissected biochemically, but can also be reactivated *in vitro*, it represents a good model system to better understand what is essential within axonemal tubulins for the expression of flagellar motility, especially with respect to the dynein-tubulin interaction.

Inhibition of Flagellar Motility by Anti-tubulin Antibodies

Generation of Specific Monoclonal Antibodies

Several anti-tubulin antibodies have been reported to inhibit flagellar motility (Asai, Brokaw 1980; Okuno et al. 1981; Asai et al. 1982). All of them reduced the flagellar beating amplitude, but not beat frequency, and these authors suggested that dynein-microtubule interactions were not affected. A general mechanism of inhibition by anti-tubulin antibodies was proposed, suggesting that they all reduce motility because of an increase in stiffness in the whole axonemal machinery, caused by the steric hindrance effect triggered by these antibodies, irrespective of the specificity of their binding site on tubulin.

However, other specific anti-tubulin antibodies have recently been shown to act via a mechanism different from the non-specific steric hindrance mechanism. Panels of monoclonal antibodies (mAbs) have been generated against total axonemal proteins from sea urchin spermatozoa. The mAbs were first screened to select those clones producing antibodies inducing inhibition of demembranated-reactivated sea urchin sperm models. The positive mAbs were further characterized by immunoblotting to select only the monospecific clones and by functional assays using different species as model systems. Finally, the antigens were characterized, and a more detailed motility study was conducted on sea urchin sperm models. Once demembranated, the 9+2 sea urchin models can be reactivated to nearly 100% with a very homogeneous population. The axonemes are about 40 μm long, thus allowing sufficient length for detailed analysis of motility parameters. So far, this approach has allowed the study of anti-dyneins, anti-tubulins and a few other anti-unknown proteins (Audebert et al. 1999; Cosson et al. 1996; Gagnon et al. 1994, 1996; Gingras et al. 1996, 1998). The corresponding epitopes turned out to be conserved during a long period of evolution (similar reactivity by immunoblotting or immunofluorescence, similar types of inhibition on the reactivated flagellar movement in distant species such as dinoflagellates, *Chlamydomonas*, distant species of sea urchins and humans). Therefore, the corresponding antigens are believed to play a crucial role in flagellar motility. According to the particular case, the antigen has been biochemically characterized, or cloned and sequenced. To date, this functional approach has allowed three groups of mAbs and their corresponding proteins to be characterized. mAb D1 allowed the properties of IC1, an intermediate chain of outer dynein arms, to be described (Gagnon et al. 1994). Two other interesting antibodies allowed new proteins that have been cloned and sequenced to be described. The first mAb, D405, is an anti-30 kD protein that shows a large similarity (74%) with p28, a light chain from inner dynein arms of *Chlamydomonas*, but also a highly basic proline-rich domain absent from p28 and similar to a domain present in dynamin (Gingras et al. 1996). The second mAb, D316, is an anti-64 kD protein that shows a limited similarity (37% and 25%, respectively) with radial spoke head proteins RSP 4 and 6 from *Chlamydomonas* but also original features such as a global pI of 4.0 due in part to three Asp/Glu-rich stretches (Gingras et al. 1998). Finally, three mAbs producing interesting inhibition patterns were obtained: B3, an anti-alpha-tubulin (Gagnon et al. 1996); and C9 and D66, anti-beta-tubulins (Cosson et al. 1996; Audebert et al. 1999).

This chapter will now focus on anti-tubulin antibodies recently developed, as well as anti-tubulin from other sources, and attempt to answer the following questions: 1) Are PTMs of tubulin important for flagellar motility? 2) Are all PTMs important for sperm motility? 3) Are primary chain epitopes also important? 4) Do alpha- and beta-tubulins play equivalent roles in flagellar motility? 5) What is so special about the beta-tubulin C-terminal domain?

Are PTMs of Tubulin Important for Flagellar Motility?

When incubated with motile sea urchin sperm models, mAb B3 completely inhibited flagellar motility at very low concentrations; B3 recognized alpha-tubulin in sea urchin sperm axonemes and also recognized brain alpha- and beta-tubulins (Gagnon et al. 1996). Subtilisin cleavage of sea urchin sperm tubulin, in conditions that remove only one part of the C-terminal sequence, prevented B3 recognition of the truncated tubulin, indicating that the epitope recognized by B3 was restricted to the last 10 to 15 amino acid residues of the C-terminal domain of tubulin. ELISA assays established that B3 reacted only with the glutamylated forms of alpha-tubulin peptides from sea urchin or mouse brain, indicating that it recognized polyglutamylated motifs in the C-terminus domain of alpha-tubulin (Fig. 1). GT335, a previously characterized anti-polyglutamylated tubulin mAb (Wolff et al. 1992), showed a similar reactivity towards sea urchin tubulin and inhibited reactivated flagellar motility with characteristics similar to those induced by B3. On the other hand, other tubulin antibodies directed against various epitopes of the C-terminal domain (YL1/2, DM1A) had no effect on motility while having binding properties similar to that of B3. B3 and GT335 both acted by decreasing the beating amplitude without affecting the flagellar beat frequency or the velocity of sliding microtubules of trypsin-treated axonemes (Table 1). B3 and GT335 were also capable of inhibiting the motility of flagella of *Oxyrrhis marina*, a 400,000,000-year-old species of dinoflagellate, and those of human sperm models. Localization of the antigens recognized by B3 and GT335 by immunofluorescence techniques revealed their presence all along the axoneme of sea urchin spermatozoa and of *Oxyrrhis* flagella. However, they did not label the distal tip of flagella nor the cortical microtubule network of the dinoflagellates, both labeled by non-inhibitory anti-tubulin mAb such as DM1A or YL1/2. This original pattern may suggest that the central pair of microtubules and/or the distal portion of outer doublets may not contain polyglutamylated tubulin (P Huitorel, D White, J Cosson and C Gagnon, unpublished results). At very low concentrations of antibodies (similar to those used in the inhibition of motility assays), B3 and GT335 labeled the axonemal structure of sea urchin spermatozoa with a gradient decreasing from the sperm head to the tip (Fig. 2). This result does not seem to be due to a trivial accessibility problem, since the same axonemes were homogeneously labeled by DM1A or YL1/2. The distribution of tubulin epitopes along flagellar length is discussed below. This information indicates that a PTM, the polyglutamylation of alpha-tubulin, plays a dynamic role in a dynein-based motility process. The biological significance of this basic mechanism is further emphasized by its presence in evolutionary distant species, from dinoflagellates to humans.

Are All PTMs of Tubulin Important for Flagellar Motility?

The results described above, obtained with B3 and GT335, showed the importance of a particular PTM of tubulin in flagellar motility. At this point, the question is whether several or all post-translational modifications are involved in flagellar motility. To answer this question, other well characterized anti-tubulin mAbs were used, among which was 6-11B-1, specific of the acetylated form of alpha-tubulin. This antibody has the unique property of recognizing an epitope (acetylated Lys 40) in the N-terminal domain of tubulin (Fig. 1). This epitope is highly abundant in all known axonemal structures. While mAb 6-11B-1 recognized alpha-tubulin from sea urchin sperm axonemes, it did not inhibit the flagellar motility of sperm models (Gagnon et al. 1996). YL1/2, a well known mAb specific for the tyrosinated form of alpha-tubulin (Fig. 1), stained axonemal structures very well and recognized alpha-tubulin from sea urchin, but again did not reduce reactivated flagellar motility. These results may be surprising because this YL1/2 epitope is the C-terminal domain of tubulin. These results were confirmed using several other anti-tubulin antibodies specific of PTMs including affinity purified polyclonal antibodies specific of tyrosinated (L141), non-tyrosinated (missing the C-terminal Tyr) (L3), non-tyrosinatable (missing the C-terminal Glu-Tyr) (L7), as seen in Figure 1. None of them induced a significant reduction of motility (Gagnon et al. 1996).

At this point, one can conclude that while polyglutamylation of tubulin is essential, several other PTMs of

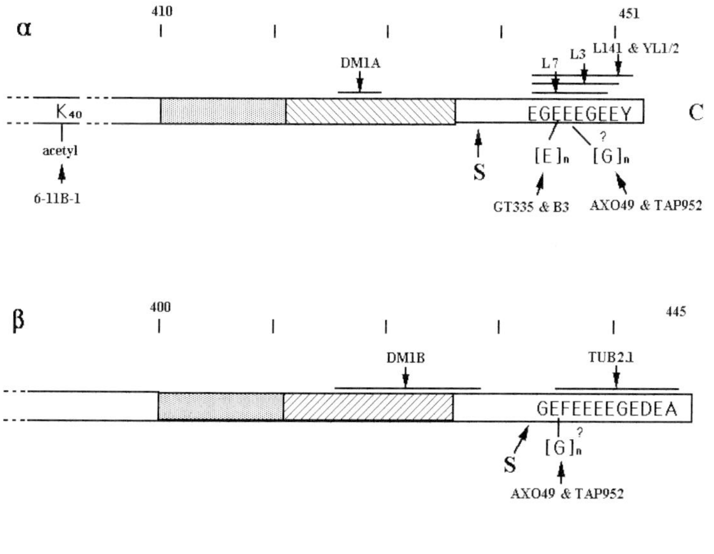

Figure 1. C-terminal domains of axonemal tubulins. Alpha- and beta-tubulin C-terminal sequences show three domains: a domain conserved within all tubulins (gray shading); a domain conserved within alpha- (top left to right bottom slanted lines) and within beta-tubulins (left bottom to right top slanted lines), identified by the mAb DM1A and DM1B, respectively; and a highly acidic and variable C-terminus (open area) subject to post-translational modifications recognized by several mAbs. A portion of the C-terminus can be removed by subtilisin cleavage (S). The site of polyglycylation is not well defined (?).

Table 1: Antibody inhibition of axonemal motility.

Antibody*	Specificity‡	Inhibition Potency°	Amplitude Inhibition	Frequency Inhibition
B3 (m)	Tub polyGlu (α)	50	+++	+
GT335 (m)	Tub polyGlu (α)	83	+++	+
YL1/2 (m)	Tub Tyr (α)	<0.6	-	-
L141 (p)	Tub Tyr (α)	0	-	-
L3 (p)	Tub Glu (α)	0.7	-	-
L7 (p)	Tub Δ2 (α)	1.3	-	-
DM1A (m)	Tub (α)	<0.3	-	-
6-11B-1 (m)	Tub acet (α)	<0.7	-	-
AXO49 (m)	Tub polyGly (α+β)	5	++	++
TAP952 (m)	Tub polyGly (α+β)	3.5	++	++
C9 (m)	Tub (β)	31	+	+++
D66 (m)	Tub (β)	14	++	+++
Tub 2.1 (m)	Tub (β)	0.1	-	-
DM1B (m)	Tub (β)	0.3	-	-

* m = monoclonal antibody; p = polyclonal antibody.
‡ Tub polyGlu (α): Tubulin polyglutamylated on alpha-tubulin only in axonemes, on alpha- and beta-tubulins in brain; Tub Tyr (α): alpha-tubulin bearing a Tyr residue at its C-terminus; Tub Glu (α): alpha-tubulin after removal of the C-terminal Tyr; Tub Δ2 (α): alpha-tubulin after removal of the penultimate Glu residue; Tub acet (α): acetylated alpha-tubulin; Tub polyGly (α+ β): tubulin polyglycylated on alpha- and beta-tubulins.
° The inhibition potency was evaluated from inhibition kinetics at different antibody concentrations. The time for half inhibition ($t_{1/2}$ min) was measured for each concentration of mAb (C, µg/ml), and the equation $P = [\ 1 / t_{1/2} \times C\ ] \times 100$ was used to obtain the relative inhibition potency (ml/min.mg).

tubulin do not seem to play a critical role in flagellar motility. Then the question is whether another modification, different from polyglutamylation, could also play such an essential role. Two monoclonal antibodies, AXO 49 and TAP 952, have been produced against carboxy-terminal peptides from *Paramecium* axonemal tubulin and probed with polyglycylated synthetic peptides. Both mAbs recognize different levels of tubulin polyglycylation, one of the most recently identified PTMs of tubulin discovered in the C-terminal domain of *Paramecium* axonemal tubulin (Redeker et al. 1994) (Fig. 1). Tubulin polyglycylation is widely distributed in organisms ranging from ciliated protozoa to mammals and is the last PTM taking place after acetylation and polyglutamylation in a proximo-distal direction at the end of the maturation process during spermatogenesis in *Drosophila* (Bressac et al. 1995) and in the flatworm *Echinostoma caproni* (Platyhelminthes), as seen in Iomini et al. (1998). The major original effect produced by AXO 49 or TAP 952 on reactivated sperm motility was the induction of an erratic type of swimming, followed in time (or at higher mAb concentrations) by a reduction of both the beating amplitude and frequency mostly in the proximal and distal sections of the axoneme (Table 1). This pattern of inhibition was quite different from the one observed in the presence of B3 or GT335. This is very striking since AXO 49 and TAP 952 recognize different levels of polyglycylation of tubulin, taking place in the C-terminal domain of both alpha- and beta-tubulins at positions very close to the polyglutamylation site recognized by B3 or GT335 (Bré et al. 1996; Gagnon et al. 1996). Immunofluorescence localization in different organisms show the existence of a graded type of labeling. AXO 49 revealed a decreasing gradient, whereas TAP 952 revealed an opposite increasing gradient from the base to the distal tip of spermatozoa (Fig. 2), suggesting a graded distribution of tubulin polyglycylation levels along the axoneme (Bré et al. 1996). Therefore, different levels of tubulin polyglycylation exist in sea urchin spermatozoa that seem to play an important role in flagellar motility. In addition, from all these results emerges the concept of a "hot spot" being present within the C-terminal sequence of tubulin, which is the site where the two PTMs (polyglutamylation and polyglycylation) occur and seem to play a key role in flagellar motility.

Are Primary Chain Epitopes Also Important for Flagellar Motility?

Not only alpha- but also beta-tubulin possesses epitopes important for flagellar motility. The mAb C9 (Cosson et al. 1996) recognized a subset of beta-tubulin isoforms. Low concentrations of C9 (0.1 µg/ml, which is about 1 mole of antibody per 33 moles of total tubulin) blocked the motility of sea urchin sperm models primarily by decreasing the sliding velocity and frequency of flagellar beating to less than 1 Hz, while later in time modifying the shear angle along the axoneme, especially in the distal end (Table 1).

Other well known anti-tubulin antibodies (such as DM1A, YL1/2 and DM1B) had no effect on motility at concentrations 100-fold higher than those effective for C9 (Cosson et al. 1996; Gagnon et al. 1996). The epitope recognized by C9 is not yet determined since all types of limited proteolysis investigated so far to generate tubulin peptides to locate the epitope lead to the loss of epitope immunoreactivity. Indirect evidence suggests that C9 epitope may reside also in the C-terminal domain, but this remains to be established. The effects of C9 on motility were not restricted to flagella of sea urchin spermatozoa. Flagellar beating of the dinoflagellate *Oxyrrhis* was completely blocked by C9 in a manner reminiscent of that of sea urchin sperm flagella. C9 also inhibited the motility of human spermatozoa and of *Chlamydomonas*. Immunofluorescence studies revealed that C9 binds to the whole axoneme of sea urchin spermatozoa and *Oxyrrhis* flagella, including their distal tips, together with the cortical network of *Oxyrrhis* cell body (unpublished results), shown in Figure 2. It interferes with the flagellar beat frequency from several distant species of sea urchin sperm, dinoflagellate, algae and human sperm flagella, suggesting that the epitope recognized by C9 is conserved over

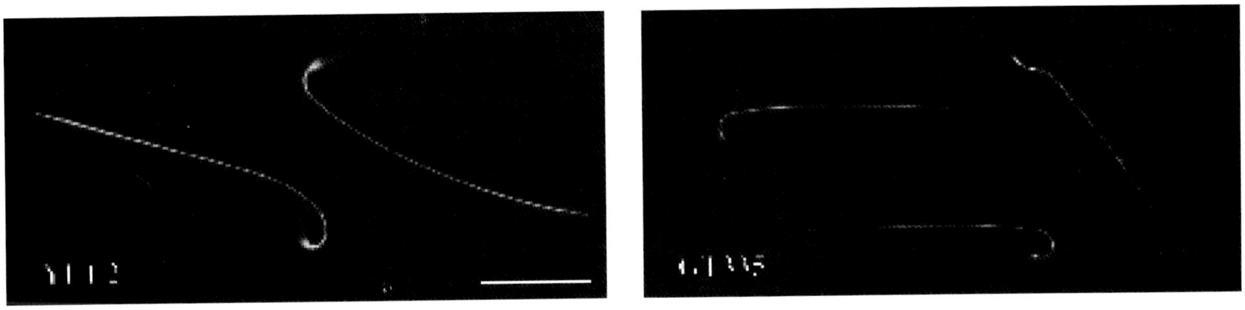

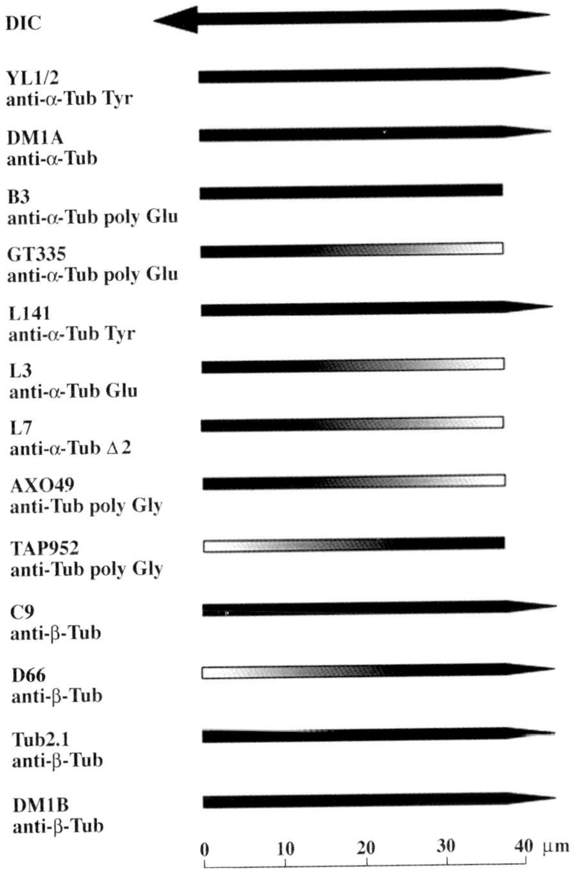

Figure 2. Immunostaining of sea urchin spermatozoa. Spermatozoa from *Paracentrotus lividus* were formaldehyde-fixed and stained with a panel of anti-tubulin mAbs. Two examples of immunofluorescence labeling are shown (scale bar: 10 μm) and a schematic representation summarizes the results.

a long period of evolution and plays an important role in sperm motility. As far as is known, C9 is the first anti-tubulin antibody reported that specifically reduces the flagellar beat frequency to very low values (down to a fraction of a hertz) in a manner reminiscent of the inhibitory effect of anti-dynein antibodies (Gibbons et al. 1976). These results suggest that the C9 anti-beta-tubulin mAb possibly interferes with the interaction between outer dynein arms and microtubules. The results also argue against the initial proposal that all anti-tubulin antibodies reduce flagellar motility via a non-specific steric hindrance effect leading to a general increase in stiffness of the axoneme (Asai et al. 1982).

Another recently produced anti-movement mAb, D66, reacted specifically with a subset of beta-tubulin isoforms and inhibited motility (Audebert et al. 1996). Limited proteolysis, high performance liquid chromatography (HPLC), peptide sequencing, mass spectroscopy and immunoblotting experiments indicated that D66 recognized an epitope localized in the primary sequence 423Q-E435 of the carboxy-terminal domain of *Lytechinus pictus* beta 2-tubulin, and that this sequence belonged to class IVb, the major beta-tubulin isotype present in axonemes (Audebert et al. 1999). D66 labeled the whole length of sea urchin sperm axonemes, including the distal tip (Fig. 2). Incubations under low antibody concentrations revealed a gradient type of labeling, with a fluorescence signal increasing from the base to the distal tip of the axoneme. D66 inhibited first flagellar beat frequency (Table 1) and later (or at higher antibody concentrations) the beating amplitude in the distal part of the axoneme. Because the primary effect of this antibody on sperm motility is to decrease the flagellar beat frequency, similarly to that of anti-beta-tubulin C9, the sequence 423Q-E435 may be involved in the transient tubulin-dynein head interaction required for axonemal motility.

Do Alpha- and Beta- Tubulins Play Equivalent Roles in Flagellar Motility?

Another anti-beta-tubulin polyclonal serum (NS20) has been described that reduces microtubule dependent motility (axonal transport, mitosis, flagellar motility) by 50% at high concentrations (Goldsmith et al. 1991). It recognizes motifs contained in the beta-tubulin sequence 400G-F436 (which is 37 amino acids long), proposed by these authors to be the primary binding site for motors of several classes onto microtubules, because the corresponding synthetic peptide itself induces the same attenuation of motility as the polyclonal anti-tubulin antiserum. This sequence is close to, but does not include, the C-terminal sequence of beta-tubulin. It comprises a consensus sequence among both alpha-

and beta-tubulins (400GEGMDEMEFTEA411), a sequence highly conserved in beta- but absent from alpha-tubulins (412ESNMNDLVSEYQQYQ426), and a portion of the variable C-terminal domain (427DATAD-EQGEF436), but not the acidic C-terminus itself (437EEEGEEDEA445), as seen in Figure 1. However, it would be important to know which parameters of flagellar motility are affected in these experimental conditions, which only measured the time required for total motility arrest. Interestingly, these authors also observed the same extent of inhibition in the presence of the alpha-tubulin equivalent sequence (411E-E447), but 80% inhibition using a combination of both peptides (Goldsmith et al. 1995). These results are best explained by the presence of functionally distinct dynein binding sites on alpha- and beta-tubulins near their respective C-terminus. This conclusion is also supported by other results (Table 1). Whereas two anti-beta-tubulin mAbs (C9, D66) primarily lower flagellar beat frequency (Cosson et al. 1996; Audebert et al. 1996), two anti-alpha-tubulin mAbs (B3, GT335) primarily decrease the beating amplitude (Gagnon et al. 1996). Moreover, antibodies specific for polyglycylation (AXO49, TAP952), a PTM present in both alpha- and beta-tubulins, induce a reduction of both motility parameters. Taken together, these results argue in favor of different roles played by alpha- and beta-tubulins in flagellar motility.

What is So Special about the Beta-Tubulin C-Terminal Domain in Flagellar Motility?

Interestingly, Goldsmith et al (1995) observed a 60% inhibition of motility using a 15 amino acids long peptide which mimicked the sequence 412E-Q426, which is highly conserved and specific of beta-tubulins. From this last result, one would predict that the mAb DM1B, specific for the conserved beta-tubulin sequence 416N-A430 (Breitling, Little 1986), should interfere with flagellar motility. However, this mAb does not affect motility (Cosson et al. 1996). Moreover, the new anti-beta-tubulin D66 mAb that inhibits flagellar beat frequency (Audebert et al. 1996, 1999) is specific of an epitope contained in the sequence 423Q-E435 previously identified as a binding site for microtubule associated proteins MAP2 and Tau, and for cytoplasmic dynein (Kotani et al. 1990; Maccioni et al. 1988). These results emphasize the importance of the beta-tubulin sequence 423Q-E435 in its interactions with other proteins critical to microtubular functions.

Recent data have also pointed out the presence in axonemal beta-tubulins of a consensus sequence EGEF present in the same position within the last 15 amino acids long C-terminal variable domain (Raff et al. 1997). This motif overlaps with the sequence recognized by the mAb D66, a strong inhibitor of human and sea urchin flagellar motility (Audebert et al. 1996, 1999).

As a conclusion, it should be emphasized that all these results, obtained with a large variety of anti-tubulin antibodies, suggest that only a small number of tubulin epitopes, located in the C-terminal domain of both tubulins, are directly involved in flagellar motility and probably in the ATP-dependent dynein-tubulin interaction.

Microtubules Motors Interactions

Tubulin Sequences Involved in MAPs and Motors Binding

To understand flagellar motility, it is of paramount importance to understand the mechanisms responsible, at the molecular level, for force production. However, very little is known on axonemal dynein-tubulin interactions, the key regulating elements. The use of tubulin synthetic peptides allowed the localization of binding sites for MAPs and cytoplasmic dyneins on both alpha- and beta-tubulins nearby residues 430 and 420, respectively (Maccioni et al. 1988). The location of these sites is in agreement with those identified as sites required for flagellar motility (Audebert et al. 1996, 1999; Bré et al. 1996; Gagnon et al. 1996). Microtubule-binding assays performed in the presence of cytoplasmic dynein, kinesin or MAPs *in vitro* also provided data. Comparison of binding properties of tubulins before or after severing their C-terminal domains by subtilisin cleavage has allowed the localization of binding sites relative to the cleavage sites. However, the presence of paclitaxel (taxol), known to induce abnormal microtubular structures, in many experiments led to different subtilisin cleavage sites, and conflicting results were reported, as seen in a review by Cleveland (1990). A partial consensus has developed which supports the concept that MAPs and motors bind to different sites on tubulin (Marya et al. 1994). It has been shown in well defined conditions that MAPs bind to cleaved tubulin (Multigner et al. 1995). Since the site of subtilisin cleavage is at D438 in alpha- and Q433 in beta- pig brain tubulin, and that the released tubulins remain competent for binding motors, these results indicate that the acidic C-terminus is not the primary binding site for MAPs on tubulin, even if it regulates this binding (Boucher et al. 1994). Recent experiments showed that kinesin can be crosslinked to the acidic C-terminus of both alpha- and beta-tubulin subunits (Tucker, Goldsmith 1997), which suggests that the C-terminus is in close proximity to the bound motor, but does not necessarily indicate that it is the primary binding site. Such data are still missing in the case of axonemal dyneins. One genetic study beautifully established that the C-terminal 15 amino acids of beta-tubulin are not necessary for microtubule polymerization during *Drosophila* spermatogenesis (Fackenthal et al. 1993). However, such tubulin built up axonemal microtubules that did not support any sort of axonemal assembly. Therefore, the C-terminus of beta-tubulin appears to be crucial for interactions between microtubules and other axonemal components necessary to build up the axonemal machinery *in vivo*.

Diversity of Axonemal Tubulins

Axonemal tubulins present a very high diversity, which is due more to PTMs than to isotype complexity. One alpha-tubulin (class II, testis-specific) and one beta-tubulin (class IVb, testis-specific) are the major gene products present in spermatozoa (Lewis, Cowan 1988; Renthal 1993). These results have been partially confirmed using antibodies specific for class II, III and IV beta-tubulins. Beta IV (testis-specific) is the major isotype present in mammalian axonemes, while two other minor beta-tubulins (class II, ubiquitous and class III, neuronal specific) are also present (Renthal 1993). In the case of alpha-tubulin, at least two different C-terminal sequences have been found in the same species of sea urchin sperm flagella (Mary et al. 1996) that seem to belong to class I (ubiquitous) and II (testis-specific). However, at least 10 alpha- and 10 beta-tubulin species can be separated by two-dimensional sodium dodecyl sulphate-polyacrylamide gel electrophoresis analysis, a finding reminiscent of the brain tubulin heterogeneity (Fouquet et al. 1994). A series of alpha isoforms is acetylated, while most alpha and beta isoforms appear polyglutamylated (at least in mammals, but only alpha isoforms are significantly polyglutamylated in sea urchin sperm flagella). Proteolytic cleavage, HPLC and mass spectrometry of C-terminal tubulin peptides allowed the identification of about 30 alpha and 15 beta isoforms in sea urchin (Mary et al. 1996) or bull (Plessmann, Weber 1997) sperm axonemes. A large portion of both alpha- and beta-tubulins are unmodified, while a combination of detyrosination, removal of the penultimate glutamate, polyglutamylation and polyglycylation is present among alpha-tubulins of both species, and polyglycylation is responsible for beta-tubulin diversity.

Localization of Tubulin Isoforms within the Axoneme

How are the different tubulin isoforms located within the axoneme? Are they evenly distributed all along the axoneme? Are they all present in all types of axonemal microtubules? Current evidence indicates that the answer to the last two questions is no.

First, the mAb GT335 produced an uneven type of labeling along mammalian axonemes (which have accessory structures around the axoneme), suggesting either an uneven distribution of glutamylated tubulin or an uneven accessibility in different portions of the axoneme (Fouquet et al. 1994). However, a gradient type of labeling has also been observed on sea urchin sperm axonemes with GT335 or D66 (Huitorel et al. 1995), or with AXO49 and TAP952, specific for glycylated tubulins (Bré et al. 1996)(Fig. 2). Since this model does not have any accessory structure, the results suggest a "real" gradient type distribution of tubulin isoforms along the axoneme which is reminiscent of the gradient of several outer and inner dynein polypeptide chains along the axoneme (Asai, Brokaw 1993; Gagnon et al. 1994). It must be remembered that axonemal structures are built up progressively during spermatogenesis by progressive distal addition of precursors which are arranged from base to tip (Dutcher 1995; Johnson, Rosenbaum 1992). Shortly after assembly, a series of tubulin PTMs occurs also from base to tip, which may in part explain their gradient type of distribution (Bré et al. 1996; Stephens 1992). Of course, the precise localization of these various tubulin epitopes remains to be determined by immunoelectron microscopy at high resolution.

Second, the axoneme contains outer doublet microtubules and central pair microtubules which are known to be biochemically and functionally different (Dutcher 1995). Moreover, sea urchin axonemes have been biochemically dissected, leading to purified A- and B-fibers of outer doublets. Direct sequence, mass spectrometry and immunological analyses showed that A microtubules are made of homogeneously unmodified alpha- and beta-tubulin gene products, whereas B microtubules are made of tubulins presenting a whole variety of PTMs described above (Multigner et al. 1996). This result is most provocative, considering that PTMs usually occur on stable microtubules and A fibers are the most stable microtubules known. The difference in composition in tubulin isoforms is not due to a specific lack of accessibility in those A fibers (Multigner et al. 1996), and certainly reflects differences in terms of function. Recently, tyrosylated tubulin was shown to be present in the whole axoneme, whereas detyrosylated tubulin was shown only in B fibers of *Chlamydomonas* outer doublets (Johnson 1998). It must be remembered that A fibers are the substrate of permanent attachment for both outer and inner dynein arms, whereas B fibers are the binding sites for the transient dynein-B microtubule ATP-dependent interactions. It is tempting to speculate that only unmodified tubulin has the proper conformation and affinity for binding dynein arms at their base, whereas only a subclass of modified tubulins in B microtubules may have the proper chemical/conformational properties in their C-terminal domain to be competent for ATP-dependent interactions with the dynein motor domains. This proper modification could possibly be the polyglutamylation of alpha-tubulin and the polyglycylation of beta-tubulin.

Conclusions

All these results point to two crucial parts of the tubulin molecules: the site of the PTMs polyglutamylation and polyglycylation in alpha-tubulin, and a portion of primary sequence in the C-terminal domain of beta-tubulin. These two classes of sites may be directly involved in the transient ATP-dependent binding of dynein heads leading to force production. The determination of the precise localization of the corresponding epitopes in the axonemal structure should remain an important tool to study the interaction between molecular motors and microtubules.

References

Afzelius BA. Role of cilia in human health. Cell Motil Cytoskel 1995; 32:95-97.

Alberts B, Bray D, Lewis J, Raff M, Roberts K, Watson JD. Molecular Biology of the Cell. 3rd ed. Garland Publishing, New York, 1994.

Alexander JE, Hunt DF, Lee MK, Shabanowitz J, Michel H, Berlin SC, McDonald TL, Sundberg RJ, Rebhun LI, Frankfurter A. Characterization of posttranslational modifications in neuron-specific class III beta tubulin by mass spectrometry. Proc Natl Acad Sci USA 1991; 98:4685-4689.

Alonso MC, van Damme J, Vandekerckhove J, Cross RA. Proteolytic mapping of kinesin/ncd-microtubule interface: nucleotide-dependent conformational changes in the loops L8 and L12. EMBO J 1998; 17:945-951.

Asai DJ. Multiple dynein hypothesis. Cell Motil Cytoskel 1995; 32:129-132.

Asai DJ, Brokaw CJ. Effects of antibodies against tubulin on the movement of reactivated sea urchin sperm flagella. J Cell Biol 1980; 87:114-123.

Asai DJ, Brokaw CJ. Dynein heavy chain isoforms and axonemal motility. Trends Cell Biol 1993; 3:398-402.

Asai DJ, Brokaw CJ, Thompson WC, Wilson L. Two different monoclonal antibodies to alpha-tubulin inhibit the bending of reactivated spermatozoa. Cell Motil 1982; 2:599-614.

Audebert S, Koulakoff A, Berwald-Netter Y, Gros F, Denoulet P, Eddé B. Developmental regulation of polyglutamylated alpha- and beta-tubulin in mouse brain neurons. J Cell Sci 1994; 107:2313-2322.

Audebert S, White D, Cosson J, Huitorel P, Eddé B, Gagnon C. Characterization of a new monoclonal antibody against beta-tubulin that blocked sperm motility. Mol Biol Cell 1996; 7:48a.

Audebert S, White D, Cosson J, Huitorel P, Eddé B, Gagnon C. The carboxy-terminal sequence 427D-E432 of beta-tubulin plays an important function in sperm motility. Eur J Biochem 1999; 261:48-56.

Barra HS, Arce CA, Argarana CE. Posttranslational tyrosination-detyrosination of tubulin. Mol Neurobiol 1988; 2:133-153.

Bernstein M, Beech P, Katz SG, Rosenbaum JL. A new kinesin-like protein (Klp1) localized to a single microtubule of the *Chlamydomonas* flagellum. J Cell Biol 1994; 125:1313-1326.

Bobinnec Y, Moudjou M, Fouquet JP, Desbruyères E, Eddé B, Bornens M. Glutamylation of centriole and cytoplasmic tubulin in proliferating non-neuronal cells. Cell Motil Cytoskel 1998; 39:223-232.

Boucher D, Larcher JC, Gros F, Denoulet P. Polyglutamylation of tubulin as a progressive regulator of *in vitro* interactions between the microtubule-associated protein tau and tubulin. Biochemistry 1994; 33:12471-12475.

Bré MH, Redeker V, Quibell M, Darmanaden-Delorme J, Bressac C, Cosson J, Huitorel P, Schmitter JM, Rossier J, Johnson T, Adoutte A, Levilliers N. Axonemal tubulin polyglycylation probed with two monoclonal antibodies: widespread evolutionary distribution, appearance during spermatozoan maturation and possible function in motility. J Cell Sci 1996; 109:727-738.

Breitling F, Little M. Carboxy-terminal regions on the surface of tubulin and microtubules. Epitope locations of YOL1/34, DM1A and DM1B. J Mol Biol 1986; 189:367-370.

Bressac C, Bré M-H, Darmanaden-Delorme J, Laurent M, Levilliers N, Fleury A. A massive new posttranslational modification occurs on axonemal tubulin at the final step of spermatogenesis in *Drosophila*. Eur J Cell Biol 1995; 67:346-355.

Brokaw C. Operation and regulation of the flagellar oscillator. In: (Warner FD, Satir P, Gibbons I, McIntosh JR, eds) Cell Movement. New York: Alan R Liss, 1989, pp267-279.

Brokaw C, Kamiya R. Bending patterns of *Chlamydomonas* flagella: IV. Mutants with defects in inner and outer dynein arms indicate differences in dynein arm function. Cell Motil Cytoskel 1987; 8:68-75.

Bulinski JC, Gundersen GG. Stabilization and post-translational modification of microtubules during cellular morphogenesis. BioEssays 1991; 13:285-293.

Cleveland DW, Joshi HC, Murphy DB. Tubulin site interpretation. Nature 1990; 344:389.

Cosson J. A moving image of flagella: news and views on the mechanisms involved in axonemal beating. Cell Biol Int 1996; 20:83-94.

Cosson J, White D, Huitorel P, Eddé B, Cibert C, Audebert S, Gagnon C. Inhibition of flagellar beat frequency by a new anti beta tubulin antibody. Cell Motil Cytoskel 1996; 35:100-112.

Curry AM, Rosenbaum JL. Flagellar radial spoke: a model molecular genetic system for studying organelle assembly. Cell Motil Cytoskel 1993; 24:224-232.

Downing KH, Nogales E. Tubulin and microtubule structure. Curr Opin Cell Biol 1998; 10:16-22.

Dustin P. Microtubules. 2nd ed. Springer-Verlag, Berlin, 1984.

Dutcher SK. Flagellar assembly in two hundred and fifty easy-to-follow steps. Trends Biochem Sci 1995; 11:398-404.

Dutcher SK, Trabuco EC. The UNI3 gene is required for assembly of basal bodies of *Chlamydomonas* and encodes delta-tubulin, a new member of the tubulin superfamily. Mol Biol Cell 1998; 9:1293-1308.

Dutcher SK, Huang B, Luck DJ. Genetic dissection of the central pair microtubules of the flagella of *Chlamydomonas reinhardtii*. J Cell Biol 1984; 98:229-236.

Eddé B, Rossier J, Le Caer JP, Desbruyères E, Gros F, Denoulet P. Posttranslational glutamylation of alpha-tubulin. Science 1990; 247:83-85.

Eddé B, Rossier J, Le Caer J-P, Berwald-Netter Y, Koulakoff A, Gros F, Denoulet P. A combination of posttranslational modifications is responsible for the production of neuronal alpha-tubulin heterogeneity. J Cell Biochem 1991; 46:134-142.

Eddé B, Rossier J, Le Caer J-P, Promé J-C, Desbruyères E, Gros F, Denoulet P. Polyglutamylated alpha-tubulin can enter the tyrosination-detyrosination cycle. Biochemistry 1992; 31:403-410.

Erickson HP. FtsZ, a tubulin homologue in prokaryote cell division. Trends Cell Biol 1997; 7:362-367.

Fackenthal JD, Turner FR, Raff E. Tissue-specific microtubule functions in *Drosophila* spermatogenesis require the beta 2-tubulin isotype-specific carboxy terminus. Dev Biol 1993; 158:213-227.

Faguy DM, Doolittle WF. The evolution of cell division. Curr Biol 1998; 8:R338-R341.

Felix M-A, Antony C, Wright M, Maro B. Centrosome assembly *in vitro*: role of gamma-tubulin recruitment in *Xenopus* sperm aster formation. J Cell Biol 1994; 124:19-31.

Fouquet J-P, Eddé B, Kann M-L, Wolff A, Desbruyères E, Denoulet P. Differential distribution of glutamylated tubulin during spermatogenesis in mammalian testis. Cell Motil Cytoskel 1994; 27:49-58.

Fox LA, Sawin KE, Sale WS. Kinesin-related proteins in eukaryotic flagella. J Cell Sci 1994; 107:1545-1550.

Fulton C, Simpson PA. Selective synthesis and utilization of flagellar tubulin. The multi-tubulin hypothesis. In: (Goldman R, Pollard T, Rosenbaum J, eds) Cell Motility Book C: Microtubules and Related Proteins. Cold Spring Harbor, MA: Cold Spring Harbor Lab, 1976, pp987-1005.

Gaertig J, Cruz MA, Bowen, J, Gu L, Pennock DG, Gorovsky MA. Acetylation of Lysine 40 in alpha-tubulin is not essential in *Tetrahymena thermophila*. J Cell Biol 1995; 129:1301-1310.

Gagnon C. Regulation of sperm motility at the axonemal level. Reprod Fertil Dev 1995; 7:847-855.

Gagnon C, White D, Huitorel P, Cosson J. A monoclonal antibody against the dynein IC1 peptide of sea urchin spermatozoa inhibits the motility of sea urchin, dinoflagellate and human flagellar axonemes. Mol Biol Cell 1994; 5:1051-1063.

Gagnon C, White D, Cosson J, Huitorel P, Eddé B, Desbruyères E, Paturle-Lafanechère L, Multigner L, Job D, Cibert C. The polyglutamyl lateral chain of alpha-tubulin plays a key role in flagellar motility. J Cell Sci 1996; 109:1545-1553.

Gard DL, Kirschner MW. A polymer-dependent increase in phosphorylation of beta-tubulin accompanies differentiation of a mouse neuroblastoma line. J Cell Biol 1985; 100:764-774.

Gardner LC, O'Toole E, Perrone CA, Giddings T, Porter ME. Components of a "dynein regulatory complex" are located at the junction between the radial spokes and the dynein arms in *Chlamydomonas* flagella. J Cell Biol 1994; 127:1311-1325.

Gentleman S, Kaiser-Kupfer MI, Scherins RJ, Caruso R, Robison WG, Lloyd-Muhammad RA, Crawford MA, Pikus A, Chader GJ. Ultrastructural and biochemical analysis of sperm flagella from an infertile man with a rod-dominant retinal degeneration. Hum Pathol 1996; 27:80-84.

Gibbons BH, Gibbons IR. The effect of partial extraction of dynein arms on the movement of reactivated sea urchin sperm. J Cell Sci 1973; 13:337-357.

Gibbons BH, Ogawa K, Gibbons IR. The effect of anti-dynein 1 serum on the movement of reactivated sea urchin sperm. J Cell Biol 1976; 71:823-831.

Gibbons IR. Cilia and flagella of eukaryotes. J Cell Biol 1981; 91:107s-124s.

Gibbons IR, Rowe AJ. Dynein: a protein with adenosine triphosphatase activity from cilia. Science 1965; 149:424-426.

Gibbons IR, Gibbons BH, Mocz G, Asai DJ. Multiple nucleotide-binding sites in the sequence of dynein beta heavy chain. Nature 1991; 352:640-643.

Gingras D, White D, Garin G, Multigner L, Job D, Cosson J, Huitorel P, Zingg H, Dumas F, Gagnon C. Purification, cloning, and sequence analysis of a Mr=30,000 protein from sea urchin axonemes that is important for sperm motility. J Biol Chem 1996; 271:12807-12813.

Gingras D, White D, Garin J, Cosson J, Huitorel P, Zingg H, Cibert C, Gagnon C. Molecular cloning and characterization of a radial spoke head protein of sea urchin sperm axonemes. Mol Biol Cell 1998; 9:513-522.

Goldsmith M, Connolly JA, Kumar N, Wu J, Yarbrough LR, van der Kooy D. Conserved beat-tubulin binding domain for the microtubule-associated motors underlying sperm motility and fast axonal transport. Cell Motil Cytoskel 1991; 20:249-262.

Goldsmith M, Yarbrough LR, van der Kooy D. Mechanics of motility: distinct dynein binding domains on alpha- and beta-tubulin. Biochem Cell Biol 1995; 73:665-671.

Gurland G, Gundersen GG. Stable, detyrosinated microtubules function to localize vimentin intermediate filaments in fibroblasts. J Cell Biol 1995; 131:1275-1290.

Haimo LT, Rosenbaum JL. Cilia, flagella and microtubules. J Cell Biol 1981; 91:125s-130s.

Hecht NB, Distel RJ, Yelick PC, Tanhauser SM, Driscoll CE, Goldberg E, Tung KSK. Localization of a highly divergent mammalian testicular alpha-tubulin that is not detectable in brain. Mol Cell Biol 1988; 8:996-1000.

Hirose K, Lockhart A, Cross RA, Amos LA. Three-dimensional cryoelectron microscopy of dimeric kinesin and ncd motor domains on microtubules. Proc Natl Acad Sci USA 1996; 93:9539-9544.

Hoenger A, Milligan RA. Motor domains of kinesin and ncd interact with microtubule protofilaments with the same binding geometry. J Mol Biol 1997; 265:553-564.

Holy J, Schatten G. Recruitment of maternal material during assembly of the zygote centrosome in fertilized sea urchin eggs. Cell Tissue Res 1997; 289:285-297.

Huang B. *Chlamydomonas reinhardtii*, a model system for the genetic analysis of flagella structure and motility. Int Rev Cytol 1986; 99:181-215.

Huang B, Ramanis Z, Luck DJ. Suppressor mutations in *Chlamydomonas* reveal a regulatory mechanism for flagellar function. Cell 1982; 28:115-124.

Huitorel P. From cilia and flagella to intracellular motility and back again: a review of a few aspects of microtubule-based motility. Biol Cell 1988; 63:249-258.

Huitorel P, Gagnon G, White D, Cosson J, Eddé B, Cibert C. Antitubulin antibodies show a non uniform distribution along the axoneme in flagella. J Cell Biochem 1995; 19B:108.

Hyams JS, Lloyd CW. Microtubules. Wiley-Liss, New York, 1994.

Iomini C, Bré M-H, Levilliers N, Justine J-L. Tubulin polyglycylation in Platyhelminthes: diversity among stable microtubule networks and very late occurrence during spermiogenesis. Cell Motil Cytoskel 1998; 39:318-330.

Jamieson BGM, Ausio J, Justine J-L. Advances in spermatozoal phylogeny and taxonomy. Mém Mus Natl Hist Nat 1995; 166:1-565.

Johnson KA. Pathway of the microtubule-dynein ATPase and the structure of dynein: a comparison with acto-myosin. Ann Rev Biophys Chem 1985; 14:161-188.

Johnson KA. The axonemal microtubules of the *Chlamydomonas* flagellum differ in tubulin isoform content. J Cell Sci 1998; 111:313-320.

Johnson KA, Rosenbaum JL. Polarity of flagellar assembly in *Chlamydomonas*. J Cell Biol 1992; 119:1605-1611.

Kamiya R, Kurimoto H, Sakakibara H, Okagaki T. A genetic approach to the function of inner and outer arm dynein. In: (Warner FD, Satir P, Gibbons I, McIntosh JR, eds) Cell Movement. New York: Alan R Liss, 1989, pp209-218.

Khan IA, Luduena RF. Phosphorylation of beta III-tubulin. Biochemistry 1996; 35:3704-3711.

King SM, Wilkerson CG, Witman GB. The Mr 78,000 intermediate chain of *Chlamydomonas* outer dynein arm dynein interacts with alpha-tubulin *in situ*. J Biol Chem 1991; 266:8401-8407.

King SM, Barbarese E, Dillman JF III, Patel-King RS, Carson JH, Pfister KK. Brain cytoplasmic and flagellar outer arm dyneins share a highly conserved Mr 8,000 light chain. J Biol Chem 1996; 271:19358-19366.

Kotani S, Kawai G, Yokoyama S, Murofushi H. Interaction mechanism between microtubule-associated proteins and microtubules. A proton magnetic resonance analysis on the binding of synthetic peptide to tubulin. Biochemistry 1990; 29:10049-10054.

Laferrière NB, MacRae TH, Brown DL. Tubulin synthesis and assembly in differentiating neurons. Biochem Cell Biol 1997; 75:103-117.

Larcher JC, Boucher D, Lazereg S, Gros F, Denoulet P. Interaction of kinesin motor domains with alpha- and beta-tubulin subunits at a tau-independent binding site. J Biol Chem 1996; 271:22117-22124.

LeDizet M, Piperno G. Identification of an acetylation site of *Chlamydomonas* alpha-tubulin. Proc Natl Acad Sci USA 1987; 54:5720-5724.

Lewis SA, Cowan NJ. Complex regulation and functional versatility of mammalian alpha- and beta-tubulin isotypes during the differentiation of testis and muscle cells. J Cell Biol 1988; 106:2023-2033.

L'Hernault SW, Rosenbaum JL. *Chlamydomonas* alpha-tubulin is posttranslationally modified by acetylation on the epsilon-amino group of a lysine. Biochemistry 1985; 24:473-478.

Liao G, Gundersen GG. Kinesin is a candidate for cross-bridging microtubules and intermediate filaments. Selective binding of kinesin to detyrosinated tubulin and vimentin. J Biol Chem 1998; 273:9797-9803.

Lindemann CB, Kanous KS. A model for flagellar motility. Int Rev Cytol 1997; 173:1-72.

Luck DJ. Genetic and biochemical dissection of the eucaryotic flagellum. J Cell Biol 1984; 98:789-794.

Luck DJ, Piperno G, Ramanis Z, Huang B. Flagellar mutants of *Chlamydomonas*: studies of radial spoke-defective strains by dicaryon and revertant analysis. Proc Natl Acad Sci USA 1977; 74:3456-3460.

Luduena RF. Are tubulin isotypes functionally significant? Mol Biol Cell 1993; 4:445-457.

Luduena RF. Multiple forms of tubulin: different gene products and covalent modifications. Int Rev Cytol 1998; 178:207-275.

Maccioni RB, Rivas CI, Vera JC. Differential interaction of synthetic peptides from the carboxyl-terminal regulatory domain of tubulin with microtubule-associated proteins. EMBO J 1988; 7:1957-1963.

MacRae TH. Tubulin post-translational modifications. Enzymes and their mechanism of action. Eur J Biochem 1997; 244:265-278.

Mary J, Redeker V, Le Caer J-P, Promé J-C, Rossier J. Class I and IVa beta-tubulin isotypes expressed in adult mouse brain are glutamylated. FEBS Lett 1994; 353:89-94.

Mary J, Redeker V, Le Caer J-P, Rossier J, Schmitter, J-M. Posttranslational modifications in the C-terminal tail of axonemal tubulin from sea urchin sperm. J Biol Chem 1996; 271:9928-9933.

Marya PK, Syed Z, Fraylich PE, Eagles PAM. Kinesin and tau bind to distinct sites on microtubules. J Cell Sci 1994; 107:339-344.

McKeithan TW, Rosenbaum JL. Multiple forms of tubulin in the cytoskeletal and flagellar microtubules of *Polytomella*. J Cell Biol 1981; 91:352-360.

Mocz G, Gibbons IR. ATP-insensitive interaction of the amino-terminal region of the beta heavy chain of dynein with microtubules. Biochemistry 1993; 32:3456-3460.

Mohri H. Amino-acid composition of "Tubulin" constituting microtubules of sperm flagella. Nature 1968; 217:1053-1054.

Multigner L, Pignot-Paintrand I, Saoudi Y, Job D, Plessmann U, Rüdiger M, Weber K. The A and B tubules of the outer doublets of sea urchin sperm axonemes are composed of different tubulin variants. Biochemistry 1996; 35:10862-10871.

Murphy DB. Functions of tubulin isoforms. Curr Opin Cell Biol 1991; 3:43-51.

Navara CS, Simerly C, Zoran S, Schatten G. The sperm centrosome during fertilization in mammals: implications for fertility and reproduction. Reprod Fertil Dev 1995; 7:747-754.

Nogales E, Wolff SG, Downing KH. Structure of the alpha-beta tubulin dimer by electron crystallography. Nature 1998; 391:199-203.

Oakley BR. Gamma-tubulin. In: (Hyams JS, Lloyd CW, eds) Microtubules. New York: Wiley-Liss, 1994, pp33-45.

Ogawa K. Four ATP-binding sites in the midregion of the β-heavy chain of dynein. Nature 1991; 352:642-645.

Okuno M, Asai DJ, Ogawa K, Brokaw CJ. Effects of antibodies against dynein and tubulin on the stiffness of flagellar axonemes. J Cell Biol 1981; 91:689-694.

Omoto CK, Kung C. The pair of central tubules rotates during ciliary beat in *Paramecium*. Nature 1979; 279:532-534.

Ozols J, Caron JM. Posttranslational modification of tubulin by palmitoylation: II. Identification of sites of palmitoylation. Mol Biol Cell 1997; 8:637-645.

Paturle-Lafanechère L, Eddé B, Denoulet P, Van Dorsselaer A, Mazarguil H, Le Caer JP, Wehland J, Job D. Characterization of a major brain tubulin variant which cannot be tyrosinated. Biochemistry 1991; 30:10523-10528.

Paturle-Lafanechère L, Manier M, Trigault N, Pirollet F, Mazarguil H, Job D. Accumulation of delta 2-tubulin, a major tubulin variant that cannot be tyrosinated, in neuronal tissues and in stable microtubule assemblies. J Cell Sci 1994; 107:1529-1543.

Piperno G, Fuller MT. Monoclonal antibodies specific for an acetylated form of alpha-tubulin recognize the antigen in cilia and flagella from a variety of organisms. J Cell Biol 1985; 101:2085-2094.

Piperno G, Ramanis Z. The proximal portion of *Chlamydomonas* flagella contains a distinct set of inner dynein arms. J Cell Biol 1991; 112:701-709.

Piperno G, Huang B, Ramanis Z, Luck DJ. Radial spokes of *Chlamydomonas* flagella: polypeptide composition and phosphorylation of stalk components. J Cell Biol 1981; 88:73-79.

Piperno G, Ramanis Z, Smith EF, Sale WS. Three distinct inner dynein arms in *Chlamydomonas* flagella: molecular composition and location in the axoneme. J Cell Biol 1990; 110:379-389.

Plessman U, Weber K. Mammalian sperm tubulin: an exceptionally large number of variants based on several posttranslational modifications. J Prot Chem 1997; 16:385-390.

Porter ME, Johnson KA. Dynein structure and function. Ann Rev Cell Biol 1989; 5:119-151.

Raff EC. The role of multiple tubulin isoforms in cellular microtubule function. In: (Hyams JS, Lloyd CW, eds) Microtubules. New York: Wiley-Liss, 1994, pp85-109.

Raff EC, Fackenthal JD, Hutchens JA, Hoyle HD, Turner FR. Microtubule architecture by a beta-tubulin isoform. Science 1997; 275:70-78.

Raybin D, Flavin M. Modification of tubulin by tyrosylation in cells and extracts and its effect on assembly *in vitro*. J Cell Biol 1977; 73:492-504.

Redeker V, Melki R, Promé D, Le Caer JP, Rossier J. Structure of tubulin C-terminal domain obtained by subtilisin treatment. The major alpha and beta tubulin isotypes from pig brain are glutamylated. FEBS Lett 1992; 313:185-192.

Redeker V, Levilliers N, Schmitter J-M, Le Caer J-P, Rossier J, Adoutte A, Bré M-H. Polyglycylation of tubulin: a posttranslational modification in axonemal microtubules. Science 1994; 266:1688-1691.

Renaud FL, Rowe AJ, Gibbons IR. Some properties of the protein forming the outer fibers of cilia. J Cell Biol 1968; 36:79-90.
Renthal R, Schneider BG, Miller MM, Luduena, RF. Beta IV is the major beta-tubulin isotype in bovine cilia. Cell Motil Cytoskel 1993; 25:19-29.
Sale WS, Satir P. Direction of active sliding of microtubules in *Tetrahymena* cilia. Proc Natl Acad Sci USA 1977; 74:2045-2049.
Satir P, Wais-Steider J, Lebduska S, Nasr A, Avolio J. The mechanochemical cycle of the dynein arm. Cell Mot 1981; 1:303-327.
Schatten G. The centrosome and its mode of inheritance: the reduction of the centrosome during gametogenesis and its restoration during fertilization. Dev Biol 1994; 165:299-335.
Shelanski ML, Taylor EW. Properties of the protein subunit of central-pair and outer-doublet microtubules of sea urchin flagella. J Cell Biol 1968; 38:304-315.
Sleigh MA. Patterns of movement of cilia and flagella. In: (Sleigh MA, ed) Cilia and Flagella. New York: Academic Press, 1974, pp79-92.
Smith EF, Lefebvre PA. The role of central apparatus components in flagellar motility and microtubule assembly. Cell Motil Cytoskel 1997; 38:1-8.
Smith EF, Sale WS. Regulation of dynein-driven microtubule sliding by the radial spokes in flagella. Science 1992; 257:1557-1559.
Smith EF, Sale WS. Mechanisms of flagellar movement: functional interactions between dynein arms and the radial spoke-central apparatus complex. In: (Hyams JS, Lloyd CW, eds) Microtubules. New York: Wiley-Liss, 1994, pp381-392.
Song Y-H, Mandelkow E. The anatomy of flagellar microtubules: polarity, seam, junctions, and lattice. J Cell Biol 1995; 128:81-94.
Stephens RE. Tubulin in sea urchin embryonic cilia: post-translational modifications during regeneration. J Cell Sci 1992; 101:837-845.
Sullivan KF. Structure and utilisation of tubulin isotypes. Ann Rev Cell Biol 1988; 4:687-716.
Sullivan KF, Cleveland DW. Identification of conserved isotype-defining variable region sequences from four vertebrate beta tubulin polypeptide classes. Proc Natl Acad Sci USA 1986; 83:4327-4331.
Summers K, Gibbons I. Adenosine triphosphate-induced sliding of tubules in trypsin-treated sperm flagella of sea urchin sperm. Proc Natl Acad Sci USA 1971; 68:3092-3096.
Thaler CD, Haimo LT. Microtubule motors: mechanisms of regulation. Int Rev Cytol 1996; 164:269-327.
Tucker C, Goldstein LSB. Probing the kinesin-microtubule interaction. J Biol Chem 1997; 272:9481-9488.
Villasante A, Wang D, Dobner P, Lewis SA, Cowan NJ. Six mouse alpha-tubulin mRNAs encode five distinct isotypes: testis-specific expression of two sister genes. Mol Cell Biol 1986; 6:2409-2419.
Wang H, Satir P. The 29 kD light chain that regulates axonemal dynein activity binds to cytoplasmic dyneins. Cell Motil Cytoskel 1998; 39:1-8.
Warner, FD. Organization of interdoublet links in *Tetrahymena* cilia. Cell motil 1983; 3:321-332.
Warner FD, McIntosh JR. Cell movement. Vol 2. Kinesin, dynein, and microtubule dynamics. Alan R Liss, New York, 1989.
Warner FD, Satir P, Gibbons IR. Cell movement. Vol 1. The dynein ATPases. Alan R Liss, New York, 1989.
Wilson PG, Borisy GG. Evolution of the multi-tubulin hypothesis. BioEssays 1997; 19:451-454.
Wilson PG, Zheng Y, Oakley CE, Oakley BR, Borisy GG, Fuller MT. Differential expression of two gamma-tubulin isoforms during gametogenesis and development in *Drosophila*. Dev Biol 1997; 184:207-221.
Witman GB. Axonemal dyneins. Curr Opin Cell Biol 1992; 4:74-79.
Witman GB, Wilkerson CG, King SM. The biochemistry, genetics, and molecular biology of flagellar dynein. In: (Hyams JS, Lloyd CW, eds) Microtubules. New York: Wiley-Liss, 1994, pp229-249.
Wolff A, de Néchaud B, Chillet D, Mazarguil H, Desbruyères E, Audebert S, Eddé B, Gros F, Denoulet P. Distribution of glutamylated alpha and beta-tubulin in mouse tissues using a specific monoclonal antibody, GT335. Eur J Cell Biol 1992; 59:425-432.
Wooley DM. Studies on the eel sperm flagellum. I. The structure of the inner dynein arm complex. J Cell Sci 1997; 110:85-94.

45

Immunocontraceptive Vaccines for the Control of Wild Animal Populations: Antigen Selection and Delivery

M. K. Holland
CSIRO
K. Beagley
University of Newcastle
C. Hardy
L. Hinds
CSIRO
R. C. Jones
University of Newcastle

For the past 150 years, Australia's unique flora and fauna have been under threat principally from three pest species, the rabbit, the fox and the cat, introduced by European settlers. In that time period, 18 species of Australian mammals—half the world total—have become extinct. Significant contributors to this have been habitat modification, through clearing of large tracts of land, and the introduction of herbivores. These factors resulted in habitat suitable for native animals existing only in small, isolated areas of low agricultural value. This is the exact situation in which predation by or competition from these pest species has greatest impact (Morton 1990). These pest species, together with the accidentally introduced house mouse, cause significant economic impact; rabbits alone cost $600 million in lost agricultural production annually. During extreme situations, such as occur during mouse plagues, loss of an entire crop can occur as well as significant human discomfort and the potential such situations hold for transmission of disease.

The traditional approach to this problem has been to use lethal based strategies, such as poisoning, shooting or trapping, which can have significant localized, short-term impacts on populations of pest species, provided the rate at which animals are removed from the population exceeds that at which they are born or recruited into it. However, there are clear limitations to lethal based methods. Most of these procedures are expensive, labor intense, are not always simple to implement and need to be constantly re-applied to ensure efficacy. When control is being considered at a continent-wide level, as is required in the case of foxes and rabbits, these become major limitations. In addition, newer concerns relating to issues of animal welfare and the precision with which only the target species can be affected by the lethal control procedure being employed demand consideration.

One of the more successful ways to achieve broad-based lethal control has been biological control through the use of species-specific pathogens. Introduction of the myxoma virus from South America into the Australian rabbit population is one of the great examples of where this approach, so successful with a number of invertebrate species, has been applied to the vertebrate kingdom (Fenner, Radcliffe 1965). The population of the Australian European rabbit (*Oryctolagus cuniculus),* in almost 100 years prior to the introduction of myxoma virus, had increased from 24 founding animals to almost 600 million. This represented, at the time, approximately 60 rabbits for every Australian man, woman and child. Within a very few years the rabbit population crashed because of the high (99.9%) lethality of the original virus. Unfortunately, within three years, attenuated strains of the virus were being isolated from the field (Fenner, Woodroofe 1953) and there was clearly strong selection for genetic resistance on the part of the rabbit. Host and virus rapidly evolved toward the state seen with the virus and its native host, the cottontail rabbit. However, myxoma still maintains a significant impact on rabbit populations, with large numbers of animals dying in annual epizootics. Unfortunately, rabbits which survive infection are solidly immune to re-infection and provide the breeding basis for the next generation. However, while rabbit numbers have steadily risen, they have never again achieved the levels encountered some 50 years ago. Nevertheless, rabbits remain the cause of an estimated $600 million annual loss in agricultural production and so there is considerable interest in the development of new methods for their control.

Lethal, species-specific pathogens have not been identified for the fox or the mouse. Agents which cause disease in foxes invariably also cause disease in dogs and probably also in the native Australian canid, the dingo. In other species, such as cats, the situation is made more complex by the fact that many cats are highly valued companion animals. There are thus clear limits to both the applicability and long-term utility of lethal biological control.

Considerations of the questionable effectiveness of some current methods used for control, including their cost and specificity, applicability across the diverse climatic zones of continental Australia and whether they are regarded as humane, have become the drivers for consideration of new ways to employ existing technologies more effectively (Williams et al. 1995) and for the development of novel technologies which address these concerns. One such technology is fertility control (Tyndale-Biscoe 1991). How this technology might be developed to meet the different reproductive strategies presented by the fox, cat, mouse and rabbit is clearly going to be very different for each species. In addition, other factors, such as species distribution, reproductive life and immunology, all differ. This chapter focuses on how fertility control might be achieved for rabbits.

Fertility Control

Fertility control can potentially be aimed at either sex. The ideal contraceptive would target both sexes to maximize efficacy. In this context, spermatozoa are attractive targets. However, the female remains the breeding unit of the population and in species such as mice and rabbits where monogamous relationships do not form, the female becomes, of necessity, the primary target. In such species, small numbers of fertile males can cover the entire female breeding population. In species where strong pair bonding forms, especially if the bonding lasts over a number of breeding seasons, either sex may be effectively targeted.

A number of steps within the reproductive process may be targets for fertility control in either or both sexes. These form several categories: 1) Effects on the hormonal control of reproduction—these can be achieved at levels from the hypothalamus to the pituitary to the gonads; 2) Effects on gametogenesis—these might be either direct effects on developing gametes or effects on the complex intragonadal regulatory mechanisms which coordinate this process; 3) Direct effects on the oocyte or spermatozoa; 4) Direct targeting of the pre-implantation embryo; 5) Interference with implantation; 6) Effects on the post-implantation embryo.

To maximize contraceptive efficacy, an antifertility agent may target more than one of these steps. A number of factors influence this selection process: how is infertility to be induced (chemical, endocrinological or immunological means?); how is the infertility agent to be delivered (immunization, oral delivery?); what side effects can be tolerated (loss of libido, effects on cholesterol levels, and so on?). The answers to these questions are clearly very different, for antifertility agents intended for animal or human use, and indeed even within the field of animal use, may differ according to whether the species targeted is a companion animal or is regarded as a pest.

To be effective, fertility control must be of long-term duration. In the case of pest species, ideally a single exposure would remove the animal from the breeding population permanently. This both minimizes cost and maximizes efficacy. However, it precludes the use of most chemical agents, which suffer the additional liability of rarely being species-specific. It addition, it must also be highly effective with large numbers of animals being rendered infertile; again, ideally after a single exposure. One approach which can potentially meet these restrictions is immunocontraception.

Immunocontraception

Immunocontraception relies on the induction of an antibody or a cellular immune response to a reproductive antigen which interferes with a key step in reproduction, resulting in infertility. For species such as the house mouse, which is a problem because of relatively localized eruptions of the population as a consequence of particular environmental conditions, it is feasible to disseminate the immunocontraceptive through a bait. Such an approach has two difficulties: 1) how to package the immunocontraceptive antigen so that it survives within the bait and after ingestion, so that 2) the antigen is presented at the gut in such a way that it provokes an immune response which is effective within the reproductive tract.

Currently, experiments are investigating a number of approaches to this question, ranging from the use of inert systems, such as microbeads and immune stimulating complex molecules (ISCOMs), to bacterial ghosts to live but genetically disabled vectors, such as vaccinia virus and Salmonella strains. Such an approach, combined with accurate prediction of when population build-ups were to occur, could potentially provide an effective means of population control.

However, for more widely dispersed species such as foxes and rabbits, a major difficulty is how to deliver immunocontraception to the target population both cheaply and effectively. As yet, no species-specific disseminating vector which could be used in foxes has been identified. However, in the case of rabbits, there is the myxoma virus, which only affects selective lagomorph species. In Australia, this means only the two introduced species, the European rabbit (*O. cuniculus)* and hare (*L. europus*).

Viral Vectored Immunocontraception

The concept of viral vectored immunocontraception relies on combining the principle of immunocontraception with one of the most recent advances in vaccine technology, the use of modified viruses to carry biologically effective antigens. In the case of rabbits, the existence of myxoma virus provides an important opportunity. Myxoma virus is a member of the family *Poxviridae* Genus *Leporipoxvirus*. It is a double stranded DNA virus of about 163 kb related to vaccinia virus. Recombinant vaccinia viruses containing foreign genes are used as routine expression vectors both *in vivo* and *in vitro* for a range of biological applications (Binns, Smith 1992). Perhaps the best known use of such a virus is as part of the bait used to immunize foxes against rabies (Bacon, MacDonald 1980). It is easy to imagine that such a bait could also carry an engineered virus which would produce an immunocontraceptive response.

Essentially the techniques which permit construction of a recombinant vaccinia virus can be applied to construct recombinant myxoma viruses (Jackson, Bults 1992). Two steps are involved. First, the foreign gene is inserted into a plasmid insertion vector behind a pox virus promoter which regulates the time of expression in the cycle of viral replication and the magnitude of expression of the foreign gene. The foreign gene is flanked by normally contiguous viral DNA sequences. Second, the plasmid construct is used to transfect virus infected cells. The site of insertion of the foreign gene into the myxoma genome is determined by homologous recombination between these viral flanking sequences and the same sequences within the viral genome.

The result is transfer of the foreign gene from the plasmid into the viral genome. This resultant recombinant virus now can replicate to produce more recombinant viruses but as it does so, the foreign gene is expressed (Fig. 1). If this foreign gene encodes a

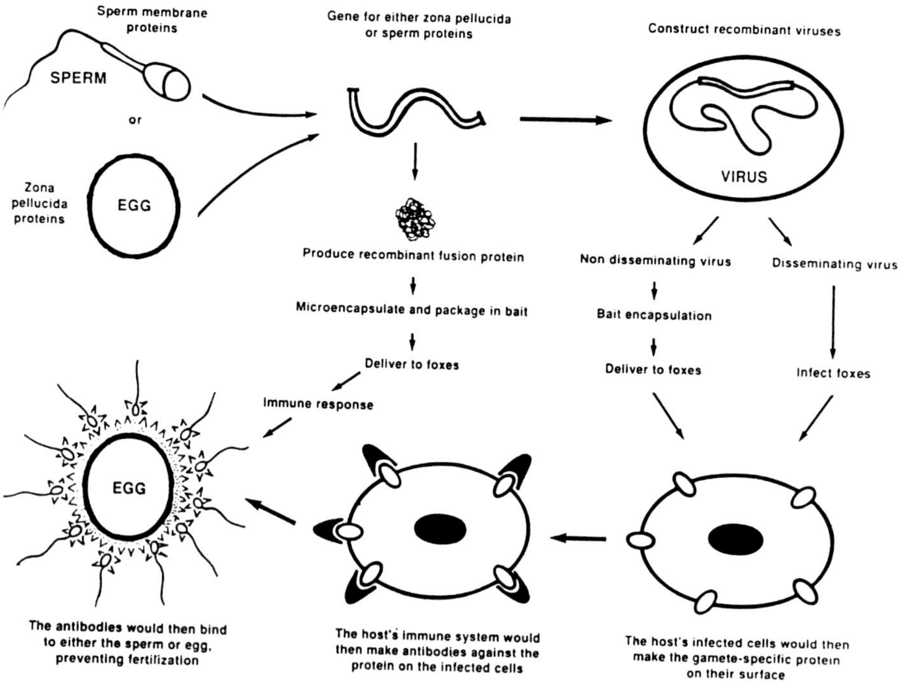

Figure 1. The concept of viral vectored immunocontraception. A gene encoding a component of the sperm or egg essential to fertilization is inserted into the myxoma virus by homologous recombination. The recombinant virus infects a rabbit and, as it replicates to produce more recombinant virions, it produces the protein encoded by the inserted gene. The rabbit mounts an immune response to the virus and also to the protein encoded by the inserted gene. This antibody binds to its antigen on the surface of sperm or egg and effectively inhibits fertilization.

component of rabbit sperm, or perhaps the zona pellucida, it will trigger an immune response which will then attack this antigen wherever it is located and in so doing produce a contraceptive response.

The success of viral vectored immunocontraception is thus determined by identification and selection of the appropriate target antigen and construction of a recombinant virus capable of delivering the antigen with appropriate immunogenicity. Indeed, these questions of selection of an antigen and immunogenicity are at the heart of any immunocontraceptive vaccine program.

Antigen Selection in Immunocontraception

A number of steps, ranging from hormones to components of gametes to the developing embryo, can potentially provide targets for immunocontraception and research on a number of these has recently been summarized in several books (Alexander et al. 1990; Talwar, Raganathy 1995). However, much of this research has been conducted with the ultimate goal of developing a vaccine for human use. This places restrictions, which do not always apply in vaccines intended for animal use, on the selection of antigens. For example, most people would regard targeting the developing fetus to produce an abortion later than the first trimester as ethically unacceptable. For many, this extends back in development to cover even the process of implantation itself. However, in the case of animals, while inducing late stage abortions would probably also be equally unacceptable, preventing implantation or attacking the early developing fetus may be quite acceptable.

In contrast, there are also restrictions which apply solely to animal use. For example, many pest species have strong social hierarchies which determine which females in a population breed. The balance of the reproductive hormones is often important in establishing and maintaining an animal's position in this hierarchy. Thus, an immunocontraceptive vaccine which targets these hormones may perturb the social hierarchy, releasing subordinate animals to breed. Potentially, this could lead to an increase in the pest animal population (Caughley et al. 1992).

Overlaying all these issues is the question of species specificity of the intended antigen. Crucial also is the question of maximizing efficacy. When all these factors are considered, gamete antigens, and particularly spermatozoa, emerge as the most likely source of species specific-antigens which are highly immunogenic, potentially effective in either sex, act through the acceptable path of preventing fertilization and which could be produced cheaply and in large quantities using current molecular technology.

Spermatozoal Antigens

Spermatozoa are regarded by the immune system as non-self antigens because they are first produced at puberty, long after immune competence has been established. Generally spermatozoa are shielded from immune surveillance in males by the blood-testis barrier (Johnson 1973) and by a degree of local immune tolerance (El Demiery et al. 1985) or immunosupression (Lehmann, Emmons 1989). However, when this shielding breaks down through disease, such as with mumps, or is deliberately breached, as in vasectomy, the production of anti-sperm antibodies results (Gubin et al. 1998). There is a highly significant correlation between the presence of anti-sperm antibodies and infertility (Heidenreich et al. 1994) in men. What prevents females from mounting an immune response to spermatozoa is less clear. However, isolation to within the female tract and some direct immunosupression seem indicated. Nevertheless, some females do produce anti-sperm antibodies which have long been associated with infertility (Escuder 1936) and, more recently, failed *in vitro* fertilization (Shibahara et al. 1996). However, attempts to use these infertile antisera to identify sperm antigens which may be used in a contraceptive vaccine have generally failed (Raghupathy et al. 1990) although there have been promising leads, especially using new molecular technologies.

An alternative approach is to test the ability of purified sperm antigens for their immunocontraceptive potential (Table 1). Both O'Rand (1995) and Naz (1995) have recently reviewed the evidence for sperm antigens which are possible candidates for an immunocontraceptive vaccine and that process will not be repeated here. Their analysis is in agreement with that of the authors of this chapter and suggests that the three sperm antigens for which there is most compelling evidence of their immunocontraceptive effectiveness are PH-20, PH-30 and LDH-C_4 (O'Rand, Lea 1997). These sperm antigens have all been shown capable of producing an immunocontraceptive response in their own species as well as in other species.

Accordingly, researchers set about obtaining the rabbit homologs of these proteins. The approach taken was to isolate and clone the encoding gene from a rabbit testicular cDNA library and produce recombinant protein using bacterial expression systems. This approach yields large amounts of protein, sufficient to conduct large scale immunization trials, rapidly and inexpensively. It is much easier and simpler than attempting a laborious and inefficient isolation from testis. However, as O'Rand has pointed out recently, "the problem of converting a successful native contraceptive antigen into an effective synthetic or recombinant gamete immunocontraceptive is at the heart of the problem" of designing a successful immunocontraceptive vaccine (O'Rand, Lea 1997). Thus, testing these recombinant rabbit sperm antigens was of general interest.

Table 1: Candidate sperm antigens for immunocontraceptive vaccine development effective by active immunization.

Antigen	Species of origin	Species in which effective
PH-20	guinea pig	guinea pig (male and female)
	monkey	
	human	
LDH-C_4	rabbit	rabbit
	mouse	mouse
	human	baboon
PH-30	guinea pig, mouse	guinea pig
	rabbit, bull	
	rat, human	
SP-10	baboon	baboon
	human	
	monkey	
RSA 2/ SP-17	rabbit	rabbit
	mouse	
	human	
FA-1	rabbit, monkey	rabbit
	mouse, human	
	bull	

PH-20

The rabbit PH-20 gene was cloned from a lambda gt11 rabbit testis library using guinea pig PH-20 cDNA (a generous gift of Dr. P. Primakoff) as probe (Holland et al. 1997). The rabbit gene was expressed in the FLAG system to produce recombinant protein (Fig. 2). This protein was used to immunize female rabbits (n=10) according to a standard protocol (Fig. 3). Blood samples were taken throughout and serum titres to PH-20 showed clear response to immunization (Fig. 4). The antisera to recombinant PH-20 reacted with the region on sperm to which PH-20 is localized. It also reacted with testis sections in a specific manner (Fig. 5) and on Western blots of sperm extracts reacted with native PH-20 protein. This antisera also had an inhibitory effect on fertilization rates in an *in vitro* fertilization assay (data not shown). These properties are consistent with the recombinant protein inducing an immune response which affects the native protein. However, no effect on the fertility of animals immunized either subcutaneously or directly into the Peyer's patch, in attempt to provoke a local mucosal response within the reproductive tract, was detectable.

While it is clear that the antiserum raised to recombinant PH-20 interacts with native protein in a highly specific manner, it is possible that the epitopes recognized on the recombinant protein differ from those seen on the native molecule. An expression system such as FLAG, in which the fusion protein carries a small (six amino acid) tag which should have minimal effects on folding of the recombinant protein, was selected in attempt to maximize the chances that the recombinant protein would resemble the native protein. In the absence of structural studies comparing native and recombinant protein, it is unclear how successful this attempt was, despite the fact that antisera to recombinant PH-20 recognized native PH-20, indicating at least some epitopes must be shared.

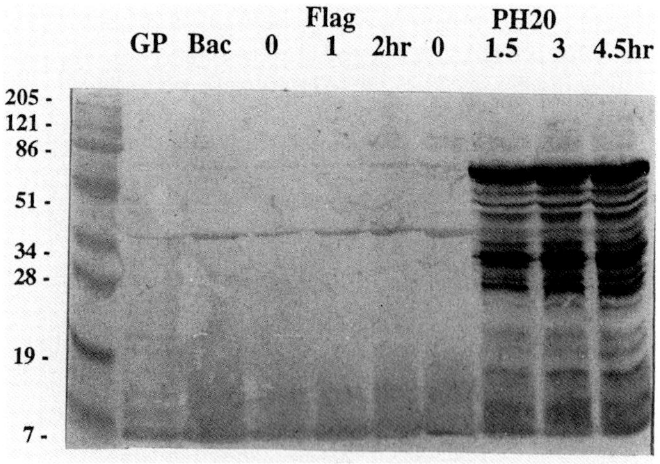

Figure 2. Production of recombinant rabbit PH-20 (see arrow) in *E. coli* using the FLAG expression system. Protein is produced after induction by ITPG. The FLAG only adds a small hexapeptide to the recombinant protein, so it is very similar in size to native PH-20.

In an attempt to further understand structure-function relationships in PH-20, sequence analysis was used to identify three peptides, each of 15 amino acids, which were likely to be exposed on the surface of native PH-20 (Holland et al. 1997). These peptides were synthesized chemically and conjugated to key hole limpet haemocyanin before being used to immunize rabbits using the protocol previously described. While there were significant differences in immunogenicity among the peptides, the resultant antisera still recognized native protein on Western blots and on testis sections (data not shown). However, again there was no effect on the fertility of animals immunized subcutaneously. This suggests that none of these peptides is involved in any immunocontraceptive response to PH-20.

After this series of negative results, it would be reassuring to know that native rabbit PH-20 could produce infertility in immunized rabbits. To this end, researchers are attempting to purify sufficient PH-20 from rabbit testes to conduct this experiment.

Antigen Immunization Protocol

Day 0	Priming Injection	400μg FCA
Day 28	First Boost	200μg FIA
Day 42	Second Boost	200μg FIA
Day 56	Fourth Boost	200μg FIA
Day 66	Mate	
Day 96	Fertility Analysis	

Figure 3. Immunization protocol and sampling protocol for female rabbits immunized with recombinant sperm antigens.

PH-30

The cloning, sequencing and expression of rabbit PH-30 alpha and beta has been described previously (Hardy, Holland 1996). The respective subunits, produced as maltose binding protein (MBP) fusion proteins, were used to immunize rabbits in a protocol similar to that previously described for PH-20. Again the resultant antisera reacted with sperm and on Western blots with native fertilin subunits (Hardy, Holland 1997) but no effect on fertility was detected with either subunit (Table 2), despite the fact that antisera to both subunits exhibited inhibitory effects in *in vitro* fertilization assays (Table 3). Again it is possible that the recombinant proteins do not display conformational epitopes seen on native protein which are essential to the immunocontraceptive response because of differences in tertiary or quaternary structure. Alternatively, it is possible that neither native PH-20 nor native PH-30 are immunocontraceptive in rabbits. Given the profound effect of both these antigens in guinea pig and assuming their homologs in rabbits perform a similarly conserved function, this explanation seems unlikely.

Another explanation might lie in the nature of the immune response within the rabbit reproductive tract after immunization. It seems clear that when the antibodies to either recombinant PH-20 or PH-30 can access the native protein, as in *in vitro* fertilization assays, antibody can adversely affect sperm-egg binding and thus fertilization rates. If antibody does not readily gain access to the protein within the reproductive tracts of immunized animals, infertility would not result.

Immune Responses within the Rabbit Reproductive Tract

The immune responses within the rat female reproductive tract have been studied in a long series of articles by Wira and colleagues (Kaushic et al. 1998). These studies essentially show that estrogen is the driving force for antibody entry. No such detailed studies exist for the rabbit. Accordingly, researchers used micropuncture to obtain samples from the vagina, uterus and oviduct of animals immunized against PH-20 using the protocol outlined earlier. The antibody levels in reproductive fluids were compared to that in blood. Anti-PH-20 IgG level in vaginal fluid was 0.016% of the levels found in blood. Uterine fluid (0.078%) and oviduct fluid (0.072%) levels were higher but still a very small fraction of the serum response.

Rabbits are an induced ovulating species and are normally locked in pro-estrus. Whether induction of ovulation through administration of hCG would change the antibody response was studied in PH-20 immunized animals. Some 12 hours after induction of ovulation

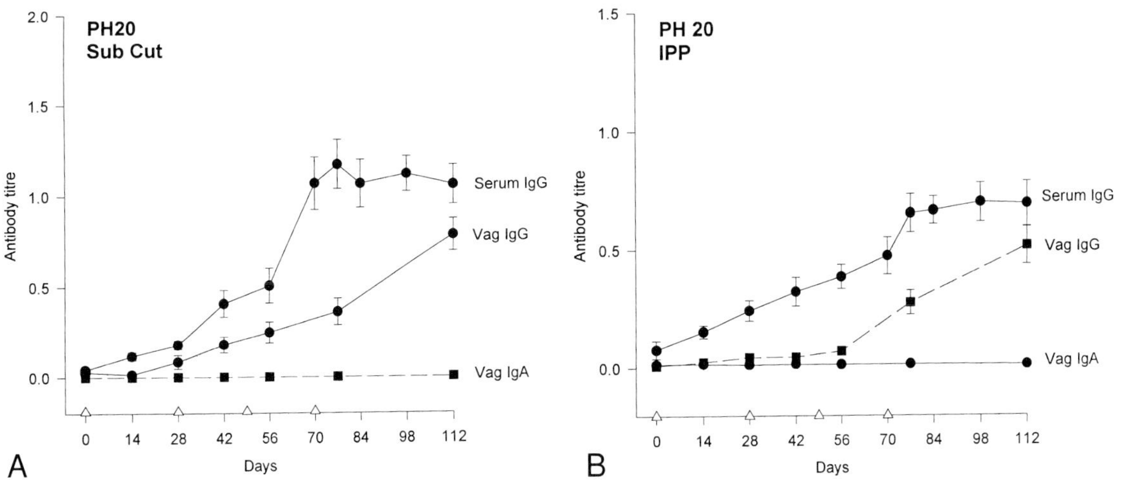

Figure 4. Serum antibody responses in female rabbits immunized a) subcutaneously (Sub Cut) and b) intra Peyer's patch (IPP) with recombinant rabbit PH-20. Protein was administered according to the schedule in Figure 3. For intra Peyer's patch injections, animals were anesthetized, their intestine exposed and Peyer's patches visualized and injected. Boosting injections were given intradermally.

(the time at which fertilization would normally occur) only vaginal antibody levels changed, increasing by an order of magnitude to be 0.16% of serum levels. The uterus and oviduct levels showed no change. It would thus seem that limited entry of antibody into the female rabbit reproductive tract is a major factor in determining whether an immunocontraceptive response can be produced. Current research is focused on understanding antibody entry into the female rabbit reproductive tract and on different approaches to inducing mucosal responses within the tract. Preliminary studies with a recombinant myxoma virus indicate that while the virus can effectively present antigen in a manner which results in a serum immune response, the response in the reproductive tract remains low. Thus, achieving immunocontraception using a recombinant myxoma virus for delivery will require selection of an antigen which can be accessed via the circulation. To this end a recombinant myxoma virus containing the rabbit zona pellucida gene encoding the ZPB protein was tested. This produces an immune response which renders a significant number of rabbits infertile. Thus, the concept of delivering a reproduction antigen via a recombinant virus and producing an immune response which affects fertility has been established. Much remains to be done to optimize both selection of antigen and to achieve the optimal immune response.

Conclusions

Fertility control provides a potential new strategy in the overall battle for the control of feral pest populations. Its applicability will largely be influenced by many factors, including the reproductive biology and

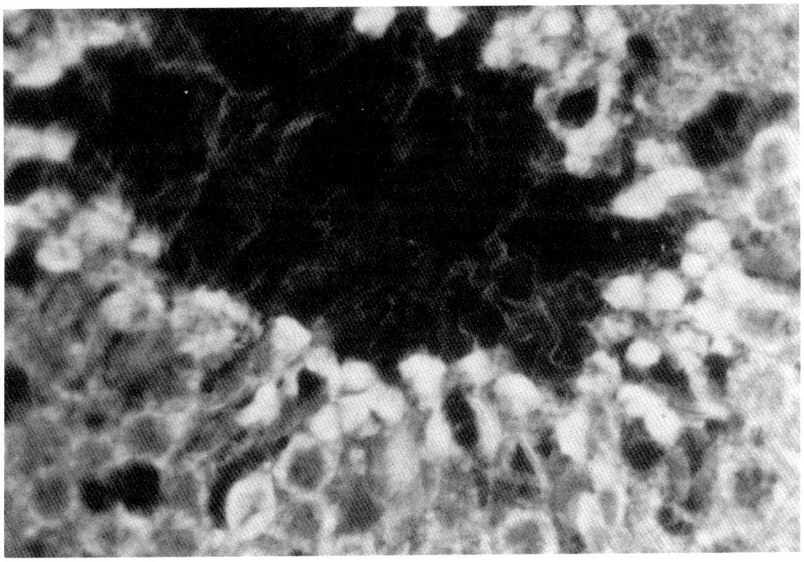

Figure 5. Localization of PH-20 to the rabbit testis. Rabbit testis was immersion fixed in Bouins solution, embedded in paraffin and sections 5µm cut and placed onto glass slides. After dewaxing, the slides were probed with mouse anti-rabbit PH-20 and antibody localization was visualized using a fluorescein conjugated anti-mouse secondary antibody.

Table 2: Fertility of female rabbits immunized subcutaneously (SC) or intra Peyer's patch (IPP) with rabbit recombinant fertilin subunits.

	Fertilin alpha		Fertilin beta		MBP*	
	SC	IPP	SC	IPP	SC	IPP
Total rabbits	6	9	8	10	6	9
Infertile rabbits	1	1	1	1	0	0
Viable fetuses**	4.5±1.0	5.2±0.8	3.8±0.7	4.5±0.6	4.3±0.5	4.7±0.6

*MBP = maltose binding protein; **mean ± SEM.

Table 3: Effect of anti-rabbit fertilin α and β on fertilization rates *in vitro*.

	Antibody treatment Fertilin alpha	Fertilin beta	MBP*
Experiments	5	4	5
Total eggs	63	77	69
% Fertilized	28 ± 2**	14 ± 4	20 ± 9
Zona bound sperm	3.0 ± 0.4**	1.5 ± 0.6	5.0 ± 0.8
Sperm in perivitelline space	0.3 ± 0.1**	0.1 ± 0.0	0.9 ± 0.1

*MBP = maltose binding protein; **mean ± SEM.

immunology of the target species. For rabbits, antigens which require high levels of antibody to be established within the reproductive tract are unlikely to be successful because of limitations to antibodies entering the female tract. That fact led researchers to consider antigens, such as components of the zona pellucida, which act outside the reproductive tract.

Acknowledgements

Financial support for these studies was through the Vertebrate Bio-control Co-operative Research Center and the Australian Research Grants Commission. The authors thank the many technicians, including Hannah Clarke, Jenny Grigg, Clayton Walton and John Wright, who contributed to this project.

References

Alexander NJ, Griffin D, Spieler JM, Waites GM. Gamete Interaction: Prospects for Immunocontraception. Wiley-Liss, New York, 1990.
Bacon PJ, MacDonald DW. To control rabies: vaccinate foxes. New Scientist 1980; 87:640-645.
Binns MM, Smith GL. Recombinant poxviruses. CRC Press, London, 1992.
Caughley G, Pech R, Grice D. Effect of fertility control on a population's productivity. Wildlife Res 1992; 19:623-627.
El-Demiry MIM, Hargreave T, Busuttil A, James K, Ritchie A, Chisholm G. Lymphocyte sub-populations in the male genital tract. Brit J Urol 1985; 57:769-774.
Escuder CJ. Temporary biologic sterility of women injected with human sperm. Arch Fac Med Montevideo 1936; 21:889-903.
Fenner F, Radcliffe FN. Myxomatosis. Cambridge University Press, Cambridge, 1965.
Fenner F, Woodroofe GN. The pathogenesis of infectious myxomatosis: the mechanism of infection and the immunological response in the European rabbit. Brit J Exptl Pathol 1953; 34:400-418.
Gubin DA, Dmochowski R, Kutteh WH. Multivariate analysis of men from infertile couples with and without antisperm antibodies. Am J Reprod Immunol 1998; 39:157-160.
Hardy CH, Holland MK. Cloning and expression of recombinant rabbit fertilin. Mol Reprod Dev 1996; 45:107-116.
Hardy CH, Clarke HG, Nixon B, Grigg JA, Hinds LA, Holland MK. Examination of the immunocontraceptive potential of recombinant rabbit fertilin sub-units in rabbit. Biol Reprod 1997; 57:879-886.
Heidenreich A, Bonfig R, Wilbert D, Strohmaier W, Engelmann U. Risk factors for antisperm antibodies in infertile men. Am J Reprod Immunol 1994; 31:69-76.
Holland MK, Andrews J, Clarke H, Walton C, Hinds L. Selection of antigens for use in a virus vectored immunocontraceptive vaccine: PH-20 as a case study. Reprod Fert Dev 1997; 9:117-124.
Jackson RJ, Bults HG. A myxoma virus intergenic transient dominant selection vector. J Gen Virol 1992; 73: 3241-3245.
Johnson MH. Physiological mechanisms for the immunological isolation of spermatozoa. Adv Reprod Physiol 6:279-324.
Kaushic C, Frauendorf E, Rossoll R, Richardson J, Wira CR. Influence of the estrous cycle on the presence and distribution of immune cells in the rat reproductive tract. Am J Reprod Immunol 1998; 39:209-216.
Lehmann D, Emmons LR. Immune phenomena observed in the testis and their possible role in infertility. Am J Reprod Immunol Microbiol 1989; 19:43-52.
Mims C. Aspects of the pathogenesis of virus disease. Bacteriol Rev 1964; 28:30-45.
Morton S. The impact of European settlement on the vertebrate animals of arid Australia: a conceptual model. Proc Ecol Soc Aust. 1990; 16:201-219.
Naz RK. Sperm-antigen based birth control vaccines. In: (Talwar GP, Raghupathy R, eds) Birth Control Vaccines. RG Landes Co, Austin, Texas, 1995.
O'Rand MG. Antigenic targets on sperm for immunological interception. In: (Kurpisz M, Fernandez N, eds) Immunology of Human reproduction. Bios Scientific Publishers, Oxford, 1995.
O'Rand MG, Lea IA. Designing an effective immunocontraceptive. J Reprod Immunol 1997; 36:51-59.
Raghupathy R, Shaha C, Gupta SK. Autoimmunity to sperm antigens. Curr Opin Immunol 1990; 2:757-760.
Shibahara H, Mitsuo M, Ikeda Y, Shigeta M, Koyama K. Effect of sperm immobilising antibodies on pregnancy outcome in infertile women treated with IVF-ET. Am J Reprod Immunol 1996; 36:96-100.
Talwar GP, Raghupathy R, eds. Birth Control Vaccines. RG Landes Co, Austin, Texas, 1995.
Tyndale-Biscoe CH. Fertility control in wildlife. Reprod Fert Dev 1991; 3:339-343.
Williams K, Parer I, Coman B, Burley J, Braysher M. Managing vertebrate pests—rabbits. Aust Gov Print Ser, Canberra, 1995.

46 *Comparative Cryobiology of Mammalian Spermatozoa*

The Procedures of Cryopreservation

Basic Aspects of Cryobiology

Mechanisms of Injury Caused by Freezing

Proposed Genetic Analysis of Freezing Injury of Spermatozoa

Conclusions

S. P. Leibo
L. Bradley
University of Guelph

It is impossible to exaggerate the influence that the freezing of mammalian spermatozoa has had on agriculture, animal husbandry and human medicine. Tens of millions of cows artificially inseminated with cryopreserved bull spermatozoa have delivered calves (Iritani 1980) and thousands of women have been successfully impregnated with frozen-thawed spermatozoa (Federation CECOS et al. 1989). Although less widely used, spermatozoa of many other mammalian species, especially of domestic animals, have also been successfully cryopreserved (Iritani 1980). As used here, "success" is defined as meaning that live young have resulted from fertilization of oocytes with frozen-thawed spermatozoa. A comprehensive summary of both the methods and the use of cryopreserved spermatozoa for artificial insemination (AI) in mammals as well as in other vertebrate species has been published by Watson (1990). The same author has also described and analyzed more recent observations and basic concepts regarding preservation of spermatozoa (Watson 1995). In addition, a detailed description of the fundamental cryobiology of mammalian spermatozoa has also been recently published (Gao et al. 1997). There have been numerous other recent reviews of sperm cryopreservation, including those with somewhat more limited focus on specific species, such as the human (Watson et al. 1992a; Royere et al. 1996), cattle (Foote, Parks 1993; Rodriguez-Martinez et al. 1997), horses (Amann, Pickett 1987), dogs (England 1993; Farstad 1996) and sheep (Salamon, Maxwell 1995a, b).

A fascinating aspect of the reviews by Watson (1990) and the two by Salamon and Maxwell (1995a, b) is their description of the fact that mammalian spermatozoa were first frozen more than 60 years ago and that motile cells were recovered after thawing. According to Salamon and Maxwell (1995a), as early as 1947 to 1950, I.V. Smirnov had achieved success in vitrifying very small volumes of spermatozoa to temperatures as low as -78°C or -183°C, even producing offspring after AI with frozen-thawed rabbit spermatozoa.

It is generally accepted, however, that it was Christopher Polge, Audrey Smith and Alan Parkes (Polge et al. 1949; Smith, Polge 1950) who were primarily responsible for devising reliable and reproducible methods to freeze spermatozoa to low temperatures. Most importantly, they showed that glycerol protected fowl and bull spermatozoa against damage caused by freezing to -79°C, the temperature of dry ice. In 1951, Stewart reported that one cow out of five that were inseminated with bull spermatozoa frozen to -79°C by the method of Smith and Polge (1950) delivered a normal bull calf. As Stewart (1951) said: "If spermatozoa can remain fertile at -79°C at all, it is reasonable to believe that at such low temperatures fertility might be maintained for very long periods, in which case the advantages to be derived from the application of this technique to specialized breeding and breeding experiments would be obvious." Shortly thereafter, Polge and Rowson (1952) demonstrated the full potential of their method by inseminating 25 cows, 15 of which produced live calves. In 1954, Bunge et al. reported the births of four children from women inseminated with cryopreserved spermatozoa. Since those early reports, sperm freezing has become a routine and widely practiced and accepted procedure.

Measured in terms of the numbers of species whose spermatozoa have been successfully cryopreserved and of the very large numbers of live young resulting from AI with cryopreserved spermatozoa, the procedures of sperm cryopreservation seem eminently successful. However, that is only partially correct. On average, about half of all human and cattle spermatozoa are damaged or destroyed by freezing and thawing, limiting the overall efficiency and efficacy of semen preservation. With other species, the results may be even poorer and more variable. Furthermore, as Critser et al. (1987) demonstrated with human spermatozoa, even those cells that have survived freezing and are motile with intact membranes do not maintain their viability and fertilizing capacity as long as do unfrozen spermatozoa. There is also evidence that pregnancies resulting from fertilization with frozen-thawed spermatozoa may not be as "healthy" as those conceived by fertilization with fresh spermatozoa. Although >85% of oocytes collected from cows that had been inseminated by AI with cryopreserved sperm were fertilized, many of these presumptive pregnancies were lost during the first two weeks after insemination (Diskin, Sreenan 1980). Six to 10 weeks later (60 to 90 days post-insemination), the non-return rate was 65 to 70%, indicating an additional 20 to 30% loss of early pregnancies. Because non-return rates have been widely used to estimate fertility of dairy cattle, considerable effort has been devoted to enhancing their predictive ability (Reurink et al. 1990; Van Doormaal 1993). Nevertheless, non-return rates have limitations when used to judge changes in sperm freezing methods. A survey of 102 Ontario dairy farms revealed that the actual pregnancy rate diagnosed by rectal palpation at about 40 to 60 days after AI was much lower than that estimated by non-return rates. On average, only 47% of 9,800 inseminations of cows with frozen-thawed semen resulted in confirmed pregnancy (Kelton et al. 1993). As mentioned above, embryonic mortality appears to be greater with frozen-thawed spermatozoa than with fresh; this loss may amount to as much as 25% in some breeds (Kummerfeld et al. 1978; Wijeratne 1973).

Finally, the ultimate calving rate of cows inseminated with frozen semen often falls well shy of that resulting from AI with fresh semen, let alone the rate achieved by natural breeding. As an example, Olds (1978) compiled data from several sources showing that 62% of 41,492 cows bred one time by natural service delivered live calves. Of 690 cows that were artificially inseminated with fresh spermatozoa and diagnosed by rectal palpation to be pregnant at 50 days, only 46 (7%) did not deliver a live calf at term (Fosgate, Smith 1954). In contrast, although the non-return rate of ~330,000 cows resulting from inseminations with cryopreserved sperm from 882 sires was 70%, the actual full-term calving rate was only ~50% (Taylor et al. 1985).

Taken together, these reviews indicate that the use of non-return rates as the primary tool to evaluate current methods to cryopreserve spermatozoa has limitations and may lead to overly optimistic assessments. What is noteworthy about these differences between fresh and frozen spermatozoa is that they have been observed with fertilization and pregnancy in cattle. It is with cryopreserved bovine spermatozoa that the most experience has been gained and the largest numbers of pregnancies have been produced. For most other mammalian species, there is considerably less information about differences between fresh and frozen cells, and thus, the influence of changes of freezing procedures on the ultimate outcome of live births.

Despite such limitations, sperm cryopreservation is widely practiced. For cattle breeding, the capability to freeze bull spermatozoa and to store it for extended times without loss of fertilizing capacity has permitted rigorous genetic selection of bulls to be exercised. For example, in Canada the use of frozen semen for artificial insemination of cattle began around 1957 (Penner 1990); parenthetically, it was recently shown that bull spermatozoa frozen by Penner in 1957 was still capable of successfully fertilizing oocytes 37 years later (Leibo et al. 1994). By 1974, a total of 2.4 million dairy cows were being bred by AI with cryopreserved spermatozoa to meet market demand for milk and dairy products.

Thereafter, however, as a result of highly selective breeding, it became possible to decrease this number of cows by almost 40% to a total of 1.4 million cows in 1988 (Penner 1990). Yet this cow herd was still large enough to provide sufficient milk for a Canadian population that had grown significantly during that 14-year period. Several years ago, Foote (1981) described analogous data for the United States. He noted that during the 25-year period from 1950 to 1975, the number of milking cows in the U.S. decreased from >22 million to <11 million; during this same interval, annual milk production almost doubled from 2,415 kg/cow in 1950 to 4,706 kg/cow in 1975. This was the time during which cryopreservation of bull spermatozoa was beginning to be used in conjunction with artificial insemination.

The derivation of extremely productive cattle has been possible because cryopreserved spermatozoa can be used to inseminate large numbers of cows to evaluate the productivity of the bulls' daughters. It is the characteristics of the sire that influence his daughters' capacity to produce large quantities of high-quality milk. This entire process to determine the genetic contribution of a bull to milk production is long and complex. It requires the following steps: 1) identification and selection of young bulls; 2) freezing of hundreds of "doses" of their semen; 3) insemination of heifers with the cryopreserved specimens to produce many dozens of calves; 4) selection of the female calves, allowing time for them to reach sexual maturity; 5) breeding of these heifers to make them pregnant so that they can be milked; and 6) evaluation of their productivity as daughters of the bulls that had been selected years before.

Methods used to cryopreserve bull spermatozoa for AI have changed relatively little in the past 40 years (Graham 1978; Pickett, Berndtson 1974; Watson 1990). The only major change is that in the 1950s and 1960s, frozen semen samples were stored in dry ice chests at -79°C, whereas now they are stored and transported in liquid nitrogen (LN_2) at -196°C. But even with modern methods, from 30 to 50% of all spermatozoa of most mammalian species are damaged or destroyed by the freezing-thawing process itself (Chen et al. 1993; Foote, Parks 1993; Rodriguez-Martinez et al. 1997; Salamon, Maxwell 1995a; Watson 1995; Watson et al. 1992a). For example, bull ejaculates that exhibit 70 to 80% progressive sperm motility before freezing suffer a decline to 30 to 40% sperm motility after thawing (Foote et al. 1993; Thomas et al. 1998). To compensate for this loss and yet maintain non-return rates of 65 to 70% by AI, more than two-thirds of all semen specimens cryopreserved in Canada, as in many other countries, are frozen at concentrations of $>10 \times 10^6$ sperm/dose. With unfrozen semen, in contrast, as few as 1×10^6 sperm/dose can yield equivalent pregnancy rates (Foote 1978; Shannon et al. 1984). This 10-fold difference between effective concentrations of fresh versus frozen sperm has very large economic consequences because it decreases the breeding units (the number of "doses" or straws) produced per sire. Furthermore, when spermatozoa are frozen at concentrations much higher than necessary for inseminations with fresh semen, the net effect is that the inherent fertility of the bull is confounded, because the contribution of high sperm concentrations to conception cannot be identified (Pace et al. 1981).

With cattle, for example, several factors determine the actual sperm concentration that is frozen in each straw. First, there are very significant differences among males with respect to the freezing sensitivity of their spermatozoa. Second, as mentioned above, almost half of all spermatozoa of most bulls is damaged or destroyed by freezing. To compensate for those losses, the sperm concentration frozen in each semen dose may be increased by a factor of >2 or 3 times the concentration of 10 x 10^6 sperm/dose that is considered to be the minimum that may be used commercially. Therefore, if the volume of an ejaculate is small and/or if the sperm concentration is low, the specimen often cannot be diluted very much. If a valuable bull happens to produce spermatozoa that are especially sensitive to freezing, it may be difficult to freeze enough doses of his semen to meet market demand. With young sires undergoing progeny testing, loss of 30 to 50% of their spermatozoa to freezing damage may mean that certain bulls are culled because they do not meet minimum criteria, and their genetic potential is lost.

Although spermatozoa of many mammalian species have been successfully cryopreserved, deriving mechanistic understanding of the reasons why spermatozoa are damaged by freezing and thawing has been difficult. Among the reasons that this has been difficult is the peculiar structure of the spermatozoon, the obvious differences in the size and shape of spermatozoa from different species, the unusual characteristics of the acrosome covering the sperm head, and the complex and heterogeneous nature of the sperm membranes (Hammerstedt et al. 1990; Parks 1997). To illustrate the diversity of these specialized cells, spermatozoa of six species are shown in Figure 1. There are obvious differences in the size and shape of the sperm heads—for example, compare those of the bull (A), the human (D) and the mouse (E)—as well as the length and structure of the principal pieces or tails—compare the stallion (C) and the mouse (E). The extent of these differences is also shown by the data in Table 1, taken from the tabulation by Cummins and Woodall (1985) of observations of many investigators. The total absolute lengths of spermatozoa as well as normalized lengths are shown. For comparison, also listed are normalized lengths of the spermatozoa shown in Figure 1. It is clear that the relative lengths of spermatozoa of different species differ, and that accurate measurements are difficult to make. Recently, Curry et al. (1996) have attempted to make a detailed analysis of sperm size by determining the surface areas (SA) and volumes (V) of ram and human spermatozoa. They found the respective values of SA and V were 135 μm^2 and 31.3 μm^3 for ram and 106 μm^2 and 22.2 μm^3 for human spermatozoa. As discussed by Curry et al. (1996), Watson (1995), and Gao et al. (1997), these characteristics, as surface area to volume ratios, have important implications for determining optimum cooling and warming procedures for cryopreservation. These dimensions yield SA/V ratios of 4.3 and 4.8 for ram and human spermatozoa, respectively. The relevance of these ratios to cryopreservation is discussed below.

Table 1: Dimensions of mammalian spermatozoa.

Species	Cummins & Woodall*		This paper
	μm	Normalized*	Normalized*
Cattle	53.5	100	100
Pig	40.0-54.6	75-102	83
Dog	62.7	117	83
Horse	60.6	113	82
Human	54.5-59.5	102-111	68
Mouse	122.9-124.3	~231	183

* Data of several observers, as collated by Cummins and Woodall, 1985; dimensions in μm. Normalized values were calculated relative to the measured length of the bovine sperm, either as specified by Cummins and Woodall, or as measured with a map measurer from enlargements of the photographs shown in Figure 1.

Many reviews have enumerated the multiple variables that influence survival of frozen-thawed spermatozoa (Coulter 1992; Foote, Parks 1993; Graham 1978; Parks, Graham 1992; Rodriguez-Martinez et al. 1997; Watson 1990, 1995). Factors that may affect survival include cooling and warming rates, sperm concentration, packaging (the container in which the semen is frozen), the extender solution itself, as well as the glycerol concentration, sugars, antibiotics, egg yolk and other supplements (original references are cited in the reviews listed above). It is notable that almost all spermatozoa of most species frozen throughout the world are cryopreserved using glycerol as the principal cryoprotective additive (CPA). This is so despite the fact that other compounds can also protect sperm against freezing damage. For example, compounds such as diothiothreitol (Richardson, Sadleir 1967) and ethylene glycol (Rao, David 1984) can protect human spermatozoa as effectively as glycerol. In a detailed study of human sperm permeability, this latter fact of the effectiveness of ethylene glycol has recently been confirmed by Gilmore et al. (1997). For cattle semen, there are recent indications that other CPAs might be as effective as glycerol (DeLeeuw et al. 1993).

Analysis of sperm cryopreservation is also complicated by the fact that, with all species for which sufficient data exist, there are large differences among different males in the freezing sensitivity of their sperm (Chen et al. 1993; Dhami et al. 1992; Iritani 1980). Furthermore, another confounding aspect of many experimental studies of sperm freezing is that semen specimens from different males are often pooled to yield a composite "typical" specimen. This may mask very significant male-to-male differences, differences that may themselves yield important information about freezing sensitivity. Such differences among males in the response of their spermatozoa to freezing have been reported for cattle (Shamsuddin et al. 1993) and humans (Glaub et al. 1976; Cohen et al. 1981; Taylor et al. 1982; McLaughlin et al. 1992), as well as for horses (Amann, Pickett 1987; Blach et al. 1989) and dogs (Thomas et al. 1993).

For all of these reasons, there are renewed efforts to attempt to understand basic aspects of the cryobiology of mammalian spermatozoa so as to improve procedures for their cryopreservation. That it might be possible to

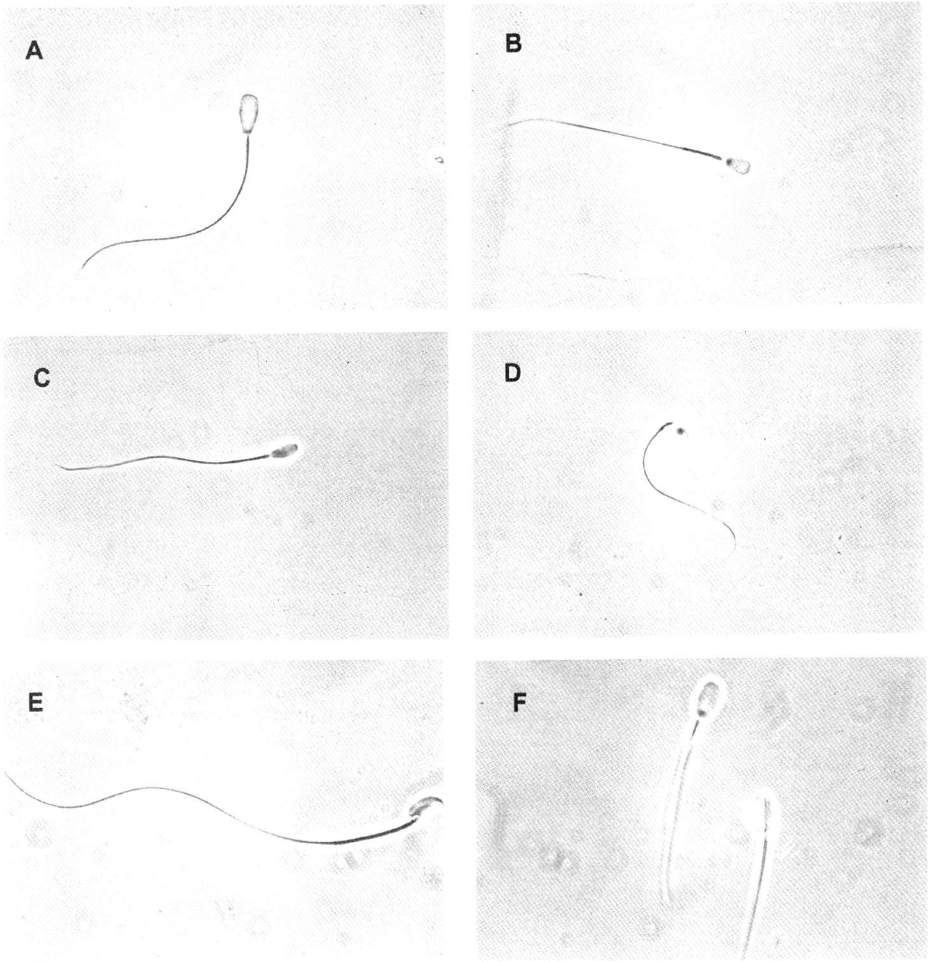

Figure 1. Phase contrast photomicrographs of spermatozoa. A: Bull; B: Dog; C: Stallion; D: Human; E: Mouse; F: Boar.

improve the current 50 to 70% survival of cryopreserved spermatozoa is suggested by the fact that the average survival of cryopreserved bovine embryos is 85 to 95% (Hasler et al. 1995, 1997; Niemann 1991). Paradoxically, bovine embryos have a volume ~40,000 times greater than that of spermatozoa; that characteristic alone would seem to render embryos much more susceptible than spermatozoa to freezing injury. Yet the opposite is true.

One important reason for the difference between the efficiency of sperm versus embryo cryopreservation is that embryos were not successfully cryopreserved until 1972 (Whittingham et al. 1972), more than 20 years after spermatozoa were first frozen (Polge et al. 1949). Success for embryo preservation depended on the considerable understanding in fundamental cryobiology of other types of cells that had emerged during the intervening years (Leibo et al. 1970; Mazur 1963; Mazur et al. 1972; Meryman 1966). Previously, because ~50% of sperm do survive freezing, and because a bull's ejaculate may contain 10 to 20 billion spermatozoa, there seemed little need to improve the efficiency of sperm freezing. Now, however, intense international competition in the cattle industry means that the loss of >30% of collected sperm is no longer economically acceptable.

With human semen, the need to improve the efficiency of sperm cryopreservation is even more critical since it may literally be a matter of life or death. It has been demonstrated that women may become infected with human immunodeficiency virus (HIV) by AI with fresh human semen. The potential for transmission of HIV, as well as of other viruses, has provided a strong impetus to improve the efficiency of sperm cryopreservation. Because of the risk of AIDS (autoimmune deficiency syndrome) and other sexually-transmitted diseases, it is now standard practice to quarantine cryopreserved human semen specimens for at least six months, to permit the sperm donor to be tested before his samples can be released for use. This drastically increases the indirect costs of semen preservation, because samples must be stored for an extended time and the sperm donor must be tested repeatedly. If the donor does not return to be tested at the end of the quarantine period, his semen may not be used. Any improvements in sperm freezing will have a large impact on the overall efficiency of human AI to treat infertility.

Mammalian spermatozoa have been successfully cryopreserved for decades. However, as mentioned above, within a species, survival after freezing of spermatozoa from different males varies widely. These differences make sperm freezing even less efficient since spermatozoa from different males may exhibit widely variable responses to the same freezing conditions. Reasons for these significant differences among different individuals remain unexplained. Examples of differences observed in four species are shown in Figure 2. Prior to freezing, sperm motility or viability may not vary much at all, as with the dog (A) and cattle (D), or it may vary significantly from 30 to 70%, as with human (C) and horse (B). However, in these examples, post-thaw motility or membrane integrity varied from 1 to 30%. Moreover, the post-thaw survival was independent of the pre-freeze value. Similar observations have also been described for spermatozoa of humans (Glaub et al. 1976; Kramer et al. 1993), bulls (Thomas et al. 1998), stallions (Amann, Pickett 1987) and boars (Larsson et al. 1976). Such differences among males of freezing sensitivity of their spermatozoa are widespread. Two detailed analyses of this phenomenon in humans have been published. In statistical analyses of multiple ejaculates of various characteristics of the semen of 315 men, Heuchel et al. (1983) found that the within-subject variance of sperm motility after freezing was much smaller than the between-subject variance. A similar observation was made by McLaughlin et al. (1992). These latter authors found that the mean percentages of survival after freezing (% motility after freezing ÷ % motility before freezing) of multiple sperm samples varied from 26 to 75%. For cattle, one consequence of this male-to-male variability in freezing sensitivity is that

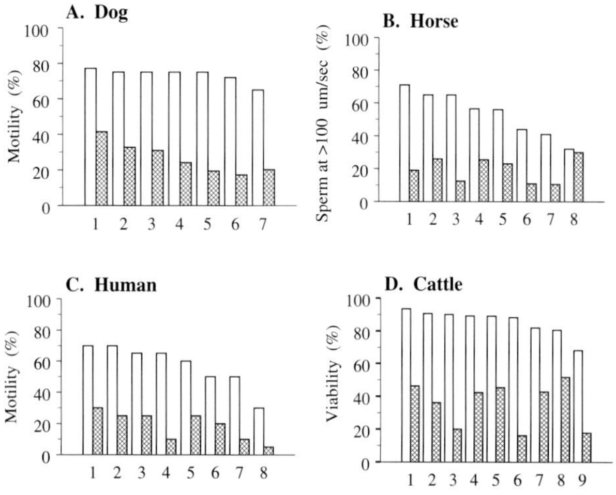

Figure 2. Comparison of pre-freeze and post-thaw assays of spermatozoa of four species. The open bars are the pre-freeze and the hatched bars are the post-thaw values. A: Dog, data of Thomas et al. (1993); B: Horse, data of Blach et al. (1989); C: Human, data of Cohen et al. (1981); D: Cattle, unpublished data of L. Bradley and S.P. Leibo.

many bulls whose genetic characteristics make them desirable cannot be used as sperm donors because their spermatozoa cannot be cryopreserved at a sufficiently high level to make them usable commercially.

The Procedures of Cryopreservation

That there are differences in semen characteristics among individuals is not surprising. Almost 45 years ago, Mann (1954) had documented large variations in the composition and characteristics of seminal plasma among different individuals of various species. Analyses of within-subject and between-subject variability of human semen specimens also reveal large differences among men with respect to the osmolality (Polak, Daunter 1984) and also sperm concentration, total spermatozoa and ejaculate volumes (Schwartz et al. 1979). However, that differences among males in sperm freezing sensitivity are found almost universally among many species and that the differences are so large is striking. At a superficial level, this phenomenon of male-to-male differences in sperm freezing sensitivity seems difficult to understand since the process of freezing and thawing is relatively simple. For spermatozoa, it usually consists of the following steps:

1. Ejaculated spermatozoa are collected, from animals commonly by use of an artificial vagina, and from humans by masturbation.
2. Either with or without being washed free of seminal plasma, cells are diluted with a buffer solution referred to as an "extender," often containing either egg yolk or milk as a supplement. The cells are then cooled very slowly to ~+5°C.
3. At that temperature, the spermatozoa are mixed with a solution of a cryoprotective additive (CPA), usually consisting of ~4 to 8% glycerol in the same milk or egg yolk extender. The spermatozoa are then distributed into plastic ampoules, or straws.
4. The samples are cooled from +5°C at about 10° to 20°C/min to a temperature of about -80°C and then placed into LN_2 for storage.
5. To restore their function, the cryopreserved spermatozoa are thawed rather rapidly and the CPA removed.

Given the simplicity of this method, it would seem reasonable for all cells of a given type to respond similarly to the same conditions of freezing and thawing. For example, most collections of cattle embryos from different cows usually exhibit high survival after cryopreservation (Hasler et al. 1995, 1997). Similarly, frozen mouse embryos also exhibit very high survival after thawing (Rall, Wood 1994; Dinnyés et al. 1995). High survival means that embryos collected from different animals or even different cattle breeds or mouse strains must have responded similarly to freezing. That is clearly not the case for spermatozoa. The question then becomes, "How can significant male-to-male differences in freezing sensitivity of spermatozoa be explained?"

Basic Aspects of Cryobiology

When cells are frozen and thawed, they undergo several cycles of osmotic dehydration and rehydration, resulting in extreme volume changes. The first occurs when cells are placed into a cryoprotective additive, such as glycerol, again when the solution freezes, then when it thaws, and finally when the CPA is removed from the cells. The theoretical responses of spermatozoa when exposed to various concentrations of several CPAs and when the CPAs are diluted have been modeled mathematically and have been summarized in detail in the recent review by Gao et al. (1997). These authors have quantitatively described the large volumetric changes that spermatozoa undergo as they equilibrate with CPAs, and have also illustrated the serious consequences that result from relatively small differences in various methods used to expose spermatozoa to CPAs and to dilute them. These volume excursions of spermatozoa result from the outflow of water when a cell is first exposed to a hypertonic solution, and from the re-entry of water when a cell that has equilibrated with a CPA is returned to isotonic solution. When spermatozoa are to be frozen, they are usually suspended in the CPA at temperatures near 0°C, a fact with important implications with respect to the fluidity of the sperm cell membrane. The role of the composition of sperm cell membranes of various species and the consequences of cell volume fluctuations at different temperatures have been described in detail by Parks (1997) and by Hammerstedt et al. (1990).

An equally important volume change occurs when cells are frozen. When a cell suspension is cooled several degrees below the freezing point of the suspending solution (about -1.5°C for 5% glycerol in saline), the solution freezes. That is, the liquid water crystallizes as ice. As the temperature is lowered, an increasing amount of water forms ice. As more ice crystals form and grow, they exclude unlike substances, such as cells and dissolved solutes. Rodriguez-Martinez and Ekwall (1998) have performed cryo-scanning electron microscopy of frozen suspensions of boar spermatozoa. Their elegant micrographs (Figs. 3, 4) of freeze-fractured specimens illustrate very dramatically the consequences of freezing solutions and the stresses to which spermatozoa are exposed during cryopreservation. In Figure 3, the large lacunae formed by ice crystals can be clearly seen, as well as the narrow channels formed by the solution and cells as they became concentrated during freezing. Even more dramatic is the higher magnification micrograph of the same specimen showing boar spermatozoa partially "embedded" in the veins of concentrated and solidified solution (Fig. 4). A "pit" from which another spermatozoon had been dislodged during the fracturing process is also evident in that figure.

When cell suspensions are frozen, they are cooled at finite rates often referred to as "slow" or "fast." But as pointed out by Mazur (1963), those expressions are relative terms. He noted that a rate that is slow for one cell type may be rapid for a second type. This is most clearly seen if one compares survival curves for different types of cells. As shown in Figure 5, the survival pattern exhibited by three very different cell types frozen in the same CPA solution is similar: survival is low (relatively or absolutely) at "slow" cooling rates, increases with "moderate" cooling, and then decreases at "fast" rates. The "optimum" cooling rate is the one at which maximum survival is observed. Based on the mathematical analysis by Mazur (1963) of the response of cells to freezing, he and his colleagues hypothesized that these "inverted U-shaped" survival curves resulted from the

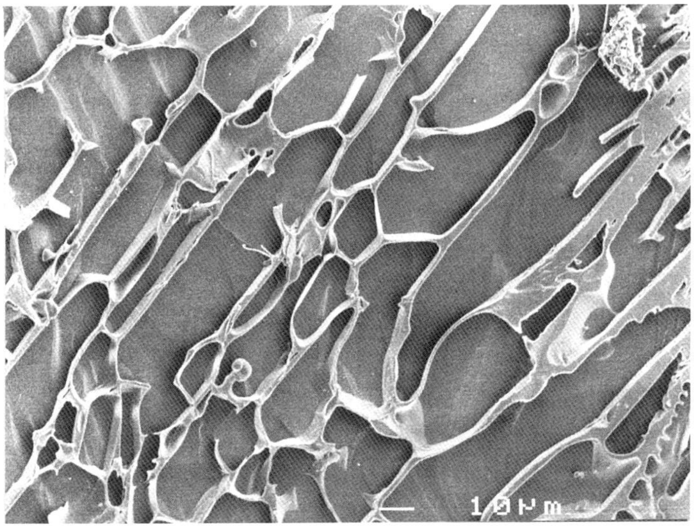

Figure 3. Scanning electron micrograph of a freeze-etched specimen of boar spermatozoa. The white bar is 10 µm long. The figure is that of Rodriguez-Martinez and Ekwall (1998) and is used with written permission of the publisher.

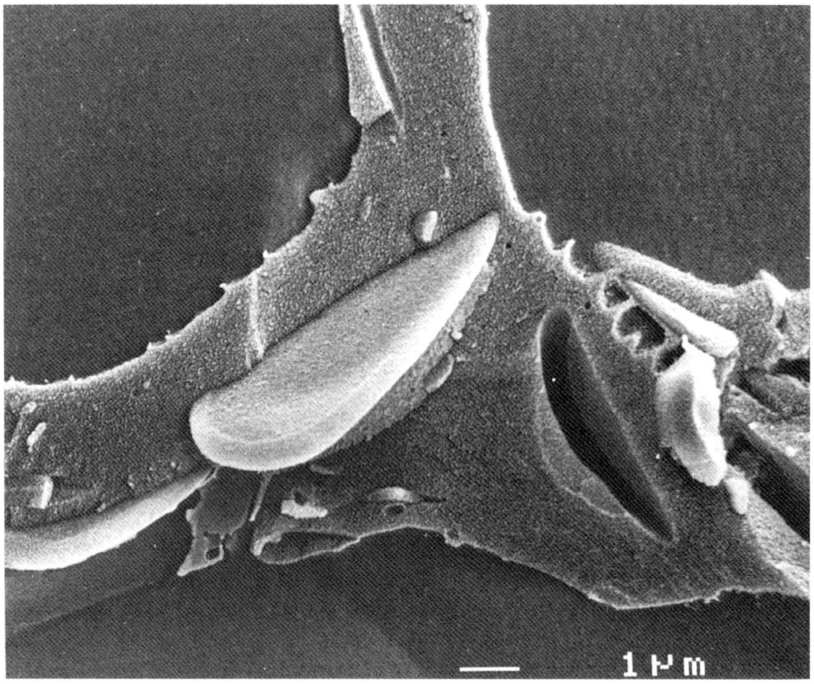

Figure 4. Higher magnification of the same specimen shown in Figure 3. The white bar is 1 μm long. The figure is that of Rodriguez-Martinez and Ekwall (1998) and is used with permission of the publisher.

interaction of two factors oppositely dependent on cooling rate (Mazur et al. 1972; Mazur 1984). This hypothesis states that the effect of Factor 1 decreases with increasing cooling rate; thus, survival increases. The effect of Factor 2 increases with increasing cooling rate; therefore, survival decreases. The mechanisms responsible for injury from Factor 1, termed "solution effects," are still largely unresolved and keenly debated among cryobiologists. The expression "solution effects" refers to the fact that the properties of aqueous salt solutions that are partially frozen at temperatures between -3° to about -35°C are drastically altered by removal of water in the form of ice. Once the first ice crystals have formed, as the temperature decreases, the solute concentration increases, some salts may crystallize, the pH increases (or decreases), the viscosity increases significantly, and the relative composition of the solutions changes. Factor 2 is generally, although not universally, interpreted to be intracellular ice formation. In the example shown, the respective optimum rates for oocytes, lymphocytes and erythrocytes are about 1°C/min, 15°C/min and 1,000°C/min. This example is for very different types of cells from three species. Another example of the same phenomenon is shown for bovine spermatozoa and embryos in Figure 6. In this case, only the right-hand leg of the embryo survival curve is evident, because the lowest rate tested was 0.3°C/min. At this rate, an apparent maximum survival of morula-stage embryos of ~55% was observed (Hochi et al. 1996). For bull spermatozoa, maximum survival was found to occur at a cooling rate of ~80°C/min (Woelders et al. 1997). The significance of this can be better appreciated if one considers the relative sizes of a bovine embryo and a bovine spermatozoon. The volume of a spherical morula is ~1.77×10^6 μm^3 with a surface area of 7.07×10^4 μm^2; thus, its SA/V ratio is 0.04. Assuming that bovine spermatozoa have dimensions similar to those of ram spermatozoa (as determined by Curry et al. 1994), then the bovine sperm SA/V ratio is ~4, or about 100-fold larger than that of bovine embryos. The SA/V ratio of a cell is one characteristic that influences the rate at which it can lose water; and the rate at which a cell can lose water is a principal determinant of its optimum cooling rate. The data in Figure 6 show that the optimum rate for spermatozoa is

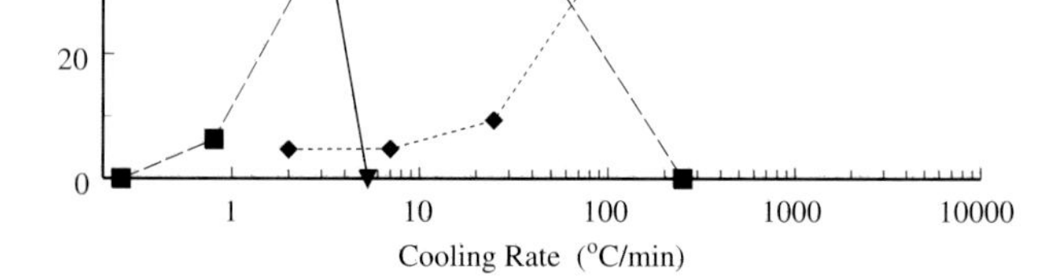

Figure 5. Survival of three types of cells suspended in 1 M dimethyl sulfoxide as a function of the cooling rate. The figure is redrawn from data collated by Leibo (1981).

>1000-fold faster than that of embryos.

For reasons briefly described above, efforts to determine mechanisms responsible for injury to spermatozoa have increased substantially during the past decade or so. As explained in recent reviews by Watson (1995) and by Gao et al. (1997), it has become apparent that empirical optimization of procedures to cryopreserve spermatozoa from various species has limits. No single CPA protects spermatozoa of all species, and there are not single optimum cooling and single optimum warming rates for each sperm species. Further evidence of the complexity of the problem is illustrated by the collation of sperm survival curves for three species shown in Figure 7. The data for boar are those of Fiser and Fairfull (1990), for human are those of Watson et al. (1992a), and for ram are those of Duncan and Watson (1992). This compilation is intended only to illustrate generalizations regarding the effect of cooling rates on sperm survival, since the procedures used in the three investigations differed in their details. Nevertheless, certain common features do emerge from this comparison. For example, survival rates of human, boar and ram spermatozoa seem somewhat less dependent on cooling rate than is true of many other types of cells. Although somewhat indistinct, the apparent optimum rates for human, boar and ram spermatozoa are about 10°C/min, 30°C/min and 50°C/min, respectively. In addition, the survival curves for all three sperm types do show the inverted U-shape described above for other types of cells. Although this clearly is not proof, it does suggest that similar factors operate to determine survival of all these types of cells.

The fact that sperm survival curves exhibit broad plateaus when plotted as a function of cooling rate has several possible explanations. As is true of other types of cells, it may result from the interaction of several factors, such as warming rate and CPA concentration. However, another explanation is that it reflects the fact that spermatozoa, even from the same species, let alone the same individual, are a much more heterogeneous population of cells. For a given species or certainly for an individual animal, most types of cells, such as erythrocytes, lymphocytes or oocytes, have very consistent and uniform properties. But even a single ejaculate of spermatozoa from one individual contains a diverse population of cells at various stages of maturation and with different capabilities of capacitation. The membranes of such spermatozoa are known to vary considerably in their composition and fluidity, characteristics that influence their permeability to water and to solutes, as well as their susceptibility to chilling injury (Parks 1997; Hammerstedt, Parks 1987; Schlegel et al. 1986). Furthermore, even with a homogeneous population of cells, the response to a single physical or chemical insult usually is variable. For example, even cultured cells (effectively, a cloned population of genetically identical cells) display a broad survival curve as a function of cooling rate (Mazur et al. 1972). Therefore, the broad survival plateaus shown in Figure 7 may simply result from the composite sensitivity of a diverse population of cells, each population exhibiting its own heterogeneous response. Regardless of the explanation of the survival plateau, it has been difficult for injuries to spermatozoa caused by freezing and thawing to be definitively identified.

Mechanisms of Injury Caused by Freezing

Most cryobiologists accept the logic that inverted U-shaped survival curves plotted as a function of

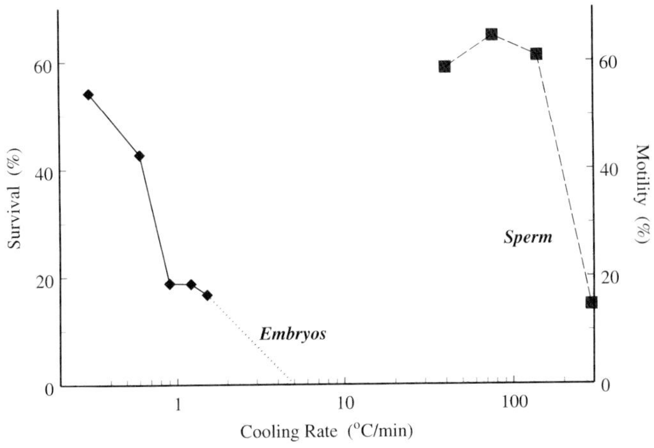

Figure 6. Survival of bovine embryos and of bovine spermatozoa as a function of cooling rate. The embryo data are those of Hochi et al. (1996) and the sperm data are those of Woelders et al. (1997).

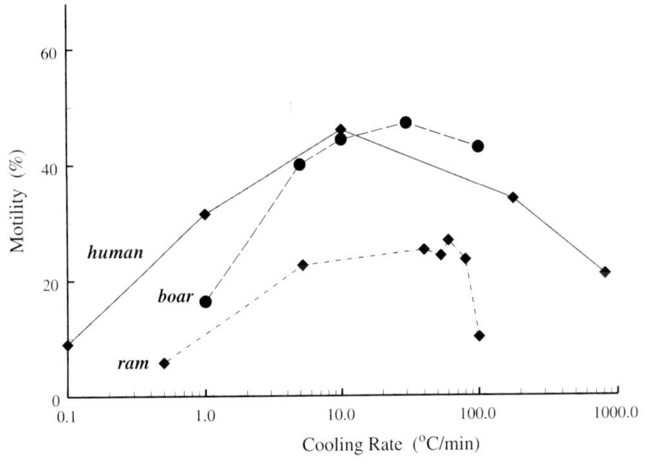

Figure 7. Post-thaw motility of spermatozoa of three species as a function of cooling rate. The data for human are those of Watson et al. (1992a), for boar are those of Fiser and Fairfull (1990), and for ram are those of Duncan and Watson (1992).

cooling rate indicate that two factors oppositely dependent on cooling rate operate to damage cells during freezing. Nevertheless, there is considerable debate as to actual causes of injury that occur during cryopreservation, especially to spermatozoa. Several possibilities are discussed in the reviews by Watson (1995) and by Gao et al. (1997). Here again, sperm survival curves plotted as a function of cooling rate provide clues. Another example is provided by the data for human spermatozoa redrawn in Figure 8 from the observations of Henry et al. (1993). They assayed post-thaw motility, as well as membrane integrity and mitochondrial function (data not shown here). They found that sperm survival varied as a function of a broad range of cooling rates, from 0.1°C/min to 800°C/min, with a distinct optimum. But the optimum cooling rate was dependent on the rate at which the frozen spermatozoa were warmed. If the samples were warmed rapidly, the optimum was 10°C/min, whereas if warmed slowly, the optimum was 1°C/min. These data reveal a phenomenon first described for embryos and other types of mammalian cells that was termed "rescue" of frozen cells (Leibo 1976b). Consider human spermatozoa cooled at 800°C/min (Fig. 8). At least 20% of them must have survived freezing to -196°C. But whether these spermatozoa ultimately survive depends on the rate at which they are warmed. If warmed slowly, the 20% survival is halved to 10%. Such reduction of survival caused by slow warming of cells that have been cooled rapidly has long been considered prima facie evidence of intracellular ice formation. Rapid warming causes metastable intracellular ice crystals to melt before they can grow to damaging size, whereas slow warming permits time for recrystallization of this intracellular ice, resulting in cell injury and death. Now consider spermatozoa cooled at 0.1°C/min (Fig. 8). During cooling from -3° to -33°C, these cells would have spent 300 min exposed to partially frozen solutions that were becoming increasingly concentrated. If warmed rapidly, then only ~5% of the cells survived. However, if warmed slowly at 1°C/min, then 18% of the spermatozoa survived. This seems paradoxical, since this slow warming would have added an additional 30 min of exposure to so-called "solution effects." The fact that slow warming "rescues" slowly frozen spermatozoa suggests that the damaging factor may be an osmotic phenomenon rather than being chemical toxicity. One possibility previously suggested is that damage is caused by a volume excursion of the cells occurring at subzero temperatures (Leibo 1976a).

Regardless of interpretation, it is generally agreed that the response to freezing is primarily influenced by a cell's permeability to water and to solutes. Therefore, to understand sperm cryobiology, considerable effort has been devoted during the past several years to determining the permeability characteristics of various species of spermatozoa (Curry et al. 1994, 1995, 1996; Duncan, Watson 1992; Gao et al. 1993, 1995; Gilmore et al. 1996, 1997; Noiles et al. 1993; Watson et al. 1992b). As reviewed by Watson (1995) and by Gao et al. (1997), the coefficients of water permeability (L_p, the hydraulic conductivity) of the spermatozoa of several species have been measured; a few examples are listed in Table 2. In addition, the activation energies (a measure of the temperature dependence of L_p) have also been calculated, and range from about 1 kcal/mole for ram spermatozoa up to 2 or 3 kcal/mole for human cells and to ~18 kcal/mole for rabbit spermatozoa. The most recent entry to the list of permeability coefficients is for mouse spermatozoa as determined by Devireddy et al. (1998b). This latter value is notable for several reasons. First, it was determined by differential scanning calorimetry (DSC) of mouse spermatozoa that were scanned at subzero temperatures. The values shown for the other species were determined by measurements made at temperatures >0°C. One purpose of these determinations is to permit mathematical modeling of the response of spermatozoa during freezing, as first described for other types of cells by Mazur (1963), so as to derive the "optimum" cooling rate. Two recent attempts have been made with human and ram spermatozoa by Noiles et al. (1993) and by Curry et al. (1994). Using the values for L_p (see Table 2) and the activation energies of L_p, these

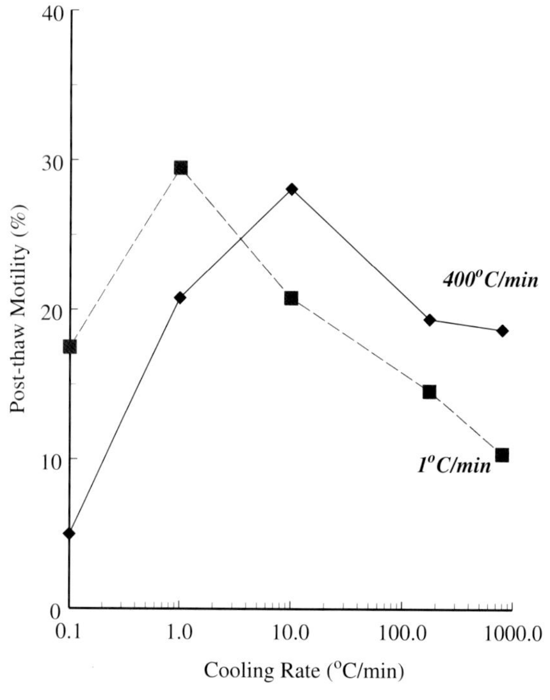

Figure 8. Post-thaw motility of human spermatozoa as a function of cooling rate. The frozen specimens were warmed either at 400°C/min or at 1°C/min. The data are those of Henry et al. (1993). For the sake of clarity, the figure has been redrawn without the error bars shown in the original figure.

investigators calculated the predicted loss of water from spermatozoa cooled to subzero temperatures at various rates. In both cases, the calculated optimum rates failed to conform with empirical observations. An example from Curry et al. (1994) for human spermatozoa is shown in Figure 9. According to calculated predictions, human spermatozoa ought to dehydrate sufficiently during cooling to -5°C so that they will not freeze intracellularly even when cooled at 1000°C/min. The prediction is that the optimum cooling rate will be >1000°C/min (curve 2 in Fig. 9). Several empirical observations are that the optimum rate for human spermatozoa is ~1°C/min to 10°C/min, not 1,000°C/min (Figs. 7, 8). This discrepancy between observation and prediction led Curry et al. (1994) to conclude that "it now seems questionable that the decline in sperm survival with increased cooling rate illustrated by the right-hand portion of the graph is a result of intracellular ice formation." These authors suggest that one possible mechanism might be damage to the sperm membrane caused by exceeding a critical gradient in osmotic pressure, by which water flux exerts a frictional force across the membrane.

An alternative explanation of the apparent discrepancy between observation and calculation has recently been suggested by Devireddy et al. (1998b). As mentioned above, they have devised a new technique using DSC to measure water transport properties of cells that is not dependent on cell shape or morphology, and that permits measurements during cooling at subzero temperatures (Devireddy et al. 1998a). As shown in Table 2, this new method yielded an L_p value for mouse spermatozoa of 0.01 μm/min per atmosphere with an activation energy of 22.5 kcal/mole. Using these values, Devireddy et al. (1998b) have calculated volumetric shrinkage of mouse spermatozoa cooled at various rates. Their calculations shown in Figure 10 indicate that mouse spermatozoa cooled at rates of 20°C/min would have dehydrated sufficiently by -12°C so as not to freeze intracellularly. However, cells cooled at rates of 50°C/min or certainly 100°C/min would be likely to freeze intracellularly. These predictions of Devireddy et al. (1998b) for optimum cooling rates agree quite well with empirical observations for spermatozoa of several species (Fig. 7). More specifically, Songsasen et al. (1997) recently described a successful procedure to freeze mouse spermatozoa by cooling them at 20°C/min before plunging them into LN_2. These investigators used spermatozoa cryopreserved this way for *in vitro* fertilization (IVF) of oocytes, and produced live young by transfer of resultant embryos. In this case, the rate predicted by mathematical modeling to be optimum agrees with experimental observation. Thus, it would seem that intracellular ice formation in spermatozoa is still an adequate explanation as a cause of injury in cells cooled at rates faster than the optimum.

Table 2: Coefficients of water permeability of mammalian spermatozoa.

Species	L_p (μm/min/atm)	References
Cattle	10.8	Watson et al. 1992b
Sheep	8.47	Curry et al. 1994
Human	2.40	Noiles et al. 1993
	2.89	Curry et al. 1994
Rabbit	0.63	Curry et al. 1995
Mouse	0.01	Devireddy et al. 1998b

Proposed Genetic Analysis of Freezing Injury of Spermatozoa

Despite intense efforts of numerous investigations conducted over several decades, many questions regarding the basic mechanisms responsible for injury to mammalian spermatozoa remain unanswered. As an alternative, researchers at the University of Guelph have recently tried to use methods of assisted reproduction and of genetics to try to answer some of these questions.

During a project to study *in vitro* production of bovine embryos, a set of monozygotic quadruplet bulls was produced from an *in vitro*-derived embryo by separating the blastomeres of a single embryo (Johnson et al. 1995). In a moment of whimsy, they were named for the Beatles. These quadruplet bull calves, shown in Figure 11, were proven by DNA analysis to be genetically identical. Just as twins share many identical characteristics,

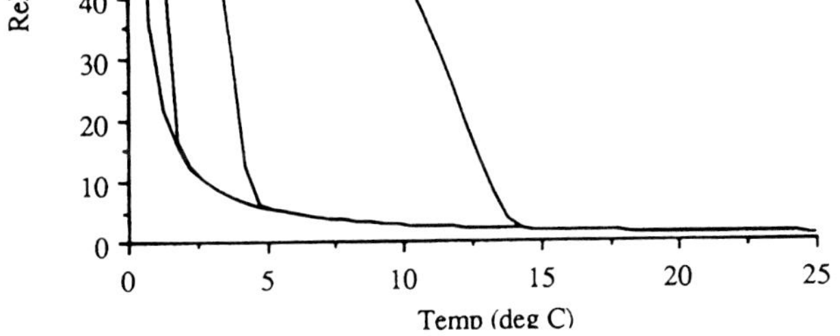

Figure 9. Calculated change in the relative volume of human spermatozoa as a function of subzero temperatures for samples cooled either infinitely slowly to remain in osmotic equilibrium (E) or at rates of 100°C/min (1), 1000°C/min (2), 10,000°C/min (3), or 100,000°C/min (4). The figure is from Curry et al. (1994) and is used with written permission from the publisher.

it seemed reasonable to hypothesize that each of the quadruplets might produce spermatozoa that were very similar to spermatozoa from their identical siblings. To test this notion, the animals were allowed to reach sexual maturity. When they were >14 months old, multiple ejaculates were collected by professional animal handlers at a commercial AI center. Various characteristics of the semen of these quads were measured, and survival of their spermatozoa after freezing was determined. Survival was assayed by motility measurements and by use of a double dye assay to determine sperm membrane integrity. Some of the semen characteristics of these genetically identical bulls are shown in Table 3. With one notable exception, the superficial characteristics of semen from these bulls are alike; the average ejaculate volumes, sperm cell concentrations, and pH and osmolalities of the seminal plasmas are similar. The one exception is the cell concentration in ejaculates of the bull named "John"; on average, the cell concentration in his specimens was only half that of the other three bulls. No explanation of that fact is available at present.

Despite this difference, preliminary measurements of the freezing sensitivity of the quads' spermatozoa have been made (Nishimura, Leibo 1995). In replicate experiments (three samples/bull) spermatozoa from the four bulls were suspended in TEST-egg yolk buffer containing 4% glycerol, and cooled from -5° to -80°C at rates of 2.5°C/min, 10°C/min or 25°C/min, and then plunged into LN_2. After being thawed rapidly, samples were washed free of the CPA solution, incubated in Sperm TALP medium at 37°C for >90 min, and sperm motility estimated by microscopic examination. One example of the mean survivals is shown in Figure 12; the average motility of the quads' spermatozoa cooled at 2.5°C/min, 10°C/min or 25 °C/min was about 35%. Survival of the spermatozoa when plotted individually for each animal varied from ~25 to 40%. Overall, the quads' spermatozoa, whether survival was assayed by motility or by measurements of membrane integrity, seemed to behave very similarly.

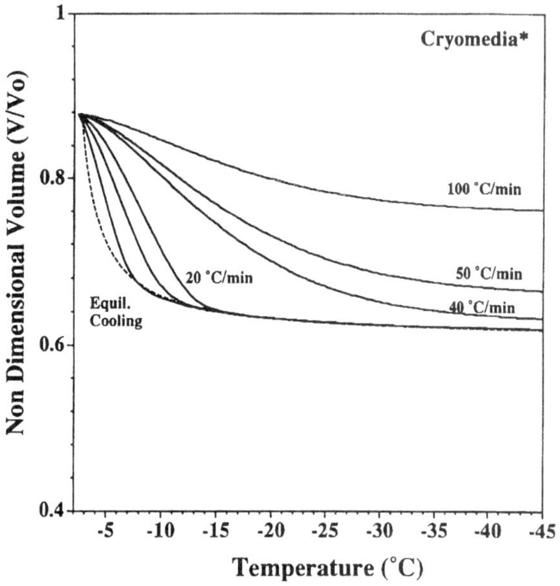

Figure 10. Calculated volumetric shrinkage of mouse spermatozoa suspended in 4% glycerol, 18% raffinose and 15% egg yolk when cooled at 5°C/min, 10°C/min, 20°C/min or faster. The dotted curve shows the response of cells cooled infinitely slowly. The figure is from Devireddy et al. (1998b) and was generously made available by the authors before publication.

Another approach to the question of male-to-male variability has begun by using genetically identical, inbred strains of mice. Beginning with the observations of Tada et al. (1990), several methods to cryopreserve mouse spermatozoa have been developed as an alternative means to "bank" valuable inbred and transgenic strains of mice. One of these methods was developed using spermatozoa from the hybrid strain B6D2F1; epididymal sperm were suspended in 0.45 M raffinose + 0.3 M glycerol in D-PBS supplemented with 25% egg yolk (Songsasen et al. 1997). This same method has also been used to cryopreserve ejaculated as well as epididymal spermatozoa, and to freeze spermatozoa from two inbred strains. The strains examined were those commonly used in molecular biology: strain 129/J is the source of most embryonic stem (ES) cell lines, and C57BL6/J provides "host blastocysts" into which ES cells are injected (Songsasen, Leibo 1997). As assayed with vital dyes by flow cytometry, sperm of the three strains exhibit the following relative survivals: B6D2F1 > 129/J > C57BL6. Furthermore, as shown by results in Table 4, when spermatozoa from the three strains were used for *in vitro* fertilization, although there was no difference in the fertilization rate for control sperm, the rates for frozen sperm varied significantly. Analogous differences in freezing sensitivity among inbred mouse strains have also been reported by Nakagata and Takeshima (1993).

After embryos produced by IVF with frozen sperm were transferred, live mouse pups were obtained from the inbred strain B6D2F1 and from 129/J mice, but none from C57BL6 mice. Measurements of the relative sensitivity of sperm from the three strains to osmotic shock revealed that those of C57BL6/J were most sensitive, those of 129/J intermediate, and those of B6D2F1 most resistant (Songsasen, Leibo 1997). Mice from inbred strains are by definition considered to be genetically identical. In one sense, therefore, the response of spermatozoa from three inbred mouse strains is analogous to that of three males of species that are genetically heterogeneous, such as cattle, dogs or humans. The inbred strains studied in these first experiments were selected only because they are commonly used in molecular genetics. Therefore, it was a fortuitous coincidence that their sperm exhibited such widely different sensitivities

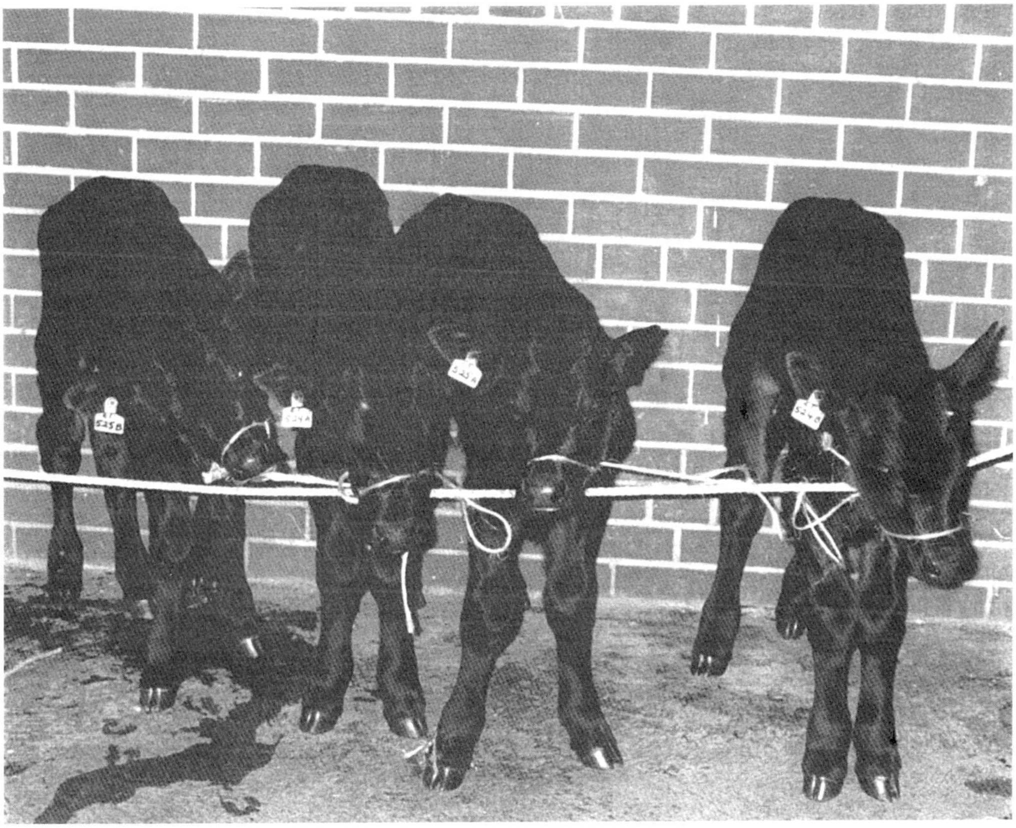

Figure 11. Photograph of monozygotic quadruplet bull calves produced by separating the blastomeres of a single four-cell bovine embryo that itself had been produced by *in vitro* maturation and *in vitro* fertilization. The methods used to produce the monozygotic embryos and the quadruplet calves were described by Johnson et al. (1995).

to freezing and to osmotic shock.

There is now strong evidence that the membrane characteristics of spermatozoa influence their response to freezing. The fact that the freezing sensitivity of spermatozoa from three different mouse strains differs significantly provides presumptive evidence for the interpretation that freezing sensitivity of sperm is genetically determined. In the future, investigations of the freezing sensitivity of spermatozoa from various inbred strains of mice, especially transgenic or gene "knock-out" strains whose basic characteristics have been deliberately altered, may provide a new approach to address problems in sperm cryobiology. There are many unanswered questions regarding the cryobiology of mammalian spermatozoa. Not the least among those questions is an explanation of why spermatozoa from different males of various species seem to vary significantly in their response to cryopreservation. The use of embryo manipulation and other methods of assisted reproduction as well as of molecular genetics may provide new ways to address these questions.

Conclusions

During the past 50 years, the ability to cryopreserve spermatozoa of many species of mammals has had a profound impact on human welfare, medicine, agriculture and science. Innumerable animals and humans have been born as a result of fertilization performed by artificial insemination or by *in vitro* fertilization with spermatozoa that were cooled and frozen to temperatures of -79° to -196°C. Thousands, if not tens of thousands, of investigations have been performed with spermatozoa of humans and of domestic, laboratory, companion and wild animals in attempts to improve methods to preserve them, as well as to understand causes of injury and

Table 3: Characteristics of semen specimens of monozygotic bulls.

Bull	Ejaculate Volume (ml)	Sperm Concentration (cells x10^9/ml)	Plasma pH	Osmolality (mOsm)
Ringo	8.3 ± 1.8 (26)	1.59 ± 0.41 (28)	6.3 ± 0.3 (18)	302.7 ± 9.0 (20)
George	10.4 ± 4.1 (25)	1.13 ± 0.38 (25)	6.2 ± 0.3 (15)	298.2 ± 12.0 (19)
Paul	8.6 ± 1.9 (24)	1.20 ± 0.47 (24)	6.3 ± 0.3 (17)	298.0 ± 10.1 (17)
John	9.9 ± 2.1 (25)	0.60 ± 0.12 (20)	6.5 ± 0.2 (16)	293.4 ± 8.8 (17)

The figures shown are the mean values ± S.D. for the number of specimens given in parentheses for the characteristics of each bull.

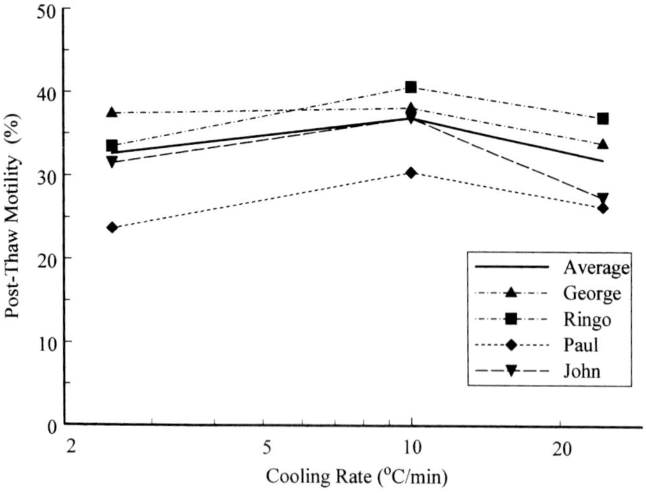

Figure 12. Post-thaw motility of bovine spermatozoa as a function of cooling rate. The spermatozoa were collected from the four bulls whose photograph is shown in Figure 10. Replicate experiments were conducted with spermatozoa of the four bulls, and the curves show the means of the replicates. Then, the means of the four means were calculated to give the curve labeled "average." The data are those of Nishimura and Leibo (1995).

Table 4: Fertilization of B6C3F1 oocytes by sperm from three mouse strains*.

Sperm Strain	Sperm Treatment	Oocytes	% 2-cell (mean SE)
B6D2F1	Control	143	89.4±5.0
	Frozen	461	61.2±4.1
129/J	Control	135	75.4±16.1
	Frozen	395	17.2±4.3
C57BL6/J	Control	116	81.9±7.6
	Frozen	299	3.0±1.5

*The data are those of Songsasen and Leibo, 1997.

death resulting from freezing and thawing. Very substantial progress has been made. There are entire billion dollar industries and human clinical practices based on the capability to successfully freeze and store millions of samples of spermatozoa, the samples themselves often containing only a few tenths of one ml of cell suspension. Despite this progress, much remains to be learned about the cryobiology of mammalian spermatozoa. As novel methods of molecular biology, analytical biochemistry, physical chemistry, reproductive biology and embryology join forces with cryobiology, it seems reasonable to predict that even greater success will be achieved in the near future. This brief review was intended to summarize a few facets of the interesting and important subject of sperm cryobiology.

References

Amann RP, Pickett BW. Principles of cryopreservation and a review of cryopreservation of stallion spermatozoa. Equine Vet Sci 1987; 7:145-173.

Blach EL, Amann RP, Bowen RA, Frantz D. Changes in quality of stallion spermatozoa during cryopreservation: plasma membrane integrity and motion characteristics. Theriogenology 1989; 31:283-298.

Bunge RG, Keettel W, Sherman J. Clinical use of frozen semen, report of 4 cases. Fertil Steril 1954; 5:520-529.

Chen Y, Foote RH, Tobback C, Zhang L, Hough S. Survival of bull spermatozoa seeded and frozen at different rates in egg yolk-tris and whole milk extenders. J Dairy Sci 1993; 76:1028-1034.

Cohen J, Felten P, Zeilmaker GH. *In vitro* fertilizing capacity of fresh and cryopreserved human spermatozoa: a comparative study of freezing and thawing procedures. Fertil Steril 1981; 36:356-362.

Coulter GH. Bovine spermatozoa *in vitro*: a review of storage, fertility estimation and manipulation. Theriogenology 1992; 38:197-207.

Critser JK, Huse-Benda AR, Aaker DV, Arneson BW, Ball GD. Cryopreservation of human spermatozoa. II. Post-thaw chronology of motility and of zona-free hamster ova penetration. Fertil Steril 1987; 47:980-984.

Cummins JM, Woodall PF. On mammalian sperm dimensions. J Reprod Fert 1985; 75:153-175.

Curry MR, Millar JD, Watson PF. Calculated optimal cooling rates for ram and human sperm cryopreservation fail to conform with empirical observations. Biol Reprod 1994; 51:1014-1021.

Curry MR, Redding BJ, Watson PF. Determination of water permeability coefficient and its activation energy for rabbit spermatozoa. Cryobiology 1995; 32:175-181.

Curry MR, Millar JD, Tamuli SM, Watson PF. Surface area and volume measurements for ram and human spermatozoa. Biol Reprod 1996; 55:1325-1332.

De Leeuw FE, de Leeuw AM, den Daas JHG, Colenbrander B, Verkleij AJ. Effects of various cryoprotective agents and membrane-stabilizing compounds on bull sperm membrane integrity after cooling and freezing. Cryobiology 1993; 30:32-44.

Devireddy RV, Raha D, Bishof JC. Measurement of water transport during freezing in cell suspensions using a differential scanning calorimeter. Cryobiology 1998a; 36:124-155.

Devireddy RV, Swanlund DJ, Bishof JC. The effect of extracellular ice on the water permeability parameters of mouse sperm plasma membrane during freezing. Am Soc Mech Eng J Biomech Engineer 1998b; 120:559-569.

Dhami AJ, Sahni KL, Mohan G. Effect of various cooling rates (from 30°C to 5°C) and thawing temperatures on the deep-freezing of *Bos taurus* and *Bos bubalis* semen. Theriogenology 1992; 38:565-574.

Dinnyés A, Wallace GA, Rall WF. Effect of genotype on the efficiency of mouse embryo cryopreservation by vitrification or slow freezing methods. Mol Reprod Dev 1995; 40:429-435.

Diskin MG, Sreenan JM. Fertilization and embryonic mortality rates in beef heifers after artificial insemination. J Reprod Fert 1980; 59:463-468.

Duncan AE, Watson PF. Predictive water loss curves for ram spermatozoa during cryopreservation: comparison with experimental observations. Cryobiology 1992; 29:95-105.

England GCW. Cryopreservation of dog semen: a review. J Reprod Fert 1993; Suppl 47:243-255.

Farstad W. Semen cryopreservation in dogs and foxes. Anim Reprod Sci 1996; 42:251-260.

Federation CECOS, Le Lannou D, Lansac J. Artificial procreation with frozen donor semen: experience of the French Federation CECOS. Hum Reprod 1989; 4:757-761.

Fiser PS, Fairfull RW. Combined effect of glycerol concentration and cooling velocity on motility and acrosomal integrity of boar spermatozoa frozen in 0.5 ml straws. Mol Reprod Dev 1990; 25:123-129.

Foote RH. Extenders and extension of unfrozen semen. In: (Salisbury GW, ed) Physiology of Reproduction and Artificial Insemination of Cattle. San Francisco: W.H. Freeman and Co, 1978, pp442-493.

Foote RH. The artificial insemination industry. In: (Brackett BG, Seidel GE, Seidel SM, eds) New Technologies in Animal Breeding. New York: Academic Press, 1981, pp13-39.

Foote RH, Parks JE. Factors affecting preservation and fertility of bull sperm: a brief review. Reprod Fertil Dev 1993; 5:665-673.

Foote RH, Chen Y, Brockett CC. Fertility of bull spermatozoa frozen in whole milk extender with trehalose, taurine, or blood serum. J Dairy Sci 1993; 76:1908-1913.

Fosgate OT, Smith VR. Prenatal mortality in the bovine between pregnancy diagnosis at 34-50 days post-insemination and parturition. J Dairy Sci 1954; 37:1071-1073.

Gao DY, Ashworth E, Watson PF, Kleinhans FW, Mazur P, Critser JK. Hyperosmotic tolerance of human spermatozoa: separate effects of glycerol, sodium chloride, and sucrose on spermolysis. Biol Reprod 1993; 49:112-123.

Gao DY, Liu J, Liu C, McGann LE, Watson PF, Kleinhans FW, Mazur P, Critser ES, Critser JK. Prevention of osmotic injury to human spermatozoa during addition and removal of glycerol. Hum Reprod 1995; 10:1109-1122.

Gao DY, Mazur P, Critser JK. Fundamental cryobiology of mammalian spermatozoa. In: (Karow A, Critser JK, eds) Reproductive Tissue Banking. San Diego, CA: Academic Press, 1997, pp263-328.

Gilmore JA, Du J, Tao J, Peter AT, Critser JK. Osmotic properties of boar spermatozoa and their relevance to cryopreservation. J Reprod Fert 1996; 107:87-95.

Gilmore JA, Liu J, Gao DY, Critser JK. Determination of optimal cryoprotectants and procedures for their addition and removal from human spermatozoa. Hum Reprod 1997; 12:112-118.

Glaub JC, Mills RN, Katz DF. Improved motility recovery of human spermatozoa after freeze preservation via a new approach. Fertil Steril 1976; 27:1283-1291.

Graham EF. Fundamentals of the preservation of spermatozoa. In: The Integrity of Frozen Spermatozoa. National Academy of Sciences, Washington, DC, 1978, pp4-44.

Hammerstedt RH, Parks JE. Changes in sperm surfaces associated with epididymal transit. J Reprod Fert 1987; Suppl 34:133-149.

Hammerstedt RH, Graham JK, Nolan JP. Cryopreservation of mammalian sperm: what we ask them to survive. J Androl 1990; 11:73-88.

Hasler JF, Henderson WB, Hurtgen PJ, Jin ZQ, McCauley AD, Mower SA, Neely B, Shuey LS, Stokes JE, Trimmer SA. Production, freezing and transfer of bovine IVF embryos and subsequent calving results. Theriogenology 1995; 43:141-152.

Hasler JF, Hurtgen PJ, Jin ZQ, Stokes JE. Survival of IVF-derived bovine embryos frozen in glycerol or ethylene glycol. Theriogenology 1997; 48:563-579.

Henry M, Noiles EE, Gao DY, Mazur P, Critser JK. Cryopreservation of human spermatozoa. IV. The effects of cooling rate and warming rate on the maintenance of motility, plasma membrane integrity, and mitochondrial function. Fertil Steril 1993; 60:911-918.

Heuchel V, Schwartz D, Czyglik F. Between and within subject correlations and variances for certain semen characteristics in fertile men. Andrologia 1983; 15:171-176.

Hochi S, Semple E, Leibo SP. Effect of cooling and warming rates during cryopreservation on survival of *in vitro*-produced bovine embryos. Theriogenology 1996; 46:837-847.

Iritani A. Problems of freezing spermatozoa of different species. In: IX Int Congr AI Anim Reprod, Madrid, 1980; pp115-132.

Johnson WH, Loskutoff NM, Plante Y, Betteridge KJ. Production of four identical calves by the separation of blastomeres from an *in vitro* derived four-cell embryo. Vet Rec 1995; 137:15-16.

Kelton DF, Martin SW, Hansen DS. Development and implementation of the Ontario Dairy Monitoring and Analysis Program. 17th World Buiatrics Congr. St Paul, MN, 1993; 285-290.

Kramer RY, Garner DL, Bruns EA, Ericsson SA, Prins GS. Comparison of motility and flow cytometric assessments of seminal quality in fresh, 24-hour extended and cryopreserved human spermatozoa. J Androl 1993; 14:374-384.

Kummerfeld HL, Oltenacu EAB, Foote RH. Embryonic mortality in dairy cows estimated by non-returns to service, estrus, and cyclic milk progesterone patterns. J Dairy Sci 1978; 61:1773-1777.

Larsson K, Einarsson S. Influence of boars on the relationship between fertility and post-thawing sperm quality of deep frozen boar spermatozoa. Acta Vet Scand 1976; 17:74-82.

Leibo SP. Freezing damage of bovine erythrocytes: simulation using glycerol concentration changes at subzero temperatures. Cryobiology 1976a; 13:587-598.

Leibo SP. Preservation of mammalian cells and embryos by freezing. In: (Simatos D, Strong DM, Turc J-M, eds) Cryoimmunology. INSERM, 1976b, pp311-324.

Leibo SP. Preservation of ova and embryos by freezing. In: (Brackett BG, Seidel GE, Seidel SM, eds) New Technologies in Animal Breeding. New York: Academic Press, 1981, pp127-139.

Leibo SP, Farrant J, Mazur P, Hanna MG Jr, Smith LH. Effects of freezing on marrow stem cell suspensions: interactions of cooling and warming rates in the presence of PVP, sucrose, or glycerol. Cryobiology 1970; 6:315-332.

Leibo SP, Semple ME, Kroetsch TM. *In vitro* fertilization of bovine oocytes with 37-year-old cryopreserved bovine spermatozoa. Theriogenology 1994; 42:1257-1262.

Mann T. Chemical and physical properties of whole ejaculated semen. The Biochemistry of Semen. John Wiley & Sons Inc, New York, 1954, pp30-35.

Mazur P. Kinetics of water loss from cells at subzero temperatures and the likelihood of intracellular freezing. J Gen Physiol 1963; 47:347-369.

Mazur P. Freezing of living cells: mechanisms and implications. Amer J Physiol 1984; 247:C125-142.

Mazur P, Leibo SP, Chu EHY. A two-factor hypothesis of freezing injury. Exp Cell Res 1972; 71:345-355.

McLaughlin EA, Ford WCL, Hull MGR. Motility characteristics and membrane integrity of cryopreserved human spermatozoa. J Reprod Fert 1992; 95:527-534.

Meryman HT. Cryobiology. Academic Press, New York, 1966.

Nakagata N, Takeshima T. Cryopreservation of mouse spermatozoa from inbred and F_1 hybrid strains. Exp Anim 1993; 42:317-320.

Niemann H. Cryopreservation of ova and embryos from livestock: current status and research needs. Theriogenology 1991; 35:109-124.

Nishimura K, Leibo SP. Effect of cooling rates on motility and viability of frozen-thawed sperm from genetically identical, monozygotic quadruplet bulls. Cryobiology 1995; 32:589.

Noiles EE, Mazur P Watson PF, Kleinhans FW, Critser JK. Determination of water permeability coefficient for human spermatozoa and its activation energy. Biol Reprod 1993; 48:99-109.

Olds D. Insemination of the cow. In: (Salisbury GW, ed) Physiology of Reproduction and Artificial Insemination of Cattle. San Francisco: WH Freeman and Co, 1978, pp579-610.

Pace MM, Sullivan JJ, Elliott FI, Graham EF, Coulter GH. Effects of thawing temperature, number of spermatozoa and spermatozoal quality on fertility of bovine spermatozoa packaged in .5 ml French straws. J Anim Sci 1981; 53:693.

Parks JE. Hypothermia and mammalian gametes. In: (Karow A, Critser JK, eds) Reproductive Tissue Banking. San Diego, CA: Academic Press, 1997, pp229-261.

Parks JE, Graham JK. Effects of cryopreservation procedures on sperm membranes. Theriogenology 1992; 38:209-222.

Penner W. Bovine Artificial Insemination Technical Manual. Canadian Association of Animal Breeders, Woodstock, ON, 1990.

Pickett BW, Berndtson WE. Preservation of bovine spermatozoa by freezing in straws: a review. J Dairy Sci 1974; 54:1287-1301.

Polak B, Daunter B. Osmolarity of human seminal plasma. Andrologia 1984; 16:224-227.

Polge C, Rowson LEA. Results with bull semen stored at -79°C. Vet Rec 1952; 64:851.

Polge C, Smith AU, Parkes AS. Revival of spermatozoa after vitrification and dehydration at low temperatures. Nature 1949; 169:626-627.

Rall WF, Wood MJ. High *in vitro* and *in vivo* survival of day 3 mouse embryos vitrified or frozen in a non-toxic solution of glycerol and albumin. J Reprod Fert 1994; 101:681-688.

Rao B, David G. Improved recovery of post-thaw motility and vitality of human spermatozoa cryopreserved in the presence of dithiothreitol. Cryobiology 1984; 21:256-541.

Reurink A, Den Daas JHG, Wilmink JBM. Effects of AI sires and technicians on non-return rates in The Netherlands. Livestock Prod Sci 1990; 26:107-118.

Richardson DW, Sadleir RMFS. The toxicity of various non-electrolytes to human spermatozoa and their protective effects during freezing. J Reprod Fert 1967; 14:439-444.

Rodriguez-Martinez H, Ekwall H. Electron microscopy in the assessment of cryopreserved spermatozoa viability. The Americas Microscopy and Analysis 1998; May, pp11-13.

Rodriguez-Martinez H, Larsson B, Pertoft H. Evaluation of sperm damage and techniques for sperm clean-up. Reprod Fertil Dev 1997; 9:297-308.

Royere D, Barthelemy C, Hamamah S, Lansac J. Cryopreservation of spermatozoa: a 1996 review. Hum Reprod 1996; 2:553-559.

Salamon S, Maxwell WMC. Frozen storage of ram semen. I. Processing, freezing, thawing and fertility after cervical insemination. Anim Reprod Sci 1995a; 37:185-249.

Salamon S, Maxwell WMC. Frozen storage of ram semen. II. Causes of low fertility after cervical insemination and methods of improvement. Anim Reprod Sci 1995b; 38:1-36.

Schlegel RA, Hammerstedt R, Cofer GP, Kozarsky, Freidus D, Williamson P. Changes in the organization of the lipid bilayer of the plasma membrane during spermatogenesis and epididymal maturation. Biol Reprod 1986; 34:379-391.

Schwartz D, Laplanche A, Jouannet P, David G. Within-subject variability of human semen in regard to sperm count, volume, total number of spermatozoa and length of abstinence. J Reprod Fert 1979; 57:391-395.

Shamsuddin M, Rodriguez-Martinez H, Larsson B. Fertilizing capacity of bovine spermatozoa selected after swim-up in hyaluronic acid-containing medium. Reprod Fertil Dev 1993; 5:307-315.

Shannon P, Curson B, Rhodes AP. Relationship between total spermatozoa per insemination and fertility of bovine semen stored in Caprogen at ambient temperature. New Zealand J Agric Res 1984; 27:35-41.

Smith A, Polge C. Storage of bull spermatozoa at low temperatures. Vet Rec 1950; 62:115-116.

Songsasen N, Leibo S. Cryopreservation of mouse spermatozoa. II. Relationship between survival after cryopreservation and osmotic tolerance of spermatozoa from three strains of mice. Cryobiology 1997; 35:255-269.

Songsasen N, Betteridge K, Leibo S. Birth of live mice resulting from oocytes fertilized *in vitro* with cryopreserved spermatozoa. Biol Reprod 1997; 56:143-152.

Stewart DL. Storage of bull spermatozoa at low temperatures. Vet Rec 1951; 63:65-66.

Tada N, Sato M, Yamanoi J, Mizorogi T, Kasai K, Ogawa S. Cryopreservation of mouse spermatozoa in the presence of raffinose and glycerol. J Reprod Fert 1990; 89:511-516.

Taylor JF, Everett RW, Bean B. Systematic environmental, direct and service sire effects on conception rate in artificially inseminated Holstein cows. J Dairy Sci 1985; 68:3004-3022.

Taylor PJ, Wilson J, Laycock R, Weger J. A comparison of freezing and thawing methods for the cryopreservation of human semen. Fertil Steril 1982; 37:100-103.

Thomas CA, Garner DL, DeJarnette JM, Marshall CE. Effect of cryopreservation on bovine sperm organelle function and viability as determined by flow cytometry. Biol Reprod 1998; 58:786-793.

Thomas PGA, Larsen RE, Burns JM, Hahn CN. A comparison of three packaging techniques using two extenders for the cryopreservation of canine semen. Theriogenology 1993; 40:1199-1205.

Van Doormaal BJ. Linear model evaluations of non-return rates for dairy and beef bulls in Canadian AI. Can J Anim Sci 1993; 73:795-804.

Watson PF. Artificial insemination and the preservation of semen. In: (Lamming GE, ed) Marshall's Physiology of Reproduction. 4th ed. London: Churchill Livingstone, 1990, pp747-869.

Watson PF. Recent developments and concepts in the cryopreservation of spermatozoa and the assessment of their post-thawing function. Reprod Fertil Dev 1995; 7:871-891.

Watson PF, Critser JK, Mazur P. Sperm preservation: fundamental cryobiology and practical implications. In: (Templeton AA, Drife JP, eds) Infertility. London: Springer-Verlag, 1992a, pp102-114.

Watson PF, Kunze E, Cramer P, Hammerstedt RH. A comparison of critical osmolality and hydraulic conductivity and its activation energy in fowl and bull spermatozoa. J Androl 1992b; 13:131-138.

Whittingham DG, Leibo SP, Mazur P. Survival of mouse embryos frozen to -196° and -269°C. Science 1972; 178:411-414.

Wijeratne WVS. A population study of apparent embryonic mortality in cattle, with special reference to genetic factors. Anim Prod 1973; 16:251-259.

Woelders H, Matthijs A, Engel B. Effects of trehalose and sucrose, osmolality of the freezing medium, and cooling rate on viability and intactness of bull sperm after freezing and thawing. Cryobiology 1997; 35:93-105.